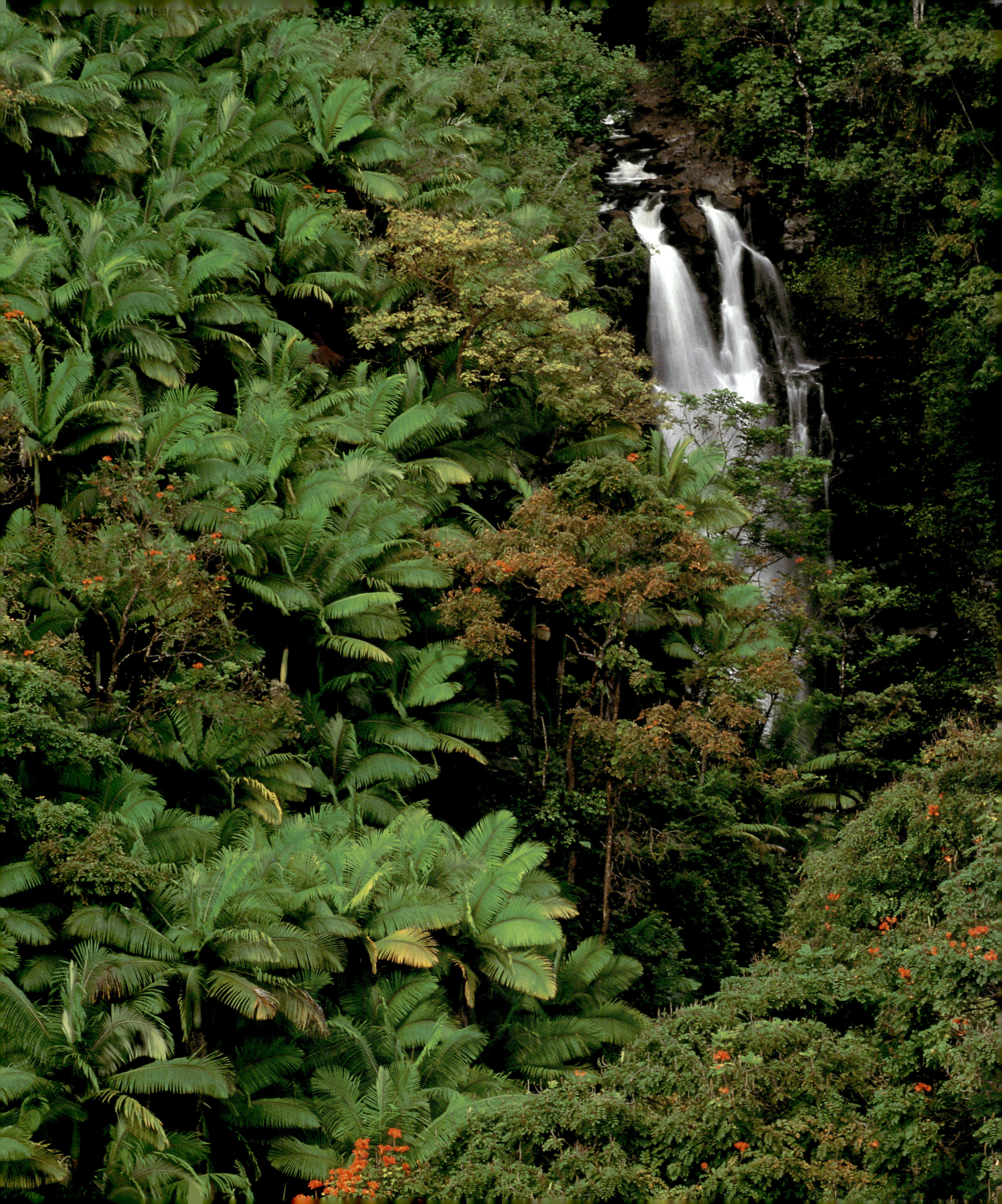

MEGADIVERSITY

EARTH'S BIOLOGICALLY WEALTHIEST NATIONS

RUSSELL A. MITTERMEIER
CRISTINA GOETTSCH MITTERMEIER

COORDINATOR
PATRICIO ROBLES GIL

FOREWORD
EDWARD O. WILSON

CEMEX

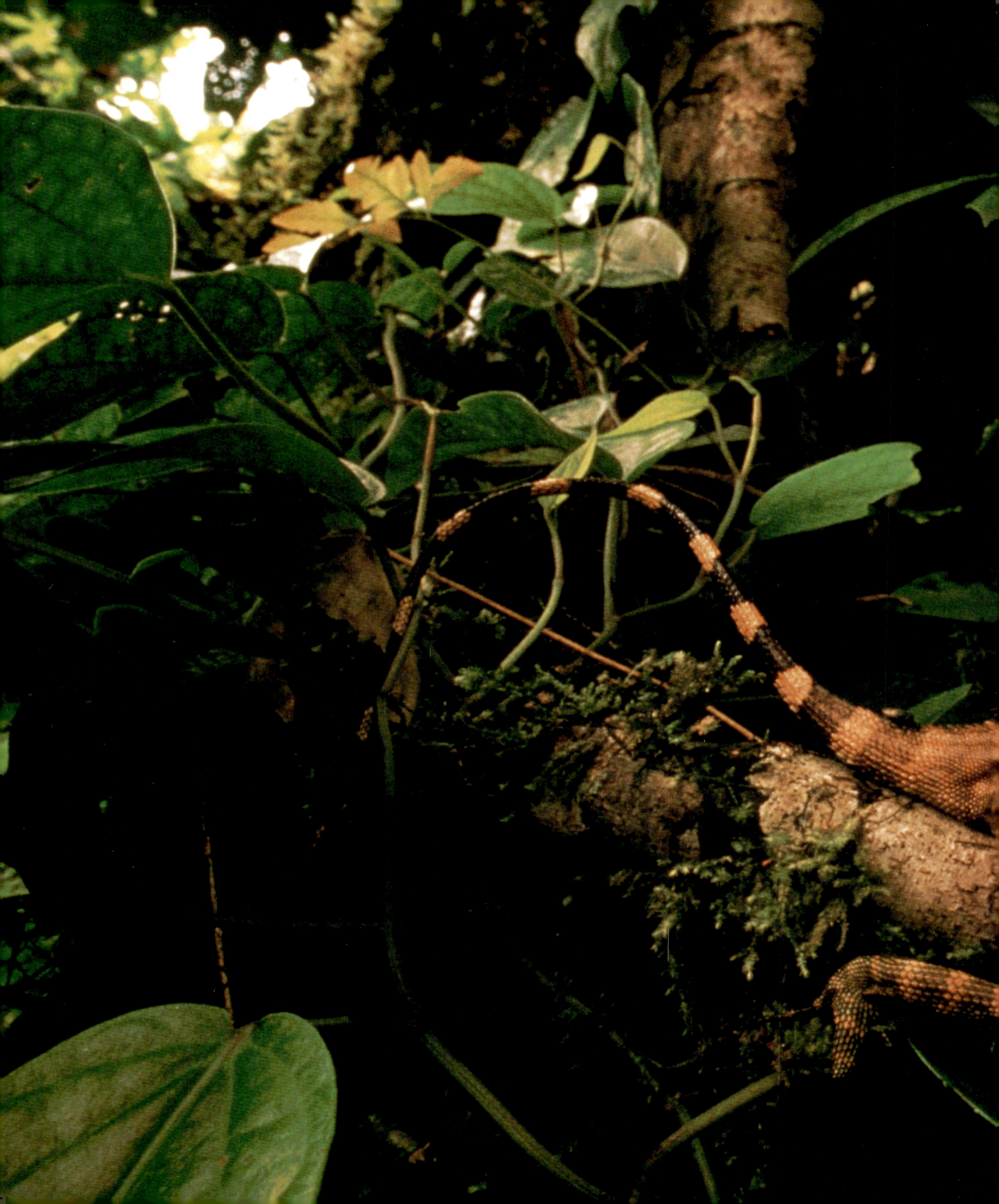

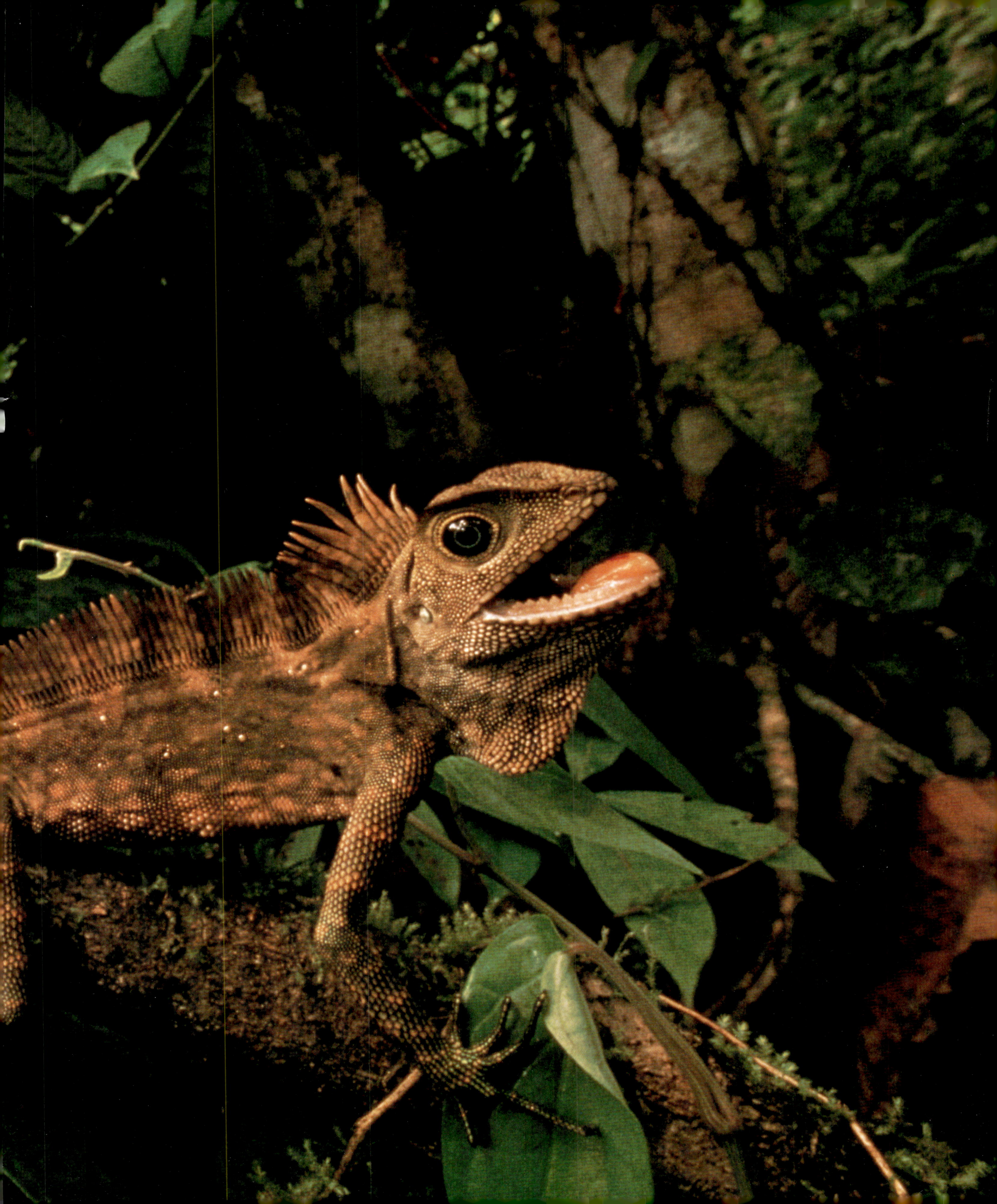

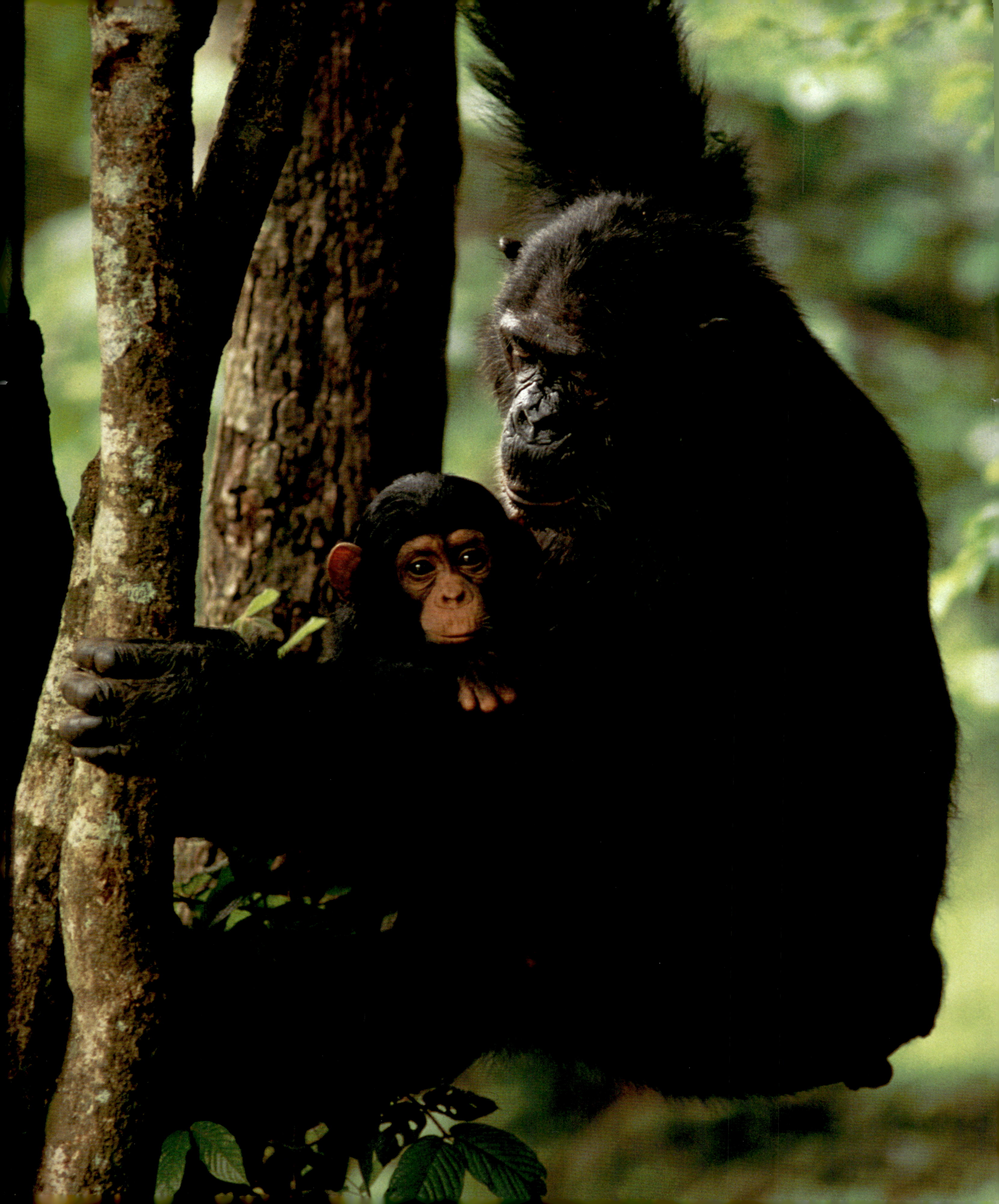

On the threshold of the coming millennium, humans still do not know the magnitude of life on Earth. Ironically, although we have precise data on other planets where no signs of life have been found, we have only identified less than 5% of the organisms with which we share this world.

In these critical times we should foster a change in values favoring respect for the life forms on our planet.

CEMEX, a multinational company, presents *Megadiversity: Earth's Biologically Wealthiest Nations*, and is proud to publicize for the first time, in a book, the concept of megadiversity, which was proposed and structured by Conservation International as part of a conservation strategy.

Knowledge of biological diversity is one of the first steps toward creating conservation strategies. Based on that knowledge arises the concept of the seventeen megadiversity countries, as they have been called in this book, and the need to offer an overview of them. These countries are privileged to have the greatest biological wealth on Earth, and while this represents a huge competitive advantage in the international sphere, it also entails an enormous responsibility, for the fate of life on Earth depends to a great extent on them.

No other world priority can be more global or more far-reaching than that of conserving life on this planet. For that reason, the responsibilities implied by this effort should be accepted and shared by all sectors of society.

Therefore, with this book CEMEX reaffirms its commitment to the conservation of natural resources, and to drawing the world's attention to this biological wealth which has yet to be fully appraised.

We hope that the 1990s will be considered the decade in which, thanks to greater scientific knowledge and a growing social awareness, humans achieved a deeper understanding of the natural world, enabling them to curb the loss of its important resources, which are the very underpinnings of human existence.

It is high time that this new conservationist ethic grew and bore fruit, so that the human beings who inhabit this diverse planet Earth never cease to marvel at its wonders.

CEMEX

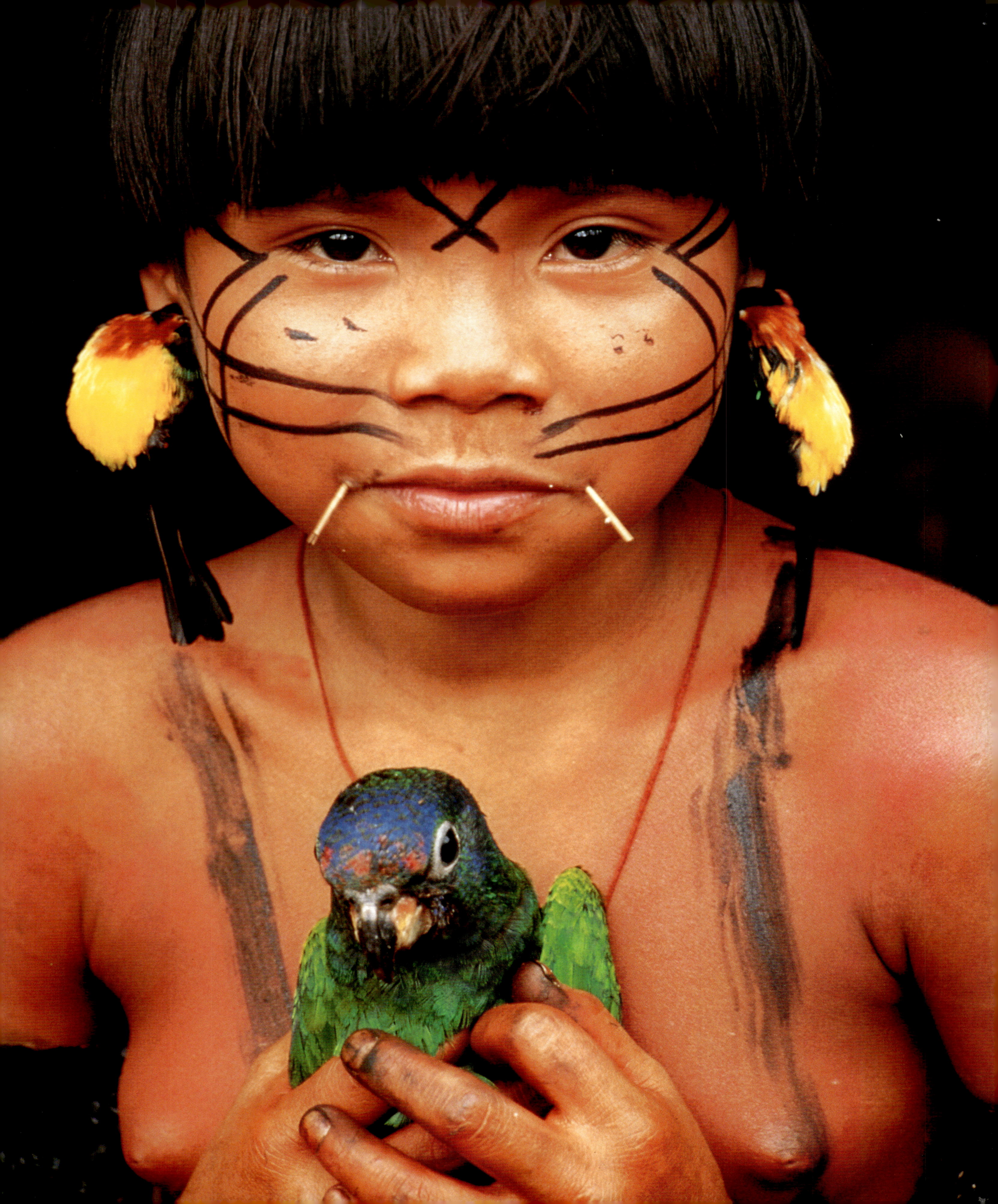

The mission of Conservation International is conservation of our planet's biological diversity, that wealth of species, ecosystems and ecological processes that make our living planet what it is, and the impending loss of which is the most critical issue we face at the end of the twentieth century. Our organization seeks to accomplish this by demonstrating that conservation can be a win-win situation in which natural systems *and* human societies' quality of life can be maintained. Although we are fundamentally rooted in the soundest scientific underpinnings and place great emphasis on increasing knowledge of species and ecosystems, we are ultimately a people-focused organization that works closely with human communities in all our target countries. We also place great emphasis on the economic and policy dimensions and seek to develop private sector partnerships to achieve our biodiversity conservation objectives. We believe strongly in the need to work closely with the business community, as well as with governments to meet what we believe are ultimately common objectives for all humanity. We are also committed to working with local communities at the grass-roots level to make sure that these people who live in close proximity to and depend more than anyone else on healthy functioning ecosystems are the first to benefit from our programs.

Our field-based conservation program is strategically focused on the highest-priority areas for biodiversity conservation in the world. Rather than trying to do everything, we decided early on that we needed to invest our limited resources in those areas that had the greatest concentrations of biodiversity and were also under the most severe threat: the *threatened biodiversity hotspots* and the *megadiversity countries*. The hotspots are an ecosystem-based approach that highlights the fact that over 50% of global terrestrial biodiversity is found in only about 2% of the land surface of the planet, and that these areas are generally the ones under greatest threat as well. The megadiversity country approach is focused on political entities rather than ecosystems, and recognizes that a very small number of countries harbor the vast majority of life on Earth. These hotspots and megadiversity countries have been our major theaters of operation, and we in fact work in 10 of the 17 that have been identified in this book. The concept of B-13 countries as a biological counterpoint to the G-7 countries helps to focus attention on the fact that living natural resources are concentrated in much the same way as economic wealth; if the linkages can be made between future economic prosperity and maintenance of biodiversity resources, then we may have effected a shift in priorities and global leadership that would have far reaching geopolitical implications.

We also believe strongly in the need for effective dissemination of knowledge on biodiversity. As a culture, we are still amazingly ignorant about the rest of life on our planet. A publication like *Megadiversity* goes a long way towards creating linkages between people and other forms of life and stressing the relevance of biodiversity to long-term sustainable development.

For these reasons and many more, we are delighted to be involved with CEMEX in the production of this important book which, for the first time, lays out the details of the megadiversity country concept and demonstrates clearly how important these megadiversity countries are to global efforts to conserve biological diversity. The fact that CEMEX is already a leader on environmental issues in one of the top five megadiversity countries is especially relevant. With their previous books, especially their three-volume series, *Mexico: Diversity of Fauna, Flora, and Cultures*, and the many field programs they have helped to support, they have made a great contribution to conservation in Mexico. With this book, they are expanding into the international arena, and we look forward to their leadership and involvement in global environmental issues as well.

Finally, we would like to congratulate Russ and Cristina Mittermeier and Patricio Robles Gil for what they have accomplished with this book. It is visually very appealing and scientifically sound; it highlights the special position these countries occupy, and also establishes the linkages between biological and cultural diversity. It should help to focus attention on the megadiversity countries, and demonstrate to people from these countries that they are indeed in a privileged position to possess such a significant portion of the Earth's living natural resources. Needless to say, the long-term economic value of these resources is also of great significance, and should be considered a major competitive advantage in the global marketplace.

In closing, we would like to again express our appreciation to CEMEX for their great vision in sponsoring this book and many other important environmental initiatives, and for serving as a model for corporations around the world. We hope that our collaboration on this important publication is the first of many joint efforts in the future.

CONSERVATION INTERNATIONAL

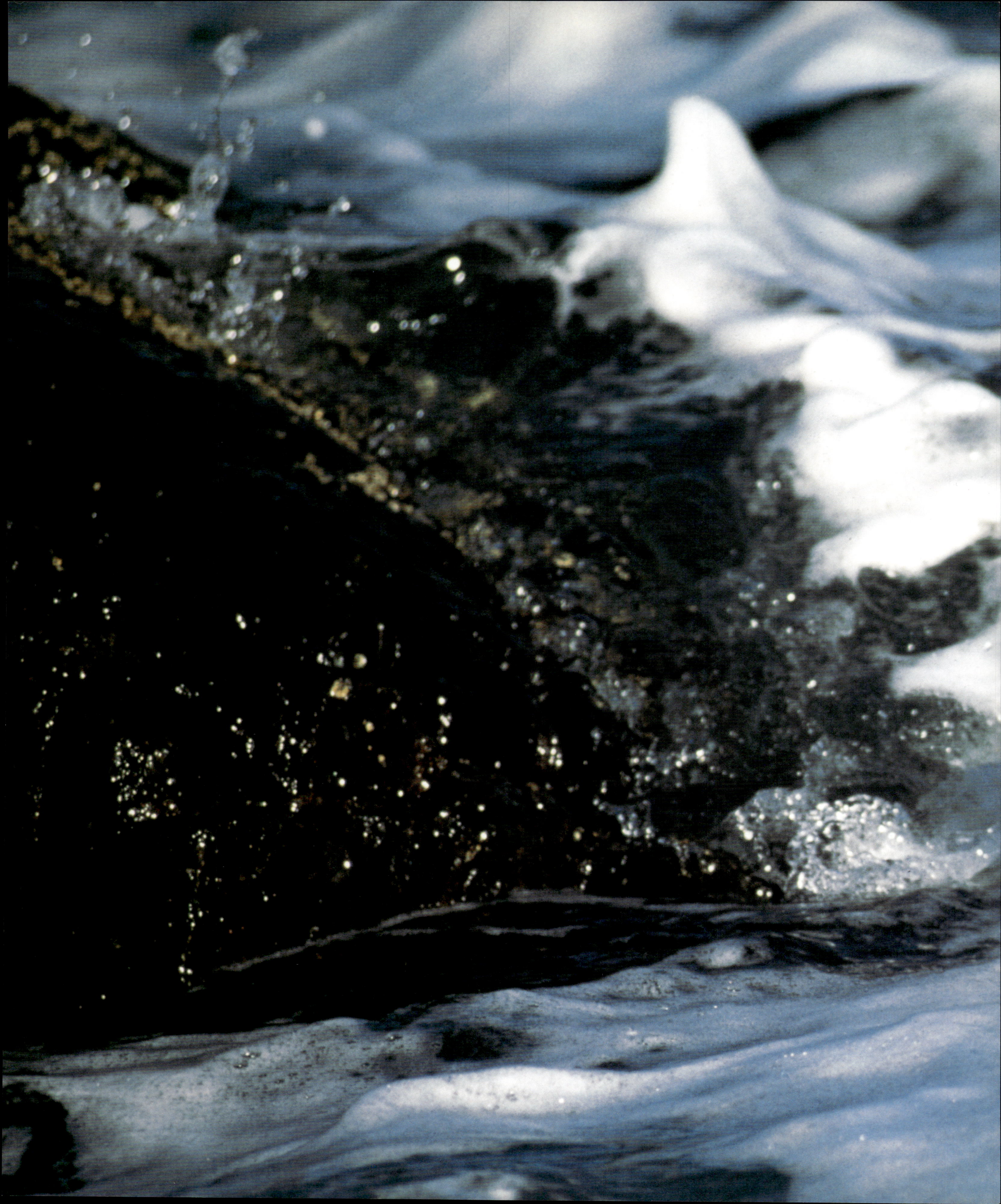

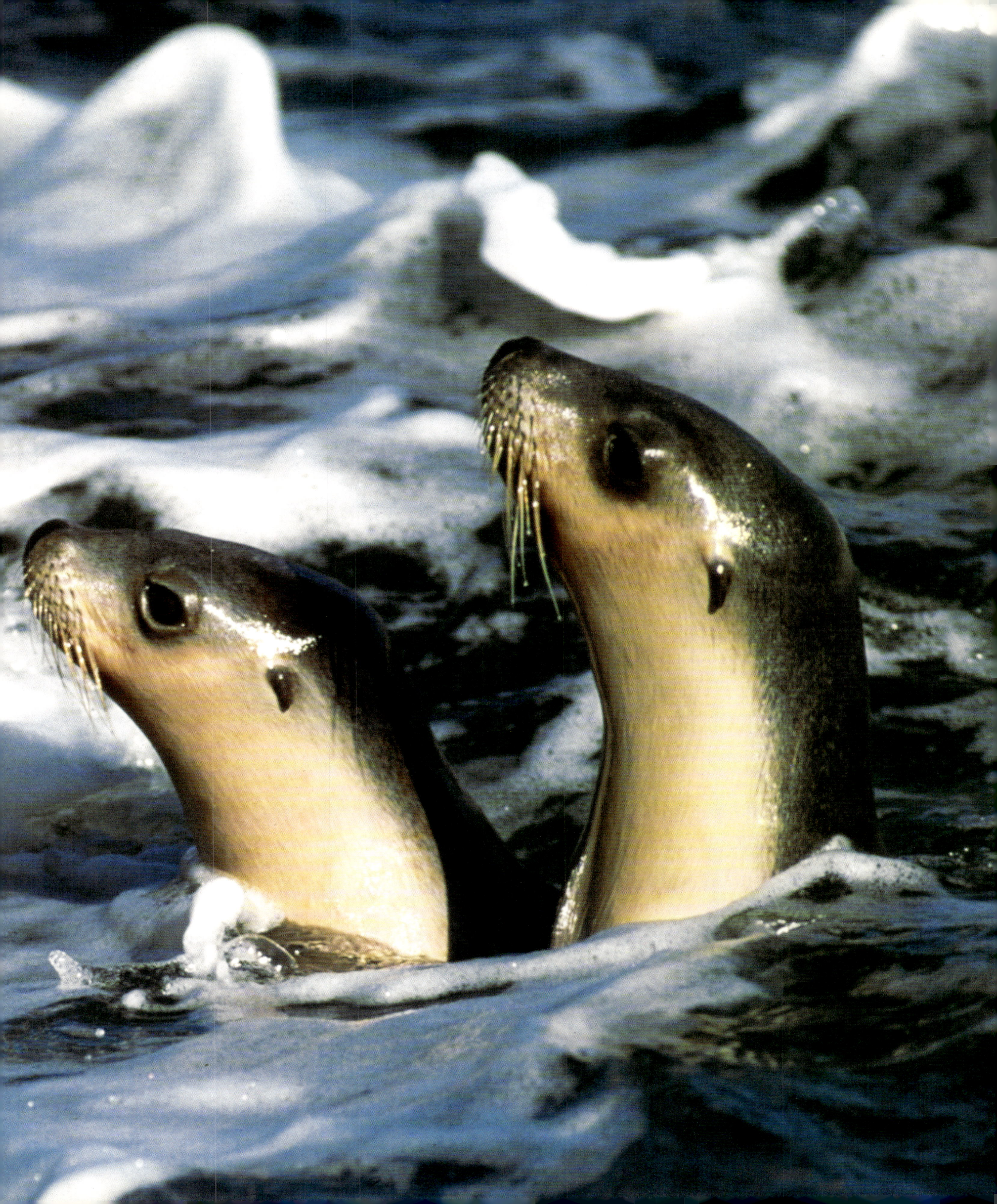

Knowledge is the power that gave humanity control of Earth, and no knowledge is more important to humanity than that concerning the biosphere on which our existence depends. It is a fact agreed to by every scientist of my acquaintance who is an expert on the subject, that species of plants and animals are disappearing at a faster rate than at any time since the end of the Mesozoic Era, 65 million years ago. The cause is not the strike of a giant meteorite, which occurred back then, but the destructiveness of human activity. Estimates of the extinction rate vary according to the method used, but mostly they fall between 100 and 1 000 times higher than the base level that existed before the global spread of the late Paleolithic *Homo sapiens*.

The first species wiped out were mostly large mammals and birds. Now, for the first time, insects and other small animals and plants are disappearing in large numbers. They are vanishing from the land, from freshwater environments, and from coral reefs. The longest scythe of destruction being swung by humanity is the devastation of the natural environment. Add to that the growing effects of pollution, overharvesting and the introduction of exotic species able to competitively exclude native ones. Unimpeded, during the next 30 years the loss of species could reach 20% of the world's existing stock.

As Russell and Cristina Mittermeier document in this important book, the process is not spread uniformly over the world. Global biodiversity is heavily concentrated in the tropics and, in particular, in a group of nations that the Mittermeiers have aptly termed the "megadiversity countries." Because poverty and population growth are so much higher in most of these same countries, it is no surprise to find that conservation problems there are also far more severe.

It follows that a million dollars spent on conservation in Madagascar or Papua New Guinea confers much more benefit on the world as a whole than does that same amount spent on conservation in Denmark or Canada. It also follows that conservation efforts in the megadiversity countries will disproportionately benefit their citizens. Native biodiversity is a source of pride for each country in turn, composing as it does a shining part of the national heritage, and it is also a future source of material wealth. The plants, animals, and microorganisms endemic to the country offer it many dimensions of unique economic opportunity.

The fauna and flora of every country in the world is, of course, priceless. But by concentrating research and aid for conservation on the biologically richest countries first, the greatest value can be obtained from whatever global resources are made available. The Mittermeiers have performed a valuable service by helping to call this important fact to the world's attention. In the present work, they have brought the megadiversity countries onto center stage, where they can secure the star status they deserve.

The outstanding layout and world-class photo selection by Patricio Robles Gil adds even more to this book. It will attract the attention of a wider audience and help to further the cause of biodiversity even more.

Edward O. Wilson

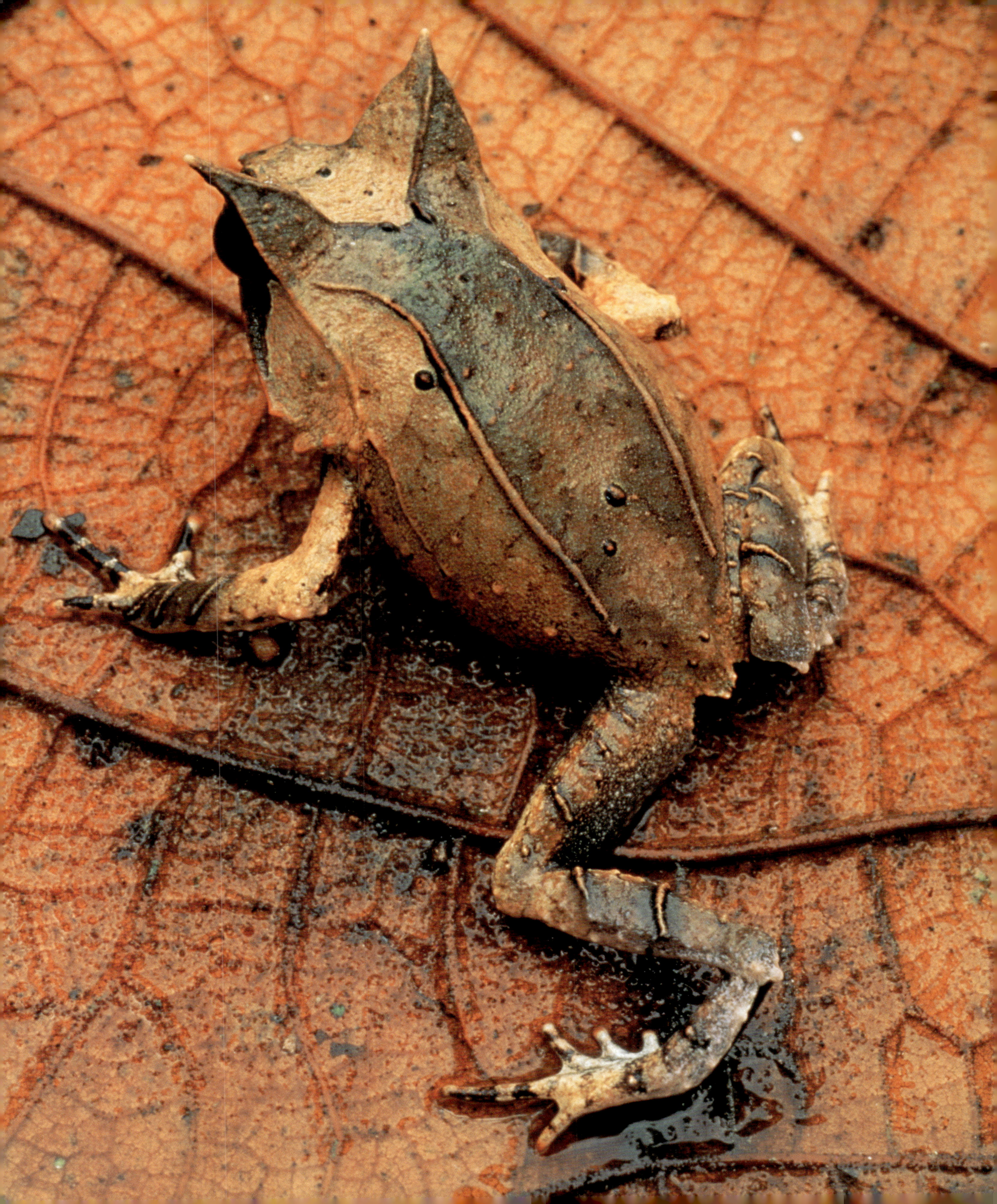

EARTH'S BIOLOGICALLY

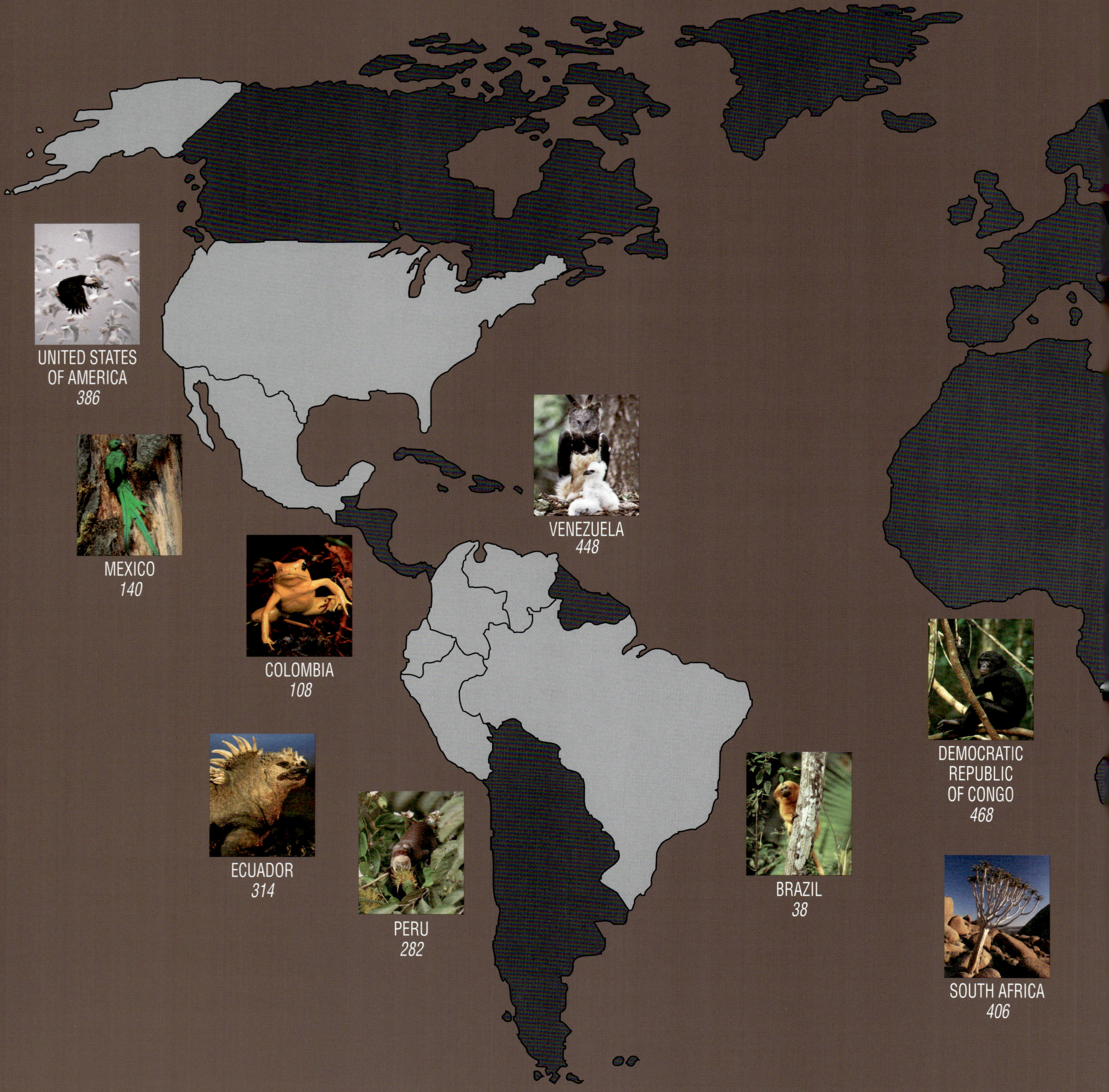

WEALTHIEST NATIONS

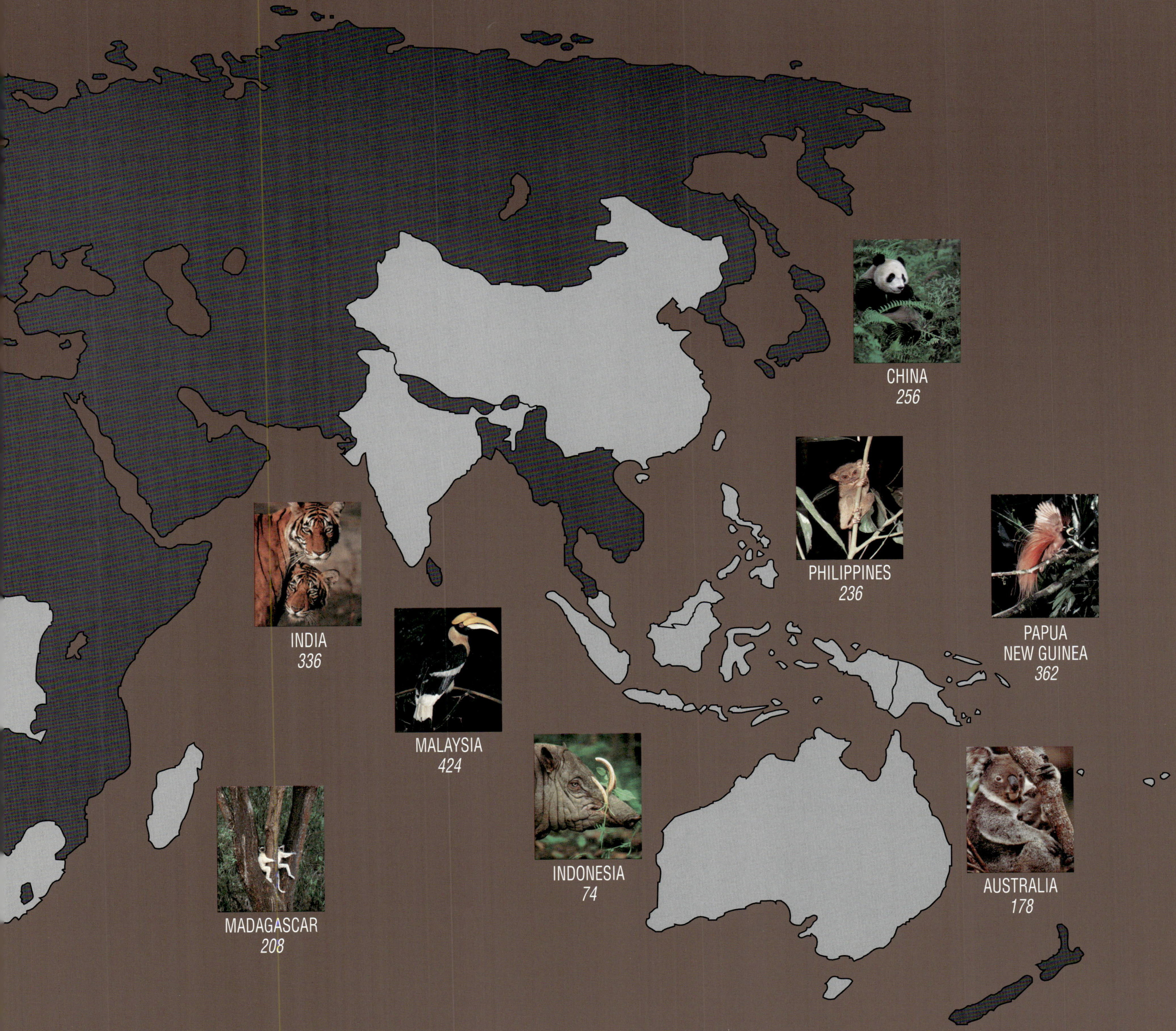

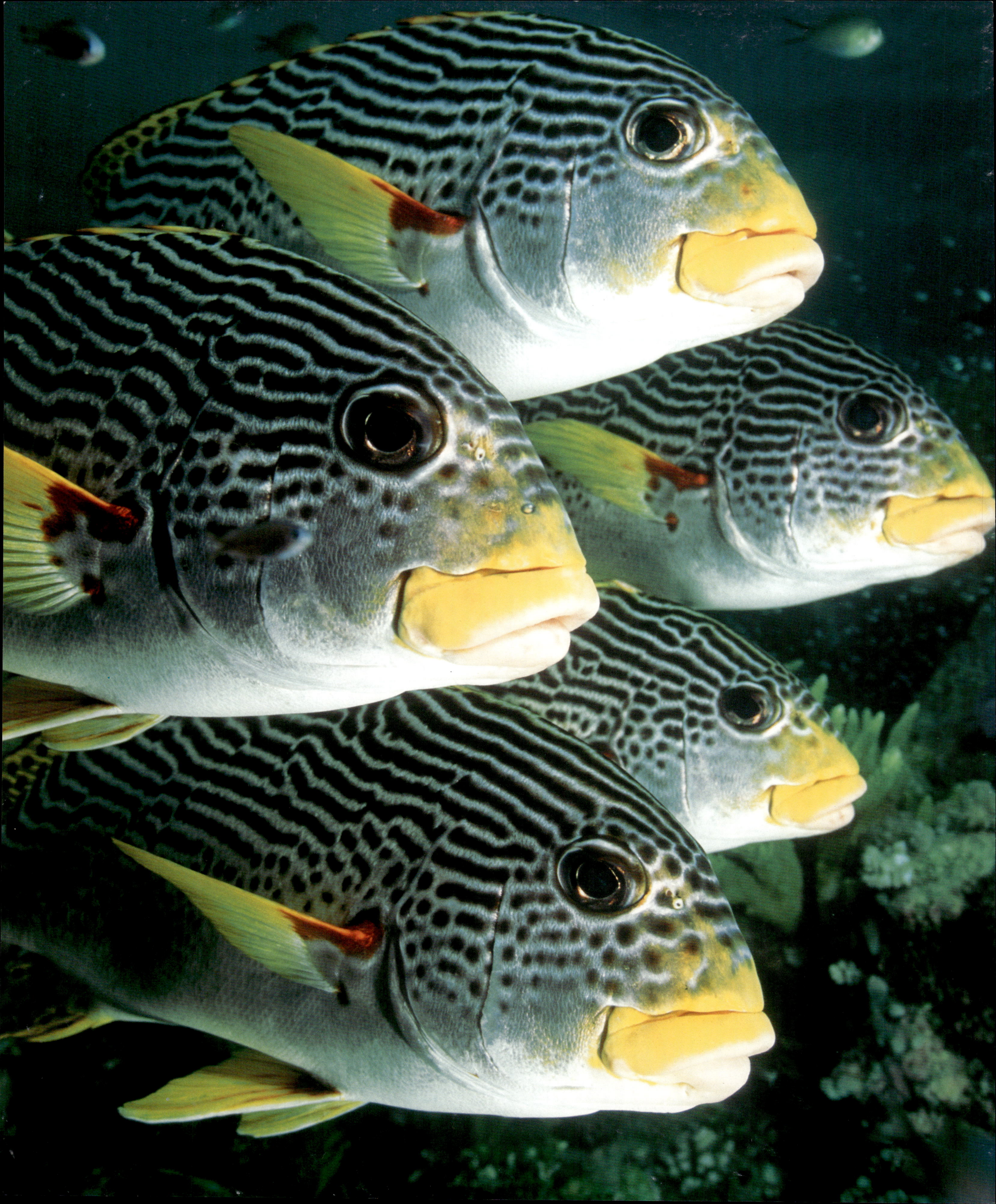

INTRODUCTION

Conservation of our planet's biological diversity, the sum total of life on Earth, is increasingly being recognized as the most fundamental of all environmental issues, and probably the single most important problem of our times. Little by little, with occasional quantum leaps (*e.g.*, the Earth Summit in Rio de Janeiro in 1992, the Biodiversity Convention), biodiversity conservation is rising to the top of the agenda of national governments, bilateral and multilateral development agencies (*e.g.*, USAID, the World Bank), and the private sector. Nonetheless, the challenges that we face in achieving biodiversity conservation are vast, in part because of the many threats to species and ecosystems, especially in the tropics, but also because of our enormous ignorance of the scope and the true value of biodiversity.

At this point in history, our measure of biodiversity is embryonic at best. Science has thus far described between 1.4 and 1.8 million species of living creatures including animals, plants, and microorganisms (E.O. Wilson, pers. comm.); however, estimates and projections made over the past few years indicate that we are probably looking at 100 million or more species in all, especially in the tropical rain forests and the deep ocean trenches, but also in the soils of our own backyards. Needless to say, our ignorance of the multitudes of ecological processes, probably at least one to two orders of magnitude more than the total species number, is even more profound. Ironically, we know more about the surface of the moon, can send spaceships to the farthest reaches of the solar system (which for all their scientific interest are simply never going to be as important to us as the healthy functioning of planet Earth), and can put millions of bits of rapidly retrievable information onto tiny computer chips, yet we do not know, to more than two orders of magnitude, how many other forms of life share this planet with us. Furthermore, our lack of understanding of the true values —the goods and services— that this biodiversity already provides and could provide in the future, is similarly staggering. Trying to achieve "sustainable" development based on the use of living, biological resources in the face of such lack of knowledge is rather like trying to construct a Boeing 747 with nothing more than a hammer, some nails, and a few sheets of scrap metal. Clearly, some reordering of our research and investment priorities is urgently needed.

At the same time, we have very clear information on the disappearance of those species and ecosystems of which we *are* aware. We have already lost more than 60% of the Earth's primary tropical rain forest, the coral reefs are being dynamited and poisoned into oblivion, and the latest *(1996) IUCN Red List of Threatened Animals* (coproduced by the International Union for Conservation of Nature and Natural Resources, or IUCN, and Conservation International) indicates that fully 25% of all mammals and 11% of all birds are already *critically endangered*, *endangered* or *vulnerable*. These are surely underestimates, with the actual numbers of threatened species almost certainly being higher than these preliminary numbers indicate.

Given the many problems we face, the biodiversity crisis sometimes seems overwhelming, especially if viewed only from the macro level. Yet there are ways to make the issue much more manageable, and the foremost among these is priority-setting. Biodiversity is by no means evenly distributed over our planet, with certain areas being far richer than others in overall diversity and in *endemism* (species that are found only in a particular place, and nowhere else). Furthermore, some of the richest areas for biodiversity, especially in the tropics, also happen to be under the most severe threat. By focusing on those that happen to have the greatest concentrations of biodiversity and the highest likelihood of losing significant portions of that biodiversity over the next few decades, we can get maximum conservation impact out of each dollar invested. Focusing heavily on the highest-priority areas under greatest threat may not always be politically convenient or politically appropriate, but it is essential if we are to have any hope of maintaining the full range of life on Earth.

Several preliminary attempts at biodiversity priority-setting were carried out in the early to mid-1980s, one of the most interesting of which was a 1983 internal priority-setting exercise for Latin America by World Wildlife Fund-U.S. This study recognized some of the most important priority areas of the planet, among them the Atlantic forest region of Brazil and the tropical Andes, and stimulated a number of other later exercises within the WWF family of organizations. Unfortunately, the results of these studies were never published. The first person to recognize the importance of priority-setting and clearly articulate its role was Dr. Norman Myers, one of the leading thinkers in the environmental movement. In a groundbreaking paper published in 1988, he recognized 10 *threatened hotspots* in the world's tropical rain forests, and estimated that they contained 13.8% of all plant diversity in just 0.2% of the total land area of the planet. In a subsequent analysis, he added several other ecosystem types and came up

TABLE 1. Land Areas of Megadiversity Countries

Country	*Area (km^2)*
China	9 561 000
USA	9 372 143
Brazil	8 511 965
Australia	7 686 810
India	3 287 782
DRC[a]	2 344 000
Mexico	1 972 544
Indonesia	1 916 600
Peru	1 285 210
South Africa	1 221 037
Colombia	1 141 748
Venezuela	912 050
Madagascar	587 045
PNG[b]	475 369
Malaysia	329 749
Philippines	300 780
Ecuador	283 561
Total	**51 189 393**

a. Democratic Republic of Congo (formerly Zaire).
b. Papua New Guinea.

On p. 2, rain forest vegetation in Nanue River Valley, island of Hawaii, U.S.A.

On pp. 4-5, Bornean crested lizard (Gonocephalus grandis) *in lowland rain forest, Sabah, Borneo.*

On p. 6, chimpanzee (Pan troglodytes) *mother and baby.*

On p. 8, Yanomami girl and blue-headed parrot (Pionus menstruus), *Parima-Tapirapaco National Park, Venezuela.*

On pp. 10-11, two Australian sea-lion pups (Neophoca cinerea).

On p. 13, Bornean horned frog (Megophrys nasuta), *Sabah, Borneo.*

On the opposite page, five lined sweetlips, Australia.

Table 2. Distribution of Animal Phyla by Habitat[a]

SYMBIOTIC = 15 (4 only symbiotic)	MARINE = 28 (13 endemic)	
Orthonectida	**Placozoa**	Porifera
Dicyemyda	**Ctenophora**	Cnidaria
Nematomorpha	**Gnathostomulida**	Platyhelminthes
Acanthocephala	**Kinorhyncha**	Nemertea
	Loricifera	Nematoda
Porifera	**Priapulida**	Rotifera
Cnidaria	**Pogonophora**	Gastrotricha
Platyhelminthes	**Echiura**	Tardigrada
Nemertea	**Chaetognatha**	Mollusca
Nematoda	**Phoronida**	Kamptozoa
Rotifera	**Brachiopoda**	Sipuncula
Mollusca	**Echinodermata**	Annelida
Kamptozoa	**Hemichordata**	Arthropoda
Annelida		Bryozoa
Arthropoda		Chordata
Chordata		

FRESHWATER = 14 (0 endemic)	TERRESTRIAL = 11 (1 endemic)
Porifera	**Onychophora**
Cnidaria	
Platyhelminthes	Platyhelminthes
Nemertea	Nemertea
Nematoda	Nematoda
Rotifera	Rotifera
Gastrotricha	Tardigrada
Tardigrada	Mollusca
Mollusca	Sipuncula
Kamptozoa	Annelida
Annelida	Arthropoda
Arthropoda	Chordata
Bryozoa	
Chordata	

a. Endemic or only symbiotic families are in boldface.

Table 3. Species Numbers for Major Groups of Vertebrates and Higher Plants Used in the Analysis of Megadiversity (Global Totals)

VERTEBRATES	
Mammals	4 629
Birds	9 040
Reptiles	6 458
Amphibians	4 222
Total Non-Fish Vertebrates	24 349[a]
Fish-Osteichthians	18 000
Fish-Chondrichtians	850
Fish-Agnathans	60
Total Fish	18 910
Total All Vertebrates	**43 259**
HIGHER PLANTS	
Angiosperms-Dicotyledons	170 000
Angiosperms-Monocotyledons	50 000
Gymnosperms	529
Ferns	10 000
Bryophytes	16 600
Other	1 300
Total Higher Plants	**248 429**[b]

a. Rounded off to 24 000 for purposes of this megadiversity analysis.
b. Rounded off to 250 000 for purposes of this megadiversity analysis.

On the opposite page,
Shoebill stork (Balaeniceps rex).

with a total of 18 areas that accounted for 20% of terrestrial plant biodiversity in just 0.5% of the land area. Beginning in 1989, both Conservation International and the MacArthur Foundation adopted Myers' *threatened hotspots* as the guiding principle for their conservation investments, with Conservation International slightly modifying Myers' list to include a few areas that had been overlooked (*Biodiversity at Risk*, Conservation International, 1990). CI's *Rain Forest Imperative* (1990) was the first strategy published by a conservation organization that actually used priority-setting approaches to define the methods and rationale for the geographic areas in which it worked.

At about the same time, two other priority-setting approaches, *major tropical wilderness areas* and *megadiversity countries*, came into being. These are two concepts, developed by R.A. Mittermeier in 1988 and 1989, that seek to complement the *hotspots* approach. The *major tropical wilderness areas* approach again emphasizes high biodiversity tropical systems, but focuses on the opposite end of the threat spectrum from the *hotspots*. Whereas the *hotspots* consist mainly of heavily exploited and often highly-fragmented ecosystems, greatly reduced in original extent (usually 1-25% of the original habitat remaining), the *major tropical wilderness areas* are still largely intact (over 75% of original vegetation cover) and have very low human population density (less than 5 inhabitants/km^2). Myers came up with a very similar concept that he called *good news areas*, but he only touched on them in the paper that first defined the threatened hotspots (Myers, 1988).

These wilderness areas have great importance as storehouses of biodiversity, as major watershed areas, and as controls against which we can measure the management of the more devastated *hotspots*. They are also the last places where indigenous people are likely to have any hope of maintaining their traditional lifestyles, and are certain to assume increasing recreational, aesthetic, and spiritual value on an ever more overcrowded planet. Among the *major tropical wilderness areas* that still exist are the southern Guianas, southern Venezuela and adjacent parts of extreme northern Brazilian Amazonia, parts of the upper Amazon in Brazil, Colombia, Ecuador, Peru and Bolivia, the Congo (=Zaire) Basin in Central Africa, and much of the island of New Guinea.

In this book we focus on the third of these priority-setting approaches, *megadiversity countries*, which looks at biodiversity priorities by political units, in this case sovereign nations. This approach was first proposed in a paper that Mittermeier gave at the Smithsonian Institution's Biodiversity Conference in 1988, the influential proceedings of which were edited by Prof. E.O. Wilson of Harvard University (Mittermeier, 1988). Mittermeier (1988) dealt exclusively with primates and their tropical forest habitats and recognized that, although 92 countries have wild primate populations within their borders, just four —Brazil, Madagascar, Indonesia, and the Democratic Republic of Congo, or DRC (=Zaire)— are responsible for fully two thirds of all primate diversity. Shortly thereafter, the concept was expanded to look at biodiversity in general (Mittermeier and Werner, 1990; McNeely *et al.*, 1990; Mittermeier and Goettsch de Mittermeier, 1992). In these early versions, we first highlighted seven countries (Brazil, Indonesia, Colombia, Mexico, Madagascar, the Democratic Republic of Congo, and Peru), then 12, and finally, with this book, 17 out of a global total of more than 230. These 17 countries are shown in the Map of Megadiversity Countries at the beginning of this chapter, and are listed in Table 1. Together they account for at least two thirds and probably three fourths or more of all biodiversity including terrestrial, freshwater and marine, and their significance in global efforts to conserve biodiversity simply cannot be overstated. Taken together with the *threatened hotspots* and *major tropical wilderness areas* to be covered in later books in this series, the megadiversity country approach enables us to achieve very significant coverage of the Earth's biological resources, and to focus on countries and ecosystems where we are likely to have the greatest overall impact.

Table 4. Higher Plant Diversity and Endemism

Country	*Total diversity*	*Endemics*	*Endemics as % of global diversity of higher plants*[a]
Brazil	~50 000-56 000	~16 500-18 500?	6.6-7.4
Indonesia	~37 000?	14 800-18 500	5.9-7.4
Colombia	45 000-51 000	15 000-17 000	6.0-6.8
Mexico	18 000-30 000	10 000-15 000	4.0-6.0
Australia	15 638	14 458	5.8
Madagascar	11 000-12 000	8 800-9 600	3.5-3.8
China	27 100-30 000	~10 000	~4.0
Philippines	8 000-12 000	3 800-6 000	1.5-2.4
India	>17 000	7 025-7 875	2.8-3.2
Peru	18 000-20 000	5 356	2.1
PNG	15 000-21 000	10 500-16 000	4.2-6.4
Ecuador	17 600-21 100	4 000-5 000	1.6-2.0
USA	18 956	4 036	1.6
Venezuela	15 000-21 070	5 000-8 000	2.0-3.2
Malaysia	15 000	6 500-8 000	2.6-3.2
South Africa	23 420	16 500	6.6
DRC	11 000	3 200	1.3
Total		**155 475-183 025**[b]	

a. 250 000.
b. The seventeen megadiversity countries have between 155 475 and 183 025 species as endemics, or 62.2-73.2% of global higher plant diversity.

The megadiversity country concept is based on four premises:

- *The biodiversity of each and every nation is critically important to that nation's survival, and must be a fundamental component of any national or regional development strategy;*
- *Nonetheless, biodiversity is by no means evenly distributed on our planet, and some countries, especially in the tropics, harbor far greater concentrations of biodiversity than others;*
- *Some of the richest and most diverse nations also have ecosystems that are under the most severe threat; and*
- *To achieve maximum impact with limited resources, we must concentrate heavily (but not exclusively) on those countries richest in diversity and endemism and most severely threatened; investment in them should be roughly proportional to their overall contribution to global biodiversity.*

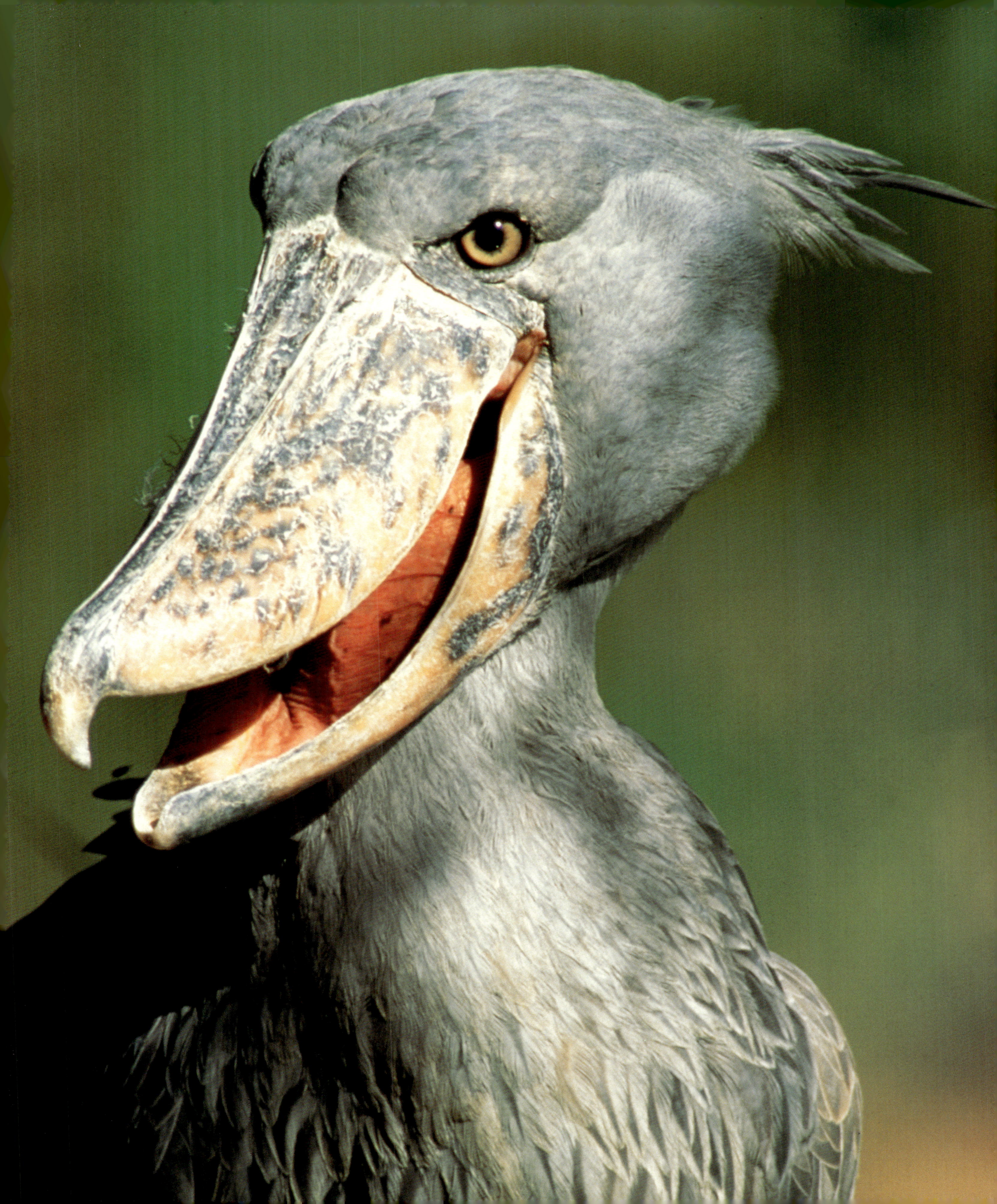

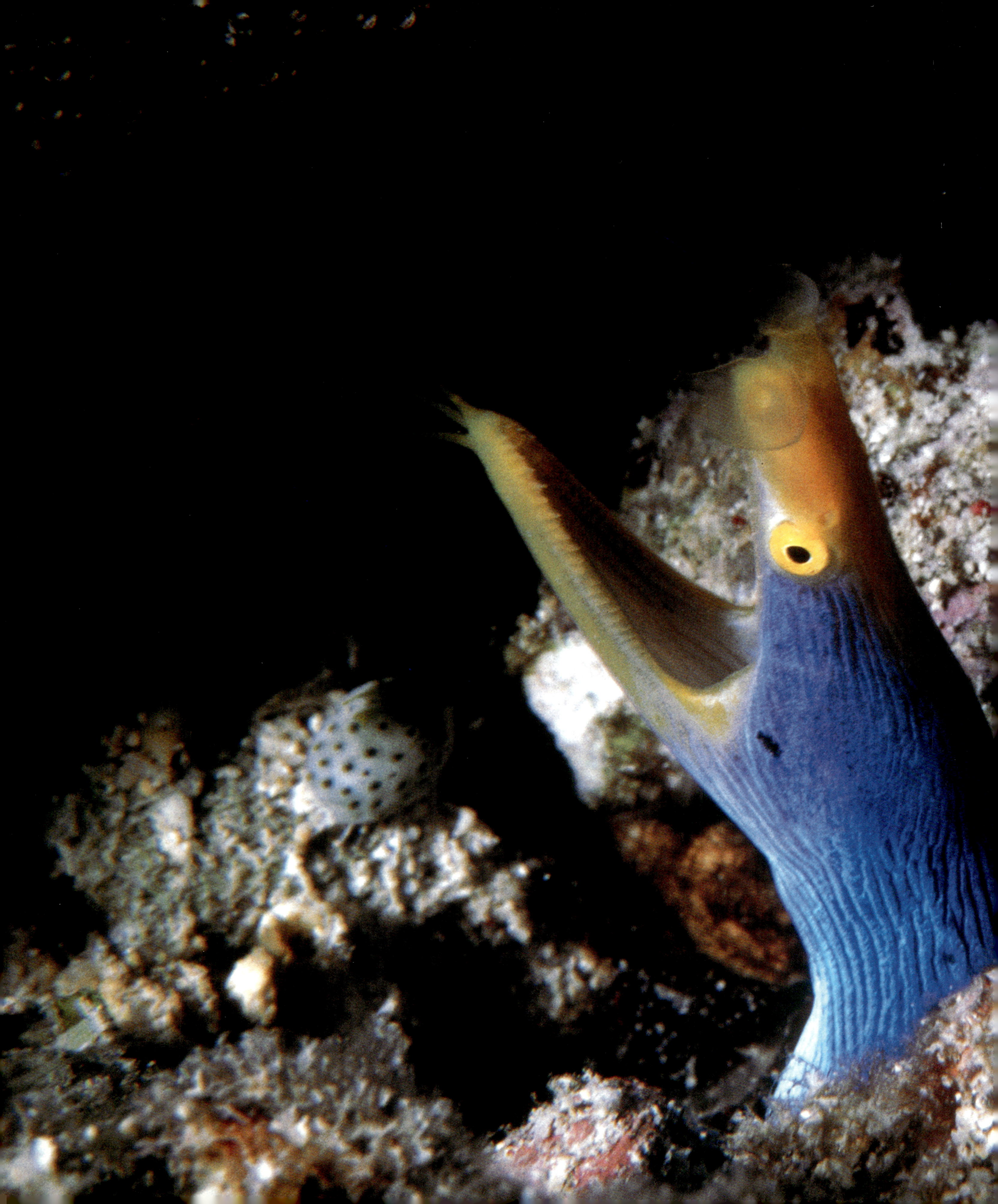

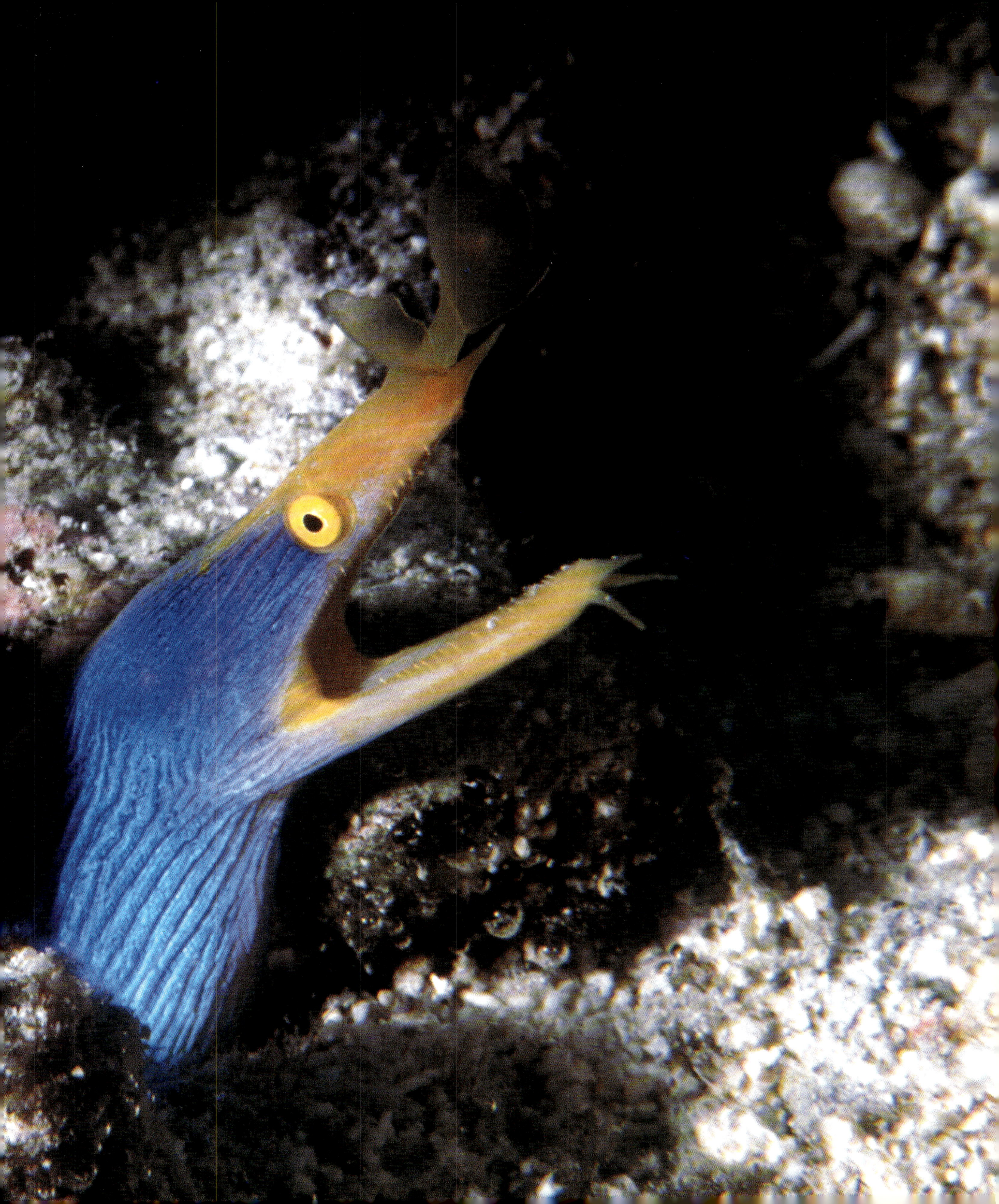

An additional premise relates to human cultural diversity and the fact that there tends to be a strong correlation between areas of high biodiversity and marked cultural diversity. That is, most of the megadiversity countries tend to be "megadiverse" in human cultural diversity as well, as measured by the number of unique languages and tribal groups contained in each.

Table 5. Vertebrate Diversity and Endemism

Country	*Mammals*	*Birds*	*Reptiles*	*Amphibians*	*Total non-fish vertebrates*	*Non-endemic: endemic*	*Endemics as % of global diversity: all non-fish vertebrates*	*Freshwater fish*[a]
Brazil	524	1 622	468	517	3 131	3.97:1		>3 000
	(131)[b]	(>191)	(172)	(294)	(788)		~3.3	
Indonesia	515	1 531	511	270	2 827	3.33:1		1 400
	(201)	(397)	(150)	(100)	(848)		3.5	
Colombia	456	1 815	520	583	3 374	5.32:1		>1 500
	(28)	(>142)	(97)	(367)	(634)		2.6	
Mexico	450	1 050	717	284	2 501	3.12:1		468
	(140)	(125)	(368)	(169)	(802)		~3.3	
Australia	282	751	755	196	1 984	1.47:1		183
	(210)	(355)	(616)	(169)	(1 350)		~5.6	
Madagascar	105	253	300	178	836	1.33:1		75
	(77)	(103)	(274)	(176)	(630)		2.6	
China	499	1 244	387	274	2 404	4.97:1		1 010
	(77)	(99)	(133)	(175)	(484)		2.0	
Philippines	201	556	193	63	1 013	2.14:1		330
	(116)	(183)	(131)	(44)	(474)		1.98	
India	350	1 258	408	206	2 222	5.65:1		750
	(44)	(52)	(187)	(110)	(393)		1.6	
Peru	344	1 703	298	241	2 586	7.56:1		855
	(46)	(109)	(98)	(>89)	(342)		1.4	
PNG	242	762	305	200	1 509	4.25:1		282
	(57)	(85)	(79)	(134)	(355)		1.5	
Ecuador	271	1 559	374	402	2 606	8.41:1		>44
	(21)	(37)	(114)	(138)	(310)		1.3	
USA	428	768	261	194	1 651	4.34:1		790
	(101)	(71)	(90)	(126)	(388)		1.6	
Venezuela	288	1 360	293	204	2 145	11.35:1		1 250
	(11)	(45)	(57)	(76)	(189)		0.8	
Malaysia	286	738	268	158	1 450	8.90:1		600
	(27)	(11)	(68)	(57)	(163)		0.7	
South Africa	247	774	299	95	1 415	9.69:1		153
	(27)	(7)	(76)	(36)	(146)		0.6	
DRC	415	1 094	268	80	1 857	13.55:1		962
	(28)	(23)	(33)	(53)	(137)		0.6	
Total					**8 433**[c]			

a. Freshwater fish data is for total species diversity only; endemism figures are not available.
b. Figures in parentheses are for endemic species.
c. The seventeen megadiversity countries have 8 433 endemic non-fish vertebrates, or 35.1% of global diversity in these groups.

Table 6. Top 12 Countries in the World for Non-Fish Vertebrate Diversity

Country	*Vertebrates*
Colombia	3 374
Brazil	3 131
Indonesia	2 827
Ecuador	2 606
Peru	2 586
Mexico	2 501
China	2 404
India	2 222
Venezuela	2 145
Australia	1 984
Bolivia[a]	1 940
DRC	1 857

a. Bolivia is not a megadiversity country, but has high enough non-fish vertebrate diversity to make the top twelve.

On pp. 20-21, two ribbon eels, Indonesia.

Given these basic premises, what then determines whether or not a country makes the megadiversity list? A number of criteria were considered, but after much discussion and analysis, it was decided that the principal criterion should be *endemism*, first at the species level and then, as an additional consideration, at higher taxonomic levels such as the genus and the family (*phyletic endemism*). Although higher-level endemism is very important, figures on it have been harder to obtain than for species-level endemism; consequently, for this analysis the species level has been used for the principal quantitative assessment.

Use of endemism as the principal criterion is in line with the IUCN's *"doctrine of ultimate responsibility,"* which holds that a nation with the only populations of a particular endangered species has *ultimate responsibility* for ensuring the survival of that particular species. The same concept applies at the ecosystem level when we are talking about habitats and vegetation types unique to a particular country. In other words, regardless of how much international interest and support there might be in saving a particular endemic species or ecosystem, the final decision and ultimate responsibility for its fate rests with the country in which that particular target organism or natural area is found.

Second, we decided that the primary criterion for inclusion on the megadiversity list had to be *plant* endemism, as most other forms of life ultimately depend on plants in some way. Furthermore, since *higher plants* are by far the most diverse and important and the best information is available on them, we have used them as representatives of the Plant Kingdom as a whole.

Nonetheless, we also felt strongly that plants could not be the only criterion, and that animals would have to be considered as well. Looking at available data that can be used for comparative purposes, we found that the best information available is for birds, and then for mammals, reptiles and amphibians. As a result, these four are the vertebrate groups that have provided the foundation for our megadiversity assessment of the Animal Kingdom. Where possible, we have also used both freshwater fish and marine fish in order to have more thorough coverage of the vertebrates. Unfortunately, data on fish is uneven and incomplete and data on endemism is particularly difficult to obtain. We were able to find sufficient comparative data on freshwater fish diversity and included the information in two tables, but we have not yet been able to do the same for marine fish. Comparative endemism data was lacking for both groups, so we were unable to fully incorporate fish endemism as a quantitative criterion for megadiversity status. This is also why the rather cumbersome term "non-fish vertebrates" is used throughout this book.

Although there is generally much less information on invertebrates, we have also included some invertebrate data for comparison. This is exclusively on two of the best studied groups of invertebrates, the butterflies (and especially the swallowtail butterflies) and the tiger beetles. Both swallowtail butterflies and tiger beetles have been analyzed globally on a country-by-country basis, with total diversity and endemism carefully calculated (Collins and Morris, 1985; Pearson and Cassola, 1992), making them ideal for this study.

Although species-level endemism is the principal quantitative criterion for megadiversity country status, a number of other criteria have been taken into consideration as well. These include the following:

• *Species Diversity*: The total number of species living within the borders of the country in question (including migratory species), with special consideration being given to countries with more than 2 000 non-fish vertebrates and more than 10 000 higher plants, even if they do not qualify on the endemism criteria (*e.g.*, the Democratic Republic of Congo).

• *Higher-Level Diversity/Phyletic Diversity*: The total number of genera and families living within the borders of the country. Special consideration was also given to countries with high generic and family diversity and, as mentioned above, especially to those with endemic families and high numbers of endemic genera (*e.g.*, Madagascar, Australia, the Philippines).

• *Ecosystem Diversity/Beta Diversity*: The number of different vegetation types, ecosystem types, major habitats, etc. within the country.

• *Presence of Marine Ecosystems*: The presence of marine ecosystems of some kind is essential to qualify for the list, since the marine realm has so many higher-level groups (*e.g.*, Phyla, Classes), especially in the Animal Kingdom, that do not occur in the terrestrial or freshwater realms (Table 2).

• *Presence of Tropical Rain Forest Ecosystems*: The presence of at least some tropical rain forest is also taken into consideration, given that these are usually the richest terrestrial ecosystems; countries with unusually large blocks of tropical rain forest (*e.g.*, the Democratic Republic of Congo) were given special attention.

Ecosystem diversity is taken into consideration because it is yet another measure of higher levels of diversity. Such *beta diversity*, as it is called, is especially noteworthy in countries like Mexico, Indonesia and China, each of which straddle two of the planet's six major biogeographic regions (the Nearctic and the Neotropical in the case of Mexico; the Indo-Malayan and the Australasian in the case of Indonesia; the Palaearctic and the Indo-Malayan in the case of China), and incorporate important transition zones between them (*e.g.*, Wallacea in Indonesia). It is also very high in countries with great topographic relief such as Peru, Ecuador, and Colombia, and those that have both great topographic relief and latitudinal diversity, such as China, India, and the United States. All of these have major mountain ranges, and terrestrial ecosystems ranging from deserts to high-altitude grasslands to rain forests, with the United States even including Arctic ecosystems and boreal forest. Countries with many islands (or entirely composed of islands) are also high in ecosystem diversity, since the interface between terrestrial and marine systems is especially rich and variable. This is clearly the case for Indonesia, the Philippines, Madagascar and Papua New Guinea.

All megadiversity countries have marine ecosystems; the list does not include any landlocked nations. This is especially important because the coastal and marine components of these countries make enormous contributions to their overall diversity, especially at the highest levels of organization (Phylum, Class) within the Animal Kingdom. Of the 33 phyla of animals, 13 are restricted to marine systems, and none are restricted entirely to freshwater systems; only one is limited to the terrestrial realm. A total of 28 phyla are found in the marine realm (including the 13 endemics), 14 are also found in freshwater, and 11 are also in the terrestrial realm. Fifteen phyla include symbiotic organisms, and four of these are found only in symbiotic situations (Table 2). Consequently, by including the presence of marine ecosystems as a criterion for megadiversity status, we guarantee that a very large and significant component of animal diversity (and certainly also marine plant and microorganism diversity) is well-represented in the megadiversity country list. Although we do not have a quantitative cutoff point for marine diversity, preferring just to use presence/absence as the determining factor, a country *must* have marine ecosystems within its borders to qualify for the megadiversity list. This is one of the reasons why certain landlocked countries with reasonably high terrestrial diversity (*e.g.*, Bolivia, Uganda) were not included in the list.

Presence of tropical rain forests of some kind is also an important consideration because of the great contribution that these ecosystems make to terrestrial and freshwater diversity. Consequently, all but one country on the megadiversity list have at least some tropical rain forest ecosystems, the one with the smallest area being the United States. Furthermore, the vast majority of remaining tropical rain forests occurs within the megadiversity countries, with Brazil, Indonesia and the Democratic Republic of Congo accounting for almost half. South Africa is the only country on the list that lacks tropical rain forests entirely, but its Mediterranean-type systems are the richest in the world, and actually exceed many tropical rain forests in overall diversity and endemism.

Use of endemism, and especially plant endemism, as the principal criterion in determining megadiversity status is consistent with the approach followed by Mittermeier, Norman Myers and Jorgen Thomsen (1997) to determine threatened biodiversity hotspot status. In the hotspots analysis, a cutoff point of *1% of total higher plant diversity endemic to the area in question* was the criterion for determining hotspot status. A total of 19 areas qualified using this criterion. When a vertebrate criterion of *1% of total non-fish vertebrate diversity (mammals, birds, reptiles and amphibians) endemic to the hot 37 37spot* was added, the list was narrowed down to 10 top-priority hotspots that had both 1% of all higher plants and 1% of all non-fish vertebrates, as endemics. An additional seven areas that met a 0.5% plant criterion were also included in a second tier of hotspots.

With the megadiversity countries, a very similar set of criteria was applied; the only difference was that for these countries the cutoff point for plants was *2% of total global diversity in higher plants endemic to a given country*. Given roughly 250 000 higher plant species worldwide (Table 3), this means that a country would have to have 2% of 250 000, or 5 000 species, endemic to it in order to qualify for megadiversity country status.

Although plant endemism was the primary determinant of megadiversity status, we also applied a set of criteria for the four non-fish vertebrate groups (mammals, birds, reptiles and amphibians), for which good comparative data on diversity and endemism by country was available. The vertebrate criterion was used to compare relative importance of the countries on the list, and to provide a broader perspective on each one's contribution to global biodiversity. However, for vertebrates, we used both a 2% criterion and a 1% criterion to determine different levels of priority among the

Table 7. Top 12 Countries in the World for Non-Fish Vertebrate Endemism

Country	*Endemic species*	*Percent endemism*[a]
Australia	1 350	5.62
Indonesia	848	3.53
Mexico	802	3.34
Brazil	788	3.28
Colombia	634	2.64
Madagascar	630	2.63
China	484	2.02
Philippines	474	1.98
India	393	1.64
USA	388	1.62
PNG	355	1.48
Peru	342	1.43

a. This figure represents the percentage of total global diversity in these four vertebrate groups that is *endemic* to the country in question.

Table 8. Top 12 Countries in the World for Total Bird Diversity

Country	*Birds*
Colombia	1 815
Peru	1 703
Brazil	1 622
Ecuador	1 559
Indonesia	1 531
Venezuela	1 360
India	1 258
Bolivia[a]	1 257
China	1 244
DRC	1 094
Kenya[a]	1 078
Mexico	1 050

a. Bolivia and Kenya are not megadiversity countries, but have high enough bird diversity to make the top twelve.

Table 9. Top 12 Countries in the World for Bird Endemism

Country	*Endemic species*	*Percent endemism*[a]
Indonesia	397	26
Australia	355	47
Brazil	>191	12
Philippines	183	33
Colombia	>142	8
Mexico	125	12
Peru	109	6
Madagascar	103	41
China	99	8
PNG	85	11
USA	71	9
India	52	4
Total	**1 912**[b]	

a. Percent endemic out of total bird species in country.
b. Total number of birds endemic to top 12 countries = 1 912.
Proportion of global bird diversity endemic to top 12 countries (1 912/9 040) = 21.2%.

TABLE 10. Top 12 Countries in the World for Total Mammal Diversity

Country	*Mammals*
Brazil	524
Indonesia	515
China	499
Colombia	456
Mexico	450
USA	428
DRC	415
India	350
Peru	344
Uganda[a]	315
Kenya[a]	309
Tanzania[a]	306

a. Uganda, Kenya and Tanzania are not megadiversity countries, but have high enough mammal diversity to make the top twelve.

TABLE 11. Top 12 Countries in the World for Mammal Endemism

Country	*Endemic species*	*Percent endemism*[a]
Australia	210	74
Indonesia	201	39
Mexico	140	31
Brazil	131	25
Philippines	116	58
USA	101	24
Madagascar	77	73
China	77	15
PNG	57	24
Argentina[b]	47	18
Peru	46	13
India	44	13
Total	**1 247**[c]	

a. Percent endemic out of total mammal species in country.
b. Argentina is not a megadiversity country, but has high enough mammal endemism to make the top twelve.
c. Total number of mammals endemic to top 12 countries: 1 247.
Proportion of global mammal diversity endemic to top 12 countries (1 247/4 629) = 26.9%.

TABLE 12. Top 12 Countries in the World for Total Reptile Diversity

Country	*Reptiles*
Australia	755
Mexico	717
Colombia	520
Indonesia	511
Brazil	468
India	408
China	387
Ecuador	374
PNG	305
Madagascar	300
South Africa	299
Peru	298

On the opposite page, mountain lion (Felis concolor), *Sonoran Desert, Mexico.*

countries that had already qualified for megadiversity status on the basis of their plant endemism. The first cutoff point for vertebrates was the same as for plants, *2% of total global diversity in non-fish vertebrates endemic to the country in question.* Given roughly 24 000 species of mammals, birds, reptiles and amphibians combined (Table 3), a country would have to have 2% of 24 000, or 480 species, endemic to it in order to qualify for the first tier of the list. Since very few countries qualified on both the 2% plant and 2% vertebrate criteria, and of those that didn't some had very high plant diversity and/or very high vertebrate diversity and met one or more of the other above-mentioned criteria as well, we also added a 1% criterion for vertebrates to better understand the relative importance of those that had already qualified on the plant criterion. To meet the 1% cutoff, a country had to have *1% of total global diversity in non-fish vertebrates endemic to it.*

Using the plant criterion, 16 countries qualified for the megadiversity list by having 5 000 or more higher plant species as endemics. Actually, three of these, the United States, Ecuador, and the Philippines, may not exceed 5 000 endemic plants, but they came close enough to warrant their inclusion on this list. This is especially the case for the two tropical countries, where estimates of plant diversity and endemism are imprecise at best, the range for the Philippines being 3 800 to 6 000 species and that for Ecuador 4 000 to 5 000 (Table 4). Brazil, Indonesia and Colombia each have between 5.9 and 7.4% of higher plant species as endemics, but these are all very rough estimates; surprisingly, South Africa is at a well-documented 6.6% and Australia is at 5.8%, also a well-documented figure; Mexico and Papua New Guinea are in the 4 to 6.4% range; China is at about 4%; Madagascar is between 3.5 and 3.8%; and India, Peru, Venezuela, and Malaysia are between 2 and 3.2%; the Philippines is somewhere between 1.5 and 2.4%, depending on which estimate one believes; Ecuador is at 1.6-2.0%, and the United States at 1.6%.

When the same 2% criterion is applied for non-fish vertebrates (mammals, birds, reptiles and amphibians), only eight countries qualify by having 480 endemics in these four groups combined. These are: Brazil, Indonesia, Colombia, Mexico, Australia, Madagascar, China and the Philippines (Table 5). Brazil, Indonesia, and Mexico actually have more than 3% of total non-fish vertebrate species as endemics; Colombia and Madagascar have between 2 and 3% each; and China has 2%. Amazingly, Australia has a staggering 5.6% of total non-fish vertebrate diversity endemic to it. The Philippines is at 1.98% with its 474 endemics (only six species away from the 2% cutoff), and was considered close enough to be included with the other seven, especially since so many new vertebrate species are still being discovered there. An additional five countries, India, Peru, Papua New Guinea, Ecuador, and the United States, made the 1% cutoff and have between 1.3% and 1.6% of non-fish vertebrates as endemics; finally, three more, Venezuela, Malaysia, and South Africa, all of which have high plant diversity and endemism, are below 1% for vertebrate endemism, in the 0.6-0.8% range.

The eight countries that qualify on the 2% criteria for both higher plants and non-fish vertebrates could be considered the "top eight"; the next five that qualify on the 2% criterion for plants and the 1% criterion for non-fish vertebrates could be considered a "second tier"; and the three that qualify on the 2% plant criterion but do not achieve the 1% cutoff for non-fish vertebrates could be considered part of a "third tier" or "honorable mention" category. However, when fish data are added, the line between the first eight and the next eight in the second and third tiers becomes somewhat blurred. Although data on *fish endemism* is simply too incomplete to be used as a quantitative criterion, there is enough information on *freshwater fish diversity* to provide some insight on the relationships between the first eight countries and those in the second and third tiers, thanks mainly to work by McAllister *et al.* (1997) and Kottelat and Whitten (1996) (Table 16). When this information is examined, it turns out that some second and third tier countries like Venezuela, Peru, the United States, and India have much more diverse freshwater fish faunas than "top eight" countries like Madagascar, Australia, the Philippines and Mexico. Although the fish faunas of Madagascar and Australia in particular are very rich in *endemics*, including endemic genera and some very unusual families, the total number of endemics in the higher *diversity* countries like Peru, India, Venezuela and the United States is likely to surpass that of the four relatively species-poor countries in the "top eight." Furthermore, what little we know of marine fish diversity indicates that second tier countries like Papua New Guinea, the United States and India are likely to rank high in this category as well, further blurring the distinction between the first eight, the second five, and the last three.

For these reasons, we have decided to recognize the top 16 countries as a single group, with the understanding that one can rank them in different ways depending on which groups of organisms are given the greatest importance (indeed, one such attempt at ranking, using what is known of both diversity and endemism for the five vertebrate groups, two invertebrate groups and higher plants, is presented in Table 23). In addition, we have decided to include the Democratic Republic of Congo as one of the megadiversity countries. It only has 1.3% of higher plants as endemics and only 0.6% of non-fish vertebrates. However, it has high non-fish vertebrate diversity, its 1 857 species being good enough for twelfth on the world list (and likely a gross underestimate, since the figure for amphibian diversity, 80 species, is almost certainly too low); it ranks sixth globally in freshwater fish diversity with 962 species, many of them endemic; and it has the second largest extent of tropical rain forest on Earth, being exceeded in this category only by Brazil. It is also the single richest country in animal diversity and endemism on the African mainland, the sole contender being Tan-

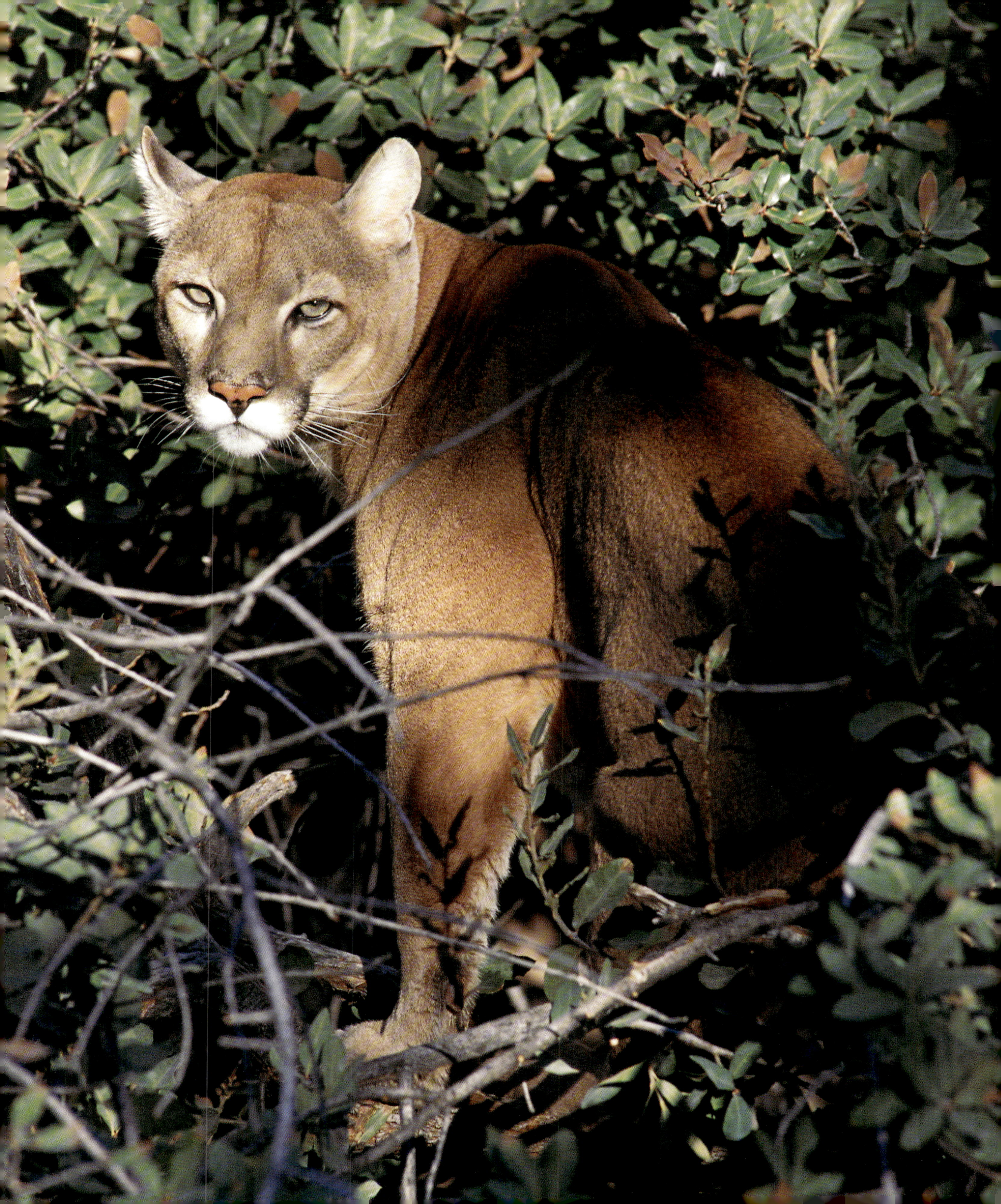

Table 13. Top 12 Countries in the World for Reptile Endemism

Country	*Endemic species*	*Percent endemism*[a]
Australia	616	82
Mexico	368	51
Madagascar	274	91
India	187	46
Brazil	172	37
Indonesia	150	29
China	133	34
Philippines	131	68
Ecuador	114	30
Peru	98	33
Colombia	97	19
USA	90	34
Total	**2 430**[b]	

a. Percent endemic out of total reptile species in country.
b. Total number of reptiles endemic to top 12 countries = 2 430.
Proportion of global reptile diversity endemic to top 12 countries (2 430/6 458) = 37.6%.

Table 14. Top 12 Countries in the World for Total Amphibian Diversity

Country	*Amphibians*
Colombia	583
Brazil	517
Ecuador	402
Mexico	284
China	274
Indonesia	270
Peru	241
India	206
Venezuela	204
PNG	200
Australia	196
USA	194

Table 15. Top 12 Countries in the World for Amphibian Endemism

Country	*Endemic species*	*Percent endemism*[a]
Colombia	367	63
Brazil	294	57
Madagascar	176	99
China	175	64
Australia	169	86
Mexico	169	60
Ecuador	138	34
PNG	134	67
USA	126	65
India	110	53
Indonesia	100	37
Peru	>89	37
Total	**2 047**[b]	

a. Percent endemic out of total amphibian diversity in country.
b. Total number of amphibians endemic to top 12 countries = 2 047.
Proportion of global amphibian diversity endemic to top 12 countries (2 047/4 222) = 48.5%.

On the opposite page, elk bull or wapiti (Cervus elaphus), *Yellowstone National Park, U.S.A.*

zania, which has only about one third as many endemic higher plants as the DRC. For these reasons, we felt that the DRC could not be left off the megadiversity list, and have included it as the seventeenth country.

We have also given the megadiversity countries the acronym "B-17," the "Biodiversity-17," to compare them to the G-7 economic powers. We feel that this comparison is quite appropriate, since the B-17 harbor an even larger share of the world's biological wealth than the G-7 have of its economic might. Perhaps in the future it will be possible to have periodic B-17 summits to help determine the future of the planet's biodiversity in much the same way that the G-7 now meet on a regular basis to discuss issues of global economic concern.

Several other countries were examined and considered for megadiversity status, but did not make the final list. These included Russia, France, Bolivia, Cameroon, Tanzania, Kenya, Uganda, Myanmar (=Burma), Thailand, and Vietnam. Russia has high plant diversity, although it is difficult to determine precise figures for the country of Russia since most of the numbers available refer to the former Soviet Union. The Asian portion of the Soviet Union had approximately 17 500 plant species, of which about 2 500 were endemic. If the small European portion of Russia is added, it is unlikely that the country would make the cutoff point for plant endemics; vertebrate diversity and endemism are also fairly low. France has relatively high plant endemism because its overseas territory, New Caledonia, by itself accounts for 2 551 endemics; otherwise France is quite low in plant and animal endemism, even when other overseas departments like French Guiana, Martinique and Guadeloupe are taken into account. Even the high plant figures for New Caledonia were not sufficient to place France on this list. Bolivia has a large area of tropical rain forest; very important tropical Andean ecosystems; higher non-fish vertebrate diversity (1 940 species) than several countries on the megadiversity list, particularly high bird and butterfly diversity; and high plant diversity and endemism (17 350 species including 4 000 endemics). However, it has low non-fish vertebrate endemism (only 47 species) and, most importantly, no marine ecosystems whatsoever. Tanzania, Cameroon, Kenya and Uganda have high plant diversity by African standards (10 000, 8 260, 6 506 and 5 406 species respectively), but plant endemism is low by global standards (1 122 plant endemics for Tanzania, 156 for Cameroon, 265 for Kenya and just 30 for Uganda). Vertebrate diversity was fairly high for all three, with 1 688 non-fish vertebrates for Tanzania, more than 1 145 for Cameroon, 1 651 for Kenya, and 1 467 for Uganda, but vertebrate endemism was lower than for the DRC or South Africa, with 113 endemics for Tanzania, 105 for Cameroon, 42 for Kenya, and only 9 for Uganda. Uganda was also excluded because, like Bolivia, it has no marine ecosystems. Finally, the three Asian countries that received consideration, Myanmar, Thailand and Vietnam, all had reasonably high plant diversity, with 14 000, 12 000 and 8 000 species respectively, but plant endemism appears to be low, with only 1 071 endemics for Myanmar, approximately 800 for Vietnam, and an uncertain number for Thailand. Non-fish vertebrate diversity was again reasonably high, with 1 445 species for Burma, 1 377 for Thailand, and 1 230 for Vietnam, but endemism was low for all three: only 50 species for Myanmar, 59 for Thailand and 43 or more for Vietnam (WCMC, 1992; Caldecott *et al.*, 1996). These countries were all given serious consideration for inclusion on the megadiversity list, and several actually surpassed some of the megadiversity countries in total non-fish vertebrate diversity (*e.g.*, Bolivia, Tanzania, and Kenya came in at eleventh, thirteenth, and fourteenth globally in this category, surpassing Papua New Guinea, Malaysia, South Africa, the Philippines and Madagascar). However, in the all-important category of endemism, the principal determinant of megadiversity status, *none* of the above surpasses any of the 17 megadiversity countries in either non-fish vertebrate endemism or plant endemism (except Bolivia, which surpasses the DRC in the latter category). Consequently, although there is no doubt that all of these countries are important in global efforts to conserve biodiversity, none of them came close enough to meeting our criteria to justify their inclusion on the megadiversity list.

Looking at the 17 megadiversity countries, we find that they have within them at least 8 433 species, or 35.1% of total non-fish vertebrates, and between 62% and 73% of higher plants as endemics (Tables 4-5). Given these very high levels of endemism, we conservatively estimate that at least two thirds of all non-fish vertebrates and more than three quarters of all higher plants are found within the borders of the megadiversity countries. Although we have not included figures for marine and freshwater endemism in this analysis, the presence of countries like Brazil, Colombia, Peru, Indonesia, the DRC, and China for freshwater systems and Australia, Indonesia, Papua New Guinea, the Philippines, Colombia, and Mexico for marine systems virtually ensures that the representation of non-terrestrial species and ecosystems is at least as high as that of terrestrial organisms, and probably over three quarters as well.

If we look at Tables 4-7, we can see how these megadiversity countries compare to one another and how important they are on an individual basis. In terms of total non-fish vertebrate diversity, Colombia ranks first with 3 374 species, followed by Brazil with 3 131, Indonesia with 2 827, Ecuador with 2 606, Peru with 2 586 and Mexico with 2 501. If non-fish vertebrate endemism is examined, the order changes somewhat, with Australia far ahead of all the others at 1 350 species, 1.59 times higher than Indonesia, which is second at 848. Mexico follows in third place with 802, Brazil with 788, Colombia with 634 and Madagascar with 630. When data on freshwater and marine fish endemism is finally added, the relative positions of these countries will change once again, but they are all likely to remain at or near the top of the global list in each category.

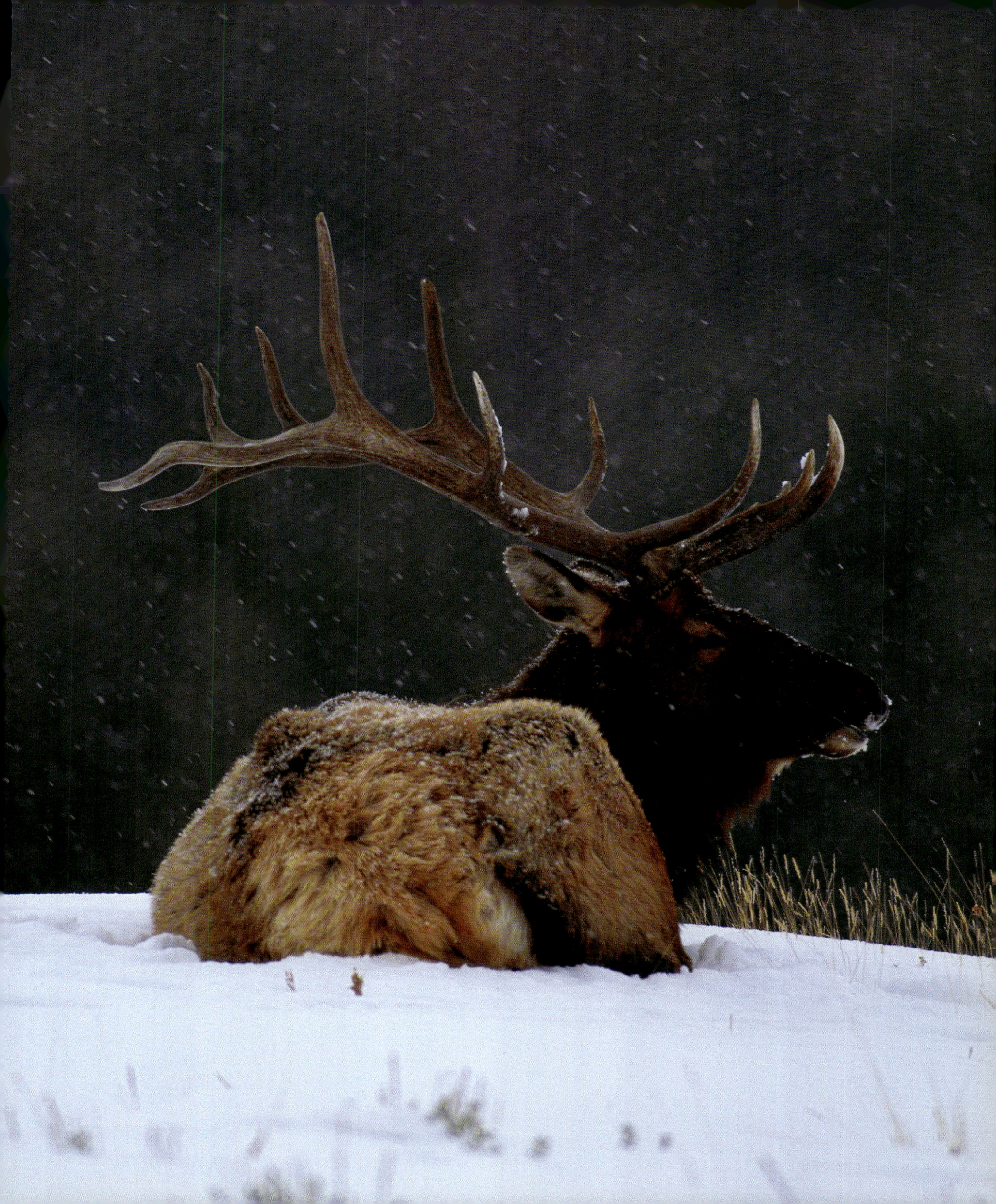

TABLE 16. Top 12 Countries in the World for Freshwater Fish Diversity

Country	*Freshwater fish*[a]
Brazil	>3 000
Colombia	>1 500[b]
Indonesia	1 400
Venezuela	1 250
China	1 010
DRC	962
Peru	855
Tanzania[c]	800
USA	790
India	750
Thailand[c]	690
Malaysia	600

a. Source: McAllister *et al.*, 1997; Kottelat and Whitten, 1996; B. Chernoff, pers. comm.
b. Estimates for Colombia run as high as 3 000 (J. Hernández-Camacho, pers. comm.)
c. Tanzania and Thailand are not megadiversity countries, but have high enough fish diversity to make the top twelve.

TABLE 17. Top 12 Countries in the World for Butterfly Diversity

Country	*Butterflies*[a]
Peru	3 532
Brazil	3 132
Colombia	3 100
Bolivia	3 000
Venezuela	2 316
Mexico	2 237
Ecuador	2 200
Indonesia	1 900
DRC	1 650
Cameroon[b]	1 550
Tanzania[b]	1 400
China	1 200

a. Source: N. Myers, pers. comm.
b. Cameroon and Tanzania are not megadiversity countries, but have high enough butterfly diversity to make the top twelve.

TABLE 18. Top 12 Countries in the World for Butterfly Endemism

Country	*Endemic species*[a]
Indonesia	700
Philippines	352
Peru	350
Colombia	300
Madagascar	250
Brazil	200
Mexico	200
Ecuador	200
China	200
Bolivia[b]	200
PNG	200
Australia	174
Total	**3 326**

a. Source: N. Myers, pers. comm.
b. Bolivia is not a megadiversity country, but has high enough butterfly endemism to make the top twelve.

TABLE 19. Top 12 Countries in the World for Swallowtail Butterfly Diversity

Country	*Swallowtail butterflies*[a]
Indonesia	121
China	104
India	77
Brazil	74
Ecuador	69
Myanmar[b]	68
Colombia	59
Peru	58-59
Malaysia	54-56
Mexico	52
Philippines	49
DRC	48

a. Source: Collins and Morris, 1985.
b. Myanmar is not a megadiversity country, but has high enough swallowtail butterfly diversity to make the top twelve.

In plant diversity, Brazil is first, if the somewhat controversial figure of 50 000-56 000 species is used, and Colombia is a close second at 45 000-51 000. Indonesia is third at roughly 37 000 (K. Kartinawata, pers. comm.). In plant endemism, Brazil, Colombia and Indonesia are all in roughly the same 15 000-18 500 species range (although it must be emphasized that the plant endemism figures for these countries are very imprecise). Yet South Africa and Australia are also in the same range with a reasonably well-documented 16 500 and 14 458 endemics respectively. Papua New Guinea with 10 500-16 000, Mexico with 10 000-15 000, and China with about 10 000 endemic plants, each have approximately 4% of global plant diversity and endemism (Table 4).

Looking at specific vertebrate and invertebrate groups (Tables 8-22), one finds that it is almost always the same handful of countries in the top five and the same in the top 12, regardless of which group of organisms one examines. Some of the top countries are richer in overall diversity; others are richer in endemics. Some have very high numbers for both. In birds and amphibians, the megadiversity countries of the Neotropics dominate. Colombia is first in both of these groups; Peru is second in birds, followed by Brazil, Ecuador and Indonesia, whereas Brazil is second in amphibians, with Ecuador a close third, and Mexico and China a good distance behind (Tables 8, 14). Brazil leads in mammals, followed by Indonesia, China, Colombia and Mexico (Table 10). Australia is the global leader in reptiles, followed closely by Mexico and then Colombia, Indonesia and Brazil (Table 12). Brazil also leads in freshwater fish, with Colombia a likely second, followed by Indonesia, Venezuela, and China (Table 16). Australia leads in marine fish because it includes a large tropical ichthyofauna in the north and a temperate ichthyofauna in the south. Mexico, with 3 000-3 500 is also quite high in marine fish, as are Brazil, Indonesia and Colombia, although precise data is not available. Peru leads in butterflies, followed by Brazil, Colombia, Bolivia and Venezuela (Table 17). As for tiger beetles, Indonesia is the leader, with India, Brazil and Madagascar not far behind, and DRC in fifth place (Table 21).

When endemism is considered, most of the same countries reappear in the top five, but two of the smallest countries, Madagascar and the Philippines, also rank high (Tables 9, 11, 13, 15, 18, 20, 22). Indonesia is first in endemic birds, with Australia a close second, and Brazil, the Philippines and Colombia further behind (Table 9). In endemic mammals, Australia is first with Indonesia a close second, and Mexico, Brazil and the Philippines in the third, fourth, and fifth positions (Table 11). In reptile endemism, as in overall reptile diversity, Australia and Mexico are again first and second (although Australia is far ahead with an incredible 616 endemics or 82% endemism); Madagascar, India and Brazil occupy the next three positions (Table 13). In amphibian endemism, Colombia and Brazil are first and second, followed by Madagascar, China and Australia (Table 15). In butterfly endemism, Indonesia is first, followed by the Philippines, Peru, Colombia and Madagascar (Table 18), and in tiger beetles, it is Madagascar at the top of the endemics list, followed by Indonesia, Brazil, India and the Philippines (Table 22).

Table 23 gives the results of a simple ranking for the top five positions in each of the groups of organisms for which data is available (birds, mammals, reptiles, amphibians, freshwater fish (diversity only), butterflies, tiger beetles, and higher plants). If a country ranks first in the world in one of these groups it is given a 5, if it ranks second it is given a 4, third gets a 3, fourth gets a 2, and fifth a 1. The top five are ranked for both diversity and endemism, except for freshwater fish, where data on endemism is lacking. Using this system, we find that Brazil comes out on top in animal and plant diversity, with Colombia a close second, Indonesia third, and Peru and Mexico a good distance behind. When endemism is considered, Indonesia comes out first, with Brazil second, Australia third, Madagascar fourth and Colombia fifth. Combining diversity and endemism, Brazil comes out first with 48 points, Indonesia second with 40 and Colombia third with 36. Australia follows behind with 21, Mexico with 15, and Madagascar with 14. Needless to say, this is a very rough way of ranking countries, but it does serve to reinforce the point that it is always the same countries that top the global lists.

The question arises as to whether one can use this ranking system and other qualitative data to determine which country is *number one*, the single richest country on Earth in biodiversity. The answer is *more or less*, with the key contenders being Brazil and Indonesia, followed by Colombia, Australia and Mexico. Starting with the plant data, Brazil appears to have the highest plant diversity, with Colombia second and Indonesia third, and all three have comparable levels of endemism. However, there is great variation in the estimates for both plant diversity and endemism for all three countries —the range for plant diversity for Brazil, for example, goes from 34 000 (P. Raven, pers. comm.) to 56 000 (WWF/IUCN, 1997)—, so the order could change as more information becomes available. Nonetheless, it is likely that all three of these countries will remain at or near the top of the global plant list. Australia and South Africa also have very high plant endemism, and could possibly exceed Brazil, Colombia and Indonesia, at least in endemism; the high ends of the estimates for Mexico and Papua New Guinea's plant endemism are in the same range, but the lower end for both is some 4 000-5 000 species lower than the other five.

In terms of non-fish vertebrate diversity, Brazil, Indonesia, Colombia and Australia all post impressive numbers. Brazil has a total of 3 131 non-fish vertebrates or 304 more than Indonesia, whereas Indonesia has 848 *endemic* non-fish vertebrates, or 60 more than Brazil. Colombia actually has the highest total for non-fish vertebrates (3 374), but only 634 endemic species, roughly the same as Madagascar. Australia, surprisingly, has far more endemic species than any other country, with 1 350, but its total non-fish vertebrate diversity figure is only 1 984. Mexico is third in endemics with 802 species, but sixth in diversity with 2 501 species. If fish are added to the vertebrate equation, some new wrinkles emerge. Brazil is by far the richest country in terms of freshwater fish diversity, with at least 3 000 species, more than double that of Colombia or Indonesia. Colombia is second with perhaps 1 500, although this is very rough estimate, and Indonesia is third with a somewhat sounder estimate of 1 400. Australia has very low freshwater fish diversity with only 183 species, but its marine fish diversity probably ranges between 4 000 to 4 500, likely the highest on Earth. Indonesia may equal or even exceed Australia's marine fish diversity, but good estimates are not available. Brazil's is much lower, and probably only some 2 500 species.

The figures for terrestrial invertebrate diversity are even less precise than for fish and plants. It is likely that Brazil, being the largest tropical country and having the greatest expanse of tropical rain forest, will emerge at the top of this list; however, in the two groups for which information is available, butterflies and tiger beetles, the data favor Indonesia. Indonesia is first in tiger beetle diversity and second to Madagascar in endemism, whereas Brazil is third in both categories. For butterflies, Peru is first and Brazil second in diversity, but Indonesia is first in endemism.

In terms of diversity and endemism per unit area, the much smaller Indonesia and Colombia clearly come out higher than Brazil and Australia, which are the fifth and sixth largest countries on Earth. But, as discussed below, species/area relationships are not a primary determinant in the megadiversity approach.

Using the ranking system discussed above, which includes data on diversity *and* endemism for birds, mammals, reptiles, amphibians, butterflies, tiger beetles, and higher plants, and data on diversity *only* for freshwater fish, Brazil comes out narrowly ahead of Indonesia, with Colombia third and Australia and Mexico way behind in fourth and fifth positions. However, this system is based on species data, and does not take into account higher levels of animal organization, such as phylum, class, order, family and genus. Indonesia (with Irian Jaya) spans two of the world's major biogeographic regions, the Indo-Malayan and the Australasian, and includes a broad transition zone, Wallacea, with its own unique faunal complement; this means that it has a very unusual mix, not only of species but of orders and families as well, surpassing even Brazil and Colombia.

Australia has a large number of endemic genera and families because of its long isolation; however, for the same reason, it is entirely lacking in many major vertebrate groups (*e.g.*, primates, carnivores, ungulates) that are major components of the fauna in many other megadiversity countries. It also owes a portion of its high vertebrate endemism to very large numbers of species in a small number of groups (*e.g.*, marsupials, skinks, geckos), making for a less balanced fauna than in a country like Indonesia. Colombia has high diversity in most groups of organisms, but its endemism is lower than that of Brazil, Indonesia, Australia and Mexico in vertebrates, and comparable to Brazil and Indonesia in plants. Mexico is strong in many categories, but places in the top three only for total non-fish vertebrate endemics, mammal endemics, and reptile diversity and endemism. Otherwise, it is usually a little behind the other two contenders.

All things taken into account, Brazil and Indonesia must be considered the top two countries for global biodiversity, although which one can actually be considered the global leader remains uncertain. Since much of the biodiversity of these two countries remains undescribed and poorly understood, the final decision as to which is number one will have to wait for the future —and since they are so close, it will always depend to some extent on how one chooses to weight the value of one group of organisms versus another. In any case, it is clear that these two large, dynamic, rapidly-emerging economic powers occupy a very special position even in the context of the wonderfully rich megadiversity countries. Provided that their conservation programs and their commitment to this issue develop as quickly as their economies, they are likely to remain the global biodiversity leaders for a long time to come.

Colombia is in a solid third position, since it leads the world in two major vertebrate groups (birds and amphibians), is a likely second or third in higher plant diversity, has medium to high endemism in some terrestrial groups, is second in freshwater fish, and is likely to rank high in marine organisms as well, since it has coasts in the two major oceans, the Atlantic and the Pacific. Diversity and endemism per unit area are quite high as well, and compare favorably with the very high diversity and endemism per unit area in the smallest megadiversity countries. Australia and Mexico vie for fourth position. Mexico generally has higher vertebrate diversity, except for reptiles, and a more balanced fauna that mixes North and South American elements; Australia is higher in non-fish vertebrate endemism, and its overall endemic vertebrate total is much higher. Both are relatively low in freshwater fish diversity, and high in marine fish diversity. Mexico has more than double the number of freshwater fish, while Australia has more marine fish. Mexico has higher plant diversity than Australia, whereas Australia probably has more endemics (although the high end of the estimate for Mexico slightly exceeds the well-documented total for Australia). Basically, it is a toss-up for fourth position, but there is no doubt that these two countries join Brazil, Indonesia and Colombia in the top five.

In any case, what really matters is not the contest to determine which country is first on the megadiversity list, but rather which of these and the other megadiversity countries will develop the most effective programs to maintain the biodiversity that they are so fortunate to possess.

Priority-setting exercises by several other institutions highlight still further the importance of the megadiver-

Table 20. Top 12 Countries in the World for Swallowtail Butterfly Endemism

Country	*Endemic species*	*Percent endemism*[a]
Indonesia	53	44
Philippines	21	43
China	15	14
Brazil	11	15
Madagascar	10	77
India	6	8
Mexico	5	10
Taiwan[b]	5	16
Malaysia	4	7
PNG	4	11
USA	4	11
Cuba[b]	4	4
Total	**142**[c]	

a. Percent endemic out of total swallowtail butterfly species in country. Source: Collins and Morris, 1985.
b. Taiwan and Cuba are not megadiversity countries, but have high enough swallowtail butterfly endemism to make the top twelve.
c. Total number of swallowtail butterflies endemic to top 12 countries = 142.
Proportion of global endemism in megadiversity countries (10) in top 12 (133/573) = 23.2%.

Table 21. Top 12 Countries in the World for Tiger Beetle Diversity

Country	*Tiger beetles*[a]
Indonesia	217
India	193
Brazil	184
Madagascar	176
DRC	134
Mexico	116
USA	111
PNG	111
Thailand[b]	102
Malaysia	95
Philippines	94
South Africa	94

a. Source: Pearson and Cassola, 1992.
b. Thailand is not a megadiversity country, but has high enough tiger beetle diversity to make the top twelve.

Table 22. Top 12 Countries in the World for Tiger Beetle Endemism

Country	*Endemic species*	*Percent endemism*[a]
Madagascar	174	99
Indonesia	103	47
Brazil	97	53
India	82	42
Philippines	74	79
Australia	72	89
Mexico	57	49
USA	45	41
PNG	45	41
South Africa	40	43
Sri Lanka[b]	32	58
DRC	27	20
Total	**848**[c]	

a. Percent endemic out of total tiger beetle species in country. Source: Pearson and Cassola, 1992.
b. Sri Lanka is not a megadiversity country, but has high enough tiger beetle endemism to make the top twelve.
c. Total number of tiger beetles endemic to top 12 countries = 848.
Proportion of global endemism in megadiversity countries (11) in top 12 (816/2 028) = 40.2%.

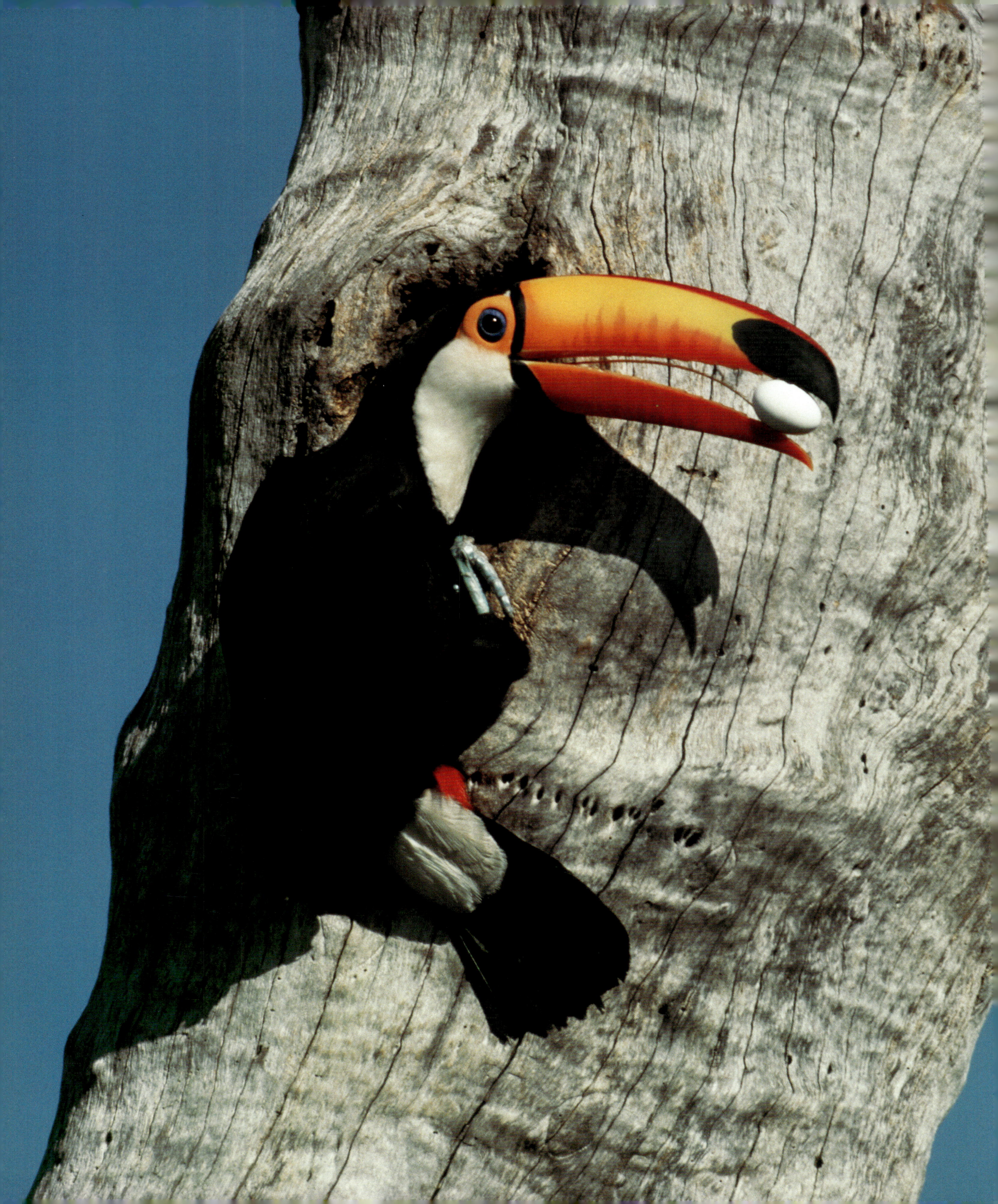

Table 23a. Ranking of Megadiversity Countries: Diversity

Country	*Birds*	*Mammals*	*Reptiles*	*Amphibians*	*Freshwater fish*	*Butterflies*	*Tiger beetles*	*Higher plants*	*Total*
Brazil	3	5	1	4	5	4	3	5	**30**
Colombia	5	2	3	5	4	3	—	4	**26**
Indonesia	1	4	2	—	3	—	5	3	**18**
Peru	4	—	—	—	—	5	—	—	**9**
Mexico	—	1	4	2	—	—	—	1	**8**
China	—	3	—	1	1	—	—	2	**7**
Australia	—	—	5	—	—	—	—	—	**5**
Ecuador	2	—	—	3	—	—	—	—	**5**
India	—	—	—	—	—	—	4	—	**4**
Venezuela	—	—	—	—	2	1	—	—	**3**
Bolivia	—	—	—	—	—	2	—	—	**2**
Madagascar	—	—	—	—	—	—	2	—	**2**
DRC	—	—	—	—	—	—	1	—	**1**
Philippines	—	—	—	—	—	—	—	—	**0**
South Africa	—	—	—	—	—	—	—	—	**0**

Table 23b. Ranking of Megadiversity Countries: Endemism

Country	*Birds*	*Mammals*	*Reptiles*	*Amphibians*	*Butterflies*	*Tiger beetles*	*Higher plants*	*Total*
Indonesia	5	4	—	—	5	4	4	**22**
Brazil	3	2	1	4	—	3	5	**18**
Australia	4	5	5	1	—	—	1	**16**
Madagascar	—	—	3	3	1	5	—	**12**
Colombia	1	—	—	5	2	—	2	**10**
Philippines	2	1	—	—	4	1	—	**8**
Mexico	—	3	4	—	—	—	—	**7**
India	—	—	2	—	—	2	—	**4**
Peru	—	—	—	—	3	—	—	**3**
South Africa	—	—	—	—	—	—	3	**3**
China	—	—	—	2	—	—	—	**2**
Bolivia	—	—	—	—	—	—	—	**0**
DRC	—	—	—	—	—	—	—	**0**
Ecuador	—	—	—	—	—	—	—	**0**
Venezuela	—	—	—	—	—	—	—	**0**

Table 23c. Top 12 Countries in the World in Combined Rankings: Diversity and Endemism[a]

Country	*Diversity*	*Endemism*	*Total*
Brazil	30	18	**48**
Indonesia	18	22	**40**
Colombia	26	10	**36**
Australia	5	16	**21**
Mexico	8	7	**15**
Madagascar	2	12	**14**
Peru	9	3	**12**
China	7	2	**9**
Philippines	0	8	**8**
India	4	4	**8**
Ecuador	5	0	**5**
Venezuela	3	0	**3**

a. Based on the top five countries for diversity in five vertebrate groups (birds, mammals, reptiles, amphibians, and freshwater fish), two invertebrate groups (butterflies and tiger beetles), and higher plants, and the top five countries for endemism in four vertebrate groups (birds, mammals, reptiles, and amphibians), two invertebrate groups (butterflies and tiger beetles), and higher plants. First place is given 5 points in each category of diversity and endemism, second place 4 points, third place 3 points, fourth place 2 points, and fifth place 1 point.

sity countries. For instance, in 1992 BirdLife International (formerly the International Council for Bird Preservation, ICBP) conducted a study of Endemic Bird Areas, those areas that are critical for the 2 609 species of birds that have had in historical times a global range of less than 50 000 km^2. This study, in effect, looked at the "hotspots" for bird diversity around the world. An examination of their publication (ICBP, 1992), shows that the largest number of EBA's is found in Indonesia (24), followed by Brazil (19), Peru (18), Colombia (14), and Mexico (14). Numbers for other EBA's are in Table 24. Of the 221 EBA's recognized, 177 or about 80% (slightly less if overlap is taken into consideration) are found in the megadiversity countries.

Another study by WWF/IUCN (1994, 1995, 1996) focused on Centers of Plant Diversity and Endemism. Its purpose was to identify those areas around the world which, if conserved, would safeguard the greatest number of plant species. The two other principal objectives were to document the many benefits, economic and scientific, that conservation of those areas would provide to society and to outline a strategy for their conservation. Criteria for selection of these centers was similar to that of our analysis, in that emphasis was placed on *species richness* and especially *species endemism*. Other characteristics considered were diversity of habitat types, presence of gene pools of plants of actual or potential value to humans, sites with species adapted to special edaphic conditions, and sites under imminent threat of large-scale devastation. Of the centers identified, roughly 269 (or slightly less if some minimal overlap between neighboring countries is taken into account) were partially or entirely in the 17 megadiversity countries; China had the largest number with 41, followed by Papua New Guinea with 39, the United States and Indonesia with 36, and Malaysia with 29 (Table 25). Worldwide, the study selected 234 of the most important centers for Data Sheet Treatment; 125 of these, or 53.4%, fell within the borders of the megadiversity countries.

If we look at diversity at greatest risk, the percentage of the most endangered diversity represented in the megadiversity list is even higher than overall diversity. Using the most up-to-date analysis, the *1996 IUCN Red List of Threatened Animals*, as a basis for analysis, we find that the megadiversity countries account for the greatest percentage of species in the *critically endangered, endangered* and *vulnerable* categories. Of the 1 107 birds considered critically endangered, endangered or vulnerable, at least 833 (75.2%) are found in the megadiversity countries (Table 26). For mammals the figure is even higher. Fifteen megadiversity countries are in the top-twenty list of countries with the most threatened mammals. These 15 account for 852 species, or 78% of the 1 096 species identified in the *1996 IUCN Red List* (Table 27). Reptiles and amphibians were only partially analyzed. However, of those assessed, at least 89.7% of the reptiles and 58.8% of the amphibians were from the megadiversity countries. It is not unreasonable to say that more than 80% and perhaps as much as 90% of the world's most endangered biodiversity occurs within the megadiversity countries. Given that not all megadiversity countries were included in the IUCN's top-twenty analysis and that threatened species data for the tropical megadiversity countries is incomplete at best, there is little doubt that the future of much of the planet's biodiversity at greatest risk depends very much on what happens in these 17 countries.

Indonesia tops the list for both mammals and birds in the three threatened categories (critically endangered, endangered, vulnerable), with China, India, Brazil and Mexico rounding out the top five in mammals; Madagascar has the highest *percentage* (44%) of its mammals endangered. As for birds, Brazil, China, and India are again in the top five with Indonesia, joined this time by the Philippines (Tables 26-27). Far

Table 24. Endemic Bird Areas (EBA) as defined by BirdLife International[a]

Country	*Total EBA's in country*	*Priority 1 EBA's*
Indonesia	24	16
Brazil	19	8
Peru	18	4
Colombia	14	4
Mexico	14	1
PNG	12	3
China	12	2
Ecuador	11	4
Philippines	9	6
USA	8	3
India	8	2
Venezuela	8	2
Australia	8	1
Madagascar	5	5
South Africa	3	0
DRC	2	2
Malaysia	2	1

a. Source: ICBP, 1992.
Total number of Endemic Bird Areas worldwide: 221.
Total in 17 megadiversity countries: 177.
Proportion in megadiversity countries: 80%.

Table 25. 12 Countries with the Highest Number of Centers of Plant Diversity and Endemism[a]

Country	*Centers of Plant Diversity and Endemism*
China	41
PNG	39
USA	37
Indonesia	36
Malaysia	29
Mexico	17
Brazil	14
Australia	14
Philippines	12
DRC	12
Colombia	10
India	10

a. Source: WWF/IUCN, 1994, 1995, 1997.

Table 26. Countries with the Most Species of Threatened (CR, EN, VU) Birds[a]

Country	*Threatened birds*
Indonesia	104
Brazil	103
China	90
Philippines	86
India	73
Colombia	64
Peru	64
Ecuador	53
USA	50
Vietnam	47
Thailand	45
Australia	45
Myanmar	44
New Zealand	44
Argentina	41
Russia	38
Mexico	36
Malaysia	34
Japan	33
PNG	31

a. Source: IUCN, 1996.
Total critically endangered (CR), endangered (EN), and vulnerable (VU) in *1996 IUCN Red List* = 1 107.
Total in megadiversity countries (13) on this list = 833.
Proportion in megadiversity countries (833/1 107) = 75.2%.

On the opposite page, Toco toucan (Ramphastos toco), *Brazil.*

TABLE 27. Countries with the Most Species of Threatened (CR, EN, VU) Mammals[a]

Country	*Threatened mammals*
Indonesia	128
China	75
India	75
Brazil	71
Mexico	64
Australia	58
PNG	57
Philippines	49
Madagascar	46
Peru	46
Kenya	43
Malaysia	42
DRC	38
Vietnam	38
Ethiopia	35
USA	35
Colombia	35
Thailand	34
South Africa	33
Tanzania	33

a. Source: IUCN, 1996.
Total critically endangered (CR), endangered (EN), and vulnerable (VU) in *1996 IUCN Red List* = 1 096.
Total in megadiversity countries (15) on this list = 852.
Proportion in megadiversity countries (852/1 096) = 77.7%.

TABLE 28. Species/Area Relationships: Megadiversity Countries with the Highest Non-Fish Vertebrate Diversity per Unit Area

Country	*Species / 1 000 km²*
Ecuador	9.2
Malaysia	4.4
Philippines	3.4
PNG	3.2
Colombia	3.0
Venezuela	2.4
Peru	2.0
Indonesia	1.5
Madagascar	1.4
Mexico	1.3
South Africa	1.2
DRC	0.8
India	0.7
Brazil	0.4
Australia	0.3
China	0.3
USA	0.2

TABLE 29. Species/Area Relationships: Megadiversity Countries with the Highest Non-Fish Vertebrate Endemism per Unit Area

Country	*Vertebrate endemic species/1 000 km²*
Philippines	1.6
Ecuador	1.1
Madagascar	1.1
PNG	0.75
Colombia	0.55
Malaysia	0.49
Indonesia	0.44
Mexico	0.41
Peru	0.27
Venezuela	0.21
Australia	0.18
India	0.12
South Africa	0.12
Brazil	0.09
DRC	0.06
China	0.05
USA	0.04

On the opposite page, mandarin duck (Aix galericulata).

less data is available for fish, and Brazil, with the greatest freshwater fish diversity on Earth, barely appears on the endangered fish list for lack of information. Once more detailed analysis of fish endangerment is available, it is certain that countries like Brazil, Indonesia, Colombia and the DRC will again account for a major portion of the most endangered species for this group of organisms. As regards invertebrates, the lack of information on the tropical megadiversity countries is even more extreme, and what holds true for fish will be even more so in this case.

Also of interest here is the issue of diversity in relation to size. One of the criticisms of the megadiversity concept is that most of the countries are large and larger countries should have more species than smaller ones. The first response to that is that the main purpose of the megadiversity country concept is not to provide a theoretical analysis of species-area relationships; rather, coming back again to the *doctrine of ultimate responsibility*, it is to demonstrate that a very small number of countries (regardless of their size) carry an inordinately large share of the planet's biodiversity. This is a *biopolitical* statement, *i.e.*, a political message to the most diverse countries, based on the best biological information available, that they have both a great global responsibility and, at the same time, an enormous competitive advantage in biological resources. As we learn more about the many values of biodiversity, this advantage translates into economic terms as well. Consequently, it is important to point out that the message of the megadiversity country concept is largely independent of size, and simply tries to highlight how special a handful of countries are and what a fundamental role they play in the efforts to conserve life on Earth. This is not intended in any way to diminish the importance of the biodiversity of other countries; it is merely meant to give added value to those countries that ultimately will be in a "make-or-break" position to determine whether or not life on Earth, as we know it, can in fact be maintained well into the next millennium.

That said, it is interesting to analyze the countries on the megadiversity list by their size, and to take a quick look at species-area relationships. It is certainly true that there are some very large countries in the megadiversity list. China, the United States, Brazil, Australia and India are in fact the third, fourth, fifth, sixth and seventh largest countries on Earth. At the same time, the world's two largest countries, Russia and Canada, do not make the list. Furthermore, there are several very small countries on the list, the smallest being Ecuador, which is about the size of the American state of Nevada, followed closely by the Philippines, Malaysia, Papua New Guinea, and Madagascar. Two of these small countries, Madagascar and the Philippines, are among the most important, so clearly it is not true that megadiversity countries are diverse just because they are large. The remaining seven countries on the megadiversity list are placed in the medium to large category, ranging in size from Venezuela at 912 050 km² to the Democratic Republic of Congo at 2 344 000 km².

To assess species diversity and endemism per unit area, we used species per 1 000 km² as a sample indicator (Tables 28-29). Looking first at the table for "diversity per unit area," it is not surprising that there is a strong correlation between the size of the country and species per unit area, with most of the smallest countries at the top of the list and most of the largest at the bottom. Colombia comes out higher in diversity per unit area than expected, and Madagascar somewhat lower. The "endemism per unit area" table is comparable. As expected, Madagascar moves up much higher on this list, and the two smallest countries, the Philippines and Ecuador, top the list. Colombia again is in fifth position, quite high given its size, and Australia moves up several places from fourteenth to tenth. There are no great surprises here; only the message that the smallest of the megadiversity countries and some of the medium-sized ones like Colombia require special attention even in the context of the megadiversity countries. The packing of diversity in each remaining patch of natural habitat in these countries is especially high, and everything that remains is of great significance. In countries like the Philippines and Madagascar, which are also under great population stress and have already lost most of their natural vegetation, the situation is especially severe.

Finally, we have the importance of human cultural diversity. Given the strong linkages that exist between human culture and biodiversity, as so clearly highlighted by Professor José Sarukhán, former Rector of the National Autonomous University of Mexico, it is not surprising that the megadiversity countries also stand out in this category. Indeed, it appears that, as with biodiversity, these countries also account for a major portion of the planet's cultural diversity, perhaps even a larger one than is the case for biodiversity. However, it is difficult to obtain precise estimates of the number of human cultures, especially in the larger tropical countries. We have tried to define cultural diversity using both the number of distinct languages and recognized tribal groups, and when we began this analysis, we expected to find a strong correlation between the two. This was true in most of the countries; however, for several others, there was a considerable difference between the number of languages identified and the number of tribal groups recognized. In some cases, there were far more languages recognized; in others, there were far more tribes than languages. As a result, our overview of human cultural diversity in the megadiversity countries is still rough, especially for the larger countries; nonetheless, in spite of the unevenness of the data, we feel it does give a good indication of the relative weight of the megadiversity countries in this important category —both in comparison with one another and with the rest of the world.

Based on our analysis, the leaders in terms of current human cultural diversity are Papua New Guinea, India, Indonesia, the United States and the Democratic Re-

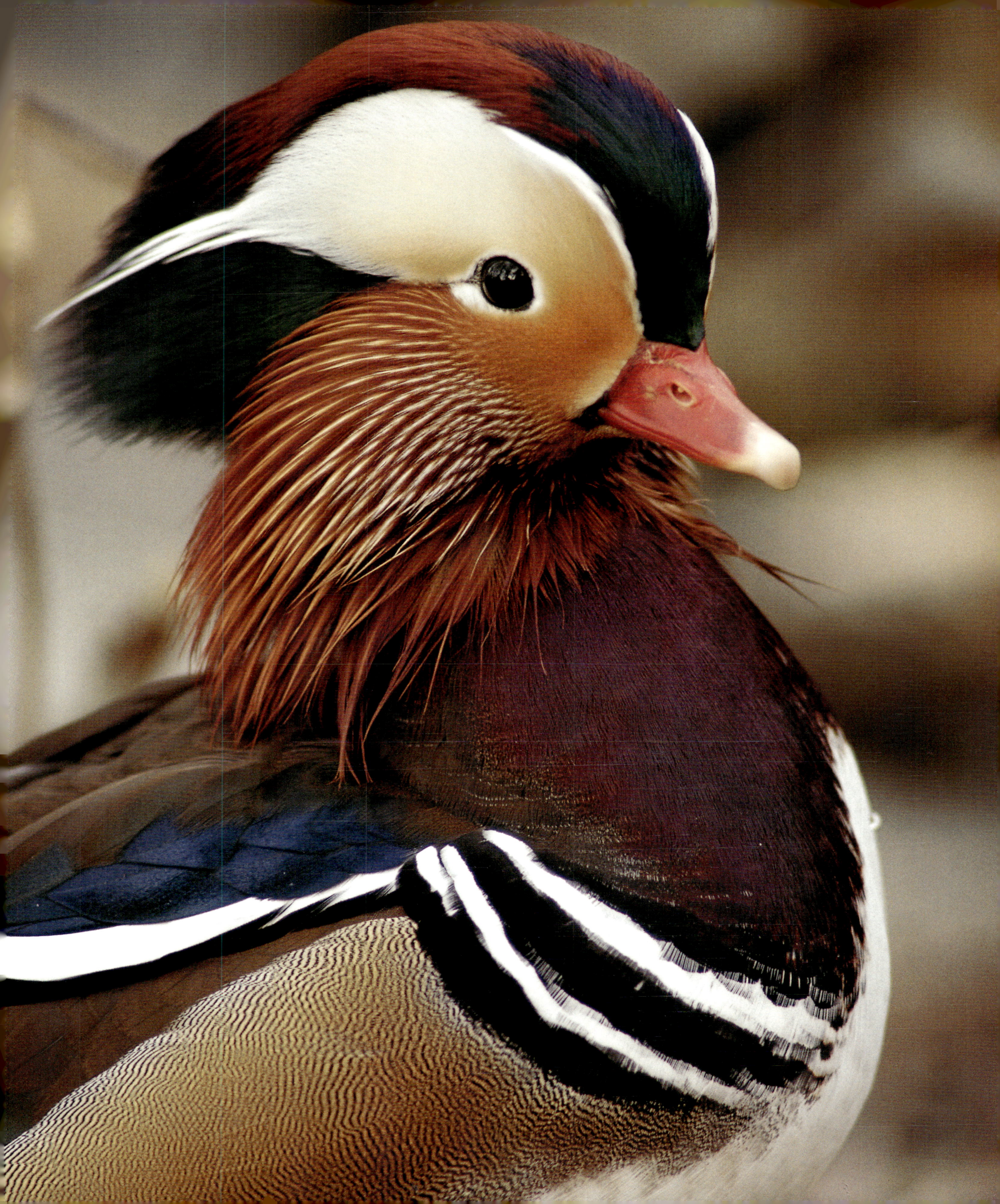

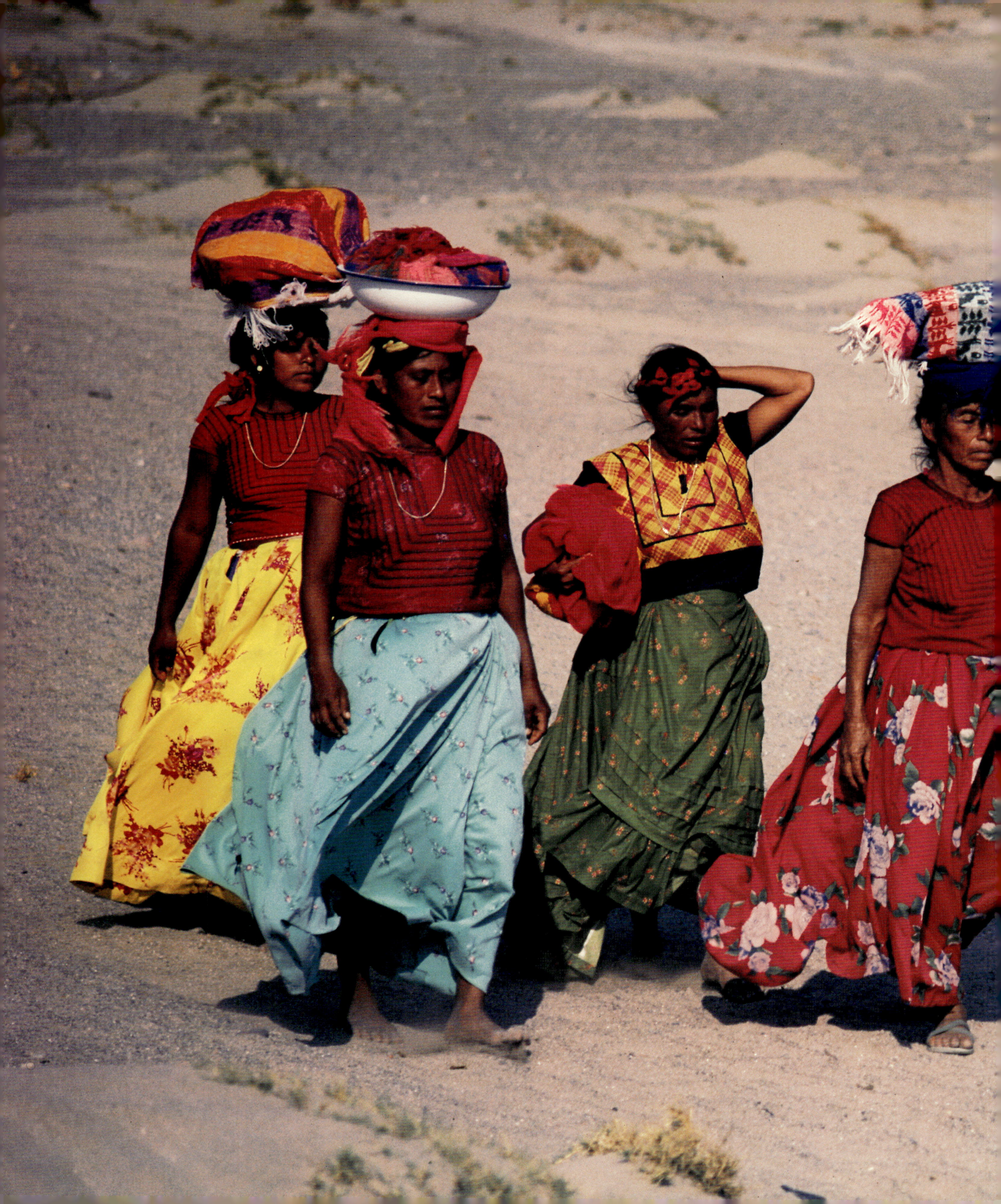

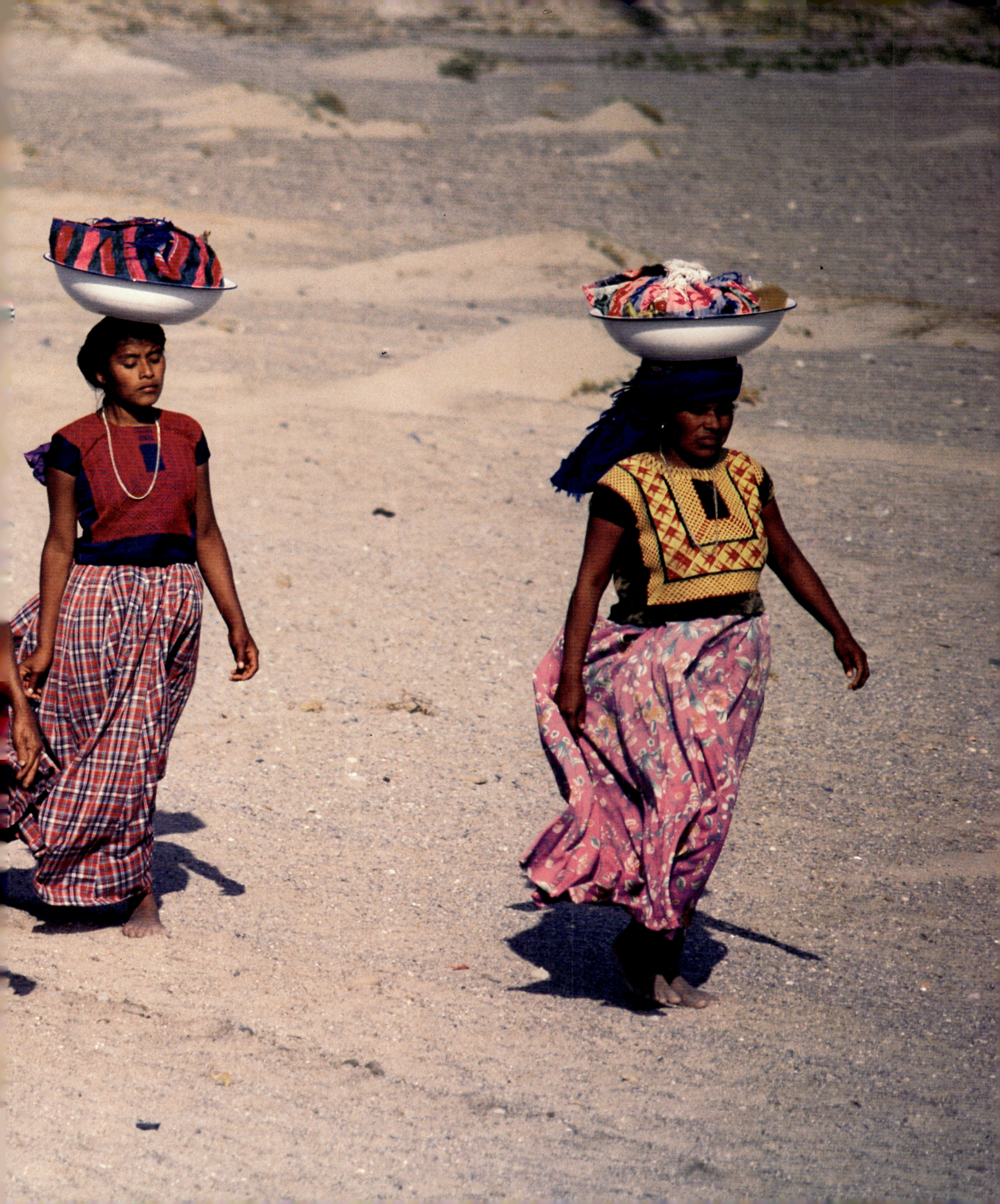

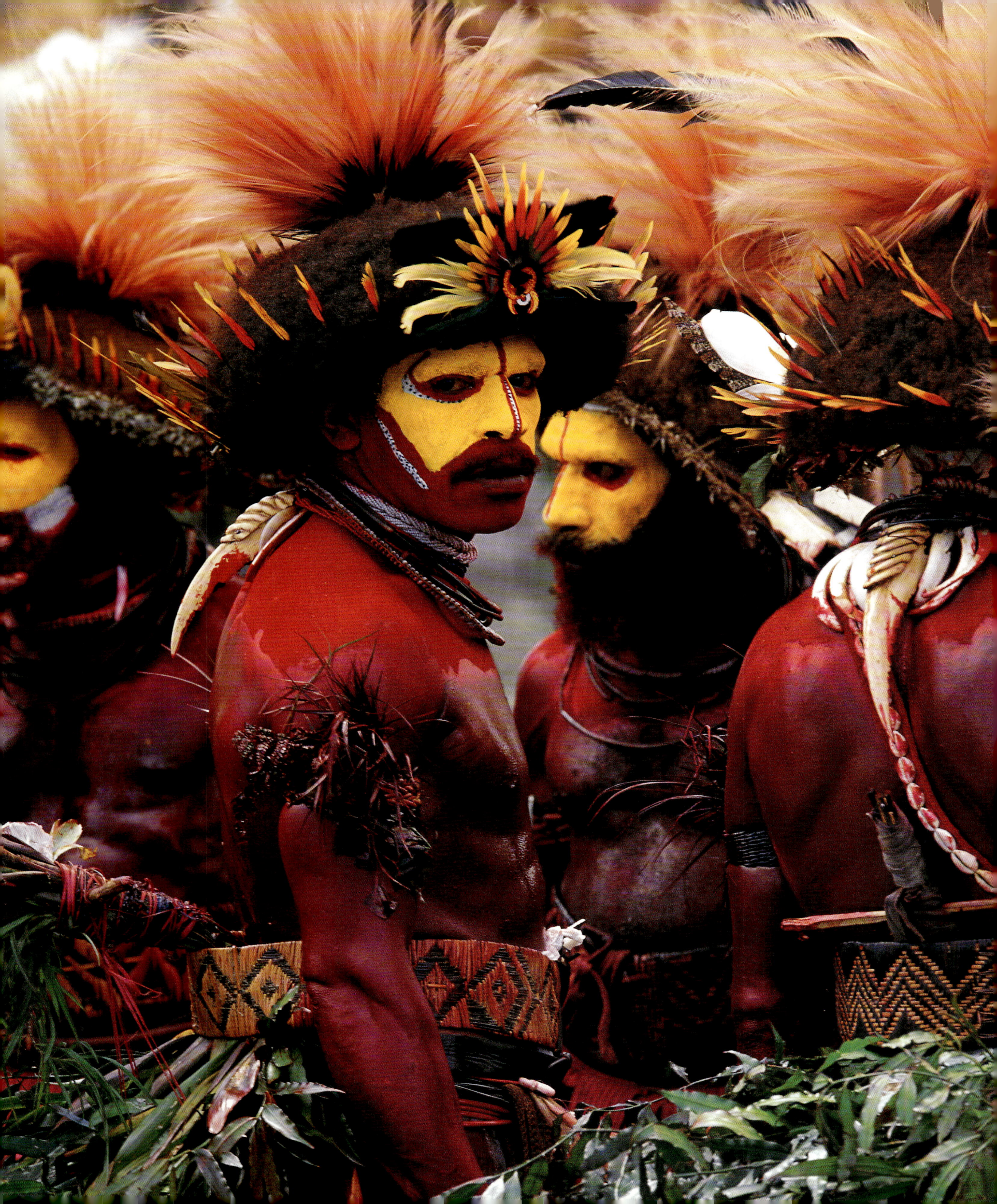

public of Congo, followed by Mexico, Brazil, Australia, the Philippines and Peru in approximate order of diversity (Table 30). Papua New Guinea has the largest number of languages at 875, and this seems to be a good approximation of the number of tribes as well. By contrast, there is a wide discrepancy between the number of languages (1 652) and the number of tribes (613) for India, leading us to believe that the estimate of 1 652 probably refers to relatively minor dialect variations as well. In the case of Australia, the situation is reversed. The number of languages is estimated at 200-250, whereas the number of tribes is given as 600-700. In both cases, we have used the lower figure to determine the position of the country in Table 30. In the case of Mexico, there is a major difference of opinion as to the number of languages spoken. In the book *Mexico: Diversity of Cultures*, published by CEMEX, Toledo (1995) cites a figure of 240 for the number of languages; in contrast, Martínez (1986; cited in Bye, 1993) believes that there were 120 spoken when the Spaniards arrived and that only 54 of these are still spoken today. In the case of Brazil, the number of tribes and languages stands at a reasonably well-documented 206, but we also include the figure of 286, which is the approximate number of tribes and languages that existed at the beginning of this century; at least 80 groups have gone extinct in the last 100 years, and many more since the arrival of the Portuguese in 1500. Also of note is that fact that some of the megadiversity countries, such as China, India, Indonesia, Papua New Guinea and the DRC are still populated mainly by their original inhabitants; others, such as the United States, Brazil, Australia, and Venezuela, are largely "immigrant" countries with the original inhabitants now being only a tiny fraction of the population. Still others, such as Peru, Colombia, Ecuador, South Africa, and Malaysia, fall somewhere in between; they continue to have large populations of the people that were there at the beginning of the colonial period, but also have significant "immigrant" populations. In this book, our estimates of human cultural diversity are based mainly on the number of indigenous groups in the country, and not on the "immigrant" cultures. Although the latter are obviously of great importance as well, especially in those countries where they are the dominant element in the population, we felt that the indigenous cultures were more appropriate as an indicator of the original cultural diversity of the country in question than the cultures which had appeared in very recent times.

The dynamic interaction between indigenous people and their natural environment is at the very essence of both conservation and the study of human cultural and biological evolution, and the coevolution of the two requires much further investigation while it is still possible. Both biodiversity and human cultural diversity are disappearing rapidly, and the window of opportunity remaining to carry out such research is already small and ever shrinking. In this book, we try to highlight the linkages between biodiversity and human culture wherever possible, with the goal of stimulating more interest in each individually, as well as in the intricate, multitudinous relationships that exist between them.

In conclusion, the main purpose of this book is to stress the enormous importance of a small number of countries in the conservation of our planet's biodiversity. Although the biodiversity of each and every country on Earth is of great importance and fundamental to long-term human survival, a handful of countries will determine the future of much of life on Earth. By documenting the importance of these countries, we hope to focus international attention on the key role they play and to instill in them a sense of pride in the heritage that they are so fortunate to possess. Ultimately, we believe that this pride and a recognition of the aesthetic and spiritual values of biodiversity will be the factors determining whether or not biodiversity is maintained. However, over the short to medium term, it is also important to demonstrate that conservation of biodiversity can be a win-win situation in economic terms as well, and that the rich biological heritage the megadiversity countries are so fortunate to possess is not a burden, but rather a major competitive advantage. This is just starting to be realized in most circles, but is almost certain to become much more evident with the rapid growth of the biotechnology industry (for which biodiversity is the raw material), the explosion of ecotourism, the growing interest in natural products of many different kinds, and an increasing recognition, documentation and quantification of the many goods and services that biodiversity already provides. The message that biodiversity is economically important is, of course, a global one, but it is a particularly strong message for the megadiversity countries, which share such a large portion of the planet's living natural resources.

Obviously, the international community will have to assist in a major way to help the megadiversity countries conserve their plants, animals, and diverse natural ecosystems, and we believe that investment in these countries should be roughly in proportion to the two thirds to three quarters of life on Earth that they possess. However, in the final analysis, it is up to the countries themselves to internalize *the doctrine of ultimate responsibility* and recognize that the survival of the majority of Earth's living creatures is first and foremostly up to them. They will certainly gain economically from doing this, but in the end, success or failure will come down to whether or not these countries accept their moral responsibility towards other forms of life. If we accomplish nothing else in this book than to begin a process of recognition of these issues within the megadiversity countries, then we will have achieved something of lasting value in global efforts to maintain the magnificent diversity of life on our planet.

RUSSELL A. MITTERMEIER
CRISTINA GOETTSCH MITTERMEIER
PATRICIO ROBLES GIL

TABLE 30. Top 12 Countries in the World for Human Cultural Diversity

Country	*Human groups*
PNG	875
India	613 (1 652)[a]
Indonesia	336
USA	300
DRC	250-300
Mexico	240? (54-120)[b]
Brazil	206 (286)[c]
Australia	200-250 (600-700)[d]
Philippines	111
Peru	96
Colombia	81
China	56

a. The estimate for tribal groups in India is 613; the estimate for languages is 1 652. Since it is believed that the higher figure includes minor dialect variations, we have chosen the lower, more conservative number to reflect human cultural diversity.
b. Estimates of cultural diversity for Mexico vary. Toledo (1995) mentions 240 languages. Martínez (1986; cited in Bye, 1993) indicates that there were 120 languages spoken at the time of conquest, but that only 54 remain today.
c. The estimate for extant tribes in Brazil is 206; it is believed that at least 80 tribes have gone extinct this century. We have therefore given the higher number in parentheses, indicating the minimum diversity that existed at the beginning of the twentieth century.
d. The case of Australia is the reverse of that for India; there, the estimate for languages is 200-250, whereas the estimate of different tribal groups is 600-700. To be conservative, we have chosen the lower figure.

On pp. 34-35, Huave Indian women, Isthmus of Tehuantepec, Oaxaca, Mexico.

On the opposite page, Huli wigmen dancers, Southern Highlands Province, Papua New Guinea.

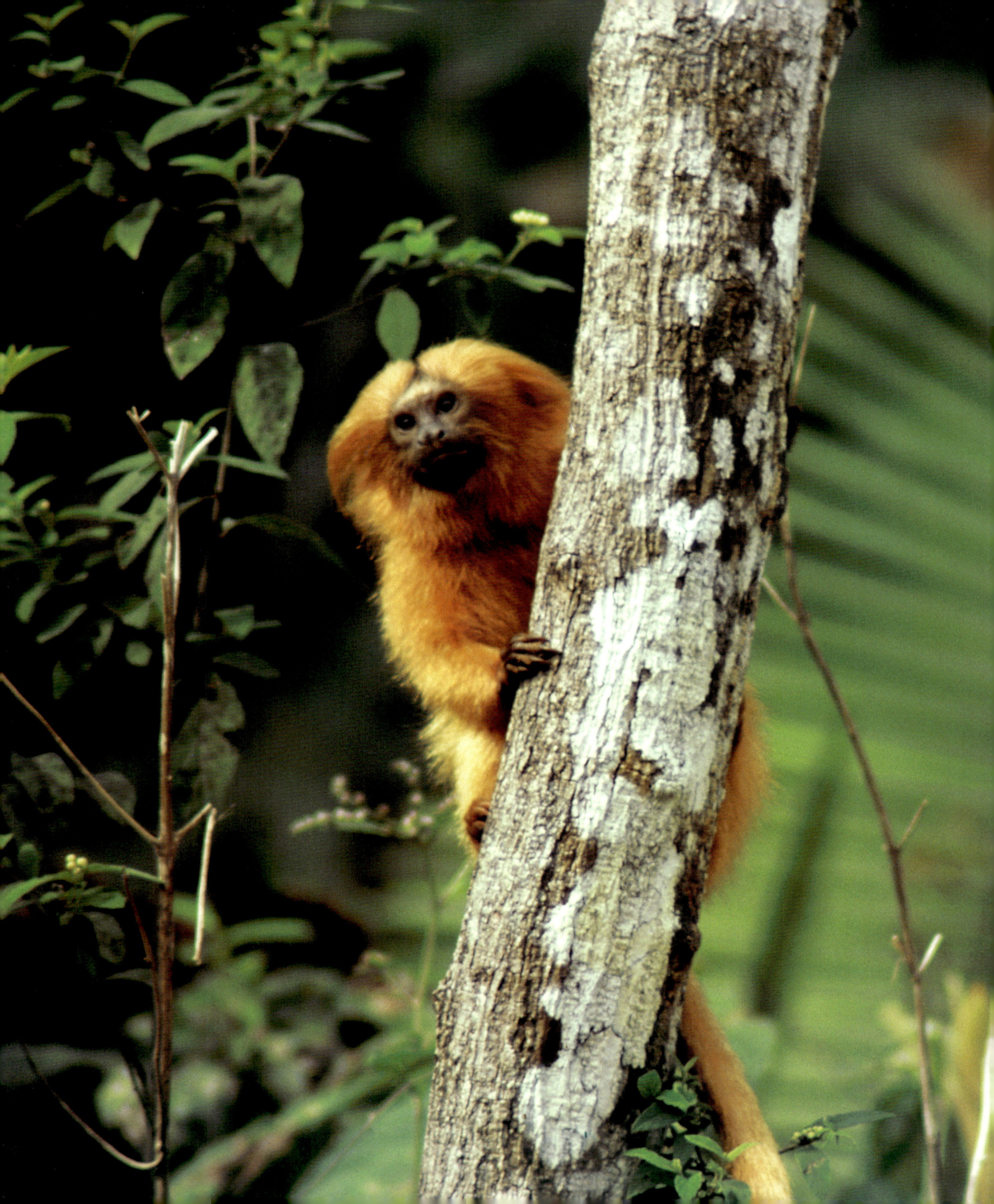

BRAZIL

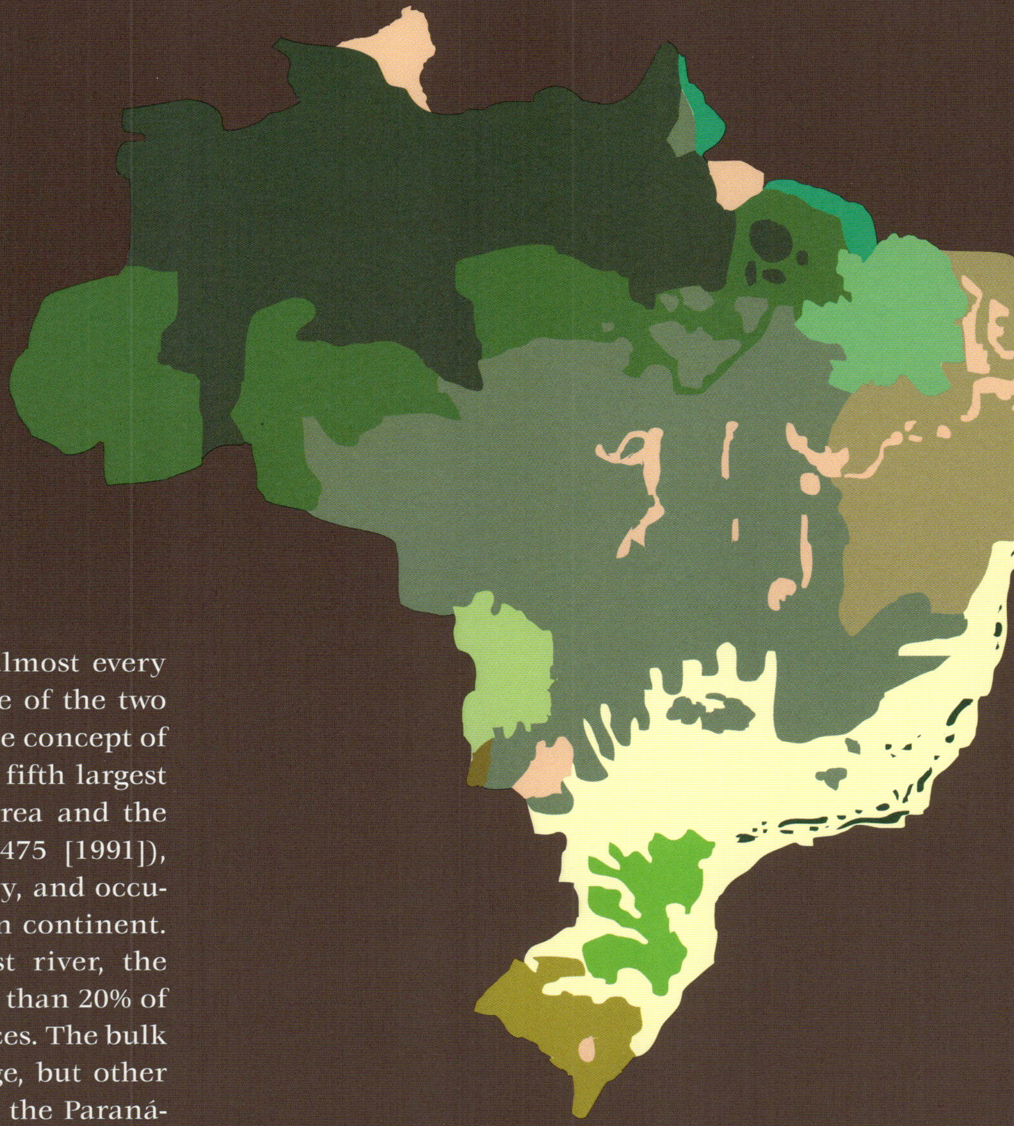

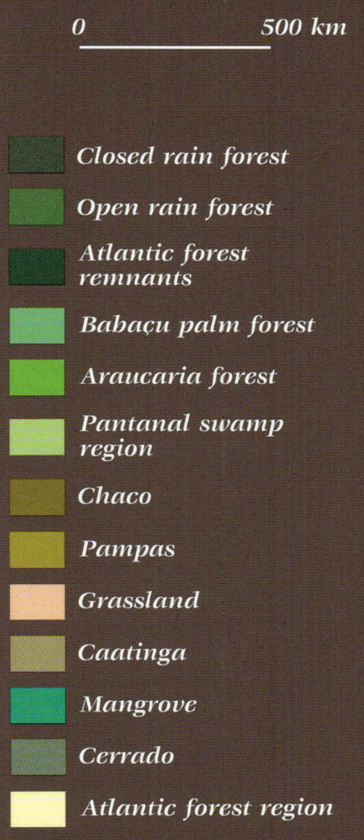

Brazil is a country of superlatives in almost every way and, along with Madagascar, is one of the two countries that originally gave birth to the concept of megadiversity (Mittermeier, 1988). The fifth largest country on Earth (8 511 965 km^2) in area and the sixth in human population (146 825 475 [1991]), Brazil is also the largest tropical country, and occupies roughly half of the South American continent. In addition, it has the world's largest river, the mighty Amazon, and accounts for more than 20% of the world's renewable freshwater resources. The bulk of this water is in the Amazon drainage, but other major rivers like the São Francisco and the Paraná-Paraguai contribute as well. Brazil is also unquestionably the richest country in the world in terms of overall terrestrial and freshwater diversity. When marine ecosystems are included, Indonesia may in fact have an advantage in total biodiversity, and its very unusual mix of Indo-Malayan and Australasian faunal elements makes its terrestrial diversity very special as well. In any case, Indonesia and Brazil are the top two megadiversity countries on Earth by virtually any measure, and which of the two is actually number one will require much more research to determine with certainty.

Taking into consideration other categories of priorities, Brazil has within its borders two of the world's top 19 *threatened biodiversity hotspots*, including the Atlantic forest region, which ranks among the top five, and the Cerrado-Pantanal complex, which was recently added to the hotspots list. It also has some of the largest and most important *major tropical wilderness areas* remaining, including extensive portions of northeastern, central and western Amazonia. Indeed, Brazil almost certainly has the largest blocks of intact tropical wilderness of any nation. Although parts of Amazonia are being rapidly modified by development, especially at the southern and southeastern fringes, the bulk of this region is still in very good condition, and some of the areas in central Amazonia that were clear-cut in the 1980s because of large-scale subsidies for cattle ranching have demonstrated remarkable regenerative capacity (*e.g.*, the region immediately north of Manaus).

Looking at the specifics for Brazil, we find that it is almost always at the top of the global list for every group of organisms used in this analysis. In diversity of non-fish vertebrates it is second to Colombia with 3 131 species, and in endemism for non-fish vertebrates it is fourth behind Australia, Indonesia and Mexico; however, it is clearly the global leader in freshwater fish diversity with at least 3 000 species (and probably many more) and has very high marine fish diversity as well (ca. 2 500 species). If fish are added to the totals, Brazil moves to the top of the global list for total vertebrate diversity. Unfortunately, good comparative data is not available for endemism of either freshwater or marine fish, so a precise determination of ranking in overall vertebrate endemism cannot be made at this time.

Needless to say, Brazil is very high on the lists for diversity and endemism in individual groups of organisms as well. It is first in the world in mammal diversity, with 524 species, and fourth in mammal

On the opposite page, the golden lion tamarin **(Leontopithecus rosalia)** *is perhaps the best known of all Brazilian flagship species. A small, brilliantly golden little monkey, it is endemic to the lowlands of the state of Rio de Janeiro. One of the world's most endangered monkey species, it is now down to about 500 individuals in the wild. This individual was photographed in the Poco das Antas Biological Reserve in Rio de Janeiro in 1980.*

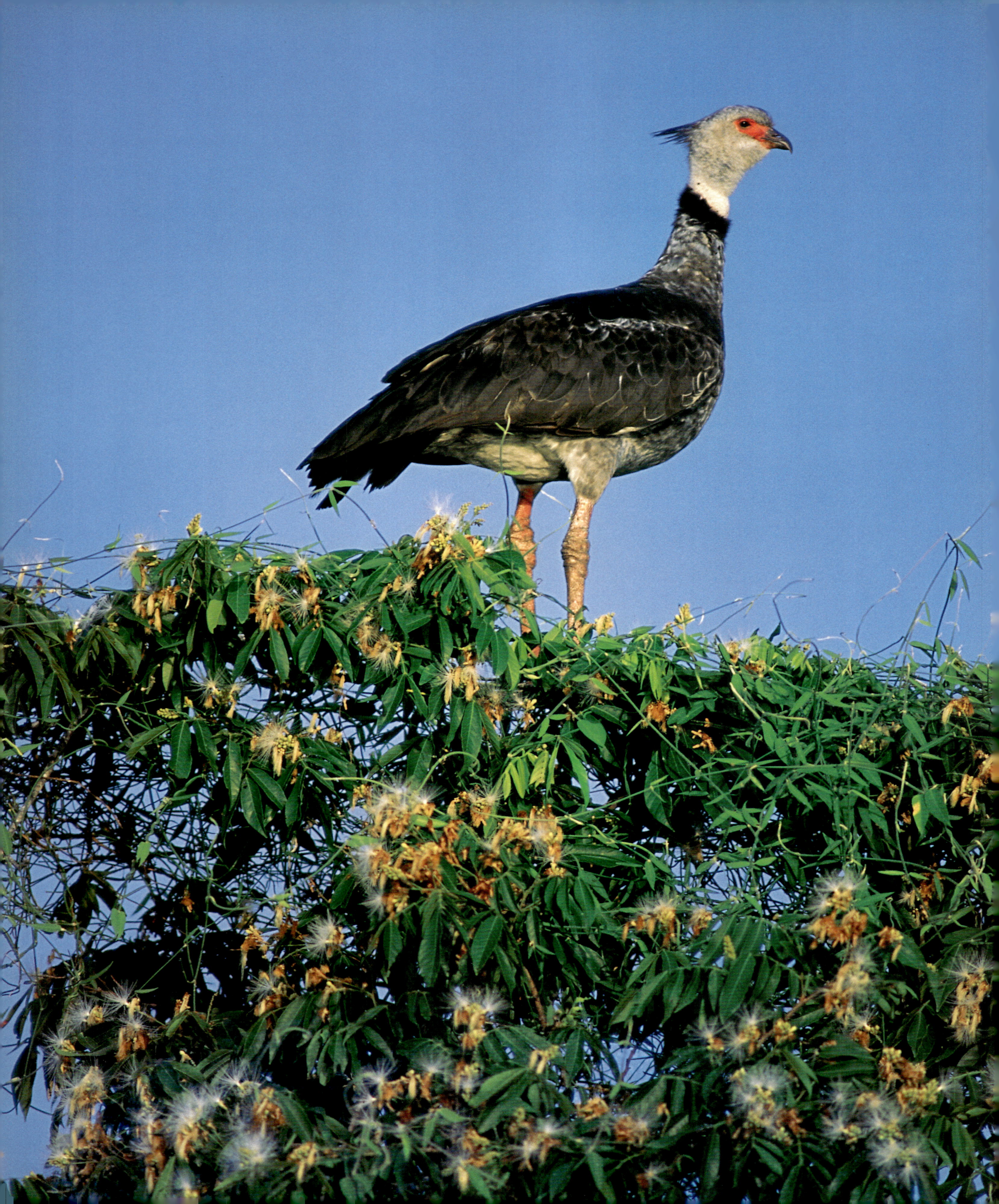

endemics with 131. In bird diversity, it is third behind Colombia and Peru with 1 622 species, and third in endemic birds (over 191 species) behind Indonesia and Australia. It also has the second highest number of BirdLife International's Endemic Bird Areas (19), second only to Indonesia's 24 areas. In reptiles, Brazil is fifth in both diversity (468 species) and endemism (172 species), and in amphibians it is second in both categories as well, with 517 species of which 294 are endemic. Within these major vertebrate groups, several are particularly outstanding in Brazil. These include the primates with 77 species or 27% of the world's total, more than double the figure for the next highest country on the list, and the psittacine birds with 70 species, again the world's highest.

Terrestrial invertebrate diversity is also very high, and may well turn out to be the highest on Earth, with perhaps 10-15 million insect species alone —although precise counts for these creatures will not be available in our lifetimes. Looking at the best-known insect groups, Brazil is third in both tiger beetle diversity and endemism with 184 species of which 97 are endemic, second in overall butterfly diversity with 3 132 species, and fourth in swallowtail butterfly diversity and endemism (74 species, 11 endemic).

Although there is much disagreement over the total number of higher plants in Brazil, with estimates ranging from 34 000-56 000, the most cited estimates are above 50 000, which clearly puts Brazil in first place in this important category. Estimates of plant endemism are even more imprecise, but an assumption of 33% endemism in this large country gives a range of 16 500-18 500 species, which vies with Indonesia, Colombia and South Africa for highest in the world. Brazil is very high in certain individual plant groups as well, ranking third, for example, in the economically important palm family, with 387 species.

In terms of threatened species, Brazil is also high on the list. It comes in second behind Indonesia in threatened birds with 103 species, and fourth in threatened mammals with 71.

Looking at the ranking system defined in the introductory chapter (Table 23), Brazil achieves the highest ranking of any country, edging out Indonesia 48 to 40. However, as indicated in the introduction, it is difficult to determine which of these two countries is the global leader in biodiversity. What is clear, however, is that Brazil and Indonesia are the top two countries in the world for biodiversity, and they both occupy crucial, pivotal roles in global efforts to save life on Earth.

Much of Brazil's biodiversity is to be found in its vast area of tropical forest, in which it is by far the world's leader. Brazil has some 357 480 000 ha of tropical rain forest, which accounts for over 30% of the world's total and is triple that of the next richest, Indonesia. Brazil has more tropical rain forest than all the rest of South and Central America combined, more than all of Asia and more than all of Africa Taking all kinds of forest into consideration (tropical, temperate and boreal), Brazil is second only to Russia in total forest cover, but the huge forest tracts of Siberia cannot compare in diversity to even a small portion of Brazil's rain forest (indeed, Russia, despite its large size and immense area of forest, does not even make the megadiversity country list). Most of Brazil's tropical rain forest is to be found in Amazonia. This immense region is comparable in size to the continental United States and is shared by 9 countries, but fully 62% occurs within the borders of Brazil.

Cultural diversity in Brazil is also quite high. Although Brazil is now largely an immigrant country like the United States, with people from many parts of the world, its indigenous population, albeit only about 0.2% of the population now, is still very significant with 206 surviving tribes. This is the seventh highest on Earth by our estimate.

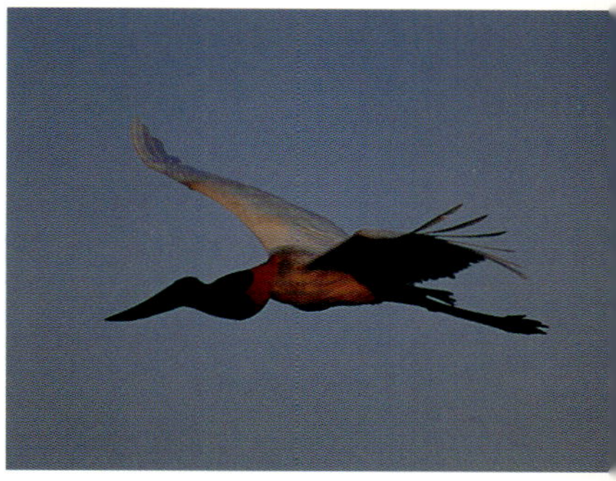

Natural habitats

The natural vegetation of Brazil can be divided into 6-7 major regions and many subregions. Amazonia is by far the largest and covers about 42% of the country, being, as indicated above, the largest block of tropical forest falling within the borders of a single nation. About 80% of Amazonia is still intact, but forest destruction since the early 1980s has been especially heavy in certain regions (*e.g.*, the state of Rondônia, the southern part of the state of Pará, the northern part of the state of Mato Grosso), and the trends are not promising. For example, in 1987 alone, it is estimated that some 8 000 000 ha of primary forest were destroyed. Although there was a decrease in annual forest destruction in the early 1990s, in part because of a large-scale recession in Brazil, it appears that conversion of Amazonian forests may in fact again be on the upswing, especially along the southern fringes of the region in the transition zone between rain forest and the drier Cerrado formations.

Although the vast Amazonian forests seem homogeneous from the air, this region is very diverse, and is composed of many different vegetation types, like the *terra firme* forests of the interfluvial regions, the *várzea* or floodplain of the white water rivers, the *igapó* or permanently flooded forests of the black water rivers, the campos or savanna enclaves in the middle of the *terra firme*, and a variety of other smaller, but very distinctive formations, such as the *campinas* and *campinaranas* growing on white sands and the montane forest growing on the sandstone *tepuis* on the Venezuela-Brazil border. This wide range of habitats and the transition zones or edges between them contribute to the great biodiversity of the Amazon forests, as do the great rivers themselves, which serve as barriers to gene flow and stimulate speciation.

The rivers of Amazonia deserve special attention, since nowhere else on Earth is there anything to compare with them. The early Portuguese explorers referred to the mainstream of the Amazon as *O Rio Mar*,

On the opposite page, the southern screamer (Chauna torquata), *a large, unusual bird that is common in the Pantanal swamplands of Mato Grosso and Mato Grosso do Sul. This one was photographed at Fazenda Kaiman, in the Pantanal.*

Above, a jabiru stork (Jabiru mycteria) *is one of the two most important flagship birds in the Pantanal region. A species that is now quite rare in most other parts of its range, the jabiru is still common in the Pantanal.*

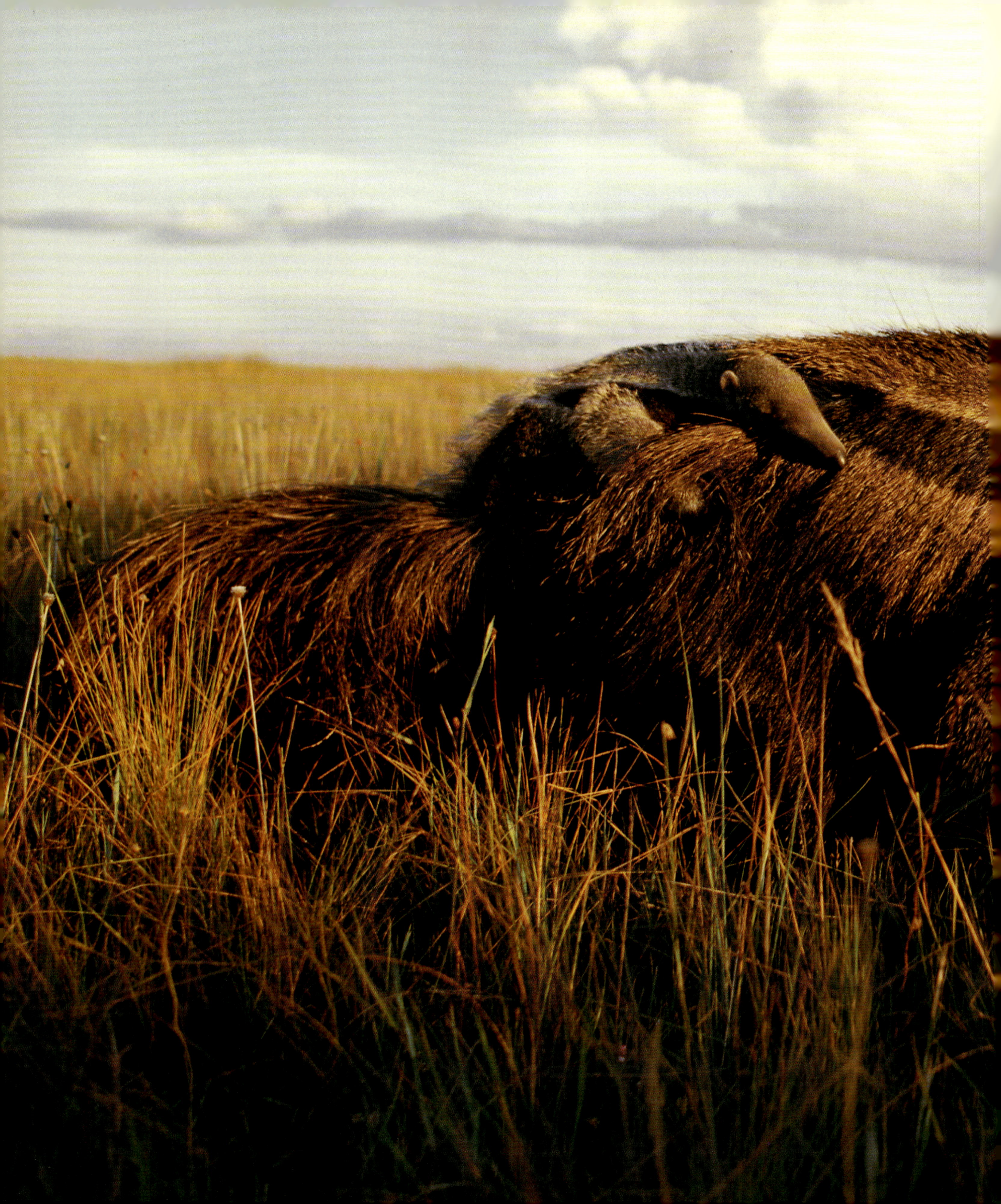

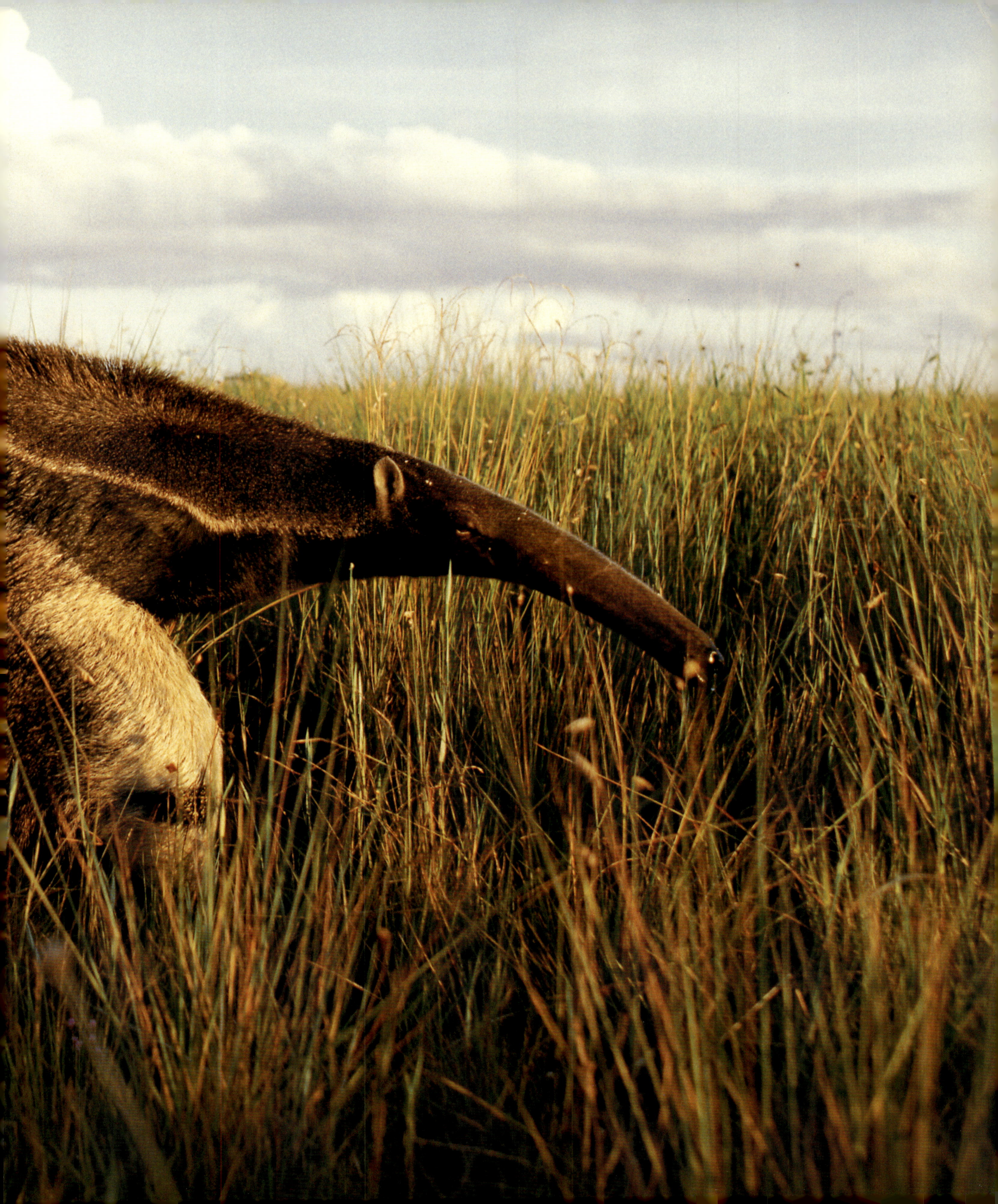

"The River Sea," and indeed the lower reaches of the Amazon itself and the mouths of some of the larger rivers such as the Rio Negro are so wide that one can sit in a boat in the middle of the stream and barely see either bank. The mainstream of the Amazon begins in the Andes, and then flows some 6 250 km to the sea, through a drainage basin of about 7 million km^2. It is surpassed in length only by the Nile (6 670 km), but is by far the largest river in the world in terms of discharge. Nearly 20% of the world's river water flows through the Amazon drainage. The flow of the Amazon mainstream at its mouth averages about 175 000 m^3 per second, incorporating the waters of thousands of smaller tributaries and also the full volume of its major affluents, some of which (for example, the Rio Negro and the Xingu, Madeira, and Tapajós Rivers) are themselves among the largest rivers on Earth. The massive scale of this truly imposing river system can be felt at Manaus, in the center of the Brazilian Amazon, where the black waters of the Rio Negro flow into the white waters of the Amazon, a spectacular sight known as the "Meeting of the Waters." The Rio Negro, draining a little under a third of the Guayana Massif, is considered the second largest river in the world in terms of discharge (but not in length, estimated at 1 700 km), slightly exceeding the Congo of Africa. The Rio Negro alone contributes about 15% of the total discharge of the Amazon, three times the volume of the Mississippi, and more than the annual flow of all the European rivers combined. The Amazon is truly a river without equal.

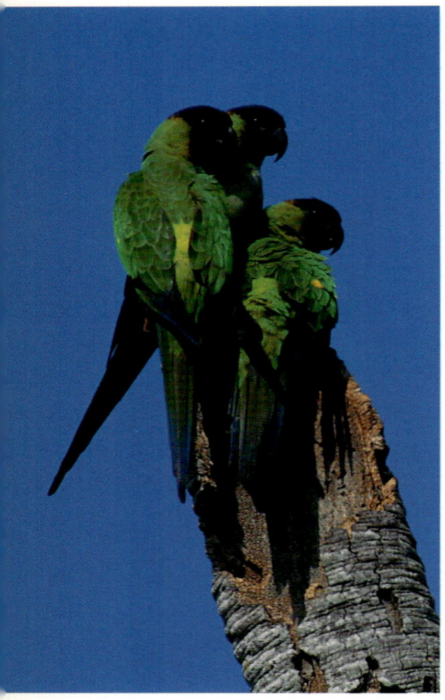

On pp. 42-43, the giant anteater (Myrmecophaga tridactyla), *seen here with an infant riding on its back, is one of the most impressive flagship species of northern South America. Although not endemic to Brazil, it is widespread in the Cerrado and Pantanal ecosystems and also occurs in rain forest, although at much lower densities.*

Above, a black-hooded parakeet (Nandayus nanday) *from the Pantanal.*

The different types of rivers in Amazonia are worthy of mention as well. Not only do they serve as geographic barriers among terrestrial species, but their diversity and the striking differences between them are a major contributor to the region's unparalleled aquatic richness. Indeed, the interplay between forest and river is perhaps more intimate here than anywhere else on Earth, even to the point of having fish that have evolved to feed on fallen fruits in areas of flooded forest. All rivers rise and fall during the year, some as much as 15 m, flooding forests along their margins. Certain kinds of rivers are rich in nutrients, while others are extremely poor. Forest composition in a particular area depends on whether or not the forest is flooded throughout the whole year or just parts of it, and forests that are flooded are affected by the composition of the river waters inundating them.

Terra firme or dryland forests are not affected by seasonal fluctuations in river levels. They make up the vast and, until very recently, largely unexplored influvial areas, or about 98% of greater Amazonia. Even the youngest soils of the *terra firme* date back to the Tertiary period, and the soils of the great plates to the north and south of the main river, the Guayana and Brazilian Shields, on which much of the *terra firme* forest grows, are among the oldest geological formations on Earth. The soils of these forests, therefore, have been exposed to leaching for millions of years, and are generally poor. Most of the nutrients in these forests are stored in the living plants, not in the soil itself, and the plants have an unmatched ability to capture and store available nutrients from the air and from whatever decaying organic matter may be available. In effect, you have an extremely rich ecosystem literally perched on top of an extremely poor substrate that barely provides support to the living structures resting on it. Indeed, unlike temperate forest trees, which have deep root systems extending vertically into the rich substrate below them, the large trees on *terra firme* forests sit almost flat on top of the ground, their root systems extending horizontally for maximum support and to access as broad an area of available nutrients as possible. This is best seen when one of the forest giants of Amazonia falls to the ground: its flat and relatively superficial root system looks like it simply couldn't hold up such a large structure. Numerous support adaptations have also evolved to assist in keeping the trees of Amazonia in an upright position, among them buttress roots stretching out from the base of large trees and stilt-roots aimed at providing extra support. Another feature of the plants of the *terra firme* forest is the chemical defense system that they have evolved to avoid predation. Since nutrients are at such a premium in this kind of forest, strong chemicals of various kinds (*e.g.*, tannins) have evolved to fend off herbivorous insects and other animals in such low primary productivity habitats.

The flooded forests of Amazonia, as distinct from those of the *terra firme*, are permanently inundated by one of the three major kinds of Amazonian rivers: "white water" rivers, "black water" rivers, and "clear water" rivers. White water rivers, like the mainstream of the Amazon, the Japurá-Caquetá, the Purus, and the Madeira, originate in the mineral-rich Andes and carry a heavy silt load. Their color can be described as *café au lait*, they have turbid opaque waters, and are slightly acidic to slightly basic. These rivers drain only about 12% of Amazonia, but carry 86% of the total dissolved salts and 82% of the suspended solids discharged each year. The floodplain or *várzea*, as it is called in Brazil, of the white water rivers can be quite wide, reaching 100 km in some areas. The forest growing on this floodplain is rejuvenated each year by mineral-rich silts left by the receding river water. It has high productivity, but a lower number of tree species than the surrounding *terra firme* forests, perhaps due to the much smaller geographic area that it occupies.

The black water rivers, of which the Rio Negro is the best example, are transparent, acidic, usually the color of "strong tea," and carry virtually no nutrient-rich silt. These rivers are characterized by low, sloping, poorly-defined banks covered with *igapó*, the Brazilian name for the periodically or permanently flooded swamp forest of these rivers. The black water rivers originate in the very poor substrate of the Guayana and Brazilian Shields, and the secondary defensive compounds, aimed at minimizing nutrient loss to the plants, leach from the fresh vegetation and the decomposing litter on the forest floor. In the flooded areas, litter falling directly into the river decays, resulting in oxygen-deprived waters. The biocidal nature of the plant secondary compounds, together with the low oxygen content and extremely poor nutrient content of the black water rivers, results in reduced animal communities. Biomass of invertebrates, fish, reptiles, amphibians, birds and mammals is much lower in these rivers than in the white water areas. Brazilian black water rivers are sometimes referred to as *rios de fome* (rivers of hunger) because of their low productivity. Unlike the white water floodplain, which often includes wide expanses of treeless, savanna-like campos and "lakes," the inundated margins of the black water rivers are almost always covered with *igapó* forest. Although these black water rivers are lower in biomass, their diversity is nonetheless very high, and includes species endemic to this kind of water and the *igapó* forests.

Clear water rivers, like the Xingu and the Tapajós, are similar to black water rivers in the absence of suspended silt and tendency toward acidity. However, the banks of these rivers are high and stable, and consequently there is less oxygen-consuming, acid-rich organic matter in the waters. Animal biomass appears to be higher than in the black water rivers, but not as high as in the white water rivers.

The sheer size of Amazonia, its tropical location, the great variety of habitat types, and the fact that much of the area is still intact, make this region one of the great storehouses of biodiversity left on the planet; estimates of the share of global diversity found there range from 10-15%. Needless to say, a major portion of Brazil's biodiversity and its ranking as the world's top country for freshwater diversity is attributable to Amazonia. Nonetheless, many other ecosystems exist within Brazil, and several of these are also high in diversity and endemism, and generally much more threatened than Amazonia.

The next most important ecoregion in Brazil is the Atlantic forest region of eastern Brazil, a unique series of rain forest ecosystems quite distinct from the much more extensive Amazonian forests to the northwest. This is a very rich area with high levels of endemism, and is home to 261 species of mammals (73 endemic), 620 of birds (160 endemic), 260 of amphibians (128 endemic), and around 20 000 plant species (more than 6,000 endemic). Indeed, as with Amazonia, if the Atlantic forest region were a separate country, it would easily qualify as a megadiversity country in its own right.

The Atlantic forest region once stretched almost continuously from the states of Rio Grande do Norte and Ceará in northeastern Brazil as far as Rio Grande do Sul, the southernmost Brazilian state. It included the typical Atlantic coastal forests on the narrow, low-lying coastal plain, the forests on the foothills of the Serra do Mar and related mountain ranges that roughly parallel the coast, and the forests on the slopes of the Serra do Mar itself, reaching up to 1 800-2 000 m. In northeastern Brazil, these forests covered only a narrow coastal strip rarely exceeding 50-100 km in width, and were replaced inland by drier Caatinga and Cerrado formations of the Brazilian interior. In the southeastern states of Brazil, the Atlantic forest covered a much broader area, extending inland in some places as far as 500-600 km. In addition, the Atlantic forest covered a good part of Brazil's south, with a large block of the unusual araucaria or Brazilian pine forests in the middle. In its primeval state, the Atlantic forest occupied between 1 000 000 and 1 200 000 km^2, and extended north to south over an area roughly equivalent to the eastern seaboard of the United States. It is the third largest major formation within Brazil (after Amazonia and the Cerrado), occupying about 12% of the country's territory, and the second most diverse biologically.

It is now generally accepted that forests extending into the eastern portion of Paraguay and extreme northern Argentina (Misiones) are also part of the Atlantic forest complex. However, regardless of how the southern limits of the Atlantic forest are defined, it is clear that it is largely a Brazilian formation, and that most of the diversity and most of the conservation problems occur within Brazil.

When the first naturalists, especially Maximilian, Prince zu Wied-Neuwied of Germany, passed through the Atlantic forest early in the last century, they found some of the richest, tallest and most impressive forests on Earth, with an especially rich orchid and bromeliad flora and an abundant and diverse fauna. Even then, however, some regions of the Atlantic forest, especially states like São Paulo and parts of the northeast, had already been colonized for more than two centuries and were starting to show the effects of large-scale habitat destruction. Indeed, this was the first part of Brazil to be colonized; it has always been the population and agricultural center of the country, and is now the industrial center as well. Two of the three largest cities in all of South America, São Paulo and Rio de Janeiro, are located only about 400 km apart in the central portion of the Atlantic forest region, and at least two thirds of Brazil's population now lives there. The result has been widespread, largely uncontrolled forest destruction for lumber and charcoal, and to make way for coffee plantations, sugar, cocoa, eucalyptus, rubber, cattle pasture and industry. The destruction has been especially rapid in the last three decades of heavy development and economic expansion, and the changes that have taken place in just the last five to ten years have been so extensive that it will take years to assess the full ecological damage.

In any event, in the Atlantic forest we are dealing with a highly endangered series of ecosystems in which only a tiny fraction of the original forest cover remains. The destruction has been especially severe in the low-lying coastal region and in the narrow strip of forest in the long-exploited northeast, where less than 1% of the original primary forest remains. However, it has also spread into the foothills and onto the slopes of the Serra do Mar, which used to be much less accessible. As a result, looking at the region as a whole, less than 8% of

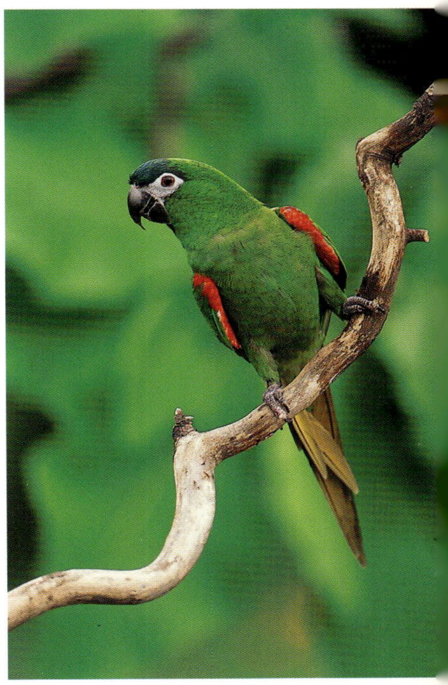

Brazil has the world's highest psittacine bird diversity, with some 70 species of parrots, macaws, and their relatives. Seen here is a red-shouldered or Hahn's macaw (Diopsittica n. nobilis).

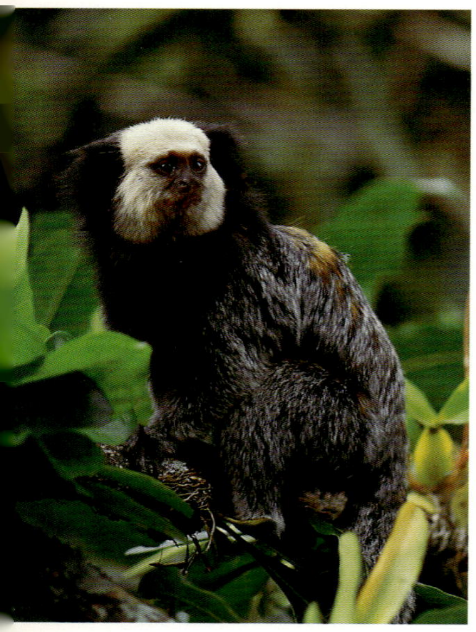

The striking white-faced marmoset (Callithrix geoffroyi) *is somewhat more abundant than its relative, but is also restricted to remaining patches of Atlantic forest in Minas Gerais, Espírito Santo and extreme southern Bahia.*

On the opposite page, a marmoset of the genus Callithrix, *a genus that has several new forms waiting to be described. All but one of these is endemic to Brazil, and six are found in the Atlantic forest region of eastern Brazil. The buff-headed marmoset* (Callithrix flaviceps) *seen here in one of its last strongholds, the Caratinga Biological Station at Fazenda Montes Claros in Minas Gerais, is the most endangered member of this genus.*

the original vegetation remains intact. Most of that is to be found in the steep-sided coastal mountains of the states of São Paulo and Paraná, with some still remaining in the mountains of Rio de Janeiro as well. Elsewhere, a large chunk of what little remains is in a series of federal- and state-protected areas, and in some large, privately-held blocks.

The widespread destruction has resulted in a tremendous decline of the native fauna and flora, a substantial percentage of which is endemic (*e.g.*, 51% of the rodents; 54% of the trees; 79% of the primates, etc.), and there is no doubt that the region represents a conservation problem of enormous proportions. Indeed, the Atlantic forest region of Brazil is listed among the top five threatened hotspots of the world, along with Madagascar, the Philippines, the tropical Andes, and Mesoamerica. Overall diversity is also high in the Atlantic forest, with per unit area diversity actually exceeding that of Amazonia in some groups. For instance, the Atlantic forest is home to 261 species of mammals, compared to roughly 353 for the Amazon, but it is only about one fifth the size of the Amazon. If one combines this high per unit area diversity with the fact that so little of the Atlantic forest remains intact, it is obvious why the region is so important in global terms. Furthermore, and not surprisingly, of the seven species considered to have gone extinct in Brazil in recent times (two birds, four butterflies and one dragonfly), all are from the Atlantic forest.

The Cerrado region of central and central-western Brazil is the largest ecoregion in the country after the Amazon, occupying about 28% of the national territory. It is composed of a variety of drier vegetation formations, ranging from open grasslands to various kinds of savanna woodland, and culminating in the *cerradão*, a forest formation similar to rain forest but with a different botanical composition, interspersed with luxuriant gallery forests. Unlike the Amazon, and rather like the American Midwest, the Cerrado has richer soils and is considered the nation's agricultural frontier. Much of Brazil's rapid agricultural expansion, for crops like soybeans in particular, is taking place in this region, which is being destroyed more rapidly than Amazonia. Furthermore, the protected area system in the Cerrado is far less representative than that of either the Amazon or the Atlantic forest. Although overall diversity and endemism are not as high as in the rain forest formations, plant and mammal diversity are quite significant, and there are a number of endemic species. One, the gallery forest rodent, *Juscelinomys candango*, endemic to the narrow strips of forest of the central Cerrado, may already be extinct.

The Pantanal region of central-western Brazil and adjacent portions of Paraguay and Bolivia is one of the internationally best known regions of Brazil. It is the world's largest wetland, covering some 110 000 km^2 and consisting mainly of the low-altitude floodplain of the Paraguai River and its tributaries, which drop off the *planalto* (inland plateau) of Central Brazil. It is basically a huge, low-altitude "sinkhole" in the center of the South American continent, and is bordered to the northwest by the semideciduous forest of the Amazonia-Cerrado transition and to the southwest by the dry Chaco formations of Bolivia and Paraguay. The vegetation includes elements from all of these formations, and is a diverse mosaic of swamp, gallery forest, and several different kinds of Cerrado and open grassland. Much of the area is flooded from December to June, and there are large, permanently-flooded areas (including lakes of up to 100 km^2), mainly in the northwestern part of the region in the state of Mato Grosso. The extent of flooding increases as one approaches the Paraguai River to the west. In essence, the Pantanal is a seasonally inundated version of the Cerrado, and the Pantanal-Cerrado complex was recently identified as one of the top threatened hotspots on Earth, joining the still higher-priority Atlantic forest region.

The Pantanal has an abundant and visible fauna, having been compared to the savannas of East Africa in terms of its wildlife-watching potential. It is probably the most important area in South America for wetland birds, and is also home to other spectacular species, such as the rhea (*Rhea americana*), the southern screamer (*Chauna torquata*), and the hyacinth macaw (*Anodorhynchus hyacinthinus*). It is also an important migratory bird stopover point and wintering ground, falling within three major migratory routes, the Central Brazilian, the Rio Negro and the Cis-Andean.

The Pantanal is also known for its abundance of mammal and reptile species. The capybara (*Hydrochoerus hydrochaeris*), the world's largest rodent, is quite common along the waterways and is easily observed in the dozens or even hundreds. The tapir (*Tapirus terrestris*), the jaguar (*Panthera onca palustris*), the puma (*Felis concolor*), and the giant anteater (*Myrmecophaga tridactyla*), which are quite difficult to see in the wild in Amazonia, are much more likely to be encountered here than anywhere else within their range. Indeed, during my first trip to the Pantanal, I [R.A.M.] managed to see all four of these animals in the wild in one week; in all the rest of the time I have spent in South and Central America (27 years), I have only seen the jaguar two other times and the puma once. The marsh deer (*Blastocerus dichotomus*), the largest deer in South America, also has its stronghold here, and several other very rare species like the maned wolf (*Chrysocyon brachyurus*) and the giant otter (*Pteronura brasiliensis*) are found in the Pantanal as well. The Pantanal jaguar is especially noteworthy since it is by far the largest form of the jaguar, with large males growing to be almost the size of a female lion and probably twice as heavy as their Amazonian cousins.

The most obvious reptile in the Pantanal is the Paraguayan caiman (*Caiman yacare*). Extremely large aggregations can be easily seen in this region, the most impressive that we saw having approximately 3 000 individuals in a pool no longer than 500 m and averaging 20 m in width. This was at the height of the dry season

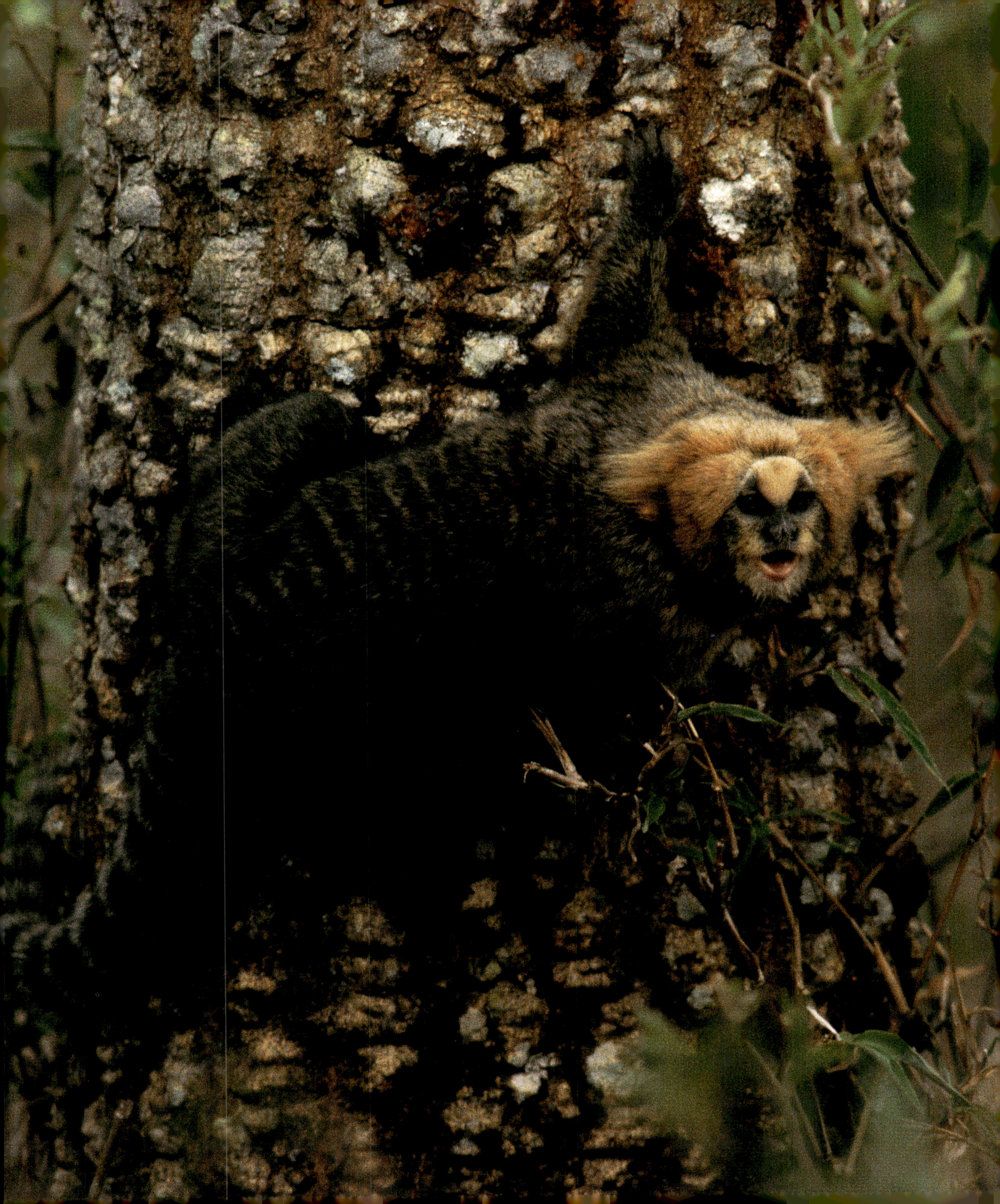

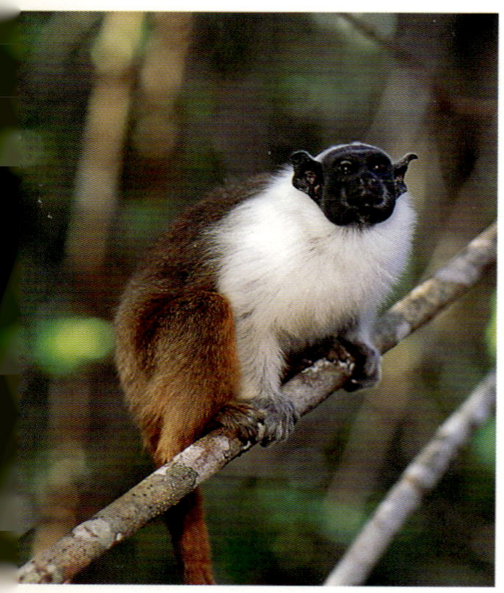

The marmoset and tamarin monkeys of the family Callitrichidae reach their greatest diversity in Brazil, with many endemic species. The pied bare-faced tamarin (Saguinus b. bicolor) *is endemic to a small area of Brazilian Amazonia centered on the city of Manaus, and is considered endangered.*

and definitely exceptional, but aggregations of dozens and even hundreds can be seen at almost any time.

The plight of this species was what originally attracted international attention to the Pantanal, since the animal was once heavily poached. In the 1970s and 1980s, for example, an estimated 1 000 000 skins per year were obtained illegally, which resulted in severe reductions in certain areas. Fortunately, however, the Paraguayan caiman, like its relative the American alligator, has been able to resist the heavy pressure and, with more protective measures in place, seems to be bouncing back in some areas.

The predicament of the Paraguayan caiman was also the original stimulus for the creation of the Pantanal Defense Society, SODEPAN. This unusual group is a nongovernmental organization composed of large landholders in the Pantanal who are interested in protecting the area for future generations to develop its economic potential through ecotourism and non-timber forest products like honey. They originally banded together in the mid–1980s because of the threat posed by caiman poachers on their land. Indeed, private landowners taking responsibility for the conservation and long-term ecological integrity of their property is probably the most viable conservation alternative for the Pantanal. The government maintains only two protected areas, the 112-km^2 Taiamã Ecological Station and the 137-km^2 Pantanal National Park, both of which are almost entirely inundated in the wet season and together cover only a very small percentage (0.2%) of the Pantanal's total area.

The principal economic activity of the Pantanal is cattle raising, which was introduced at the end of the last century, and most land is in the hands of large *fazendeiros* (ranch owners), a number of whom have tracts exceeding 1 000 km^2. Cattle raising and associated burning of grasslands in the dry season have affected the vegetation, and it is believed that little of the Pantanal's vegetation remains undisturbed. Nonetheless, cattle raising in the Pantanal is far less destructive than in Amazonia, since there are large areas of natural grassland and relatively little felling of forest. Much of the region has low human and cattle population densities, so the opportunity still exists to maintain a large portion in good condition. Other economic activities in the Pantanal include fishing, which was once sustainable when carried out by local people but is now becoming increasingly commercial and destructive, and a growing tourism industry. Given the great abundance of fauna in the Pantanal and the ease with which it can be observed, this region could become one of the great tourist destinations of the continent. Fortunately, a number of landowners have recognized this and have started to create hotel-fazendas, which are attracting an increasing flow of national and international tourists.

The Caatinga of northeastern Brazil is composed mainly of dry thornscrub formations, and is by far the driest region of Brazil. Some experts believe that this region was much more forested in the recent past, but that it has been completely modified and turned into a semidesertic region by human activities. In any case, this somewhat inhospitable region also has one of the highest poverty levels in Brazil. Species endemism is low in the Caatinga, but there are some interesting endemic species, among them the *tatu-bola* or three-banded armadillo (*Tolypeutes tricinctus*) and two spectacular macaws that rank among the most endangered birds in the world, Lear's macaw (*Anodorhynchus leari*) and Spix's macaw (*Cyanopsitta spixii*).

The last and least significant of the major Brazilian ecosystems is the Pampas, the grasslands of extreme southern Brazil that are contiguous with similar formations in Argentina and Uruguay. These are low in diversity and endemism compared with the tropical Brazilian ecosystems, and are better represented in the other two countries, so we won't go into further detail on them here.

Flagship species

As would be expected of the country that has the highest terrestrial biodiversity on Earth, Brazil has a large number of flagship species. Without a doubt, the two most important of these are the nonhuman primates and the Psittacinae birds, the parrots and macaws and their relatives. In terms of primate diversity, Brazil is unmatched on the planet, with fully 16 genera, 77 species and some 128 taxa, 54% of which are endemic to the country and found nowhere else. As if this great diversity were not enough, new species continue to be discovered, especially in the Amazon region. Since 1990, there have been seven new species described, and several others are awaiting description at this time. The bulk of Brazil's primates occur in the two main forest regions of the country, the Amazon and the Atlantic forest. In the Brazilian portion of the Amazon, there are 60 monkey species belonging to 14 genera and in the Atlantic forest there are six genera and 24 species and subspecies, 19 (79%) of which are endemic.

Given the degree of destruction in the Atlantic forest, it is there that we face the greatest conservation challenges, and two monkey genera from this region have been the most visible symbols for conservation during the past two decades, *Leontopithecus* and *Brachyteles*. The lion tamarins of the genus *Leontopithecus* and the muriquis of the genus *Brachyteles* are endemic to the Atlantic forest, and are among the most endangered primates in the world. Interest in the Atlantic forest as a whole first grew out of concern for the future of one of the lion tamarins, the famous golden lion tamarin (*Leontopithecus rosalia*), and began at an international level in 1970. As more and more research on these animals was conducted, especially beginning in 1979 with a World Wildlife Fund–supported program on primates and protected areas in the Atlantic forest, it became obvious that this region was both among the most diverse and the most endangered on Earth. This initial interest in primates has now blossomed into a number of internationally and nationally supported conservation projects for the region as a whole —an excellent example of the kind of leveraging effect that interest in flagship species can provide.

Four species of lion tamarins are currently recognized, the golden lion tamarin (*Leontopithecus rosalia*) restricted to low-lying forest remnants in the state of Rio de Janeiro and now down to an estimated 470-631 individuals in the wild (supplemented by a roughly equal number in captivity in Brazil, the United States and Europe); the golden-headed lion tamarin (*Leontopithecus chrysomelas*), from the southern part of the state of Bahia and numbering between 6 000-15 000 (supplemented by some 600 in captivity); the black lion tamarin (*Leontopithecus chrysopygus*), from five forest remnants in the heavily devastated interior of the state of São Paulo and numbering perhaps 1 000 in the wild; and finally the recently discovered black-faced lion tamarin (*Leontopithecus caissara*), from the coastal area on the Paraná-São Paulo border and numbering about 400 individuals. The black-faced lion tamarin was unknown to science until 1990, when it was discovered by two Brazilian researchers in Paraná State, a real surprise to all of us working on New World primates. It is the only one of the lion tamarins not yet being held in captivity. All of these monkeys are in grave danger of extinction, with *L. rosalia*, *L. chrysopygus*, and *L. caissara* all being considered critically endangered, and *L. chrysomelas* still slightly more abundant than its relatives, being placed in the endangered category.

The muriquis, genus *Brachyteles*, are the largest of the New World monkeys, and the largest mammals entirely endemic to Brazil. Very poorly known until the late 1970s, these monkeys are restricted to forest remnants in the states of Bahia, Espírito Santo, Minas Gerais, São Paulo, and possibly still Rio de Janeiro and Paraná. Once very abundant, they were one of the main sources of protein for early explorers traveling through the region, and one estimate, by the Brazilian scientist Alvaro Aguirre, puts the original population at 400 000. Now, however, with widespread forest destruction and hunting, the total population is 1 158 documented animals. To further complicate the matter, research conducted in the late 1980s indicated that what we thought was just one species, *Brachyteles arachnoides*, is in fact two distinct species, *Brachyteles arachnoides* from the mountains of São Paulo and *Brachyteles hypoxanthus* from regions further north. The northern populations number 234 individuals, whereas those in the south are more numerous, totalling some 924 individuals. Fortunately, conservation efforts now in place should make it possible to ensure the survival of these critically endangered animals.

In any case, the lion tamarins and the muriquis are really the equivalent of the giant panda in China, animals that have helped to focus global attention on Brazil and its unique biodiversity. They have become symbols, within Brazil and internationally, of the unique fauna and flora of the Atlantic forest, and now appear on T-shirts, buttons, posters, and a wide variety of educational materials in schools and universities. Indeed, the golden lion tamarin has even been featured in floats in Rio de Janeiro's famous carnival. Quite a change for animals that were virtually unknown even within Brazil just two decades ago.

The lion tamarins are also demonstrative of the kind of international cooperation that is so critical to the survival of endangered species. There are now international cooperative agreements and management committees for all four species of lion tamarins, and all the animals held in American and European institutions are officially the property of the Brazilian government. International conservation organizations and key zoos have, in turn, helped conservation efforts in Brazil with technical assistance and financial support, complementing that being provided by the Brazilian government itself. This lion tamarin program just celebrated its 25th anniversary in May, 1997. With the muriquis the situation is comparable, except that there is not yet a self-sustaining captive population for these larger monkeys. Only about 7-8 are being kept at the Rio de Janeiro Primate Center, and most of these are hybrids between the two species.

Among the Psittacinae birds, several species also stand out as excellent symbols of conservation in Brazil. The large blue macaws have been particularly important, although conservation efforts on their behalf have not yet been as successful as those for the primates. One of the most endangered birds in the world, Spix's macaw (*Cyanopsitta spixii*), is an example. This beautiful animal, a monotypic genus, is now down to a single individual in the wild in the state of Piauí in Brazil's northeast, and perhaps 40 individuals in captivity, mostly outside of Brazil. The single individual has been monitored daily for the past three years. Recently, a captive-born mate was released in the area, in the hope the pair would breed. So far, there has been no success in this attempt.

Another highly endangered macaw is Lear's macaw (*Anodorhynchus leari*), restricted to a small area in the Raso da Catarina region of the state of Bahia, one of the driest parts of Brazil's driest ecosystems, the Caatinga. This large, spectacular species is down to perhaps 100 individuals in the wild, and has also been the subject of a major conservation effort over the past 15 years. The second of the large blue macaws, the well-known hyacinth macaw (*Anodorhynchus hyacinthinus*) is somewhat more abundant than its Bahian relative, but still numbers only in the low thousands. It is found mainly in the Pantanal region, with outlying populations in the southern Amazon and western Cerrado. Long persecuted for the pet trade, where it can bring exorbitant prices in the tens of thousands of dollars, it too has declined precipitously in recent years. However, once again, Brazilian and internationally supported efforts are helping to document the status of this species and save it from extinction.

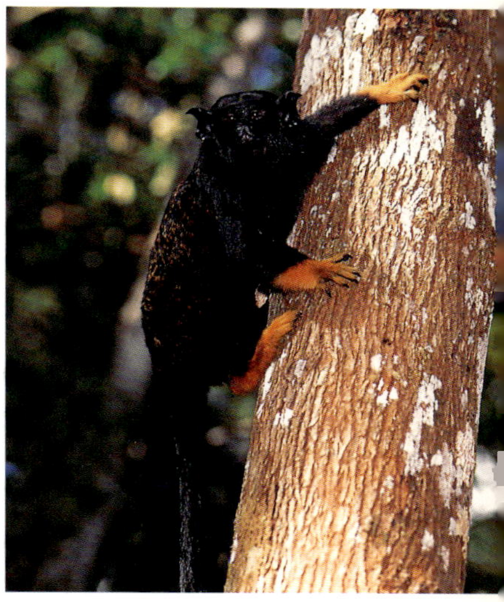

The golden-handed tamarin (Saguinus m. midas), *by contrast, is widespread in the northeastern quadrant of Amazonia and occurs in Brazil, southern Venezuela and the Guianas. It is probably the most abundant of all callitrichid species.*

On pp. 50-51, Iguaçu Falls in the state of Paraná in southern Brazil. Iguaçu Falls, which is on the Argentina-Brazil border, is arguably the most impressive series of waterfalls in the world. A major international tourist attraction, it is protected in national parks on both sides of the border.

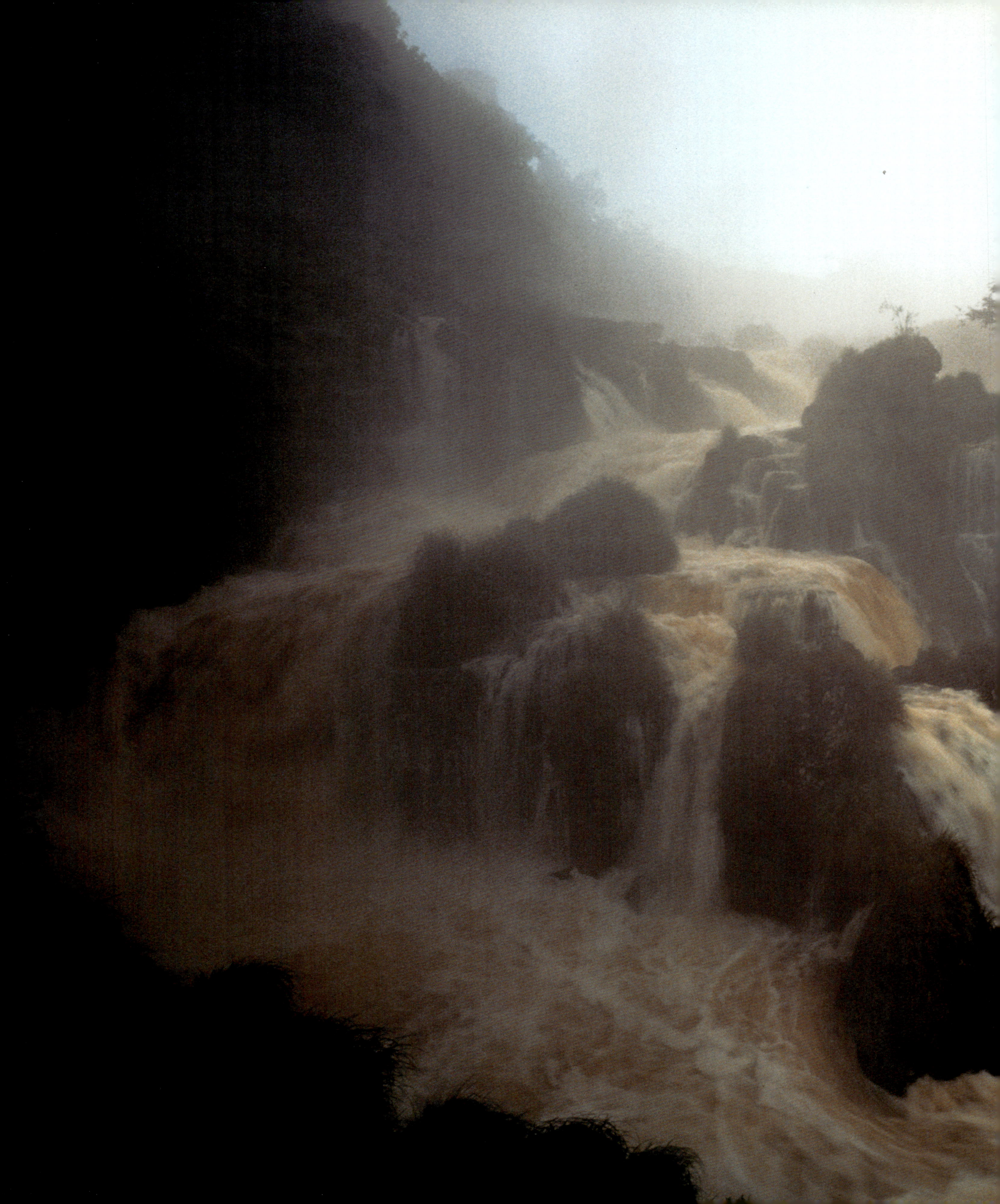

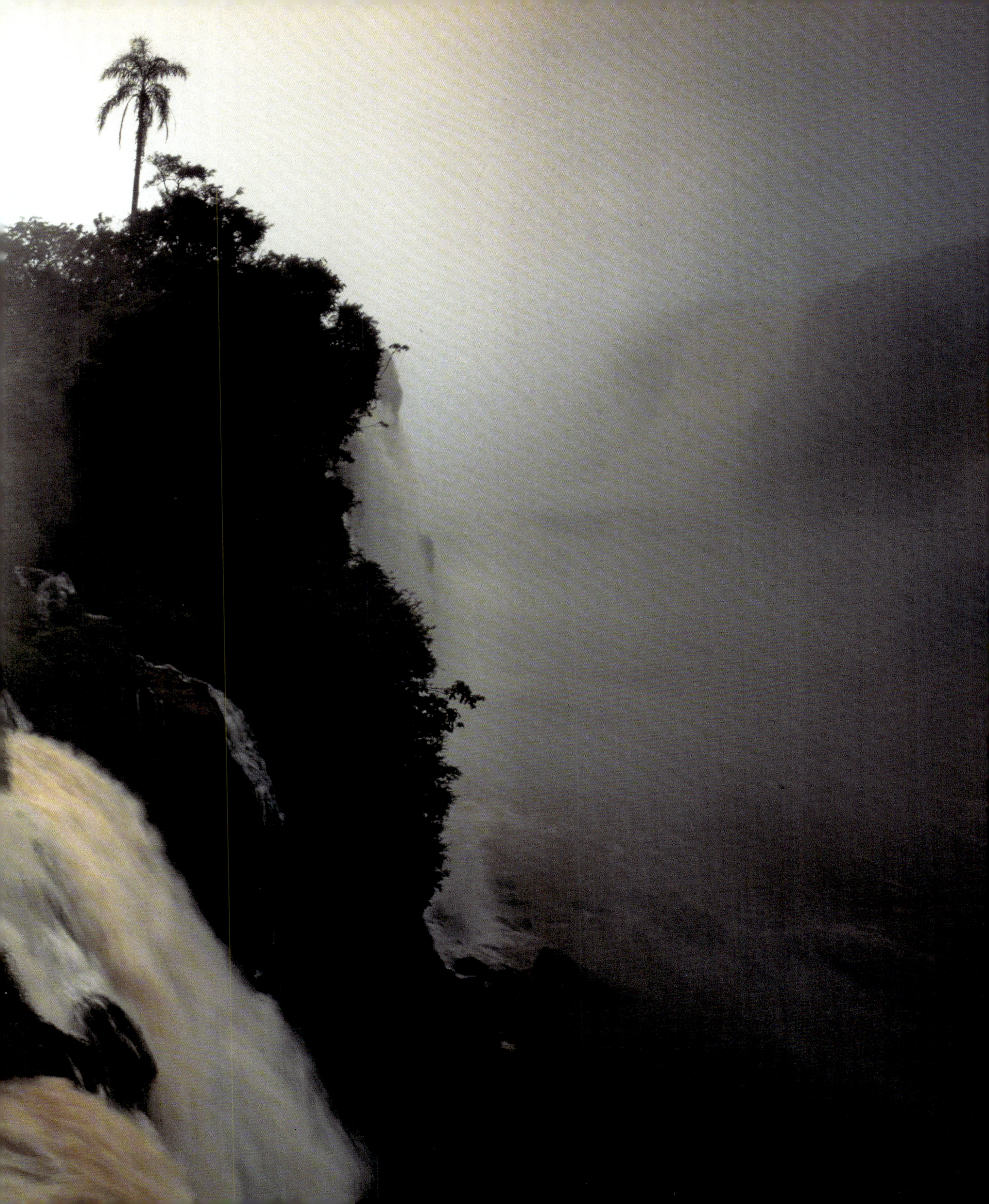

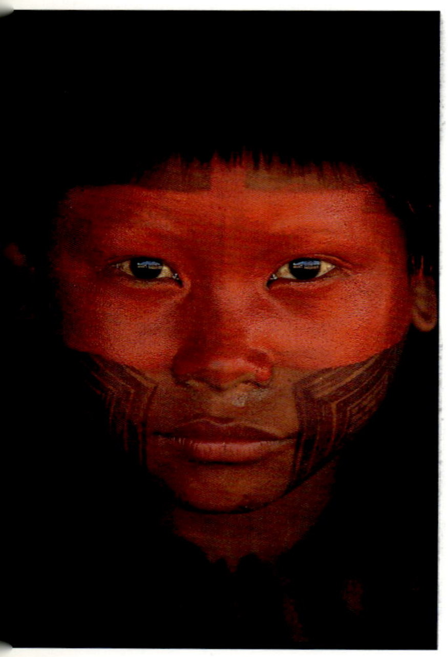

Brazil's surviving indigenous people are divided into 206 distinct tribal groups. Of these, the Kayapó Indians of the middle Xingú River region in southeastern Brazilian Amazonia are one of the most distinctive and best known. A Kayapó child, above, decorated with red face paint made from red urucum dye (Bixa orellana) *and temporary body tattoos made from genipapo fruit* (Genipa americana).

On the opposite page, a Kayapó man wearing a headdress made of yellow oropendola feathers, bead armbands, and a necklace made of beads and freshwater mollusk shells.

The third of the large blue macaws of the genus *Anodorhynchus* is the glaucous macaw (*Anodorhynchus glaucus*), which has not been seen in this century. However, since it comes from a remote and little explored region of the country, which includes some of the rugged formations bordering on the southern Pantanal, there is a slim chance that it may still survive.

Aside from the primates and the psittacines, there are several more poorly-known endemic species that also rate as flagships. Among these are the thin-spined porcupine (*Chaetomys subspinosus*), an odd, porcupine-like animal that appears to be intermediate between the tree porcupines and the spiny rats, and is restricted to a small area of southern Bahia and Espírito Santo in the Atlantic forest region; the three-banded armadillo (*Tolypeutes tricinctus*) from the Caatinga of Bahia, which rolls into a tight ball for protection and moves like a miniature Volkswagen beetle; and the maned sloth (*Bradypus torquatus*), an unusual relative of the wide-ranging three-toed sloth that is restricted to remaining forest areas from Rio de Janeiro to southern Bahia.

Aside from the endemics, there are a number of Brazilian flagship species that are shared with other South American countries. These include the jaguar, and especially the Pantanal jaguar (*Panthera onca palustris*); the giant otter (*Pteronura brasiliensis*), largest of the otters and highly endangered; the Amazonian manatee (*Trichechus inunguis*), a freshwater relative of the better-known salt water manatee; the maned wolf (*Chrysocyon brachyurus*), an elegant and shy dweller of the Cerrado that looks like a large-eared fox with the legs of an antelope; the giant anteater (*Myrmecophaga tridactyla*); the giant armadillo (*Priodontes giganteus*); the immense pirarucu (*Arapaima gigas*) of Amazonia, one of the largest fish in the world; and the giant Amazon river turtle (*Podocnemis expansa*) which, in spite of centuries of massive overexploitation, still survives on some of the more remote tributaries of the Amazon.

Some Brazilian species appear to have already gone extinct in recent times, although very few compared to what has been lost in places like Australia, Madagascar and Hawaii. One species of curassow bird from northeastern Brazil seems to have disappeared, and another, the razor-billed curassow (*Mitu mitu mitu*) from the heavily deforested state of Alagoas in northeastern Brazil, appears to be on the verge of extinction. How many invertebrates and plants disappeared from the more than 90% of the Atlantic forest that has already been destroyed will never be known, but certainly there must have been losses of species never even described by science. However, as threatened hotspot ecosystems go, the Atlantic forest has still lost relatively little of its biodiversity.

Cultural diversity

We will never know for certain how many indigenous human cultures existed in Brazil when Europeans first arrived there in 1500, but there is no doubt that it was very rich indeed. It was originally thought that humans arrived in Brazil only about 20 000 years ago, following the last glaciation and coming over the Bering Strait. However, recent archaeological studies in northeastern Brazil indicate that human presence dates back to at least 50 000 years, and that colonization may well have taken place by sea. The enormous linguistic diversity of the indigenous populations throughout Brazil also indicates a complex colonization process that will probably never be fully understood.

Estimates on indigenous population at the time of European contact vary widely, usually from 500 000 to 5 million, and no one is sure how many tribes existed at that time. However, we do know that there are now 206 existing tribes, and that an additional 80 were destroyed just between 1900 and 1957 (Ribeiro, 1957), mostly in areas where rubber extraction, Brazil nut gathering, and mining were taking place. During that period, the indigenous population collapsed from about one million to only 200 000. Given that many other tribes disappeared in the first four centuries of European colonization, among them the Oti-Xavante and the Carijó in the south, the warlike Tupinambá, who dominated the Brazilian coast from the mouth of the Amazon to Cananeia in São Paulo state, and the large and powerful Omagua and Tapajós tribes living along the Amazon mainstream, it is likely that many other tribes also existed in 1500, perhaps on the order of several hundred more than those actually documented. In any case, the diversity of Brazil's human cultures was, and, with all its losses, still is one of the richest in the world.

According to the latest study (Ricardo, 1996), Brazil's Indians have rebounded somewhat from the situation in the 1950s and now number about 280 000, most of them living on designated Indian lands. Of the 206 documented groups, 71 (34%) have a population of less than 200 individuals, 49 (24%) have a population of 201-500, 32 (15%) a population of 501-1 000 and 44 groups (22%) populations of 1 001-5 000. In other words, 95% of all remaining Indian groups number less than 5 000 individuals. Just four groups, the Sateré-Mawé, the Potiguara, the Xavante and the Yanomami, range between 5 000 and 10 000; three number between 10 000 and 20 000 including the Guajajara, the Terena and the Macuxi, and two, the Kaingang and the Ticuna number between 20 000 and 30 000. Only one group, the Guarani, exceeds 30 000 individuals. All in all, the original indigenous inhabitants of Brazil number only about 0.2% of the population of the country.

A number of Brazilian Indian tribes still maintain significant elements of their traditional cultures. Like most other indigenous people, they live basically from slash-and-burn agriculture and from fishing and hunting, and have a rich spiritual life. Brazilian Indians are especially renowned for their elaborate featherwork, which compares in intricacy and beauty to the magnificent woodworking of the Sepik River cultures of Papua New Guinea and the elaborate masks and other art

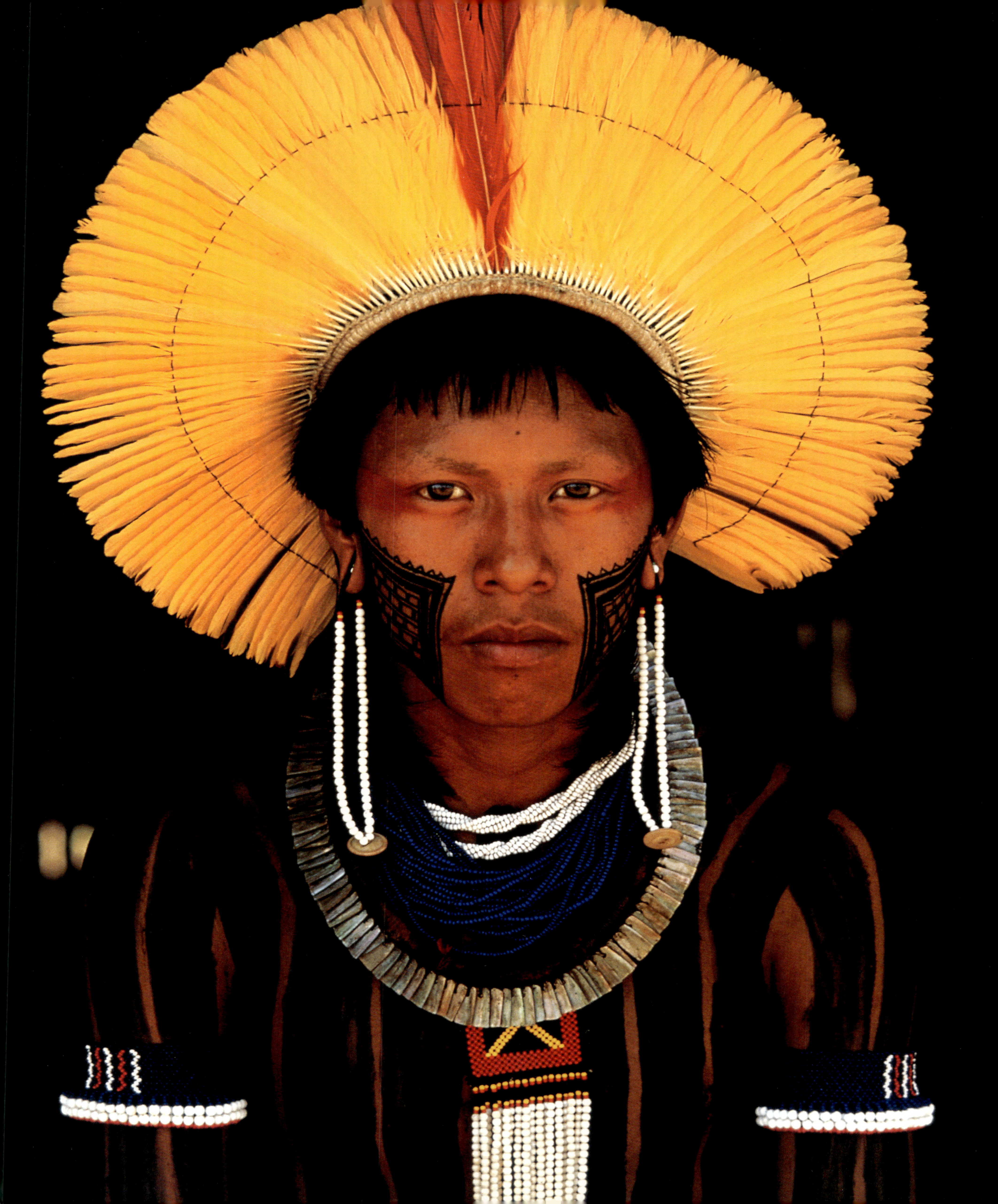

found in Melanesia. Parrots, macaws, toucans and a wide variety of other birds provide the main raw material for this art, which is also the basis for much of the religion and spiritual world of the Amazonian Indians, once again demonstrating a close linkage between human cultures and biodiversity.

Among the best known of the Amazonian Indian groups are the Kayapó, a medium-sized tribe that numbers between 3 500 and 4 500 individuals living in about 20 villages in the middle reaches of the Xingu River region of the southeastern Amazon. A politically astute and warlike group that has fought strongly for its rights, the Kayapó now have areas of some 10 million ha demarcated as their domain. They have been in regular contact with mainstream Brazilian society for only about 30 years, but are very visible in any political activities involving Indians, such as the 1992 Earth Summit in Rio de Janeiro, where they were better represented than any other indigenous culture. The Kayapó also have one of the richest art traditions of any Amazonian tribe, with a wide variety of featherwork and other art used in their religious ceremonies and everyday life. They are part of the Gé (or Jé) linguistic group, and refer to themselves as the Mebêngokrê, or "people of the water source." The Xikrin, the Mekragnoti, and the Gorotire are subgroups of the Kayapó.

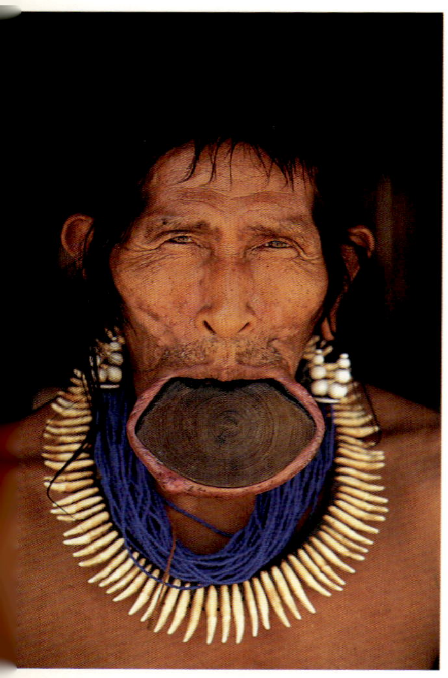

Another Kayapó man with the traditional lip disk and a necklace made of monkey teeth.

Traditional enemies of the Kayapó, the Karajá live on the largest fluvial island in the world, the 500 km-long and 120 km-wide Bananal Island, or Ilha do Bananal, in the Araguaia River, another of the Amazon's larger southern tributaries. Another isolated language family, the Karajá have lived on this island since pre-Columbian times, and now number some 1 900 people. They have a very distinctive art form, of which the circular *laheto* headdress is the most distinctive element. They are also well-known for their clay dolls, or *litjoko*, which now appear on a regular basis in Brazilian tourist shops and depict people and animals engaged in a variety of different activities.

Living near the Karajá north of the Ilha do Bananal are the Tapirapé. Although they are a Tupi-speaking group, they are thought to have appeared in the Bananal region at roughly the same time as the Karajá, and share with them a number of cultural elements. Through warfare with the Kayapó and introduced diseases, they are now down to about 380 individuals living in one reservation on the left-bank tributary of the Araguaia that bears their name. They are best known for their large, semicircular *cara grande* or *upé* masks, made of wood decorated with numerous small feathers and surmounted by a band of larger feathers. The *cara grande* mask is one of the four or five most spectacular and well-known pieces of Amazonian Indian art.

Also in the Xingu region, but this time in the upper reaches of the Xingu, is the 2 600 000-ha Xingu National Park in northern Mato Grosso state, actually a large indigenous reserve rather than a true national park. This reserve was created in 1961 after more than a decade of efforts by the Villas Boas brothers, tireless champions of Brazil's indigenous people. Although the tribes represented in the Xingu National Park are a mix of several completely different cultures and linguistic groups, they have been brought together through their common problems and a united front against an ever-encroaching modern society. This is another artistically rich region, with a wide variety of beautiful featherwork and masks used in traditional ceremonies. A particularly interesting common element in the region was the keeping of live harpy eagles in large cages in the center of villages. This enabled the people to periodically pluck feathers from these birds for headdresses and other ornaments, but also had strong spiritual meaning. The harpy was seen as the chief of the air, whereas the human chief occupied the same role in the terrestrial realm. When a tribal chief died, the harpy was often ritually killed as well.

In all, there are 10 villages in the Xingu Park, representing four different language families, Tupi, Carib, Arawak and an isolated language group. The Arawak family is composed of the Mehinako (149 people), the Waura (226 people), and the Yawalapiti (196 people). Carib speakers include the Kalapalo (326 people), the Kuikuro (343 people), the Nafuqua (64 people), and the Txicão (214 people). The Tupi-Guarani speaking family is represented by the Aweti (93 people), the Kamayura (303 people), and the Kayabi (1 200 people). The Trumai, who originate from a distant region of the southeast, speak a language not related to the other three families and number only 89 people. As should be obvious from the very small numbers of people in each of these tribes, their survival both as cultural and biological entities hangs in the balance.

Another well-known group are the Xavante, who call themselves Akwé and, like the Kayapó, are Gé-speakers. They once occupied a large area of southwestern Mato Grosso state in the upper reaches of the Araguaia and Xingu Rivers. A very proud, warlike group, all contacts with them up until the 1940s usually resulted in bloodshed. The Xavante now live in some 10 reserves and, as of 1989, were estimated to number 7 100 people.

Also in Mato Grosso state live the Rikbaktsa (also known as Canoeiros or Erigpaktsa), a group that numbers about 700. These people are largely hunter-gatherers with some agriculture, and represent yet another isolated language group with no close affinities to the larger language families of South America. The Rikbaktsa also possess a rich and diverse art, with some affinities to the Mundurucú, who live further to the north. Their most impressive work of art is a headdress known as the *myhara*, which consists of a crown with long macaw feathers coming down the back and flanked by long, thick strands of human hair.

The Mundurucú and the related Sateré-Mawé of the region between the Madeira and Tapajós Rivers in the states of Amazonas and Pará have been largely acculturated and integrated in the dominant *caboclo*/neo-Amazonian society of Brazilian Amazonia. The Mundurucú number some 3 000 people and the Sateré-Mawé

5 825. The Mundurucú were renowned as fierce warriors and headhunters, and used a dried skull with the skin still attached and elaborately decorated with macaw feathers as a trophy, one of the more striking artifacts of Amazonian Indian cultures. The Sateré-Mawé use an unusual "ant-glove" in their initiation ceremonies. Like the ant board of the Wayana-Apalai, these woven gloves have stinging ants inserted between the weave. The young initiate has to stick his hand into the glove and resist the painful stings of the ants without complaint.

The Urubú-Ka'apor Indians live in the eastern corner of Amazonia in the state of Maranhão. This Tupi-speaking group numbers some 500 people and lives in a 500 000-ha reserve in the Gurupi River region. Urubú-Ka'apor art differs considerably from that of other Amazonian tribes, and involves what is probably the finest handiwork of any culture in the region. Rather than emphasizing large and spectacular pieces, they make small, delicate, intricately-woven pieces in which the brilliant blue and purple feathers of the cotingas dominate. The Urubú-Ka'apors' exquisite workmanship and use of color are unmatched in Amazonia.

Moving up to the extreme northern portion of Brazilian Amazonia, on the border with Suriname and Guyana, one finds several distinctive and related Carib-speaking tribes, the Waiwai, the Wayana-Apalai, and the Tirio. Each of these straddle the border with the neighboring Guianas, with the Waiwai also living in Guyana and, to a lesser extent, in Suriname; the Wayana-Apalai in Suriname and French Guiana; and the Tirio mainly in Suriname, with one village in Brazil. The Waiwai number some 1 300-1 400 people; the Wayana and Apalai (actually two closely-related tribes speaking very similar languages), some 450 in Brazil, perhaps 250 in French Guiana and Suriname; and the Tirio between 700 and 1 000 spread over three main villages in Suriname and Brazil.

The Wayana-Apalai are best known for their elaborate initiation ceremonies, featuring the spectacular *orokó* headdress, arguably the most spectacular of all Amazonian Indian artwork, and the ritual ant boards (*kunana*) with their interwoven stinging ants and wasps. These ant boards, which may have 50 or more extremely painful *Paraponera* ants woven into them, are applied to the bodies of both boys and girls as part of their passage into adulthood. The round meeting houses of the Wayana are also decorated with the circular wooden *mariwana* boards painted with various animal and spirit figures.

Finally, the Yanomami, who live in one of the world's least disturbed wilderness areas on the border between extreme northern Brazil and southern Venezuela. As a result of their isolation, they are one of the few groups that managed to come into the 1970s with a mostly intact culture, and with a larger population than any other Amazonian group. As of 1988, they numbered 9 975 in Brazil and a 1992 census in Venezuela put their numbers there at 15 193. Unfortunately, in the latter part of the 1970s, with the Brazilian Perimetral Norte road and the discovery of gold in the region, some 50 000 *garimpeiros* or gold prospectors invaded the region, bringing with them the same problems, diseases, violence, etc. that have plagued other Amazonian Indian tribes since 1500. Entire villages were wiped out by disease, and the Brazilian prospectors did not stop at the Venezuelan border. To control the Brazilians, the Venezuelans posted larger military patrols in the Yanomami area, but the soldiers brought with them some of the same threats as the gold miners. Reacting to heavy international protests against mistreatment of the Yanomami and claims of genocide, the Brazilian government finally responded in 1990, and by 1992 had demarcated some 20 000 000 ha as Yanomami land and removed some of the gold miners. Nonetheless, as recently as July, 1993, a group of gold miners massacred Yanomami in the village of Hwaximeutheri, spurring yet another round of international protests. Clearly, the relationship between Brazilian Indians and the rest of the country's populace is still being defined, and what the eventual outcome will be for the Indians is difficult to ascertain.

In any case, beginning in the late 1970s, Brazil's Indians have come to terms with the need to fight for their own rights instead of depending on the government to look after them. In 1980, the first Indian organization, the Union of Indigenous Nations (UNIND) was founded to deal with issues such as legislation and demarcation of indigenous lands. In Brazilian Amazonia, lands designated for Indians now account for over 100 million ha, or about one fifth of Legal Amazonia (MINTER-FUNAI, 1989), so progress is being made. Nonetheless, much more needs to be done, and further articulation of the linkages between biodiversity conservation and survival of indigenous cultures in Brazil, and especially in Amazonia, is likely to provide positive results.

RUSSELL A. MITTERMEIER
GUSTAVO A.B. DA FONSECA
ANTHONY B. RYLANDS
CRISTINA GOETTSCH MITTERMEIER

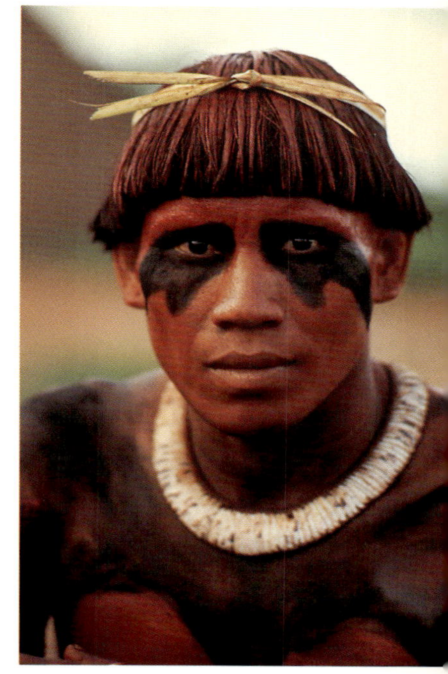

A Kuikuro man in the Xingú National Park in the southeastern quadrant of Brazilian Amazonia, with a necklace made of freshwater mussel shells.

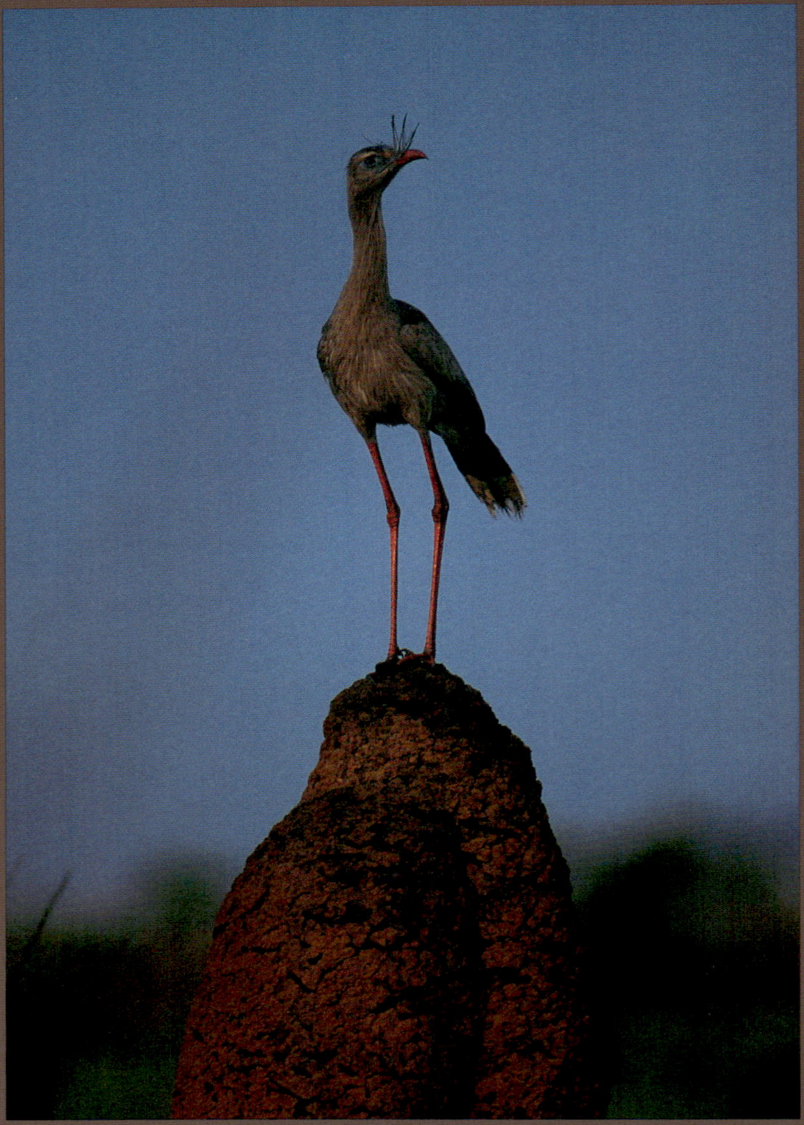

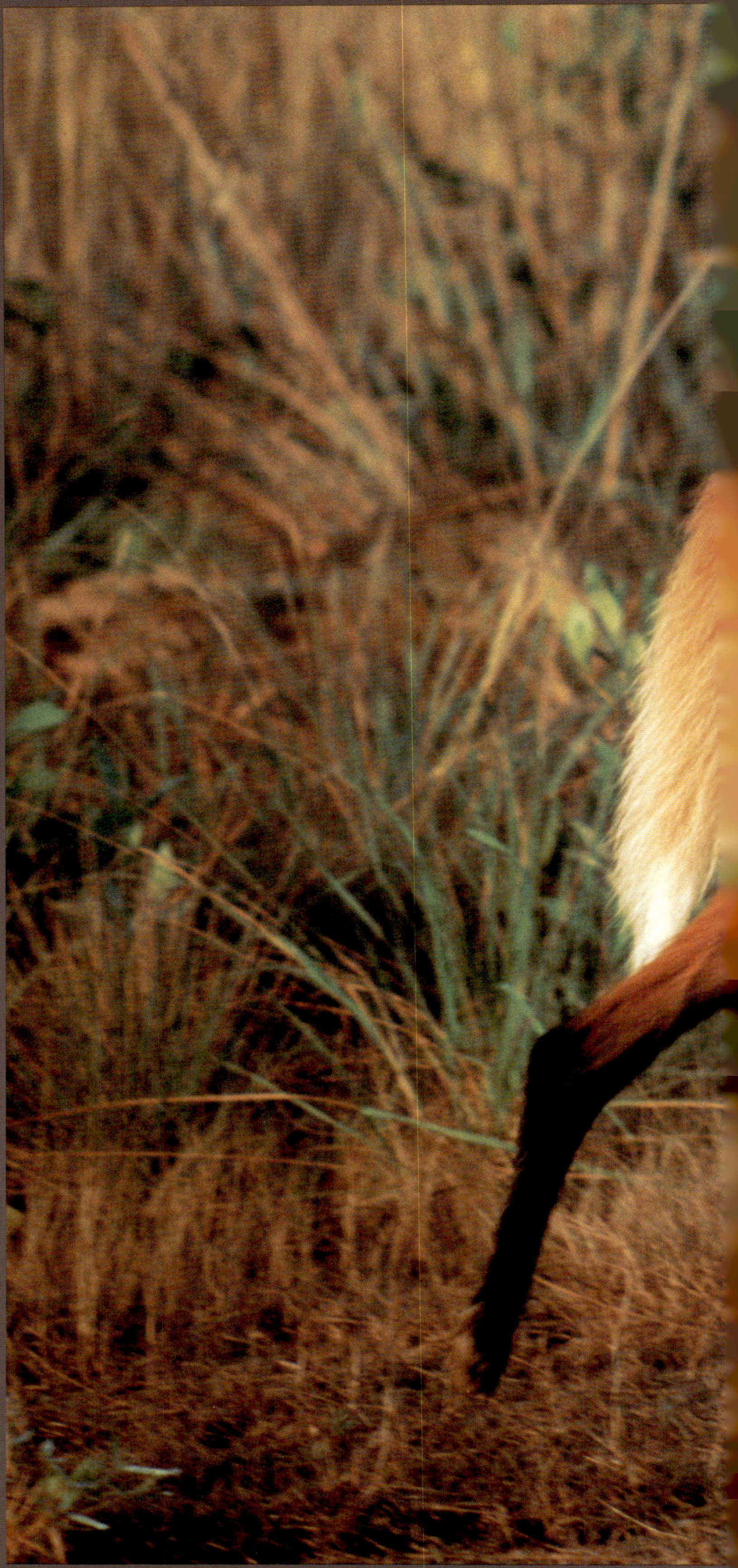

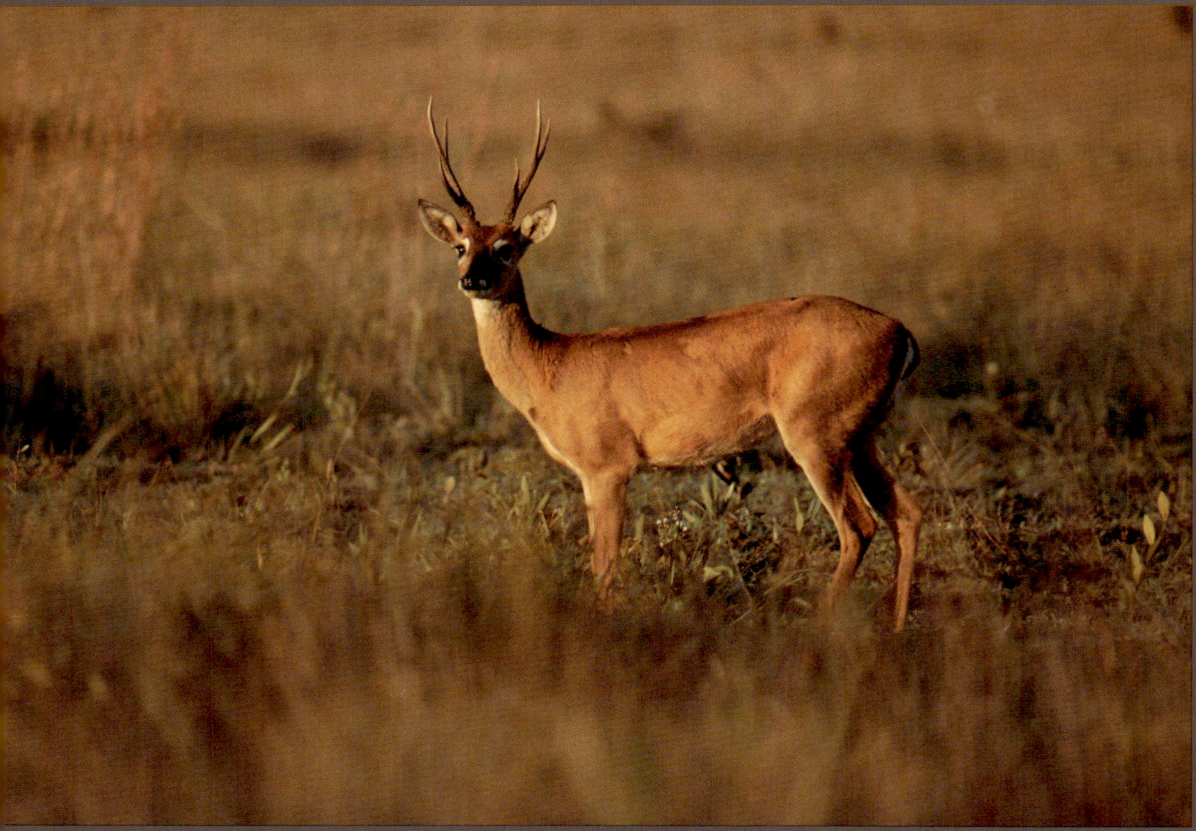

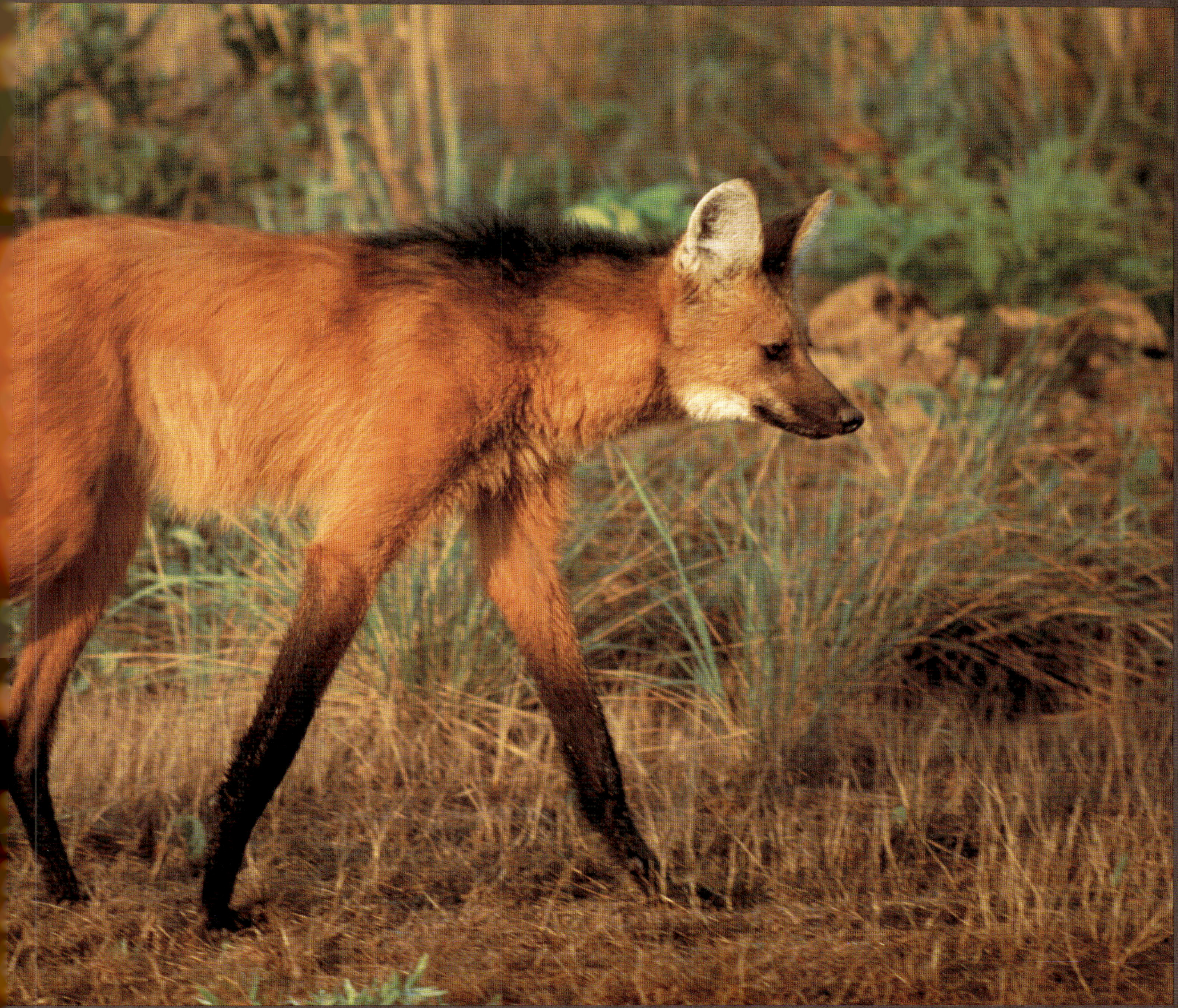

Fauna of the Cerrado. The Cerrado is a complex of grasslands, dry forests, and a variety of different transitional formations, and is the second largest ecoregion in Brazil after Amazonia. Seen here are several typical Cerrado species. Upper left, the red-legged seriema (Cariama cristata), *an ecological equivalent of the African secretary bird, is a long-legged savanna dweller that feeds on small vertebrates; this individual was photographed in Emas National Park, largest of the Cerrado protected areas. Lower left, the Pampas deer* (Ozotoceros bezoarticus) *is a deer species common in undisturbed parts of the Cerrado and the Pantanal. Above, the maned wolf* (Chrysocyon brachyurus) *looks like a cross between a large fox and an antelope; widespread, but sparsely distributed and secretive, it is rarely seen.*

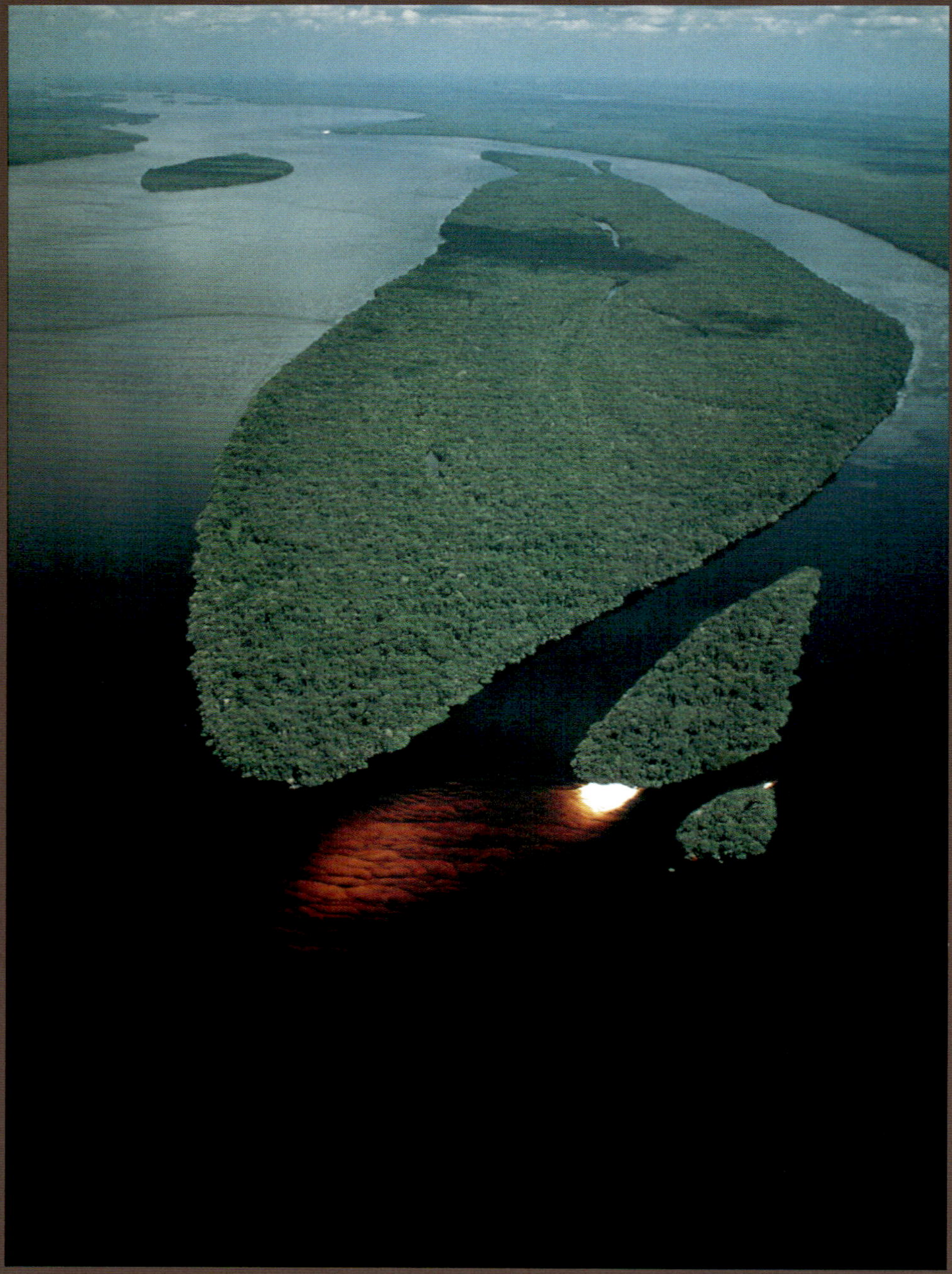

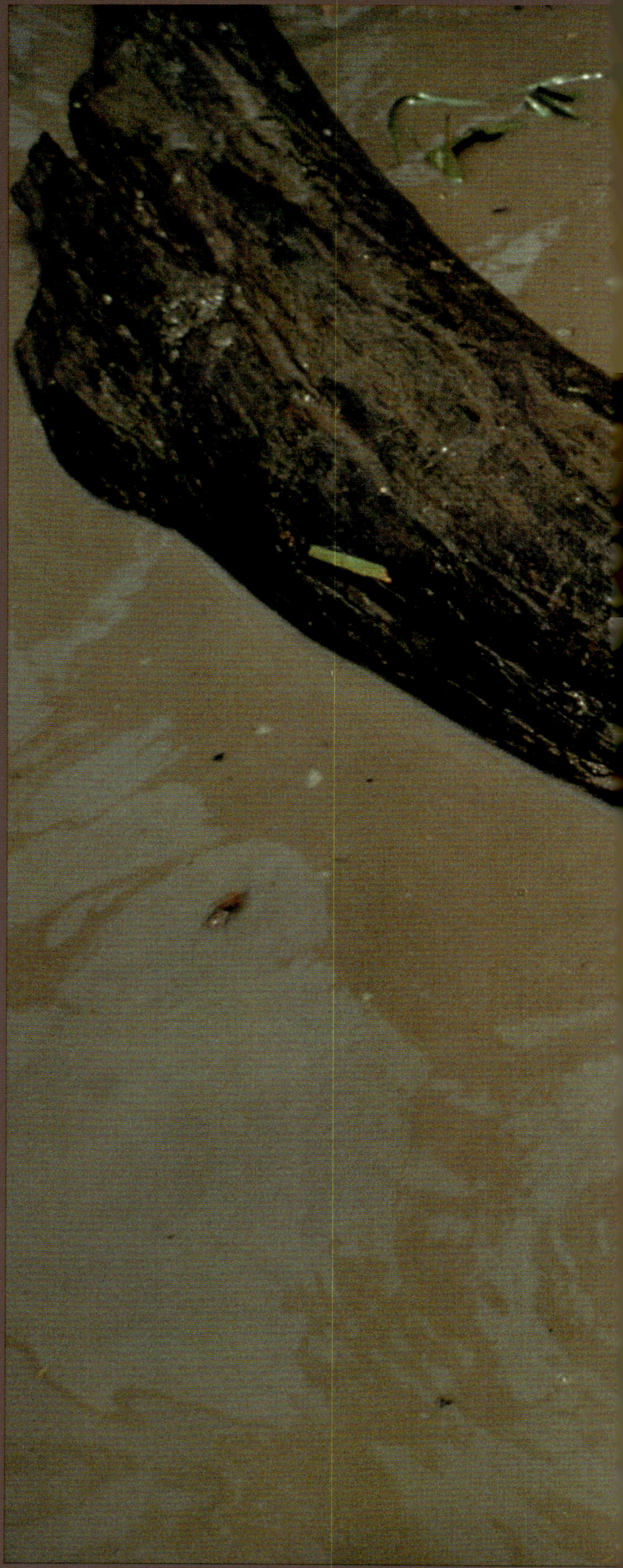

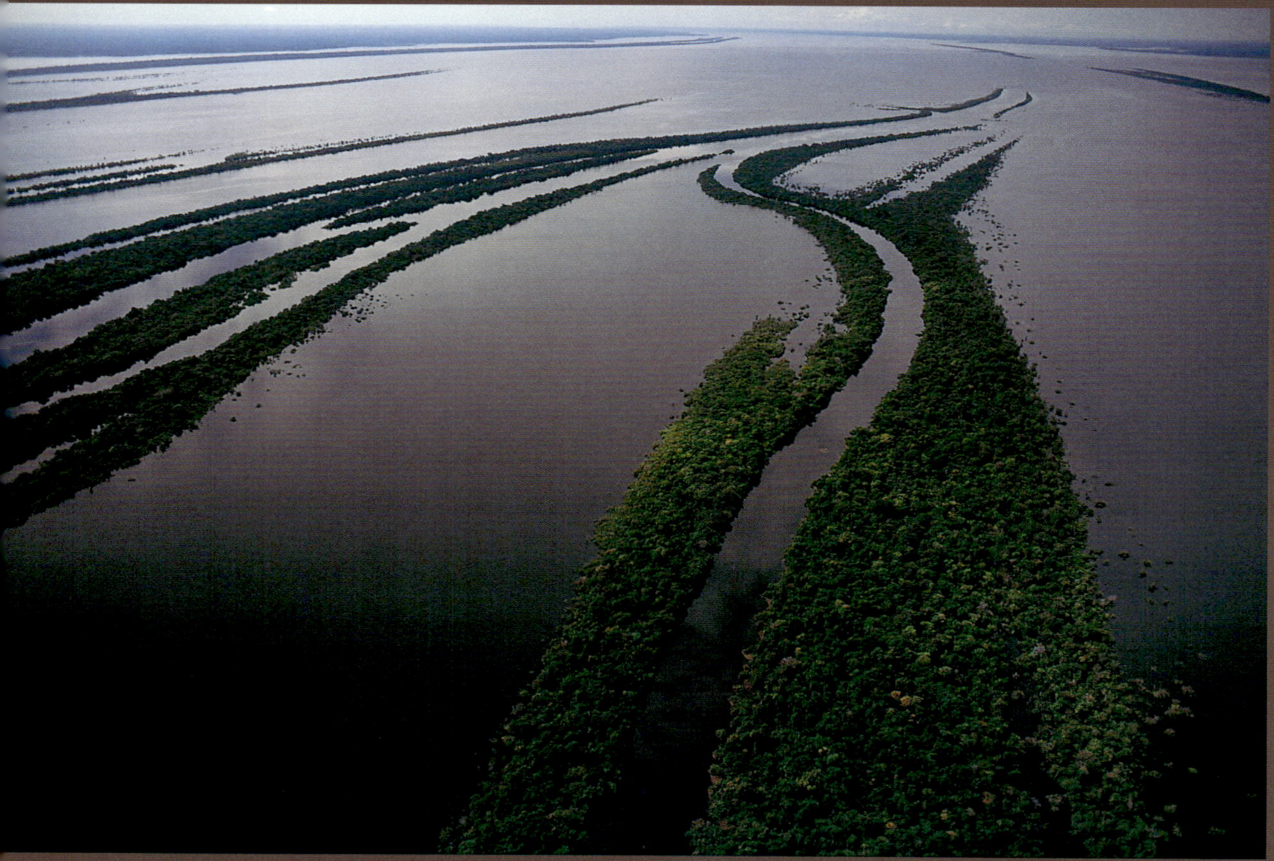

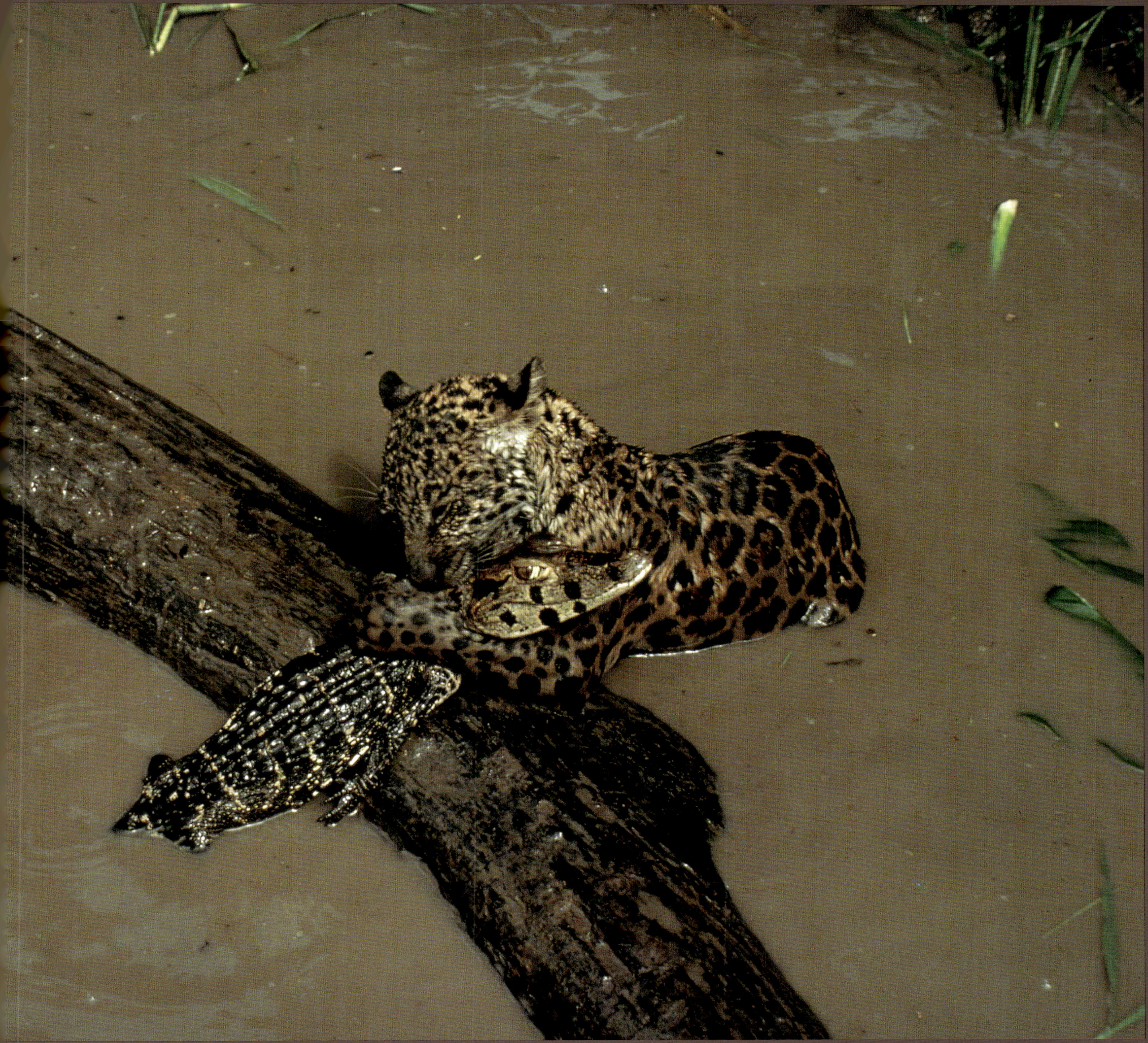

*Amazonia is by far the world's largest remaining tropical rain forest region, and 62% of it lies within the borders of Brazil. The Rio Negro, upper left, is an enormous black water tributary that enters the mainstream in the central portion of Amazonia, just below the city of Manaus. Lower left, the Negro's waters, stained by organic materials to the color of dark tea, flow over a white sand bank. The Anavilhanas archipelago near the mouth of the Rio Negro consists of a series of long, narrow fluvial islands that are essentially a delta of this huge river; the islands are protected in the Anavilhanas Ecological Station shown here, lower left. Above, one flagship species consumes another, as a young black caiman (*Melanosuchus niger*), largest of the South American crocodilians, falls prey to a juvenile jaguar (*Panthera onca*).*

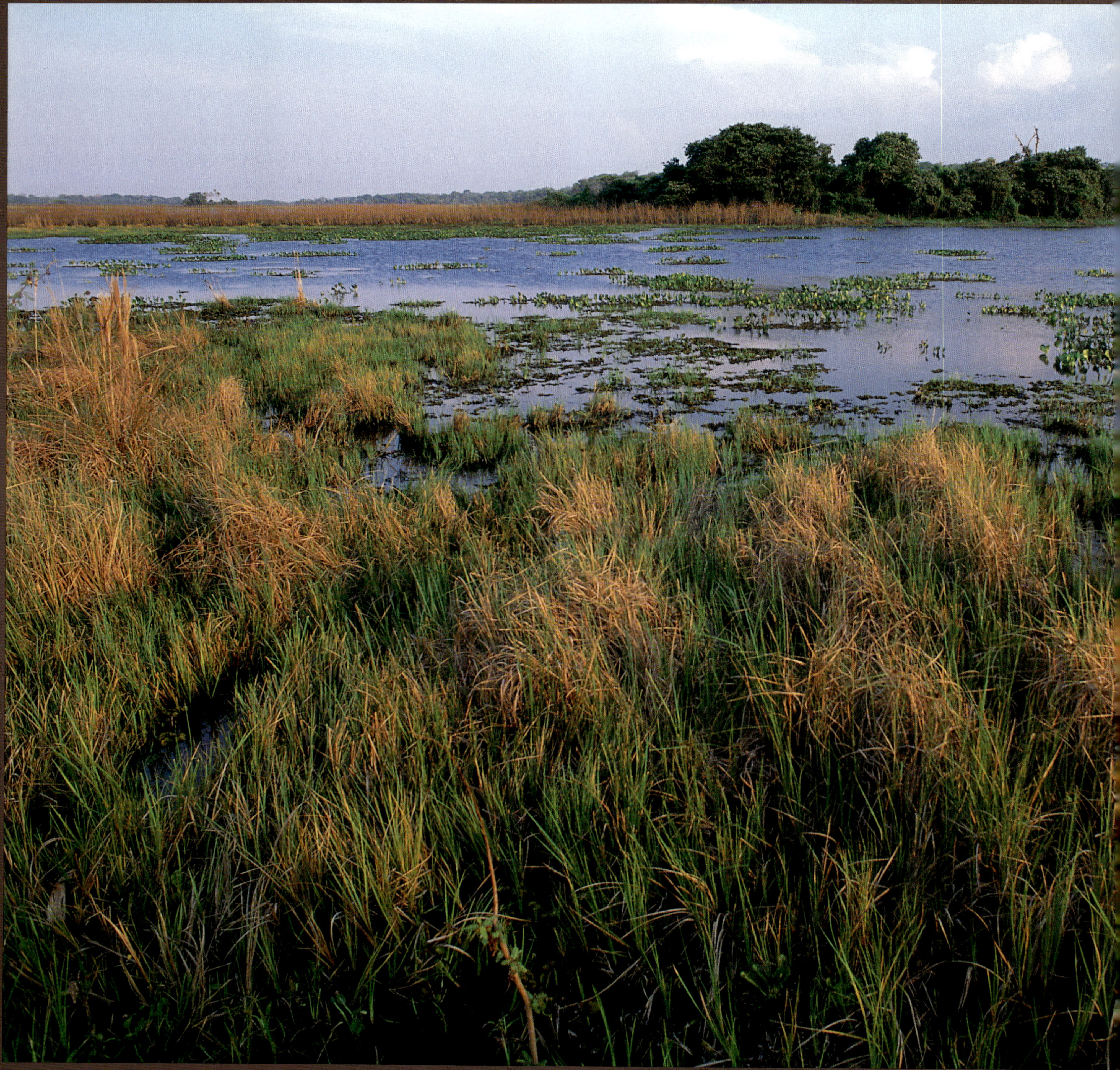

*The Pantanal of Mato Grosso is the world's largest wetland, and much of it is flooded during a good part of the year. Above, an island of dry land, known locally as a capão, serves as a refuge for terrestrial species in the midst of a flooded grassland. Upper right, the hyacinth macaw (*Anodorhynchus hyacinthinus*), is the largest of the macaws, and one of the two most important avian flagship species of the Pantanal, the other being the jabiru. Although this magnificent macaw ranges beyond the Pantanal, this region is its last major stronghold. Lower right, the red brocket deer (*Mazama americana*) is a widespread South American deer species that occurs both in the rain forests of Amazonia, and in forest islands and more open areas in the Pantanal.*

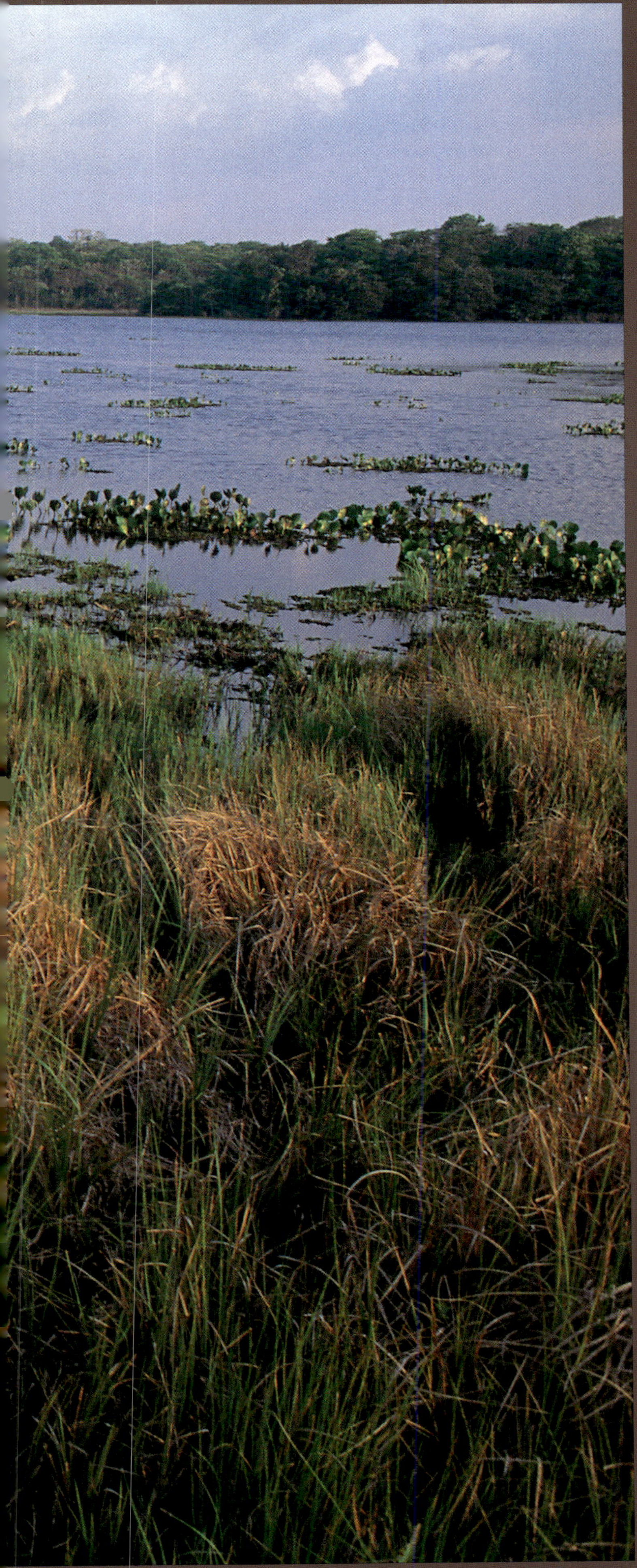
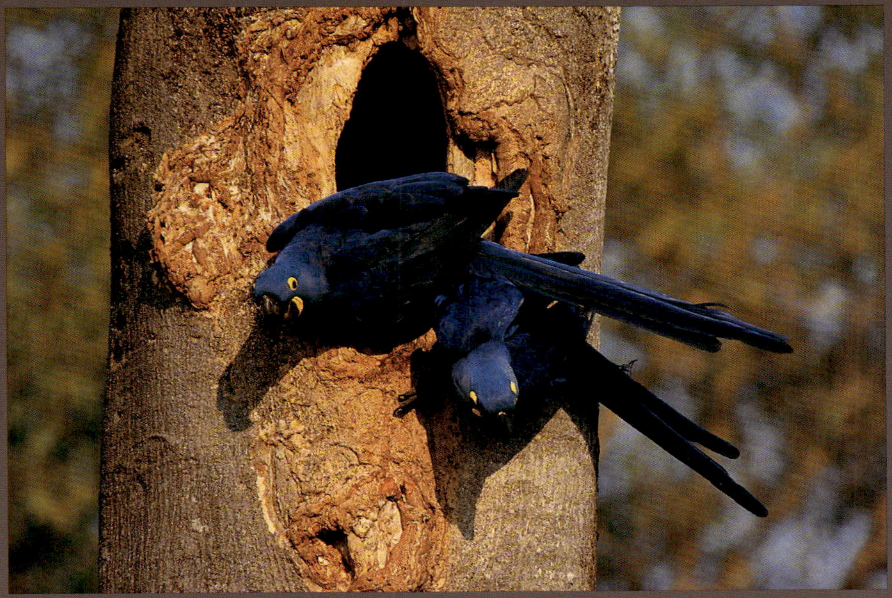
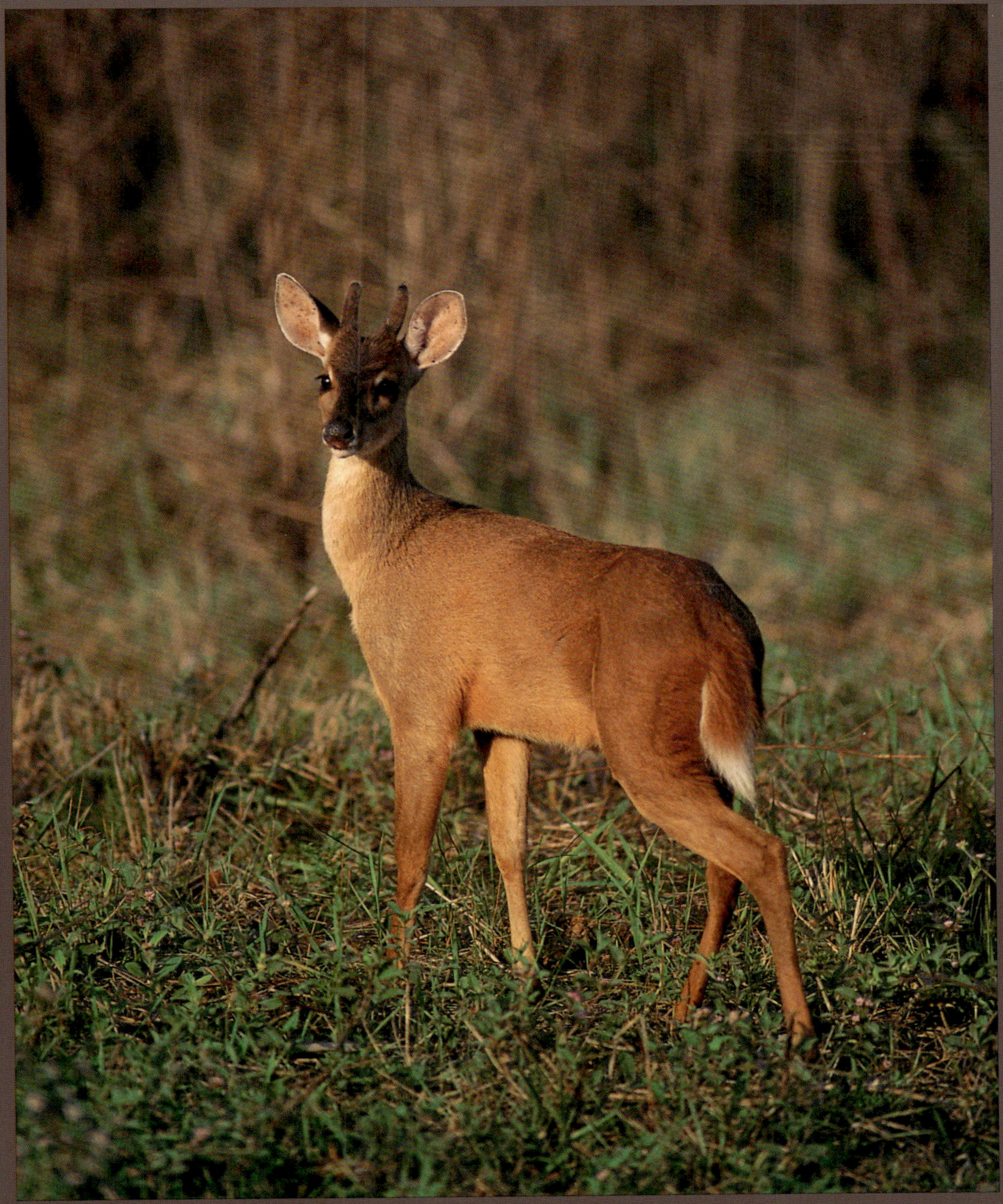

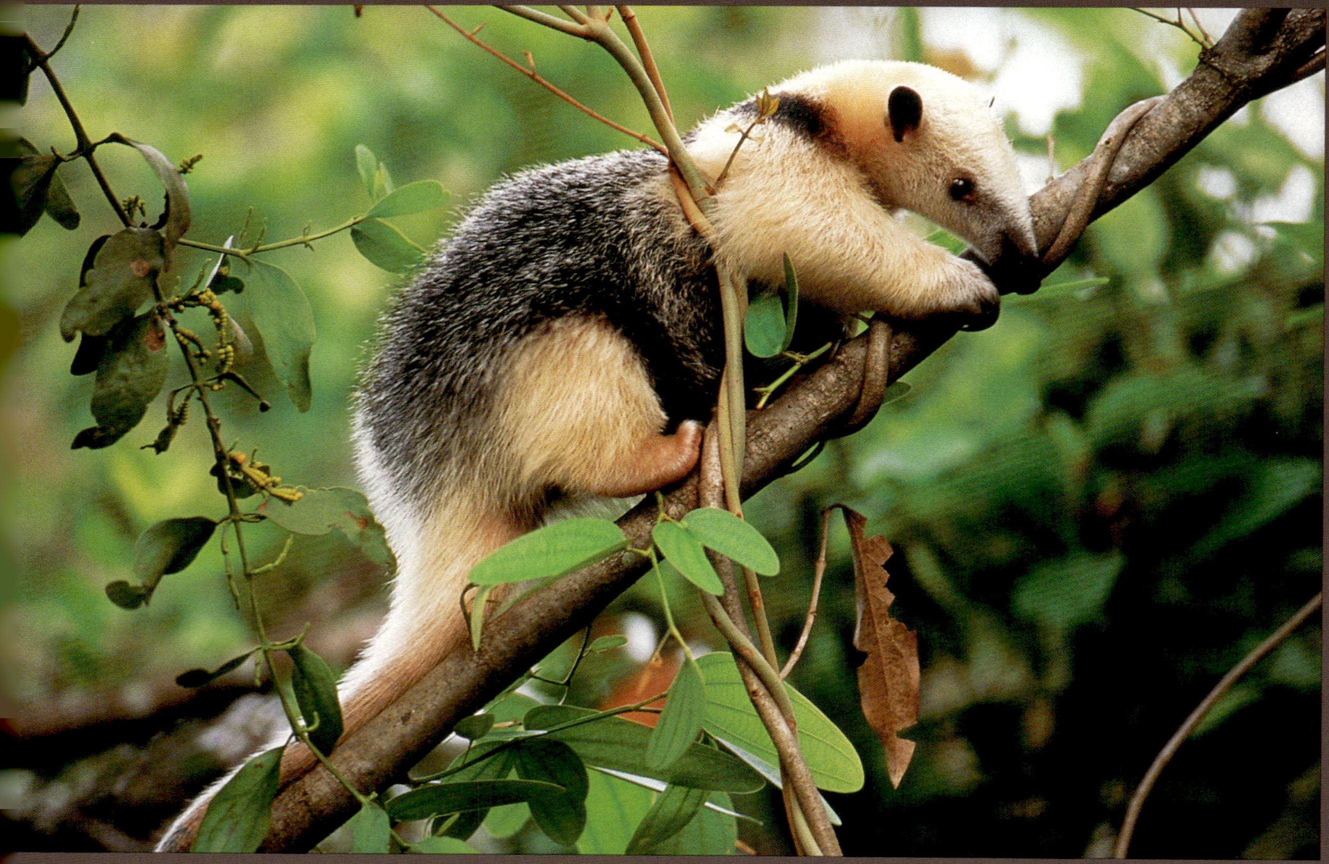

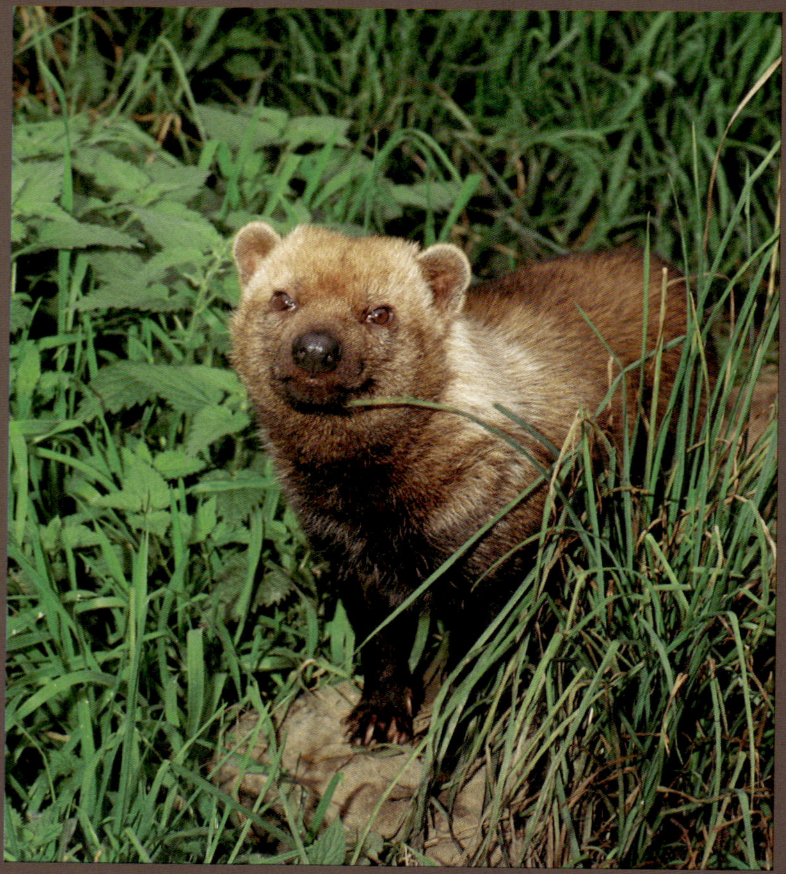

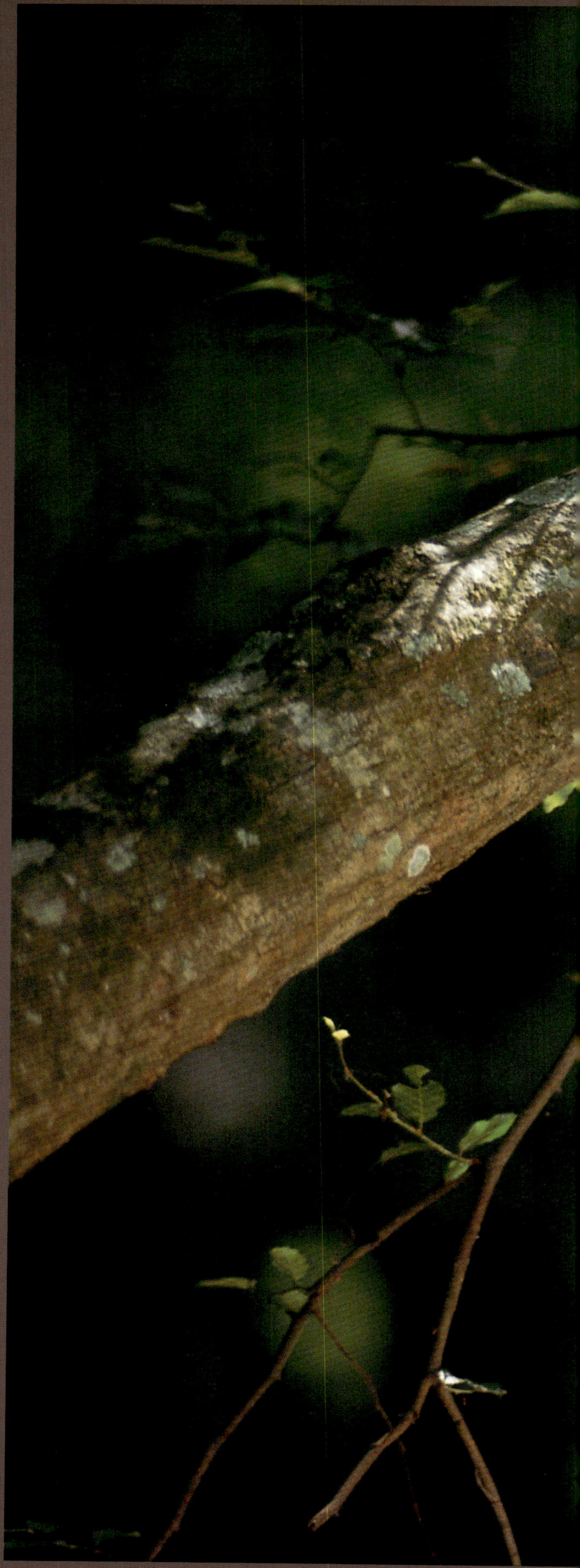

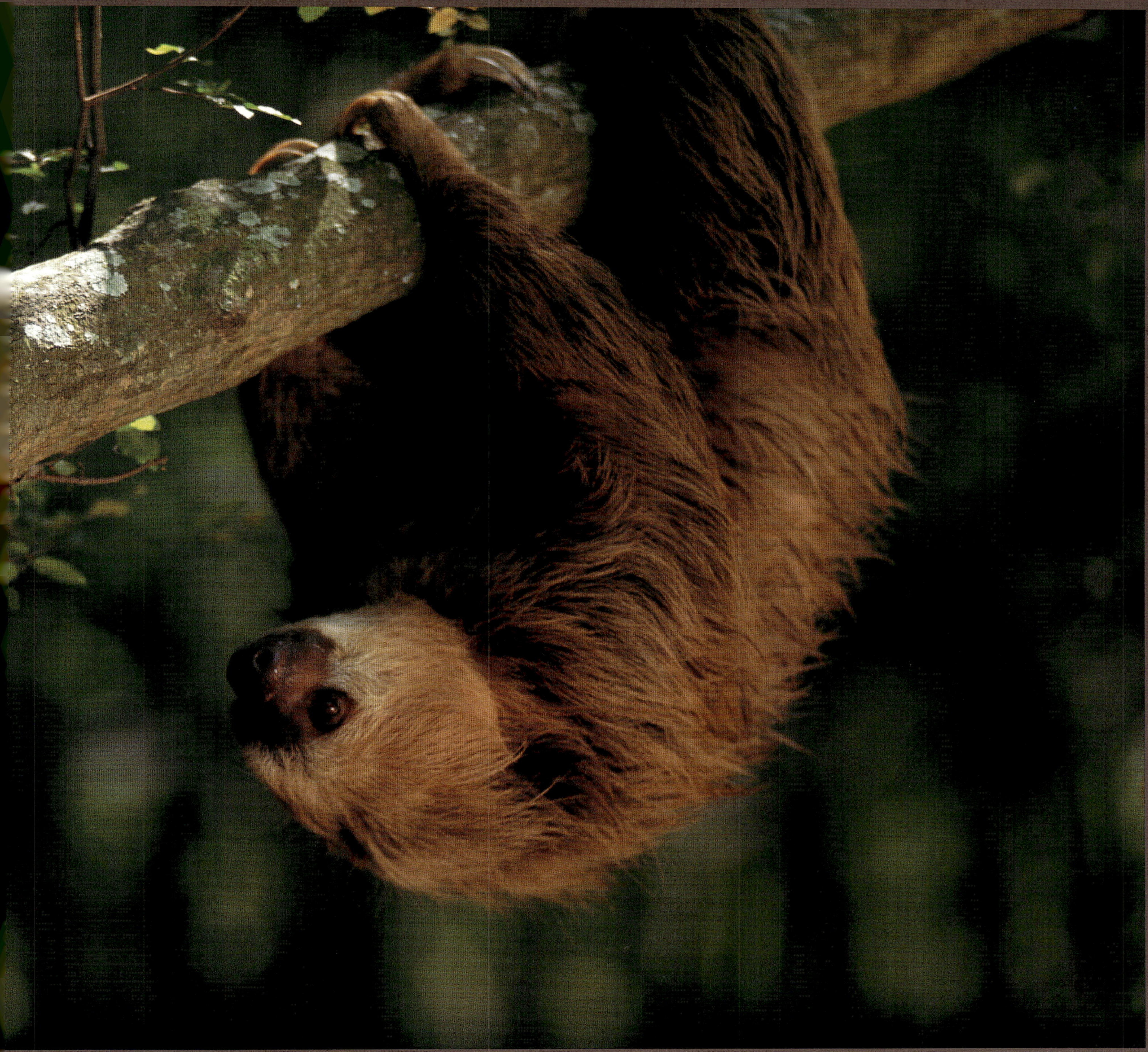

Brazil has the world's highest mammal diversity, with 524 described species. Shown here are some of the more interesting non-endemic species. Upper left, the tamandua, or collared anteater (Tamandua tetradactyla), *is the medium-sized member of the anteater radiation in South America, which also includes the giant anteater, and the tiny silky or pygmy anteater* (Cyclopes didactylus); *this tamandua was photographed in the Pantanal. Above, the two-toed sloth* (Choloepus didactylus) *is the larger and more aggressive of the two living South American sloth genera. Lower left, the bush dog* (Speothos venaticus) *is a small wild dog that is very common in northern South America, but is sparsely distributed and rarely seen.*

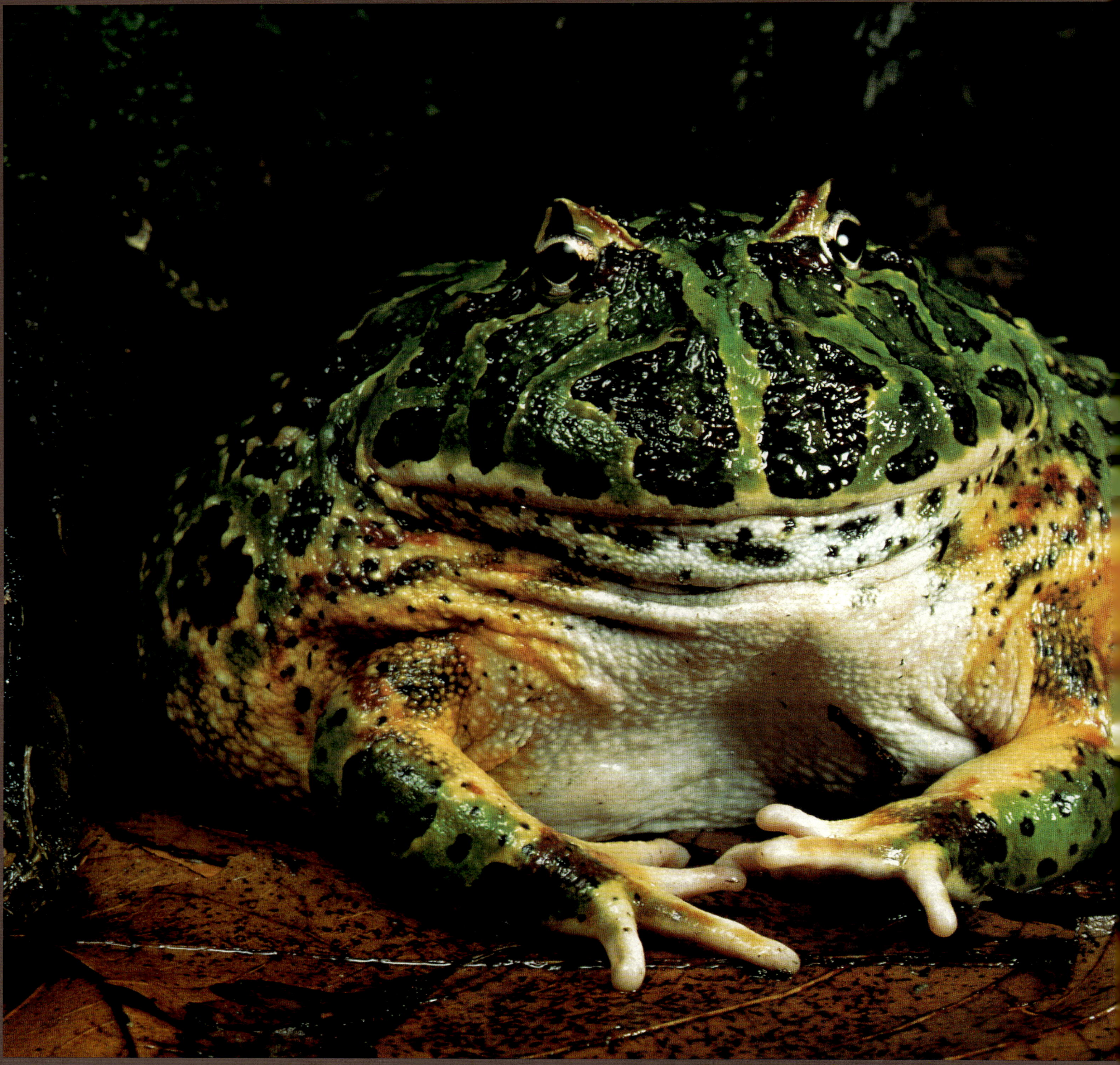

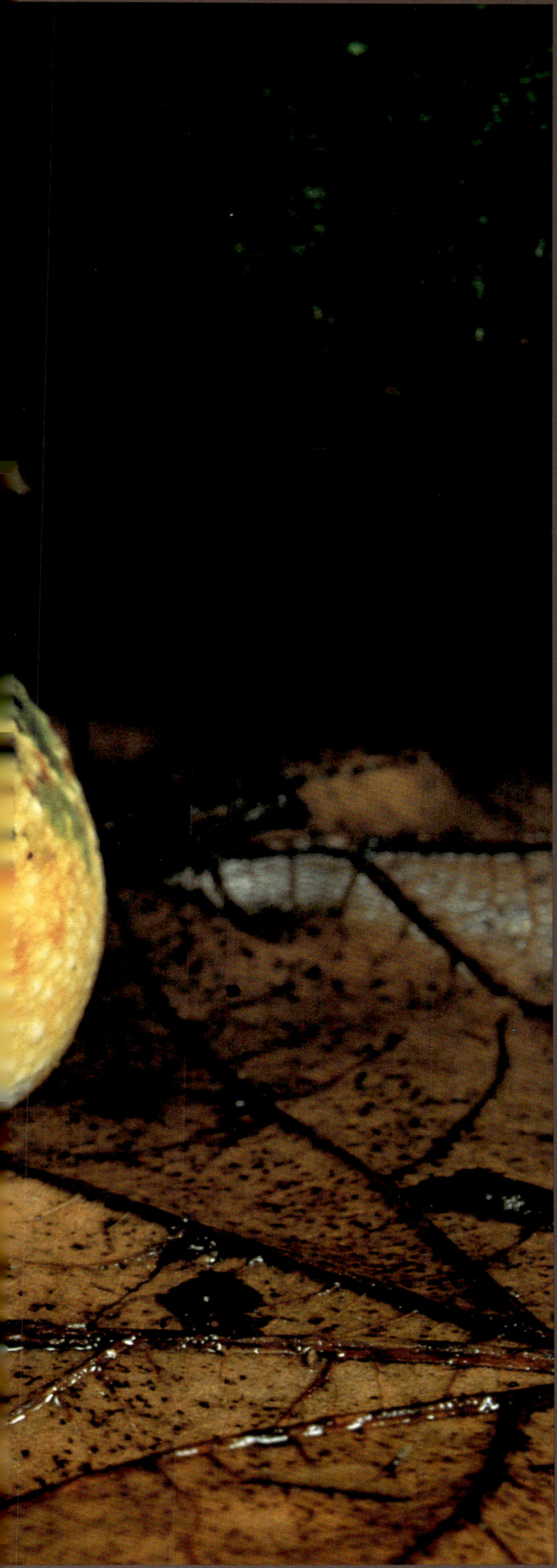

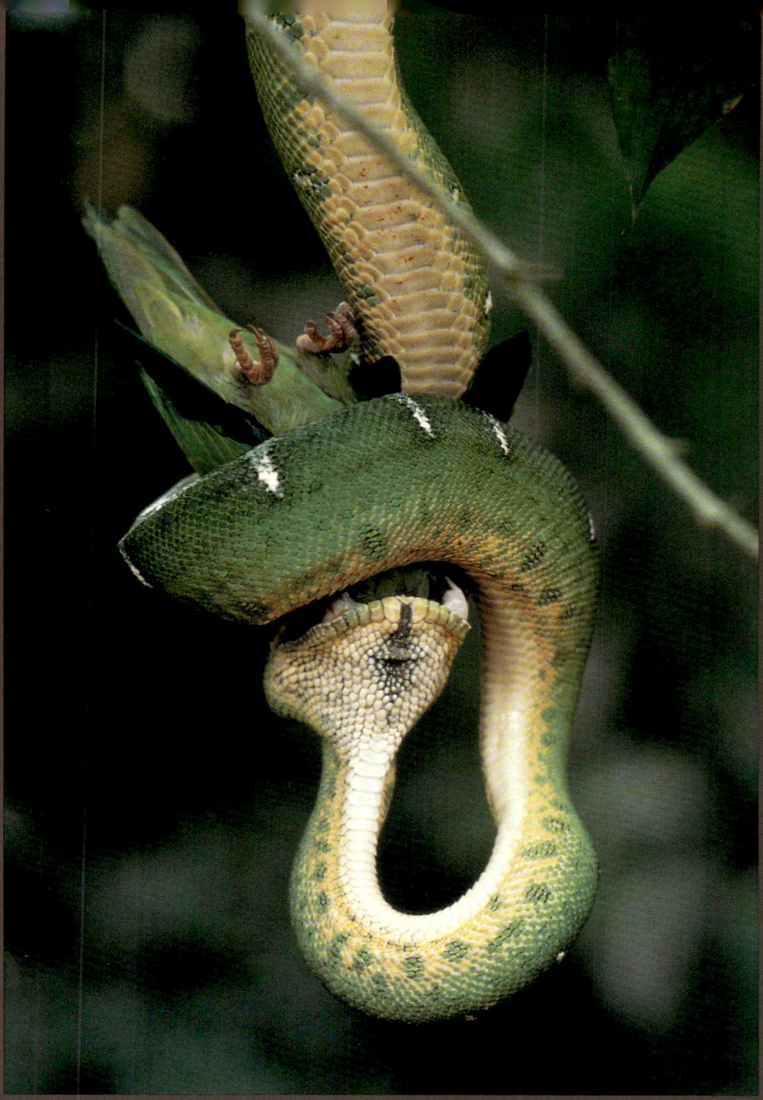

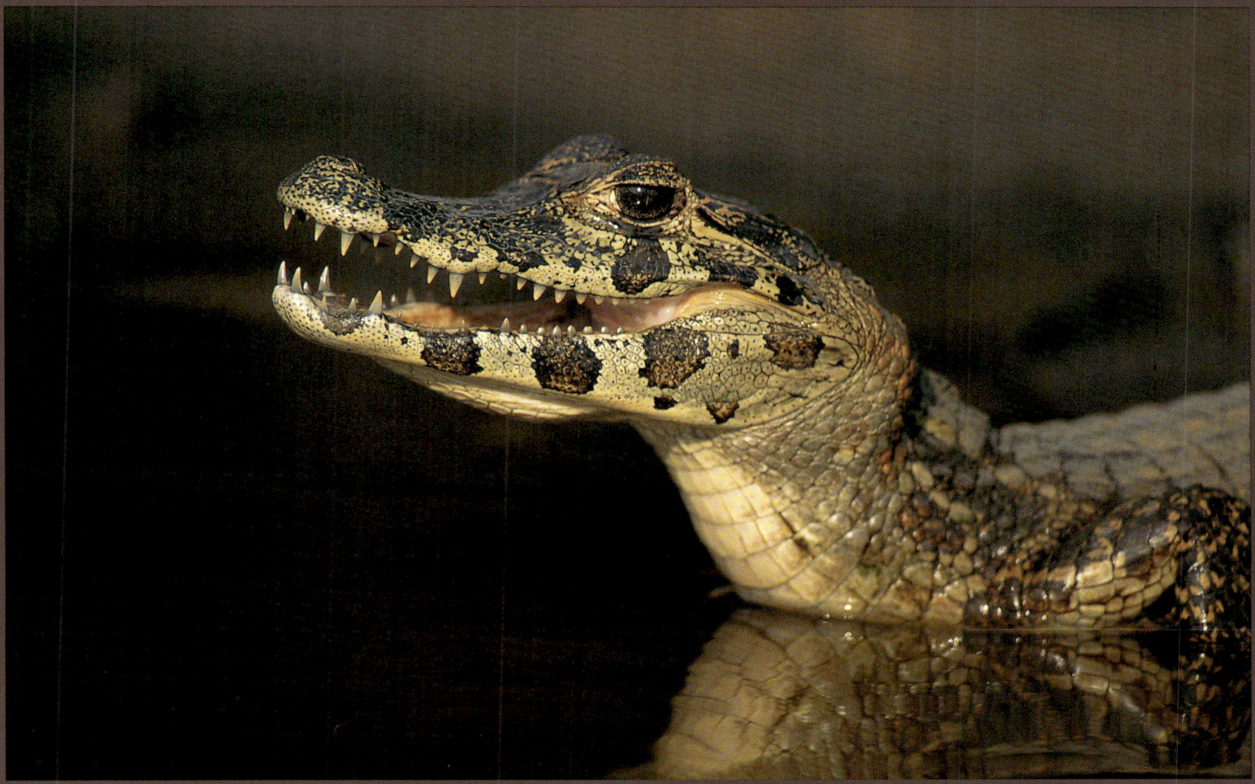

Brazil is the world's second richest country in amphibians and fifth richest in reptiles. Left, the huge frog-eating frog (Ceratophrys ornata) *is a sit-and-wait predator that hides in leaf litter on the forest floor, and pounces on any animal that passes by, including mice, rats, and other small mammals. Upper right, the emerald tree boa* (Corallus caninus) *is an arboreal predator widespread in the rain forests of Amazonia; here, one is swallowing a parrot that it snatched out of the air with its long, curved teeth. Lower right, a juvenile black caiman* (Melanosuchus niger) *will grow into one of the world's three largest reptiles; it can reach 6-7 m in length and vies with the saltwater crocodile of the Asia-Pacific region and the gharial of South Asia for the title of largest living reptile.*

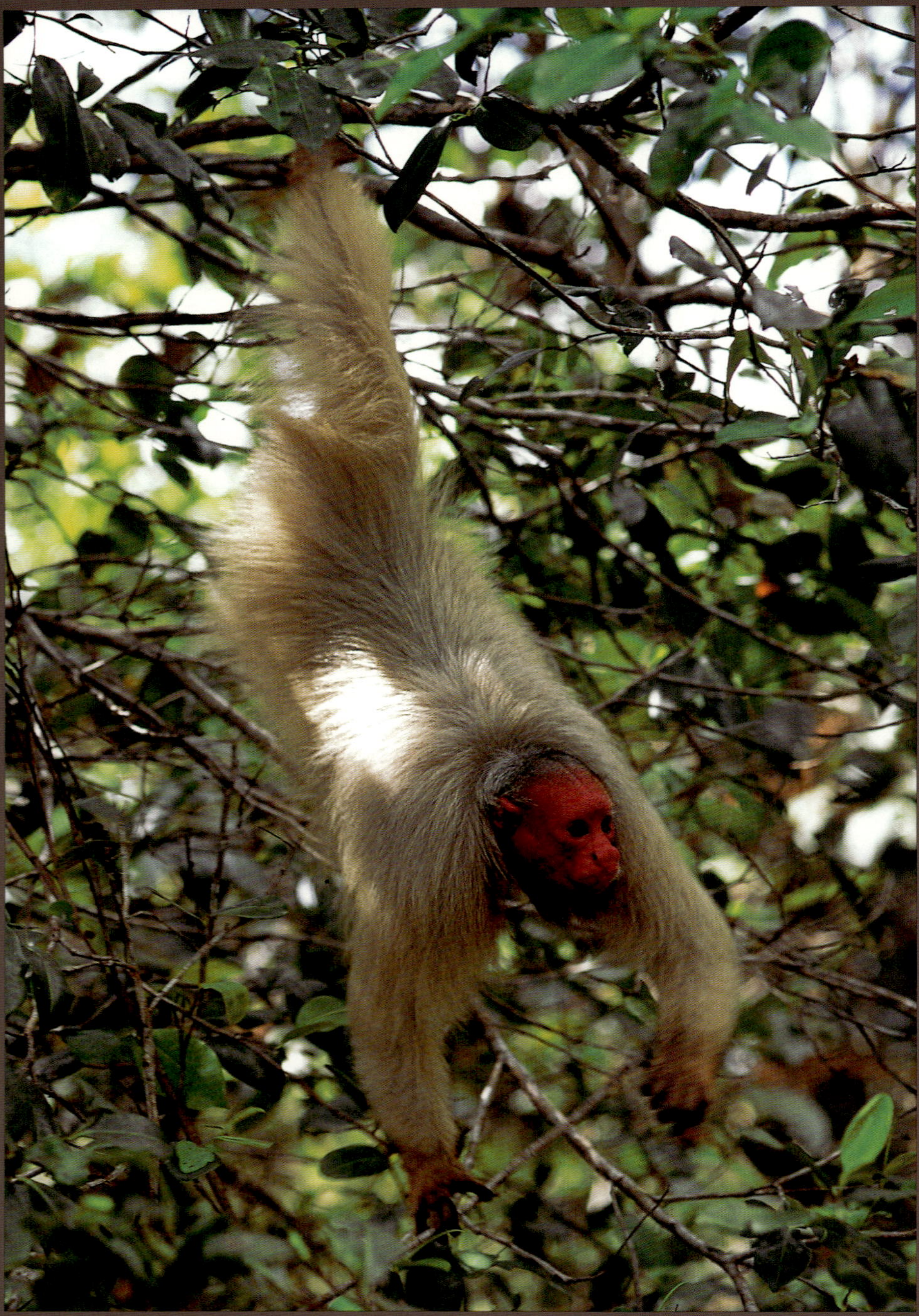

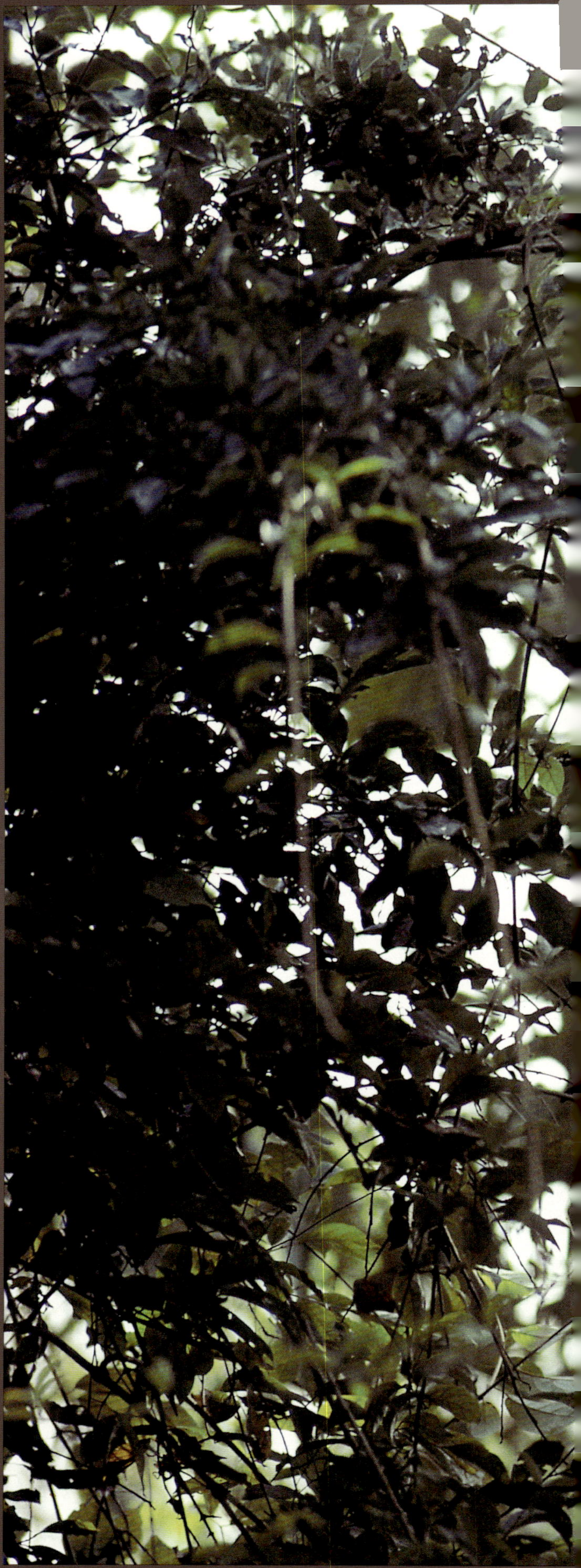

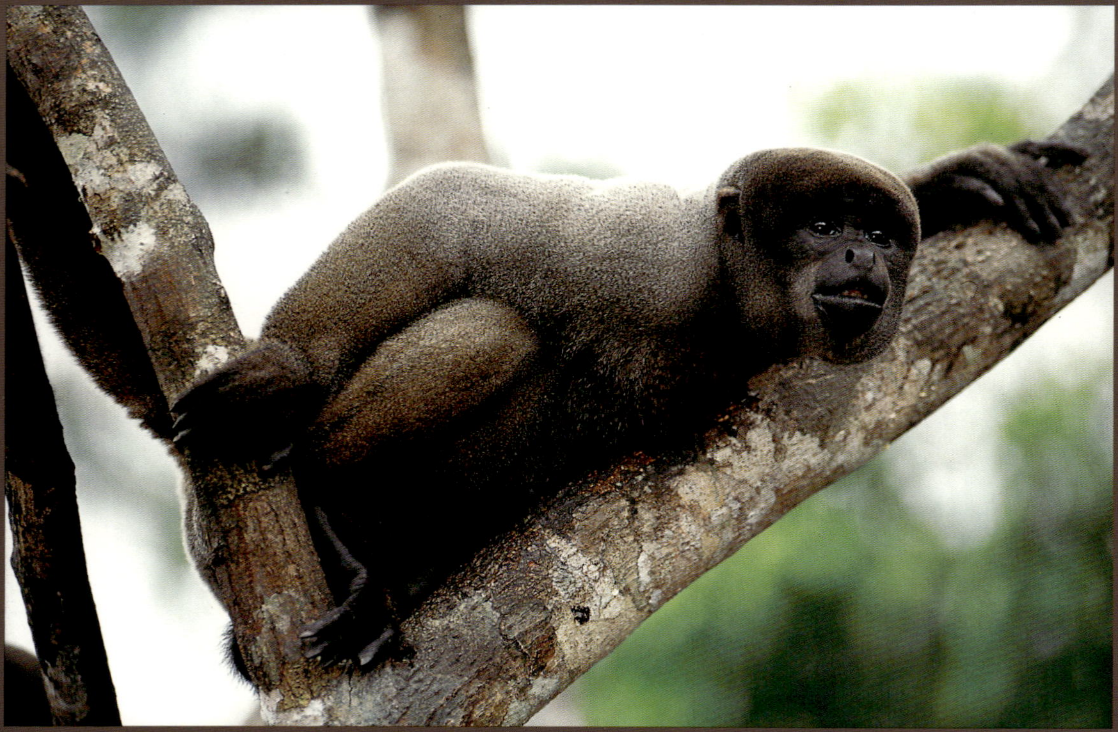

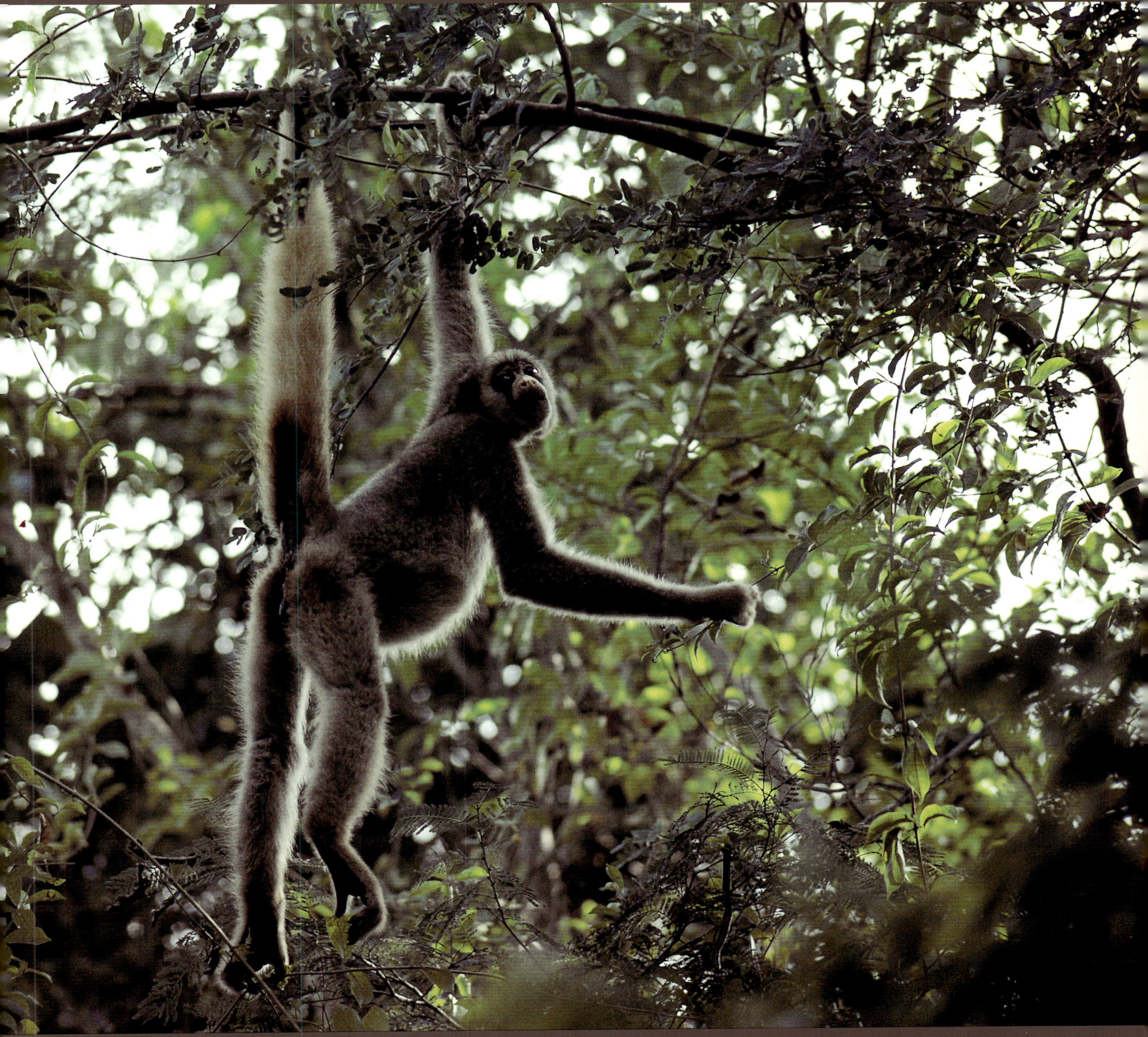

The muriqui, above, is the largest of the South American monkeys and also the largest mammal endemic to Brazil. Once thought to consist of only one species, it has now been divided into two distinct species. The one shown here, Brachyteles hypoxanthus, *is found in remaining forest patches in the states of Minas Gerais and Espírito Santo; this one was photographed in the single most important area for this species, Fazenda Montes Claros in Minas Gerais. Upper left, the white uakari* (Cacajao calvus calvus) *is another Brazilian endemic and is found only in a small area of várzea forest, in the floodplain of the upper Brazilian Amazon. The uakaris are the only South American monkeys with short tails. This one was photographed in the Mamirauá Ecological Station. Lower left, the woolly monkeys* (genus Lagothrix) *are among the largest of the Amazonian monkeys, and prefer undisturbed primary forest habitat. They are heavily impacted by hunting pressure, being one of the preferred monkey food species.*

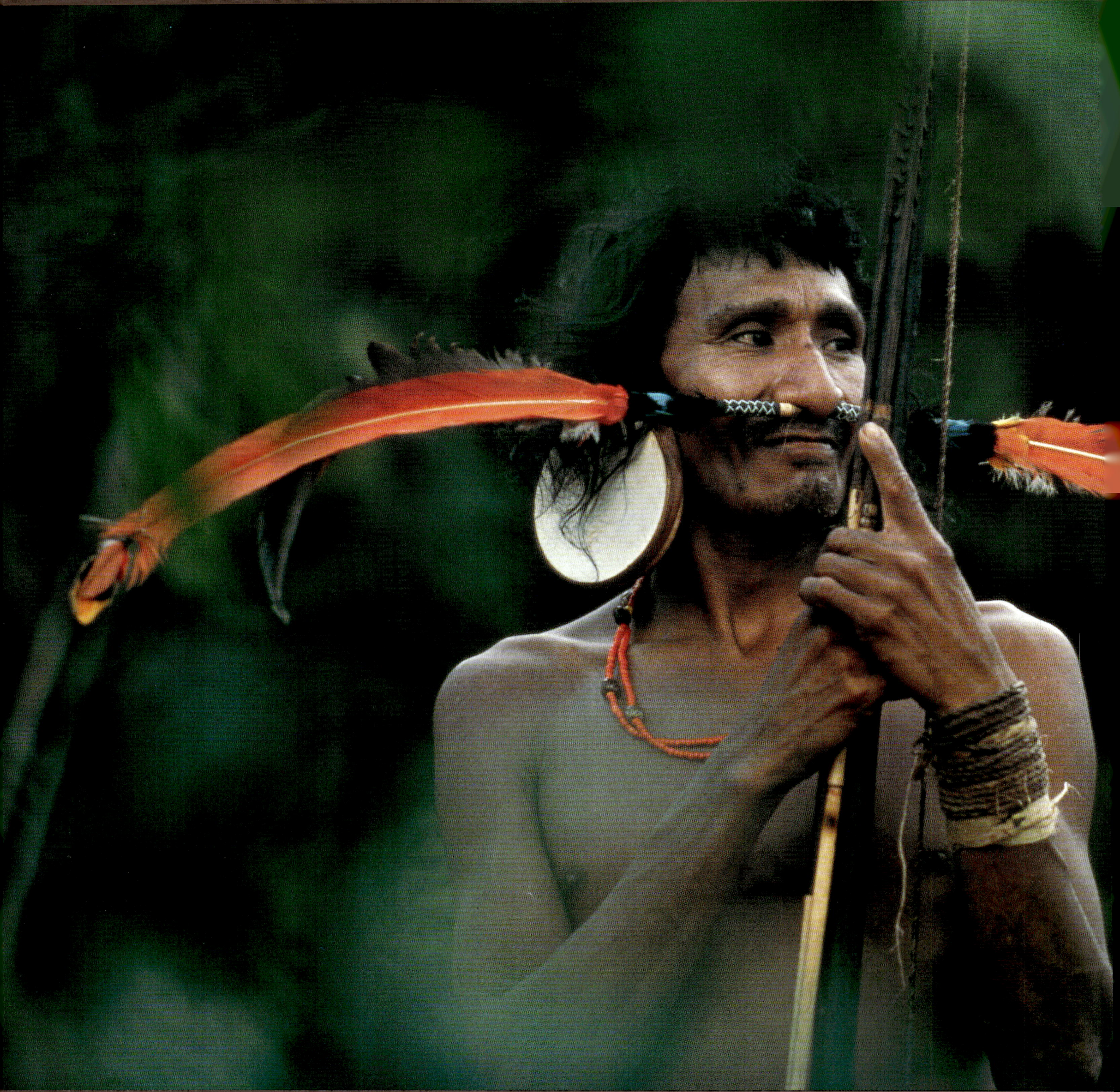

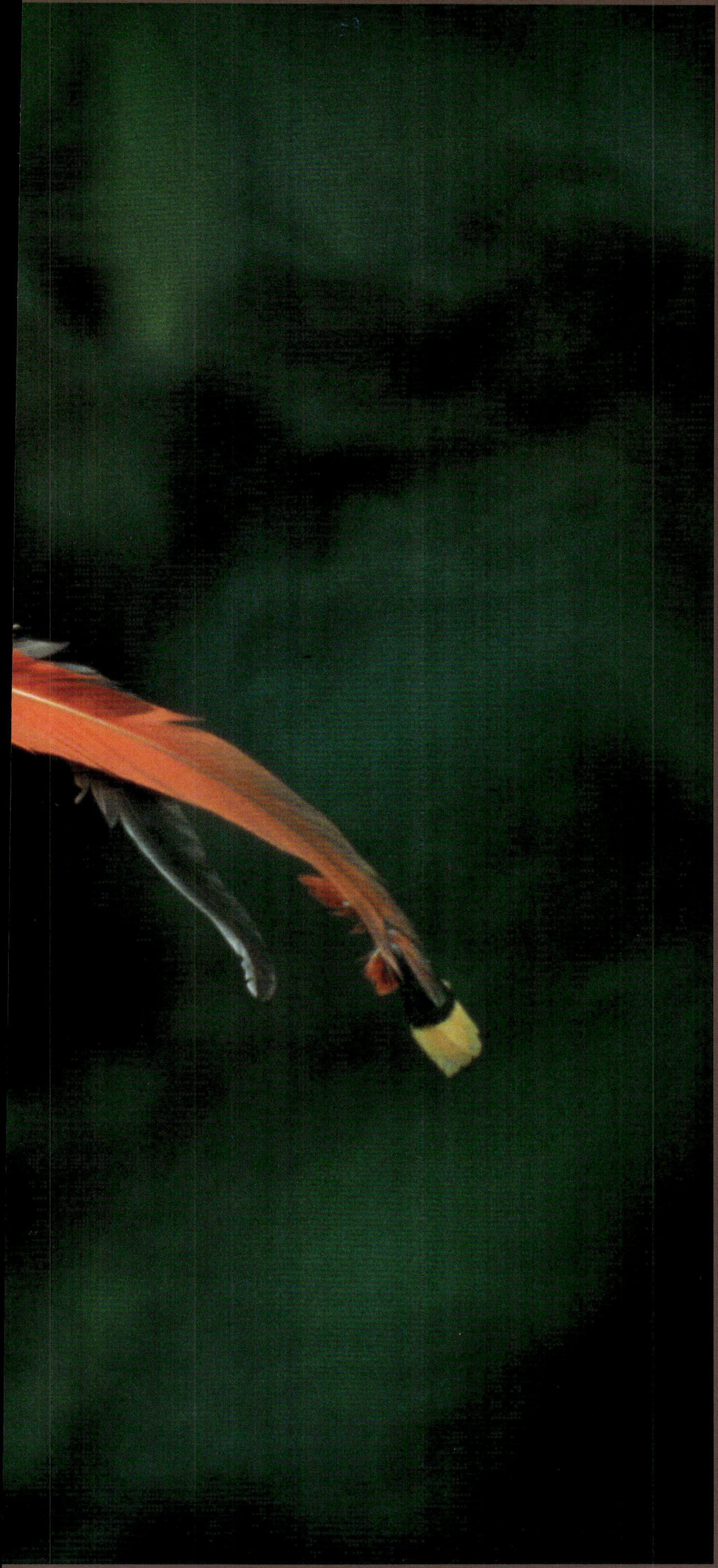

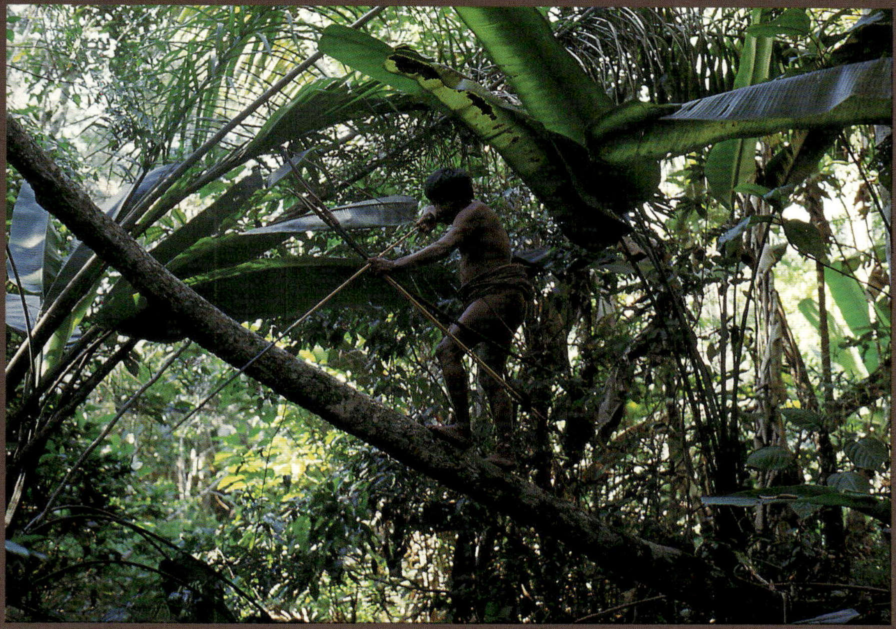

Most of Brazil's remaining Indian tribes are very small, with many of those that are left numbering less than 1 000 individuals. The Rikbaktsa and the Uru-eu-wau-wau are among the smaller tribes. Left, this Rikbaktsa warrior from Mato Grosso has a 15-cm balsa wood disk in his earlobe and two long macaw feathers piercing his nasal septum. Above, the Uru-eu-wau-wau warrior is hunting with an arrow, with his typical cane belt (Rondônia).

The Yanomami Indians are one of the largest remaining Amazonian tribes, and live in a very remote region on the Brazil-Venezuela border. Unfortunately, their previously little-disturbed home is now being invaded by garimpeiros, *the infamous Brazilian gold prospectors, and the Yanomami are starting to suffer many of the same problems of contact that have plagued Amazonian Indians for centuries. Below, a group of Yanomami cut up a giant anteater that has been killed for food. On the right, a group of Yanomami in their traditional* shabono *drink plantain soup from a large trough made of a hollowed-out log. In the photo, the individual on the right wears a traditional arm ornament made of macaw feathers.*

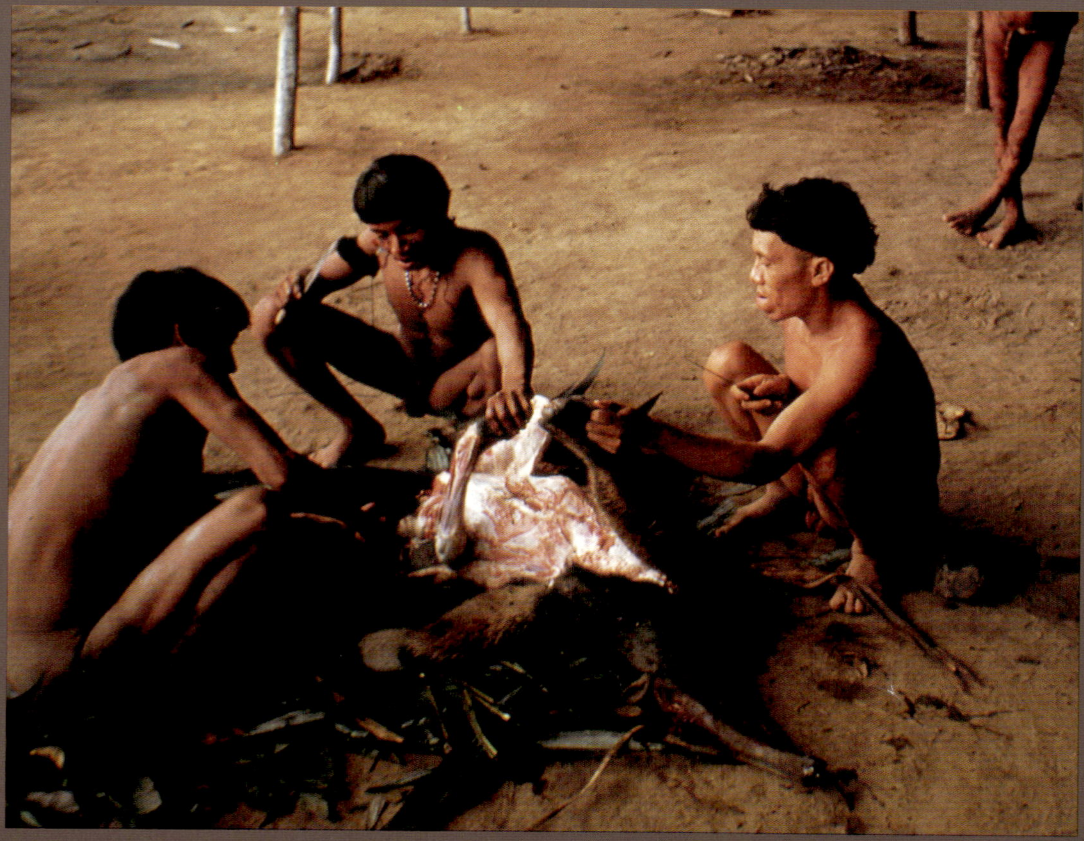

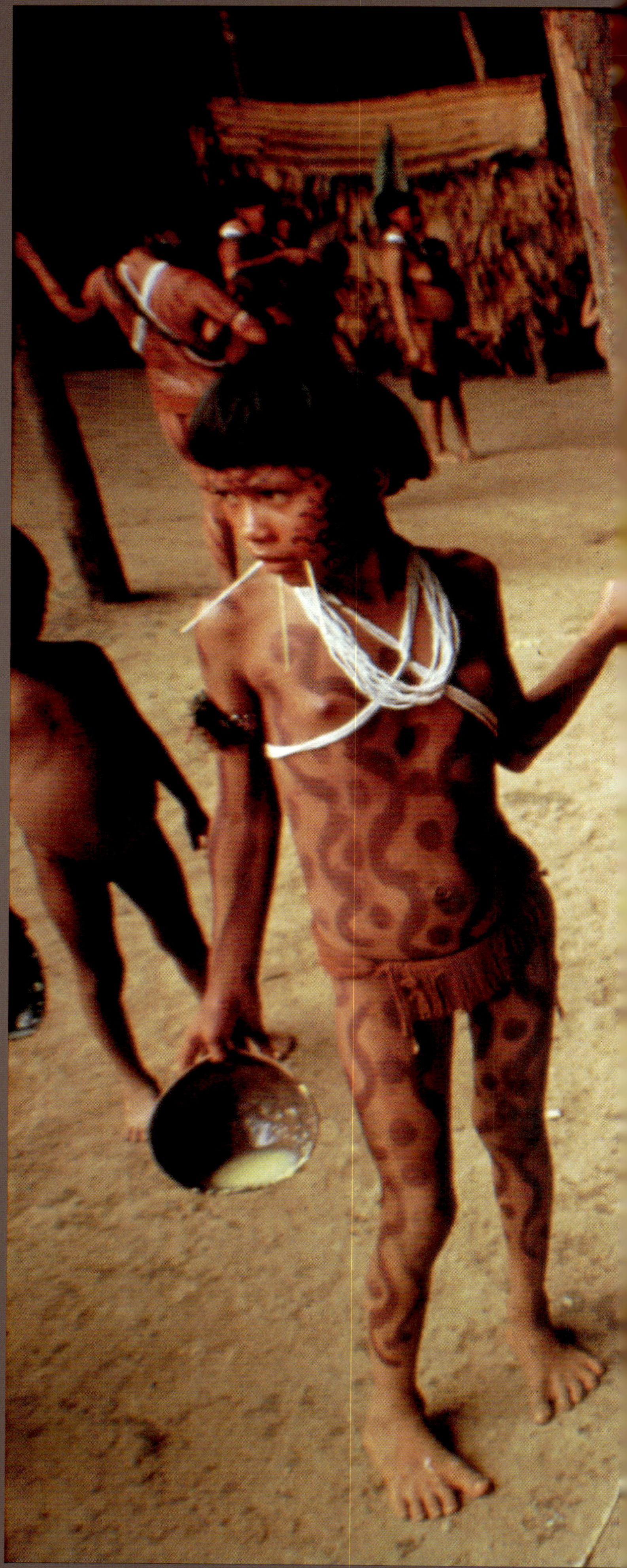

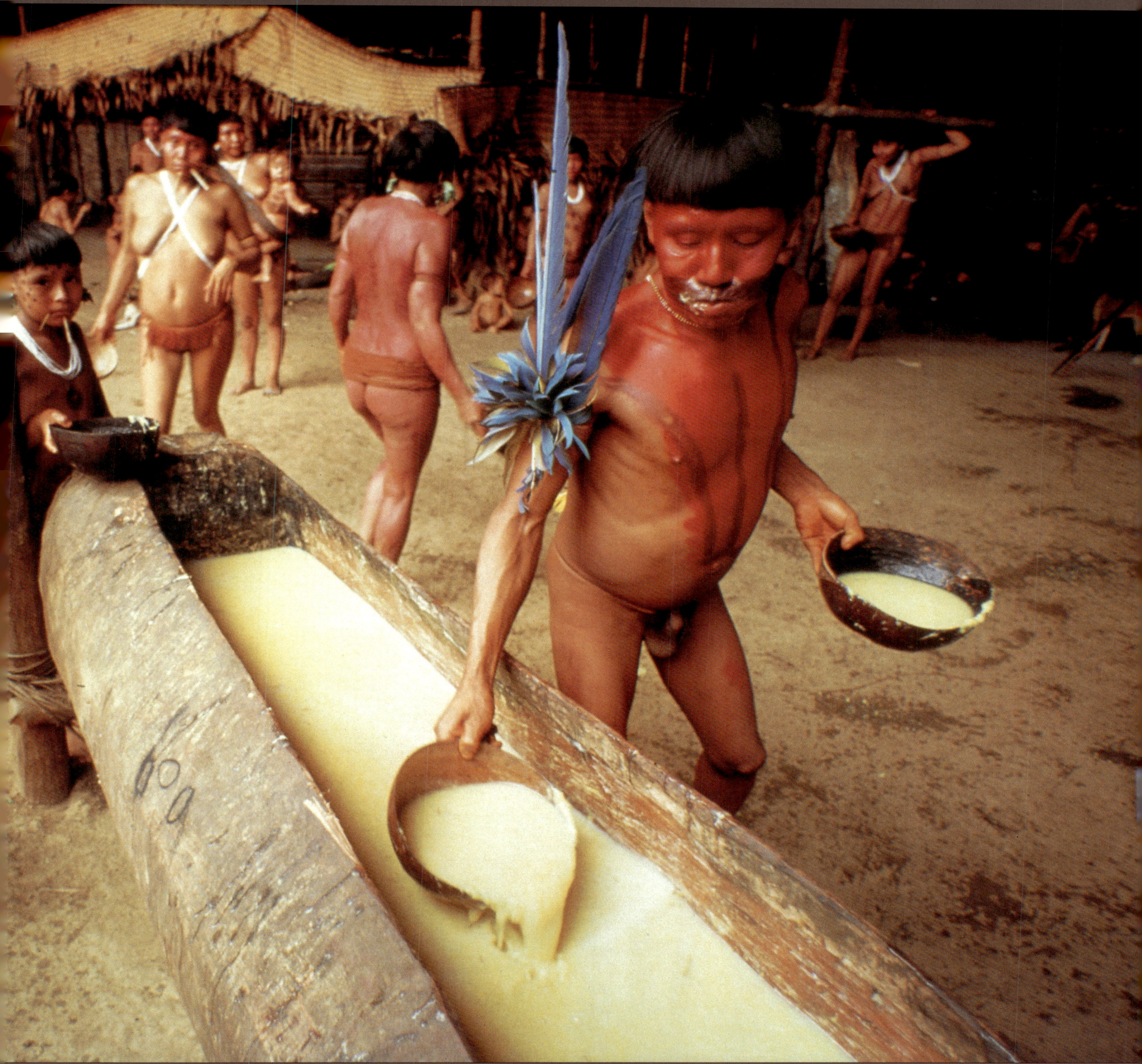

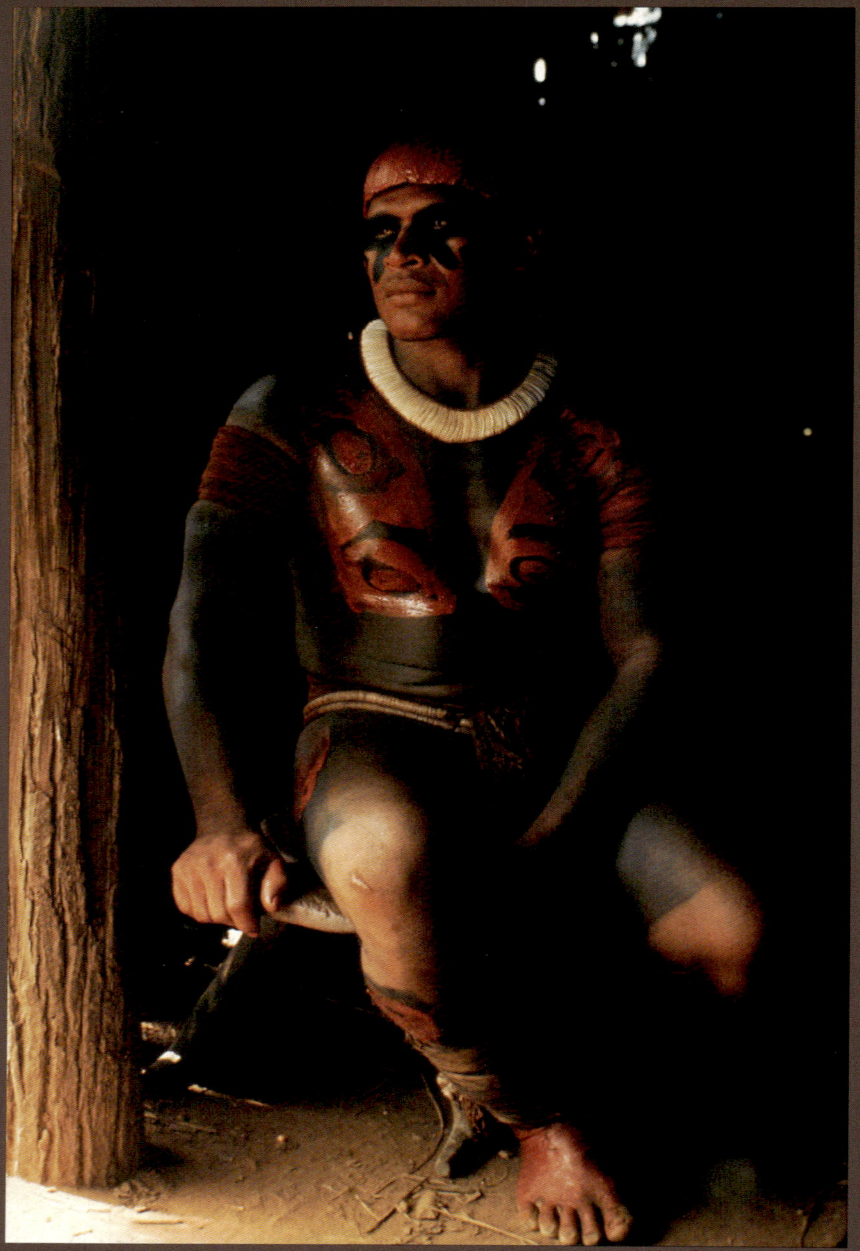

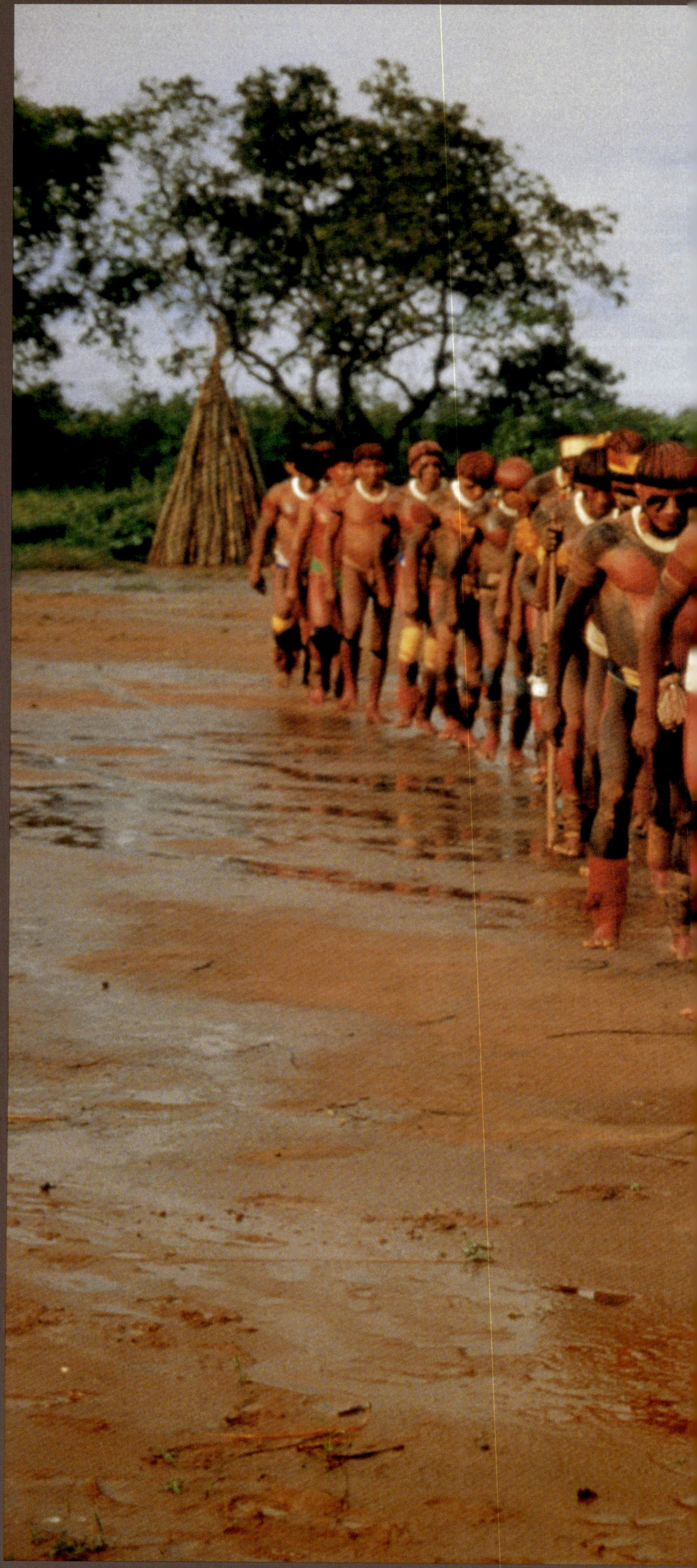

The Xingú National Park in southeastern Brazilian Amazonia has what is probably the greatest concentration of Indian tribes left in Amazonia. Above, a Waura man with red body paint and a necklace of white riverine mussel shells and a belt of terrestrial snail shells waits to participate in a ceremony. On the right, Kuikuro men dressed in their traditional outfits take part in a dance. The Waura now number only about 187 individuals, the Kuikuro 277.

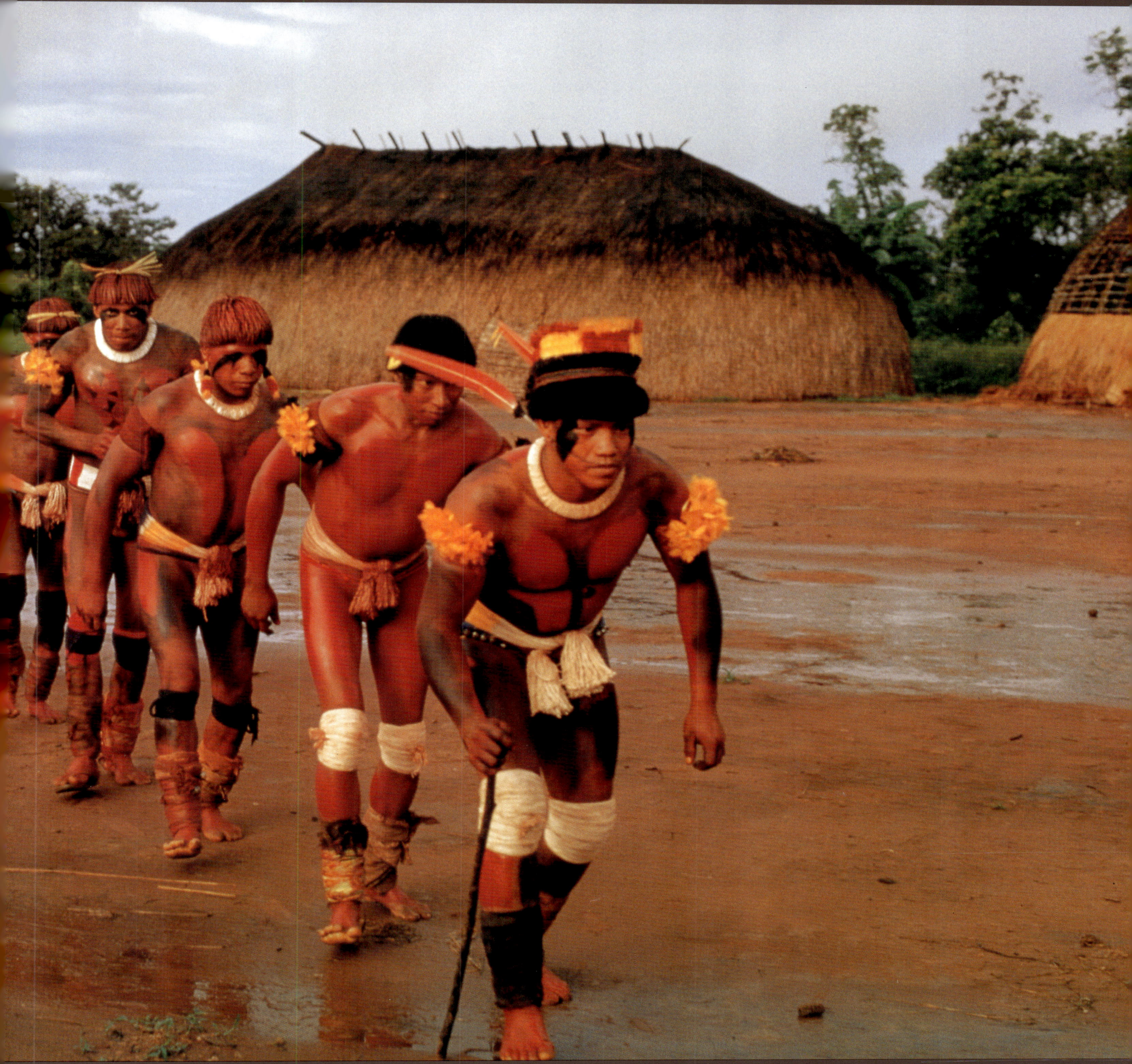

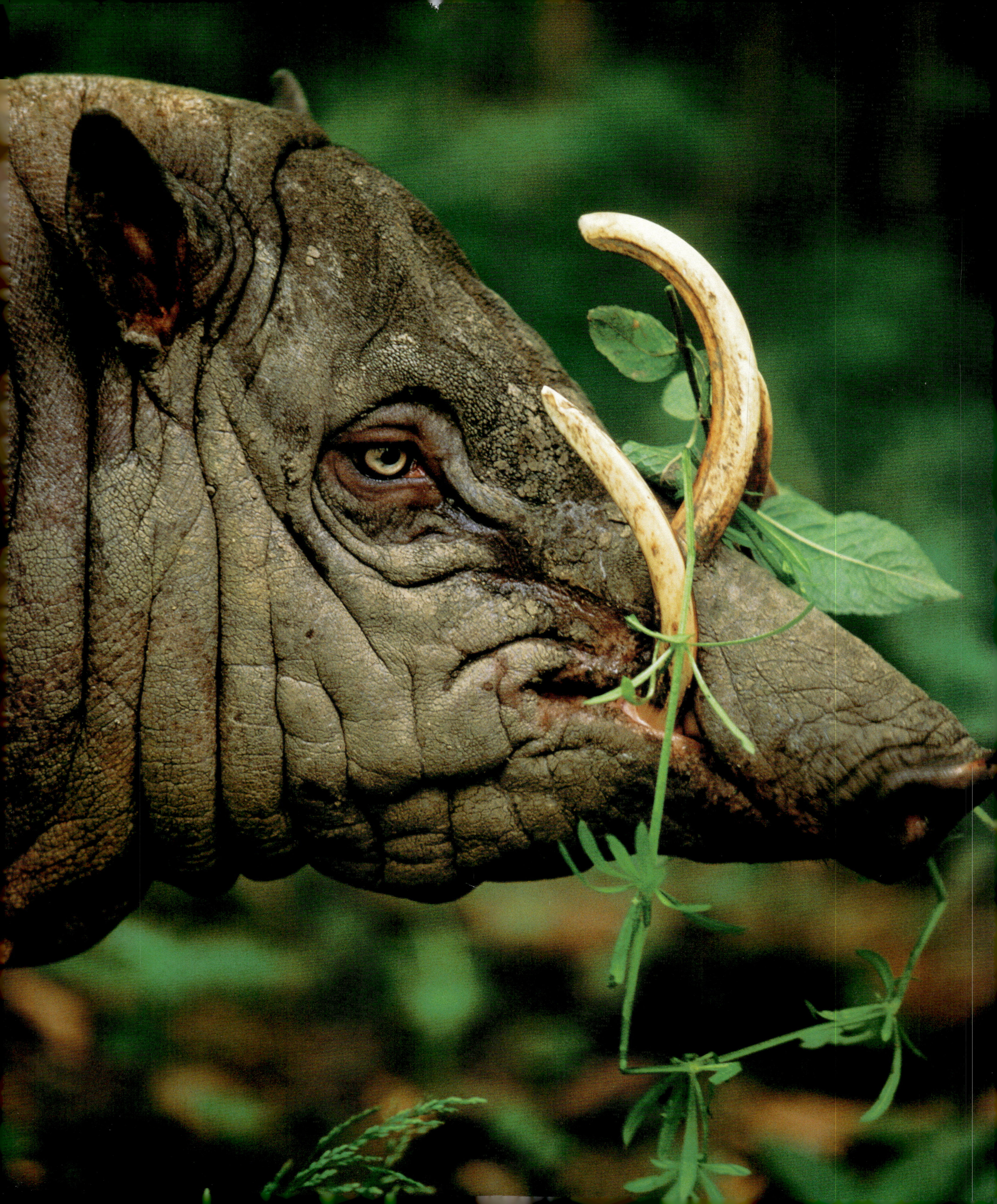

INDONESIA

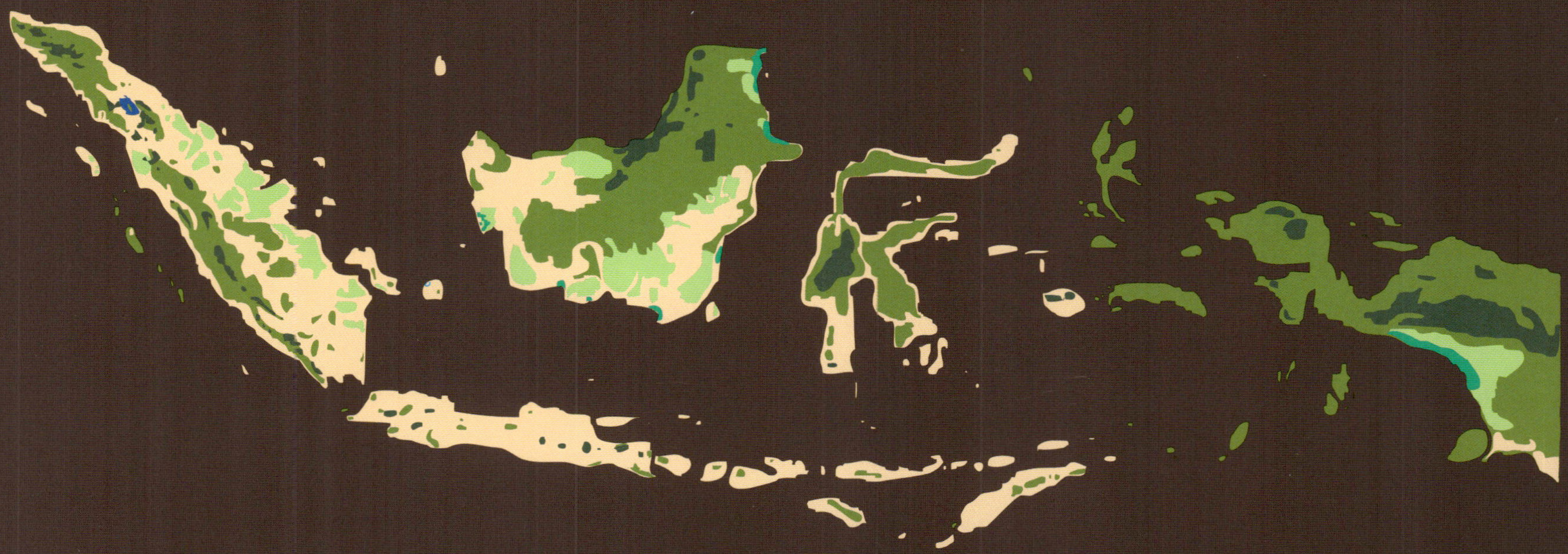

0 500 km

Cultivated area

Mangrove

Montane rain forest

Lowland rain forest

Swamp forest

Bodies of water

Indonesia vies with Brazil for the title of *richest country on Earth* in terms of biological diversity. While Brazil surpasses Indonesia in terrestrial and freshwater species diversity, Indonesia unquestionably leads in marine diversity and has very impressive terrestrial and freshwater diversity as well. Furthermore, Indonesia, along with Mexico, is one of only three megadiversity countries that straddles two of the world's six major biogeographic regions, in this case the Australasian and Indo-Malayan realms. Of additional interest is the large transition zone between these two major regions, known as Wallacea, which itself has high diversity and endemism. Indonesia's two major biogeographic regions and its fragmentation into over 17 000 islands, including major portions of the second and third largest islands on Earth, New Guinea and Borneo, may give it an advantage over Brazil in total generic- and familial-level diversity, but the limited knowledge that we have of biodiversity in general makes it difficult to know with certainty. In any case, Indonesia is unquestionably one of the top two countries on Earth for biodiversity and the likely leader in marine biodiversity.

Looking at the number of species in key groups of organisms, Indonesia is second on the world mammal list behind Brazil, with 515 species, 39% of them endemic; fourth in reptile diversity with 511 species, 150 of them endemic; fifth in bird diversity with 1 531 species, 397 of them endemic; and sixth in amphibian diversity with 270 species, 100 of them endemic. It is also first in psittacine bird diversity, with 75 species, 38 of which are endemic, and in the top 4 globally for primate diversity, with 35 species. Indonesia is also in the top five in plant diversity with an estimated 38 000 higher plant species; heads the world list in palm diversity with 477 species, 225 of them endemic; and has over half of the 350 species of dipterocarp trees, a very important group economically (with 155 species being endemic to Borneo alone). Indonesia also ranks behind only Brazil and possibly Colombia in freshwater fish diversity (about 1 400 species). The country's land area (1 916 600 km^2) makes it the fifteenth largest country on Earth, between Libya and Mexico, but its area —including the sea within the archipelago— is far more extensive. It has only 1.3% of the world's land surface, but has roughly 12% of the world's mammals, 7.3% of the world's reptiles and amphibians, and 17% of the world's birds. These are extremely high figures which only Brazil and a handful of other countries can match. Nevertheless, as more information becomes available, the species totals will almost certainly increase among the plants, fishes, reptiles, amphibians and invertebrates, and possibly among the mammals and birds as well.

In addition to being a megadiversity country, all of Indonesia is also included in two threatened biodiversity hotspots, Western Sunda and Wallacea, and one major tropical wilderness area, Melanesia, which includes Irian Jaya. It also has the world's highest number, 24, of Endemic Bird Areas, as identified by BirdLife International, and 16 of these are considered "first priority." In WWF/IUCN Centers of Plant Diversity and Endemism, it ranks in the top 4, together with China, Papua New Guinea, and the United States.

On the opposite page, the babirusa (Babyrousa babyrussa) *is one of the world's most bizarre mammals, and certain the most extraordinary pig. Its name in Bahasa Indonesia means "pig-deer," and it is endemic to Sulawesi. The upper canines of the male emerge vertically from the upper jaw, penetrate through the skin of the nose, and then curve backwards over the front of the face towards the forehead.*

The coming together of two major biogeographic regions in Indonesia is so impressive that it stimulated the great British explorer and naturalist Alfred Russel Wallace to co-develop with Darwin the Theory of Evolution and to create the science of biogeography. During Wallace's long stay in the country in the middle of the last century, he was struck by the great differences in the faunas of eastern and western Indonesia, and there seemed to him to be a divide between Borneo and Sulawesi, and between Bali and Lombok. This later became known as Wallace's Line, with a fauna to the west characterized by Asian animals such as pheasants, barbets, large cats, rhinoceroses, deer, apes, and tree shrews found throughout the Sunda Region of Peninsular Malaysia, Sumatra, Java, Bali and Borneo, and an Australian fauna to the east characterized by cockatoos, friarbirds, birds of paradise, and marsupials.

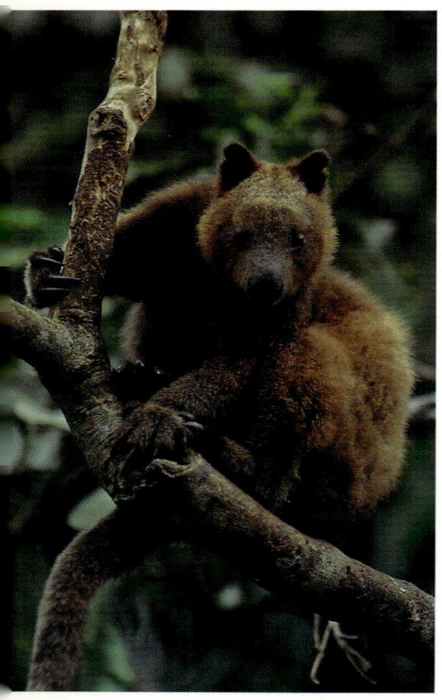

Doria's tree kangaroo (Dendrolagus dorianus), *largest of the tree kangaroos, is found both in Irian Jaya in Indonesia and in Papua New Guinea.*

The area between Wallace's Line and New Guinea is regarded as a transitional biogeographic subregion known as Wallacea, and comprises the islands of Sulawesi, the Moluccas, and the smaller islands that form Nusa Tenggara. In Sulawesi, for example, one can see monkeys and marsupials and cockatoos and hornbills in the same trees, making it one of the great zoogeographic laboratories on Earth.

The islands of western Indonesia share many biological and cultural similarities and, together with Peninsular Malaysia, the rest of Borneo, the Philippine Island of Palawan and the southern part of the Thailand isthmus, form the area known as Sundaland, which lies on the Sunda Shelf, an extension of the Asian continent where the sea is shallow (40 m or less). These areas have been connected with one another on and off during Indonesia's geological past as the sea level has risen and fallen due to glaciations and ice melting, alternately allowing and preventing the dispersion of animals and plants from the Asian continent and among different islands. In contrast with western Indonesia, all the islands of eastern Indonesia, except for Irian Jaya, have never been connected to a major landmass. Irian Jaya, on the other hand, sits on the Sahul Shelf, an extension of the northern part of Australia, with which it has had connections during past glaciations.

Stretching as it does some 5 000 km from east to west (a distance greater than the continental United States) and due to its complex geological past and unique biogeographic position, Indonesia has enormous ecosystem diversity. Topographically, it is also very varied, with the result that almost every kind of ecosystem can be found, from lowland rain forest and swamps to coastal and marine environments and some of the longest caves on Earth, from arid regions all the way up to the snow-clad peaks. Some 47 distinct ecosystems have been recognized, and among the megadiversity countries it ranks with the highest in this category, even when compared to other biogeographically very varied countries like Peru, Colombia, Mexico, China, India, and the United States.

In human diversity, Indonesia is also a global leader, with at least 336 distinct cultures recognized, a large portion of which (around 250) are in Irian Jaya. This figure puts Indonesia in the top three countries on Earth in this category, with only Papua New Guinea and India being higher in total numbers, but probably not in the range of different cultures present.

All in all, Indonesia, which is a major economic force in today's world, occupies a very privileged position in global biodiversity conservation, and therefore needs to incorporate biodiversity protection and use into its future development strategies in very significant ways.

Natural habitats

Indonesia has a wide range of natural vegetation types. Different types of lowland rain forests are found in most corners of the archipelago, and include peat swamps and freshwater swamps (both of which are extensive in Sumatra, Kalimantan and Irian Jaya); heath forests (Kalimantan supports the largest expanse of this vegetation type in Southeast Asia); forests on limestone; and forests on ultrabasic rocks (Sulawesi has the largest ultrabasic area on Earth). In the drier areas of the southeast, seasonal deciduous and monsoon forests and savanna grasslands are found. Many parts of Indonesia are mountainous and these slopes are covered with distinctive vegetation types which change with the altitude, and on the highest peaks of Irian Jaya, Sulawesi, Sumatra and Java there are even alpine ecosystems.

The dominant natural vegetation type found in Indonesia is lowland evergreen rain forest, and the country still has about 500 000 km^2 remaining (26% of the total area of the archipelago), which represents about 10% of the global total for this type of forest. More than 80% of Indonesia's tropical lowland rain forest is found in Sumatra, Kalimantan, and Irian Jaya. These forests are characterized by abundant plant growth and high biomass, as well as by the presence of large, woody climbing plants, large buttressed trees and the prevalence of trees with tall, smooth-barked trunks.

As one would expect, there is considerable variation across the country. For example, in western Indonesia these rain forests are dominated by a single family of trees, the dipterocarps or Dipterocarpaceae, many of which can be extremely large. This is one of the largest families of forest trees anywhere in the world. They are found mainly in southeast Asia with a single species in South America and about forty species in West Africa, but some 470 species from the Seychelles to New Guinea. Borneo has by far the most species with 267, of which 155 are endemic, while Sumatra has 106 species, only eleven of which are endemic. Further to the east, the number of species per island decreases, with only ten found in Java, seven in Sulawesi, six in the Moluccas, and fifteen on the island of New Guinea (relatively few given its large area). The size and wood characteristics of many of the species have made them among the most favored timber trees in the region. Ecologically, they are of singular interest because they have large, woody, oil-rich fruit which is produced in great abundance only every five to seven years. The fruits are produced more or less simultaneously by different dipterocarp species over large areas. This strategy is thought to be a means by which seed predation is reduced.

A tree with a similar strategy which forms another type of lowland evergreen rain forest is ironwood (*Eusideroxylon zwageri*) of Sumatra and Kalimantan, which is highly gregarious and in some areas has formed virtually single species stands. Its timber is of excellent quality and durability; thus, only about 3 500 km^2 of this formation, or one third of the original area, remains.

In limestone areas, the lowland evergreen rain forests are in many ways similar to other types of lowland forests, but when they occur on steeper slopes and rocky hilltops the shallow soil conditions can seriously affect tree growth. Forests on limestone generally have relatively few tree species, although the total number of plant species is probably not dissimilar. The relative paucity of tree species is probably due to the fact that some trees of lowland forest cannot tolerate the high calcium levels in the soil. The different tolerance of trees to calcium and the unique physical habitat of limestone makes the composition of these forests rather different from other lowland forests, giving rise to a specific community of trees. On steep limestone cliffs with bare rock faces, clefts and shelves, a distinctive herbaceous flora occurs. When there is a severe dry season, the tolerance of these plants may be exceeded and they may be forced to survive as annual plants; that is, they must complete their life cycle during the wet season so that their seeds can germinate when the following drought ends.

An interesting lowland forest formation is heath forest or *kerangas*, which in the Iban language of Kalimantan means "land that if cleared will not grow rice." This type of forest is characterized by fairly short trees that form a single-story canopy. Kalimantan has the most extensive areas of heath forest in Southeast Asia, while only a few small patches remain in Sumatra, where it was never common to begin with. Heath forests usually grow on ancient silica-rich, nutrient-poor sandstone terraces. Water drains easily through this type of soil, which is often covered by a thin layer of dead leaves and organic matter. Less than a meter below the surface, just beneath a layer of white sand, there is a hard, iron-rich layer called a *podzol pan*. This layer acts as a barrier that can stop the free flow of water causing the forest to flood after heavy rain. The water flowing out of a heath forest is usually black water, as in peat swamp forests, which is full of dark humic acids and very few nutrients. The heath forest itself has a low, uniform, single-story canopy formed by large saplings and pole trees which are so dense that walking through the forest can be quite difficult. There are fewer tree species in this kind of forest, and abundant carnivorous plants such as pitcher plants and sundews, often found in forests growing on poor, acidic soils. Where heath forests have been burned or felled, a form of slow-growing, stunted *padang* or secondary vegetation forms. Once a heath forest is felled, it cannot regrow because the organic layer of the soil is quickly oxidized, and the few nutrients that existed are washed away rapidly through the sandy soil.

There are two main types of inland swamps in Indonesia, freshwater swamps and peat swamps, which together cover about 150 000 km^2. Many of these grow along the low-lying eastern coast of Sumatra, the western and southern coasts of Borneo, and some parts of southwestern New Guinea. Freshwater swamp forests grow on mineral soils which are often waterlogged. As an adaptation to this, many trees have buttress roots to support them and roots which grow out of the soil to "breathe" air. Palms are especially abundant in freshwater swamps, and there is very little undergrowth. Few areas of freshwater swamp remain in Sumatra and Kalimantan because they are drained to create high-quality agricultural fields; the largest single remaining area is Way Kambas National Park in southern Sumatra.

Peat swamp forests differ from freshwater swamp forests because they grow in organic peat soils that overlay a dense mixture of water, decaying plants, and timber. These forests are characterized by poor soils and many slender trees with abundant undergrowth. Palms are again a common element. The surface of most of the extensive peat swamps of Sumatra and Kalimantan is above that of the surrounding land so that no nutrients enter the swamps from rain draining through or over the adjacent land. The swamps are therefore a closed system and the plants have to use nutrients from within themselves, the peat, or the rain that falls directly on them. As a result, the peat is very acid and poor in nutrients. The peat swamp forest generally occurs in up to four concentric rings of different types of vegetation. Trees become progressively shorter and more stunted towards the middle of the swamp, where the peat is deepest and the nutrients in shortest supply.

In relatively sheltered estuaries and coasts are found mangrove forests, another highly-specialized natural habitat. The trees found here can grow in freshwater but, unlike most plants, they can withstand the hostile environment with its high salt concentrations and the generally anoxic and poor soils. An estimated 24 000 km^2 of mangrove forests remain in Indonesia (1.3% of the land area). Most mangrove forests are dominated by trees from a single family, the Rhizophoraceae, which characteristically produce spear-shaped fruits which can drop into the soft mud below and begin growing. These trees have different ways of dealing with the anoxic condition of the soil. *Rhizophora* trees produce long, curved stilt-roots with large lenticels, or bark holes, which allow exchange of gases, and *Bruguiera* produce uplifted loops of root which snake in and out of the soil. These too have large lenticels. Other types of mangrove tree, *Sonneratia* and *Avicennia*, produce slim, pencil-like aerial roots.

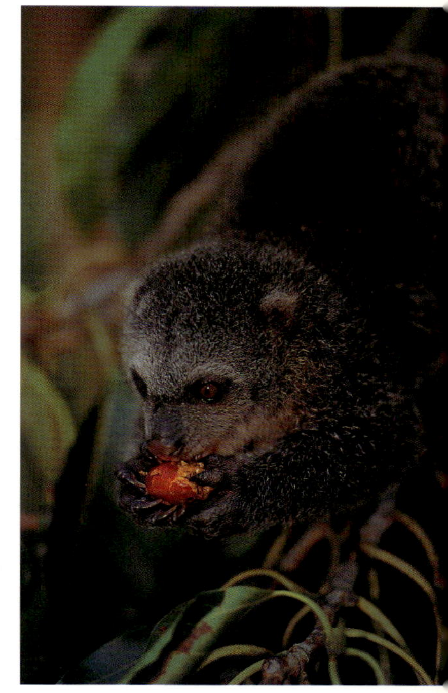

The bear cuscus (Ailurops ursinus) *is endemic to Sulawesi. This orphan is learning to eat natural foods, in this case the ripe fruit of a strangler fig* (Ficus altissima)*, at the research station in the Tangkoko-Batuangus-Duasaudara Nature Reserve in north Sulawesi.*

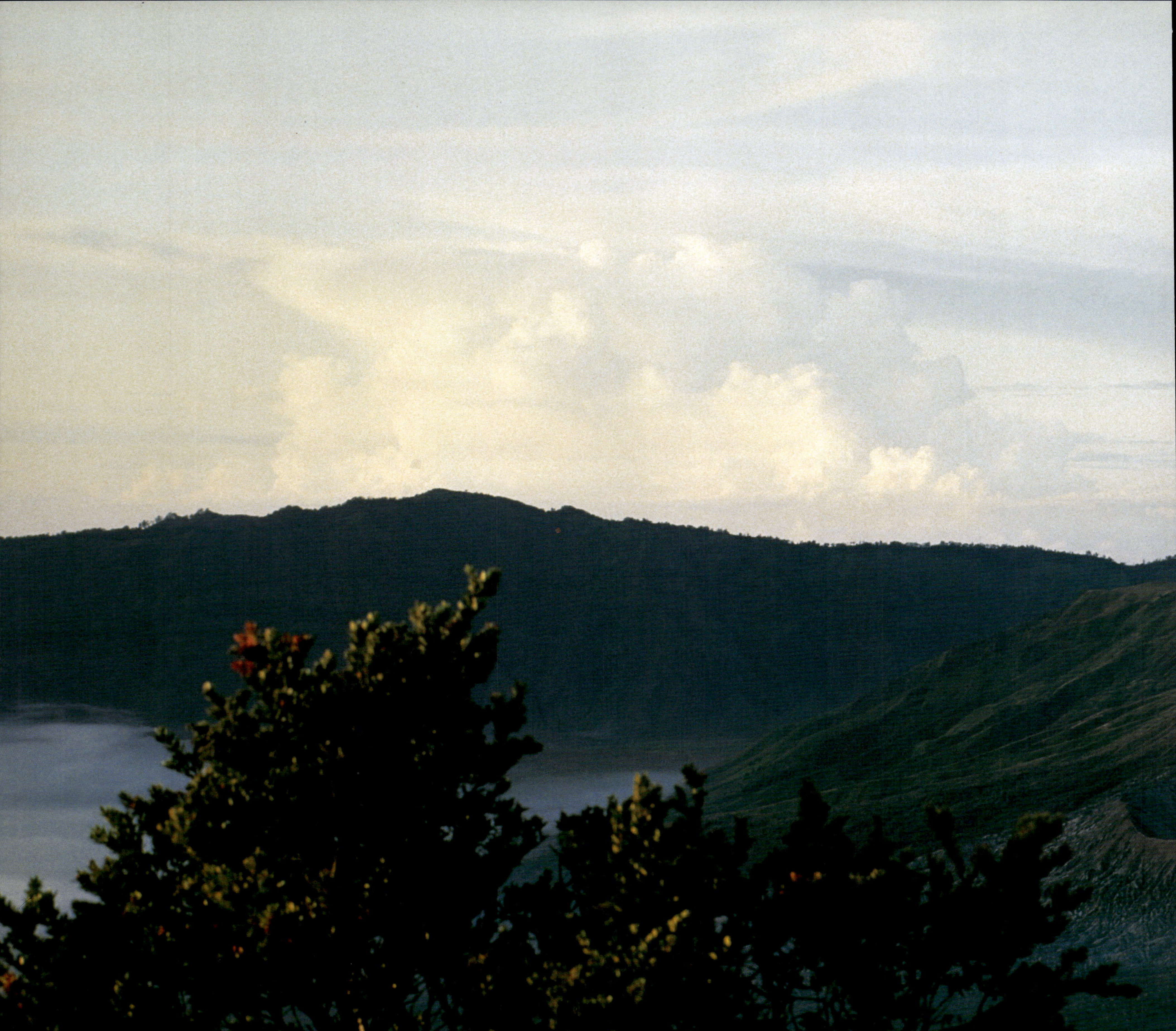

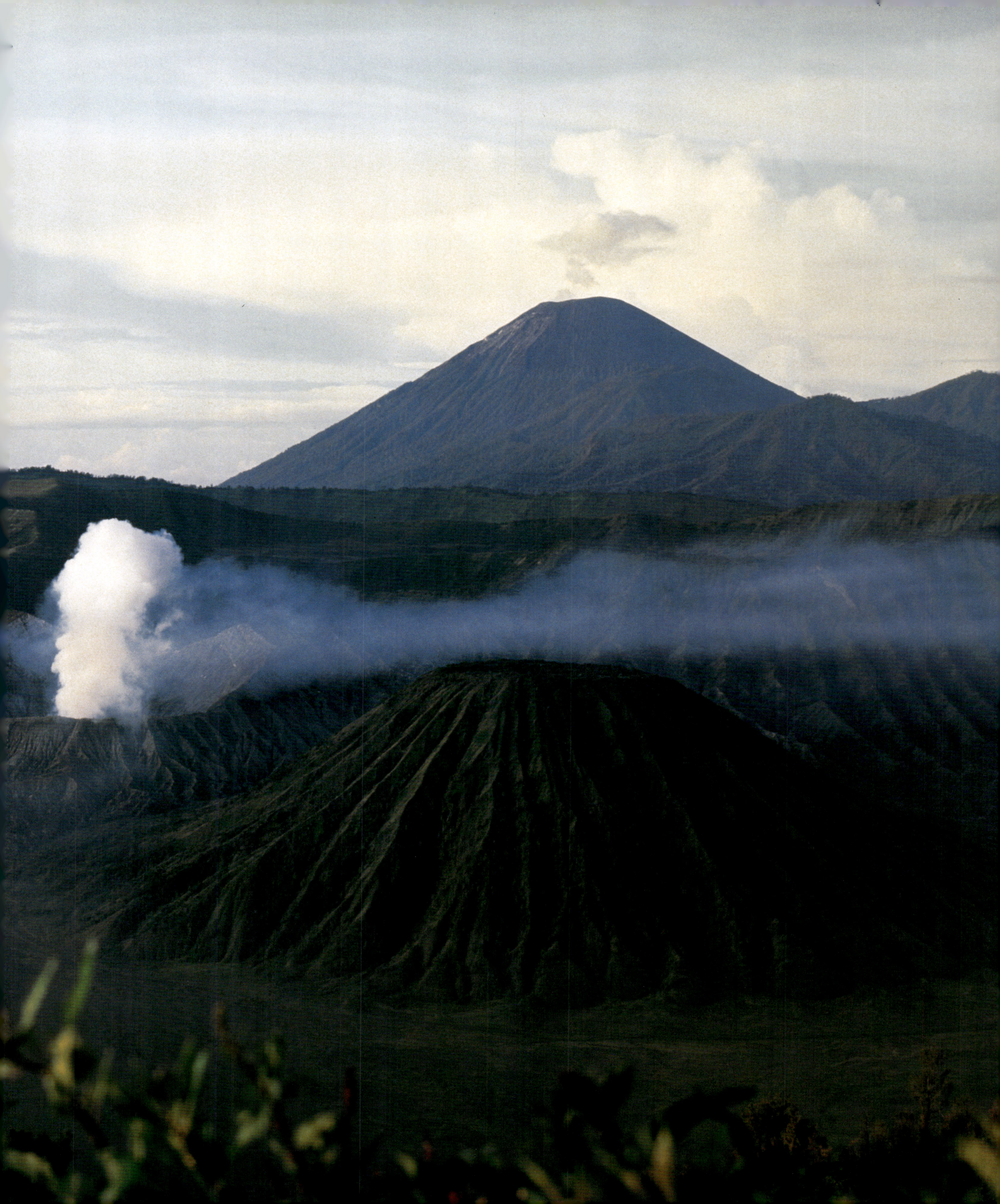

Different types of forests are found in the drier parts of the country such as much of East Java, Nusa Tenggara, parts of Sulawesi and the southern parts of New Guinea, where the influence of the dry southeastern monsoon winds is most strongly felt and where there are four or more (generally consecutive) really dry months. During the dry periods, the soils dry out, deep cracks develop, and evaporation exceeds precipitation. In contrast to trees in the less seasonal areas, many of the trees here are deciduous and have developed very deep root systems. Typically the forest comprises a simple, lightly-closed community of trees, few of which ever exceed 25 m in height, with relatively few of the species found in the lowland rain forest. Annual plants and those with underground tubers are common. Those of us who have walked through these types of forest will well remember that many of the plants are thorns, but the apparent dominance of these may be caused by people and grazing animals cutting or eating those species that are less unpleasant to deal with. Where fire and herbivores are major ecological forces, dry deciduous forest has formed a savanna vegetation with a more or less continuous grass layer and a discontinuous tree layer. These forests are mixed in composition, but not particularly rich in species. Fire is quite common in some of these areas and in certain cases may be required to stimulate seed germination by cracking open the tough outer casing of the seed.

Probably the most conspicuous plant indicators of a climate where deciduous forest would be found are the two large palms. One of these, *Borassus flabellifer*, has blue-green or grayish leaves, a smooth trunk with a diameter like that of a coconut palm, 18-cm round fruits, and no large spines on the leaf stalks. In contrast, *Corypha utan* is generally more massive, its leaves are nearly 2 m in diameter, twice as large as *Borassus*, and it regularly produces small fruits up to 5 cm in diameter. *Corypha* has the second largest inflorescence of any flowering plant, with the flowering branches projecting nearly 4.5 m above the top of the palm and setting hundreds of thousands of fruits. After fruiting, the tree dies.

The deciduous vegetation shows very clear seasonality and the contrasts are best seen at the end of the dry season. Some dormant trees such as certain *Acacia* trees start to come into leaf before rain falls on the forest itself, possibly as a result of a rise in the water table caused by rains in the hills. Bees become active as they are attracted to the sweet smells of tree flowers among their leafless branches. Once the rains fall, within a week there is a riot of leaf and flower color from the ground to the canopy. On the forest floor, lilies use up reserves stored in their deep underground bulbs.

An interesting type of dry forest is that dominated by the *sawo kecik or Manilkara kauki*, such as is found in western Bali. This dominance of a single species runs counter to the normal perception of tropical forest as being seemingly infinitely diverse. The tree is an efficiently-dispersing and rapidly-growing species, and the near single-species stands are probably the result of disturbance occurring years ago during early forest succession. In these situations, however, the dominance lasts only a single generation.

Above 1 000 m there are significant changes in the forest, the trees become rather shorter and less massive, and epiphytes such as orchids are more common. This is lower montane forest. Further up, there is abundant moss, the tree canopy gradually becomes more uniform, the trees become shorter, eventually looking squat and gnarled, and leaves are small and relatively thick. This is upper montane forest. The thicker leaves characteristic of montane habitats may be an adaptation to resist undesirable penetration by fungi, bacteria, mosses and lichens, some of which grow on leaves. The high humidity provides ideal conditions for the growth of these plants and they probably represent a greater threat to leaf life in the frequently cloud-swathed forests of high mountains.

Above the upper montane forest is the subalpine forest of yet smaller trees with smaller leaves, and branches covered with epiphytic lichens but virtually no orchids. Valleys and other depressions in the subalpine zone are generally devoid of trees, and it appears that no trees have adapted to both high altitude and waterlogged conditions, particularly where frosts sometimes occur. Lack of trees on many high mountain slopes, however, is the result of hunters having set fires in order to facilitate the hunting of deer and other quarry. Here and in the highest alpine areas of Indonesian New Guinea (Irian Jaya), there are many shrubs, tough grasses, and herbaceous plants. Some of these have woolly hairs on their leaves and stems which have been variously attributed to the ability to protect the plants against high temperatures, intense ultraviolet radiation and frost/freezing. Other high-altitude plants protect their growing buds and perennial parts with thermal insulation by retaining old leaves, and by having tufted branches and persistent scales. Such protection is necessary because it is more of an energy drain to a high-altitude plant to replace a leaf than it is to plants growing at lower altitudes.

Montane forests cover approximately 190 000 km^2 (9.9%) of Indonesia and are among the least threatened natural habitats, largely because the climate does not encourage human habitation, and because there are few crops which can be grown there.

The marine environment of Indonesia is dominated by two extensive shelf areas: the shallow Sunda Shelf to the west (the world's largest continental shelf), with a maximum depth of 200 m at its edge; and the Sahul Shelf in the east. During the last ice age these shelves were above water, forming land bridges with the neighboring continents. Today, some areas are still extremely shallow, like the sea between Kalimantan and Java, where depth ranges from 6 to 60 m. Between the two shelves there are a number of complex environments, including deep ocean trenches (some of which descend to more than 7 000 m); basins that range from 2 000-6 000 m in depth; steep-sided atolls and seamounts;

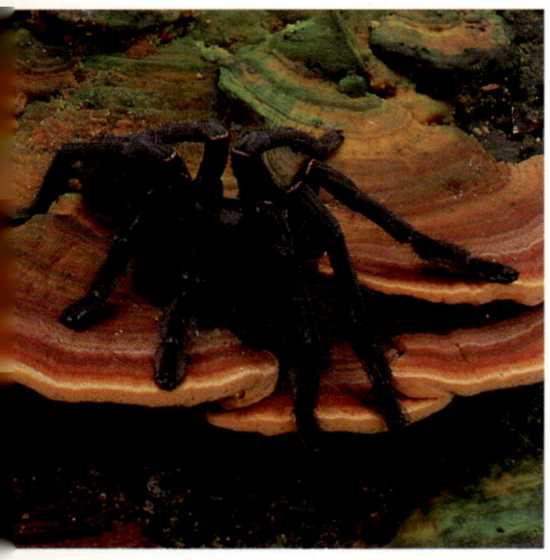

The island of Java is volcanic in origin, and still has a number of active volcanoes. Mt. Bromo, on pp. 78-79, is one of the best known, and is a popular tourist destination.

Above, a large female tarantula, or "earth tiger" walking on a bracket fungus in the Tangkoko-Batuangus-Duasaudara Nature Reserve in north Sulawesi.

submarine banks, and even submerged volcanoes. The more visible coastal and marine ecosystems span a wide range of habitats, including the most extensive mangrove forests, sea grass beds and spectacular coral reefs in Asia. The country's colorful reefs in the deep and clear seas off Sulawesi and the Moluccas are some of the world's richest ecosystems in terms of the species of corals, fishes and other reef organisms —indeed Indonesia is the world's center of diversity for corals. One of Indonesia's most spectacular marine features is the Taka Bone Rate atoll, in the Flores Sea, the third largest atoll in the world.

Flagship species

The wildlife of Indonesia has changed considerably since the first humans arrived in the archipelago. Hunting, habitat loss and population pressure are some of the factors that have accelerated the rate of transformation. Sadly, these changes have already resulted in the extinction of several species. Among the most spectacular creatures that have been lost are the Balinese and Javanese subspecies of the tiger (*Panthera tigris balica* and *P. t. sondaica*) and two species of birds: the Javan wattled lapwing (*Hoplopterus macropterus*), which has not been seen since before the Second World War, and the cerulean paradise flycatcher (*Eutrichomyias rowleyi*), which is known from a single specimen caught in 1873 on the island of Sangihe north of Sulawesi. There is a recent report of the flycatcher having been seen, but its area of remaining habitat is so small that it is likely functionally or imminently extinct, if not actually extinct. Unfortunately, the list is likely to increase in the future if strong conservation action is not taken. Indonesia holds the dubious distinction of having the largest number of vertebrate species threatened with extinction. In this section we describe a few of the most important flagship species of Indonesia.

The Greater Sunda Islands —Sumatra, Borneo (or Kalimantan), Java, and Sulawesi (or Celebes)— are the westernmost islands of Indonesia, and have a great deal of their biological wealth in common by virtue of their similar geological history and position on the Sunda Shelf. The most tectonically-active part of the shelf is on the western and southern rim. Thus, Sumatra has volcanoes down much of its western mountain chain and frequent earthquakes, whereas Borneo has no volcanoes and relatively few earthquakes. The most striking feature of Java and Bali (the latter being the westernmost of the Lesser Sunda Islands and is included here because Java and Bali are closely linked biologically) is that they have the largest number of active volcanoes in the world. On Java, volcanoes run like a spine down the back of the island, and on Bali they form a focus in the north of the island. As a result of volcanic activity both islands are extremely fertile and productive, and therefore, intensively farmed. Java is one of the most densely-populated places in the world, with some 112 million people living in its 138 204 km^2. Most of these people live in cities, however, and there are still some rather large areas of forest, although it covers only 10% of the island. This, combined with very intensive agriculture and high population, make it the area with the most acute environmental problems in the country. Java retains only 10% of its natural forest, which has resulted and is continuing to result in the inevitable loss of species and genetic resources. While this forest is relatively stable (in comparison with much of the country), it is undergoing constant attrition. Attrition is also the story on Borneo and Sumatra, although the former still has a very small human population relative to its size.

Most of these islands' animals and plants came from mainland Asia during the last Ice Age when the Sunda Shelf was exposed. Both Sumatra and Kalimantan (that part of Borneo under Indonesian administration) still have extensive forests, and many of the species within them can be found on both islands and on the much more disturbed Java and Bali to the south. Sumatra is less species-rich than Borneo, and only 12% of its plant species and 17 of its plant genera are endemic. Java and Bali are less species-rich than Sumatra and have no endemic plant genera. One of Indonesia's most striking flagship species is the Rafflesia flower (*Rafflesia arnoldi*), the world's largest flower, growing to almost one meter wide and weighing about 9 kg. It is restricted to southern Sumatra and Borneo, although there are 10 to 15 related species distributed all over Southeast Asia. They are enigmatic plants not only because they are parasitic on a few species of vine, but because they are in some respects more like a fungus than a flowering plant. They have no roots, no stem and no leaves; just a flower that looks akin to and stinks of rotting meat, which is how it attracts its pollinators, blowflies.

Primates are perhaps the most diverse group of flagship species in Indonesia. The country boasts more than 35 of the world's 275 primate species and has representatives from five families and nine genera, with 21 endemic species; Borneo can boast thirteen species. Primates are native to the Greater Sunda Islands although they can be found on a few other islands. For example, long-tailed macaques (*Macaca fascicularis*) can be found on the islands to the east of Bali, but these were almost certainly introduced by people who carried the young macaques as pets as they traveled —much as they do today. The Greater Sunda Islands and the Asian mainland share many primate species, although this is less clear on Java because the orangutan (*Pongo pygmaeus*), siamang (*Hylobates syndactylus*), and pig-tailed macaque (*Macaca nemestrina*) used to be found on Java, before they became extinct as a result of human predation.

Since the Greater Sunda Islands separated from the mainland for the last time about 10 000 years ago, some of the primates have radiated into a range of species. Among Sumatra's endemic primates are the beautiful Thomas leaf monkey (*Presbytis thomasi*) of northern

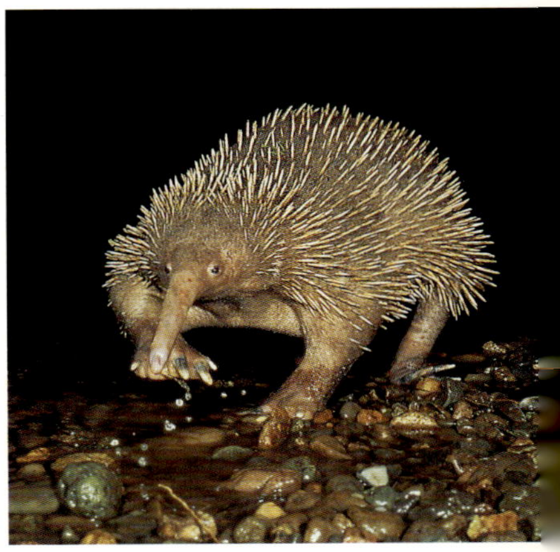

The long-beaked echidna (Zaglossus bruijnii), *one of two monotreme species found on the island of New Guinea.*

Sumatra and the varied mitered leaf monkey (*P. melalophos*), which have a number of colorful subspecies. In Kalimantan there is a different batch of endemic species, such as the red leaf monkey (*Presbytis rubicunda*), Hose's leaf monkey (*P. hosei*), and the white-fronted leaf monkey (*P. frontata*); the Bornean gibbon (*Hylobates muelleri*); and the proboscis monkey (*Nasalis larvatus*). With its pendulous nose, full belly and reddish fur, the proboscis monkey is certainly one of the world's oddest-looking monkeys. An interesting feature of this monkey is its adaptability to water. Proboscis monkeys inhabit tall coastal swampy regions and riverine forests of Borneo, and are excellent swimmers.

Common to both Sumatra and Borneo are pig-tailed macaques and the tiny, owl-eyed, gremlin-like tarsiers (*Tarsius bancanus*). These are the smallest primates found in Indonesia and among the smallest primates in the world, with adult males weighing only 75-120 g. The enormous eyes and large ears are indications of their nocturnal habits. They are efficient predators of insects and small vertebrates such as lizards, frogs and even birds. Another shared nocturnal primate is the slow loris (*Nycticebus coucang*), a solitary creeper with an omnivorous diet. It has a slow, chameleon-like creep that it uses to approach its victims before quickly grabbing them. Like the tarsier, the loris gives birth to only one young each year.

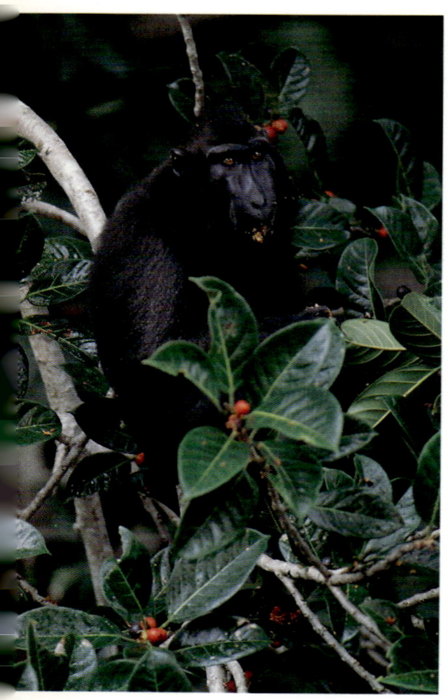

The Sulawesi crested black macaque (Macaca nigra), *formerly known as the "Celebes black ape," is one of seven different kinds of macaques endemic to Sulawesi. Here one feeds on fig fruit* (Ficus forstini) *35 m above the forest floor in the Tangkoko-Batuangus-Duasaudara Nature Reserve in north Sulawesi.*

The orang-utan (*Pongo pygmaeus*), or "man of the forest," is one of the two largest nonhuman primates, the largest arboreal animal in the world, and one of our closest living relatives. It is also one of Indonesia's most prominent and well-known flagship species. The orang-utan can weigh up to 90 kg and measure as much as 140 cm in height. A tremendously powerful creature, it is nevertheless reclusive and shy. There are a number of threats to the survival of orang-utans. Current estimates of their numbers have ranged from as low as 2 000 to 4 000, and hundreds of them are losing, or are likely to lose, their habitat as a result of forest clearing for plantations. Forest conversion in this area is so extensive that there is very little opportunity for these apes to move to alternative forest areas. In Indonesia, orang-utans were subjected to continuous illegal hunting pressures until a rehabilitation program for confiscated orang-utans was established in 1971 at Ketambe, in what is now the Mount Leuser National Park. Illegal traffic in infant and juvenile orang-utans was also suppressed through the eventual refusal of foreign zoos and transshipment countries such as Singapore and Hong Kong to accept them. In 1977, the orang-utan was listed in Appendix I of the Convention on International Trade in Endangered Species. However, the species is still considered *vulnerable* throughout its range.

The most remarkable area for primates in Asia is the Mentawai Islands to the west of Sumatra, which have been isolated from the rest of the Sunda Shelf for perhaps a million years. Despite the islands' relatively small size, all four of its primate species are endemic. The long and complex song of the female of the totally black Kloss' gibbon (*Hylobates klossi*) has been said to produce the most beautiful sound of any land mammal, while the Mentawai langur (*Presbytis potenziani*) is unique among leaf monkeys in living exclusively in monogamous groups, like gibbons. Strangely, the pig-tailed langur (*Simias concolor*) also lives in monogamous groups in some areas of the islands, but in the more typical multi-female groups in other areas. The fourth endemic species is the Mentawai macaque (*Macaca pagensis*), a swashbuckling, dog-like monkey which is as at home on the ground as it is in the trees. Two distinct species are sometimes recognized.

A final primate species, as yet undescribed but known by the people living near and working in the forests of southern Sumatra, is the *orang pendek*. Stories of this large ape have persisted in the same way as the Yeti of Nepal, but only recently have scientists had fleeting glimpses of it. Some hairs, feces, and footprints have been obtained and an analysis of DNA is eagerly awaited. Automatic cameras have been set up along trails, and soon we may be able to see what would be one of the greatest flagship species in the country.

Other important mammal flagship species include the critically-endangered Sumatran rhinoceros (*Dicerorhinos sumatrensis*) and the much larger Javan rhinoceros (*Rhinoceros sondaicus*). Both are found outside Indonesia (the Sumatran in a few pockets of the Asian mainland and the Javan in southern Vietnam), but their only long-term hope for survival is in Indonesia. The Javan rhino weighs up to 2 tons and is the largest Indonesian animal after the elephant. It has a rather inappropriate name since it was once found as far north as the Brahmaputra River in Bangladesh, and was first described from specimens caught in Sumatra. The last twelve Javan rhinos in Sumatra were shot between 1925 and 1930. There may still be some of these rhinos left in southern Vietnam. For approximately the last sixty years, the only population in Indonesia has been confined to Ujung Kulon National Park in western Java. Ujung Kulon National Park is noteworthy because the peninsula it sits on took the full force of the massive Krakatau volcano explosion in 1883. This killed much of the wildlife and destroyed the human settlements on the peninsula —the area has never been repopulated. The Javan rhino is probably the rarest large mammal on Earth today, with none held in captivity anywhere in the world and only about 50 animals left in the wild. The possibility of a second eruption killing the last surviving rhinos worries scientists, but due to the high risk involved in relocating these large animals, they are perhaps best left where they are now. Indonesia recognizes the importance of the Javan rhinoceros' survival and is actively working on *in situ* conservation projects. The country has already won awards for these efforts and has used this species as a logo for major international events, further underlining its importance as a flagship species.

The Sumatran rhinoceros has been declared one of the world's 12 most endangered species by the World

Conservation Union's Species Survival Commission (IUCN/SSC), and it is officially protected in Indonesia, where there is a declining population of perhaps only 400 of these rhinoceroses scattered in the larger forest blocks in Sumatra, and a handful in Kalimantan. It requires large areas of territory to survive (about 100 km^2 per rhinoceros), preferably of primary forest, although it eats certain species of secondary growth vegetation. The Sumatran rhinoceros is hunted illegally for its horn and other body parts, since they are highly sought after for Oriental traditional medicine. Equally threatening to its survival is the continuing loss of its habitat to agriculture, industry, and human settlement. The pressure has driven these animals uphill to rugged mountainous areas, despite their physiological need for swampy mud wallows to coat themselves against insects and cool and protect their skin.

Also found in Kalimantan, Java and Bali is the banteng (*Bos javanicus*) or wild cattle, which used to be a royal hunting quarry for Hindu kings in Java 600 years ago. These beasts can reach 1.8 m or more at the shoulder and weigh up to 900 kg. Banteng are very handsome cattle with the male and female very distinct: the male is predominantly black and the female mid-brown, but both have a contrasting white rump patch and stockings. Banteng are found in forested areas but their "preferred" habitat is not clear, since they are found in grasslands, scrub and forest. It may be that they favor mosaics of different habitats. They live in loose herds with the cow and calf as the fundamental unit, with generally only one adult male in each herd. The surplus males group together to form bachelor herds.

Among the large mammals not shared by all the Greater Sunda Islands is the tiger, though precisely why it has never lived on Borneo is unclear (it is extinct on Java and Bali). This king of the forests is really the archetypal flagship species, since its size, grace and hunting skills inspire awe and attention from humans. Conservationists estimate that there are about 400 tigers left in Indonesia. Although the Sumatran tiger is one of the smallest tigers (about 2.5 m long and 185 kg in weight), it is a powerful and feared predator. Sumatran tigers face acute space problems; being predators, they live at very low densities —typically one every 50 km^2 in relatively undisturbed habitats with adequate prey, such as pigs and deer. They can be found in many forested areas, from lowland swamps to the high-altitude plateau in Mount Leuser National Park, but they are being hunted ruthlessly to supply the demand for body parts used in Chinese medicines.

The tiger and banteng are regarded by the Javanese as epitomes of important but opposite human traits; the tiger represents base human impulses, while the banteng represents the Javanese ideal of steadfastness and restraint. The two traits are both necessary but must be kept in balance for internal harmony to be maintained. This balance was demonstrated as a political allegory during the Dutch colonial period in the form of gladiatorial fights between tigers and bantengs in large cages. To the Javanese, the tiger represented the Dutch who were deadly in firepower and quick to act, but lacked stamina. The banteng, meanwhile, represented the Javanese, being slow to anger but deadly when aroused. The result of the battles was unexpected (at least to the Dutch) because the banteng was generally victorious.

Other mammalian flagship species of the Greater Sunda Islands are the sun bear (*Helarctos malayanus*) and the pesut or freshwater dolphin (*Orcaella brevirostris*). The sun bear is the smallest of the world's bears, and is very aggressive and unpredictable, feared more by villagers and hunters than any other animal in the forest. The freshwater dolphin is one of Kalimantan's largest and most secretive mammals, it grows to be about 2 m and weighs 100 kg. This dolphin inhabits the mighty Mahakam River in East Kalimantan, as well as estuaries and inshore waters in India and Australia, and is threatened by being killed in fishing nets, overfishing of its food, and disturbance by motorboats.

The rivers of the Greater Sunda Islands harbor nearly a thousand species of fish, many of them popular in aquaria around the world; these include tiger barbs, harlequin fish, scissortails, gouramis, snakeheads, bumblebee fish, archerfish, pipefish, halfbeaks, walking catfish, glass catfish, kuhli loaches, clown loaches, tinfoil barbs, silver sharks, and featherbacks. Most exceptional of all, however, and the world's most expensive aquarium fish, is the Asian bonytongue, sometimes known as arowana (*Scleropages formosus*). A virtual mythology has evolved around this magnificent fish, especially the "super red" race from the Kapuas River in West Kalimantan. It is supposed to bring its owner good luck, especially if a cup full of its tank water is drunk before going to work in the morning.

If viewed as a single unit, the Greater Sunda Islands have an impressive 135 bird species found nowhere else in the world. Sumatra and Kalimantan both have about 350-400 species of resident land birds, with 22 of those in Sumatra (including surrounding islands) and 37 of those in Kalimantan being endemic. The smaller island of Java has 289 resident species, 20 of them endemic. The even smaller island of Bali has 204 resident bird species, but only one of those is endemic, the stunning Bali starling (*Leucopsar rothschildi*). This is a pretty white bird with a long crest, bare blue skin around the eyes and black tips to the wings and tail. It inhabits only one small area of dry lowland forests in northwest Bali, and is in imminent danger of extinction in the wild. Over a thousand are held in captivity around the world, but to lose it on Bali would be a sorry testament to Indonesian conservation.

Among the Greater Sunda birds which can act as flagships for conservation efforts, one would count the great argus pheasant (*Argusianus argus*), a bird which leads a solitary life and is rarely seen. Its *ki-au* call is, however, one of the most characteristic calls of the great western jungles of Indonesia. During the short breeding season the flamboyant male, with his incredi-

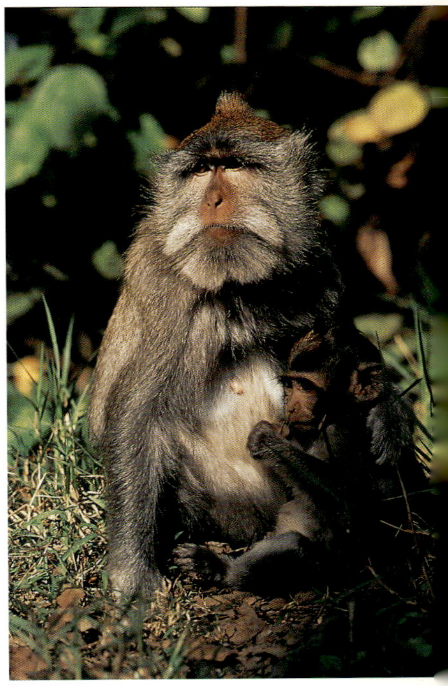

The wide-ranging crab-eating macaque (Macaca fascicularis), *or long-tailed macaque as it is sometimes called, occurs from mainland southeast Asia as far east as Halmahera in the Moluccas. It is a very adaptable species that often occurs in close proximity to man.*

On pp. 84-85, a red-knobbed hornbill (Rhyticeros cassidix) *is one of the many bird species endemic to Sulawesi.*

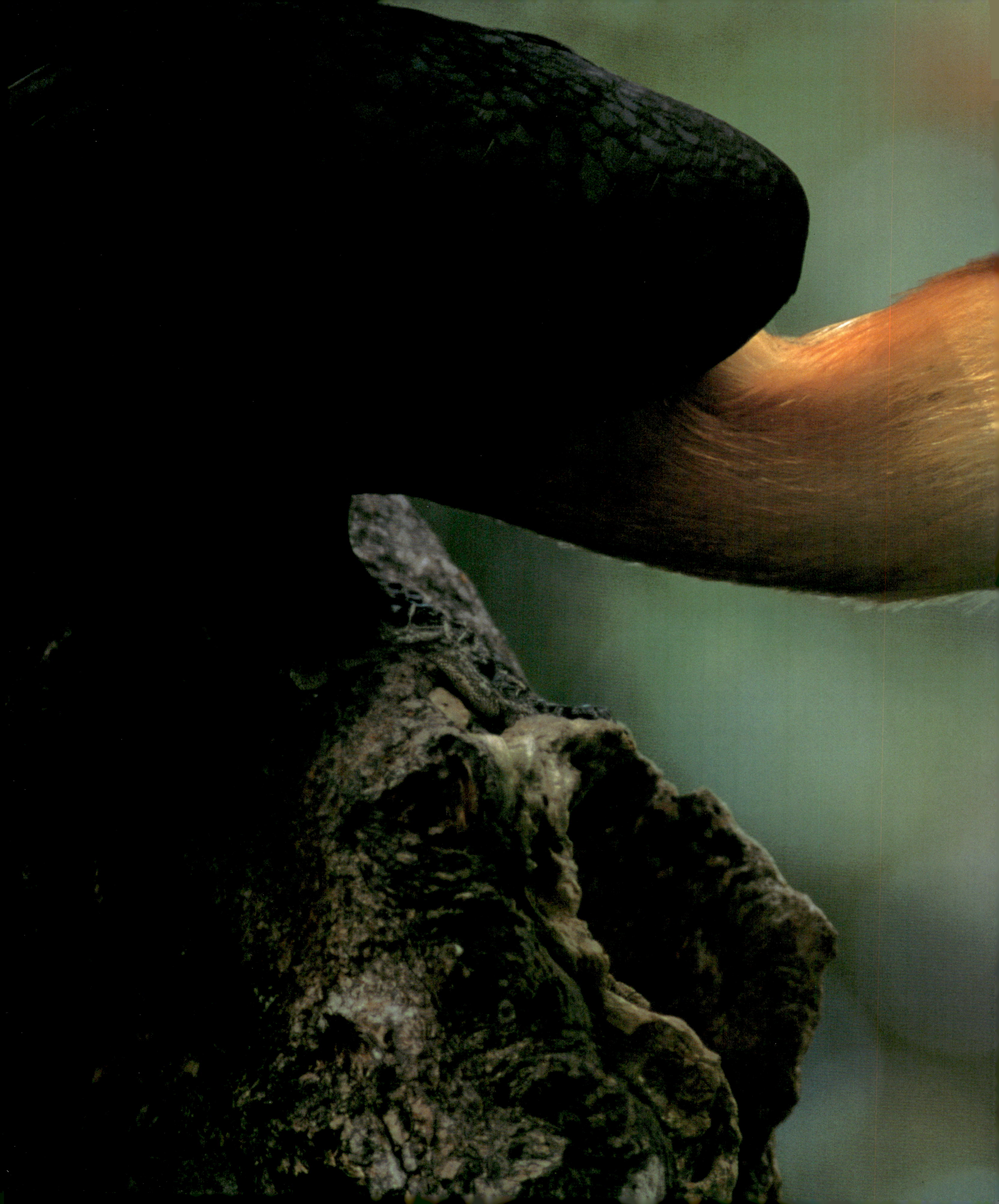

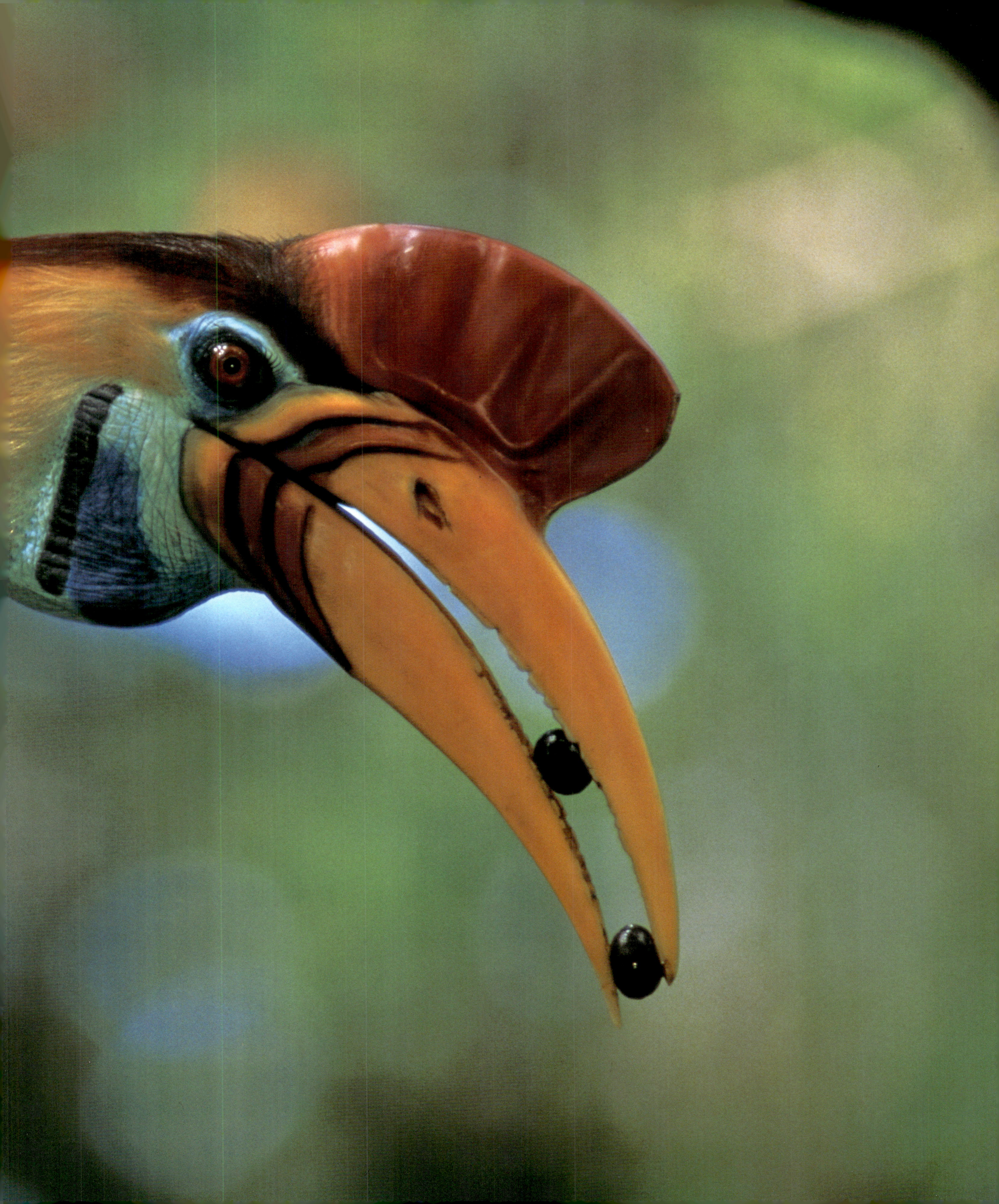

bly long central tail feathers and secondary wing feathers, courts the rather dull female by performing a strutting and seemingly choreographed dance in the deep forest.

The hornbills are a popular flagship group. These large forest birds reach sizes of up to a meter in length. They are mostly frugivores, but will not hesitate to eat any small creature that comes their way. They have distinctive, large horns or casques on the top of their heads, which give them their unique appearance. There are eight species of hornbill in Kalimantan and ten in Sumatra.

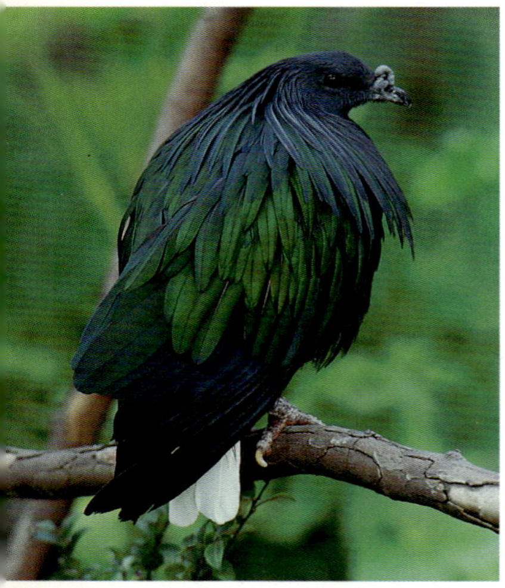

Another flagship bird is the extravagant and well-known green peafowl (*Pavo muticus*), which is found in Assam, southwestern China, and Indochina but in the Greater Sundas only on Java. It feeds by scraping in the ground for seeds, insects, and small vertebrates. Its morning calls echo through the woodlands, especially from August to October, which is the breeding season. The fact that such a large bird as this peafowl has survived until now on Java is remarkable. However, despite being a protected species, hunting has become a problem, in part for its tail feathers which are used for certain types of traditional dances in which a single mask can use 1 000 feathers. In some areas, it used to be believed that peafowl groups would follow behind hunting tigers, waiting for them to defecate and then picking the intestinal worms out of their stool. Feasting on peafowl therefore had no attraction.

As regards Sulawesi, the first and most striking feature of this island is the fact that it was not formed by volcanic activity, like Java and Bali, but rather by tectonic folding and uplifting. One fifth of this oddly-shaped island lies above 1 000 m, making it the most mountainous island in the archipelago. The towns and cities are limited to the few coastal plains found on the four peninsulas that form the island. Sulawesi has some 15 million people, one half of which live in the fertile plains and valleys of South Sulawesi.

Sulawesi and its offshore islands cover an area of 186 145 km^2. About half of Sulawesi is still forested, most of it (as with much of the archipelago) in the mountainous regions. Its varied and rugged geology is reflected in the number of forest types found there. Notable among these are the forests that grow on infertile ultrabasic rocks. These occur mostly on the east and southeast peninsulas, which harbor the world's largest tracts of ultrabasic rocks. The soils have high levels of heavy metals such as nickel and chromium, and are uncommonly infertile. The vegetation is dominated by trees of the myrtle family such as *Eugenia*, *Kjellbergiodendron* and *Metrosideros*, and the animal life in this lowland environment is very poor, with the exception of ants, which have developed a mutualistic relationship with epiphytic plants. The eastern side of Palu Bay in Central Sulawesi is the driest place in Indonesia, with a rainfall of only 500-600 mm each year. Its natural vegetation is monsoon forest, but centuries of using fire to encourage the type of forage favored by sheep, goats and cattle has turned most of the once-forested landscape into a dull grassland.

The striking Nicobar pigeon (Caloenas nicobarica) *from Irian Jaya, one of a wide variety of pigeons and doves on the island of New Guinea.*

On the opposite page, the flying lizard (Draco volans) *is a spectacular little lizard that doesn't actually fly, but rather glides from tree to tree on a broad membrane of skin.*

The lowland rain forests support a wealth of plants and a number of valuable timber species. These include the ebonies, like the darkest and hardest species *Diospyros celebica*, found in the northern and central regions. Eastern Indonesia is also home to the only species of *Eucalyptus* to make its home in rain forests, *Eucalyptus deglupta*, which has peeling bark giving the trunk an array of rich pastel shades. This tree can reach an awe-inspiring 78 m in height, and to think of these giants starting as seeds weighing just one hundredth of a gram seems incredible. They tend to grow on former bare ground such as landslips and old riverbanks.

The fauna of Sulawesi is one of the most distinctive in all Indonesia, particularly the mammals in which almost all the terrestrial species are endemic. The babirusa (*Babyrousa babyrussa*) is, without a doubt, one of the most bizarre mammals in the world, and is certainly the most extraordinary pig. It is known only from the island of Sulawesi (*Babyrousa babyrussa celebensis*), some of the Penju Islands such as Togian (*B. b. togaenensis*), and the Sula Islands and Buru (*Babyrousa b. babyrussa*). These three subspecies differ in size, hairiness and characteristics of the skull and teeth. Most striking among the babirusa's many peculiarities are the upper canines of the male, which emerge vertically from the upper jaw, penetrate through the skin of the nose, and then curve backwards over the front of the face towards the forehead, a unique feature in mammals. The function of these tusks remains unclear. They are quite brittle and therefore easily broken, and are rarely used in combat between males. The babirusa inhabits tropical rain forest on the banks of rivers and ponds abounding in water plants. It appears that babirusas are social, with groups or troops of up to eight individuals that usually share communal wallowing areas and salt licks. In recent years, large-scale commercial logging operations have posed a major and increasingly serious threat to this species, which is classed as *vulnerable* in the *IUCN Red List*; it has also been included in Appendix I of the Convention on International Trade in Endangered Species since 1982. Sadly, babirusa are among the first animals to become locally extinct after logging and land opening. Such activities not only deprive them of their moist forest habitat, but also increase their exposure to hunting pressure, especially to serve the demand for bush meat from the Minahasan people of the northern tip of Sulawesi.

The seven different kinds of Sulawesi macaques are among the most striking mammals on the island. The Moor macaque (*Macaca maura*) is found in the southern part of the southwest; the Tonkean macaque (*Macaca tonkeana*), from the mountains around Enrekang in South Sulawesi, north to Gorontalo in North Sulawesi; the black macaque (*Macaca nigra*) on the northern peninsula; and the booted macaque (*M. ochreata*) in the southeast peninsula. In addition to these, there are at least three other taxa, *Macaca nigriscens*, *M. hecki*, and

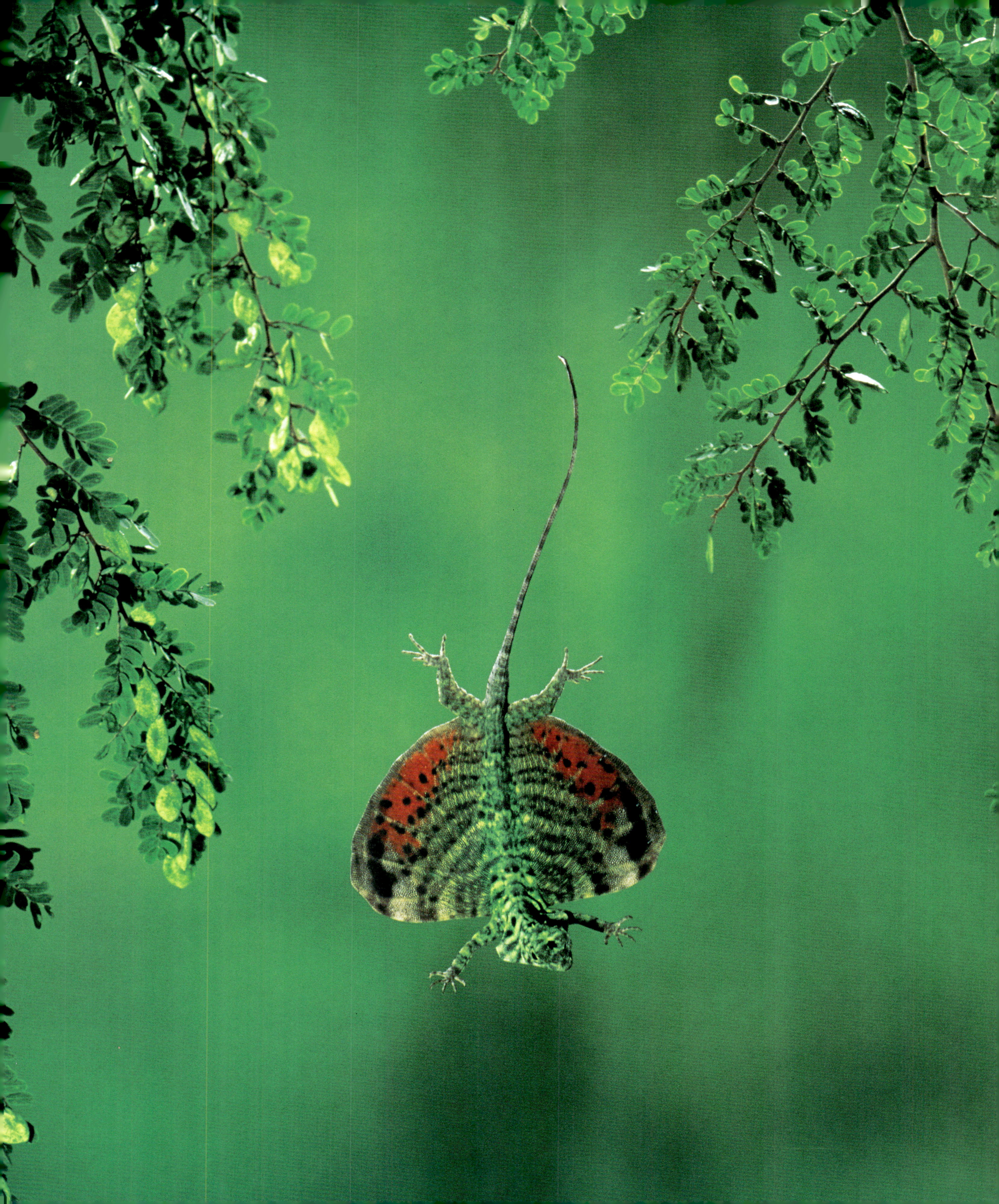

M. brunnescens, which are sometimes recognized as species and sometimes as subspecies.

While there is only one tarsier species known from all of Borneo and Sumatra, Sulawesi has no less than three; the spectral tarsier (*Tarsius spectrum*), which is found throughout the island in the lowlands, even in towns, secondary forest, and mangrove; and two species in the mountains, the smaller of which, the lesser Sulawesi tarsier (*Tarsius pumillus*) has a head and body length of less than 10 cm, making it the smallest primate in Asia. Diana's tarsier (*Tarsius dianae*) is another small species known only from Lore Lindu National Park in Central Sulawesi.

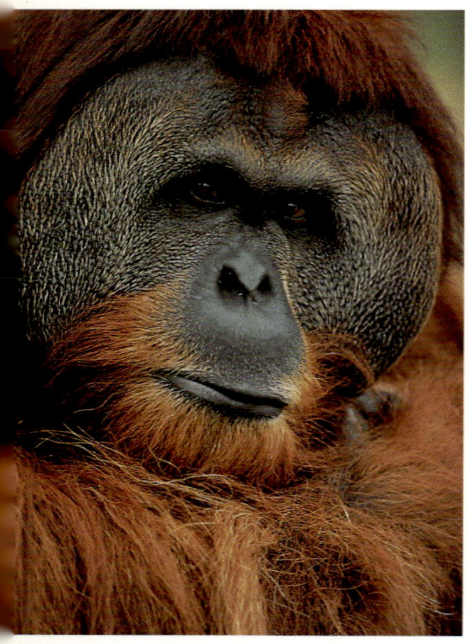

Sulawesi marks the western boundary of the great Australian mammal subclass Marsupialia. It has three very different endemic species of tree-dwelling cuscuses. One is the bear cuscus (*Ailurops ursinus*), a dark brown animal measuring up to 1 m from snout to tail. It is mostly diurnal, moving slowly through the canopy in search of leaves to eat. The dwarf cuscus (*Strigocuscus celebensis*), on the other hand, is nocturnal and dwells primarily in the lower canopy where it eats mostly insects and fruit. A third endemic species (*Phalanger pelengensis*) is found on the eastern island of Peleng.

Sulawesi has two small, endemic wild buffalo known as anoa, which stand about 75 cm at the shoulder. One (*Bubalis depressicornis*) inhabits the lowlands, while the other (*B. quarlesi*) inhabits the mountains, although the ranges of the two species appear to overlap. Anoa inhabit primary forest and are unpredictable and aggressive, rendering fruitless attempts to domesticate them. Their short, sharp horns make them the most feared animals in the region. Hunting pressure and habitat loss have led to a decline; however, the actual number of anoa in Sulawesi is not known.

Sulawesi's bird fauna is as impressive as its mammals. There are 250 species, 36% of them are endemic, and these include seventeen endemic genera. There is an abundance of pigeons and doves, which include the very rare dusky pigeon (*Cryptophaps poecilorrhoa*) and some more elusive and secretive birds such as the Sulawesi woodcock (*Scolopax celebensis*). Other avian flagship species include the endemic crowned myna (*Basilornis celebensis*), the purple-bearded bee-eater (*Meropogon forsteni*), the red-knobbed hornbill (*Rhyticeros cassidix*), and the famous maleo (*Macrocephalon maleo*). The maleo is a member of a group of birds known as megapodes or mound builders, chicken-like incubator birds. There are 22 recognized species of megapodes in seven genera and their distribution extends to Australia, New Guinea and surrounding islands, eastern Indonesia, and the Philippines. However, the maleo is restricted to Sulawesi and possibly some offshore islands. This bird gets its name from its large feet. About as big as a domestic hen, the maleo is almost entirely black, except for the white chest. It has a perkily erect tail and an unusual and distinctive black-knobbed head. This bird incubates its eggs in the ground: the male digs a pit in ash or sand that can be as much as one meter deep. The female deposits the eggs, and then the eggs are buried and the soil packed around the pit, a task that is accomplished with the use of the large feet of this bird. The egg is as unusual as its parents. Colored pink to pale red, it is about 11 cm long, and weighs more than 250 g, making it twice as long and five times heavier than the common hen's egg. Although nests are usually found in the open sun, megapodes live in shady forests, sometimes far from their nesting grounds. This species is considered endangered.

A male orang-utan (Pongo pygmaeus).

On the opposite page, in the forest in Kalimantan, this ex-captive was rehabilitated at Camp Leakey in the Tanjung Puting National Park, and then released into the wild.

Sulawesi also has a number of flagship species among the freshwater fish and invertebrates. An entire family, the duck-billed fishes (Adrianichthyidae), is restricted to the Lindu and Poso Lakes in Central Sulawesi, and consists of only four species. The female of one of these four species carries her eggs attached by thin threads to her belly until they hatch. The Malili Lakes at the northeast corner of South Sulawesi are even more remarkable, for within this complex of lakes, rapids and rivers have evolved no less than 60 species of crabs, shrimps, snails and fishes, many of which are extremely colorful. Strangely, only one species is common to all four lakes, and that is a shrimp.

Among the insects, swallowtail butterflies are very abundant in Sulawesi, with 38 species most of which are extremely beautiful, and eleven of which are endemic. As for other invertebrates, the coconut crab (*Birgus latro*), found in the remotest islands of Sulawesi and east through the Moluccas to the Pacific, is the world's largest species of terrestrial arthropod. These enormous crabs, weighing up to 5 kg, can measure 90 cm between outstretched claws. Nowadays, however, a large number of these nocturnal crabs are consumed locally and exported to other islands, and this has taken its toll; despite the fact that it is considered an internationally threatened species and is protected legally in Indonesia, it is still heavily exploited.

To the east of Sulawesi lie the Moluccas, a group of about 1 000 islands, spanning 1 300 km. The two largest islands in the group are Seram and Halmahera, with an area of about 35 000 km^2, while some of the smallest islands can't even be seen on most maps. They belong to a single province comprising 74 505 km^2 of land set in an area of 1 000 000 km^2 of sea. The capital is Ambon, located on the small island of the same name just south of Seram.

While the Moluccan forests cannot compete with the larger islands to the east and west in terms of the number of plant species, they are notable as the center of origin of plants over which countries engaged in wars due to which fortunes were won and lost. These are nutmeg and mace, which come from the same tree species (*Myristica fragrans*), and cloves (*Eugenia aromatica*), and they are the reason why the Moluccas have long been known as the Spice Islands.

While there are mammals on these islands, they are neither particularly species-rich nor really the stuff of flagship species. The marsupial rufous spiny bandicoot (*Echymera rufescens*) is found in the limestone Kai Is-

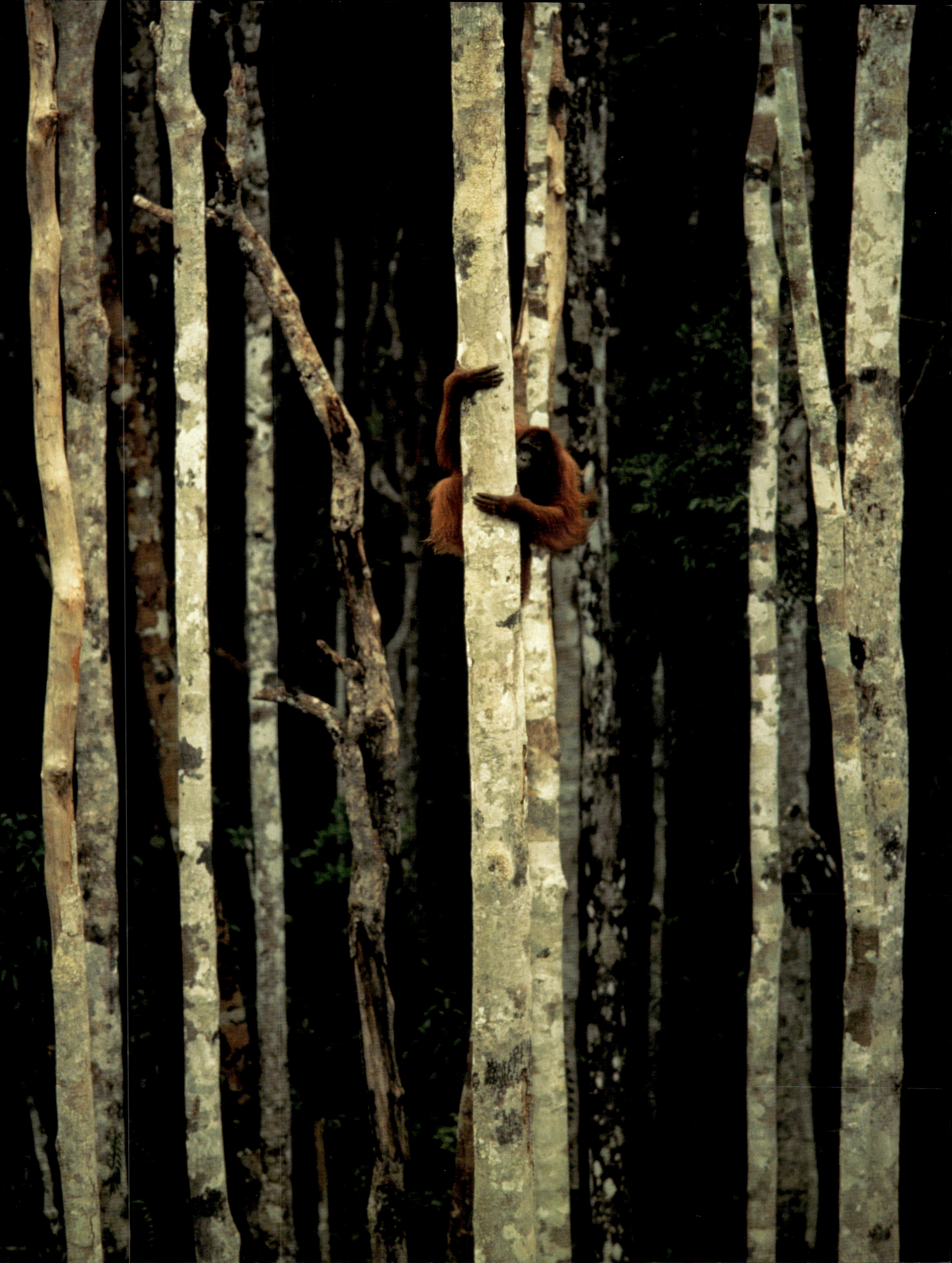

lands and low-lying Aru islands in the southeast, and the Seram bandicoot (*Rhynchomeles prattorum*) is endemic to Seram. This island also has what could be the closest thing to a flagship species in the form of the orange rat (*Melomys fulgens*). This is a remarkably brightly-colored animal with a 15-cm head and body and a proportionately very long 20-cm tail. The fine, velvety hair of its upper parts is bright orange, and its belly, white. Interestingly, it is most often seen in trees used as roosts by large *Pteropus* fruit bats which have a mantle the same color as the rat. Since birds of prey generally leave fruit bats alone, it may be that this rat is mimicking the bats and thereby gaining protection. If so, this is of exceptional interest because cases of mimicry among mammals are exceedingly rare.

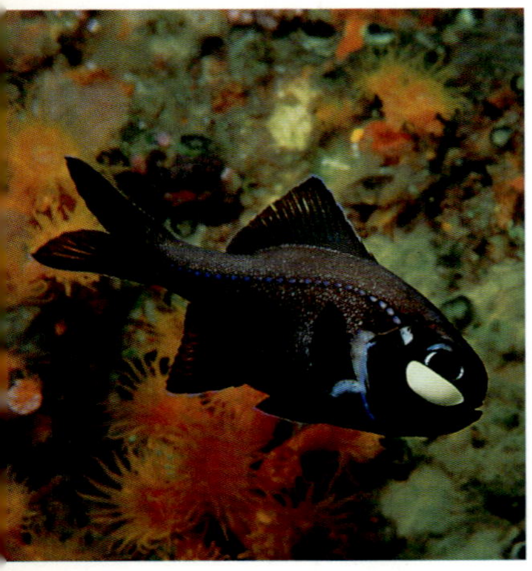

A flashlight fish (Photoblepharon palpebratus) *in the waters off Borneo. The light organ under the eye is used to attract mates and prey.*

On the opposite page, a green turtle (Chelonia mydas) *swims along a reef off the oceanic island of Sipaden, off Borneo. Sea turtles come to this island to mate.*

The more frequently seen mammals such as long-tailed macaques, civet cats, pigs, deer and even babirusas were introduced to the islands by people. Around the coasts where hunting is not too intense and large areas of sea grass occur, the dugong (*Dugong dugon*) can be found. This marine mammal is related to the elephant and is slightly larger than a man. It is the only marine mammal to feed exclusively on plants and, like the elephant, it has a very inefficient digestive system that requires large amounts of food to satisfy its nutritional demands.

The Moluccas boast some 350 species of resident birds, 89 of which are endemic to the islands. Good flagship species can be found among them, such as the cockatoos and parrots. Of the thirty-some species of parrots, perhaps the most striking is the salmon-crested cockatoo (*Cacatua moluccensis*), which lives mostly on the island of Seram and a few neighboring islands. Another striking species is the black-lored parrot (*Tanygnathus gramineus*), a green parrot with a black line running from its forehead to its eyes. There are also the hanging parrots, such as *Loriculus amabilis*, which hang upside down, like little green bats. Most parrots inhabit the lowland forests of this island group, where there is an abundance of flowering trees for them to feed on. The largest bird in the province is the southern cassowary (*Casuarius casuarius*), a bird more generally associated with New Guinea. The most beautiful is arguably the male king bird of paradise (*Cicinnurus regius*) from the Aru Islands; it has a whiter-than-white belly, blue legs, a ruby-red head and back, and two thin wires leading from the edges of the tail which end in iridescent green discs. Another contender for the title, however, is Wallace's standardwing, a bird of paradise endemic to Halmahera, which has "flags" on its wings that are used in displays when many males dance to potential mates in the top of an exposed tree.

In the absence of mammals, there is a rich reptile fauna occupying the niche of the large predators, including crocodiles, large pythons and monitor lizards. One of the Moluccas' reptilian flagship species is the sail-fin lizard (*Hydrosaurus amboinensis*), the largest lizard in these islands. It is called a sail-fin lizard because of the large fin that runs the length of its tail, which makes up two thirds of its total length, of up to one meter. These toothless lizards perforate leaves and then pull off pieces. Although they are also found in Sulawesi, New Guinea and the Philippines, they are more commonly associated with the Moluccas.

The Moluccas also have many endemic butterfly species, among which are four large bird-wing butterflies. The islands of Halmahera, Bacan, and Tidore are also home to the largest bee in the world (*Chalocodoma pluto*), the females of which can reach 4 cm in length.

East of Java and south of Sulawesi is the southernmost group of islands of Indonesia. Known as the Lesser Sundas or Nusa Tenggara, this group of islands is spread across 900 km from the islands of Bali and Lombok in the west to Timor in the east, with Sumbawa, Flores, Sumba, Komodo and Alor in between.

The Lesser Sundas are the driest part of the entire archipelago, with an average annual rainfall of about 1 500 mm. The predominantly dry climate, the relatively small size of the islands, and the large distance between them and the two nearest continents, Asia and Australia, combine to make the flora of the Lesser Sundas very poor in species. There is only 12% endemism, which appears to occur mostly on the islands of Lombok and Timor. The natural vegetation of the Lesser Sundas would be deciduous forests and savanna woodlands, with moist evergreen forests in the mountains. However, little remains of the drier forest types. The practice of starting fires in the dry season to encourage new growth for cattle fodder has resulted in a huge loss of natural forest.

The largest block of forest remaining in the Lesser Sundas is on Sumba, but there are a few remaining areas on the other islands. The characteristic trees in these mostly deciduous forest pockets are *Acacia*, *Cassia* and *Pterocarpus*. Timor has very few patches of forest left, but some of these contain the last natural stands of very important trees such as the sandalwood (*Santalum album*), source of the aromatic and highly-priced sandalwood oil and perhaps one of the most important botanical flagship species in the area. The most significant remaining stands of sandalwood are found on Sumba, Flores and Timor.

Like the flora, the fauna of the Lesser Sundas is also rather poor in species, and there is a marked and gradual impoverishment as one moves to the east. There are 242 resident bird species, 66 of which are endemic. A variety of parrots, cockatoos and honey eaters is well dispersed over the islands. Among the mammals there are eight endemic species, including four fruit bats, two insect-eating bats, one shrew, and the Flores giant tree rat (*Papagomys armandvillei*) with a head and body length of about 40 cm. Many of the relatively large terrestrial mammals, such as long-tailed macaques, civets and pigs have been introduced by people.

The most striking flagship species of the Lesser Sundas is the amazing Komodo dragon (*Varanus komodoensis*), the largest lizard in the world and native to the small island of Komodo and adjacent islands of Rinca,

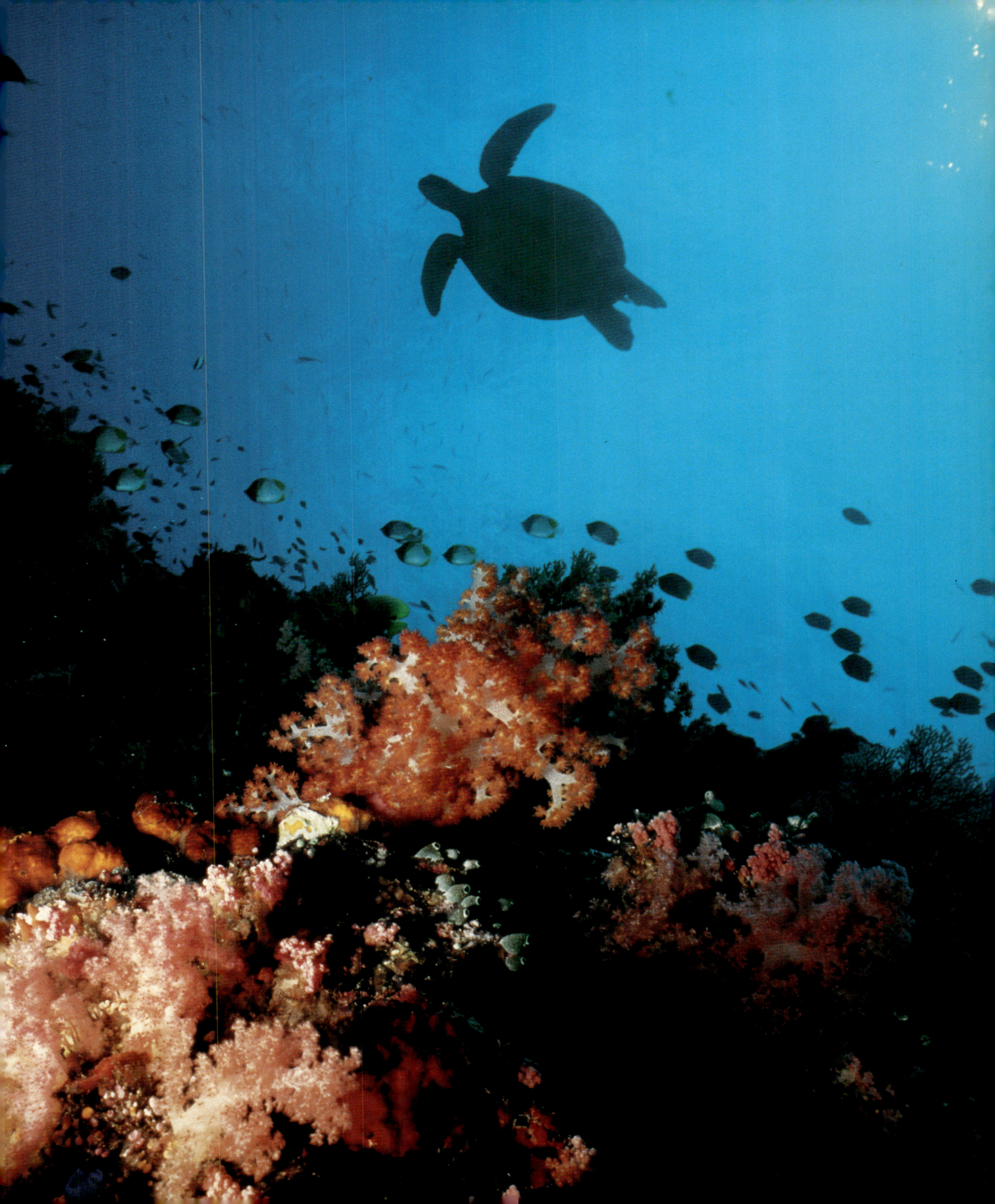

Padar and West Flores. This large predator and scavenger, also known as "the dragon" or *ora* by the local people, may grow to an impressive 2.8 m in length and weigh 150 kg. Its size, ferocious appearance, and unrestrained behavior when feeding have made it the protagonist of many early mariner's tales of fire-breathing creatures prowling the islands of the archipelago. The dragon feeds on meat, dead or alive, and is known to cannibalize younger or sick dragons. The mouths of these lizards carry a deadly load of bacteria, and once a victim has been wounded, infection will take over, weaken the victim, and eventually kill it. The dragon is armed with an acute sense of smell, which it uses to follow its prey or to detect a decaying body from as far away as eight km. Adults are capable of killing deer, wild boars, and even water buffalo. Special protected areas, which have become major tourist attractions on Komodo and some of the other islands, have been established to conserve the home of these giant reptiles.

Finally, we have Irian Jaya, the easternmost part of Indonesia, and the western part of the island of New Guinea. Irian Jaya is a land of vast tropical swamps, lush rain forests, remote highlands, and enormous mountains. The vegetation of Irian Jaya has affinities with both Asia and Australia. Among the more than 16 000 plant species, there are at least 24 endemic genera, and one third of the total plant species are orchids. Perhaps most noteworthy in Irian Jaya are the high mountains that plateau above 3 000 m and harbor a wide variety of *Rhododendron* species, some of which have become popular in cultivation.

Along the southern coast of Irian Jaya lie vast swampy areas, although there are also extensive swamps on the north coast around the delta of the Mamberamo River. Other characteristic vegetation types include the paperbark woodlands, comprised almost exclusively of the fire-resistant paperbark tree or river tea tree (*Melaleuca leucodendron*). Mangroves are abundant and vast on Irian Jaya's coastlines, and their composition is fairly similar to other mangrove forests of India and the Pacific. The southeastern corner of Irian Jaya is reminiscent of Australia, since it harbors vast expanses of grasses with very few trees in between. There is a marked dry season in this area, and dry savanna forests with species such as *Melaleuca*, *Eucalyptus*, *Casuarina* and *Acacia* have developed there.

Of the 174 non-marine mammal species recorded in Irian Jaya, roughly a third are marsupials, a third are rats and mice, and a third are bats. A total of about 100 are endemic. There are also two noteworthy species of egg-laying echidnas, representative of the monotremes, subclass Prototheria. However, the true mammalian flagship species for Irian Jaya are the tree kangaroos, which have taken the ecological role of primates. Some species are known to inhabit lowland areas, like the grizzled tree kangaroo (*Dendrolagus inustus inustus*), which lives in lowland and mid-elevation rain forest on the Vogelkop (Doberai) and Fakfak Peninsulas of northwestern Irian Jaya. Most of the rest are almost entirely restricted to tropical rain forests at higher elevations, like Seri's tree kangaroo (*Dendrolagus dorianus stellarum*), which lives in dense, mossy upper montane forest in central and western New Guinea's central mountain chain, where it is found from 2 600 to 3 200 m elevation, extending almost to the margins of the subalpine grasslands.

To date, 643 species of birds have been recorded in Irian Jaya, including 269 species endemic to the island of New Guinea. No other island or major island group in the country can match this level of distinctiveness. Although its affinities are mostly Australian, there are a few decidedly Asian birds, like Blyth's hornbill (*Rhyticeros plicatus*). There are some 42 species of parrot, of which the vulturine parrot (*Psittrichas fulgidus*), with its black body and red belly, rump and wing patch is certainly a flagship species. One of the most extraordinary endemic birds is the ground-dwelling pheasant pigeon (*Otidiphaps nobilis*), about the size of a small chicken but with a small crest at the back of the head and long legs which make it look like a pheasant. Another relatively little-known large bird is the massively-beaked, black palm cockatoo (*Probosciger aterrimus*), whose bright red skin below the eye is shielded by small black feathers when the bird is not excited.

New Guinea is perhaps best known for its birds of paradise. These include the greater bird of paradise (*Paradisaea apoda*), which got its scientific name because the specimens that first arrived in Europe were legless, these having been cut off by the hunters because the chief value of the birds was in their feathers. Other dramatic birds of paradise include the twelve-wired bird of paradise (*Seleucides melanoleuca*) and King of Saxony bird of paradise (*Pteridophora alberti*).

Striking birds which are close relatives to the birds of paradise are the bowerbirds, with eleven species in Irian Jaya, mostly in the mountains. Like the birds of paradise, they fit neatly into two groups, the relatively dull, monogamous species and the polygamous species with rather drab females and flamboyantly-adorned males. This group of birds is known for the polygamous species habit of building elaborate bowers on the ground from sticks and other materials. Some species ornament the basic structure with carefully chosen brightly-colored objects. The most outstanding species in this group are the golden bowerbird (*Sericulus aureus*), which builds bowers called "avenues" that include a pair of high fences built very strongly with interwoven twigs and grasses, and the Vogelkop bowerbird (*Amblyornis inornatus*), which distinguishes itself by building the most elaborate bowers.

The last birds deserving mention are the huge, flightless cassowaries, by far the largest animals on the island of New Guinea. All three existing species are found in Irian Jaya: the southern cassowary (*Casuarius casuarius*), the northern cassowary (*C. unappendiculatus*), and the dwarf or Bennett's cassowary (*C. bennetti*). The plumage of these birds resembles fur and is very sought after by the local people for use in their cere-

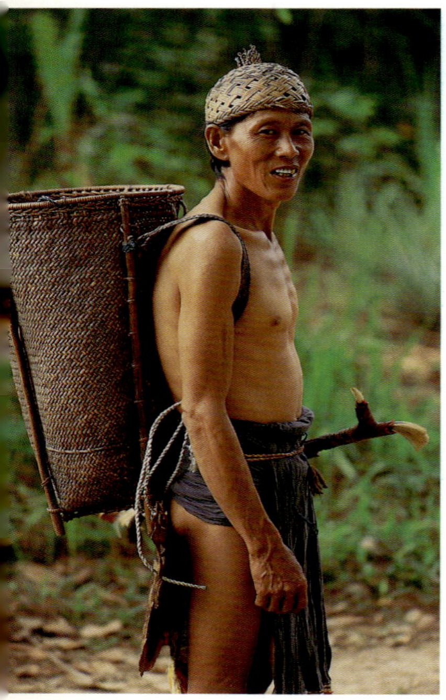

Human cultural diversity in Indonesia is among the three highest in the world. Above, a Kenyah Dayak from East Kalimantan with his traditional parang or bush knife and a backpack and hat made of rattan.

monial regalia. The wing feathers have no vanes and are reduced to long, black quills, also used for ornamentation by the local people.

The island of New Guinea is very rich in reptiles and amphibians. There are about 100 species of snakes and some 200 species of lizards, many of them endemic. In addition, there are two species of crocodile, the saltwater crocodile (*Crocodylus porosus*) and the New Guinea crocodile (*C. novaguineae*), which lives primarily in the Mamberamo river system in northern Irian Jaya. There are also six freshwater turtles, six marine turtles and well over 200 species of frogs and toads. The vast majority of Irian Jaya's frogs and toads are found in the lowlands, but some occur up to nearly 4 000 m. With so many species, it is not surprising that they adopt a wealth of life strategies. There are burrowing species, strictly aquatic species, and terrestrial species, and some whose eggs develop directly into miniature forms of the adult, such as the nine endemic *Xenobatrachus* toads. The largest frog in Irian Jaya is the Arfak frog (*Rana arfaki*), which reaches 16 cm in length.

One of the most outstandingly beautiful snakes in the world is the New Guinea green tree python (*Morelia viridis*), whose bright green skin is the color of fields of young rice sprinkled with small yellow jewels. Strangely, the young are pink, red, brown, or yellow in color. This species is very popular with hobbyists and zoos and may be threatened by the trade. It is, however, beginning to be bred in captivity and it is to be hoped that the demand for pets can be met from captive-bred specimens. Irian Jaya also has two of the world's most deadly snakes, the death adder (*Acanthophis antarcticus*), which is also found in Australia and west to Seram in the Moluccas, and the taipan (*Oxyuranus scutellatus*), from the southern half of the island and also found in Australia. The death adder is extremely aggressive, making little attempt to escape when threatened or disturbed; despite its relatively small size (up to 80 cm), it can deliver a fatal bite to humans. In contrast, the taipan tends to slither into holes when approached, but if cornered, it will strike and bite, delivering a potentially fatal dose of venom.

Finally, among the invertebrate fauna of Irian Jaya are at least 5 000 species of butterflies and moths, of which the most spectacular are surely the bird-wing butterflies. These are the largest butterflies in Southeast Asia, with wingspans of up to 33 cm.

Cultural diversity

Indonesia is the fifth most populous country in the world, with more than 200 million inhabitants. Its motto is *Unity in Diversity*, and quite appropriately, this country not only has very high biodiversity, but its human cultural diversity is very high as well, with some 336 cultures recognized. These cultures speak at least 250 different languages, but the country's lingua franca, Bahasa Indonesia, adopted only 45 years ago, is now spoken throughout the country and greatly facilitates communication.

Over 60% of all Indonesians live on the island of Java, which itself occupies just 7% of the land area, and most of them speak Javanese as their mother tongue, the thirteenth largest language group in the world ahead of Bihari, Italian and Korean. Other dominant groups are the Sundanese of West Java, the Acehnese, Minang and Batak of Sumatra, the Balinese, the Dayak of Kalimantan, the Ambonese of the Moluccas, the Buginese, Minahasan and Toraja of Sulawesi, and the Dani of Irian Jaya. A considerable number of small groups also exist, and some of these in the more remote areas have retained part of their ancestral traditions.

There are many tribes considered "uncivilized" by the more westernized inhabitants of the great urban centers of Indonesia. These include Sumatra's elusive Kubu and Talang Mamak tribes, the Penan of Kalimantan, the Alfur (Naulu) of the Moluccas, and the Asmat of Irian Jaya, who are all descendants of the Austronesian groups that drifted into the archipelago between 8000 and 7000 B.C. The policy of the central government is to bring these tribal peoples into the mainstream of Indonesian life. There are thought to be about 1.2 million of these people, 50% of them on Irian Jaya.

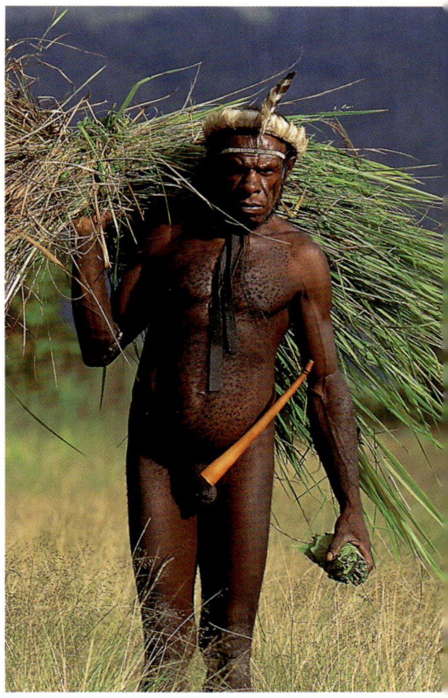

A Dani tribesman from the highlands of Irian Jaya with a feather headdress and the traditional penis gourd gets ready for a feast.

Another important element in a country composed of islands is the seafaring communities. The sea gypsies, or Bajau, are one of the last groups living almost exclusively on their *lepa*, a boat hand-hewn from meranti wood. These skilled mariners live on their boats from birth through marriage to death. Families as large as eight live together with all their possessions, which consist mostly of fishing implements, cooking utensils, clothes and even chickens, under conditions that to us may seem crowded, but that to Bajau are apparently quite comfortable. Although a large number of Bajau have now forsaken their traditional lifestyles for more settled situations, many of them continue to live on the sea. Other groups with the sea in their blood are the Buginese and the Makasarese from southern Sulawesi. These groups of people have a close relationship with the sea, and the Buginese have long been recognized as being the most skillful boatbuilders and sailors in Indonesia. They are major traders and a popular tourist stop in Jakarta is to see their high-prowed colorful boats being loaded and unloaded in the old Sunda Kelapa port. They were feared pirates in days gone by, and the term "boogieman," used to frighten small children in English-speaking countries, is thought to have originated from stories about the Buginese.

Among the important groups on Sumatra are the Batak, a family of six cultural and language groups centered inland on Lake Toba in northern Sumatra. They are a swashbuckling or "aggressive" group, as implied by their name, which may have originated as a abusive nickname used by coastal traders. When reports reached Europe of the Batak and their ritual cannibalism, that caused amazement because they were also said to have their own written language, an aristocracy,

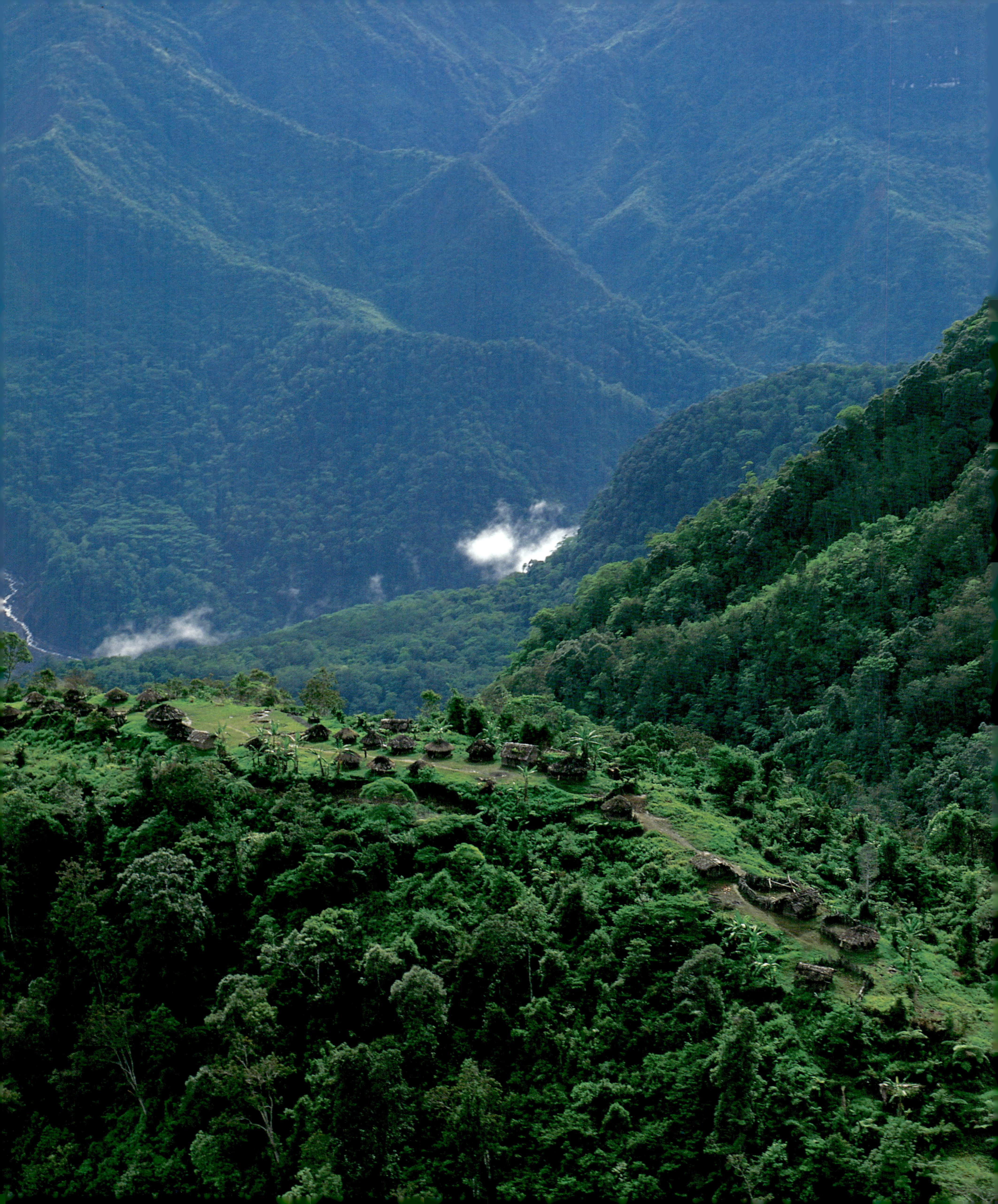

and a refined culture rather than being the expected raw savages. Each group consists of clans which are revealed in surnames, and marriage within a clan is forbidden. Irrigated terraced rice dominates the agriculture, but the highlands allow a large range of vegetables to be grown. The majority of the Batak groups are Christian, but the southern groups follow Islam.

The Minang are a group found mainly in West Sumatra and are well-known as intellectuals (a disproportionate number of them have held high positions in government). They are also known as commercial doyens, as creators of a spicy-hot cuisine which can take the skin off the mouth of the unwary, as paragons of the Islamic faith, and as members of one of the few matrilineal groups on Sumatra. Minang men traditionally travel the archipelago and beyond in search of opportunity, work and profit, and it is said that there are now more Minang in Jakarta than in their homeland.

Among the forest-dwelling people of Sumatra are the Kubu, the largest extant forest group, but even their numbers are steadily declining. While the exact origin of the Kubu remains unclear, they are physically similar to the surrounding Malay people, and their speech is a dialect of Malay. Most Kubu groups practice a combination of hunting and gathering and swidden agriculture, which may consist of rice and maize, tubers and bananas. However, there are some that practice no agriculture and simply buy rice from the proceeds of sale of forest products such as rattan and resins. Wild game and fish are an important part of their diet, but the bulk of their food is derived from either wild or cultivated plants. The nomadic existence of the Kubu permits few possessions. Apart from basic hunting tools, the most important of which are the multi-purpose bush knife and the 2.5-meter long spear, the only possessions of an average Kubu family are pots and pans, and rattan mats to sleep on. Once proud hunters of the Sumatran jungle, in recent years the Kubu have had to endure increasing pressures from the outside world. Their native forest territory has been significantly reduced as a result of logging operations and the expansion of agricultural activities by lowland farmers and corporations. The Kubu's claims to their traditional forest lands are often not recognized, and their protests have met with little success to date. Their future is far from clear. An estimated 5 000 Kubu live in Jambi and South Sumatra provinces, clinging to their traditional way of life, and they remain dependent on the forest for their survival. Unfortunately, entrepreneurs are exploiting the Kubu for their hunting skills by encouraging them to obtain protected species, which are butchered and sold on the Chinese medicine market.

The Mentawai Islands lie in the middle of the chain of islands located to the west of Sumatra. The people there are descended from some of the earliest groups to have migrated to Indonesia several thousand years ago. They have retained elements of a Neolithic culture with some Bronze Age elements, and were unaffected by the waves of Hindu, Buddhist and Islamic cultures on the mainland. Old stone axes are still occasionally found. They are organized into clans of about ten families which center their lives on the longhouse, although they have smaller field houses within an hour or two's walk. They do not have chiefs, but rather hold ceremonies called in response to the augury performed by "medicine men" and the ceremonies are overseen by a form of "master of ceremonies." Their staple is the sago palm (*Metroxylon sagu*), which is much more common in eastern Indonesia. Unlike eastern Indonesians and the people of Papua New Guinea, they never boil the sago but roast it over a fire. Feral pigs and chickens are a major source of protein, as are prawns caught in the rivers. They will also hunt primates (excluding the gibbon), deer, and other animals using a sugar-palm bow and poisoned arrows. They are among the most unusual of Indonesia's people, with the most traditional group being found on the largest island of Siberut.

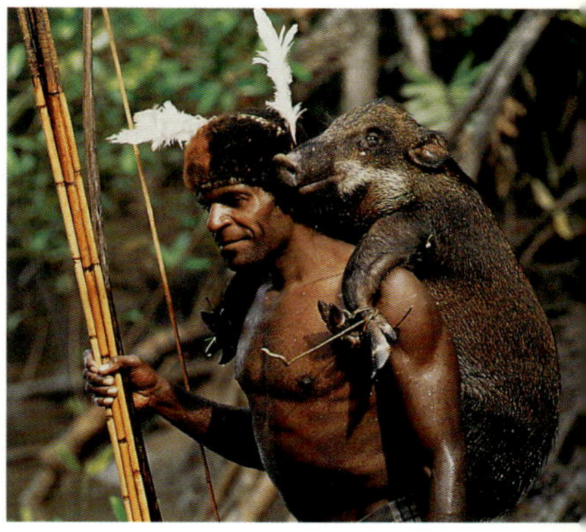

The Javanese can divided into three groups. Those of the central area live in what most people regard as the heartland of "real" Javanese culture. Here, surrounded by beautiful rice fields made fertile by a series of active volcanoes are the homes of aristocratic dynasties that have developed and encouraged a wide range of artistic endeavors for over 1 000 years. The royal domination has generated almost an obsession with status and rank, and different levels of language are used to elevate, put down, or befriend the person being spoken to. The religion of the majority is Islam, although two distinct strands are encountered: the orthodox, and those who incorporate Javanese customs, rites and mysticism. The second group is found across the northern plain, comprising people who are urban rather than rural, and perhaps not as refined or genteel as their southern kin, having had much more to do with trade than with the arts. They are uniformly orthodox in the form of Islam practiced, and the small traders have taken this with them both through Java and to the rest of the archipelago. The third group occupies the extreme east of Java, but has been largely incorporated into the other groups.

The Sundanese live almost exclusively in West Java and are the second most numerous ethnic group in the country. Their language is more closely related to the trading languages of Minang (West Sumatra) and Malay than to the neighboring Javanese, but they have adopted the levels of speech that reflect actual or perceived rank or respect. Other Javanese "inventions" such as gamelan music, shadow plays, certain dance types, and terraced irrigated rice cultivation have also been absorbed to the west and adapted. Today the Sunda culture differs from the Javanese by being more clearly Islamic and by paying less attention to social hierarchy. Sundanese are perhaps more independent by nature; even last century, they were predominantly dry-field shifting cultivators in the mountainous, forested interior, living in small villages. By being so dispersed they were less susceptible to authoritarian, central control.

The Baduy of West Java number only about 5 000

On the opposite page, a village in the "South Gap," near the Baliem Valley in Irian Jaya.

Above, an Asmat hunter wears a feather headdress and carries a pig that he has shot with his bow and arrow.

people living in the rugged and mountainous area 150 km west of Jakarta. Their population size in no way reflects their significance, the interest taken in them, or the degree of frustration felt by government officials charged with encouraging them to develop. The Sundanese living nearby hold them in great respect and indeed fear because the Baduy are said to be masters of magic and to be watched over by phantom tigers. A Baduy village has barely changed since written observations were first made, for the people believe that what was good enough for their ancestors should be good enough for them. Thus, modern accoutrements of living are banned and their agriculture is classic swidden cultivation, which has all but disappeared elsewhere on Java. Given the land pressures from within and outside their area, this type of agriculture is now practiced on much steeper slopes than it was in the past. The Baduy people probably arose from one of the ascetic religious communities which were charged with maintaining the imperiled balance between the earthly and spiritual worlds. They had to lead a very regimented and pure life, and their importance and independence were reflected in their freedom from taxation. Because of their crucial task, they maintained close ties with the princes who depended on harmony, peace, and tributes for their existence.

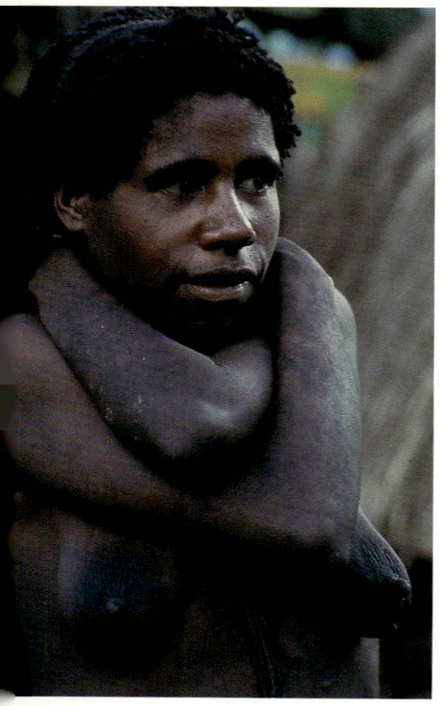

A Dani woman from the Baliem Valley in Irian Jaya embraces herself against the cold.

The heavy burden felt by the Baduy for the integrity of a principality has now been transferred to a feeling of responsibility for the entire nation. This is reflected in their brazen behavior of walking barefoot to Jakarta to make representation directly to Cabinet Ministers. After making direct representation to the President (their reigning "prince"), their 5 000-ha (just 7 x 7 km) territory has now been marked with concrete boundary posts to protect it against the encroaching modern world. In this respect, the Baduy have achieved what no other minority ethnic group has even been allowed to consider elsewhere in the archipelago. However, their current territory is a fraction of the area the Baduy used to occupy, so the population and land use must be kept in check to avoid a spiral of degradation.

The history of the Balinese is related, as one might expect, to that of the Javanese. There has been contact for over 1 000 years, and periods of peace, war, colonial subjugation, and colonial domination have taken their various turns. Some groups of native Balinese, known as the Bali Aga, objected to these changes and have kept themselves separate from mainstream Bali culture ever since. The unique Hindu culture of the Balinese has changed over time, but the adherence to the religion and its practices is still great. Music, dance, drama, shadow plays, weaving, painting, and carving as art forms are strong, and proficiency in at least one area is still pursued by the young.

Moving on to Borneo, one encounters the Dayak. "Dayak" is a catchall term for the inland peoples of Borneo, but actually covers at least eight distinct groups such as the Iban, Kayan and Kenyah. Indeed, it no more describes a cultural group than does the term "American Indians." Some are well-known because of their enormous longhouses, their stylish and curvaceous art, and their practice of headhunting. The latter was an essential part of their religious life, since the spirit within the head, when brought back and used in ceremonies, could be encouraged to help its captors. The tradition of carving was also part of the manipulation of spirits, which might otherwise bring bad fortune to the longhouse. Slash-and-burn is the traditional form of agriculture. This was no problem in the days when the available land was seemingly limitless, the population small, and the forest clearings used on a long, cyclical basis. However, today much of the good agricultural land has been converted to industrial plantations for tree crops or timber, and human population density is much higher. As a result, the interval between use of fields has been reduced, and the capability of the soil to recover has been compromised.

The Penan used to be a nomadic hunter-gatherer people who followed the cycles of the forest in the interior of Borneo. They tended to keep to themselves since the dominant Dayak and Malay were wont to enslave or simply kill them. Most are now relatively settled, inhabiting the upper watersheds of the dwindling rain forests of northern Kalimantan and neighboring Sarawak, Sabah and Brunei, whose borders they generally disregard. Yet some Penan have retained their traditional ways of exploiting Borneo's forests, the most important of which was the processing of wild forest sago (*Eugeissona utilis*), a single trunk of which could yield 4 kg of pure starch. For hunting, the men use long blowguns with poisoned darts which can be shot accurately over 100 m. These were used to hunt virtually anything that could not be caught by hand, but primarily for the bearded pig (*Sus barbatus*), which performs remarkable mass migrations through the Bornean forest following the fruiting of the dipterocarp trees.

In the Sa'dan highlands of northern South Sulawesi live the Toraja, one of Indonesia's most interesting human cultures and a largely Christian group of rice farmers with persistent animistic cultural practices. The population was divided into nobles, commoners and slaves, with the nobles having marriage links with the kingdoms in the south of South Sulawesi. Toraja land was an independent area, but never particularly rich. About a hundred years ago, however, the area became famous for its coffee. Yet the Toraja culture is best known to the outside world for its houses and its funerals. The houses have long projecting, upsweeping roofs covered with overlapping pieces of bamboo cut in half lengthwise, and the fronts are generally decorated with the horns of buffaloes sacrificed in important ceremonies. The houses are not just dwellings, but rather points of reference through which genealogies can be traced. The funerals, the most visible part of a complex cycle of ceremonies, are notable because they are so expensive that the deceased are kept in the houses for months or years until the necessary funds have been raised. At this time a large number of people gather, there are buffalo fights, kick-fighting competitions among the young men, and other ceremonies to assist the passage to heaven.

In Seram are found the Naulu people. An adult Naulu man is easily recognized by his habitual red cloth turban meant to symbolize the blood spilled by this once headhunting group in central Seram. They were famous for their fighting skills, and alliances were made with them to help other groups fight off invaders such as the Dutch. Many still practice animistic rites, making offerings to the spirits in separate spirit houses, and conducting coming-of-age initiation rites. Other groups are somewhat fearful of them because they are said to have strong magical powers. The more remote the group, the stronger the magic. Their staples are sago and cassava, supplemented by fish and game such as deer, pig and cuscus. They earn cash from selling dried cloves, damar resin, dried coconut, and nutmeg.

Indonesia has many unique cultures, but the cultural diversity of Irian Jaya, the western half of the incredibly diverse island of New Guinea, is hard to match. Irian Jaya by itself is said to have some 250 tribes, and accounts for a significant portion of Indonesia's cultural diversity. Among the largest groups in Irian Jaya are the Dani. It was in 1938 when an aircraft flew over the central range of Irian Jaya and the existence of the stone-age Dani and other related tribes was discovered by the outside world. Although they looked "savage," with men wearing little but penis gourds and women wearing small grass skirts, their broad montane valley was a landscape of organized agriculture with complex irrigation canals feeding terraced fields of sweet potato. Their lives were fearful, however, with frequent raids between the tribes, and an almost constant state of warfare. This has decreased in the last few decades and they have adopted some modern conveniences, such as metal tools, but their culture has proved remarkably resilient, even including their polygamy in which a man simply has as many wives as he can afford (his wealth being measured in pigs). Their lack of dress is all the more notable given the high altitude at which they live, but they cover their bodies with pig fat and soot which is said to reduce the effect of the cold.

Probably the best known human group of Irian Jaya, and certainly one of the most famous cultures of Indonesia, is the Asmat. Unlike the high-land dwelling Dani, the Asmat live in the swamp forests of southwestern New Guinea, not far from the southern section of the border between Irian Jaya and Papua New Guinea. Their land is bordered on the north by the towering central highlands of New Guinea and on the south by the Arafura Sea, and even today the area is very difficult to reach. Tides in this area reach 5 m, and at low tide there are wide mud flats along the coast. The only way of moving through this roadless region is by waterway, and all transport takes place by means of dugout canoe. Sago palm is the staple food of the Asmat, and larvae of the Capricorn beetle are considered a great delicacy and also have some ritual significance. Traditionally headhunters, the Asmat live in villages varying in size between 35 and 2 000 people and their total population is currently thought to be between 50 000 and 70 000. The Asmat achieved some notoriety because the young anthropologist Michael Rockefeller disappeared in their region in the early 1960s, but they are best known for their magnificent art and wood carving. This wood-carving tradition revolves around the ceremonial life of these people, which is mainly aimed at appeasing the spirits of the dead and focuses on the basic themes of Asmat culture, headhunting and ancestor worship. Many objects are carved by the Asmat, but the symbolically most important depictions of the ancestors and the best known elements of Asmat art are the shields, the ancestor sculptures, and the large and spectacular *bisj* poles. These poles are often more than five meters in height, and consist of several human figures, one above the other, culminating in a large, open-worked, pennant-shaped protrusion called the *tsjemen* or penis. The cutting of the tree for a *bisj* pole imitates the killing of an enemy, and is rich in symbolism and ritual importance. Asmat art is now being produced for commercial purposes and for sale to tourists, but it remains one of the more impressive and distinctive tribal art forms on Earth.

One of the most unusual human groups anywhere is the Korewai, inland relatives of the famous art-creating Asmat, who live in a region between the Brazza and Eilanden Rivers in Irian Jaya. They build houses of wood and sago-leaf fronds in the tops of trees as high as 30 m above the ground and even carry piglets, dogs and chickens into these arboreal dwellings. Their houses are an important part of their culture, possibly a response to raids by rival groups, including headhunting forays by the Asmat, or a reaction to the frequent inundation and mosquito infestation of the swampy land below. They plant crops such as taro, cassava, tobacco, and bananas, and still live in a largely traditional fashion in their remote region.

As is evident from the preceding, both biological and cultural diversity reach spectacular heights in Indonesia, placing this very important country in a most privileged position among the world's nations. With its growing economic power, Indonesia has the capacity to effectively incorporate biodiversity conservation and maintenance of human cultural diversity into its future development strategies if it chooses to do so. Hopefully, recognition of Indonesia as one of the world's top two megadiversity countries will help to increase awareness of the truly unique position in which this country finds itself, and will lead to more concerted action to conserve the wide array of species, ecosystems and cultures that can still be found in this island nation.

ANTHONY WHITTEN
JANE WHITTEN
CRISTINA GOETTSCH MITTERMEIER
JATNA SUPRIATNA
RUSSELL A. MITTERMEIER

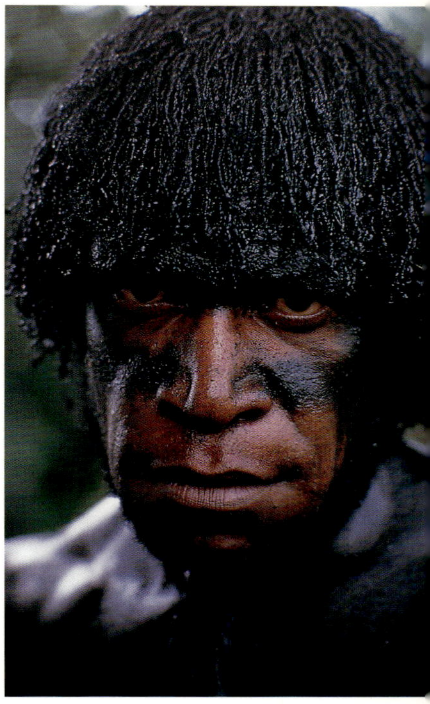

A Dani man smeared with pig fat and charcoal, Baliem Valley, Irian Jaya.

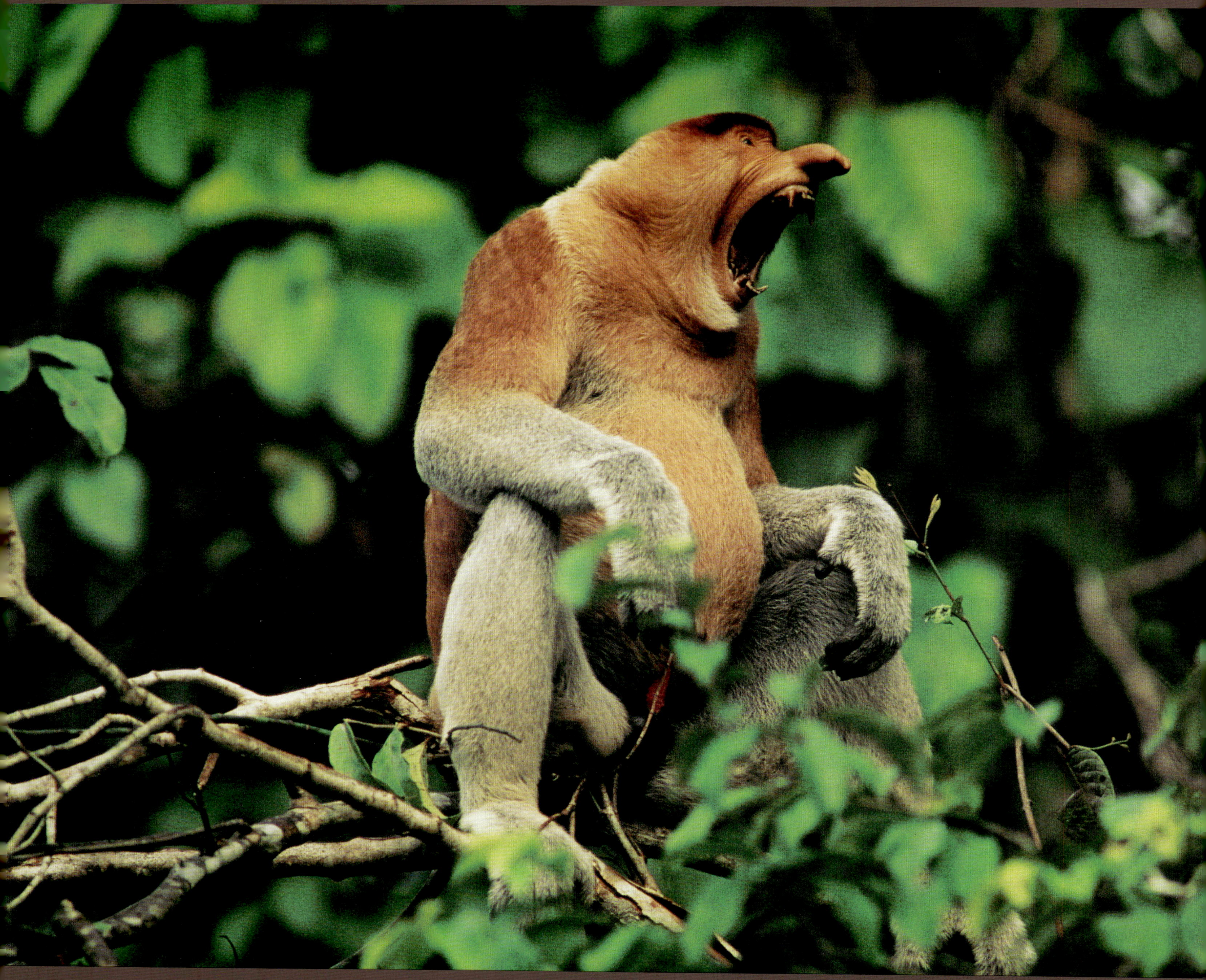

*The island of Borneo, third largest island in the world, is fringed with mangrove swamps. The amazing proboscis monkey (*Nasalis larvatus*) is endemic to Borneo and found mainly in these mangrove forests, although its range also extends inland in a number of areas. Above, a large male in a threat display. Upper right, a proboscis monkey leaps from tree to tree; although a large and bulky animal, this monkey is a good leaper and also a strong swimmer. Lower right, riverside vegetation in a peat swamp forest along the Sekonyer River, in the Tanjung Puting National Park.*

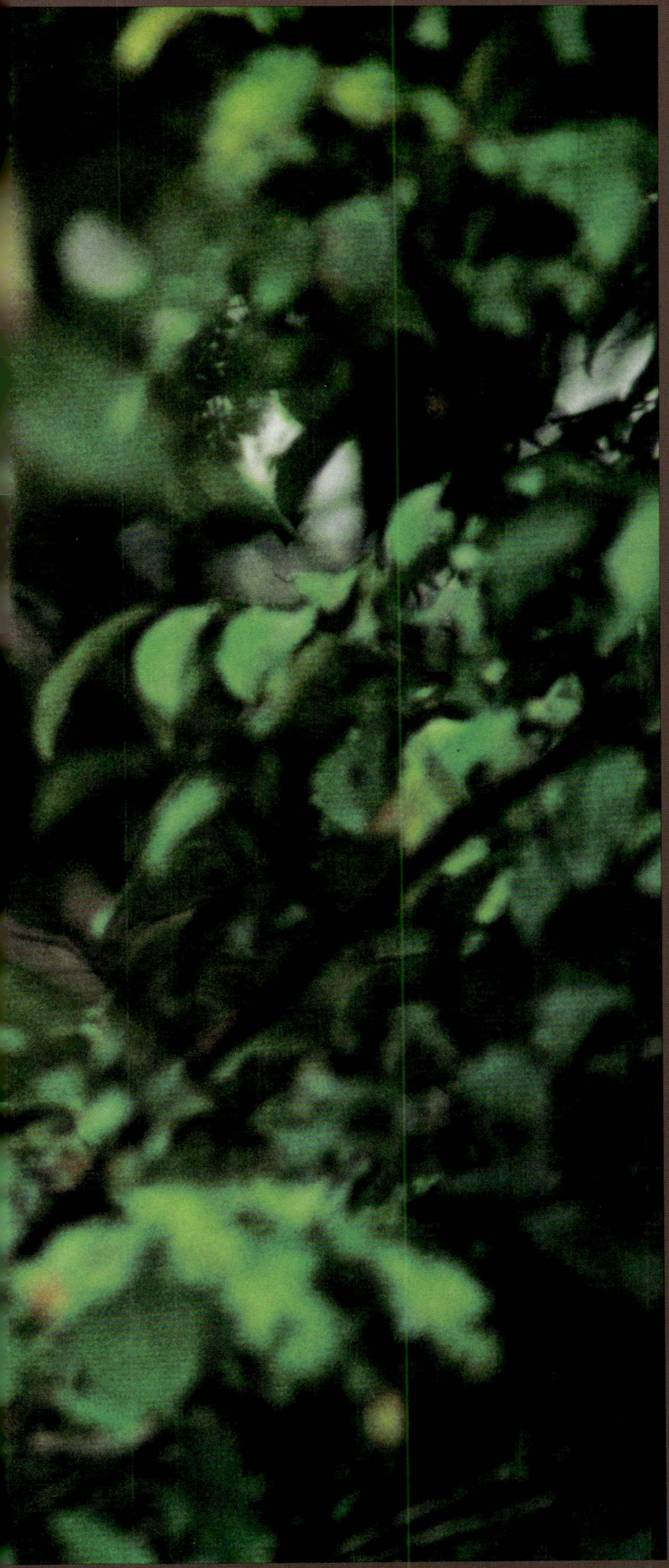

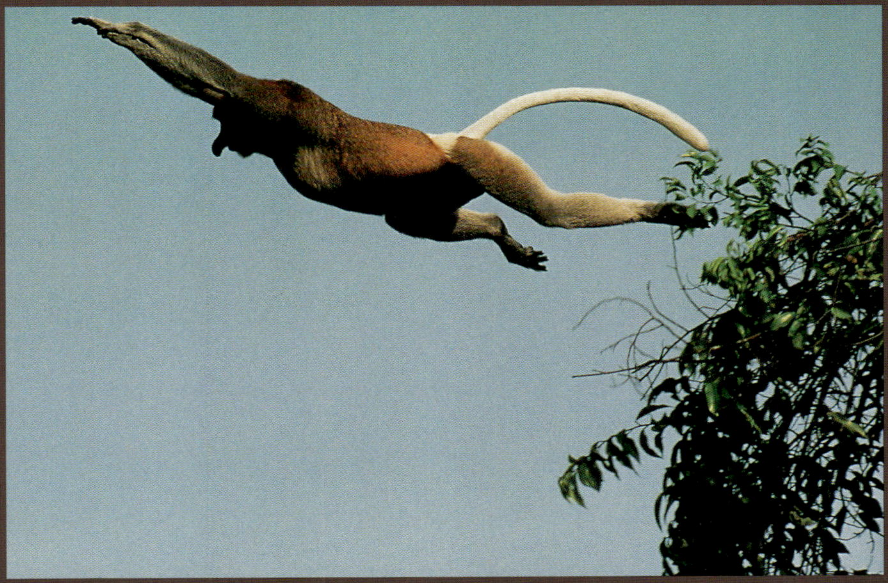

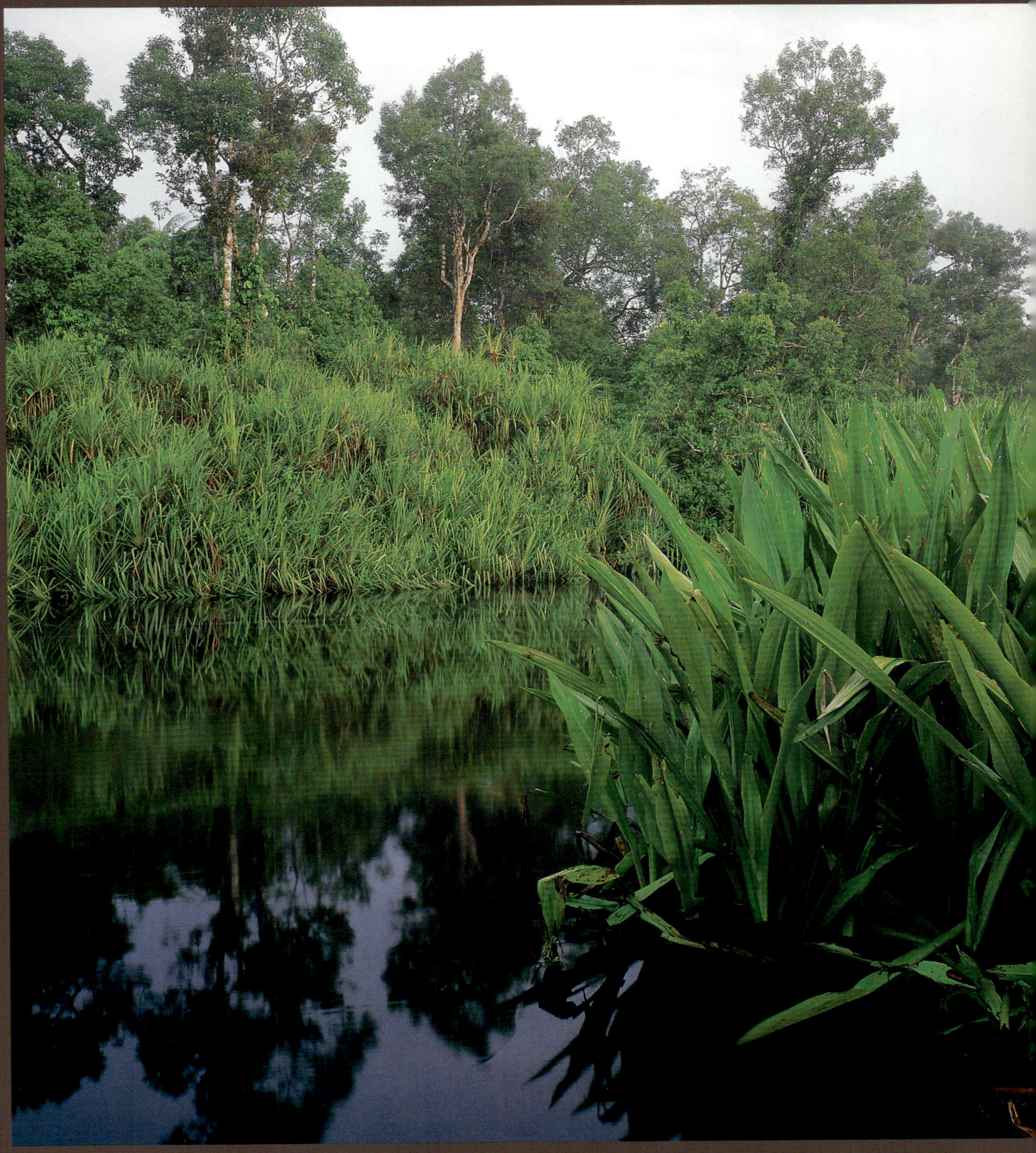

Indonesia has the third largest extent of tropical rain forest on Earth, after Brazil and the DRC, and most of its important flagship species occur in this ecosystem. Right, the Sumatran rhino (Dicerorhinus sumatrensis)*, smallest of the living rhino species, is a rain forest dweller now found mainly in Sumatra and Peninsular Malaysia, with tiny populations in Sabah in East Malaysia and perhaps still in Kalimantan.*
Below, the Rafflesia flower (Rafflesia arnoldi) *is the world's largest flower, growing to almost one meter in width and weighing about 9 kg.*

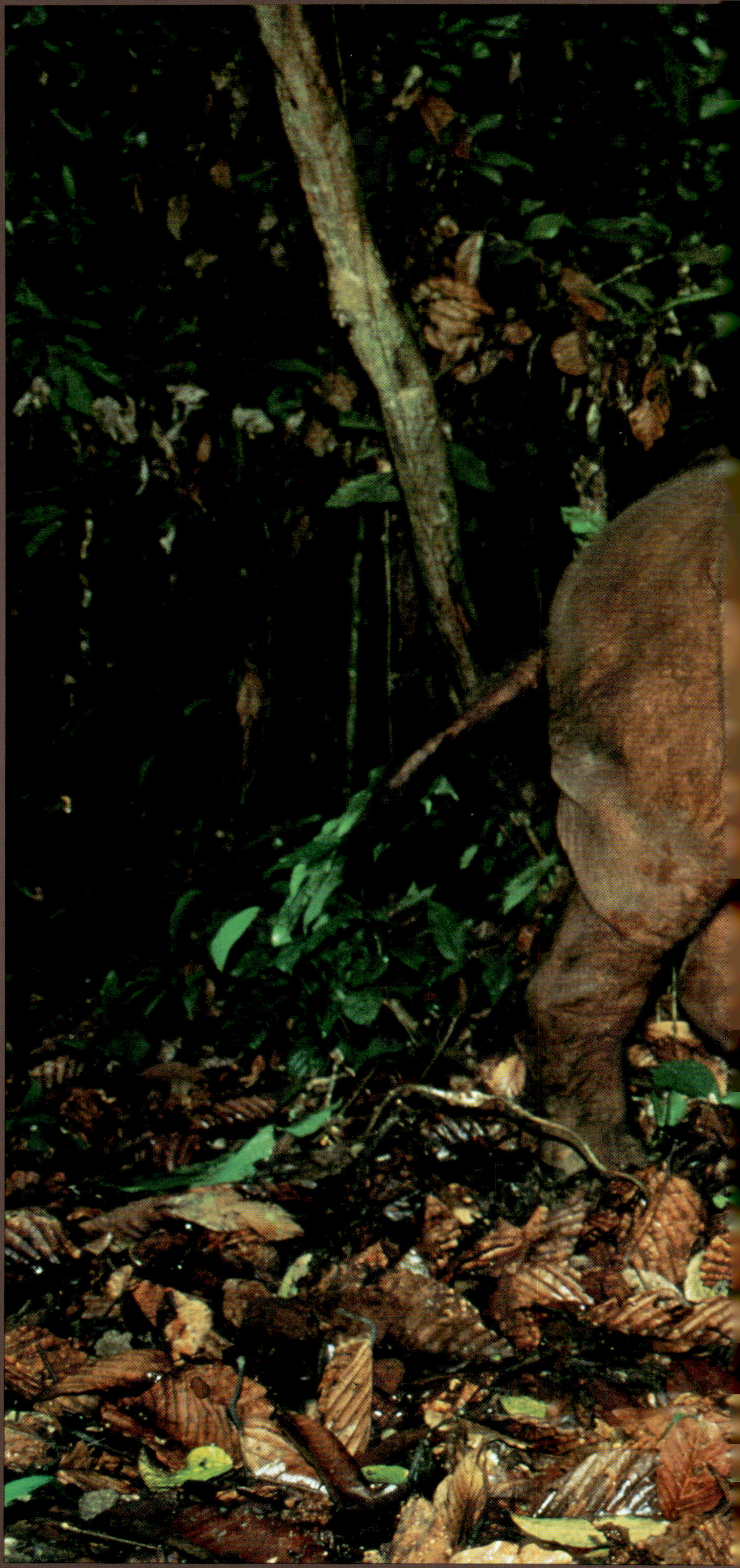

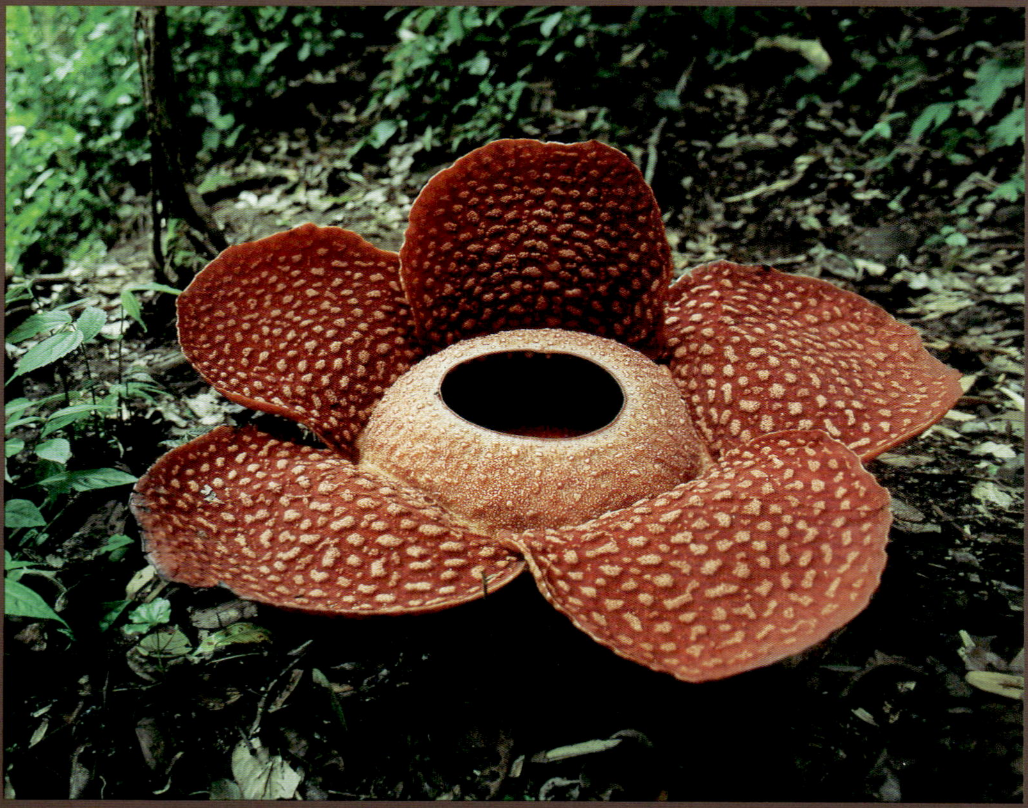

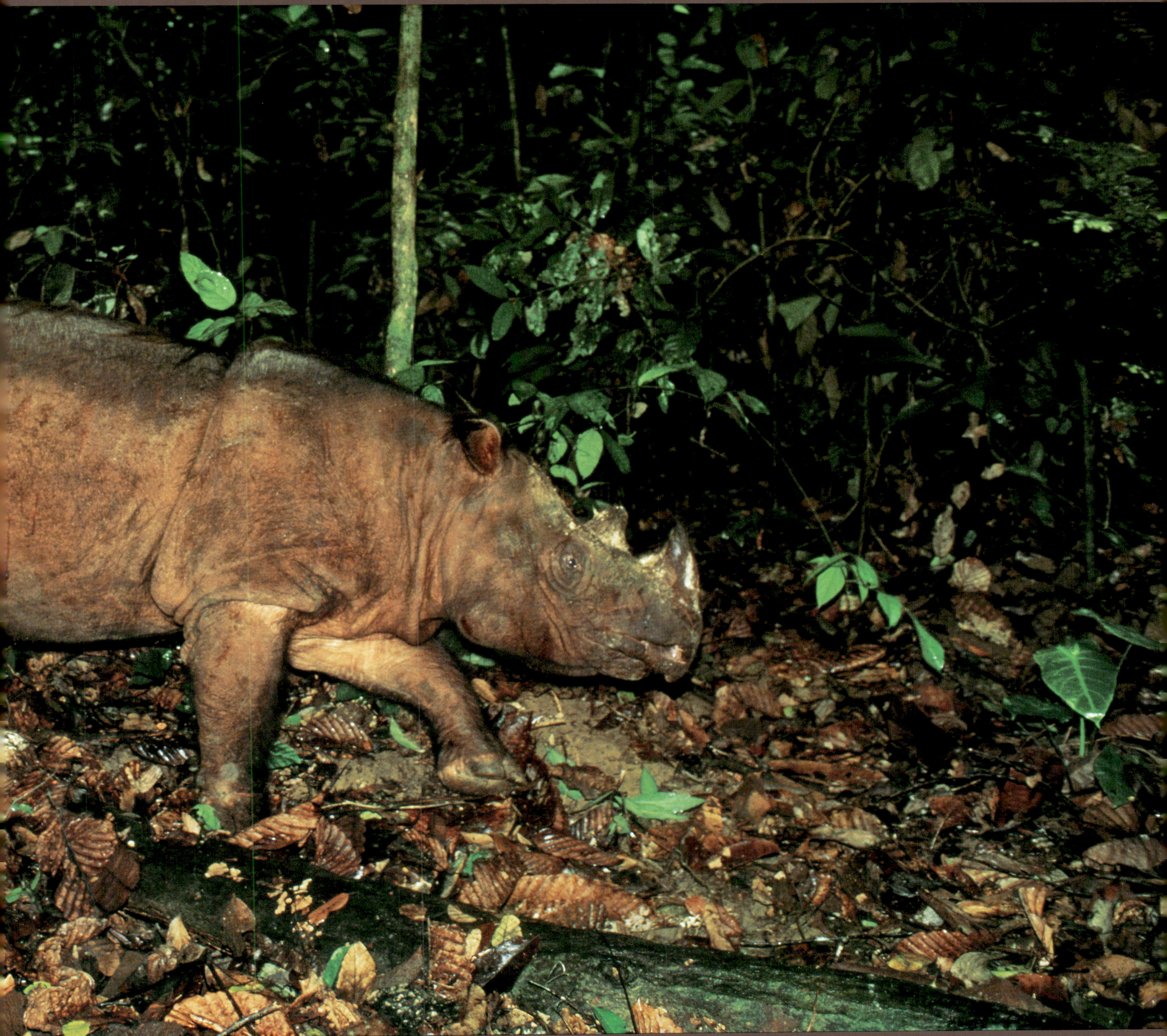

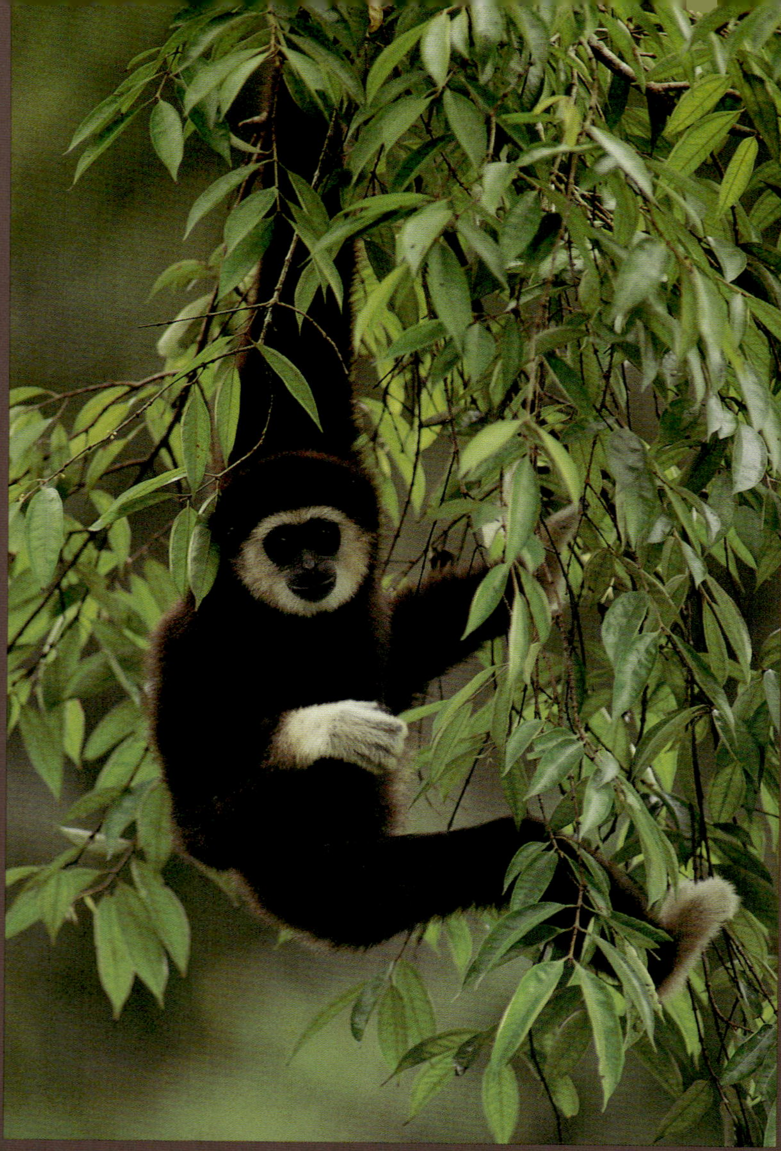

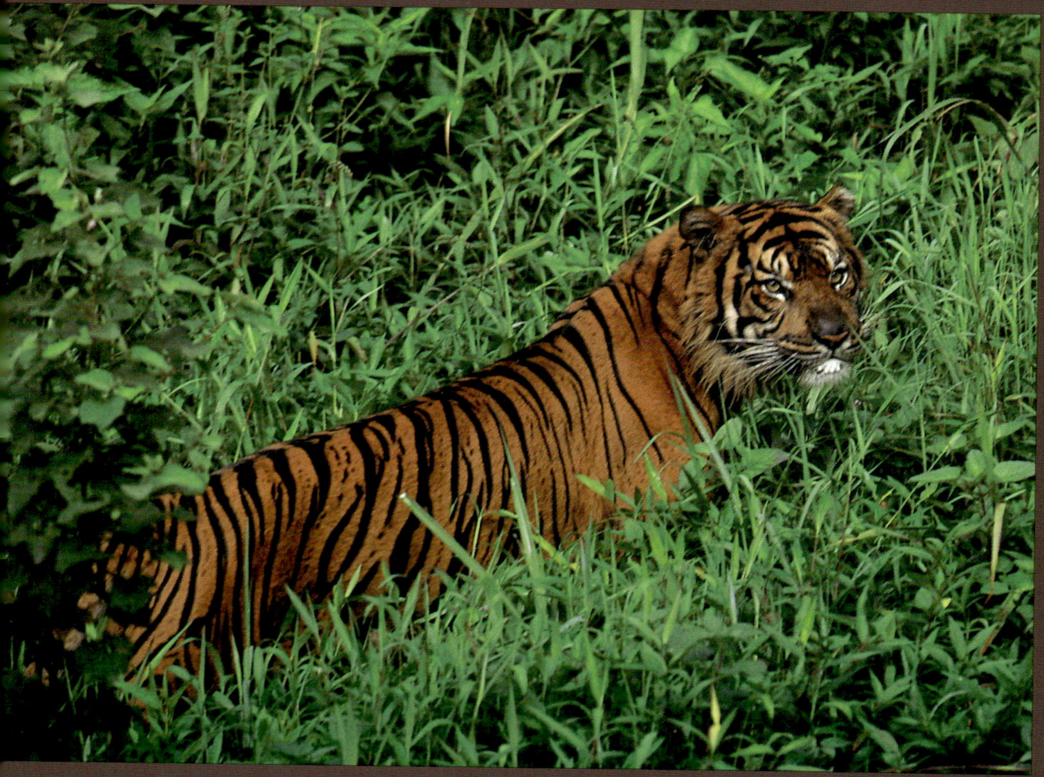

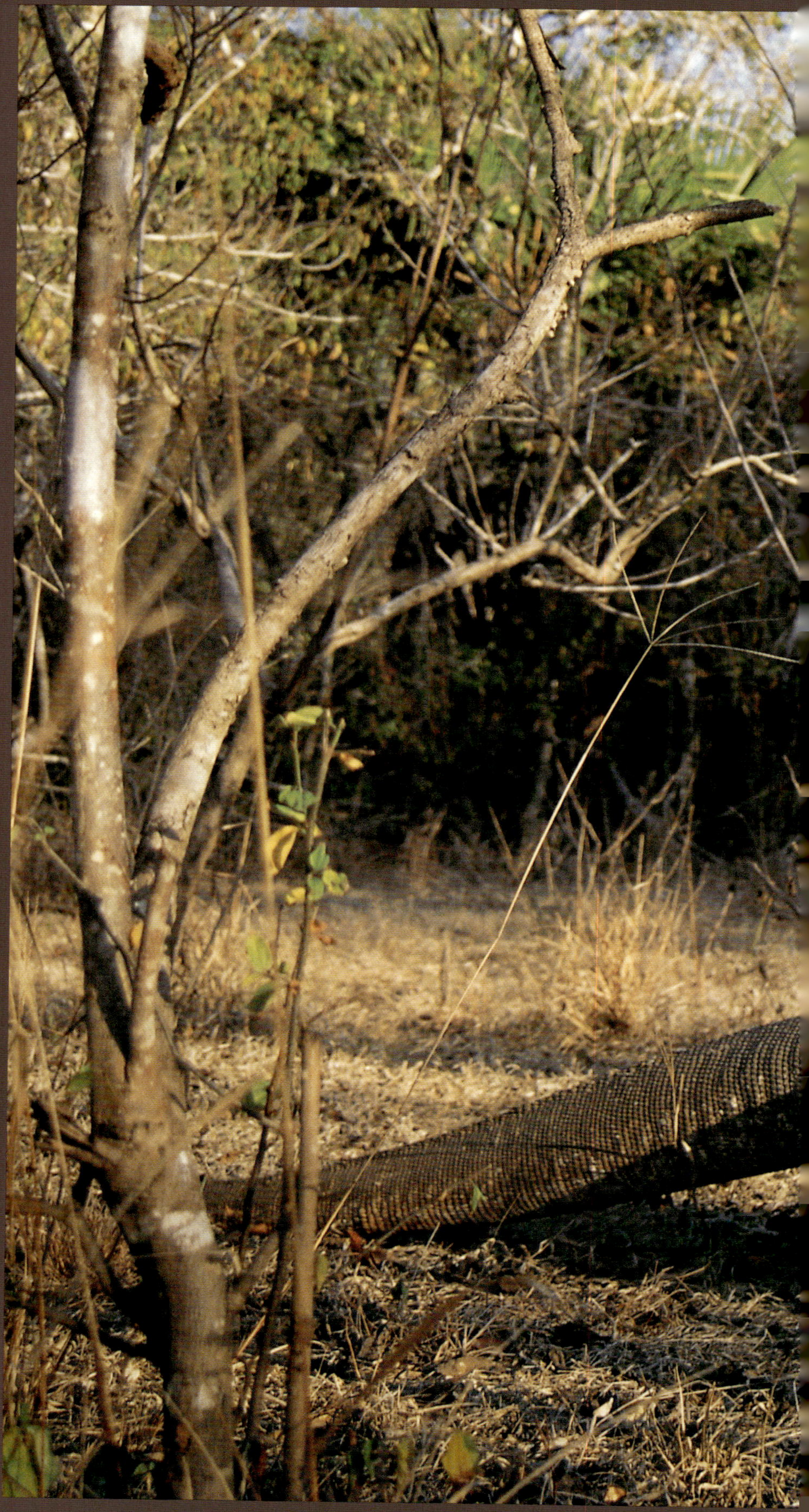

Upper left, the white-handed gibbon (Hylobates lar) *is one of six gibbon species found in Indonesia, site of the world's highest concentration of these lesser apes. Lower left, the Sumatran tiger* (Panthera tigris sumatrae) *is the only surviving race of the tiger in Indonesia, the Javan and Balinese subspecies having gone extinct in this century. Below, the imposing Komodo dragon* (Varanus komodoensis) *is the largest living lizard. It is restricted to the island of Komodo, the western end of Flores and two smaller islands, where it has become a major tourist attraction.*

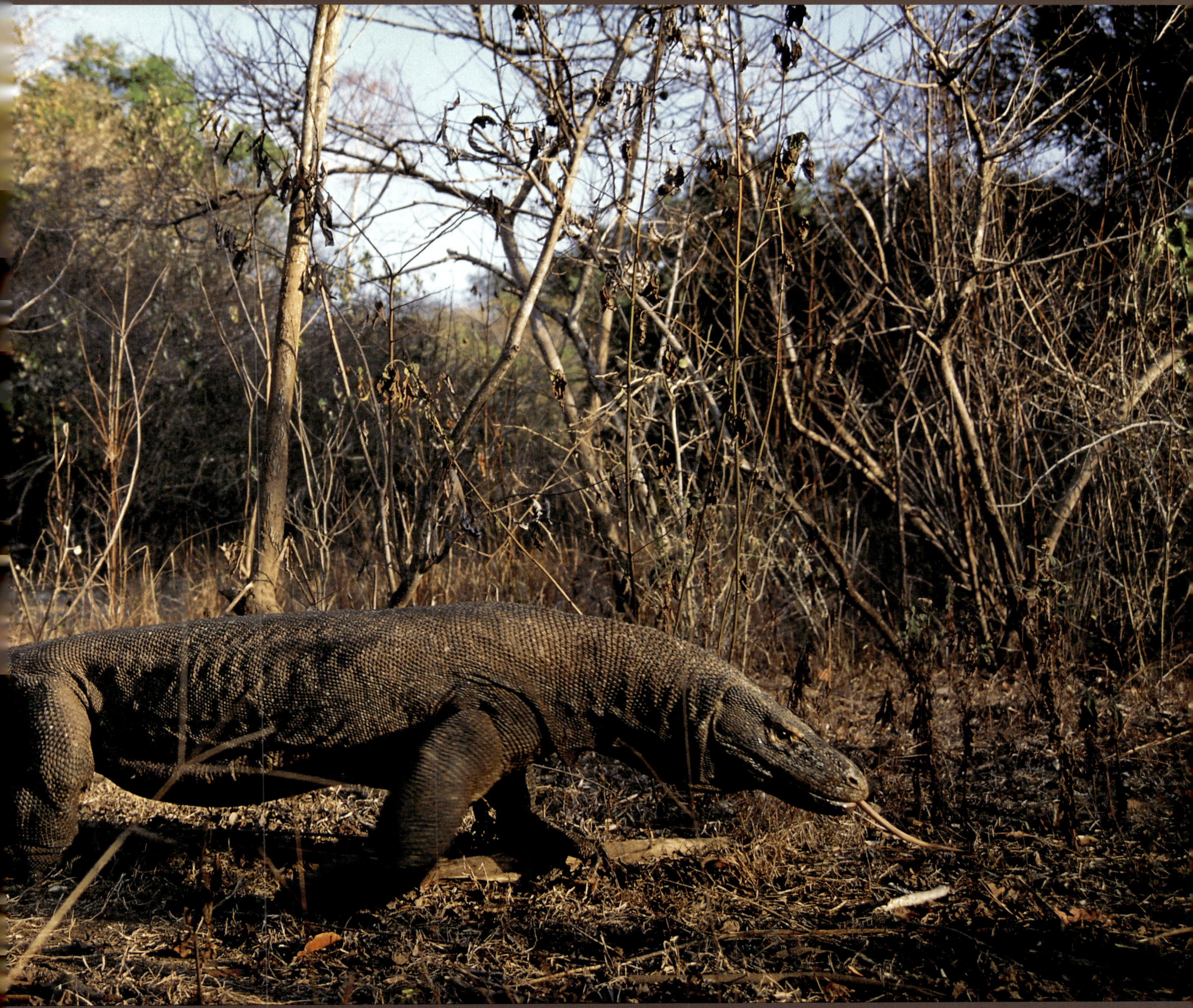

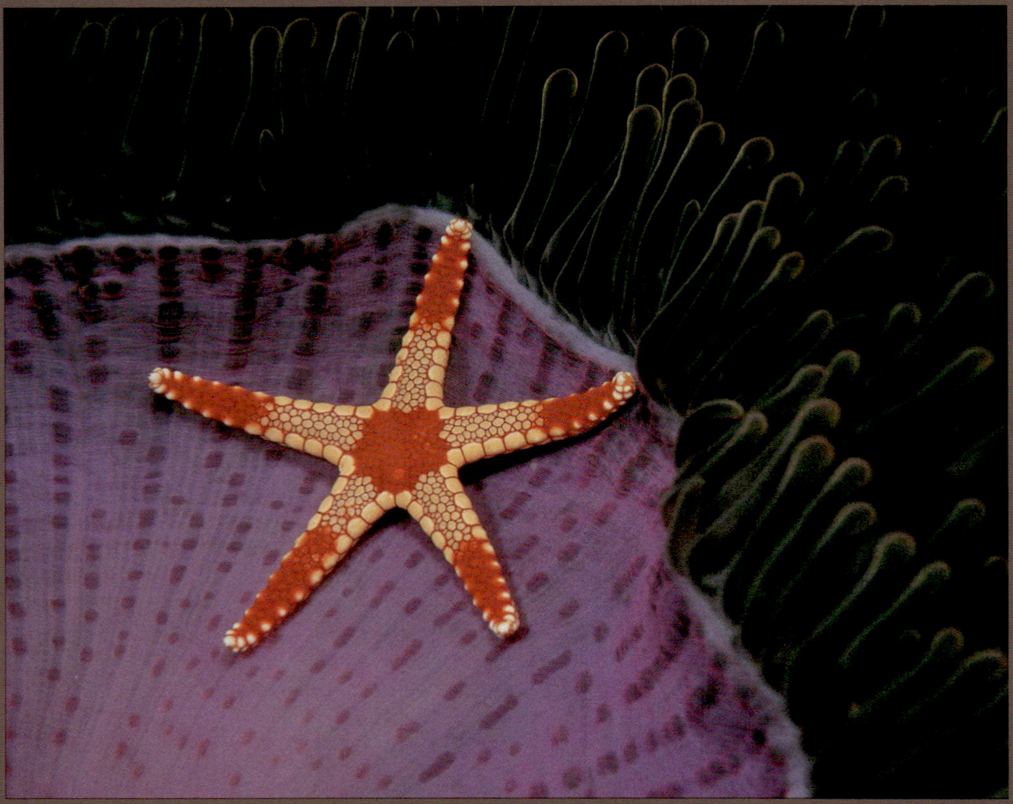

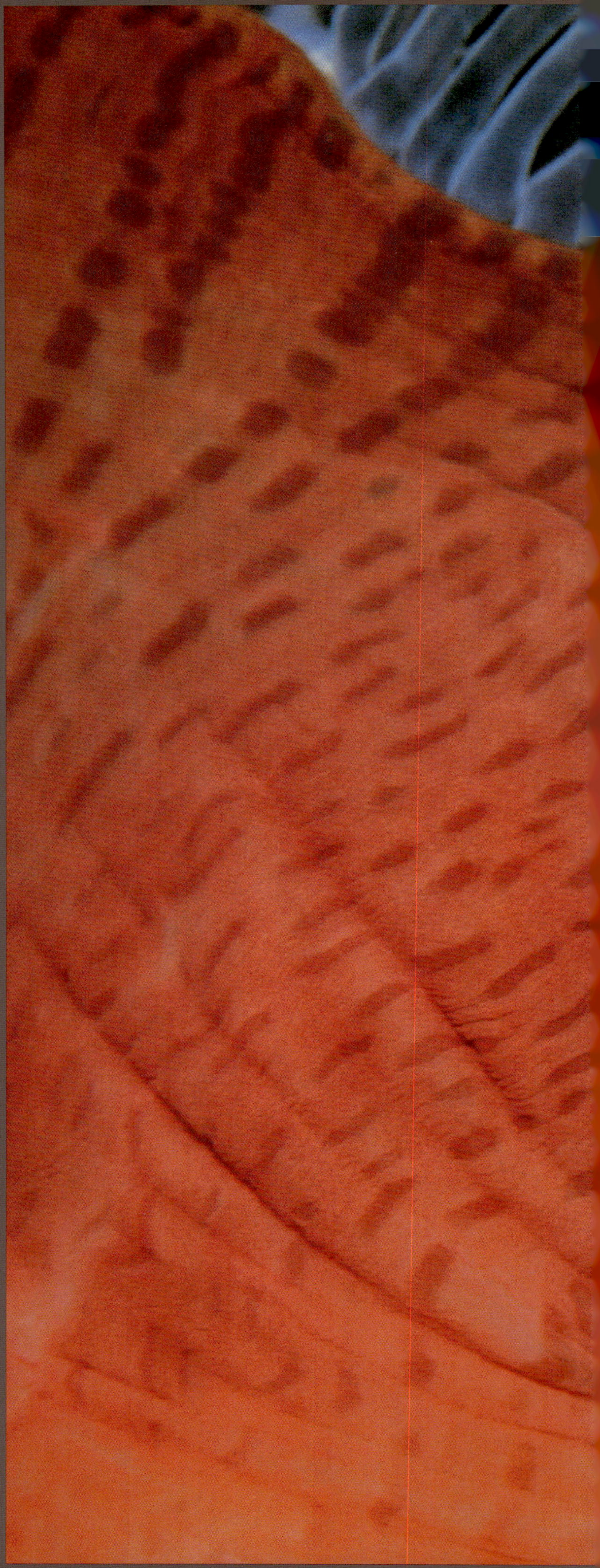

Indonesia's marine diversity may be the highest on Earth, and its coral reefs are especially rich. Above, a candy cane starfish on a sea anemone foot, Sipaden Island, Borneo. Right, one of Indonesia's many coral species, with clownfish and sea anemone.

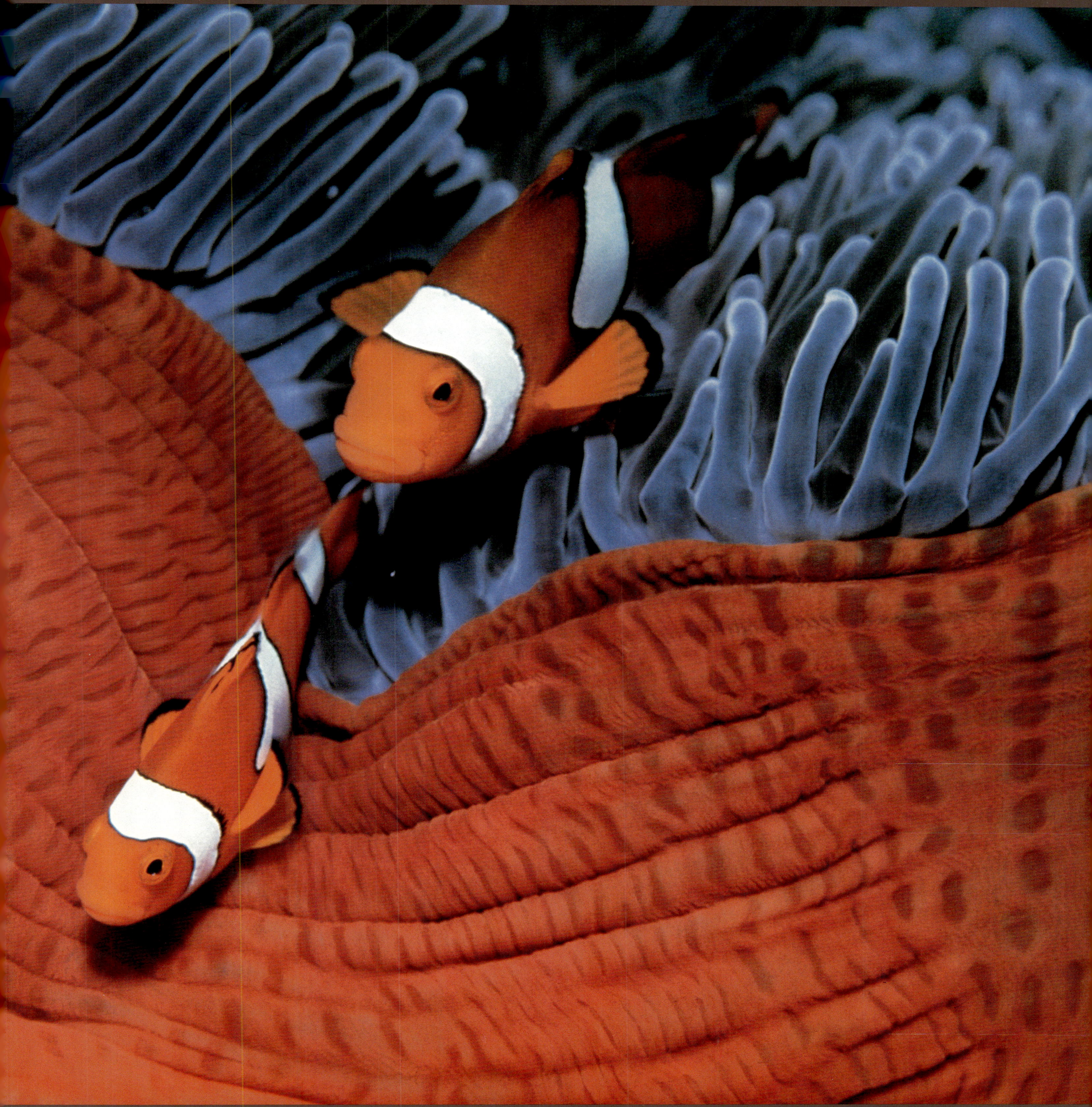

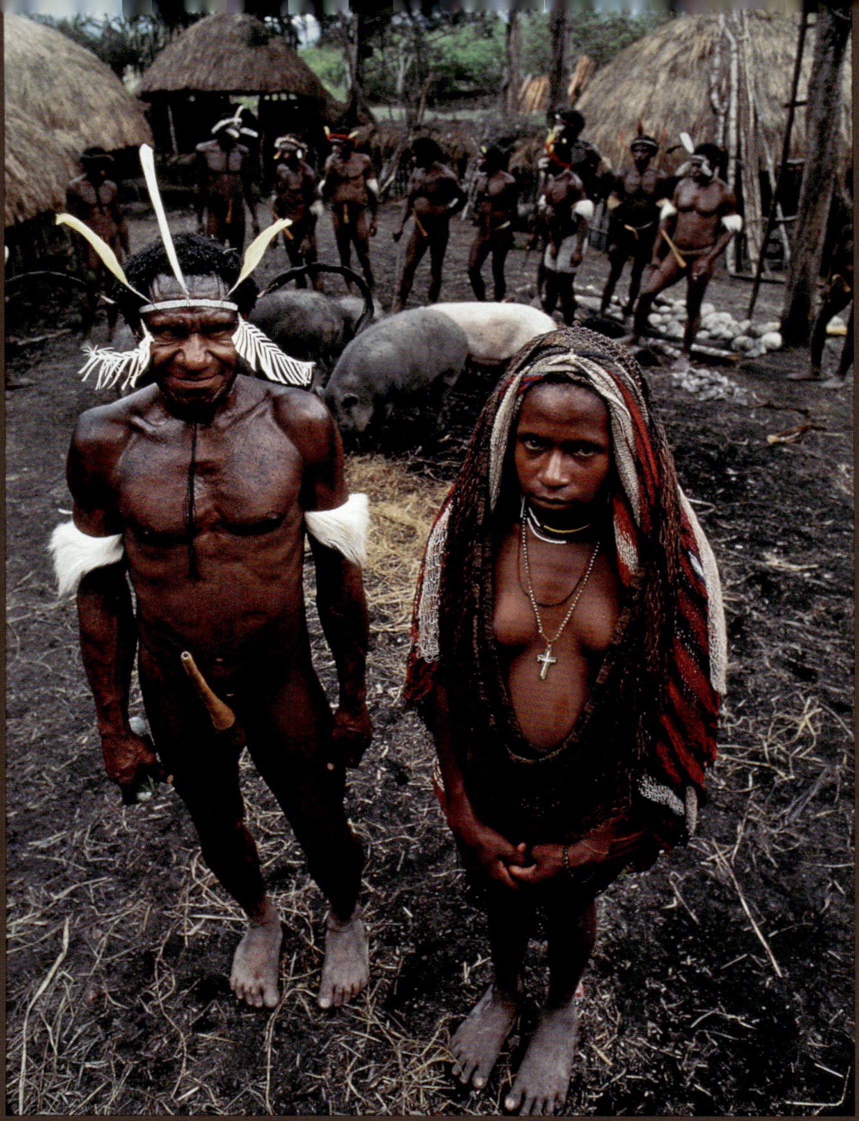

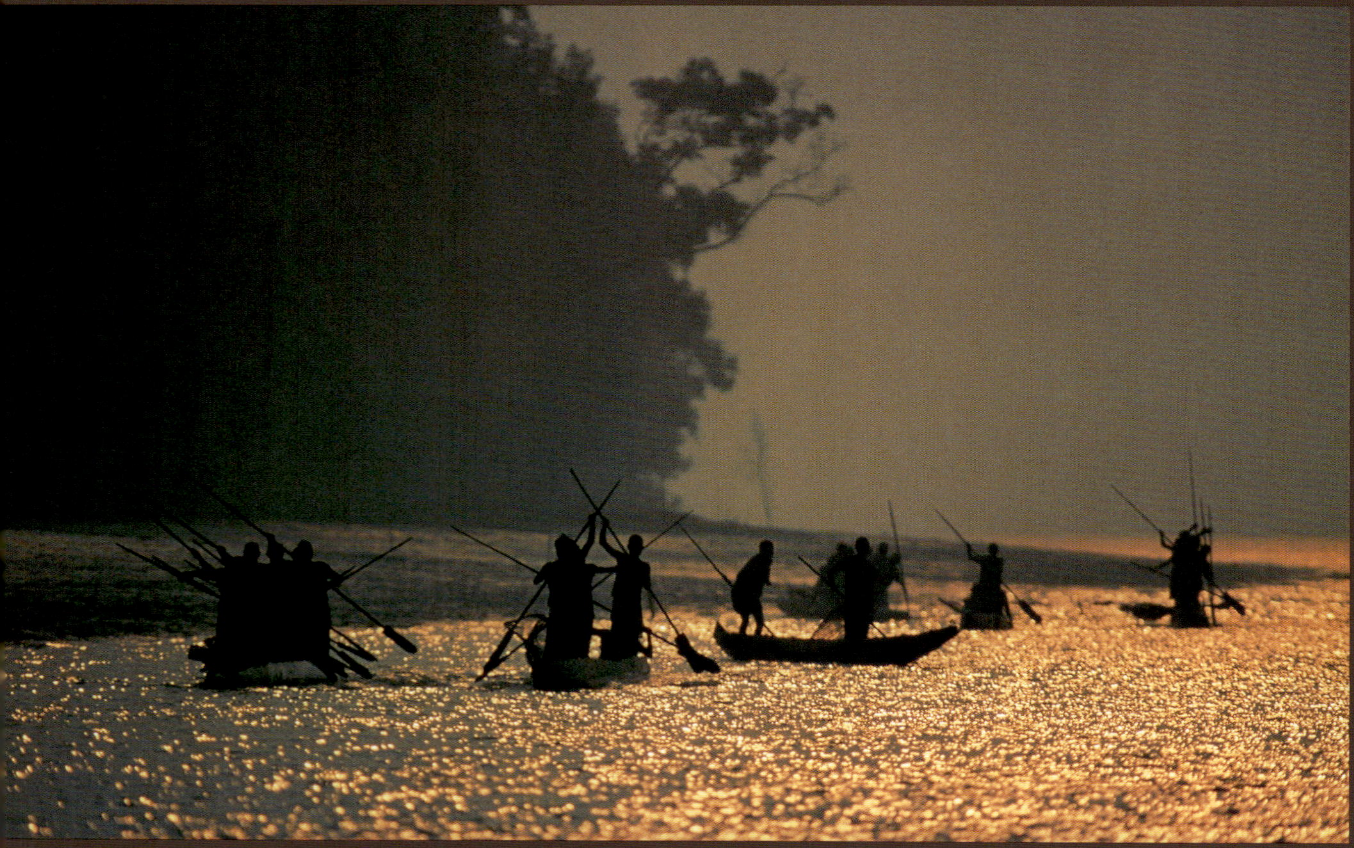

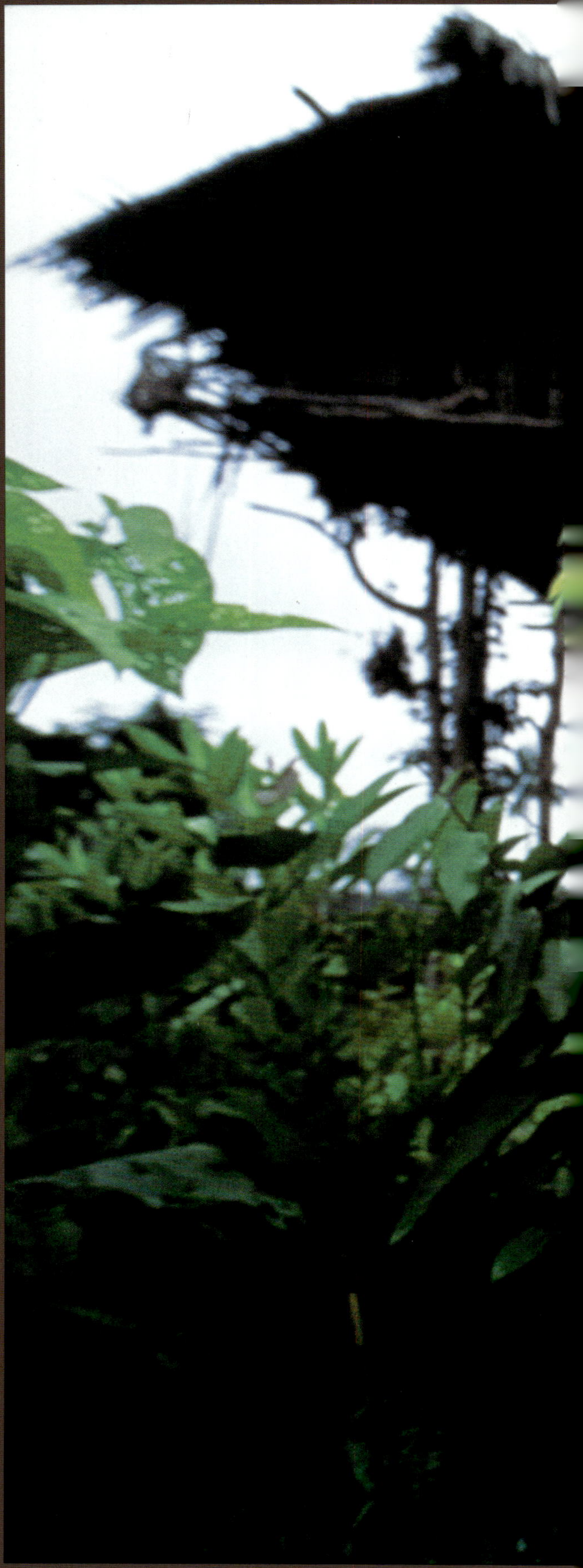

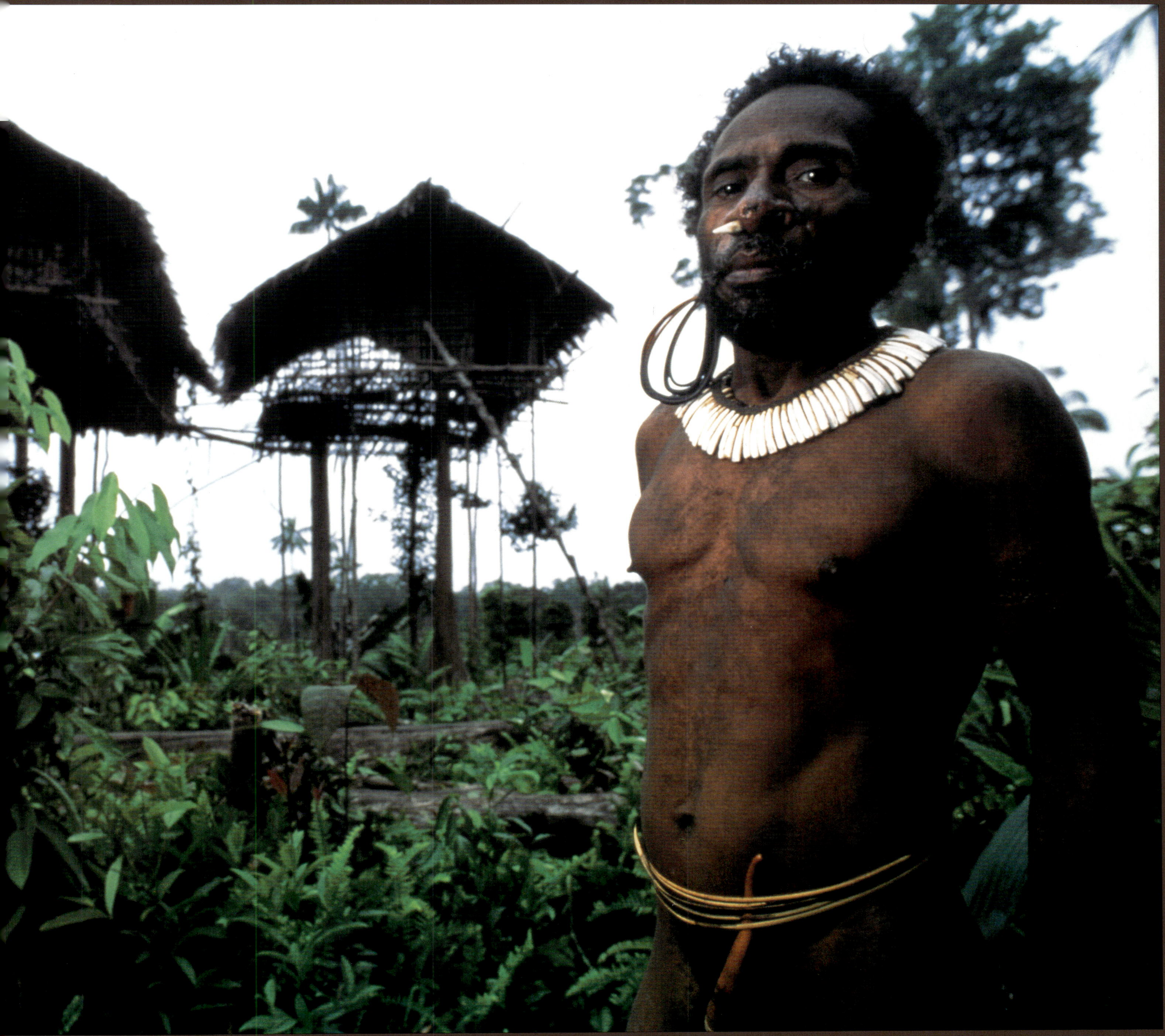

A large portion of Indonesia's very high human cultural diversity is found in Indonesia's half of the island of New Guinea, Irian Jaya. Upper left, a bartered 12-year old bride stands in front of pigs used in prenuptial negotiations. Lower left, Asmat fishermen returning with their nets, Flamingo Bay, near Agats in southeastern Irian Jaya. Above, a clan elder of the Korewai tree-house people wearing a nose bone, a bone necklace and a penis gourd. The Korewai occur inland from the coastal Asmat in southeastern Irian Jaya.

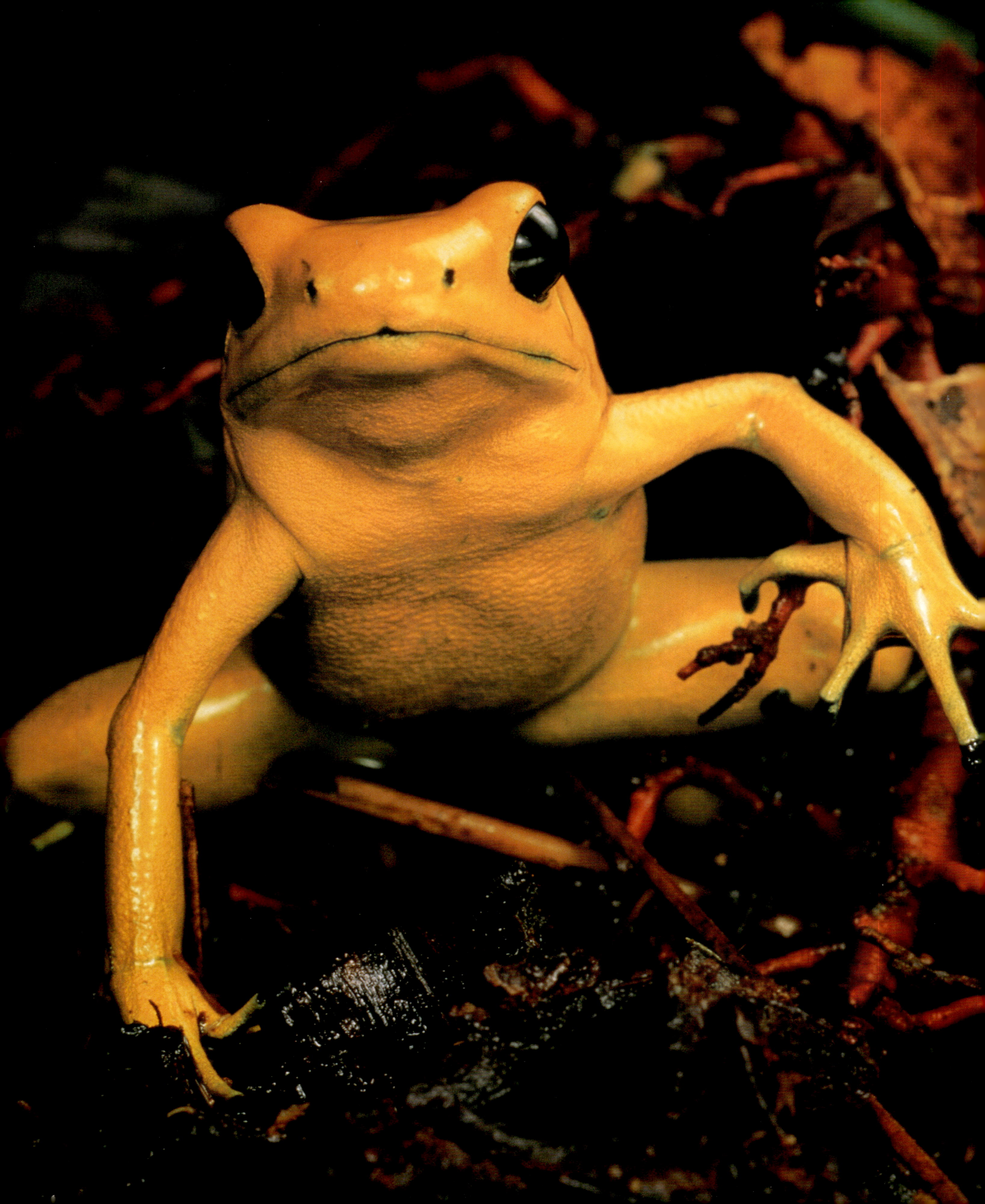

COLOMBIA

Colombia is one of the top three megadiversity countries. It is exceeded in overall diversity only by Brazil and Indonesia, and although its diversity compares very closely to those two global leaders, it is the smallest of the three. At 1 141 748 km^2, it is only 13.4% the size of Brazil and 59.6% the size of Indonesia; yet, in several categories, it has higher diversity than either of the two global leaders. As a result, it has very high per unit area biodiversity, being surpassed in this category only by Papua New Guinea, the Philippines, Malaysia, and Ecuador, which is the smallest of all the megadiversity countries and only 24.8% the size of Colombia. Looking at Colombia in another way, it occupies only 0.77% of the land surface of the planet, but has an estimated 14-15% of all terrestrial biodiversity.

Colombia also has within its borders the most important portions of two of the world's top threatened biodiversity hotspots, the tropical Andes, which ranks number one on the world list, and the Chocó/Darién/Western Ecuador hotspot. The bulk of the Chocó/Darién/Western Ecuador hotspot is in Colombia, and some of the most diverse and certainly the most endangered portions of the tropical Andean hotspot are in the three Andean cordilleras of Colombia and two isolated massifs, the Sierra Nevada de Santa Marta, and the Serranía de la Macarena, also included in this hotspot. The lowland Amazonian portion of Colombia also falls within the major tropical wilderness area of upper Amazonia, and includes some of the least disturbed portions of this rain forest wilderness. In addition, Colombia has 10 Centers of Plant Diversity and Endemism as identified by WWF/IUCN, and 14 Endemic Bird Areas as defined by BirdLife International (ICBP).

In terms of species numbers for different groups, Colombia is the world leader in two vertebrate groups, birds and amphibians. In terms of birds, Colombia has an astounding 1 815 species, of which at least 142 are endemic, and as regards amphibians, there are from 583 to possibly as many as 700 species, of which at least 367 are endemic. These figures represent very high portions of the global totals for these groups (20% for birds and 14% for amphibians), especially when measured against the fact that Colombia occupies only 0.77% of the Earth's land surface.

Plant diversity is also extremely high, with estimates ranging from 45 000-51 000 species of higher plants, of which roughly 33% are thought to be endemic. Only Brazil surpasses Colombia in this category, and does so in a much larger land area. Several of the biogeographic regions of Colombia make a major contribution to the plant total, most notably the Chocó, which stretches along the Pacific coast from Panama to Ecuador and has some of the wettest rain forests on Earth. The Chocó has some 9 000-12 000 species of plants, 25% of them endemic, in an area of just 131 246 km^2. Another very rich area is the foothills of the Andes, especially the western range near the Ecuadorian border. Colombia appears to be the world leader in orchid diversity as well, with between 3 000-5 000 of the Earth's estimated 25 000-35 000 species, and holds the record for the highest number of orchids at a single site.

Colombia ranks third in butterfly diversity behind Peru and Brazil, with approximately 3 100 species (and this is likely a vast underestimate), of which 300 are endemic; third in reptiles behind Australia and Mexico, with 520 species and 97 endemics; and fourth in mammal diversity behind Brazil, Indonesia and China, with 456 species, of which at least 28 are endemic.

Freshwater fish and other aquatic diversity appear to be quite high as well, since the country includes large portions of both the Amazon and Orinoco drainages, and all of the Magdalena River drainage, as well as several other large rivers like the Sinú, Atrato, San Jorge and San Juan. Exact figures for

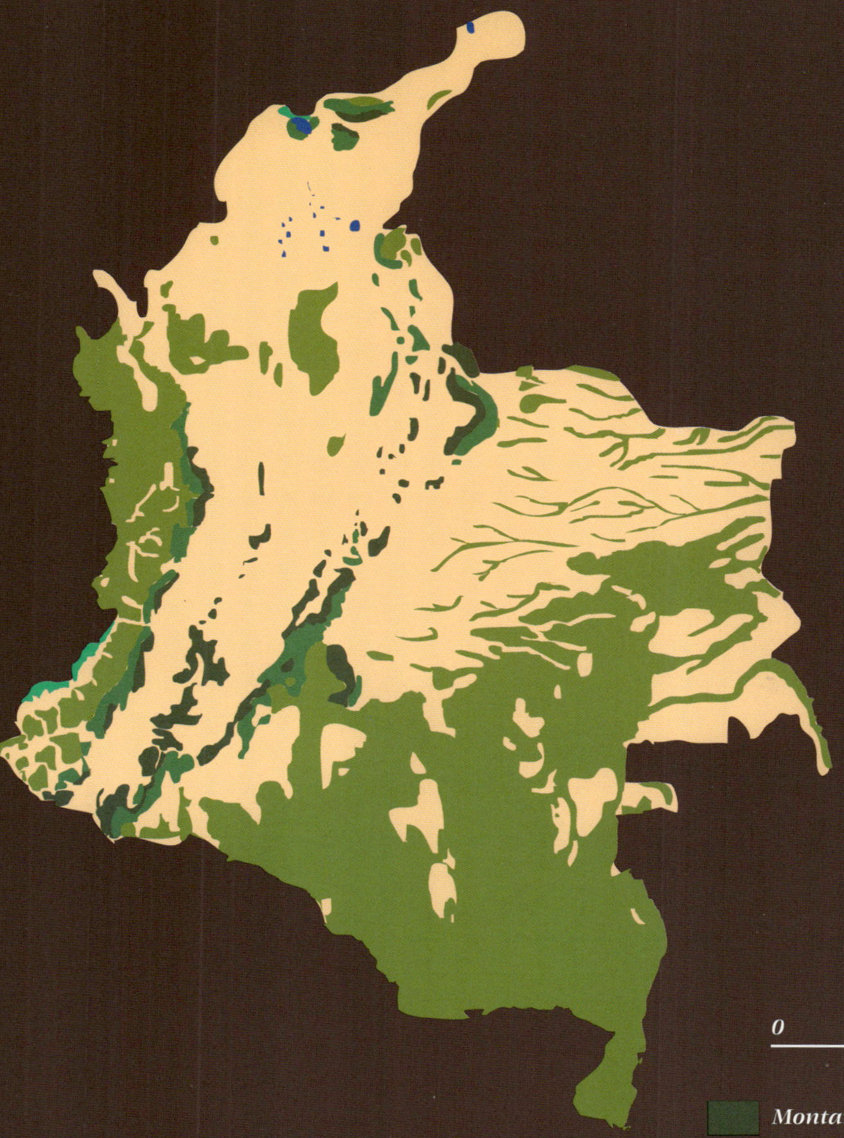

On the opposite page, the yellow poison dart frog (**Phyllobates terribilis**), *the most toxic member of the poison frog family and one of the world's most poisonous animals. Local Emberá Indians poison their blowgun darts by simply rubbing them across this animal's back. The poison remains effective for six months and is strong enough to kill a jaguar. This amazing animal is found only along the middle to upper reaches of the Saija River in the southern Chocó, and is symbolic of Colombia's rich amphibian diversity, at 583 species the highest in the world.*

Colombia's freshwater diversity are not available at this time, but one estimate places freshwater fish diversity at roughly 3 000 species, second only to Brazil. Marine diversity is also quite high because Colombia is one of few megadiversity countries possessing coasts on two oceans, the others being Mexico and the U.S.A. It has significant reef systems, especially in the Caribbean, and a number of important island groups, including Gorgona and Malpelo in the Pacific and San Andrés, Providencia and the San Bernardo Islands in the Caribbean. Extensive mangroves on both the Caribbean and Pacific coasts make Colombia one of the most important countries in the Americas for this type of ecosystem, which serves an important ecological function as nurseries for many species of marine fishes.

Ecosystem diversity in Colombia is also strikingly high, perhaps the highest on Earth, with no less than 99 biogeographic units currently identified. The country's tropical location, the variety of climates, and the geological history that gave rise to the three Andean cordilleras have resulted in a large number of distinct ecosystem types, and the country's position as the contact point between the Isthmus of Panama and the South American continent has made it central to the biological exchanges that took place between North and South America with the emergence of the land bridge. The triple cordilleras of the Colombian Andes are particularly important, since they serve as a series of barriers to gene flow and have been an enormous stimulus to evolution, even more so than in other Andean nations where there is less fragmentation of the Andes chain. Isolated mountain enclaves like the Serranía de la Macarena and the Sierra Nevada de Santa Marta are also centers of endemism in their own right. The lowland areas of Colombia, represented by two completely distinct rain forest systems, upper Amazonia and the Chocó, and the vast plains or Llanos of Orinoquia, also make major contributions to Colombia's overall ecosystem diversity. In summary, the many different river drainages, the coastlines in two oceans, the three Andean cordilleras and their outlying massifs, the range of lowland wet forest and grassland ecosystems, plus the country's position as contact between South America and the Isthmus of Panama, all combine to place Colombia in a privileged position worldwide in terms of ecosystem diversity and overall biological importance.

Cultural diversity is also high, with 81 well-differentiated and surviving ethnic groups. As with Brazil, it is impossible to determine how many indigenous groups may have originally existed at the time of European contact, but what remains certainly puts Colombia in one of the top categories for human diversity.

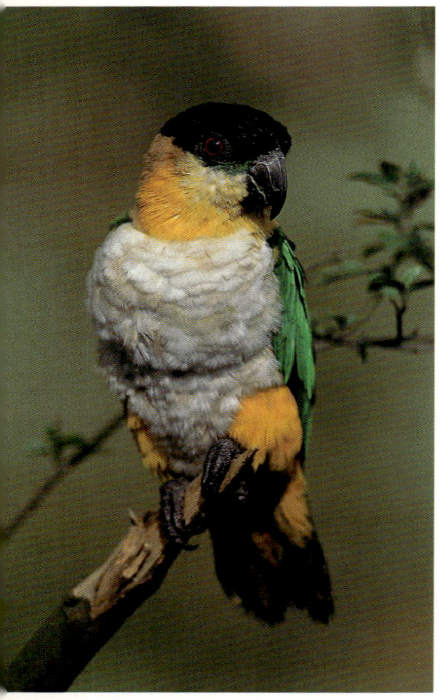

The parrot family is well-represented in Colombia, with 50 species. Above, a black-headed parrot (Pionites melanocephala).

Natural habitats

According to the most recent biogeographic analysis of Colombia (Hernández *et al.*, 1992), based on the vegetation, landscape, climate, and wildlife composition, the country can be divided into as many as 99 biogeographic units within nine different biogeographic provinces, showing the vast diversity of Colombia's natural habitats. A simpler breakdown divides the country into five major continental regions and two insular regions: the Caribbean Ocean Island Territories, the Pacific island of Malpelo, the Caribbean Arid Belt (including the Sierra Nevada de Santa Marta mountain cluster), the Chocó, the Andes, Orinoquia, and the Amazon.

The Caribbean Ocean Island Territories are comprised of the islands of San Andrés and Providencia and their adjacent cays and islets isolated in the Caribbean approximately 180 km from the Central American mainland (Nicaragua), 480 km north of the Colombian coast and 400 km southeast of Jamaica. Despite their distance from Colombia and the cultural affinities of the residents being predominantly West Indian, the islands have been part of Colombia since falling under Spanish domination in 1786. Bathymetry indicates that the two islands are most likely to be truly oceanic, never connected either to each other nor to the Central American continent. San Andrés is the largest island of the region, a low-lying carbonate island about 12.8 km long north to south and 3-5 km wide, with its highest point being approximately 90 m asl. Providencia lies approximately 70 km to the north and is visible on a clear day from San Andrés. The latter is of volcanic origin and much higher, with its tallest peak reaching 390 m. Providencia is flanked by a smaller island to the north, Santa Catalina, and both San Andrés and Providencia are surrounded by abundant coral reefs and low-lying sand cays and islets, some large and stable enough to support human inhabitants, including Roncador, Quitasueño, Serrana, Serranía and Crab Cays.

San Andrés and Providencia both possess a well-defined biota that includes a mix of elements from Central America and some typically Antillean elements. Endemic plants are scarce, though noteworthy, as they include some true Antillean taxa such as *Paurotis* and *Coccothrinax*, both palm genera of the Arecaceae family (known locally as *pacta*), and *Myrica mexiana*, a shrub restricted to the highest peak on the island of Providencia and one of only two species of this genus to be found in the New World, the remaining nearly 30 species having exclusively Old World distributions. The fauna in this province also harbors some noteworthy endemics. Mammals are solely represented by bats, many of which are endemic, and the adjacent coral cays once harbored populations of the now extinct Caribbean monk seal (*Monachus tropicalis*). Birds are represented by 94 species and 14 endemic taxa. There are also two endemic terrestrial reptile taxa, and the archipelago was once an important site for sea turtle nesting and foraging, particularly for hawksbill (*Eretmochelys imbricata*) and green turtles (*Chelonia mydas*); these species have seen dramatic declines in recent years due to overhunting and habitat loss. Chroniclers have documented an active trade in green-turtle meat, hawksbill shell and monk seal hides over the past two centuries throughout the Caribbean, frequently mentioning the islands of San Andrés and Providencia, Roncador Cay, Serranía Cay, and other islands of this group. Two centuries of human presence have also resulted in a considerable number of introduced species, some of which have proliferated with consequent adverse effects on the native fauna. Among these, the introduction of the spectacled caiman or *babilla* (*Caiman crocodilus fuscus*) has been especially damaging.

The Pacific Ocean island of Malpelo, a small volcanic rock island lying 330 km west of Colombia's Pacific coast, is the country's westernmost geographic extension. Malpelo's 3.5-km^2 surface area is punctuated by a central peak reaching an altitude of 390 m at its highest point and a number of steep promontories. As is typical of islands of this sort, the overall floral and faunal diversity is low compared to mainland numbers, yet the relative endemism is high. The island's vegetation is limited to a few grass species, algae, ferns and lichens of varieties easily established by airborne seeds or seeds transported in the guano of seabirds. The animals that inhabit the island are mostly oceanic seabirds seeking temporary resting and breeding sites; these include seasonal aggregations of as many as 25 000 boobies of the genus *Sula* (masked and red-footed boobies), red-billed tropic birds, frigate birds and shearwaters. Other macrofauna on Malpelo include three endemic lizard species (*Anolis agassizi*, *Diploglossus millepunctatus* and *Phyllodactylus transversalis*), and an endemic subspecies of terrestrial crab (*Gecarcinus johngarthia malpilensis*). Abundant marine life is found associated with the coral reefs of Malpelo, representing elements of both continental shores and that of the other nearby eastern Pacific oceanic islands and island archipelagos of Cocos, Clipperton, and Galápagos. While it is still poorly studied, scientists concur that Malpelo's marine fauna is of great interest given the island's position at the confluence of the Northern Equatorial Countercurrent with smaller coastal currents, and the importance its reefs may have as a stepping-stone in the dispersion of corals and associated reef organisms in the Pacific.

The Colombian Pacific islands of Gorgona and Gorgonilla lie close to shore adjacent to the coastal town of Guapi, department of Cauca. Due to their proximity to the mainland, these islands bear strong biological affinities to the Chocó, and have a much higher diversity overall than Malpelo. Despite the similarities to the mainland biota, however, Gorgona and Gorgonilla possess some unique endemisms, including three lizard species, some bat species and a subspecies of white-faced capuchin monkey (*Cebus capucinus*). These two islands also harbor some of the most important coral reefs of Pacific Colombia, and both humpback whales and whale sharks are seen at this island at certain times of year. Gorgona was a prison island until 1984, at which time it was declared a national park (terrestrial and marine), and is presently under the strict protection of the Colombian Ministry of the Environment.

The Caribbean Arid Belt is a region dominated by dry vegetation that stretches east from the Sinú River across the northern Colombian departments of Córdoba, Sucre, Bolívar, Atlántico, Magdalena, Cesar and La Guajira from the terminus of the wet Chocoan forests, eastward through increasingly arid habitats ending in La Guajira Desert, one of the continent's most xerophytic biomes. The gently rolling, low-lying region is bordered to the north by the Caribbean sea throughout its extent, and gives way to the northern foothills of the Andes and the major Andean valleys of the Cauca and Magdalena Rivers to the south. The coastal areas and littoral zones are dominated by mangroves (an estimated 650 km^2 on the Caribbean coast), interspersed with beaches, rocky shores and the dry decumbent vegetation typical of most Caribbean coastal biomes. The Sierra Nevada de Santa Marta mountain cluster lies as a biogeographic island within this province, and is described in greater detail below. Another biogeographic island, the Serranía de Macuira and Serranía de Jarara in the north of La Guajira Peninsula, harbors a dense vegetation completely different from the surrounding desert. This unique formation is protected as a national park.

This coastal belt was the first part of Colombia to be colonized by the Europeans 500 years ago, and was also the site of extensive pre-Columbian indigenous settlement (Sinú and Tayrona cultures principally) prior to that. As such, it is one of the most altered habitats in all of Colombia. Much of the original vegetation has been converted for agriculture, and other anthropogenic alterations such as the Canal del Dique that connects the Magdalena River to Cartagena Bay. Coastal roads that cut off many natural estuaries from the sea (*e.g.*, the Ciénaga Grande and Ciénaga de la Virgen) have done much to change the original landscape. Some of Colombia's most extensive banana plantations, for instance, are found to the east of the Sierra Nevada, the dry forests that were once typical of this region have been reduced to perhaps 5% of their original cover, and one of the Caribbean's historically largest mangrove areas in the Ciénaga Grande between the coastal cities of Barranquilla and Santa Marta, has all but disappeared.

The upper portion of La Guajira Peninsula is a true tropical desert with mobile dunes and sporadic xeric vegetation, including a number of cactus species and other succulents. The lower part of the peninsula and the upper portion of the department of Cesar have a predominance of medium-sized subxerophilous vegetation with typically Central American elements, such as the wild nispero and chicozapote (*Manilkara zapota*) that came originally from the Yucatán Peninsula, where it is used to make chewing gum. Especially along the coast, one can see columnar cacti that reach a height of 10 m or more, and bulbous-trunked trees of the genera *Bombacopsis* and *Pseudobombax*. The largest intact remnants of the dry forests that once dominated large portions of this region can be found in the Montes de María Reserve and Los Colorados Sanctuary in the departments of Sucre and Bolívar, respectively. It is in these and other riparian remnants of the northern Co-

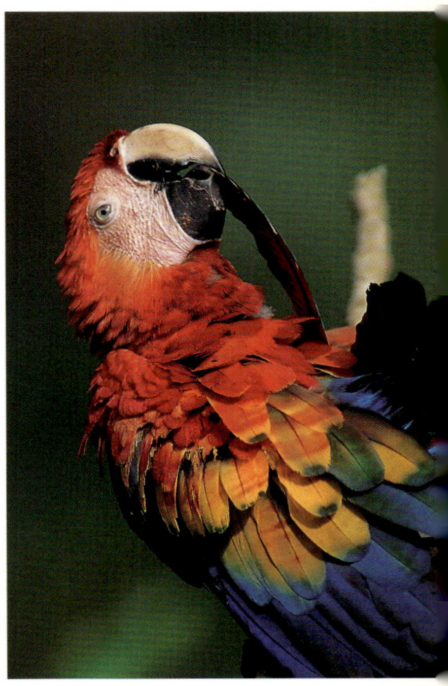

The well-known scarlet macaw (Ara macao) *is one of five large macaw species found in Colombia.*

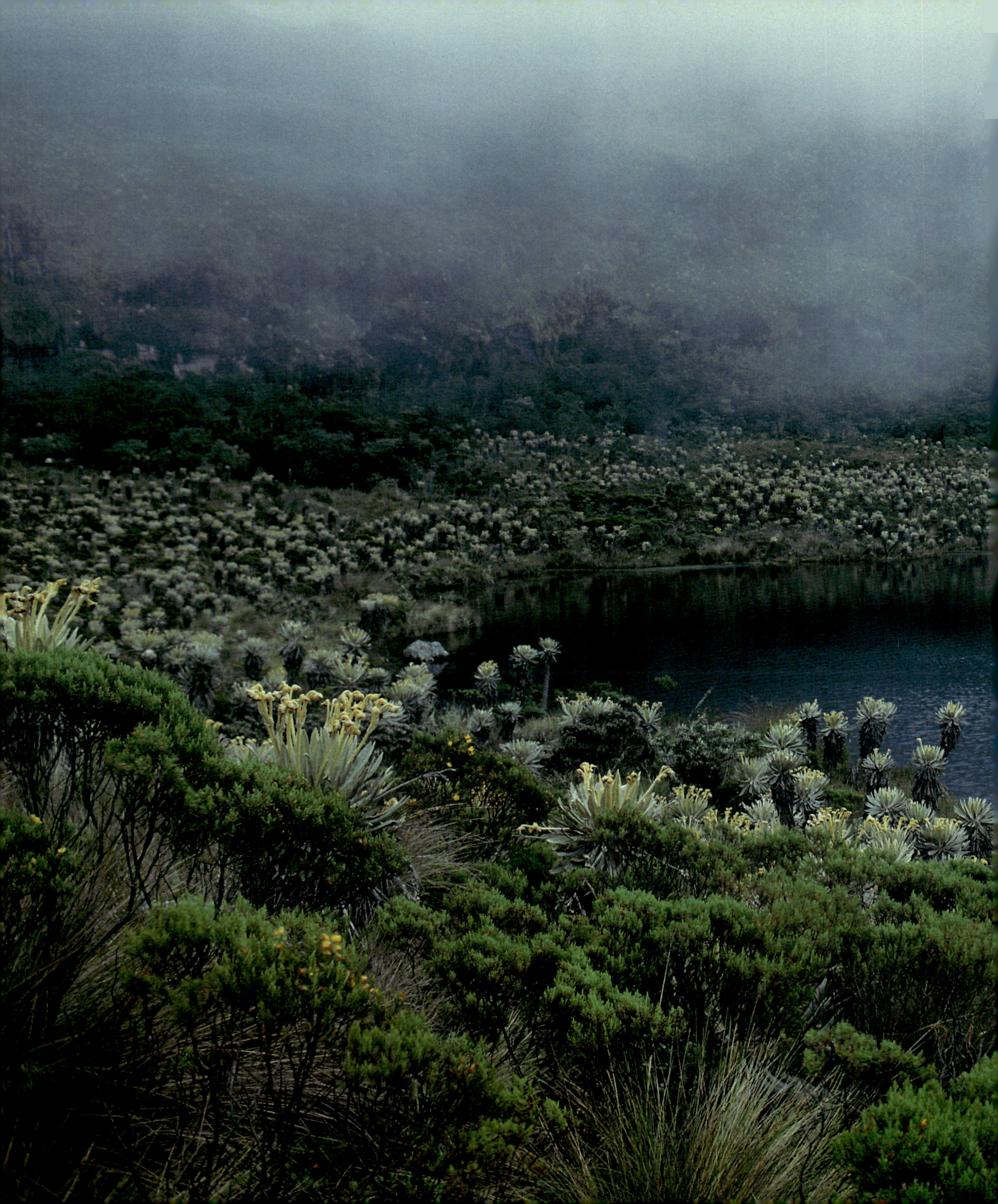

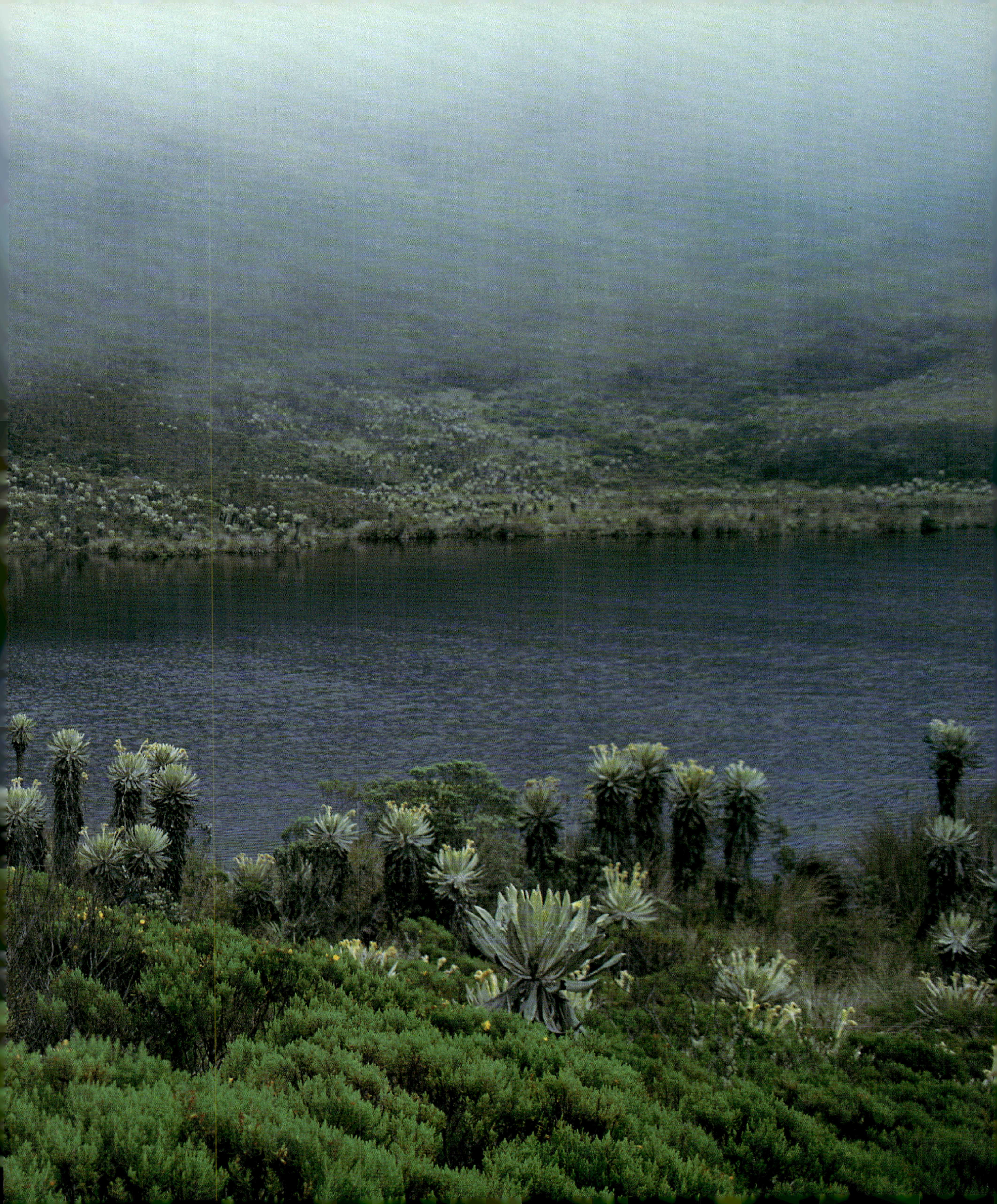

lombian dry forests that one can find the last remaining natural populations of the cotton-top tamarin (*Saguinus oedipus*), a small endemic monkey traded heavily up until the 1960s for use in the biomedical industry. Another endemic Colombian monkey, perhaps even rarer, is the white-footed tamarin (*Saguinus leucopus*). This primate can also be found in the southern parts of the Caribbean Arid Belt east of the Magdalena River.

The Sierra Nevada de Santa Marta mountain cluster, a massif separate and independent from the nearby Andes, covers an area of approximately 23 km^2 adjacent to the Caribbean coast near the city of Santa Marta. Reaching an altitude of 5 775 m (the Colón and Bolívar peaks) in just 42 linear km from the sea, the Sierra Nevada de Santa Marta is considered the tallest coastal mountain on Earth. The vast range of altitudes and climates within such a restricted area has given rise to nearly the full spectrum of Colombian ecosystems at a single site, from the aforementioned xerophytic coastal growth at lower elevations, through sub-Andean, Andean, high Andean and *páramo* vegetation (classification by Cuatrecasas, 1958), and beyond the tree line to perpetual snows, all in a mere 1.4% (some 16 000 km^2) of the country's territory. There is very strong endemism in all groups of vertebrates found in the Sierra Nevada. Fifteen of the 356 bird species occurring in the area are endemic, as are 20 of the 42 species of amphibians, and 16 of the 96 species of reptiles. There are also 190 species of mammals and 101 fish species.

The Sierra Nevada de Santa Marta is the water factory for the arid northern coast of the country and the source of over one hundred rivers that feed the more than one million Colombians residing in the region from the Ciénaga Grande east to La Guajira Desert. The waters from the Sierra support enormous banana plantations to the west, one of the country's most fertile agricultural regions to the south (Valledupar), and the western hemisphere's largest low-sulfur fossil coal operation to the east in La Guajira Peninsula. The Sierra's fresh waters spill into the Ciénaga Grande creating the estuarine conditions that support the nurseries that feed northern Colombia's important coastal fisheries.

The Sierra Nevada is also home to the last vestiges of the pre-Columbian Tayrona culture, now the Kogui, Arhuaco and Arsario tribes; the Kogui still live in traditional fashion in isolated villages in the upper reaches of the Sierra. The Tayrona culture was formerly a great nation unto itself rivaling that of the Maya in Central America in infrastructure and cultural advancement. The Tayrona demonstrated a level of respect for the land and nature (their *Madre Sierra*) so strong that although at one time the area was home to between 500 000 and 1 million Tayrona living in a complex series of more than 500 interconnected lithic villages, temples and trails throughout the massif, today virtually no negative environmental impacts are visible. Now only a few thousand Kogui are all that remain, yet they continue to preserve their rich and vibrant culture and society, largely unknown to outsiders, guarding the knowledge of how to truly manage the Sierra Nevada and its resources in a sustainable and respectful manner. The native peoples and environments of the Sierra Nevada have suffered all manner of threats since the arrival of the Spanish half a millennium ago, including slavery and the ravages of exotic illnesses, the encroachment of small-scale and then mechanized agriculture from the lowlands, the drug traders that started arriving in the 1960s and converted major areas first for marijuana and now for coca and opium poppy cultivation, and the influx of guerrilla domination and the subsequent social disorder wrought by the war between dissidents and the State. Nonetheless, through it all, the Sierra Nevada has remained a relative wilderness, albeit a threatened one, and must continue to be a critical priority for conservation in Colombia. There are three indigenous reserves, two national parks (Sierra Nevada de Santa Marta and Tayrona), plus a biosphere reserve overlapping different areas of the Sierra, and a strong conservation presence spearheaded by the Fundación Pro-Sierra Nevada de Santa Marta, a Colombian NGO founded in 1986 that has become the country's strongest nongovernmental conservation entity.

The Chocó Region comprises the Colombian portion of the Chocó biogeographic zone (Chocó/Darién/Western Ecuador hotspot), bounded by the Pacific ocean to the west, the Panamanian and Ecuadorian borders to the north and south, respectively, and the Andes mountains to the east, and traverses the Colombian departments of Antioquia, Chocó, Valle, Cauca and Nariño. Chocoan forests in the northeastern portion of this region extend slightly east around the northern Andean terminal to the forest of the lower Sinú River drainage. Covering some 131 246 km^2 (11.5% of Colombia's territory), the Chocó region has the highest rainfall in the country and perhaps the world, with 5 000-12 000 mm/year, giving rise to a complex hydrological network dominated by the San Juan and Atrato river systems and a number of other noteworthy tributaries.

Plant diversity in the Chocó is unparalleled, reaching an estimated 12 000 species, nearly one quarter of the country's total, with an estimated endemism of as much as 25-30%. The forests are very different from the Amazon lowlands in many respects, one being the abundance of palms both large and small, including *Jessenia*, *Socratea*, *Iriartea*, and *Wettinia*, and indeed the Chocó forests have a higher proportion of palms than any other forest biome on Earth.

This region is also unique in that it possesses unusually high endemism when compared to other lowland humid forest areas in the Neotropics, due most likely to its separation from the greater Amazon Basin by the Andes, to the excessively high rainfall, and to the absence of a dry season. This high endemism has been documented in many groups, most notably plants, birds and butterflies. Chocó forests near the Colombia/Ecuador border are also considered to be the area of highest bird endemism in all of South America and probably the world (Terborgh and Winter, 1980).

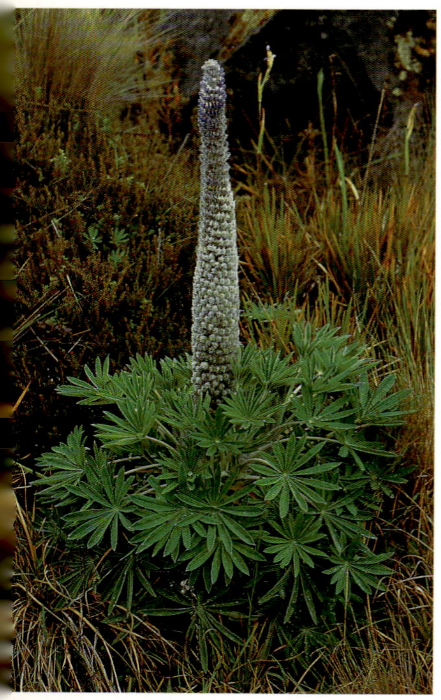

On pp. 112-113, frailejones (Espeletia *sp.*) *formations surrounding a high-altitude lake in the Santuario de Fauna y Flora Iguaque, in the department of Boyacá.*

Above, Chocó (Lupinus alopecuroides), *a characteristic species of the high-altitude* páramo.

In much of the region, forests extend all the way to the rocky shorelines of the Pacific, and the arid coastal plain characteristic of Colombia's Caribbean shores is almost entirely absent. In other areas of the Chocó region, humid forests give way to extensive mangrove ecosystems, and indeed Colombian mangroves attain their maximum biomass on the Pacific coast, covering an almost continuous belt from Cape Corrientes (department of Chocó) south to the border with Ecuador. It is estimated that, all told, mangroves in the Chocó region cover more than 2 813 km^2, and represent one of the most important expressions of this biome in all of the American Pacific. In addition to the red, white, button and black mangrove species found in the Caribbean, the Pacific mangroves have a more heterogeneous species mix that includes also *Rhizophora harrisonii* and *R. brevistyla*. Mangrove ecosystems are under threat throughout this region due to their overexploitation for timber and fuel wood, and because mangrove areas are being cleared for the development of the pond shrimp industry; this latter threat has already severely decimated most Ecuadorian mangroves and has taken a severe toll in the southern Chocó region as well, near the city of Tumaco. However, mangroves receive official protection in the region within the 80 000-ha Sanquianga National Park.

In terms of traditional inhabitants, the region is home to perhaps as many as 100 000 Indians of the Emberá, Waunana and Tule groups. Emberá Indians of four distinct ethnic groups reside in typically small communities alongside rivers in forested areas, and to a large extent have preserved their traditional lifeways. The Pacific coastal region is largely inhabited by communities of Afro-Colombians, descendants of escaped slaves brought to the region some 300 years ago to mine gold and extract other resources. These Afro-Colombian residents are mostly dedicated to artisanal fishing and small-scale agriculture. There are three cities of noteworthy size in the Chocó Region: Quibdó, the capital of the department of Chocó, lying along the Atrato River; and the two coastal ports of Buenaventura and Tumaco, each of which is connected to the interior of the country by a road that cuts latitudinally across the biome.

In contrast to other biogeographic territories in Colombia, the Chocó is still largely intact, due to a paucity of major roads, ports and other infrastructure that would otherwise allow for the easy penetration of the forest. Nonetheless, the threat of large-scale development looms large on the horizon for this region, with major infrastructure projects including a number of roads, an interoceanic canal, train routes and hydroelectric dams currently being considered by government and private sector interests. Small- and medium-scale gold mining, agricultural expansion and timber harvesting are also notable threats in the region.

The Colombian Andes split into three separate, roughly parallel branches aptly named the Western, Central, and Eastern Cordilleras, lying along north-south axes and separated from each other by the long, broad valleys of the Magdalena and Cauca Rivers. It is these cordilleras more than any other geographic feature, that make Colombia a megadiversity country, since the plethora of minute variations in altitude, climate, soils, humidity and rainfall that are found in their countless valleys, peaks and isolated pockets create an infinite variety of microhabitats that have led to high levels of diversity and endemism in the flora and fauna.

The Andean Region is the most biologically important zone in Colombia in terms of its sheer diversity, its levels of endemism, and the threat posed by five centuries of severe human impact. It is comprised of 45 distinct vegetational districts (Hernández *et al.*, 1992) ranging from sub-Andean humid forest to *páramo* and to arid, xerophytic sites. One such arid district is Chicamocha Canyon, which harbors a variety of cacti and some Central American elements, including palms of the family Zamiaceae. The other arid Andean site is Dagua Canyon, which interestingly enough is surrounded by areas with an average rainfall of 6 000 mm, although it only receives 800 mm a year itself, due to the unusual topography of the region. The Dagua is home to several endemic species such as the *Frailea* cactus, a genus whose closest representative species occur in Brazil, Uruguay, Paraguay, and the Bolivian Chaco.

There are several forest types in the Andes, beginning in the lowlands with the sub-Andean forests that extend up to 2 000 m. Sub-Andean forests are similar to those of the hot lowlands, but have fewer palm species, fewer lianas, and fewer buttress root species. The canopy can reach a height of 40 m, and at these altitudes there is constant fog and temperatures ranging between 5° and 14°C (41° and 57°F). Within the sub-Andean forest belt, vegetation begins to transition at around the 1 500 m-elevation where the Lauraceae family becomes the dominant element (Cuatrecasas, 1958; Gentry, 1991). Montane forests begin at an altitude of approximately 2 000 m, and are characterized by shorter trees and more abundant epiphytes such as mosses, lichens, ferns and algae. Palms of the genera *Aiphanes* and *Euterpe* occur commonly up to 2 500 m, and the genus *Ceroxylon* becomes more prominent at higher altitudes up to 3 200 m. Montane forest then gives way to *páramos* that begin between 3 200-3 800 m altitude, and extend up to the snow line.

The *páramo* is an ecosystem type unique to the Andes (Costa Rica's Páramo de la Muerte, despite its name, is not true *páramo*), and finds its greatest expression in Colombia. *Páramos* are characterized by a thick mat of sponge-like, highly absorbent mosses and grasses (such as *Camalagrostis*) and shrubs of the genus *Ericacea*, from which rise giant frailejones (genus *Espeletia*), some of which can reach a height of 10-12 m (*Espeletia uribeii*). The climate is cold and humid at these dizzying altitudes in the Andes, and the *páramo* landscape is typically blanketed in fog and heavy mist, giving it an eerie, otherworldly sensation. Small flowers of red and yellow with upturned petals to capture the moisture adorn much of the *páramo* vegetation, and on those

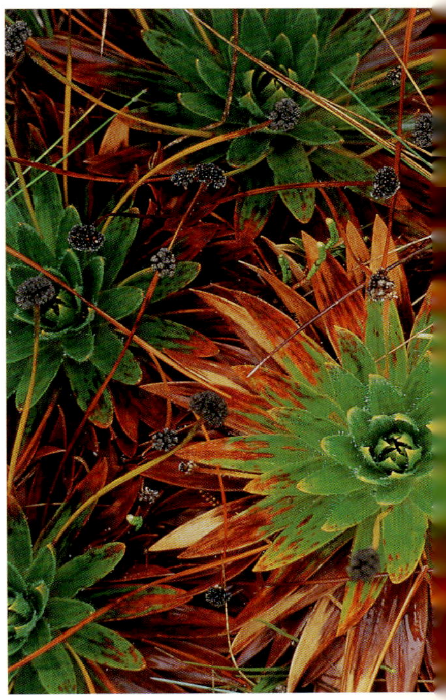

A frailejoncillo (Espeletia *sp.*) *on the Puracé Volcano, near Bogota.*

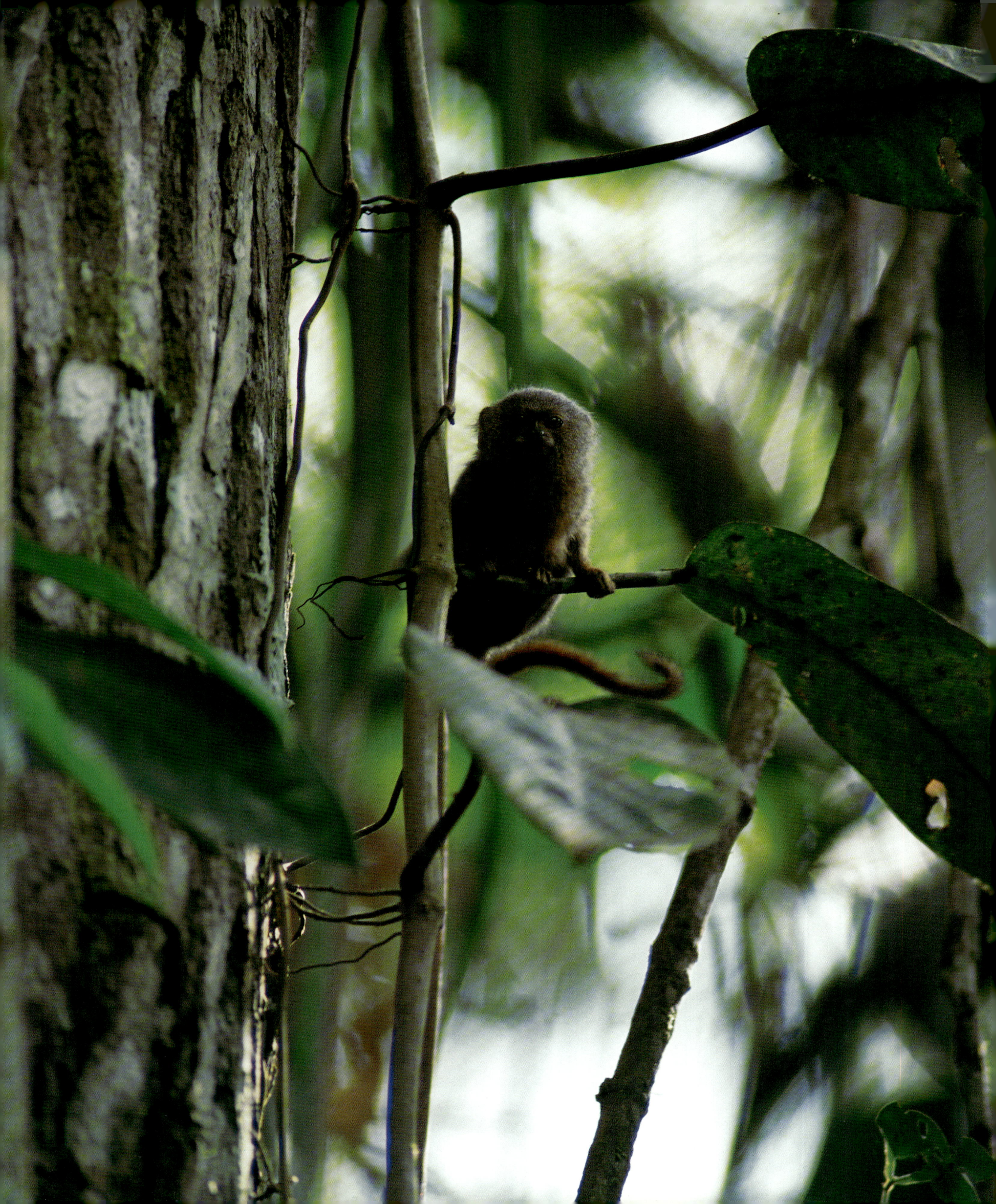

rare occasions when the sun comes out, the *páramo* presents a scene that is as brilliant and colorful as it is gray and foreboding under normal conditions.

Andean fauna is the most diverse in the country, particularly with respect to birds and amphibians. Indeed, the tropical Andean region is considered to be the most biologically diverse on Earth, and has been referred to by Myers (1988) as "the epicenter of global biodiversity." The tropical Andean hotspot extends from Bolivia north to Venezuela, but the three cordilleras of Colombia are unquestionably the richest and most diverse of all. Among the typically Andean animal species, the best known are the mountain tapir (*Tapirus pinchaque*), the Andean condor (*Vultur gryphus*), the spectacled bear (*Tremarctos ornatus*), and the dwarf deer (*Pudu mephistopheles*). There are also a large variety of amphibians such as salamanders and harlequin frogs, and scientists are discovering new amphibians in the Colombian Andes at the pace of several per year.

The Colombian Andes are of great concern to conservationists because a huge portion of the natural vegetation cover has already been lost as cities originally settled by the Spanish, such as Santa Fe de Bogotá (now over 5 million inhabitants and growing), Medellín and other major Andean population centers, continue to expand. The valley of the Cauca River, for instance, which separates the Western and Central Cordilleras, is almost entirely converted to large-scale mechanized agriculture (sugarcane, sorghum and other crops), and once extensive wetlands have all but disappeared. Similar conditions can be seen in the Magdalena River valley. The Andes is the largest geographic region of the several discussed herein, covering portions of 21 different Colombian departments, and is the region suffering the greatest degree of human disturbance at present.

Lying to the east of the Andes and forming part of the western edge of Orinoquia is the Serranía de la Macarena, an isolated elongate massif about 241 km long by 30 km wide, that exhibits some unique biogeographic features. With sheer escarpments on its western flank and more gentle slopes to the east, this *tepui*-like formation is surrounded on all sides by the Duda (northwest), Guayabero (southwest and south), Yarumales (southeast and east) and Guéjar (northeast) Rivers, and is indeed the source of a number of other smaller tributaries that feed into these. The massif itself has a transitional flora and fauna, with elements from the nearby Andes, Amazon and Orinoquia, including some noteworthy endemics such as the bromeliad *Vellozia macarenensis*. Eight species of primates can also be found in the Macarena, and a team of Colombian and Japanese primatologists from the Universidad de los Andes have undertaken in this region one of the continent's longest-term continuous research initiatives to monitor these animals. The Macarena is threatened by agricultural expansion and by indiscriminate and uncontrolled logging, especially in the south and southwest; despite the fact that this area has had national park status since 1948 (expanded in 1989), it remains difficult to manage given that it is also a stronghold of guerrilla activity and illegal coca cultivation, and has been the site of armed conflict between guerrilla and paramilitary forces for many years.

The Orinoco region, or Orinoquia, is a mosaic dominated by natural savannas and gallery forests, wetlands, and a narrow rain forest strip where it contacts the foothills of the Andes. It lies east of the Andes, north of the Amazon, is bounded by Venezuela to the north and east, and overlaps with the departments of Arauca, Casanare, Vichada and Meta. Here one finds the largest grasslands in northern South America, the Llanos, a biome that extends into Venezuela (states of Apure and Barinas) with a total area of over 50 million hectares.

Colombian Orinoquia lies atop a Guayana Shield sedimentary plain. The northern area is flat and has a few depressions that generally flood during the rainy season. The southern portion has a slightly wavy profile and a rather complex structure formed by microvalleys which are created by either permanent or temporary rivers. Orinoquía drains into the Orinoco river and its most important tributaries are the Arauca, Meta, Vichada and Guaviare Rivers. The Guaviare is 947 km in length and drains 6 400 m^3 per second, making it the second largest tributary of the Orinoco.

The vegetation in the savannas of Orinoquia is mainly grasses. There are more than 200 species of these, among which the one known as *rabo de zorro*, or foxtail (*Andropogon bicornis*), is the most common, since it has a natural resistance to fire and to the acidic soils common in this region. This grass is very hard and contains silicates in its structure which cause premature wear on the teeth of the herbivores that feed on it. Some savannas have short, bushy vegetation in which the dominant species is the *chaparro* (*Curatella americana*), a tree with rough and semirigid leaves. Others have different dominant species such as *chaparro manteco* (*Byrsonima crassifolia*), *chaparro alcornoque* (*Bowdichia virgiloides*) and *saladillo* (*Caraipa llanorum*). Other vegetation associations typical of the floodplains along rivers are stands of *Mauritia* palm called *morichales*, dominated by *Mauritia flexuosa*, the fruit of which serves as an important food source for several large vertebrates such as tapirs (*Tapirus terrestris*) and the two species of peccaries (*Tayassu tajacu* and *Tayassu pecari*). Gallery forests are another important feature of Orinoquia. They are found alongside rivers with permanent water flow and can reach several kilometers in width. The canopy of these forests can be up to 25 m in height.

Animal diversity in Orinoquia is low and, in contrast to African savannas, there are very few herbivores and very few large animals. The largest are the tapir (*Tapirus terrestris*) and the Orinoco crocodile (*Crocodylus intermedius*), a species endemic to Colombian and Venezuelan Orinoquia. The mammal species with the largest biomass in this region is certainly the capybara (*Hydrochoerus hydrochaeris*), the world's largest rodent, as it can be found by the thousands in this region. Although it also lives in other parts of Colombia and in

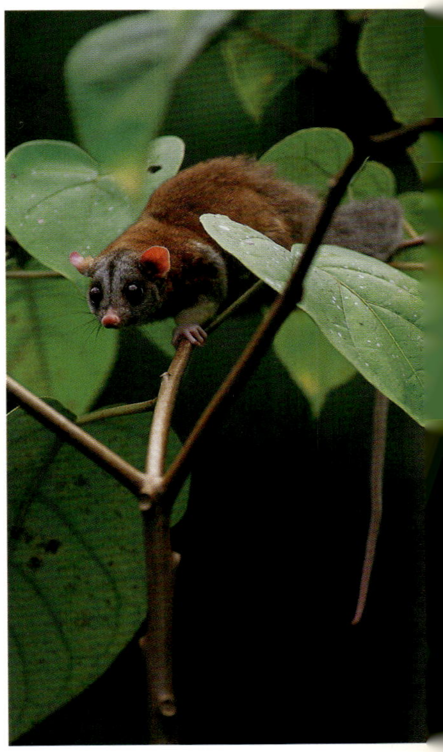

On the opposite page, a pygmy marmoset (Cebuella pygmaea) *stares at the photographer from its perch in the forest in the Amacayacu National Park near Leticia. Widespread in upper Amazonia, this marmoset is the smallest of all monkey species.*

Above, Colombia has one of the world's most diverse mammal faunas, including 33 marsupial species. Above, a western woolly opossum (Caluromys lunatus) *from the Niambi River.*

other countries, it is seldom as abundant as it is in these little-disturbed savannas. Other flagship species of Orinoquia include the giant otter (*Pteroneura brasiliensis*), the giant anteater (*Myrmecophaga tridactyla*) and the puma (*Felis concolor*). Most of the animal species found in Orinoquia are closely related to those found in Amazonia, but certain large aquatic species haven't been able to travel through the natural barriers formed by the rapids of some rivers. Thus, the Orinoco crocodile (*Crocodylus intermedius*) and the Caribbean manatee (*Trichechus manatus*) are found in Orinoquia, whereas the black caiman (*Melanosuchus niger*), the Amazonian manatee (*T. inunguis*), and the pink river dolphin (*Inia geoffrensis*), are only in the Amazon.

Perhaps the greatest threats to Orinoquia come from large-scale development, particularly hydrocarbon extraction along the Andean slopes, and road construction that will open up these remote areas to colonization. Extensive cattle ranches also pose a threat to natural vegetation and wildlife. Much of Orinoquia is still quite remote, however, reachable only by light plane into numerous small airstrips. The Tuparro National Park protects approximately 5.5 million ha of representative Orinoquia ecosystems in the far east of the biome, along the border with Venezuela.

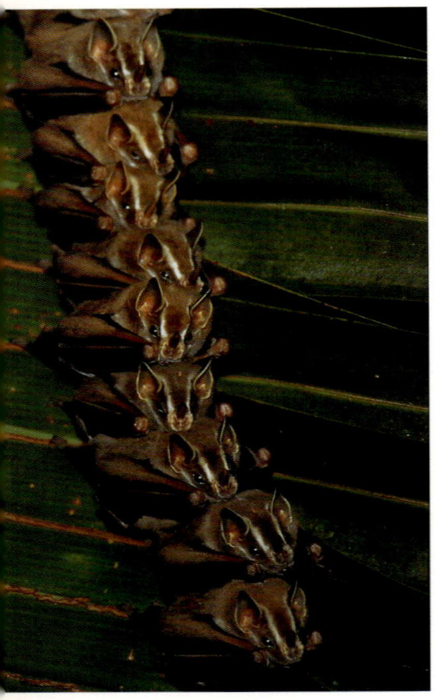

The tent-making bat (Uroderma bilobatum) *is a very appealing bat species found in Colombia.*

The Colombian Amazon occupies fully 35.4% of the southeastern quadrant of the country, bounded by the Andes to the west (departments of Putumayo, Caquetá and Meta), to the south by Ecuador and Peru, and to the east by Venezuela and Brazil. The Guaviare River, which flows from west to east, roughly marks the boundary between the Amazon (to the south) and Orinoquia (to the north). The Colombian Amazon region is almost entirely covered by rain forests, with some small islets of natural savannas such as the Yarí region, as well as areas where the rocky substrate is exposed either in the form of tables or islands (inselbergs) which emerge abruptly from the savanna. The highest of these formations, reaching up to 1 300 m above the surrounding grasslands, are the Serranía del Chiribiquete and the Mesas del Iguaje. The fauna and flora of these formations show a close affinity to the *tepuis* of southern Venezuela. Colombia is the only Amazon Basin country that possesses both the species-poor vegetation of the white sand Guayana Shield soils, and the more species-rich soils of the Andean foothills.

Colombian Amazonia is extraordinary both for its high biodiversity and the relative integrity of its ecosystems and indigenous communities. Indeed, it is perhaps the most intact of all the Amazonian nations, given the near complete absence of roads into the region and the fact that even the rivers are largely unnavigable due to rapids. The only city of any size in Amazonas is Leticia, the capital of this commissary, located at the southern tip of the Colombian Trapezium (the southernmost, trapezium-shaped extension of Colombia that touches the Amazon River; it is only reachable by air). Virtually all other access to the Amazon region comes from the few navigable rivers such as the Amazon itself, by air to remote airstrips, or by foot or canoe from nearby towns in Brazil.

Amazonia overall is the largest wilderness area on the planet, shared among nine separate South American nations; Brazil tops the list with fully 60% of Amazon forests within its territory, and Colombia is fourth with approximately 8% of the total coverage, amounting to approximately 403 000 km^2. The Colombian Amazon lies in that portion of the greater Amazon Basin known as Upper Amazonia, the least-studied, though probably the most biologically important sector of this immense wilderness. Both Schultes (1988) and Gentry (1990) considered this to be the floristically richest part of Amazonia, and in fact the latter claimed that the forests defining the Amazon/Andes interface in this region of Colombia are "the most diverse forests on Earth." Eight of the ten most dominant plant families found in these forests are also among the top ten most species-rich plant families on Earth, and this region is also packed with endemics.

Fauna in the Amazon is also diverse, and includes at least 300 mammal and 868 bird species. Some of the more conspicuous flagships are the jaguar (*Panthera onca*) and the giant otter, both of which suffered from overhunting for an active skin trade earlier in this century, but are now recovering. Fifteen of Colombia's 29 primate species are native to the Amazon, including woolly monkeys (*Lagothrix lagotricha*), spider monkeys (*Ateles belzebuth*), and black-headed uakaris (*Cacajao melanocephalus*).

Human population density in the Amazon is lower than in any other major region of Colombia (approximately 0.16 inhabitants/km^2), consisting of Indians (22%) of some 50 different tribes and 10 linguistic families that live within the forest or alongside rivers, and colonists (78%), the majority of whom live in a small handful of population centers (Puerto Inírida, Mitú, Leticia); when these latter population centers are excluded, the population density for the Colombian Amazon drops to just 0.08 persons/km^2, placing this zone amongst the least densely inhabited on Earth. No less than a dozen tribes of Indians in the Colombian Amazon have had little or no contact with the outside world. This is one of the remotest corners of the globe, one of a handful of sites on Earth where natural evolutionary processes can still take place largely unhindered by the influence of modern man and where aboriginal human communities can continue to live in a traditional way.

However, this integrity remains tenuous because of outside influences and socioeconomic pressures, many of which have resulted in extensive deforestation and social dislocation of indigenous inhabitants in other parts of Amazonia. Volumes have been written about man's attempts to penetrate and tame this vast wilderness —there was the rubber boom, the gold boom, and the wildlife trade boom, all of which have come and gone, leaving little trace. The recent history of the Colombian Amazon has been characterized by a resurgence of extractive economies and political marginalization. The presence of the federal government in the region became more evident in the sixties, as official policies of that era stimulated migration from the interior of the country and encouraged exploitation of the region's natural resources. The resulting colonization has transformed the ecosystems of the northwestern and western sectors of Amazonia. Nevertheless, recent quantitative analyses by Defler (1995) have shown that only 15% of the entire region of the Colombian Amazon has undergone conversion. Deforested areas are mostly concentrated in the foothills of the Andes and south of the Guayabero River, while extensive areas further from the Andes are almost entirely intact.

Evidence of forest loss in the Amazon has generated increasing national and international attention, particularly in the past decade, thus resulting in some positive legal and administrative changes in Colombia's Amazonian policy. Yet, entrenched exploitative attitudes are slow to change, and central government policies often have little bearing on what takes place in remote outposts of the country like the Amazon.

Surprisingly, nearly 60% of the Colombian Amazon has already been declared as either indigenous reserves or protected areas, although many of these are "paper parks" that receive little support to enforce protection. Fortunately, in some cases there is still very little pressure. Such is the case, for example, of Chiribiquete National Park (1.3 million ha) or Nukak National Park (860 000 ha), enormous areas under little threat at present (for sake of comparison, the largest protected area in Central America, the Laguna del Tigre National Park in Guatemala, comprises only 330 000 ha). Yet other parks such as Tinigua, Picachos and La Paya National Parks are threatened by colonization pressure from all sides, and have already become isolated forest islands. Despite their large size, these parks will eventually suffer the ills that derive from geographic isolation, *e.g.*, limited gene pools. All told, ten existing or proposed national parks and reserves in the Colombian Amazon account for 15.2% of its territory (61 179 km^2). Indigenous reserves, while positive for biodiversity protection, may in some cases provide only limited protection, since traditional attitudes towards conserving natural ecosystems are often based on custom and myth, and these underpinnings are increasingly being threatened by acculturation and the often rapid loss of cultural identity.

It is hoped that even more parks and protected areas will be established. An Amazonian conservation workshop (Workshop '90) that was convened in Manaus, Brazil in January, 1990, brought together most of the recognized global experts on Amazonian biodiversity to derive consensus on priorities for conservation in the nine-country region. Several areas of the Colombian Amazon currently outside parks were identified as high priorities, including the areas between the Caquetá and Putumayo Rivers, the forests adjacent to the Jirijirimo Rapids of the Apaporis River, eastern Vichada and Guainía. The results of Workshop '90 were published as a map that has already had a noteworthy impact in urging Amazon Basin countries to establish parks and reserves in high-priority areas.

One of the more impressive long-term research and conservation efforts in the Colombian Amazon is that headed by Drs. Thomas and Sara Defler with support from Conservation International and the Fundación Natura. Since 1982, this team has overseen an ecosystem conservation effort at the Caparú Reserve (Caparó is the name in a local indigenous language for woolly monkey) in the southern Colombian Amazon. The effort has focused on maintaining a scientific presence in the region, on research into key aspects of tropical forest ecology, on long-term research and monitoring of eight species of primates, on education and training for Colombian primatologists, on the establishment of a permanent protected area (National Park), and on broad surveys and detailed mapping of the entire Colombian Amazon region. The centerpiece of the park is the Estación Biológica Caparú (EBC), located on the lower Apaporis River, which has a grid of over 100 km of trails covering 8-10 km^2 of pristine rain forest.

Today more than 30 Colombian students have undertaken field research projects at the EBC, and the proposed Caparú National Park is pending presidential declaration; the park will protect some 330 000 ha of lowland tropical forest, including a small range of hills, the Serranía de Taraira, home to many endemic plants. The heart of the park is the longest oxbow lake in the Colombian Amazon (24 km). Once officially declared, Caparú National Park will also protect several endangered Amazonian animals, such as the black-headed uakari, the woolly monkey and six other primate species, plus the black caiman (*Melanosuchus niger*), the giant otter, and the pirarucu fish (*Arapaima gigas*). The Deflers' research (continuous since 1982) makes Caparú by far the longest consistently monitored site in the Colombian Amazon, and has provided scientists with valuable information for enhancing our understanding of the ecology of this poorly studied biome.

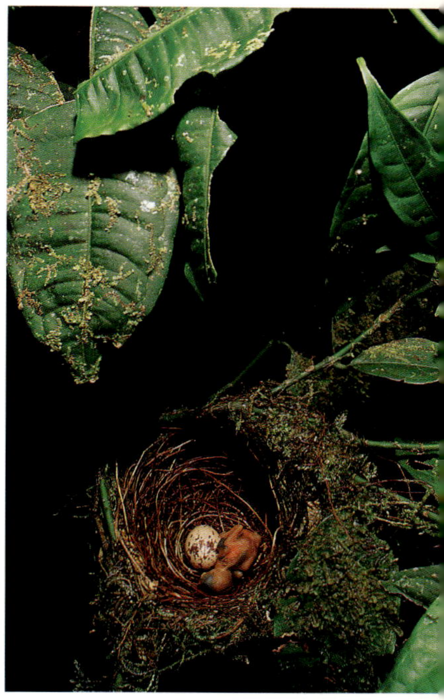

A nest of the unicolored tapaculo (Scytalopus senilis) *in the La Planada Reserve in the southern Colombian Andes.*

Flagship species

Colombia is known worldwide as the richest country in bird diversity. The current number of bird species recorded for Colombia is 1 815, including 203 migratory species and 142 endemic ones. Colombia is particularly rich in certain bird groups. For instance, there are 335 species of hummingbirds (family Trochilidae) in the Americas and fully 151, or 45% are found in Colombia. These highly attractive birds fill the very special niche of pollinating a wide variety of plants, and are the fastest and lightest of all birds. With 53 species of psittacine birds, Colombia ranks behind only Brazil and is tied with Australia for second place in the world list. Notable endemics in this group include the yellow-eared parrot (*Ognorhynchus icterotis*) and Fuerte's parrot (*Hapalopsittaca fuertesi*).

Distribution of bird species varies according to alti-

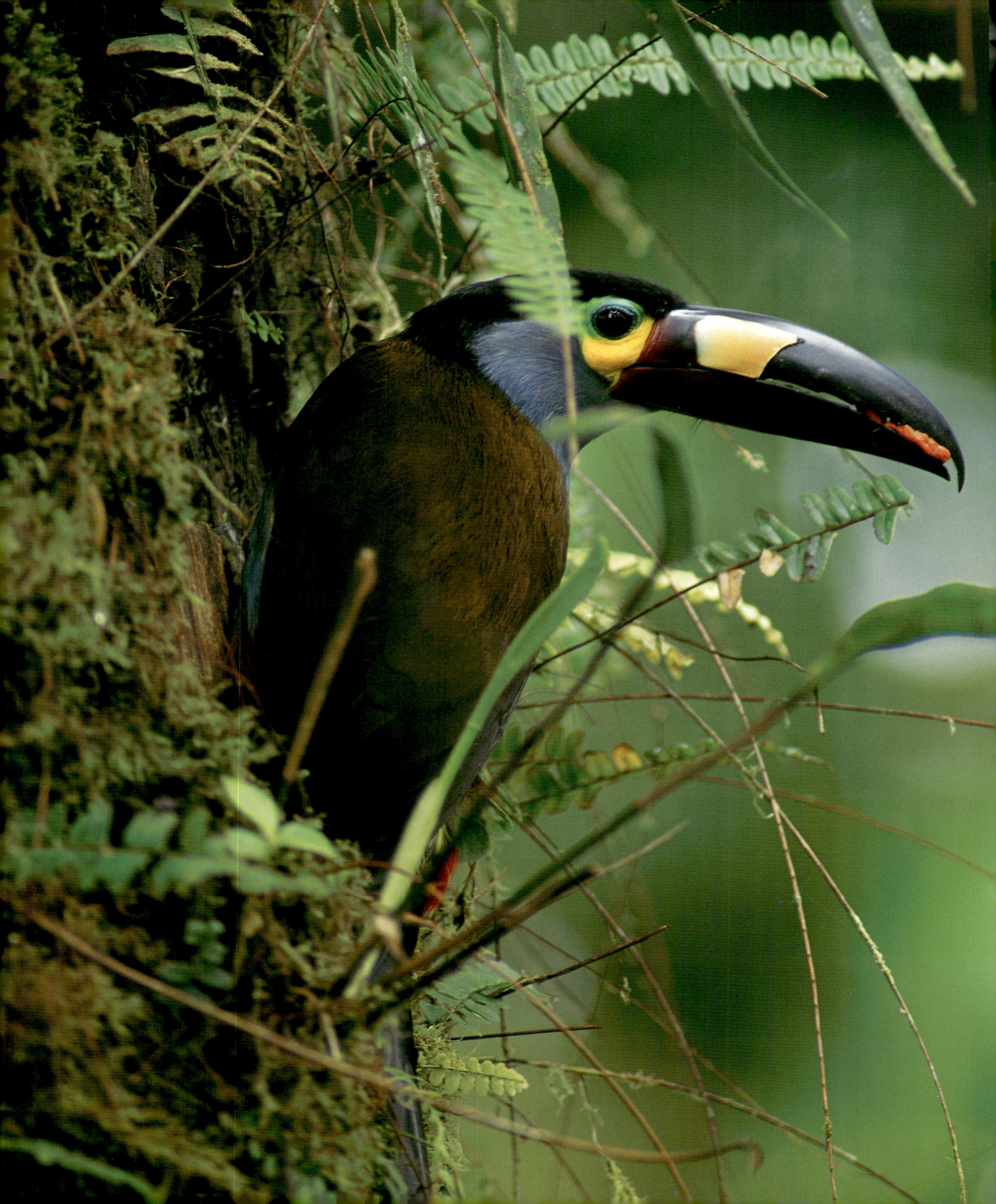

tude, but the largest concentration is found on the Andean slopes. Birds make the most of every niche available in the tropical forests of Colombia —from the open skies which are frequented by swifts and swallows (families Apodidae and Hirundinidae) through all the forest strata to the ground below, every layer is inhabited by a unique avifaunal community. Among the more spectacular of these forest dwellers are the northern helmeted curassow (*Pauxi pauxi*), the endemic blue-billed curassow (*Crax alberti*), the endemic mustached and brown-banded antpittas (genus *Grallaria*) and the recurved-billed bushbird (*Clytoctantes alixii*). Both of the aforementioned curassows, as well as the other members of the family Cracidae, are among the first to disappear in the face of hunting pressure, since they are large and very good to eat.

One of Colombia's most spectacular large birds (up to 13 kg) is the Andean condor (*Vultur gryphus*). An inhabitant of the high Andes, this carrion feeder was the target of relentless persecution earlier in the century when it was hunted nearly to extinction as an alleged cattle pest. This systematic extermination was worsened by a simultaneous decline in its natural prey as a result of habitat loss and hunting, and the Colombian populations of Andean condor are believed to have dropped into the low hundreds. For the past several years this animal has been the focus of an ambitious program to reintroduce it to several sites throughout the Colombian Andes. The highly successful reintroduction program spearheaded by the Colombian Instituto von Humboldt and Conservation International, combined with a hunting ban and government-sponsored education programs, have helped to recover wild populations from the brink of extinction. The recovery program has received good coverage in the national press and other communications media in Colombia, and has become a national conservation success story; efforts have also been made to replicate and expand the program to neighboring Venezuela and Ecuador.

Mammals are represented by 456 species in Colombia (and possibly another 40), 28 of which are endemic. One of the most spectacular is the spectacled bear (*Tremarctos ornatus*), found in the Andes and the only member of the bear group represented in South America. Its large size (175 kg) and the fact that there are many myths among the local people regarding healing and aphrodisiac powers of the meat and fat of this bear, have made it one of the most threatened species throughout its range. The largest member of the cat family in the Neotropics is the jaguar (*Panthera onca*). It has more than 100 names in Indian languages, as well as 18 in Spanish, and plays an important role in Indian mythology and folklore (Rodríguez *et al.*, 1996).

There are twenty-eight species of freshwater and marine mammals in Colombia; two of them are sirenids, the Caribbean manatee (*Trichechus manatus*), inhabitant of the great Andean rivers and tributaries of the Orinoco drainage, and the Amazonian manatee (*Trichechus inunguis*), which lives in the lower portions of the tributaries of the Amazon River itself.

The most diverse mammalian orders are the bats and the rodents. There are 173 species of bats in 9 families, and 131 species of rodents in 11 families. Bats play a crucial role in tropical rain forest ecology. They have a wide variety of feeding niches, from nectar feeders (like hummingbirds), fruit eaters (like many birds and rodents), to blood eaters and meat eaters specializing in frogs, insects, fish and birds. Bats also play an important role in pollinating flowering plants such as *Bombacopsis quinata*, in the regulation of insect populations, and as seed dispersers. Most of the rodents in the Colombian rain forests are small (rat-, mouse-, and squirrel-sized), though they range to as large as the capybara, which can weigh up to nearly 32 kg. Generally, any rodent in excess of 100 g is considered game, but the one with the best flavor is the paca (*Agouti paca*), which can reach up to 12 kg. This animal is considered a delicacy among the local people, as well as in some of the larger cities.

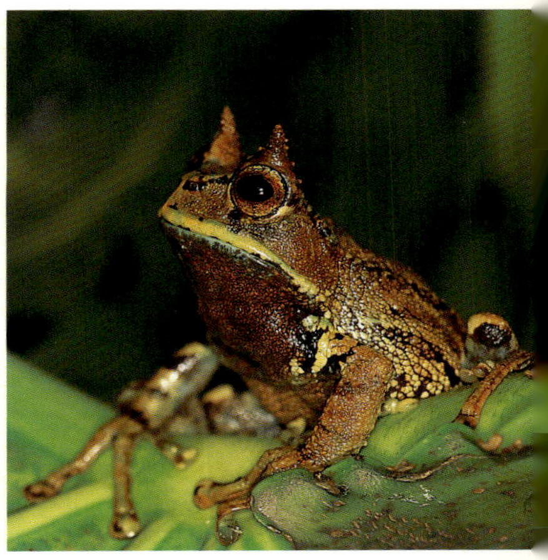

Colombia is one of the world's top countries for primate diversity, with 29 species in 3 families and fully 59 taxa, comprising about one eighth of the world's primate species. There are at least four endemic species, including two of night monkey (*Aotus brumbacki* and *A. hershkovitzi*) and two tamarins, the white-footed tamarin (*Saguinus leucopus*) and the cotton-top tamarin (*S. oedipus*).The cotton-top tamarin is particularly noteworthy because of its beauty and charisma, and has been termed "the panda of Colombia" because of its importance as a flagship species. A large mane of cottony white fur gives the cotton-top its most appropriate name. This fluffy top knot is also the source of its other vernacular name, used mostly in Germany, the "Lisztasschen" or "Franz Liszt monkey," for the Hungarian piano virtuoso and composer (1811-1886) also known for his pompadour. Due to its unique natural history and physiology (it is the only nonhuman mammal to develop colitis preceding the onset of cancer), the cotton-top tamarin is one of the most important primate models for biomedical research; indeed, it could well be the key to understanding one of the leading causes of human death in Europe and North America.

Although found abundantly in both zoological and research colonies outside Colombia, the cotton-top is highly endangered in its wild habitat in the Caribbean Arid Belt, where most of its native forests have disappeared. It has suffered heavily from live capture, especially in the 1960s and 1970s; as many as 40 000 were legally exported from Colombia for biomedical purposes and the pet trade during the 15 years prior to their 1976 inclusion on the U.S. Endangered Species List. All legal capture and export of this species has ended, yet it still suffers from an active, though substantially diminished national pet trade within Colombia. Its numbers are now severely reduced in the wild, where it is found only in a few protected forests and riparian habitats in northern Colombia west of the Magdalena, and one introduced population in Tayrona National Park.

On the opposite page, the plate-billed mountain toucan (Andigena laminirostris) *in the La Planada Reserve, one of at least 20 species of the toucan family in Colombia.*

Above, the very unusual truncated poison frog (Dendrobates truncatus).

Captive populations in foreign countries are still larger than the remaining populations in the wild. These remain strong and this species is very common in zoo collections; sadly, however, more than 30% of captive individuals die before they reach six years of age due to the aforementioned propensity for colon cancer. Interestingly, recent studies supported by Conservation International show that cotton-tops seldom if ever develop this illness in the wild.

An active research, conservation and education program headed by Dr. Anne Savage of the Roger Williams Park Zoo in Rhode Island, with support from the Colombian Ministry of the Environment, has been underway for nearly a decade at the Montes de María Reserve near Colosó, department of Sucre, one of the cotton-top's last forest strongholds. This tamarin has also been used by Conservation International in a public awareness campaign featuring T-shirts and stickers depicting the animal and bearing conservation messages.

Other important primates are the night monkeys, genus *Aotus*, the only nocturnal Neotropical primates. These animals have also felt the pressure of live capture for the biomedical industry, and have been of key importance to Colombian researchers seeking to develop a synthetic malaria vaccine and for eye research.

Other highly endangered Colombian primates are the larger ones, those first to succumb to hunting pressure by local inhabitants of Amazonia due to their attractiveness as food items. The populations of woolly monkeys (*Lagothrix*) and spider monkeys (*Ateles*) have been seriously depleted. An endemic taxon, the highlands woolly monkey (*Lagothrix lagotricha lugens*), is perhaps the most endangered of the large-bodied Colombian primates, yet little is known of its distribution, ecology and behavior. Adults are hunted throughout the year and infants are kept or sold as pets. Other smaller species, such as the squirrel monkeys (*Saimiri* spp.) and the tamarins and marmosets (*Saguinus* and *Cebuella*) are greatly sought after for the pet trade.

Colombian amphibians are the group with the highest levels of endemism in the country, especially along the Pacific coastal plain and in the Andes. There are currently 583 amphibian species described for Colombia, more than any other country on Earth. Of these, 17 are salamanders and 25 caecilians (Ruiz C. *et al.*, 1996). In addition, there is an amazingly high number of new species not yet described, more than 120 in all. The most diverse group among amphibians are the frogs of the family Leptodactylidae, with 212 species described to date; one widespread leptodactylid genus, *Eleutherodactylus*, is represented in Colombia by 169 species, over 42% of the 400 species in this genus.

Among the best flagship species for Colombia are the 33 species of poison dart frogs of the family Dendrobatidae, most notably the three frog species of the genus *Phyllobates* that are found in the Chocó. These and other species of dendrobatids produce toxic alkaloids that are secreted through the skin. Alkaloids are among the most promising natural compounds in the search for cures for cancer and other illnesses, and many of these frog poisons have already been tested in the laboratory and have yielded interesting results. One frog alkaloid has anesthetic properties 200 times stronger than morphine. Another can generate immediate and lasting erections in laboratory rats. The flagship value of these animals derives not only from their beauty and potential economic value, but also from the fact that they cannot produce the toxins outside their natural surroundings; if they are removed from the forest or if the forest is destroyed around them, their ability to synthesize these valuable alkaloids disappears. This is because the frogs do not generate the poisons themselves, but rather piece together their species-specific alkaloids from chemical building blocks that are carefully selected from their insect food items and combined through a complex, rapid and entirely unknown process. These frogs are an unusual parallel to the human "shamans" with whom conservationists work to preserve and rescue ethnobotanical knowledge. They too are the only ones who know the ingredients and methods for preparing their potent jungle medicines, and thus shortcut the process of locating these compounds for the bioprospector.

The most toxic and one of the most beautiful of these frogs is *Phyllobates terribilis*, found only at a restricted site in the Saija River Basin of the rain forests of the southern Chocó. This bright yellow frog is one of the most poisonous vertebrates in the world, and its poison can easily be absorbed through the skin. Toxicity of this rain forest jewel is such that a full-grown man can easily die by merely touching it. The Emberá Indians of the Chocó use the skin secretions of this frog to poison their blowgun darts, and to do so they don't even have to kill the animal. They merely rub the tip of the dart on the frog's back and the poison is strong enough to kill a jaguar in a few minutes; furthermore, the poison remains active for as long as six months. A conservation and research project highlighting this unique frog is currently being initiated by the Colombian National University's Natural Sciences Institute, ProSelva and Conservation International, and the frog has been featured in public awareness materials produced by these partners to draw attention to the critical threats facing this animal and its forest habitat.

Amphibians' success at diversifying lies in their ability to use different niches in the rain forest, as well as a wide range of reproductive strategies. Some of them, like the centrolinelids, lay their eggs on the vegetation overhanging water, so that when the eggs hatch, the tadpoles fall directly into the water. Others lay their eggs among decaying forest floor debris, and then carry their young on their backs and put them in the pools formed at the base of bromeliad plants. The female feeds the tadpoles with unfertilized eggs that she regularly deposits in the pool. One of the most curious forms of parental care is exhibited by the marsupial frog, which keeps its eggs between the skin folds on its back until the tadpoles are completely developed.

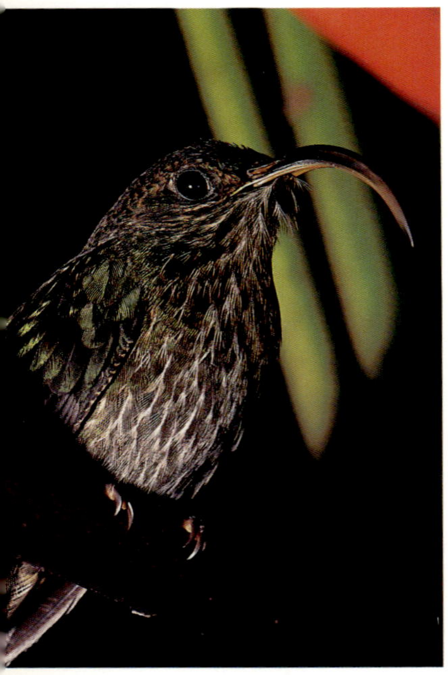

Colombia has the richest bird fauna on Earth. Above, a white-tipped sicklebill (Eutoxeres aquila) *from the Chocó region.*

On the opposite page, a male black-capped tanager (Tangara heinei) *from La Planada Reserve.*

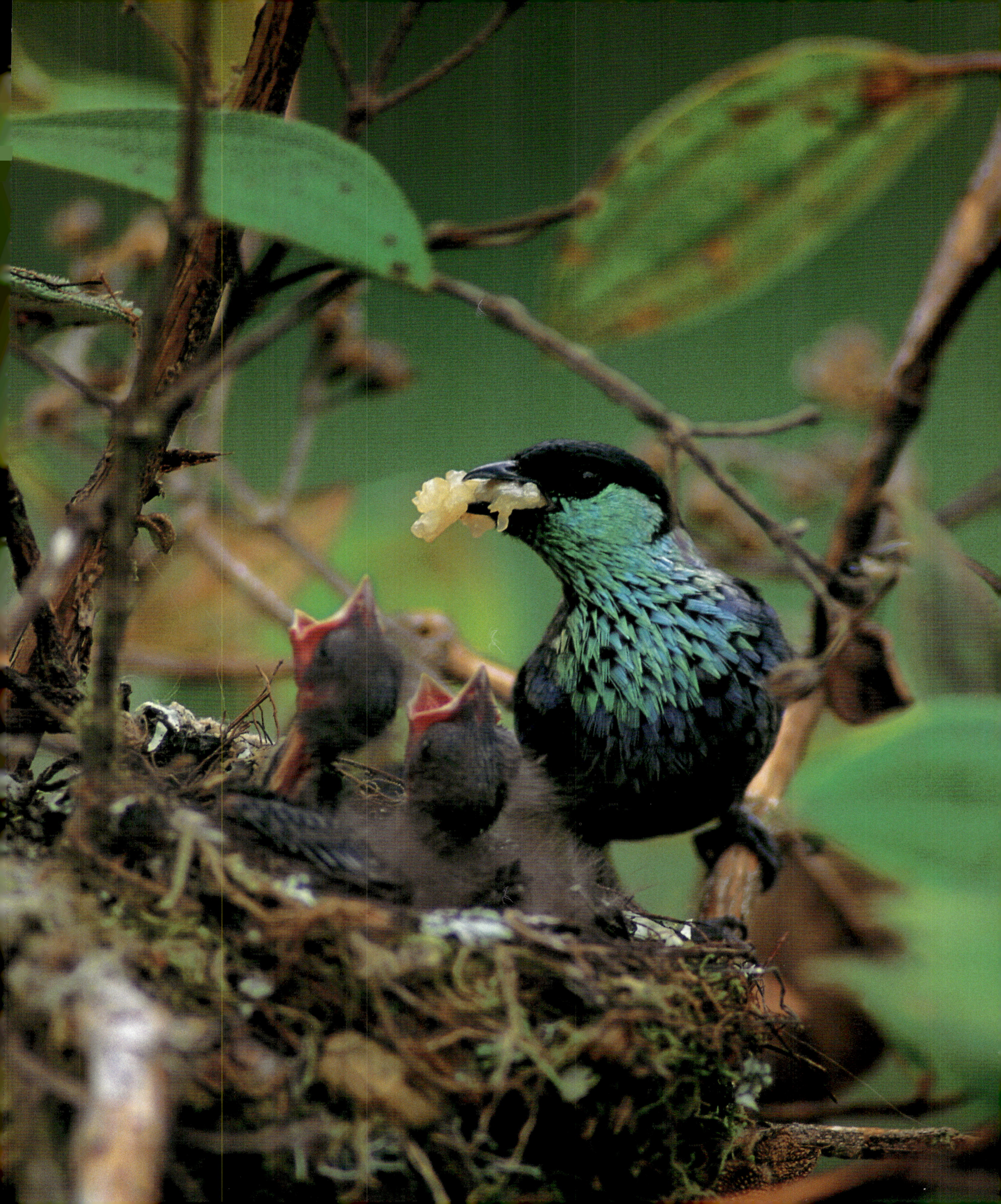

There are 520 species of reptiles in Colombia, 97 of them endemic; a total of 236 are lizards. However, as with the frogs mentioned above, there are many species yet to be described, so the numbers for this group are going to increase dramatically as scientists continue to explore new areas. There are 37 species of turtles, including 5 marine turtles that nest on Colombian beaches. The rapid development of tourism, as well as the widespread use of the meat, shell and eggs of these animals, places both marine and freshwater turtles among the most threatened animals in Colombia.

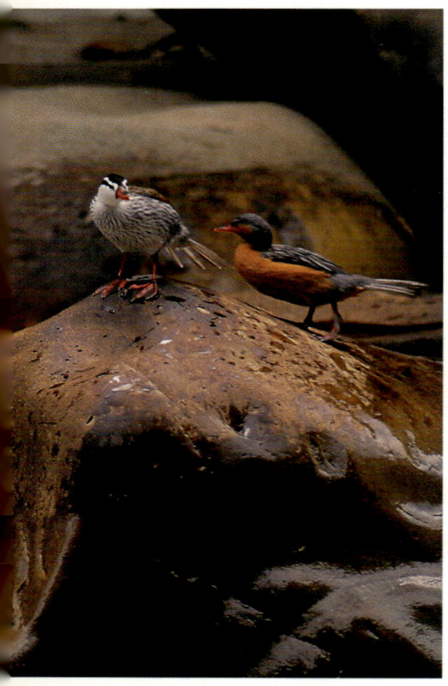

Perhaps the place where the pressure on the chelonians is most evident is in La Guajira Peninsula, on the northern coast of Colombia, where as part of local tradition and culture, hundreds of turtles are captured, slaughtered, and then distributed to be sold in street markets and restaurants. This market is encouraged by the popular belief in the aphrodisiac power of turtle meat and eggs, and the use of turtle shell in the manufacture of handicrafts. The species most affected are the green turtle (*Chelonia mydas*) and the hawksbill turtle (*Eretmochelys imbricata*). Three other marine turtles —the loggerhead (*Caretta caretta*), the olive ridley (*Lepidochelys olivacea*) and the leatherback turtle (*Dermochelys coriacea*)— are not popular as food items, but are victims of egg harvest by local peoples. The most recent estimates (1988) showed that approximately 6 000 turtles are slaughtered every year in La Guajira alone. One of the most endangered freshwater turtles is the Magdalena River turtle (*Podocnemis lewyana*), which has long been persecuted for its meat and eggs in this devastated part of Colombia, to the point that it is now considered endangered.

Of the 23 species of crocodilians known in the world, six species and nine subspecies are found in Colombia, making it the richest country on Earth for these animals. Of these, the Apaporis subspecies of the caiman (*Caiman crocodilus apaporiensis*) is endemic to the Colombian Amazon; this taxon may indeed be a full species, a question that is currently under review by taxonomists. The others include two species of crocodiles (*Crocodylus americanus, Crocodylus intermedius*), the spectacled caiman (*Caiman crocodilus fuscus, Caiman crocodilus crocodilus*), the two dwarf caimans (*Paleosuchus trigonatus, Paleosuchus palpebrosus*) and the black caiman (*Melanosuchus niger*). Crocodilians have long been subjected to heavy hunting pressure in Colombia, putting both *Crocodylus* species, and *Melanosuchus niger* in danger of extinction. The government of Colombia started a highly successful commercial captive-breeding program during the mid-1980s, and this experiment gave way to a new production concept under controlled conditions and with a sound scientific basis. This has discouraged illegal trade and has encouraged new ways to conserve these important animals.

The torrent duck (Merganetta arnata) *inhabits fast-flowing mountain streams in the Andes.*

On the opposite page, wet tropical forest of the Niambi River area, near La Planada Reserve in the southern Chocó. This area receives some 6 000 mm of rainfall.

Palms are the best plant flagships in Colombia. There are 226 species, making the country second in the world behind Indonesia. Colombia's palms play an important role in human societies as sources of food and other products. One of the most important protein sources for inhabitants of the Chocó is the fruit of the *chontaduro* palm; also, palm oils such as that extracted from the fruit of the *mil pesos* palm are important in food production (*mil pesos* oil is chemically indistinguishable from olive oil). Palm thatch is used as a resistant roofing material, palm fiber is used in the production of crafts and household materials from furniture to flooring, and coconuts are an important subsistence and export crop on both the Caribbean and Pacific coasts. Colombia's national plant is the wax palm (*Ceroxylon quinduence*), which is widely used by one of the most threatened parrots in South America, the yellow-eared parrot (*Ognorhynchus icterotis*), a poorly-studied endemic to Colombia and Ecuador.

Cultural diversity

Colombia has a rich diversity of human cultures, paralleling its globally important biodiversity. As in the case of Brazil, it is difficult to determine how many different indigenous cultures existed in pre-Columbian times, and how many were lost as a result of European colonization and disease, but the fact is that today there are some 81 surviving indigenous groups in Colombia. They are dispersed throughout the country in about 450 settlements, mostly in Amazonia, Orinoquia, the Pacific coast, the Andean Cordillera, La Guajira Peninsula and the Sierra Nevada de Santa Marta. Among the most important ethnic groups are the: Tukano, Witoto, Yagua, Yukuna and Tikuna in Amazonia; Achagua, Saliba, Tunebo and Amoura in Orinoquia; Chocó, Cuna, Emberá and Waunana in the Pacific; Zenu, Muisca, Paez in the Andes; Wayu in La Guajira, and Kogui, Arsario and Arhuaco in the Sierra Nevada de Santa Marta.

The Amerindian cultures in Colombia are very distinct, and their different origins and cultural values give the country a unique and intriguing cultural diversity. This wealth of cultures has been recognized by the Colombian Constitution and has been studied thoroughly in the last few years, thanks to the development of more serious research in anthropology and linguistics. Many indigenous communities are undergoing a process of acculturation through contact with the mestizo peasant community, thereby losing some of their traditional Amerindian legacy, and even their languages. Fully one quarter of the 450 000 Amerindians in Colombia have already lost some of their indigenous heritage. The groups that have remained more traditional still practice subsistence agriculture and hunt, fish and gather forest products. In the high Andes, some communities still depend on subsistence production of corn and tubers; in the savanna, others live a seminomadic existence and survive by hunting, fishing and gathering; in the rain forest regions of the Chocó and Amazonia, the prevalent economic activities are hunting, fishing, and some slash-and-burn agriculture;

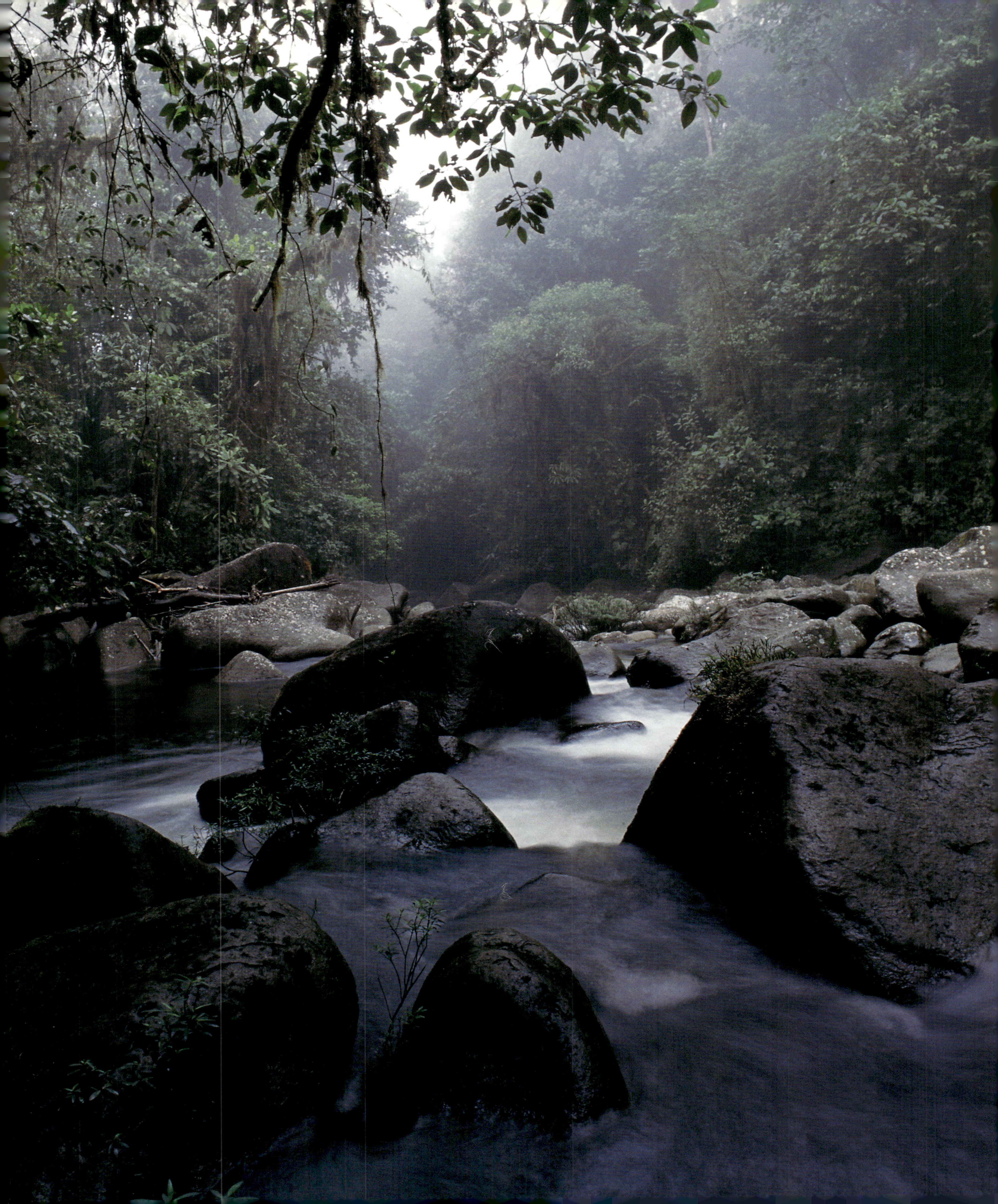

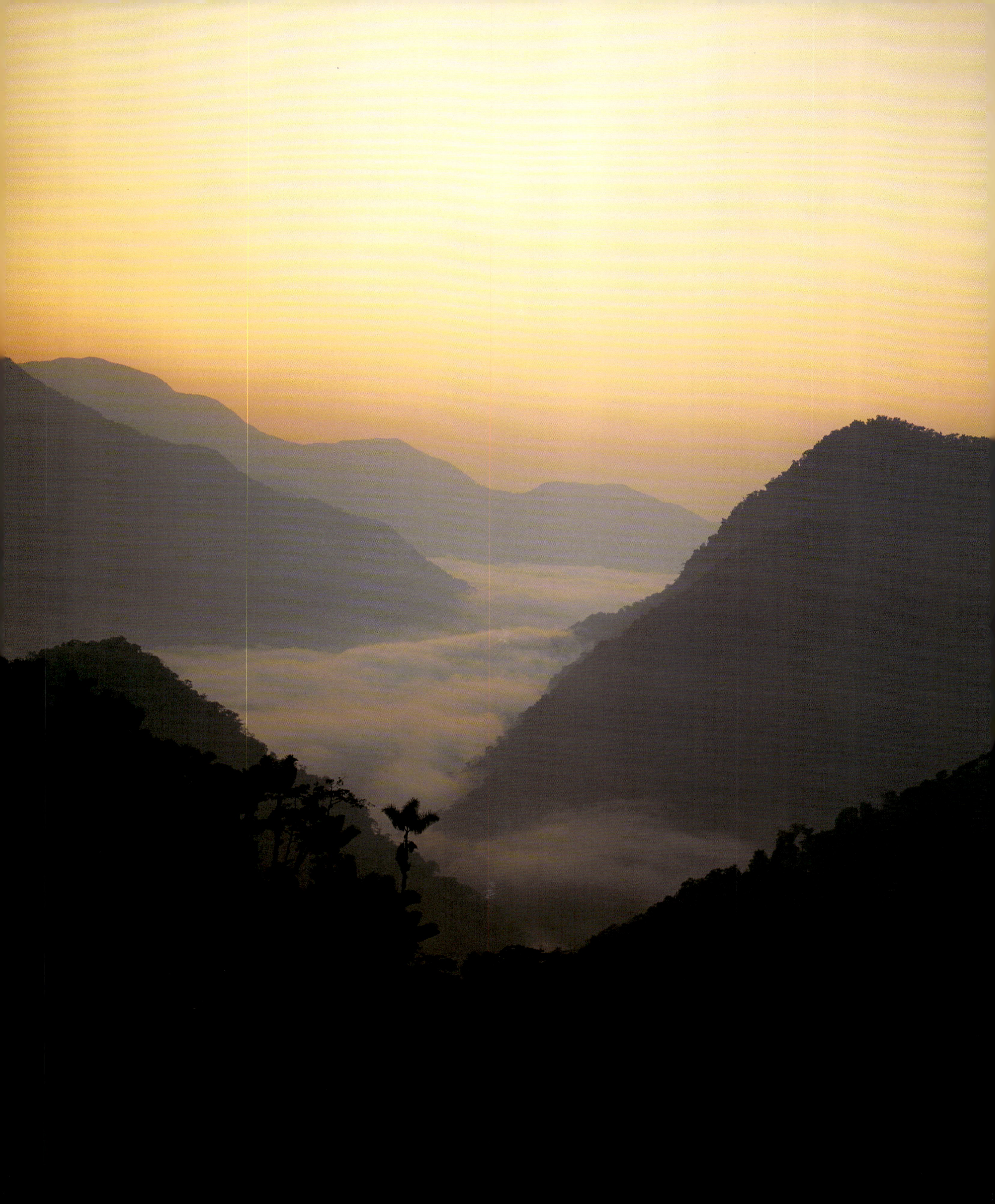

and in La Guajira, the Wayu, which are the second largest indigenous group after the Paez, are mostly sheep- and goat-herders.

In recent years, some indigenous groups have become involved with guerrilla movements. This is evident, for example, in some of the Emberá communities from the northern part of the Chocó. In other areas, indigenous communities have chosen a more democratic path, and have formed political parties and chosen representatives to the Colombian Senate through popular elections. The judiciary process to amend the constitutional rights that grant autonomy to both Indian and black populations as stipulated in the 1991 Constitution is currently being revised to aid in the preservation of indigenous cultures. The government has been setting aside indigenous reserves for the last ten years, and there are now more than 18 million ha legally recognized as indigenous reserves, the largest system of its kind in South America.

One of the most interesting groups is the Kogui of the Sierra Nevada de Santa Marta. There are about 7 000 individuals living in traditional settlements in the upper mountain reaches of this region. Their appearance gives the impression of extreme poverty, but their regal demeanor and their disinterest in mundane objects speaks of a higher understanding of the world. To the Kogui, the origin of all things comes from a maternal deity who managed to impregnate herself and hence produce all the things that exist in this world. Their supernatural world makes them less susceptible to outside influences, and they remain proud of their lifeways and origins. Chewing the leaves of the coca plant (*Erythroxylon coca*) for its mild stimulant effect is a basic element of Kogui culture. Only adult men are allowed to chew coca, and achieving adulthood requires a series of ceremonial rites and rituals. They mix the coca with lime made from the ashes of mollusk shells, which is kept in a gourd known as a *poporo*. The lime is extracted from the narrow-necked *poporo* with a stick moistened with saliva. The lime releases the alkaloids in the coca leaves and produces the desired results.

The Wayu, or Guajiros, live in La Guajira Desert and are the second largest ethnic group in the country, with a population of almost 100 000 in both Colombia and Venezuela. They conserve their original language, and do not regard the desert as an inhospitable, arid place, but rather as the center of their existence. They are highly adapted to make the most of what few resources this harsh environment provides. They live in settlements called *rancherías* and their most important economic activity is sheep- and goat-herding. They also gather wild cactus fruits to supplement their diet.

Four distinct groups of Emberá inhabit the Chocó region, numbering as many as 100 000 in small, scattered communities. The Emberá, despite having borne the brunt of Spanish colonization and subsequently the growth of a substantial Afro-Colombian culture in their region, have managed to largely maintain their isolation and most of their traditions. They typically live in villages of as many as two dozen elevated huts on the banks of both major and minor rivers throughout the Chocó, inland from the coast. They subsist by hunting and gathering, as well as some small-scale agriculture, and frequently trade for fish and other items in coastal towns that are mainly inhabited by Afro-Colombians dedicated principally to ocean fishing. Emberá customs include body painting and adornment, often with elaborately crafted bead necklaces. They are also excellent basket weavers; using locally available fibers and natural colorings, the women produce intricate and beautiful designs on all manner of baskets so tightly woven that they can effectively be used to carry water without leaking. The Emberá also practice traditional medicine which combines plant remedies with shamanistic incantations; the *jaibana* or doctor, uses songs and elaborately carved medicine sticks called *bastones* to effect his cures, and is respected as one of the community's most powerful members. As mentioned earlier, the Emberá are blowgun hunters, and employ the skin alkaloids of poison dart frogs of the genus *Phyllobates*; however, use of the blowgun is rapidly disappearing in the Chocó, being replaced by the shotgun in most areas.

The Pacific coast is inhabited mainly by Afro-Colombians, descendants of slaves brought by the Spanish some 300 years ago. People of African origin were first brought to this region by slavers from Popayán, and were employed in the gold mines and to harvest tagua, the ivory-like nut of the tagua palm. The 1991 Colombian Constitution recognizes the rights of these people to community land ownership, in much the same way as Indian tribes have reserves established in their traditional lands. The second most important black population in the country is located in the cities along the Atlantic coast, especially in Cartagena and Santa Marta, which were major slave distribution centers during the era of the slave trade, and in the small town of Palenque. The Afro-Colombian communities of the Chocó have a culture that exhibits both African roots (particularly in music and art) and elements of the local Emberá culture (in their knowledge of natural history, plant medicines, and household uses of biodiversity), thus representing an unusual mix. The local nongovernmental conservation groups, Fundación Natura and Fundación Inguedé, have been working with the support of Conservation International over the past few years to create micro-businesses around the sustainable use of some of the forest products utilized by these Afro-Colombian communities in the Gulf of Tibugá, including huina oil, wax from the leaves of a local plant, *Callathea lutea*, pita fiber from the leaves of a locally common, pineapple-like bromeliad, and tagua nuts, which are crafted by local artisans into buttons and carvings to be sold to tourists.

RODERIC B. MAST
JOSÉ VICENTE RODRÍGUEZ
RUSSELL A. MITTERMEIER
CRISTINA GOETTSCH MITTERMEIER

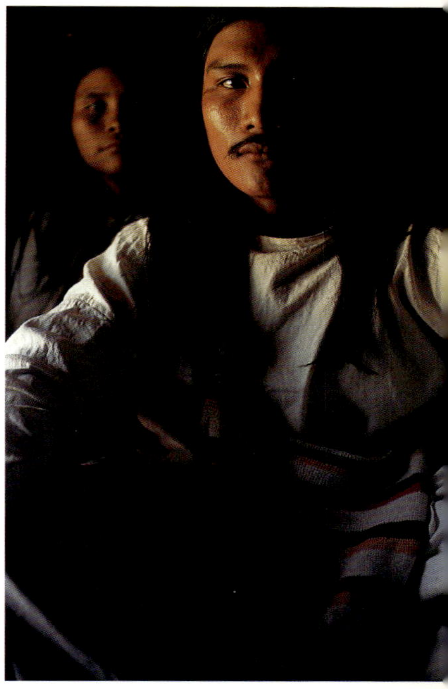

On the opposite page, cloud forest at dawn in the Sierra Nevada de Santa Marta. This isolated mountain massif in northern Colombia is rich in endemic species.

Above, a Kogui Indian, one of the traditional inhabitants of the Sierra Nevada de Santa Marta.

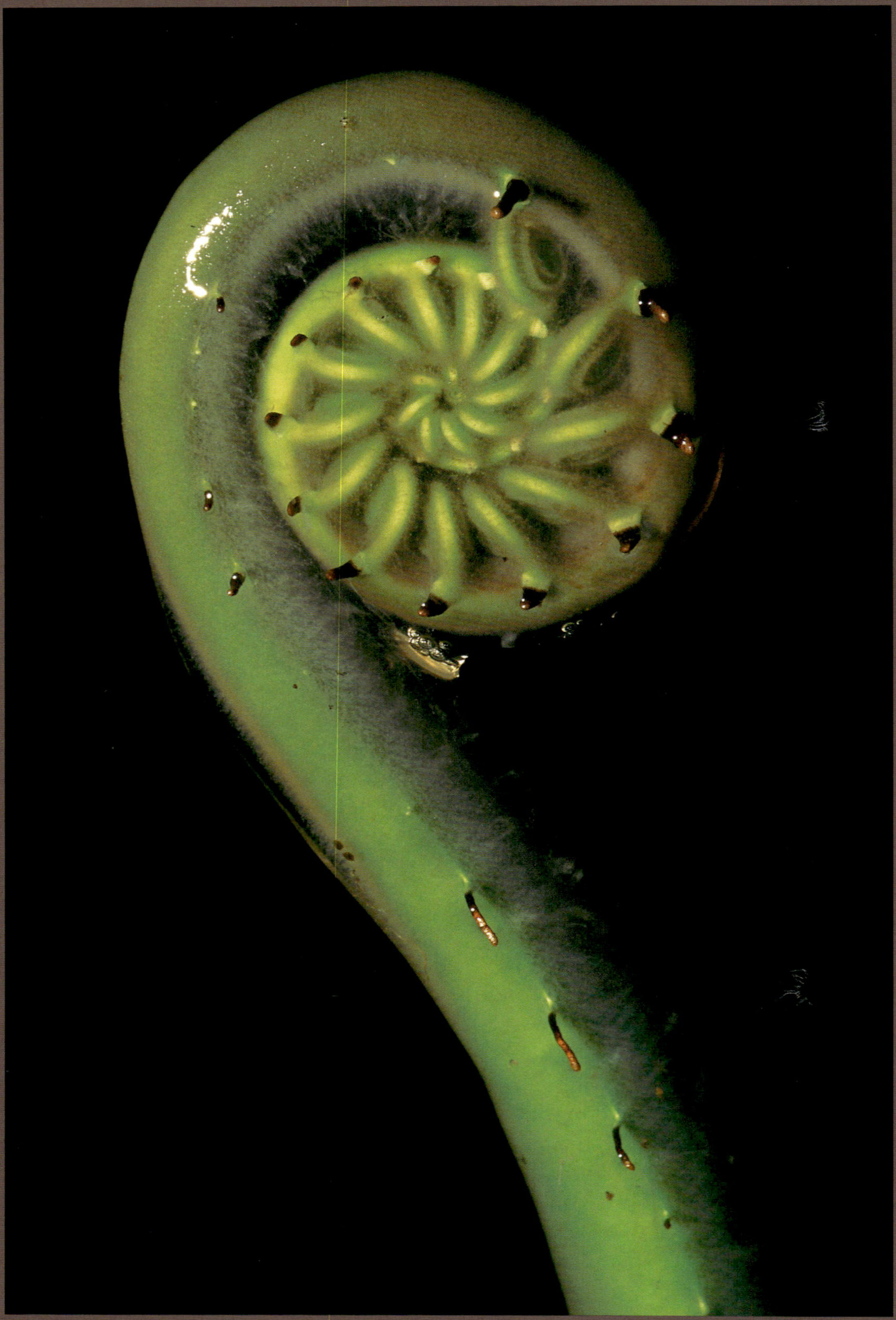

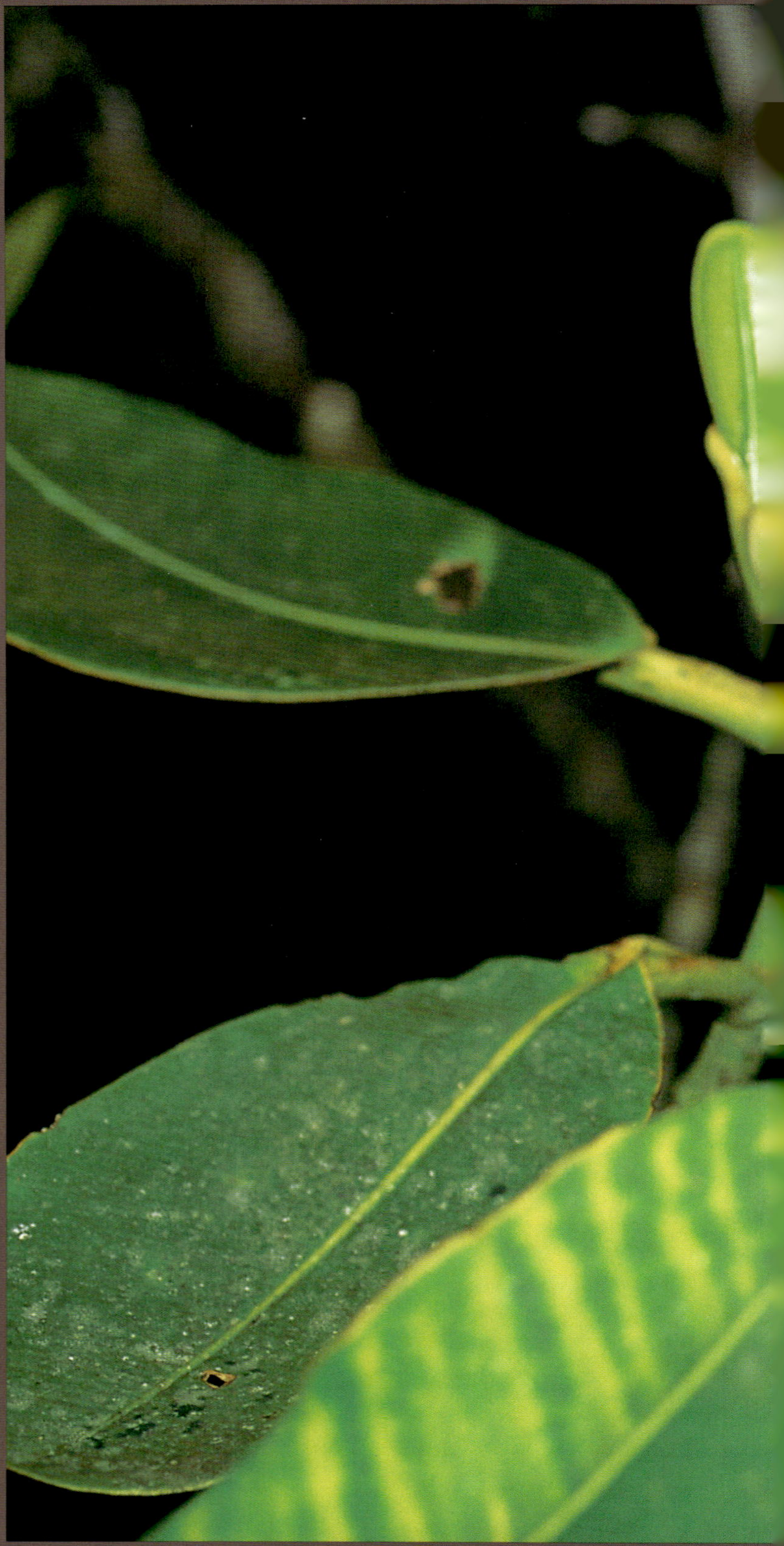

Above, a fern unfolds in the La Planada Reserve.
Upper right, a violet-tailed hummingbird (Agaiocercus celestis),
one of at least 136 hummingbirds in Colombia, the highest number for any country.
Lower right, the flower of a bromeliad growing in the Niambi River.

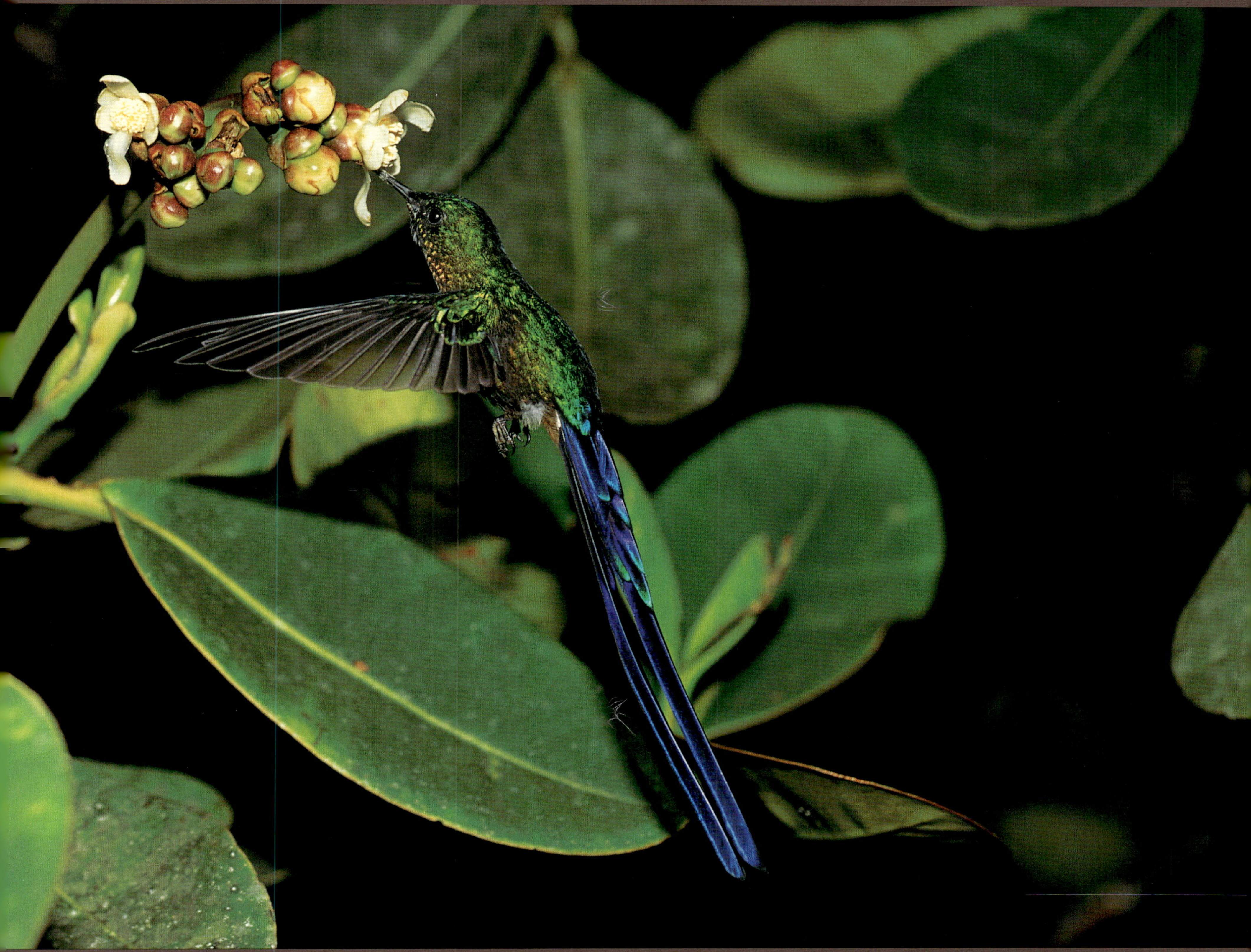

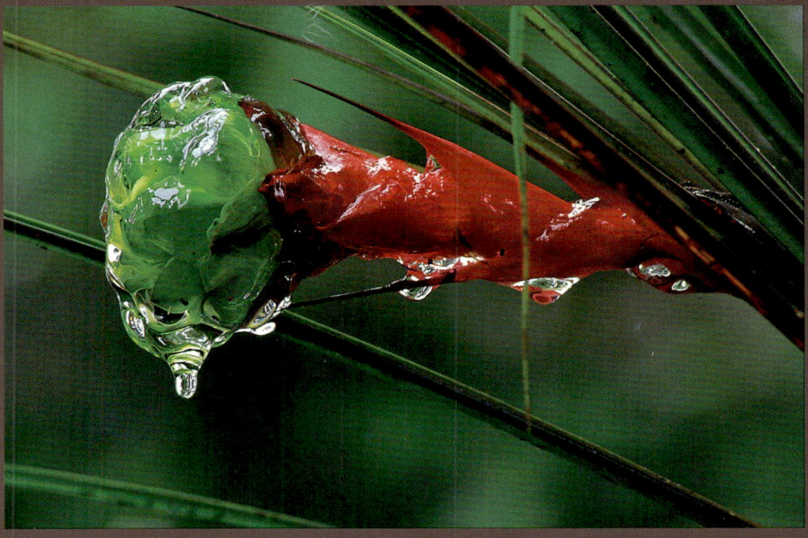

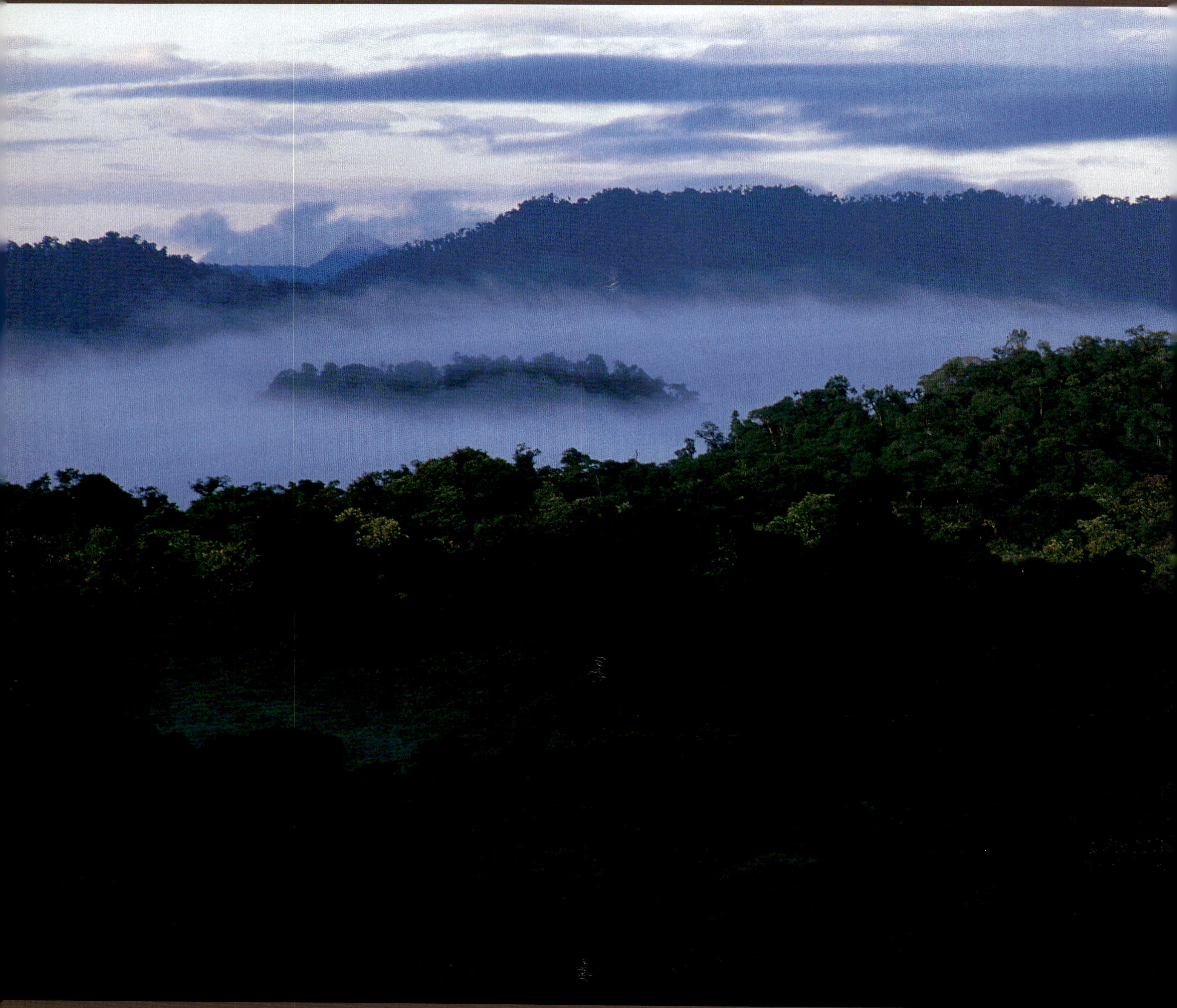

Above, the La Planada Reserve, an important protected area on the Pacific slope of the Eastern Cordillera in the department of Nariño near the border with Ecuador. This 3 200-ha privately protected reserve is mainly cloud forest habitat that reaches as high as 2 100 m and is very diverse for its small size.

Upper right, the spectacled bear (Tremarctos ornatus), *the only South American bear and one of the most important Andean flagship species. This individual was photographed in La Planada Reserve. Lower right, an interior view of the forest in La Planada, rich in ferns, bromeliads and palms.*

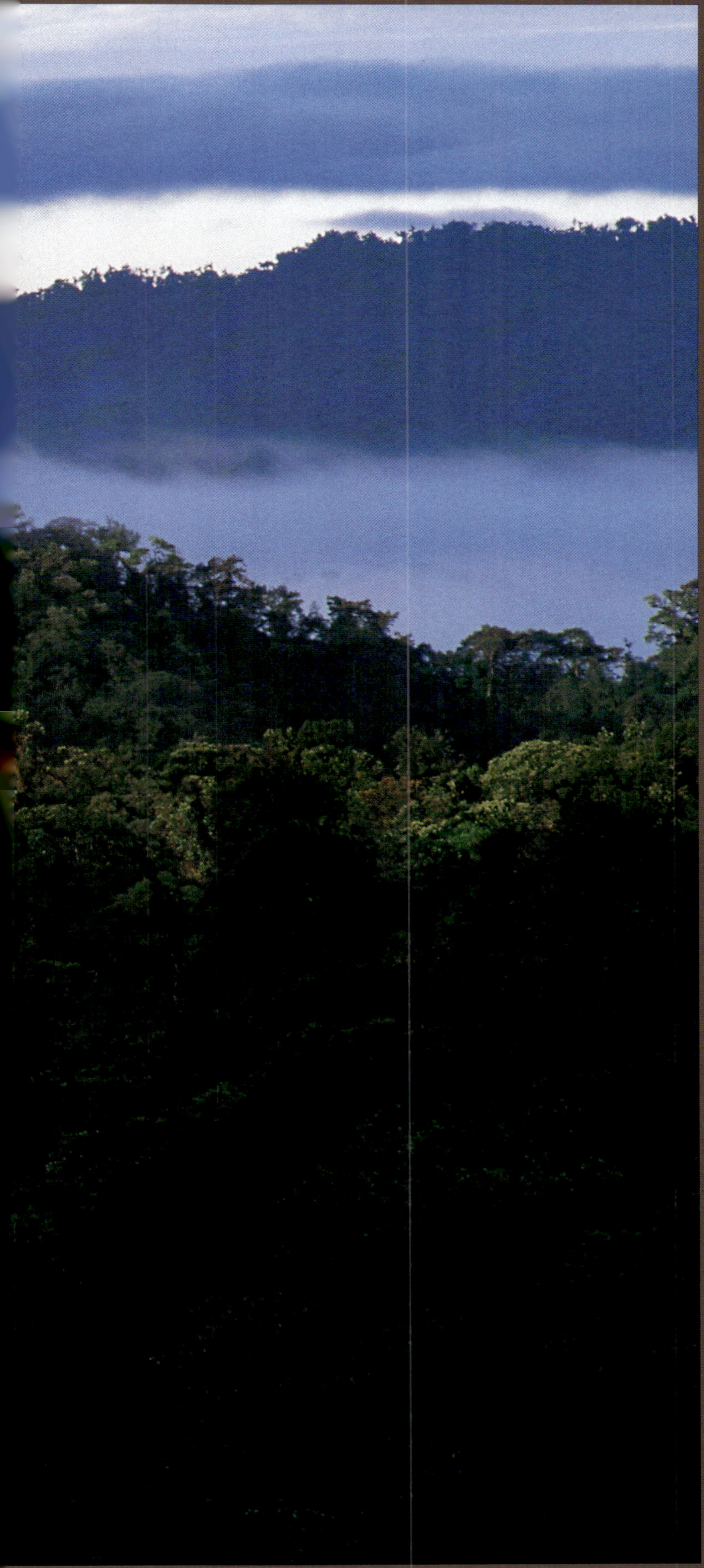

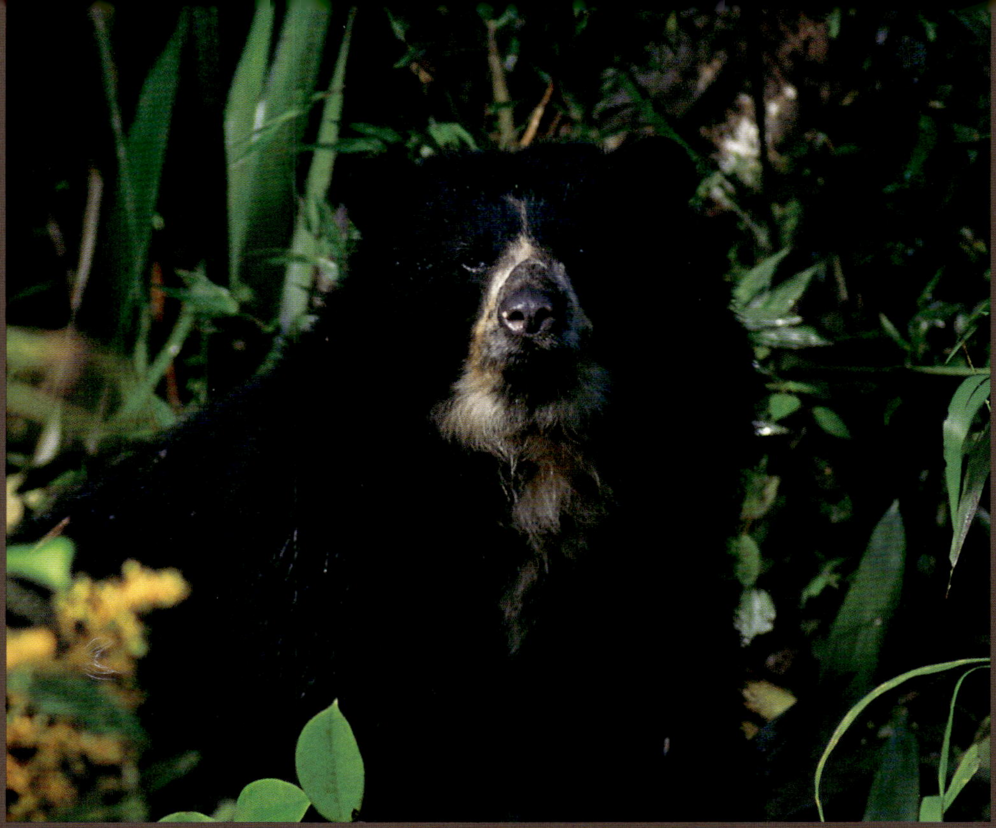

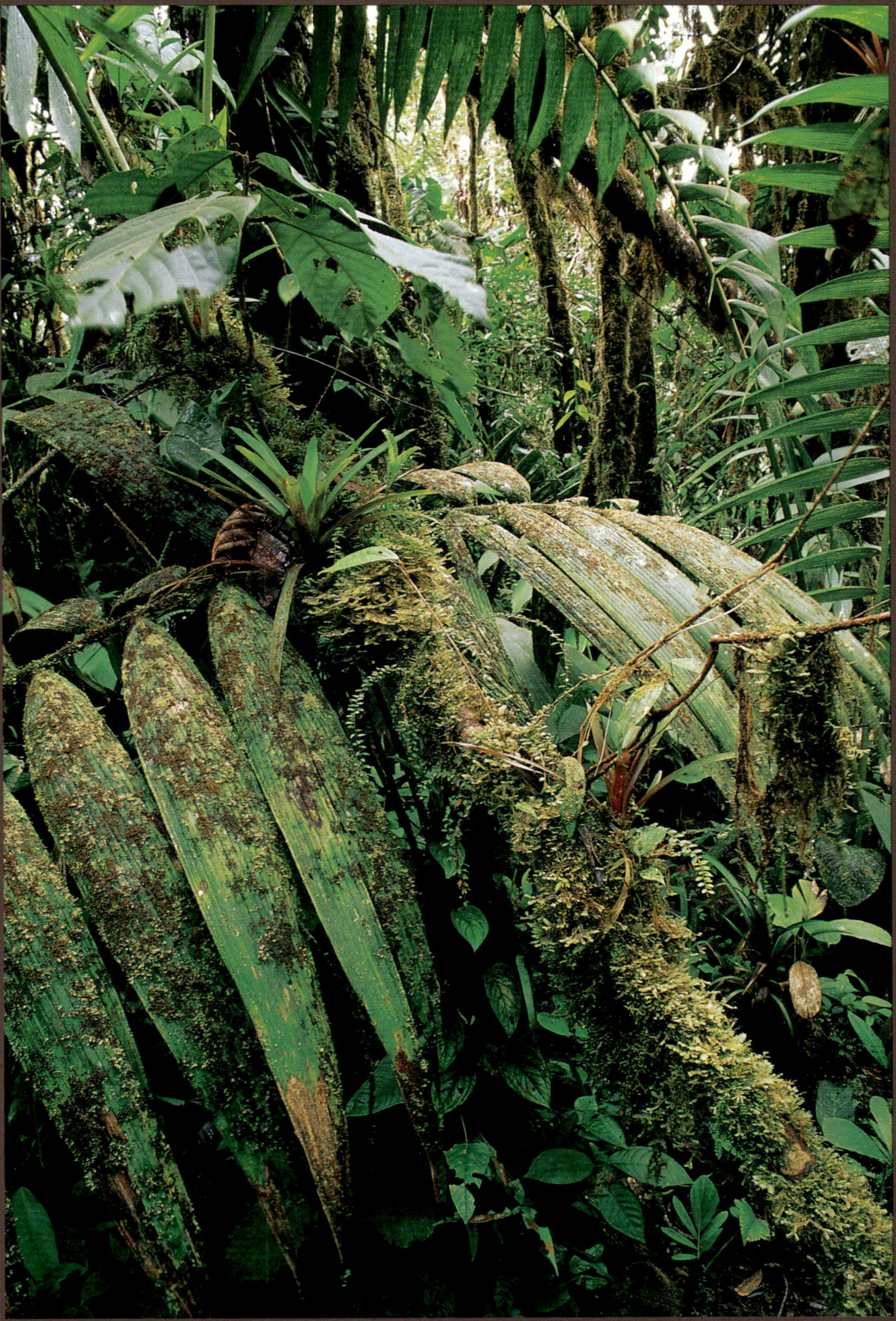

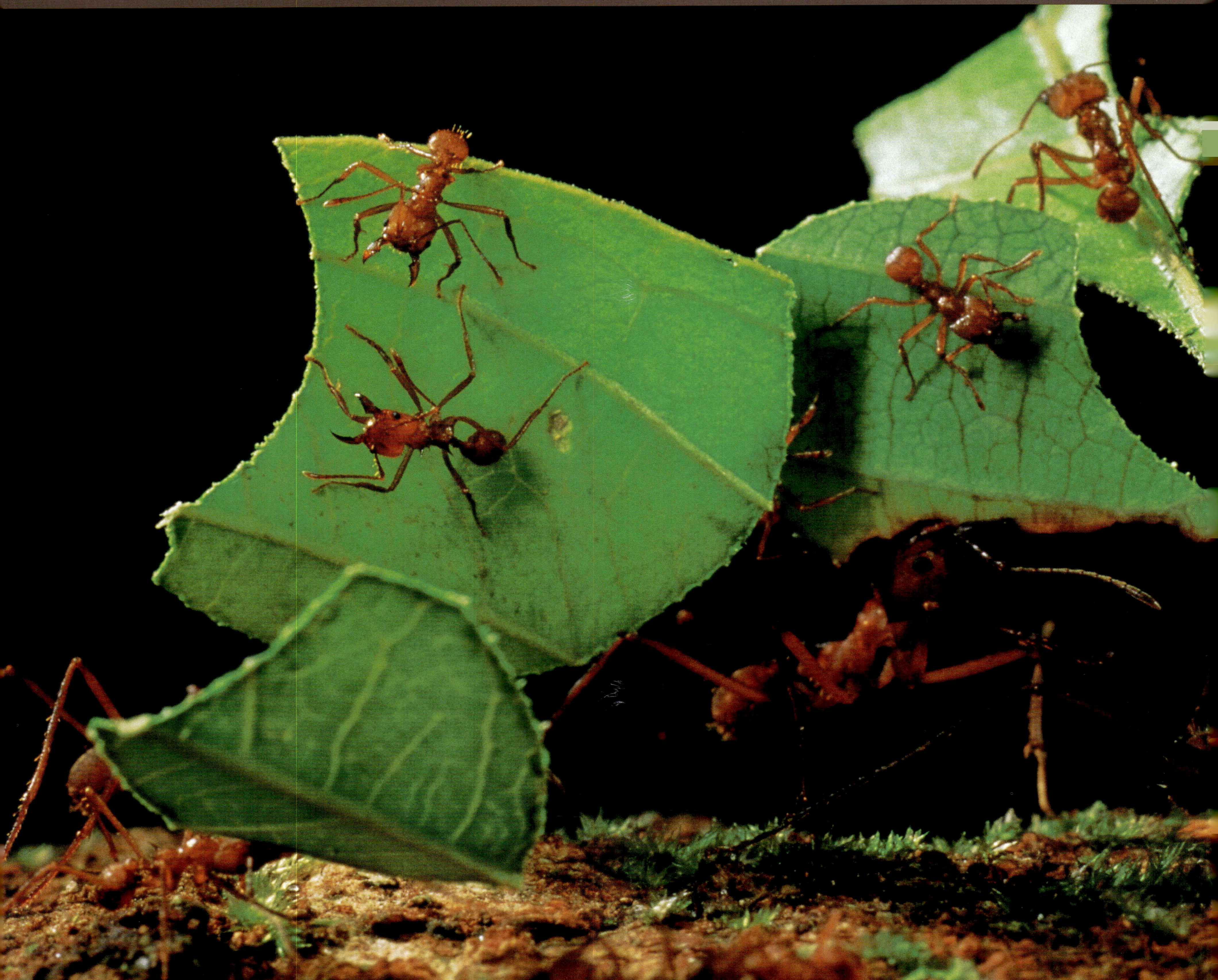

Above, hard-working leaf-cutting ants (Atta cephalotes) *are a prominent feature of the Neotropical rain forest, endlessly carrying leaves to grow their underground fungus gardens. Colombia has the world's highest amphibian diversity, most of it consisting of frogs. Upper right, a specimen of* Hemiphractus johnsoni, *a species that includes other frogs on its list of prey items. Lower right, an orthopteran moss mimic in the La Planada Reserve.*

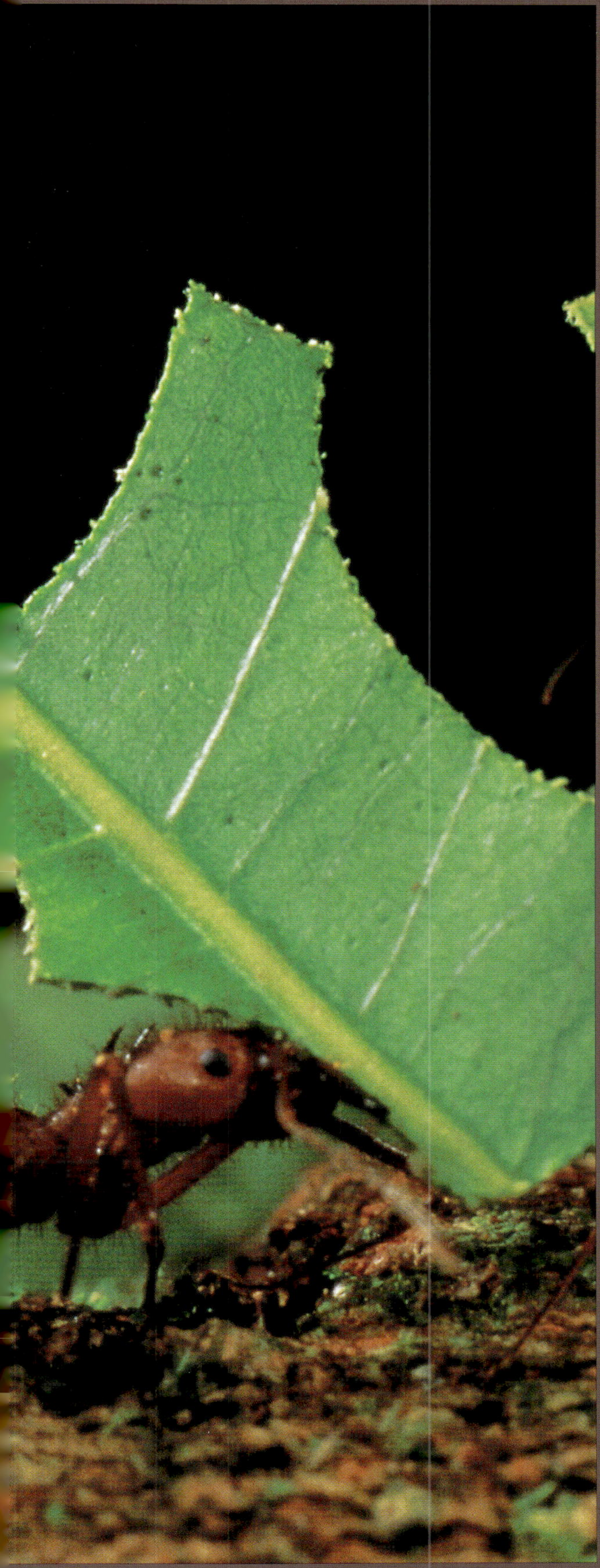
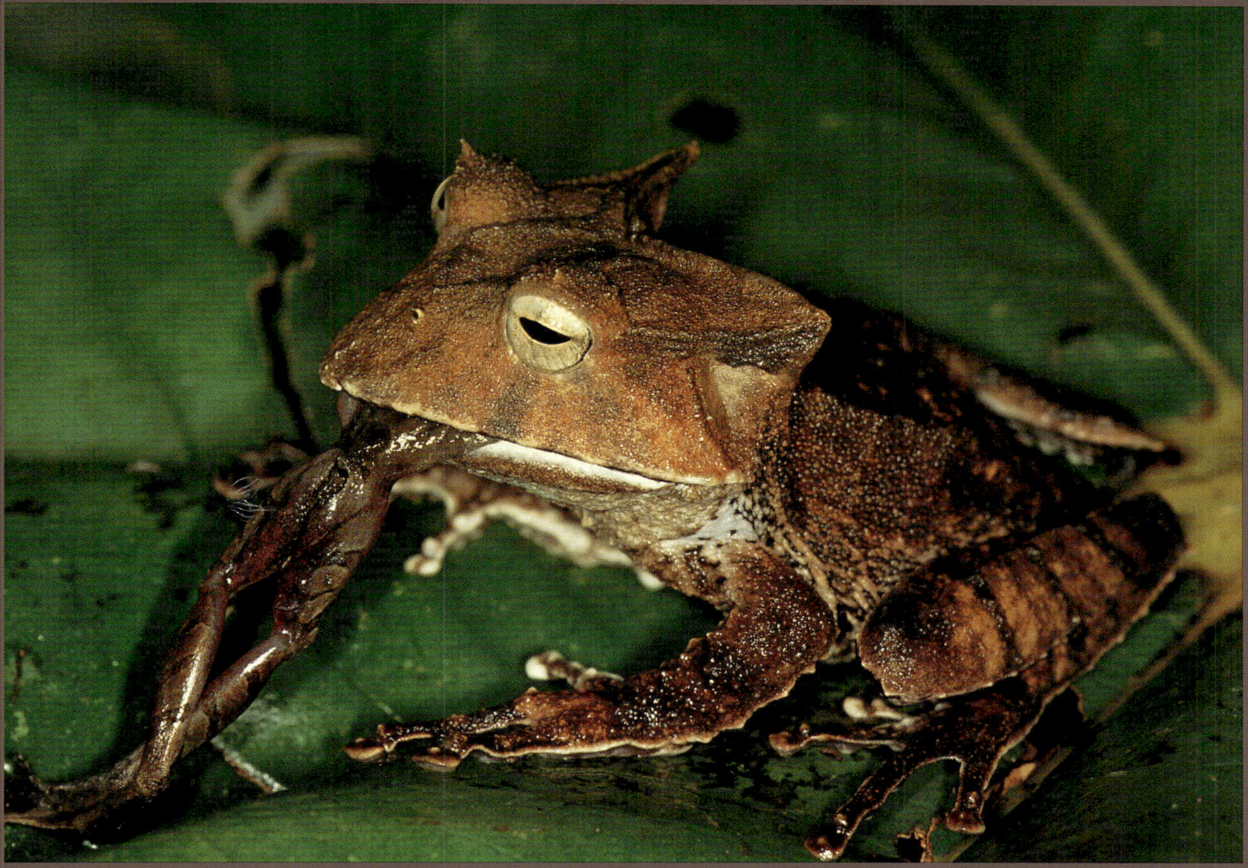
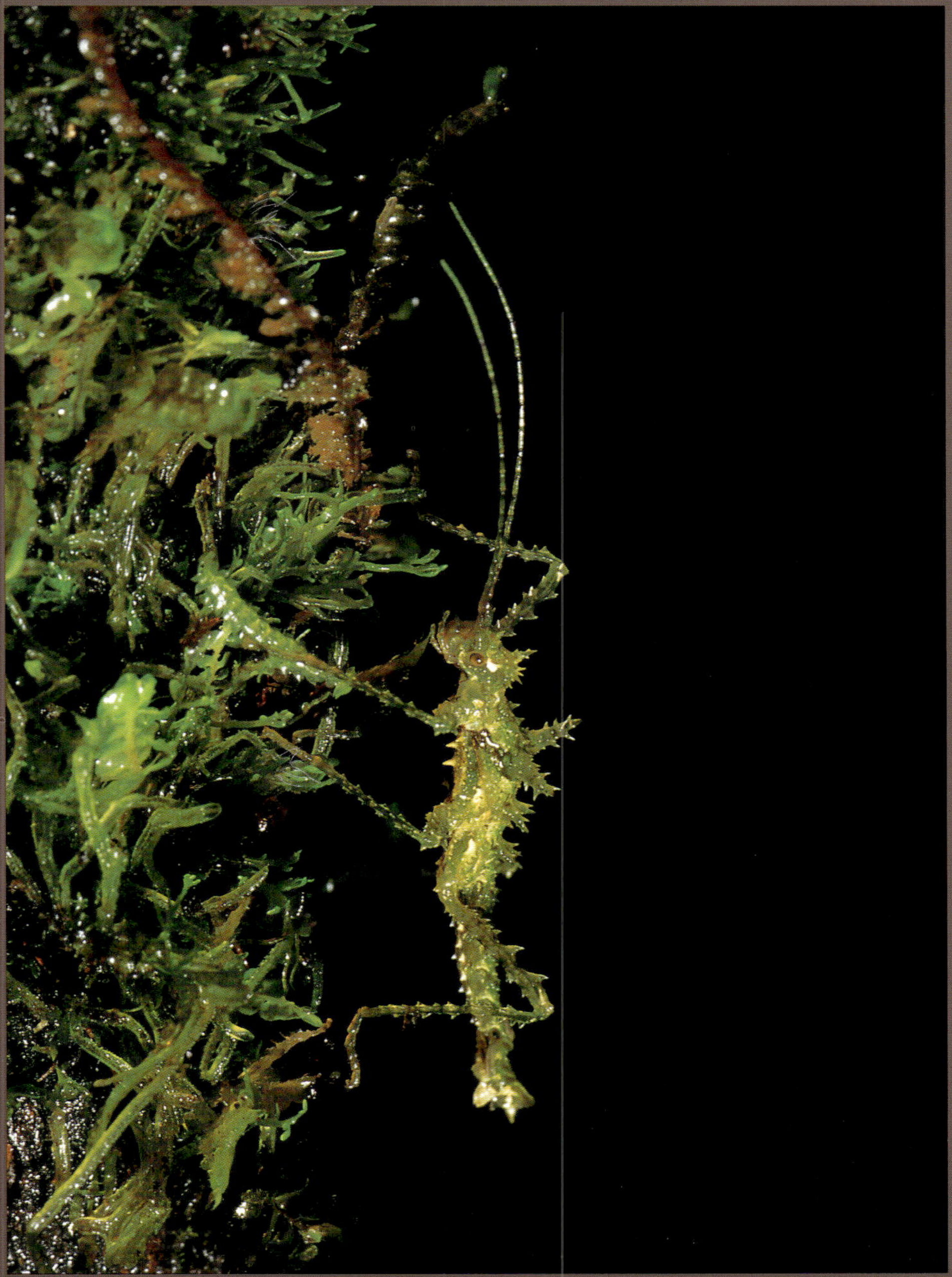

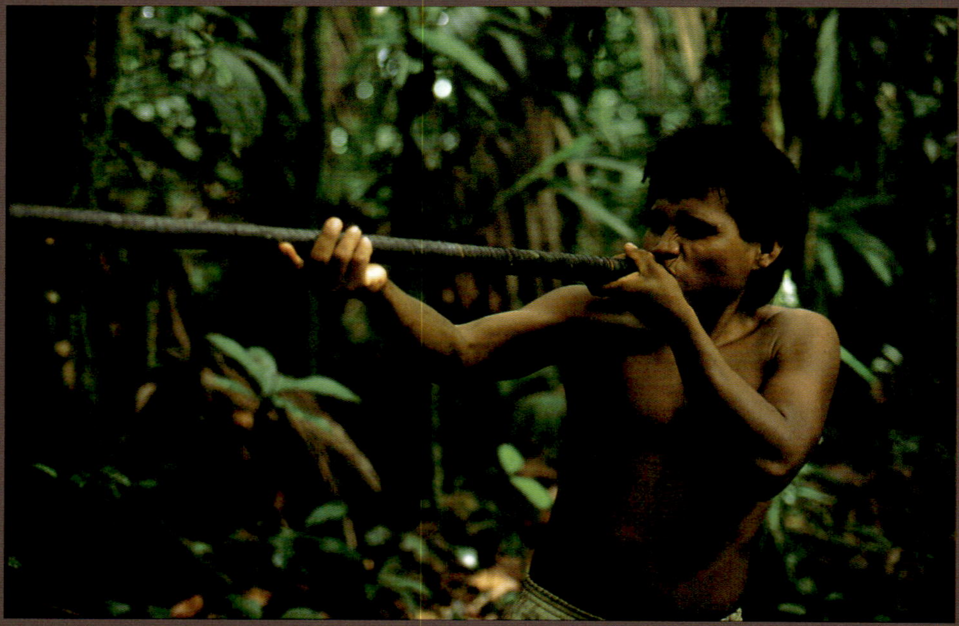

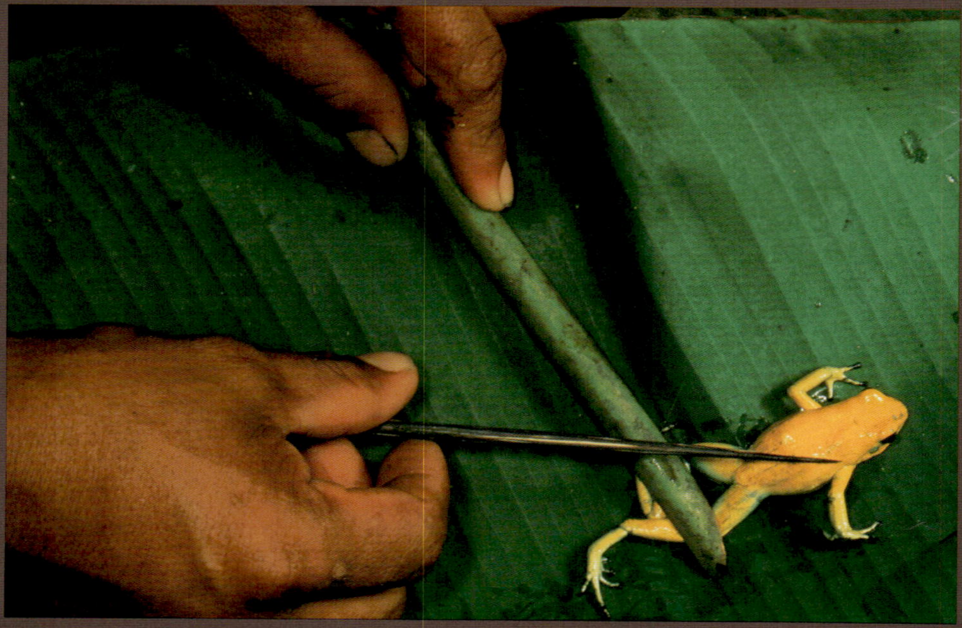

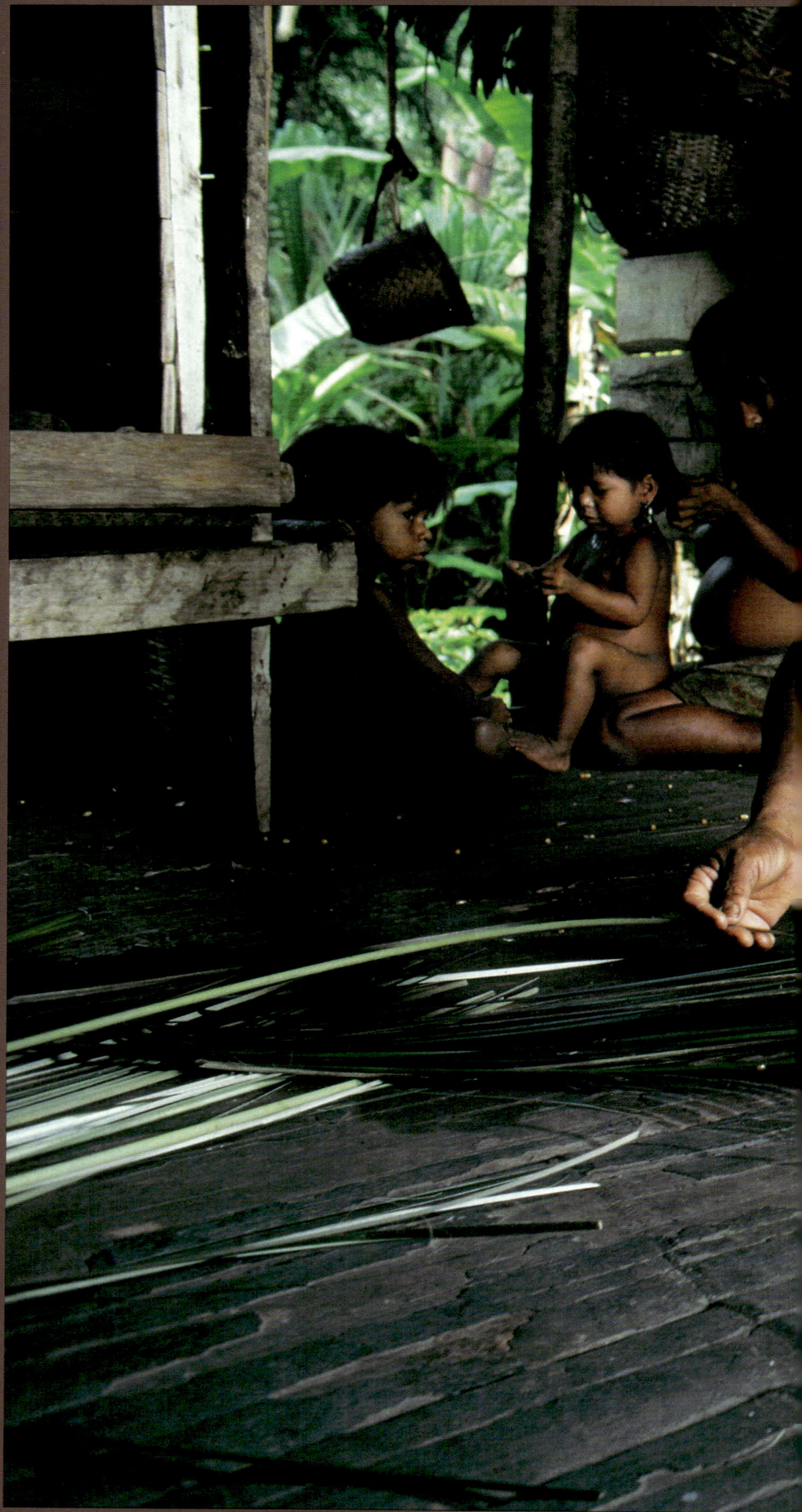

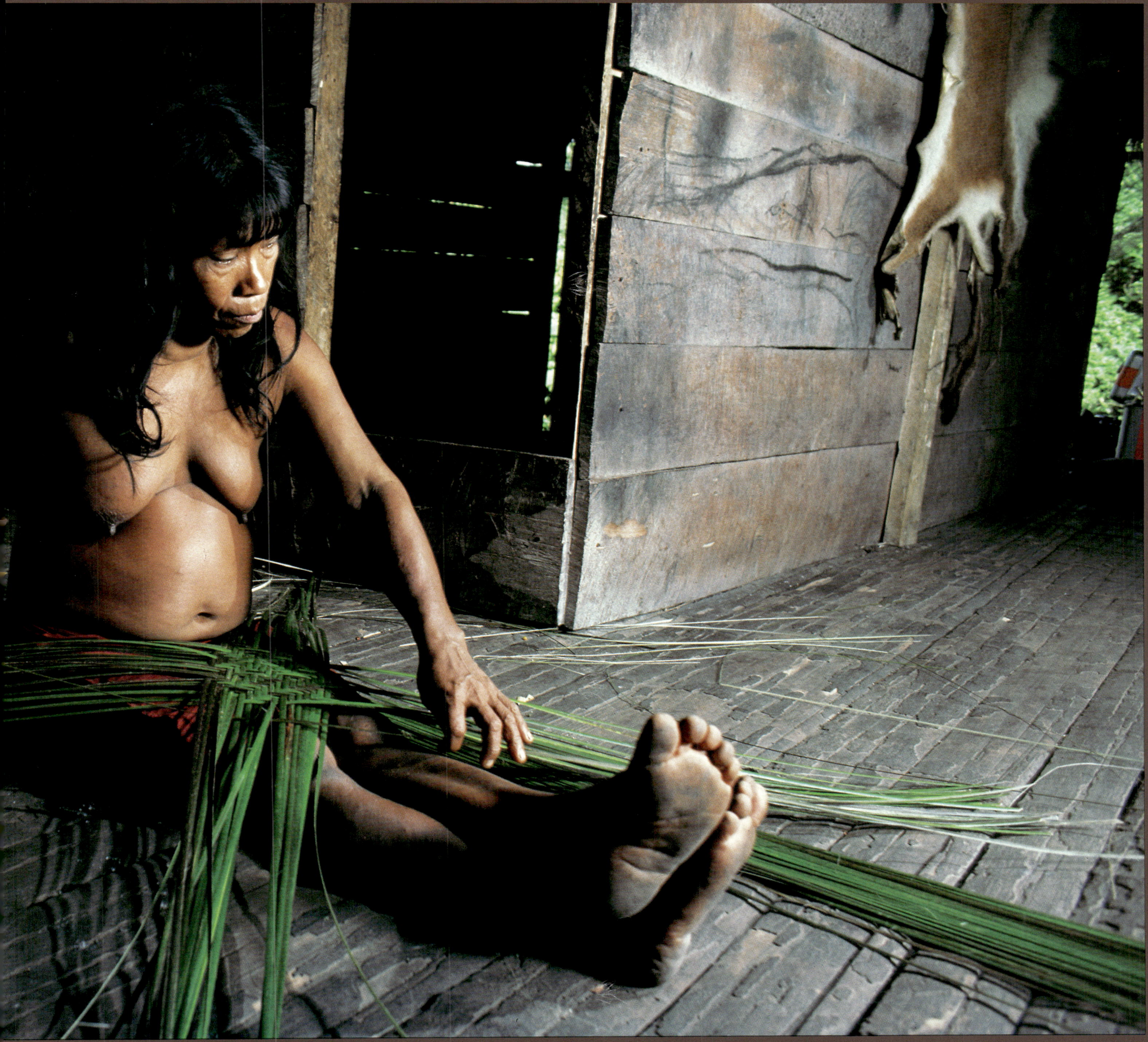

The Emberá Indians are a large group living inland in the Chocó region on the Pacific side of Colombia. Traditionally blowgun hunters, most have now adopted shotguns. Nonetheless, the blowgun does persist in several remote areas.

Upper left, an Emberá in the upper Saija River prepares to shoot a blowgun dart, which has been poisoned by rubbing it on the back of a yellow poison dart frog, **Phyllobates terribilis**, *lower left. Above, an Emberá woman making a basket out of strips of palm fiber.*

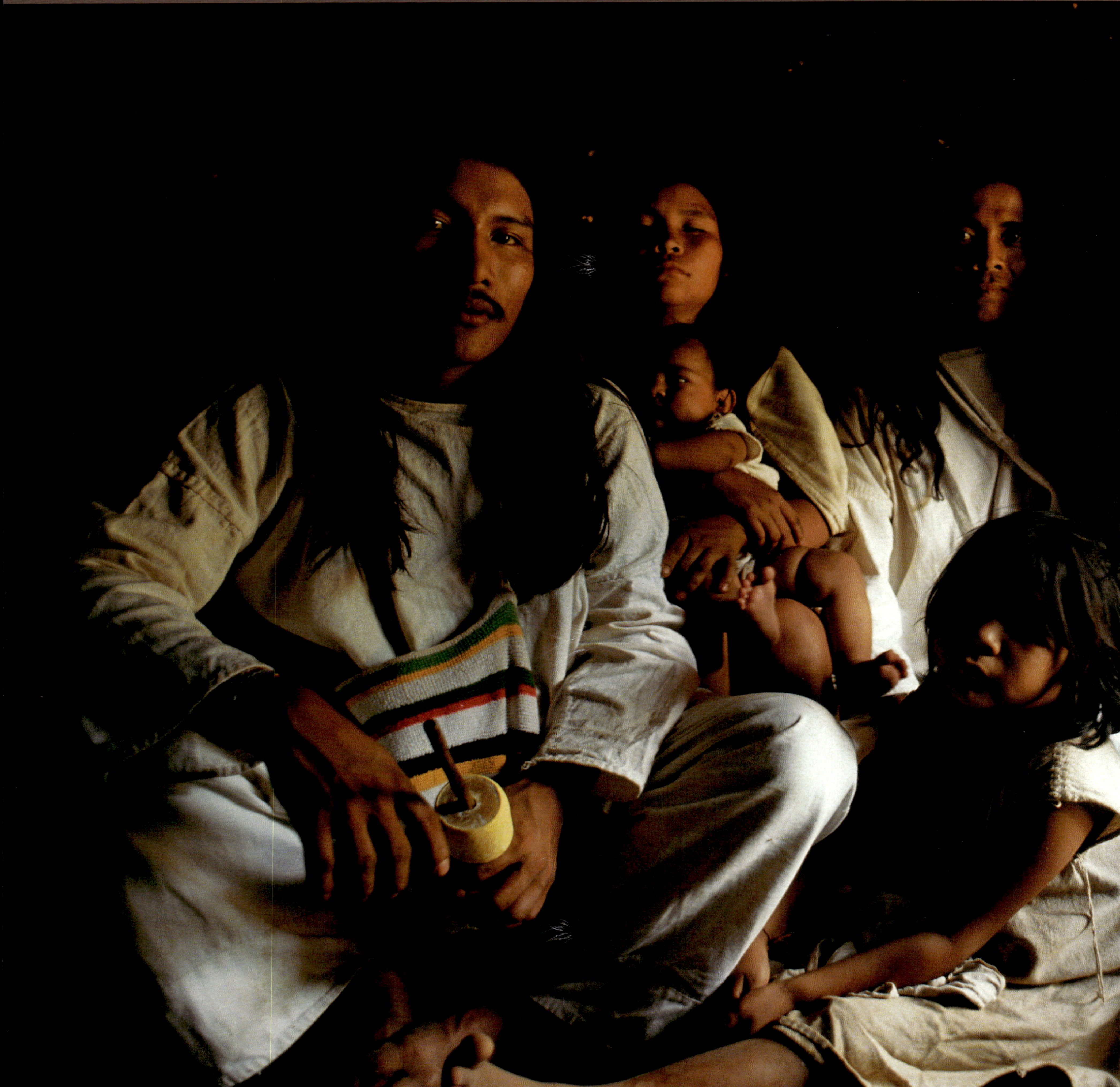

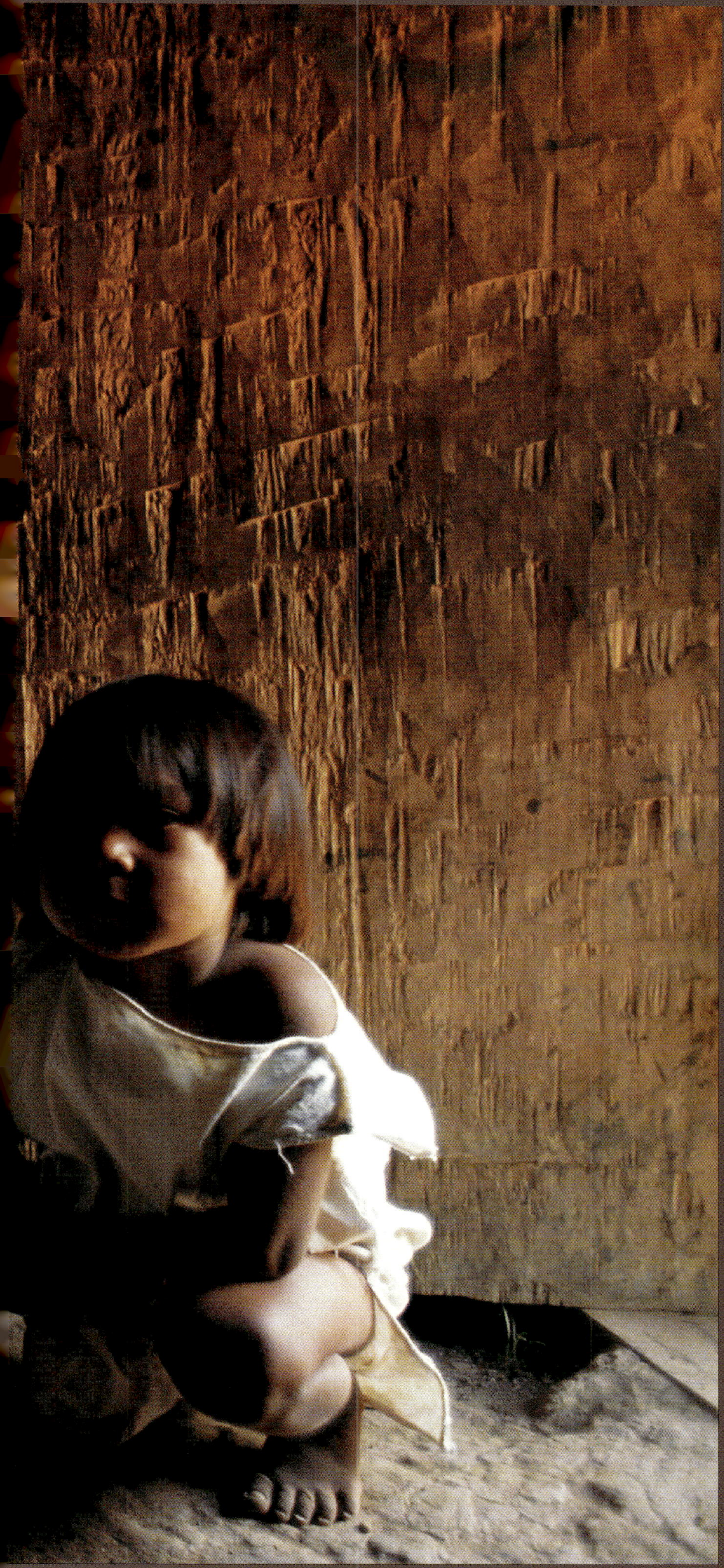

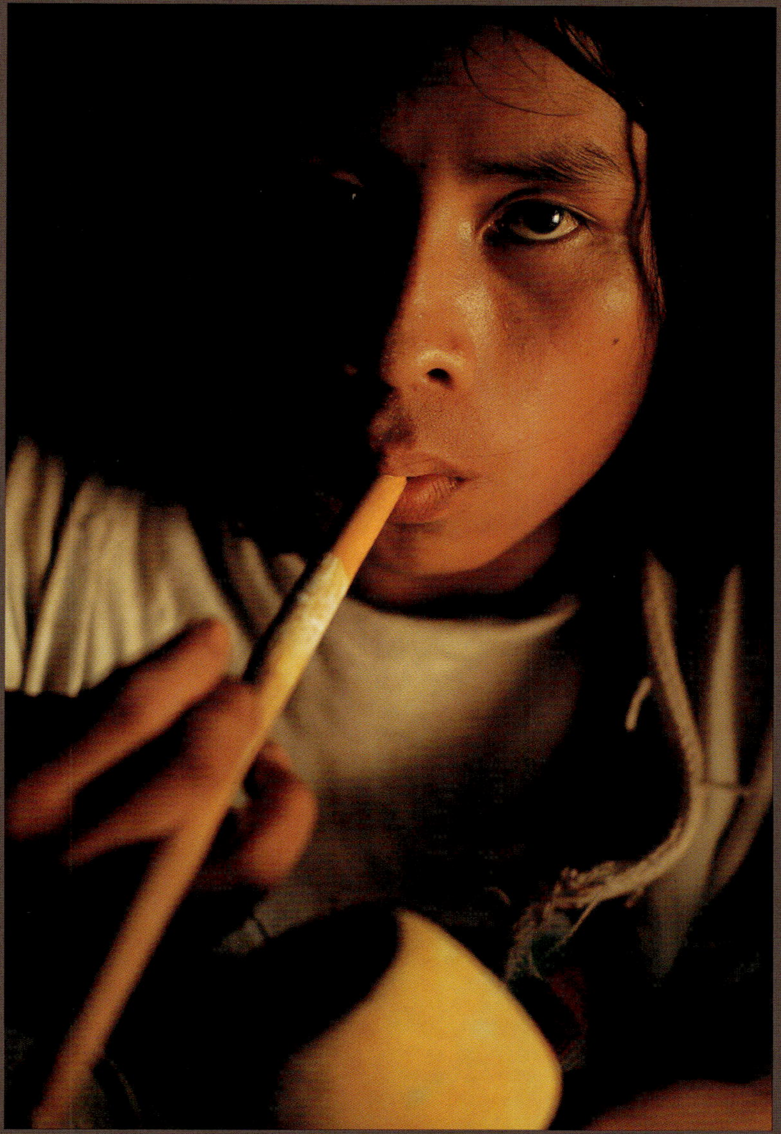

Left, a Kogui Indian family
in the Sierra Nevada de Santa Marta.
The Kogui are descendants of the Tayrona culture,
and are a very traditional people that largely shun
contact with the outside world.
The poporo *or traditional lime container, above, is used*
in coca chewing, and is the prized possession
of adult Kogui men.

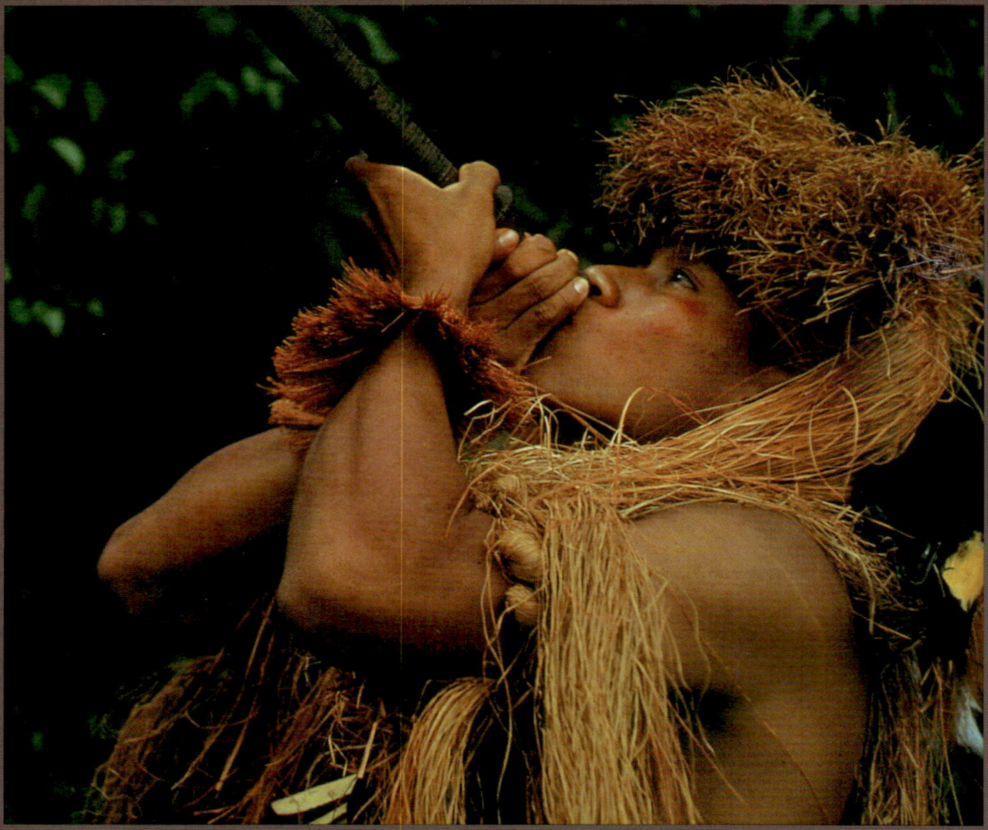

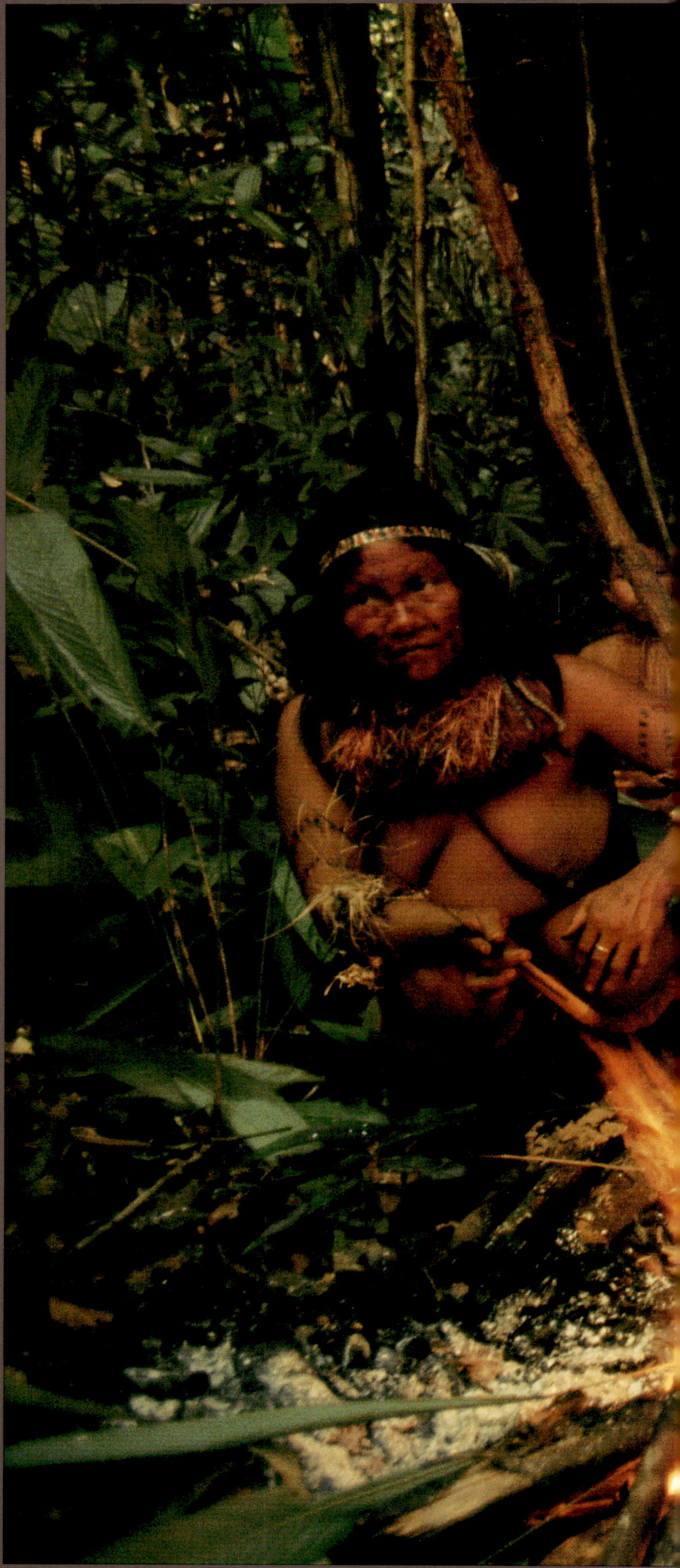

Above, a Yagua Indian in traditional palm-fiber costume, shooting a blowgun.
Right, Yagua Indian family around a campfire with their blowguns. The Yagua live on and near the upper Amazon on the Peruvian-Colombian border.
They still manufacture and use blowguns, an unusual style of hunting that is disappearing throughout Amazonia.

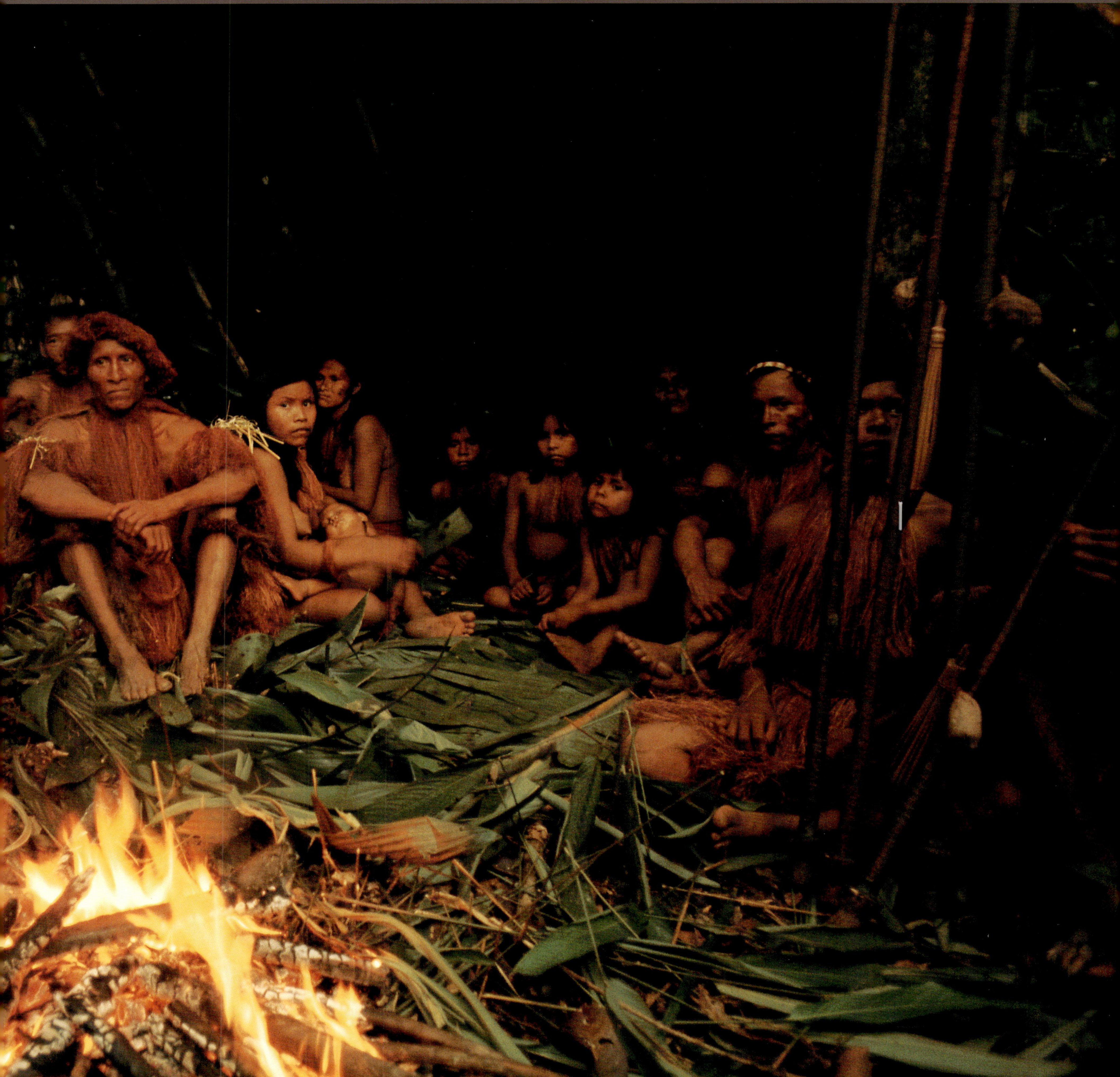

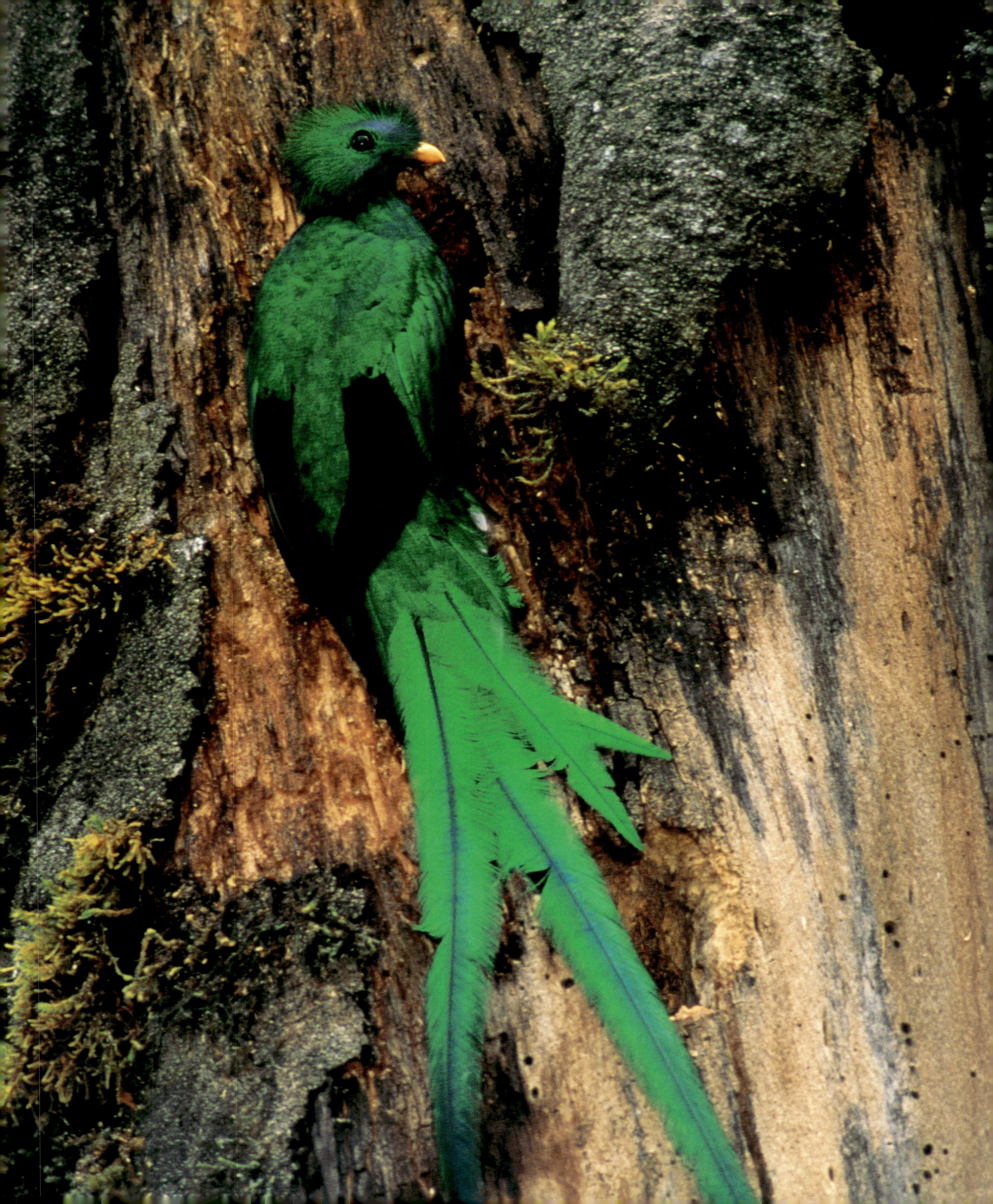

MEXICO

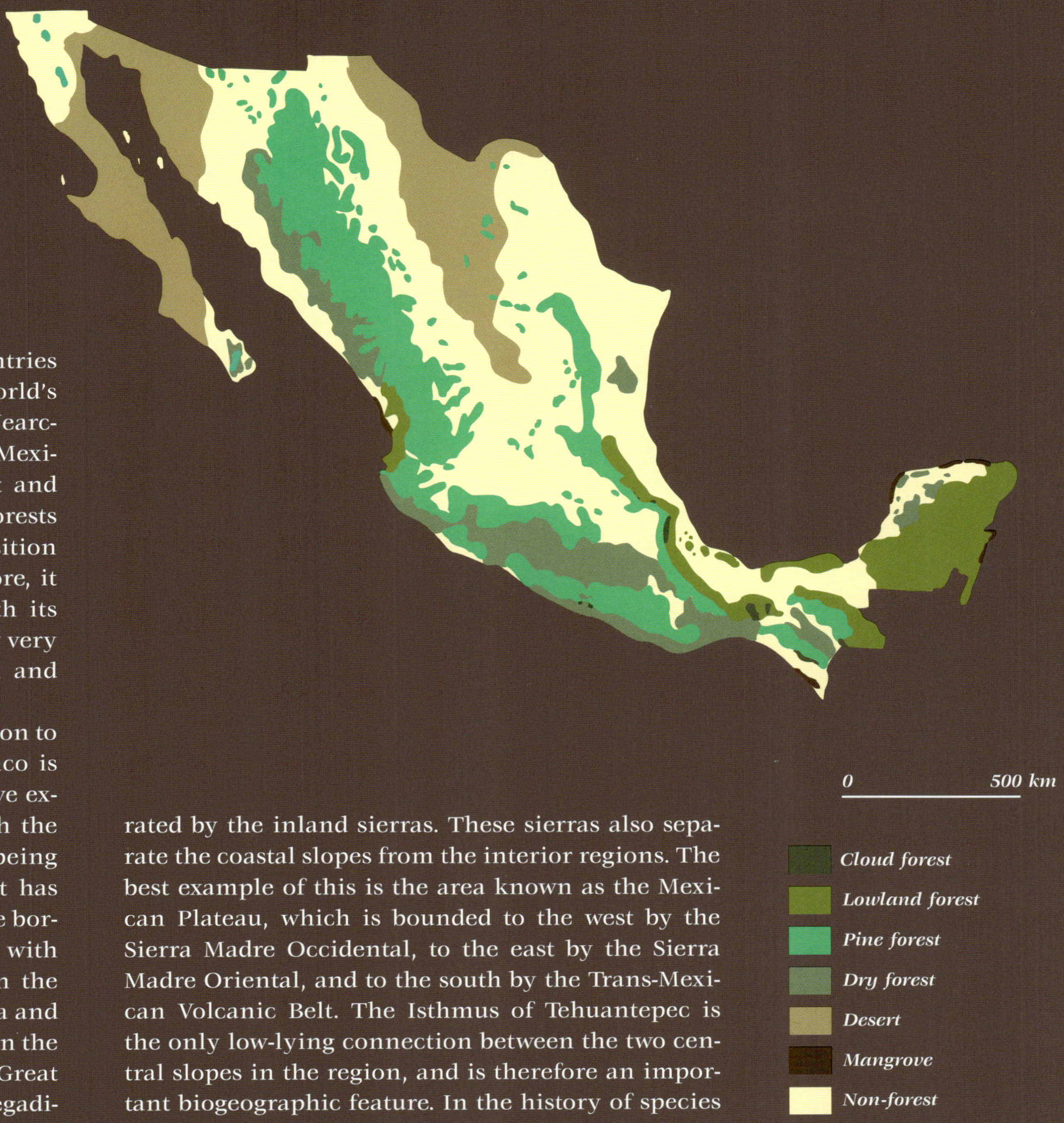

Mexico is one of the top five megadiversity countries and one of only three to straddle two of the world's major biogeographic regions, in this case the Nearctic and the Neotropical regions. Consequently, Mexico mixes elements of North American desert and pine-oak forests with the rain forests and dry forests of Central America, and also includes a transition area with high levels of endemism. Furthermore, it has a varied topography which, together with its great biogeographic diversity, gives this country very high species diversity, high generic diversity, and very high endemism in certain groups.

Marine systems also make a major contribution to Mexico's overall biodiversity. First of all, Mexico is one of the three megadiversity countries to have extensive coastlines and marine systems in both the Atlantic and Pacific Oceans (the other two being Colombia and the United States). Secondly, it has one of the few marginal seas entirely within the borders of a single country, the Gulf of California, with its high levels of diversity and endemism. On the Caribbean side, it shares with Belize, Guatemala and Honduras the second longest coral reef system in the world, an extensive area exceeded only by the Great Barrier Reef of Australia. Indeed, of all the megadiversity countries in the western hemisphere, there is no doubt that Mexico is the richest in marine diversity, and is surpassed only by Asia-Pacific nations like Indonesia, Australia and perhaps Papua New Guinea. Its marine fish fauna is indicative of this; estimates vary from between 2 000 and 2 500 (L.T. Findley, pers. comm.) to as many as 3 000 to 3 500 marine fish species in all (P. Almada-Villela, pers. comm.).

A medium-sized megadiversity country, Mexico is the seventh largest on the list, with an area of 1 972 544 km^2. Mexico's human population of 95 million is quite high, especially when considered per unit area, and this has put a lot of pressure on its natural areas. Many of these have already been severely degraded, which is why much of southern, southeastern, and central Mexico falls within the large and globally very important Mesoamerican Forests Threatened Biodiversity Hotspot.

Mexico has a highly intricate topography which often results in extremely localized species distributions. There are two coastal plains (Gulf of Mexico and Pacific coast) of varying width, which are separated by the inland sierras. These sierras also separate the coastal slopes from the interior regions. The best example of this is the area known as the Mexican Plateau, which is bounded to the west by the Sierra Madre Occidental, to the east by the Sierra Madre Oriental, and to the south by the Trans-Mexican Volcanic Belt. The Isthmus of Tehuantepec is the only low-lying connection between the two central slopes in the region, and is therefore an important biogeographic feature. In the history of species dispersal, this isthmus has acted both as a barrier to highland species to the north and south, and as a bridge for species from the Pacific and the Atlantic slopes.

Glaciations have also played a major role in the endemisms seen today in Mexico. When the climate experienced overall cooling, much water became locked up as ice, resulting in a drier climate. Humid forest remained only in the wettest areas, such as the Selva Lacandona. An example of how this influenced endemism and distribution is provided by the birds of the arid regions, which were forced into the few remaining warm refuges, such as the Balsas drainage or the Pacific side of the Isthmus of Tehuantepec. On the other hand, species of temperate areas ranged more widely during glaciations. However, with a warming of the climate and retreat of the glaciers, the temperate species were forced to find refuges of suitable habitat, usually on higher slopes, such as the Sierra Madre del Sur of Guerrero and Oaxaca, or Los Tuxtlas in southern Veracruz. Species isolated in these various refuges evolved and diverged. In addition, for long periods in the geological past, higher

On the opposite page, the resplendent quetzal (**Pharomachrus mocinno**) *is considered by some to be the most beautiful bird of the Americas. It also had great significance to the Aztec and Maya cultures. The headdresses of Aztec and Maya kings were adorned with quetzal feathers, and killing a quetzal in these cultures was punishable by death.*

sea levels separated Central America from the vast areas of both North and South America, allowing species to develop in isolation.

In terms of species diversity and endemism, Mexico usually ranks close to the top of the global list. It has the highest diversity of both pines (55 species) and oaks (138 species), the highest cactus diversity (834 species), and the highest numbers of marine mammals (33). It is sixth in total non-fish vertebrate diversity with 2 501 species. Its reptile diversity (717 species) is second only to Australia, and nearly 200 species higher than the third country on the list, Colombia. Its amphibian diversity is fourth worldwide (284 species), while its tiger beetle diversity (116 species) and butterfly diversity (2 237 species) are each sixth on the global list. Only in birds does it fail to make the top ten, its 1 050 species placing it twelfth. Freshwater fish diversity is reasonably high, with 468 species in 33 families (P. Almada-Villela, pers. comm).

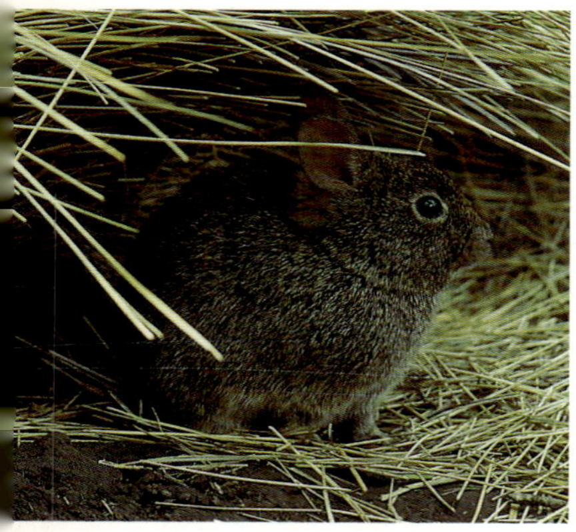

The volcano rabbit or teporingo (Romerolagus diazi) *is an unusual endemic species that can be found at the high volcano tips of the Trans-Mexican Volcanic Belt* (at the Popocatépetl and Iztaccíhuatl), *and is one of the smallest rabbits in the world.*

Endemism is even more impressive than overall diversity. Mexico ranks third on the world list in non-fish vertebrate endemism with 802 endemics. It is second in reptile endemism with 368 species (51%), again ranking behind only Australia; third in mammal endemism with 140 species (31%), behind only Australia and Indonesia; sixth in amphibian endemism with 169 endemics (60%); sixth in bird endemism with 125 species (12%), and seventh in tiger beetle endemism with 57 species (49%). Mexico also comes in fifth in Endemic Bird Areas as identified by BirdLife International (ICBP), with a total of fourteen.

Plant diversity also ranks very high, although precise data for the Mexican flora is not available. Estimates range from 18 000 to 30 000 species, with endemism at roughly 10 000 to 15 000 species. The high end of the diversity figure (30 000) would put Mexico in the top five with Brazil, Colombia, Indonesia and China. The high end of the endemism estimate (15 000) would put it in the top seven in this category. Mexico also has a large number of Centers of Plant Diversity and Endemism as identified by WWF/IUCN, its 17 centers placing it sixth on the world list.

Looking at its overall diversity, Mexico's combination of total species numbers, endemic species, endemic genera and families, great ecosystem diversity, major marine systems, and the mixing of two major biogeographic regions, puts it in a tight race with Australia for the title of fourth richest megadiversity country, after Brazil, Indonesia, and Colombia.

Mexico's human cultural diversity is also quite high, although estimates of total diversity vary. According to Martínez (1986; cited in Bye, 1993), there were 120 languages spoken at the time of conquest, and 54 of these remain today. Another estimate puts the total at 240 languages (Toledo, 1995), which may reflect dialect variations within languages. The higher figure places Mexico sixth on the list of megadiversity countries in this category, comparable to Brazil, the U.S.A., Australia, and the Democratic Republic of Congo, and clearly surpassed only by Papua New Guinea, India and Indonesia.

Natural habitats

Mexico is a country with a complex and intense geological history, one that has shaped its territory and people, and it is one of the countries in the world with the largest variety of natural habitats. From arid sand dunes to snowy peaks, its appearance changes from north to south and from east to west. Due to its geographic location, its vast coastal areas, and the two large cordilleras that divide the country, it has a wide variety of climates, and these have helped generate its globally significant biodiversity.

Six major areas of natural habitat can be identified in Mexico, and they are classified according to temperature and humidity, with altitude and latitude being important determinants as well. The first type of environment is tropical rain forest, which is found in the hottest and most humid regions of the country. This formation occurs as high and medium-high evergreen forests where the levels of moisture in the environment are so high that the vegetation never loses its foliage, and it includes many different kinds of humid forests. Mexico's tropical forests are found from southern San Luis Potosí to the Guatemalan border, including northern portions of the state of Hidalgo, southeastern Veracruz, and parts of Campeche, Quintana Roo, Oaxaca, Chiapas, and Tabasco. The northernmost extensions of rain forest are Los Tuxtlas in Veracruz, and the foothills of northern Oaxaca in the vicinity of Tuxtepec. Los Tuxtlas, rising from the coastal plain to 1 600 m, is also a minor center of endemism.

One of Mexico's most important regions of biological diversity, and one of the country's top conservation priorities, is the Selva Lacandona of eastern Chiapas. This montane tropical forest once covered 13 000 km^2 in the western watershed of the Usumacinta River. Today, less than 5 000 km^2 of this forest remain, yet it continues to provide habitat for a wide variety of Mexico's natural resources. Mexican ecologist Rodrigo Medellín points out that the Selva Lacandona is probably the country's most diverse ecosystem.

"It is," he writes, "the largest remnant of tropical rain forest in Mexico and is part of the largest expanse of that vegetation type in Central America" (1994). The Selva Lacandona has long been noted for its diversity of arthropods and plants; more recently, Medellín has stated that the Selva Lacandona's mammal fauna represents 25% of the country's total. Bats, with 64 species, account for many of these mammal species. Combined with high numbers of marsupials, rodents, and carnivores, the bats make the Selva Lacandona second only to Cuzco Amazónico, Perú in terms of the number of mammal species in a single Neotropical site (1994).

Most of these species are protected today in the 3 300-km^2 Montes Azules Biosphere Reserve, established by the Government of Mexico in 1978. In 1992, a

Presidential Decree added an additional 550 km^2 to the reserve, and the Lacandón, Tzeltal, and Chol Maya inhabitants of the Comunidad Lacandona recently created a community reserve, La Cojolita, which geographically connects the Montes Azules Biosphere Reserve to a similar one, the Maya Biosphere Reserve, in neighboring Guatemala.

Unfortunately, most of the forests of the humid southeast have now been largely cleared, mainly to provide grazing land for cattle. Today, species such as the cattle egret (*Bubulcus ibis ibis*), white-collared seedeater (*Sporophila torqueola*), and great-tailed grackle (*Quiscalus mexicanus*) are usually associated with an area that some fifty years ago supported harpy eagles (*Harpia harpyja*) and scarlet macaws (*Ara macao*).

The second form of vegetation type is tropical dry forest, which in Mexico is the largest in surface area and owes its presence to a well-defined dry season that can last 5-8 months. Here, dry forest extends from altitudes of sea level up to 1 500 m, with the dominant formation being low deciduous forest; it is lower in height than rain forest and loses its leaves during the driest months. This type of forest is located along the Pacific coast from southern Sonora to Chiapas, close to the Guatemalan border, with a notable extension into the center of the country in the Balsas River Basin; it is also found in parts of Baja California Sur and in the north of the Yucatán Peninsula. Along the Gulf Coast, it is spottily distributed in the states of Tamaulipas, San Luis Potosí, and northern Veracruz.

The vegetation in tropical dry forest is green and luxuriant during the rainy season and gray and leafless in the dry season. The tallest trees in the dry forest do not exceed 15 m, and the dominant plants are shrubs and low-branching trees. Another noteworthy characteristic of this type of vegetation is the large number of lichens attached to the tree trunks. One of the most spectacular species in this forest is the corpulent guanacaste tree (*Enterolobium cyclocarpum*), typically found in the driest areas of evergreen forests, in sub-evergreen forests, and in the most humid parts of dry forests.

In Latin America, the only regions where one can still find large areas covered by dry forests are located in Mexico and Costa Rica. In Mexico, the Chamela-Cuixmala Biosphere Reserve was created in 1993. The reserve protects over 13 000 ha of this ecosystem and a series of adjoining wetlands on the coast of Jalisco.

The next two environments are found in the temperate setting, with a cooler climate, lower plant diversity, and simpler structure. In the more humid areas of these regions, the forests are so rich in species that they sometimes resemble rain forests; in the cooler and drier areas, pine and oak forests predominate. The first type of temperate forest is known as *bosque de niebla* (cloud forest), which occurs in humid, temperate climates with greater moisture, especially in the form of fog. These cloud forests reach their northernmost distribution in Mexico, where they occur along the Atlantic slope in parts of the mountainous regions of Tamaulipas, Veracruz, and Oaxaca. On the Pacific coast, they are found intermittently from Sinaloa to Chiapas.

The trees in this humid environment are usually enormous, exceeding 50 m in height, with trunks more than two meters thick. Although some shed their leaves during the driest part of the year, the community as a whole is considered evergreen. The understory of this forest is characterized by a great profusion of palms and ferns, notably the majestic tree fern, which can reach heights of 15 to 20 m and is the most outstanding feature of this stratum. The exuberant profusion of epiphytes and semiepiphytes is another characteristic of this forest. Because of their latitude, and in sharp contrast to those of South America, Mexico's cloud forests are characterized by the presence of many northern species, such as the sweet gum, certain oaks, the *palmillo* and the *talauma*; others are clearly tropical, like dwarf avocado and trumpetwood; still others are endemic to Mexico, such as the *árbol de las manitas*. This type of ecosystem occupies only about 1% of the country's land area, yet has a rich flora and plays a crucial role in protecting watersheds. It clearly deserves greater attention from the scientific community and more formal protection before it is all turned into grazing land and coffee plantations.

The second important temperate environment is the pine-oak forest. These communities, typical of the mountainous regions of Mexico, are found throughout the country, with the exception of Yucatán and Quintana Roo. Within this category are a number of different plant associations that range from pure stands of pine, to mixtures of pine and fir, to mixtures of conifers and oaks as well as other broad-leaved plants, to more or less pure stands of oaks. The forests with oaks as a dominant element are distributed from sea level to altitudes of around 3 000 m. Coniferous forests can be found even higher, but not at sea level. At higher altitudes, pines become deformed and dwarfed due to the effects of wind and cold temperatures. Pine-oak forests appear in their physiognomy to be typical temperate forests. They can be evergreen when the dominant species are pines, or deciduous when the dominant vegetation is composed of broad-leaved elements. As indicated earlier, both pines and oaks reach their highest global diversity in these Mexican forests.

Unlike the more humid rain forests and the dry forests, the pine-oak forests are rather poor in epiphytes. The few representatives of this group include bromeliads and lichens, but their abundance and diversity cannot be compared to that of the more humid ecosystems. There is, however, a group of plants that lives on the trunks and branches of pines, but with a totally different lifestyle from that of the epiphytes. These are the mistletoes, a group of parasitic plants that are quite common in these forests.

Humid pine-oak forests and cloud forests once covered much of the Chiapas-Guatemala highlands, but today they are restricted to the steeper, higher and more remote slopes, since much of the land has been cleared

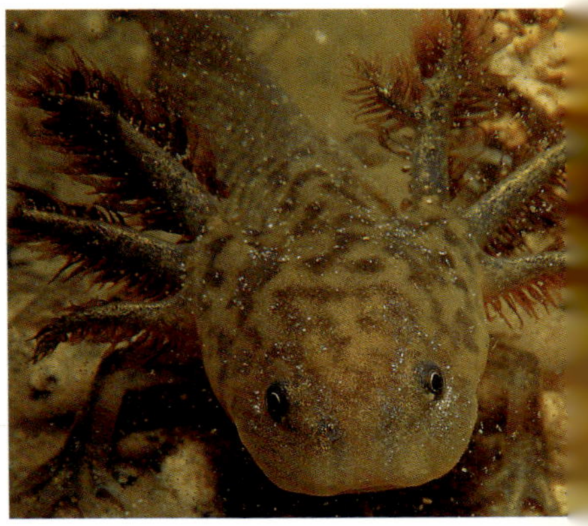

The unusual axolotl or ajolote (Ambystoma mexicanum) *is considered by some to be one of Mexico's most famous species. This salamander has a unique feature of neoteny, that is, it does not metamorphose from the larval stage to an adult salamander. In this species, the larval stage is the adult stage.*

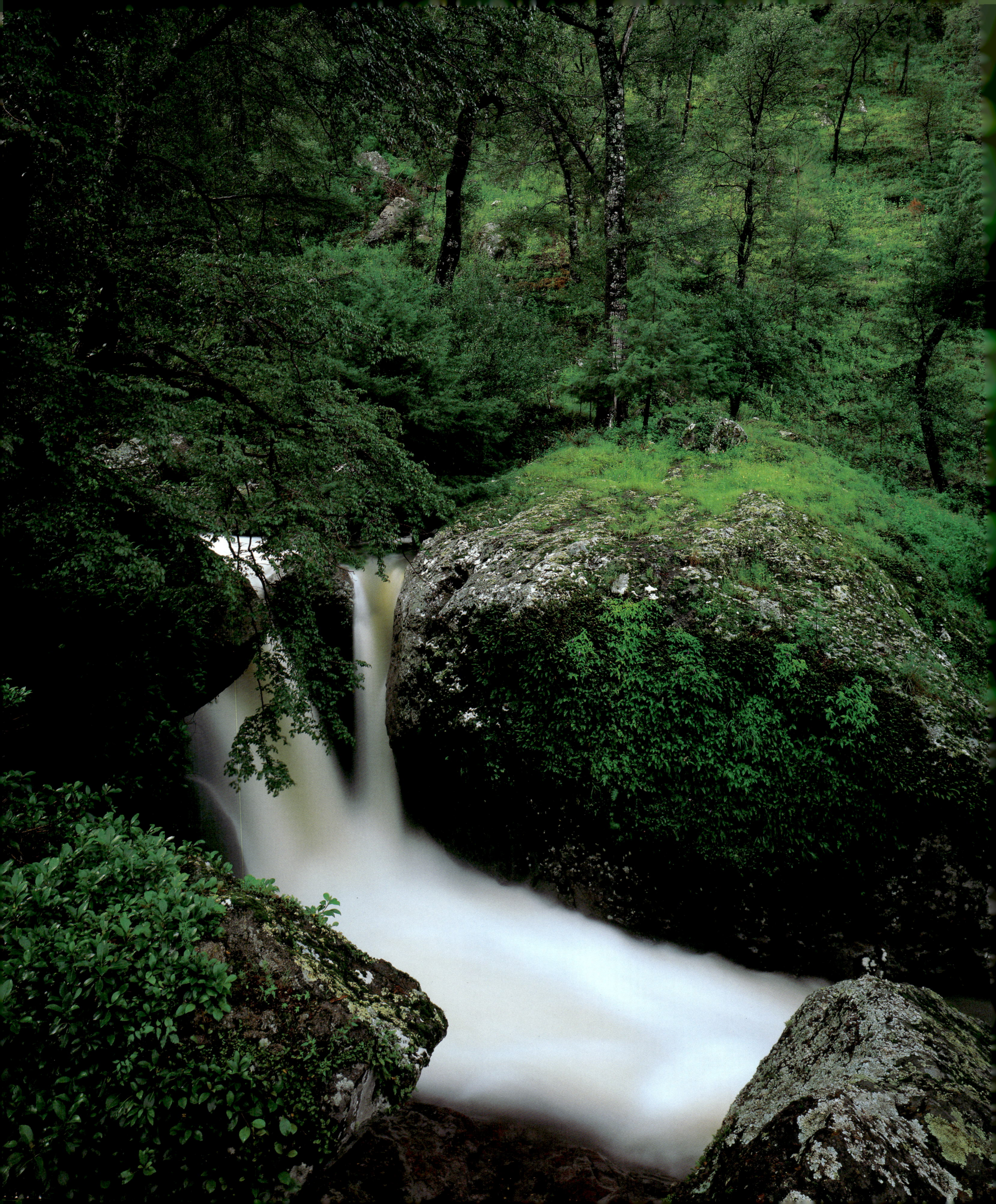

for agriculture. Here the thick vegetation consists mostly of evergreen trees, from whose branches cling a number of mosses, orchids, and bromeliads, a kind of hanging garden surrounded by tree ferns. The high diversity of these forests is partly due to the steep slopes (some of which reach angles of 60° or more) surrounded by deep and abrupt gorges that are barriers to gene flow. Other important factors that have made this high biological diversity possible are the convergence of the two major hydrographic basins of Chiapas, the Pacific coast and Usumacinta-Grijalva drainages, as well as the different types of soils found here, representing the peculiar geological history that turned this place into a refugium during Pleistocene glaciations. The Chiapas-Guatemala highlands harbor several endemic species of birds, such as the horned guan (*Oreophasis derbianus*) and the bearded screech-owl (*Otus barbatus*). Among the most endangered mammals of this region are the jaguar (*Panthera onca*), the tapir (*Tapirus bairdii*), and the spider monkey (*Ateles geoffroy vellerosus*).

Another noteworthy region where temperate forests occur is on the peaks of the Trans-Mexican Volcanic Belt, from the Pico de Orizaba in the east to the Colima Volcano in the west. These crests rise as forested islands from the dusty agricultural plains at the southern end of the Mexican Plateau. Oaks are the dominant element on the lower slopes, but as one goes higher, pine-oak and then pine forest with firs takes over on the cooler and wetter slopes. The peaks of the highest volcanoes are barren and rocky, covered with snow throughout the year. Both the highland forests and the scattered lakes and marshes of the Volcanic Belt are important centers of endemism, although many of the central highland species also range north into the Sierra Madre Occidental and south into the sierras of Guerrero and Oaxaca.

In the area near the Valley of Mexico one finds the famous pair of volcanoes called Popocatépetl and Iztaccíhuatl. Volcanic activity in this region started in the Tertiary period, but it was not until the Miocene that Iztaccíhuatl appeared. Nowadays its upper limits, reaching a height of 5 386 m, are permanently covered with snow. There are still twelve glaciers on this volcano. In the Pliocene, Popocatépetl, the other important volcano of the area, was formed. It is the youngest volcano in the area and the second highest (5 452 m). This volcano, along with the Iztaccíhuatl and the Pico de Orizaba (the highest volcano, at 5 747 m), are the only three mountains in Mexico that still have glaciers. This area, which has been declared a national park, has two types of environments: a subhumid temperate climate, in which high temperatures vary between 6.5° and 22°C and there is a rainy season during the summer from June to November; and a cold climate, which occurs above 4 000 m. Vegetation is clearly distributed according to these two climates. The *oyamel* forests (*Abies religiosa*) are found from 2 700 to 3 300 m above sea level, and are associated with two species of pine (*Pinus montezumae* and *P. ayacahuite*). At approximately 3 500 m, the *oyamel* forest begins to merge with the pine forest dominated by *Pinus hartwegii*, a very cold-resistant species that in the higher areas coexists with juniper (*Juniperus monticola*) and several species of grass, the most dominant being *Festuca tolucensis*.

The northern sierras of the Baja California Peninsula are closely related to the mountains of southern California and support several species that occur nowhere else in Mexico. Among the species that once were found in this region was the California condor (*Gymnogyps californianus*), which is now extinct in this part of its range.

Deserts and grasslands are two natural habitats characterized by an absence of trees; these two types of environments cover a vast portion of Mexico. At higher altitudes where the cold climate doesn't allow for tree growth is a vegetation type dominated by grasslands known as *zacatonales*, with scattered small bushes. In these grassy areas, the most notable mammals include several species of moles, squirrels, mice (*Microtus* spp., *Peromyscus* spp.), and other rodents. Populations of coyote (*Canis latrans*), white-tailed deer (*Odocoileus virginianus*) and mountain lion (*Felis concolor*) have been drastically reduced, but still persist in some areas. A particularly noteworthy species is the volcano rabbit or *teporingo* (*Romerolagus diazi*), an endemic monotypic genus found only in the high mountain grasslands of the Volcanic Belt. This small lagomorph is one of Mexico's most important flagship species.

A wide variety of microhabitats exist in these arid and semiarid vegetation associations, but all share one common feature, the low average height of the vegetation (less than 4 m); nonetheless, in some instances there may be taller individual plants such as the saguaros and yuccas in the desert, and occasional pines and oaks in the grasslands.

The dry and hot conditions of the desert provide the perfect environment for a profusion of leafless species and species with small leaves. These typical desert plants, including the Cactaceae, Crassulaceae, Agavaceae, and other related families, are characterized by thorns and succulent leaves or stems, and come in an astounding array of sizes, shapes and colors.

The deserts and grasslands of Mexico cover an area of approximately 50% of the country, and can be found in a wide variety of ecological conditions determined by the amount of rainfall, the degree of soil drainage, the temperature and its seasonal variations, and the local topography. Generally, these types of environments, which range from scrublands to grasslands and low thorny forests, occur in the northern half of the country, but they can also be found under special conditions in some southern states like Oaxaca and Puebla. In areas of higher humidity, the semiarid landscape merges into deciduous tropical forest.

The most noteworthy desert regions of Mexico are the Baja California Peninsula, the Sonoran Desert and the Chihuahuan Desert in the northwest. Most of Baja California is covered by rugged, inhospitable deserts.

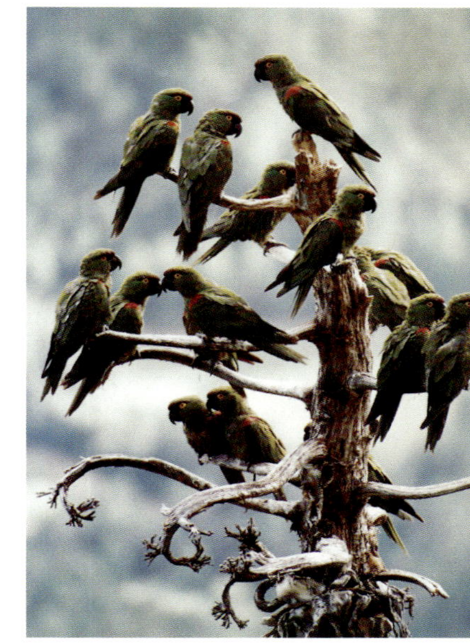

On the opposite page, pine-oak forest at Cerocahui River Falls in the Urique Canyon complex, Barranca del Cobre National Park, in the Sierra Madre Occidental in Chihuahua. Mexico has the highest diversity of pines and oaks in the world.

Above, the maroon-fronted parrot (Rhynchopsitta terrisi) *is a species of the Sierra Madre Oriental and is endemic to Mexico.*

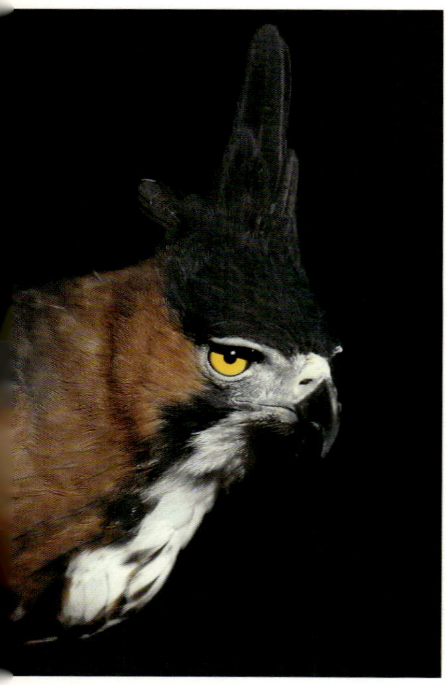

Bird diversity in Mexico is relatively high with almost 300 species more than the United States. Among the most spectacular species are the ornate hawk eagle (Spizaetus ornatus) *and the King vulture* (Sarcoramphus papa). *Above, the ornate hawk eagle is a powerful predator and ranges from Mexico to Argentina, mostly in rain forest habitats.*

The birds are usually the most interesting representatives of the fauna, but some large and important mammals like the mountain lion (*Felis concolor*), the bobcat (*Lynx rufus*), and the pronghorn (*Antilocapra americana peninsularis*) occur as well. This Baja subspecies of the pronghorn, locally known as *berrendo*, is considered to be very endangered in Mexico, and is another important flagship for the country. Some of the desert islands in the Gulf of California are an important breeding area for seabirds, such as the least storm petrel (*Oceanodroma microsoma*), Heermann's gull (*Larus heermanni*), and the elegant tern (*Sterna elegans*), which have their most significant breeding grounds on these islands.

Some of the offshore Pacific Islands are also centers of endemism. These include Guadalupe Island, which harbors the only colony of Guadalupe fur seals (*Arctocephalus townsendii*) known in the world; the Revillagigedo Islands (San Benedicto, Socorro, Roca Partida, and Clarión, and other smaller islands), and Clipperton Atoll. Like many offshore islands, Mexican islands have a high percentage of endemic species and subspecies.

The Sonoran Desert covers an area of 126 256 km^2, and is one of the most interesting and diverse deserts in the world. It harbors a variety of plant and animal communities, and is especially rich in cactus diversity. Large portions of this desert are covered by gobernadora (*Larrea divaricata*) and burrobush, or bur sage (*Franseria dumosa*). These plants have adapted to the extremely arid conditions by means of a large root system that allows the plant to absorb moisture from a wide area. Fauna associated with these plant communities include many reptiles, such as the leopard lizard (*Crotaphytus wislizenii*) and the desert iguana (*Dipsosaurus dorsalis*), as well as small mammals like the antelope jack rabbit (*Lepus alleni*) and the desert rabbit (*Sylvilagus audubonii*). Another interesting plant community is formed by the paloverde tree (*Cercidium microphyllum*) and the giant saguaro cactus (*Carnegiea gigantea*), typical of this region. The saguaro is the one of largest cacti in the world, and can reach a height of 15 m and a weight of 6 tons. As one moves south and inland, the desert gives way to deciduous thorn scrub and woodland. In southern Sinaloa and Nayarit, the thorn forest may grade into taller, tropical semideciduous forest, and into oak scrub, pine-oak, and pine forest in the Sierra Madre Occidental. Fingers of tropical forest extend north and inland along river valleys to southern Sonora and southwestern Chihuahua.

The northeastern desert plant associations are known as chaparral. This arid region is covered mostly by short, thorny bushes and has many species in common with the Sonoran Desert. It covers parts of the Sierra Madre Oriental and ranges from Coahuila to southern Nuevo León. On the northeastern portion of the Atlantic slope, where the Trans-Mexican Volcanic Belt nearly squeezes out the coastal plain, and in the north, where the plains bordering the Rio Grande merge into the Mexican Plateau, desert scrub gives way to deciduous woodland. This may grade into tropical semideciduous forest as one moves south and inland.

Two other categories of terrestrial habitats also exist, the wetlands, which are areas that are permanently or seasonally flooded, and the coastal regions, where the ocean meets the land, forming environments with very specific types of vegetation such as mangrove swamps and coastal lagoons.

A particularly noteworthy wetland is the oasis of Cuatro Ciénegas ("Four Marshes"), one of the world's few inland desert wetlands. Situated in a small valley in the Chihuahuan Desert, Cuatro Ciénegas covers only about 1 200 km^2 and is fed by underground springs. The origin of its water remains poorly understood, but it appears to emerge from deep aquifers. Also, since it comes from so far underground, it lacks sediments and is crystal clear, making for wonderful visibility. Not surprisingly, given its isolated location, it is particularly rich in endemic species. More than 60 endemics have now been cataloged from this tiny area, including cacti, snails, scorpions, fish, and reptiles. Perhaps the best known is the Coahuila box turtle (*Terrapene coahuila*), a relative of the wide-ranging North American box turtles. The Cuatro Ciénegas endemic represents an interesting stage in box turtle evolution in that it is intermediate in habits between its terrestrial relatives and its aquatic ancestors and, though dome-shaped like other box turtles, it is largely aquatic. Whether this is a retention of earlier aquatic behaviors or a secondary adaptation to the desert ecosystem in which it lives remains to be determined. Fortunately, this locality, which was under many different threats, was declared a Natural Protected Area in 1994, which will hopefully guarantee its survival over the long term.

On the Caribbean side, one finds the low-lying coast of the Yucatán Peninsula, which is characterized by long, sandy beaches. Along the north coast of the Yucatán, however, extensive lagoon systems support large numbers of nesting and migrant waterbirds. This northern area is one of great ecological and faunistic value. It includes the estuaries of Celestún, Dzilam de Bravo, and Lagartos. The low mangrove and deciduous forests found there create an intricate network of channels that provide refuge for a wide variety of animals. The formations known as *rías* are important as well, with the *Ría de Lagartos* being particularly noteworthy. It is located on the Gulf of Mexico and covers an area of 47 340 ha. The lack of human activities and its isolation make it an ideal habitat for several important species of birds, including a sizable population of flamingos (*Phoenicopterus ruber*). Mangrove forests in this area are composed of four species: the red mangrove (*Rhizophora mangle*), the black mangrove (*Avicennia nitida*), the white mangrove (*Laguncularia racemosa*), and the button mangrove (*Conocarpus erectus*).

In the area of the Caribbean there are also other wetlands of significant size such as Sian Ka'an in Quintana Roo, which is a Biosphere Reserve with marine, mangrove, and tropical rain forest ecosystems. Other such areas are the Centla swamps in Tabasco, and the Agua

Brava wetlands in Nayarit, bordering the Pacific Ocean.

Among the marine systems of Mexico, two stand out in particular, the Yucatán Peninsula Barrier Reef and the Sea of Cortés, or Gulf of California. The Yucatán Peninsula Barrier Reef is the longest reef system in the western hemisphere and the second longest in the world. It consists of over 400 mangrove, sand and coral islands and islets, and extends from Yucatán, along the Belize coast and down to Honduras. The northernmost portion of this barrier reef lies across from the lower tip of the island of Cancún, a site known as Punta Nizuc. The diversity of this system is among the highest for coral reefs in the western hemisphere.

The most important marine system in Mexico and one of the most important in the world is the Gulf of California. It extends between the country's northwestern mainland coast and the Baja California Peninsula. It is 1 200 km long, has an average width of 150 km, and extends south from the mouth of the Colorado River to an imaginary line drawn between Cabo San Lucas, Baja California Sur and Cabo Corrientes in Jalisco. A combination of circumstances, such as its rugged submarine topography, its orientation from northeast to southeast, the more than 200 islands and islets found scattered over its 160 000-km^2 surface, and its relative isolation from the Pacific Ocean, have allowed the development of a very biologically and oceanographically distinct environment. The Gulf of California is one of the richest seas in the world in terms of biodiversity, even though it only covers 0.044% of the world's ocean surface. Recent studies have found that it has 14 of the 32 marine phyla known, 11 of which have endemic species in this sea. Up to date 5 047 species of marine and coastal fauna have been recorded, of which 383 are endemic to the Gulf. Among the best represented groups of organisms in the Gulf are the macroinvertebrates, with 3 943 species. There are also 866 fish species (more than 55% of all the families of marine fish), 164 species of aquatic birds, and 32 species of marine mammals (82% of all marine mammals found in the northeastern Pacific). In the more than 200 islands and islets found throughout the Gulf, over 600 plant species, 65 of reptiles, 40 of mammals, 134 of terrestrial birds, and 45 of migratory and resident birds have been recorded. Six bird species have 60-95% of their populations nesting in the Gulf. Of the 887 plant and animal species recorded there, 90 are endemic, and 60 of those are reptiles.

This incredible wealth is due in part to its subtropical location, with cool to warm waters in the winter and warm to hot waters in the summer. Also contributing is its rugged submarine topography which ranges in depth to 4 000 m. Its rich variety of ecosystems include 356 413 ha of mangrove forests, 600 000 ha of coastal lagoons, large extensions of beaches, and several types of bottoms including sandy, rocky and loamy substrates. Especially noteworthy are the hydrothermal vents found in the abyssal basin across from Guaymas.

The Gulf of California is characterized by its highly productive waters, which are the result of two upsurging areas that bring the nutrient-rich waters from the bottom to the surface. These upsurges are caused by the shifting winds during the winter and summer and they cause a microalgae bloom that sustains important populations of fish, such as sardines, which in turn are consumed by sailfishes, swordfishes, dorados, seabirds, and marine mammals. Another important feature are the sheltered coastal ecosystems that harbor the larval stages of many marine organisms.

Flagship species

Mexico has great species diversity and the third highest number of endemic non-fish vertebrate species on Earth. Several of these stand out as flagship species symbolic of this country's rich natural heritage. These include a number of endemics and several other spectacular species with ranges that extend beyond the borders of this country.

One of the most unusual endemic species is the volcano rabbit or *teporingo* (*Romerolagus diazi*), which inhabits the snow-covered slopes of the Popocatépetl Volcano. Also found in the Ajusco Volcano, this rabbit is a remnant of remote geological eras. It is also one of the smallest rabbits in the world, not surpassing 30 cm in height and 600 g in weight. With its short and abundant yellow and brown fur, it blends in perfectly with the surrounding grass. It lives in groups of four or five individuals and frequently uses loud vocalizations to keep in touch with other members of its group. These animals feed mostly on grasses, herbs, and seeds. This species has been listed as endangered by the IUCN for over 25 years. The main threats to its survival are habitat fragmentation due to human encroachment, forest fires and shifting agriculture.

Another important Mexican flagship mammal is the pronghorn or *berrendo*. In Mexico there are three subspecies of *berrendo*: *Antilocapra americana peninsularis*, *Antilocapra a. sonorensis*, and *Antilocapra a. mexicana*, and all three are critically endangered. *Peninsularis* only occurs in the central section of Baja California Peninsula, with less than 200 individuals. *Sonorensis* is distributed in the northwestern part of Sonora and as far north as a small area in Arizona, and also numbers no more than 200 individuals. Lastly, the *mexicana* subspecies is found in the northern part of Chihuahua; its populations (600 individuals) are distributed in small groups within the fences dividing the different ranches in the area. In 1996, 65 individuals of this subspecies were reintroduced in Coahuila, in a valley where these *berrendos* had been extinct for over 50 years. This effort was made possible thanks to the coordination of the Mexican organization Unidos para la Conservación and also thanks to the support of the New Mexico Department of Wildlife. The long-term objective of the program is to use this first reintroduced herd in order to repopulate the *berrendo* in northern Mexico.

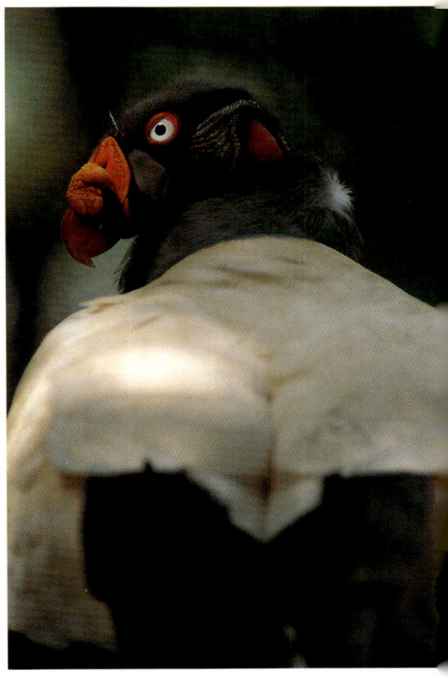

The king vulture is by far the most attractive of the Neotropical vultures, and one of the most imposing and appealing bird species in general. Like the ornate hawk eagle, it is widespread from Mexico to Argentina and lives in the tropical lowlands.

On pp. 148-149, desert of central Baja California with Boojum trees (Idria columnaris), *cardon cactus, old-man cactus, chollas and other vegetation characteristic of the Peninsula. Mexico has what are probably the most diverse deserts on Earth.*

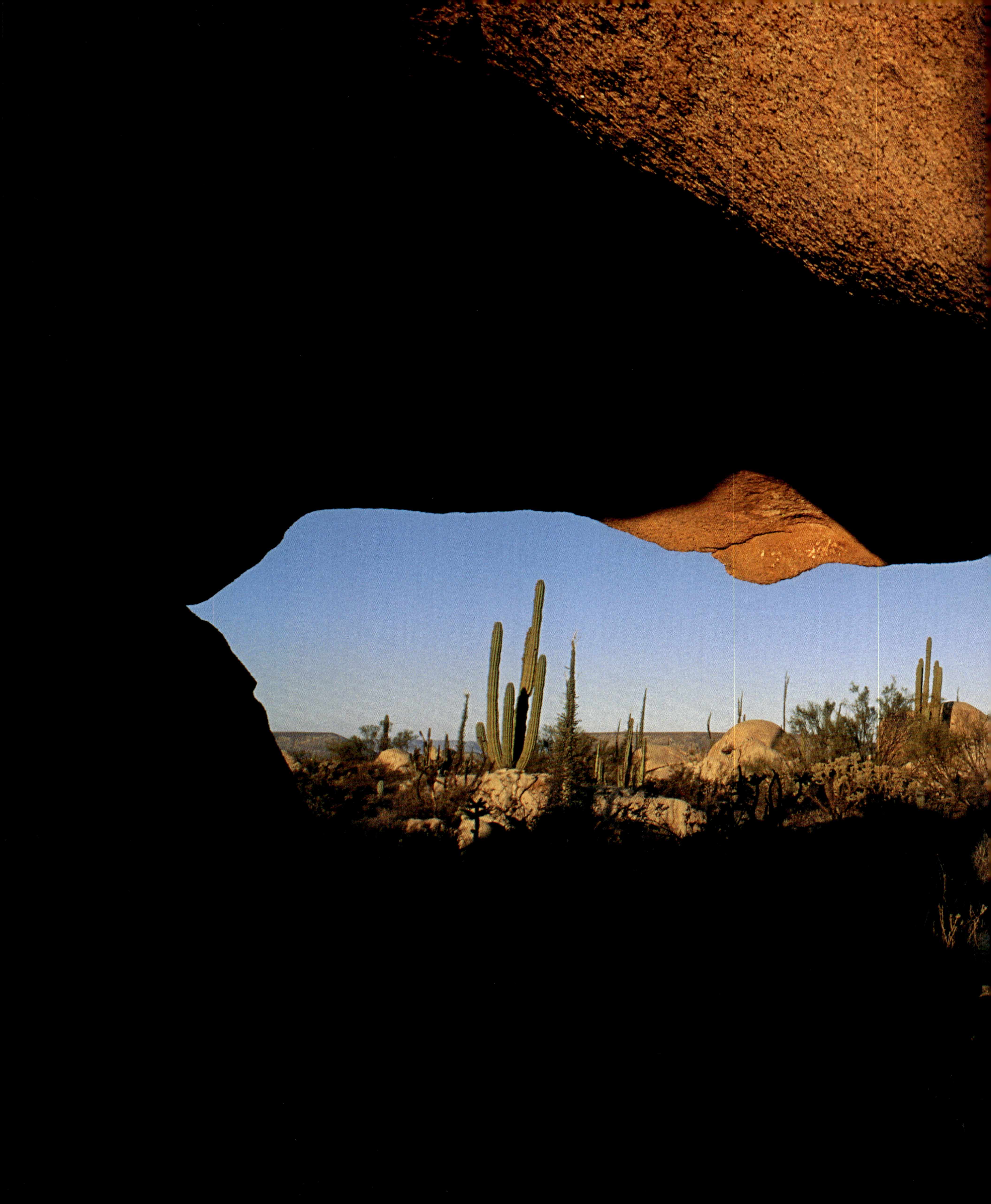

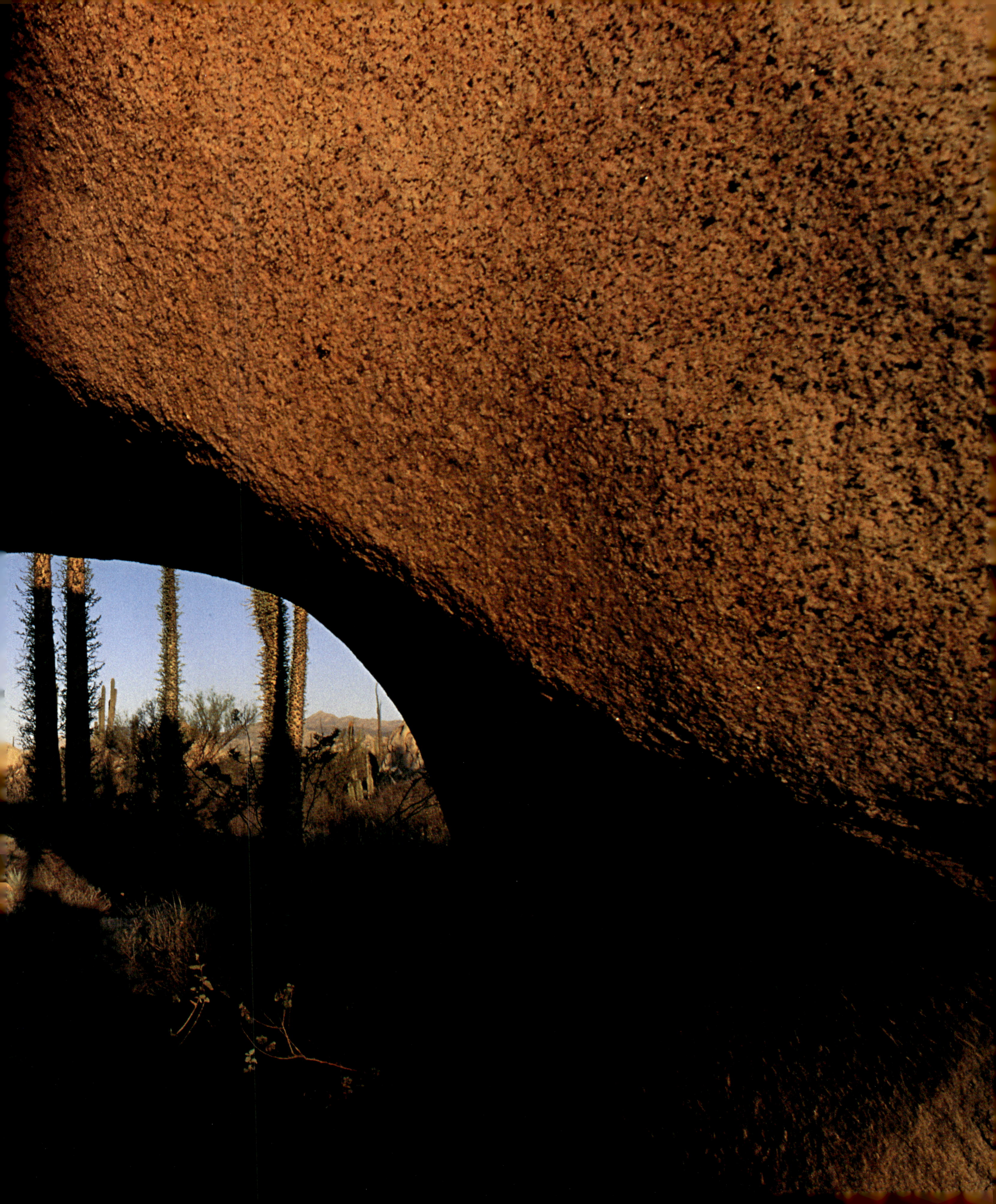

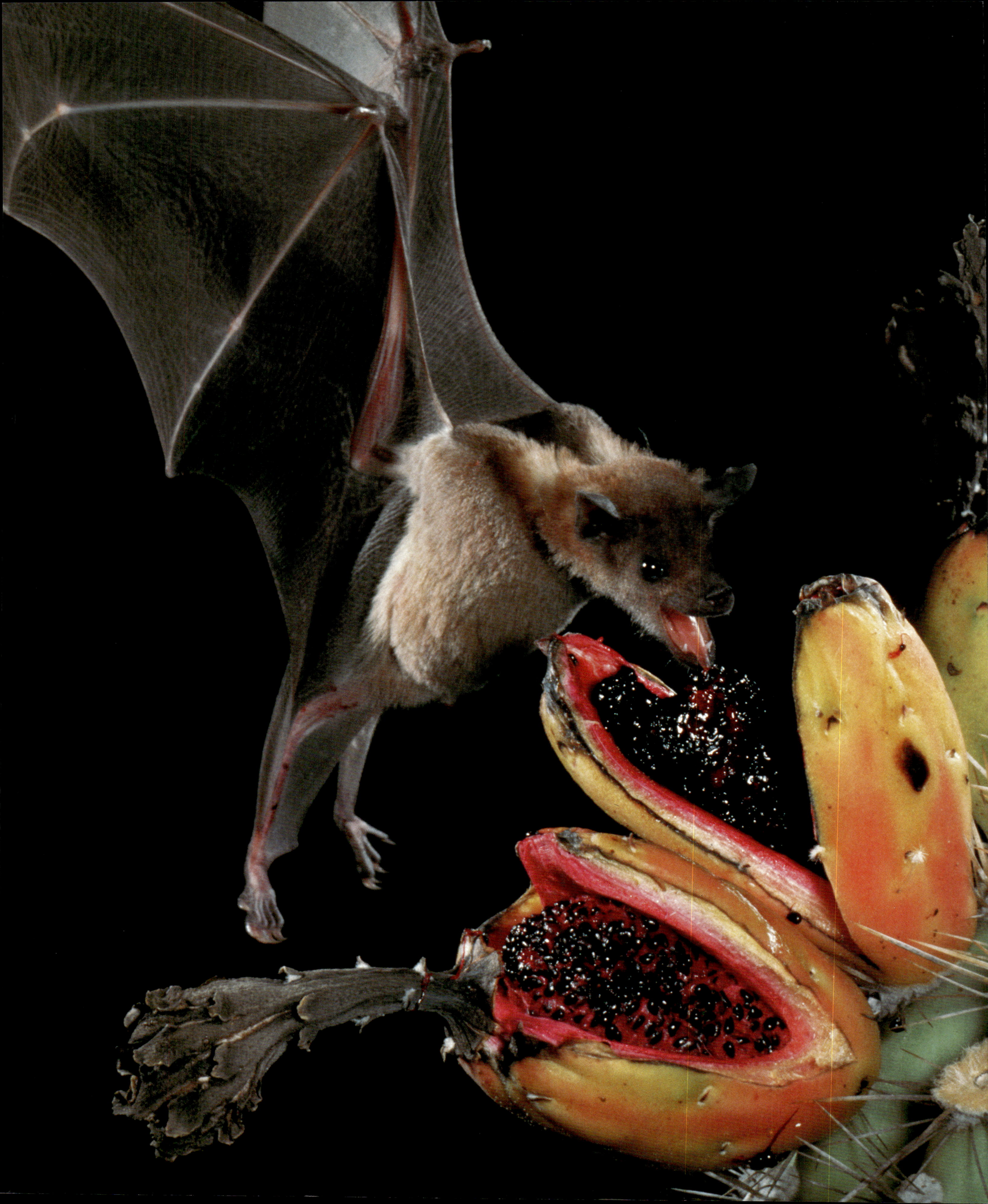

Mexico's rich cetacean fauna includes a number of important flagship species such as the *vaquita*, the gray whale, and the fin whale. One of the smallest and most endangered cetaceans in the world is the *vaquita* (*Phocaena sinus*). A tiny cetacean found only in the upper Gulf of California, it was not described by scientists until 1958. Unfortunately, it is in great danger of extinction due to changes to its habitat and incidental capture in sardine fishing nets. The remaining population, estimated to be no more than 400 individuals, is now protected in the Upper Gulf of California and Colorado River Delta Biosphere Reserve.

The gray whale (*Eschrichtius robustus*) is another important cetacean flagship. Although not restricted to Mexico, it comes to the sheltered lagoons of the Baja California Peninsula to breed every year. In the spring, the gray whales travel to the Bering Sea and the Arctic Ocean to feed on the abundance of krill. This allows them to store enough food for their long journey to the Baja California coast. The largest concentrations of these creatures are to be found in the famous Ojo de Liebre Lagoon; others go into the Gulf of California, but they generally seek shallow, sheltered lagoons to mate, give birth and tend to their calves. They rarely feed during their stay in Mexican waters, and in February they start their migration north, with the males leaving first and the females and calves traveling later.

In 1949, the Mexican government became one of the first in the world to protect these marine mammals. Thanks to this sustained effort, the gray whale population in the Pacific has increased, to a point where today it has over 27 000 individuals. This represents a major success story for the conservation of the world's whales.

The largest Mexican flagship species is the fin whale (*Balaenoptera physalus*), also the second largest animal on Earth, being exceeded in size only by the mighty blue whale (*Balaenoptera musculus*). It has two distinct populations, one in the northern hemisphere and one in the southern hemisphere. The fin whales in the northern hemisphere attain 24 m in length, while their southern counterparts grow to 26.8 m, with females being slightly larger than males. Fin whales are sometimes found singly or in pairs, but more often in pods of three to seven individuals. Several pods, totaling as many as fifty animals, may be concentrated in a small area. Although they are distributed worldwide and make yearly migrations from their wintering grounds to their breeding grounds, there is one population considered resident in the Gulf of California.

Moving on to the reptiles, an important group of flagship species are the Helodermatidae, the world's only poisonous lizards. Included in this family are two species, the Gila monster (*Heloderma suspectum*) and the Mexican beaded lizard (*Heloderma horridum*), and a total of six different taxa. Although one subspecies of Gila monster, *Heloderma suspectum cinctum*, is endemic to the southwestern U.S. and one subspecies of beaded lizard, *Heloderma horridum charlesbogerti*, is found only in a small area of Guatemala, the family is mainly Mexican in distribution. Four of the six taxa occur in Mexico, and three of the subspecies of the beaded lizard, *Heloderma h. alvarezi*, *Heloderma h. horridum*, and *Heloderma h. exasperatum*, are endemic to the country. These highly distinctive, very attractively patterned lizards are desert dwellers. They have grooved rather than hollow teeth, and their venom is introduced into the body of prey or potential predator by chewing, rather than the more efficient injection of the kind employed, for instance, by the rattlesnakes and their allies.

Mexico even has a globally-renowned amphibian flagship, the highly unusual axolotl or *ajolote*. This animal is considered by some to be one of the country's most famous species, and if scientific literature is any indication, this must be the case. In an exhaustive analysis of the herpetofauna of Mexico published in seven volumes in the 1970s, American herpetologist Hobart M. Smith noted that nearly forty percent of the total literature on Mexican reptiles and amphibians deals with the axolotl. The axolotl is a salamander of the wide-ranging family Ambystomatidae, which includes well-known North American species like the tiger salamander (*Ambystoma tigrinum*) and the spotted salamander (*Ambystoma maculatum*). However, the axolotl, which is a distinct species, *Ambystoma mexicanum*, has the unique feature of neoteny, meaning that it does not metamorphose from the larval stage to an adult salamander. Rather, the larval stage is the adult stage. This unusual feature attracted the attention of the scientific community early on; thus, this species has become the most widely used non-mammalian laboratory animal. A shipment of 34 individuals sent to Paris in 1863 from Lake Xochimilco in the Valley of Mexico, supplemented by a second shipment in 1866, became the foundation of a breeding stock that has supplied almost all of the laboratories that have worked on this animal over the past 130 years —almost certainly one of the purest lines of experimental animals.

Although the name *ajolote* is used in Mexico for a number of different species including *Ambystoma tigrinum*, *A. lacustris*, *A. lermaensis*, *Bathysiredon dumerili*, and *Rhyacosiredon* spp., the true axolotl, and the one that is known around the world, is *Ambystoma mexicanum* from Lake Xochimilco. In spite of its wide use in the laboratory, the axolotl has been little studied in nature and some of the most fundamental questions about its ecology, behavior, and even taxonomy remain to be answered. Given the current condition of its habitat, it is questionable whether this is even still possible.

Moreover, the true axolotl (*Ambystoma mexicanum*) and the other species referred to as *ajolotes* in Mexico have long been popular as a source of food, and in some areas are even thought to have medicinal properties. The name *ajolote* is of Aztec origin and has been interpreted in various ways, but it is most likely that it derives from *atl*, meaning water, and *xolotl*, meaning dog —in other words, "water dog." Interestingly, this is precisely the name used in parts of the United States for the larvae of the closely-related tiger salamander and

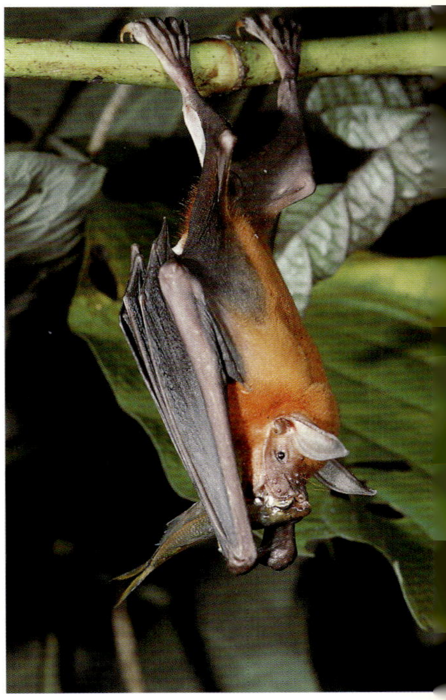

On the opposite page, Mexican long-nosed bat (Leptonyctens curasoae) *feeding on the fruit of the saguaro cactus* (Carnegiea gigantea).

Above, fishing bat (Noctilio leporinus) *in the Chajul region of the Lacandón rain forest, feeding on a small fish it has just captured.*

also for mud puppies (*Necturus* spp.), another group of American salamanders.

With the second highest reptile diversity on Earth, Mexico not surprisingly also has several reptile flagship species. Noteworthy among these are the marine turtles. With seven of the world's eight species of marine turtles found on its beaches and territorial seas, Mexico is the leading country worldwide for these reptiles. The marine turtles found in Mexican waters are the leatherback or *laud* (*Dermochelys coriacea*), the Atlantic green turtle or *tortuga verde* (*Chelonia mydas*), the Pacific green turtle or *tortuga prieta* or *tortuga negra* (*Chelonia agassizi*), the hawksbill or *carey* (*Eretmochelys imbricata*), the loggerhead or *caguama* (*Caretta caretta*), the olive ridley or *golfina* (*Lepidochelys olivacea*) and Kemp's ridley or lora (*Lepidochelys kempi*). The most important for Mexico is Kemp's ridley, a flagship species whose entire global population nests on a single site, Rancho Nuevo on the coast of the northeastern state of Tamaulipas. Kemp's ridley is also the most endangered of all sea turtles, with a total world population now estimated to be as few as 1 000 individuals. Named for Richard Kemp of Key West, Florida, who sent the first type specimen to Samuel Garman of Harvard in 1880, this species is also called the bastard turtle, or ridler turtle; these names derive from the fact that up until the early 1960s virtually nothing was known of its ecology, where it foraged and reproduced, and especially, where it nested. Since no one had seen it nesting in the Caribbean, it was believed by fishermen to be a hybrid, and not a distinct species. It was not until the early 1960s, after a twenty-year search, that the famous turtle biologist Archie Carr and some of his colleagues actually discovered its nesting beaches about 360 km south of the United States border in Tamaulipas. The key to locating its nesting site was a home movie filmed by a wealthy landowner from Tampico, Don Andrés Herrera, who had seen the daylight mass-nesting phenomenon of the Kemp's ridley in 1947, but had shelved the footage, being unaware of its value to science. The film showed an estimated 40 000 turtles nesting in that year; by the time of its discovery in the 1960s, numbers had already dropped dramatically due to overharvesting of adults and eggs. By the late 1960s, the Mexican government began to provide protection for Kemp's ridley, and this grew into a massive international rescue operation run jointly by the Mexican Departamento de Pesca and the U.S. Fish and Wildlife Service. This included protection of eggs and adult females, antipoaching controls, scientific research and head-starting programs for neonate turtles, as well as an effort to establish a nesting colony in southern Texas, where the species was thought to have nested decades earlier. Today the Kemp's ridley population remains small, but the decades of conservation efforts seem to have paid off. Small numbers of adult females are returning to nest in southern Texas, and nesting females are steadily increasing in numbers on the Tamaulipas beaches.

Among the birds, the horned guan (*Oreophasis derbianus*) deserves special mention. It is a bizarre cracid, endemic to the remote cloud forests of Chiapas and Guatemala where habitat destruction has made it vulnerable of extinction. Its most striking feature is a red horn of bare skin on top of its head, strongly contrasting with an otherwise black and white body. A dweller of the upper canopy, it is most often located by the unusual humming song of the male. And in these same forests, the most spectacular and beautiful bird in the Americas, the resplendent quetzal (*Pharomacrus mocinno*), cuts through the sky like a waving serpent. The Aztec and Maya civilizations worshipped this extraordinary bird whose hindfeathers were used to adorn the headdresses of kings and nobles.

Finally, we have a non-endemic species that has become one of Mexico's best known flagship species, the monarch butterfly (*Danaus plexippus*). This country shares the distribution of this migrant species with the United States and Canada, but the concentrations that it reaches in a few small areas in Mexico are one of the great biodiversity phenomena on our planet. Each year the two different populations of monarch butterflies that breed in Canada make their journey south. The western population goes to California to overwinter, and the eastern population makes its way down the Sierra Madre to spend the winter in the *oyamel* forests of the Mexican mountains of the states of México and Michoacán, an area that has recently been declared a Special Biosphere Reserve for the Monarch Butterfly. There, they carpet the forest in staggering densities. They rest high up in their cold mountain shelter and use up almost none of the valuable energy that they will need to make the journey back. In February, they fly north again, some of them stopping on the Gulf coast of Texas and breeding there. Their offspring then continue the trip north to Canada.

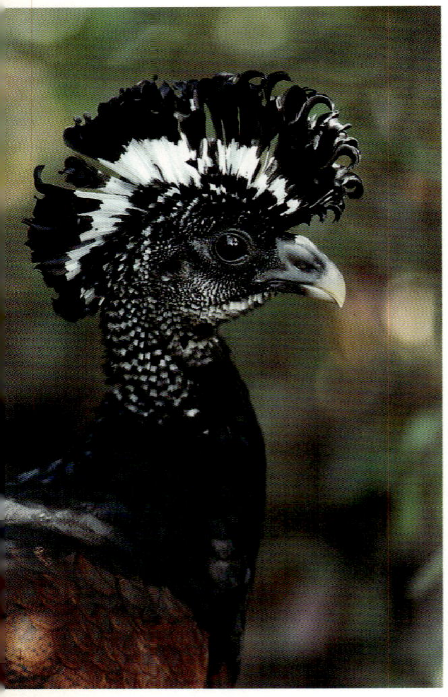

Mexico has seven species of the large cracid birds, which include the curassows, the guans, and the chachalacas. Pictured above is a female great curassow (Crax rubra).

Mexico is the only country with both species of wild turkeys. The male ocellated turkey (Agriocharis ocellata) *shown on the opposite page is a large, spectacularly-colored bird with distinctive red warts on the head. The female is duller in coloration and lacks the warts. The ocellated turkey is found in the Yucatán Peninsula and adjacent parts of Guatemala and Belize.*

Cultural diversity

In the early sixteenth century, when Hernán Cortés touched shore on the territory that would later become modern Mexico, over 17 million indigenous people called the region home (Denevan, 1992). They hunted, fished, and farmed in scores of ecosystems, ranging from the Sierra Tarahumara of northwestern Mexico to the lowland rain forest of the Selva Lacandona in the extreme southeast. The region was also host to two of the New World's most developed civilizations, the Aztec civilization which was still in its heyday, and the Maya, which already lay in ruins by the time Cortés arrived.

The earliest inhabitants of Mexico were hunters who entered the area by at least 10 000-20 000 BC, following migrating bison, mammoths, and mastodons from the colder northern reaches of North America. By 7 000 BC, some of these families had begun to gather in villages, where they cultivated corn, squash, avocados, beans, peppers, and tomatoes. Gradually, a select few of these villages evolved into permanent settlements, and by

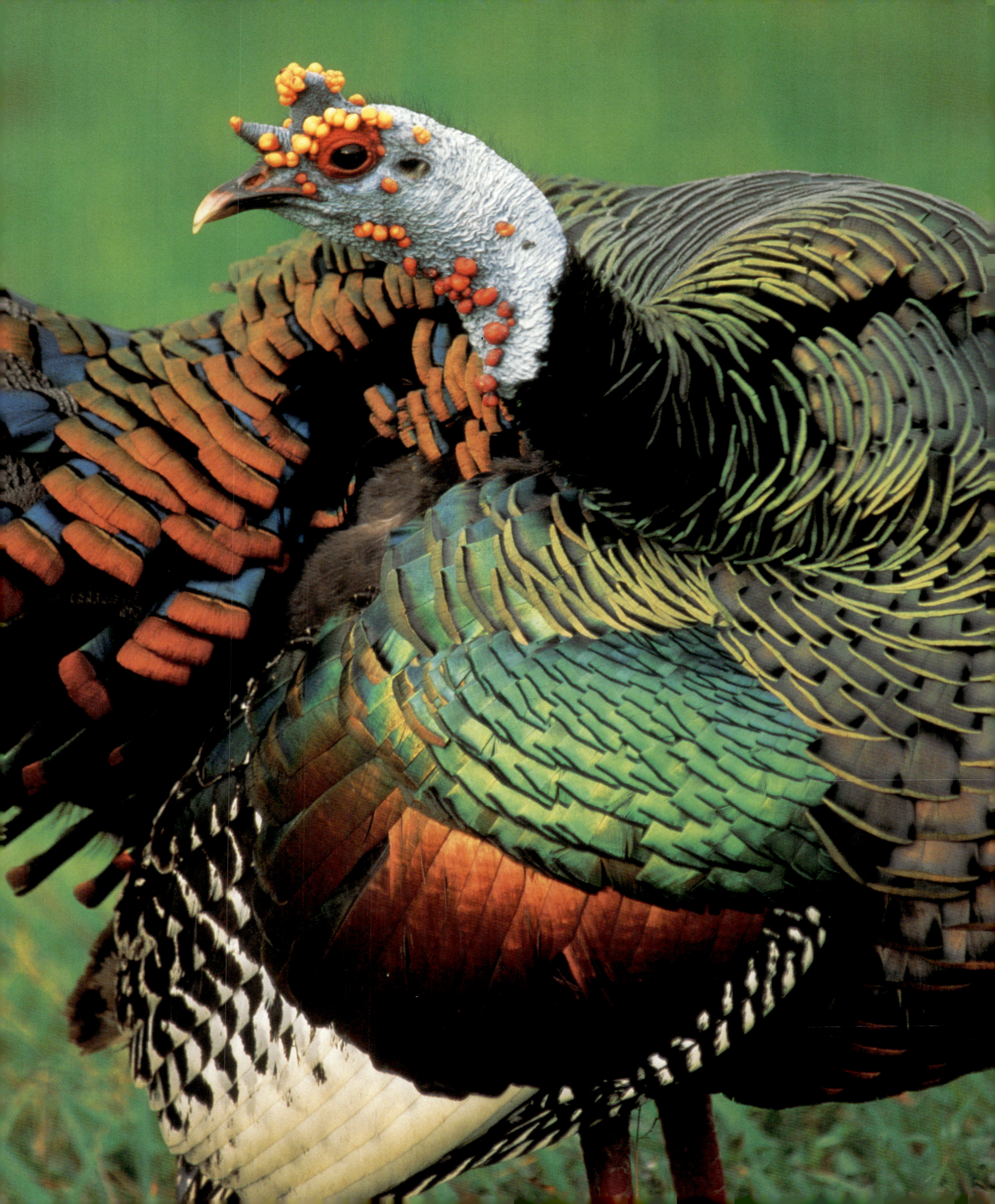

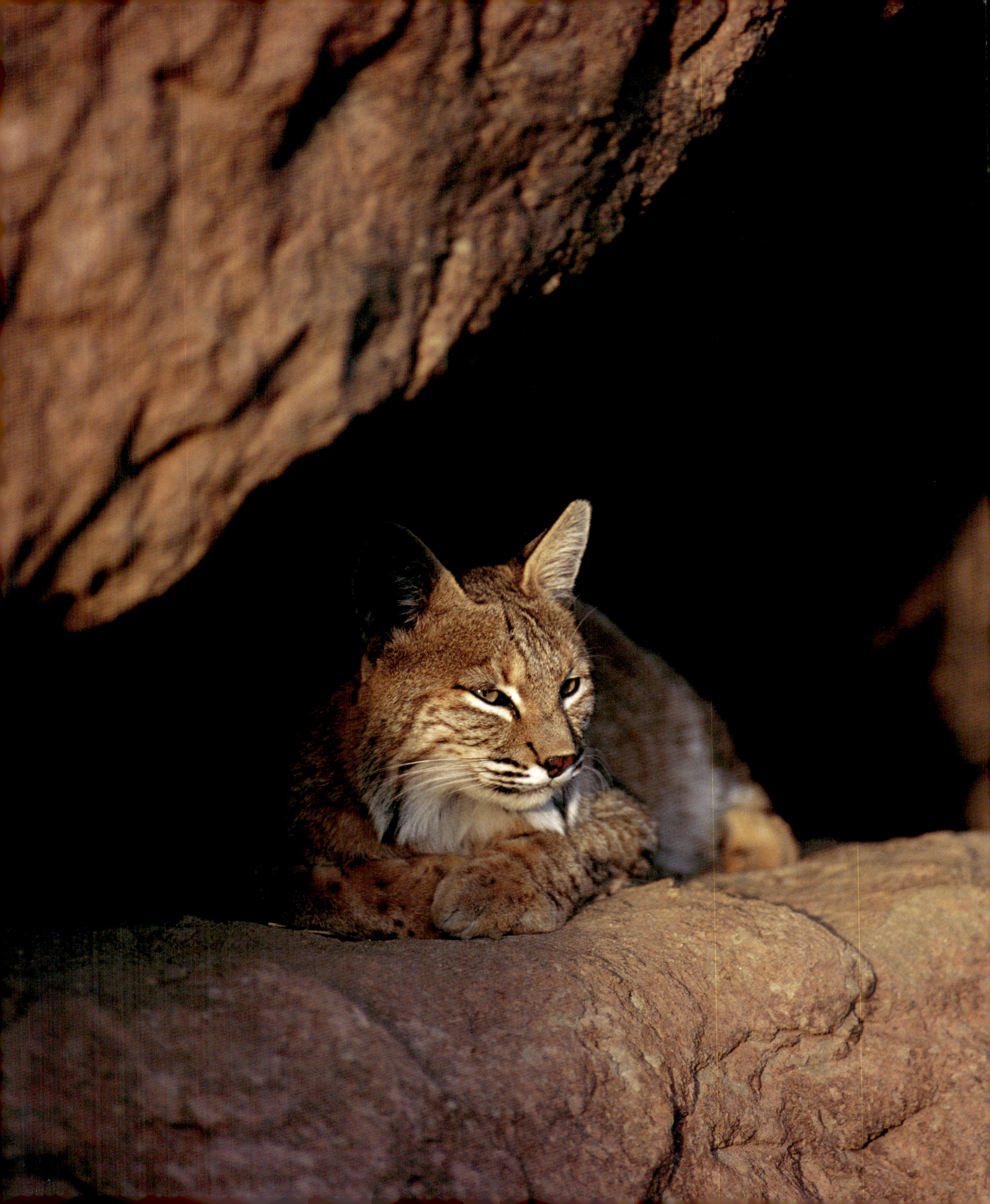

1 000 BC the first flat-topped earthen structures were built as sites for religious ceremonies.

In the Maya region of southeastern Mexico, the indigenous inhabitants began to build these early earthen platforms around open courtyards to form religious centers, and by 500 BC these villages had evolved into vibrant centers of economic and ceremonial activities, with temples as high as 20 m. Over the next 1 000 years, the Maya turned the biological diversity of the forest-covered highlands and lowland rain forests into the foundation for the most advanced civilization in the Americas, a civilization that flourished for more than ten centuries. In cities such as Palenque, Bonampak, and Yaxchilán, the Maya developed sophisticated mathematics, astronomy, and writing, as well as a calendar system that measures time more accurately than the modern Gregorian calendar.

For reasons that are still fuel for scholarly debate, the Maya civilization disintegrated between AD 700 and 900, leaving behind the stone ruins of dozens of ancient cities, and descendants who comprise the diversity of Mexico's Maya peoples today. Combined with the Maya peoples of neighboring Guatemala, Belize, and Honduras, the Maya number more than 11 million people, speaking 30 languages and dialects. After the Quechua speakers of the Andean countries of South America, they are the second largest group of indigenous people in the western hemisphere.

Although the highly-advanced civilization of the Maya had changed drastically by the time Cortés reached Mexico in 1519, the early Spanish conquistadores found a later civilization, the Aztecs, in a large, fertile valley with a lake dominated by the elegant city of Tenochtitlan. With over 100 000 inhabitants, the Aztec capital was larger than any Spanish city of its time. The Spanish conquest of the Aztec capital and its surrounding empire is the story of the birth of Mexico, and the admixture of people that resulted from the subsequent intermarriage of Spaniards with Aztecs, Maya, and scores of other indigenous groups, was the foundation of the country's growing mestizo population.

At the close of the twentieth century, Mexico is still home to a diverse family of indigenous peoples. Of the 120 languages spoken at the time of the conquest, 54 are still spoken by Indians in Mexico today (Martínez 1986; cited in Bye 1993), although there is some controversy as to the total number of languages spoken. Toledo (1995), for example, cites 240. Ten percent of the nation's 95 million are considered descendants of indigenous tribes with little or no European blood.

The total number of indigenous speakers in Mexico has been increasing since 1950. Speakers of Chol, Tzeltal, Tzotzil, Mazatec, Amuzgo, and Tlapanec are growing in number, while Huichol, Huastec, Mixe, Mixtec, and Chinantec speakers are remaining constant. A few linguistic groups are declining in number, including the Chontal, Nahuatl, Zapotec, Yucatec Maya, and Tarahumara (Bye, 1993). Thirteen of the 32 Mexican states have been classified by the Mexican census as "eminently indigenous": Oaxaca, Chiapas, Quintana Roo, Campeche, Guerrero, Hidalgo, Nayarit, Puebla, San Luis Potosí, Veracruz, Yucatán, México, and Michoacán (Barry, 1992). Together, Mexico's indigenous groups have legal control of 160 000 km^2, or 8% of the country's territory (Durning, 1993).

One of the largest of these indigenous reserves is that of the Comunidad Lacandona in eastern Chiapas. Created in 1971, the Comunidad Lacandona designates 641 000 ha of lowland forest as the homeland of the 500 Lacandón Maya, 5 000 Tzeltal Maya, and 3 000 Chol Maya who live inside the reserve. Simultaneously, this indigenous reserve overlaps 85 percent with the Montes Azules Biosphere Reserve, which was declared by Mexico and recognized by UNESCO in 1978.

The most traditional of the reserve's inhabitants are the Lacandón. Until the mid-1960s, the immensity of the lowland tropical forest protected the Lacandón Maya from many of the changes that have characterized other indigenous groups in Mexico. As a result, the Lacandón preserved environmental adaptations that are especially appropriate to the forest ecosystem. From the biological diversity of eastern Chiapas, the Lacandón gather fruit, wild animals, natural insecticides, fish poisons, fiber for rope, incense for religious ceremonies, wood for houses, furniture, and canoes, and medicinal plants to cure a toothache or a snakebite.

The Lacandón Maya also practice a system of traditional agroforestry that produces food crops, trees, and animals on the same plot of land simultaneously. The Lacandón orchard-garden (*milpa*) combines up to 79 varieties of food and fiber crops on single-hectare plots cleared from the tropical forest. After burning the forest to clean the plot of insects and weeds, the farmer plants fast-growing tree crops and root crops in the nutrient-rich ash —papayas, bananas, chayotes, maniocs, sweet potatoes, and plantains. Then, during the course of the year, the farmer and his family add corn, chiles, limes, watermelons, squash, tomatoes, cotton, tobacco, rice, beans, sugar cane, cocoa, and onions. Planting times for some of these crops are keyed to seasonal signals, like the flowering of specific forest tree species. When the flowers of the mahogany tree fall, for example, the Lacandón farmer knows that it is time to plant the spring corn crop. The falling of wild tamarind flowers indicates the time to plant tobacco.

The biological diversity of the Lacandón *milpa* has a corollary benefit: pacas, peccaries, brocket deer, and other edible mammals are drawn to the young shoots of the agricultural plot, and Lacandón hunters lie in wait for them to add vital protein to their diet. Studies by forest ecologists indicate that the presence of *milpa* plots in the tropical forest can actually increase the populations of some wildlife species.

Over the past dozen years, most Lacandón have become fully integrated into the economy of modern Mexico. Paved roads now pass near all three of the major Lacandón communities of Naja, Mensabak, and Lacanjá Chan Sayab, and most children attend bilingual

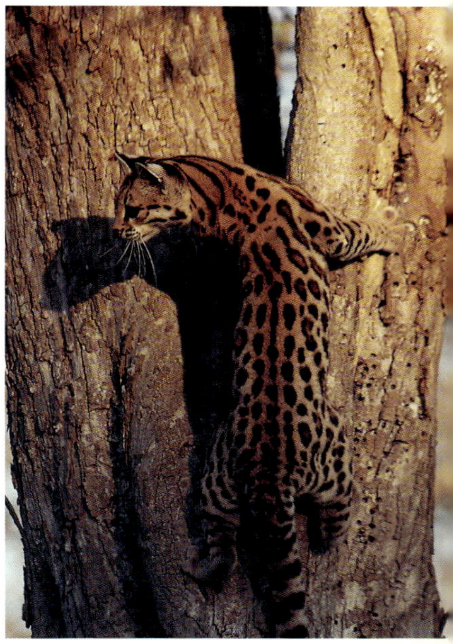

Six cat species occur in Mexico. On the opposite page, the bobcat (Lynx rufus) *is a North American species that extends into Mexico and occupies a wide variety of habitats.*

Above, the margay (Felis wiedi) *is a Neotropical species that extends north into Mexico and just into the southwestern United States. It is an adept tree climber and probably the most arboreal of all the American cats.*

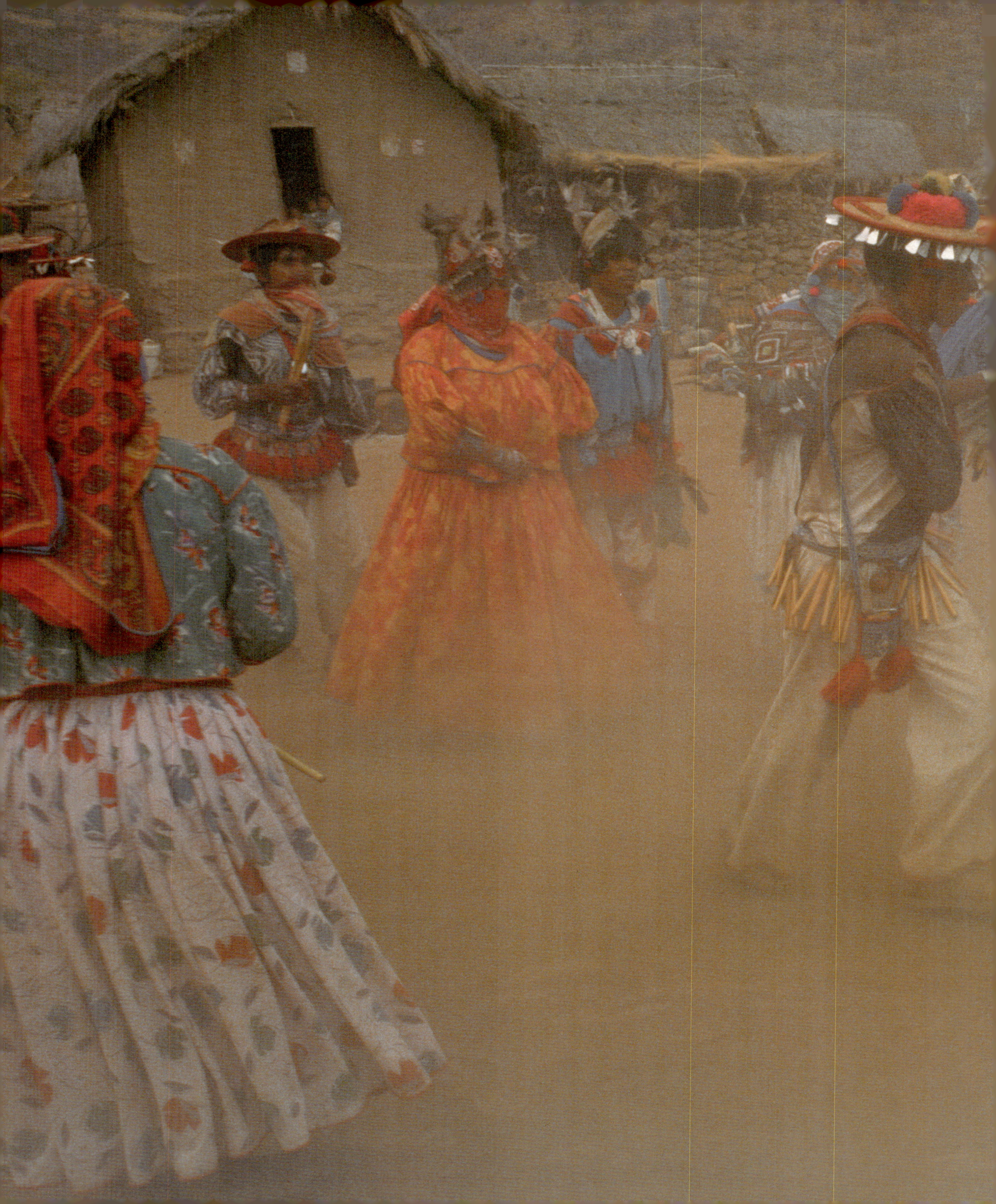

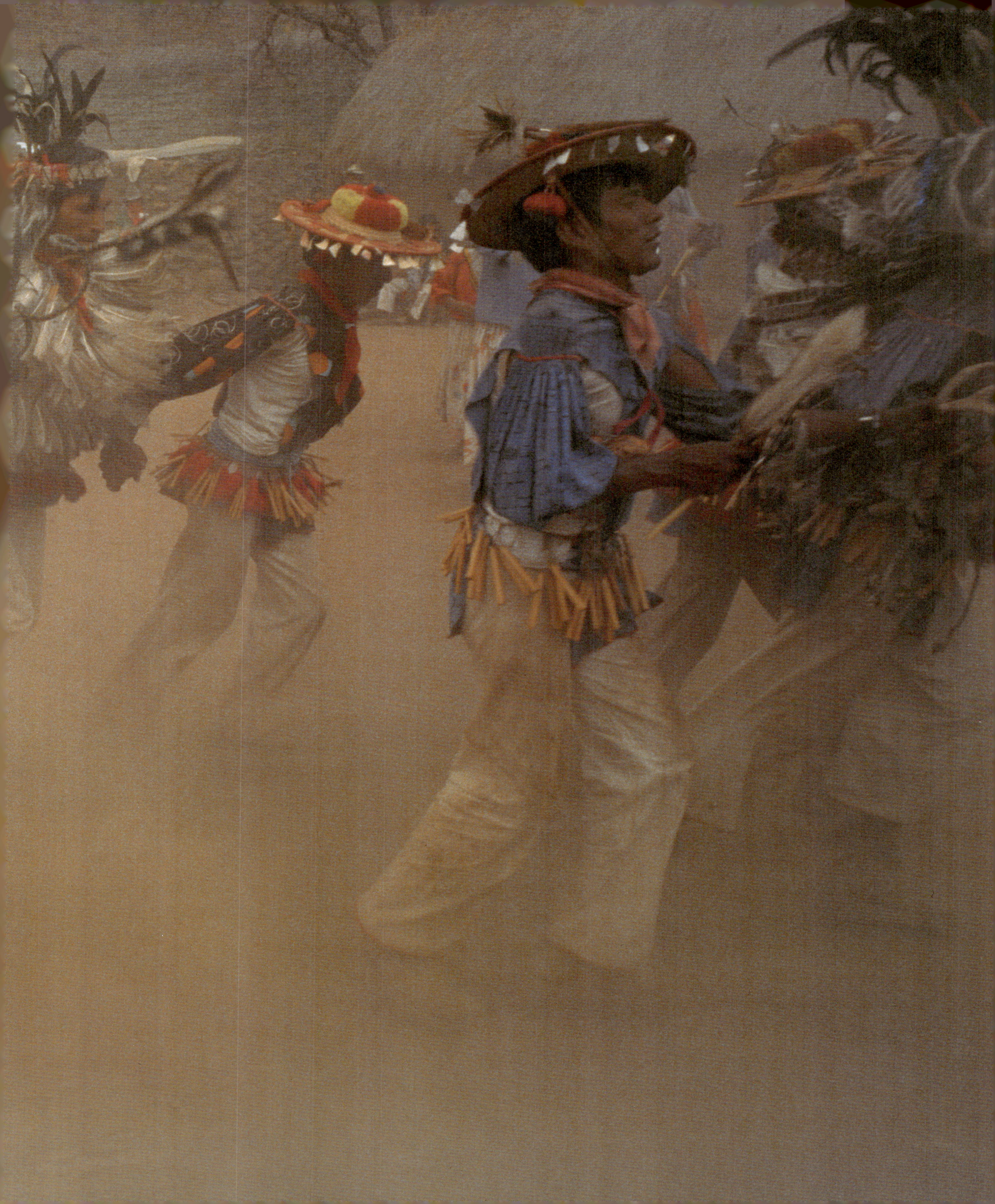

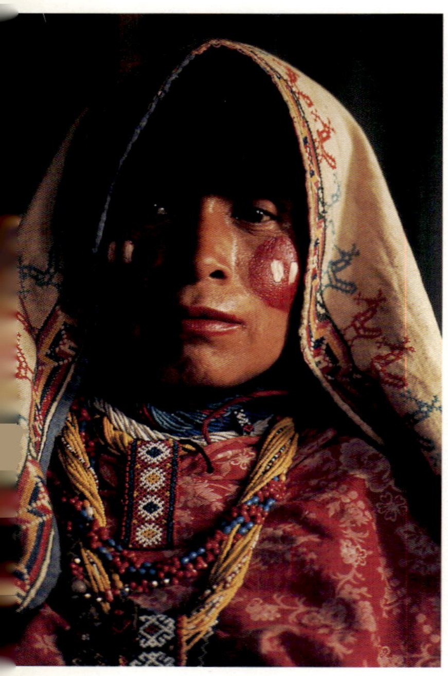

schools. Increasingly, only the older adults keep alive the cultural heritage of understanding the ecological intricacies of their tropical forest home.

Another noteworthy indigenous group are the Tarahumara from Chihuahua, who are world-renowned for their stamina and long-distance running ability. When first encountered in the early seventeenth century by Jesuits, the Tarahumara occupied a vast area of the Sierra Madre on both sides of the Papigóchic River. Today, the 28th parallel marks their northernmost boundary and they have lost all but the headwaters of the Papigóchic. In the process of colonization, they lost their land to logging and mining companies, modified their traditional productive activities, and became more dependent on introduced goods. Today there are federal, missionary and private forces at work in the Sierra Tarahumara, continuing to change Tarahumaran society.

Nearly every adult Tarahumara is a craftsperson. They weave woolen blankets, make pottery, and the women are excellent embroiderers. Most Tarahumara yarns are undyed, natural wool, meaning the predominant colors are black or white. For splashes of color and design, weavers buy or trade for commercially spun and dyed yarns or, more rarely today, they use indigo or other native dyes to color white yarns.

The main staples of the Tarahumara diet are corn, squash and beans. Corn is eaten in different ways, including *pinole* (ground and sweetened), *esquites* (seasoned), *atole* (a thick drink made with corn and sugar), *tamales*, and simply boiled or roasted. One of their most famous contributions to corn cuisine is their famous beer or *tesgüino*, a fermented drink made year round from sprouted corn. This *tesgüino* is a very important part of Tarahumara culture. Given the scattered and isolated nature of their settlements, the *tesgüinada* is a valuable opportunity for these people to come together. It is not only a religious or social gathering, but one that brings the people together in a *ranchería* or household. During such parties, certain essential tasks that cannot be performed by a single family, such as heavy agricultural or construction work, are often carried out.

Another mountain-dwelling group that has maintained much of its traditional lifestyle is the Huichol. The Huichol, who call themselves *wisrrarikas*, meaning "people who live in places with thorny plants," are outstanding because they have been among the most successful indigenous groups in preserving their own special vision of the world and their environment. Living in the farthest reaches of the Sierra Madre Occidental in the states of Jalisco and Nayarit, the Huichol have isolated themselves in order to maintain their spiritual integrity. They are also world-famous for their exquisite beadwork, which they sell to obtain those western goods that are of interest to them. Indeed, the Huichol culture is one of Mexico's most interesting ones due to its vision of the natural world. Every year the Huichol make pilgrimages to the deserts of San Luis Potosí in search of peyote (*Lophophora williamsii*), a hallucinogenic cactus which they use to come into contact with the spirit of their ancestors and with the cosmos.

The coastal regions of Mexico have been the home of several indigenous groups. Today, there are mainly four cultures that base their survival entirely or partially on resources from the sea: the Mayo from the Sonora-Sinaloa border, the Seri from the arid coast of the Sonoran Gulf of California, the Nahua from Michoacán, and the Huave in the state of Oaxaca. Of these, the Seri are perhaps the most interesting and the most traditional. In their harsh and highly variable environment, they remain extremely independent, nonconforming, and very adaptable to the world's changing circumstances. Most Seri have historically had access to marine resources for their sustenance, and also depend on the desert for subsistence. The Seri were able to adapt to centuries of European pressure without radically changing their social structure or economy, and on the verge of the twenty-first century, they remain as one of last nonagricultural hunter-gatherer tribes on Earth. The Seri are familiar with and use more than 400 species of desert plants. One of their main activities is fishing; however, with the increase in tourism in the Seri communities of Desemboque, Punta Chueca and Kino Bay, and with the decline of the marine resources in their area, the Seri now rely heavily on the sale of their ironwood carvings. This craft has been so successful with tourists that is now copied by other non-Seri carvers as well.

Although much traditional knowledge has vanished as their elders have died out, the Seri's cultural identity remains strong. Their language is basically intact and their population is increasing. The Seri have met the challenges of the modern world in much the same way that they reacted to earlier Spanish-Mexican influences —by taking advantage of new resources, new markets, and new opportunities, without radically altering their independent and flexible way of life.

All told, the indigenous people of Mexico have made many important contributions to the world, including 5 000 useful species of plants (both wild and cultivated), as well as 3 000 medicinal species used by different cultures. Their conscientious use of their natural environment as evidenced by their consumption of nontraditional protein sources is also noteworthy. Indeed, as an indicator of this, Mexico is the number one entomophagous country in the world, with 230 species of insects included in the diet of the Mexican people. Thanks to their ingenuity, the ancient Mexicans created the *chinampa*, an agricultural system built above the water on platforms made of mud and plant remains, which fed thousands of people in pre-Hispanic times and still subsists today.

CRISTINA GOETTSCH MITTERMEIER
RUSSELL A. MITTERMEIER
JAMES NATIONS
ALEJANDRO ROBLES
MARÍA DE LOS ÁNGELES CARVAJAL
PATRICIO ROBLES GIL

Huichol Indians (pages 156-157) live mainly in the highlands of Nayarit and Jalisco. Here they are performing a ceremonial dance. Notice the use of various kinds of feathers as ornamentation.

Above, Huichol Indian woman with face paint and other traditional ornaments.

On the opposite page, a Huichol Indian boy from the Sierra Madre Occidental, painted traditionally in preparation for a religious celebration.

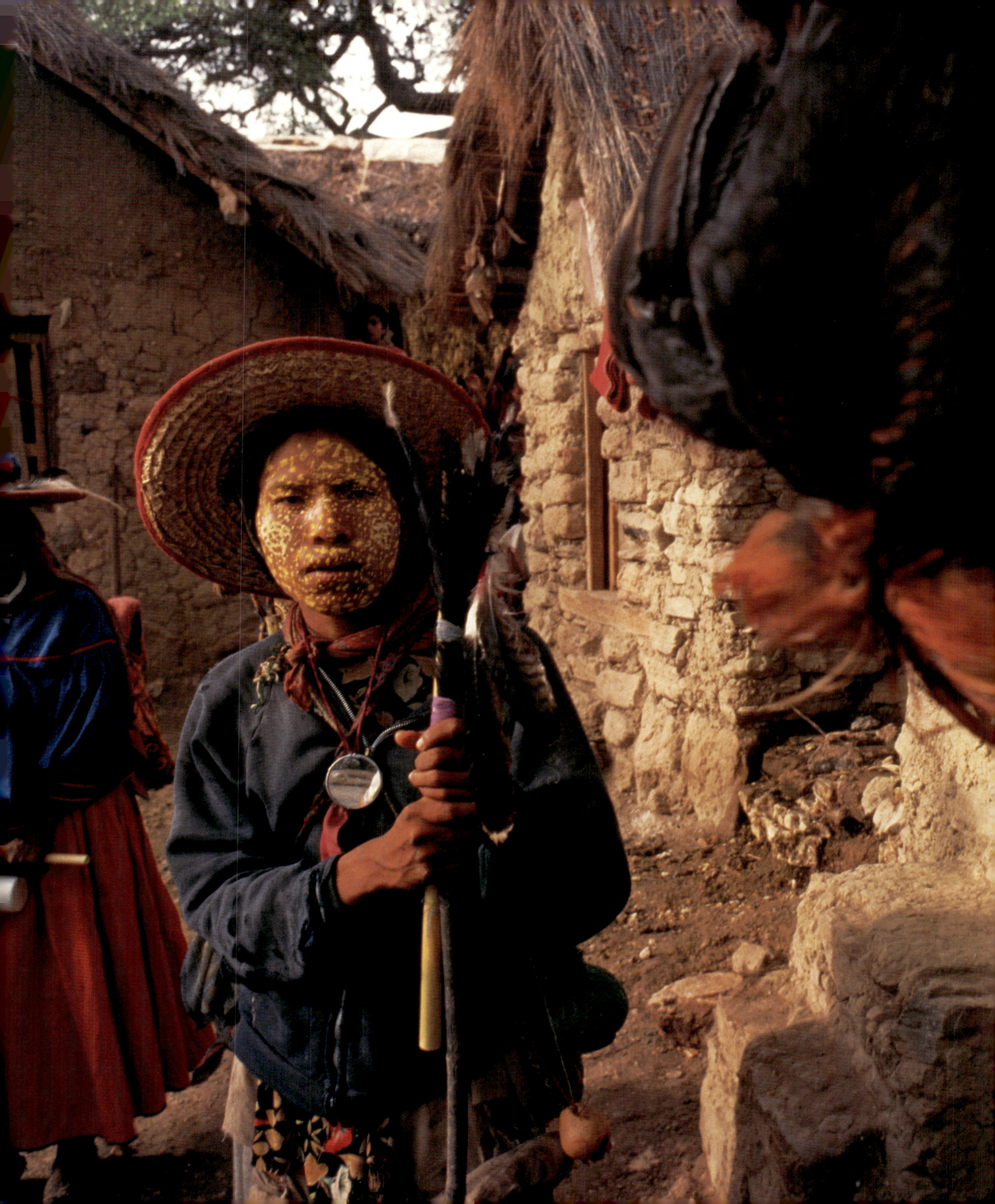

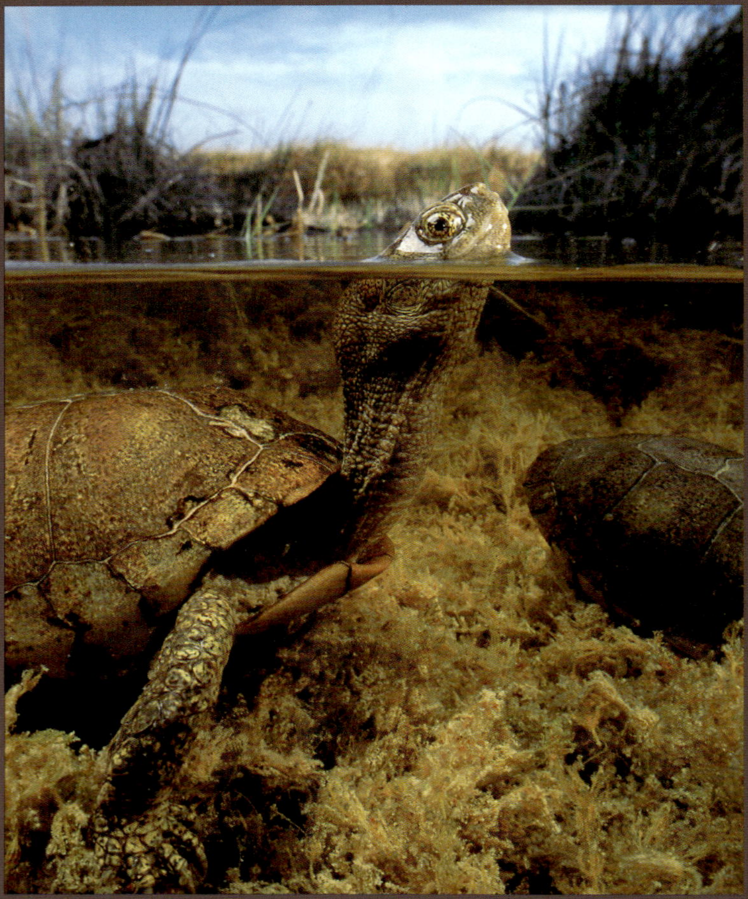

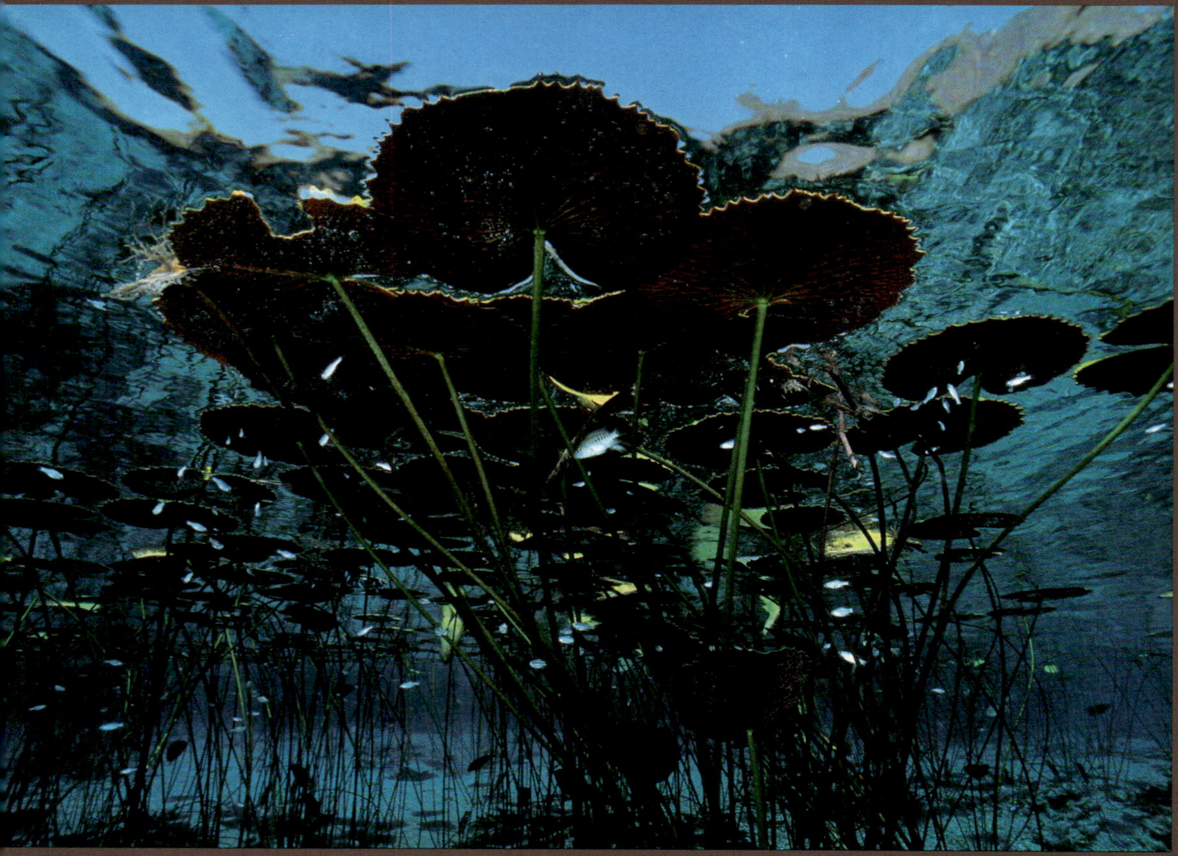

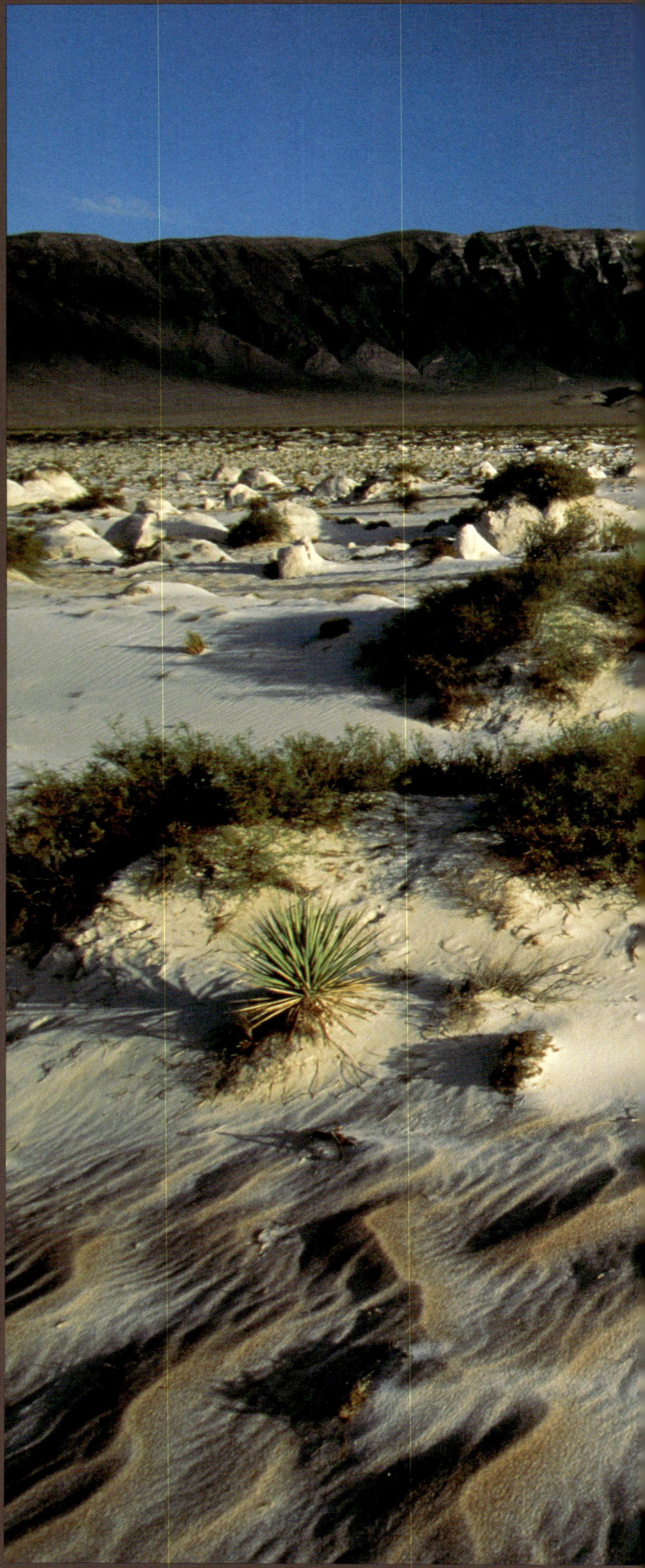

Upper left, at the Laguna de los Burros, a Coahuilan box turtle spends much of its time underwater, while other box turtles evolved as land creatures. Lower left, long-stemmed water lilies provide shade and refuge for schools of pupfishes, tetras and cichlids. Since the springs of Cuatro Ciénegas flow from deep underground, there is no sediment to mar visibility in these crystal-clear waters.

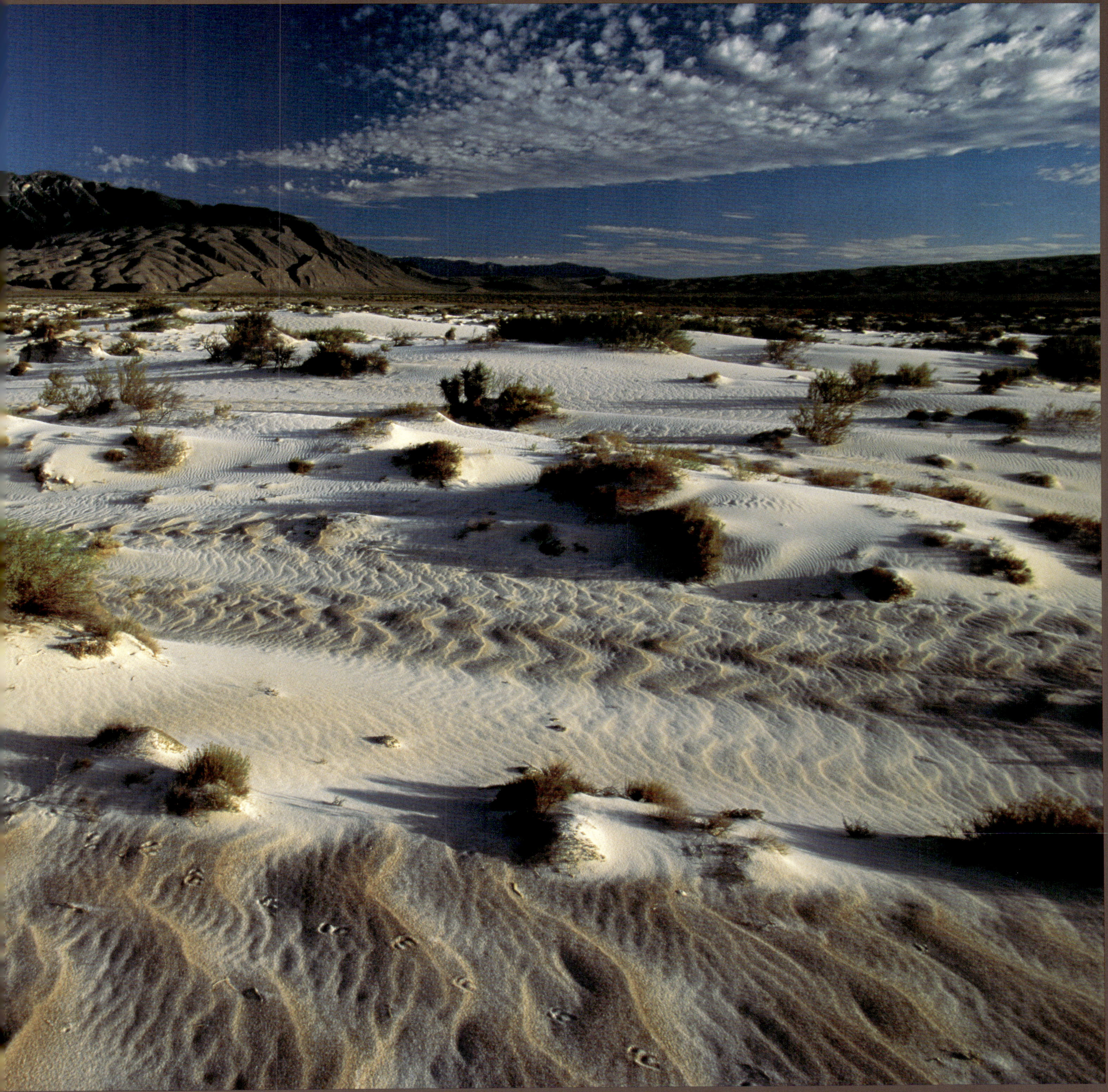

Cuatro Ciénegas (Four Marshes) is a small valley in the Chihuahuan Desert in the state of Coahuila, where underground water flows to the surface in a system of subterranean channels, probably from deep aquifers. It is a mini center of endemism, with more than 60 endemic species, including fishes, turtles, snails, scorpions and cacti in an area of only about 1 200 km^2. The photo above shows a valley of gypsum dune fields, the only ones of their kind in Mexico. These are near the Churince Lagoon.

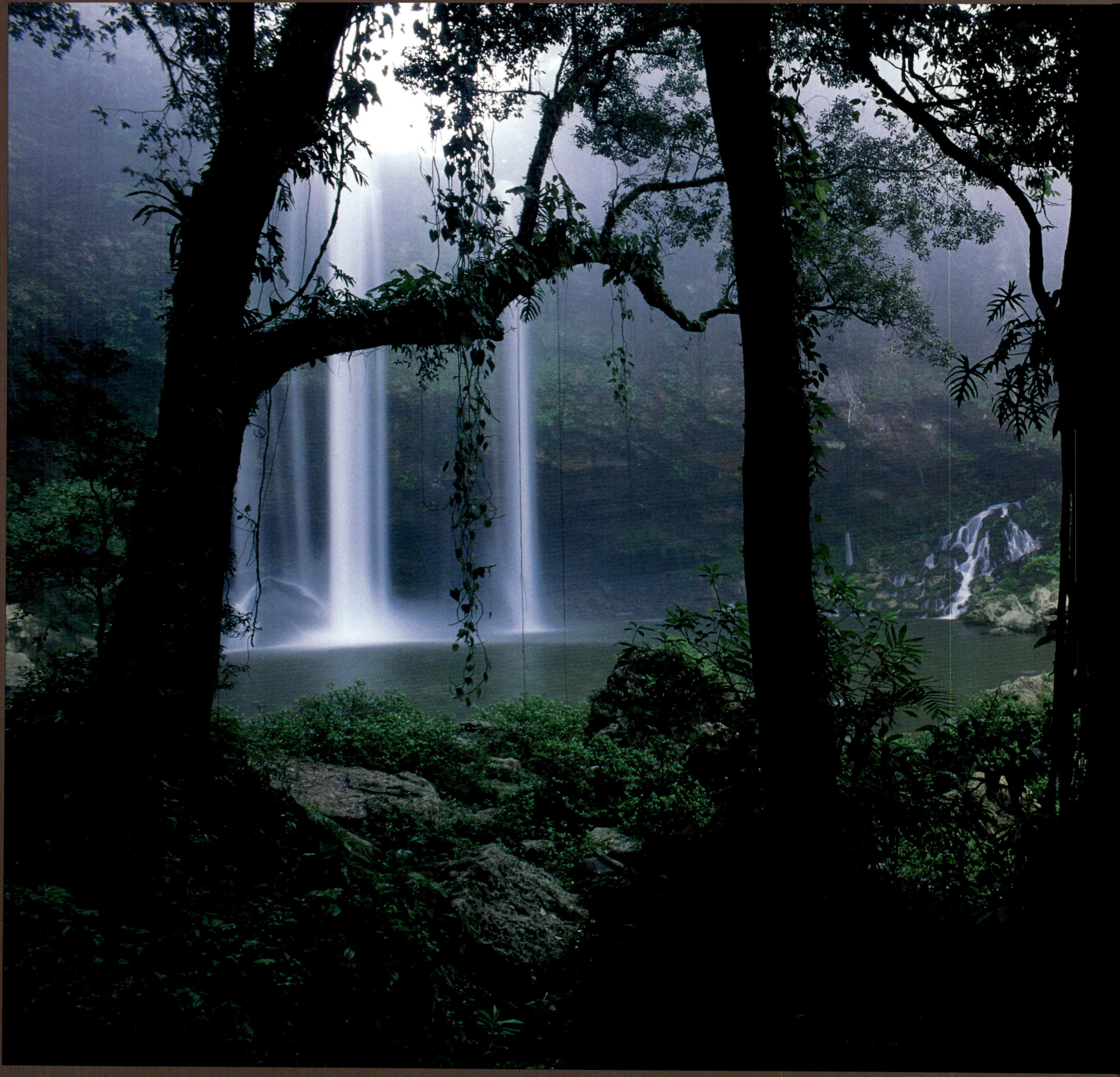

Mexico has the most important tracts of rain forest in North America. The Selva Lacandona occupies only 0.25% of the country's land surface and yet has more than 20% of its diversity. Above, the Misol-ha Waterfall is located in the northern part of the Selva. Upper right, the Mexican mantled howler monkey (Alouatta palliata mexicana) *is one of the two howler monkey species in the country. This mother and infant were photographed in Tabasco. Lower right, the jaguar* (Panthera onca) *is the largest mammalian predator in the Neotropics and still occurs in reasonable numbers in Mexico. This young female was photographed in the wild in Calakmul Biosphere Reserve.*

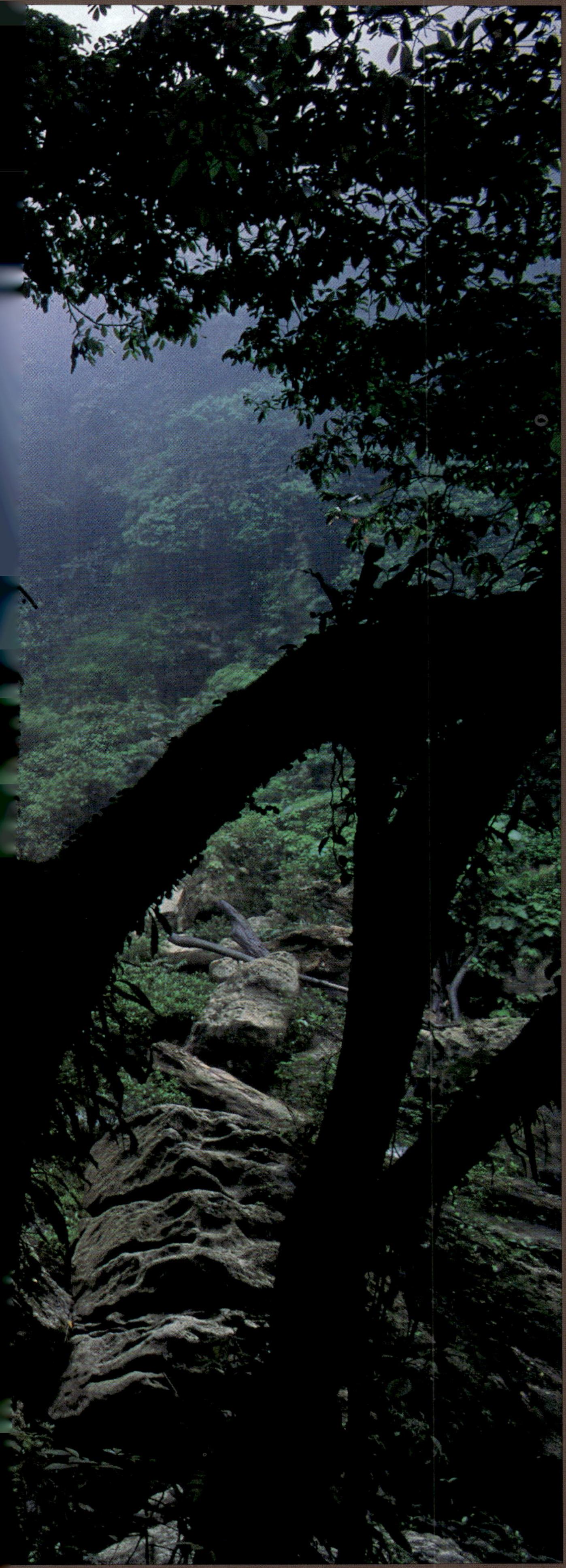

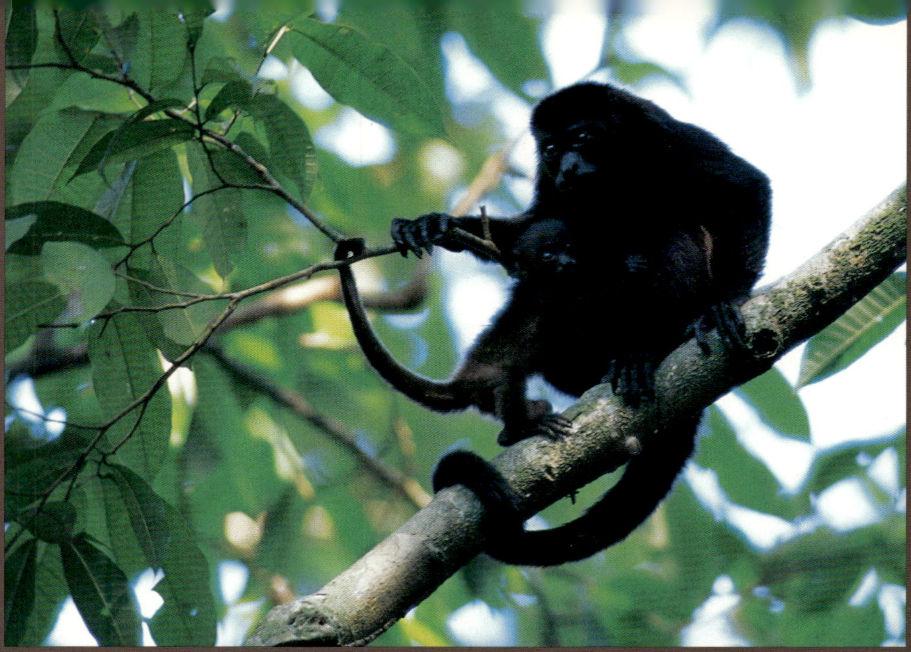

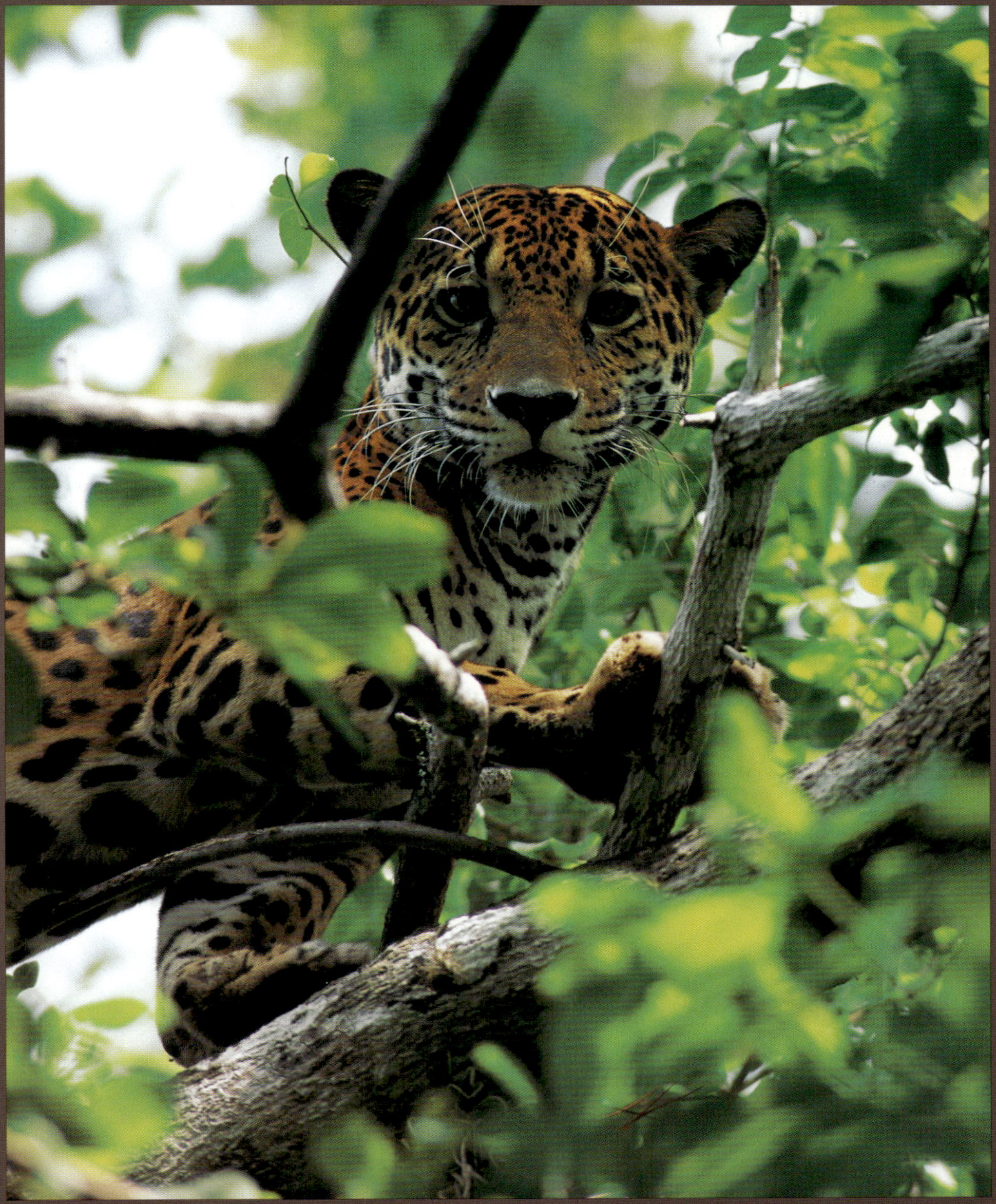

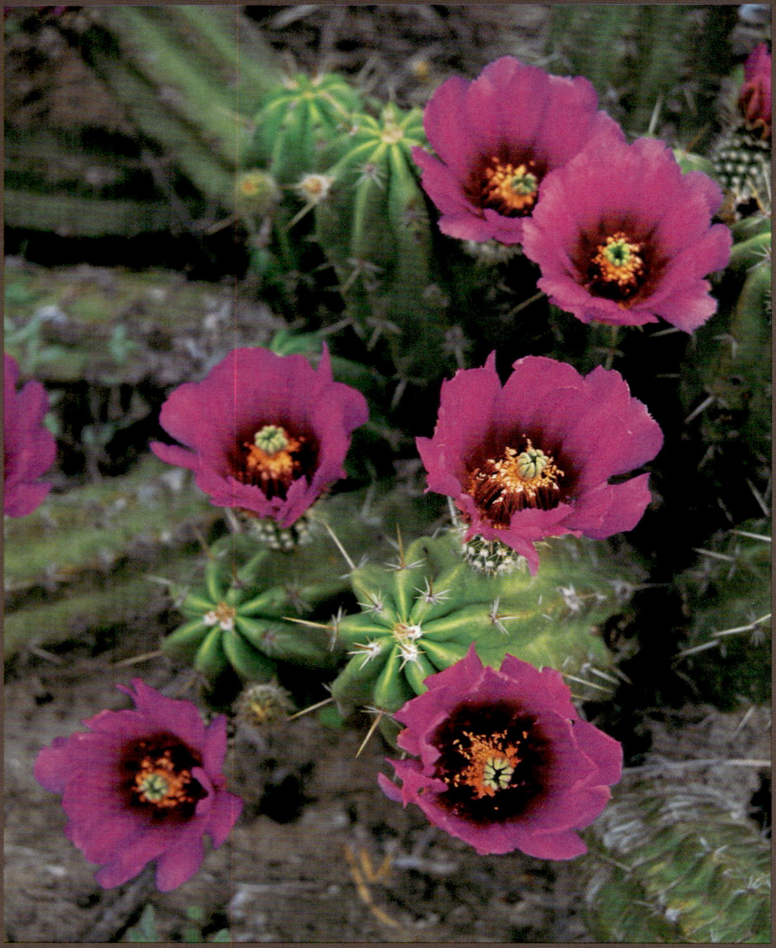

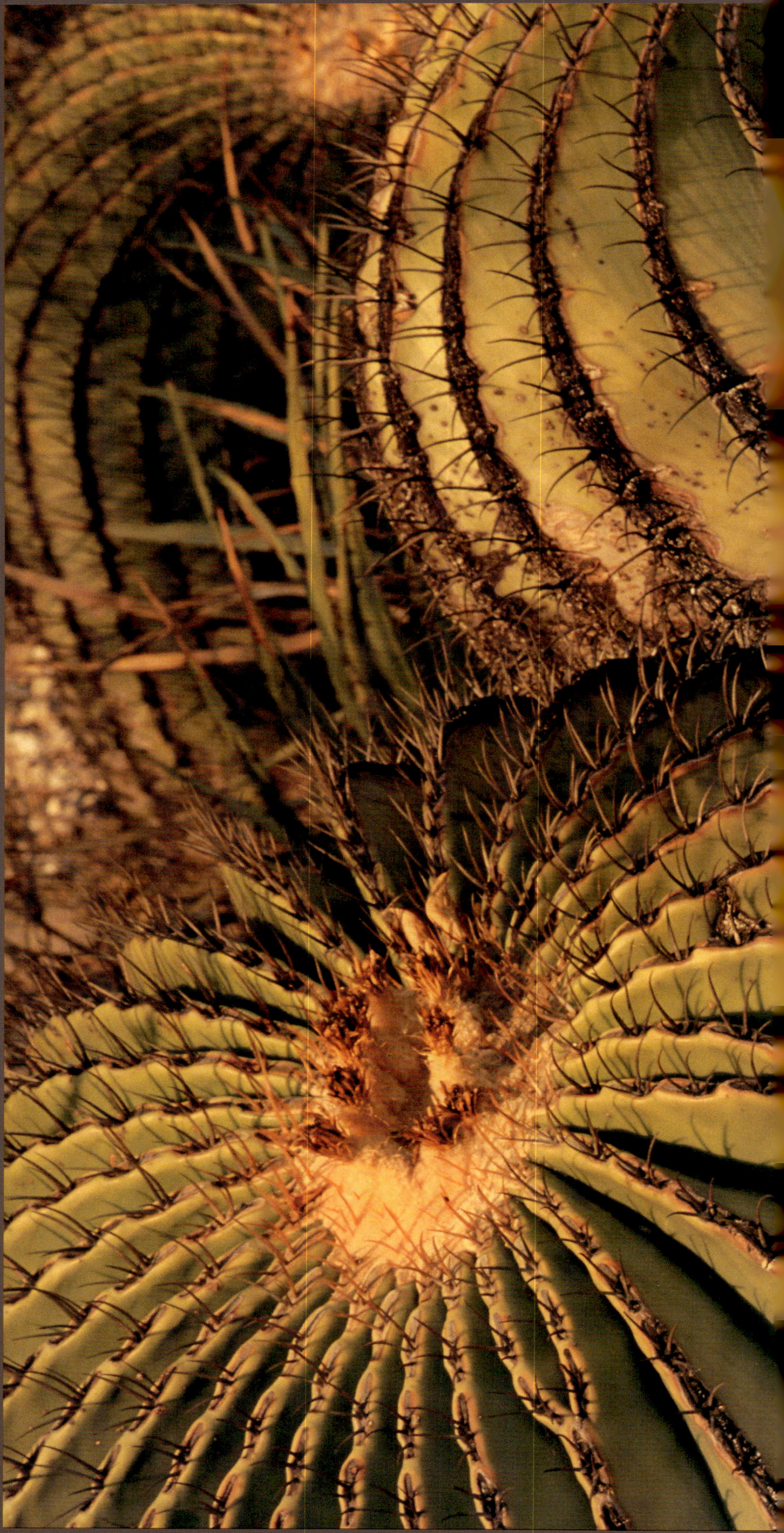

Mexico has the highest diversity of Cactaceae in the world, with 832 species. Above, a little organ-pipe cactus (Echinocereus spp.) *in Santa Maria, Coahuila. Right, a great barrel cactus* (Echinocactus platyacanthus) *in Tehuacán, Puebla.*

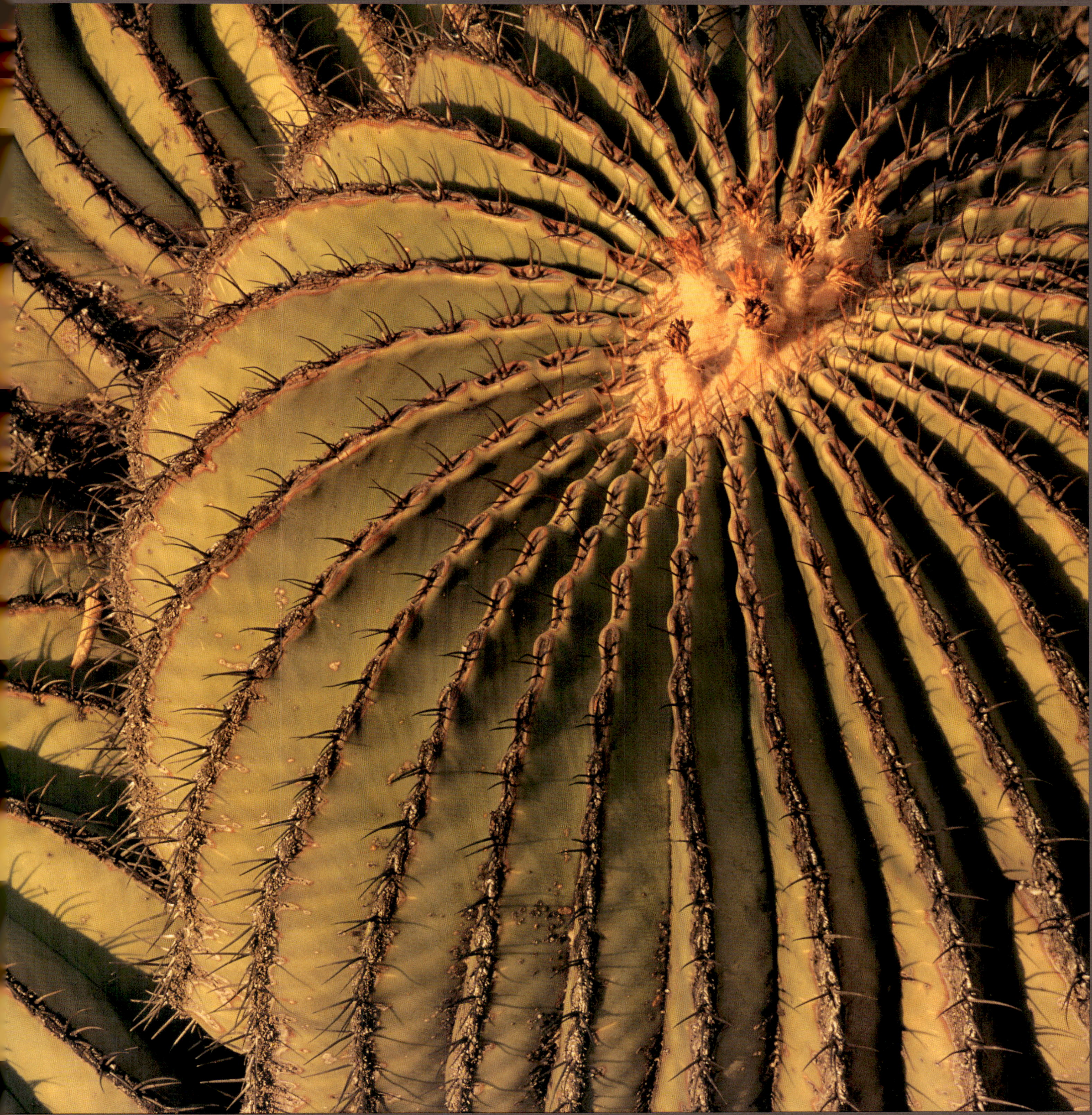

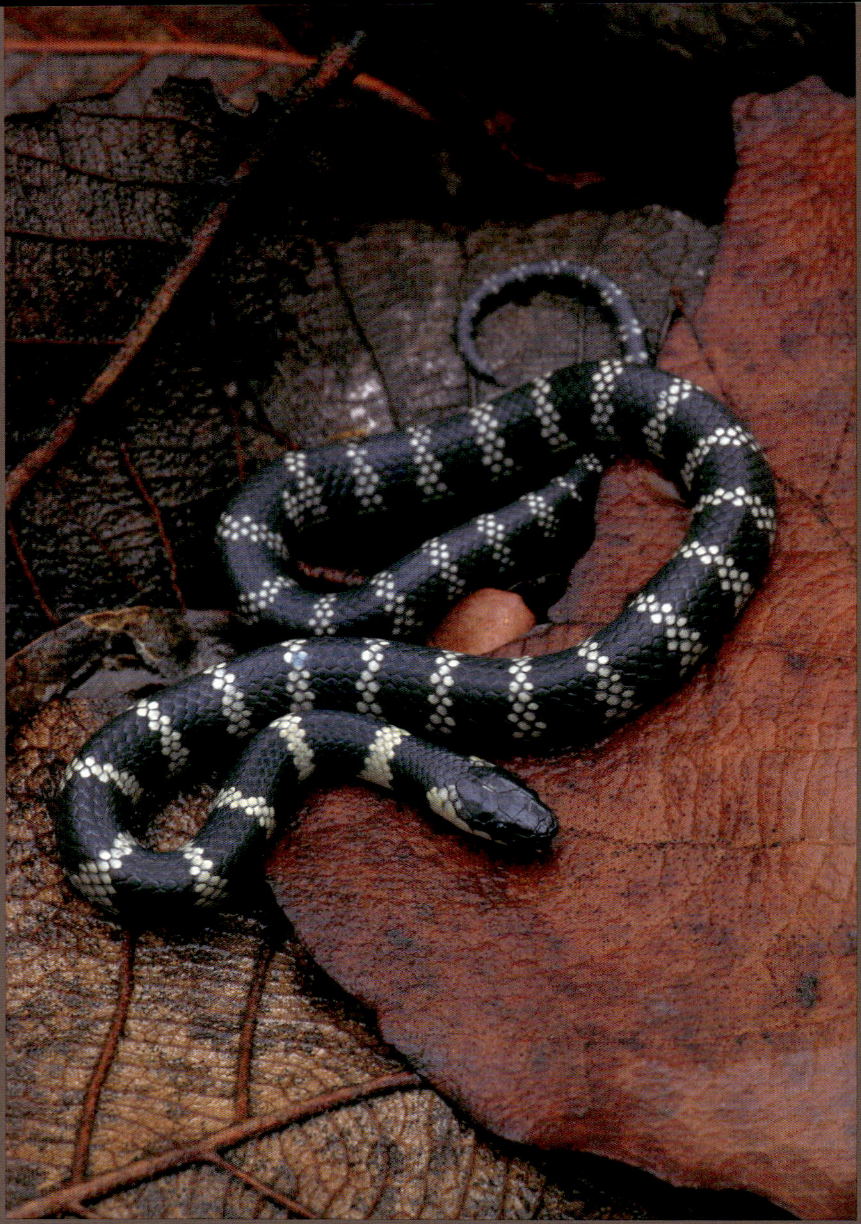

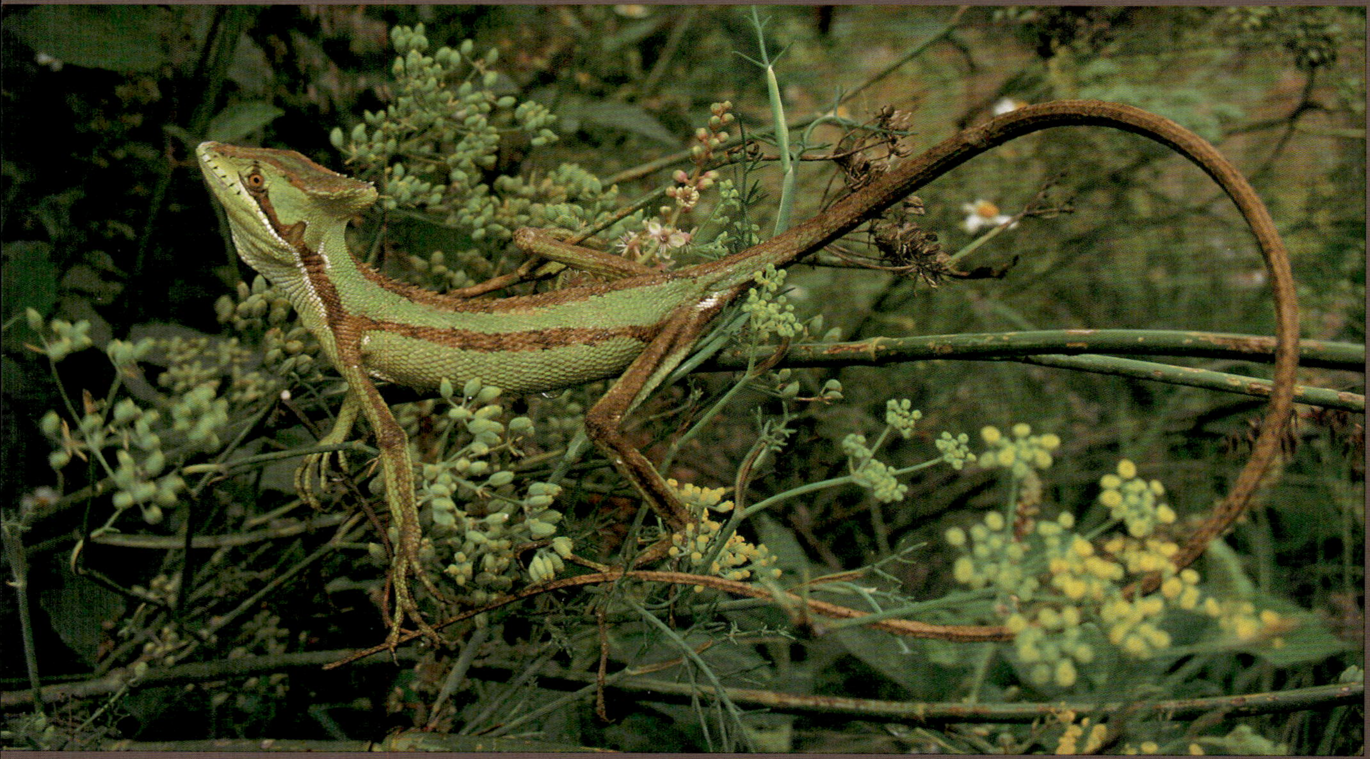

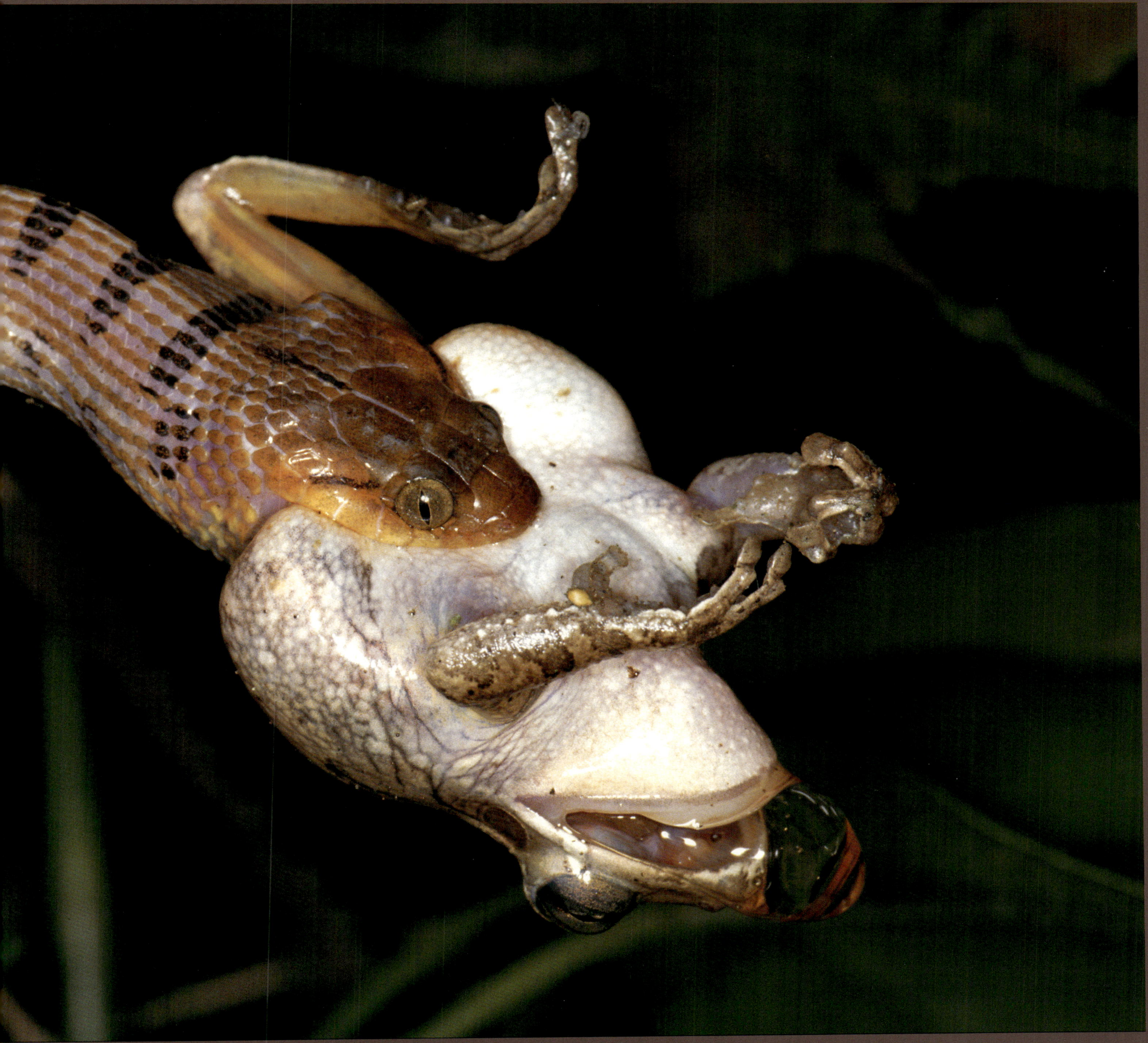

In Mexico, the diversity of environments influences the variety of habits. Here we have three species belonging to the reptile world: upper left, a banded snail-sucker snake (Sibon faciata), *on the floor of a tropical forest; lower left, a serrated casque-headed lizard* (Laemanctus serratus), *mimetically green on a tree branch and, finally, above, a northern cat-eyed snake* (Leptodeira septentrionalis), *a nocturnal reptile that feeds on amphibians, eating a Mexican tree frog* (Smilisca baudini).

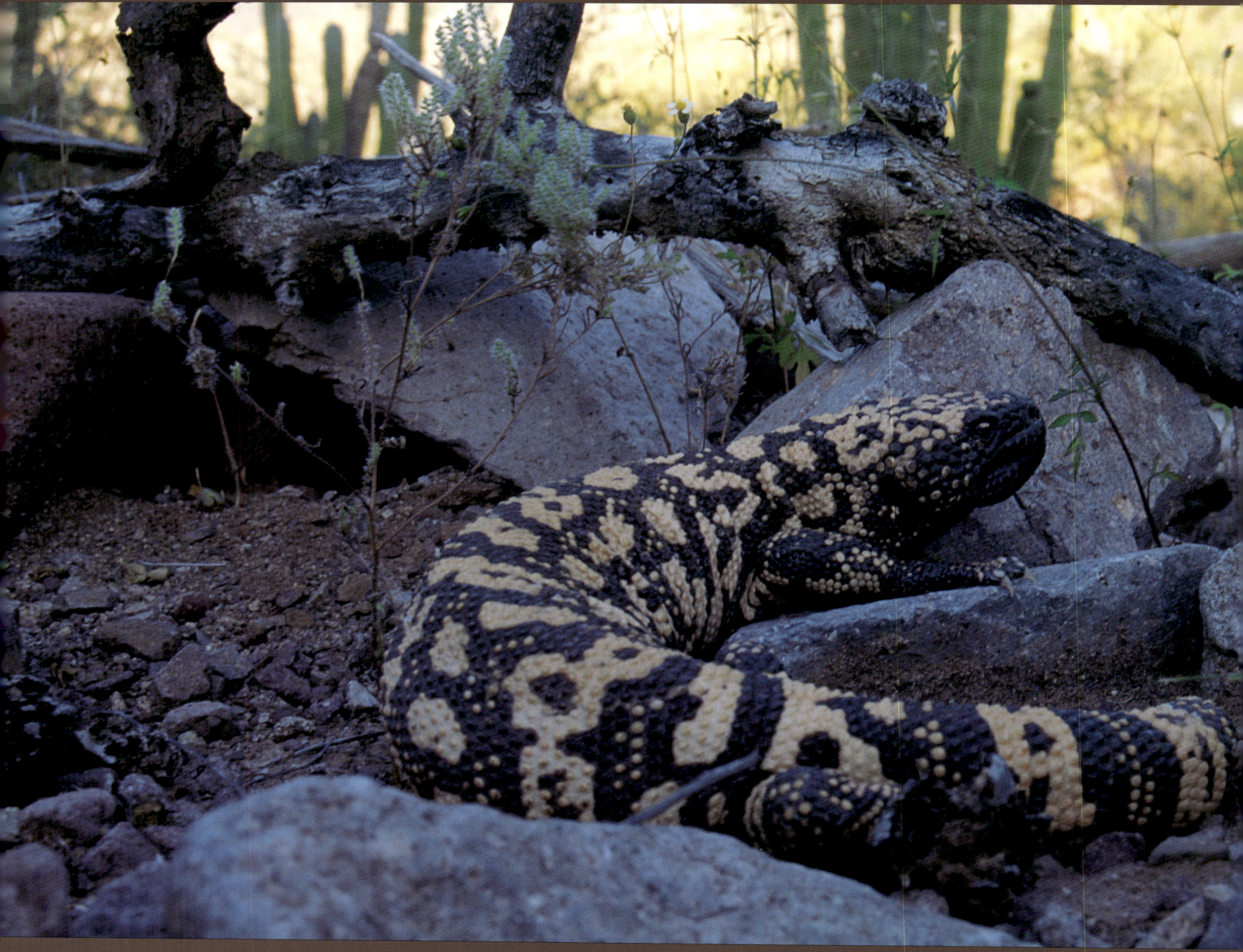

Reptiles tend to peak in diversity in drier ecosystems, although they are quite diverse in rain forest ecosystems as well. Much of Mexico's high reptile diversity is found in the vast desert regions of the country. The famous Gila monster (Heloderma suspectum) *shown above, is one of the only two poisonous lizards in the world. Two species of tortoises occur in Mexico: the widespread desert tortoise* (Gopherus agassizii) *and, upper right, the large Bolson's tortoise* (Gopherus flavomarginatus). *A Mexican endemic, this tortoise is in the Mapimí Biosphere Reserve. Mexico has the world's highest rattlesnake diversity, with 24 species. Lower right is a speckled rattlesnake* (Crotalus mitchelli), *which occurs in both Mexico and the United States.*

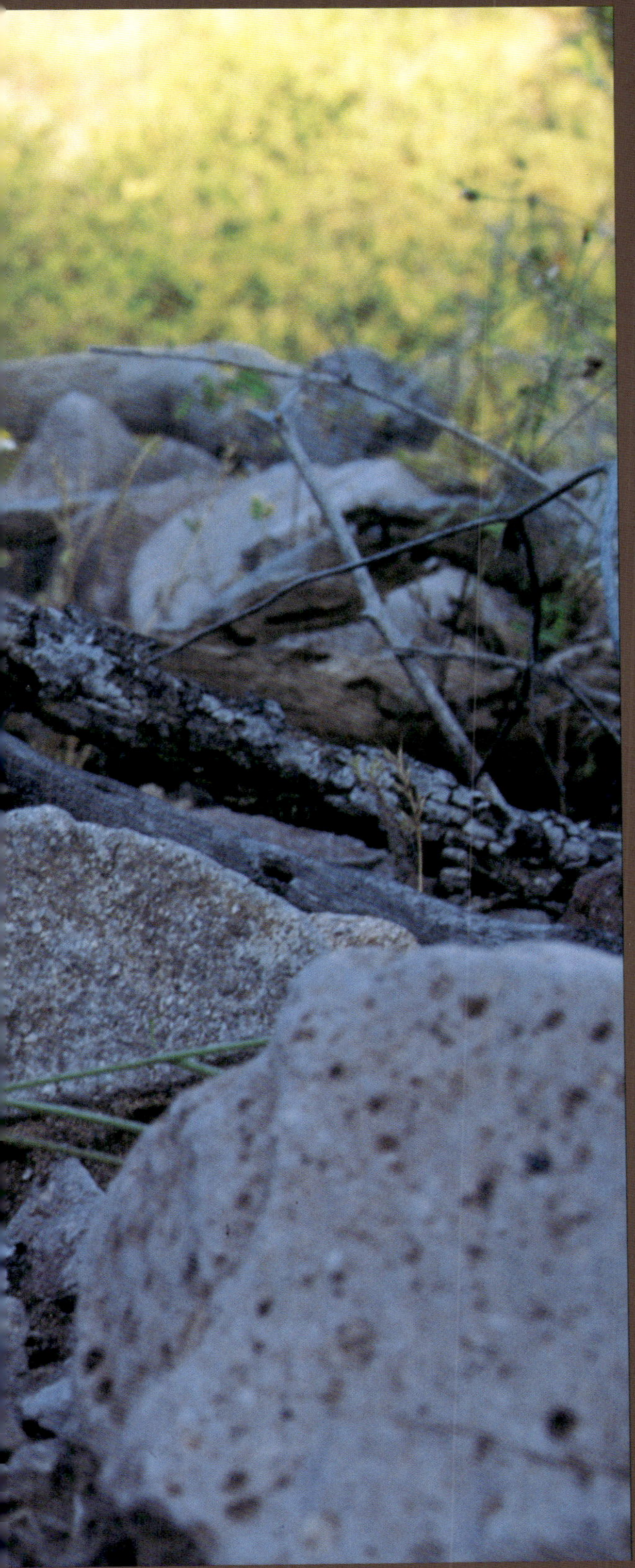

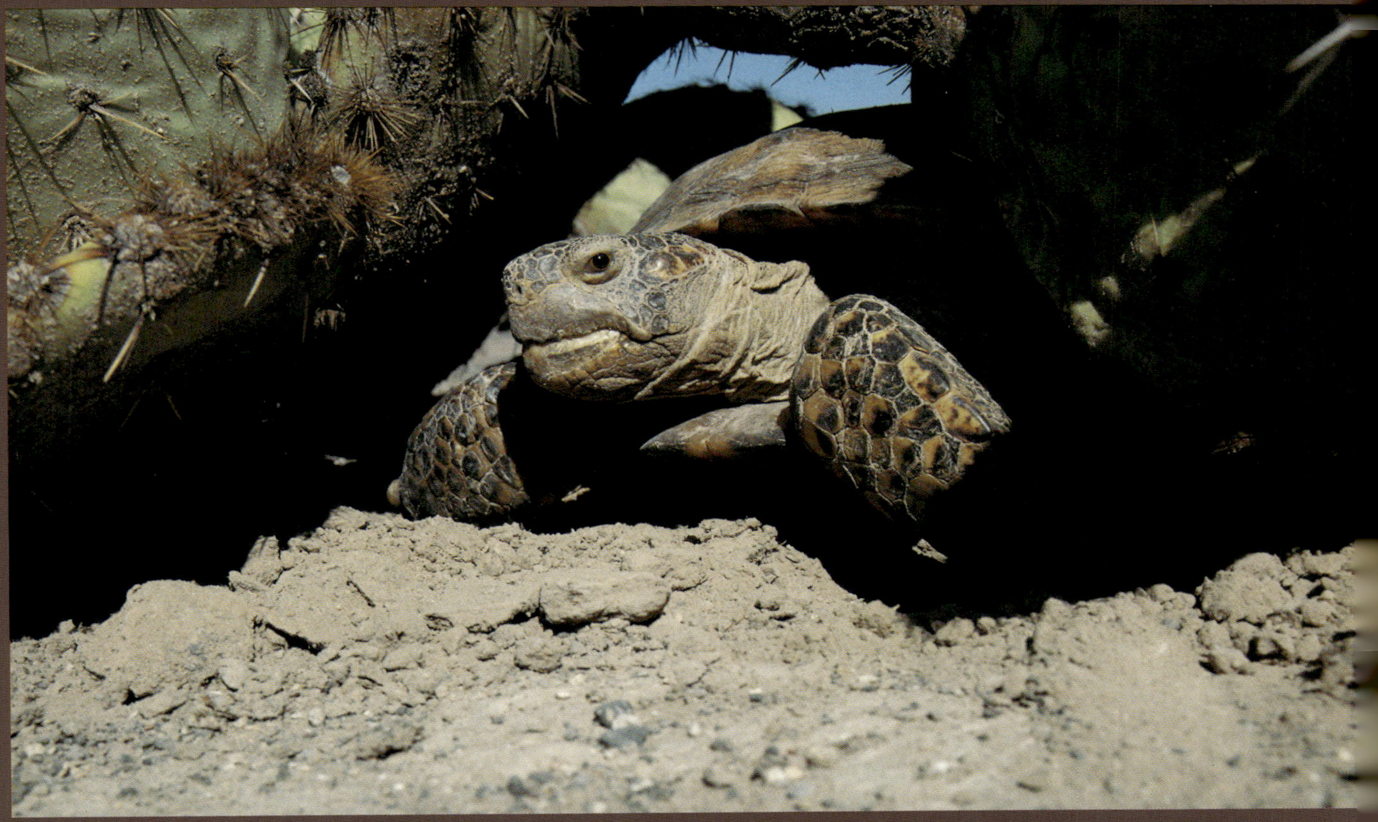

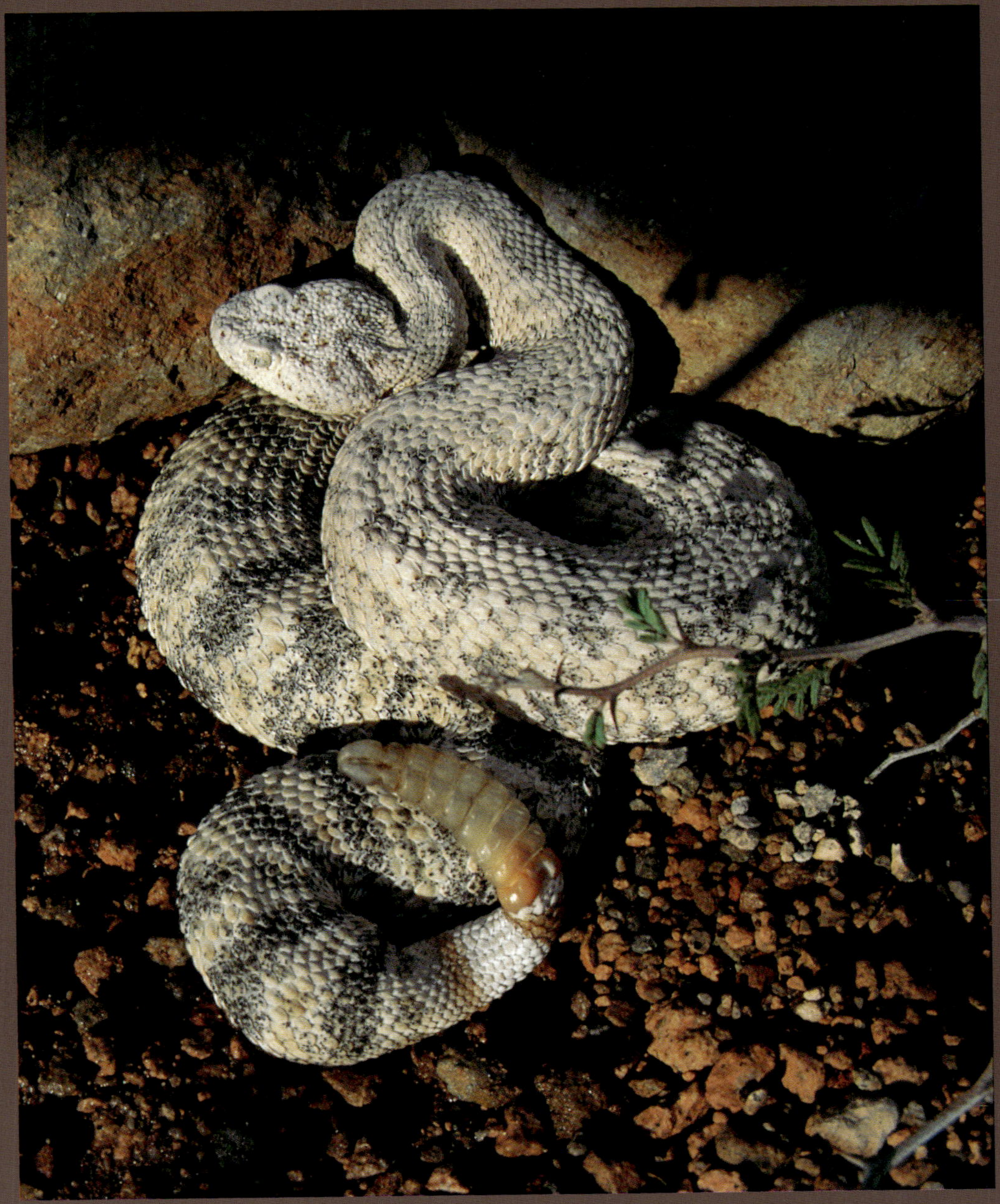

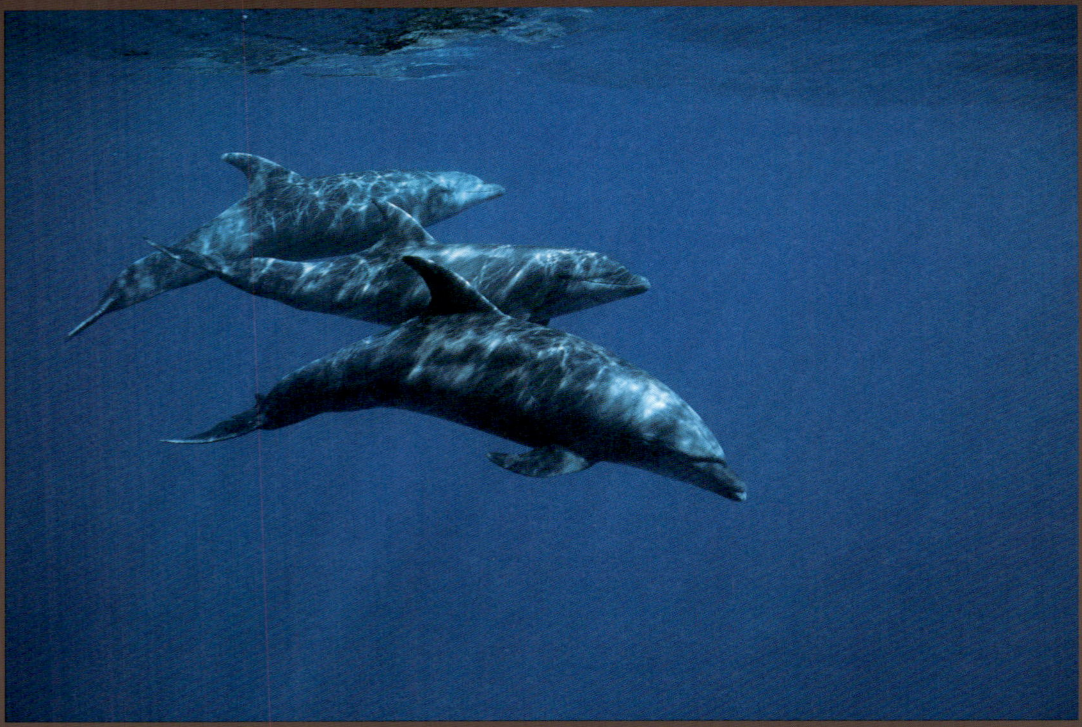

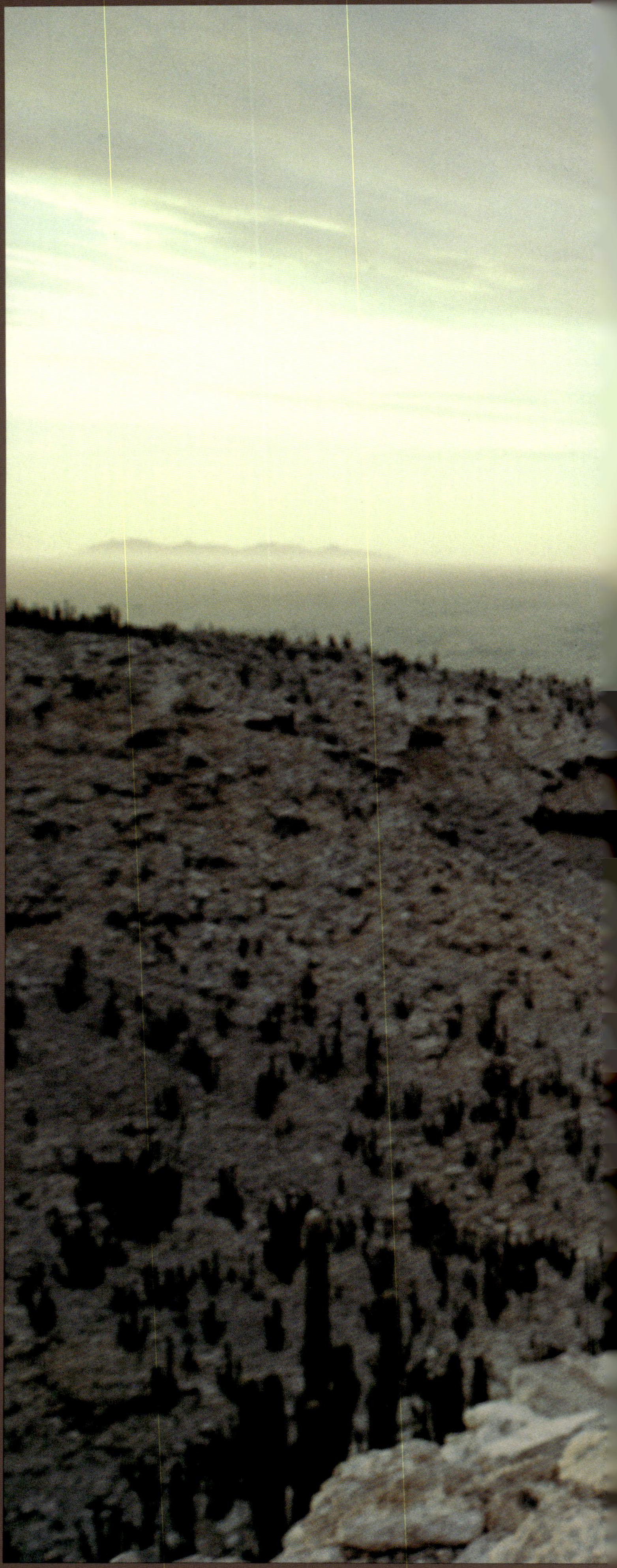

The Gulf of California is one of the most important marine systems on Earth, with 33 of the world's 77 cetaceans found within its boundaries. Above, a group of bottlenose dolphins (Tursiops truncatus), *one of the more common species. The islands in the Gulf of California are of great importance as well, and are in many respects a mini version of the Galápagos Islands in Ecuador in terms of their diversity and endemism. On the right, a blue-footed booby* (Sula nebouxii) *surveys its domain on the island of San Pedro Mártir.*

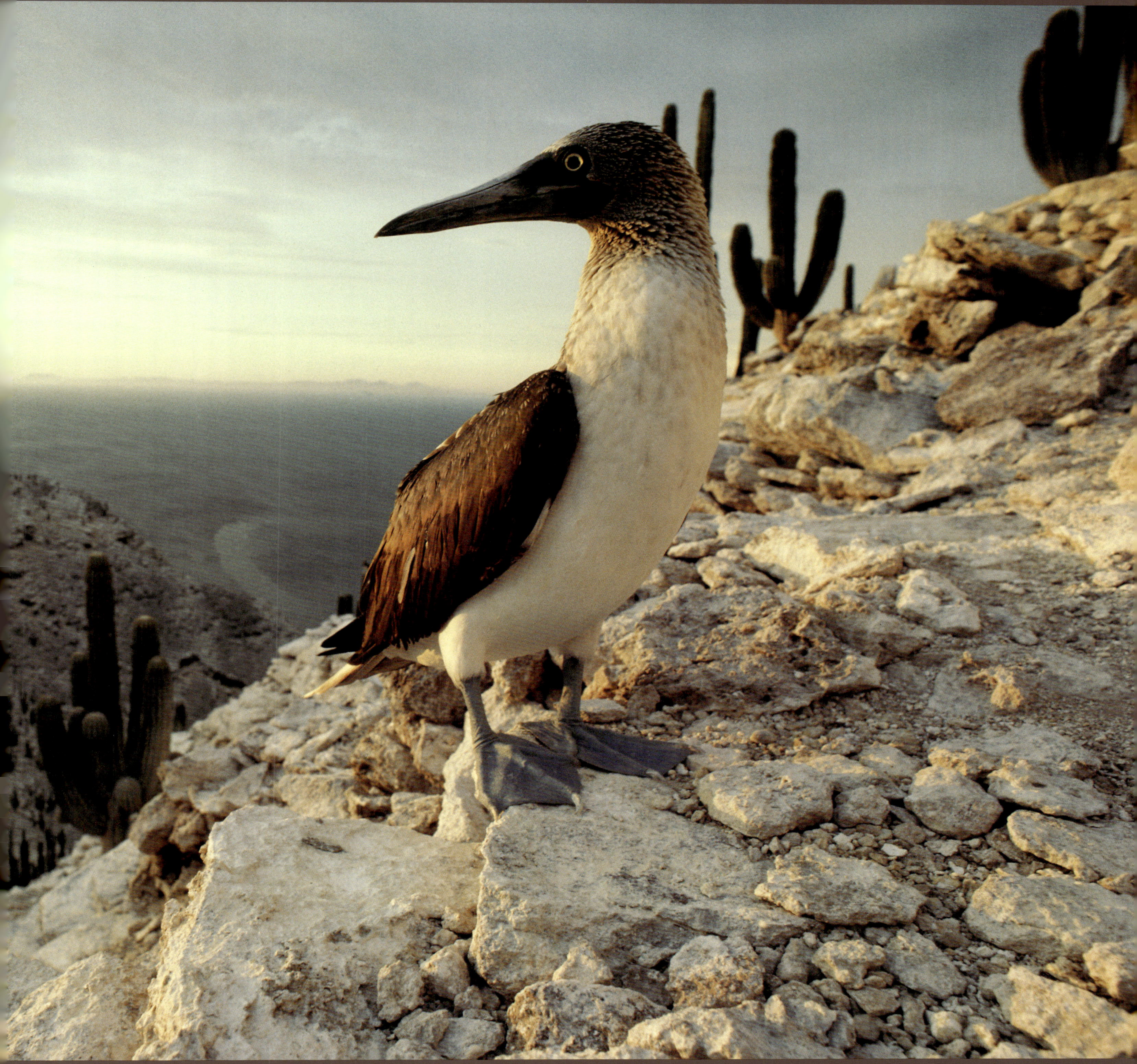

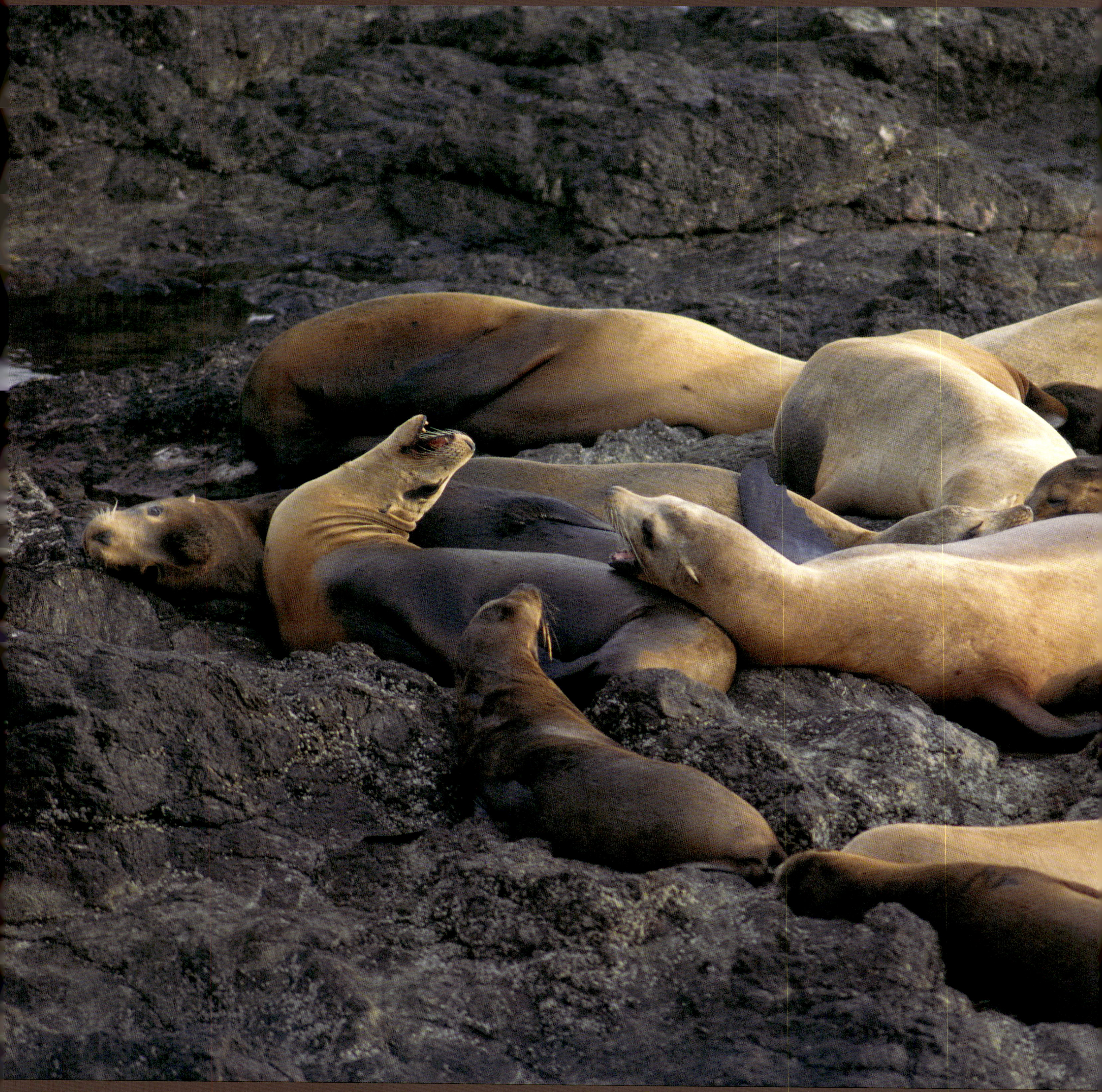

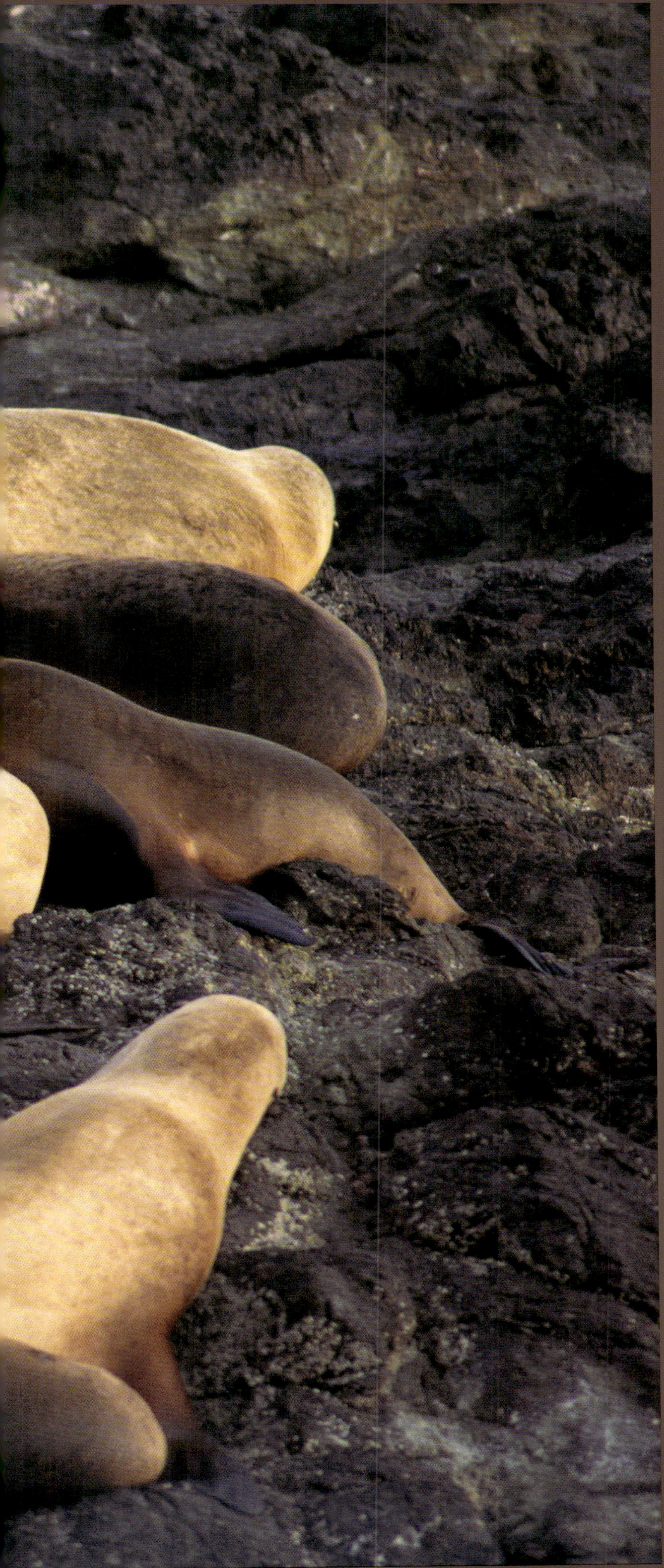

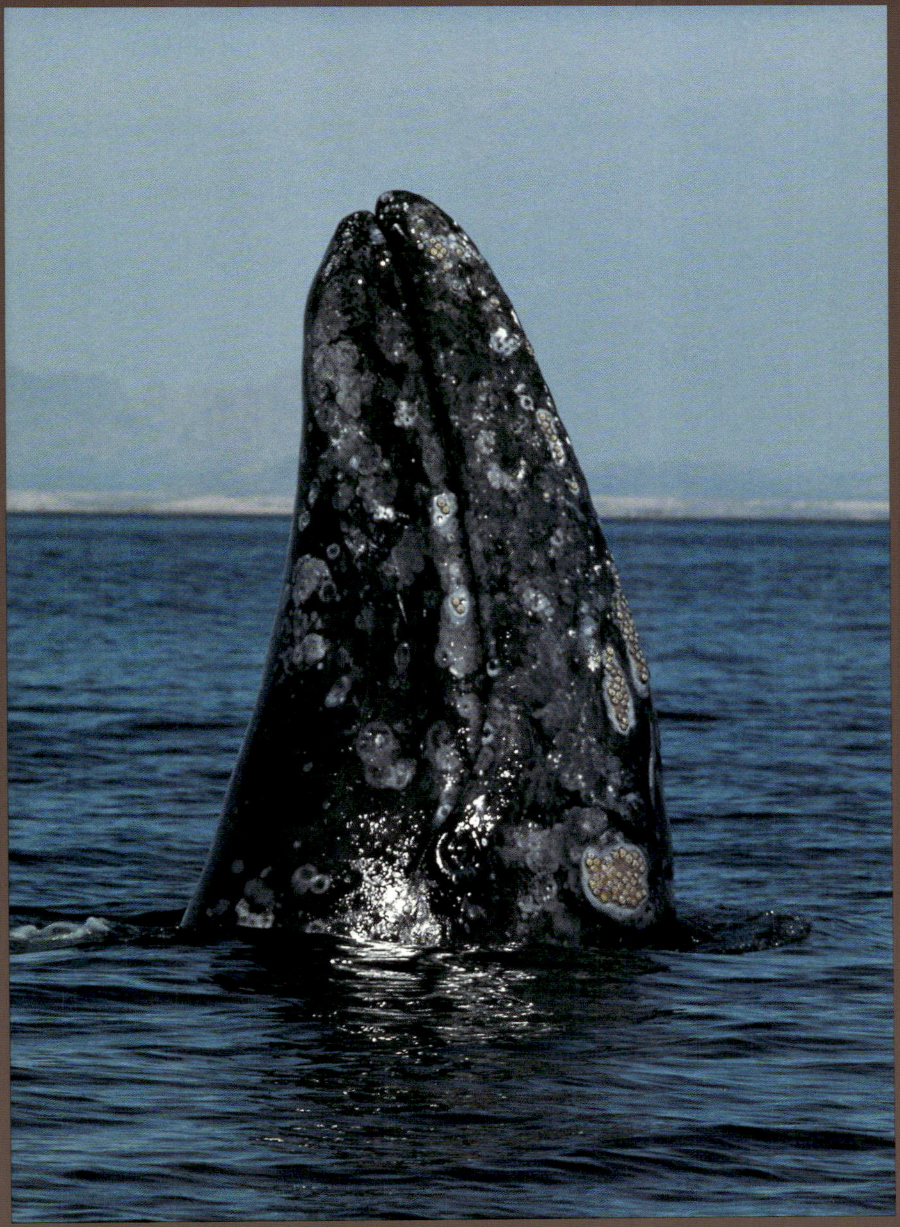

Above, a gray whale (Eschrichtius robustus) *breeds in the coastal lagoons in Baja California, like the San Ignacio Lagoon, and is one of the 13 whale species found in the Gulf of California. Left, a group of California sea lions* (Zalophus californianus) *a species common in the Gulf of California, that breeds on many of the islands.*

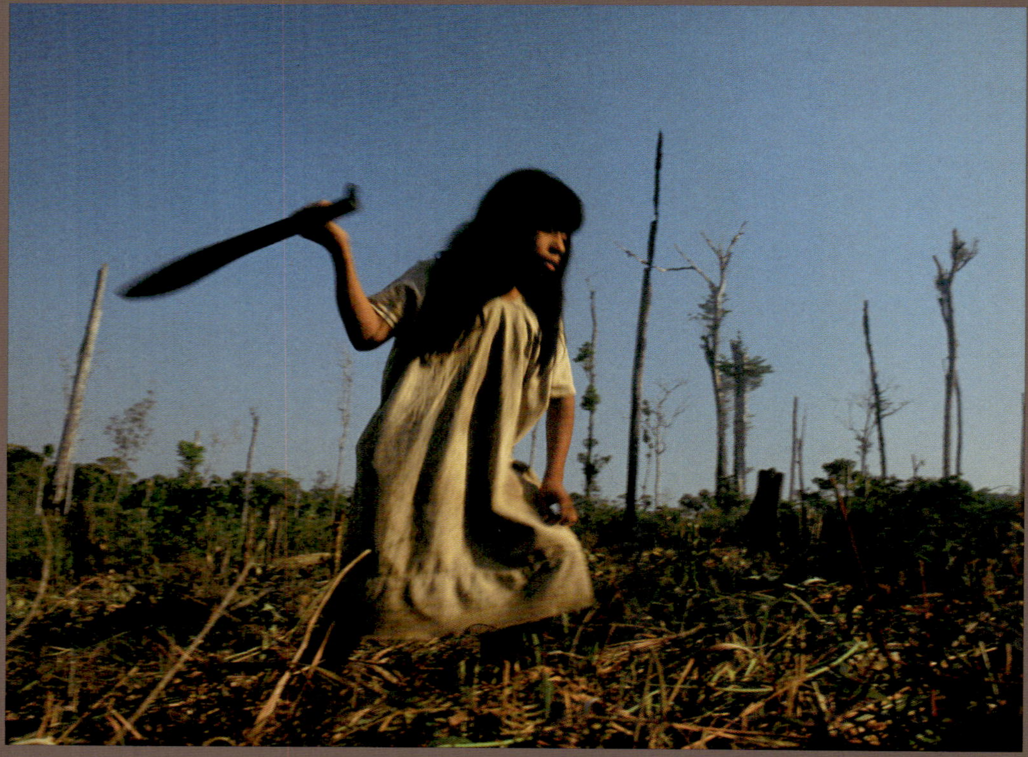

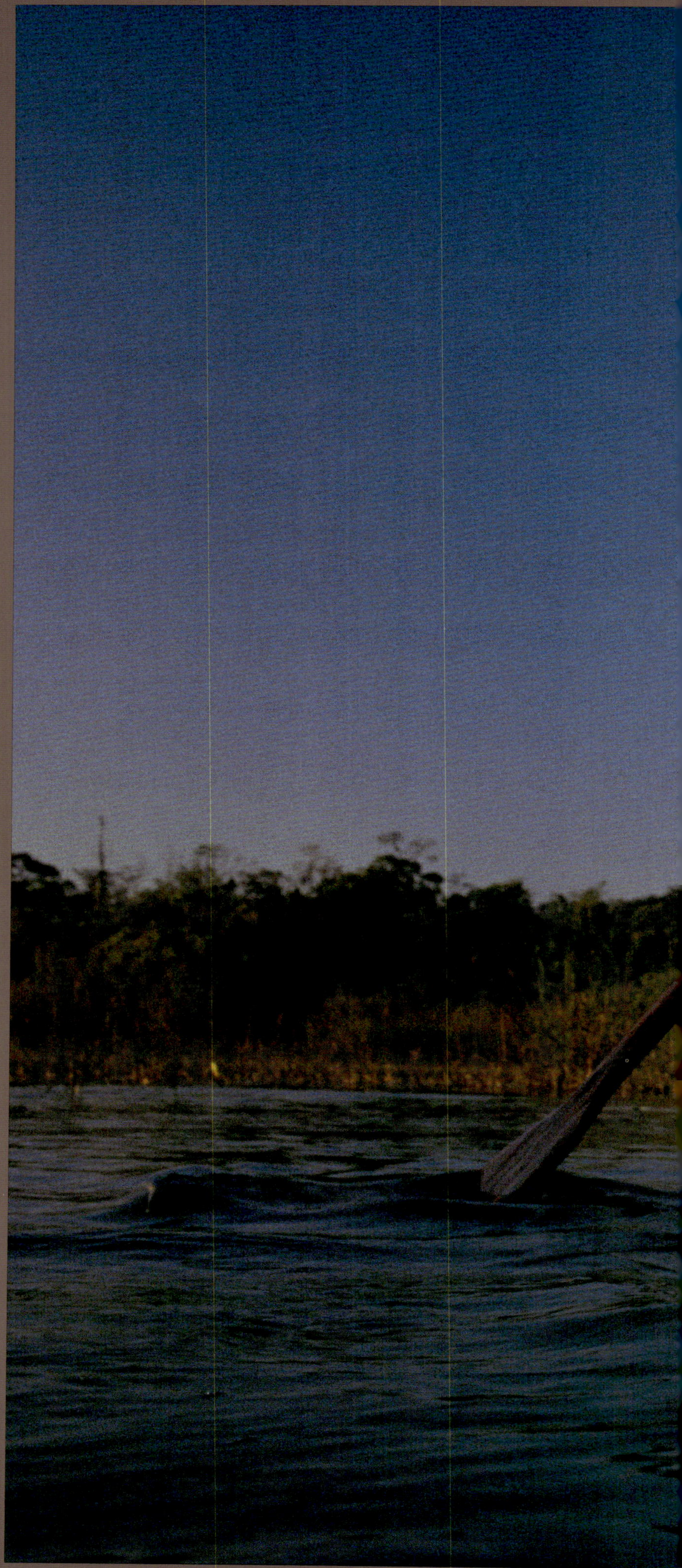

The Lacandón Indians are one of the most traditional Indian groups in Mexico and inhabit the Selva Lacandona, which is named after them, in Chiapas. They are very autonomous and independent, and produce for their own consumption, raise birds and bees for trade, and also resort to fishing and hunting to survive. This ethnic group has traditionally lived in isolation, which helped them to maintain their rites and beliefs. Three of their main gods are Hachakyum, the creator; Kisin, god of the underworld; and Nojoch-Yum Chac, god of rain. Above, man with a machete in a field. Right, Indians in a canoe on a lagoon in the Selva Lacandona.

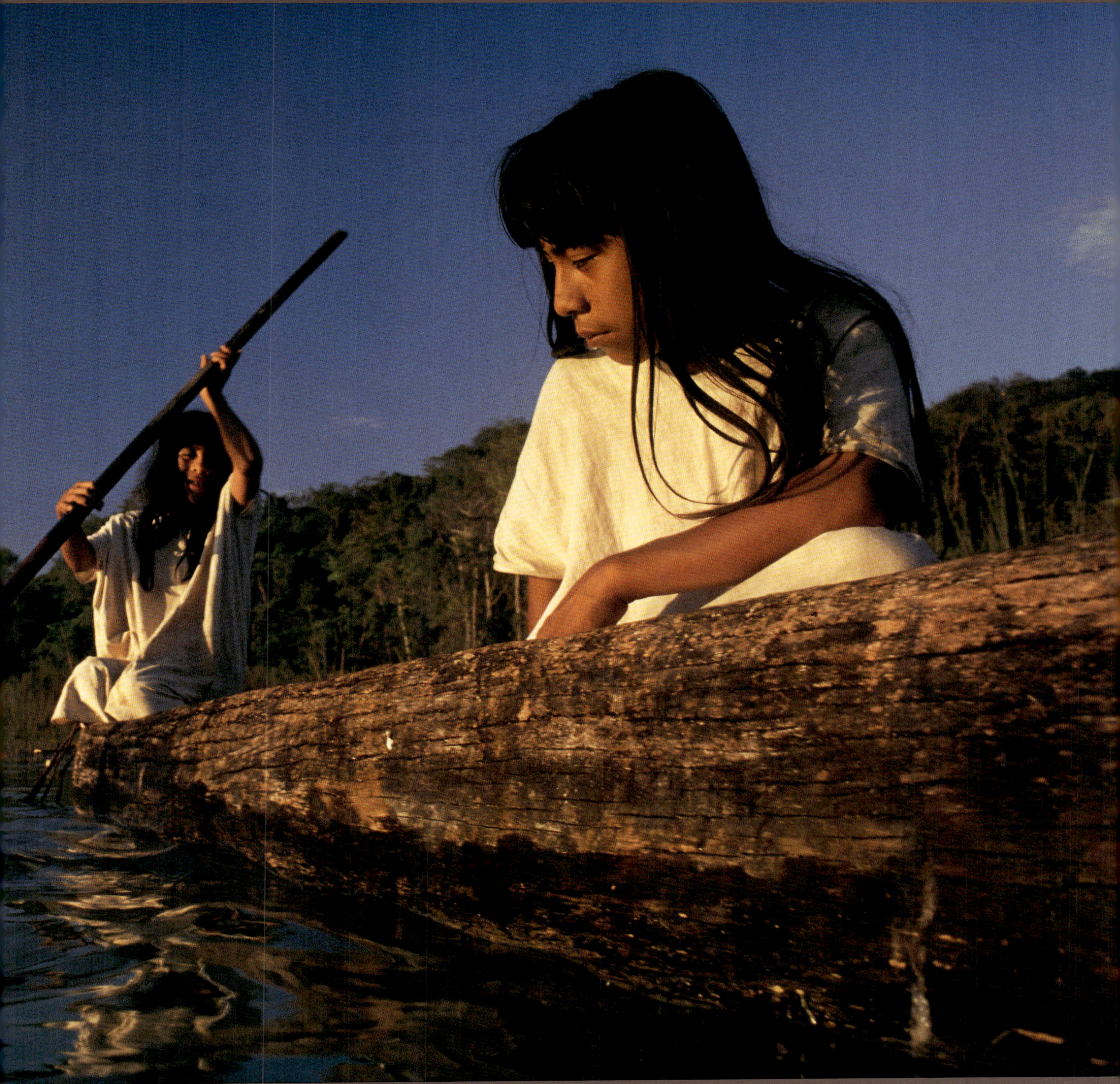

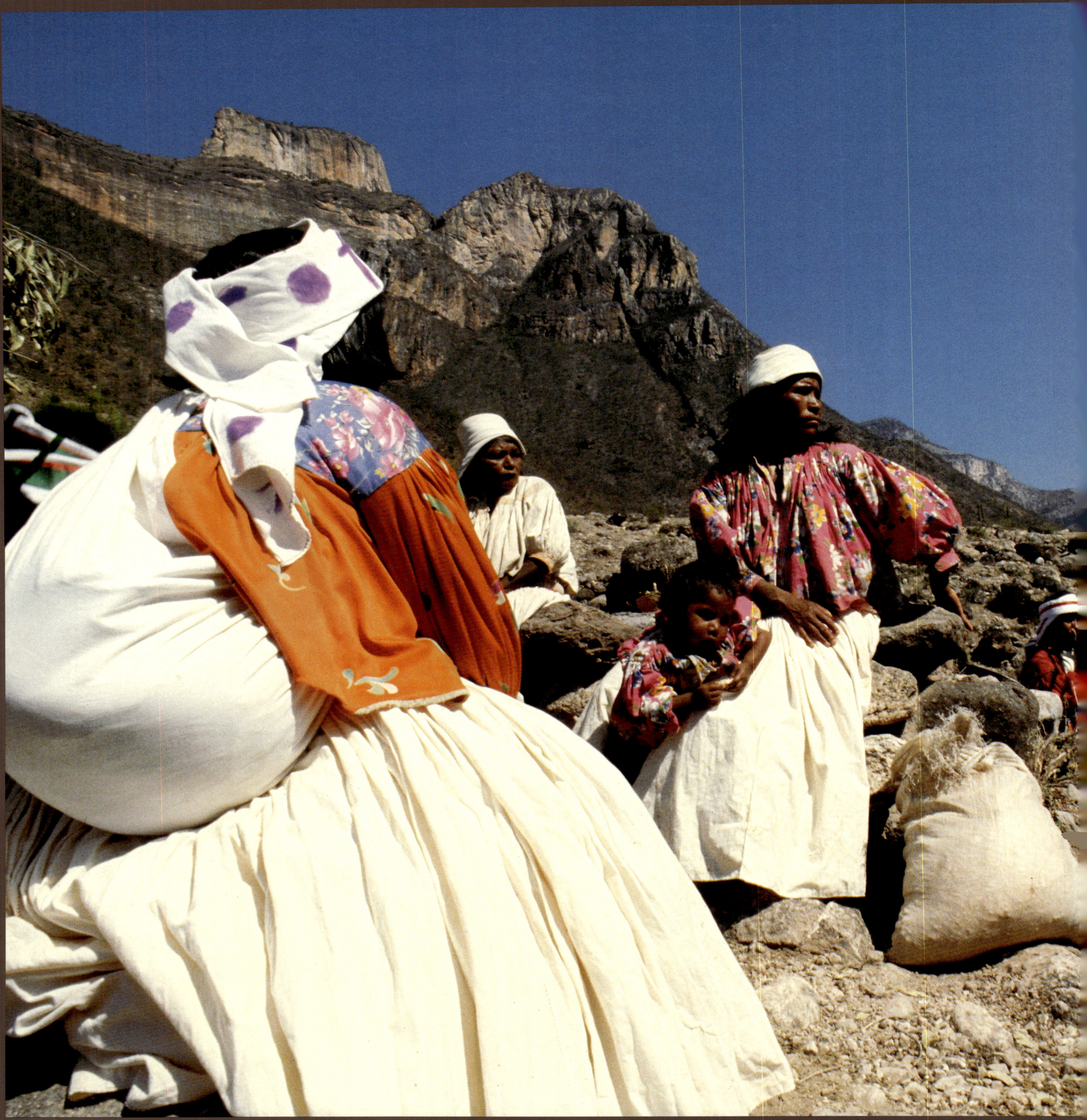

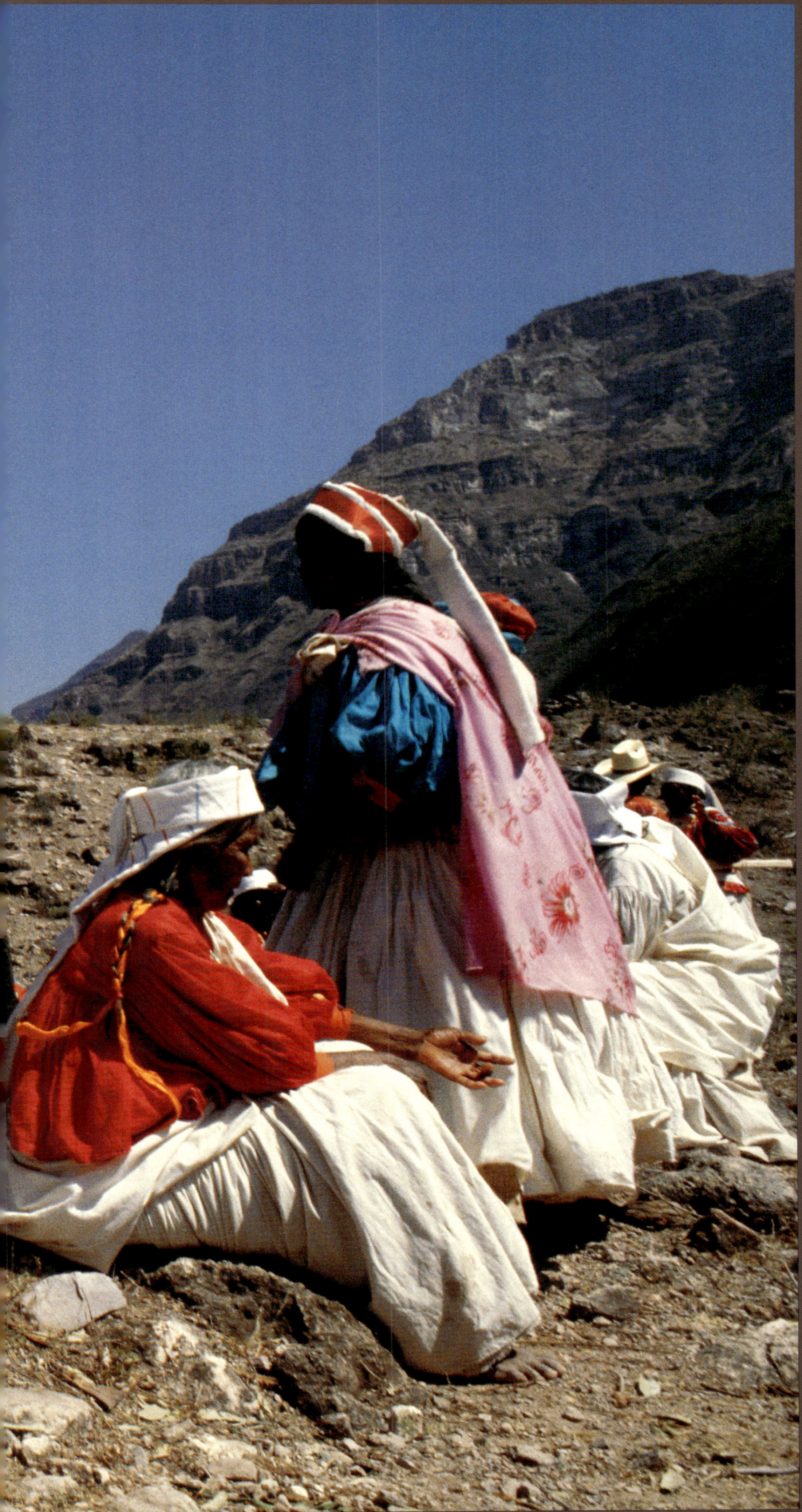

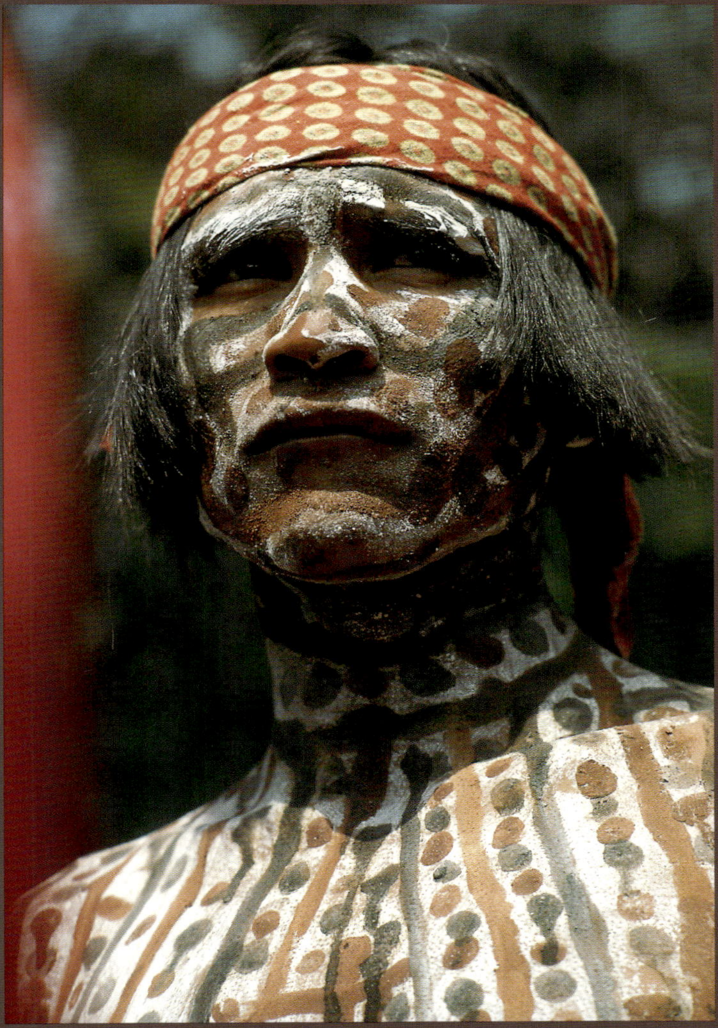

The Tarahumara Indians, who inhabit the mountain zone of Chihuahua, have experienced very difficult problems to survive since the Europeans arrived in the area in the sixteenth century. They fought for their independence but were defeated and confined to increasingly smaller areas. Today they still live in isolation, struggling very hard to live. Above, man painted traditionally in preparation for Holy Week celebration. On the left, another Holy Week celebration at Munerachi, Chihuahua.

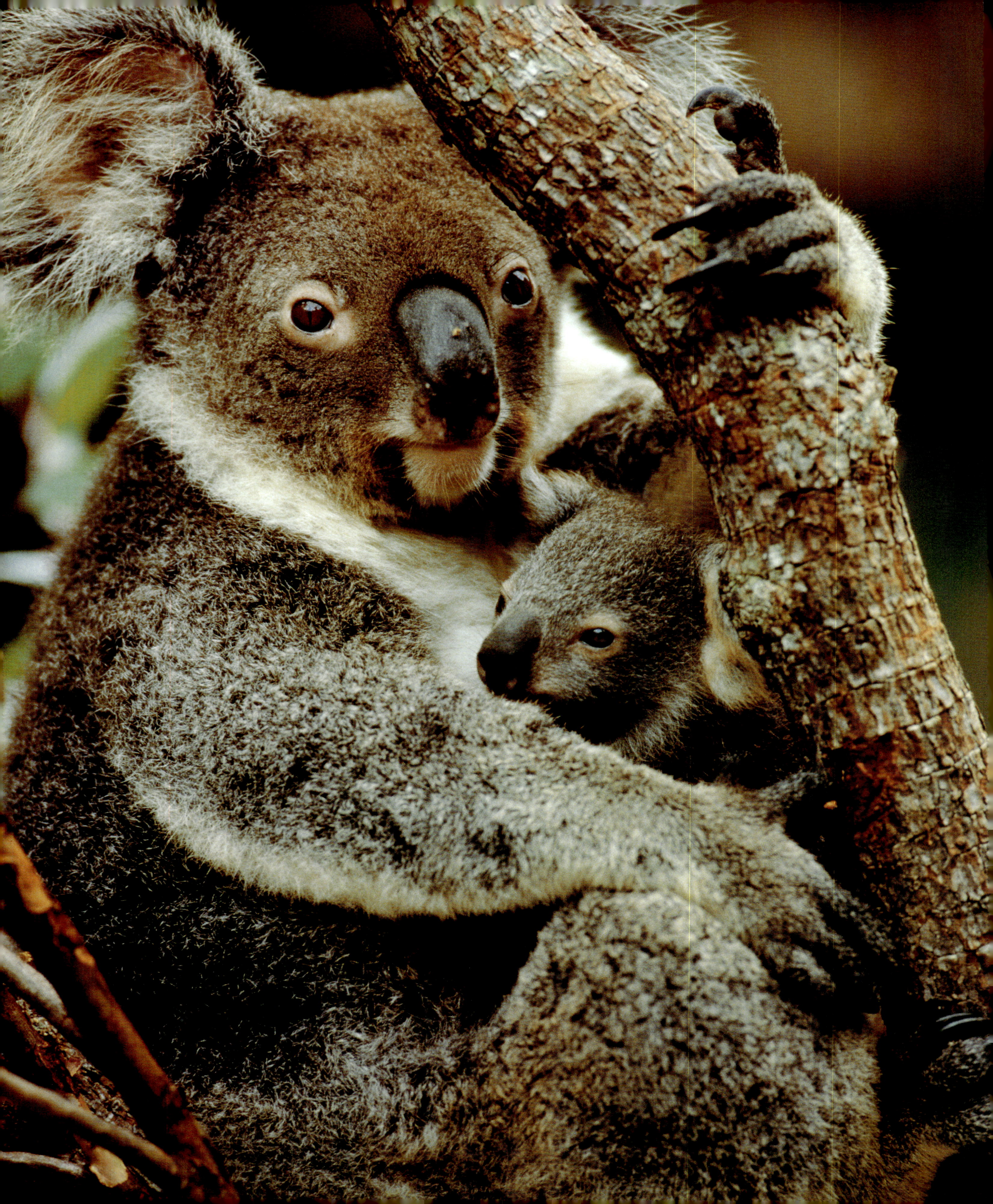

AUSTRALIA

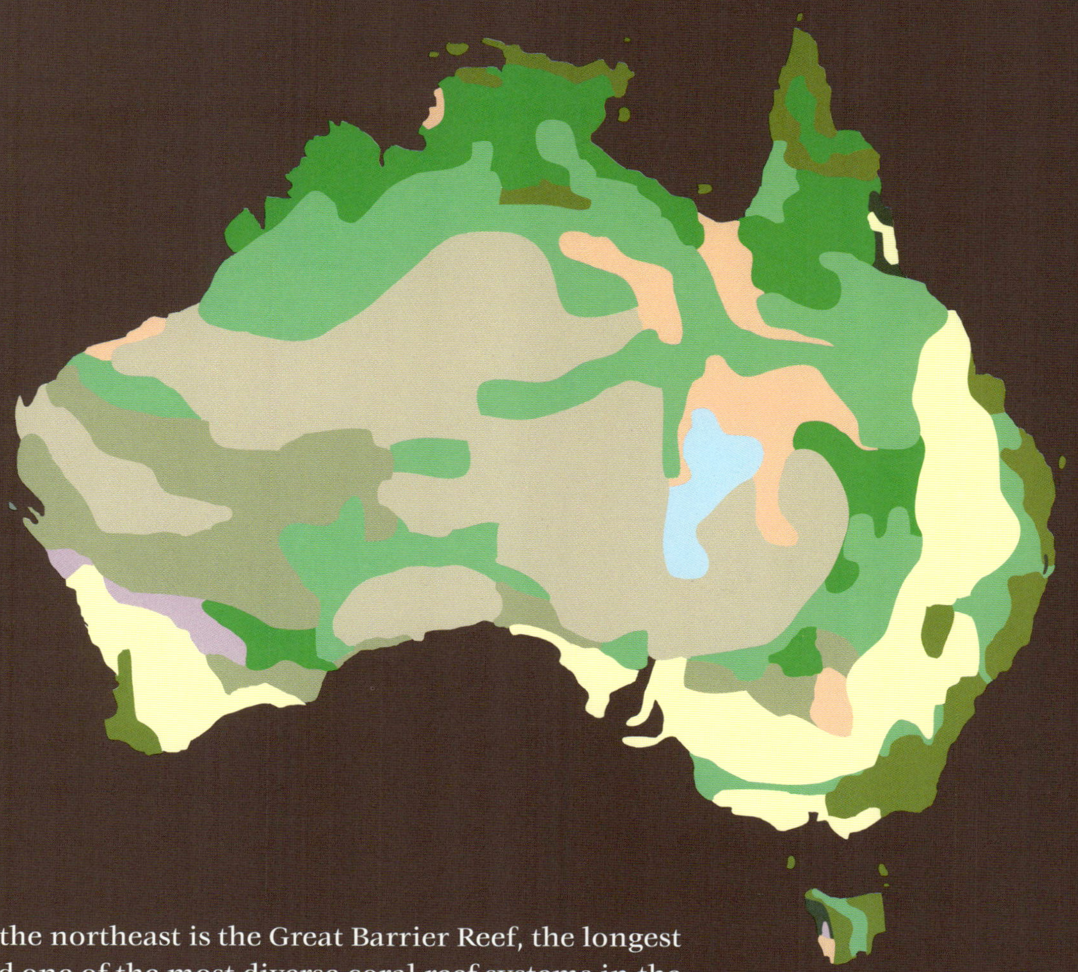

Australia is both the smallest inhabited continent and the only continent that is just one country. It is also the driest and the flattest continent, with the highest peak, Mt. Kosciusko, reaching only 2 230 m. Mean elevation is only about 305 m, by far the lowest of any continent. Since there are very few mountains to stimulate the production of rain clouds or to provide runoff, the continent is generally quite dry, although rainfall gradually increases towards the coasts. All of Australia's rivers together drain less than half the area of the continent, and the Lake Eyre Basin in the northeastern part of South Australia is one of the largest areas of internal drainage in the world, occupying 1 243 194 km^2. The country's major mountain range runs more or less parallel to the continent's eastern coast; it is called the Great Dividing Range because on its eastern side its rivers flow into the Pacific, while on the west they go into the Murray River system or other systems of internal drainage that for the most part flow into the ground in sloping, porous layers of sandstone.

The sixth largest country on Earth (7 686 810 km^2), Australia has one of the lowest human population densities, with only about 18 million people in this vast land area, mostly concentrated in several large cities (Sydney, Melbourne, Adelaide, and Brisbane) on the eastern and southern coasts. Just as an example of Australia's vast, largely uninhabited extent, the state of Western Australia is almost as large as the three largest American states (Alaska, Texas and California) combined and larger than all but one other country behind Australia (India); indeed, if Western Australia were a country, it would be the tenth largest in the world. In spite of this, it has only 1.5 million people, of which 1.4 million live in the city of Perth. Several other states (Queensland, Northern Territory) are also enormous and sparsely inhabited.

Australia also has an interesting mix of tropical and temperate ecosystems. In the northeast, the Cape York Peninsula and the tropical rain forests of northern Queensland share much with New Guinea, whereas in the far south, especially in southeastern Australia and in Tasmania, one finds giant *Eucalyptus* forests and some of the most important stands of temperate rain forest left on Earth. In the central part of the country are some of the largest and driest deserts on Earth, notably the Simpson Desert, and in the northeast is the Great Barrier Reef, the longest and one of the most diverse coral reef systems in the world. Needless to say, this mix of temperate and tropical ecosystems leads to great biodiversity, which is perhaps best reflected in the country's marine life. Marine fish, for example, number some 4 300 species, almost double that of Brazil and probably the highest on Earth, representing roughly 20% of total global diversity in this group. Such wealth is achieved because of a combination of a temperate fauna and a tropical fauna, and because of the marine systems entirely surrounding this island continent. The fact that Australia is so dry has also led to a considerable level of diversity in groups that tend to adapt well to dry systems, especially the reptiles; not surprisingly, Australia has by far the highest reptile diversity on Earth, with 755 species, 616 of which are endemic. On the other hand, groups that tend to be more diverse in wetter habitats have correspondingly fewer representatives in Australia; hence, amphibian diversity is relatively low at 196 species (less than half that of tiny Ecuador, smallest of the megadiversity countries), although 169 of these are endemic.

Given that it is a large, isolated island, Australia has very high levels of endemism in addition to its marked diversity in certain groups and, as in Madagascar, this endemism extends to higher levels of taxonomic organization (family, genus). For instance, Australia's total plant diversity is 15 638 higher plants, which is fairly considerable and certainly puts it in the top 12 countries, but its endemism, at 14 458 species, is extremely high and is comparable to plant endemism for Brazil, Colombia, Indonesia, and South Africa. Furthermore, at the family level,

On the opposite page, koala mother and joey. The koala (**Phascolarctos cinereus**) *is one of Australia's best known and most popular flagship species, and indeed is one of its best loved creatures.*

Australia is unmatched, with 12 endemic plant families. Brazil and Colombia, in spite of their very high plant diversity, have no endemic families, and only Madagascar and New Caledonia, with five endemic families each, can compare in this category. Vertebrate diversity is also quite high, with the combined number for mammals, birds, reptiles and amphibians totalling 1 984, or tenth on the world list. However, in terms of endemism, Australia, with 1 350 endemics in these four vertebrate groups, is by far the highest, with Indonesia a distant second at 848 endemics.

Indeed, Australia tops the world in two of the four vertebrate groups, with 210 out of 282 mammals and 616 out of 755 reptiles being entirely confined to Australia. In bird endemism it is second to Indonesia with 355 out of 751 species restricted to the country. Even Colombia, with an amazing 1 815 bird species compared to Australia's 751, has less than half the number of endemic bird species (142). In amphibian endemism, Australia comes in fifth behind Colombia, Brazil, China and Madagascar and tied with Mexico at 169 endemics out of a total of 196 species.

Australia also has within its borders two of the world's top-priority global biodiversity hotspots, southwestern Australia and southeastern Australia/Tasmania, both of which are described in more detail below. It also has 14 Centers of Plant Diversity and Endemism as identified by WWF/IUCN and eight Endemic Bird Areas as defined by BirdLife International (ICBP). When Australia's high diversity and very high species endemism at the family and genus level are taken into consideration, this biologically very unusual country clearly ranks in the top five nations on Earth, and vies with Mexico for the fourth position overall.

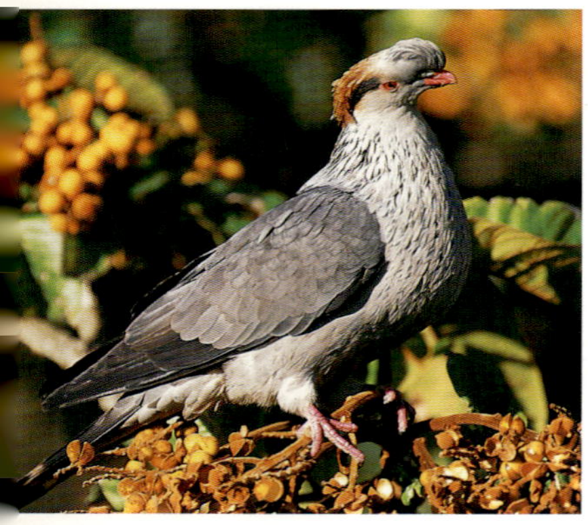

The topknot pigeon (Lopholaimus antarcticus) *from the east coast of Australia. Pigeons and doves reach their greatest diversity in Indonesia, New Guinea and Australia.*

Natural habitats

Australia is located in the southwestern part of the Pacific Ocean about halfway between southeast Asia and Antarctica. The continent straddles the Tropic of Capricorn, with some 40% of it being on the northern tropical side, and the rest in the temperate zone. All of Australia's natural environments divide neatly between these two climatic regions. The tropical portion of Australia includes several different types of environments, including the wet tropical region, the humid and subhumid tropical regions, and the arid and semiarid tropical regions; the temperate portion also includes humid and subhumid, and arid and semiarid zones.

The major vegetation types of Australia are rain forest, tall open forest, woodland, scrub and shrubland, heathland, grassland, salt-marsh and mangrove forests, alpine and subalpine vegetation, and coastal dune vegetation. Rain forests in Australia occupy only about 20 000 km^2, or 0.26% of the land surface, and occur within three climatic ranges, namely, tropical, subtropical and temperate. The transition zone from tropical rain forest to temperate rain forest also harbors a wide variety of life. Rain forests spread in a broken chain from the Cape York Peninsula south to Tasmania and occur also on the northernmost tip of the Northern Territory. Within tropical Australia, the major rain forest massifs occur in clusters separated by dry corridors, which act as effective barriers to gene flow. Australia's southern temperate rain forests have a spotty distribution across a latitudinal belt between 28°S and 44°S.

Rain forests in the most humid region of Australia, which is known as the Wet Tropics, are classified as closed rain forests because the canopy is closed and the trees are densely spaced, in contrast to the open and generally scattered sclerophyllous dry forest vegetation that covers most of the forested area of the continent. All the remaining rain forests of Australia outside of the Wet Tropics are classified as open rain forests due to the discontinuous nature of their canopy.

The Wet Tropics region is located in the southeastern corner of the Cape York Peninsula, in the northern part of the state of Queensland, and it gets its name because it receives the most rainfall in all of Australia, from 1 200 to 4 000 mm a year. This area is mainly covered with closed rain forest but other formations can be found there as well, such as sclerophyllous forest and more open savanna woodland. Closed rain forests have three or more tree layers with or without emergents. More than 800 species of trees occur in the Wet Tropics rain forest. Despite the fact that they represent only 0.1% of Australia's land area, the Wet Tropics are home to about 50% of all Australian bird species, 30% of marsupial species, 27% of frog species, 23% of the reptile species, and 60% of butterfly species. The Wet Tropics also contain the most diverse assemblage of primitive angiosperm plants in the world. Of the 19 known families of primitive angiosperms, 13 occur in this area. These living fossils provide unique insights into the early evolution and diversification of flowering plants.

The rest of the Cape York Peninsula falls into the humid tropical climate category, receiving less rain than the Wet Tropics. Most of the peninsula is affected by the seasonal nature of the climate, and the vegetation includes open *Eucalyptus* woodlands, open shrubland and monsoon rain forests. The heathlands in the far northeast are the result of the sandstone soils, but the main formation in the region is open rain forest, as is also the case on the northernmost tip of the Northern Territory.

The Cape York Peninsula area is large even by Australian standards, and rich in diversity. It is the part of Australia nearest to New Guinea, and over millions of years there has been considerable interchange of fauna and flora with that island. The richest areas are the mid-peninsula rain forests of the McIlwraith and Iron Ranges. There are many species shared with New Guinea, like the saltwater crocodile (*Crocodylus porosus*), which still has a secure home in the rivers of the peninsula; the green tree python (*Morelia viridis*), one of the most attractive snakes in the world; and the spotted cuscus (*Phalanger maculatus*), as well as a variety of other endemic possums. There are also several other endemic species, including the spiny bandicoot (*Echymipera rufescens*); the bare-backed fruit bat (*Dobsonia moluccensis*); and two of the four Australian birds of paradise, the magnificent riflebird (*Ptiloris magnificus*), and the trumpet manucode (*Manucodia keraudrenii*).

The other humid tropical area is found in the northernmost portion of the Northern Territory. This rocky area is formed of Lower Proterozoic sedimentaries deposited about 2 billion years ago. These rocks form the base of the region, but the most important geological feature is a rock formation known as Kombolgie, which is made up of horizontal layers of sandstone that constitute the massive Arnhem Land plateau. The Alligator River cuts its way through much of the plateau and forms a very impressive gorge. The plateau has a general elevation of 250-350 m, but in places it can be much higher. About one third of the plateau is bare rock, and another third shallow sand, supporting heath and spinifex. The remaining third is deeper sand with oases of evergreen myrtle (*Allosyncarpia ternata*). Drainage is mostly through extensive systems of billabongs and lagoons that flow towards the Alligator River. Vegetation includes *Eucalyptus* woodlands, shrublands and grasslands. Paperbark forest dominated by melaleucas (*Melaleuca argentea, M. leucadendron*) fringes the rivers and swamps. There are also pockets of monsoon rain forest of considerable scientific interest.

Species richness in this area is paralleled only by the Wet Tropics of Queensland. Kakadu National Park, with an area of 1 755 200 ha, is in this region. Because of its great diversity of physical environments, it harbors at least 58 species of mammals, 275 species of birds, 100 species of amphibians and reptiles, and 50 species of fish, as well as an enormous diversity of plants.

The subhumid tropical region of the country occupies a large area in the northeastern portion of the continent, exclusive of the coastal regions of the Cape York Peninsula and the northern tip of the Northern Territory. It is the predominant environment in a good part of northern Queensland, the northernmost part of the Northern Territory, as well as the northernmost tip of Western Australia, where it can be found in the area known as The Kimberleys.

The main types of environments found in the subhumid tropical region are woodland, grassland, and sedgeland, all typical of drier climates. Woodlands are a structural plant formation that usually has a grassy component, dominated by perennial woody plants over 2 m tall. Because of their structural similarity, the terms "savanna" and "woodland" are often interchangeable. Aside from having a very high proportion of endemic species, these woodlands also have two large groups of sclerophyllous trees and shrubs, namely *Eucalyptus* and *Acacia*; between them they dominate almost every plant association in the continent. These *Eucalyptus* and *Acacia* forest types, better known as sclerophyllous forests, are a truly Australian formation, although they can also be found in New Guinea. Nevertheless, the species-rich genera *Eucalyptus* and *Acacia* may be regarded as the most typical genera of the Australian woody flora. Both are widespread, both have high levels of species endemism, and both are conspicuous dominants of major plant formations. In drier areas, *Acacia* is often the community dominant, while in more humid areas *Eucalyptus* dominates, reaching an extreme in the open *Eucalyptus* forests in southern Australia; however, even there, *Acacia* remains a strong representation in the understory.

In temperate Australia, the most important formation is temperate rain forest, which occurs predominantly in Tasmania, in disjointed areas in southcentral Victoria and along the Great Dividing Range in New South Wales and far southeastern Queensland. The island of Tasmania is the last great temperate wilderness remaining in Australia, and one of the last in the world. Western Tasmania, in particular, is second only to the Wet Tropics in average annual rainfall. All of this area has at least 200 days of rain annually. What sets it apart from most of Australia, except for the Wet Tropics, is not just its ruggedness and beauty, but the presence of the mesophyllous temperate rain forest. This formation, also known as cloud forest, is dominated by myrtle beech (*Nothofagus cunninghamii*), and includes three of Tasmania's endemic pines, the celery-top pine (*Phyllocladus asplennifolius*), the slow-growing Huon pine (*Lagarostrobos franklinii*), and the King Billy pine (*Athrotaxis selaginoides*).

Around 96% of the 765 000 ha of Tasmanian rain forests and most of the mainland rain forests are on crown land. Some 60% of Tasmania's forests are in dedicated reserves or in reserves recommended by the government. Another 15% has been allocated to areas for evaluation of conservation potential. Most of the larger mainland forest patches are in Nature Conservation Reserves. The fauna of this region is not as rich as that of eastern Tasmania; this wilderness area is the last stronghold for species that were more widely distributed in the past. For many years, there was hope that the Tasmanian tiger or thylacine (*Thylacinus cynocephalus*) would find refuge in the western wilderness of Tasmania. However, in spite of several supposed sightings, this species is thought to be extinct.

Alpine and subalpine habitats in mainland Australia and Tasmania occur over a relatively small area. On the mainland they are confined to the southeast of the continent at elevations of 1 370-1 525 m, and their area of 5 180 km^2 is only 0.07% of the country's land area. In Tasmania, alpine habitats extend down to 915 m, and their area of 6 480 km^2 is about 10% of the total Tasmanian territory.

The temperate eastern highlands of Victoria State consist of hills, mountains, and the country's highest plateaus, also known as the "high plains." Temperatures drop with increasing altitude, and rainfall varies from about 650 mm to more than 1 500 mm a year, with occasional heavy snowfalls occurring in the higher country. At the highest elevations, a variety of alpine and subalpine communities can be found, the most com-

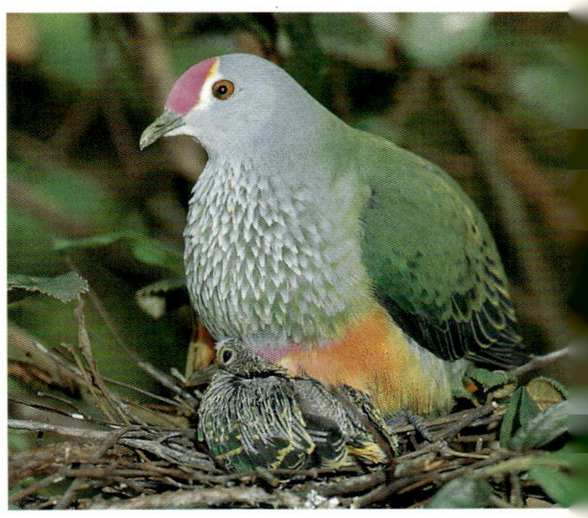

Rose-crowned fruit pigeon (Ptilinopus regina), *another pigeon species from the subtropical rain forest of Australia's east coast.*

mon of which are *fjaldmark*, herbfields, grassland, bog and woodland. Tall evergreen forests occur at elevations of 750-1 400 m, while woodlands occur on drier sites with 650-900 mm rainfall. The open forests occur at lower elevations and give way to woodlands at still lower elevations.

The arid and semiarid regions occupy almost the entire western portion of the continent and include the tropical part of Western Australia, most of the Northern Territory, except for its northernmost tip, and parts of western Queensland. The eastern arid region of central Australia covers approximately 400 000 km^2 and includes the corners of three adjoining states, South Australia, Queensland, and the Northern Territory. This area is very interesting and comprises some of Australia's most outstanding features, such as Lake Eyre, the Great Artesian Basin, the Tirari Desert and the Simpson Desert. The most outstanding vegetation formations in the dry tropical region are open woodland, shrubland and open shrubland, as well as small patches of grassland in the westernmost part of Western Australia; the eastern arid region includes grassland and sedgeland, open shrubland, open woodland and a patch of herbland.

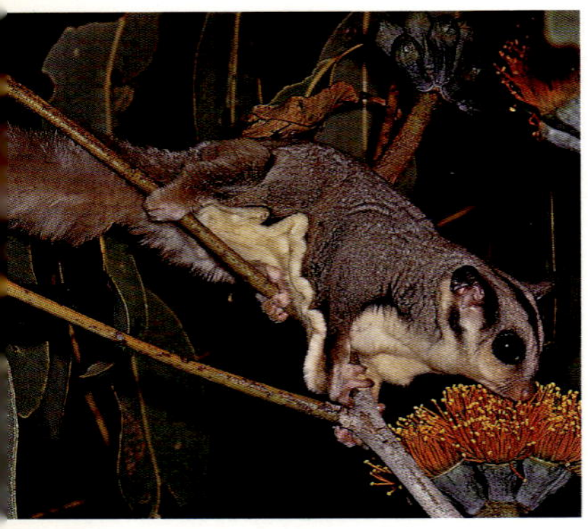

A sugar glider (Petaurus breviceps) *is intermediate in size between the tiny feathertail and the greater glider. It can volplane through the air for more than 50 m.*

On the opposite page, tall, open eucalyptus forest with rain forest understory, known in Australia as Wet Eucalyptus Forest. The genus Eucalyptus *reaches its greatest diversity in Australia. This photo was taken in Errinundra National Park in East Gippsland; the eucalyptus species is shining gum* (Eucalyptus denticulata) *and the tree fern in the understory is* Dicksonia antarctica.

The three major areas of sand-ridge deserts are largely uninhabited by man. These are the Simpson Desert, north of Lake Eyre; the Great Victoria Desert, north of the Nullarbor Plain; and the Great Sandy Desert on the inland northwest. Altogether there are approximately 1 942 491 km^2 of sandy deserts, about one fourth of the continent (an area almost as big as Mexico).

The Simpson is Australia's harshest desert, although occasional floods hit the region causing water to gush into, and even fill, the usually dry Lake Eyre. This lake, the bed of which covers 9 300 km^2, is rarely completely filled. Up to 90% of the lake bed is dry most of the time, but the sight of the vast expanse of water in the desert when the lake is flooded is a remarkable one. The Tirari Desert, a 12 000-km^2 sand-ridge desert, is located on the eastern side of Lake Eyre. This extension of the Simpson Desert contains some of the most interesting fossil remnants of vertebrate fauna in Australia.

The Copper River is a good example of the peculiar floodplains that form when the river level rises and water spreads out over the immense plains moving through an intricate maze of channels that fill out as the water flows into the desert. The diversity of flora and fauna of the eastern region is a reflection of the variety of habitats found there: sandy desert, stony desert, salt lake, river channels and floodplains. *Eucalyptus* woodland, open woodland, scrub and heath, shrubland, and herbland are the most prominent types of natural vegetation found in this harsh environment.

Finally, there is the subhumid temperate region of the southwestern part of the state of Western Australia. This area covers almost 310 000 km^2 and encompasses a wealth of environments that go from high forest, woodlands, heathland, and shrubland to the wild coastline. The rainy season takes place during the winter, and the region has the highest rainfall in the far southwest, with 1 400 mm per year. There is a broad inland band that is somewhat drier, with rainfalls between 350 and 600 mm and the driest region, which extends from Shark Bay to the curve of the Great Australian Bight, gets only 300 mm of rain per year. This region is recognized as one of the world's richest botanical areas, with 4 000 species of plants recorded, 83% of which are endemic. Indeed, given the very high levels of diversity and endemism and the degree of threat to what remains of this region, it has qualified as one of the world's top 19 *threatened biodiversity hotspots*.

This area is also where the Stirling Range is located, just northeast of the whaling town of Albany. This mountain range erupts suddenly from the surrounding plain into a profile of ragged peaks, the highest of which, Bluff Knoll, towers 1 073 m. These peaks attract rainfall and occasionally winter snow. The moisture enables this region to be the habitat for a remarkable variety of plants, creating a series of microhabitats that have an unusually high number of restricted endemics. Of the nearly 1 000 plants recorded in the range, over 50 are endemic. The Stirling Range is a national park surrounded by agricultural land, and is an ecological island where many of the area's remarkable species have taken their last refuge.

There are two types of forest found in this region that occur nowhere else in Australia. The first is known as jarrah because the dominant species is the jarrah (*Eucalyptus marginata*) and the second is the karri forest, where the dominant species is the karri (*Eucalyptus diversicolor*), one of the world's largest trees. Jarrah forest covers approximately 1.4 million ha, and all the forest trees found in it are endemic. Karri forest, on the other hand, is very threatened by the woodchip industry. It has been almost entirely logged, and only about 20 000 ha of mature karri forest survive within the Shannon Basin National Park and in pockets of pure karri stand and mixed forest between Cape Naturaliste and Walpole-Nornalup National Park.

Heathlands are another formation that occurs in this area, since they are particularly well adapted to nutrient-deficient sandstone soils. The number of plant species that can be found in Australian heathlands varies from as few as 33 to as many as 131 species/ha. Heathlands are dominated by low shrubs up to 2 m in height, with small, evergreen, sclerophyllous leaves and extensive root systems. Several species may be codominant within them, including the uniquely southwest Australian *Banksias* with their beautiful flowers, the grass tree (*Xanthorrhoea* spp.), which has the greatest biomass, and *Allocasuarina*. Although heathland vegetation is widespread across Australia in all humid and subhumid areas, the greatest concentration is found on the sand plain and lateritic soils of southwestern Australia, where 50% of the 3 700 Australian heathland species occur.

Aside from all the terrestrial environments discussed above, there are two more natural habitats that con-

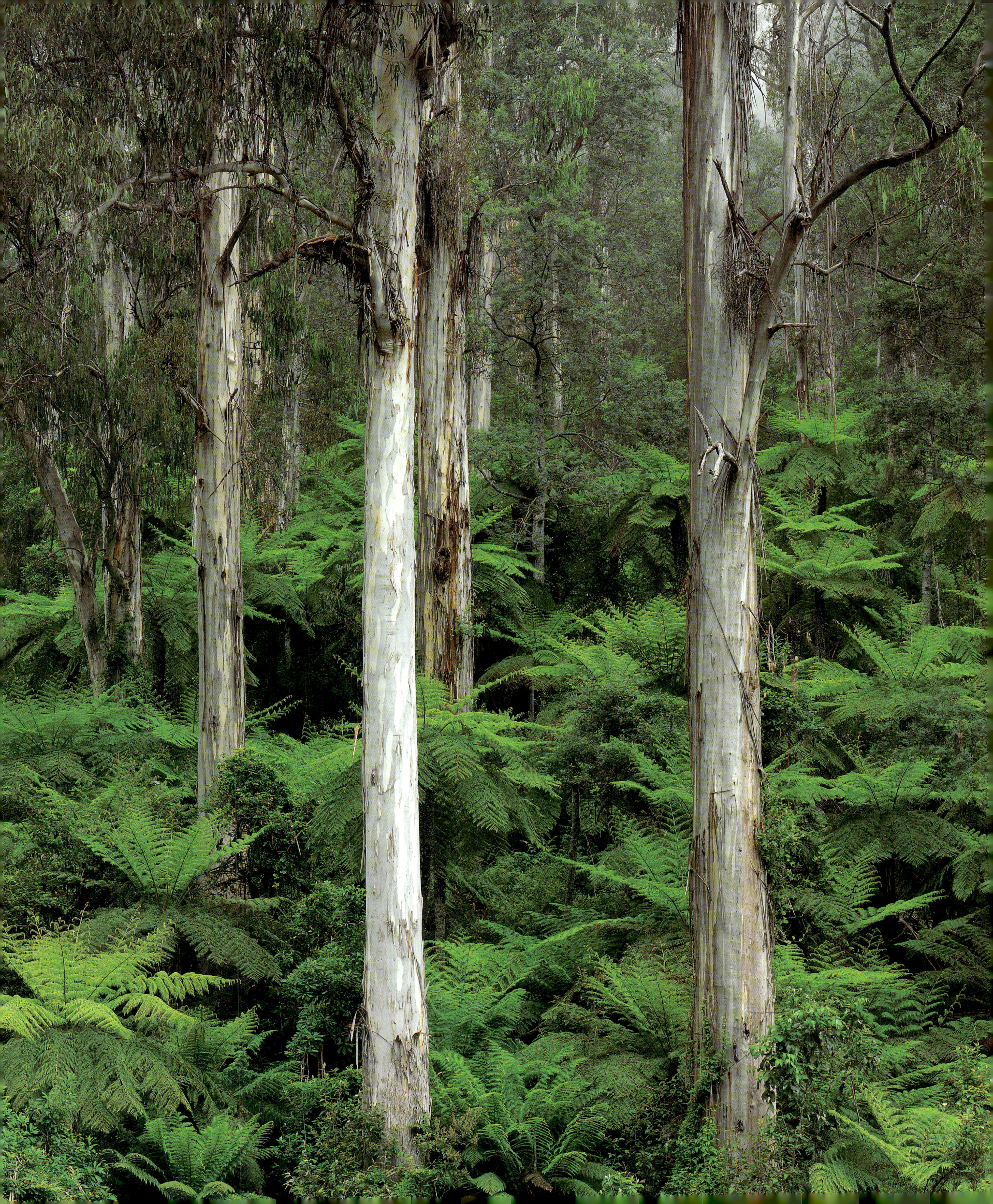

tribute to making Australia one of the most diverse countries in the world. These are the Great Barrier Reef off the coast of Queensland and the Subantarctic islands located in the Southern Sea. The barrier reef, extending for almost 2 000 km along the coast of north Queensland, is the site of a bewildering array of marine life-forms that together constitute one of the world's richest and most diverse marine ecosystems. The Great Barrier Reef, declared a World Heritage Site in 1981, begins near the mouth of the Fly River in the Gulf of Papua and ends at Lady Elliot Island, just north of Fraser Island. It encompasses a huge marine province comprised of many ecosystems aside from coral reefs. These include 540 high continental islands, sandy cays, mangrove forests, sea-grass beds, and the soft, sandy sea bottoms between the reefs. It is considered one of the richest marine habitats on Earth, supporting an extraordinary diversity of life-forms which both create and depend on the reef. Many of the species found there exist nowhere else in the world, and almost every marine organism is richly represented. There are over 1 500 fish species, 400 species of coral and 4 000 species of mollusks. A similar diversity is found among other major marine groups, like crustaceans, echinoderms, anemones and marine worms. Some rare and endangered species like the dugong (*Dugong dugon*) have major strongholds in this region.

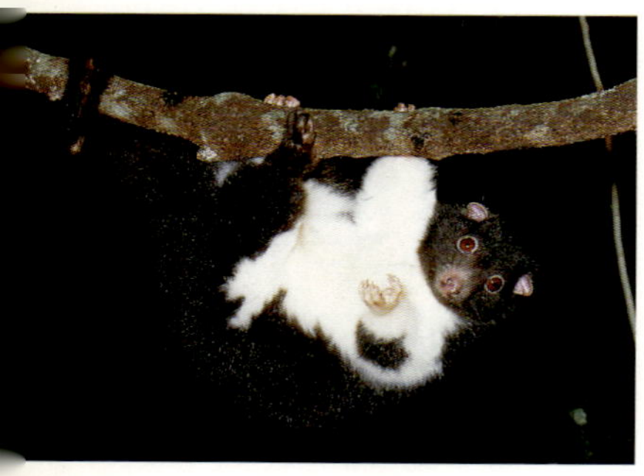

A Herbert River ringtail possum (Pseudocheirus herbertensis), *photographed in Queensland. This is one of the endemic marsupial species of the Queensland tropical rain forest.*

The last Australian environment is the one found in the Subantarctic islands. Macquarie Island is notable as the habitat of the world's largest colony of macaroni or royal penguins (*Eudyptes chrysolophus*). It also has a large number of king penguins (*Aptenodytes patagonicus*), as well as gentoo penguins (*Pygoscelis papua*) and rockhopper penguins (*Eudyptes chrysocome*). Similarly conspicuous on the beaches and contributing to an incomparable wildlife spectacle are huge numbers of southern elephant seals (*Mirounga leonina*), and three species of fur seal, the New Zealand fur seal (*Arctocephalus forsteri*), Antarctic fur seal (*A. gazella*), and the Subantarctic fur seal (*A. tropicalis*). The seals and penguins are sustained by the abundant marine life that inhabits the submarine shelf around the island. So too are the many seabirds that include albatrosses, cormorants, shags and the great skua (*Catharacta skua*). Vegetation on the islands is sparse. There are no trees and the tallest vegetation is the tussock grass that provides protection for nesting seabirds. In the Heard-McDonald group, the poorly-developed soils support tussock only in the lower areas. Above 200 m, mosses and lichens constitute the sole vegetation. The more northerly Heard and McDonald islands have similar fauna to that found on Macquarie Island.

Flagship species

Given its isolation and very high levels of endemism, Australia has many unique species found nowhere elsewhere in the world. Some really stand out as flagships that have long been identified with the island continent; the best known are the kangaroo and the koala, followed closely by the platypus and the wombat. The kangaroos and the koalas are the true icons of Australia (along with the boomerang), and are usually the first things schoolchildren around the world learn about when the geography lesson turns to this country. The kangaroo even appears on the Australian coat of arms.

The mammal fauna of Australia is dominated by marsupials and the primitive monotremes, which together account for about 56% of all species; the two other terrestrial mammalian orders that are represented naturally are the rodents and bats. As is the case in Madagascar, Australia's mammal fauna reaches very high levels of diversity in certain groups, especially the marsupials (119 species), whereas many others are entirely absent (*e.g.*, primates, ungulates, placental carnivores, etc.). The monotremes, the world's only egg-laying mammals, are entirely restricted to Australia and New Guinea, with the platypus (*Ornithorhynchus anatinus*) being endemic to eastern Australia and the two genera of echidnas being found in both Australia and New Guinea. Their name, Monotremata, means "one hole," and refers to the fact that reproduction and excretory functions use the same orifice, unlike the placental mammals.

The platypus is probably the most unusual mammal on Earth, with its duck-like bill, its flat, beaver-like tail, and the sharp poison spur on the ankle of the male, which makes it the only poisonous mammal in the world. These features, combined with its egg-laying behavior, shared only with the echidnas, combine to create a truly unique creature. The platypus is found from the high altitudes and winter snows of Tasmania and the Australian Alps to the tropical rain forest lowlands and plateaus of Queensland. It is aquatic and eats a wide variety of invertebrates, but will occasionally take small vertebrates as well. It is a solitary animal, but several individuals may use the same body of water, indicating that they may not be territorial. The platypus has a limited capacity for recolonization, and many of the bodies of water in which it once lived have been modified by dams, agricultural activities, and pollution, leading to local extinctions. Nonetheless the animal is still relatively abundant throughout much of its original range, and should be considered vulnerable but not endangered.

The short-beaked echidna (*Tachyglossus aculeatus*) is the only member of its family in Australia, and it occurs in New Guinea as well. It is also an egg-laying mammal and is covered with long spines. This echidna is distributed over most of Australia, from regions of winter snow to the driest deserts, and its only habitat requirement appears to be a supply of ants and termites upon which to feed. Although sparsely distributed, it does not appear to be in any immediate danger.

The 119 species of marsupials in Australia can be divided into two main groups: the carnivorous and insectivorous Dasyuroidea, with 42 species, and the herbivorous marsupials, with some 75 species. The dasyuroids include tiny to medium-sized predators with unlikely names like phascogale, planigale, dunnart, kowari, kultarr, ningaui, and antechinus. The best known are the quolls, or native cats (*Dasyurus* spp.), aggressive predators about the size of the American opossum and with distinctively spotted coats, and the Tasmanian devil (*Sarcophilus harrisii*), largest of the living dasyuroids, and an animal made famous by an American cartoon character of the same name. The most spectacular of all the marsupial carnivores, the thylacine (*Thylacinus cynocephalus*), also known as the Tasmanian wolf or Tasmanian tiger, was eliminated on the Australian mainland by the dingo tens of thousands of years ago, and probably became extinct in Tasmania in the 1930s, the last known individual dying in the Hobart Zoo in 1936. Although there is still a slim chance that it survives in remote corners of Tasmania, much attention has been devoted to finding it over the past few decades without any success. The likely loss of this striped, dog-like creature with a huge, gaping mouth is most unfortunate, since it was a truly unique species and one of the most impressive of the marsupials.

The herbivorous marsupials number roughly 75 species, of which the kangaroos or Macropodoidea, with 48 species, are the most diverse. The best known are the largest members of the family, the eastern gray kangaroo (*Macropus giganteus*), which can reach 2.3 m in head-body length and 66 kg in weight and was the first species collected by Europeans on Cook's voyage in 1770; the western gray kangaroo (*Macropus fuliginosus*); and the red kangaroo (*Macropus rufus*), which occurs mainly in the drier regions of the country. However, kangaroos come in many different sizes and varieties, from the tiny musky rat kangaroo (*Hypsiprymnodon moschatus*), which is the most primitive kangaroo, to the bettongs, the potoroos, the pademelons, the wallabies, the hare wallabies, the rock wallabies, the nail-tailed wallabies, the wallaroos, the monjons, the nabarleks, the quokkas, and the arboreal tree kangaroos. Five members of the kangaroo family have already gone extinct: the desert rat kangaroo (*Caloprymnus campestris*) from a very restricted, hot dry area of central Australia; the broad-faced potoroo (*Potorous platyops*) from southern and southwestern Australia; the eastern hare wallaby (*Lagorchestes leporides*) from southeastern Australia; the toolach wallaby (*Macropus greyi*) from the southeastern corner of South Australia State; and the crescent nail-tailed wallaby (*Onychogalea lunata*) from a large area of western and central Australia. The loss of 5 out of 48 species in this very special group of animals since the arrival of Europeans represents one of the most significant extinction spasms in recent times.

The success of kangaroos as a group seems to be largely due to their ability to use fibrous plant food, particularly grasses, and the larger species occupy a grazing niche that is filled by ungulates in other parts of the world. European settlement and hunting has had little effect on the larger kangaroos and wallabies; indeed, it may have benefited them through an increase in grasslands brought about by land-clearing and through development of watering holes for cattle and sheep in dry areas. However, many of the smaller species from woodlands and open country have fared much worse because of loss of habitat, competition from introduced herbivores like the rabbit, and direct predation by introduced placental carnivores like the fox.

The other great flagship species of Australia is the koala (*Phascolarctos cinereus*). The koala must rate among the cutest and cuddliest of all animals, and has been popular around the world for decades. Divided into three subspecies, this delightful creature occupies a large but greatly fragmented range in eastern Australia. While the harsh winter excludes it from the montane forests along the backbone of the Great Dividing Range, it is widely distributed on either side. The koala feeds almost exclusively on *Eucalyptus* leaves, so its distribution is closely associated with this genus of tree. However, how it manages to survive on *Eucalyptus* is one of the most interesting issues in koala biology, since *Eucalyptus* foliage is poor-quality food, low in protein and with high concentrations of indigestible compounds such as fiber and lignins, and toxins like phenolics and terpenes. The koala seems to manage on this diet through a combination of behavioral, anatomical, and physiological adaptations. It is inactive for up to 20 hours a day and has very low energy requirements for a mammal. Its food is very efficiently processed, its sharp-edged teeth grinding the leaves to a fine paste and releasing plant cell contents quickly for immediate absorption in the stomach; toxic compounds in the leaves are inactivated in the liver and excreted. The very long caecum has one of the largest capacities of all mammals, and fermentation products add to the koala's energy supply. Although a relatively slow-moving creature, it is an excellent climber and can also bound along the ground with surprising, if somewhat amusing, agility when necessary.

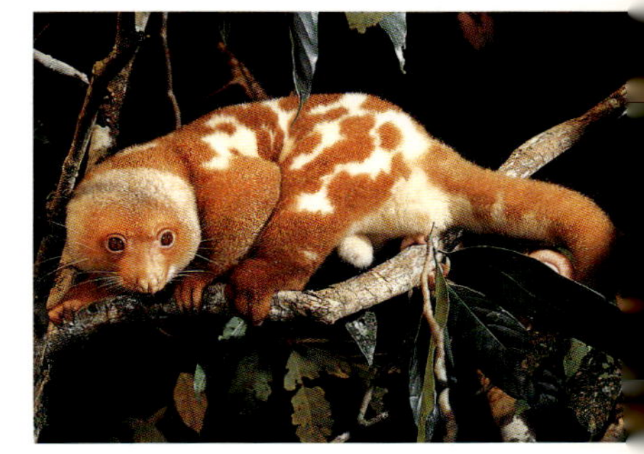

*A spotted cuscus (*Phalanger maculatus*), one of the more wide-ranging cuscus species that occurs both on the Cape York Peninsula of northern Queensland and on the island of New Guinea.*

Although the animal has been the focus of much attention and even has nongovernmental organizations dedicated exclusively to its conservation, there are still many management problems to overcome. Remnant populations living at high densities in isolated habitat fragments are at greatest risk. Eye and reproductive tract diseases resulting from the bacterium *Chlamydia psittaci* are particularly severe in some overcrowded populations, although reproduction remains high in most areas. Since the koala occurs in Australia's most populous area, habitat conservation is the most important issue, and the need to link isolated habitats through corridors is especially critical.

The three wombat species, rather like terrestrial versions of the koala, are short, stocky, powerful grazers that live in burrows. The two hairy-nosed wombats —the northern hairy-nosed wombat (*Lasiorhinus krefftii*) and the southern hairy-nosed wombat (*Lasiorhinus latifrons*)— live in grasslands and savanna and are very restricted in range. The southern species is still rela-

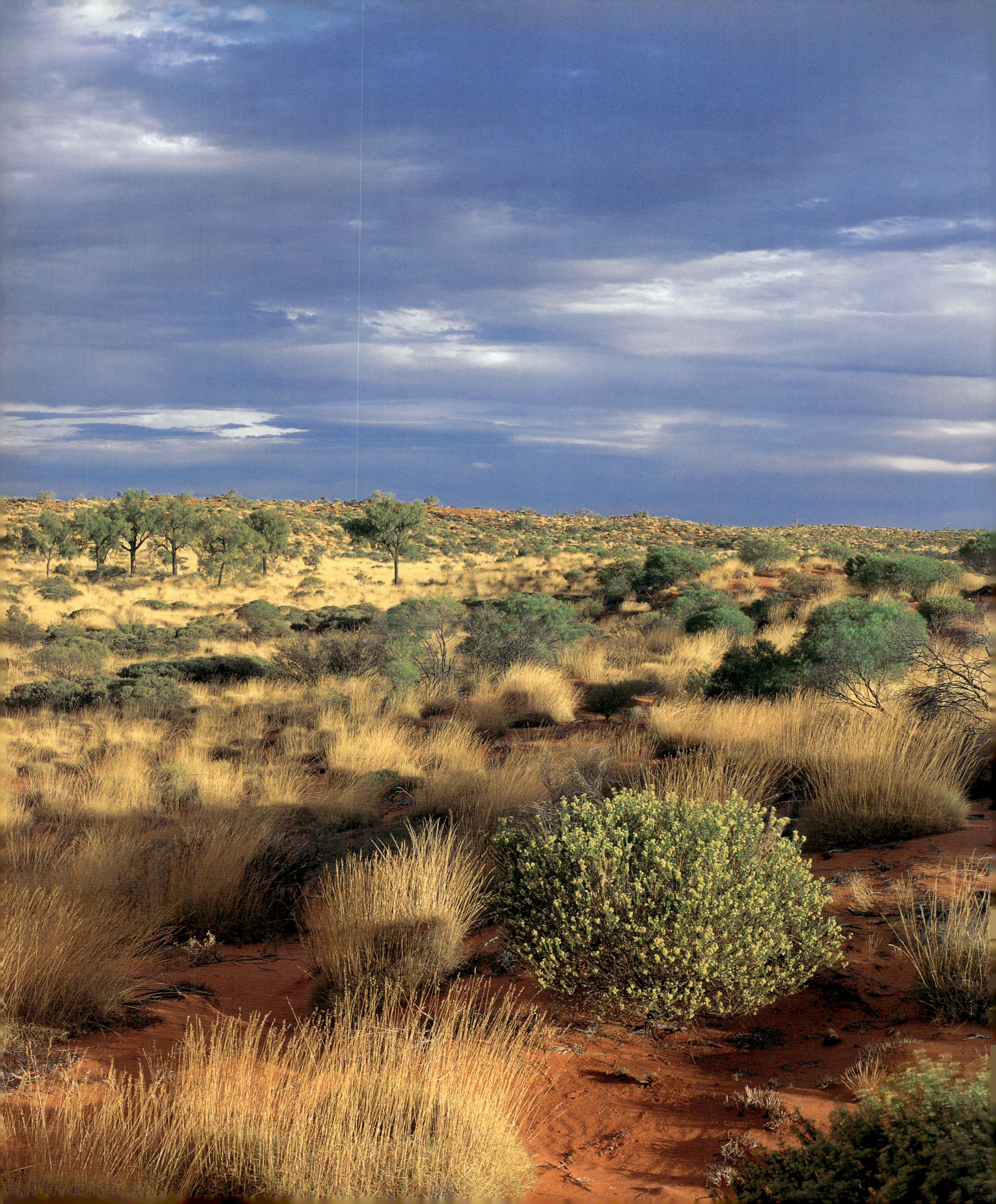

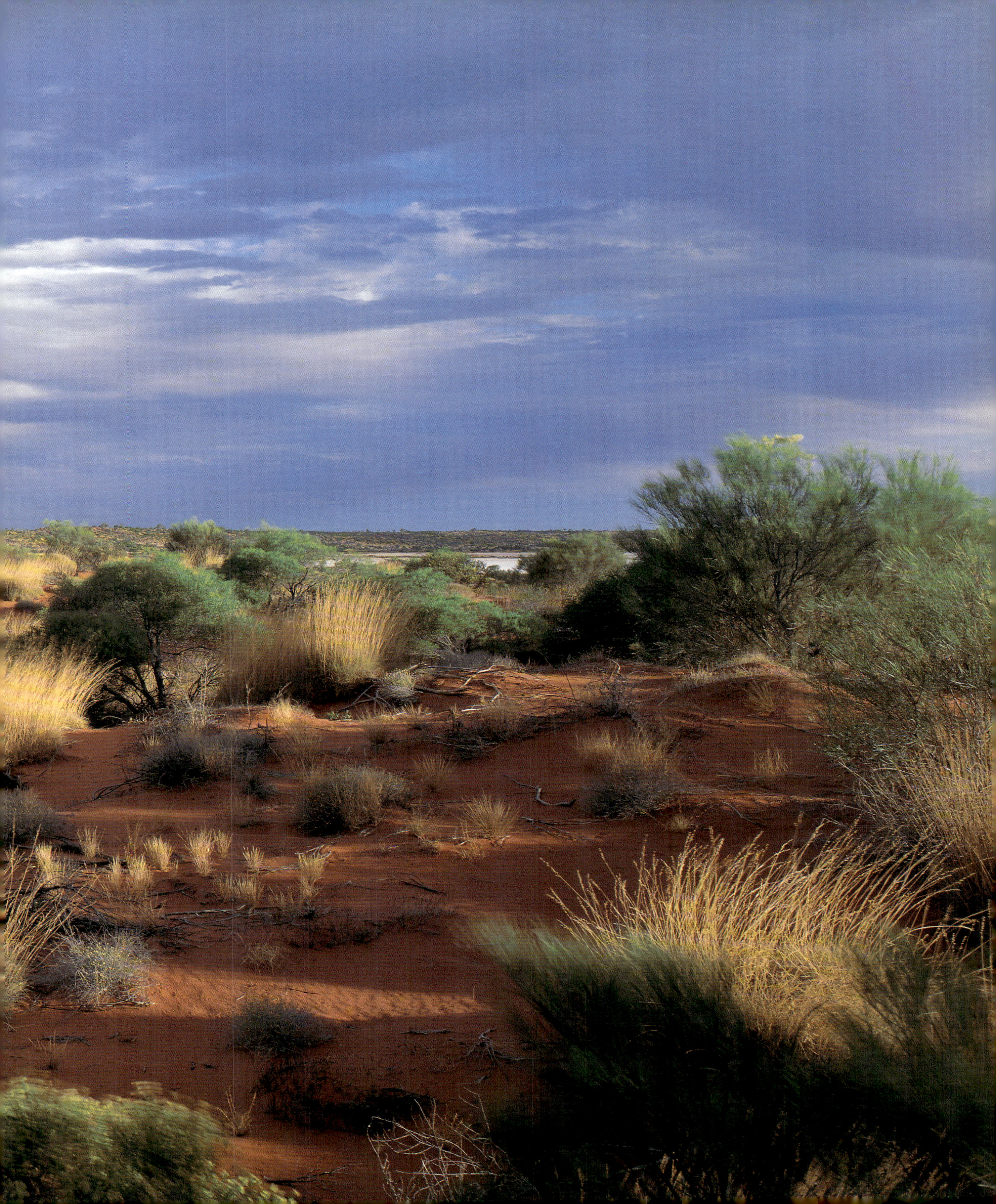

tively common in its limited distribution, whereas the northern species is considered endangered. The common wombat (*Vombatus ursinus*) is found in the forested, mountainous areas of southeastern Australia, where it grazes in forest clearings. It digs elaborate burrows up to 20 m long, and is mainly nocturnal in behavior. It is protected in all states in which it occurs except Victoria, where it is classified as vermin mainly because of the damage it causes to rabbit-proof fences.

Other important marsupials in Australia include the Phalangeroidea or possums, which include 11 species of ringtails and gliders, six species of bushtails and cuscuses, and six species of pygmy possums and feathertail gliders, and the Peramelemorphia or bandicoots and bilbies, a group of small to medium-sized, long-nosed terrestrial omnivores. Three of the eleven species of bandicoots and bilbies have already gone extinct, including the lesser bilby (*Macrotis leucura*) from central and western Australia, the desert bandicoot (*Perameles eremiana*), and the distinctive pig-footed bandicoot (*Chaeropus ecaudatus*), which had a large range from western to central and southern Australia. As with the kangaroos, the loss of 3 out of 11 bandicoots and bilbies (27%) over the past 200 years represents a significant recent extinction spasm.

Finally, we should mention three very distinctive, taxonomically unique marsupials that are little known to the outside world, but have the potential to be very significant flagship species. The first is the numbat (*Myrmecobius fasciatus*), a delightful, squirrel-sized marsupial anteater with a long, pointy nose and white stripes on a brownish background. It is the only member of a monotypic family, Myrmecobiidae, within the Dasyuroidea. A diurnal species now restricted to southwestern Western Australia, it is quite rare but can still be seen in a few protected areas such as Dryandra National Park south of Perth. Although it has natural predators, it appears that the introduced fox has been particularly damaging to it, and programs to control the fox have resulted in population increases in areas like Dryandra. Strangely, although termites are abundant in Australia, a large marsupial anteater equivalent to the giant anteater of South America, the pangolins of Africa and Asia, and the aardvark and aardwolf of southern Africa never evolved in Australia; the numbat is the best that the marsupials could come up with. A delicate, very attractive little creature that is at great risk of extinction, the numbat is a delightful flagship species.

The second of the unusual marsupials is the honey possum (*Tarsipes rostratus*), usually considered to be in its own family, Tarsipedidae, within the suborder Phalangerida. However, in the opinion of some experts, it is different enough to warrant even higher distinction within the marsupials. One of the few truly nectarivorous mammals, this tiny, mouse-sized creature, weighing only 7-12 g, has a prehensile tail, a long, pointy snout, and an elongate, protrusible, brush-tipped tongue especially adapted for probing flowers for nectar and pollen. It does not use claws for climbing, but rather grips branches with the expanded tips of its fingers and toes rather like a small primate; its generic name, *Tarsipes* —"foot like the tarsier"— refers to a supposed resemblance to the primate tarsiers (*Tarsius* spp.), found in southeast Asia. It shares characters with several other major marsupial groups, but is apparently the sole survivor of a long-extinct group of marsupials, having been able to survive because of its very specialized diet and lifestyle. The honey possum is found only in southwestern Australia, where it occurs in the coastal plain heaths. Given this distribution, which is largely coincident with the newly-identified southwestern Australian biodiversity hotspot, it would be an excellent flagship for this globally important region.

The last of the extremely unusual marsupials is the marsupial mole, genus *Notoryctes*, which is so distinctive that it is placed in its own order, the Notoryctemorphia. Although it superficially resembles the placental golden mole of South Africa, a striking case of convergent evolution, this strange, golden-furred, blind burrower does not show a close relationship to any other marsupials and remains something of an enigma. Like the golden mole, but unlike the moles of Eurasia and America, the marsupial mole digs with two flattened claws and its burrow fills in behind it, making it appear as if it were "swimming" through the desert sand in which it lives. Two closely-related forms are known, *Notoryctes caurinus*, from the northern parts of the range, and an animal very poorly known, and *Notoryctes typhlops*, which occurs in the central portion of Western Australia, parts of southwestern Northern Territory and the northwestern corner of South Australia; both species are considered endangered.

The platypus, the numbat, the honey possum and the marsupial mole are excellent examples of monotypic, higher-level endemic taxa (orders, families) restricted to Australia and for which that country has ultimate responsibility. This is one of the many reasons for Australia's globally very high ranking among the megadiversity countries. Just as an indication of the importance of these animals, the marsupial mole's order, Notoryctemorphia, is the *only mammalian order endemic to a single nation*.

Moving on to the reptiles, Australia has a number of spectacular species, not surprisingly, given the fact that it has the highest diversity of reptiles on Earth. Among these, perhaps the most conspicuous are the crocodiles. There are two species of crocodiles in Australia, the saltwater or estuarine crocodile (*Crocodylus porosus*) and the endemic Australian freshwater crocodile (*C. johnstoni*). Both occur exclusively in the tropical portion of the continent, from about Rockhampton on the east coast of Queensland to Broome on the northwest coast of Western Australia. Because the interior of this region is arid, crocodiles are restricted to the coastal fringe and the rivers that drain into it. Saltwater crocodiles occupy the immediate coastal areas, including both tidal and nontidal rivers, saline and freshwater swamps, and an assortment of billabongs. The smaller freshwater croc-

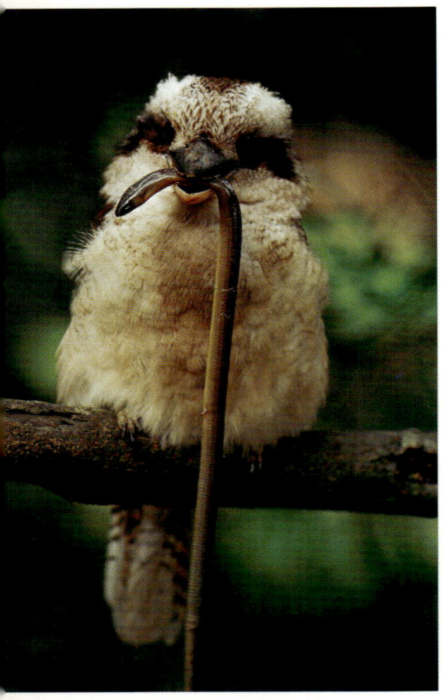

On pp. 186-187, the Little Sandy Desert of Western Australia, near Lake Aerodrome. Australia is the driest of the continents, with a large portion of its area covered with desert.

Above, the kookaburra (Dacelo *spp.*) *is one of Australia's best known flagship birds, and is world-renowned for its raucous laughing call. Here one is seen eating a legless lizard.*

odiles are usually found inland in the upper reaches of rivers and in billabongs and swamps.

The saltwater crocodile is probably the most aggressive and dangerous of all crocodilians, and one of the largest of all reptiles, with only the Indian gharial and the black caiman of South America as serious competitors. Although it averages 4.6-5.2 m in length, 6-m individuals are not unusual and there are old, unconfirmed records of specimens exceeding 8 m. This length, combined with the animal's enormous bulk, would make the saltwater crocodile *the* largest of all living reptiles. Made famous by the *Crocodile Dundee* movies, the saltwater crocodile does in fact eat an occasional human, be it a local person or an unwary tourist, and almost any body of water in tropical Australia, regardless of its size, has signs warning people of the presence of dangerous crocodiles. One insurance company in Cairns even has a special crocodile insurance policy that pays a victim's family $50 000 if the policy holder is eaten by a crocodile. Aside from humans, which are definitely within the size range of saltwater crocodile prey, this species feeds on a wide variety of fish and other aquatic organisms and will also take any animal that can be caught from the water's edge. It also has spectacular leaping ability, with individuals up to 2 m in length being able to leap vertically out of the water almost their entire body length. This behavior has even become a tourist attraction in the Adelaide River in the Northern Territory, with habituated crocodiles performing this leaping behavior in their natural environment in response to bait suspended from long poles held out from the safety of a tourist boat.

Aside from its occasional appearances in the international press because of its man-eating behavior, the crocodile's most interesting relationship with man is the one it has with the aborigines. The role that crocodiles play in traditional aboriginal life in Australia varies between different groups. However, some groups believe that crocodiles are members of their extended families and thus hold them in special regard. There are aboriginal dances and stories about crocodiles, the beginnings of which may date back thousands of years. There are beliefs that the spirits of dead people are contained within the bodies of some large crocodiles, and there may be widespread mourning when a particularly large crocodile is killed. Crocodiles are also a dominant theme in both rock art and bark paintings, and gastroliths, the rocks found in crocodile stomachs, have special significance in certain ceremonies.

Turtle diversity in Australia is high as well, with 22 species, of which two genera and 13 species are endemic. The two endemic genera are the recently described *Rheodytes leukops*, known only from the Fitzroy River in Queensland and not described until 1980, and the western swamp turtle (*Pseudemydura unbrina*), known only from one or two small swamps at Bullsbrook, near Perth. The latter is probably the most endangered freshwater chelonian in the world. Australia also has six species of marine turtles, including the unusual flatback turtle (*Natator depressus*), a regional endemic that is found mainly in northern Australia.

Australia is also renowned for its snake fauna, which is both very diverse and unusually rich in venomous species. Of the 169 snake species recorded for the country, fully 110 (65%) are poisonous, including 77 members of the cobra family (Elapidae) and an additional 33 species of the two sea snake families, Hydrophiidae and Laticaudidae. This is a far higher percentage of venomous reptiles than any other country (or continent) on Earth, and makes an encounter with a snake in Australia much more interesting and potentially dangerous than anywhere else. Several of the venomous species are among the most toxic in the world, including the taipan (*Oxyuranus scutellatus*), which can reach 2 m and is the largest venomous species in Australia, and the fierce snake or western taipan (*Oxyuranus microlepidotus*). Other very dangerous species are the black tiger snake (*Notechis ater*), the brown snakes (*Pseudonaja* spp.), the black snakes (*Pseudechis* spp.), the king brown snake (*Pseudechis australis*), and the death adders (*Acanthophis* spp.).

One of the most unusual of Australian reptiles is the thorny devil or moloch (*Moloch horridus*). Named for the Semitic god, Moloch, for whom children were burned in sacrifice, this little lizard of the large family Agamidae measures only about 9 cm and does not live up to the evil reputation of its namesake; indeed, it is actually quite cute. It is covered with spectacular thorny spines over its entire body, including one large, curved, horn-like spine above each eye and a thick, spiny nuchal hump on the back of its neck. A desert dweller occurring over a wide area of dry western and central Australia, it feeds only on ants. Although the spines may serve as a deterrent to some predators, their main function appears to be water collection in the arid environment inhabited by this lizard. On cool desert nights, dew forming on the spines is carried by fine channels across the moloch's body and into its mouth. The moloch is the only member of a monotypic genus endemic to Australia. Interestingly, the unrelated horned "toads" of the southwestern deserts of the United States and Mexico, a group of iguanid lizards of the genus *Phrynosoma*, have evolved a similar body shape and thorny skin in adaptation to a desert environment, but none of them have carried it to such an extreme as has the moloch.

Several other agamid lizards are also worthy of note. These include the bearded dragons and the frilled lizard. The frilled lizard (*Chlamydosaurus kingii*) is a monotypic genus characterized by the presence of a large, loose ruff or frill around its neck. When the lizard is at rest, this spectacular frill lies in folds on the neck and shoulders, but it can be erected suddenly and with good effect when the lizard is alarmed and opens its mouth. The bearded dragons of the genus *Pogona* include seven species, of which the *Pogona barbata*, the typical bearded dragon, is the best known. The animal's name derives from a well-developed pouch or "beard"

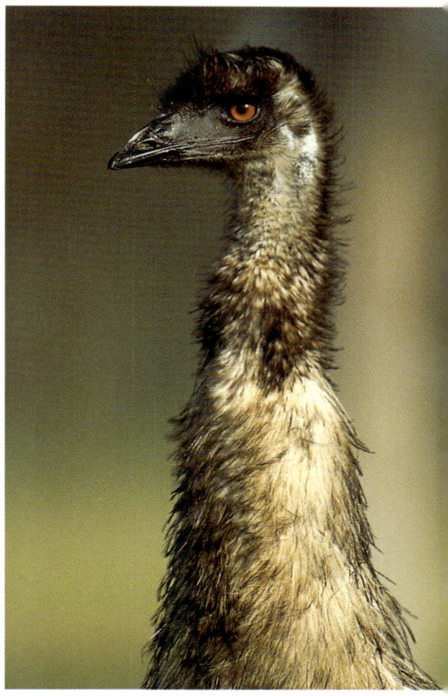

The emu (Dromaius novaehollandiae), *a ratite bird related to the ostrich, the rhea and the cassowary, is Australia's largest bird. It is widespread on the continent.*

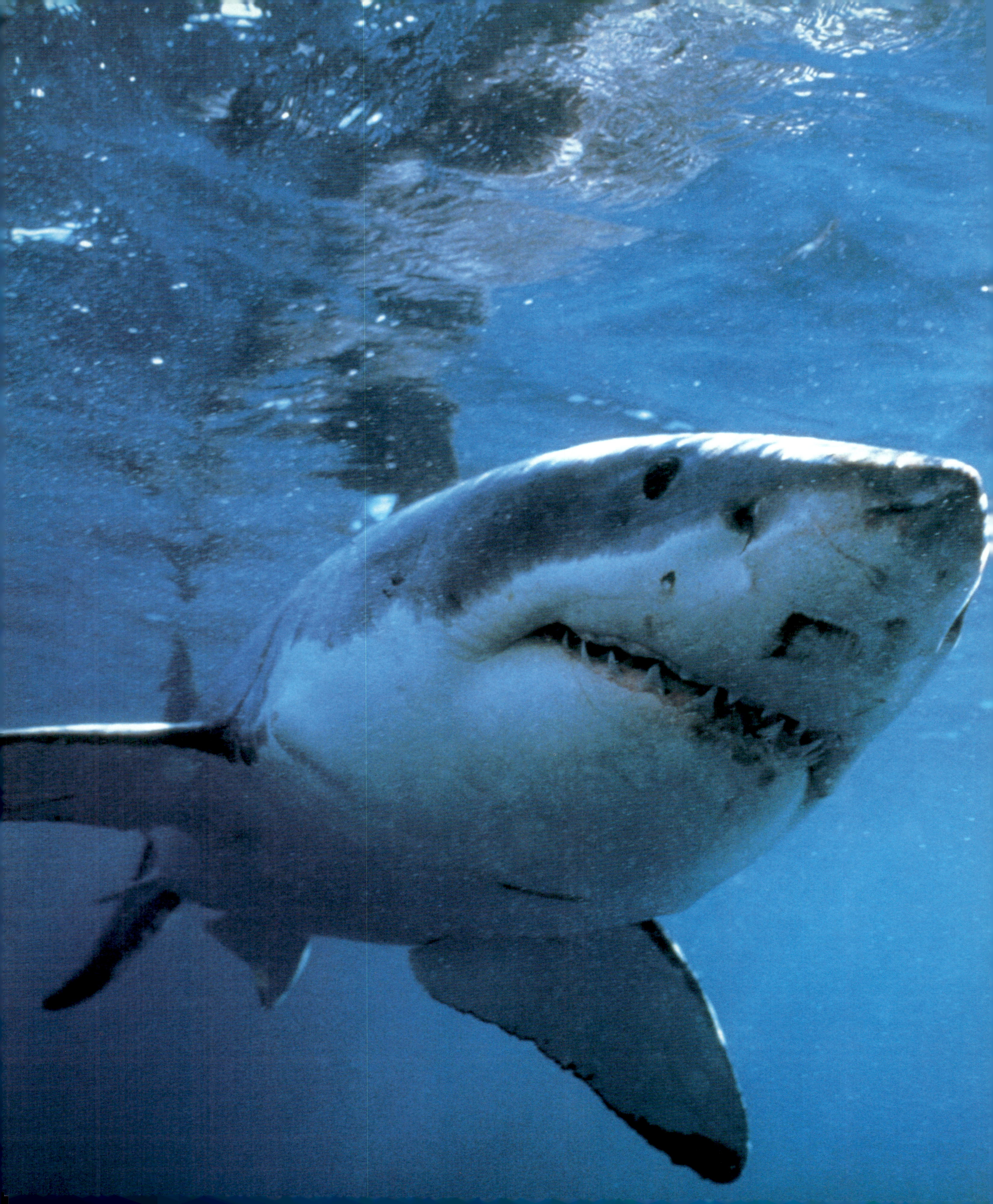

on the throat in adults, the hind edge of which is bordered by long, spiny scales. As with the frilled lizard, these spines flare out when the animal is excited and opens its mouth, again enhancing the size of the mouth to impress potential predators or competing members of the same species.

The skinks reach enormous diversity in Australia, with 31 genera and at least 317 species, the highest in the world for this widespread and diverse group, and also making them the single most diverse reptile family in the country; indeed, the skinks account for 42% (317/755) of all reptile species in Australia which, as already noted, is the most diverse country on Earth for reptiles. The most appealing of the skinks are the blue-tongued skinks of the genus *Tiliqua* and the unusual shingle-back (*Trachydosaurus rugosus*). There are six species of *Tiliqua*, all but one of which are endemic to Australia, with the eastern blue-tongued skink (*Tiliqua scincoides*) being the best known. Large and heavy-bodied with relatively short limbs and short tails, they have long distinctive blue tongues. Easily kept in captivity, very docile and omnivorous, they are popular pets among reptile fanciers and breed well in captivity. The shingle-back is a very scaly version of the blue-tongued skink to which it is closely related. It has very rough dorsal scalation and its short, thick tail is shaped like a second head, presumably to confuse predators. The blue-tongued skinks reach about 30 cm in head-body length and the shingle-back 25 cm, making them the largest of all Australian skinks.

Other important Australian lizards include the monitors and the geckos. The monitors reach their greatest diversity in Australia, with 25 species, including the perentie (*Varanus giganteus*), which can exceed 2 m and is the second largest lizard in terms of overall size, being exceeded only by the Komodo dragon (*Varanus komodoensis*) of Indonesia. The geckos are represented by more than 100 species, again a global high.

Moving on to the birds, perhaps the most spectacular group is the parrot family, including the cockatoos, the rosellas, the cockatiels, the budgerigars, the lorikeets and their relatives. Fifty-three species occur in Australia, tied with Colombia for second highest diversity in the world after Brazil, which has 70 species. Among the most striking species are the palm cockatoo (*Probosciger aterrimus*), which is a large, black, solitary bird with an enormous bill. It is found in the Cape York region of Australia, and also in New Guinea. Other spectacular species include the other black cockatoos, the red-tailed black cockatoo (*Calyptorhynchus banksii*), the glossy black cockatoo (*C. lathami*), the yellow-tailed black cockatoo (*C. funereus*), the long-tailed black cockatoo (*C. baudinii*), and the white or pink-colored cockatoos, among them the sulfur-crested cockatoo (*Cacatua galerita*), the pink or Leadbeater's cockatoo (*C. leadbeateri*), the long-billed corella (Cacatua tenuirostris), the little corella (*C. pastinator*), and the widespread galah (*C. roseicapilla*). The Australian parrots, lorikeets and rosellas are all very noteworthy as well, with some of the most spectacularly colored being the Australian king parrot (*Alisterus scapularis*), the rainbow lorikeet (*Trichoglossus haematodus*), the red-collared lorikeet (*Trichoglossus rubritorquis*), and the crimson rosella (*Platycercus elegans*). The night parrot (*Pezoporus occidentalis*), and the ground parrot (*Pezoporus* spp.) are unusual in their nocturnal behavior, since almost all other psittacines are diurnal. Finally, we have the budgerigar (*Melopsittacus undulatus*) and the cockatiel (*Leptolophus hollandicus*), widespread Australian natives that have become world famous as some of the most popular cage birds on Earth.

Another important flagship bird and the country's national bird, is the emu (*Dromaius novaehollandiae*), largest of the Australian birds. When Europeans arrived, Australia was inhabited by at least four isolated populations of the endemic emu, one on King Island, one on Kangaroo Island, one in Tasmania, and one on the mainland. Nowadays, the emu only exists on the mainland, where it is widely distributed. In the arid interior, it is nomadic and to a certain extent migratory. In Western Australia it is considered a pest and therefore an extensive fence system has been installed to keep it away from crop areas. This flightless bird is very mobile and needs to move constantly in search of food. It eats young green shoots, insects, fruit and seeds, and also swallows stones to help mechanically digest its food by triturating it in the stomach. The male, which is smaller than the female, spends eight weeks incubating the eggs; after the chicks have hatched, the male takes care of the young. The female lays up to 20 large and glossy green eggs. These are eaten by goannas, dingoes, feral pigs and other nest predators.

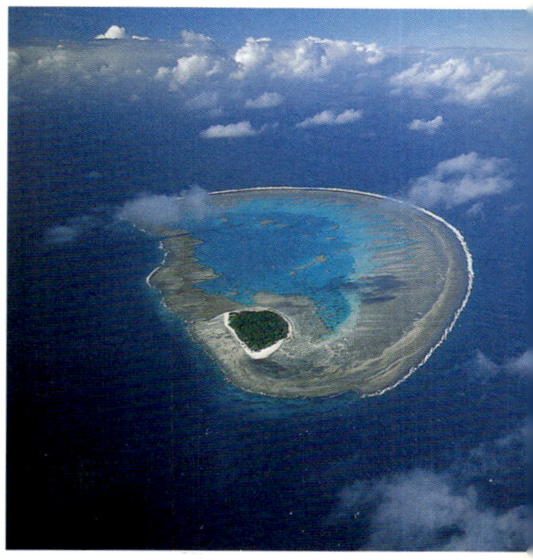

Cultural diversity

It is still not certain when humans first arrived in Australia. Until recently, radiocarbon dating had put the first arrival at around 40 000 years ago. However, evidence has come to light in the past few years indicating that it may have been as long as 60 000 years ago or perhaps even more. In any case, by the time the English buccaneer William Dampier first arrived in Australia in 1688, there were at least 300 000 Australian aborigines living on the continent, speaking at least 200 to 250 distinct languages and dividing themselves into no less than 600 to 700 different tribes, a very high concentration of human cultural diversity.

The aborigines were well-adapted to the rigors of the Australian continent, and had spread out over the millennia to occupy almost every part of it. Their material culture, though appearing simple and primitive to the Europeans, worked very well in their environment. Coastal aborigines obtained a substantial portion of their food from the sea, while those living in the rain forests and the deserts developed weapons that enabled them to hunt the wide variety of game species that existed in these regions. Weapons like the club, the spear,

On the opposite page, the great white shark (Carcharodon carcharias) *is one of the most ferocious predators on Earth, and appears to be particularly abundant off the Australian coast.*

Above, Lady Musgrave Island, part of Australia's Great Barrier Reef, the most extensive reef system on Earth.

and the spear-thrower predominated, while the boomerang, a kind of modified throwing club, became the best known contribution of aboriginal society to the outside world. Although boomerang-like weapons were also used by other peoples, the returning variety was invented by the aborigines, although apparently it was used mostly in play rather than in hunting. Larger boomerangs for hunting and for fighting were often much heavier and less curved than the returning varieties. Interestingly, neither the bow and arrow nor the blowgun ever made it to Australia, nor were they developed independently on this continent. Gathering of plant foods was also an important part of aboriginal culture, and the vast knowledge that aborigines have of their natural environment and the uses to which it can be put is still impressive today, in spite of all the upheavals that their cultures have undergone over the past two hundred years.

Aborigines lived mainly in family groups, bands, or small communities, with the area occupied by such groups being dependent on the kind of environment in which they lived; territories being smaller in richer coastal or forest environments compared to the poorer, arid desert environments. However, since they never developed sedentary agriculture, they never lived in large settlements such as towns or cities.

The aborigines also had (and still have) a very strong spiritual life and religious beliefs deeply rooted in the idea of the *Dreamtime*. The *Dreamtime* to the aborigine was the period at the beginning of time when the Earth was not formed as it is now, but was rather more like a great plain stretching to the horizon. This was a period of creation, when great spirit-heroes rose from the plain where they had been sleeping, took the form of animals, wandered the Earth, and behaved something like humans. Then the *Dreamtime* suddenly ended, and the spirit-heroes went into the sky or into holes in the ground, leaving behind signs of their existence such as hills, rivers, caves, rocks and other features of the landscape. They also left behind the spirits that were to bring about life in human, animal and plant form, as well as fire, laws to be followed, and the first weapons and other implements to enable man to live on Earth. All aborigines and other living creatures are descendants of these original beings, who didn't really die but rather carried on in different forms. Consequently, in aboriginal belief, death is merely a physical end; the spirit survives and the past is thus strongly linked to the present and remains very real. Aborigines everywhere have a strong belief in the *Dreamtime* and its general pattern of creation, and it is a fundamental part of their social system and world view.

Yet as was so often the case when Europeans and indigenous cultures came into contact, the European discovery of Australia had disastrous impacts on aboriginal culture. The situation of the Tasmanian aborigines is one of the most indicative and perhaps the saddest of all. Following an initial friendly encounter with Tasmanians of the D'Entrecasteaux Channel in southeastern Tasmania in 1802, the first British settlers left Port Jackson in 1803 intent on establishing a prison camp in Tasmania, then known as Van Diemen's Land, after the Dutch explorer who first saw the island. Shortly after the arrival of the British, the first massacre of Tasmania took place and there ensued a war between Tasmanians and Europeans that lasted almost 30 years. By 1830, only 300 Tasmanians out of an original population of about 5 000 remained, and by 1847 all but 47 of these had died. This warfare, together with the introduction of European diseases, in effect exterminated the Tasmanians, with the last pure individual of the race, an old woman named Trucanini, dying in 1876 (although Tasmanians of mixed blood survive to the present day).

Unfortunately, this pattern repeated itself on the Australian mainland during the colonial period and well into this century. The result today is that people who claim aboriginal descent now number only about 265 000 out of a population of some 18 million, and the vast majority of these are of mixed genetic heritage. Of the original territory of Australia that was entirely theirs, they now own only about 13% —a high percentage compared to what is now in the hands of indigenous people in countries like the United States and Brazil, but not very much in a sparsely-populated country like Australia. Furthermore, land still in the hands of aborigines is usually in the least fertile and most undesirable areas.

In December, 1993, 305 years after the English and the aborigines first met, Australia's then Prime Minister celebrated the passing of legislation that recognized the legality of native title in Australia. This was, in effect, a declaration of peace, and a recognition of the fact that aboriginal tribal law has a place in modern Australian society, and it heralded a new era in relations among Australia's people. Nonetheless, the past cannot be altered or denied. A war of two centuries' duration was fought, and a whole array of human cultures was drastically modified or wiped out entirely, with the culture of the victorious race largely supplanting that of the vanquished (Flannery, 1996a).

In today's Australia, aborigines hold title to some significant chunks of land, the largest of which is Arnhem Land in the Northern Territory, where they have an Aboriginal Reserve of some 8 million ha. Aborigines have succeeded in maintaining elements of their culture, especially in some of the more remote desert areas, where they continue to eat a wide variety of bush foods, practice their ceremonies and produce their arts and crafts, although most of these are now aimed at a growing tourist market. However, very few people still live a truly traditional lifestyle, and many of the languages and traditions of some of the smaller groups are unfortunately on the verge of extinction, with several languages now being spoken by very small numbers of older people.

Cristina Goettsch Mittermeier
Russell A. Mittermeier

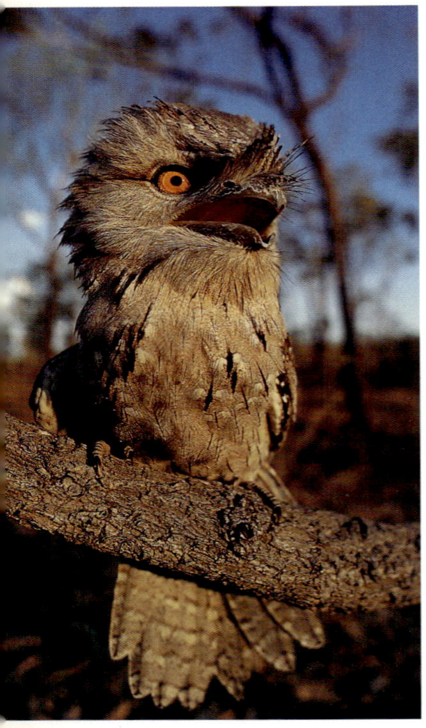

The tawny frogmouth (Podargus strigoides), *a large and bizarre relative of the nighthawks and nightjars that is endemic to Australia.*

On the opposite page, the Australia brush turkey (Alectura lathami), *a large megapode bird that is common in the forests of Queensland. Megapodes are known for their very large eggs which are incubated in decomposing piles of vegetation or in warm volcanic sands.*

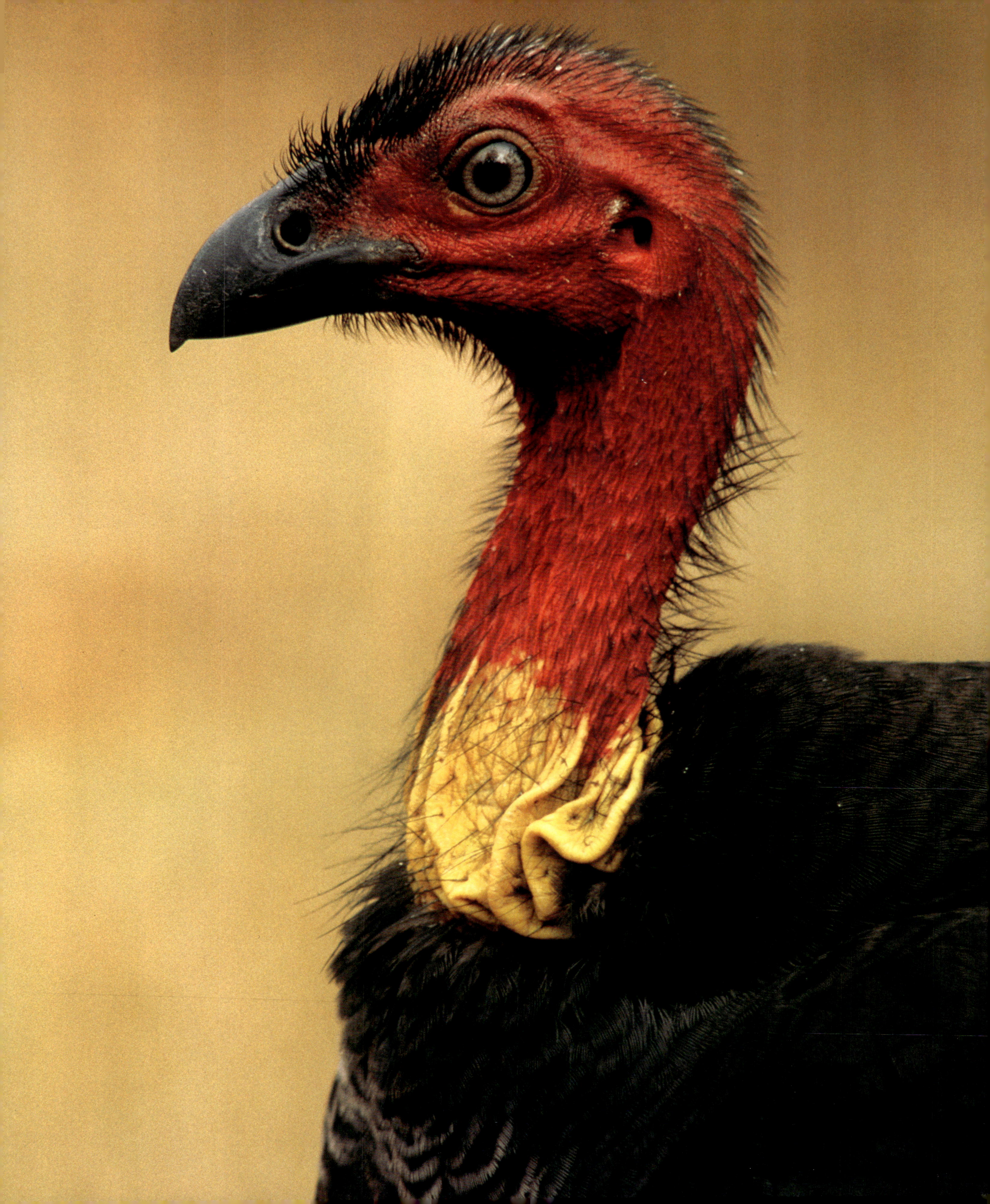

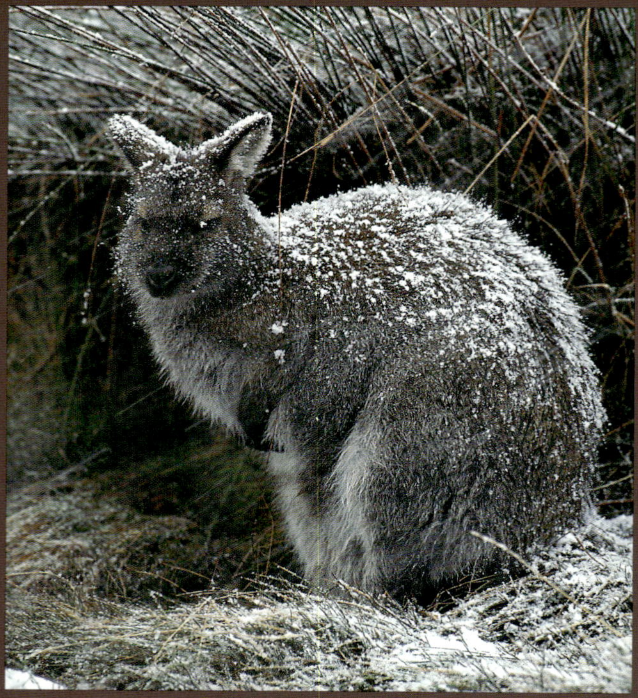

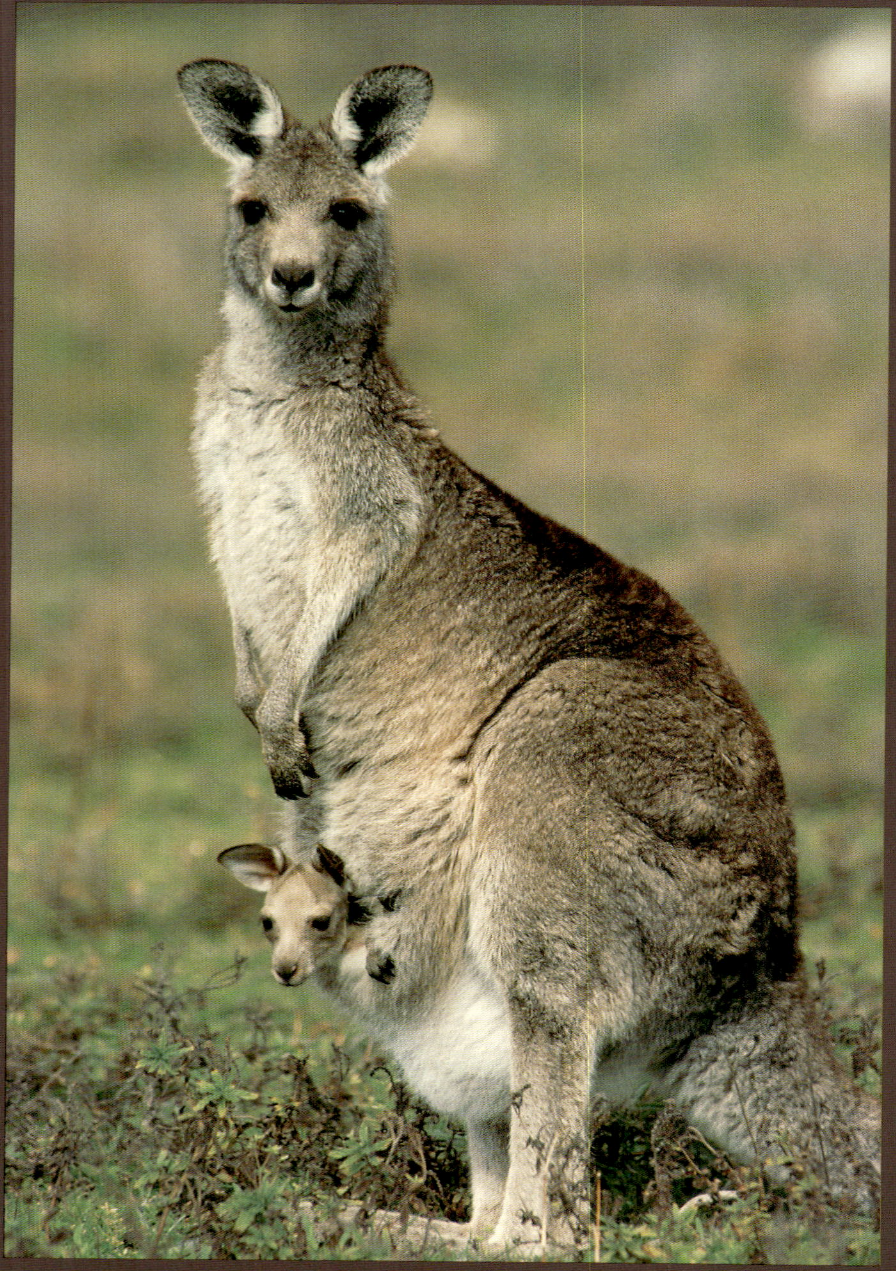

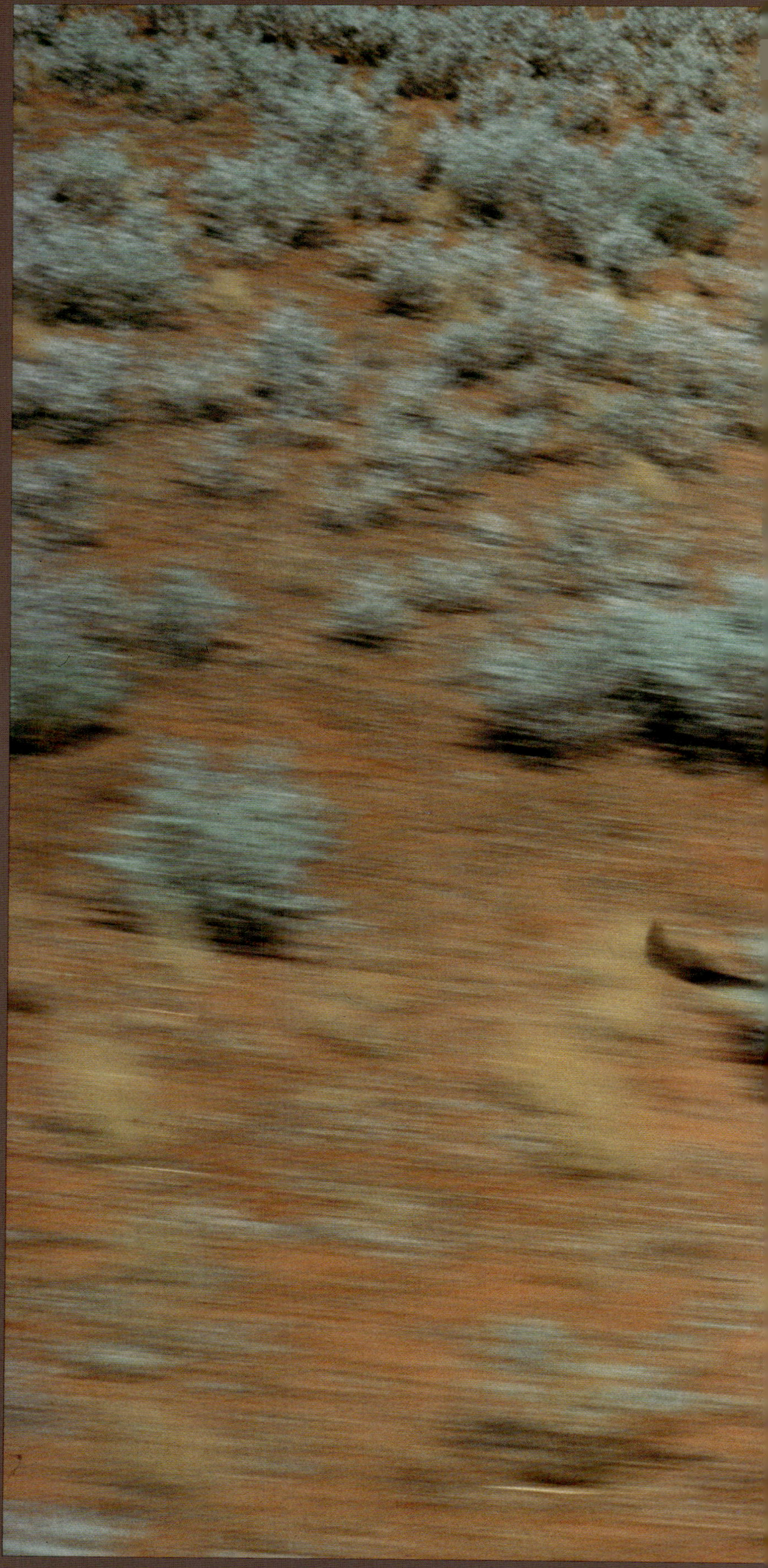

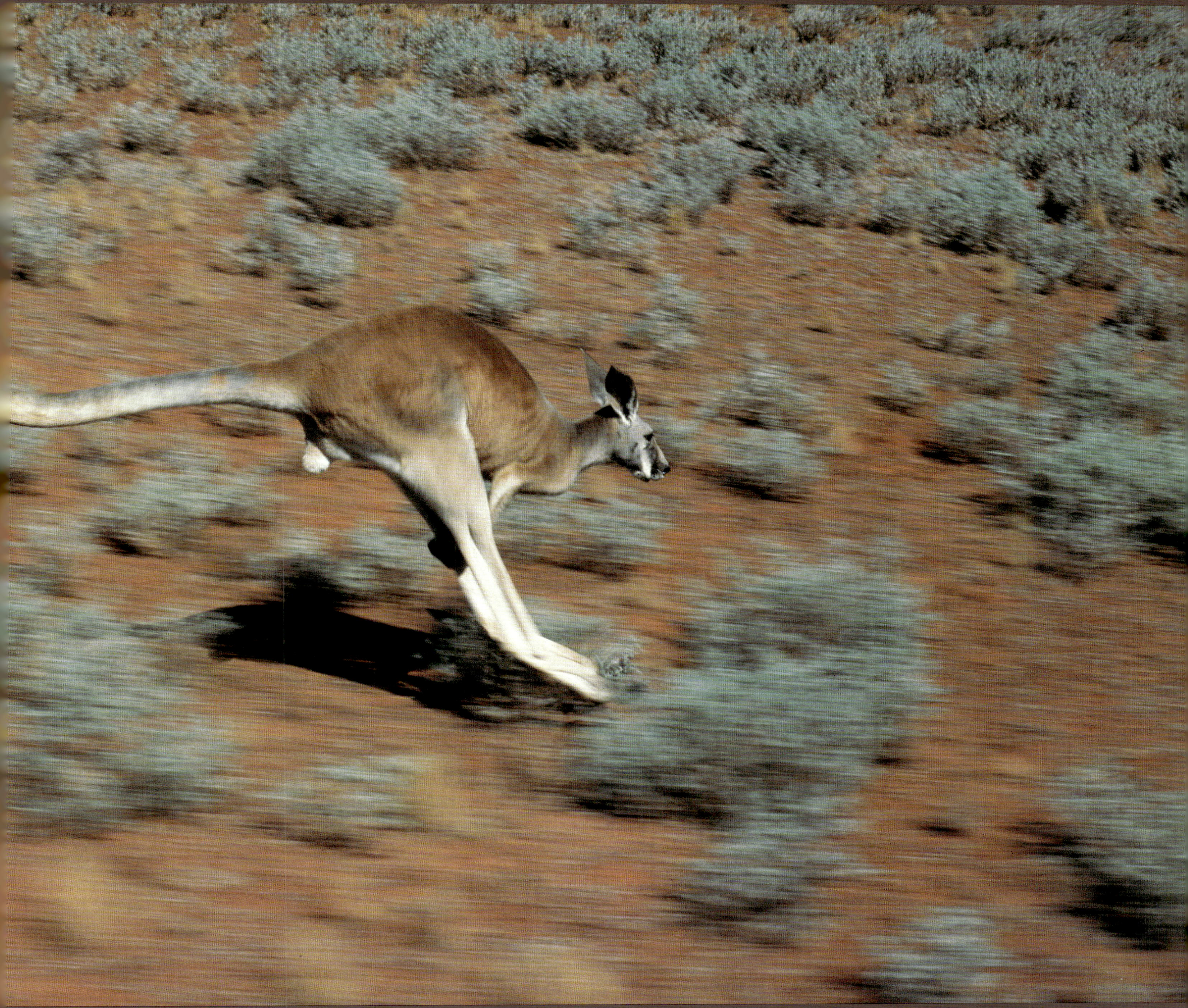

The kangaroos are Australia's best known animal group, and come in many different shapes and sizes. Three of the 48 are shown here. The red-necked or Bennett's wallaby (Macropus rufogriseus), *upper left, is a medium-sized species from Tasmania.*

Lower left, the eastern gray kangaroo (Macropus giganteus), *shown here with its joey, is the largest of all kangaroos. Above, the red kangaroo* (Macropus rufus) *is surpassed only in size by the two species of gray kangaroo, and is characteristic of drier regions.*

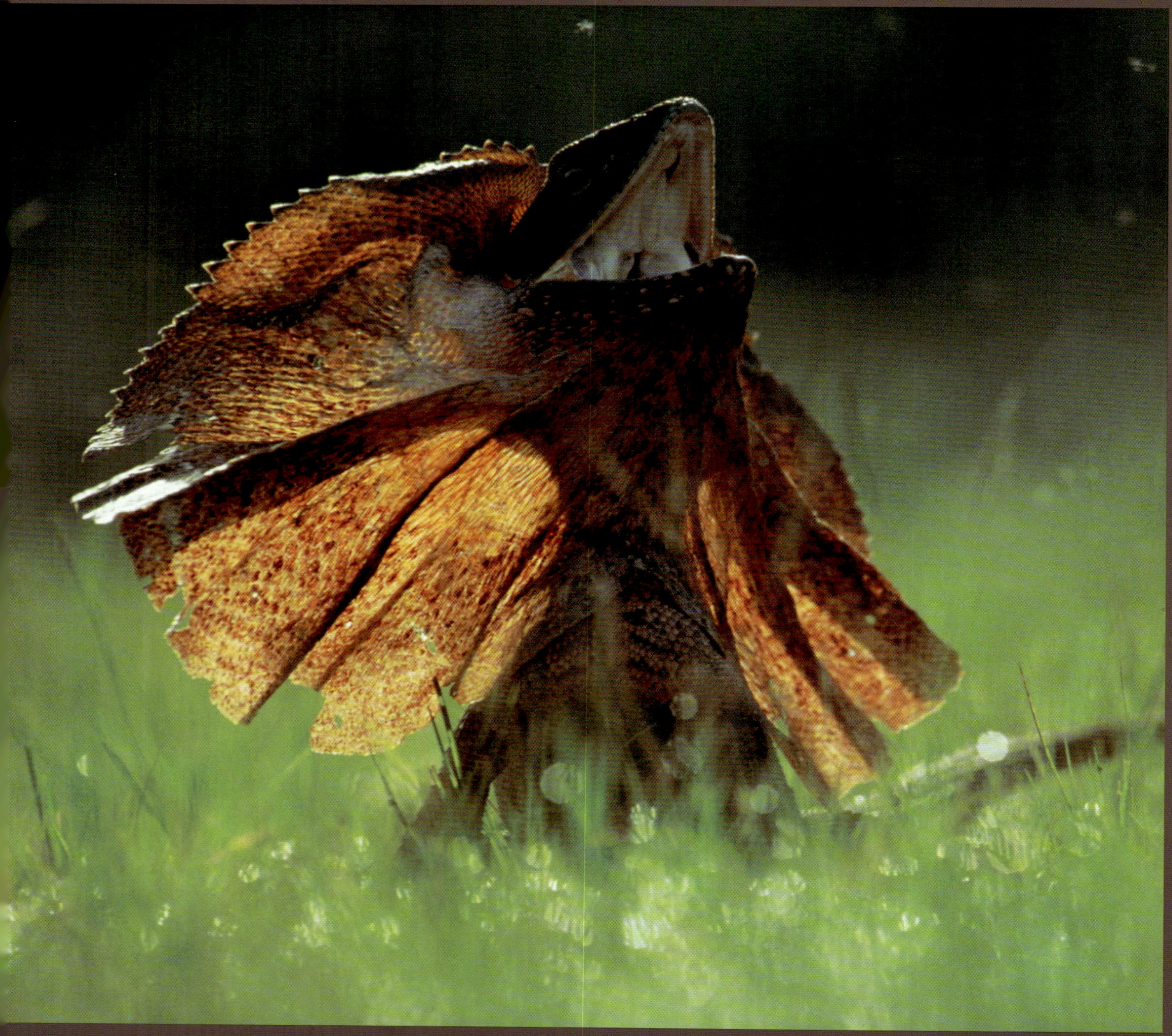

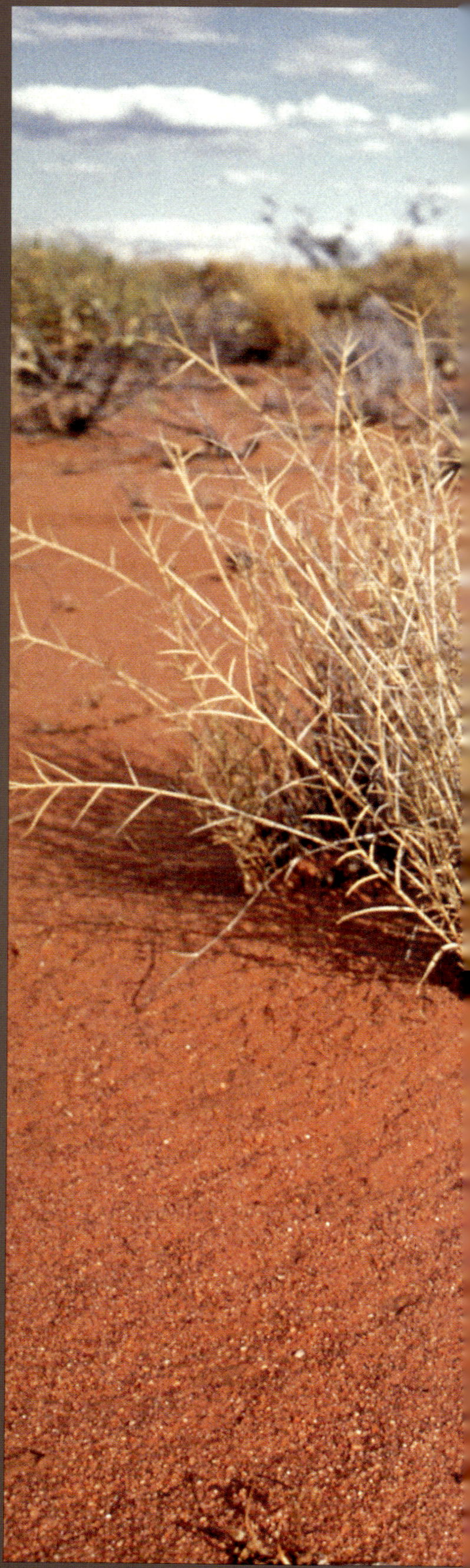

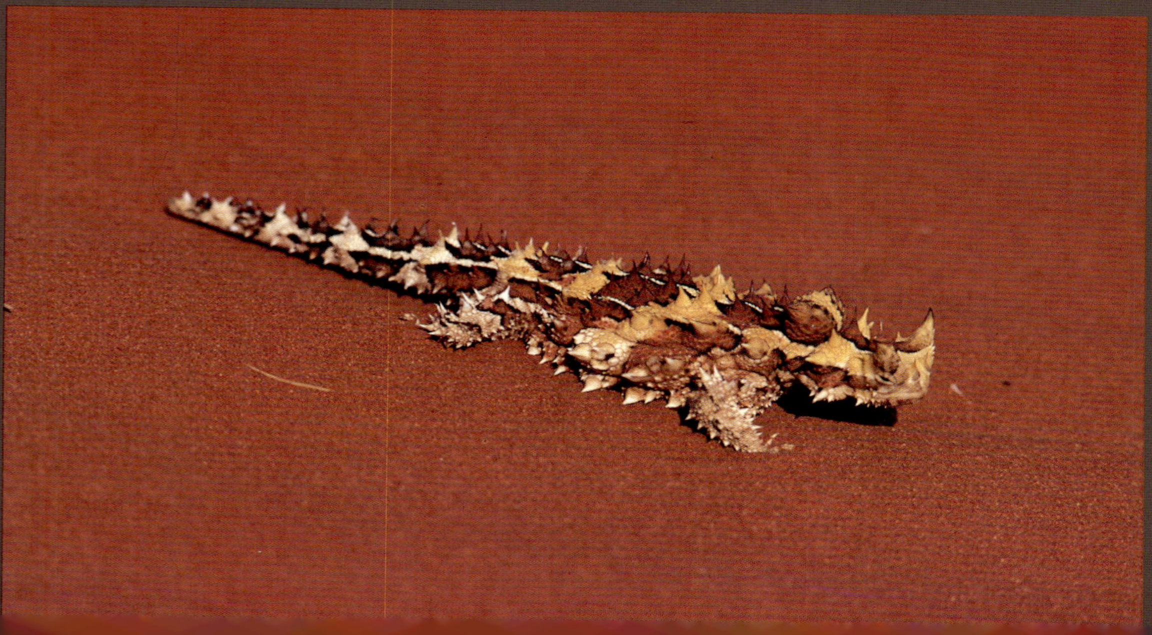

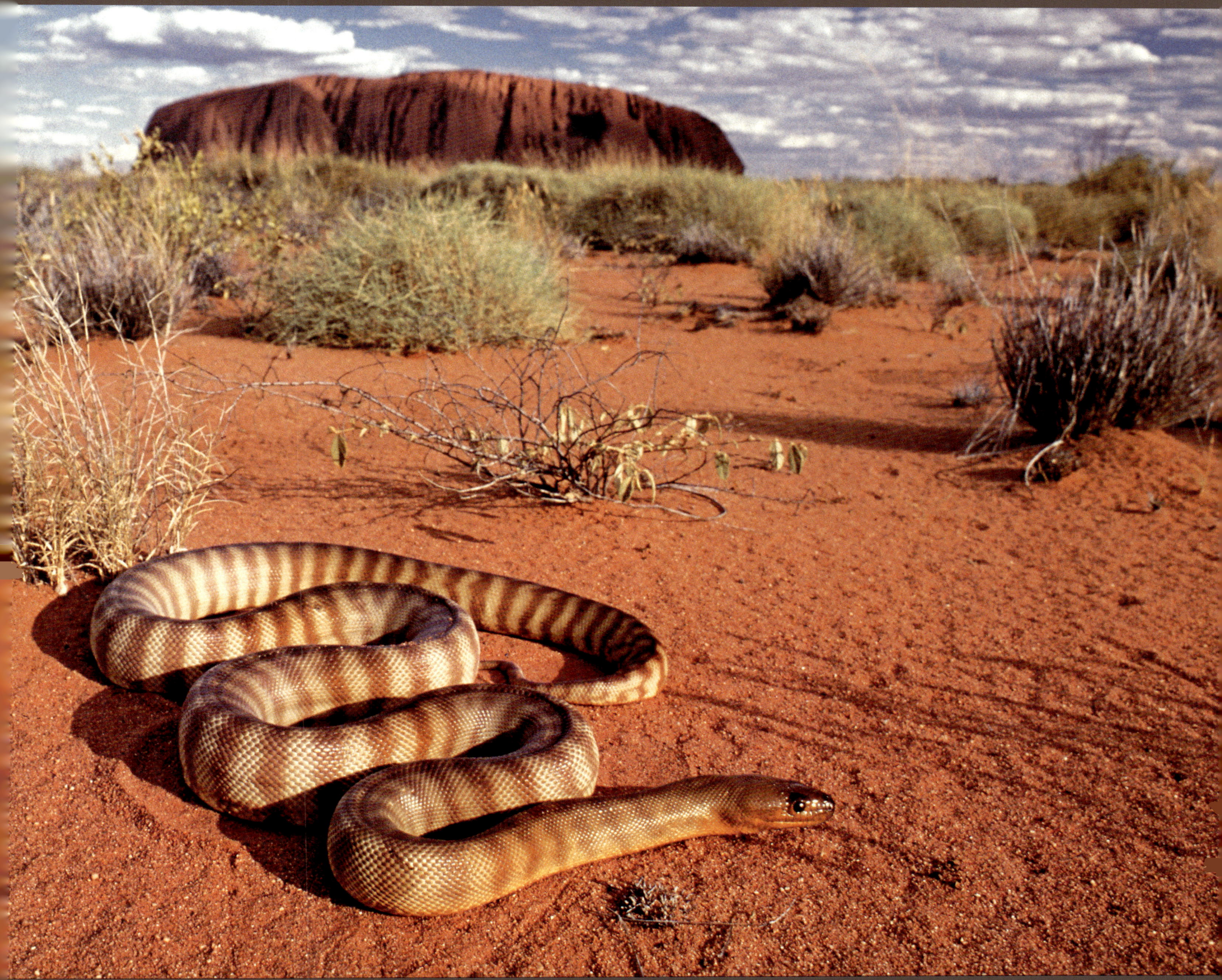

Australia has the highest percentage of poisonous snakes of any country on Earth, with some 65% of all snakes found there being venomous, including some of the world's most dangerous species.
*Above is one of the harmless species, the woma python (*Aspidites ramsayi*), photographed in the desert area near Ayers Rock, Uluru National Park, Northern Territory. Australia also has the richest reptile fauna on Earth, and the highest number of endemic species.*

*Upper left, the frilled lizard (*Chlamydosaurus kingii*) is a striking species characterized by the large, loose ruff or frill of skin around its neck, which can be erected when the lizard is alarmed, greatly enhancing the apparent size of the animal to frighten predators.*
*Lower left, the moloch or thorny devil (*Moloch horridus*) is perhaps the most unusual and appealing of all of Australia's flagship reptiles. It is a gentle, harmless creature that feeds exclusively on ants.*

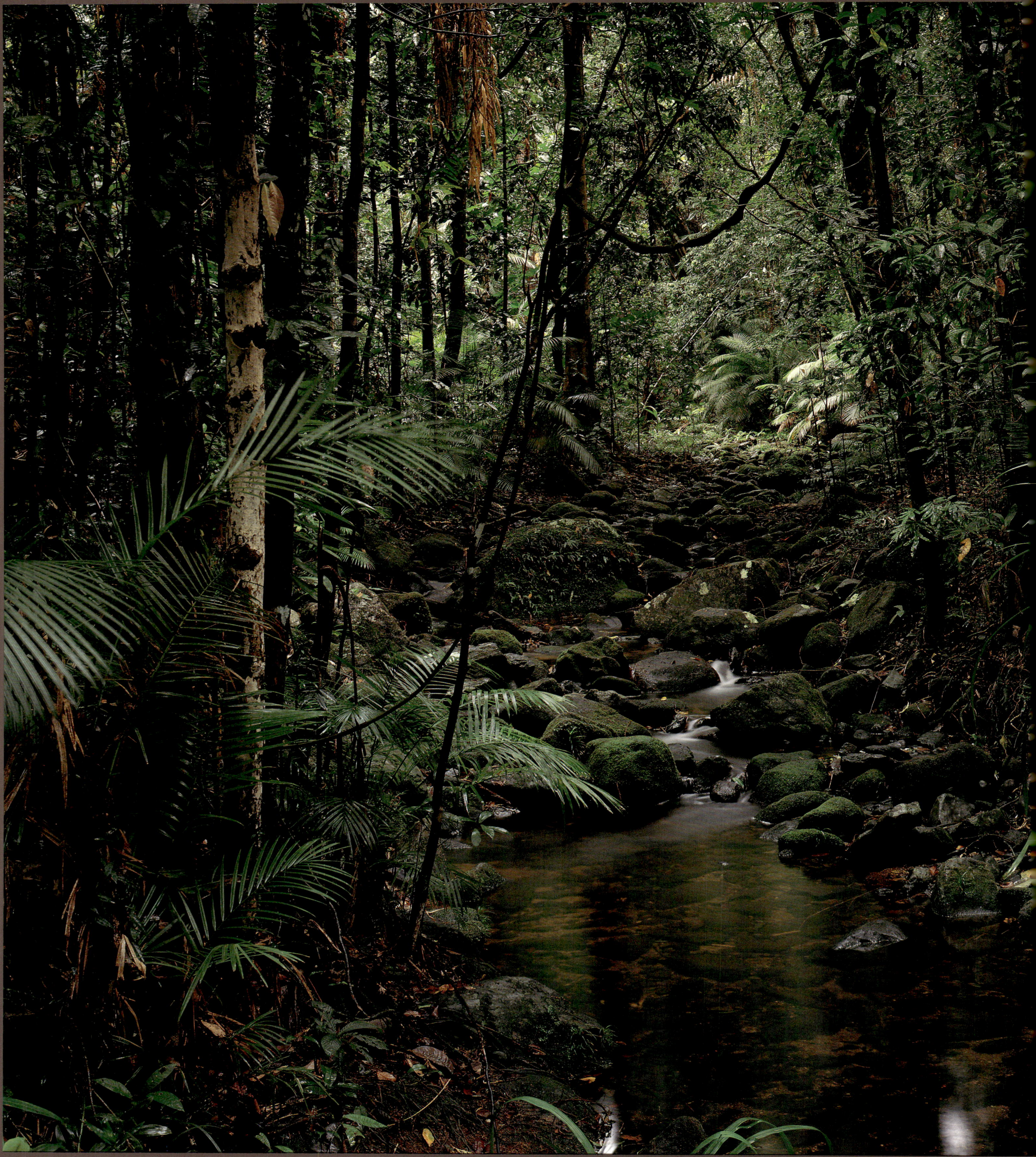

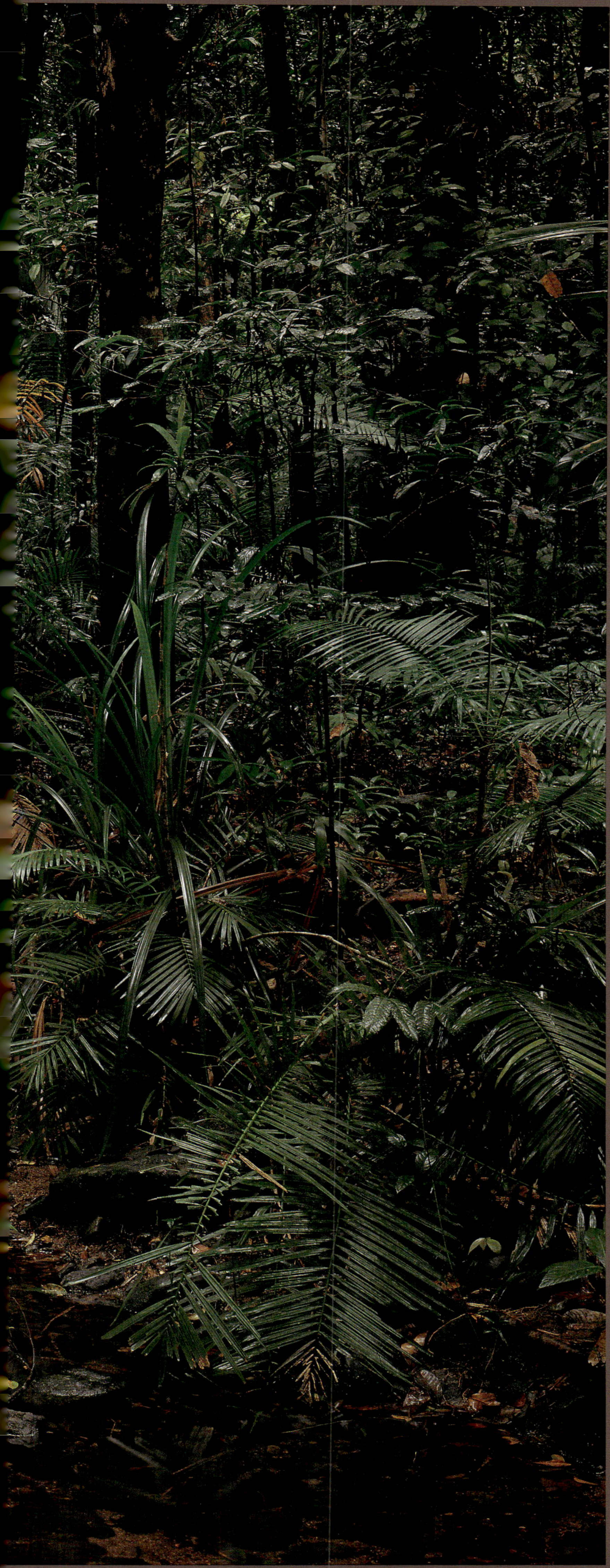

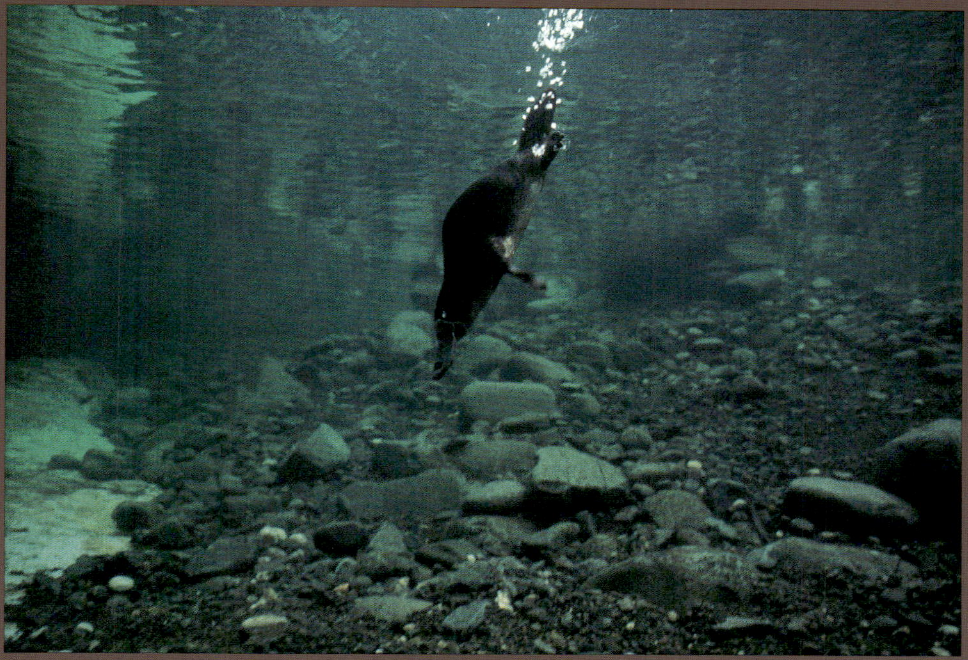

Although largely a dry continent, Australia has a small but very important strip of tropical rainforest in north Queensland. The most important protected area in the Queensland rain forest is Daintree National Park. Seen here (on the left) is dense rain forest in the Mossman Gorge section of this park, which has perhaps the highest diversity of any protected area in Australia. Above, the platypus (Ornithorhynchus anatinus) *is one of only three monotremes, and perhaps the most unusual mammal on Earth. The platypus is an aquatic species found in appropriate habitats along the east coast of Australia from Tasmania to Queensland.*

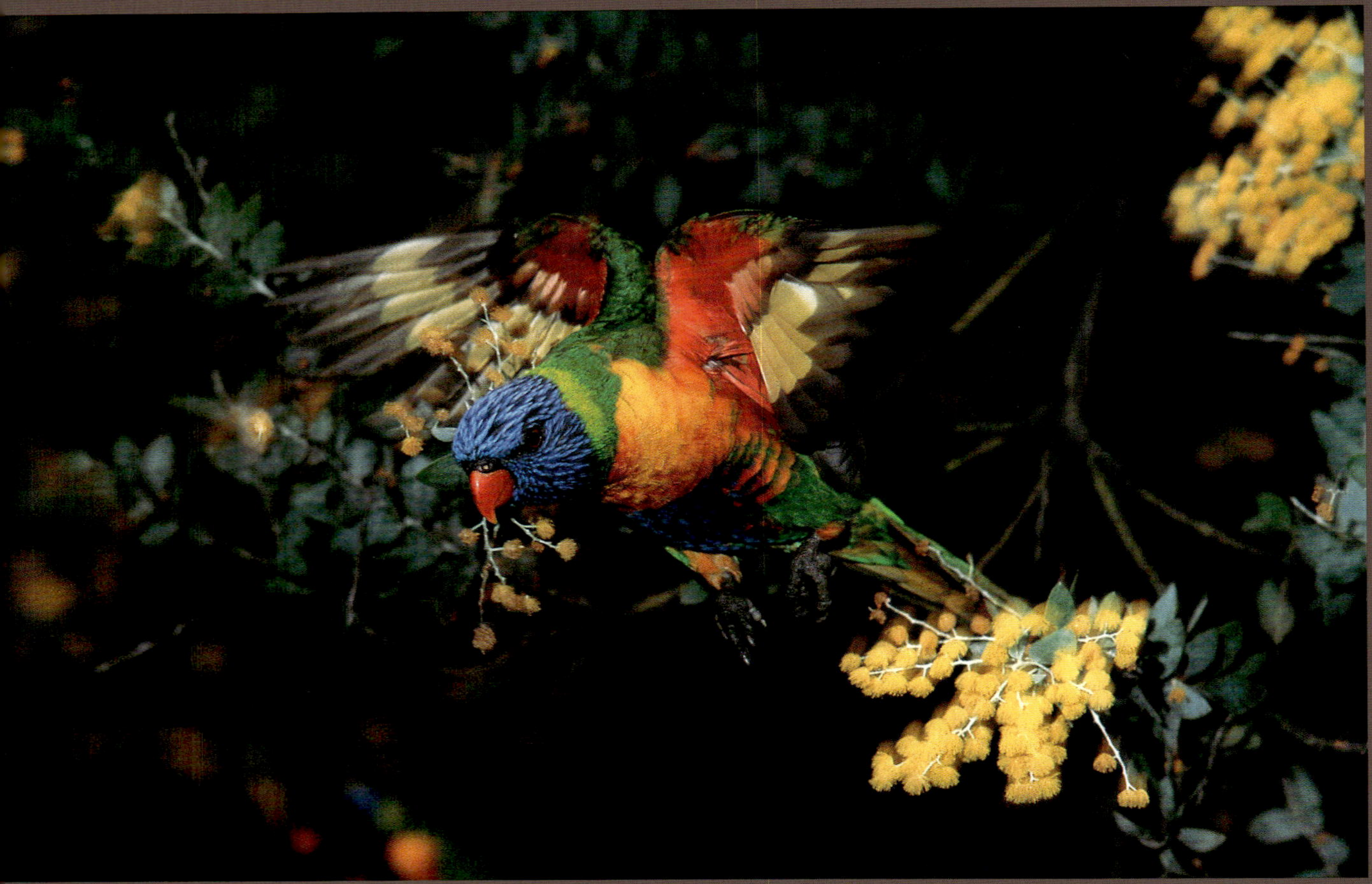

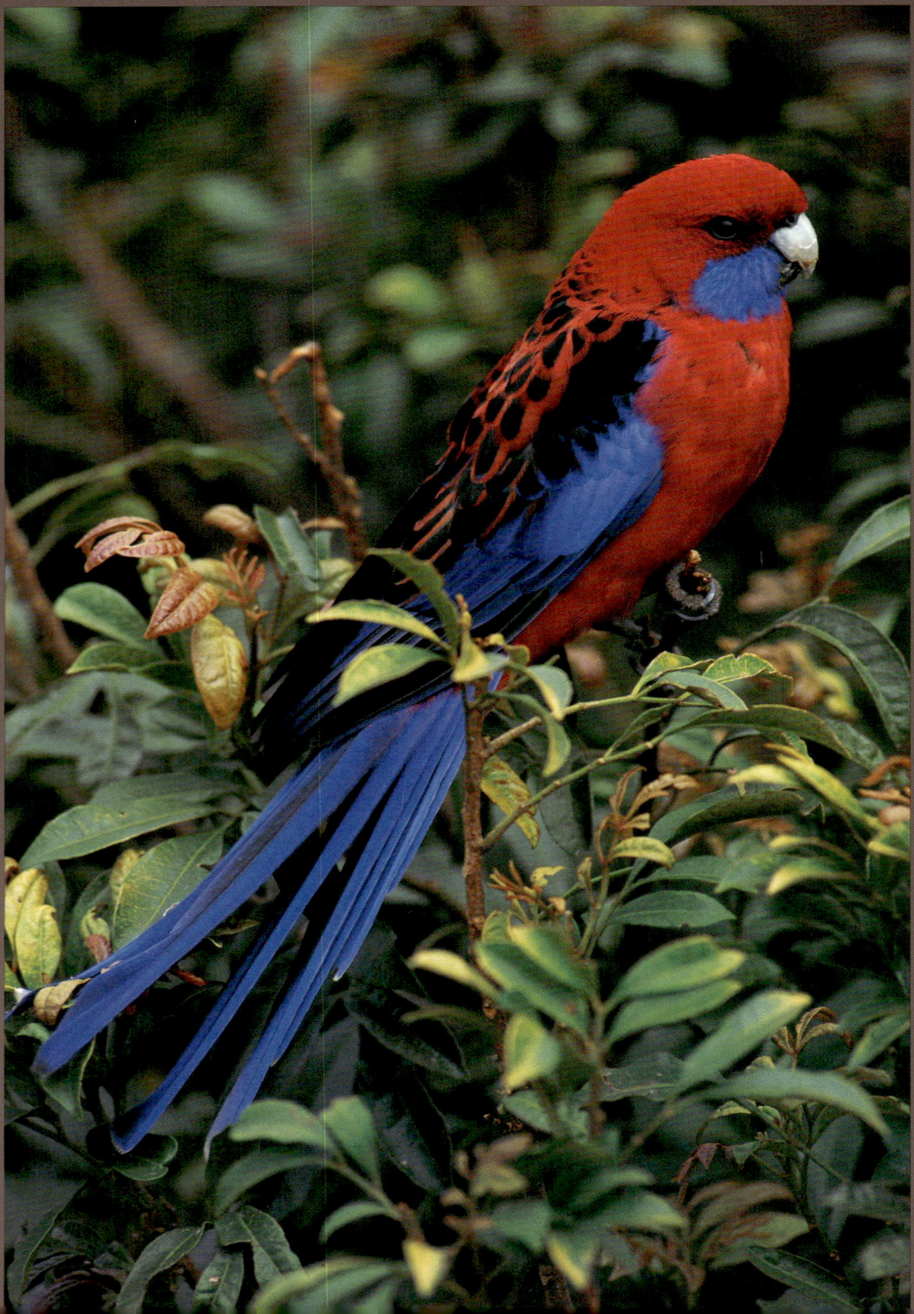

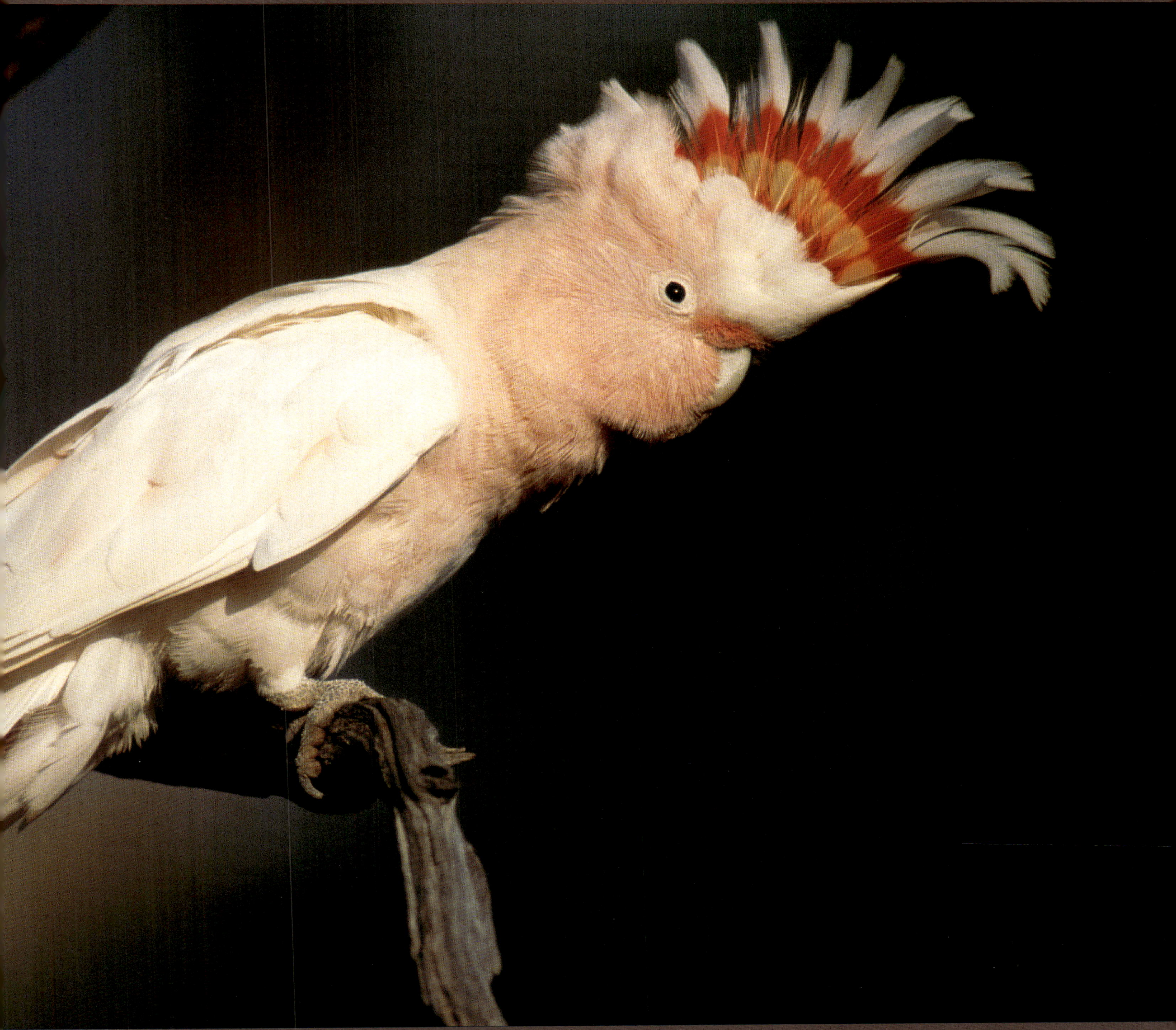

Australia is one of the three richest countries on Earth for psittacine birds, which are represented there by the parrots, the cockatoos, the lorikeets and a number of other species. Many of them are strikingly beautiful. Upper left, the rainbow lorikeet (Trichoglossus haematodus) *is a common species that occurs even in urban areas, and ranges along the entire east coast of Australia and into New Guinea as well.*

Lower left, the crimson rosella (Platycercus elegans) *is also common in eastern Australia; the individual here was photographed in Lamington National Park in southern Queensland. Above, the pink or Leadbeater's cockatoo* (Cacatua leadbeateri) *is characterized by its distinctive crest. It occurs in southern and western Australia.*

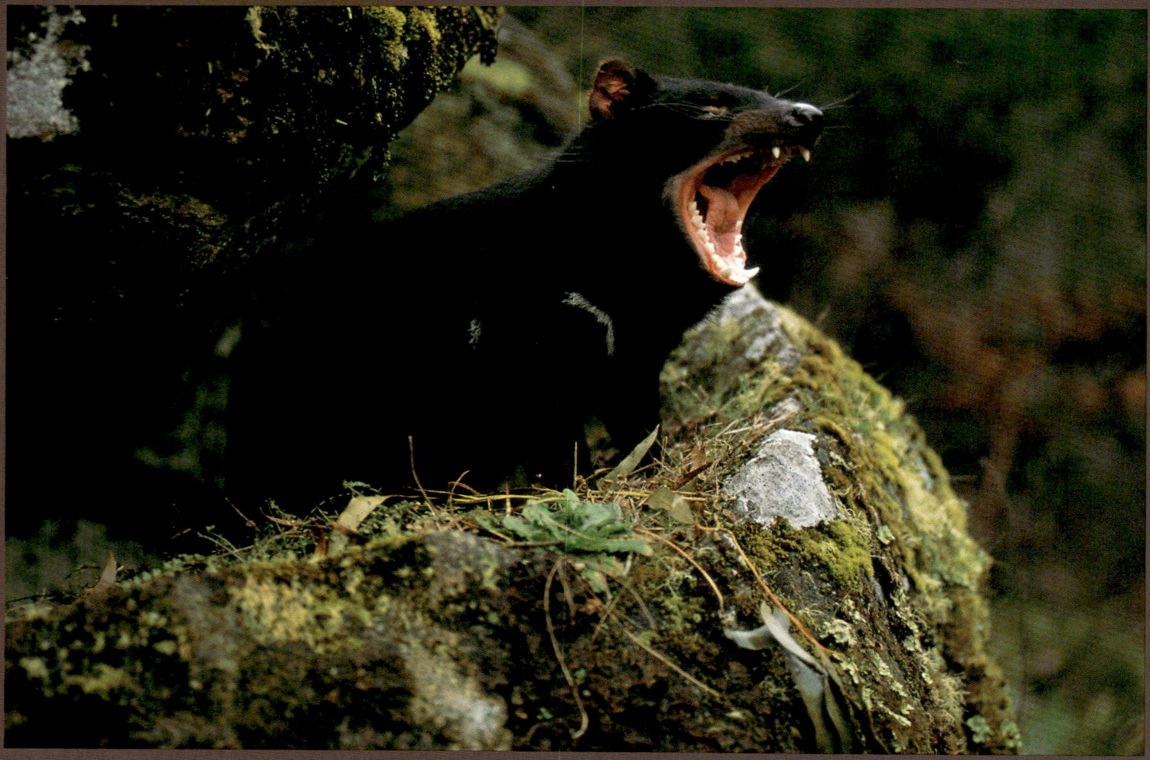

Above, the Tasmanian devil (Sarcophilus harrisii) *is a ferocious predator and the largest of the living carnivorous marsupials. Once found on the mainland as well, it is now restricted to the island of Tasmania.*
Right, the saltwater crocodile (Crocodylus porosus) *is one of the two or three largest of all reptiles, and can reach 6-7 m, and possibly even 8 m in length. It is one of the most aggressive of the crocodilians, and among the world's most dangerous large animals. To the great credit of the Australians, it has been very effectively protected, and is not uncommon in many areas of Queensland and the Northern Territory.*

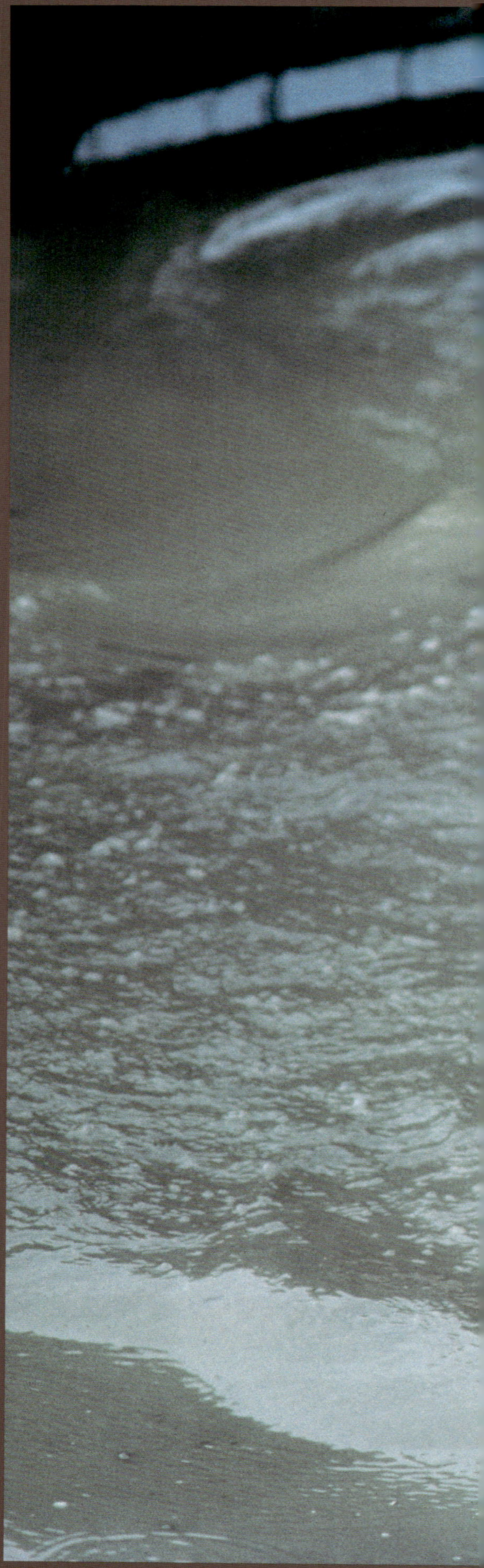

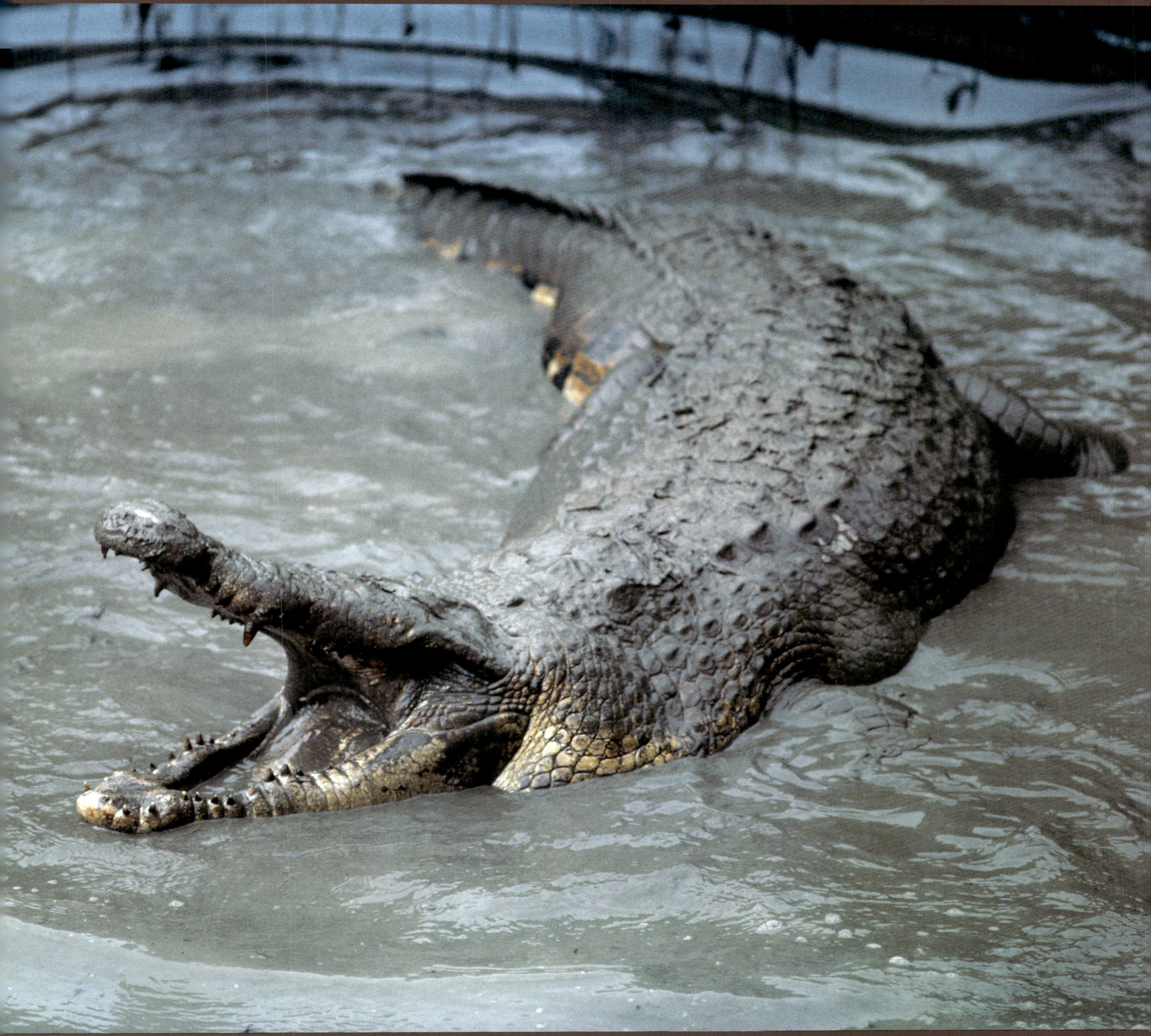

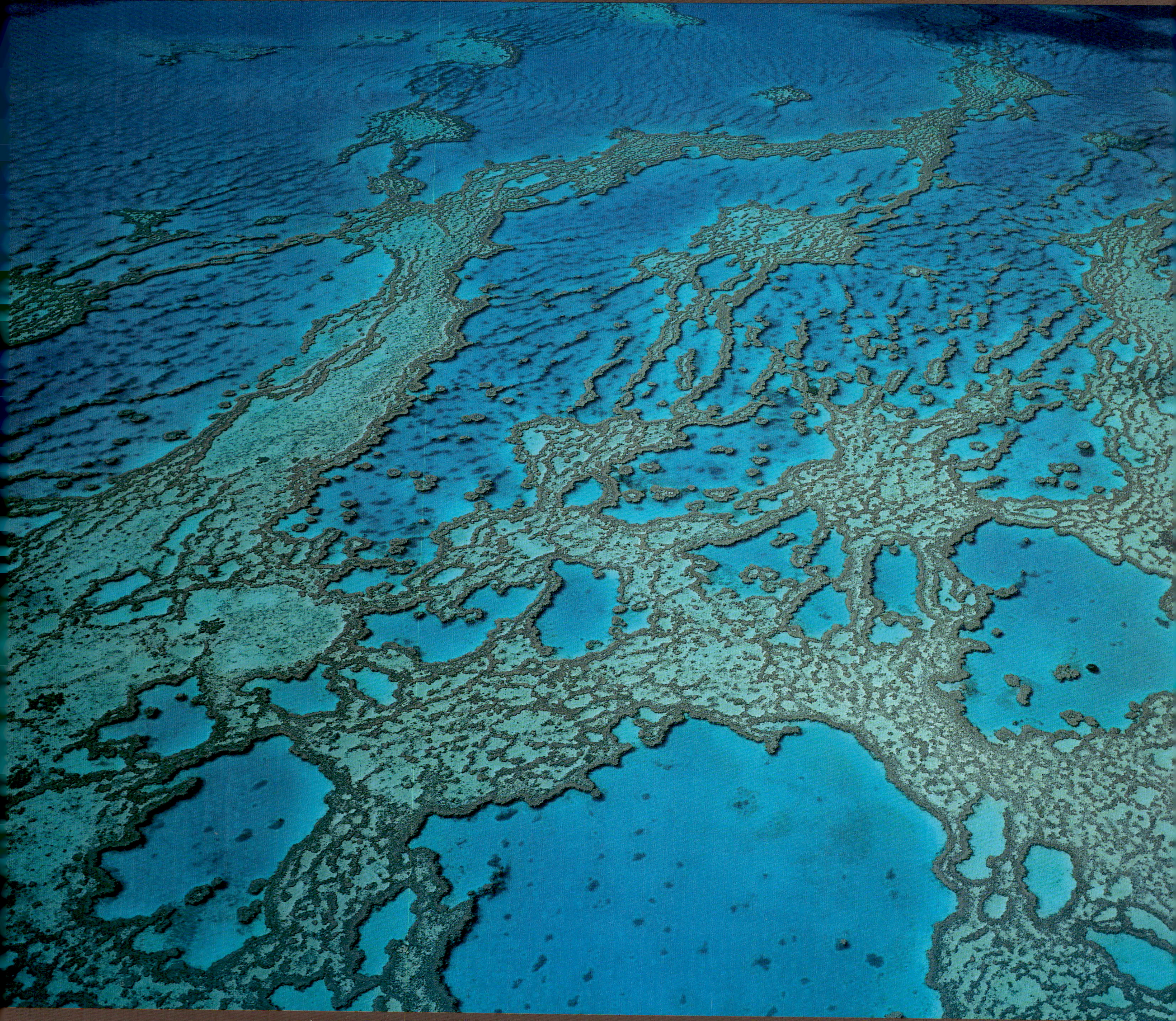

Australia has what is probably the most diverse marine fish fauna on Earth, with a diversity that exceeds 4 000 species. This is in part because it combines two major faunas, a temperate one and a tropical one that includes the largest reef system in the world, the Great Barrier Reef.

Above, the Hardy Reef in the Great Barrier Reef Marine Park in Queensland. Upper right is a leafy sea dragon from Kangaroo Island, a very unusual species of sea horse. Lower right is a stand of soft coral trees of the family Siphonogorgiidae; these are found in the Coral Sea, and come in a great variety of colors.

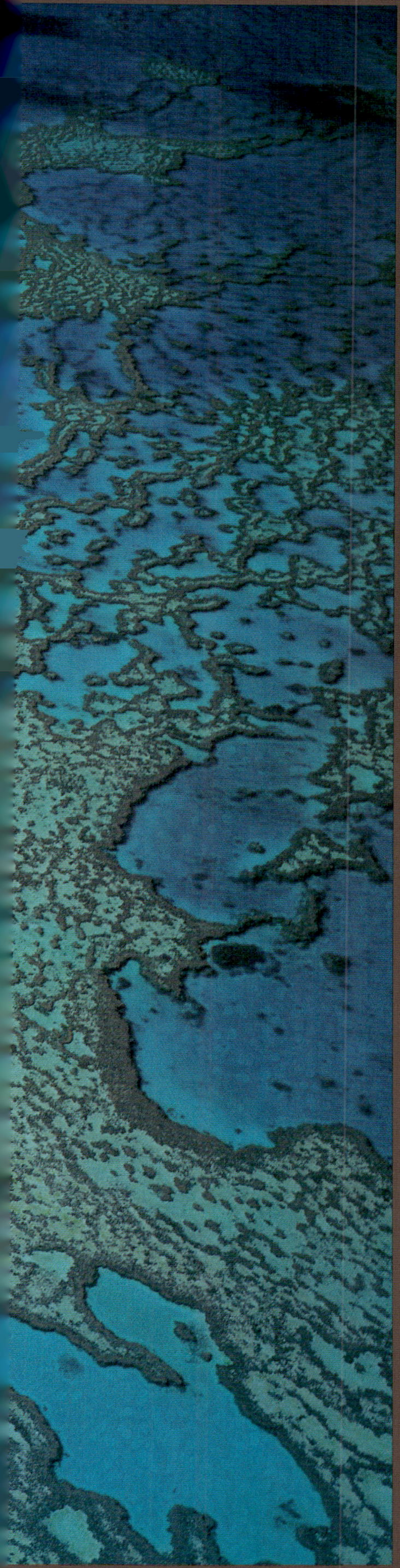

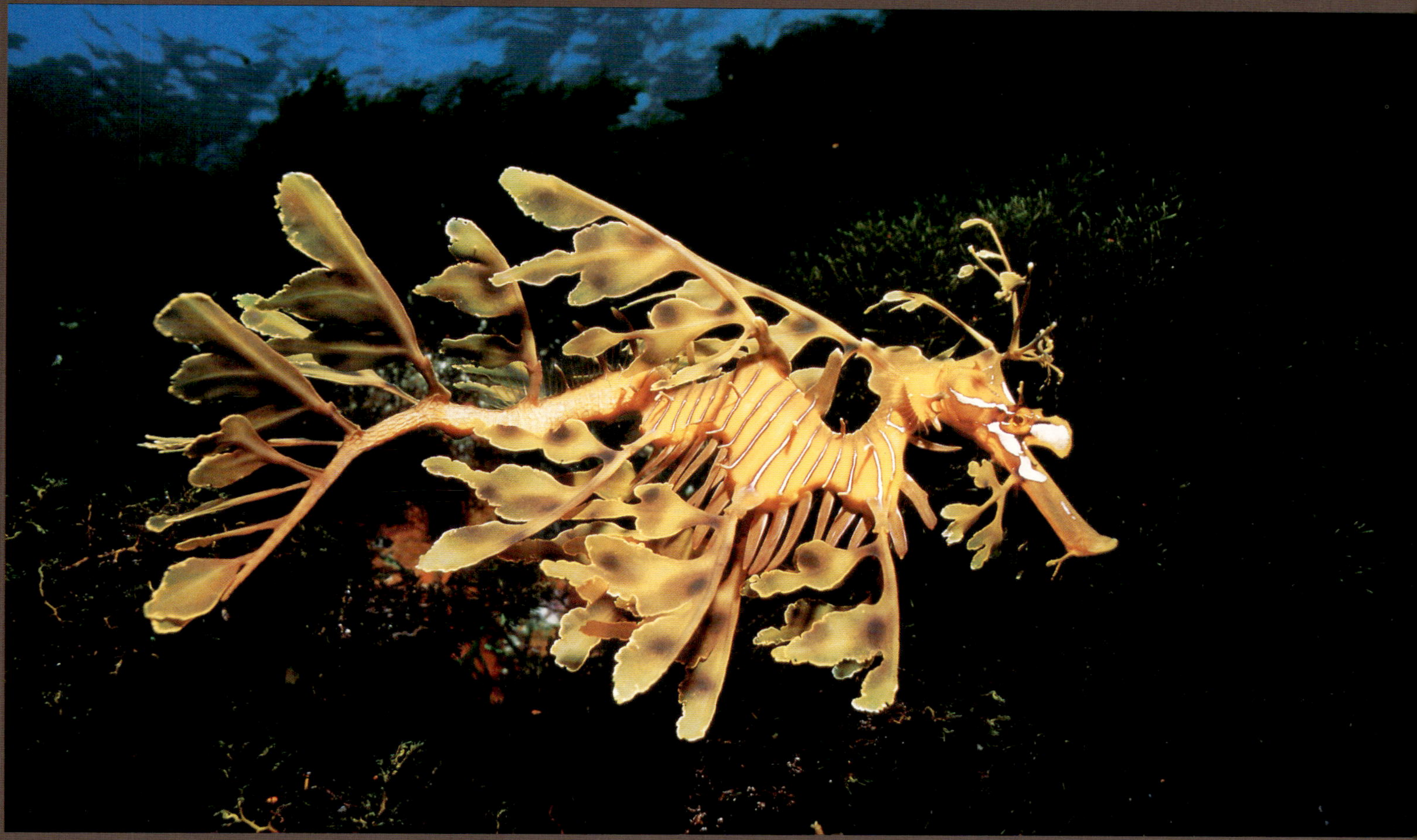

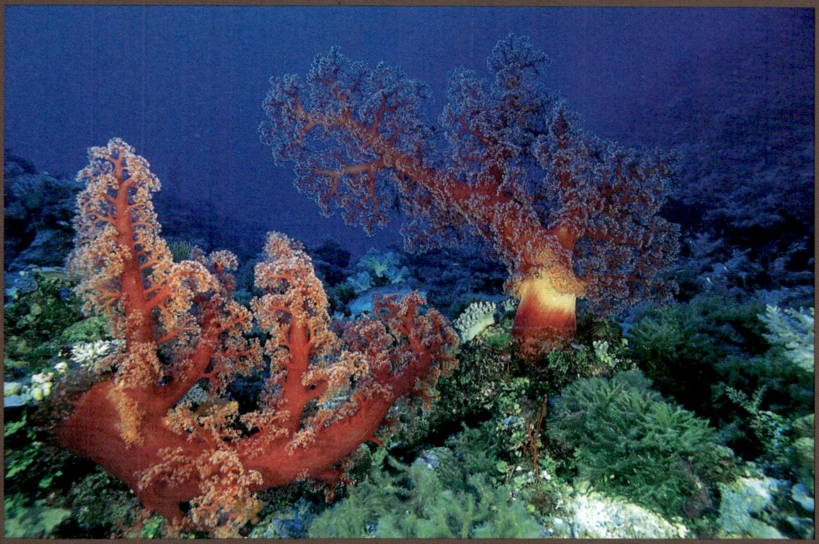

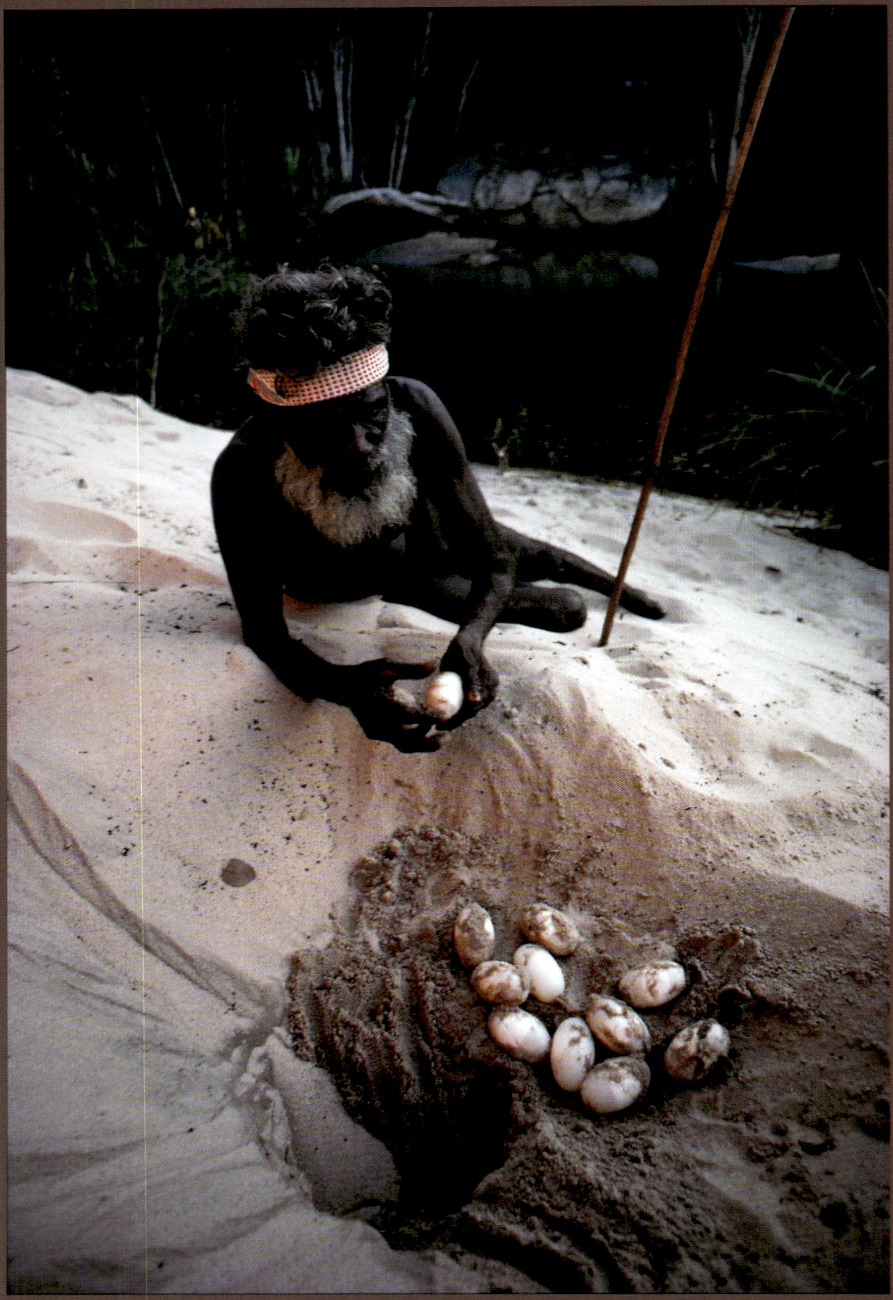

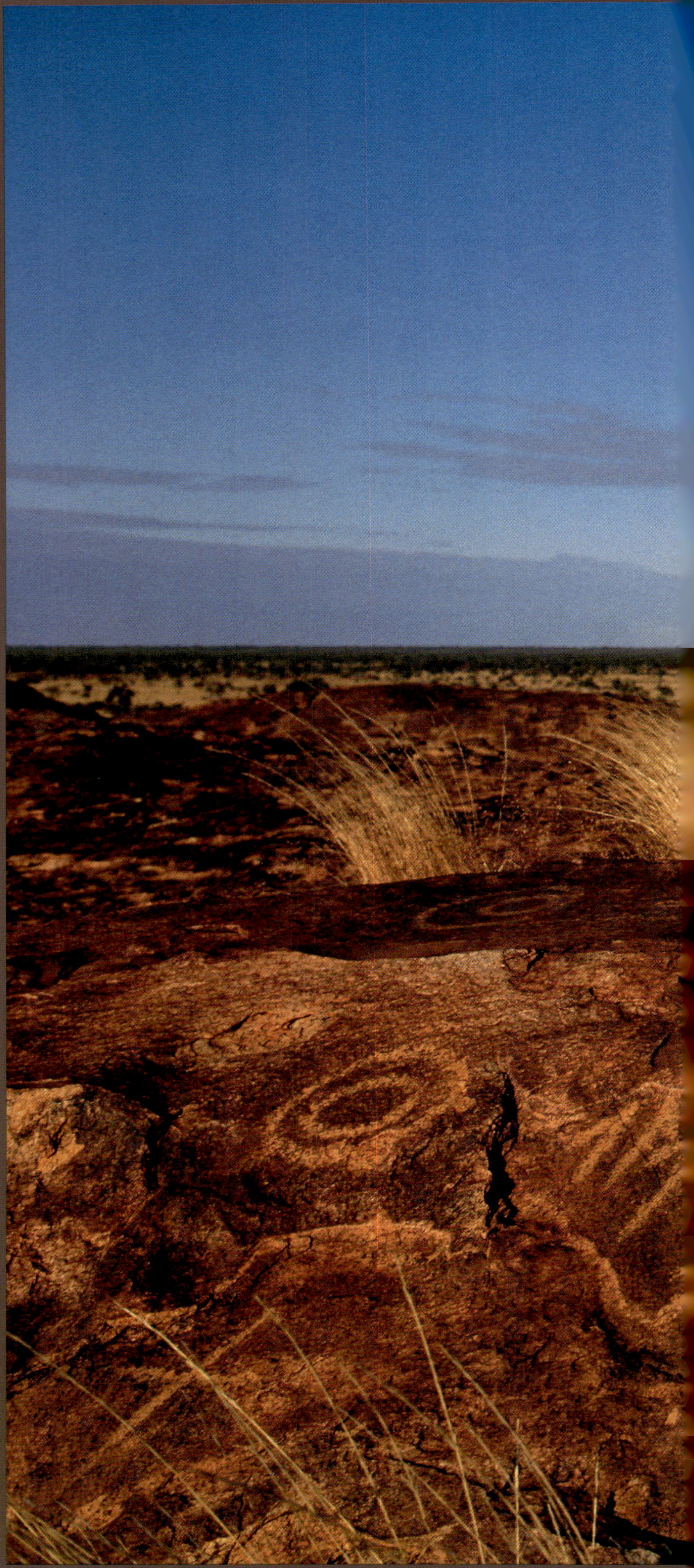

Although the Australian aborigines have suffered greatly at the hands of Europeans over the past two centuries, many groups still survive.
In the image above at the Kakadu National Park, a man with edible freshwater crocodile eggs.
He is a Gagudju elder.
On the right, a member of the Eastern Arrernte people of northern Australia, ponders the carvings of circles and tracts made by ancient predecessors on Corroboree Rock in Simpson Desert.

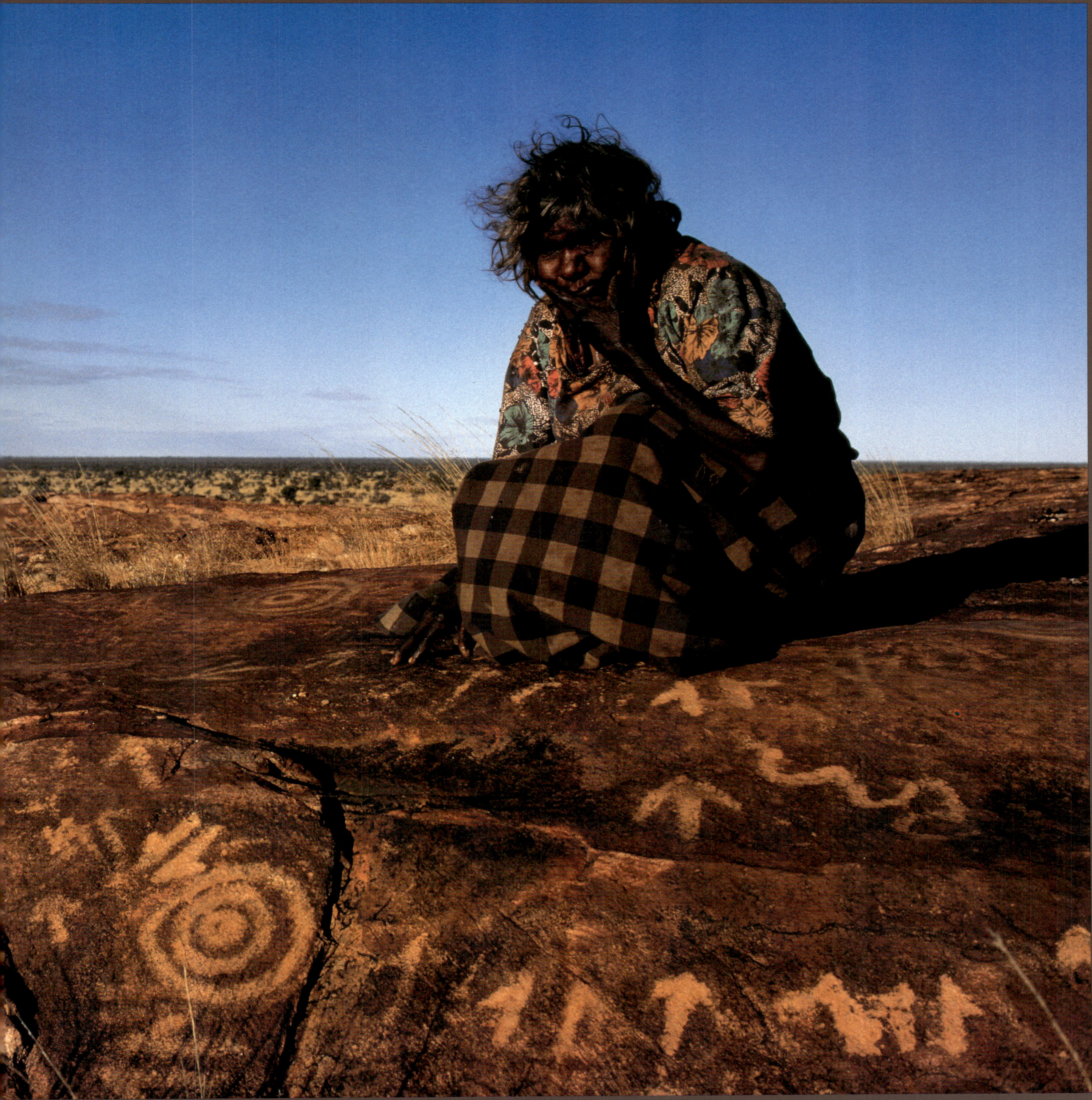

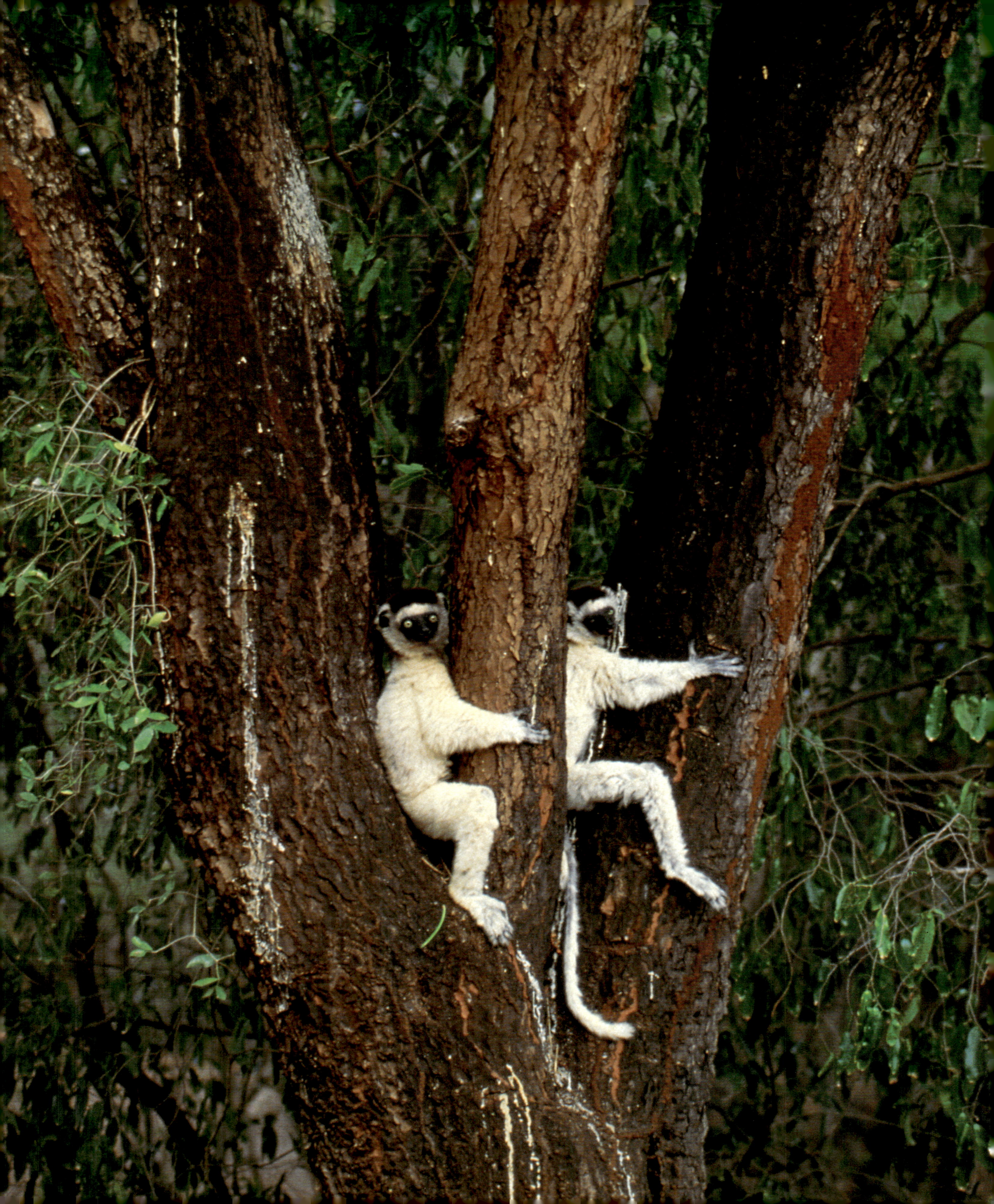

MADAGASCAR

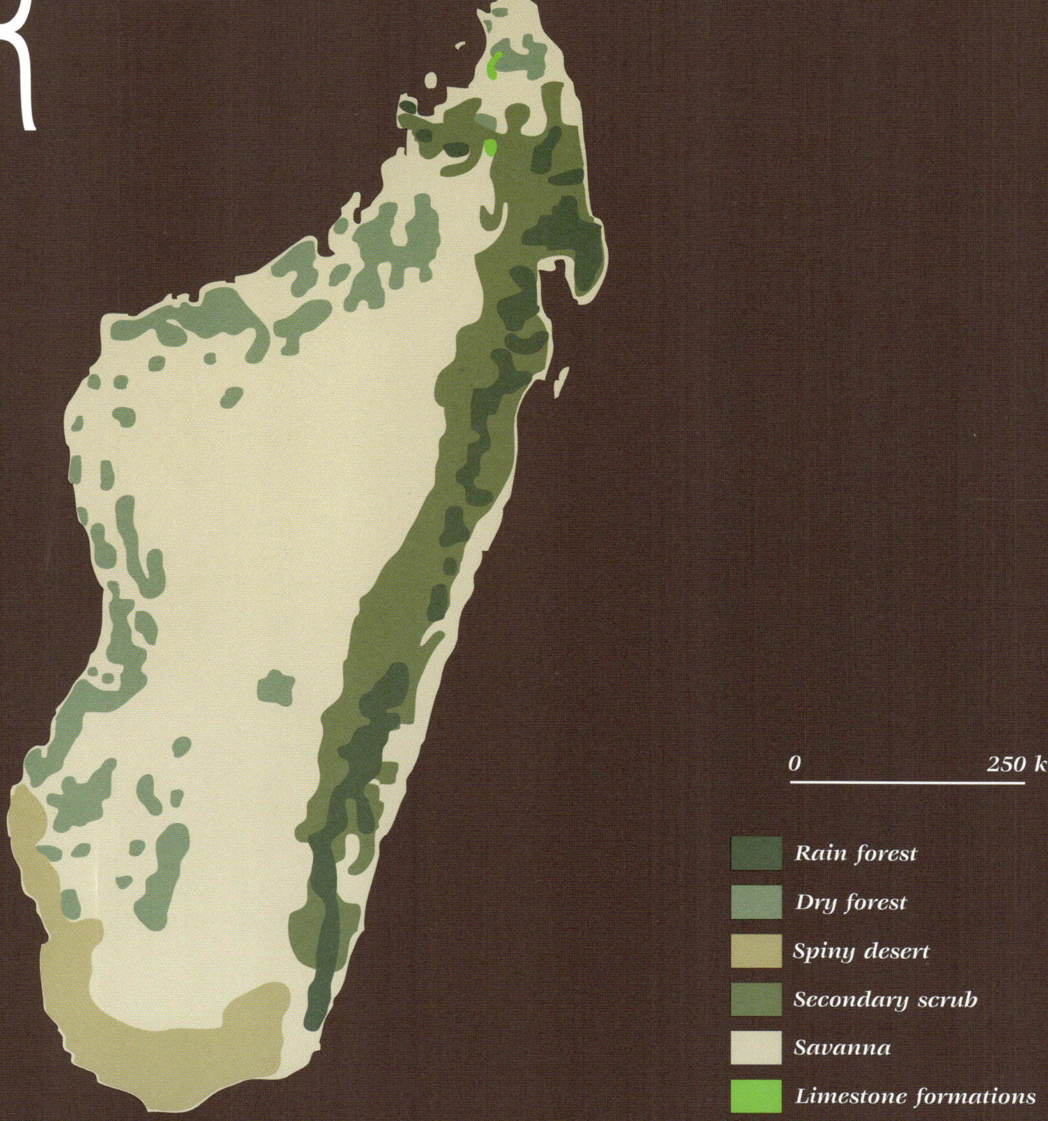

Madagascar is a very special place, even in the context of the wonderfully diverse megadiversity countries. The fourth largest island on Earth (and the largest oceanic island), it is about the size of France or the state of Texas, covering 587 045 km^2 or 0.4% of the land surface of the planet. Although one of the smallest megadiversity countries, in biological terms it is a mini-continent. In spite of being located only about 400 km from the eastern coast of Africa, Madagascar, once part of Gondwanaland, has been isolated from other landmasses for more than 160 million years. Consequently, most of the plant and animal species found there evolved apart from the rest of the world and unique to the island. The resultant very high levels of endemism, not just at the species level but at the generic and family levels as well, really set Madagascar apart from all other megadiversity countries except perhaps for Australia, which has comparable endemism at the generic and familial level, but is 13 times larger than Madagascar. Indeed, in many ways, Madagascar is a unique evolutionary experiment, a living laboratory unlike anywhere else on Earth.

Since it is situated largely within the tropics, Madagascar also has very high species diversity in certain groups of organisms, especially given its relatively small size. For instance, although it occupies only about 1.9% of the total land area of the African region, it has more orchid species than all of the African mainland and indeed is home to about 25% of all African plants (Jolly *et al.*, 1984). Reptile and amphibian diversity is also quite high, with 300 species of reptiles and 178 species of frogs being recorded to date, and more being discovered every year. The diversity of our closest relatives, the non-human primates, represented in Madagascar by lemurs, is extremely high as well, with 32 species and 50 different taxa of lemur found in Madagascar, representing 12.8% of all living primates.

On the other hand, Madagascar is entirely lacking in certain groups of animals, including many that are typical of the African mainland, so close and yet so far away in evolutionary terms. For instance, there are no elephants, no rhinos, no giraffes or other antelopes, no lions, no hyenas, no wild dogs, no monkeys (in spite of the high primate diversity represented by the lemurs) and no poisonous snakes. Still others, like the rodents, the bats, the mongoose family (the only group of carnivores represented on Madagascar) and the birds are quite low in species diversity, but exhibit very high levels of endemism. Only 105 species of mammals have been recorded, but 77 are endemic. Furthermore, 37 of 52 genera and 6 of 15 families are also endemic, a remarkably high level.

Indeed, it is endemism rather than total diversity that puts Madagascar on the megadiversity list, and especially endemism at higher taxonomic levels like the genus and the family. The figures are truly striking. Of the estimated 11 000-12 000 species of flowering plants, more than 80% are thought to be endemic (the actual total remains to be determined, and the estimate varies from expert to expert), and there are five plant families restricted to Madagascar. Compare this to Brazil, which is the world's leader in total plant species diversity at 55 000 but has no endemic families. Only Australia, with 12 endemic families, and tiny New Caledonia (another chip of ancient Gondwanaland), with five endemic families, can compare. Even within Madagascar, the levels of regional endemism are striking. The plants of the southern spiny desert, for example, are 95% endemic just to that region at the species level, and 48% at the generic level; they do not even occur anywhere else in Madagascar. Among the birds, which exhibit

The lemurs are Madagascar's most famous flagship species. Some 50 different kinds exist, all of them endemic to the country. Among the most spectacular are the beautiful sifakas, shown on the opposite page, of which Verreaux's sifaka (Propithecus verreauxi verreauxi) *is the most common. These sifakas are resting in a large tamarind tree in the famous Berenty Reserve in southeastern Madagascar, a private reserve that has long been a popular tourist destination.*

low levels of species diversity, with only 253 species, 103 (41%) are endemic, as are 35 out of 162 genera; furthermore, there are a full five families found only on Madagascar.

Though they have no endemic families, the reptiles and amphibians also exhibit very high levels of endemism at the species and generic levels. A total of 274 out of 300 reptile species are endemic (91%) and 36 out of 64 are endemic genera (56%). These all represent a degree of higher-level endemism that few other places can match. With the frogs (the only amphibians represented on the island), endemism is almost complete. The 178 species of frog include only one or two introduced species and the number of endemic genera stands at 16 or 17 out of a total of 18.

However, once again, the most striking example is that of the lemurs, the great flagship species of the country. The 32 species and 50 taxa mentioned above are 100% endemic to Madagascar (with only two found on the neighboring Comoros, where they were almost certainly introduced by man); moreover, the 32 species are divided into 14 endemic genera and five endemic families, a degree of primate endemism at higher taxonomic levels that no other country can approach. Even Brazil, the world's leader in total primate species diversity with 77, has only two endemic genera and no endemic families. Looking at Madagascar's great primate diversity in another way, we see that this small country has great global responsibility, a feature shared with the other megadiversity countries. Although it is only one of 92 countries with wild primate populations, it alone is responsible for 21% (14/65) of all primate genera and 36% (5/14) of all primate families, making it the single highest-priority area on Earth for the conservation of these, our closest living relatives.

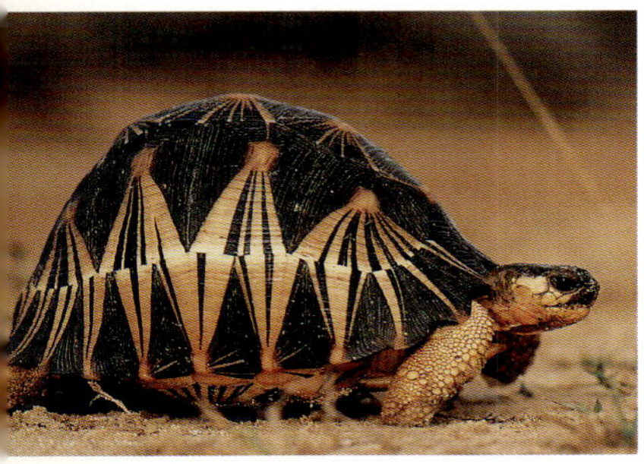

The radiated tortoise (Geochelone radiata) *is perhaps the most beautiful chelonian in the world. Restricted to the southern spiny desert region of Madagascar, it is still common in areas where it is considered* fady *or taboo; in other areas, it is frequently eaten by local people and declining rapidly.*

Natural habitats

The natural habitats of Madagascar are mainly forest formations, ranging from rain forest in the eastern part of the country to dry forest in the west and desert-like formations in the far south. In biogeographic terms, the country can be divided into two major regions: Eastern and Western (White, 1983). The Eastern region includes four floristic domains: Eastern, Central, High Mountain and Sambirano. The Western region is divided into just two, the Western and the Southern. Each is characterized by specific types of vegetation.

Within the Eastern region, the Eastern Domain includes the lowland rain forests between sea level and 800 m altitude, extending along the east coast from Sambava in the north to Fort-Dauphin (=Tolanaro) in the south. Average annual precipitation is between 2 000-3 000 mm and there is no dry season. The height of the evergreen forest canopy averages between 20-30 m with no emergent trees. This region is characterized by very high species diversity and endemism, and also comprises one of the most endangered vegetation types, most of it having already being cleared (only the much more restricted Sambirano and High Mountain Domains are more at risk). The largest remaining tracts of this rain forest formation are found in the northeastern and southeastern parts of the island. The forests of the Eastern Domain and what remains of the Central Domain higher up the eastern escarpment constitute the area usually referred to in the popular literature as the *Eastern Rain Forest Region*.

The forests of the Central Domain parallel those of the Eastern Domain and extend westward to Madagascar's central plateau above 800 m and to altitudes of 1 300 m and occasionally as high as 2 000 m. The average rainfall exceeds 1 500 mm and there is no dry season. Species diversity is as high as in the Eastern Domain, but endemism is actually higher. The forest canopy is lower, averaging between 20-25 m. Epiphytic vegetation is more plentiful and the herbaceous stratum more developed. Although the Central Domain is the most extensive biogeographic formation in the country, most of it has already disappeared, especially on the central plateau itself, where almost nothing remains. Most of what is left of the Central Domain forests are on the eastern slope heading down toward the lowlands of the Eastern Domain.

The High Mountain Domain comprises five distinct, geographically separated montane regions, Tsaratanana in the northwest, Marojejy in the northeast, Ankaratra in the central east, Andringitra in the central southwest, and Andohahela in the southeast. Forests in this formation grow at altitudes of 2 000-2 867 m and rainfall is substantial throughout the year. Species diversity is lower than in the two previously discussed formations, but endemism is very high. The two largest intact blocks remaining are Marojejy and Andohahela.

The Sambirano Domain consists of a small enclave of seasonal moist forest in the northwest. Together with the Tsaratanana Massif, it constitutes the northern end of Madagascar's central mountain range. Annual rainfall exceeds 2 000 mm. The Sambirano is characterized by high species diversity and a high level of endemism, and represents something of a transition zone between Madagascar's Eastern and Western Regions. Forest canopy height is about 30 m with some emergents.

Moving on to the Western Region, this part of the country has been divided into the Western Domain and the Southern Domain. The Western Domain is the area referred to in the popular literature as the *Western Dry Forest Region* or the *Western Dry Deciduous Forest Region*. The vegetation of the Western Domain now consists of discontinuous patches of dry deciduous forest on the coastal plains and limestone plateaus, ranging from sea level to 800 m. It covers the area from Antsiranana in the north to Morombe in the southwest. Annual rainfall ranges from 500-2 000 mm, lightest in the south and heaviest in the north. There is a marked dry season of almost seven months during which many of the canopy trees shed their leaves. The shrub layer is well-developed and vines are common. Plant species diversity is lower than in the Eastern Region, but the level of endemism is higher. Together with the lowland rain forests of the Eastern Domain, the lowland dry forests are the most endangered, the principal threats being slash-and-burn cultivation, fire, and uncontrolled use by cattle, which are heavily concentrated here.

The Southern Domain is the driest part of the country, and the part usually referred to as the *Southern Spiny Desert Region*. It is characterized by deciduous thicket or thorn shrub, and is perhaps the least disturbed of Madagascar's major biogeographic domains. It runs southward from Morombe on the west coast, covering much of the island's southern tip from sea level to 400 m. Rainfall in this area is sparse and irregular, ranging from 300-800 mm. The dry season is marked and very long. Forest height is low and the formations are usually impenetrable due to a high incidence of thorny vegetation. The Didiereaceae, a family superficially resembling the Cactaceae and endemic to the Southern Domain, and various species of *Euphoria* are the dominant plant groups. Species diversity and levels of endemism, especially in plants, are very high. The primary threats to forests of the Southern Domain are charcoal and firewood production, the uncontrolled collection of ornamental and medicinal plants, and overexploitation of the land by livestock, especially cattle and goats. Slash-and-burn agriculture is not as significant a threat due to the poor quality of the soils.

Flagship species

Madagascar has a number of striking flagship species, the best known of which are the lemurs, an entire radiation of nonhuman primates unique to the island. Of the 50 different kinds that exist, several stand out as being especially interesting. Among these are virtually all members of the family Indriidae, the indri, the sifakas and avahi. The indri (*Indri indri*) is the largest of the living lemurs; it stands about one meter high and weighs approximately 7 kg. It is one of the truly spectacular, yet little known, species on the planet. It looks like a cross between a koala and a giant panda and leaps from tree to tree like an arboreal kangaroo. Its eerie morning call, an extended wailing sound used to identify the territory of the small family groups in which these animals live, is one of the signature sounds of the Malagasy rain forest.

Living partly sympatrically with the indri is the diademed sifaka (*Propithecus diadema diadema*), roughly equal in size to the indri but with a long tail. This white, brown and honey-colored animal is considered by many to be the most beautiful of all the living lemurs. Another very interesting lemur is Tattersall's sifaka (*Propithecus tattersalli*), described by scientists in 1938. Living in a tiny area in northeastern Madagascar, it hangs on in tiny forest fragments, now under heavy pressure from gold miners, and is one of the three most endangered lemurs. Another seven sifaka species exist, ranging in color from pure white to jet black, and rank among our favorites in the entire Animal Kingdom.

The avahis are the third genus of the indriid family. About half the size of the smallest sifakas, they are woolly little creatures with large owl-like eyes. Unlike their diurnal relatives, they are active only at night. Two species exist, one in the eastern rain forest and the other in the western dry deciduous forest.

The best known of the Malagasy lemurs is the ring-tailed lemur (*Lemur catta*), whose distinctive raccoon-like tail immediately identifies it. It is a frequent symbol for Madagascar and an animal often kept in captivity. A number of other typical lemurs exist (genus *Eulemur*), and occupy most of the remaining forest habitats on the island.

The bamboo lemurs are also an interesting group. Two of them, the greater bamboo lemur (*Hapalemur simus*) and the golden bamboo lemur (*Hapalemur aureus*), rank among the most endangered along with Tattersall's sifaka, and the golden bamboo lemur was not discovered by scientists until 1985. Indeed, a project that I helped to support to search for the greater bamboo lemur wound up not just rediscovering that species, which hadn't been seen for 20 years, but finding the previously unknown golden bamboo lemur as well. Another member of the genus, the Lac Alaotra bamboo lemur (*Hapalemur griseus alaotrensis*), has one of the most unusual habitats of any primate on Earth. It lives only in the reed beds fringing Lac Alaotra, Madagascar's largest lake and its most important rice-growing area. As more and more of this lake is turned into rice paddy, the survival of this lemur becomes increasingly unlikely. On top of this, it is still hunted for food and is one of the few lemur species kept regularly as pets by local people.

Among the small lemurs, several stand out as well. The pygmy mouse lemur (*Microcebus myoxinus*) is the smallest living primate, weighing in at only about 30 g. It was first described in 1858, then synonymized with the wide-ranging gray mouse lemur (*Microcebus murinus*) and forgotten for more than 130 years. In 1992, a group of German scientists conducted a detailed study of this animal and the sympatric gray mouse lemur and discovered that it was a very distinct species. We had actually photographed the pygmy mouse lemur in the wild in 1985 without knowing what it was. It was not until the work of the German scientists eight years later that we found out that it was in fact the long-overlooked pygmy mouse lemur.

Other interesting nocturnal species include the fat-tailed dwarf lemur (*Cheirogaleus medius*), which stores a large amount of fat in its tail to use over a six-month season of inactivity, and the fork-marked lemur (*Phaner furcifer*), an attractive, very active nocturnal species with a loud barking call that stirs up the Malagasy night and seems far too loud to come from an animal that weighs just half a kilogram.

Finally, the mysterious aye-aye, unquestionably the strangest and most unusual of all primates and one of

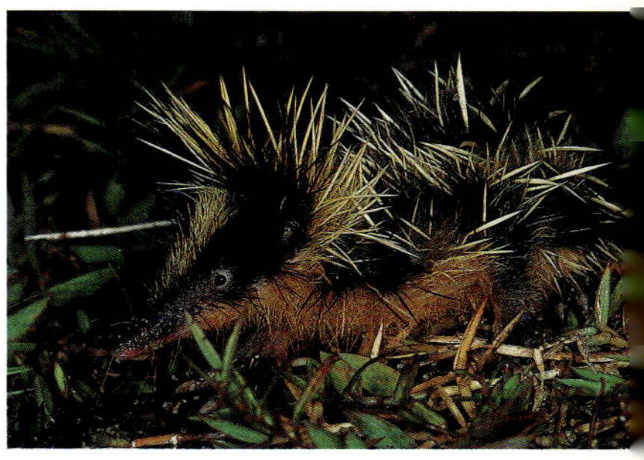

The striped tenrec (Hemicentetes semispinosus) *is one of about 30 species of tenrec, an almost exclusively Malagasy group of hedgehog-like insectivores. When disturbed, this attractive and unusual little creature has the habit of vibrating its spines to further discourage predators that are not sufficiently deterred by the spines themselves.*

On pp. 212-213, orchids (Angraecum *sp.*) *in Ranomafana National Park, one of the most important protected areas in the eastern rain forest region of Madagascar.*

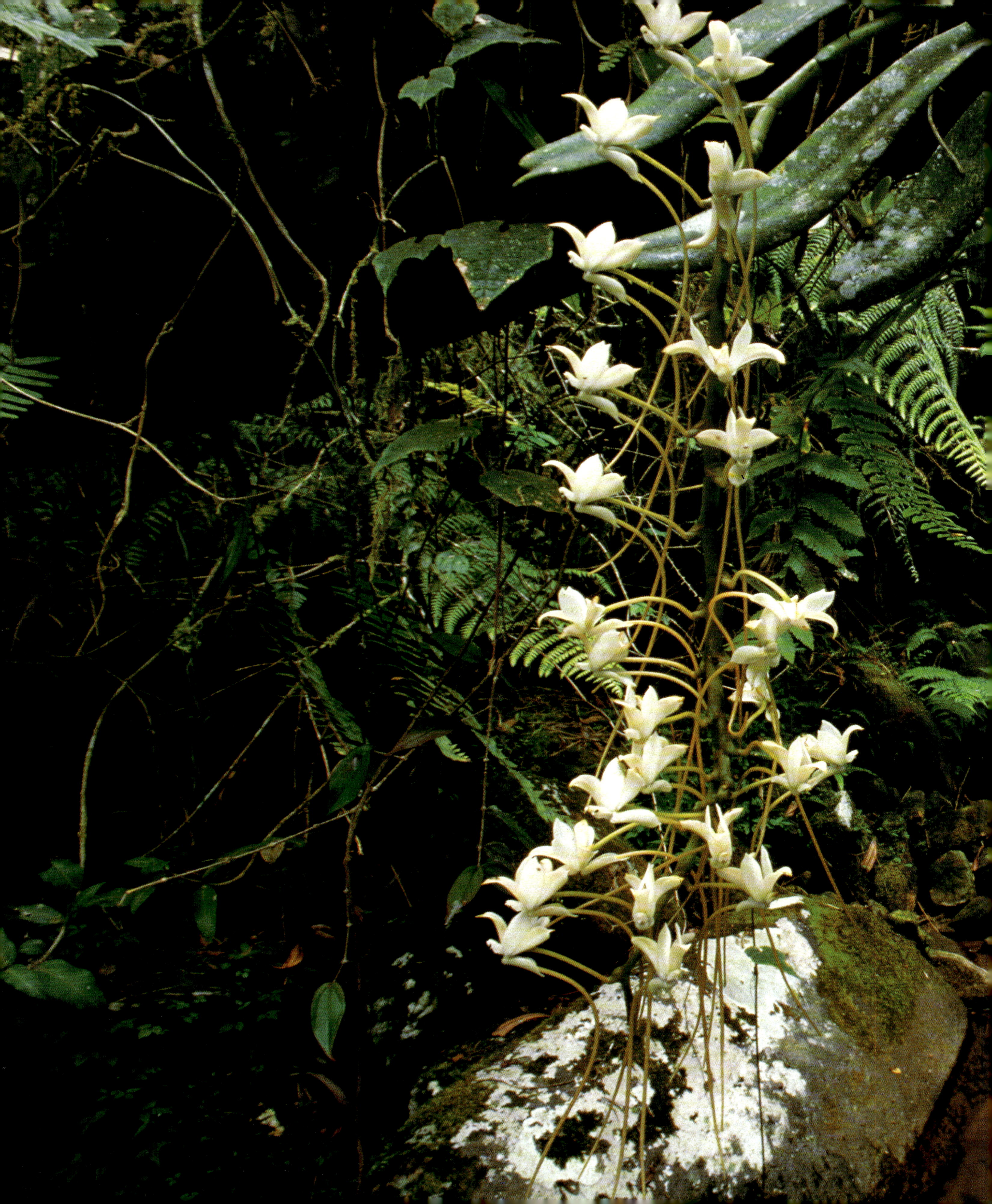

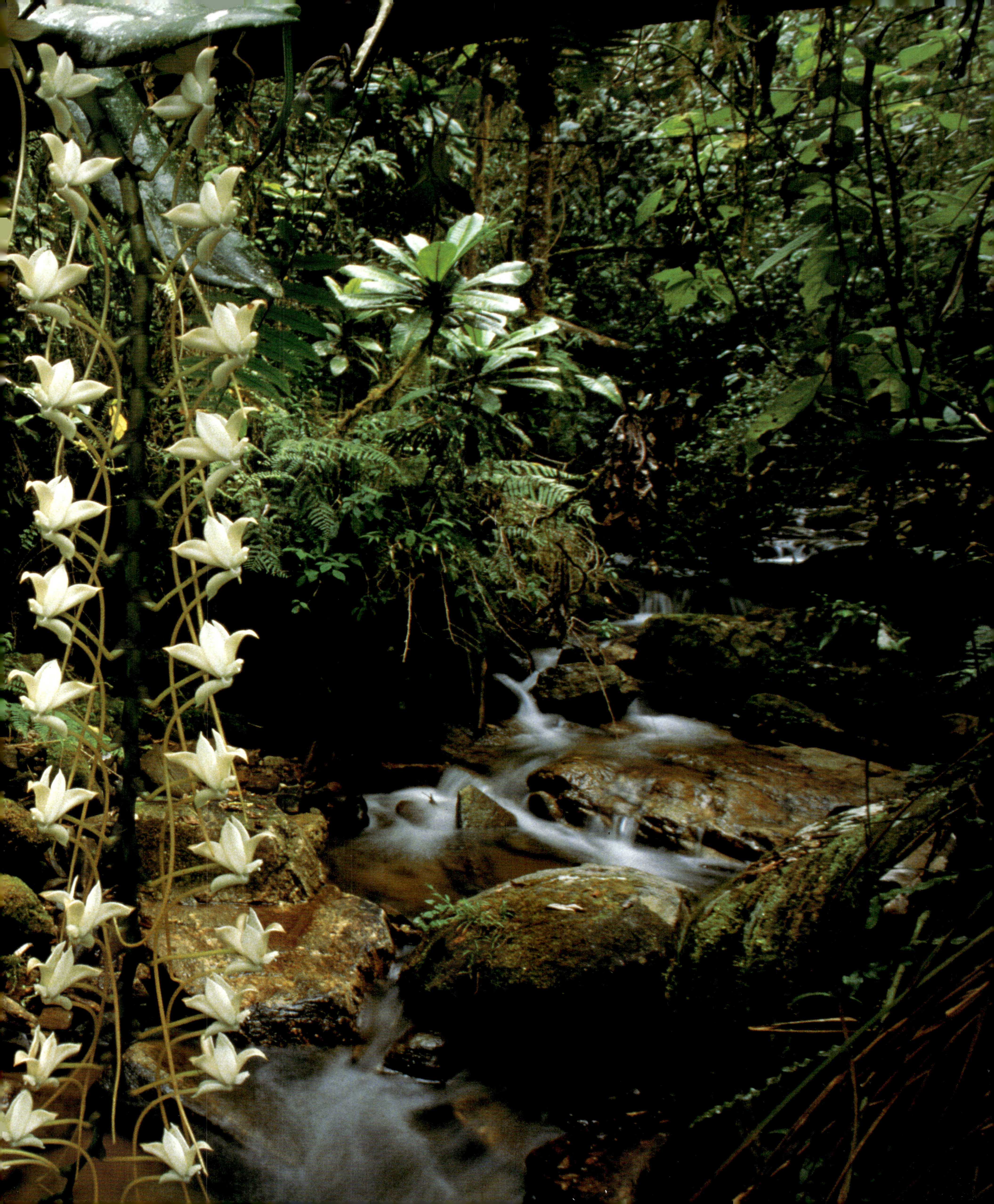

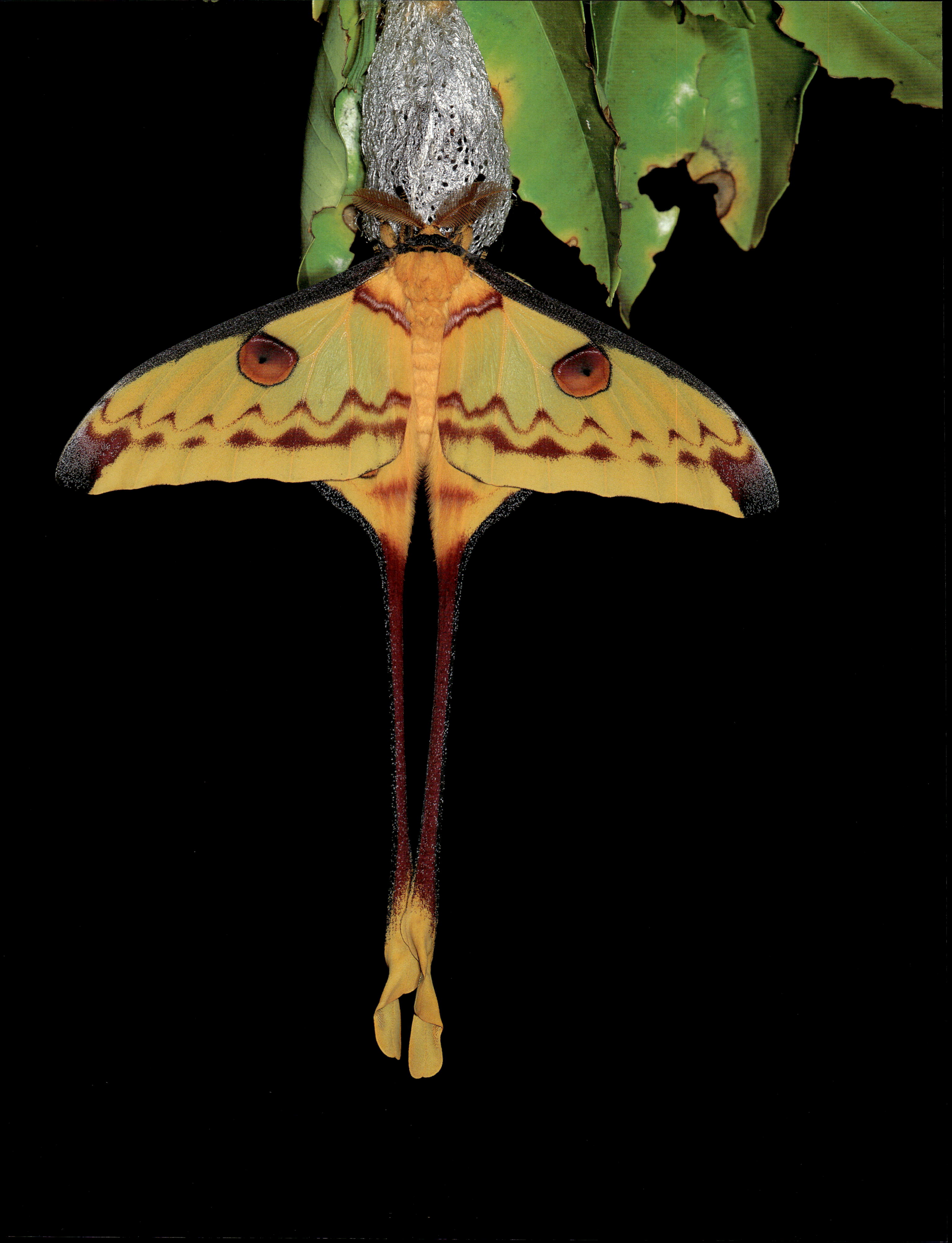

the most unusual mammals on Earth. About the size of a large house cat, this highly distinctive lemur has huge, fox-like ears, a shaggy coat of fur, continuously growing incisors like a rodent (the only primate with this feature), and a long skeletal middle finger on each hand. Its very sensitive ears enable it to hear grubs moving about in rotten wood, its incisors enable it to gnaw a hole into the wood, and its long middle finger reaches in and plucks the grub from the hole. It is able to use the same procedure on a ripe coconut, leading to its being considered an agricultural pest in some areas. In many ways it occupies a woodpecker-like niche in Madagascar, the avian woodpecker being absent from the island. It also shares unusual parallels with the striped possum (*Dactylopsila*) from Queensland, Australia and Papua New Guinea, another mammalian equivalent of the woodpecker. The aye-aye is also part and parcel of the mythology of Madagascar's people. Many Malagasy are truly frightened of the animal, and consider it a harbinger of evil. Indeed, it is often killed on sight because it is such a sign of bad luck, and in some areas a village may be abandoned if one appears. Until recently, the aye-aye was considered extremely endangered and on the verge of extinction. Yet with the substantial increase in lemur field work in the late 1980s and early 1990s, researchers developed a search image for this secretive primate and its nests, and it appears to be somewhat more abundant than previously believed. However, it is still considered endangered, and as the only representative of an entire primate family, the Daubentoniidae, this animal remains one of the highest global primate conservation priorities.

As remarkable as Madagascar's living lemur fauna currently is, it pales by comparison with what existed as recently as a thousand years ago. Since the arrival of humans on Madagascar, fully eight genera and at least 15 lemur species have become extinct, representing almost a third of all known species (living and extinct) and fully 36% (8/22) of known genera. Furthermore, virtually all the extinct species were diurnal and larger than any of those surviving today. They included some spectacular creatures like *Megaladapis*, which resembled a large Australian koala and grew to the size of a calf; *Babakotia*, which resembled a large, primate version of the South American tree sloth; and the truly amazing *Archaeoindris*, which was larger than an adult male gorilla and probably occupied a niche similar to that of the now extinct North American ground sloth. And the lemurs were not the only animals to disappear. Along with them went several other wonderful creatures, including at least one species of giant tortoise, a pygmy hippopotamus, and an aardvark, not to mention the colossal elephant bird referred to below. Indeed, the loss of the giant lemurs, the elephant birds and other species of Madagascar represents one of the three or four very striking extinction spasms in recent times, the others being the loss of a large portion of Hawaii's native species since the arrival of Polynesians and then Europeans, the disappearance of many of Australia's endemic mammals since colonization by Europeans, and the comparable losses of the giant moa birds and other endemics of New Zealand. All of these episodes have occurred within historical times, all are at least in part attributable to human impact on native ecosystems, and all demonstrate that the much-discussed potential for major extinction spasms in our lifetimes is not just a figment of the conservationist's fertile imagination. They are a very real possibility, and could well occur again in other parts of the world.

The other mammals of Madagascar are not nearly as spectacular as the lemurs, but several are worthy of note nonetheless. There are no native moles, shrews or hedgehogs in Madagascar, but their niche is partly occupied by the uniquely Malagasy tenrecs, family Tenrecidae. The tenrecs are a primitive group of insectivores that are divided into 30 species and eight genera. They range in size from the tiny shrew-like *Microgale cowani* (6 cm in length) to the common tenrec (*Tenrec ecaudatus*), a spiny hedgehog-like creature that weighs up to 1.5 kg. This highly adaptable creature holds the mammalian record for offspring, having up to 32 young at a time feeding on as many as 29 nipples. It is also considered a delicacy among the Malagasy people and is commonly eaten; in spite of this, it remains widely distributed and common. The tenrecs also include some very unusual creatures like the aquatic tenrec, *Limnogale mergulus*, about 20 cm long excluding the tail and resembling a very small long-nosed otter. This endangered animal lives in fast-flowing streams and is known from only a handful of localities. It is rarely seen, and an individual caught in 1990 was the first to be seen alive in 25 years. Subsequently, several others were captured for research and observation, then released back into the wild in Ranomafana National Park.

Among the rodents, which are very low in diversity in Madagascar (only 11 species), the most spectacular is the giant jumping rat (*Hypogeomys antimena*), a monotypic genus restricted to a tiny area (approximately 100 km^2) of western dry deciduous forest north of Morondava. It is a large species weighing up to 1.2 kg, and looks like a giant gerbil with long rabbit ears.

Madagascar is also surprisingly lacking in carnivores, with none of the larger species so prevalent in mainland Africa, no lions or leopards or any other members of the cat family, no hyenas, and no hunting dogs. The only Malagasy carnivores are members of the civet, genet and mongoose family, the Viverridae, but these include some very interesting animals divided into three subfamilies (Fossinae, Galidiinae and Cryptoproctinae), eight genera and nine species, of which all but one (*Viverricula indica*) are endemic to the island. The most striking of these is the fossa (*Cryptoprocta ferox*), largest of the Malagasy carnivores, and the fanaloka (*Eupleres goudoti*), a very rare earthworm eater with a long, pointy nose and a thick, bushy tail. The fossa is still widespread in Madagascar and apparently quite adaptable, but rarely seen. In spite of its large size, up to 1.5 m, including the tail, it is able to climb trees in

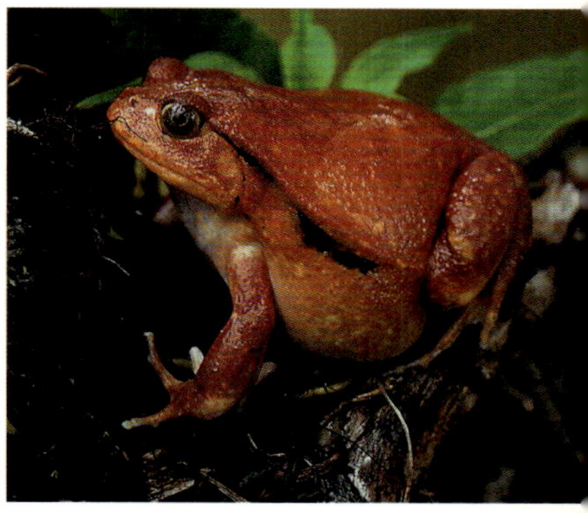

On the opposite page, insect diversity in Madagascar is still poorly understood and there are undoubtedly many undescribed species. Moths are particularly diverse, and include some large and spectacular species like the one shown here.

Above, the tomato frog (Dyscophus antongili) *is a large, striking species that is perhaps the best amphibian flagship for Madagascar. Restricted to a small geographic area in the vicinity of Maroantsetra in northeastern Madagascar, it is now bred outside the country for the herpetocultural trade.*

search of small lemurs and other prey. It is an unusual animal that reminds me of a cross between Australia's now-extinct predatory marsupial, the thylacine (minus the stripes), and a small puma, integrating both dog- and cat-like features.

After the lemurs, the most interesting creatures of Madagascar are the reptiles, which achieve a very high diversity and incomparable levels of endemism on this island. The reptile total is now 300 and growing almost monthly, as new species are uncovered in every expedition sent out to explore the remaining natural habitats of the country. Madagascar's reptile diversity is on par with that of many much larger megadiversity countries, and none can match its endemism. Of the different groups of reptiles, the lizards are the most diverse, with more than 180 species, and the turtles and tortoises are especially noteworthy. The snakes are represented by only three families, the Colubridae, the Boidae and the Typhlopidae, with the terrestrial poisonous snakes so prevalent in Africa being absent from Madagascar. Madagascar separated from the African mainland before the evolution of poisonous snakes (as was the case with higher primates as well) and none ever made their way to the island over the ocean barrier. Two species of poisonous marine snakes are found along the Malagasy coast, but one can walk through a Malagasy forest, day or night, without worrying about stepping on a venomous serpent. Particularly noteworthy in Madagascar are three species of boas (*Boinae*), a subfamily found mainly in Central and South America, but entirely absent on the African mainland where pythons occupy the niche for large constricting snakes. This unusual pattern of distribution is repeated with the iguanid lizards, which are dominant in the New World, and also occur in Madagascar and on Fiji but are absent from the African mainland, where the agamid lizards are the common lizard family.

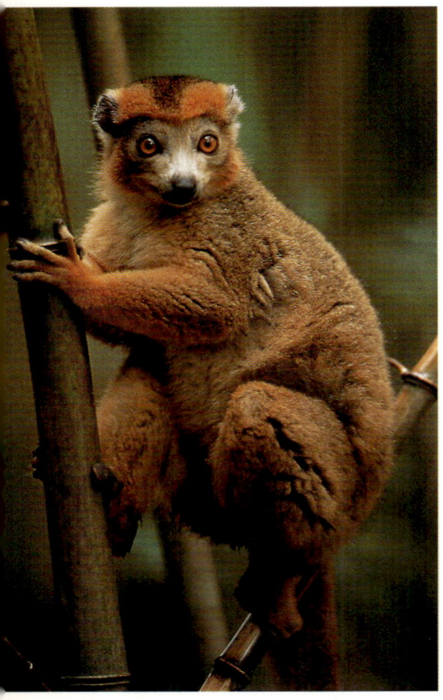

The crowned lemur (Eulemur coronatus) *is restricted to the northernmost tip of Madagascar, and is found in both humid forest and dry forest habitats. It is sexually dimorphic in color, the individual here being male.*

Probably the most spectacular snake in Madagascar is the twig-nosed snake (*Langaha madagascariensis*). A long, pencil-thin, brown species, this snake is distinguished by the long, leaf-shaped projection on the nose in the female and the pointy, thorn-like extension in the male. These snakes occupy a niche very similar to that of the vine snakes (*Dryophis* and *Oxybelis*) of South and Central America.

Of the many lizards on Madagascar, the most popular are undoubtedly the chameleons. Indeed, if one asks the visitor to Madagascar what he or she would most like to see, lemurs are usually the first priority, but chameleons are a close second. Madagascar is the main center of diversity for the chameleons with about half of the species (57/131) thus far described and new ones being discovered on a regular basis. Madagascar includes the largest of the chameleons, Oustalet's chameleon (*Furcifer oustaleti*), which can reach 68.5 cm, and the smallest, the pygmy stump-tailed chameleon (*Brookesia minima*), which measures only 3.4 mm total length and is one of the smallest vertebrates. The chameleons are best known for their unmatched ability to change color, and they are among the most wondrously-colored creatures in the Animal Kingdom even without considering their color-changing abilities. They are also characterized by several amazing adaptations to arboreal life, including the opposable finger and toes, with two lateral fingers opposing three inner ones on the hands, and the reverse being true (three lateral toes opposing two inner ones) on the feet. Add to this the very sharp claws and the prehensile tail and you have a very effective climbing machine. The diurnal chameleons also have very specialized eyes which are set in turrets and can move independently from one another, thus monitoring the environment in all directions. These eyes have given rise to a Malagasy proverb, "Be like the chameleon, keep one eye focused on the future and the other on the past." Its eyeset also gives the chameleon very precise binocular vision, enabling it to use its other famous adaptation, the long sticky tongue. This tongue, which can be equal in length to the chameleon's entire body, can be shot out with great speed and drag the unsuspecting prey right into the chameleon's waiting jaws.

The brilliant colors for which the chameleons are best known are characteristic of the larger species. The small, stump-tailed chameleons are usually drab brown in color, but are adorned with spectacular spines, tubercles and horns. Indeed, some of the most heavily adorned species like *Brookesia perarmata*, a "mini-dinosaur," would look great in a science-fiction movie, especially if magnified several times. Unlike their highly arboreal larger relatives, these little creatures are mostly terrestrial and move about either in the leaf litter or very low on mossy tree trunks. Their best defense is their drab coloration, but they are also able to rapidly vibrate their bodies, a striking defensive behavior that must startle potential predators. In addition, they can play dead, looking like nothing more than a clump of dirt. Indeed, the first wild *Brookesia* we ever saw, on the island of Nosy Mangabe, behaved in precisely this manner. We found it in the middle of a rainstorm as we were climbing a hill. It had pulled in its skinny little legs and fallen over sideways, and we only noticed it because it began rolling down the steep slope. *Brookesia* are truly remarkable creatures, and among our favorites in the animal world; they are not easy to see, but the patient observer should be able to locate them, especially with the assistance of local Malagasy guides.

Two other noteworthy flagship lizard groups are the day geckos (*Phelsuma* spp.) and the leaf-tailed geckos (*Uroplatus* spp.) Although most members of the wide-ranging gecko family are nocturnal and cryptic in coloration, the day geckos are brightly colored and as their name implies, diurnal. They are quite beautiful, with many being strikingly colored with emerald green, orange, red, and blue patterns. Twenty species of day gecko occur on Madagascar, 16 of them endemic, with the most impressive being the large *Phelsuma madagascariensis*. The leaf-tailed geckos keep to the more typical gecko pattern of nocturnality, and are as inconspicuous as the day geckos are conspicuous. The nine species of *Uroplatus* are characterized by large eyes, a large triangular head, and a short flattened tail that comes in many different sizes and shapes depending on the species. *Uroplatus* lives mainly in the rain forest and is among the best camouflaged of all lizards.

Madagascar also has great amphibian diversity, but all of it is in the frog group, the salamanders and caecilians being entirely absent from the island. With 178 species and counting, all but one or two of them endemic, Madagascar has one of the richest and most unique amphibian faunas in the world. As with the lemurs and many other groups, this diversity is not just at the species level but at the generic level as well, these 178 species being divided into 18 genera of which 16 or 17 are endemic. The best known of the Malagasy frogs are the tomato frogs (*Dyscophus* spp.) and the mantellas (*Mantella* spp.). The mantellas are small, very brightly-colored forest floor frogs that show amazing convergence with the equally brilliant poison dart frogs (*Dendrobates*, *Phyllobates*) of the Neotropics. They also appear to be mildly poisonous, but their toxicity does not reach the extreme levels of certain South American species like *Phyllobates terribilis*. Perhaps the most striking and beautiful of all the mantellas is the orange-gold *Mantella aurantiaca*, which is sold from time to time in the international pet trade.

The three members of the genus *Dyscophus* are large terrestrial frogs and the only representatives of the family Microhylidae in Madagascar. The largest of the three is the bullfrog-sized tomato frog (*Dyscophus antongili*) from the Maroantsetra area of northeastern Madagascar. It is bright red in color, and probably the best known of the country's amphibians.

Last but not least among the flagship reptiles of Madagascar are the chelonians and the crocodiles. The single species of crocodile on Madagascar, *Crocodylus niloticus*, is not endemic, but rather the same one that is widespread on the African mainland. Although not as interesting biologically as most of the other reptiles in the country, it is of importance to some of Madagascar's human cultures. For example, in Lake Anivorano between Diégo-Suarez and Ambilobe, it is held sacred by local villagers and protected by a *fady* (taboo). Elsewhere, in the Ankarana Massif, it lives in subterranean rivers, an unusual habitat for a crocodilian.

Of the 14 turtles and tortoises, nine, including the five marine turtles, are found elsewhere. The remaining five, four tortoises and one freshwater turtle, are endemic, and of considerable international interest. The radiated tortoise (*Geochelone radiata*) of the southern spiny desert is arguably the most beautiful chelonian on Earth. Its brilliant, radiating star yellow pattern on a black background is particularly stunning, especially when seen in the context of the stark Didiereaceae forest of the far south, its natural habitat. It looks as if someone had planted it there, and that it could not possibly have originated in such forbidding terrain. A large individual (up to 40 cm) with a well-defined pattern must rank among the most beautiful creatures on Earth.

There are two small species in the southern part of the country, the Madagascar spider tortoise (*Pyxis arachnoides*) and the Madagascar flat-tailed tortoise (*Pyxis planicauda*). The spider tortoise is fairly widely distributed in the south and southwest. The flat-tailed tortoise, on the other hand, is extremely restricted in range and critically endangered. It only occurs in a region of some 100 km^2 to the north of Morondava in the southwest, roughly the same area occupied by the giant jumping rat, *Hypogeomys antimena*.

The fourth of the endemic tortoises and the largest of the non-marine chelonians of Madagascar is the angonoka or plowshare tortoise (*Geochelone yniphora*), generally referred to as "the most endangered tortoise on Earth" (although several others, including some of the Galápagos tortoises and perhaps even the Madagascar spider tortoise, are equally deserving of this title). Reaching a length of 45 cm, this large striking animal is not quite as beautifully patterned as the radiated tortoise, but impressive in its own right. It is restricted to a tiny, dry deciduous forest peninsula in the Soalala region of the Bay of Baly in northwestern Madagascar, and has a surviving wild population of 100-400 animals. Fortunately, a captive breeding program was established almost a decade ago by the Jersey Wildlife Preservation Trust and based at the Forest Station of Ampijoroa in the Ankarafantsika Natural Reserve near Majunga in western Madagascar. This program has proven quite successful, and has bred several dozens of these tortoises for possible future reintroduction.

Finally, the Madagascan big-headed side-neck turtle (*Erymnochelys madagascariensis*) is a large freshwater species found in lakes and rivers in the west. This animal is closely related to the *Podocnemis* turtles of northern South America, and is quite endangered because of overexploitation. A captive-breeding program for it is soon to be initiated at the Ampijoroa Station alongside the current program for the angonoka tortoise.

Moving on to the birds, the most impressive flagship bird species of Madagascar are unfortunately now extinct. These were the elephant birds, which included the giant elephant bird (*Aepyornis maximus*), at 450-500 kg the heaviest bird that ever lived and exceeded in height only by the largest moa of New Zealand (now also extinct). This huge creature, which grew to be 3 m tall, had eggs that weighed 45 kg. In fact, the egg of the elephant bird is considered the largest *cell* of the Animal Kingdom, with a volume of 7-9 liters or seven times that of an ostrich egg. Little is known of its natural history, but it may have persisted in remote areas up until the arrival of the first Europeans in Madagascar, who heard reports of a *vorn be* in the south of the country. The elephant bird is also thought to have given rise to the legendary *roc* of Sinbad the Sailor, a distinct possibility since early Arab sailors visiting Madagascar undoubtedly heard stories about the elephant bird as well.

Of course, none of the living birds of Madagascar can compare to the extinct elephant birds. Nonetheless, the

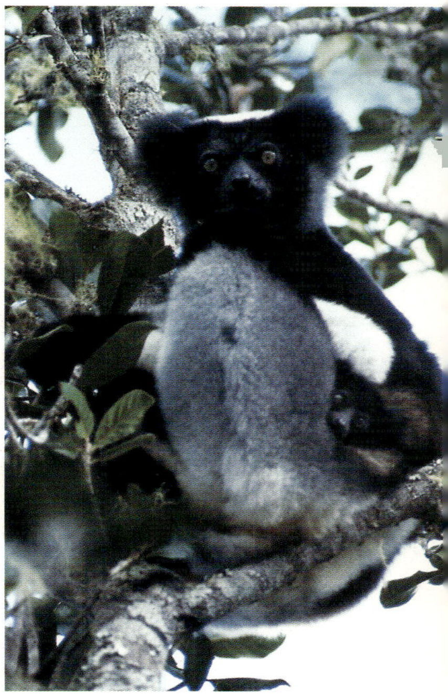

The indri (Indri indri) *is one of the two largest living lemurs, and one of the most spectacular creatures in the Animal Kingdom. The ultimate flagship species for Madagascar, it looks like a cross between a koala and a giant panda, and bounds from tree to tree like an arboreal kangaroo. Its eerie, wailing morning call is one of the most characteristic sounds of the eastern rain forest.*

On pp. 218-219, fire-resistant palm trees (Mademia nobilis) *growing in open savanna in Isalo National Park, in the arid region of southern Madagascar.*

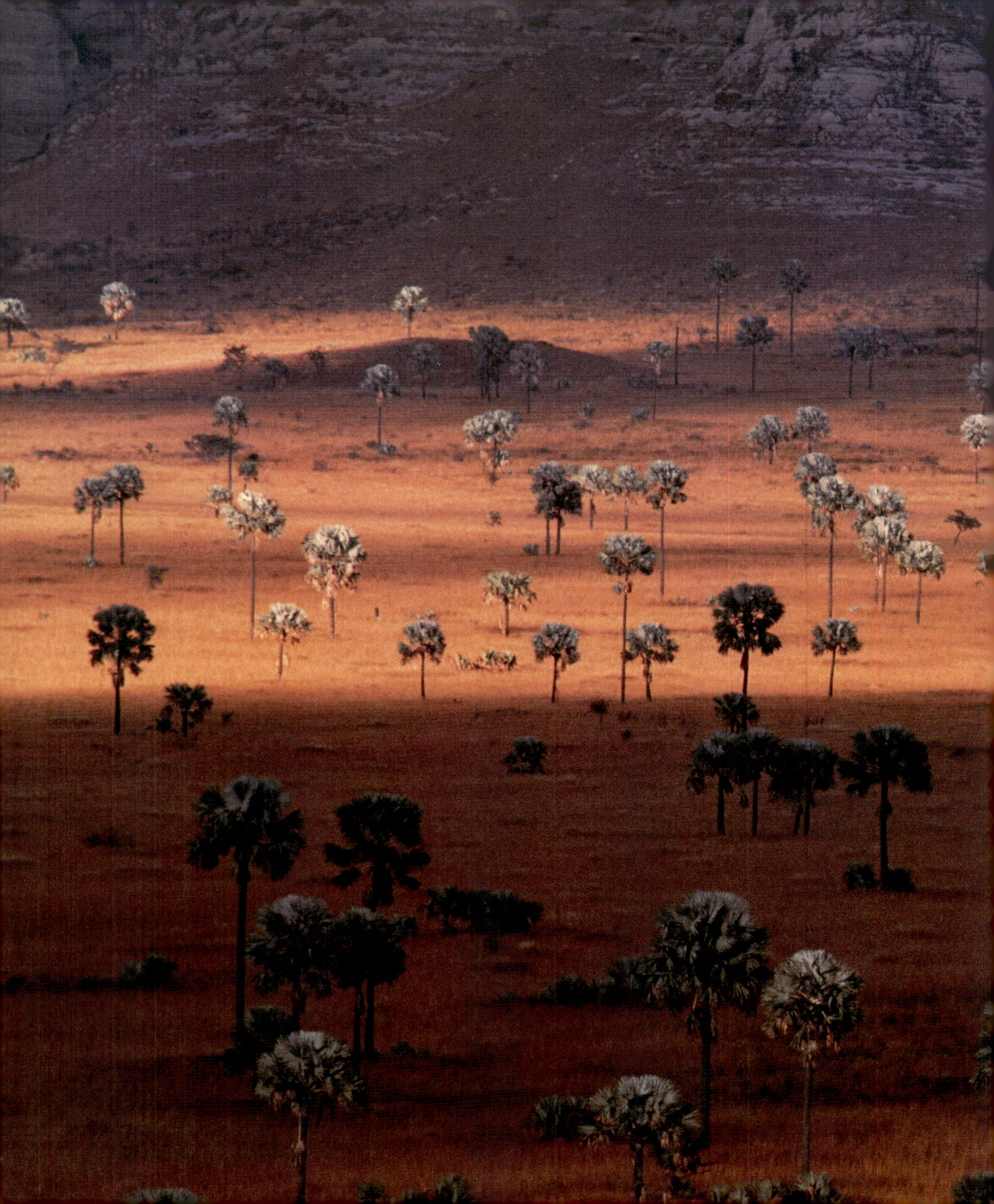

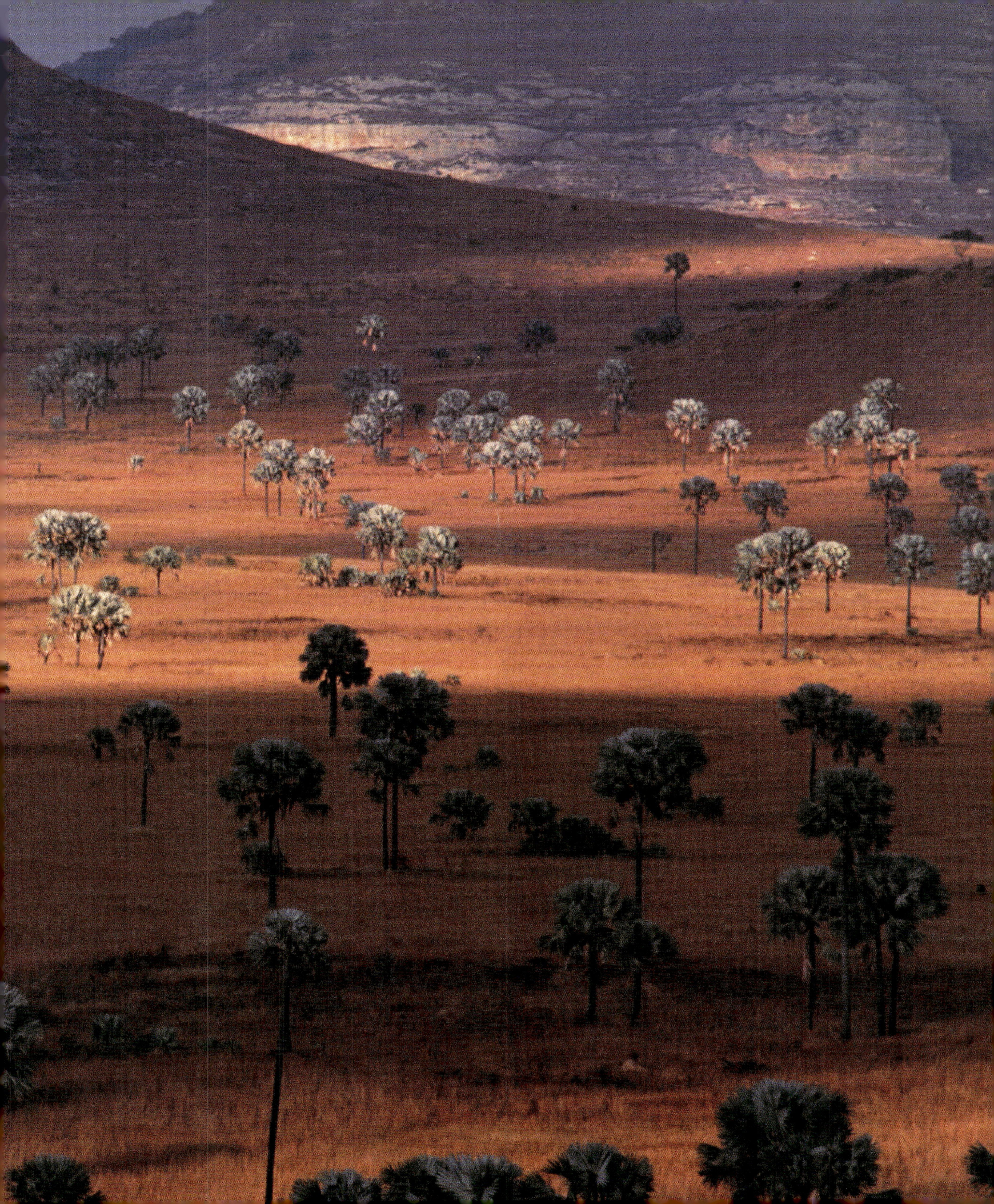

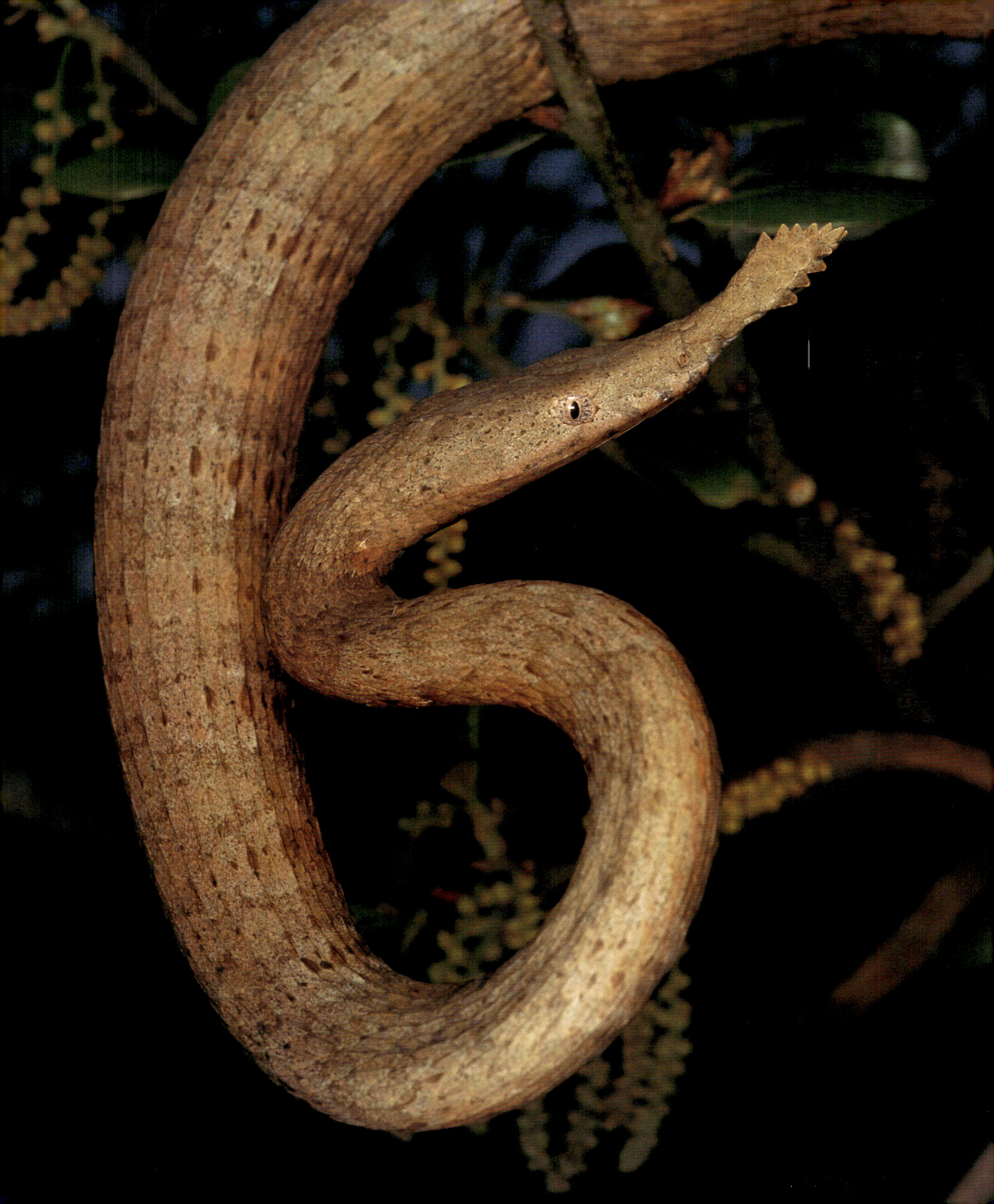

country does have a very high level of endemism, including three families found only on Madagascar itself and an additional two endemic to the Malagasy region (Madagascar, the Comoros, Mauritius, Reunion and small satellite islands). These include the mesites (family Mesitornithidae), a taxonomic enigma since they have features common to several other bird orders, the ground rollers (family Brachypteracidae), the cuckoo rollers (family Leptosomatidae), the asities and sunbird asities (family Philepittidae), and the vangas (family Vangidae). The 14 species of vangas display spectacular evolutionary diversification of beak types leading to their being divided into 11 distinct genera. Their adaptive radiation has been compared in evolutionary interest to that of the Galápagos finches (Geospizinae), and the honeycreepers (Drepanidinae) of the Hawaian islands. The couas are another important group, endemic at the subfamily level, the Couinae, and include ten species that are restricted to Madagascar.

Moving on to the Plant Kingdom, Madagascar is unusual in that it has a number of flagship species of plants as well. The best known of these are the baobabs, which are very distinctive trees with thick trunks that often look in the dry season as if they had been planted upside-down. The baobabs are found only in Australia, Madagascar, and mainland Africa and, of the eight species, six are endemic to Madagascar (*Adansonia madagascariensis*, *A. za*, *A. grandidieri*, *A. perrieri*, *A. suarezi*, *A. rubrostipa*). The African species *A. digitata* is also found in Madagascar and may be native, but its close association with villages indicates that it was probably brought in by Arab traders. These unique trees, each individual with its own distinctive "personality," are as much a part of the Malagasy landscape as are the lemurs, and are found mainly in the western dry deciduous forest, the southern spiny desert and the far north. Given their striking appearance, they are often the focus of superstitions and folklore. They are also used to produce rope, and unfortunately are sometimes cut down in times of drought so that the moist wood can provide fodder for cattle.

After the baobabs, and perhaps even better known within Madagascar, is the traveler's palm (*Ravenala madagascariensis*), not a palm at all but rather a representative of the banana family (Musaceae). This statuesque, fan-shaped tree is named for the large quantity of water stored in its trunk, which can be cut open to provide sustenance for thirsty travelers. So well-known is this tree that it serves as the symbol of Madagascar's national airline. Furthermore, the hotel rating system for the country doesn't use stars but rather *Ravenala*, a five *Ravenala* hotel being the top of the chart.

Most of the other flagship plants of the country are succulents, plants of the drier regions with water-storing tissues, usually in the form of flattened stems or leaves. Among the most interesting of these are the cactus-like Didiereaceae, a family endemic to the southern spiny desert. These cactus-like trees are the most striking feature of Madagascar's dry south, and include ten species divided into four genera (*Alluaudia*, *Alluaudiopsis*, *Decarya* and *Didierea*). The most impressive of these is *Alluaudia ascendens*, which can reach a height of almost 15 m, and is somewhat reminiscent of the saguaro cactus of the southeastern United States. The octopus tree (*Didierea madagascariensis*) is another unusual member of this family. Didiereaceae have stiff, very sharp spines, and one of the many wildlife wonders of Madagascar is to see the fluffy white Verreaux's sifaka (*Propithecus verreauxi*) jumping from bole to bole of these intimidating trees. How it manages to accomplish this without being impaled remains a mystery.

Other distinctive succulents include the Euphorbiaceae which, along with the Didiereaceae, is one of the dominant families of the southern spiny desert, the aloes of the family Liliaceae, and the squat bottle-shaped *Pachypodium* of the family Apocynaceae, which to the casual observer look like mini-baobabs.

Finally, it is important to note that there are a number of Malagasy plants of international economic value. The best example of these, and probably the most-cited example of the economic importance of plants is the rosy periwinkle, *Catharanthus roseus*. This species, which has now been introduced all around the tropical world, contains 75 different alkaloids, two of which are critically important in the treatment of childhood leukemia and Hodgkin's disease. Madagascar is also a center of diversity for wild relatives of coffee, with at least 50 species known to date. Some of these are naturally very low in caffeine and their potential in the global coffee market, the world's largest agricultural commodity, is starting to be investigated by Conservation International.

Cultural diversity

The human cultures of Madagascar have very unusual origins, and are truly unique. Also, it is important to note that, of all the megadiversity countries, Madagascar is the one to have been occupied by humans for the shortest period of time. All of the other megadiversity countries have been occupied by human populations for tens of thousands, if not hundreds of thousands, of years; in contrast, our own species did not arrive on this isolated island until as recently as 1 500-2 000 years ago and, in some cases, as recently as 500-600 years ago. Consequently, there has not been the same kind of coevolution of people and natural ecosystems as in other parts of the world. Some of the land use practices brought from Asia and Africa, where the Malagasy people originated, have been particularly damaging to the fragile natural ecosystems of the island.

Current research indicates that the first people to arrive on Madagascar came from the region that is now Indonesia and Malaysia. The Malagasy language, though there are many dialects, is basically Malayo-Polynesian in origin and most resembles some of the languages of central Borneo. In spite of the close prox-

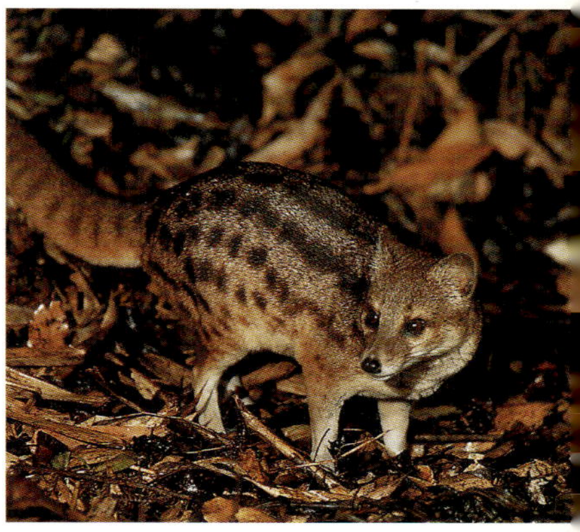

Madagascar is one of the few larger land areas in the tropics that is devoid of poisonous snakes. However, a wide variety of harmless species is found on the island, of which the leaf-nosed snake (Langaha nasuta) *is perhaps the most unusual. This species is sexually dimorphic, with the male having a twig-shaped nasal appendage and the female the leaf-shaped structure shown here (on the opposite page).*

Madagascar has very few native carnivore species compared to mainland Africa. All nine that are found in Madagascar are members of the mongoose family Viverridae. Above, one of the better known species is the fanaloka (Fossa fossana)*, which is widespread in remaining natural habitats of the country.*

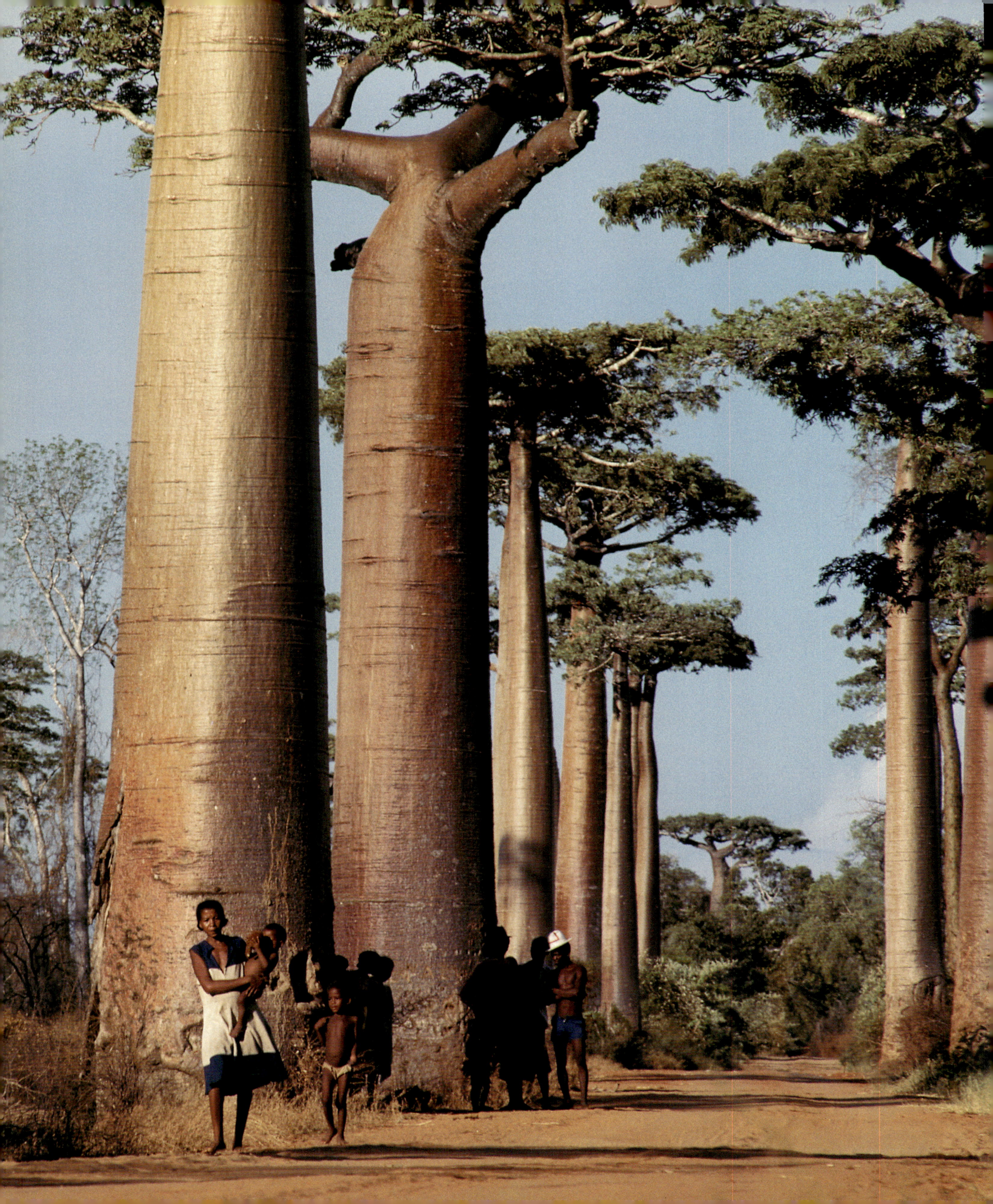

imity of Africa, Bantu/Swahili elements in the language are minimal and largely restricted to the northwest, where there is now regular contact with the Comoros. These Asian origins of the language speak strongly for the heavy early influence of Asia in Malagasy culture, and lesser involvement from Africa.

How the first Malagasy actually arrived is still not entirely certain. However, given the long distances, it appears that some came in their outrigger canoes via southern India and East Africa, where they established small colonies, and that the actual migration may have occurred over several generations. It is also possible, if not likely, that some of the early arrivals covered the distance from Indonesia (ca. 6 400 km) in a single voyage, since some of the Malagasy are racially very similar to the Indonesians. This is especially true of the Merina people of the High Plateau, living in the area around the capital city of Antananarivo.

Given the diverse origin of the Malagasy people, their culture is an unusual mix of African and Asian elements, including rice-growing from Indonesia, cattle-raising from Africa, and a strong belief in the power of the ancestors. The various cults of the dead vary from tribe to tribe, and are manifested in different ways, from the elaborate tombs of the Antandroy and the Mahafaly tribes to the *famadihana* (turning of the bones) ceremony of the Merina, in which the bones of dead relatives are periodically disinterred, wrapped in a fresh shroud, shown around the village, and returned to the family tomb. The *famadihana* in particular is reminiscent of some of the elaborate ceremonies of Indonesian tribes like the Toraja of Central Sulawesi.

From the slaughter of many cattle upon the death of an important Antandroy to the Merina *famadihana*, such events are an occasion for much feasting and celebration. Indeed, the Antandroy and Mahafaly rarely eat their cattle on other than ceremonial occasions, much as is the case among many East African tribes.

In any event, the modern Malagasy are divided into 18 tribes that culturally and racially display varying degrees of Asian and African origins. These range from the Merina, among whom the Indonesian racial and cultural influence is very strong, to the Makoa, who descend from African slaves and are almost entirely African. Several of the tribes are particularly worthy of note, including the Merina, the Sakalava, the Betsimisaraka, the Antandroy, and the Vezo.

The Merina, whose name means "People of the Highlands," live in the capital city of Antananarivo. They are the most Asian of all the Malagasy, and the largest of the tribes in terms of population. It was under their kings, Andrianampoinimerina and Radama I, that Madagascar was first united as a country in the nineteenth century. The Merina continue to be the dominant political force in the country, although recently there have been attempts at decentralization. Along with the Betsileo, who also live on the central plateau, they are the premier rice-growers of Madagascar.

The Sakalava, or "People of the Long Valleys," occupy the largest geographic area and live in the western part of the country between Tulear and Majunga. They are dark-skinned with Polynesian features and short, curly hair. Once a powerful tribe with their own kings and queens, they were conquered by the Merina in the nineteenth century. The Sakalava are cattle-raisers; a person's wealth is measured by the number of cattle owned, much as in parts of East Africa.

The Betsimisaraka, whose name means the "The Many Inseparables," occupy a large area of the eastern rain forest and are the second largest of the tribes in population. They cultivate rice, both in small paddies in the valleys, and dry rice on the forested slopes. They also produce vanilla and a variety of forest products. Racially, they are a mix of Indonesian and African elements, with the occasional influence from European pirates who frequented this part of Madagascar in centuries gone by.

The Antandroy, whose name appropriately translates as "People of the Thorns," inhabitat the southern spiny desert. They are a dark-skinned people who live in simple dwellings often made out of wood of the Didiereaceae. Their small villages are usually surrounded by fences of introduced *Opuntia* cactus or native Didiereaceae. They are cattle-raisers, and subsist mainly on millet, maize and cassava. The men wear distinctive conical hats, and carry metal-tipped spears. They are famous for their huge, rectangular tombs, the walls of which are often brightly painted with scenes from the dead person's life. They are among the most interesting, and least changed of the Malagasy, having never been completely conquered either by the Merina or the French.

Finally, the Vezo, who are usually recognized as just a subtribe of the Sakalava, rather than a separate tribe. They live on the coast in the Morondava area, and are fishermen rather than agriculturists. They are particularly interesting in that they still live a lifestyle that must be very much like the earliest arrivals on the island, using outrigger canoes and other primitive fishing techniques. Most other Malagasy have largely lost their maritime skills in the millennium or so that they have been living in Madagascar, but the Vezo still retain them.

Resolving the conflict between Madagascar's fragile natural ecosystems and destructive land use practices such as cattle raising and dry rice production in rain forest areas is one of the great challenges for this unique country. The needs of the people have to be addressed but current practices are clearly unsustainable. The great challenges faced by conservationists in the years to come will be to find solutions that improve the quality of life of the people of Madagascar, while at the same time maintaining the great biodiversity upon which they ultimately depend and which is of such great interest to the world at large.

RUSSELL A. MITTERMEIER
CRISTINA GOETTSCH MITTERMEIER

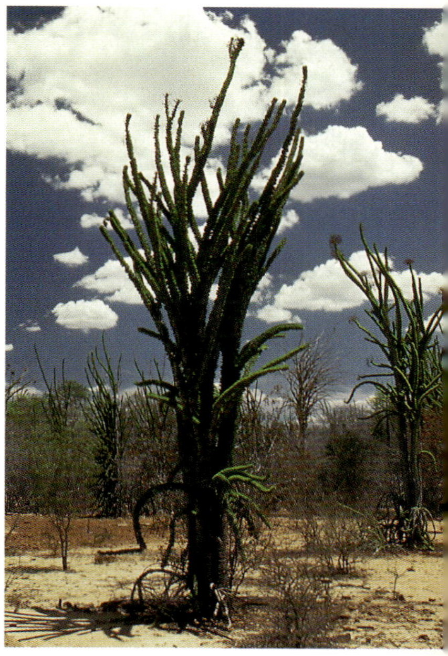

Plants are also important flagship species for Madagascar. The bizarre and wonderfully appealing baobabs reach their greatest diversity in this country, with seven of the world's eight species occurring there, six of them endemic. On the opposite page, the famous "highway of the baobabs" is perhaps the most impressive concentration of these trees anywhere in the country. It is located just outside Morondava, in the western dry deciduous forest of southwestern Madagascar, and features the tallest and most impressive of them all, Grandidiers baobab (Adansonia grandidieri).

Above, the Didiereaceae of the southern spiny desert superficially resemble the Cactaceae of the New World, but are actually an endemic family exclusive to southern Madagascar. The species shown here is Alluaudia ascendens.

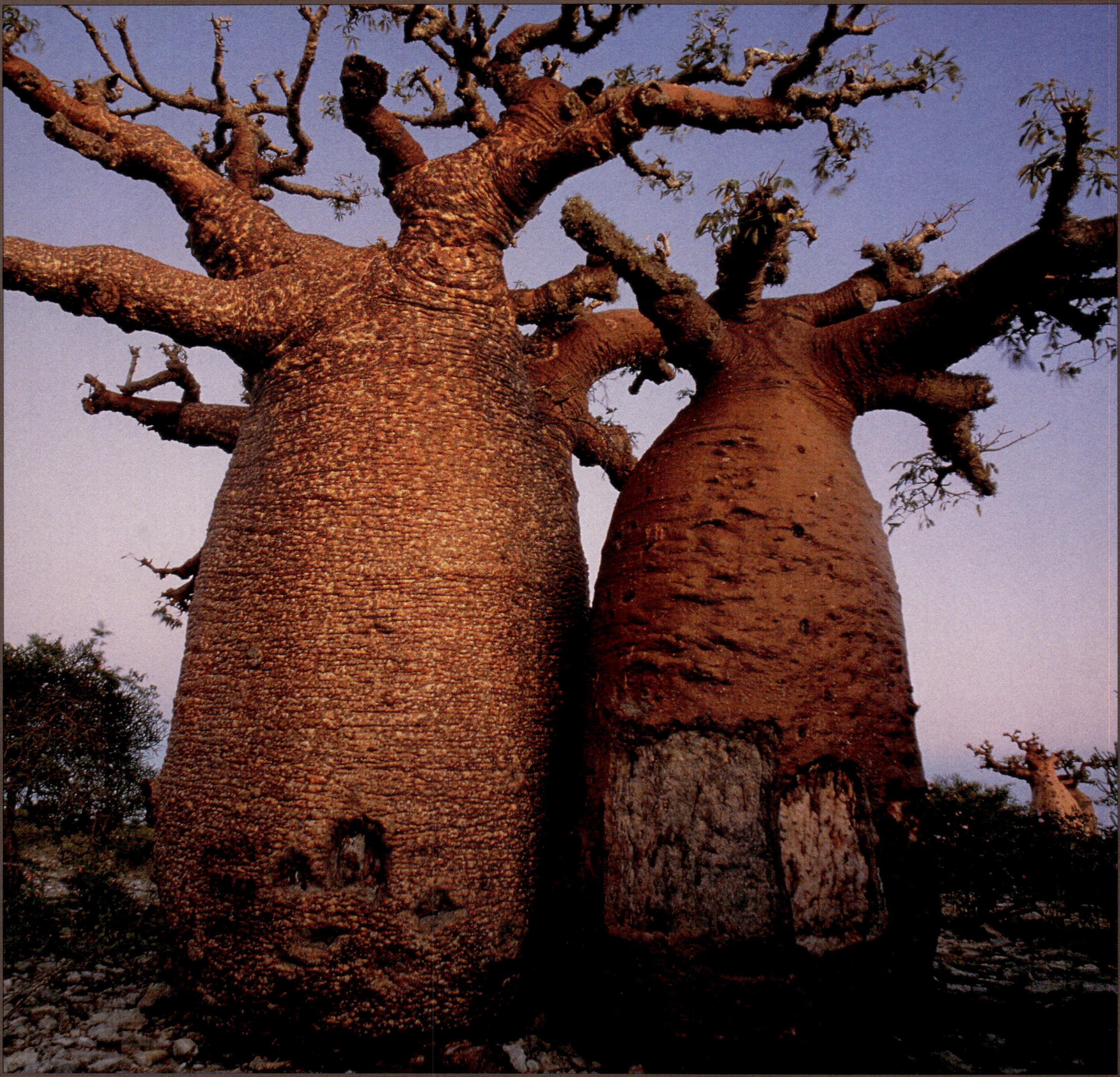

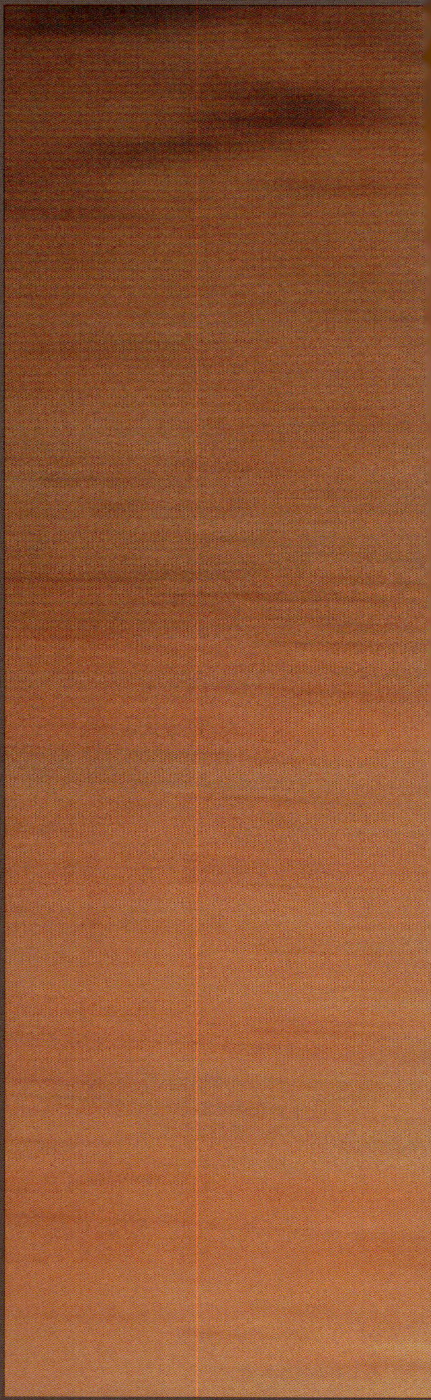

Above, baobabs often achieve imposing bulk, and in the dry season look as if they had been planted upside down. Local people sometimes strip the bark from baobabs to provide fodder for cattle, thatch roofs or for medicinal purposes, as is the case of this individual.
Upper right, the best known of Madagascar's lemurs is the ring-tailed lemur (Lemur catta), *which has long been kept in zoos around the world.*
Restricted to southern and southwestern Madagascar, it is the most terrestrial living lemur (lower right), and moves about in groups that can exceed 20 individuals.

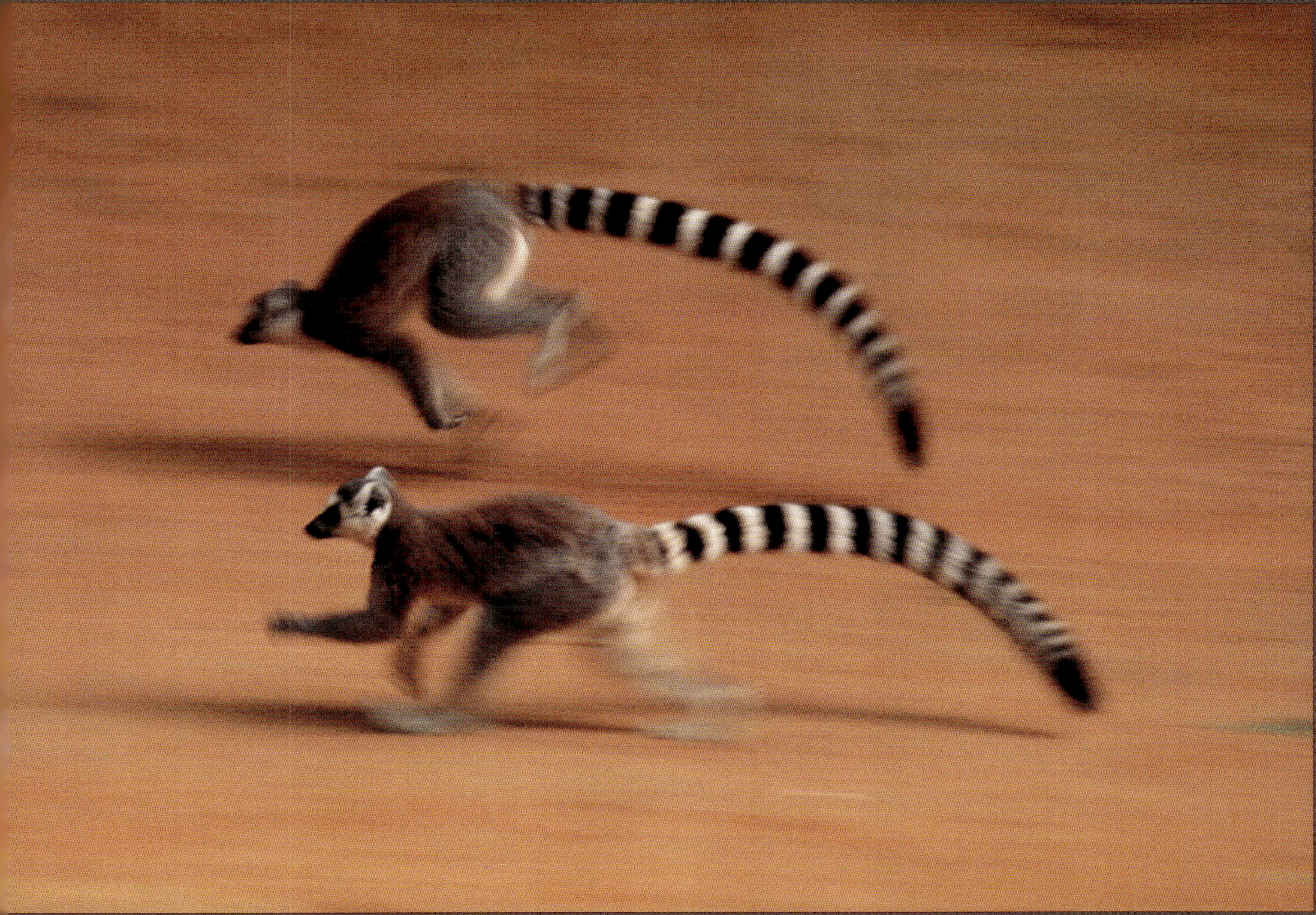

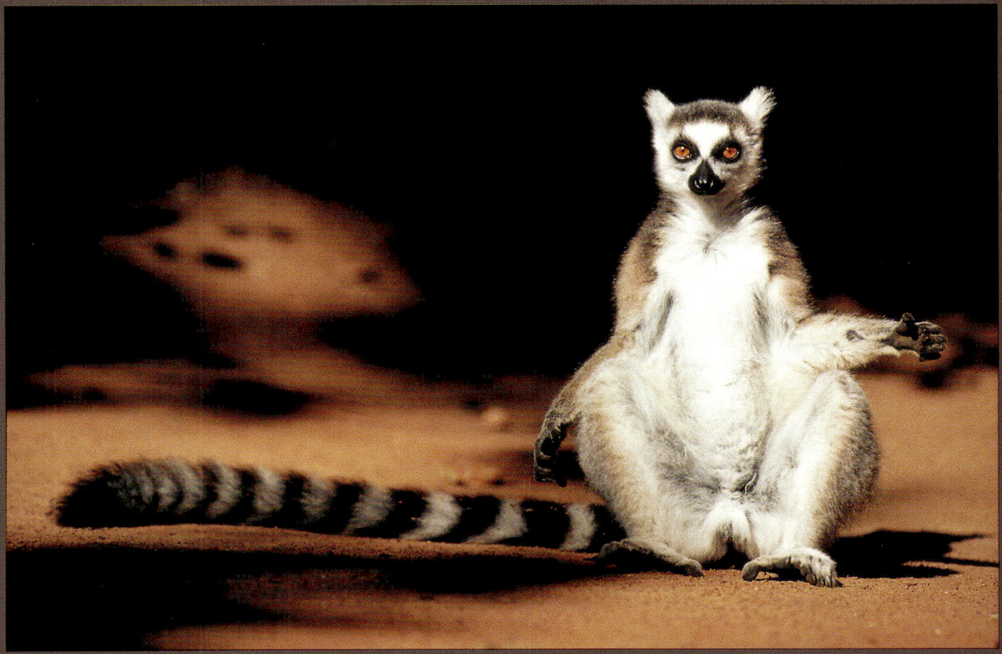

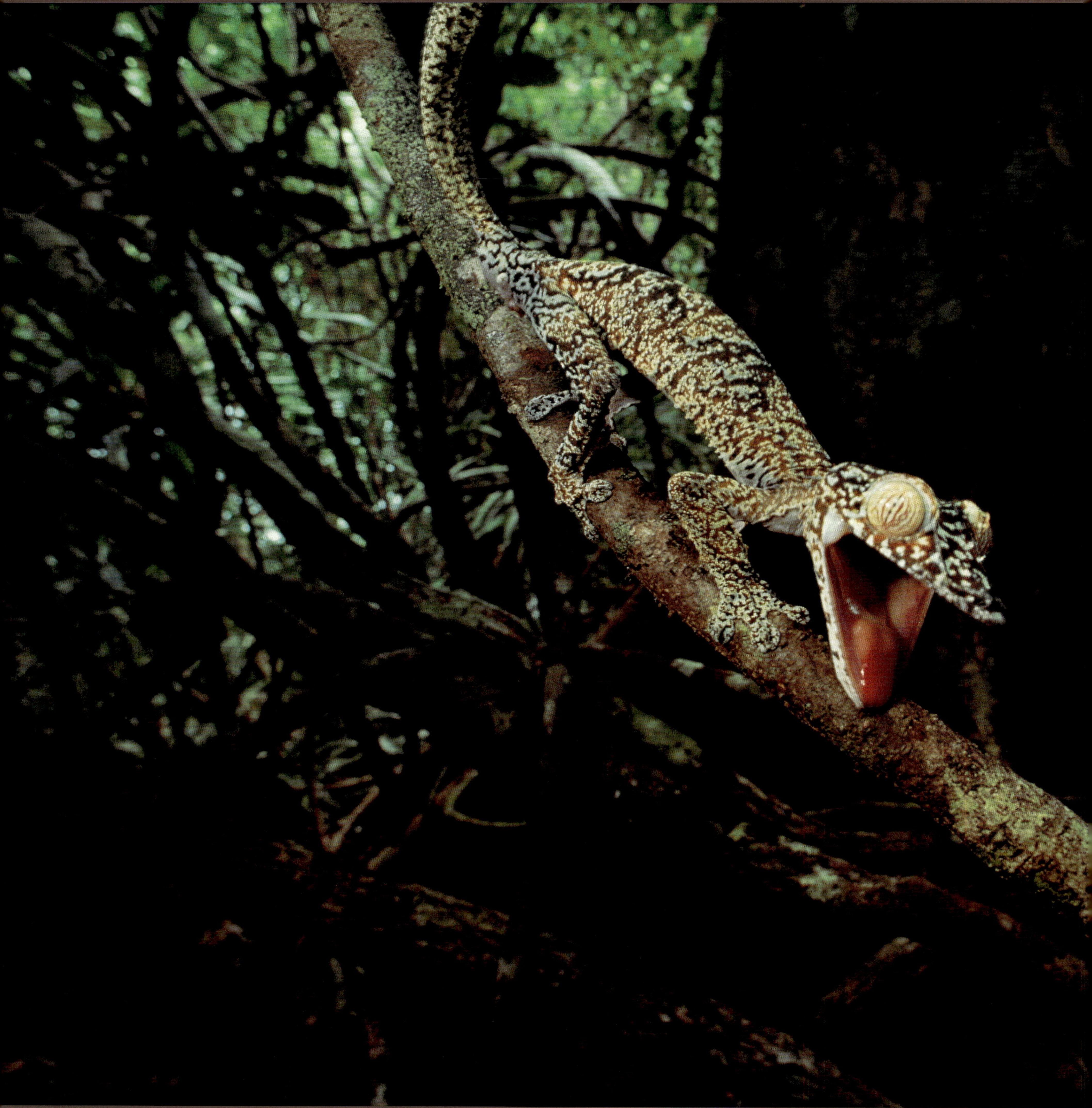

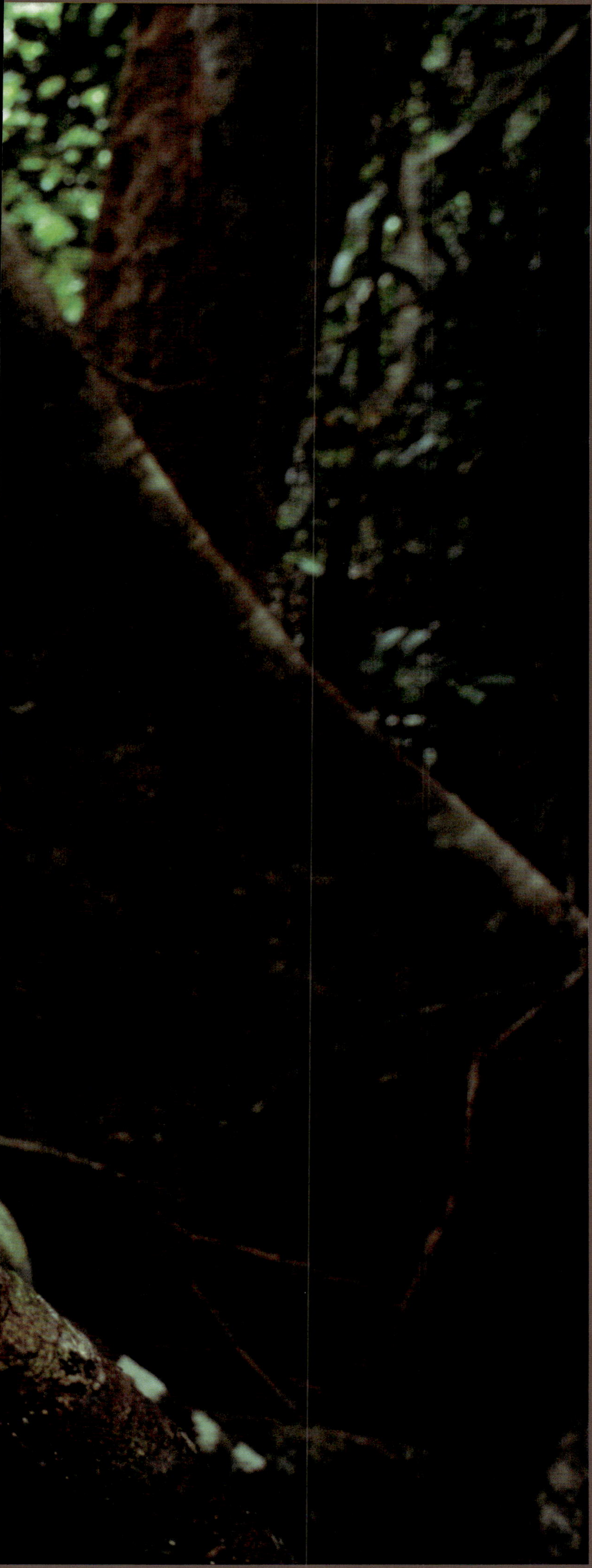

Aside from the chameleons, Madagascar includes a number of other unusual lizard groups. The geckos are well represented (left), and include at least nine species of leaf-tailed geckos of the genus Uroplatus, *characterized by the unusual shapes of their tails. This species* (Uroplatus fimbratus) *is the best known, and is a nocturnal inhabitant of the eastern rain forest. Below, this Urania moth* (Urania riphoeus), *a diurnal moth, is among the most beautiful species in Madagascar, a country with the fifth highest number of endemic butterflies on Earth.*

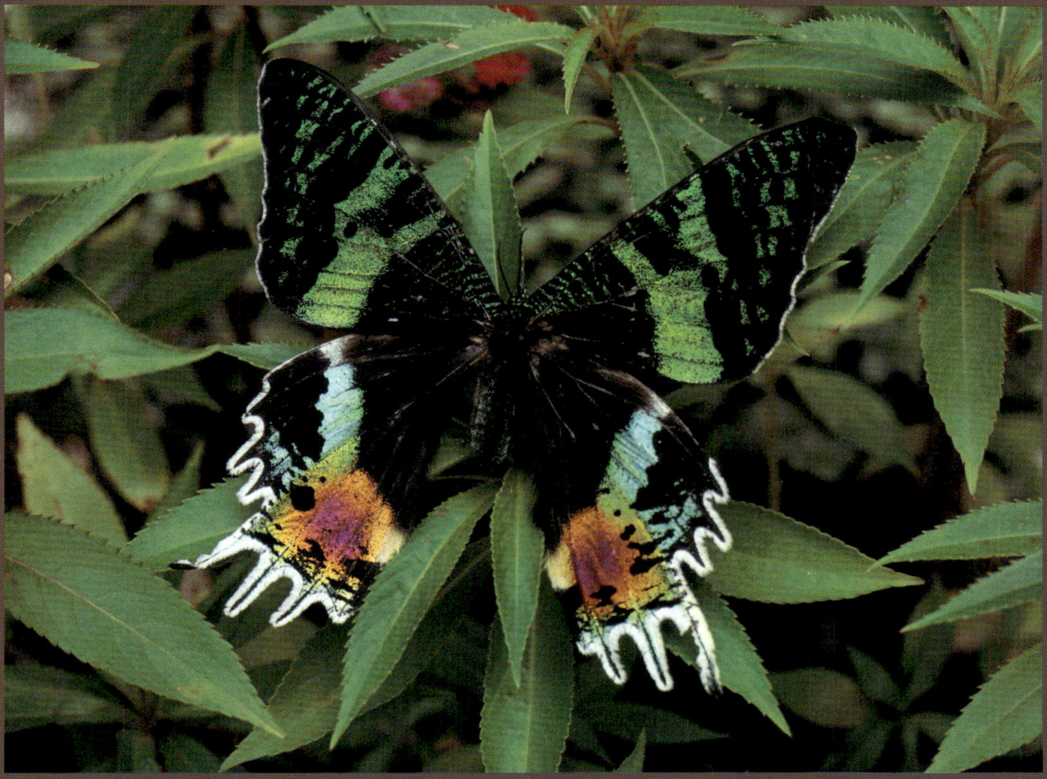

Lemurs come in many different shapes and sizes. Below, the mysterious aye-aye (Daubentonia madagascariensis), *is certainly the most unusual.*
The only living representative of its own primate family, Daubentoniidae, it is the only primate that has rodent-like, continuously-growing incisors.
It is considered a bad omen in many parts of its range and is feared by local people.
Right, the red-fronted brown lemur (Eulemur fulvus rufus) *is one of the more typical lemur species.*
Lower right, the pygmy mouse lemur (Microcebus myoxinus) *was first described in 1858, then synonymized with the more common gray mouse lemur* (Microcebus murinus) *and forgotten for more than 130 years.*
It was finally rediscovered by a group of German scientists in 1992.
It is the smallest of all living primates and weighs only about 30 g.

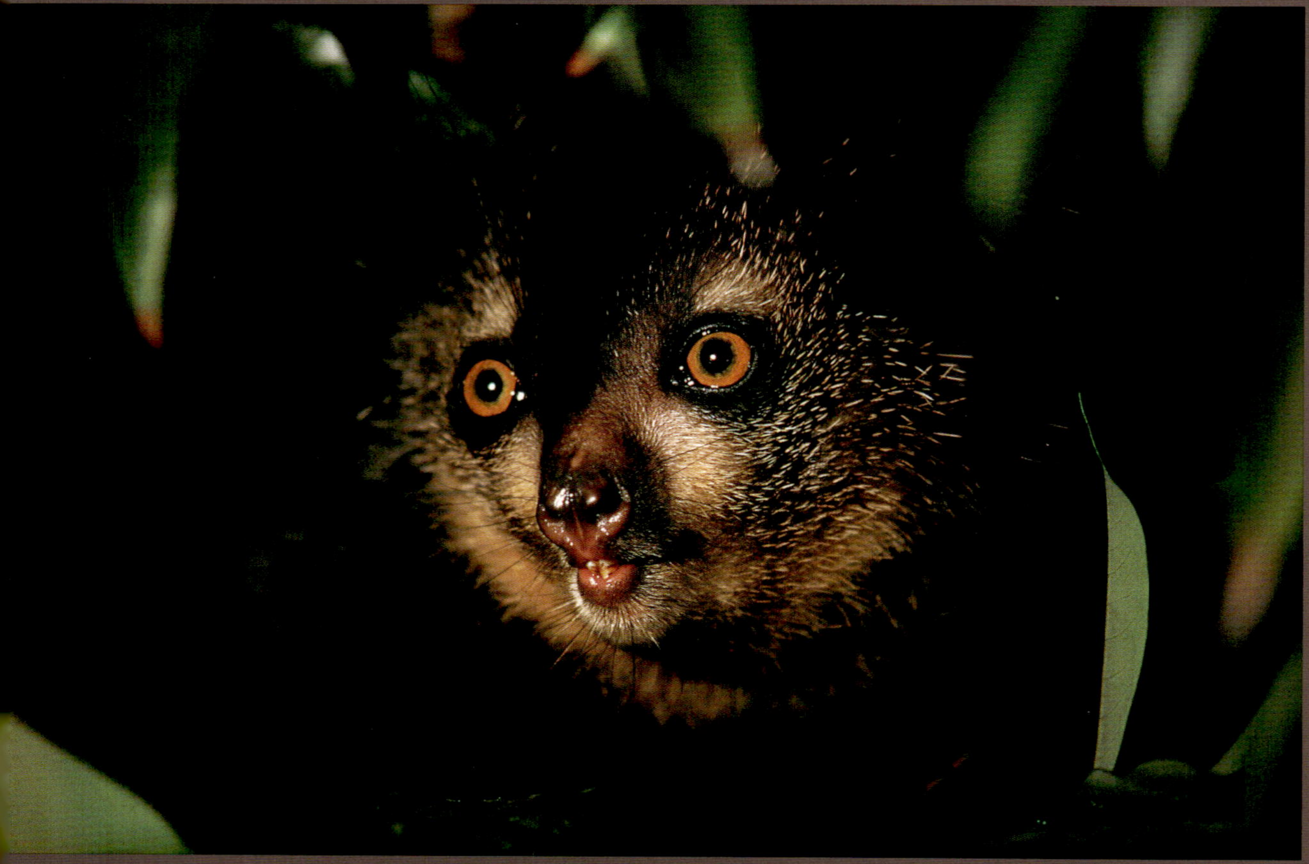

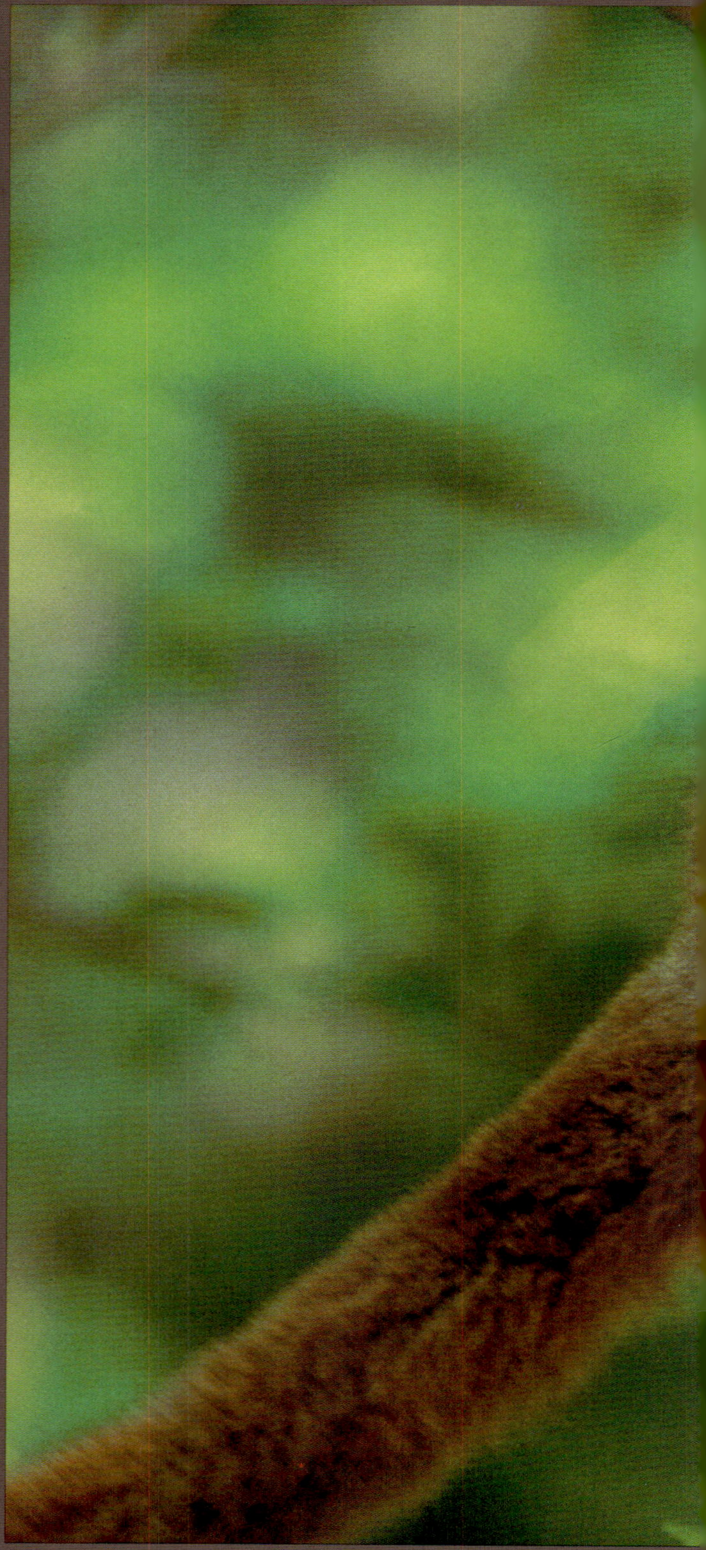

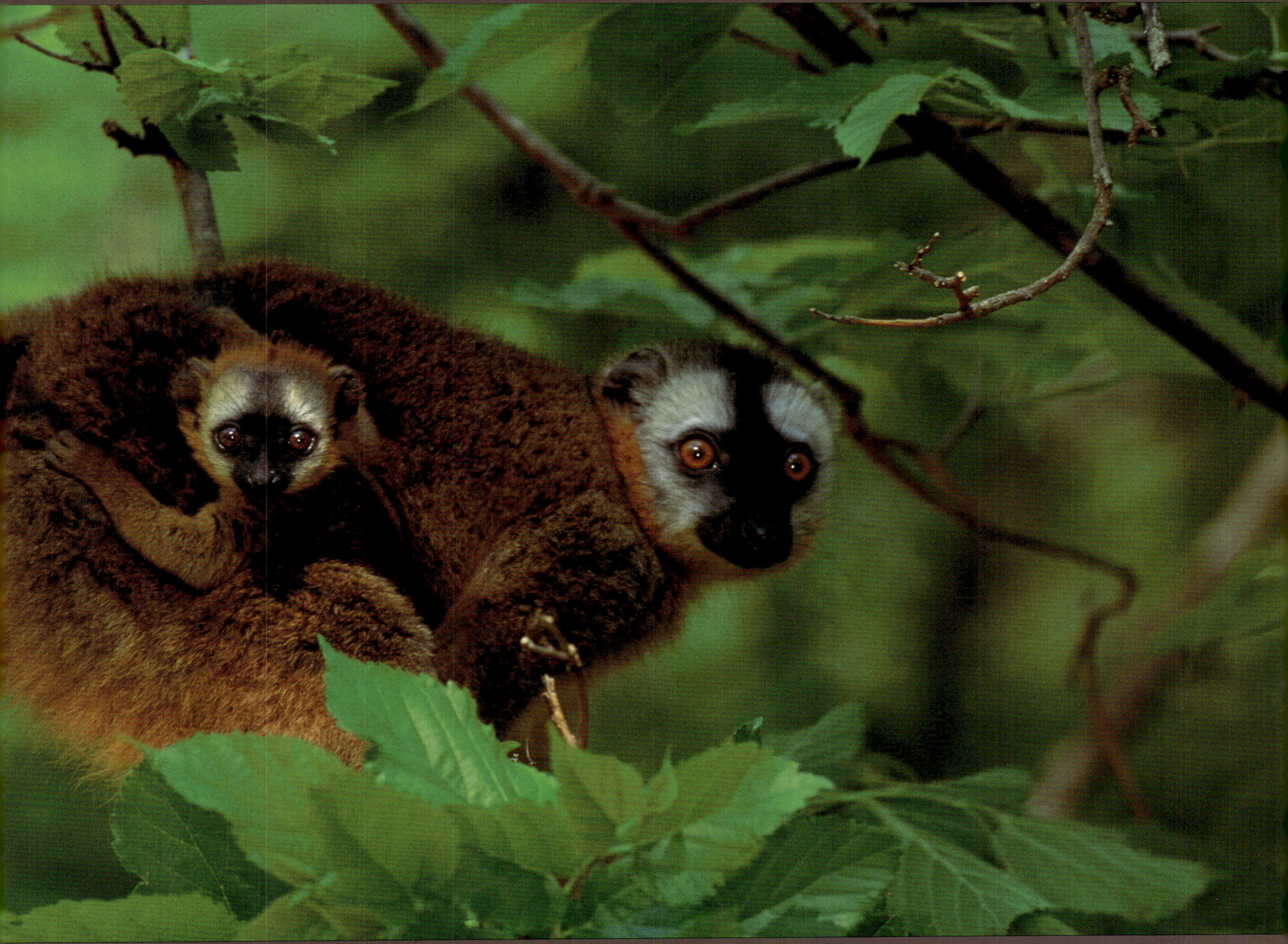

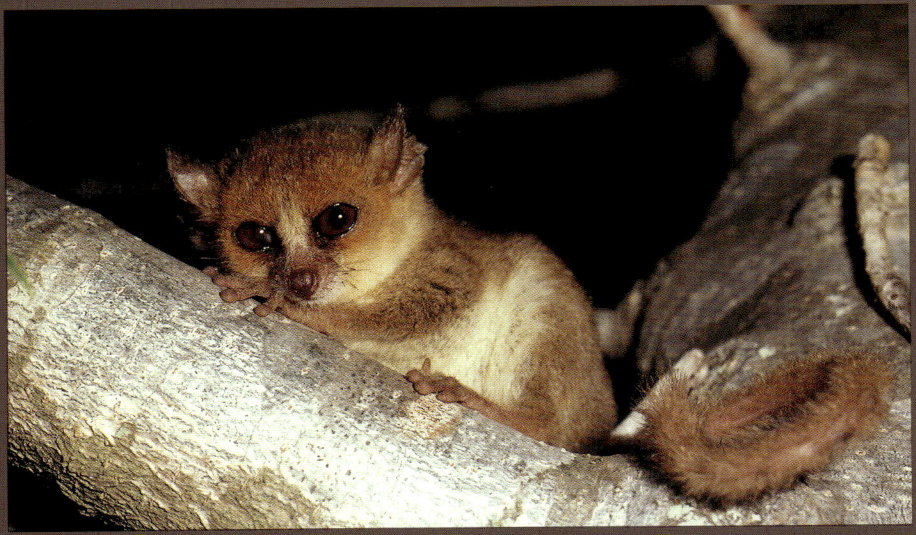

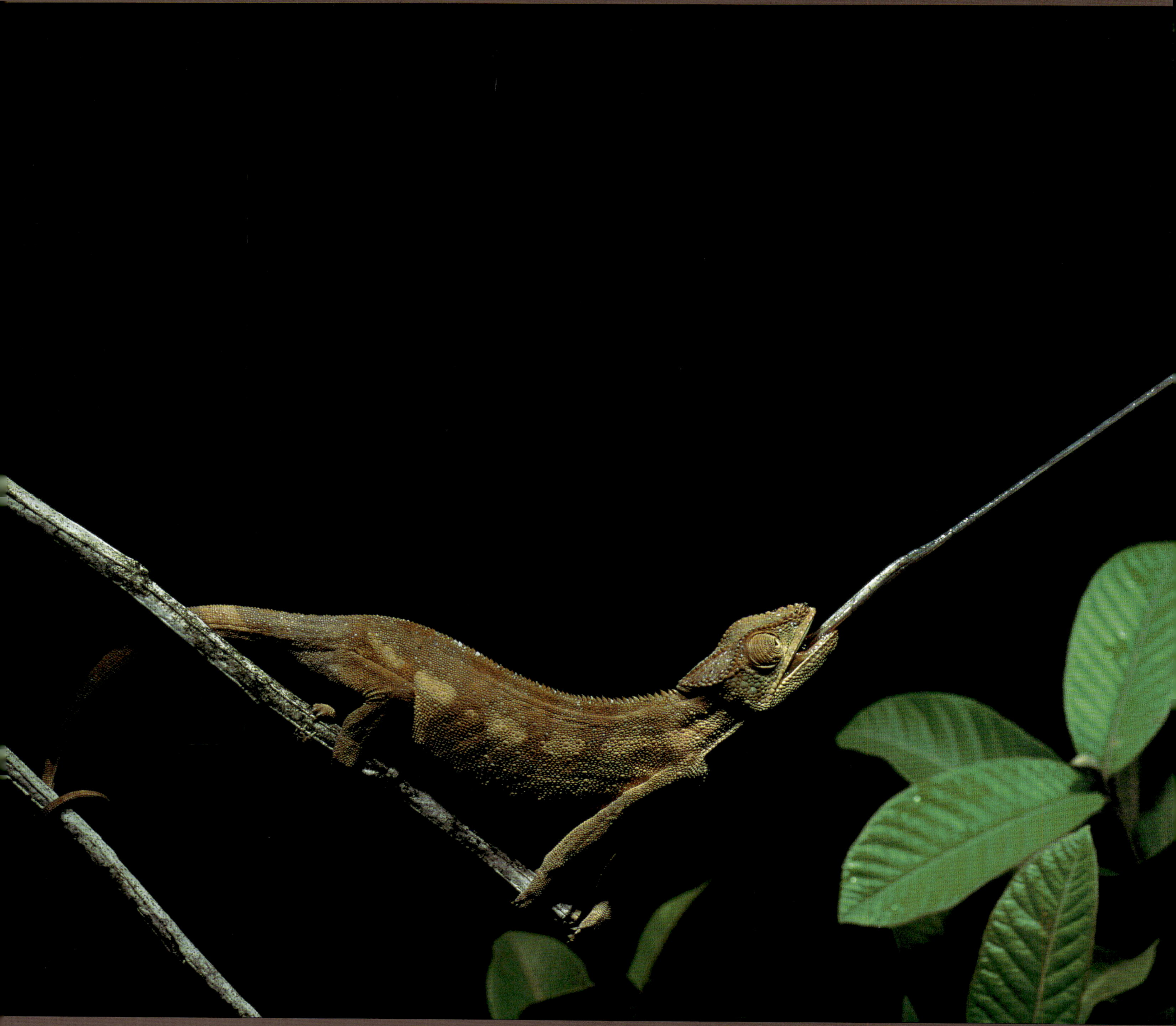

The chameleons are the most famous reptiles of Madagascar and are surpassed as flagships only by the lemurs. The chameleons reach their greatest diversity in Madagascar, with about half of the known species occurring only there. These include the Furcifer oustaleti *(above) and the world's largest, Parson's chameleon* (Chameleo parsoni), *which can reach 70 cm (upper right).*

Lower right, a stump-tailed chameleon of the genus Brookesia. *Unlike the larger species, the stump-tailed chameleons are generally drab brown in color and live on the forest floor or in very low vegetation. However, they are adorned with a grotesque armor formed of spines, tubercles and horns. Some like* Brookesia perarmata *depicted here look like mini-dinosaurs.*

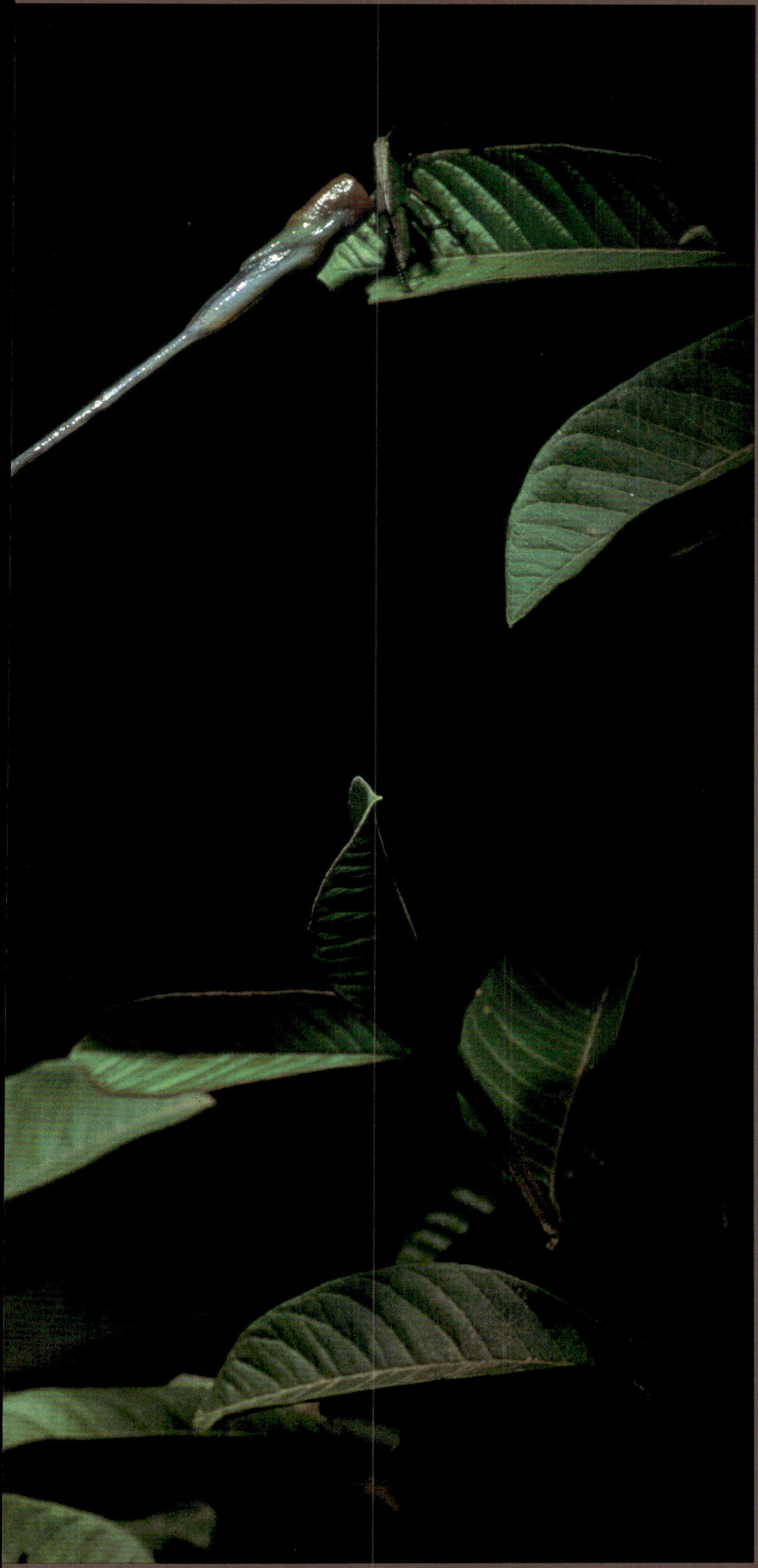

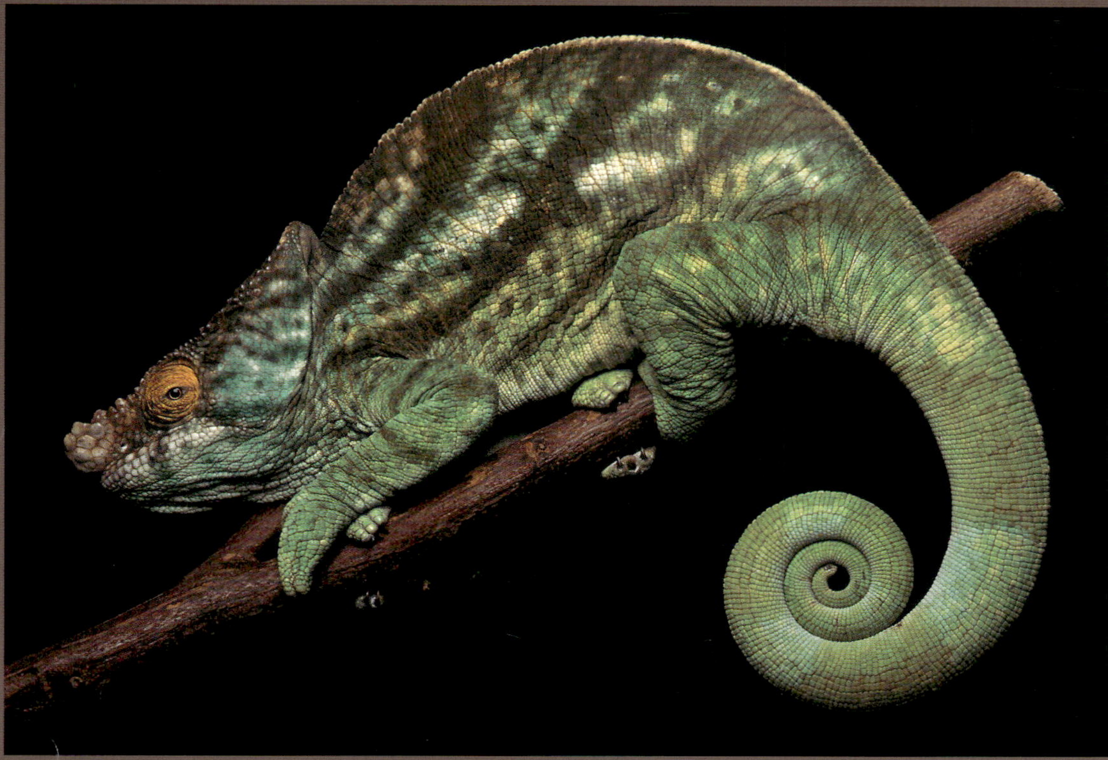

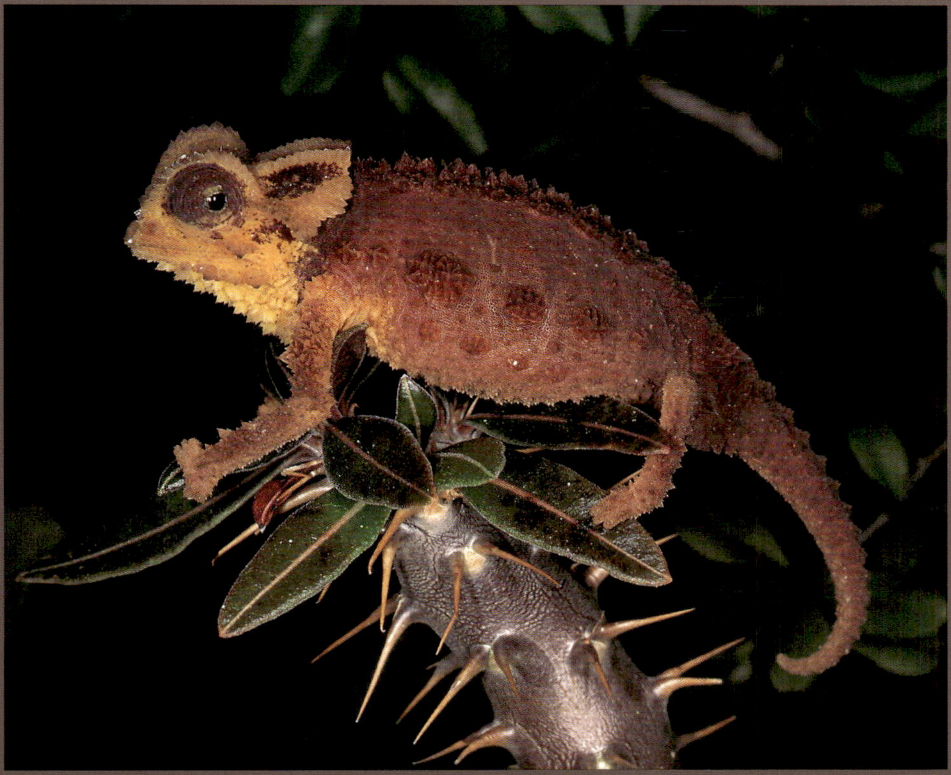

Without appropriate national and international attention, Madagascar's unique biodiversity may be approaching its twilight. Much has already been lost in the past 2 000 years, and much more could disappear in the near future. Below, the wonderful octopus tree (Didierea madagascariensis), *one of the Didiereaceae endemic to the southern spiny desert, is at risk. These slow-growing trees are being cut down for charcoal and firewood over large areas of their restricted range. On the right, these giant baobabs in the sunset stand as a mute testimony to Madagascar's global importance and to its fragility, since they are not in a protected area and could be cut down for cattle fodder at any time.*

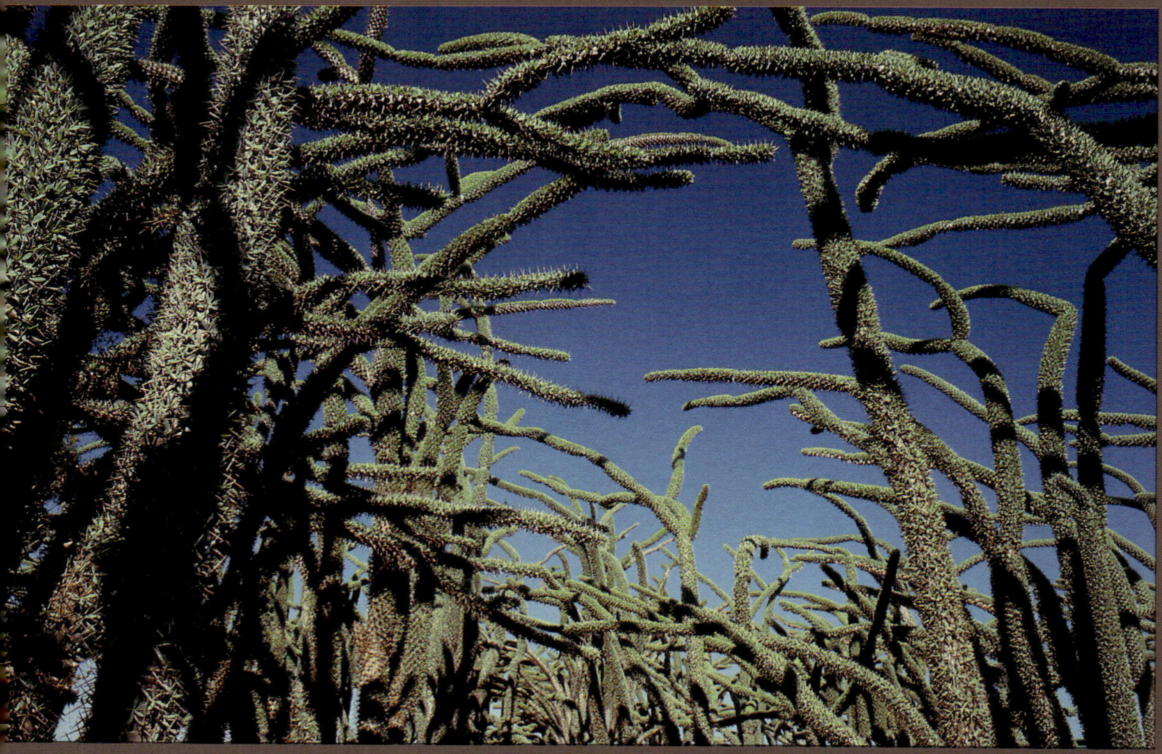

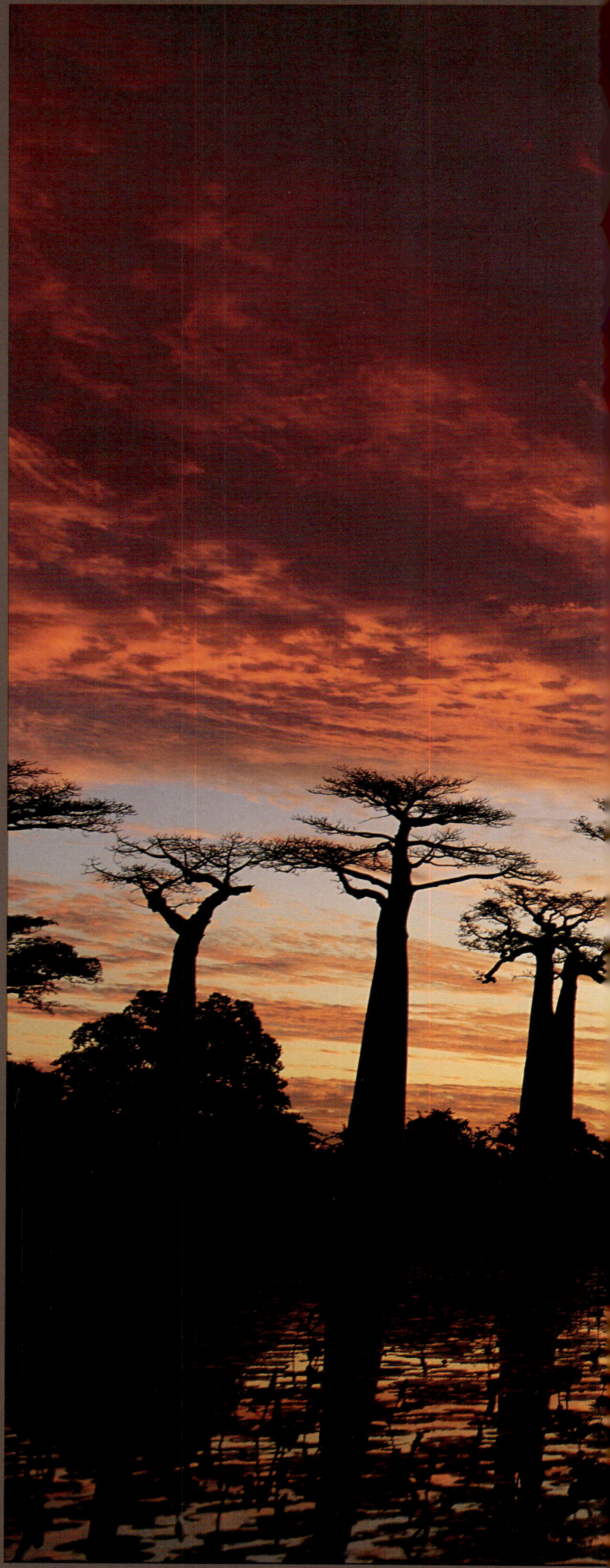

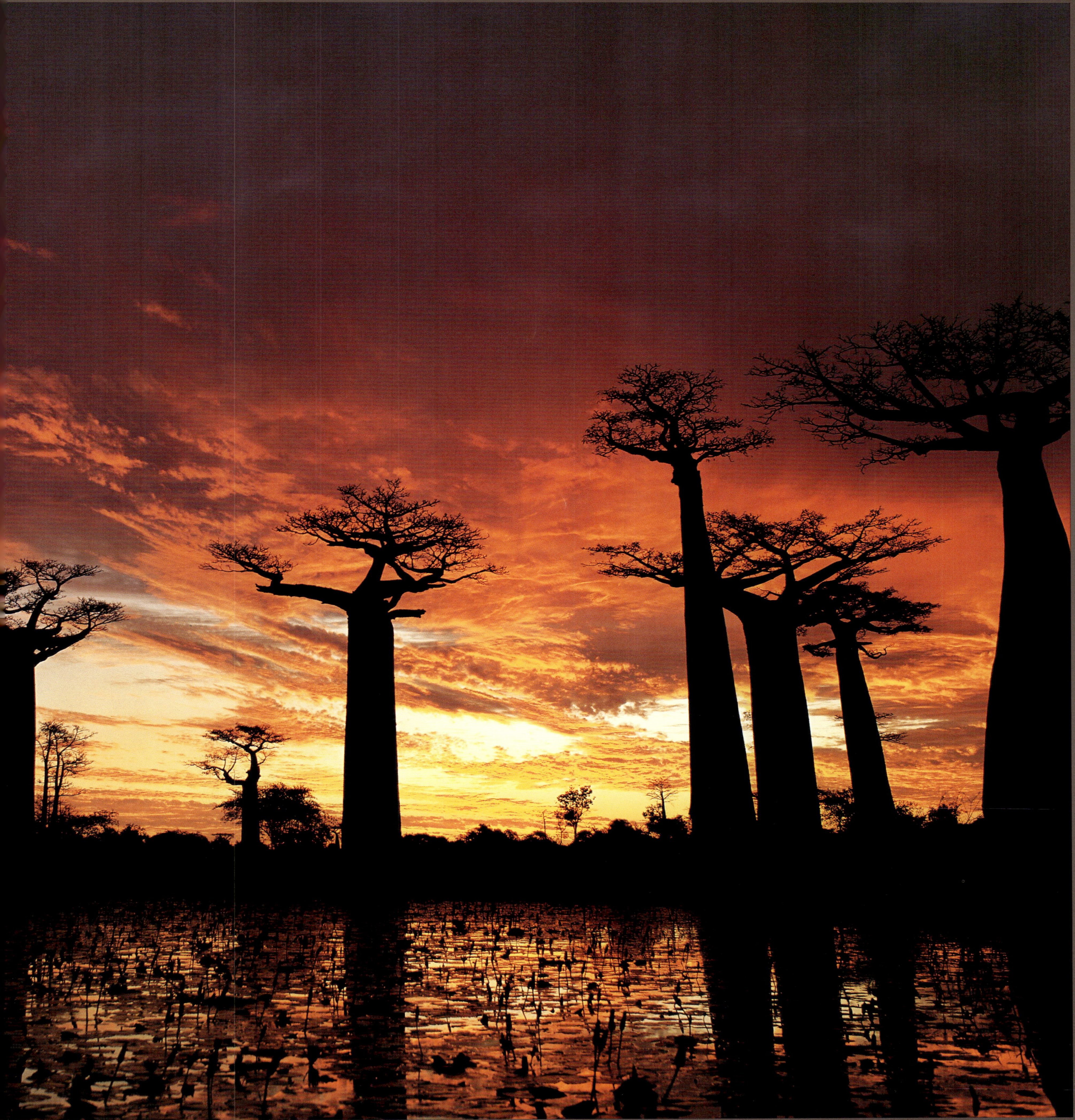

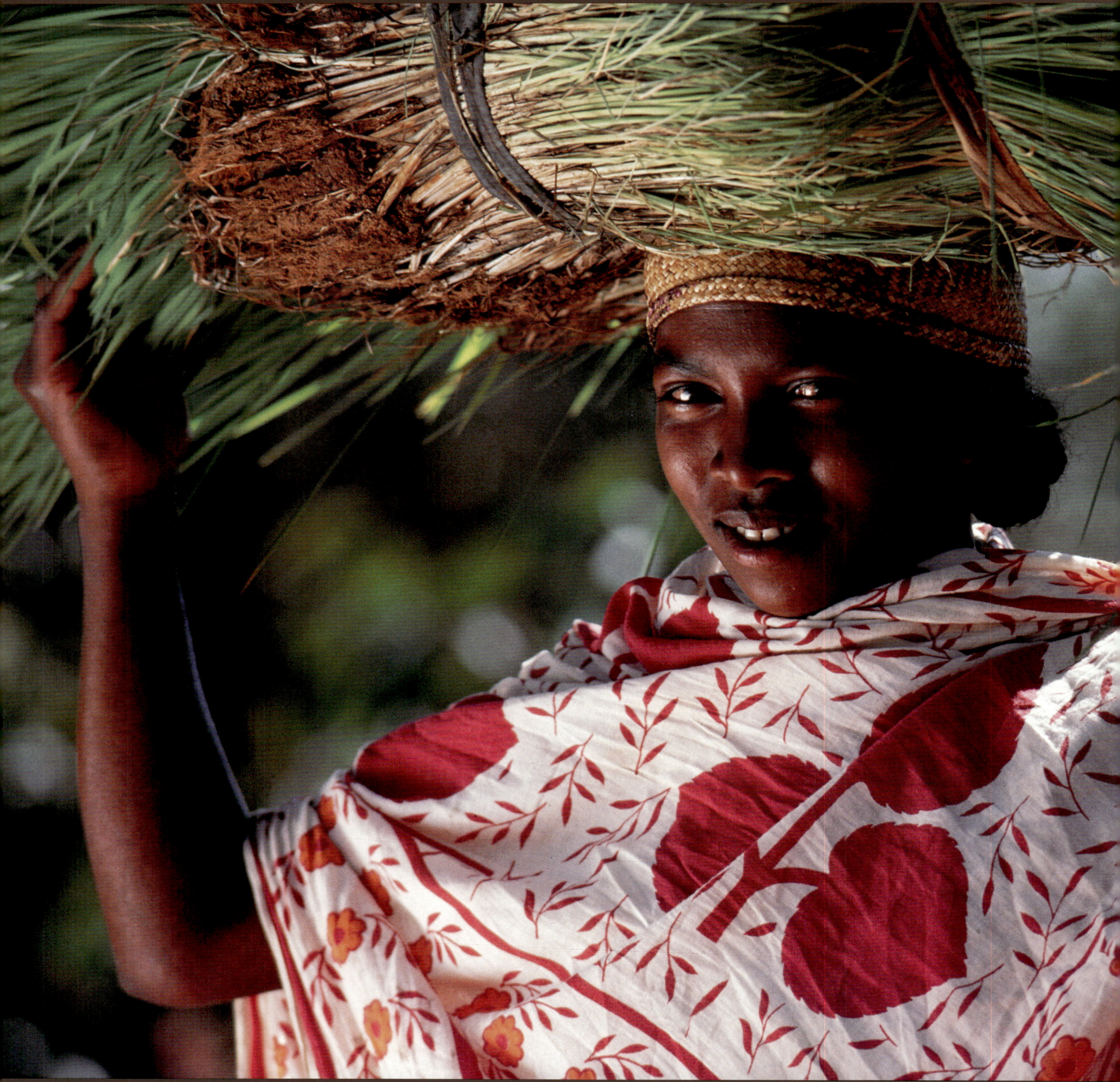

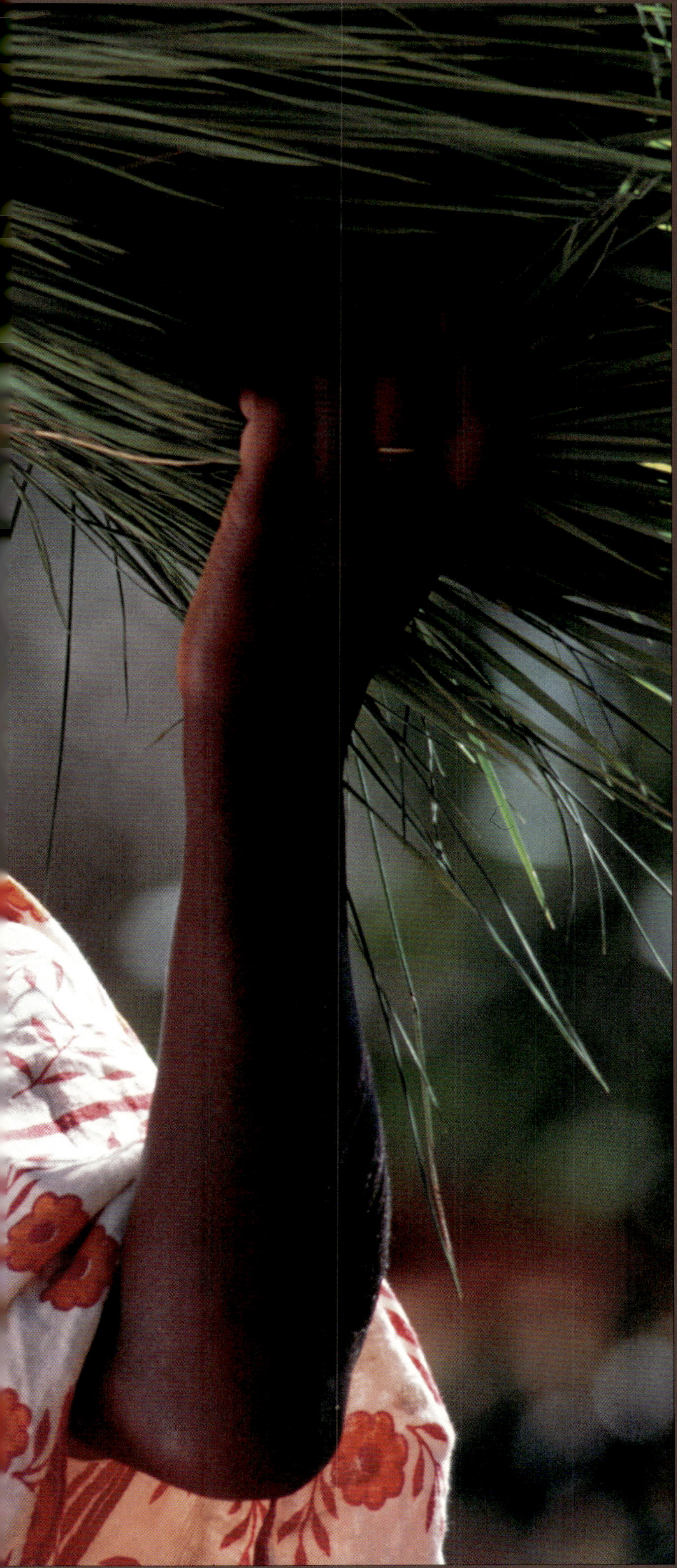

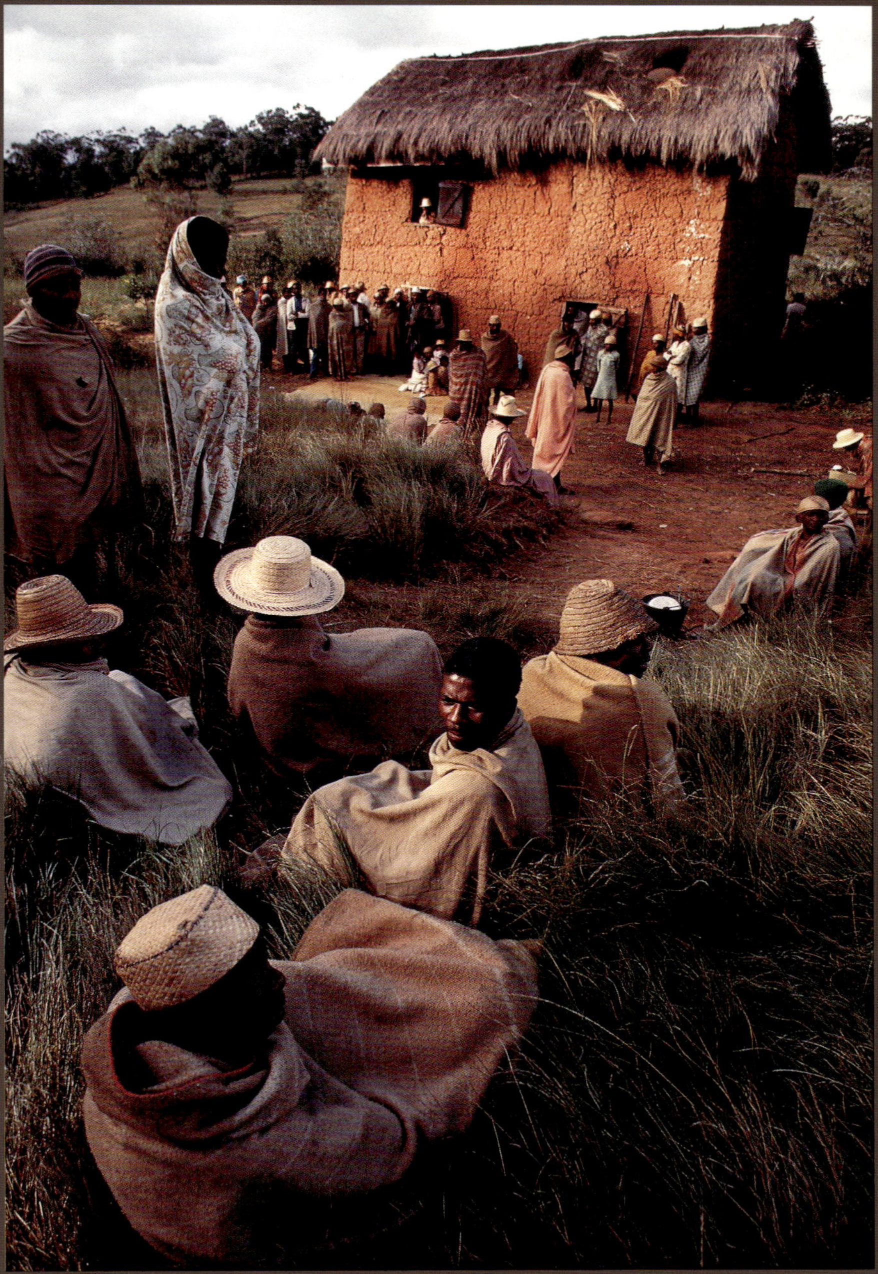

Madagascar's people originated both in Asia and Africa, and arrived on the island only about 1 500-2 000 years ago. Their culture is a unique mix of elements from both continents, together with their own distinct traditions that have evolved over the past two millennia. Left, a Betsileo girl from Fianarantsoa carries rice shoots to a paddy, one of the traditions inherited from Southeast Asia. Traditions of ancestor worship are also important in Madagascar. In the photo above, villagers in Ambalavao in the central plateau region gather for a wake. Treatment of ancestors by some of the people on the high plateau shares many features with Indonesian cultures like the Toraja of Sulawesi.

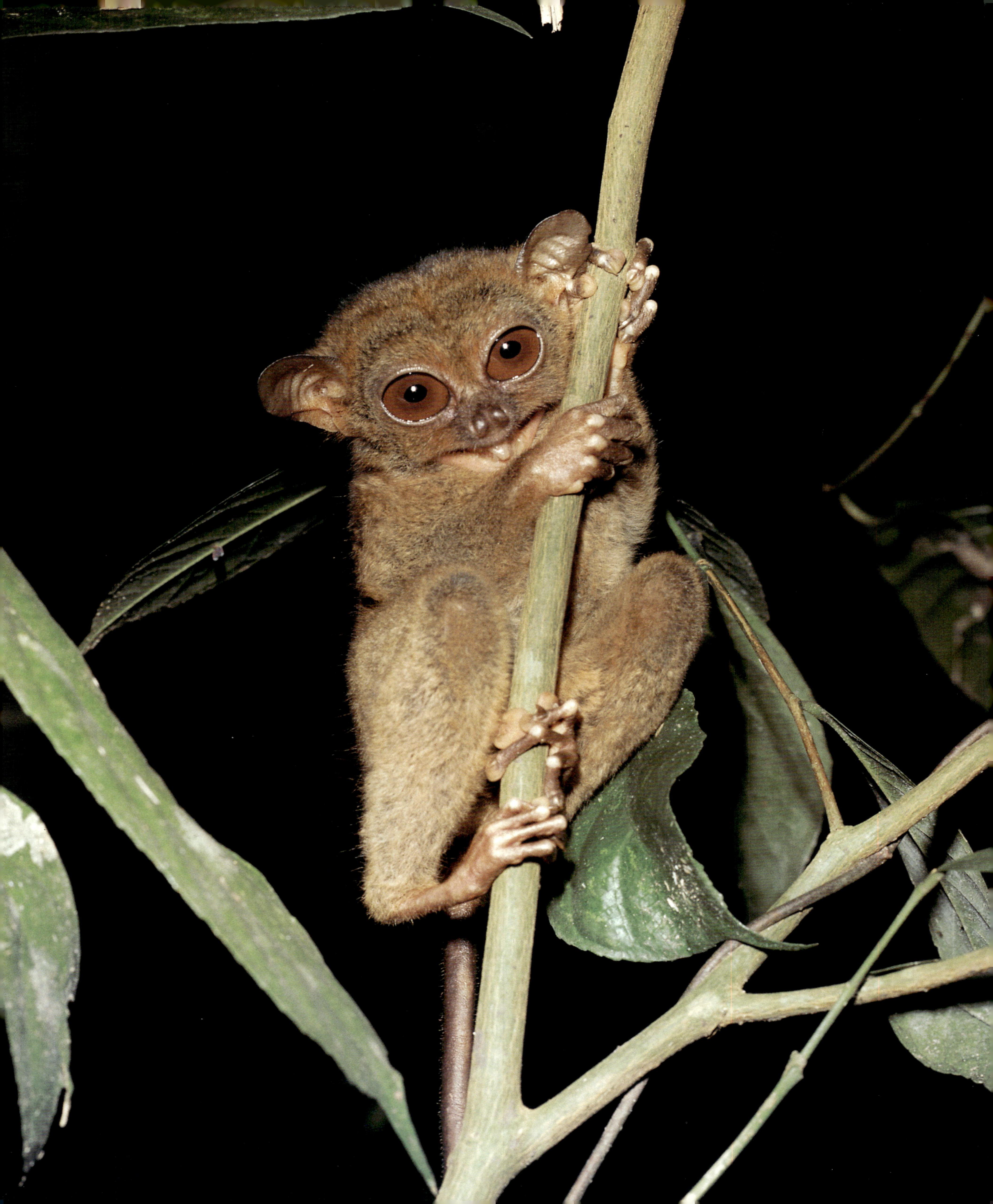

PHILIPPINES

The island nation of the Philippines is located in the westernmost portion of the Pacific Ocean, north of Borneo and Sulawesi (which are parts of Malaysia and Indonesia), and due east of Vietnam across the South China Sea. Covering only 300 780 km^2, it is the second smallest of the megadiversity countries after Ecuador, and also one of the highest-priority threatened hotspots on Earth. It is a place of great beauty, with rugged volcanic mountains reaching down to the shore, coral reefs extending into bright blue waters, and fertile plains now supporting vast areas of cropland for rice.

The more than 7 000 islands that make up the Philippines (over 700 of which are inhabited by humans) lie entirely in the tropics, between 5°N and 21°N. The archipelago is formed by a series of isolated geological fragments that have long and complex geological histories, most dating back 30 to 50 million years, originating far out into the Pacific Ocean. These islands are part of the "Ring of Fire" of the Pacific Basin which extends from Indonesia to Japan and eastern Russia, and around the western edge of both American continents. With at least 17 active volcanoes, earthquakes and volcanic activity in the Philippines are a part of daily life. The Philippine archipelago stretches over 1 810 km from north to south, and measures 1 104 km at its widest point. It is only 241 km from Taiwan, and its nearest islands are just 40 km from the Malaysian portion of Borneo. Rainfall is high through most of the country, typically 2 000 to 3 000 mm per year in the lowlands and reaching up to 12 000 mm per year on some high mountain peaks. The Philippines sits on one of the world's primary typhoon tracks, with up to 35 typhoons passing through the archipelago each year.

The varied geological histories of the different parts of the Philippines, combined with the high degree of topographic diversity, result in an exceptionally diverse biota. Indeed, the Philippines has been designated as one of the world's top 19 biodiversity hotspots, and in fact is in the top five in this elite group. The country actually consists of a group of fairly independent hotspots. There are five major and at least five minor Centers of Endemism, each of which supports its own unique set of mammals, amphibians, plants, birds, reptiles, and butterflies, and each of them geographically isolated from the others. The five larger centers are Luzon, Greater Mindoro, Greater Palawan, Greater Mindanao and Greater Sulu. In addition to these mammal-based centers of endemism, the Philippines also has within its borders 12 of the Centers of Plant Diversity and Endemism recognized by WWF/IUCN. These include Mount Apo, Mindanao; Mount Kitanglad, Mindanao; Mount Pulog, Luzon; the Palanan wilderness area, Luzon; Palawan; Sibuyan Island; Romblon Island; southern Samar Island; Mount Baloy, central Panay Island; Mount Isarog, Camarines Sur Province; Mount Talinas and Lake Balinsasayao; and Mount Makiling and Mount Banahaw. Looking at Endemic Bird Areas as identified by BirdLife International (ICBP), there are 9 in the Philippines, one of the highest totals for any country. These include the Luzon Mountains, the Luzon lowlands and foothills, Mindoro, Negros and Panay, Cebu, Palawan, Samar-Leyte-Bohol, the Mindanao lowlands and the Sulu Archipelago. These areas of endemism range in size from Luzon, the largest island at 103 000 km^2, which has 31 endemic species of mammals, to tiny Camiguin Island, a minute 265-km^2 speck north of Mindanao which has at least two species of endemic mammals and one endemic frog species.

The patchwork aggregate of islands has resulted in high species diversity in certain groups and very high endemism. There are at least 201 mammal spe-

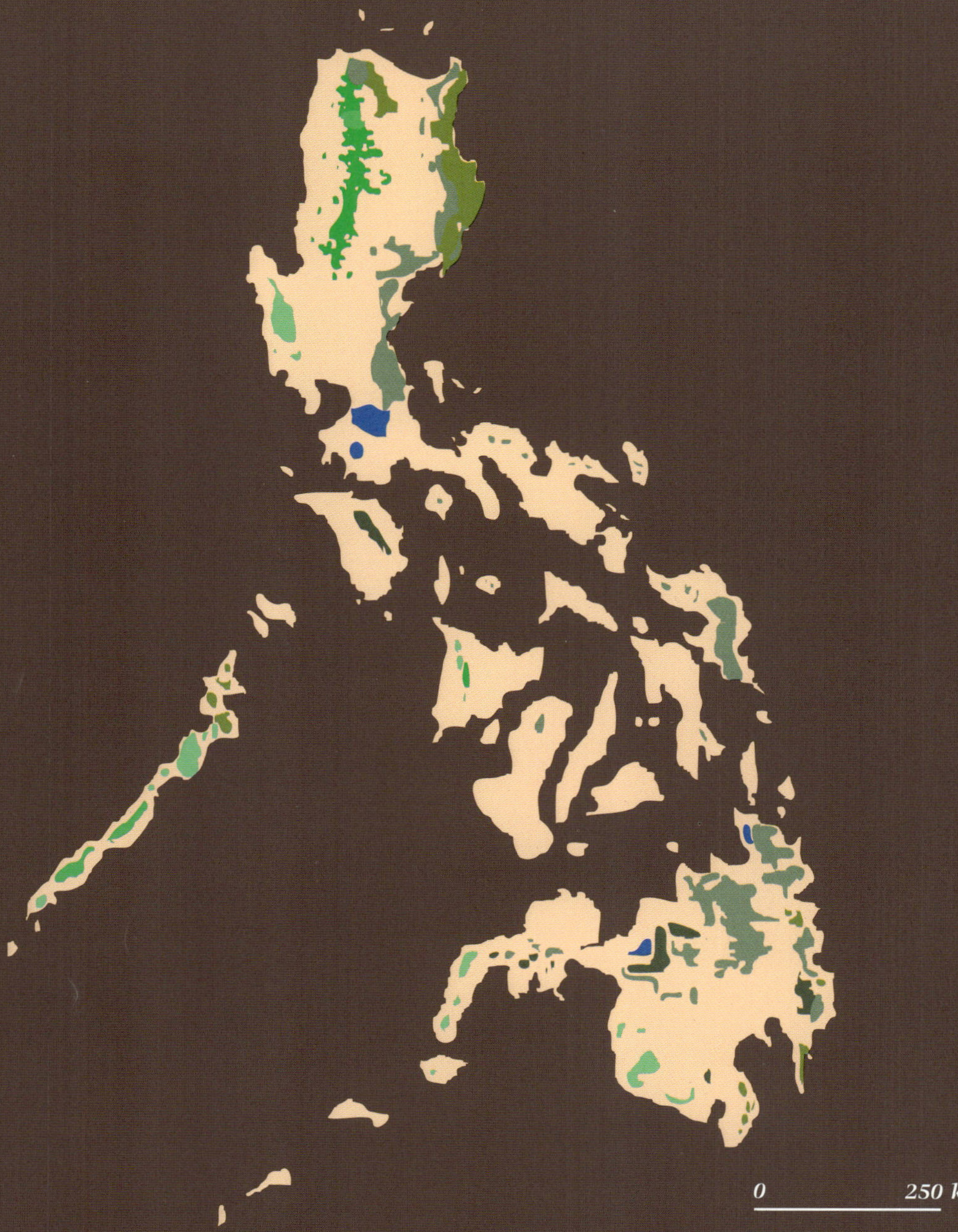

On the opposite page, the very interesting and unusual Philippine tarsier (Tarsius syrichta) is one of the few primate species found in this country. A very distinctive lineage, the tarsiers go back to the earliest days of the primates. Tiny creatures that look like mammalian versions of tree frogs, they are aggressive little predators that feed on a variety of insects and small vertebrates. This one is from Malagos, on Mindanao.

cies in the country, 180 terrestrial and 21 marine; of these 116 are endemic, good enough for fifth place on the global list in this group. Bird diversity is relatively low at 556 species, but endemism is fourth in the world at 183 species. Reptile diversity is moderate at 193 species, but endemism is eighth at 131 species. Amphibians have low diversity with 63 species, but most (44) are endemic; freshwater fish diversity, estimated at 330 species, is high given the size of the country.

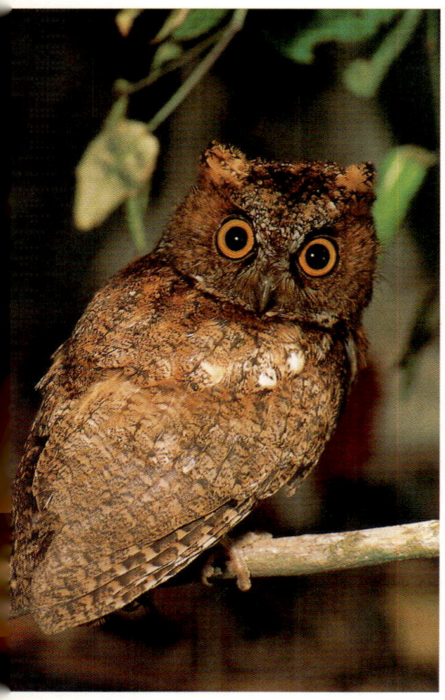

The Philippines has high bird diversity for its size, and very high endemism. Above, one of the endemics, a Luzon scops owl (Otus longicornis).

Based on butterflies and tiger beetles, the Philippines has very high invertebrate diversity as well. For all butterflies there are 895 species, of which a very high 352 are endemic, the second highest number of endemics in the world after Indonesia. For the swallowtail butterflies the pattern is the same, 49 species and 21 endemics, with the number of endemics placing it second behind Indonesia. In tiger beetles, the Philippines is an amazing eleventh on the global diversity list, with 94 species, the smallest country in the top twelve, and fifth in endemism with 74 species. As with the vertebrates, all of these figures are very impressive when the small size of the country is considered.

There is some uncertainty as to the number of plants in the Philippines. Early estimates were around 8 000 higher plant species with perhaps 50% endemism; most recent estimates are in the order of 10 000-12 000 species (S. Sohmer, pers. comm.), again with up to 50% endemism. Using the lower end of the recent estimates, the number of endemic higher plant species in the Philippines would be 5 000, good enough to make the 2% plant cutoff for the top 12 megadiversity countries. Palm diversity and endemism are especially high with 157 species, of which 109 (69%) are endemic.

Given the vast number of islands in the Philippines (second only to Indonesia), its tropical location, and a coastline twice as long as that of the United States, it is not surprising that its marine diversity should also be remarkably high. Diversity of coral species in the Philippines may be the highest on Earth, and total coastal marine diversity probably ranks among the top three or four countries in the world, together with Indonesia and Papua New Guinea.

The biodiversity of the Philippines is truly impressive in global terms. This is certainly true of overall diversity, but especially so as regards endemism. In absolute numbers, its endemic species in groups like mammals, birds, reptiles, butterflies, tiger beetles and plants are comparable to and often exceed much larger megadiversity countries.

Indeed, the total number of non-fish vertebrate endemics is 474, good enough for eighth on the world list. Of the seven countries above it, only Madagascar is comparable in size and even it is almost double the size of the Philippines. All the others are much bigger, Brazil, for example, being 28 times its size, and Australia 26 times larger. When per-unit-area endemism is considered, the Philippines is truly amazing.

Unfortunately, the Philippines is also the world leader in degree of endangerment, with only 6-8% of natural vegetation remaining and far less than that in primary condition —leading to the dubious distinction of generally being considered the *hottest of the hotspots* and the most severely endangered of the megadiversity countries. A comparison with Madagascar, another country that is both a top megadiversity country and a top-five global biodiversity hotspot, is particularly informative. Although Madagascar is known as a country that has lost most of its natural vegetation, with only 10-15% thought to remain, its situation is still better than that of the Philippines. Madagascar, at 587 045 km^2, is nearly twice the size of the Philippines, and its human population only about 20% of that of the smaller country, some 14 million compared to over 73 million. Consumption patterns are much higher in the Philippines than in Madagascar, and growing more rapidly as well. When all of these factors are coupled with the high concentrations of diversity and endemism, it is easy to see why many experts feel that the Philippines stands at the brink of a serious biodiversity crisis, and that it could be the site of the first major extinction spasm of the twenty-first century. Only concerted local, national and international efforts can prevent this, and it is hoped that recognition of this country as one of the leading megadiversity nations will help to focus attention in the most effective ways.

Fortunately, there have been some improvements in recent years. The recently-developed Integrated Protected Areas System (IPAS) was funded by the Global Environmental Facility of the World Bank, and is now managed out of the Philippines Department of Environment and Natural Resources (DENR). It has chosen ten top-priority sites for increased management and protection. Of the IPAS sites, one of the most important is in the Palanan Wilderness Area, also known as the Northern Sierra Madre Wilderness Area, the largest such area left in the Philippines. It has very high plant and wildlife diversity, including the last Philippine eagle population outside of Mindanao and is also home to the Agta, one of the very unusual Negrito groups of this country. This area was very poorly known scientifically prior to a survey conducted in 1991 by Conservation International, with BirdLife International, Leiden University (The Netherlands), Isabela State University (the Philippines), and the DENR. The team's findings focused national and international interest on Palanan, and began the process of protecting it. At 200 000 ha, the national park that has been established there is by far the largest in the country, and covers perhaps 10% of the remaining primary rain forest of the Philippines.

As part of the IPAS program, involvement of local communities in resource protection and management is being handled by a group called NGOs for Integrated Protected Areas, Inc. (NIPA). In 1996, NIPA selected the Northern Sierra Madre Wilderness Foundation as the host NGO to manage the Palanan project. This is a consortium of several local organizations and Conservation International's Philippines office, which was instrumental in the development of the Northern Sierra

Madre Wilderness Foundation and its ongoing local capacity-building efforts. The program now underway involves an integrated approach to conservation that includes scientific surveys of forest and marine resources, sustainable management of these resources, social development and capacity building , enterprise development, and addressing concerns critical to indigenous people living in the area, especially the Agta.

Conservation International is also working in the Tubbataha Reef in Palawan. This is the most important coral reef system in the Philippines, and is unequaled in size and diversity elsewhere in the country. This area has already been declared as the first national marine park in the Philippines, and is important for biodiversity conservation, ecotourism, and as a source of fish larvae for the entire Sulu Sea, which sustains sizable reef fisheries in the southern part of the country. The Tubbataha Reef represents an important source of foreign exchange from tourism, as a well as a source of income for subsistence fishermen from the Cagayancillo Islands and elsewhere in the country. The reef is being destroyed and productivity diminished by fishermen who use destructive practices such as dynamite fishing. Legal protection began in 1988, with the establishment of a national park, and the site was subsequently declared a World Heritage Site. Following a visit by President Fidel V. Ramos in 1995, a Presidential Task Force was created to develop and implement a management plan for Tubbataha. Conservation International, one of three NGO members of this Task Force, has been primarily involved in livelihood and enterprise development, including projects such as seaweed cultivation and ecotourism.

Another major need involved further control of logging. This activity has been banned in 64 of 73 provinces, being permitted only in the nine with significant forests remaining. This benefits the 64, but concentrates the destruction in the few places where significant habitat for flagship species like the Philippine eagle still exists. Timber concessions should be tightly controlled, and logging operations limited to areas where damage to biodiversity will be minimal. And logging companies should be required to replant natural forest after logging is completed. Replanting of deforested areas that have been converted to grassland (now 25% of the country) should also be a high priority, although such programs have been plagued by especially high levels of corruption in the past. All things considered, and given the very high biodiversity of everything natural that remains in the Philippines, the best alternative for the country would be to ban all further logging in natural forest areas, and to declare them all protected zones. In view of what has already been lost, this would amount to a small percentage of the country, but would have enormous global biodiversity implications. This alternative certainly represents both the moral high ground on this issue, as well as the best long-term economic solution for the country.

Human cultural diversity in the Philippines is also quite high, although lower than that in Papua New Guinea and Indonesia. At latest count, there are 111 ethnolinguistic groups in the country, with some languages subdivided into a great many local dialects. Nearly all ethnic groups are Malayan in origin, although there is some evidence that the Negrito groups may represent a proto-Malayan origin. There are strong elements of Chinese, Spanish, and American culture overlain on the basic Malayan culture; trade and intermarriage with Chinese merchants began by at least AD 1200, and the Philippines was governed by the Spanish as a colony from about 1560 to 1898, and by the United States from 1898 to 1946. A substantial portion of the people of the country are of thoroughly mixed ancestry, and most of its modern history is one of control by strong external powers. Christianity is the dominant religion, but Islam is practiced by the Moros of the southern Philippines, and animism is practiced by some isolated groups (and remains an element of religious beliefs in a significant portion of the population).

All things considered, the Philippines is one of the biologically richest and most severely threatened of all the megadiversity countries, and one worthy of very special attention for both the terrestrial and marine realms. Nowhere else have expanding human populations pushed so many species so close to the brink of extinction; failure to take rapid and effective action in this country could result in one of the first significant major extinction spasms of the twenty-first century.

Natural habitats

Prior to disturbance by humans, virtually the entire area of the Philippines (about 94%) was covered by some form of rain forest; only a few seasonal forests existed in the south-central part of the country, a small area of mixed forest and savanna occurred in the Cagayan Valley of northern Luzon, and some pine-dominated cloud forest in north-central Luzon. Trees of the family Dipterocarpaceae were diverse (at least 50 species) and probably most numerous; these trees, sometimes referred to as Philippine mahogany, have dense, beautifully lustrous hard wood much sought after for use in furniture in developed countries, as well as for a wider range of local uses. Because the country is generally well-watered, and soils are largely composed of fairly homogeneous volcanic materials, most variation in habitats is associated with the changes in temperature and rainfall that occur along elevational gradients as one proceeds from the shoreline and beaches to the hills and mountains dominating the Philippines.

Philippine coral reefs are estimated to cover about 30 000 km², equal to nearly 10% of the land area, and are often described as being among the most colorful and productive in the world. With 488 out of the 500 species of corals that occur worldwide, the Philippines has one of the richest coral faunas. Moreover, the roughly 1 400 marine fishes that occur in Philippine

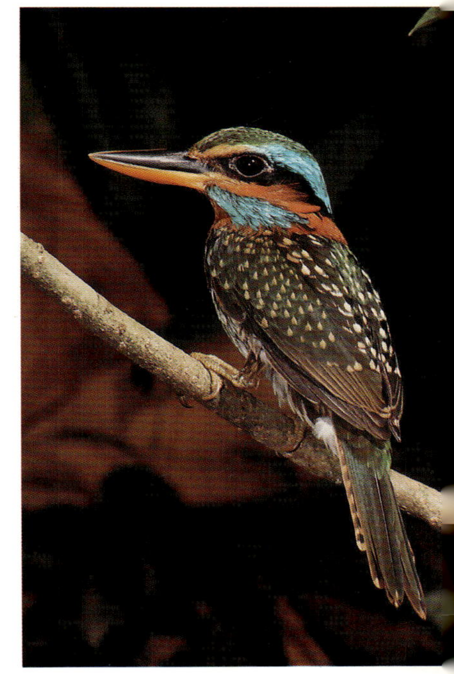

The spotted wood kingfisher (Actenoides lindsayi), *another of the endemic birds of the Philippines.*

On pp. 240-241, the flying gecko (Ptychozoon intermedium) *does not actually fly, but rather glides on extended membranes. This one is from Mount Apo.*

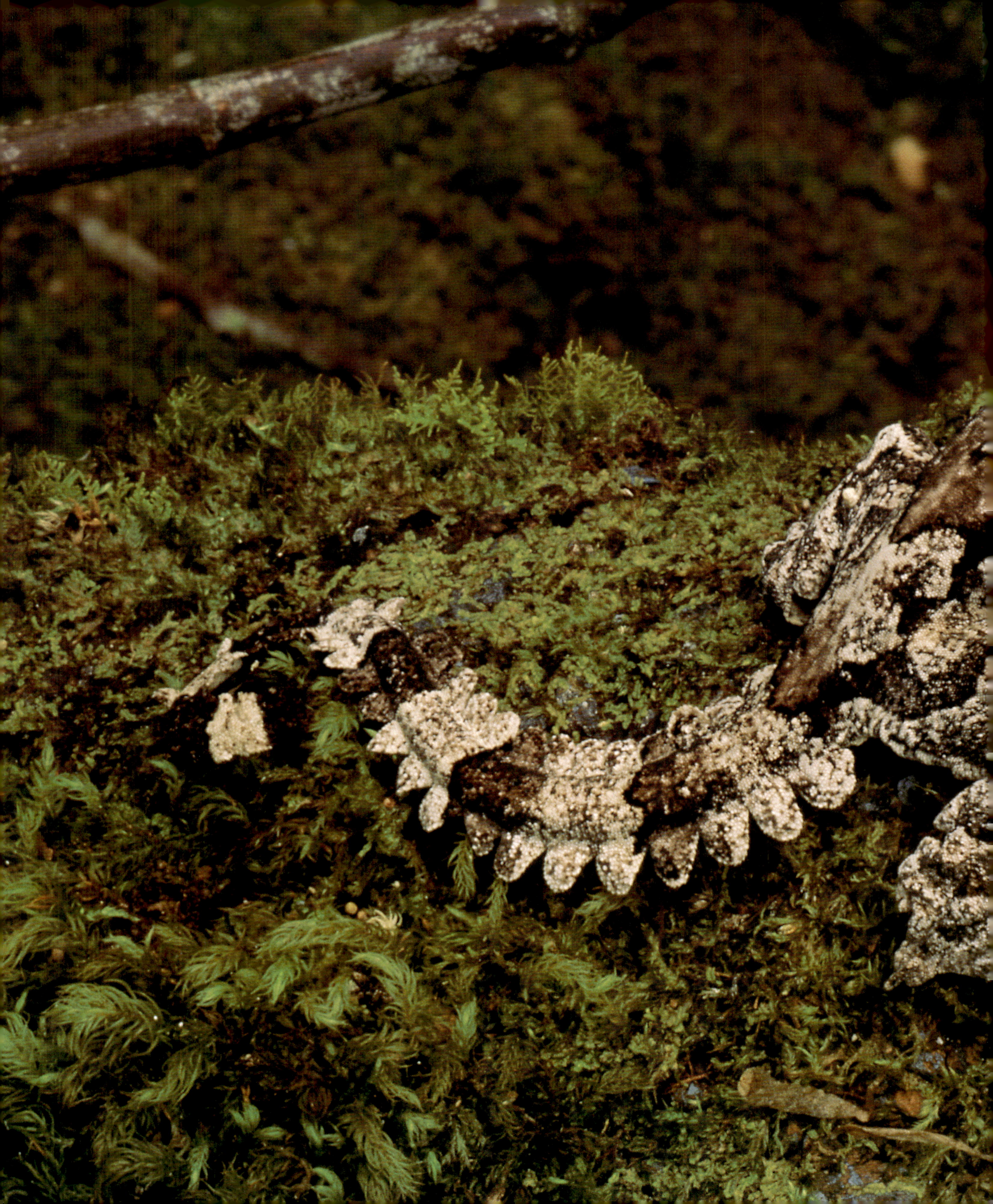

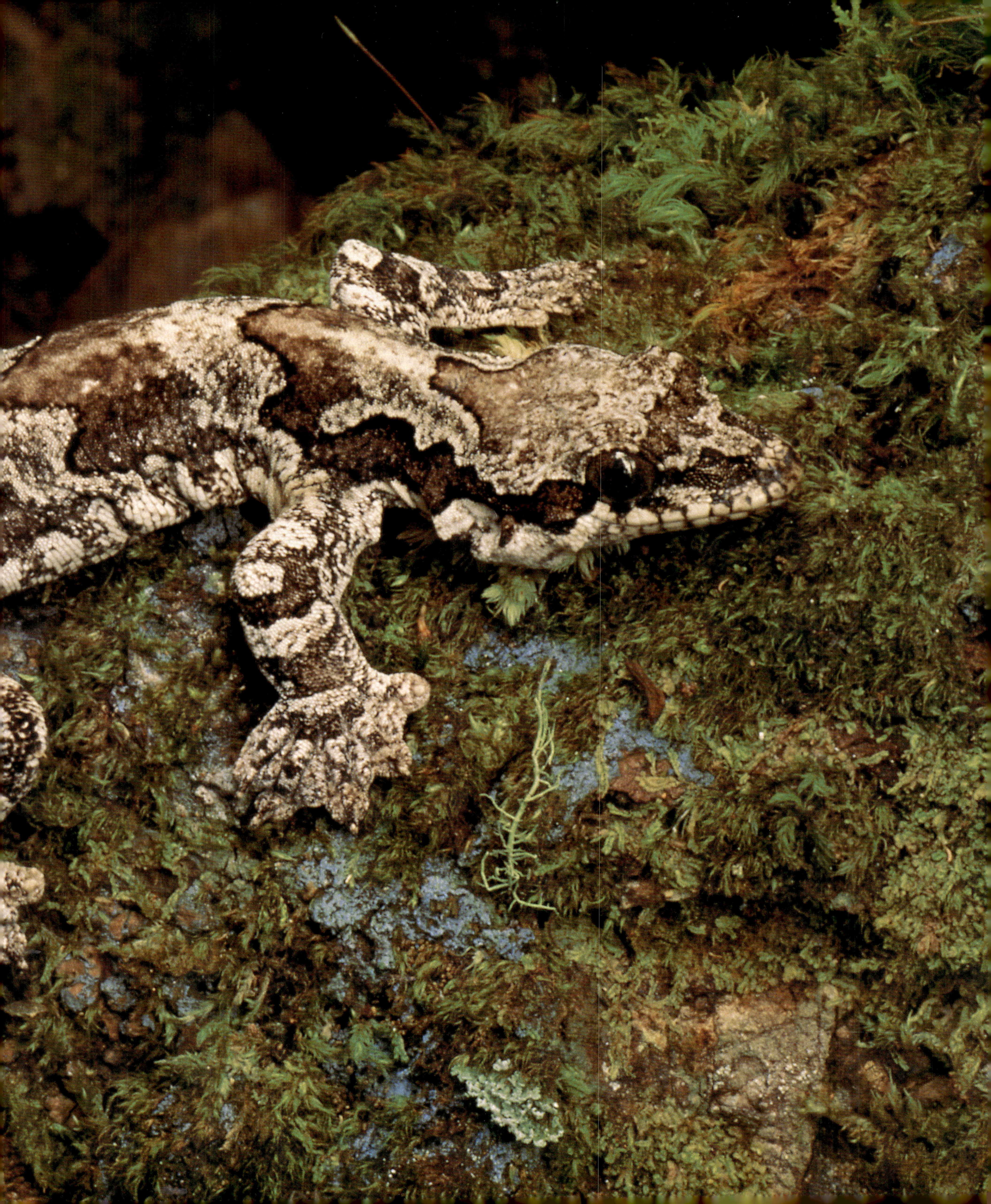

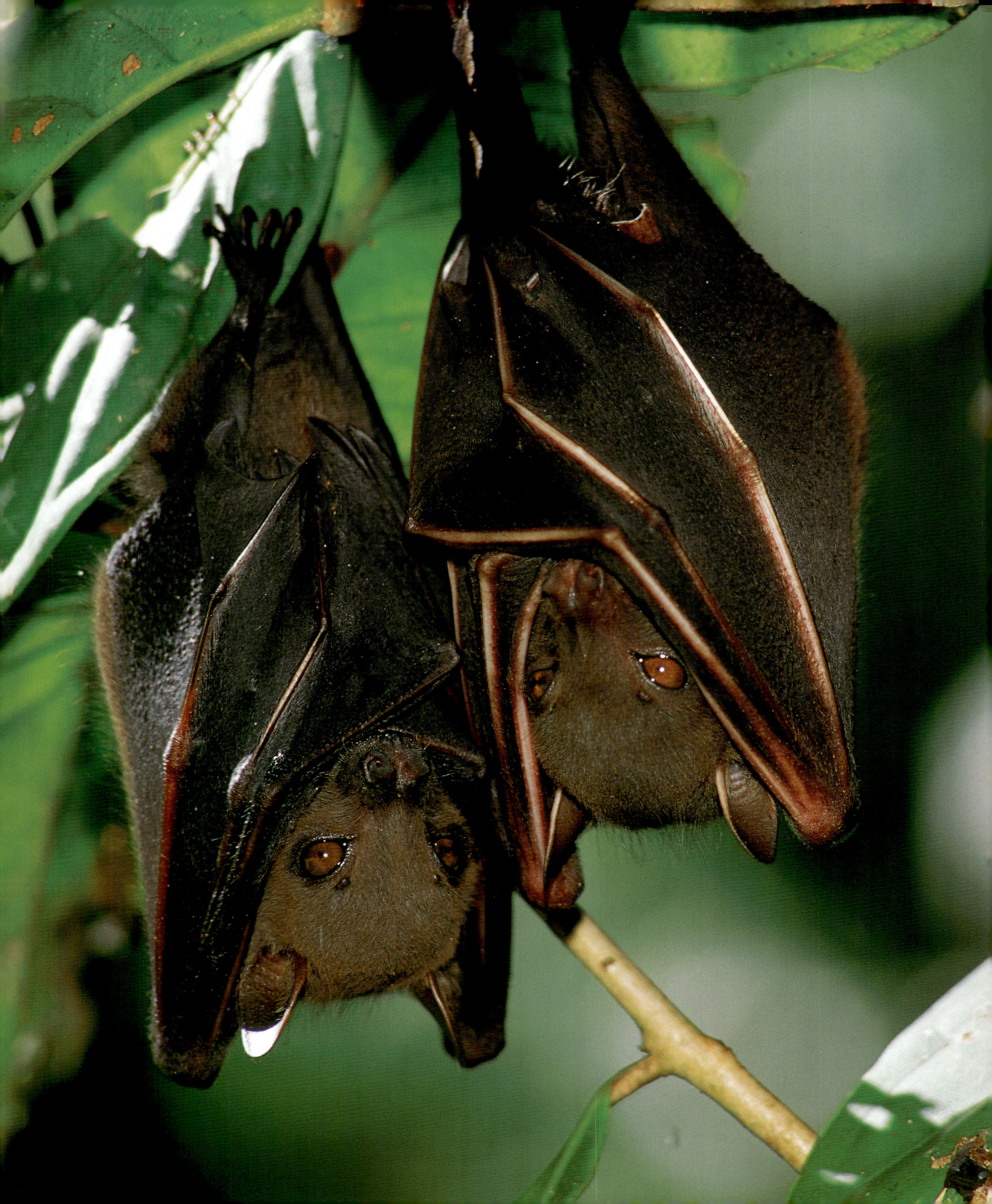

coral reefs include about 240 endemics, clearly establishing its place as one of the top marine megadiversity countries. There are at least 21 species of marine mammals, including sperm whales, Bryde's whale (a baleen whale that reaches 15 m in length), a large number of porpoises, and dugongs. Unfortunately, destructive fishing methods (including dynamite, poisons, and dragging), combined with overfishing by both subsistence and commercial fishermen, have badly damaged the coral reefs; a study in 1981 by the University of the Philippines found that only 5.5% of the nation's coral reefs could be described as being in *excellent condition*, and more than 70% in *fair or poor condition*.

Mud flats near the mouths of rivers and at the heads of bays were originally covered by mangrove, dominated by the genera *Brugiera*, *Ceriops*, and *Rhizophora*, with at least 83 additional species of plants as part of the community. This was a crucial breeding habitat for many species of marine fishes, where the influx of organic material and fresh water from the rivers, combined with protection from predators and the waves of the open sea, provided ideal conditions for the development of fish-fry, and also provided excellent habitat for some shorebirds. Virtually all of this habitat is now gone, the wood cut for charcoal and the land converted to fish or prawn ponds, or taken over by urban sprawl.

Once covered by tangles of pandans (*Pandanus*), fast-growing trees, and grasses before giving way to succession to forest, the beaches have now been almost entirely settled by humans. Given the enormous coastline of the country, this habitat was once quite extensive.

Portions of the Philippines that lie on limestone outcrops sometimes support a distinctive type of forest dominated by the molave tree (*Vitex parviflora*). Since leguminous trees are also often present, this type of forest is also called *leguminous forest*, as well as *limestone forest*. These limestone regions are often extremely rugged, giving rise to karst topography —a seemingly endless series of treacherously sharp limestone ridges, cliffs, and small caves. Because the limestone is porous, rainfall drains away quickly, so that the molave forest is often quite dry during the two- to four-month dry season. The molave trees tend to grow in pockets of deep soil and to be widely separated, with the intervening spaces taken up by the leguminous trees and bamboos. Although the rugged nature of the terrain tends to help protect this type of forest, the molave trees are much sought after for lumber, and so the habitat has gradually been greatly reduced in area.

On the high plateau and valleys of north-central Luzon at 900-1 500 m, a vast pine forest now grows, dominated by the species *Pinus insularis* and *P. merkusii*. The pines are usually widely spaced, with areas between them dominated by grass. Although pine forest is native to the region, cutting of native broad-leaved trees and annual burning of the grasses has favored the spread of the pine forest. This part of the country has a marked dry season, due to its relative isolation from the sea, but heavy rains fall at times, and fog is common during much of the year. This habitat is home not only to one of the giant cloud rats mentioned below, but also to such seemingly out-of-place birds as bullfinches (*Pyrrhula leucogenys*) and crossbills (*Loxia curvirostra*), as well as the Ifugao and Bontoc peoples who built the world-famous rice terraces of northern Luzon.

Rain forest dominated by trees of the family Dipterocarpaceae once covered most of the Philippines from sea level to about 1 000-1 400 m. Massive trees with huge buttresses, reaching over 50 m in height and 2 to 3 m in diameter, dominated a community that once provided habitat for most of the species of the country. The weathered volcanic soils that support this type of forest are fertile, leading to a profusion of species and high biomass above ground. This forest is usually evergreen; only in a few seasonal regions do any of the trees drop their leaves synchronously. In addition to dipterocarps, giant fig trees are conspicuous and provide much of the fruit that supports the giant bats, parrots, monkeys, and dozens of other species. This forest type reached its best development on the lowland plains of the larger islands, where heights of 30-40 m were common; in the low hills, many of the same species were present, but the trees tended to be smaller, often reaching only 25-30 m. Due to the economic value of dipterocarps, they were cut as lumber for export as early as the mid-1800s, and became an economic mainstay of the American colonial government in the first half of the twentieth century; they were especially heavily plundered during the Marcos administration. And the fertile plains were of great value for agriculture, and so were not allowed to regenerate to forest. Today virtually all the forested plains are gone, and most of the forest on the hills as well. Primary lowland forest probably now covers only 3-4% of its original area, although larger tracts of secondary growth are present.

Beginning between 1 000 and 1 400 m on most mountains, lowland forest grades into montane forest (called lower montane forest by some botanists) dominated by oaks. Trees may reach 25 m in height, but 12-20 m is more common; few have buttresses. Rainfall at this elevation is usually about two to three times that at sea level, leading to an increase in epiphytes such as orchids and moss. Leaf litter, which is almost absent in lowland forest, is common, and often the canopy is somewhat broken by the steepness of the terrain, so that ground plants are much more common than in the lowland forest. Although montane forest originally covered only a small portion of the country —perhaps 10-12%—, its limited economic value has resulted in less being destroyed, and it may now represent half of the primary forest remaining in the Philippines.

At elevations from about 1 200 to as high as 1 700 m, the forest undergoes another dramatic change. As one reaches higher into the mountains, the temperature drops, annual rainfall continues to rise, and usually the steepness of the terrain increases as well. Tree height declines to as little as 2-4 m, moss grows so thick that it covers not only stones and the trunks of trees but twigs

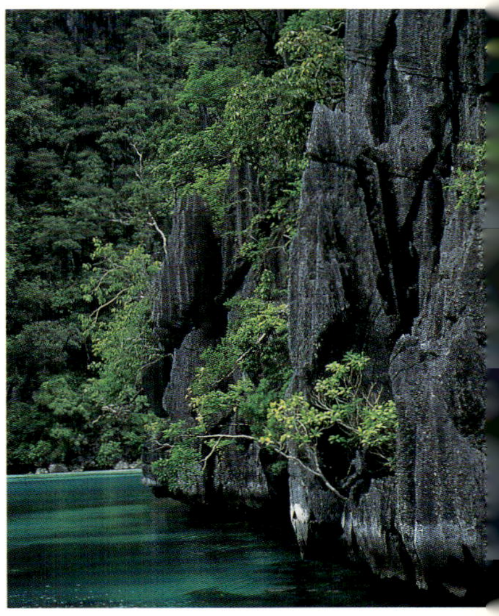

On the opposite page, a short-nosed fruit bat (Cynopterus brachyotis) *found at 1 100 m on Mount Apo, Mindanao.*

Above, a Palawan peacock pheasant (Polyplectron emphanum).

and parts of leaves as well, orchids appear in tremendous profusion, and pitcher plants become one of the most common vines. At its best, mossy forest is a miniature world of deep jade green, with tree trunks covered by moss twice as thick as the trunk, and where roots and slowly decaying vegetation can accumulate several meters above the ground —leading to great uncertainty as to where the "ground" really is. Rainfall can be as high as 12 000 mm per year, and even during the dry season mist and fog are so thick that over a centimeter of water can fall each night from the strands of moss that hang from the trees and become oversaturated. It is here that levels of endemism are the highest, at least proportionately: in a recent survey of mossy forest on Mt. Kitanglad on Mindanao, all of the mammals were endemic to the Philippines, and most to Mindanao (including two previously unknown species); all 16 of the bird species endemic to Mindanao were present, and many thought to be rare were actually quite common. In addition, there were 10 species of earthworms, all of them endemic and all undescribed species.

Perhaps 6-8% of the Philippines was originally covered by this type of vegetation. Since mossy forest is of no direct economic value (although hugely valuable for watershed protection), it is probably the least disturbed habitat in the country, accounting for about a third of the primary forest that now remains.

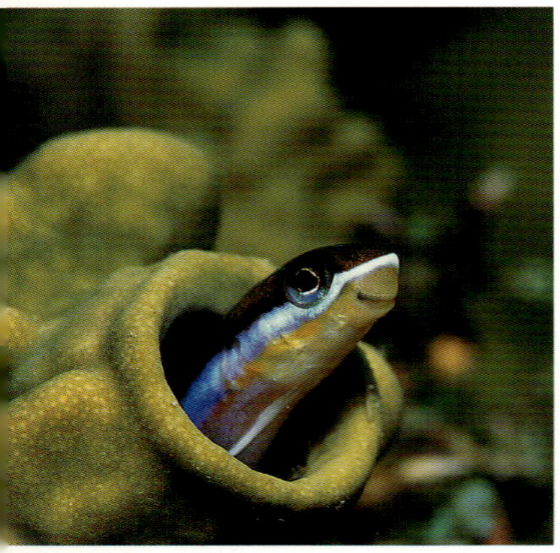

A neon-cheeked blenny (Plagiotreinus rhinorhynchos) *from the Tubbataha Reef off Palawan, probably the most important reef system in the Philippines.*

On the opposite page, a green turtle (Chelonia mydas) *glides gracefully through the water off Balicascag Island.*

Flagship species

Isolated island groups throughout the world typically lack large mammals, and often have dwarfed versions of the few large mammals that are present. Additionally, isolated islands usually have few carnivores, which tend to be small. The Philippine fauna is typical in both of these respects: large, flashy species are few and far between. What makes the Philippine biota so special is the tremendous diversity of small species, each living its life in a unique fashion. Some are quite unusual, and a few verge on the bizarre and attract a good deal of attention. Nevertheless, there are some large, even spectacular species that are present, and these are often the species best known to the people of the Philippines. Following are described some of both groups: the large and dramatic, and the small and remarkable.

Perhaps the most widely-known wild animal within the Philippines is the Philippine eagle (*Pithecophaga jefferyi*) which, as the second largest eagle in the world, forms a dramatic exception to the rule of predators on islands being small. These huge birds live only in primary lowland rain forest, where they feed on flying lemurs or colugos (an odd group of gliding, leaf-eating animals confined to Southeast Asia), snakes, flying squirrels, and occasional monkeys. From what little is known of them, they are monogamous and very long-lived, breeding only once a year. At times they are remarkably inconspicuous, sitting on a perch at the head of a valley while they watch for prey, only occasionally soaring long distances. The territory of a given pair is thought to be 20-50 km^2, with a nest usually built in a high treetop where visibility is very good. Incubation lasts about two months, and brooding for nearly four.

With such a low reproductive rate and such a large home range, it is not surprising that the destruction of lowland rain forest has hit the Philippine eagles especially hard. Once found on most of the large islands of the country, they are now believed to be extinct on all but Luzon and Mindanao, where the only large tracts of forest remain. Even worse, it is the lowland forest that the eagles require, for their prey live only there, and lowland forest has been the most devastated habitat in the country. The total population of these eagles in 1996 is thought to be no more than 30 breeding pairs, most in inaccessible mountains on Mindanao, and only a few in the Palanan Wilderness Area of the Northern Sierra Madre Mountains of northern Luzon. Despite prolonged effort, attempts to breed them in captivity have been largely unsuccessful. Protection of the remaining forest habitat is clearly their only hope. Fortunately, their remaining habitats are excellent places for the conservation of other species, and these places also typically form critically important watersheds for coastal cities. Since the Philippine eagle is the nation's top carnivore and most dramatic and widely-known species, its survival or extinction is watched as one of the most important indicators of the beginning of environmental collapse in this country.

Hornbills are one of the more conspicuous groups of birds in the Philippines because of their often large size, boisterous calls, and prominent place in traditional folk legends. Their English name comes from the hollow, inflated, and usually brightly-colored structures on the top of the huge bill that serve as resonating chambers for their calls. Their feathers are preferred by the few remaining hunter-gatherer groups in the country for fletching their arrows, and often adorned traditional headdresses. Hornbill females are walled with mud into a hollow cavity in a tree by their mates while they incubate the eggs, seemingly as a strategy to exclude predators; the females are then fed through a narrow opening by the males. The largest species is the size of a small turkey. There are five species of hornbills in the country, all listed as endangered or critically endangered because of habitat loss and overhunting. Many islands have quite distinctive populations, and some of these may be different species.

The twelve species of parrots in the Philippines once filled the skies over forests and traditional villages with bright, noisy flocks. Most spectacular is the Philippine cockatoo (*Cacatua haematuropygia*), a bright white bird with vermilion pink under the tail coverts and a splash of yellow on the inner tail and flight feathers. As noisy raiders of crops, especially corn, these birds also figure prominently as troublemakers in traditional folktales. Once present in large flocks on nearly every island, they have been vastly reduced by hunters seeking birds for the pet trade, with thousands leaving the country

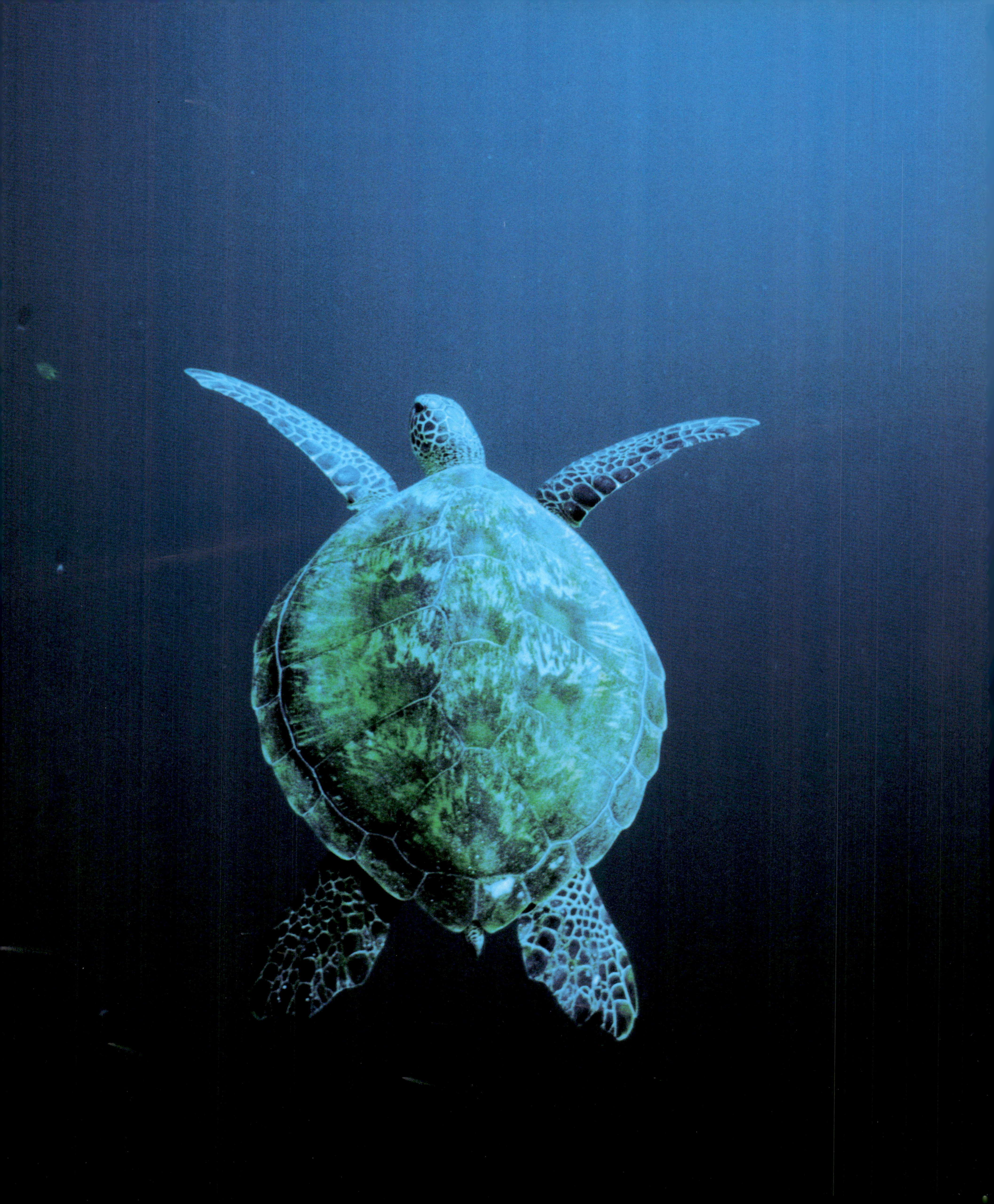

every year during the 1970s and most of the 1980s, to the extent that, in over 15 years of fieldwork, Lawrence Heaney has yet to see one in the wild. The smaller parrots are nearly as actively sought, but their smaller size and higher rate of reproduction have made them slightly less vulnerable. The acrobatic antics of the bright blue and green racquet-tails in flying through wreaths of mist in a cloud forest is one of the most unforgettable sights in the country.

The largest and most impressive of the flagship mammal species in the Philippines is the tamaraw (*Bubalus mindorensis*), the dwarf water buffalo of Mindoro. Although they are indeed dwarfs, standing only about half as tall as a domestic water buffalo, an adult may weigh 180 to 230 kg. They were traditionally viewed as being very fierce, turning to charge on hunters, or even to circle around and take hunters from behind. This reputation served them poorly in this century, as hunters attempting to demonstrate their own courage by killing tamaraws reduced their numbers from roughly 10 000 at the turn of the century to only 100-300 today. Habitat destruction also played a major role in this decline. Once entirely covered by rain forest, Mindoro is now one of the most severely deforested islands in the country. Since this island is one of the nation's Centers of Endemism, efforts to protect the tamaraw could help to protect habitat for all of the Mindoro endemic species, but habitat-protection and captive-breeding programs have been entirely unsuccessful so far.

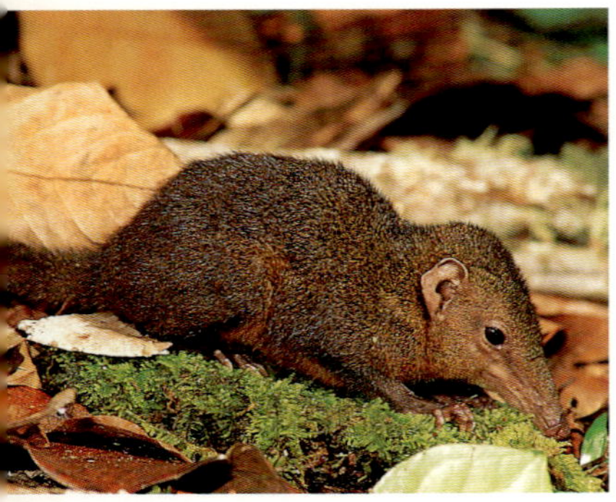

Once considered primitive primates, the tree shrews are now placed in their own order, the Scandentia. This species, Urogale everetti, *was photographed at 1 100 m on Mount Apo, on the island of Mindanao.*

One of the most beautiful species in the country is the Visayan spotted deer (*Cervus alfredi*). Related to the sambhar deer of India and Southeast Asia, the Visayan spotted deer is also a dwarf, with an adult male standing only 71-76 cm at the shoulder and the female substantially smaller. Their English name reflects their unusual trait of retaining as adults the spots present on fawns —only a few other species of deer share this feature. Recognized as a valid species only a decade ago because the deer of the Philippines had been so poorly studied, this deer was immediately placed on nearly every international list of endangered species because of its precarious status. Once found on four moderately large islands and one smaller island in the center of the country (in the Negros-Panay faunal region), they are now extinct on three of the islands and barely surviving in the wild on the other two, with a total population in the wild of probably only a few hundred. Fortunately for the deer, a captive-breeding program begun at Silliman University on Negros Island has been quite successful, sparking similar programs at four other institutions, so that the total captive herd is now up to nearly 100. Unfortunately, little effort has been made to either protect habitat or end hunting; Negros and Panay, where the few wild deer remain, are among the poorest islands in the country, with huge numbers of impoverished subsistence farmers living in steep, upland areas that formerly supported populations of the deer. Protecting the deer requires some meaningful improvement in conditions of rural poverty.

The Philippines is one of only two countries in the world with three species of wild pigs (the other being Indonesia), and two of these are endemic. One of the endemics, the Visayan warty pig (*Sus cebifrons*), is found in the same area as the Visayan spotted deer, and has also been severely impacted by deforestation and hunting. Their higher rate of reproduction makes these animals better able to survive, but they suffer from an additional problem: when populations drop low enough, they begin to hybridize with free-ranging domestic pigs, and rapidly lose their characteristic features and genetic distinctiveness. Protecting habitat, reducing hunting pressure, lessening rural poverty, and managing feral domestic pigs all become intertwined in a complex fashion, demonstrating the need for coordinated social and conservation programs.

Some of the most dramatic species in the Philippines are the fruit bats; with 26 species, this country has one of the largest fruit bat faunas, with eight species of giant flying foxes among them. The largest of these is becoming the new "poster child" for Philippine conservation; the golden-crowned flying fox (*Acerodon jubatus*) has a wingspan up to about 1.7 m, and weighs up to approximately 1.4 kg. Although some species elsewhere have slightly larger wingspans, these are more heavily-built bats, and are probably the world's largest flying mammals. They often fly 20-30 km in a night, seeking out fruit (especially wild figs) in lowland forest. Reports from the late 1800s and early 1900s document the presence of fruit bat colonies of the two largest species of 100 000 to 125 000 on nearly every large island in the country; their slow, steadily stroking wings carried them over every village and town, giving rise to folk legends of spirits that carry away children and impregnate careless maidens. Today the larger species are virtually gone, represented by scattered colonies of a few hundred —only the few largest wilderness areas have colonies that number in the thousands. Hunting and destruction of lowland forest have reduced the endemic species in particular since they appear to survive only in primary forest, although some of the non-endemic species have been able to survive by raiding orchards. One species from the island of Mindoro, discovered only a few years ago and now in the process of being formally described, may be the most severely endangered species of bat in the world.

The Philippine islands have surprisingly few species of primates —only three species are known, and two of them (the slow loris and the crab-eating macaque) are widespread elsewhere in Southeast Asia. The one endemic species is an appealing little animal, locally called "mago," which bears a striking resemblance to the cartoon character "Mr. Magoo." This is the Philippine tarsier (*Tarsius syrichta*), with large, protruding eyes that it blinks and squints in bright sunlight. These tarsiers are found throughout the Mindanao faunal region in primary forest, secondary forest, and scrubby agricultural areas; this flexibility in habitat use stands them in good stead, for they are currently more widespread and common than most endemic Philippine mammals. They feed on large insects and small lizards, and are able to leap with great accuracy in their efforts to catch their prey. Because they are widespread and well-known (and often kept as pets), they are one of the traditional flagship species in the Philippines, helping to provide popular support for habitat and wildlife protection in the southern portion of the country.

The seemingly most unlikely flagship group of mammals for the Philippines are the rodents. Although there are some tree squirrels and chipmunk-like animals, most would be considered by a taxonomist to be "true rats" —members of the subfamily Murinae. However, to the eye of most people, they would be bewildering —from tiny, long-snouted animals with minute eyes to something that looks vaguely like a tree-living woodchuck with long, flowing black hair. The diversification of native, forest-dwelling rodents in the Philippines is comparable to or even more impressive than the classic examples of evolutionary biology, Darwin's finches in the Galápagos Islands and the honey-creepers of Hawaii. An astonishing 15 new species of these rodents have been discovered in the last ten years, a figure which, combined with the new bats found, makes the Philippines one of the world's primary sites for discovery of previously-unknown species of mammals.

The few rodents in this group that conform to the classical notion of a flagship species are the giant cloud rats of the genera *Crateromys* and *Phloeomys*. This includes the "woodchuck with flowing black hair," more properly called the Luzon bushy-tailed cloud rat. These animals are apparently endemic to the pine forests of upland Luzon, where they feed on the buds and bark of pine trees. Their relatives, the Luzon giant cloud rats, which feed on tender young leaves, weigh up to 2.5 kg, making them the largest "true rats" in the world —although they look far more like a large, shaggy squirrel. Even these large rodents are still poorly known, with new species described in 1981, 1985, and 1996. Because of heavy hunting pressure and habitat destruction, all of the cloud rats and Luzon giant rats are considered to be rare and declining, and one is feared to be extinct.

One of the most intriguing rodents is a species found only on Mt. Isarog, an extinct volcano that supports the largest patch of primary forest remaining in southern Luzon —an area of less than 50 km^2. On the upper half of the mountain, in montane and mossy forest that receives up to 12 m of rain each year lives a species known as the Isarog shrew-rat (*Rhynchomys isarogensis*). It was formally named only in 1981, on the basis of a single specimen with no habitat information, but was recognized as having an extremely long snout, hefty hind legs, and short, stout forelegs with sturdy claws on the forefeet. When Heaney and colleagues visited the mountain seven years later, they chanced to capture one, and took it back to camp where they tried to feed it food from the forest. Seeds, fruit, and large insects were all rejected after being sniffed by the increasingly hungry animal. Finally someone happened to kick over a stone, and underneath it were some small, squirming earthworms. When one of these was held up for the shrew-rat, it leaped on the worm with gusto; it grasped one end of the worm with its incisors, and used a forepaw to strip the dirt from the outside of the worm. It then swallowed the worm like a long piece of spaghetti. We now know that this species is a specialist on earthworms, hopping on its haunches along trails that it clears through the mossy forest, pouncing on earthworms with its long, narrow snout and using its robust foreclaws to dig out the worms that attempt to escape.

At least another eight species of rodents in the Philippines also rely heavily on earthworms as a food source; some are burrowers, some dig through dead leaves at the surface of the ground, but only one other species pursues its food in the same way as the Isarog shrew-rat. Several of these eight, and many other different, but equally unusual Philippine forest rodents, are like the Isarog shrew-rat in that they now live in tiny remnant patches of forest surrounded by vast, and to them, totally inhospitable agricultural lands. Although these species may have no great impact on the world, they are biologically unique, and serve as good indicators of the health of the rain forest ecosystems. The fact that 19 rodents are listed in the Philippine Red Data List as endangered species indicates that even small-bodied species, which seemingly should be tolerant of habitat disturbance, have been pushed to the brink, and that the ecosystems which support them are in need of rapid and effective protection.

Reptiles are remarkably diverse in the Philippines, with 193 species currently known. About 131 or 68% of these are endemic to the country, making this one of the richest endemic reptile faunas in the world, and several new species are described each year. The ecology and conservation status of most species is poorly known. Among those flagship species known to be endangered is Gray's monitor (*Varanus olivaceus*), which has the distinction of being the only lizard in the world to specialize on fruit and snails. Its range includes only southern Luzon, where it lives in places dominated by rugged limestone cliffs and ridges; the only monitor lizard that occupies a smaller area is the Komodo dragon (*Varanus komodoensis*) in Indonesia. Another large flagship lizard is the sail-fin lizard (*Hydrosaurus pustulatus*), an endemic species that reaches lengths of nearly a meter and weights of over 500 g. The brightly-colored sails and frills of males are used in displays to court females and intimidate other males. They live along unpolluted, shaded mountain streams, feeding on a variety of flowers, young leaves, and a few insects. Both of these lizards are heavily hunted for their delicate flesh, and the sail-fin lizard is exported in large numbers to Europe as pets. Habitat destruction is a serious problem for both the monitor and the sail-fin lizard.

Probably the most severely endangered reptile in the Philippines is the endemic freshwater crocodile (*Crocodylus mindorensis*). It is a relatively small, little-known

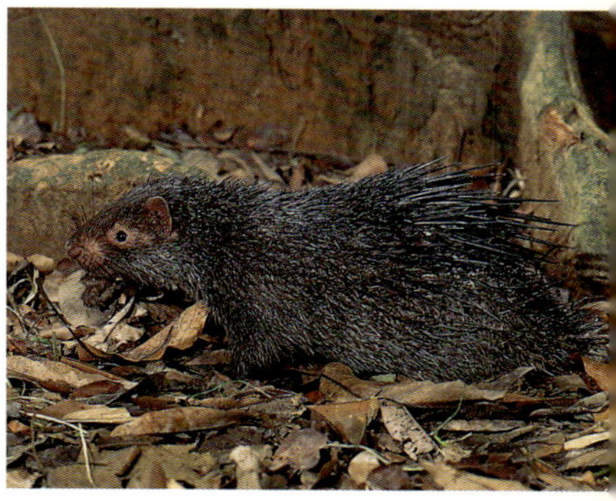

The Philippines has very high mammal endemism, with 115 of 206 mammal species restricted to the country. This is one of the endemics, the Philippine porcupine (Thecurus pumilis).

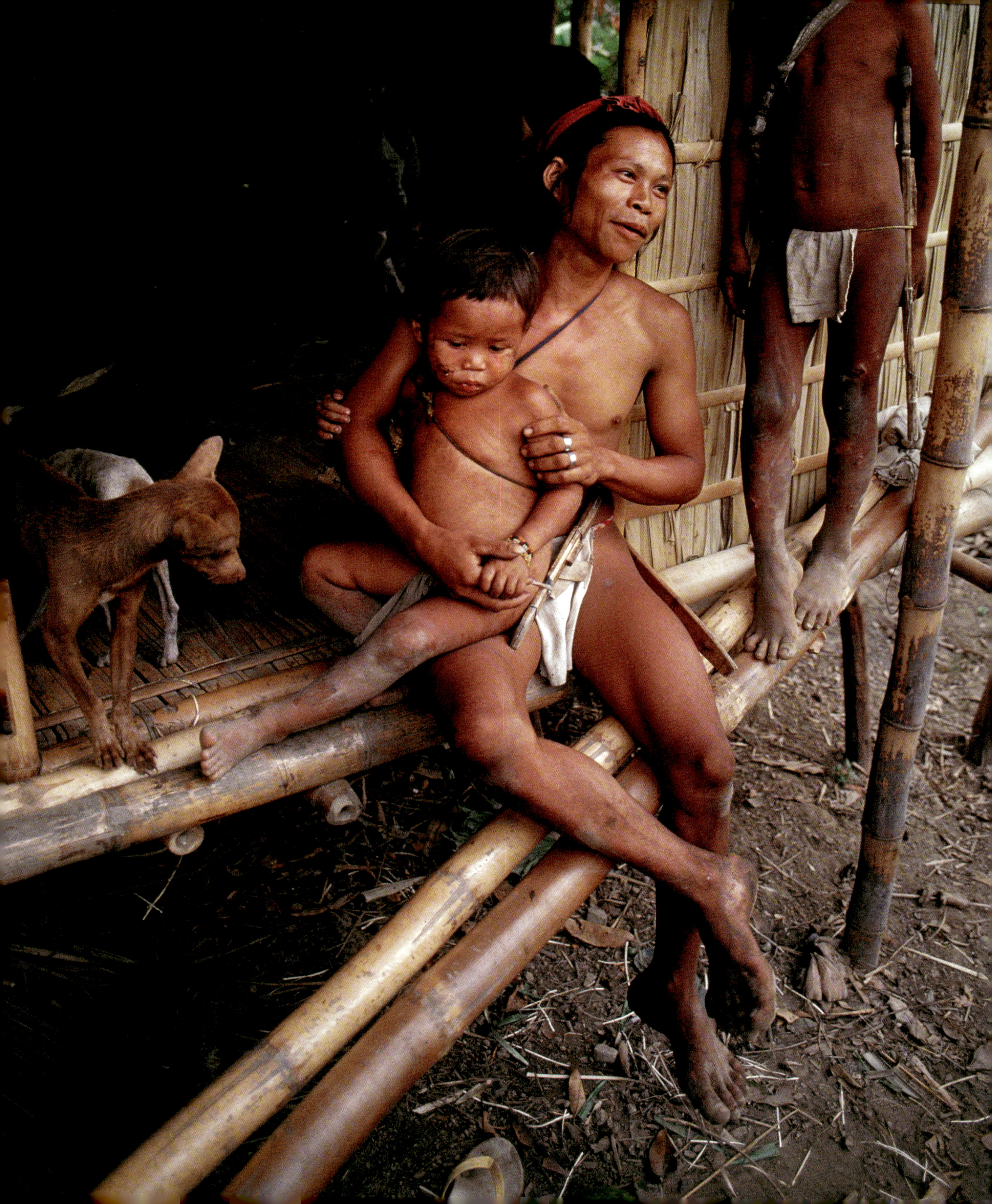

freshwater reptile. Males seldom exceed 3 m, and until recently this crocodile was considered a subspecies of the New Guinea crocodile (*Crocodylus novaeguineae*). At one time it was widely distributed throughout the archipelago, but is now restricted to Luzon, Mindoro, Masbate, Samar, Negros, Busuanga, and Mindanao. Its preferred habitat includes freshwater lakes, marshes, and the tributaries of large rivers and ponds, where it feeds mostly on fish, frogs, snakes and small water birds. The combination of heavy hunting and conversion of habitat to agriculture and aquaculture reduced wild populations to 500-1 000 by 1982, and to only about 100 in 1993. Now the principal threats are habitat loss and killing by local people. There is a large, well-funded Japanese captive breeding project, but thus far it has been ineffective. The human pressures, limited resources for captive breeding, and lack of government concern for crocodile habitat conservation programs are all cause for grave concern about the future of this species.

Cultural diversity

The roughly 73 million Filipinos are composed of a large number of ethnic and linguistic groups that maintain a surprising degree of cohesion, even in the face of Westernization and economic change. About 88% of the Filipinos can read, making this one of the most literate nations in the world, and a large percentage speak English —it is claimed that the Philippines is the third-largest English-speaking nation in the world, following only the U.S.A. and U.K.

Most —about 66 million people— are "lowlanders," people whose ancestors were part of the core of the Spanish colony and have embraced Western values and dress to a large degree. In spite of the Spanish colonial domination, very few people speak Spanish, although some Spanish words have been incorporated into most languages. Virtually all are Christian (Catholic and Protestant), although their beliefs often include aspects of traditional animism. The majority speak English, many from childhood, and often grow up knowing more about African wildlife from T.V. shows than they do about Philippine wildlife. In spite of their Westernization, however, most also speak at least one, and often three or four, Filipino languages, and identify strongly with their family and ethnic group. Cuisine varies between these groups, with Bicolano food known for its hot spices, Tagalog food for its strong Spanish influence, and Cebuano food for its exceptionally fine use of seafood. Understanding Philippine culture and politics requires good familiarity with these inconspicuous but fundamental ethnic units.

Within this society are people of partial Chinese, Spanish, and American ancestry, and these tend to be the best-educated; they often are descendants of the hacienda-owners who formed the elite of Spanish colonial society. They are also the most economically successful members of society; although forming only about 2% of the total population, "mestizos" earn about 55% of the nation's income (as of 1990). Ethnic Chinese are also present, many having arrived within the last 100 years, and often retain the traditions of their homeland.

In addition, over four million people are Muslims; they live in the southern portions of the country and are generally known as Moros. These people were never conquered by the Spanish, and maintain a strong independence movement that occasionally flares up into warfare. They are ethnically closest to several groups in Borneo, and indeed their own territory once included part of northern Borneo. The Moros actually belong to about 12 ethnic groups, each maintaining its own traditions and leadership. The Yakan of Basilan Island are among the best-known because of the beautiful textiles they produce in vibrant colors with geometric designs.

More than two million Filipinos are members of legally-recognized cultural minorities. Composed of dozens of small groups, each with its own language and traditions, these cultural minorities live in the more isolated parts of the country, especially in mountainous regions. Their isolation, and their often fierce independence, largely prevented them from being absorbed into the Spanish colonial system. Most are animists, often with beliefs in spirits of the forest that traditionally led them to protect forest and wildlife. The Ifugao and Bontoc of the Central Cordillera in north-central Luzon produced the magnificent rice terraces, some of which cover entire 600 m-high hillsides that are often listed as one of the "wonders of the world." They also produce beautiful textiles, using black and bright red natural plant dyes. The Mangyans of Mindoro (actually comprised of 12 subgroups) are both farmers (mostly sweet potatoes) and cattle-grazers, whose burning of forest to create grassland is one of the factors that has led to the decline in tamaraw populations.

A diverse set of peoples are grouped as Negritos; all tend to be shorter, darker, and curlier-haired than most Filipinos, and are thought to be the remnant population of the first wave of Asians to reach the Philippines. They are all traditionally hunter-gatherers, often living in close partnership with nearby farmers, with whom they trade meat (mostly wild pig and deer) and forest products for rice and other crops. These people go by many names, Aeta, Ati, Ita, and Eta. The Aeta live in the mountains of Negros, Samar and Leyte, and in the forest of northern Luzon, especially in the Palanan wilderness, while the blowgun-using Bataks are on Palawan Island. It is estimated that as few as 20 000 Negritos survive today, impoverished victims of loss of traditional lands to logging and agriculture who face cultural extinction in the near future if nothing is done.

LAWRENCE HEANEY
RUSSELL A. MITTERMEIER

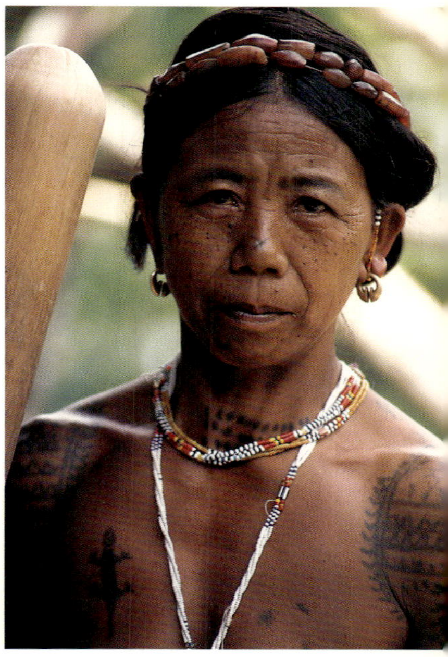

The Philippines has very high human cultural diversity, especially for a country its size. On the opposite page, holding a child, a Mangyan tribesman of Mindoro wears the same sort of loincloth worn when colonists arrived.

Above, a woman from Basao village.

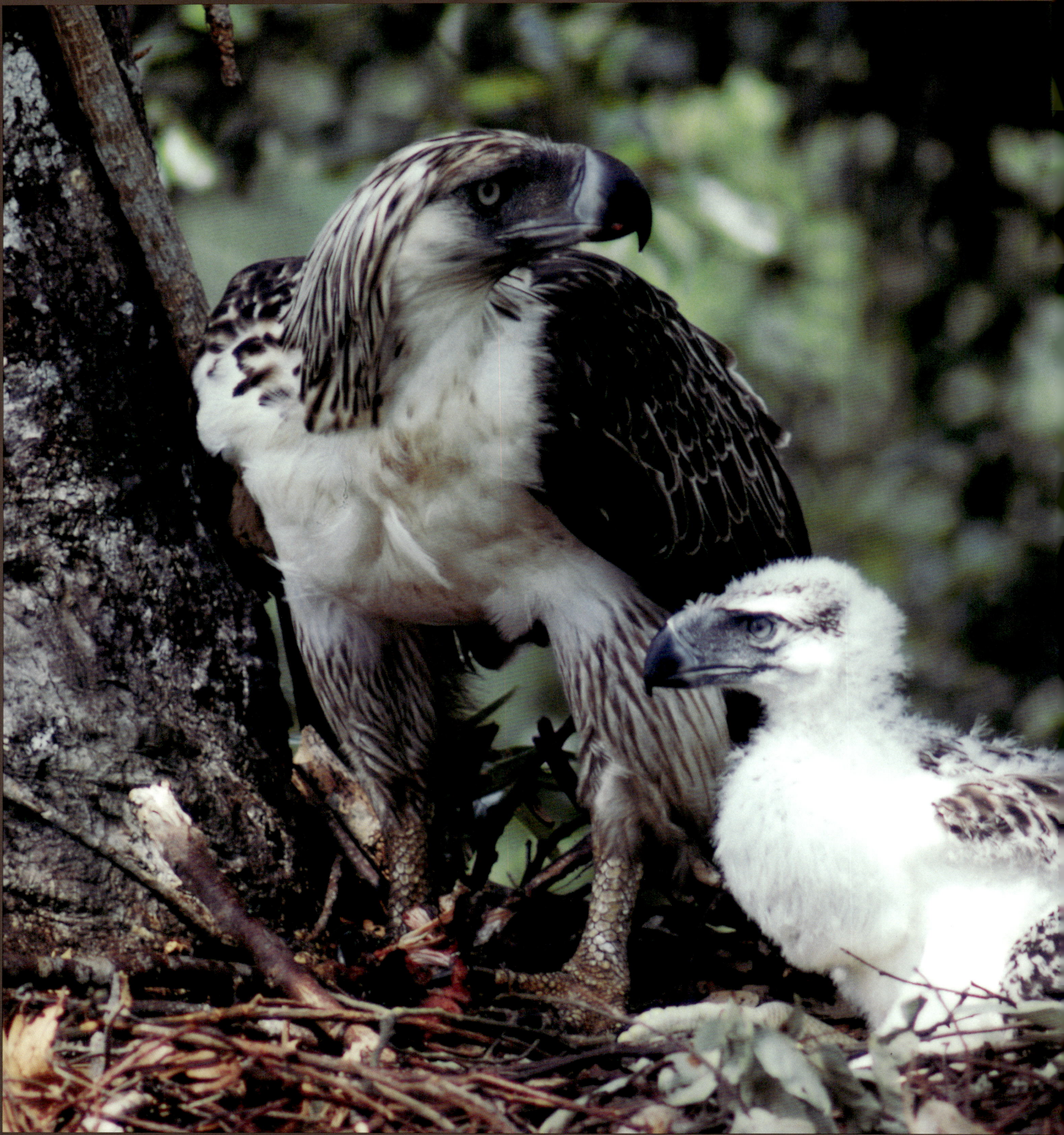

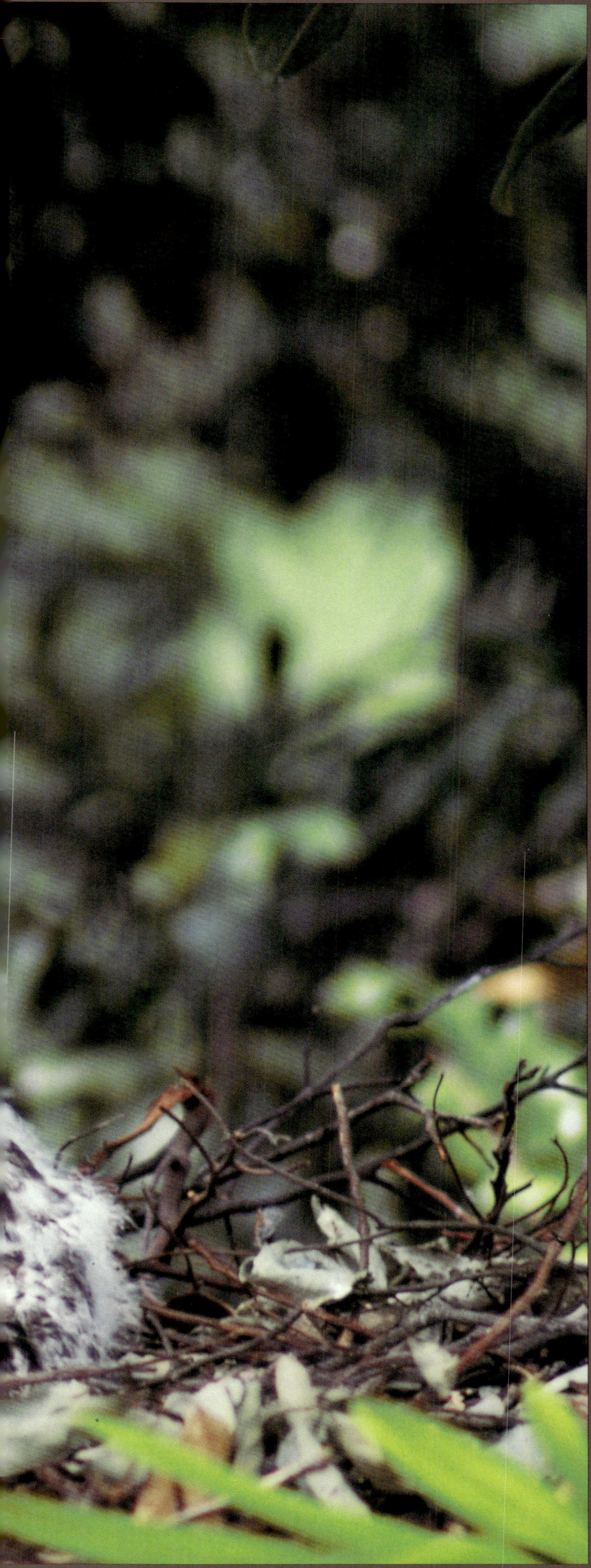

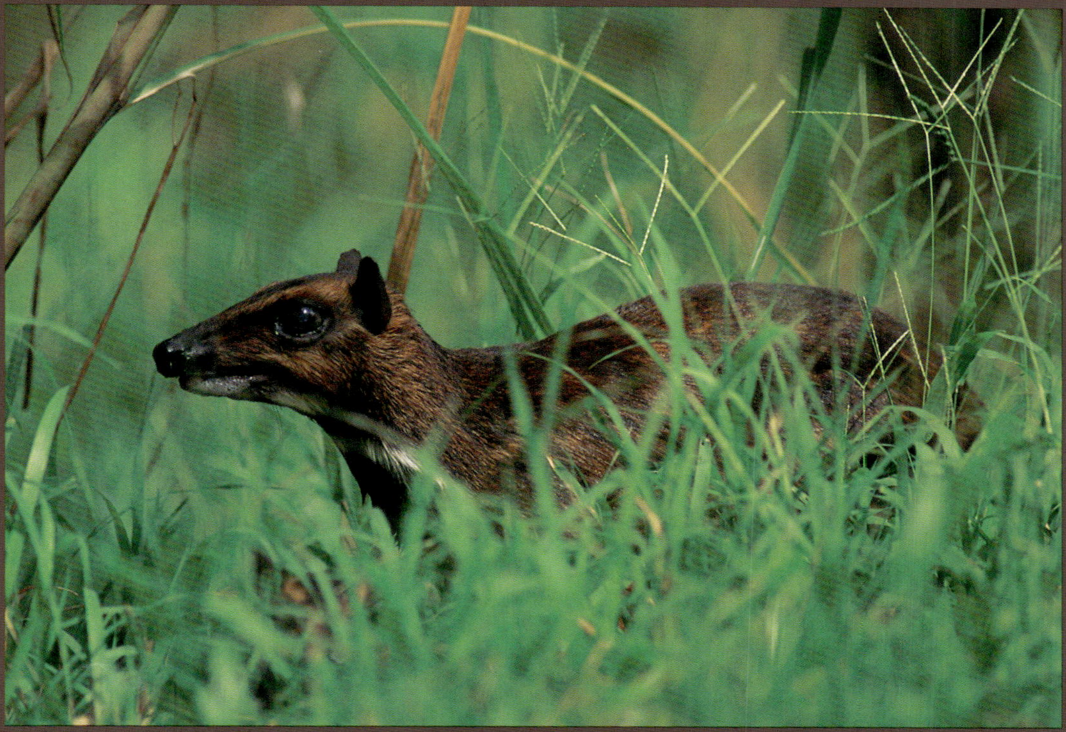

The Philippine eagle (Pithecophaga jefferyi) *is the premier flagship species of the Philippines. The tallest of the eagles, this magnificent species is now critically endangered. Left, a female on a nest with an 8-week-old chick. Above, the Philippine mouse deer* (Tragulus nigricans) *is one of the smallest of all deer.*

The Philippines has a rich bird fauna of 556 species,
of which 183 are endemic, fourth highest in bird endemics.
Below, a buff-banded rail (Rallus philippinensis).
On the right, a Palawan peacock pheasant (Polyplectron emphanum).
Both the rail and the peacock are considered endangered.

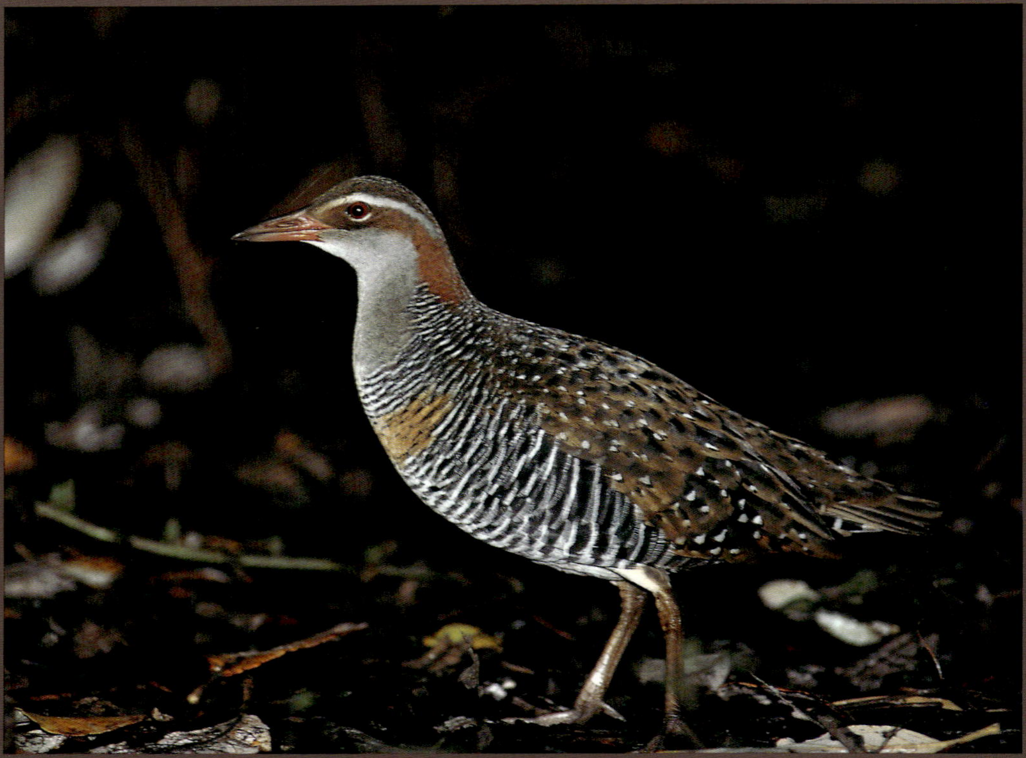

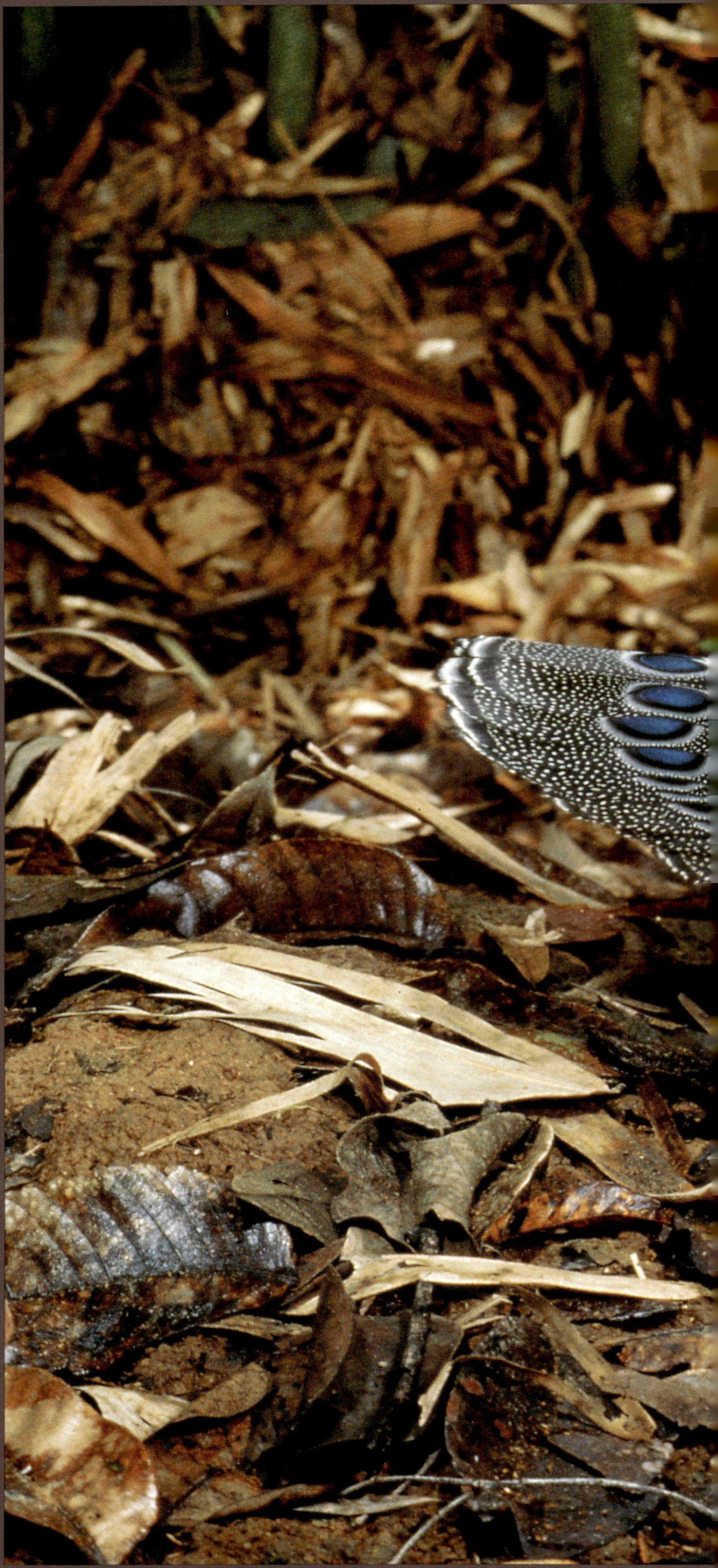

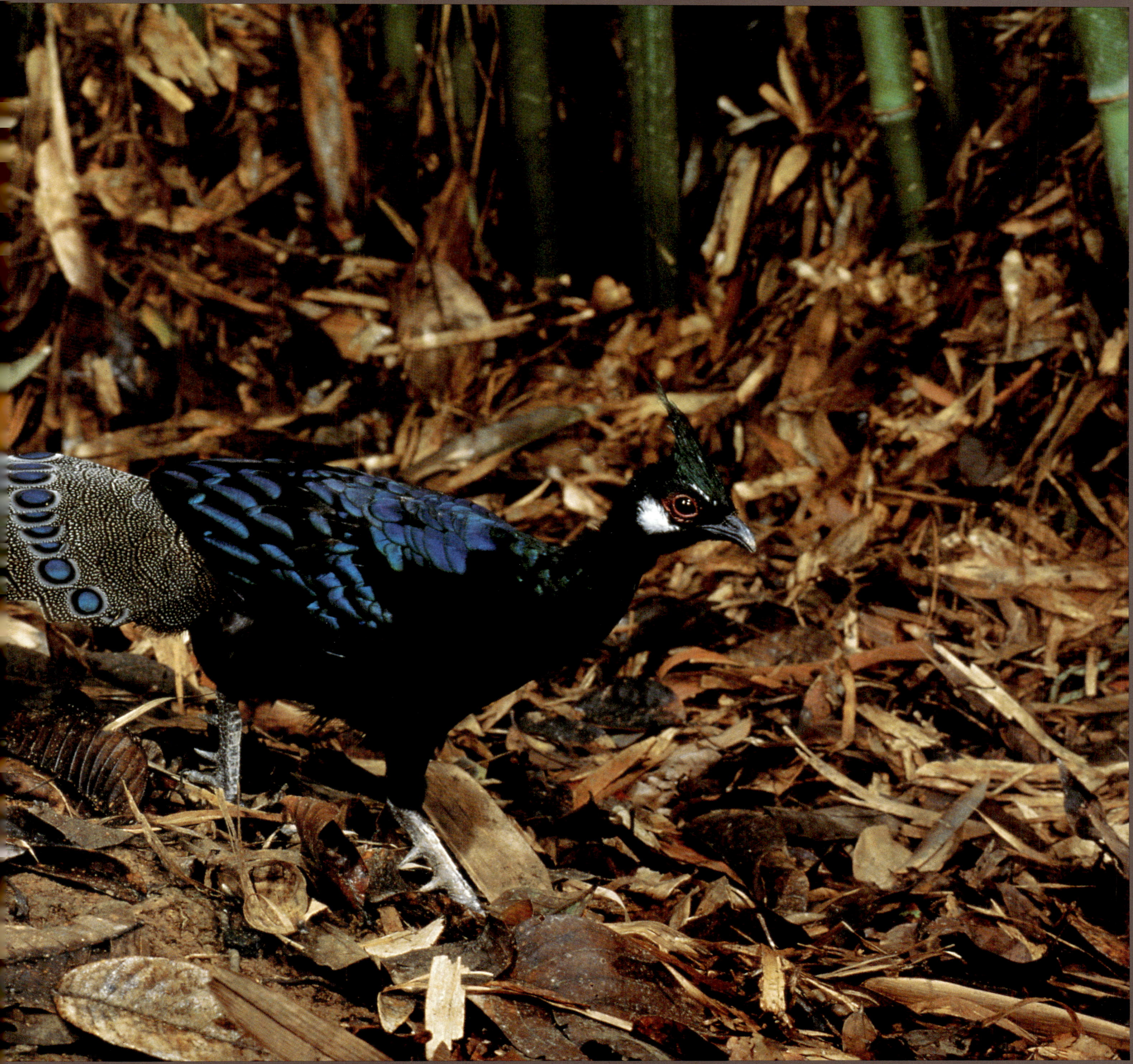

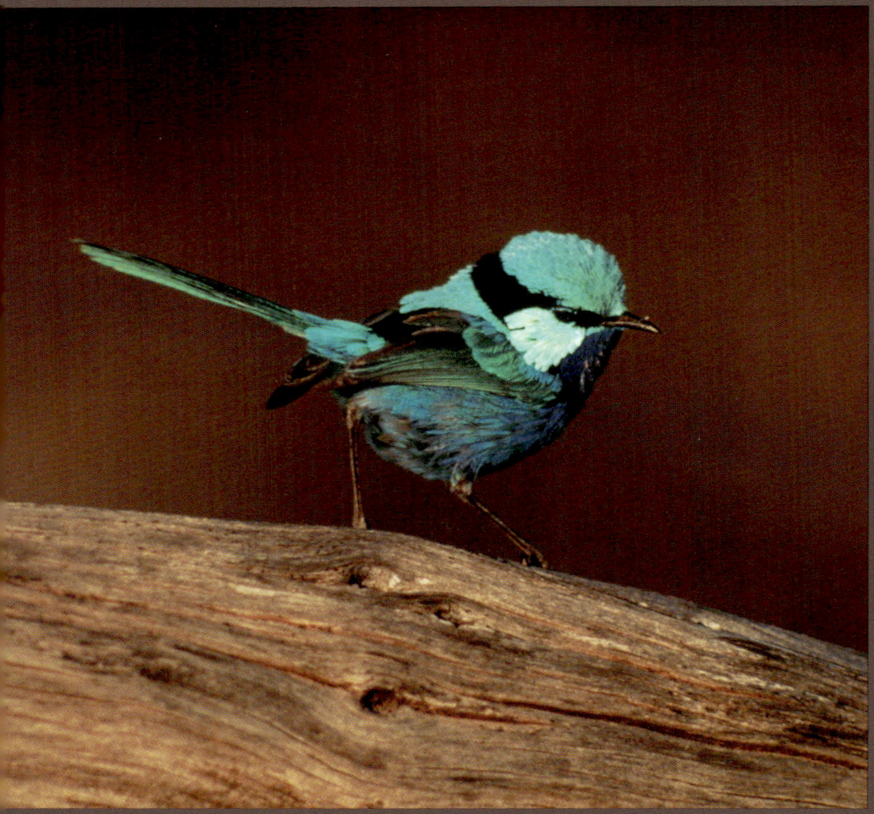

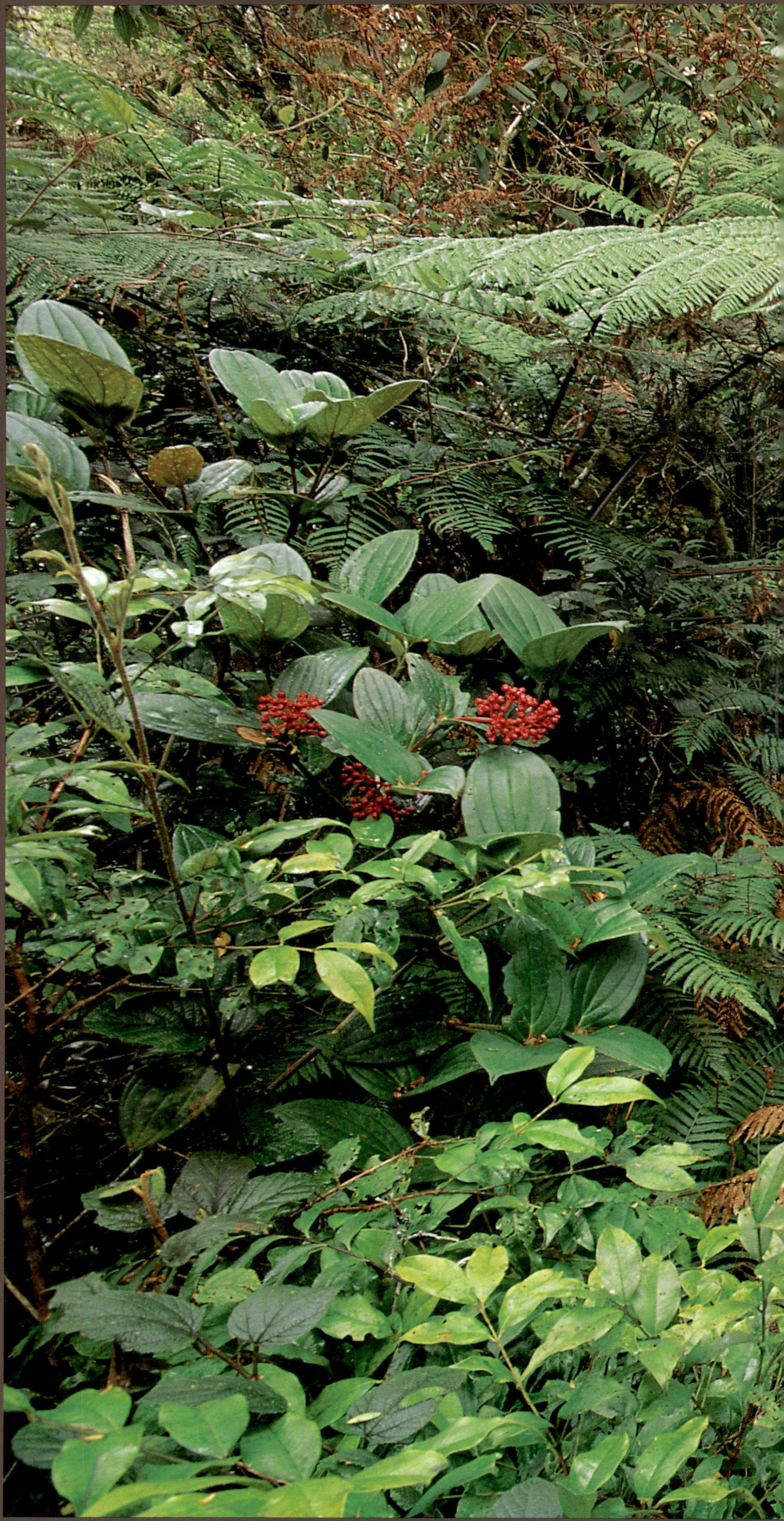

The Philippines is also the fourth highest in threatened birds. Above, a fairy-wren (Malurus splendens callainus); *right, rain forest on the island of Mindanao.*

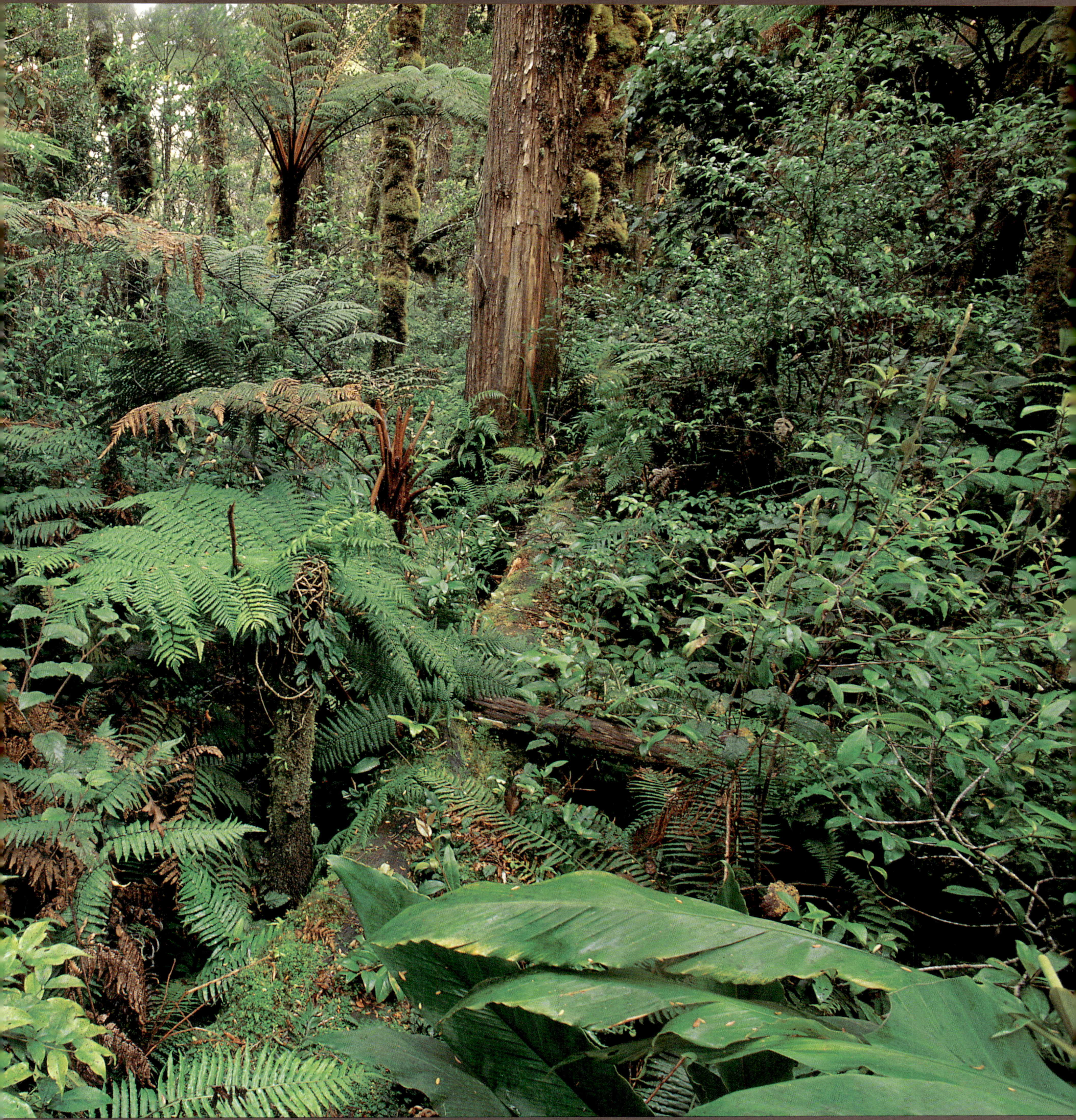

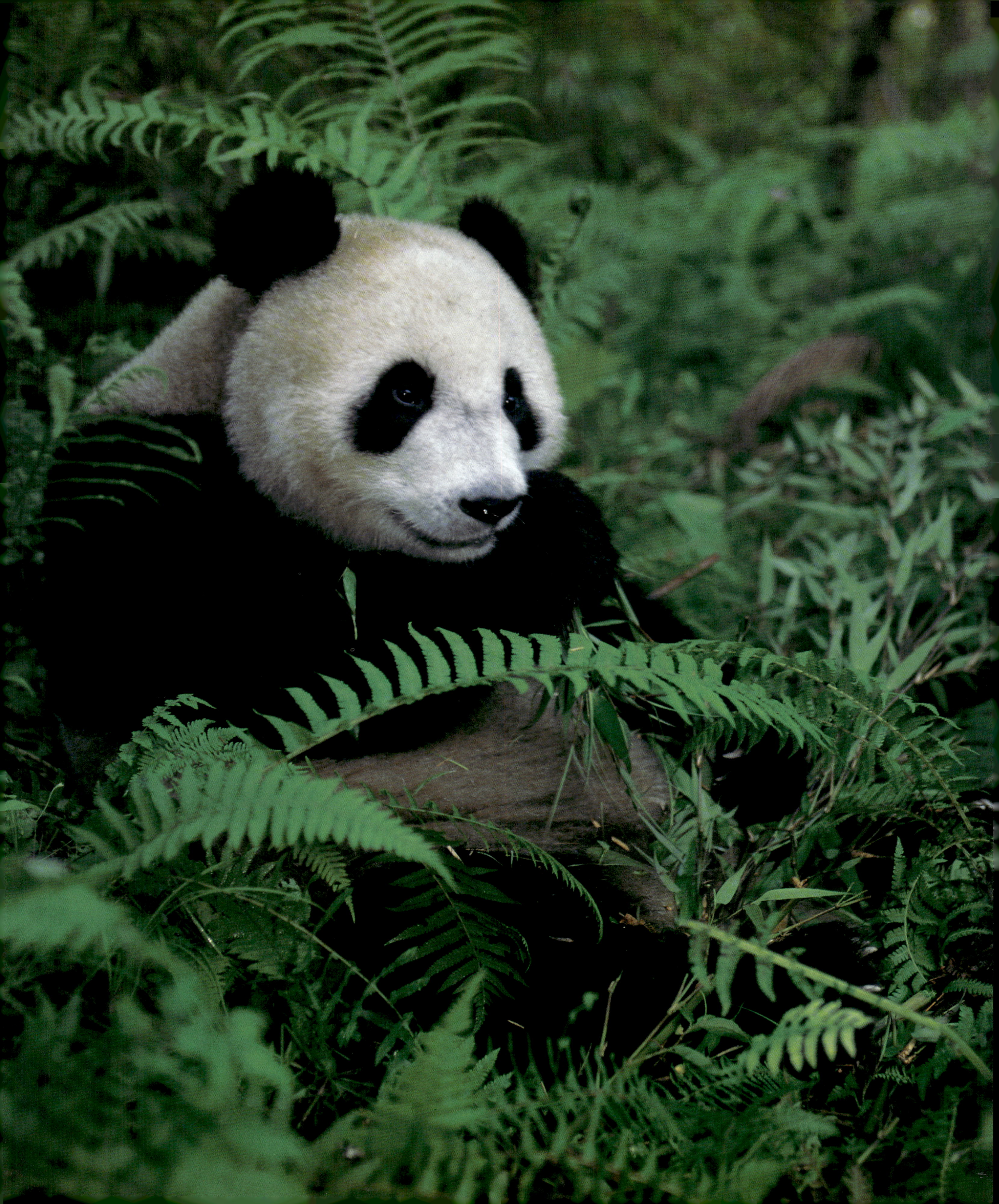

CHINA

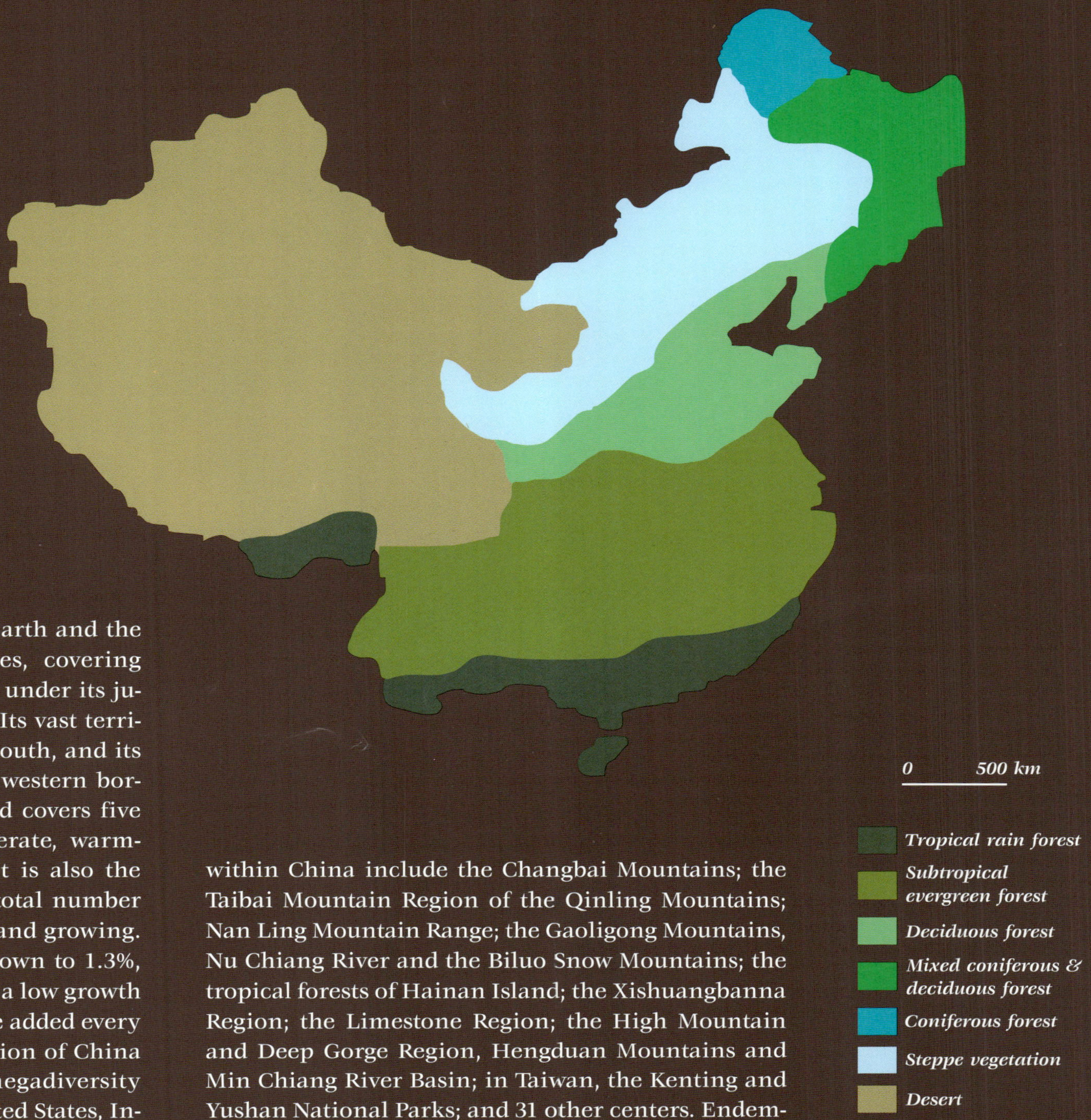

China is the third largest country on Earth and the largest of the megadiversity countries, covering 9 561 000 km^2, with a total marine area under its jurisdiction of more than 3 000 000 km^2. Its vast territory stretches 5 500 km from north to south, and its eastern coastline is 5 200 km from its western border; it spans 50 degrees of latitude, and covers five climatic zones: cold-temperate, temperate, warm-temperate, subtropical, and tropical. It is also the world's most populous nation, with a total number of inhabitants approaching 1.2 billion and growing. Although China's growth rate is now down to 1.3%, with such an immense population even a low growth rate means many millions of people are added every year. If one excludes India, the population of China alone exceeds that of all the other megadiversity countries combined, including the United States, Indonesia and Brazil, themselves sixth, seventh and eighth on the world list of most populous nations.

China is also one of only three megadiversity countries that span two major biogeographic regions, in this case the Palaearctic and the Indo-Malayan realms. As with the other two, Mexico and Indonesia, this results in an incredible mixture of completely different faunal and floral complements and an unusual transition zone between the two regions. Furthermore, there are a number of areas that served as Pleistocene refugia during the glaciations, with a number of rare and unusual species surviving as relicts.

China also includes part of one of the world's top-priority threatened hotspots, the eastern Himalayas, 41 Centers of Plant Diversity and Endemism as identified by IUCN/WWF, and 12 Endemic Bird Areas as defined by BirdLife International (ICBP). The eastern Himalayas threatened hotspot is especially rich in plant endemics, and also includes portions of neighboring Nepal, Bhutan and Myanmar. Globally important Centers of Plant Diversity and Endemism within China include the Changbai Mountains; the Taibai Mountain Region of the Qinling Mountains; Nan Ling Mountain Range; the Gaoligong Mountains, Nu Chiang River and the Biluo Snow Mountains; the tropical forests of Hainan Island; the Xishuangbanna Region; the Limestone Region; the High Mountain and Deep Gorge Region, Hengduan Mountains and Min Chiang River Basin; in Taiwan, the Kenting and Yushan National Parks; and 31 other centers. Endemic Bird Areas within China are the Tibetan Valleys, South Tibet, the Qinghai Mountains, the Central Sichuan Mountains, the West Sichuan Mountains, the South Chinese Forests, the Yunnan Mountains, the Shanxi Mountains, the Fujian Mountains, Hainan Island, West China, and the island of Taiwan.

In terms of species numbers, China ranks very high on the world list, coming out in the top eight countries on Earth in overall diversity. It is especially rich in higher plant diversity, with estimates of the total number of species ranging from 27 100 to 30 000, and approximately 10 000 endemics. Taiwan alone has 3 577 plant species, of which 1 075 are endemic. China also has 301 vascular plant families, three of them endemic (Bretschneideraceae, Davidiaceae and Eucommiaceae), and 3 116 plant genera, of which 243 are endemic. Its total number of higher plant species places China fourth on the world list, surpassed only by Brazil, Colombia and Indonesia, and in plant endemism it ranks seventh.

Vertebrate diversity is also high. In mammals,

On the opposite page, the giant panda **(Ailuropoda melanoleuca)** *is the world's single most popular flagship species, and is endemic to China. Unfortunately, it is also critically endangered, with perhaps less than 1 000 left in the wild in a number of small, fragmented, widely-separated populations. Its survival hangs in the balance.*

China has at least 499 species, good enough for third place after Brazil and Indonesia; of these, 77 are endemic, eighth highest in the world in this category. Bird diversity is 1 244 species, of which 99 are endemic, putting it in ninth position globally in both diversity and endemism. Reptile and amphibian diversity is also very significant, with 387 reptile species and 274 for amphibians, or seventh and fifth, respectively, on the world list. Endemism is comparable, with China's 133 endemic reptiles (34% of the total) again coming in seventh, and its 175 endemic amphibians (64% of the total) ranking fourth behind only Colombia, Brazil and Madagascar. Marine and freshwater diversity is also quite high in China. For example, there are 2 804 species of fish recorded from the country, 440 of them endemic; 1 794 of these are marine and 1 010 freshwater.

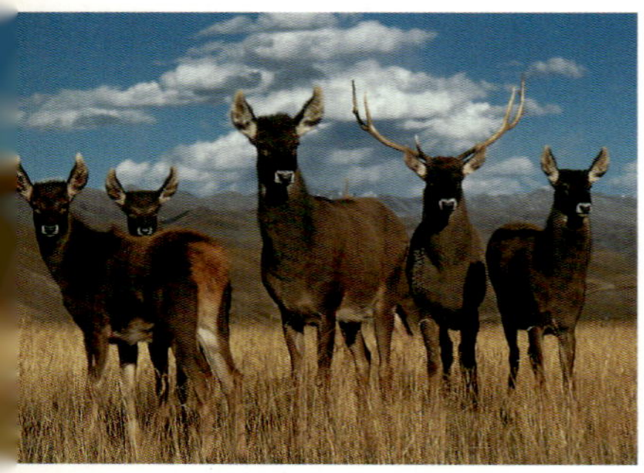

China has high mammal diversity, with 499 species in all, 77 of them endemic. Ungulates are especially diverse. Above, a white-lipped deer (Cervus albirostrus) *from Tibet.*

Not much information is available on invertebrate diversity, but what little is known places China in a very high position globally. In butterfly diversity, its 1 200 species are twelfth on the world list, and its roughly 200 endemics place it in a tie for sixth with five other countries (Brazil, Mexico, Ecuador, Bolivia and Papua New Guinea). Swallowtail butterfly diversity is especially high, with China's 104 species ranking second behind Indonesia and its 15 endemics placing it third after Indonesia and the Philippines. In tiger beetles, China's 94 species put it in a three-way tie for tenth place with South Africa and the Philippines, and its 16 endemic tiger beetles rank nineteenth. Little information is available for other invertebrate groups, but it is reasonable to assume that many species still await discovery.

Not surprisingly for such a large country spanning two different major biogeographic regions, China has tremendous ecosystem diversity, ranging from tropical seasonal rain forest and evergreen broad-leaved forest to snowcapped peaks, and including monsoon forest, semideciduous broad-leaved forest, mixed coniferous and broad-leaved forest, subalpine coniferous forest, elfin forest, boreal coniferous forest, alpine scrub, alpine meadow, alpine tundra, steppe, desert, savanna, and mangroves. Altitude ranges from the Chinese side of Mt. Everest, the highest mountain in the world, to the Turfan Depression, at 154 m below sea level, one of the lowest points on Earth.

With the world's oldest continuous civilization dating back some 5 000 years, China also has a long history of agricultural development, and is one of the eight original centers of crop diversity. Of the roughly 1 200 species of cultivated crops in the world, about 200 originated in China, accounting for about 17% of the total. There are more than 600 species of domestic plants and over 100 000 varieties, or cultivars. China also has many varieties of domesticated animals and birds. According to 1989 statistics, there were over 590 varieties of domesticated mammals and poultry, including 66 kinds of horses, 20 asses, 73 cattle, 20 water buffalo, 50 yaks, 4 camels, 79 sheep, 43 goats, 113 pigs, 109 chickens, 35 ducks, 21 geese, and 3 turkeys.

Human cultural diversity is also fairly high within the context of the megadiversity countries. Although the dominant group are the Han Chinese, which account for an astounding 93% of the population and comprise the single largest nationality in the world, there are also 55 other "minority nationalities." Although they account only for 7% of the total population, they still number over 80 million people. Indeed, fourteen of these minority groups have populations of one million people or more, and the largest, the Zhuang nationality of the Guangxi Autonomous Region, has over 13 million people.

Given its tremendously high ecosystem diversity, its high biodiversity and endemism in almost all groups of organisms, its huge human population, and its immense geographic area, not to mention its growing economic might, China is clearly one of the most important countries on Earth in virtually every category including the biological. Its position as one of the top eight most important nations in terms of biodiversity should help to further focus this country's attention on the great natural resources it is so fortunate to possess, and hopefully encourage further action on their behalf.

Natural habitats

China has a range of natural habitats that few countries can match. For convenience, we have grouped them into several broad categories, including forests, steppes and grasslands, deserts, and various montane systems, and freshwater and marine aquatic systems.

Forests vary from coniferous in the temperate zone to tropical evergreen forest in the tropical areas of the far south. The area of natural closed forest of all kinds in China is 978 470 km^2 or about 10% of the land area, and open forest covers a further 172 000 km^2 or about 2% of the country, for a total of roughly 12% still in forest. In all, there are some 212 different categories of forest, depending on the dominant or characteristic species. A separate category of bamboo forests is also quite diverse, and includes some 36 different types.

Cool-temperate coniferous forest is found at the southern end of the coniferous forest belt lying across northern Eurasia, including Siberia. The Chinese portion is mainly in the northern mountains, extending as far south as the slopes of the Da Hinggan Mountains. This habitat is very cold, and the winter lasts for nine months; in the northernmost portions in China, there is permafrost. Dahurian larch (*Larix gmelinii*) is the most common species; it occupies about 50% of the cool-temperate forest area, and can reach 20-30 m. In areas where it has been destroyed by human or natural causes, broad-leaved species such as Asian white birch (*Betula platyphylla*), David's poplar (*Populus davidiana*), Dahurian birch (*Betula dahurica*), and Mongolian oak (*Quercus mongolica*) take over and form extensive secondary forests; these occupy about 24% of the extent of this forest type.

Temperate coniferous and deciduous broad-leaved

mixed forest is found mainly in the mountains of northeastern China, including the Xiao Hinggan and Changbai Mountains. In some areas, forest still covers over 40% of the land area, making these forest blocks some of the most extensive ones left in the country. Winter temperatures are very cold, but the proximity to the Pacific Ocean has a moderating effect, resulting in warm summer temperatures. Precipitation ranges between 500 and 1 000 mm. The main coniferous species in these forests is the Korean pine (*Pinus koraiensis*), which can reach 30-40 m in height; further south the number of broad-leaved trees increases, among them the Mongolian oak and the Manchurian ash (*Fraxinus mandshurica*). The Changbai Mountains are home to a variety of medicinal plants, including the famous ginseng (*Panax schinseng*).

Warm-temperate deciduous broad-leaved forest occurs over a vast area of northern China, but is sparsely distributed in the southern part of Liaoning Province, Shandong Province, the North China Plain, and the Loess Plateau. Rainfall is usually in excess of 800 mm, although on the western border it is only 400 mm. Average annual temperature is 8-16°C, and summers are hot and wet, while winters are dry. This area was once largely covered by dense forest and forest-steppe. However, it has long been used for agricultural purposes, and there is a high human population, with the result that forest covers only about 8% of the landscape. Due to habitat destruction and modification, the main species are now typically secondary ones, including oaks (*Quercus* spp.) and Chinese pine (*Pinus tabulaeformis*).

The subtropical evergreen broad-leaved forests of China occur from the Qin Liong mountains and the Huai River in the north, south almost to the Tropic of Cancer and from the eastern coast to the Qinghai-Tibet Plateau. The largely subtropical monsoon climate of this region has high temperatures and abundant rainfall; it has not been altered much by climatic changes and glaciation for millions of years, so the flora has become very diverse under such stable conditions. There are some 1 700 genera of seed plants in this region, or 57% of the total for the country; of these, 77 are endemic monotypic genera of great antiquity, including many relict species. Angiosperm diversity is estimated at 14 500 species, or about half the total for the entire country. The oak family (Fagaceae) is the most important, and its evergreen species occupy a large percentage of the forest area. Fifteen genera and over 100 species of the laurel family (Lauraceae) occur here as well, including the world-famous cinnamon (*Cinnamomum*). At least 11 genera and over 100 species of the tea family (Theaceae) are common in this region. Tea-producing plants of the genus *Camellia* are grown commercially. The tea plant itself, *Camellia sinensis*, has long been cultivated in China for medicinal purposes and as a beverage. Bamboo is also abundant in this type of forest, with some 26 genera and 300 species; the most valuable is meso bamboo (*Phyllostachys pubescens*). The climate is warm in summer and moderate in winter, with heavy monsoon rains in summer and a little rainfall in winter as well; annual precipitation ranges from 1 000 to 2 000 mm. As a result of human disturbance, very little natural vegetation remains in this region; Masson pine (*Pinus massonica*) and Chinese fir (*Cunninghamia lanceolata*) forests are widespread in this area, but are mainly secondary or planted.

Tropical rain forests and tropical monsoon forests are found south of the Tropic of Cancer. The main difference between them is that the monsoon forest experiences a dry season in which the trees lose their leaves, while the tropical rain forest keeps its leaves all year long. Only 3% of China falls within the tropics, including the southern part of Yunnan, Guangdong and Fujian Provinces, the Zhuang Autonomous Region of Guangxi, Hainan Dao Island, Taiwan Island and the South Sea Islands. Only a small part of this area is forested. Temperature and rainfall are high, and although this is the smallest part of the country, it is the richest biologically. True rain forest in this region is limited to the wettest and warmest areas, including the southern parts of Taiwan, Guangdong and Yunnan, the Zhuang Autonomous Region of Guangxi, and southeastern Tibet; it is similar in structure to the rain forests of Southeast Asia, with trees averaging 30-40 m and some growing to more than 60 m. Tropical forests have largely been destroyed in China. In all, tropical rain forests cover only 7 150 km^2 in mainland China and 1 660 km^2 on Taiwan, with tropical monsoon forests covering an additional 17 050 km^2. Consequently, these two forest types cover no more than 0.26% of mainland China's land area and 4.58% of Taiwan, and even these figures may be overestimates. Nonetheless, these tropical areas harbor 25% or more of all of China's globally important biodiversity; Xishuangbanna alone has some 4 000-5 000 species of higher plants and over 500 vertebrate species. As an example of the economic importance of Xishuangbanna's biodiversity, there are over 300 species of medicinal plants, 200 of edible plants, 100 of timber trees, 100 of oil-producing trees, and over 50 of bamboo in this highly-endangered region alone. Hainan Island is another example; it has four endemic vertebrate species and 24 endemic subspecies, most of them endangered. These include the Hainan black gibbon, the Hainan moonrat (*Neohylomys hainensis*), the Hainan flying squirrel (*Petinomys electilis*), and the Hainan partridge (*Arborophila ardens*).

Bamboos are common and diverse in this region as well. China has the highest diversity of bamboo species in the world, with 30 genera and roughly 300 species; the area covered by bamboo groves in China is 3.2 million ha, or about one fifth of all bamboo groves on Earth. Most of China's bamboos grow in the tropical and subtropical zone, and this unusual group of grasses has long had enormous economic value.

In addition to forests, China has a rich diversity of grasslands and steppes. The grassland region in the country covers a vast area of some 3 million km^2, mainly in the western part of the northeastern region, the

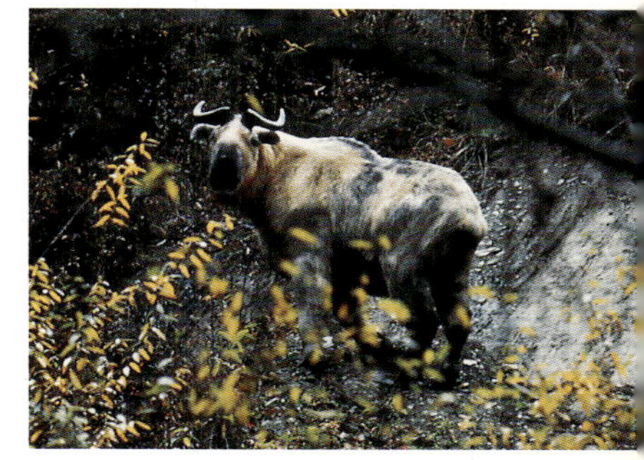

The takin (Budorcas taxicolor) *is a very large, distinctive goat antelope from alpine and subalpine bamboo forests in western China, Bhutan and Myanmar. This male is from Sichuan.*

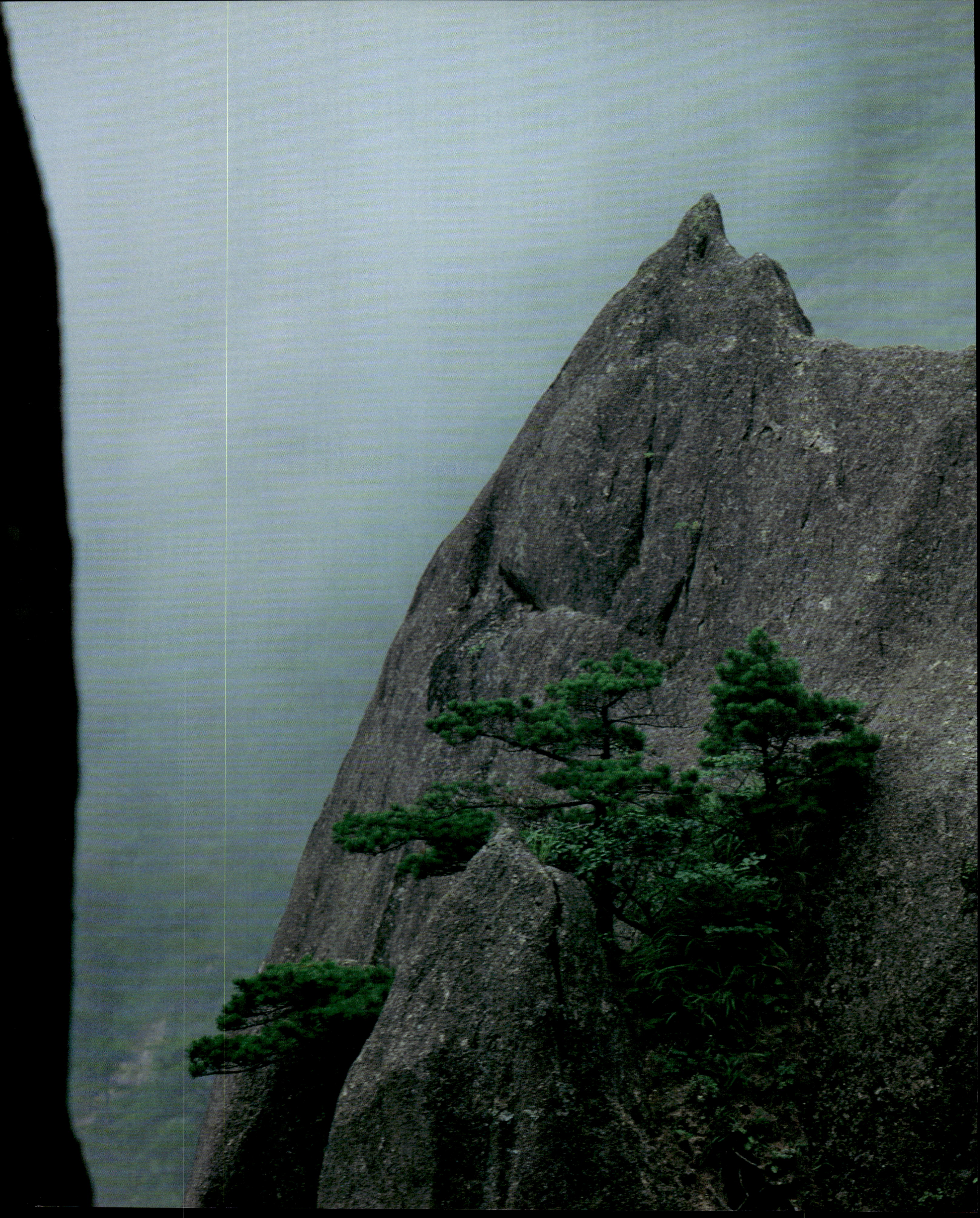

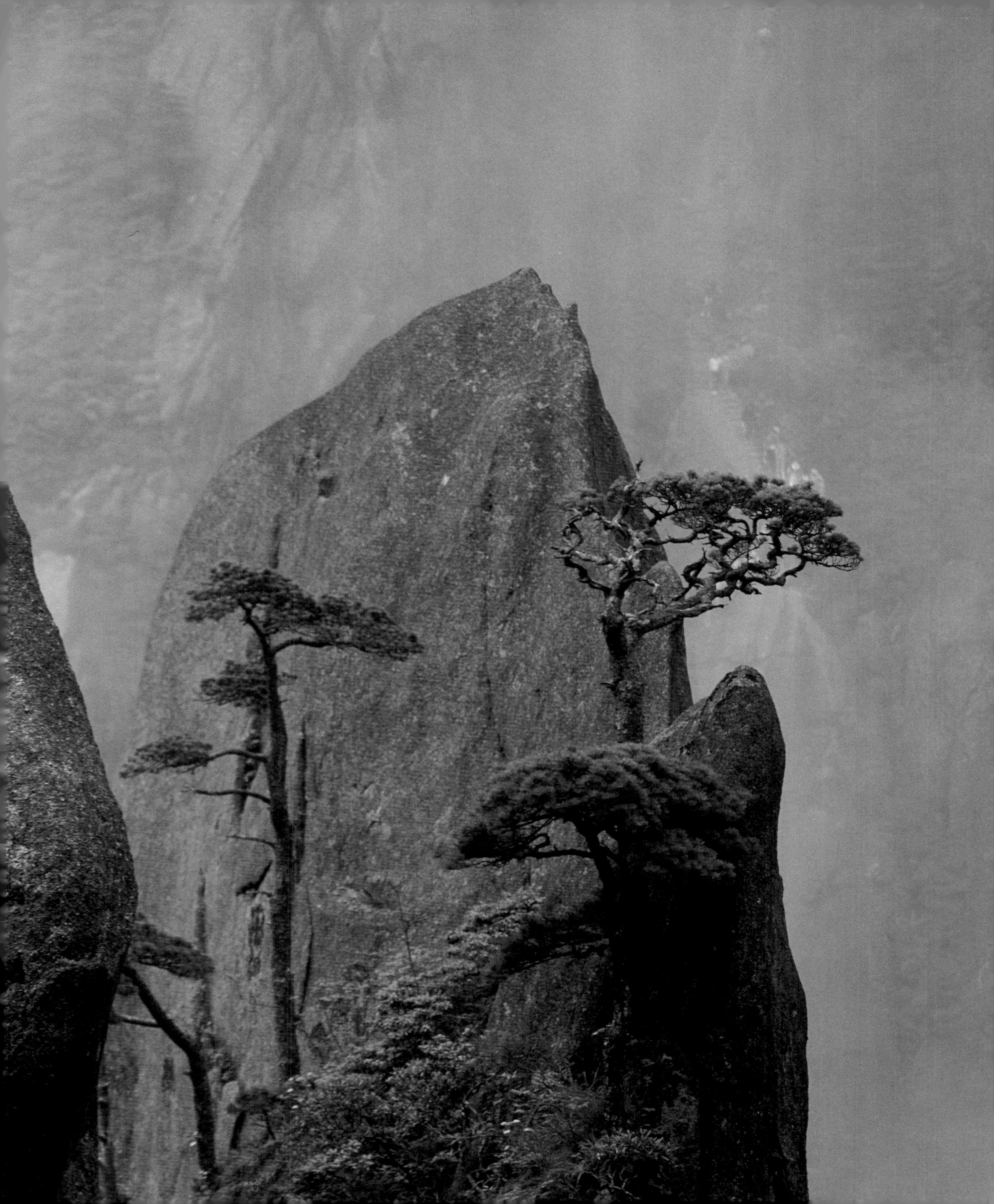

Inner Mongolian Plateau, the northern part of the Loess Plateau, and the central part of the Qinghai-Tibet Plateau. This type of grassland is dependent on the amount of rainfall and the temperature; the wetter the area, the taller the grass. In China, there are four major types of grasslands, including meadow steppe, typical steppe, desert steppe, and highland cold steppe. Meadow steppe is found on the Northeastern Plain and in Inner Mongolia, and is a transition belt between forest and steppe proper. Rainfall is between 350 and 500 mm, and grass grows luxuriantly, reaching 40-60 cm. The dominant plant group is the grass family (Gramineae) and the pea family (Leguminosae); on the Northeastern Plain and in eastern Inner Mongolia, the principal species are feather grass or Baikal needle grass (*Stipa baicalensis*), Chinese aneurolepidium (*Aneurolepidium sinensis*), and Siberian filifolium (*Filifolium sibericum*). The meadow steppe is used mainly for grazing.

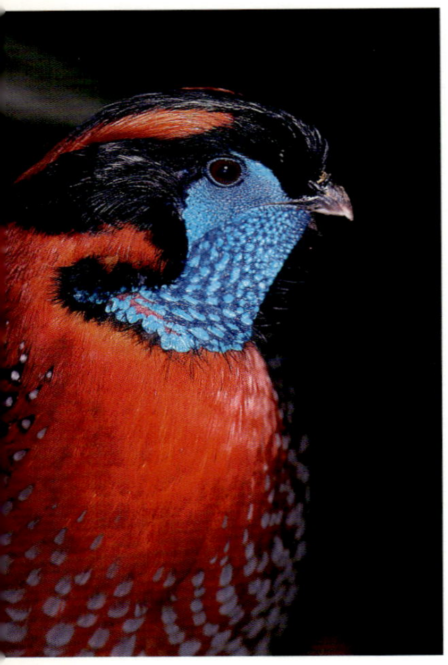

Typical steppe, or arid steppe, is drier than meadow steppe, and is characteristic of the middle part of the Inner Mongolian Plateau, the southwestern part of the Northeastern Plain, the northern Loess Plateau, and below the forest belt in the Northwestern mountain areas. True steppe exists in a semiarid transitional area between meadow steppe and desert steppe, where annual rainfall is between 250 and 400 mm. It is composed mainly of grasses, composites, and various peas and vetches, with Krylov's needle grass (*Stipa krylovii*), greater needle grass (*Stipa grandias*), June grass (*Koeleria cristata*) and crested wheat grass (*Agropyron cristatum*) being the main species. This area is also used mainly for grazing in China.

Desert steppe is found in the middle and western parts of Inner Mongolia and in the lower levels of the northwestern mountains. It stretches between the true steppe and the barren desert, in areas where rainfall ranges between 150 and 250 mm. Plants are mainly perennial herbs and some shrubs and bushes, all adapted to long, dry periods. The main species are Gobi needle grass (*Stipa gobica*), short-flower needle grass (*Stipa breviflora*) and sandy needle grass (*Stipa glareosa*). Sandy needle grass is well adapted to this climate, and forms vast areas of almost uniform desert steppe vegetation in some areas of very sandy and gritty soils. This region is used mainly for grazing of goats and camels.

Finally, there is highland steppe, found above 4 000 m on the Qinghai-Tibet Plateau and above 3 000 m in northwestern Sichuan. Plants are resistant to drought and to very low temperatures. The vegetation is dominated by low tussocks of grasses, sedges, and dwarf alpine shrubs, including purple-flowered needle grass (*Stipa purpurea*), Krylov's fescue (*Festuca kryloviana*) and Moorcroft's sedge (*Carex moorcroftii*). This area is used as pasture for cold-adapted, resistant species like Tibetan yaks, Tibetan sheep and Tibetan goats.

On pp. 260-261, China has many mountain ranges, which are very important refuges for the country's wildlife. Here a classic scene in the Huangshan Mountains of Anhui Province, with pines in the foreground.

Above, the pheasants and their relatives reach their greatest diversity in China, with 25 species and 10 endemics. The male Temminck's tragopan (Tragopan temminckii) *is one of the most spectacularly-colored of all birds.*

Deserts also occupy a large area of China, covering 1 095 000 km^2 in the arid northwest or about 11.5% of the country as a whole. Some 637 000 km^2 of this area is sandy desert, and the rest is stony desert. Deserts occur in China where annual rainfall is less than 200 mm, although there is no place in the country that does not receive at least some rainfall. Vegetation is sparse and short and there is little variety of plants; there are also large areas completely devoid of vegetation. The main desert areas in China are the Tarim Basin (906 500 km^2), the Junggar Basin, the Alxa Desert, the Ordos Desert, and the deserts of the Qaidam Basin.

Mountain ecosystems are also a major feature of the Chinese landscape, covering about 43% of the land area, and are especially important for wildlife. Most of the plains and other low-lying areas in China have been cultivated for centuries, so the mountains are where most of the forests remain. The orientation of mountains in China is also important because of its effect on climate and plant and animal distribution. Most Chinese mountain ranges run from northeast to southwest and from west to east, with fewer that run northwest to southeast, and only a handful running predominantly north to south. Three groups of mountains run west to east, and those are the Tien and Yin mountain ranges in the north, the Kunlun, Qilian and Dabie ranges in the middle, and the Nan Ling group in the south, all of them serving as important geographical boundaries. The mountains that run northeast to southwest are medium to low ranges in the east of China; they include three major groups, the Da Hinggan and Taihang Mountains; the Xuefeng Mountains, and the Changbai Mountains, Shandong Hills and Wuyi Mountains. Those that run north to south are mainly in the middle of the country, and include the Helan, Liupan and Hengduan ranges. The northwest to southeast ranges are the highest in the country, and are located in the west. These include the Altai Mountains in the northwest, and the Himalayas, the highest mountains in the world, in the southwest.

Marine and freshwater diversity is also of great importance in China. The country has more than 50 000 rivers that drain a catchment area larger than 100 km^2; these include the third longest river on Earth, the 5 550-km Yangtze (Chang), and several other major rivers, among them the 4 830-km Yellow (Huang) River, and the Amur (Heilong) River, which starts in China and ends up in Russia. There are also some 2 800 lakes of one km^2 or larger, and they cover an area of some 80 000 km^2. The coastal waters of China are comprised of four seas, the Bo Hai, the Yellow, the East China Sea and the South China Sea. The Bo Hai is one of the few seas entirely enclosed within a single country, and covers some 97 000 km^2. The coastline of China extends some 18 900 km, and includes a wide variety of coastal ecosystems, including coral reefs in the south and off the southeast coast of Taiwan and neighboring islands.

Flagship species

China is home to many different and globally important flagship species, including the world's single most popular flagship animal, the giant panda (*Ailuropoda melanoleuca*). This unique creature has been the object of more affection and publicity than any other wild animal, in spite of the fact that it did not become known to the Western world until 1869. Although it had been known to the Chinese for millennia, the first Westerner to describe it was Father Armand David, a French Jesuit naturalist who studied two skins of pandas killed by Chinese hunters in the Mu-ping Mountains. The first giant panda to reach the West alive was a female cub named Su-lin, which was obtained by American Ruth Harkness in an expedition to China in 1936 and sold to the Brookfield Zoo in Chicago for 8 700 U.S. dollars. The giant panda is found only in three places in China, and is critically endangered. It lives mainly in bamboo groves of mountainous areas near the boundary of Sichuan, Shaanxi, and Gansu provinces at altitudes of between 3 000 and 4 000 m. It eats mainly the stems and leaves of various bamboo species, but will also take fruits, bamboo rats, and carrion. The structure of its teeth and the fact that it is still attracted to meat indicates that it had a more carnivorous ancestry, but it has now become very specialized on bamboo —to the point that many die when there are periodic mass flowerings and die-offs of key bamboo species. An enigmatic animal that has been moved back and forth from the bear family to the raccoon family and back to the bear family, giant pandas superficially look and behave like bears. They occur in very low densities, between 1 per 2.9 km^2 and 1 per 9.7 km^2, and reproduce slowly. Newborn infants are tiny and poorly developed, weighing only about 100 g, compared to the 100 kg or so that an adult weighs at maturity. In all, there are only about 1 000 pandas left in the wild, and they unfortunately live in a number of small, badly-fragmented populations within their remaining range. Although they have been a major focus of international attention and investment, and are the symbol both of China's conservation efforts and of the world's largest international conservation organization, the World Wide Fund for Nature or World Wildlife Fund (WWF), their survival in the wild is by no means assured. To the contrary, they must be considered one of the world's most endangered larger mammals.

Pandas have long been kept in captivity in China and occasionally internationally as well. Following a die-off of fountain bamboo (*Sinarundinaria*) in 1983 and the death of a number of wild pandas, an effort was made to bring many of them into captivity. By 1986, a total of 64 had been taken, of which 46 were saved. The Beijing and Chengdu zoos have succeeded in breeding giant pandas, as has the zoo in Mexico City, but many other efforts, including one in the National Zoo in Washington, D.C., have been notably unsuccessful. During the 1980s, there was a controversial program that consisted of loaning pandas to a number of American zoos in return for large cash payments; this program was much criticized by the international community and finally ended. In 1996, a pair of animals was made available on long-term loan to the San Diego Zoo in California.

The red or lesser panda (*Ailurus fulgens*) is actually quite distinct from the giant panda, except that it also has a facial mask. It is much smaller, has a long tail, and striking reddish-brown fur. It belongs to the raccoon family (Procyonidae), and also eats bamboo stems and leaves, as well as fruit and birds' eggs.

China is the second most diverse country in Asia for primates, after Indonesia, and has one of the greatest concentrations of endangered primate species. The most spectacular flagship primates in China are the snub-nosed monkeys of the genus *Rhinopithecus*. These large, leaf-eating monkeys of the subfamily Colobinae include four species, of which three are endemic to China and one to Vietnam. The best known member of the group is the golden monkey (*Rhinopithecus roxellanae*), which is second only to the giant panda in terms of its global importance as a flagship for China. It has the widest distribution of any Rhinopithecus, being found in the Qionglai, Min Shan, Qinling and Daba Mountains of central China, at altitudes ranging from 1 200 m to over 3 000 m. It inhabits deciduous broad-leaved forest, coniferous forest, and a forest type that includes both broad-leaved trees and conifers. Three subspecies are recognized, *R. r. roxellanae* from the Qionglai and Min Shan Mountains of Sichuan and Gansu, with a population of perhaps 2 000-4 000; *R. r. qinlingensis* from the Qinling Mountains of Shaanxi, numbering about 4 000; and *R. r. hubiensis* of the Daba Mountains of Hubei, with a total of only about 800 individuals. The world population for this species would therefore be roughly 6 800-8 800 animals. This monkey is both arboreal and terrestrial, and lives in enormous groups of 100 to 300 individuals. It is brilliantly colored, with long, bright golden hairs extending from the shoulders like a cape. In addition to the turned-up snub nose, males of *R. roxellanae* have enlarged projections on the corners of the upper lip, the purpose of which is unknown. Males in this species average 16.4 kg, females 9.4 kg.

The second *Rhinopithecus* species, the Guizhou snub-nosed monkey, *R. brelichi*, is found only in a very small region in the center of Mt. Fanjing in the Wuling Mountains of Guizhou, with the bulk of the population being in the Fanjingshan Nature Reserve. They live in areas of mixed deciduous and evergreen broad-leaved forest at altitudes of 1 500 to 2 200 m, with a high rainfall of 2 000 mm or more. Social organization seems to be based on small, one-male groups of 5-10 individuals which can coalesce and range together in bands of up to 400 individuals. The total population of this critically endangered species is less than 1 200 and perhaps even less than 800, all in a single population. Coloration of this species is brown above, grading to gray on the lower body and with a bright patch of white between the shoulder blades. The distal limbs, the neck, and the head are black with the exception of a distinct golden brow. The chest and inner parts of the upper arms are golden as well. Size is comparable to *R. roxellanae*.

The third species, the Yunnan snub-nosed monkey,

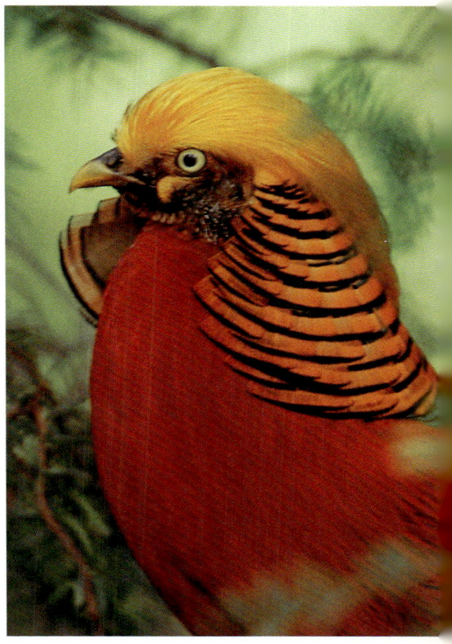

One of the pheasant species endemic to China, the golden pheasant (Chrysolophus pictus), *is now bred widely in captivity. It, too, is a spectacularly-colored creature.*

On pp. 264-265, *a wild snow leopard* (Panthera uncia). *This spectacular flagship species is found at altitudes of 3 000 to 6 000 m in the high mountains of Tibet, Qinghai, Xinjiang, Gansu and Sichuan Provinces of China and in neighboring countries as well.*

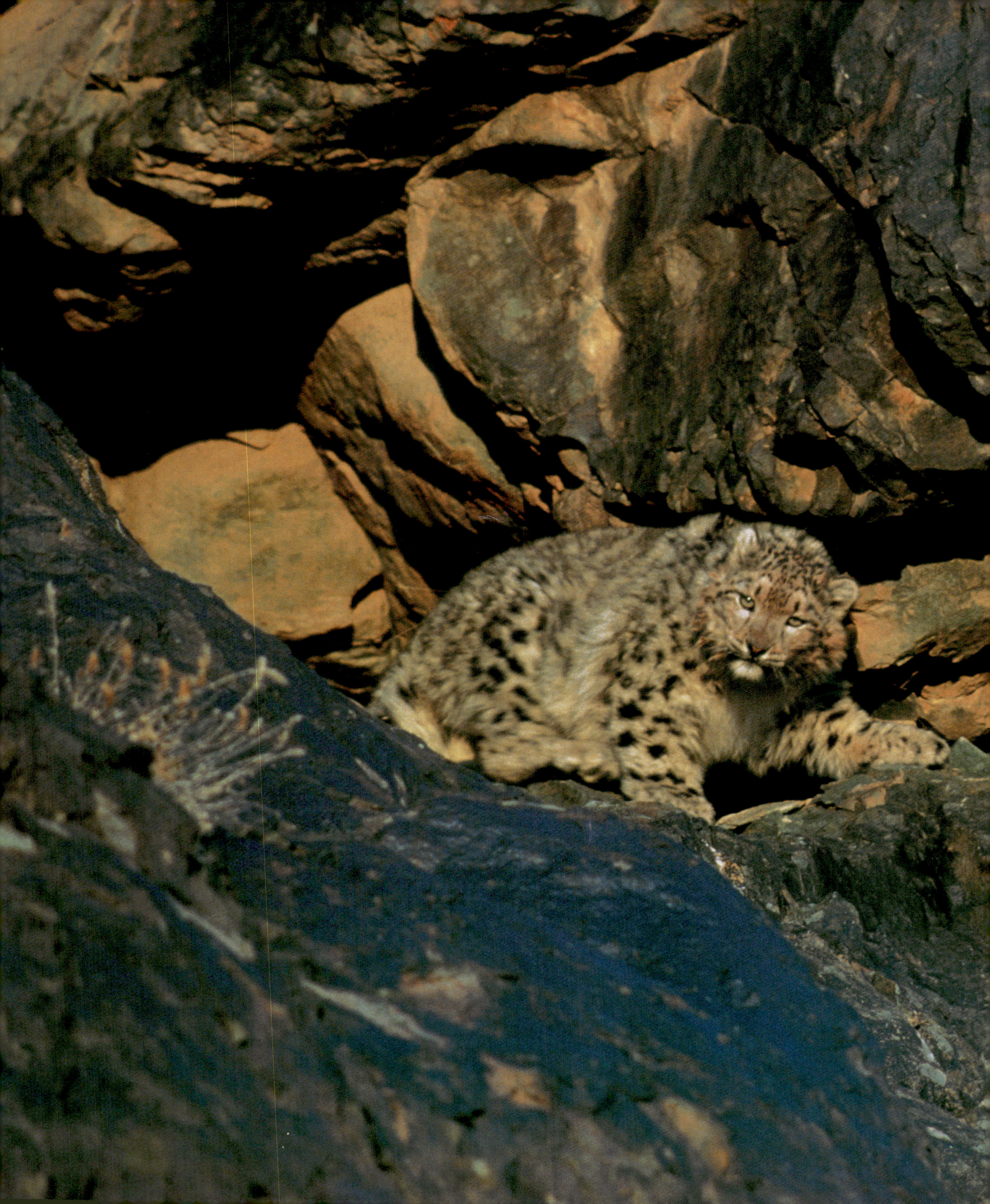

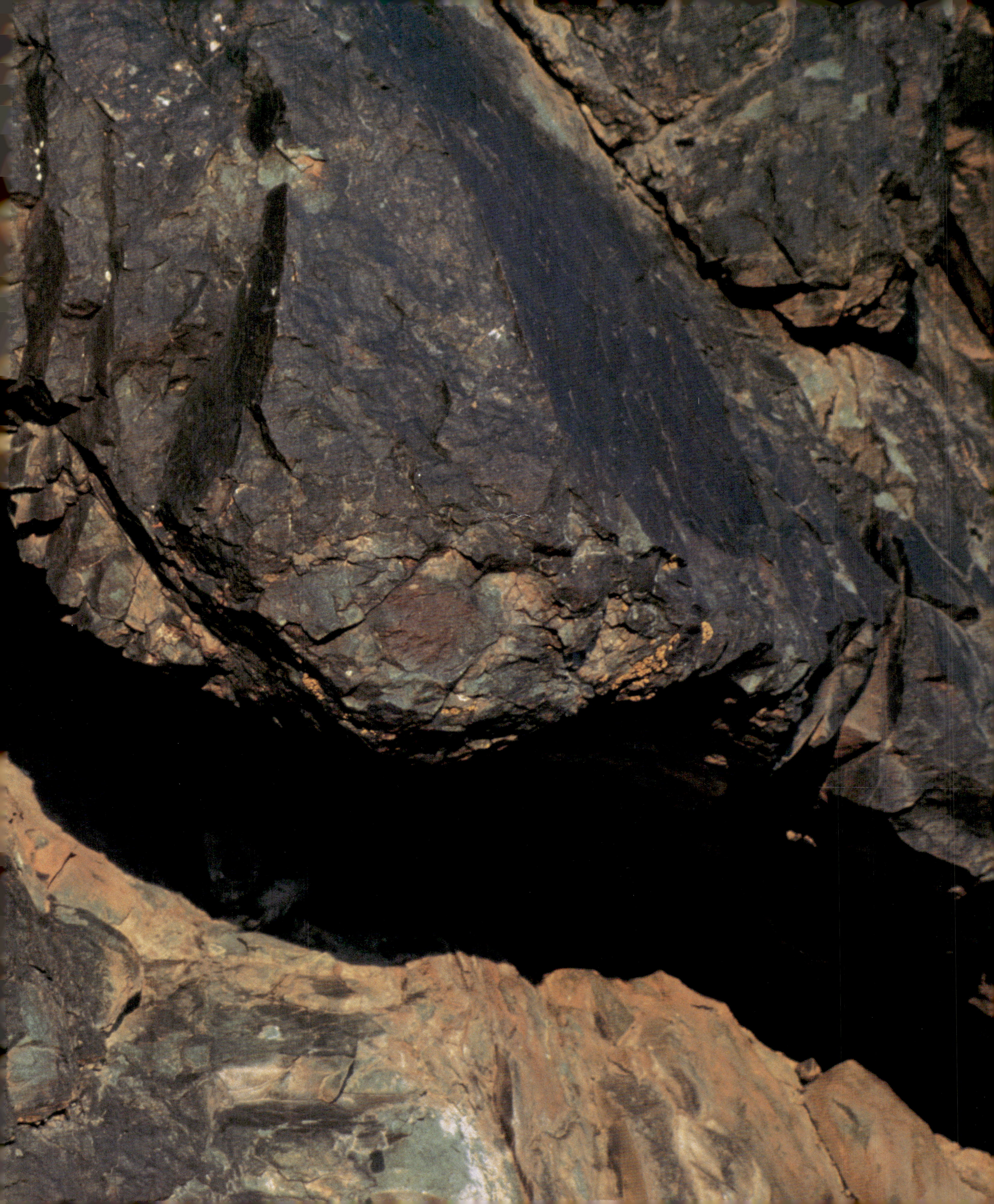

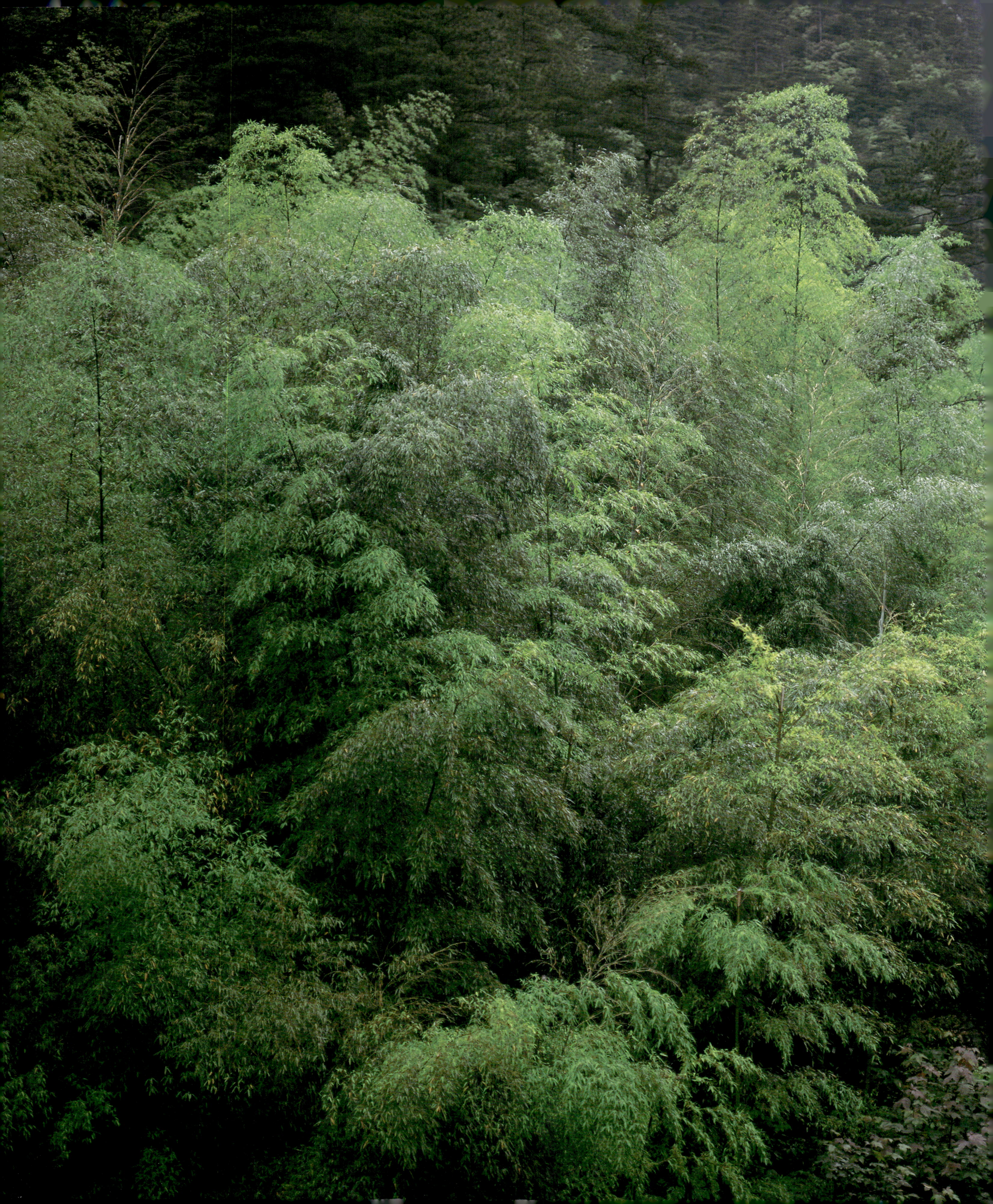

R. bieti, is found at the highest altitudes occupied by any nonhuman primate. It is only known from the Yunling Mountains of southwestern China, in high-altitude evergreen forest with a canopy composed mainly of fir, evergreen oak and rhododendron species. Its range is between 3 000 and 4 500 m. Temperatures average below 0°C for several months of the year and snow is common. There appear to be 13 distinct subpopulations of this animal in five counties of Yunnan and Tibet, and the total population is between 1 000 and 1 500 animals. This unusual species is also largely terrestrial and lives in groups ranging from under 50 to over 200 individuals. It apparently eats arboreal lichens, supplemented by leaves and occasionally fruits and seeds. Adult coloration is grayish-black to black on the back, head, and distal limbs, while the tail, flanks, chest, and front of the neck are white. Infants are white, and become light gray over the course of several months.

The Hainan gibbon (*Hylobates concolor hainanus*) is a distinct subspecies of the black gibbon that was once found in most of the natural forests on Hainan Island and numbered some 2 000 individuals as recently as the 1950s. However, there are only two isolated populations remaining, one in Bawangling Nature Reserve with four groups totaling 21 individuals prior to 1989, later decreasing to about 15 individuals because of forest destruction and poaching in the reserve. The other is on Limu mountain, where the situation is unknown and the animal is possibly already extirpated. If current trends continue, the Hainan gibbon could become the first primate taxon to go extinct in the wild in this century, and thus deserves special attention.

The South China tiger (*Panthera tigris amoyensis*) is in a critical state. Forty years ago, it was thought to number more than 4 000, but it was declared a pest by the government and hunted mercilessly. As a result, no tigers have been seen anywhere in the past 20 years, when the last one known was brought into captivity; it could well be on the verge of extinction or already gone in the wild. The captive population of 50 or so South China tigers is only in Chinese zoos, and these animals are descended from just six wild-caught individuals. It is undoubtedly the most endangered of the surviving subspecies of the tiger. The Siberian tiger (*Panthera tigris altaica*) also occurs in China, mainly in several mountainous areas in the eastern part of Heilongjiang Province and in the southeastern part of Jilin Province; perhaps 100 of these animals remain in the Chinese part of their range, but they are also found in Russia.

The snow leopard (*Panthera uncia*) is not endemic to China, but is a spectacular flagship species nonetheless. Found at altitudes of 3 000 to 6 000 m in the high mountains, it is similar to the common leopard in size and can weigh up to 200 kg. It has a beautiful thick, two-layered, grayish-white coat with circular clusters of black spots. It is now quite rare because it has long been hunted for its fur, and is now found only in small numbers in the mountains of Tibet, Qinghai, Xinjiang, Gansu and Sichuan, and in neighboring countries as well.

Père David's deer (*Elaphurus davidianus*) is a large deer and the only representative of the genus *Elaphurus*. It is especially interesting because it demonstrates the value of captive breeding in saving an endangered species from extinction. In 1865, the French missionary Father Armand David was heading to Peking when he heard of a nature park that belonged to the emperor at Na-Hai-Tzu. An outstanding zoologist (who also discovered the giant panda and had the Chinese giant salamander named after him), Father David visited the emperor's park and found an unusual deer living there. By talking to local people, he realized that almost nothing was known of this animal. He managed to obtain two skins and two live animals and sent them back to Paris. The live animals never made it but the skins did, and the famous French zoologist Milne Edwards used them to name a new species, which he called *Elaphurus davidianus* in honor of Father David. As it turned out, the animals that Father David saw in the emperor's garden were the only ones left in the world, since the species had already gone extinct in the wild. In 1869 and 1883, four of these deer were sent to the Royal Zoological Society in London, and another to Berlin; later, the Duke of Bedford obtained 18 for the park at Woburn Abbey. These turned out to be the salvation of the species because in 1895 the Hun Hoo River flooded its banks, destroying the walls around the emperor's park; the escaping deer were killed and eaten by the starving population. The few animals that stayed inside the park and didn't get killed during the flood, were later eaten by troops occupying Peking during the Boxer Rebellion in the early years of the twentieth century. One female survived, and she died in Peking in 1920, leaving the animals at Woburn Abbey as the last representatives of this distinctive genus. Since that time, many more have been bred in captivity and some have been returned to China, where a reintroduced population of some 250 animals now lives in seminatural conditions.

The blue sheep (*Pseudois nayaur*) is found in mountain valleys in Shaanxi, Sichuan, Inner Mongolia, and on the Qinghai-Tibet Plateau. It lives on exposed rock and mountain grasslands at altitudes of 2 400 m to as high as 6 000 m. Another wild sheep, the argali (*Ovis ammon*), is a large sturdy sheep with big, spiraling horns that grow up to 90 cm. It is found at altitudes of 1 000-3 000 m in mountainous cold desert or semi-desert areas. Both the argalis and the blue sheep have long been a popular target of big game hunters.

The Yangtze river dolphin (*Lipotes vexillifer*) or *baijitun*, as it is called in China, is a unique cetacean species that lives exclusively in the middle and upper reaches of the Yangtze River. This dolphin has been heavily impacted by increased vessel traffic, incidental kills in fisheries, explosions during river bank construction, and pollution, to the point that it has decreased from about 400 in the 1980s to 150 or less in the early 1990s. It is the most endangered cetacean and one of the most endangered mammals in the world. A seminatural reserve for the baijitun has been proposed at Shi Shou,

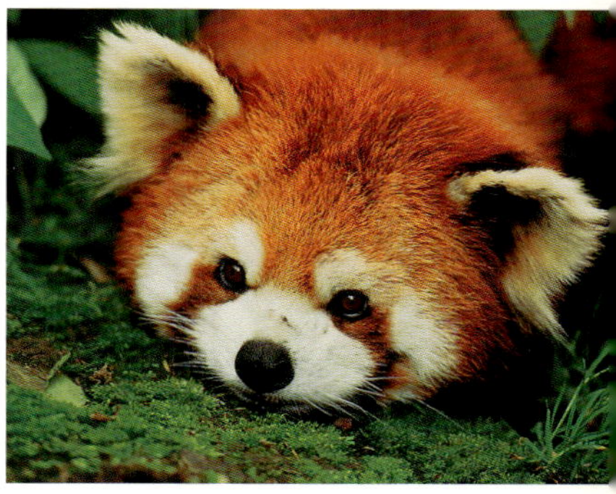

China has the highest bamboo diversity in the world, with roughly 30 genera and 300 species. The country's bamboo forests represent about one fifth of all bamboo formations on Earth. On the opposite page, a bamboo forest in the Huangshan Mountains.

Above, the red or lesser panda (Ailurus fulgens), *in spite of its name and similar facial mask, is actually quite distinct from the giant panda. It, too, feeds on bamboo, as well as fruit and bird's eggs.*

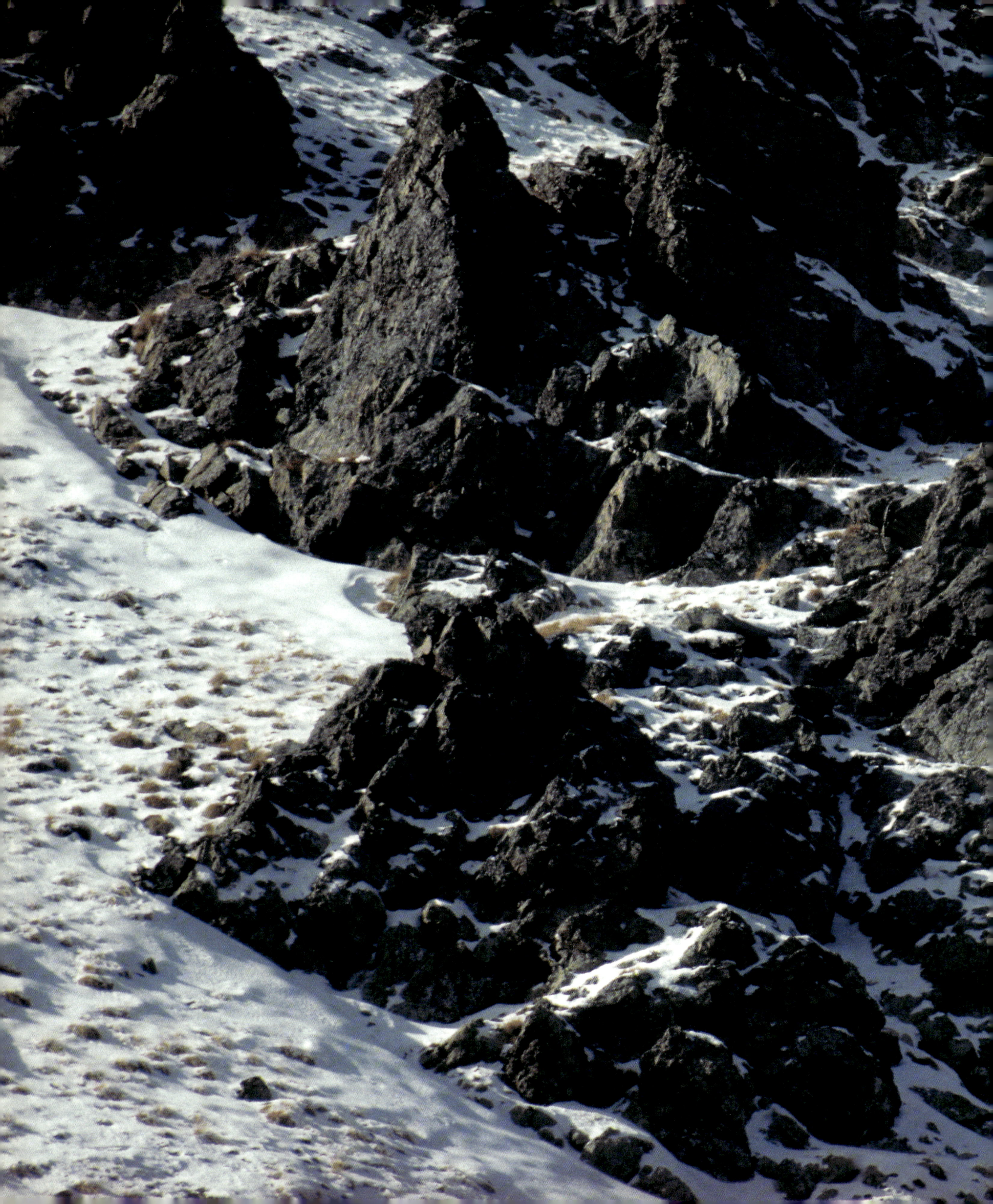

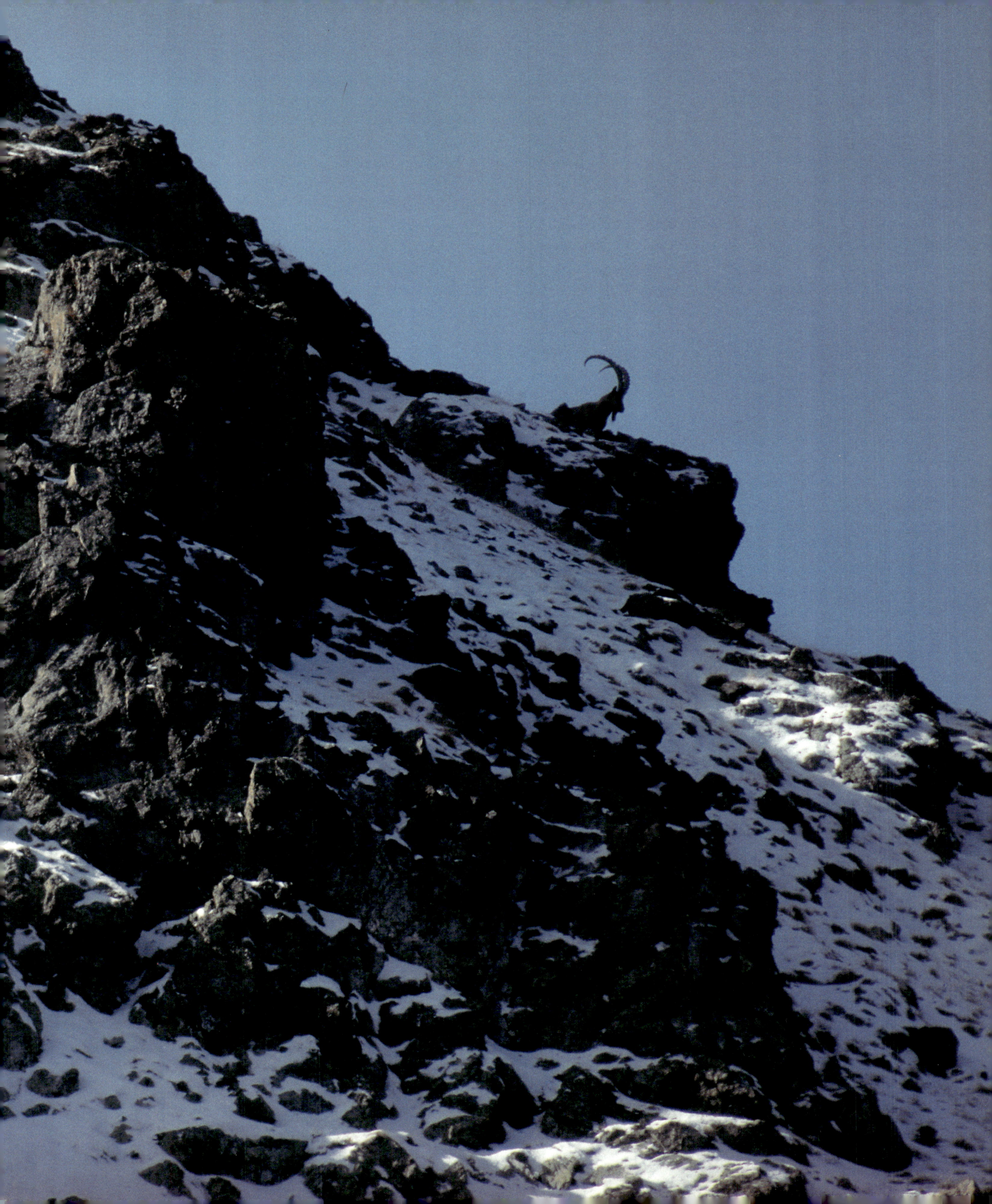

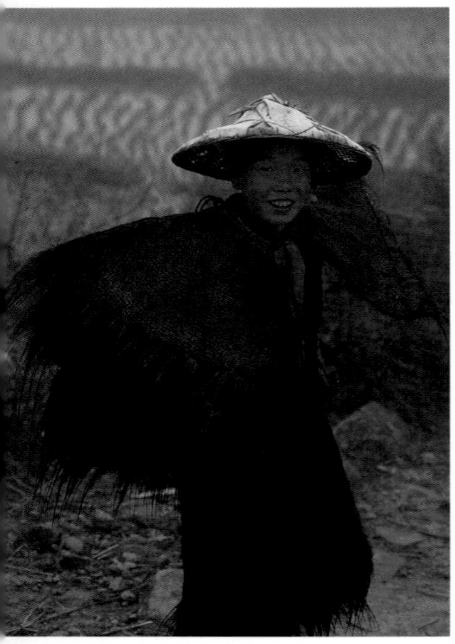

On pp. 268-269, an Asian ibex (Capra ibex siberica) *from the Tianshang Mountains. This widespread wild goat species is found in montane habitats from Central Europe to Central China.*

Above, peasant woman from Fuzhou with a cape of palm bark fiber.

but it is uncertain whether such plans are actually feasible.

Among the birds, probably the most interesting group in China is that of the pheasant and its relatives. An almost exclusively Asian group with the exception of the Congo peacock from the DRC, the pheasants reach their greatest diversity in China. Of the 50 Asian species, 25 occur in China (including Taiwan) and 10 are endemic (including two endemic to Taiwan). Some truly magnificent creatures like Temminck's tragopan (*Tragopan temminckii*), Lady Amherst's pheasant (*Chrysolophus amherstiae*), Elliot's pheasant (*Syrmaticus ellioti*), and the golden pheasant (*Chrysolophus pictus*) are included in this group. The latter two are endemic, as are the blue-eared pheasant (*Crossoptilon auritum*), the brown-eared pheasant (*C. mantschuricum*), the white-eared pheasant (*C. crossoptilon*), Reeve's pheasant (*Syrmaticus reevesii*), the Chinese monal (*Lophophorus lhuysii*), and Cabot's tragopan (*Tragopan caboti*), while the Taiwan Mikado pheasant (*Syrmaticus mikado*), and the Swinhoe's pheasant (*Lophura swinhoii*) are endemic to the island of Taiwan.

The reptiles and amphibians have several Chinese flagship species as well, including the Chinese alligator (*Alligator sinensis*), the Chinese big-headed turtle (*Platysternon megacephalum*), the Chinese xenosaur (*Shinisaurus crocodilurus*), and the Chinese giant salamander (*Andrias davidianus*).

The Chinese alligator is a smaller version of the American alligator (*Alligator mississippiensis*), and is the closest living relative of the American species. It reaches 1.5-2 m in length and is endemic to the country. Although it was at one time more widely distributed in China, the Chinese alligator is currently found in part of the lower Yangtze River, principally in the provinces of Anhui, Zhejiang, and Jiangsu. Its three main habitat types are riverine and swampy areas, low-elevation agricultural communes, and tree-farm communes up to 100 m above sea level. Because they occur in the high northern latitude, Chinese alligators spend a large portion of the year hibernating in subterranean burrows. These burrows can be quite complex, with above- and below-ground pools and numerous airholes. This animal's extensive use of these burrows and its secretive behavior has allowed it to inhabit wetland habitats in areas with dense human populations.

The Chinese alligator is without a doubt one of the world's most endangered crocodilians, with its current distribution being one tenth of its former range. The single greatest problem facing these animals is habitat destruction related to the intense human population pressures in the region. Very little natural wetland remains, and what does exist contains very few alligators. Estimates in the early 1980s suggested that no more than 500 wild individuals remained. Although they are legally protected, and a large number of communes have been classified as alligator preserves, many animals are still killed or collected for sale to zoos or government-sponsored farms.

A special Chinese Alligator Nature Reserve and Reproduction Research Center has been established in Xuancheng, Anhui, and has succeeded in captive breeding to the point that reintroduction has become possible and a second alligator farm in Hainan Province has been established. At present there are 123 Chinese alligators in zoos outside China.

The Chinese big-headed turtle is a very unusual species placed in its own family, the Platysternidae. It is a small animal that lives in mountain streams and is characterized by its unusually large head. It is found in southern and eastern China, including Hainan Island and Hong Kong, westward to Yunnan and also into Vietnam, Thailand and Myanmar.

The Chinese xenosaur or crocodile lizard (*Shinisaurus crocodilurus*) is a monotypic family endemic to Guangxi in China, where it occurs mainly in the Yaoshan Mountains. It is about 50 cm long, with a flat tail that is longer than the body. It lives on trees near streams and looks superficially like a little crocodile, hence the common and scientific names.

The Chinese giant salamander (*Andrias davidianus*) competes with its Japanese cousin for the title of largest salamander species in the world. It can reach 180 cm in length. It lives in the clear waters of high mountain streams above 1 000 m in central and southwest China, and in rocky caves where it feeds on fish and frogs. Unfortunately, it is considered a delicacy and is heavily persecuted, so that it is now endangered.

As for fish, the Chinese sturgeon (*Acipenser sinensis*) is an endemic species confined to the Yangtze River, although it may have also occurred in the Pearl River and the Yellow River. When mature, it swims back against the current to spawn, often in sections of the river with steep cliffs. The young descend back down the river to feed until they reach maturity. This species has become threatened during the last two decades, mainly after the Gezhouba Reservoir dammed the Yangtze and blocked the migration route for the sturgeon. Artificial propagation and release programs for this species have thus far succeeded. A large spawning site has been set up in the Dujiang section of the river below the dam, which hopefully will help to ensure the survival of this large and impressive fish.

Conifers are also a flagship species in China. Of the roughly 30 genera of coniferous trees in the world, about 20 genera, with roughly 200 species, grow in China. Eight of these, the dawn redwood (*Metasequoia*), the Cathay silver fir or yinshan (*Cathaya*), the golden larch (*Pseudolarix*), the taiwania (*Taiwania hayata*), the Chinese deciduous cypress (*Glyptostrobus*), the keteleeria (*Keetleria*), the Fukien cypress (*Fokienia*) and the Chinese fir (*Cunninghamia*), are endemic to the country. The yinshan (*Cathaya argyrophylla*) is a monotypic genus that dates back to the Tertiary. It is of great interest in phylogenetic studies of the pine family, and occurs in very scattered colonies of one to several dozen individuals in about 30 different sites in Guangxi, Hunan, Sichuan and Guizhou Provinces.

Of particular note is the dawn redwood or water larch (*Metasequoia glyptostroboides*), first found in Lichuan County, Hubei Province, in 1943. This resident of the subtropical evergreen broad-leaved forest zone is a living fossil that originated 100 million years ago in the Arctic regions, and spread southwards to Europe, Asia and North America. It was considered extinct, but some 600 km^2 of forests of this tree have now been found. They grow to an average height of 30 m, can live to be 300 years old, and are a significant tourist attraction.

Cultural diversity

China has the largest human population in the world, its more than 1.2 billion people accounting for over 20% of the human population of the planet. The dominant nationality is the Han, which makes up more than 93% of the total population, and is the single largest human group on Earth. In addition, there are 55 other "minority nationalities" accounting for 7% of the population. This is a small percentage, but still accounts for more than 80 million people. The minorities range in size from the Luoba of Tibet, who number only 2 100, and the Hezhe of Heilongjiang Province, of which there are only 1 500, all the way to the Zhuang of the Guangxi Autonomous Region, which have a population of 13 380 000. Fourteen of the 55 minority nationalities number more than a million people each. Although the minorities used to occupy their own areas, there has been considerable mixing over the course of history. However, certain areas still have sizable contained communities of the minority groups. These include the following Autonomous Regions: Inner Mongolia in the north, the Xinjiang Uygur and the Hui Autonomous Region of Ningxia in the northwest, the Tibet in the southwest, and the Zhuang Autonomous Region of Guangxi in the south. There are also many minority communities in the provinces of Yunnan, Guizhou, Qinghai, Sichuan and Gansu, with particularly large concentrations of more than 20 minority nationalities in the southern part of Yunnan Province. A number of these Yunnan minorities live in the tropical forest region of Xishuangbanna. As a state policy, most of the minority nationalities are not subjected to the strict birth control regulations that are applied to the Han Chinese.

The Han live mainly in the eastern and lower-lying provinces, and have long been adapted to high-density living. They are rice farmers, with paddy rice being their mainstay, supplemented by dryland farming on hill soils. Almost all natural forests have been cleared in the region of the Han, except for small areas that may have been maintained for religious or other purposes. Wildlife is one of the main protein resources, especially in the more remote areas, and traditional Chinese medicine, in use for thousands of years, is also based on wild plants and animals. There is a huge demand for such traditional medicines, and hundreds of species of roots, bark, bulbs, and leaves are gathered daily from forests and other ecosystems for personal use or for sale to drug companies and hospitals. Many species are quite endangered due to the medicinal trade, which also includes export to overseas Chinese communities.

The Mongolians are the dominant people in the vast area of Inner Mongolia in the north. They number more than 3 million and are pastoralists, moving their herds of cattle, horses, sheep and camels over hundreds of kilometers every year.

In extreme northwestern China, minority groups include the Uighurs and the Kazakhs. Although some of them are settled in towns and villages, most retain their nomadic, pastoral lifestyle, living in tents in scattered areas of Xinjiang. They are Muslims, rear goats and sheep, and farm the poor mountain soils of this very arid region. In the desert areas, they have domesticated camels, and also use horses and donkeys for transportation.

In Yunnan, the dominant minority is the Dai, who live in the autonomous prefecture of Xishuangbanna and who have practiced Buddhism for several centuries. They live in tidy villages each with its own monastery. Each village has a *longshan* or holy hill, or shares at least another larger one with a neighboring village. The *longshan* is a forested area where it is forbidden to cut trees or hunt animals, but where collecting of medicinal plants, mushrooms and other minor forest products is allowed. These areas are considered the home of spirits and it is believed that bad luck will strike the village if the *longshan* is cut down. Because of these beliefs, the *longshans* have served as mini-nature reserves in this area. The Dai also have for centuries planted fuelwood trees, *Cassia siamensis*, rather than cutting down the natural forest. Large clumps of bamboo are also planted around villages to provide shelter, wood, and other useful items. Dai men also hunt and fish, and use a wide variety of traditional medicines collected from the forest and from the *longshans*.

The Tibetans live in the harsh conditions of the high Qinghai-Tibet Plateau. The eastern part of this plateau fans out into a series of mountain chains that go down into Yunnan and Sichuan Provinces. In this region, the Tibetan tribes have adapted to upland farming. In the western part of the plateau, there is little arable land, so the people live nomadically; the domestic yak is their most important animal, although horses, sheep and goats are used as well. Tibetans use yak wool to make their homes, yak dung for fuel, yak milk to make yogurt, cheese and butter, and yak meat. Yaks are also the main draft animals for carrying supplies between campsites. Tibetans are also hunters, tracking down blue sheep, gazelles and other game with dogs and primitive muskets. Recently, with the development of the local economy, the plateau has opened to more and more hunting and its wildlife is being threatened as never before.

WANG SONG
XIE YAN
RUSSELL A. MITTERMEIER

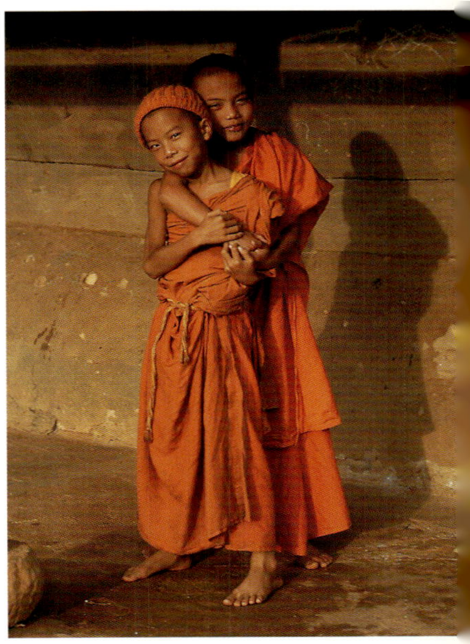

Young Buddhist monk in a temple near Jinghong, in the rain forest region of Xishuangbanna in southern China.

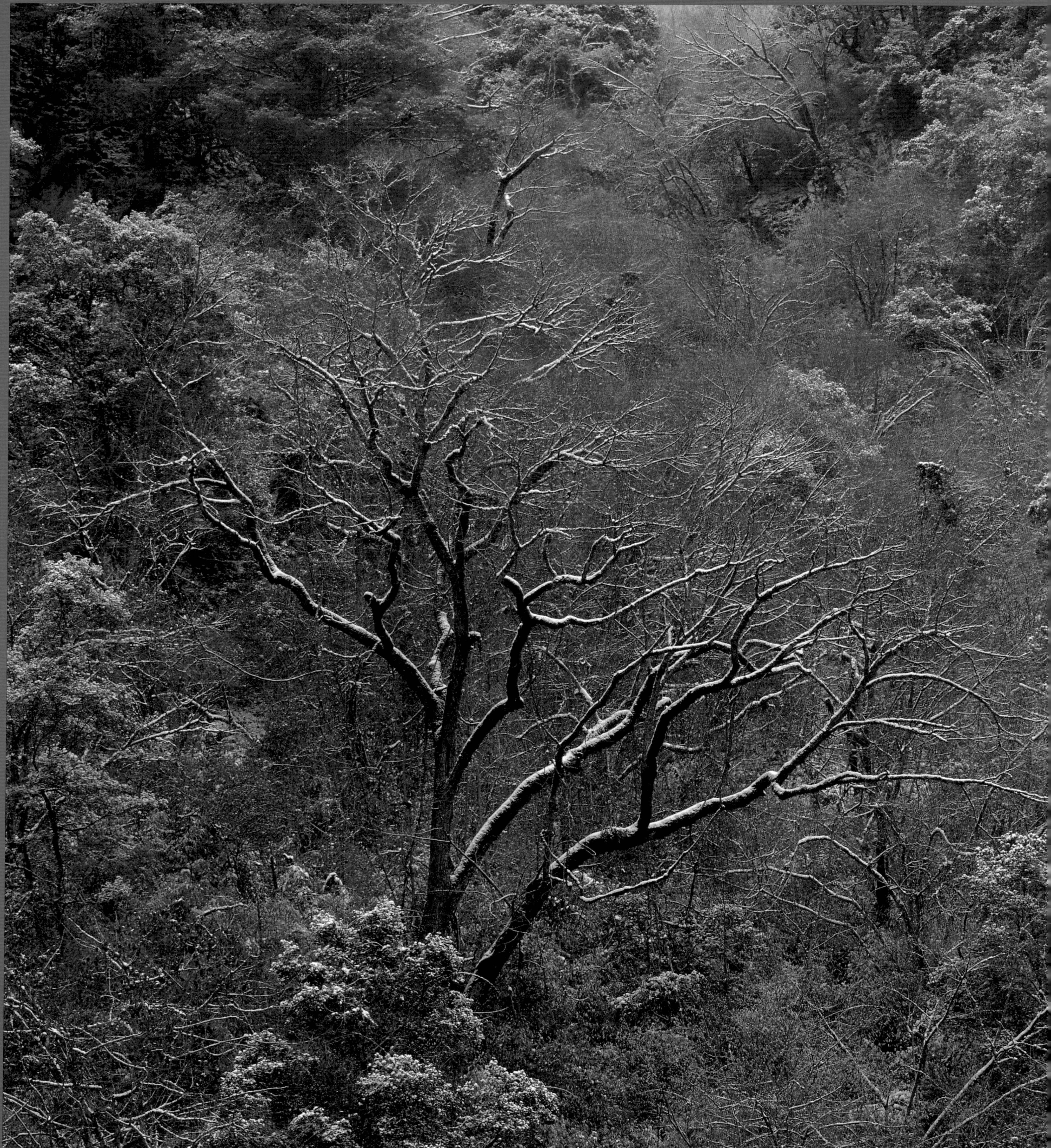

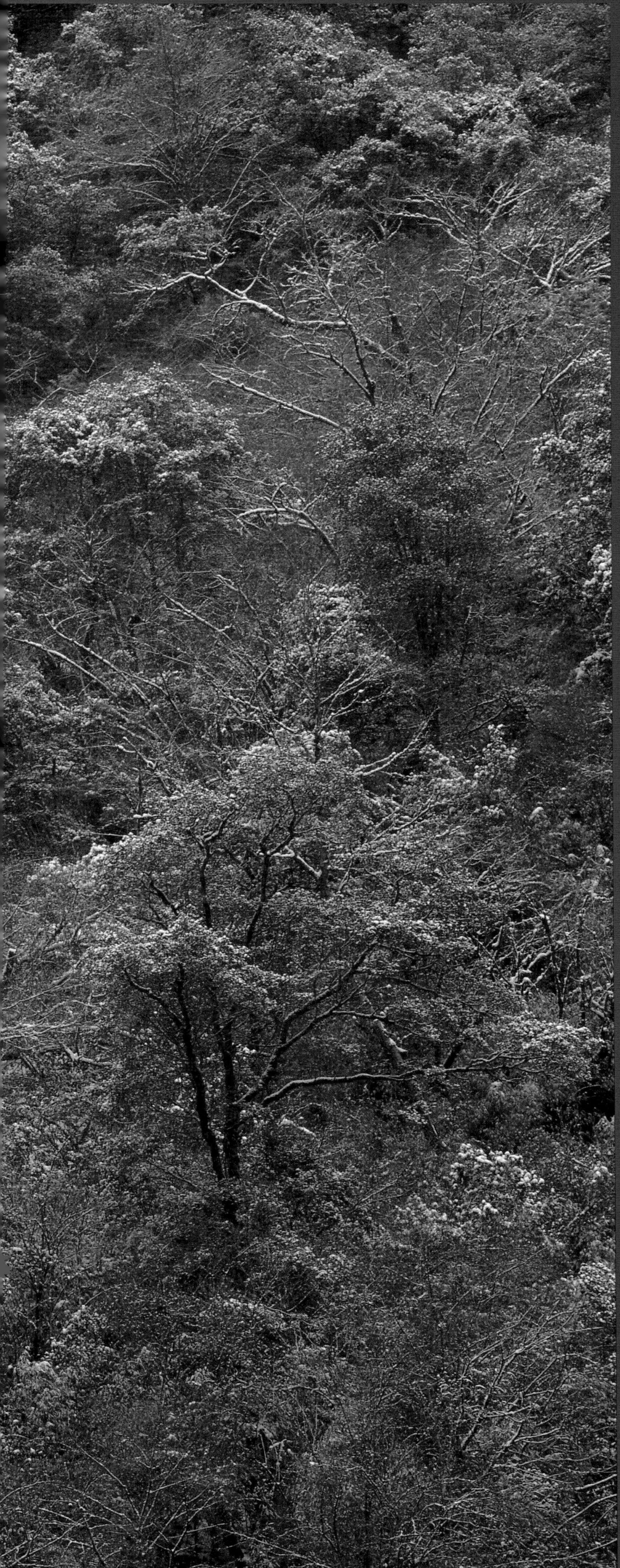

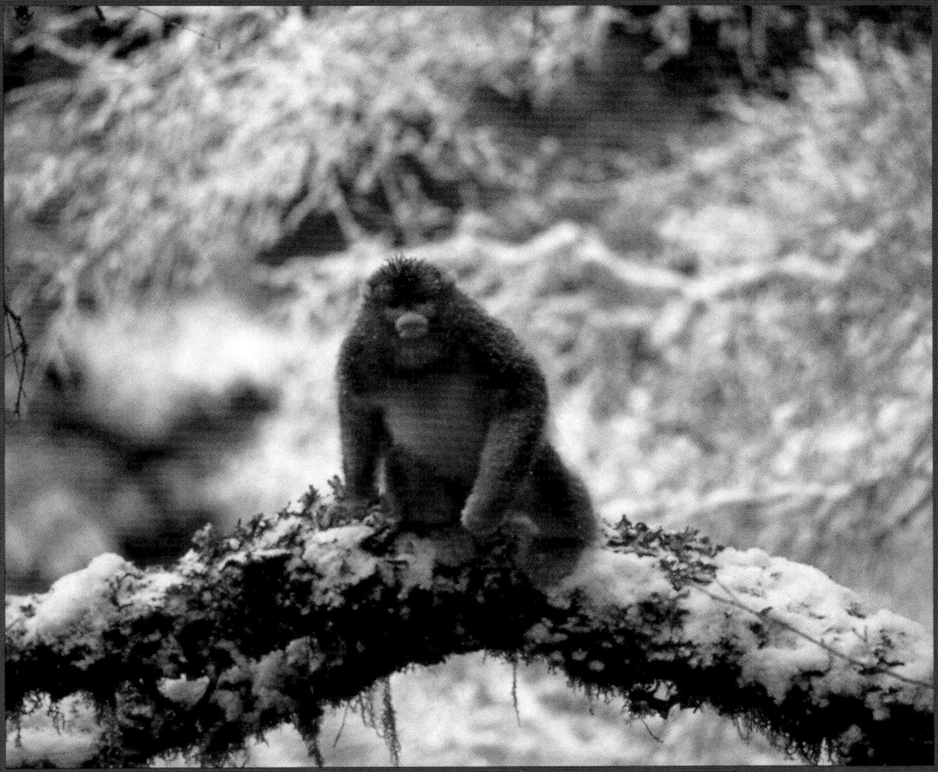

Left, mixed woodland in the Wolong Nature Reserve in Sichuan Province, home of the giant panda and the golden monkey. Above, a male golden monkey (Rhinopithecus roxellanae) *from the Wolong Reserve. This striking species is the second most important flagship in China after the panda. It lives in temperate forests and moves about in large groups of 100 to 300 individuals.*

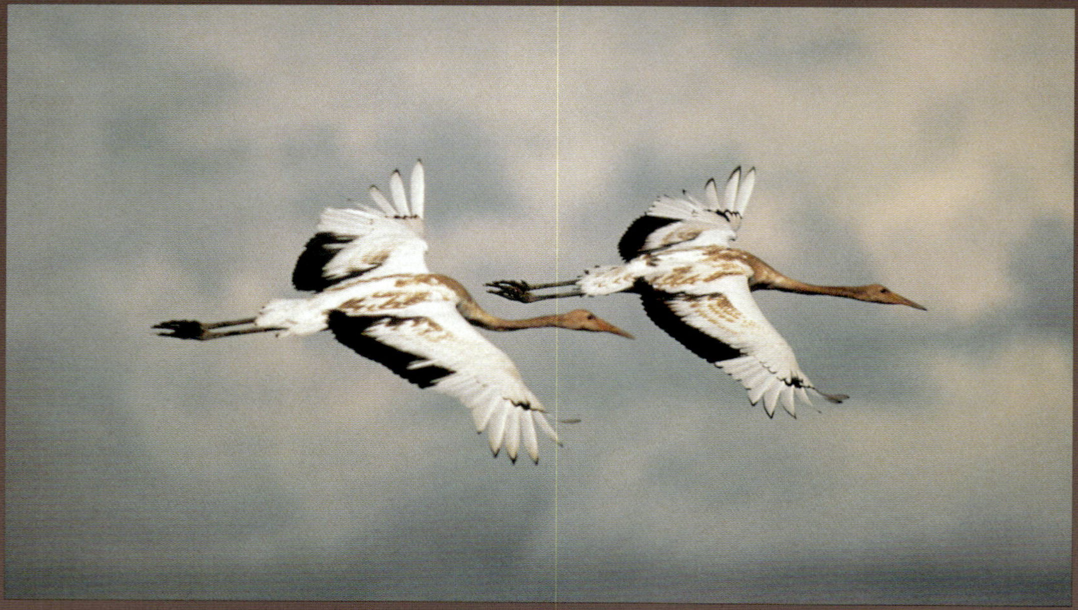

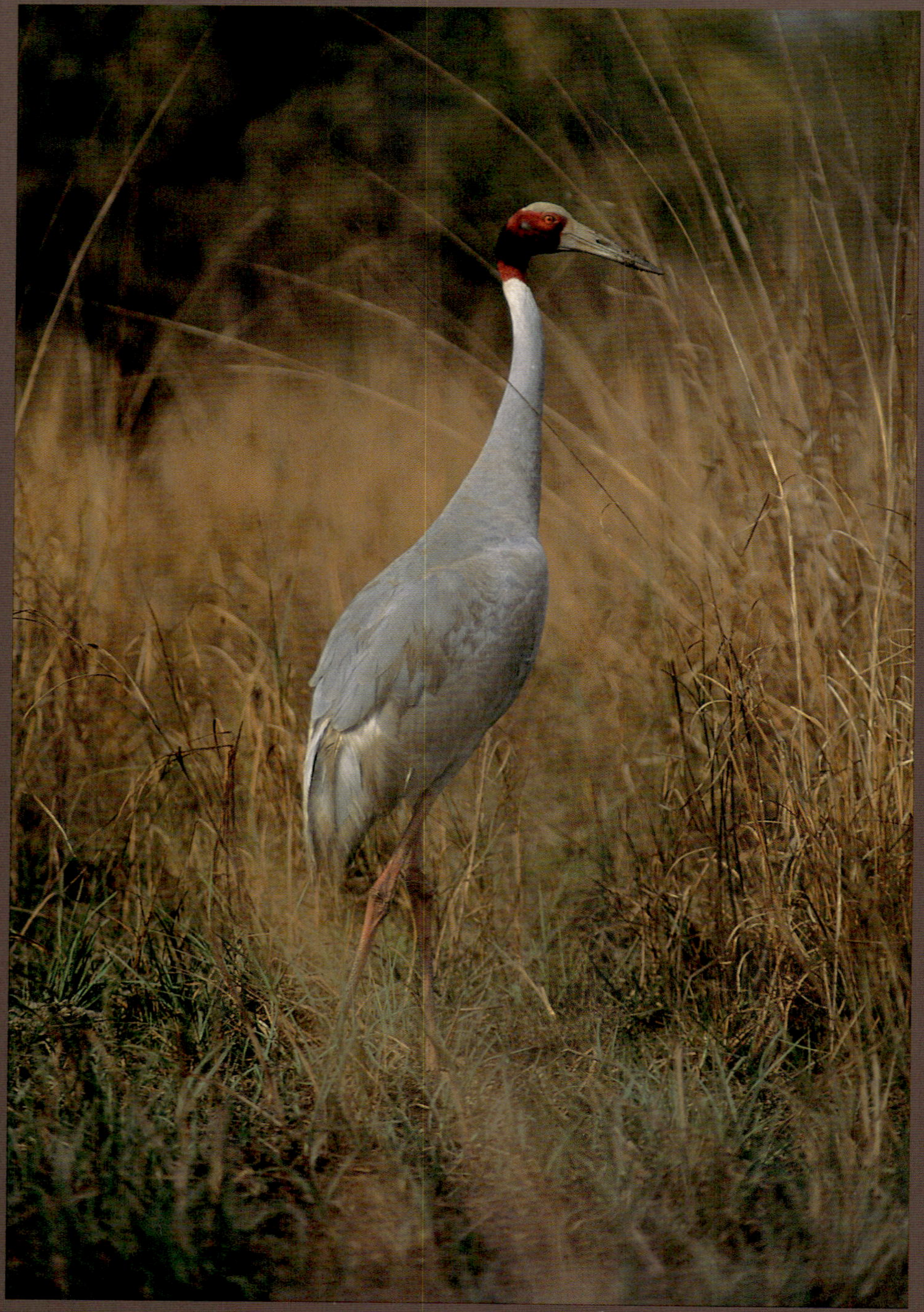

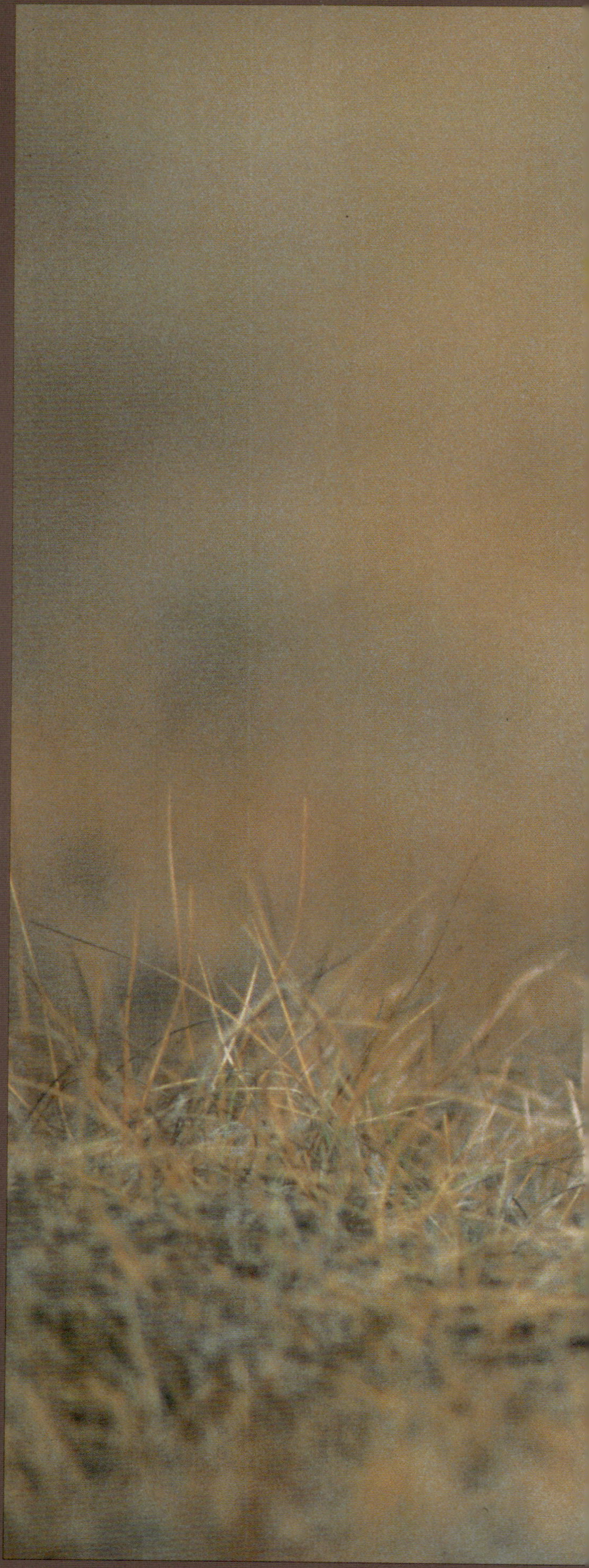

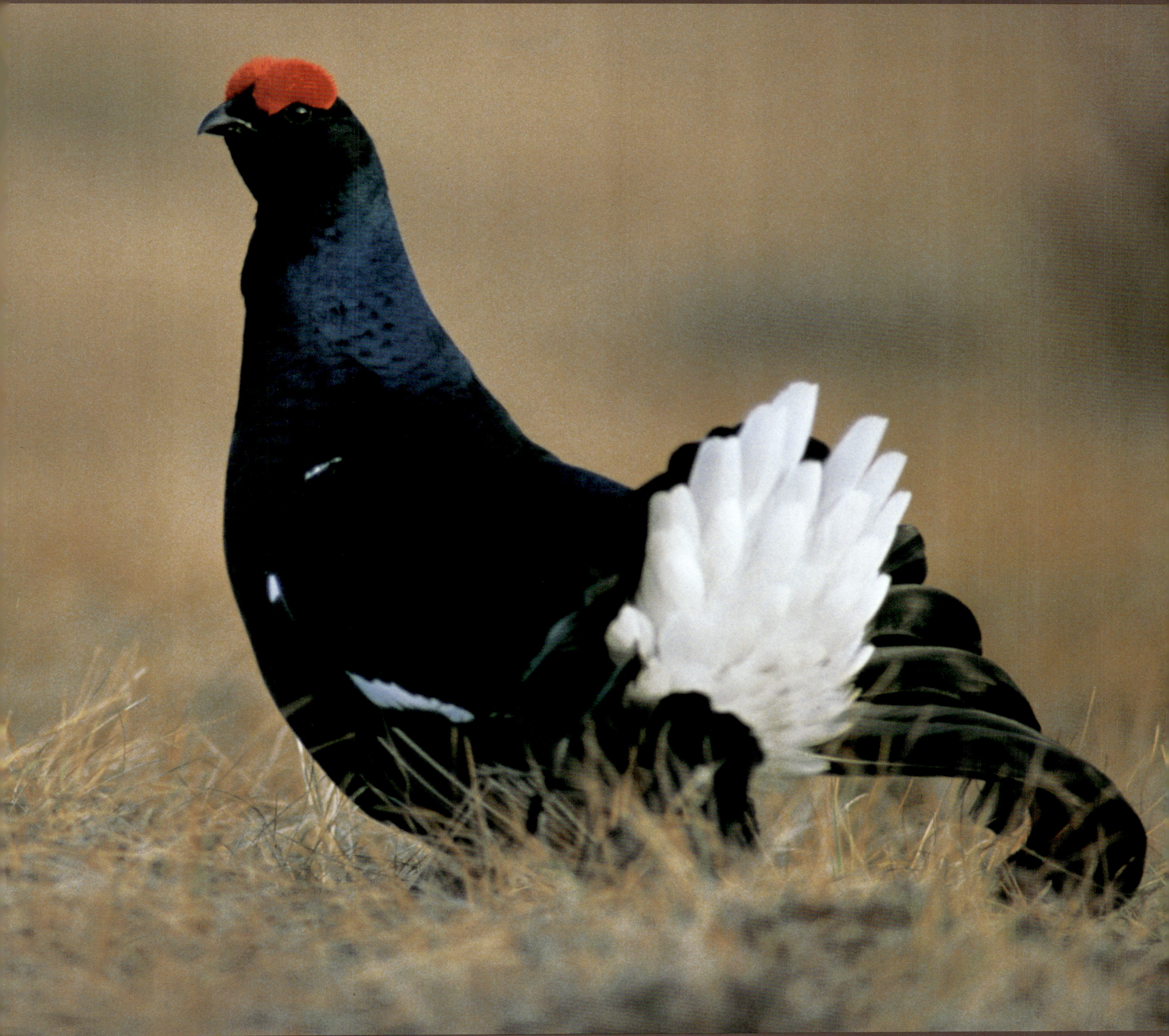

China has 1 244 bird species, some of which are depicted here. Upper left, the Manchurian red-crowned crane (Grus japonensis)*; lower left, the imposing Sarus crane* (Grus antigone) *in marshy habitat; above, a male Eurasian black grouse* (Lyrurus tetrix) *on its lek.*

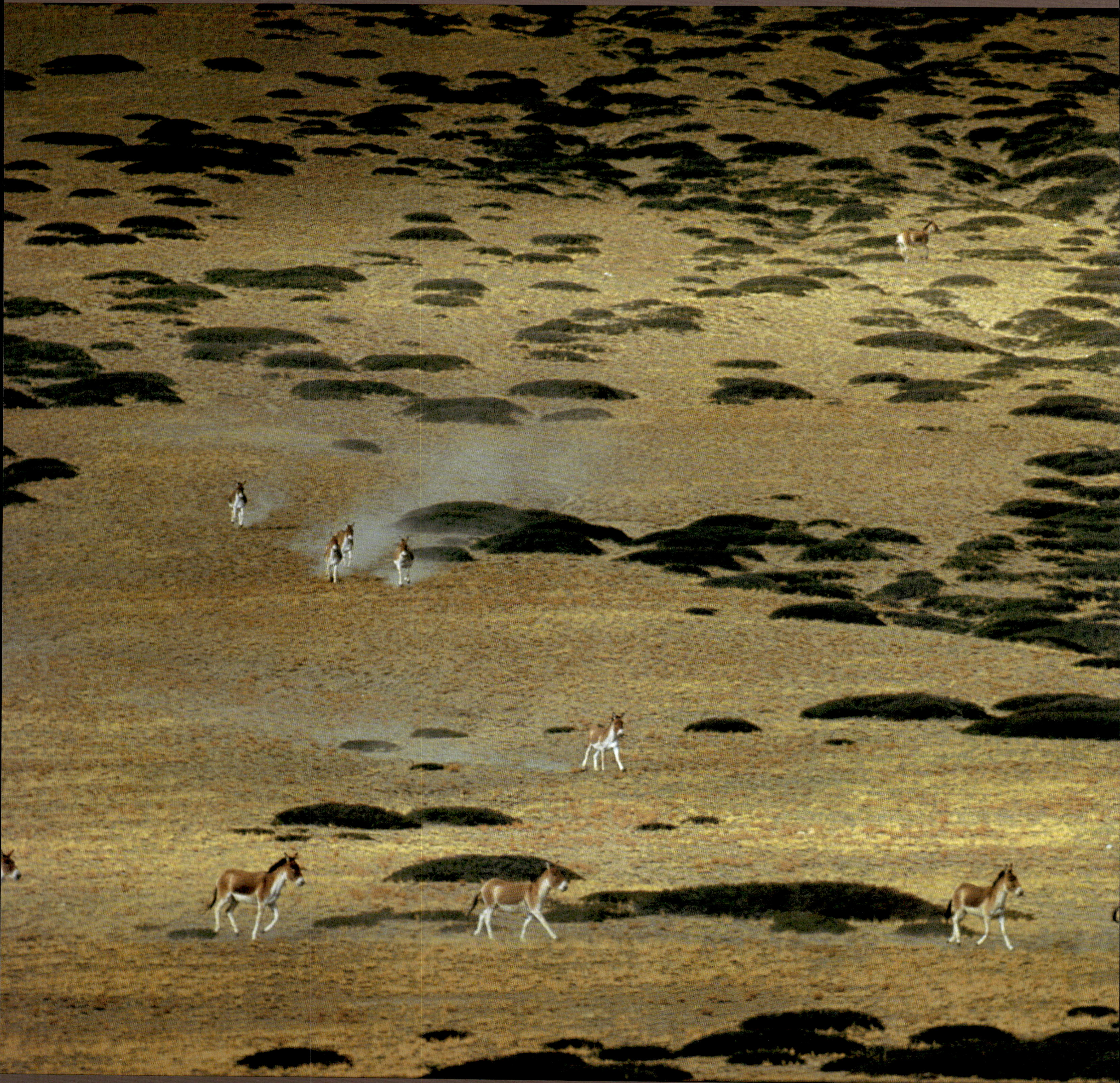

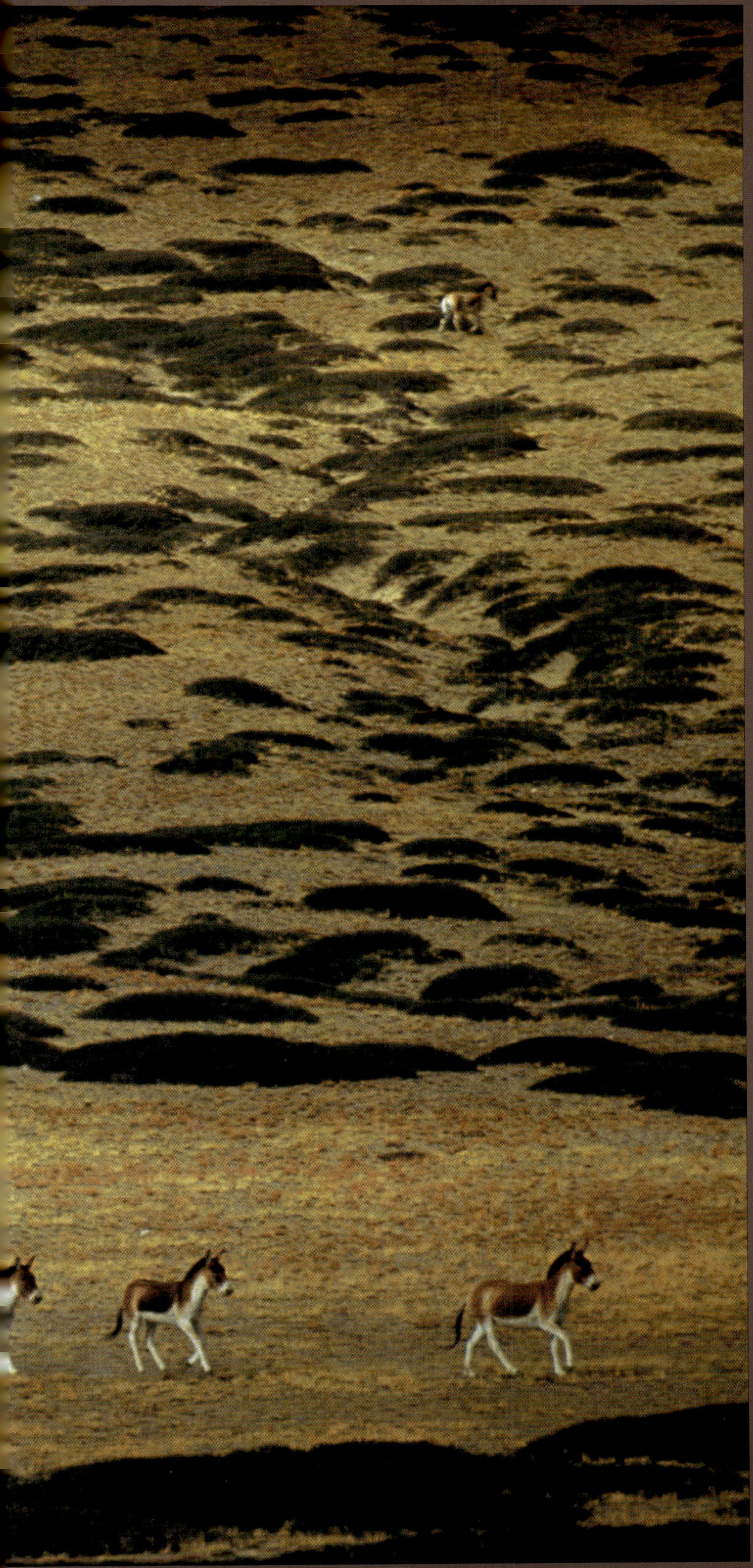

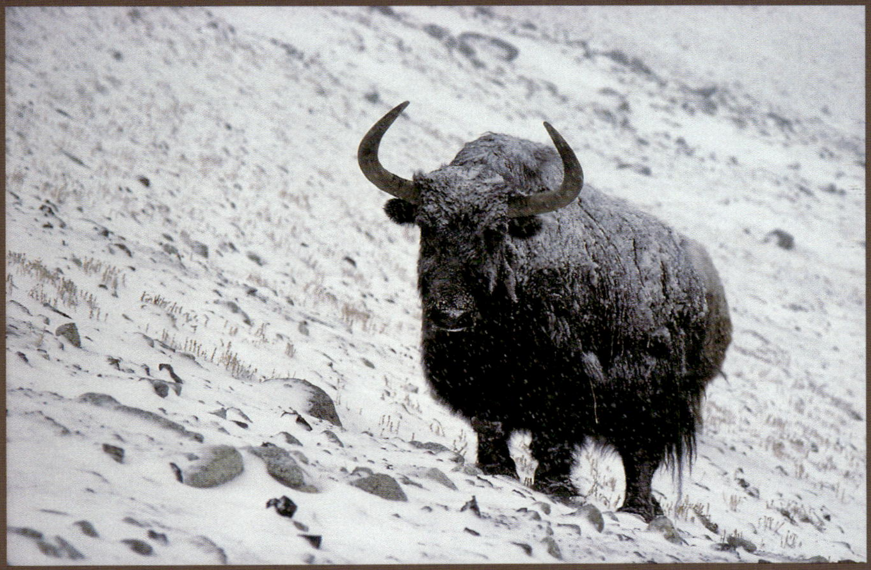

Left, a herd of kiang or Tibetan wild ass (Equus hemionus kiang) *from the Paryang Valley of western Tibet. This species is the most horse-like of the asses. Above, the wild yak* (Bos grunniens) *lives at altitudes of 4 000 to 6 000 m on alpine tundra and ice deserts on the Tibetan plateau. It is the ancestor of the domesticated yak that is so important to the indigenous people of this region.*

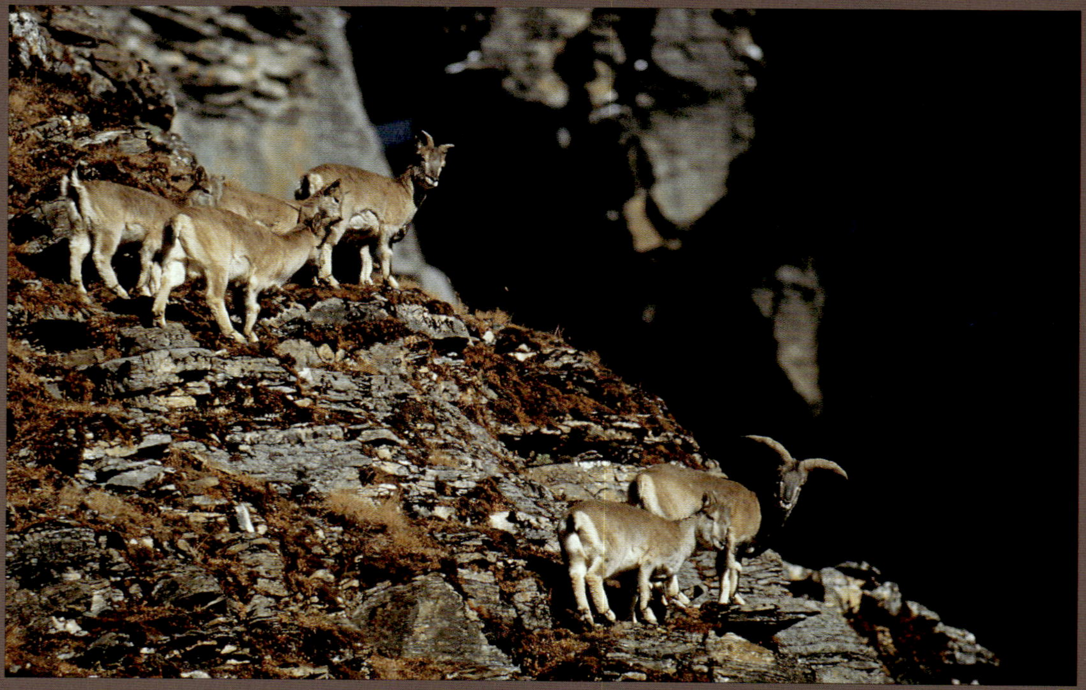

Above, a group of Tibetan blue sheep (Pseudois nayaur). *This species lives on exposed rock and mountain grasslands at altitudes of 2 400 to 6 000 m in Shaanxi, Sichuan, Inner Mongolia, and on the Qinghai-Tibet Plateau. Right, a lammergeier* (Gypaetus barbatus) *landing on a rock outcrop.*

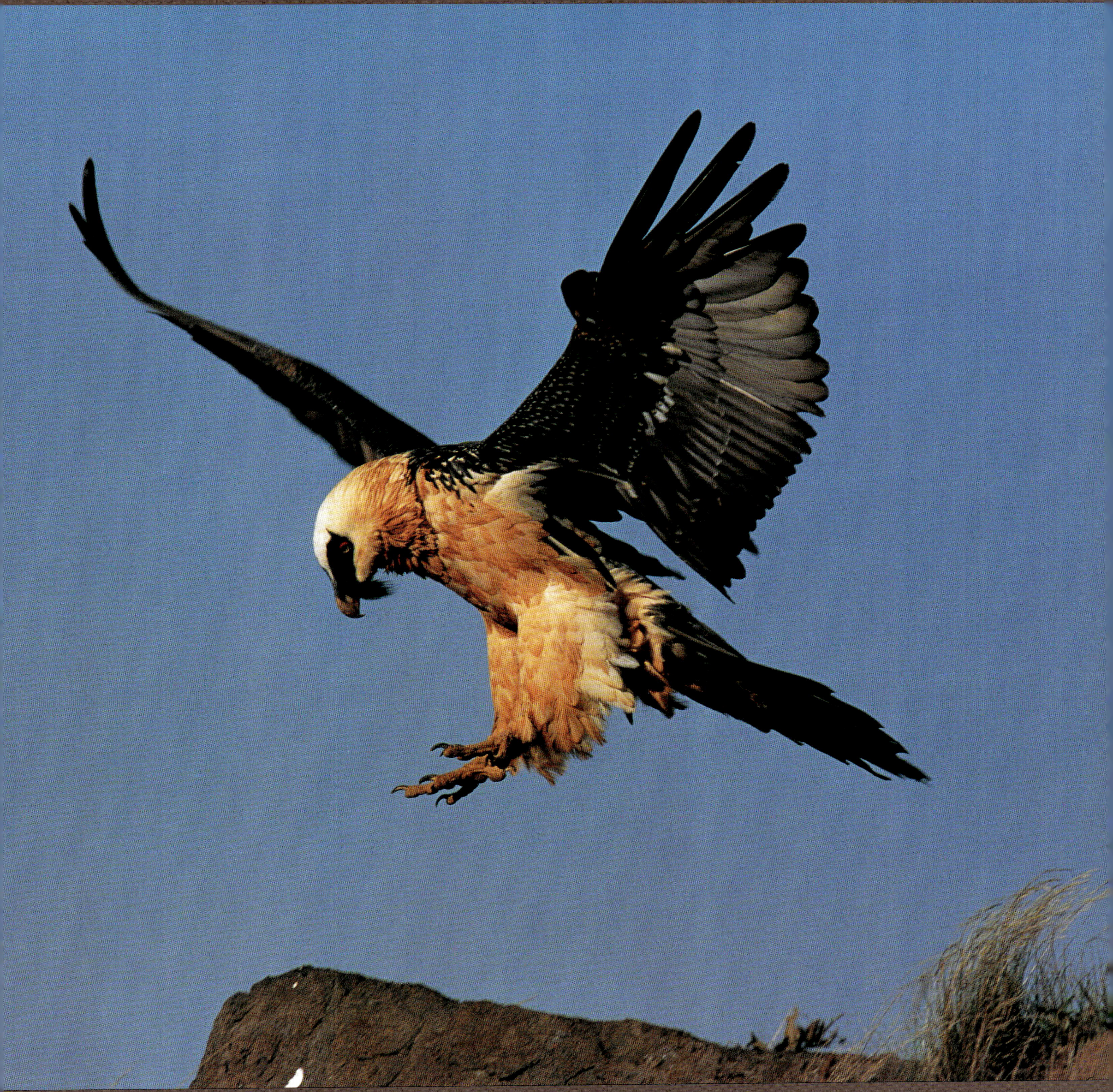

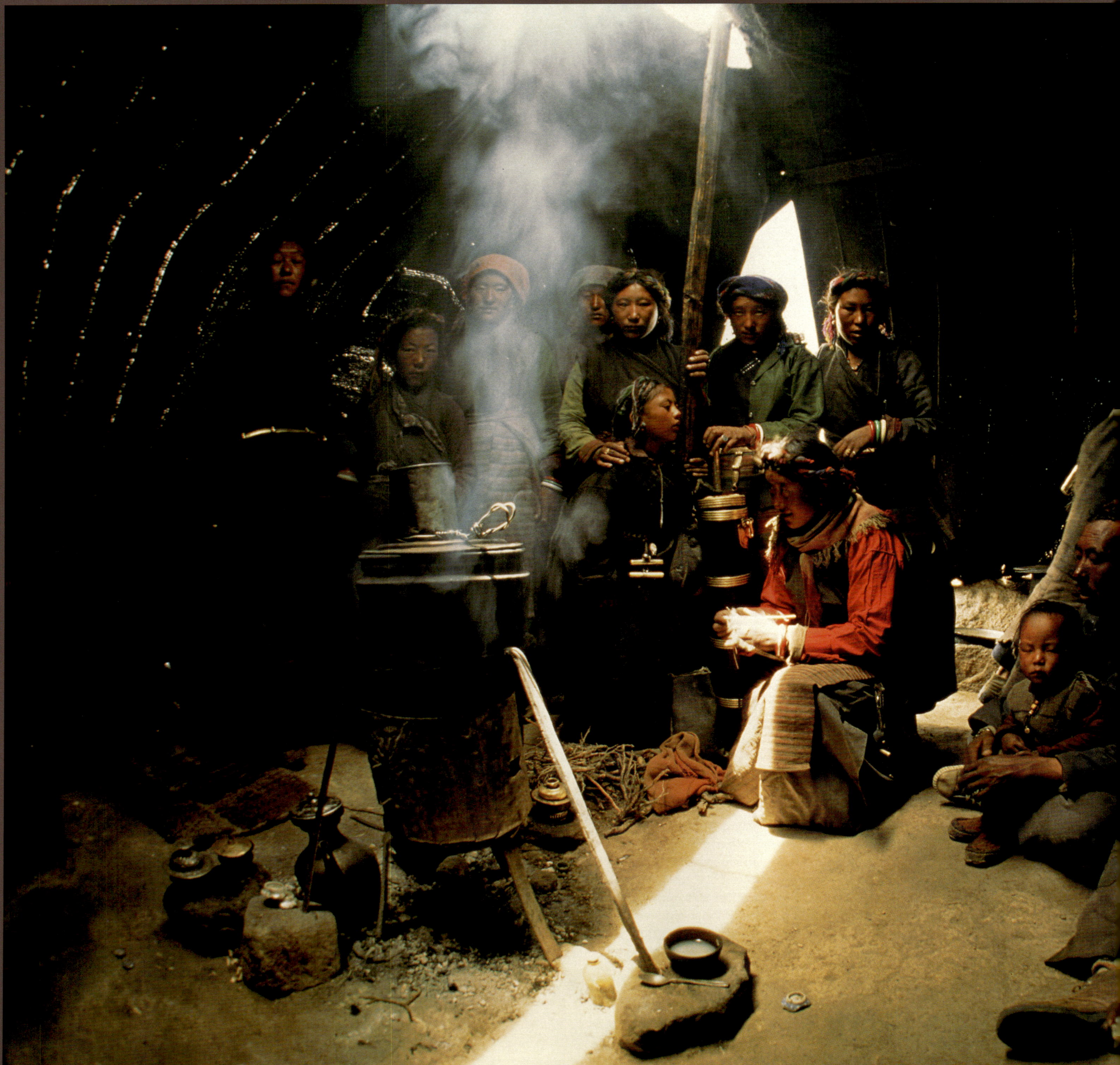

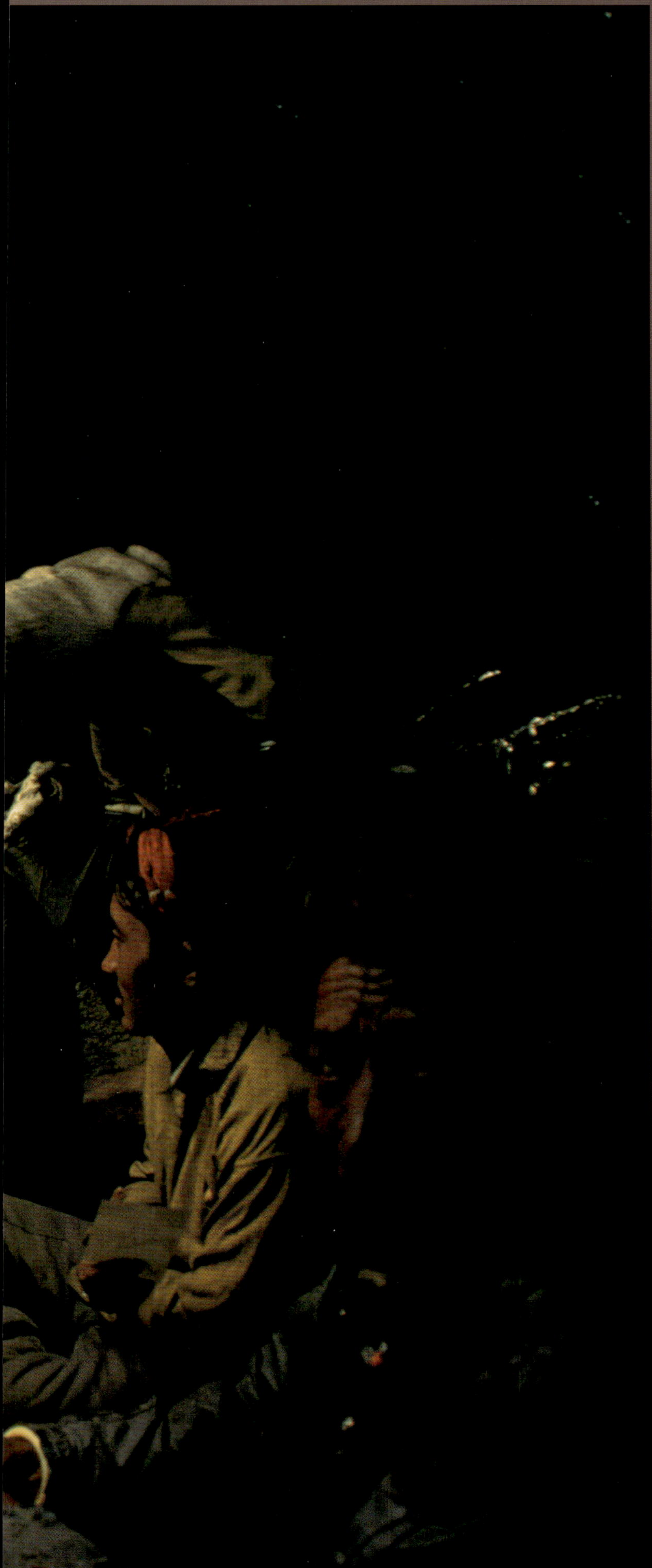

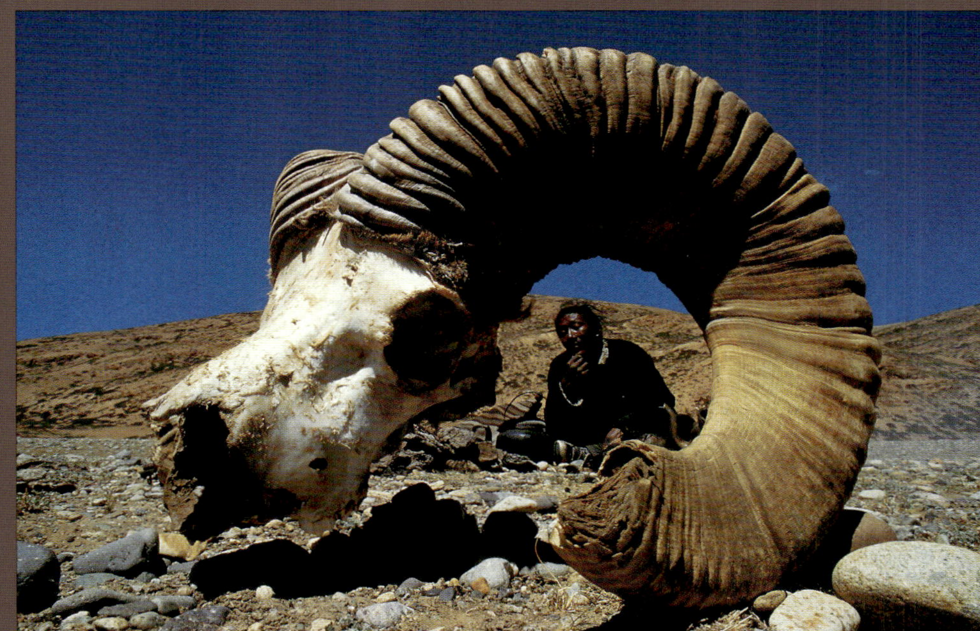

Left, a nomad family in a yak hair tent at Pekhu Tso, in western Tibet. Above, a Tibetan argali skull (Ovis ammon) *framing Tsewang Tsambu, a Tibetan nomad on pilgrimage to Mount Kailas.*

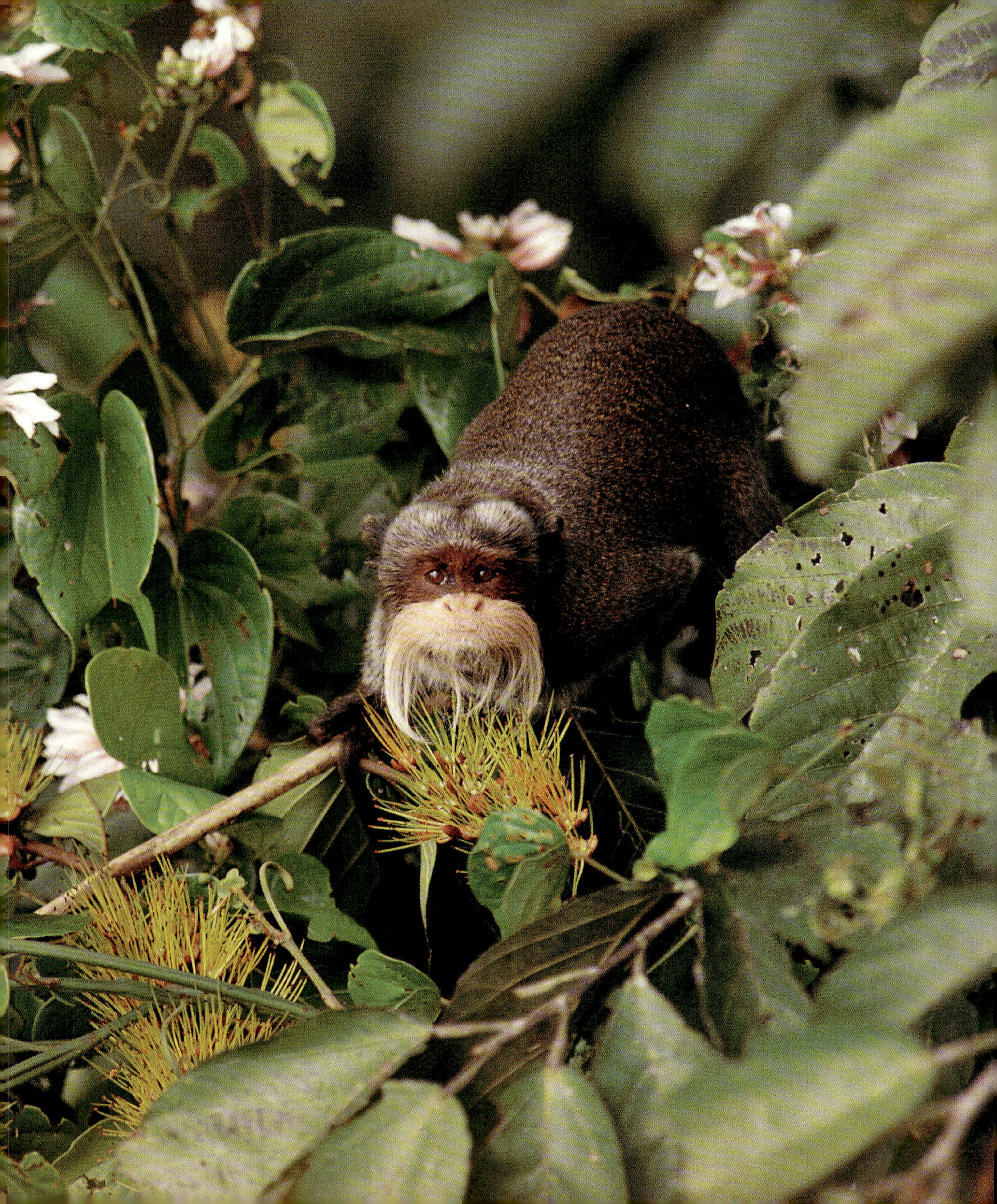

PERU

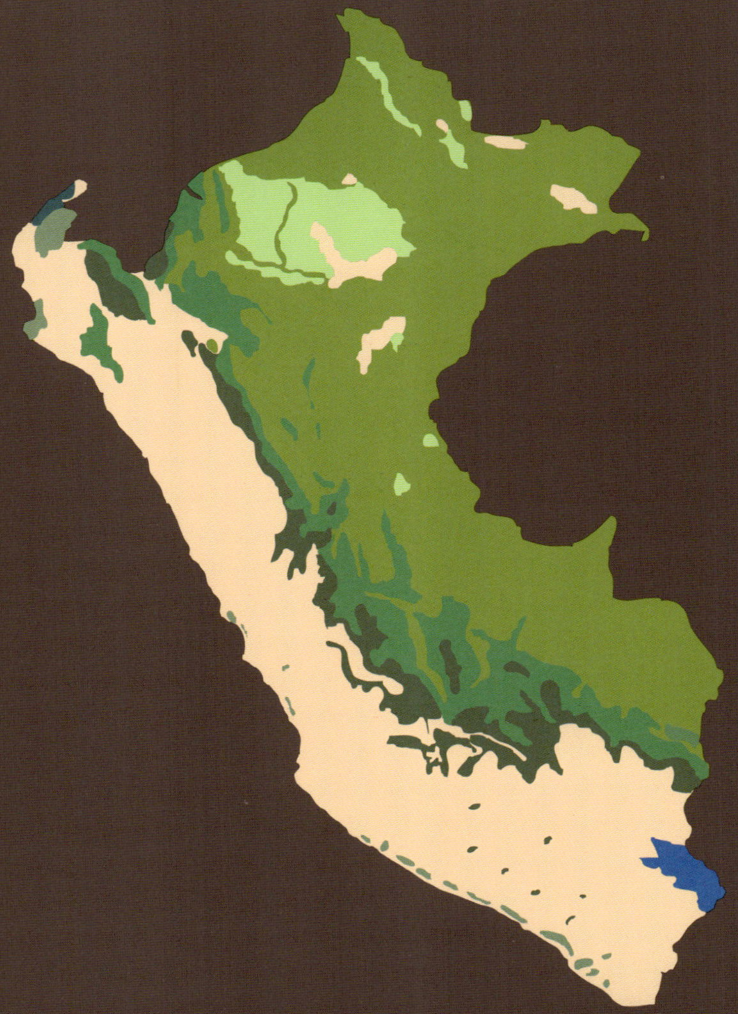

Peru is a medium-sized megadiversity country with an area of 1 285 210 km^2, between Indonesia and South Africa in size and towards the lower end of the medium-sized megadiversity country list. With ecosystems ranging from one of the driest deserts on Earth on the Pacific coast to the snowcapped peaks of the high Andes to the forest-covered eastern Andes and the broad Amazonian lowlands, Peru has one of the greatest concentrations of ecosystem diversity on Earth. It includes a portion of the world's highest-priority threatened biodiversity hotspot, the Tropical Andes, and a significant portion of one of the most important remaining tropical wilderness areas, upper Amazonia. Peru's area of rain forest is one of the most extensive on Earth, its portion of Amazonia (almost 700 000 km^2) ranking fourth in the world. It has very high levels of biodiversity and endemism in certain groups of organisms, especially birds and butterflies.

In terms of species numbers, Peru ranks a very high fifth in total non-fish vertebrates with 2 586 species, and twelfth in endemic vertebrates with 342. Its highest position in the vertebrates is in the birds, for which it ranks second in total diversity, with 1 703 species, behind only Colombia, and third in Endemic Bird Areas (as defined by BirdLife International, ICBP), its total of 18 being exceeded only by Indonesia (24) and Brazil (19). Among the Neotropical countries, it also has the second highest number of Areas for Threatened Birds (89) after Brazil, and the second highest number of threatened bird species (64), again being surpassed only by Brazil.

In mammals, Peru ranks ninth with 344 species and eleventh in endemic mammals with 46. In reptile diversity, it is twelfth with 298 species and tenth in endemic reptiles with 98. Amphibian diversity is seventh with 241 species and amphibian endemism twelfth with at least 89 species.

Freshwater fish diversity is also quite high and largely attributable to Peru's portion of Amazonia. The current estimate is 855 species, good enough for seventh on the world list, but this number is certain to increase as more information becomes available.

Peru is also quite rich in certain groups of invertebrates. It is first in the world in overall butterfly diversity with 3 532 species, and third in endemics with 350, yet it does not even rank in the top 12 in tiger beetles. Unfortunately, good comparative information for most other invertebrate groups is not available.

In plant diversity, Peru is better documented than any of the other Neotropical countries on the megadiversity list, having been the subject of a detailed study by the Missouri Botanical Garden and Peruvian institutions. Its total of 18 000-20 000 higher plant species puts it in the top nine globally in this category, and its relatively low 5 356 endemics still puts it in the thirteenth position in this category.

Cultural diversity in Peru is high given its size, and includes at least 96 ethnic groups. The country's population of 24 370 000 is moderately high for its land area, and includes a mix of Caucasian, mestizo and Asian groups, in addition to the native Amerindians.

Natural habitats

Peru can be divided into four major geographic regions from west to east: 1) the Pacific marine and insular region; 2) the coastal plains; 3) the Andean Sierra; and 4) the Amazon Basin (Selva). This simple division belies a remarkable array of habitat types and landscapes that, depending on the system employed, total 11 ecoregions, 15 biogeographic provinces, or 84 of the 101 described Holdridge life zones, the greatest number of Holdridge zones recorded within the borders of a single country.

Peru's oceans are characterized by cold Antarctic waters that travel north with the Humboldt Current for the entire length of the country; their nutrients make possible highly productive fisheries, perhaps

On the opposite page, the emperor tamarin is one of the most spectacular of the tamarin species, which are characterized by striking facial features like mustaches, pompadours, baldness and other unusual characters. Named after Emperor Franz Josef of Austria, who sported impressive mustachios, this subspecies, **Saguinus imperator subgrisescens,** *is one of two subspecies of the emperor tamarin, and is protected in Manú National Park. Here it is feeding on the flowers of the* **Combretum** *vine.*

the richest in the hemisphere, supporting a large export industry, especially for anchovies (*Engraulis ringens*). The maritime area of Peru occupies about 863 000 km^2 and 3 080 km of coastline; these waters and the numerous coastal islands and peninsulas are home to a variety of fish, invertebrates, seabirds and marine mammals.

Despite its enormous extension, the marine and insular region of Peru has only one protected area, the Paracas National Reserve, on the coast some 250 km south of Lima. Declared in 1975, this reserve occupies approximately 335 000 ha (117 000 ha on land, 218 000 ha marine) between the cities of Pisco and Ica, including the offshore islands of San Gallán in the north and Independencia in the south. Fur seals (*Obryonia* sp. and *Arctocephalus australis*), otters (*Lutra felina*), Humboldt penguins (*Spheniscus humboldti*), a wide array of seabirds and other interesting species can be found in the coastal areas and adjacent islets. The principal species are the Peruvian guanay cormorant (*Phalacrocorax bougainvillii*), the Peruvian booby (*Sula variegata*), and the pelican (*Pelecanus thagus*); they served as the basis of a large guano industry that earlier in this century supplied enormous quantities of this product for the phosphates market (fertilizers and cleaning products). The industry still maintains an active presence through the fertilizer division of the Peruvian company Pesca-Perú.

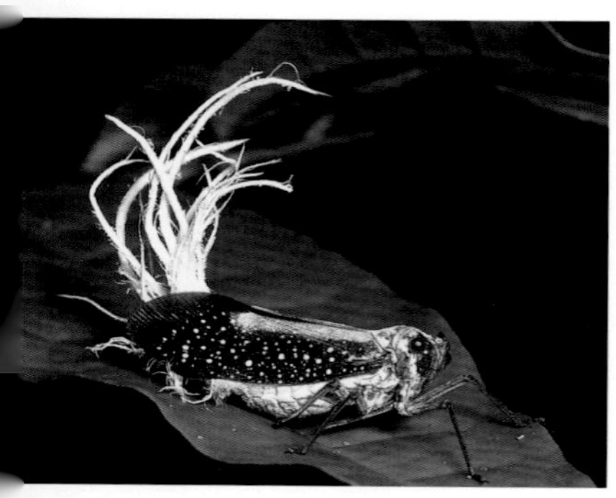

*A lantern bug (*Phanax *sp., family Fulgoridae), with a waxy tail. This family of insects has a variety of species with very unusual shapes and appendages.*

A curious legacy of the once booming, though presently diminished, guano industry is a system of 25 islands and 15 peninsulas equipped with concrete protective walls, docks, buildings, manned guard stations, and other infrastructure necessary for the exploitation of guano. Given urban expansion in coastal Peru over the past half century these areas, which had been so carefully protected for their economic importance to the guano industry, are now the last remaining refuges for many important marine species, especially seals, penguins, and the large aggregations of seabirds for which the areas were initially set aside. Given their conservation importance, the diminished economic importance of the guano industry, plus the presence of existing infrastructure, it has been proposed that a series of changes be made so that these areas can enter into the Peruvian protected areas network through the establishment of research facilities and ecotourism in conjunction with more carefully-managed and reduced guano exploitation. Indeed, the guano industry itself is historic and a potential ecotourism attraction.

Given the poor representation of the marine realm in Peru's protected areas network, the Peruvian Sociedad Pachamama undertook a study in 1990-1991 and identified and proposed two additional marine areas for protected status in northern Peru: the 1 877-km^2 Mancora Reserve for its importance as a nursery for a high diversity of demersal fishes, and the 6 073-km^2 Llescas/Lobos Islands Reserve, which has high biodiversity.

In the coastal plain region, the cold surface water of the Humboldt Current keeps the prevailing easterly winds dry, creating arid conditions for nearly the entire length of Peru's Pacific coast. Mangrove ecosystems occur in the department of Tumbes, but the remainder of the coast is desert with sparse vegetation and localized pockets of dry tropical forest in the north.

The dry forests in the Grau region of the Tumbes and Piura departments of far northwestern Peru represent the southernmost extension of the highly-endangered coastal dry forest ecoregion of southwestern Ecuador. They occupy only 23 342 km^2 (1.8%) of Peru's territory and possess low diversity overall, yet are noteworthy for their endemism and highly-endangered status. The Noroeste Peruano Biosphere Reserve protects the largest block of the dry forest biome in three contiguous areas: the Tumbes National Forest, the El Angolo Game Reserve, and the Cerros de Amotape National Park; Cerros de Amotape is a low-lying coastal mountain formation whose highest point (1 534 m) is within the park.

Resources in the dry forest have been threatened by overgrazing, and in the past by logging, with the result that endemic tree species such as *Ocotea piurensis* and *Zizyphus piurensis* (ebony) are on the verge of extinction. Several animal species have been reduced by hunting, among them the American crocodile (*Crocodylus acutus*), which reaches its southernmost distribution in this area, the mantled howler monkey (*Alouatta palliata*), also at the southern limit of its range, and the southern river otter (*Lontra longicaudis*). Adjacent to the coast and near these dry forests, one finds Peru's only mangrove forests, most of which lie within the Manglares de Tumbes National Sanctuary.

South of the mangrove and dry forest of Tumbes and Piura, the remainder of Peru's coastal zone is desert that stretches all the way to Chile (the Peruvian and Atacama Deserts). This region is dominated by sandy plains and active dunes, with large expanses that support little or no vegetation. All told, these deserts cover 140 000 km^2 (11%) of Peru, have surprisingly homogeneous temperatures year round, with occasional periods of rainfall that are influenced by the El Niño phenomenon that periodically causes great changes in oceanographic and climatic conditions.

In areas of slightly higher altitude in the desert region, one finds a curious "fog zone" effect that creates the conditions for a unique vegetation type known as *lomas*. This unusual formation occurs only in small pockets, and is unique to Peru. Found at just 40 known sites in the coastal zone and on the adjacent islands of Las Viejas, San Gallán and San Lorenzo, these *lomas* cover only 2 000 km^2, contain only 577 plant species, yet have species endemism levels that reach 40%. One of the better known (and protected) *lomas* formations is the Lomas de Lachay National Reserve, a 51-km^2 area just 60 km north of Lima. This and other *lomas* are oases of green in an otherwise bleak landscape. Since their capacity to support life derives almost entirely from their ability to gather water from the surrounding mists, one finds a predominance of epiphytes and especially bromeliads and lichens. A cactus, *Trichocereus lachayensis*, found only in Lomas de Lachay, is one of many range-restricted endemics that have resulted from the unusual combinations of altitude, climate and moisture that make the *lomas* formations possible. *Lomas* are found in two other reserves, Paracas at 1 170 km^2 and the Mejía National Sanctuary at 7 km^2, and several other *lomas* formations have been proposed for inclusion in the Peruvian Conservation Units System.

Another highlight of Peru's coastal desert are the Plains of Nazca or Pampa Colorado. An immense desert plateau covering some 1 500 km^2, the Plains of Nazca are crisscrossed with lines that from the ground appear to be random, but when viewed from an aircraft are revealed to be intricately-designed glyphs depicting animals like monkeys and hummingbirds. Some of these glyphs are over 3 km long and despite years of interest and theories regarding how works such as these could have been produced by early civilizations, their true origin remains a mystery. Indeed, Peru is full of such mysteries, including how the Inca or pre-Incan civilizations could have built (without the wheel) such amazing lithic structures as Saqsaywaman outside Cuzco, which is over 500 m long and constructed of more than 200 000 carefully-sculptured stones weighing up to 75 tons each, yet fitting so precisely with neighboring stones that not even a knife blade can be slipped between them.

Moving eastward one encounters the Andes, one of the highest mountain ranges on Earth. The Andes and their multiplicity of microhabitats, especially on the eastern slopes, are the richest region on Earth for biodiversity. In Peru, the Andean region occupies about 27% of the country's area and stretches from north to south, bounded by the dry coastal plains to the west and the lowland wet forests of the Amazon Basin to the east. The country's highest point, Huascarán at 6 773 m, is in Peru's continental divide, the Cordillera Blanca. The Peruvian Andes and parts of adjoining Bolivia are among the key Vavilov Centers (of origin and diversity of crop species) and are also important for economically valuable plants, particularly tubers such as the potato (*Solanum tuberosum*) and its hundreds of wild relatives; grains like amaranth (genus *Amaranthus*), which is the world's most nutritious plant; and a variety of edible fruits and medicinal plants. As in Colombia and Ecuador, the Peruvian Andes have been continuously occupied by humans for thousands of years. It is said that Peru harbors the densest high-altitude population in the world, and the human presence manifests itself in deforested slopes and eroded valleys. The Peruvian Andes can be divided into a number of different habitat types including the *yungas*, the *puna*, and the eastern slopes that descend into the Amazon Basin.

In the far northern portion of Andean Peru (department of Piura) and extending into Ecuador, one finds the Marañón Gap region, or Huancabamban depression; this is a low area that goes down to 500 m, and is considered to be a natural barrier dividing the northern from the southern Andes. It is one of the most important barriers to plant and animal dispersal in the Andes, especially for birds, rodents, reptiles, amphibians, and butterflies, and also serves as an east-west corridor between the Amazon and the Pacific (Gentry, 1982). The southern Andes below the Marañón Gap are older and narrower than the northern, and their biodiversity is distinct from that found to the north in Ecuador, Colombia and Venezuela.

The Peruvian *yungas* is a montane cloud forest zone including the Huancabamban depression, and extends south along the slopes of the Andes generally between 600-3 500 m altitude, covering an area of approximately 250 000 km^2. These Andean slope forests are among the richest in terms of diversity and endemism on Earth, and are home to the spectacled bear (*Tremarctos ornatus*), the pudu or *sacha-cabra* (*Pudu mephistopheles*), the yellow-tailed woolly monkey (*Lagothrix flavicauda*) and spectacular birds like the horned guan (*Crax unicornis koepckeae*). The greatest threats to the *yungas* come from the expanding agricultural frontier and the illegal cultivation of coca which, in addition to deforestation, is responsible for dumping large quantities of chemicals into the watercourses. The Río Abiseo, Yanachaga-Chemillen, Cutervo, Tingo María and Manú National Parks provide protection for *yungas* ecosystems, as do nine other Peruvian conservation units.

Yungas gives way to *puna* at higher altitudes. The *puna* is a distinct vegetation type found predominantly in Andean Peru, but also in adjacent areas of Bolivia. It is a high-altitude (above 3 300 m), alpine-like vegetation below the snow line that can be found covering most of the departments of Huancavelica, Ayacucho, Apurimac, and Puno, plus smaller portions of several others. Dominant plant species in these cold, relatively dry habitats are bunchgrasses (*Calamagrostis*, *Festuca*, *Stipa*), supplemented by a variety of herbs, grasses, sedges, lichens, mosses and ferns. Animal diversity is low overall in the *puna*, though there are at least 30 species of birds restricted to this zone, including the puna rhea (*Pterocnemia tarapacensis*), the ornate tinamou (*Nothoprocta ornata*) and the Andean condor (*Vultur gryphus*), the latter being an important national flagship species symbolic of this habitat. The *puna* is also prime habitat for the vicuña (*Vicugna vicugna*), the smallest of the South American camelids. It is also a great genetic storehouse for the wild relatives of commercially valuable crops, such as the wild tomato (*Lycopersicon hirsutum* and *L. chilensis*), and has been used as a source of traditional medicinal plants and foods for millennia.

Today, as many as 6 million Peruvians have a direct impact on the *puna* ecosystem in the cities of Cajamarca, Cerro de Pasco, Cuzco and Puno, making it one of the country's more threatened ecosystems. *Puna* is impacted by the mining industry, with toxic runoff and contamination of waters being a major concern. In addition, the *puna* is important grazing land for domesticated cattle, sheep and both wild and domesticated camelids, constituting perhaps the greatest threat to the integrity of this Andean ecosystem. Fortunately, puna habitats are protected in five national reserves (Calipuy, Junín, Pampa-Galeras, Salinas-Aguada Blanca, and Titicaca), three national sanctuaries (Huayllay, Calipuy, and Ampay), and three historical sanctuaries (Chacra-

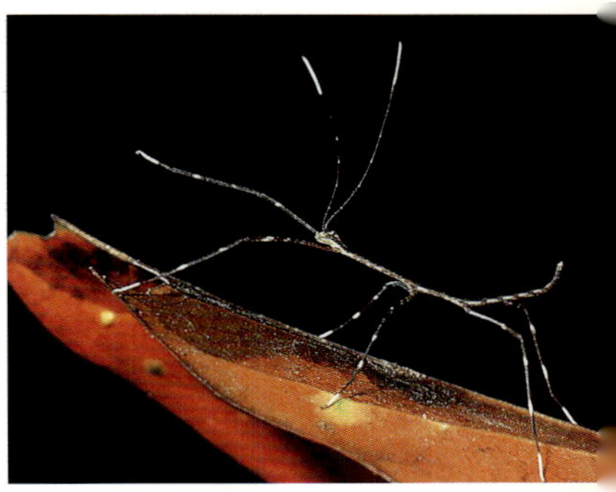

A stick insect (family Phasmidae) on the forest floor in Manú National Park, Peruvian Amazonia.

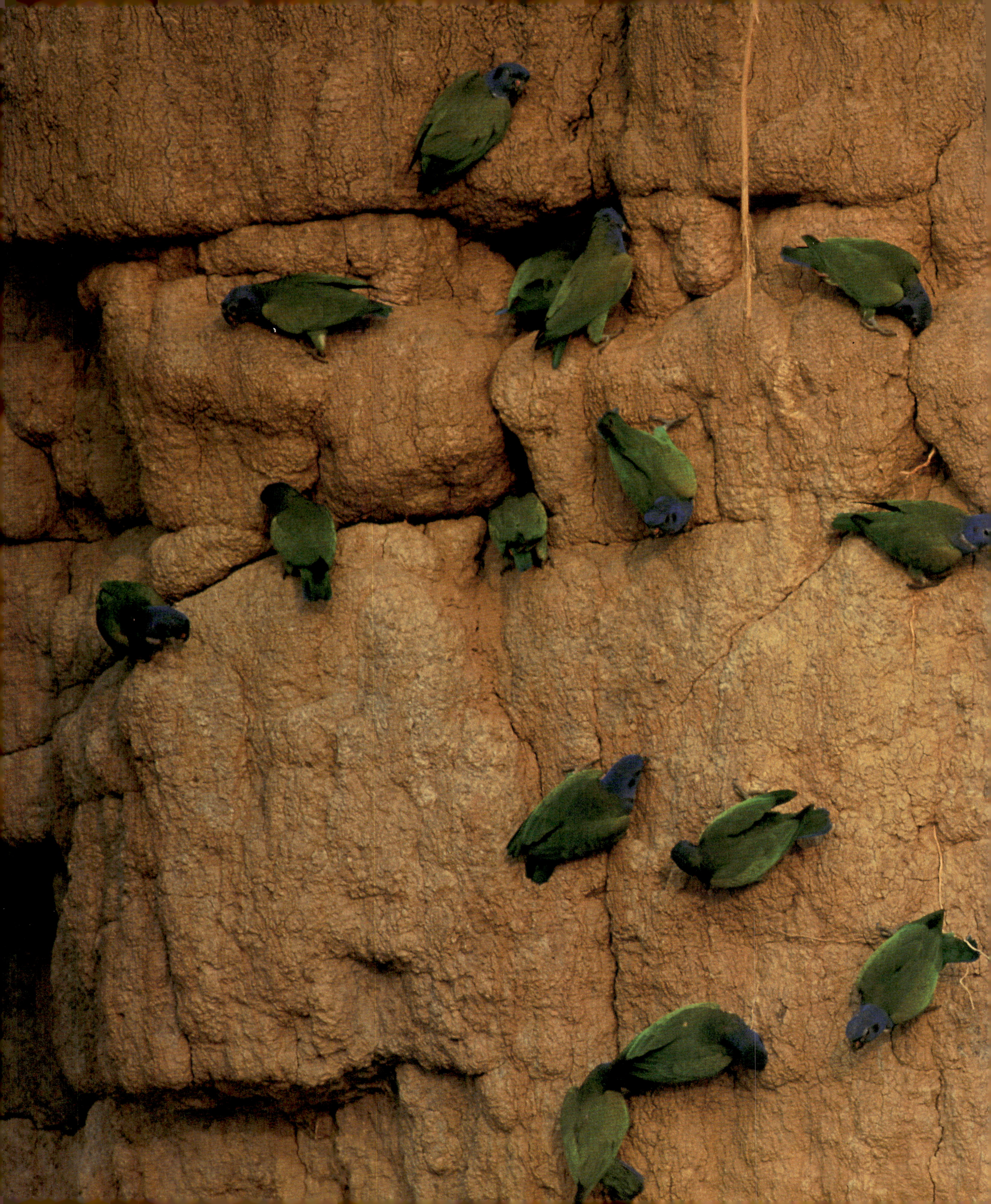

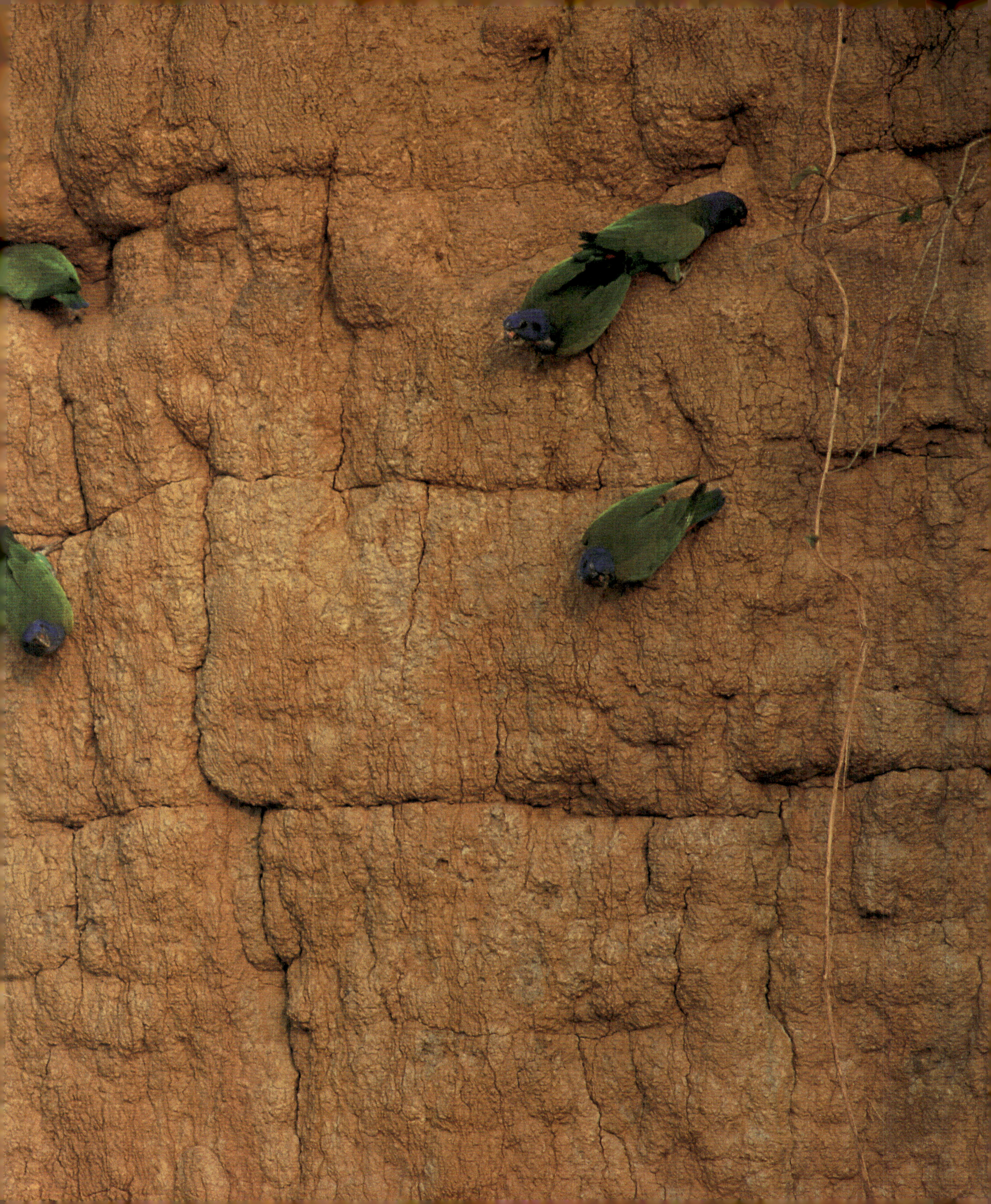

marca, Pampas de Ayacucho, and Machu Picchu), plus a Biosphere Reserve (Huascarán), which protects fully 3 400 km^2 of *puna* ecosystem. The Río Abiseo and Manú National Parks also protect a portion of *puna* habitat.

At the far eastern extent of the Peruvian *puna* biome is Lake Titicaca, which straddles the border between Peru and Bolivia. At about 3 800 m altitude, it is the highest navigable body of water in the world. It was an important cultural center for the Incas and pre-Incas and has a most interesting history; to this day, indigenous communities of the Uru people live on islands in the lake, and navigate using handmade rafts crafted from the large expanses of papyrus-like totora reed that fringe the edges of the lake. Titicaca is an important nesting area for several shorebird species, and is home to various endemic species, one of the most noteworthy being a giant frog (*Telmatobius* sp.) that inhabits the lake bottom.

Finally, the Peruvian portion of Amazonia, known as the *Selva*. This is by far the largest portion of the country, covering about 54% of the land area, or 698 521 km^2. It is a rain forest region of great global significance, and places Peru fourth on the world list of total remaining tropical rain forest cover, behind only Brazil, Zaire and Indonesia. It is also particularly important because it lies wholly within the portion of Amazonia known as upper Amazonia, which has some of the most intact remaining wilderness areas and the highest concentrations of diversity in the entire region. Finally, Peruvian Amazonia transitions into the eastern slope of the tropical Andes, generally agreed to be the richest and most diverse biodiversity hotspot on Earth. Not surprisingly, several sites in Peruvian Amazonia and especially the transitional areas between the Andes and the lowland Amazon, hold world records for single-site species diversity. In the Tambopata area in the department of Madre de Dios, for example, scientists have registered 545 bird species, 91 species of mammals, 127 species of reptiles and amphibians, and 94 fish, 29 dragonfly, 40 termite, 600 beetle, 73 horsefly, 39 bee and 1 207 butterfly species in a single 550-ha site; the butterfly count is the highest thus far recorded anywhere on Earth. In the 15 238 000-ha Manú National Park to the northwest of Tambopata, over 900 bird species have been found —over 200 species more than the United States and Canada combined.

As one descends from the high Andes to lowland Peruvian Amazonia, one passes through a series of different ecosystems, with enormous concentrations of globally-important diversity packed into a very narrow belt of land. Cloud forests occur at 3 500 m down to 1 500 m, and are characterized by high humidity and rainfall. They are rich in epiphytes and tree ferns, and have especially high orchid diversity. Known in Peru as the *Ceja de la Selva*, or "Edge of the Forest," this region is at the heart of the Tropical Andes biodiversity hotspot, often referred to as the "epicenter of global biodiversity." Among the tree species in these cloud forests are cascarilla (*Cinchona* spp.), carapacho (*Weinmannia* sp.), *Alnus jorullensis*, *Chusquea* spp., and *Gynoxys* spp. *Podocarpus*, the only genus native to Peru, is also found in these forests. These cloud forests occupy roughly 139 250 km^2 of the Selva region, or about 19.9% of this area.

Moving down to the low Selva, the dominant formation is closed mixed forest, with species like Brazil nut (*Bertholletia excelsa*) and rubber (*Hevea brasiliensis*), and commercially valuable species like mahogany (*Swietenia macrophylla*), cedars (*Cedrela* spp.), *ishpingo* (*Amburana cearensis*) and *tornillo* (*Cedrelinga catenaeformis*). These forests are rich in wildlife, especially species like primates that reach their greatest densities in these lowland rain forest formations. Indeed, Peru with 32 primate species, most of them found in this lowland rain forest area, ranks fifth in this important group of mammals. Closed mixed forest covers approximately 480 037 km^2 or about 68.7% of the Selva area.

Other smaller formations in the Selva include swamp forests, marsh forests, *pacales* and *Podocarpus* forests. Swamp forests include the famous *aguajales*, vast palm swamps dominated by species of the genus *Mauritia*, mainly *Mauritia flexuosa*. This is a valuable species that produces thatch for houses and a fruit rich in oil and vitamin B. Also found in these forests are members of the palm genera *Euterpe* and *Jessenia*. *Euterpe* palms, known as *huasai*, are the source of *chonta* or heart of palm, and *Jessenia*, known as *hungurahui*, produce an oil similar to olive oil and nuts that can be used to make soy meal. These swamps account for 16 033 km^2 of the Selva region, and marsh forests, which are seasonally-flooded formations, cover 46 740 km^2. *Pacales* are bamboo-dominated open forests. The bamboos form large clumps which reach up into the canopy at 30 m, and spread over the interspersed trees. The bamboo species are commonly *Merostachys* spp. and *Bambusa* spp.; tree genera in this unusual formation include *Schizolobium*, *Triplaris*, and *Perebea*. These forests cover about 14 933 km^2. The remainder of the Selva region is made up of *Podocarpus* forest, which covers only 1 528 km^2.

Although there is timber exploitation and some 540 sawmills in Peruvian Amazonia, most of the timber harvesting is carried out by small-scale operators working at low productivity and depending on rivers for log transportation. Most timber extracted is used within Peru, with less than 1% being exported. Several experimental forest management and silviculture systems have been tested in the country, notably the Jenaro Herrera Project, a collaborative project between Swiss Technical Cooperation and the Peruvian Amazon Research Institute (IIAP), based in Iquitos, and the Yanesh Forest Cooperative in the Palcazú Valley of the department of Pasco, a joint effort of USAID, the Peruvian Foundation for the Conservation of Nature, and World Wildlife Fund-U.S. While producing interesting results, these projects have been focused on sustained timber harvesting, with biodiversity conservation being at best a secondary consideration. Nonetheless, both projects rank among the most significant efforts at sustainable timber management any-where in the tropical world.

Nontimber forest products have great potential as

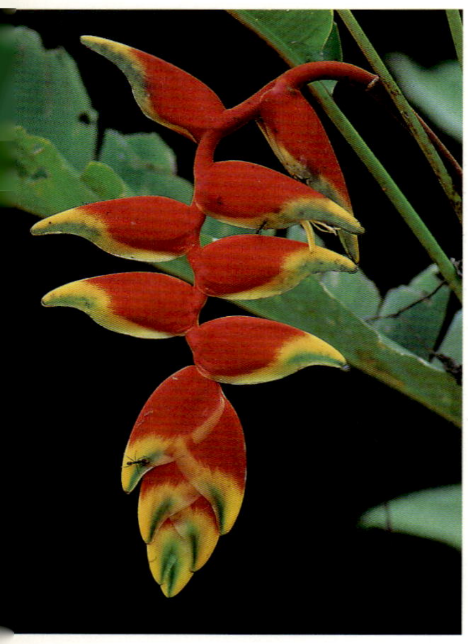

On pp. 286-287, the blue-headed parrot (Pionus menstruus) *on a collpa in Manú National Park. These collpas, which are visited by hundreds of macaws and parrots every day, are a unique feature of parts of upper Peruvian Amazonia, and are particularly impressive in Manú and in the Tambopata-Candamo region further to the south.*

Above, heliconia flower (Heliconia sp.) *in Peruvian Amazonia. The genus* Heliconia, *also known as bird-of-paradise, is a diverse and popular group belonging to the banana family, Musaceae.*

well, and include a wide variety of materials such as medicinal plants, edible fruits and ornamental species, dyes, resins, latexes, and construction materials like thatch. By far the most important of these are Brazil nuts and rubber, both of which are collected by local communities. The estimated potential production of Brazil nuts is 2 100 metric tons, worth about $ 32 million U.S. per year; for rubber it is 500 metric tons, worth about $ 700 000 U.S. annually.

However, the most sustainable and probably the best long-term source of income from the use of these forests is ecotourism. On a site-specific basis, Peru has what are probably the richest and most diverse rich-soil sites in Amazonia, characterized by unusually high densities of forest wildlife, in addition to high diversities of species. Manú National Park, created in 1973, is probably the single most impressive tropical rain forest site on Earth, with a density and diversity of wildlife without equal. Indeed, in the experience of the authors of this book, which spans all the tropical continents and covers all the major rain forest areas on Earth, Manú National Park, and especially the Cocha Cashu site, is simply unmatched. In a brief visit, one can see giant otter, black caiman, a dozen species of primates, including large ones like the black spider monkey (*Ateles chamek*) and the woolly monkey (*Lagothrix lagotricha*), an amazing diversity of birds, many different reptile and amphibian species and, with considerable luck, perhaps even a jaguar or ocelot. Many of these animals, especially the giant otter, the black caiman and the large primates, are now locally extinct or extremely hard to find elsewhere in Amazonia, yet are still easy to find in Manú.

The Tambopata-Candamo area, south of Manú, is comparable in importance. A large area, at 1 200 000 ha roughly equivalent to Manú in size, it too has been proposed for protection, with Conservation International and local partners taking the lead for the past seven years in trying to have it set aside as a national park. The process of park creation has already begun, with the establishment of the 325 000-km^2 Bahuaja-Sonene National Park in 1996. This was the first new national park declared in Peru in eight years, and the first in President Fujimori's administration. Although the declared park area is about one third of that originally proposed, it is an important first step and represents what might be a unique and precedent-setting policy innovation in that it is perhaps the world's first *expanding* park.

The reason for this unusual process is that before the park could be officially declared, an oil concession was granted by the Peruvian Ministry of Energy and Mines in March, 1996 to a consortium headed by Mobil Oil. The concession overlaps with a substantial portion of the original park proposal's core protected zone, posing a potentially severe conflict over land-use rights and the potential loss of one of the world's biologically richest ecosystems to oil development. The signing of Mobil's concession had been expected, so Conservation International had already been in direct dialogue with Mobil and Peruvian authorities about possible alternatives. Before the concession was even signed, Conservation International had been in close contact with Mobil to ensure that their seismic exploration in the region would have minimal impact, and that it would take into consideration the long-term integrity of the region's biodiversity and indigenous communities. The result is that Conservation International is now conducting on-the-ground monitoring of Mobil's exploration activities as they take place, in order to carefully document all biological and social impacts and to redirect Mobil's efforts should they threaten either the region's biodiversity or its human communities. National and international efforts to protect this region, while at the same time allowing carefully conducted oil exploration, led the Peruvian government to declare the Bahuaja-Sonene National Park, the supreme decree of which allows for the expansion of the park over time into areas relinquished by Mobil from its concession area. The Mobil contract stipulates periodic mandatory relinquishments over the seven-year period of the concession and also allows for voluntary relinquishments. Under normal circumstances, these relinquished lands would be resold as concessions to other oil companies, but in this case they should go into the park. This represents an important and hopefully precedent-setting model for a win-win situation in which park establishment goes hand in hand with oil or mineral exploitation.

There is no doubt that the Peruvian Selva, Peru's portion of greater Amazonia, is one of the most significant and diverse rain forest regions on Earth. The fact that it is still largely intact offers many options and opportunities for this to become one of the few regions where natural resource-based sustainable development can actually be achieved.

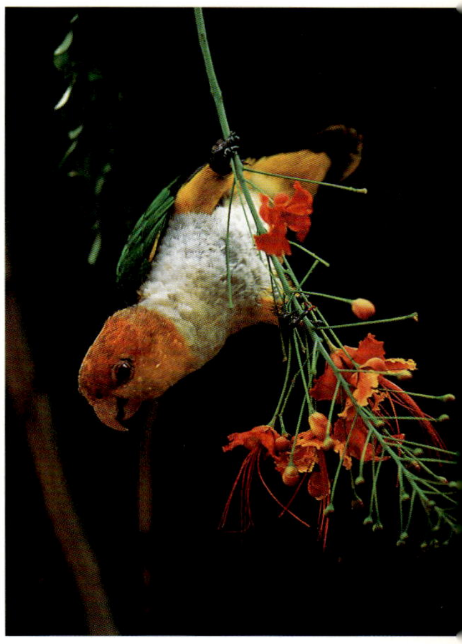

A white-bellied parrot (Pionites leucogaster), *one of the many psittacine species found in Peru. This individual was photographed in the Tambopata-Candamo region.*

Flagship species

With one of the world's most diverse vertebrate faunas, Peru has a number of globally important flagship species. Many of these are mammals, but there are a number of important bird and reptile flagships and several plant species that deserve this status as well. Perhaps the most important Peruvian flagship species is the Peruvian yellow-tailed woolly monkey or *mono choro de cola amarilla* (*Lagothrix flavicauda*). The largest mammal endemic to Peru, this uniquely Peruvian and uniquely Andean primate is of global importance, and is one of only two or three Neotropical primate species that appear to be exclusively Andean in distribution. As such it is not only a good flagship species for Peru, it is also excellent as a flagship for the Tropical Andes hotspot.

The Peruvian yellow-tailed woolly monkey was first discovered by the famous German explorer-naturalist Alexander von Humboldt in 1802 and described by him in 1812 as *Simia flavicauda*, *"le choro de la Province de Jaén."* He named the animal for the yellow ventral band on the distal tip of its tail, but he never actually saw a live specimen. His description was based on nothing more than flat, trimmed skins used as saddle covers by Peruvian muleteers near Jaén, in the department of Cajamarca. No type specimen was preserved, and Humboldt knew so little about this animal that he thought it was a species of howler monkey, rather than a woolly monkey. For over 100 years, nothing more was seen of this animal, until 1925 and 1926, when five specimens were collected by two expeditions; two of these animals wound up in the American Museum of Natural History in New York and the other three, in the British Museum of Natural History. In 1927, the specimens in the British Museum were used by the British zoologist Oldfield Thomas to describe a new species and new genus of woolly monkey, *Oreonax hendeei*. In 1963, the American zoologist Jack Fooden noted that Humboldt's *Simia flavicauda* and Thomas' *Oreonax hendeei* were one and the same and that the animal was not a distinct genus, but rather a very unusual species of woolly monkey, the correct name for which was therefore *Lagothrix flavicauda*.

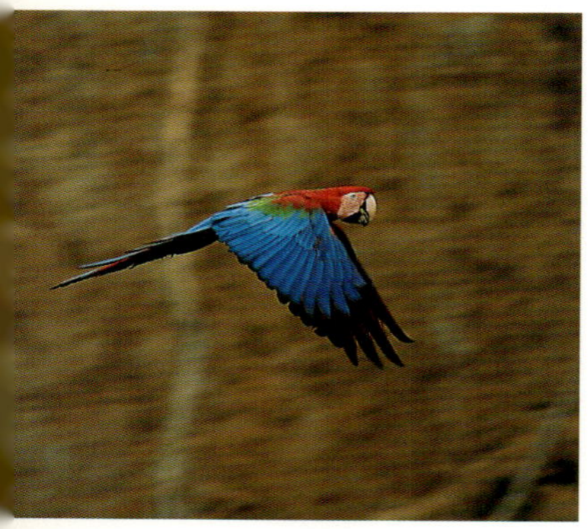

A red-and-green macaw (Ara chloroptera) *flying into a* collpa *in Manú National Park, Peruvian Amazonia.*

On the opposite page, the Manú River meandering through lowland rain forest, Manú National Park, Peruvian Amazonia. Information gathered to date indicates that Manú is the single most diverse protected area on Earth.

For almost 50 years after the collection of the specimens in 1925 and 1926, nothing was heard about this species in the wild, and it was thought that it might in fact be extinct. Then, in 1974, Russell A. Mittermeier and Hernando de Macedo-Ruiz, then Curator of Mammals and Birds at the Natural History Museum in Lima, organized an expedition to see if this animal still existed, and were accompanied on this trip by American zoologist Anthony Luscombe. Their 12-day expedition was successful in reconfirming the existence of this animal and in obtaining the first live specimen ever seen by the outside world. They also determined that it was restricted to a small area of montane and cloud forest between 500 and 2 500 m in the departments of Amazonas and San Martín in the northern Peruvian Andes. Considerable national and international press attention surrounded the rediscovery of this very attractive species, the value of which became immediately evident to the Peruvian conservation community. A subsequent expedition in 1978, again by Mittermeier and Macedo-Ruiz, succeeded in obtaining a second animal. Both of the animals were first maintained in captivity at the Natural History Museum, with the second one later being transferred to the Parque Las Leyendas, Lima's zoo. Aside from one other individual currently being maintained in the Parque Las Leyendas, these two animals are the only members of their species ever to be exhibited in captivity, and none have ever left Peru. Although it remains uncertain how many yellow-tailed woolly monkeys still survive in the wild, we estimate that there are no more than a few thousand at most.

The original rediscovery of the Peruvian yellow-tailed woolly monkey stimulated a range of efforts to conserve the habitat of this endangered species and learn more about its behavior and ecology. Many of these are still underway. Among the various efforts undertaken was a research project on the species by Peruvian zoologist Mariella Leo Luna; a program to create new protected areas in the region, notably the Abiseo National Park, which is also the site of Pre-Incan archaeological ruins of international significance; and a host of conservation education projects, including a film on the species, its placement on a Peruvian postage stamp, and the development of educational materials. Work on the Peruvian yellow-tailed woolly monkey was also instrumental in stimulating the creation of the Asociación Peruana para la Conservación (APECO), one of the oldest and most respected conservation NGOs in the country, which was founded by woolly monkey researcher Mariella Leo Luna and several of her colleagues.

The region in which the Peruvian yellow-tailed woolly monkey occurs is important not only for this species, but appears to be a center of endemism in this part of the Andes. Other interesting species found there include the marvelous spatuletail hummingbird (*Loddigesia mirabilis*), even more restricted in range than the monkey, and various endemic frog species, some of which (*Colostethus mittermeieri*, *C. idiomelus*, *C. poecilanotus*) were collected for the first time on the expeditions looking for the woolly monkey in 1974 and 1978. Important non-endemics that also occur in the area include the spectacled bear (*Tremarctos ornatus*) and the white-bellied spider monkey (*Ateles belzebuth*). The very poorly-known night monkey, *Aotus miconax*, may also overlap partly or entirely with the range of the woolly monkey; it is another Peruvian endemic and very limited in range, but almost nothing is known of its ecology or distribution, except that it was not described by science until 1983.

Similar in importance to the Peruvian yellow-tailed woolly monkey is the white-winged guan (*Penelope albipennis*), an endemic Peruvian species of the family Cracidae. Like the woolly monkey, this species was known from only three skins collected in 1876 and 1877, had not been seen in a hundred years, and was thought to be extinct. Then, in September, 1977, it was rediscovered in the dry forests at the foothills of the western Andes in the department of Tumbes by Peruvian sportsman Gustavo del Solar and American ornithologist John O'Neill. On a subsequent expedition, Hernando de Macedo-Ruiz, co-rediscoverer of the yellow-tailed woolly monkey, accompanied Del Solar to the region and also saw the species in the wild. He then announced its rediscovery to the outside world, and succeeded in again attracting considerable attention to the uniqueness and importance of Peru's biodiversity. As with the woolly monkey, this event stimulated action on behalf of the species and its habitat, including both efforts to curb hunting and preserve a portion of the remaining habitat and also a program to breed the guan in captivity and reintroduce it into its natural habitat. This species is even rarer than the woolly monkey, and is thought to number only a few hundred in the wild.

Other important Peruvian flagships are the four species of the camel family living in South America, the domesticated llama (*Lama glama*) and alpaca (*L. pacos*) and the wild guanaco (*L. guanicoe*) and vicuña (*Vicugna vicugna*). The guanaco is found mainly in Argentina (550 000 individuals), with a population of unknown size in Bolivia, a small population (19 836) in Chile, and

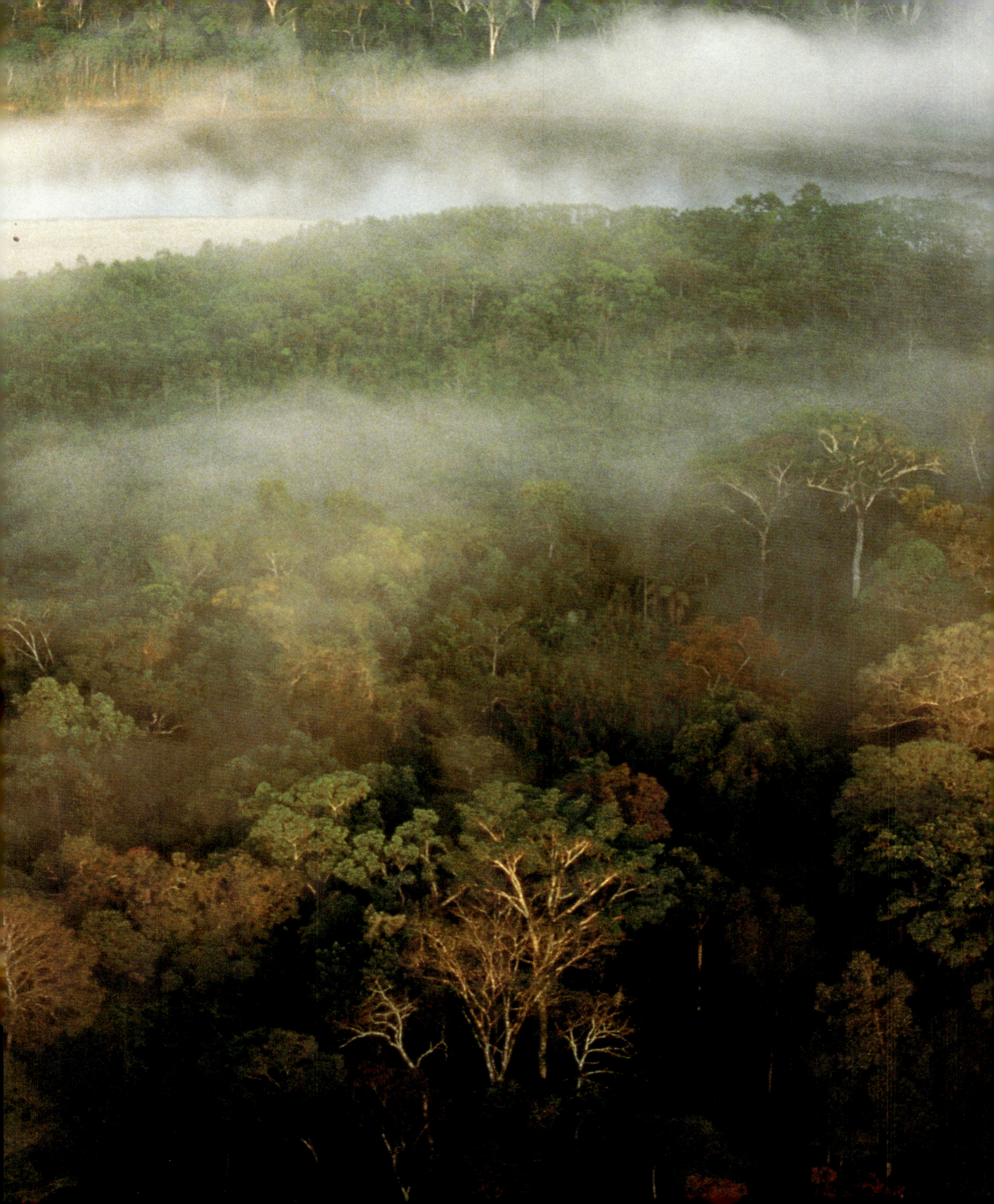

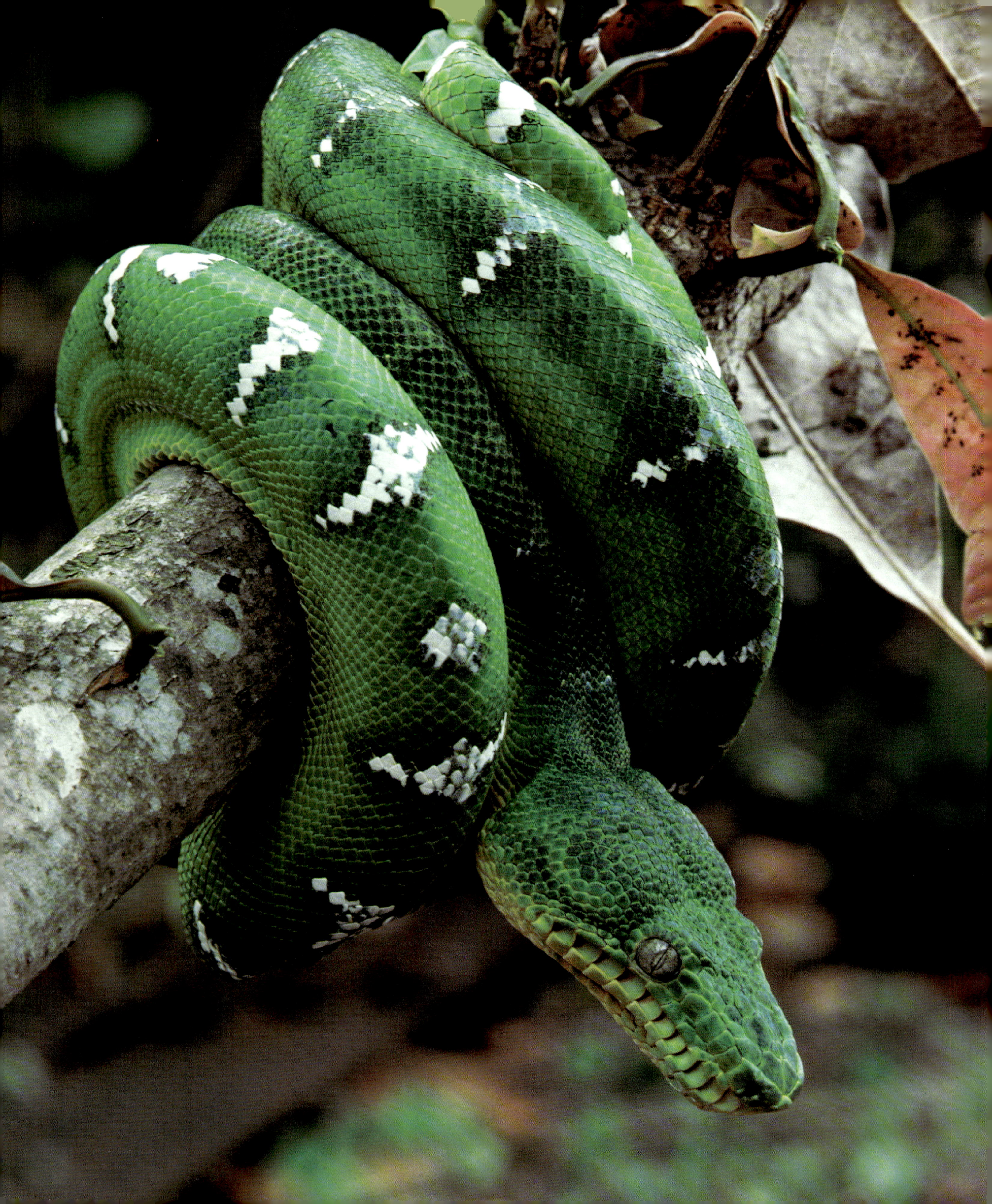

a very small one (1 347) in five departments of Peru. The vicuña, on the other hand, is most abundant in Peru, with 97 670 of the estimated 160 000 individuals of this species found in Peruvian territory. This beautiful species is the smallest of the camelids, and inhabits high-altitude regions of the Andes between 3 000 and 4 800 m, mainly in the *puna* ecosystem. It weighs only 40-50 kg, and is a social animal that lives in family groups consisting of a dominant male, several females and young less than one year old. Males are territorial, and males without territories live in bachelor groups.

The wool of the vicuña is perhaps the finest in the world, and is much in demand. During the Inca Empire, use of the vicuña was strictly controlled. When vicuña were needed for ritual uses, a special capture method called *chaku* was employed. This consisted of rounding up the vicuñas toward the end of a ravine or mountain slope where stone corrals had been constructed to contain a large number of individuals without hunting them. Several hundred people, led by highly-specialized individuals, would then capture the animals and shear them; except for a handful that would be sacrificed in rituals, the rest would be released back into the wild.

With the arrival of the Spanish and the destruction of the Inca Empire, there was no central authority able or willing to enforce conservation measures on behalf of the vicuña, and hunting of vicuña become much more intensive —to the point that Simón Bolívar prohibited it as soon as Peru became independent from Spain in 1824. Unfortunately, his decree was not enforced, and by the mid-twentieth century, the vicuña population had declined to about 10 000. Then, as of the 1970s, Peru in particular began development of a program of sustainable utilization of the vicuña, based on the premise that local communities had to be fully involved and benefit economically from the animal's management and conservation. This program, now known as the Special Project for Rational Utilization of the Vicuña, is part of the government's Ministry of Agriculture. Originally developed by renowned Peruvian conservationist Antonio Brack Egg, it defines a long-term management program that includes six subprojects in Huaraz, Huancayo, Pampa Galeras, Cuzco, Arequipa and Puno, and involves three critical protected areas: Huascarán National Park and the Pampa-Galeras and Salinas-Aguada Blanca National Reserves, of which Pampa-Galeras has the largest population and is the most important.

Although this project suffered a number of setbacks during the 1980s because of the strong presence of Shining Path (*Sendero Luminoso*) terrorists in some of the key departments for vicuñas, the long-term perspectives now appear to be quite good. The species is no longer in danger of extinction and knowledge of management techniques is well-advanced after more than two decades of research, putting the country in a position to consider sustainable utilization of the species over the long term. Peru's vicuña population is now relatively stable and constitutes about 70% of the world population, up to 97 670 as of 1992. Furthermore, the CITES Conference of 1987 agreed to allow international trade in industrial fabrics produced from vicuña wool from the Pampa-Galeras population, opening the door for creation of an international market. Problems still exist, mainly in the form of poaching and illegal trade of vicuña products, as well as the high costs of controlling illegal activities in such remote areas. Nonetheless, conservation of the vicuña is poised to become one of the success stories of biodiversity conservation, and should receive whatever assistance is necessary from the international community to help make this possible.

In addition to these key Peruvian flagships, a number of other very interesting species are found in Peru. Among these are the *taruca* or Andean huemul (*Hippocamelus antisiensis*), a medium-sized deer species living at high altitudes (3 000-4 000 m) in an area that overlaps with the vicuña; the *pudu* (*Pudu mephistophiles*), a very small deer that reaches only 32-36 cm at the shoulder, weighs about 9 kg and occurs in Andean valleys at 2 000-3 000 m; the giant otter (*Pteronura brasiliensis*), a widespread but highly-endangered species that has some of its most important remaining populations in Peru, especially in Manú National Park; the spectacled bear (*Tremarctos ornatus*), still reasonably well-represented in the Peruvian Andes; the mysterious Andean mountain cat (*Felis* [*Oreailurus*] *jacobita*), a very rare, cryptic, high-altitude cat species about which very little is known; the mountain tapir (*Tapirus pinchaque*), found in the extreme northern tip of the Peruvian Andes; the chinchilla (*Chinchilla laniger*), world-renowned for its fine fur and still found in the wild in parts of the high Andes; the giant black caiman (*Melanosuchus niger*), which is found in good numbers in some of the more remote corners of Amazonia; and the American crocodile (*Crocodylus acutus*), which reaches the southern limit of its vast range in the Peruvian department of Tumbes in northern Peru. Another Peruvian monkey species worthy of note is the Peruvian equatorial or Peruvian white-faced saki (*Pithecia aequatorialis*), an unusual species that was not described by scientists until 1987. Found from the north bank of the Marañón River in Peruvian Amazonia up as far as the Napo River area in Ecuador, it is known from only three localities. It is characterized by a striking white face in the male, superficially resembling that of the Guianan white-faced saki (*Pithecia pithecia*) of northeastern Amazonia and the Guianas. Although these two geographically widely-separated species can be confused at first glance, they are actually quite distinct. The equatorial saki remains very poorly known and has never been studied in the wild.

Peru also has several internationally-known wildlife phenomena. One of these is that of the guano birds on the islands of the Pacific, where these birds and nearby fur seal colonies combine to provide one of the more spectacular wildlife viewing opportunities on the Pacific coast of the Americas. Another, and probably the most spectacular of all, is that of the *collpas* or mineral "licks" for macaws and parrots in certain parts of Amazonia. The most impressive of these *collpas* are in Manú

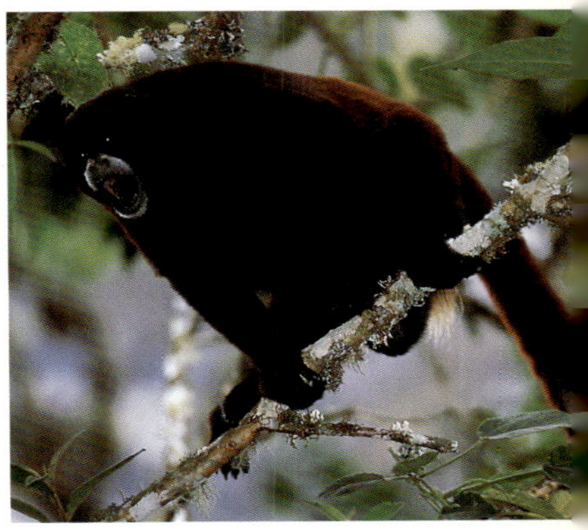

On the opposite page, the emerald tree boa (Corallus caninus) *is a very attractive arboreal species that, though rarely seen, is widespread in Amazonia.*

Above, the Peruvian yellow-tailed woolly monkey (Lagothrix flavicauda) *is the largest mammal endemic to Peru and the country's single most important flagship species. It was rediscovered after not having been seen in almost 50 years.*

National Park and the Tambopata-Candamo area, and appear not to occur elsewhere in Amazonia, at least not on the scale at which they occur in these Peruvian sites. In a handful of areas where riverbanks have been exposed by natural erosion, large mixed flocks of parrots and macaws gather en masse every morning to feed on clay of the riverbank. This takes place only in very restricted areas, although there are many riverbanks that superficially seem to be indistinguishable from the areas of intensive use. As many as 15 different species will use these selected banks every morning, with different combinations of species descending at different times, in what appears to be something of a predetermined order, to feed on these substances. The exact reason (or reasons) still remains to be determined, but it may be that these clays provide critical trace elements or that they assist in the detoxification and digestion of plant toxins in the seeds and other foods that the macaws and parrots eat —or some combination of both.

Needless to say, the ecotourism potential of seeing hundreds, if not thousands, of brilliantly-colored, raucous macaws and parrots landing on a riverbank, circling overhead, perching in trees awaiting their turn at the feeding site, and squawking endlessly is hard to match in the natural world. One tour company, called Rain Forest Expeditions and run by several young Peruvian macaw biologists turned entrepreneurs, has been leading tours to the most impressive *collpa* in the Tambopata-Candamo area for several years. They maintain a team of researchers that monitor the birds on a daily basis to ensure that the impacts of tourism are minimal, and have built an excellent, low-impact lodge that is a model of how such enterprises should be carried out.

Peru also has several plant species that fall into the flagship species category. Perhaps the most striking is *Puya raimondii*, a bromeliad of the Andes. It may take as long as a century to reach maturity, and has the largest inflorescence of any plant on Earth. Its flower spike can reach 6 m, and is composed of thousands of tiny flowers. The arrayán tree (*Myrcianthes ferreyrae*) is a flagship tree species that is represented by only 200 individuals in the Taimara River Basin in the *lomas* habitat of Atiquipa. Peru is also the center of origin of some of the world's most important crop species, among them the potato (of which there are eight species and some 3 000 varieties) and tobacco. Several other Andean crops, such as the highly nutritious grain quinoa (*Chenopodium quinoa*) and the legume *tarwi* (*Lupinus mutabilis*), though less known, are becoming increasingly popular outside Peru as well. Among the medicinal plants, perhaps the most famous is cinchona, the source of quinine, which has been of fundamental importance in the fight against malaria around the world.

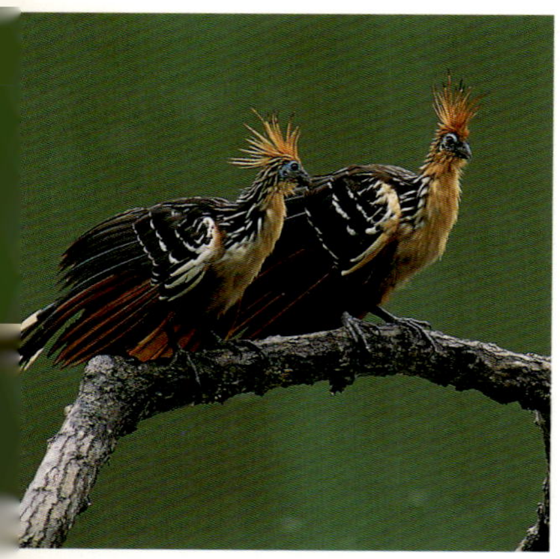

The hoatzin (Opisthocomus hoazin) is a very unusual bird species that is the only member of its family. It retains several reptilian features, including the claws on the wings of hatchlings that enable them to maneuver on branches around their nest. This species is always found at the edge of lakes and rivers.

On the opposite page, the blue and yellow macaw (Ara ararauna) is one of the more widespread of the large macaw species of South America. Here a pair rests on a palm stump in Manú National Park in Peru.

Cultural diversity

Peru today is a pluricultural, multilingual and multiethnic country, with a population of 24 370 000 divided into 15% Caucasian, 37% mestizo (of mixed European and Amerindian descent), 45% Amerindian, and 3% other, including Afro-Peruvians and Asians, among them 300 000 people of Chinese descent. The diversity of indigenous people, the original Amerindian inhabitants of the country, is quite high, especially in relation to the country's size. Although only about 15% the size of Brazil, Peru has 96 spoken languages, compared to 206 spoken in the much larger Brazil.

Peru's archaeological treasures reflect some of the world's oldest and most cultured civilizations. Amerindian groups began to settle there over 14 000 years ago. The oldest known settlement is the Pikimachay cave in the coastal region, which dates back to 12 000 BC. By 500 BC, Andean groups had developed sophisticated civilizations which cultivated potatoes, gourds, cotton, squash, beans, chili peppers, maize, avocados, and peanuts, and had even developed a process to fuse metals in a plating process. The pre-Incan Moche domesticated llamas and alpacas, utilized fertilizer, developed innovative irrigation techniques, and prevented erosion through the use of terraced slopes more than 6 000 years ago. At Sipán, a Moche ruin discovered by Dr. Walter Avila, archaeologists are today unearthing hidden treasures of a once extensive culture that lived in the Lambayeque Valley from AD 100-700. The Moche of Sipán had no written language, but developed a highly sophisticated craft culture that produced some of the country's finest ceramics, as well as enormous clay and wood structures over 70 m high. Dr. Avila's discovery of the tomb of the "Lord of Sipán" has revealed some of the finest gold pieces yet found in the Americas, and much more remains to be discovered in what appears to be an elaborate complex of tombs.

By the fifteenth century, the Incan Empire controlled the entire Andean region from Colombia to Chile. The Inca ruled this empire from the capital city of Cuzco, and practiced systems of forced labor and cultural transplantation. Their empire officially ended with the death of Atahualpa at the hands of the Spaniard Pizarro in 1533. The remnants of this important culture can be seen throughout Peru to this day in the form of ruins of lithic roads, cities, temples and other structures. The most famous of these is the remarkably intact Machu Picchu, discovered by Hiram Bingham in the early 1900s and probably the most important religious center of the Inca. Fully explored, but not yet entirely understood, Machu Picchu lies well hidden atop a mountain along the valley of the Urubamba River and consists of a maze of buildings, rooms, steps and passages with religious structures like the famous "Hitching Post of the Sun." This and other sites of archaeological importance like Saqsaywaman in Cuzco, the famous lines of Nazca, the Toro Muerto and Pajaten sites, and many others combine with Peru's rich biodiversity to make this country one of the world's premier ecotourism destinations.

The dominant ethnic group represented in Peru today is the Quechua. Descendants of the Incas, there are now up to 13 million Quechua living in Ecuador,

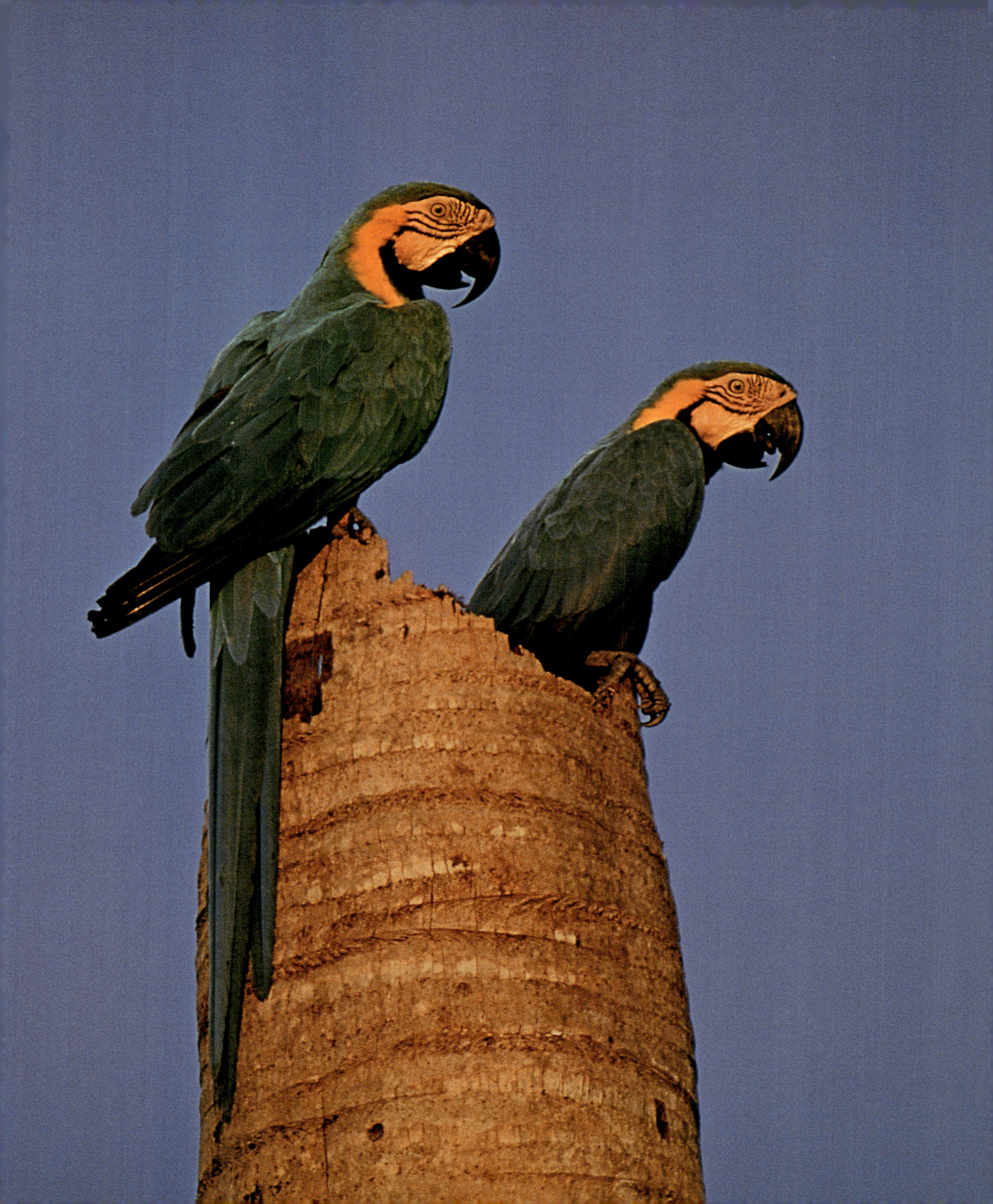

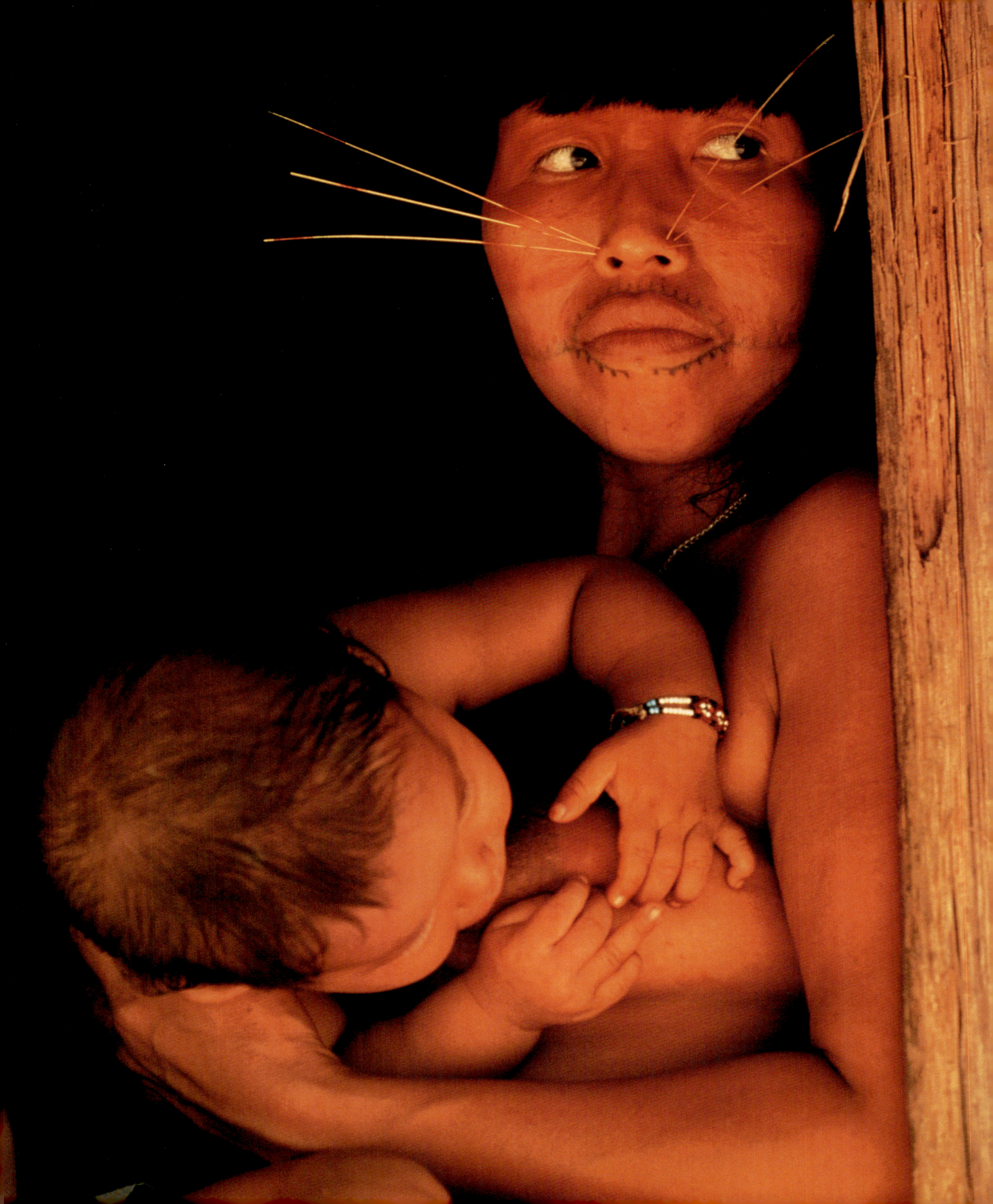

Bolivia, Peru, Colombia, Chile, and Argentina. Five million of these Quechua live in Peru, speaking over thirty separate dialects, in addition to two forms of lowland Quichua. Many of these dialects, such as that of the 300 000 Huaylas Quechua of central Ancash Department, are so different from each other that native speakers of different dialects must converse in Spanish. The most widespread dialect (and the country's joint national language with Spanish) is Cuzco Quechua, with 1 500 000 speakers. Quechua lifestyle is characterized by small highland farms and pastoral communities. These people remain a shy, highly traditional, indigenous group that has adapted well to harsh Andean environments and has survived 500 years of oppression.

After Quechua, the most widely-spoken language in Peru is Aymara. There are 350 320 Aymara and related Jaqaru inhabiting the Lake Titicaca area in the department of Puno, the departments of Moquegua and Tacna, and the Yauyos Province in the department of Lima.

The greatest cultural diversity in Peru is found on the eastern Andean slopes and in the Amazonia. Many of these groups are quite small, and some, such as the Morunahua in the department of Ucayali and the Pisabo in the department of Loreto, have still had only minimal contact with the outside world. The various tribal groups of this region are best organized by language families. These families are the Jivaroan (found within the departments of Loreto and Amazonas, and including the Aguarana, Huambisa, Achual, Candoshi and Shapra ethnic groups); the Panoan (found in Loreto, Ucayali, and Madre de Dios, and including the Shipibo, Conibo, Cashibo, Capanahua, Amahuaca, Marinahua, and Cashinahua ethnic groups); Zaparoan (Loreto; Arabela and Iquitos tribes); the widespread Arawakan (found within Madre de Dios, Pasco, Ucayali, Apurimac, Ayacucho, Cuzco, Junín, and Loreto and represented by the Ashanica, Machiguenga, Amuesha, Piro, and Culina); the Witotoan (Loreto); the Cahuapanan (represented by the Chayhuita of Loreto); the Tupí (Cocama and Cocamilla groups of Loreto); the Tacanan (the Ese'Eja of Madre de Dios); the Tucanoan (Secoya and Maijuna of Loreto); and the Peba-Yaguan (the Yagua Indians of Loreto and adjacent parts of Colombia). In addition, there are other groups for which the family affiliation is uncertain, such as the Ticuna of the department of Loreto, and the Urarina found along the Pucayacu, Chambiru, and Urituyacuri Rivers.

According to a recent census, the Indian population of the Amazon is almost 300 000 people. The eight most numerous groups are the Aguarunas, with 45 000 (15%), the Ashanica (or Campa) with some 40 000 (13%), the Lamas-Chachapoyas with 22 000 (7%), the Shipibo-Conibo with 20 000 (6.7%), the Chayahuita with 14 000 (4.7%), the Cocamilla with 11 000 (3.7%) the Quichua del Napo-Pastaza-Tigre with 11 000 (3.7%), and the Machiguenga with 9 000 (3%).

One of the better known Amazonian Indian groups is the Machiguenga, some 6 000-8 000 of whom live in the vicinity of the Manú National Park along the Urubamba and Mishagua Rivers, Cuzco region. They have come into contact with the outside world over the past twenty years as a result of scientific presence in the park and due to the presence also of oil workers in the nearby Camisea block which lies to the west of Manú and is one of the largest unexploited natural gas deposits in the Americas. While many Machiguenga have become increasingly acculturated, many more still live in a traditional fashion despite regular contact with outsiders. They are exceptional hunters with bow and arrow and fashion a wide variety of arrows, each with its own specific design for fish, fowl or mammal, made from the reed, *Gynerium sagittatum*, adorned with feathers from captive game birds that are kept for these purposes, and armed with points of bamboo or palm wood. They also gather forest products and perform small-scale agriculture for their subsistence needs. Traditional dress is the *cushma*, a tunic which they have woven for centuries from cotton that is cultivated for this purpose. The Machiguenga also have a rich pharmacopoeia of medicinal plants from the forest and a tradition of healing done by a shaman who is referred to as *seripegari* or "he who uses tobacco." The Machiguenga maintain regular contact with staff of Manú National Park in the communities of Tayakome, Yamuibato and Santa Rosa de Huacaria, while others choose to live in more remote villages that seldom are exposed to the outside world.

Peru arguably still has several uncontacted Indian groups of uncertain origin. Some of these may still occur in the Manú area; others are found in the Río de las Piedras region, which is only now starting to be opened to oil exploration.

As in the other Amazon Basin countries, the ethnic groups of Peru, particularly those of the Amazon Basin, have formed federations to represent their traditional rights within the country and continent-wide; several of these have been effective in bringing concerns about land tenure, discrimination, and other issues to the attention of the world. One of the most important is the Indigenous Federation of Madre de Dios (FENAMAD), representing several of the indigenous communities in that sparsely-populated department, considered to be one of the most biologically diverse regions on Earth. Working both with FENAMAD and the *campesino* group FADEMAD (the Agrarian Federation of Madre de Dios), Conservation International has strived to encourage a community-based planning and sustainable resource management scheme for the above-mentioned Tambopata-Candamo Reserved Zone, which is inhabited by *campesino* colonists and at least 250 Ese'Eja Indians living along the banks of the middle and lower Tambopata River. This effort complements the national park program now underway, and has been instrumental in giving full support for the park from local communities.

RUSSELL A. MITTERMEIER
RODERIC B. MAST
CARLOS PONCE DEL PRADO
CRISTINA GOETTSCH MITTERMEIER

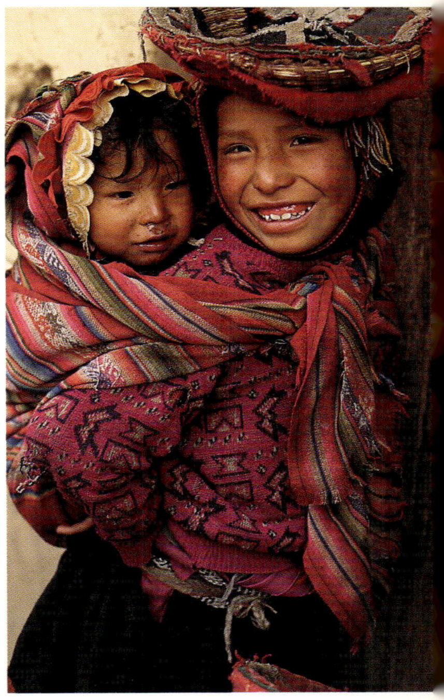

On the opposite page, a Mayoruna Indian, one of many lowland Amazonian Indian tribes in Peru.

Above, Quechua Indian children in Cuzco. The Quechua are the largest indigenous group in Peru, and are descendants of the Incas.

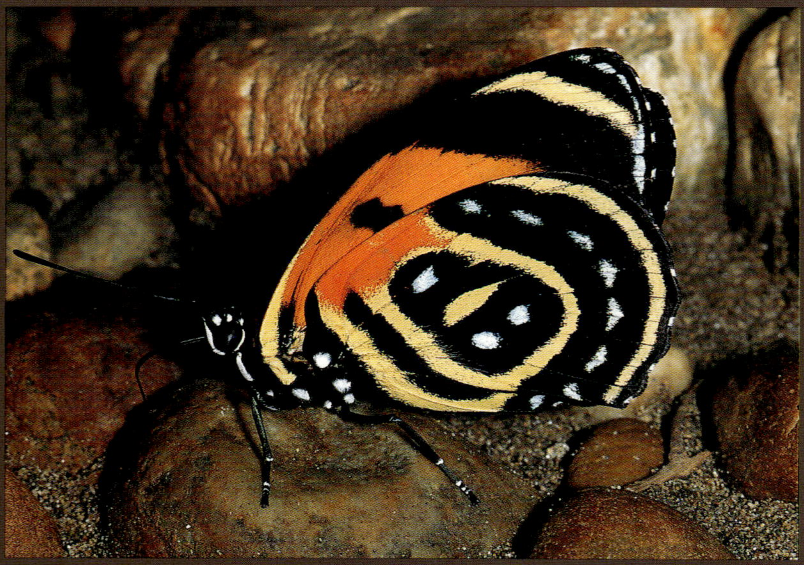

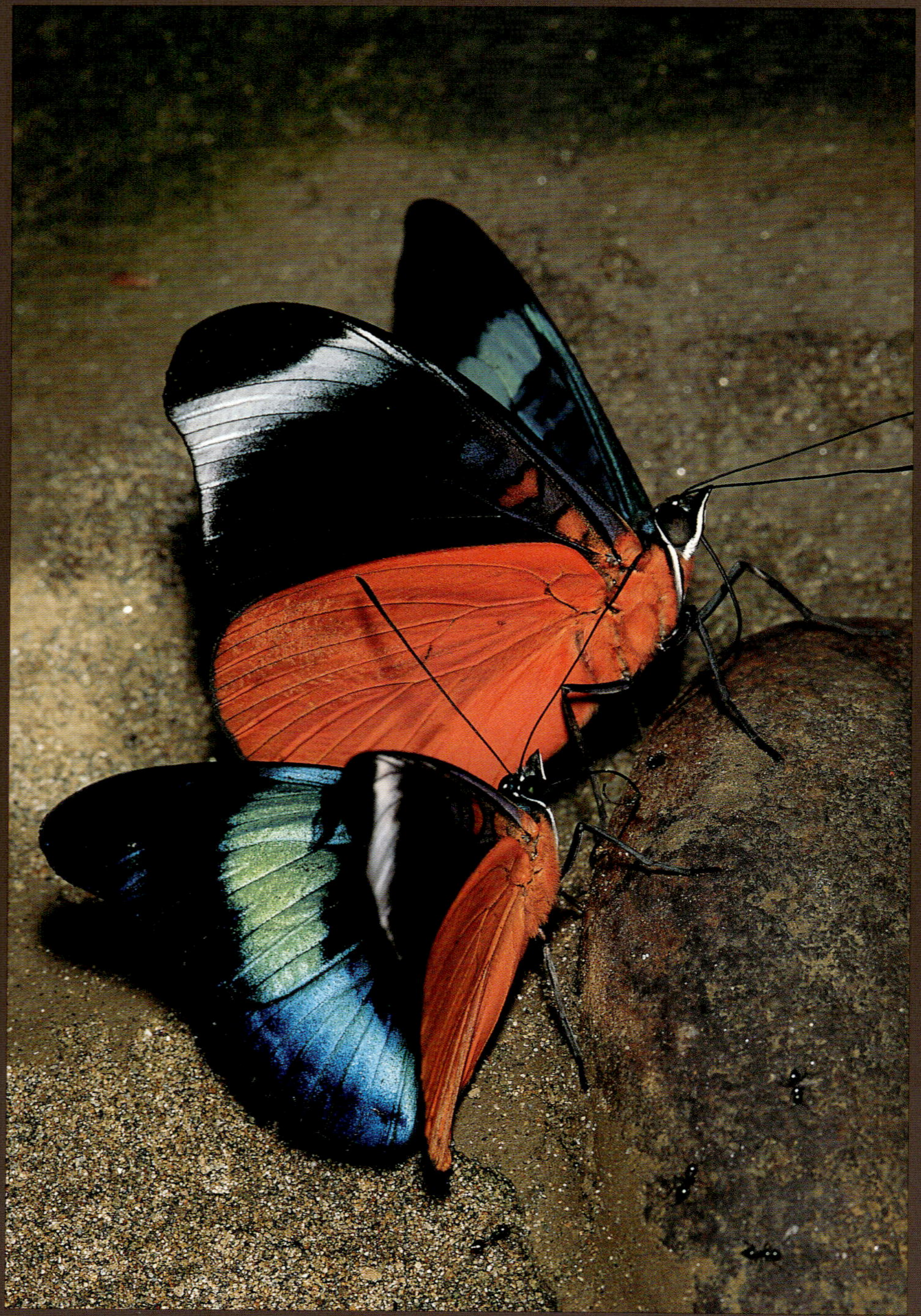

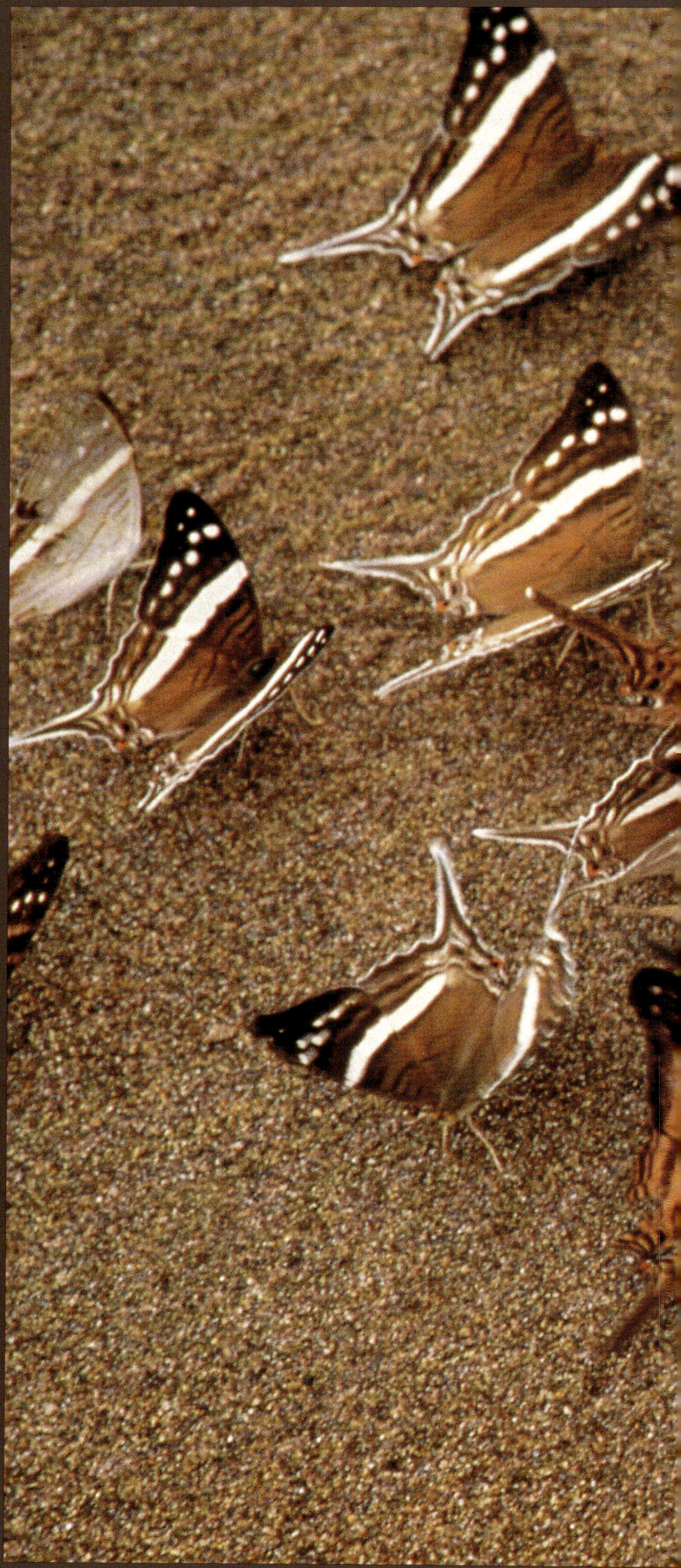

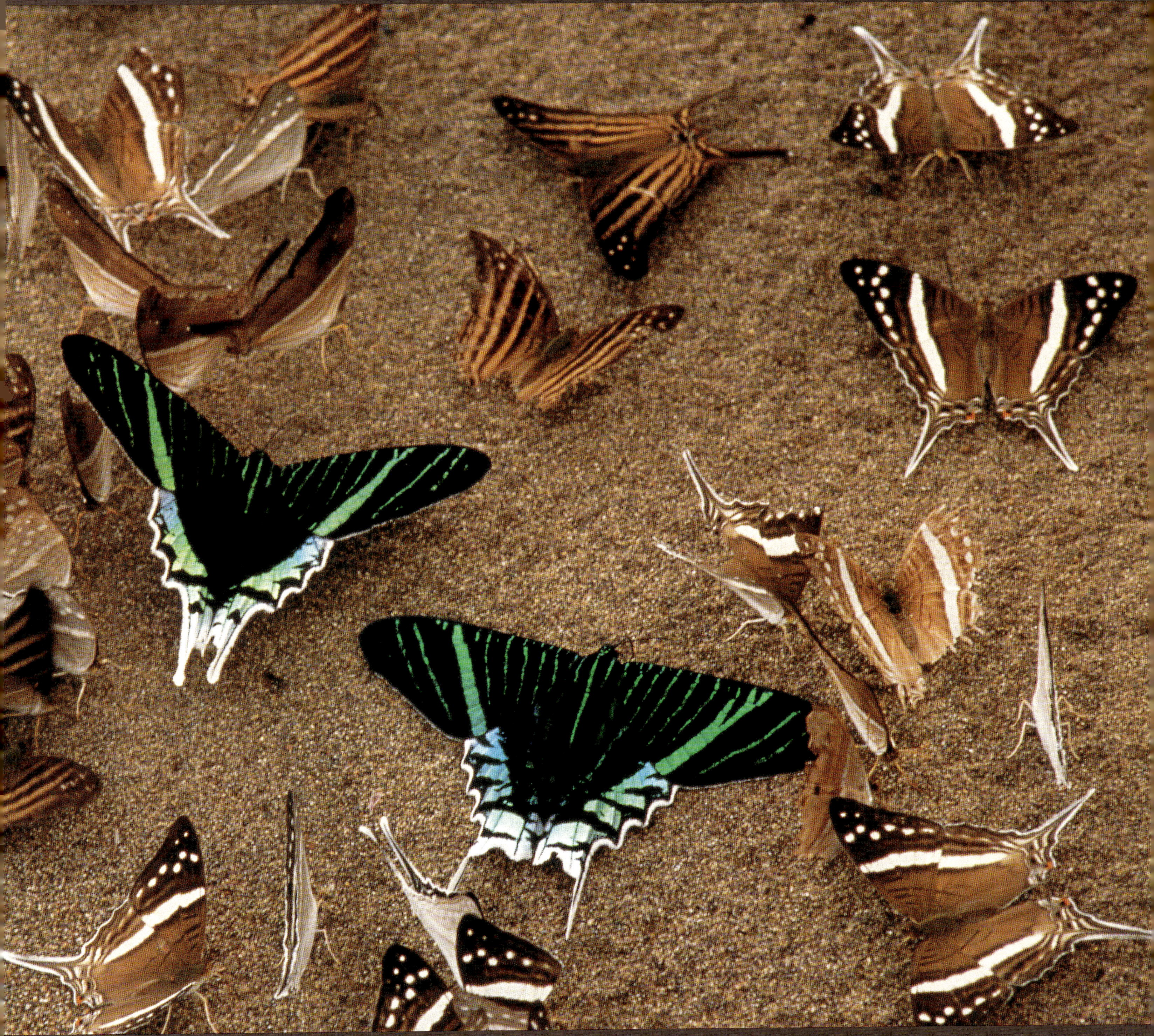

Peru has the highest recorded butterfly diversity on Earth. Here is just a small sampling of the incredible wealth of this group, which is especially rich on the slopes of the tropical Andes. Upper left, Callicore synosura*; lower left,* Panacea *aff.* prola *and above, a reunion of butterflies and moths:* Urania *sp. (moth),* Marpesia chiron *and* Marpesia crethon *(butterflies).*

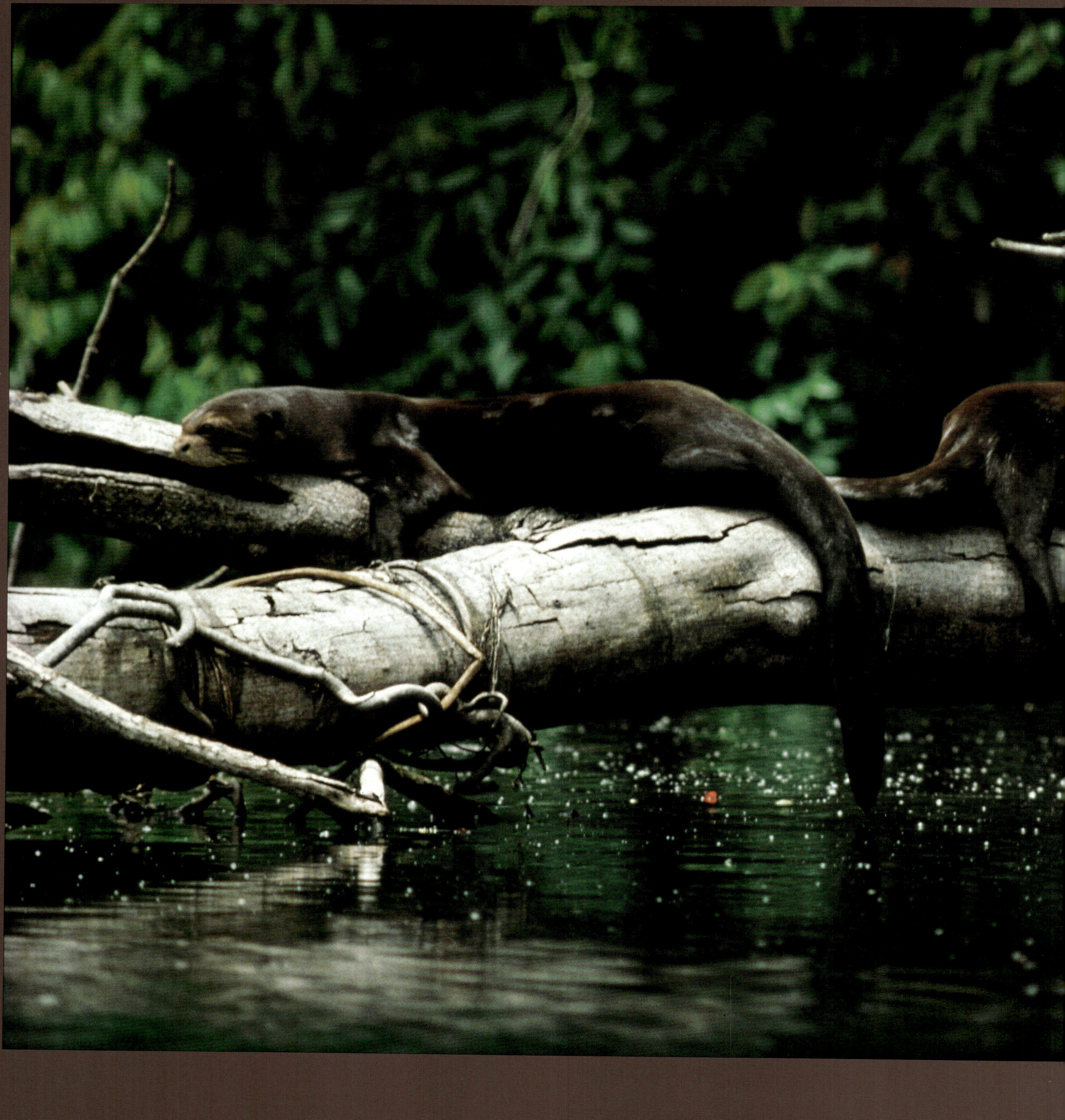

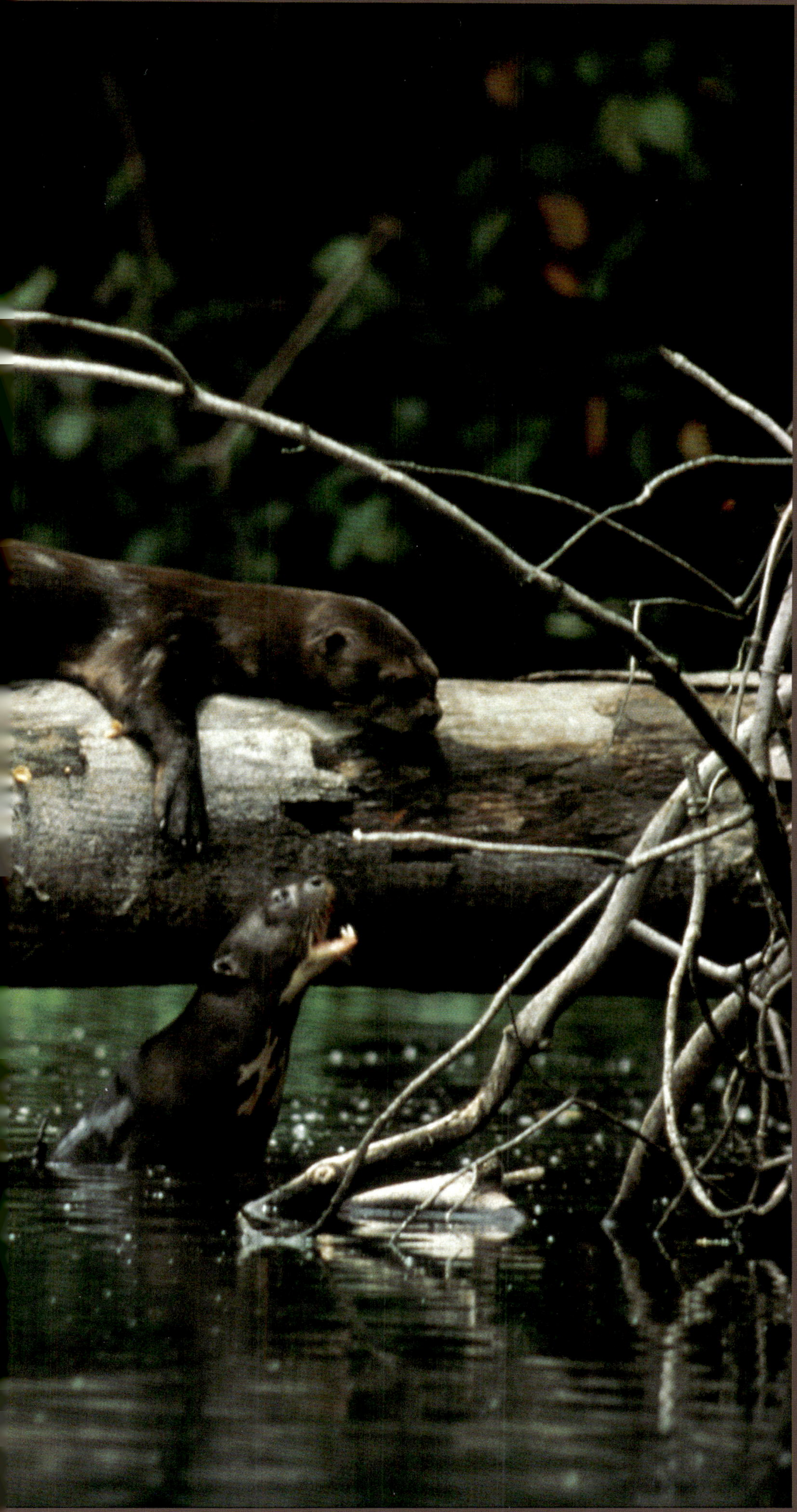

Left, the giant otter (Pteronura brasiliensis)
is one of the most endangered large mammals
in all of Amazonia, and one of the most important
flagship species in the region.
The largest of the otters in overall length, only the sea otter
of the Pacific Coast of North America compares in size.
A diurnal, vocal, highly sociable animal with a very valuable
pelt, it was long persecuted for the skin trade.
It is now protected throughout its range, but is quite rare.
These individuals were photographed
in Manú National Park, one of their last major strongholds.
Below, the pirañas (Serrasalmus *spp.*) *are fierce predatory*
fishes that are widespread and abundant throughout
Amazonia. Although they are armed with razor-sharp teeth
that can nip off the tip of a finger if not handled
with care, their reputation for savagery has been much
exaggerated by the popular press. This one
was photographed in an oxbow lake in Manú National Park.

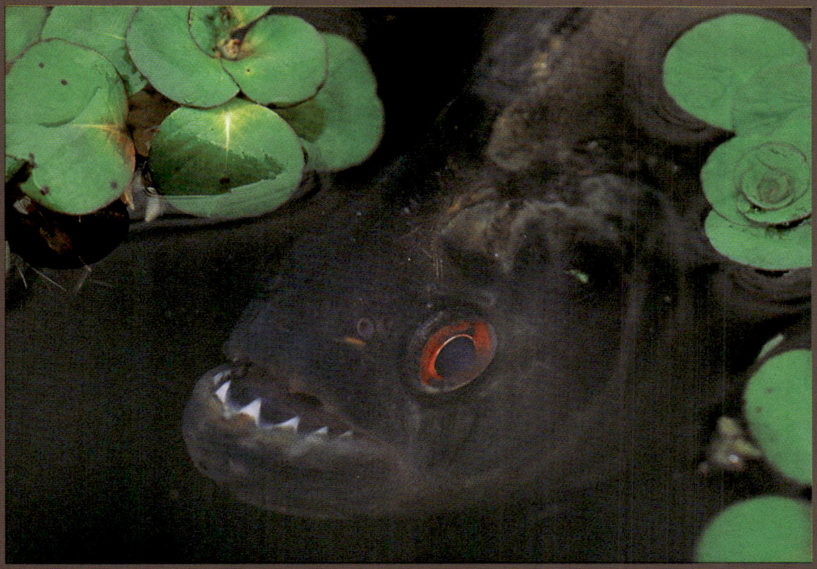

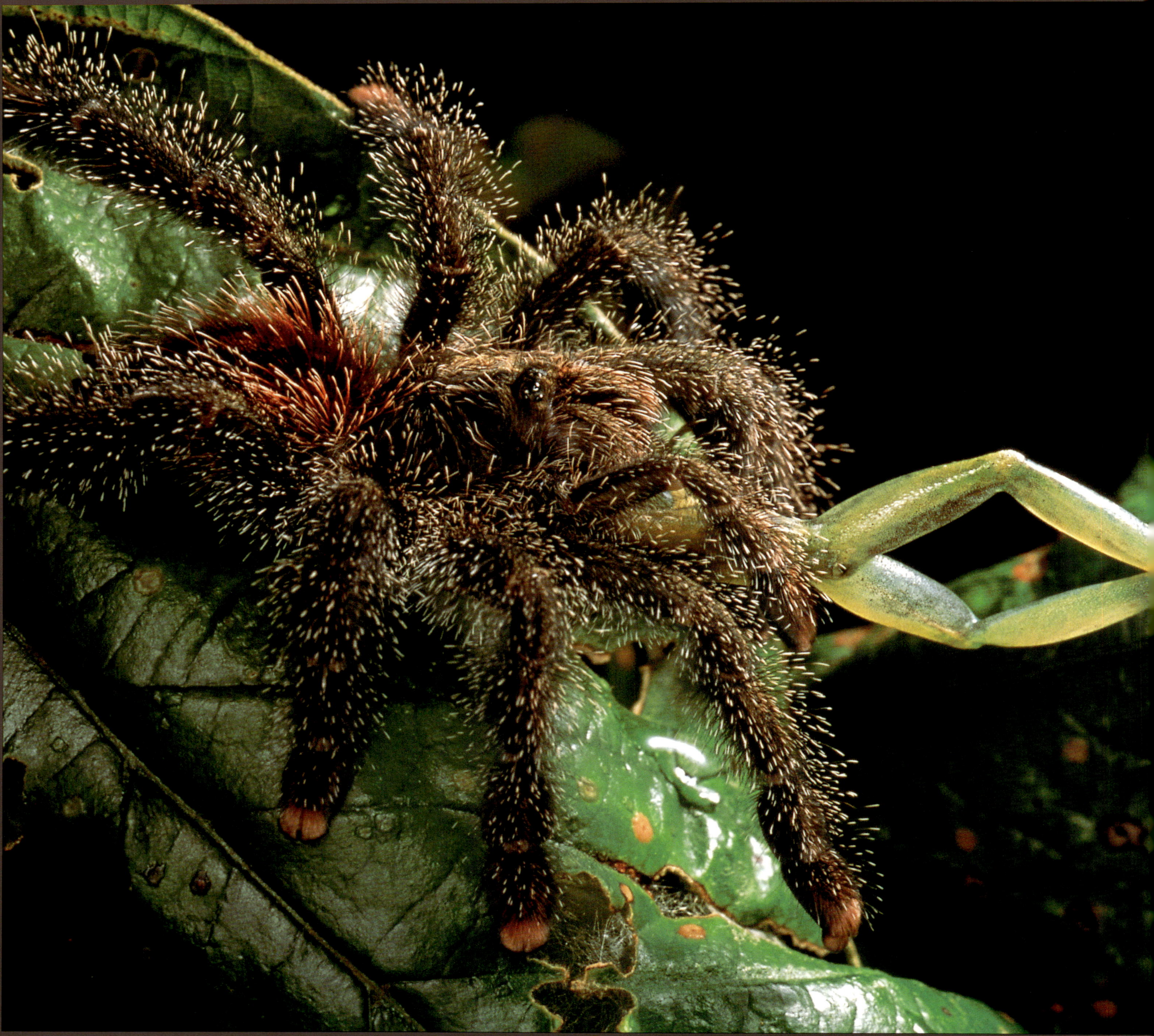

Above, a tarantula eating a tree frog. Like the piraña, these large spiders are part of the folklore of Amazonia. Although they often prey on small vertebrates, they pose little danger to humans.
Upper right, an attractive tree frog (Hyla geographica) *in amplexus, the technical term for the mating grip in frogs. Tree frogs of the family Hylidae are a very diverse group in Peruvian Amazonia that almost certainly includes many species as yet undescribed by scientists.*
Lower right, a juvenile matamata (Chelus fimbriatus) *from Amazonia. This turtle, a member of the side-necked turtle family Chelidae, is characterized by its long, snorkel-like snout, the ridges on its shell, and its broad, flat head adorned with fleshy appendages. It is clearly one of the world's most bizarre and unusual chelonians.*

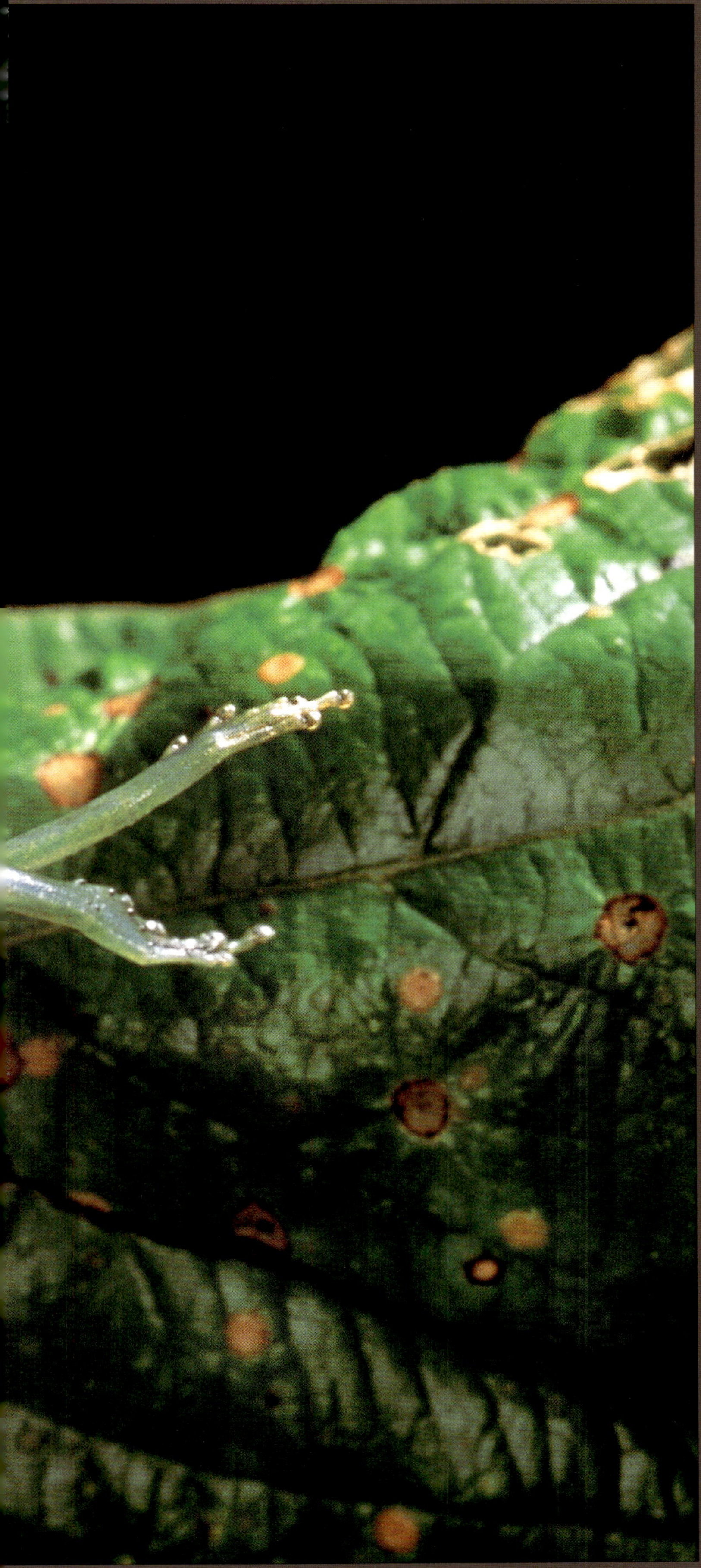

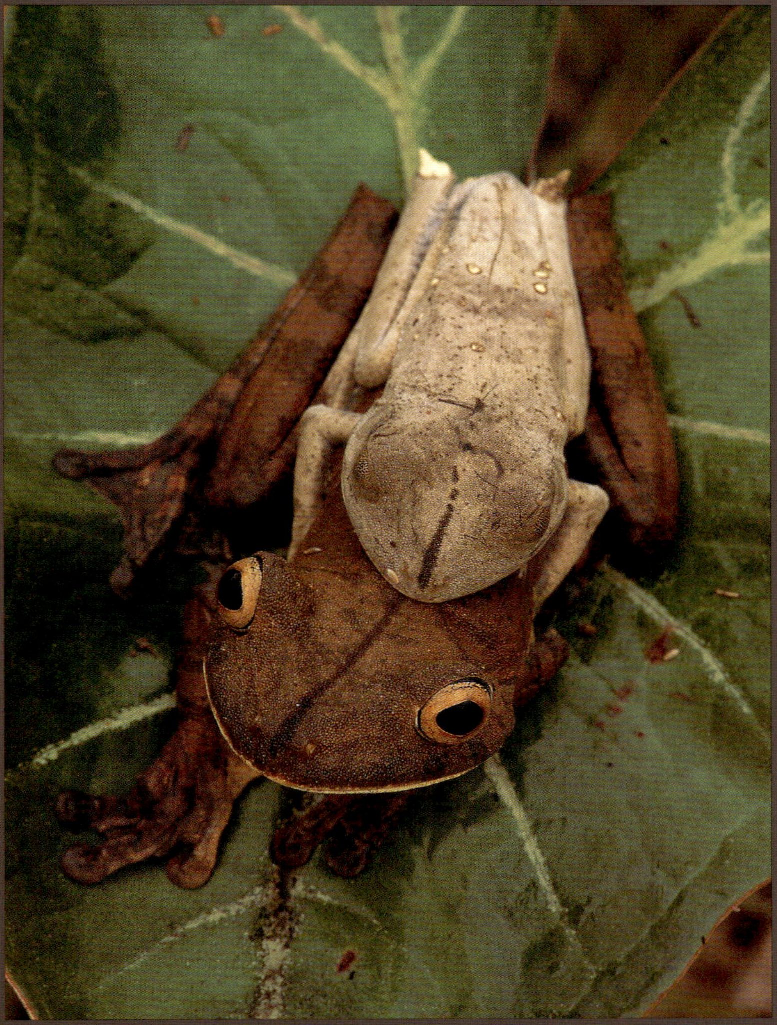

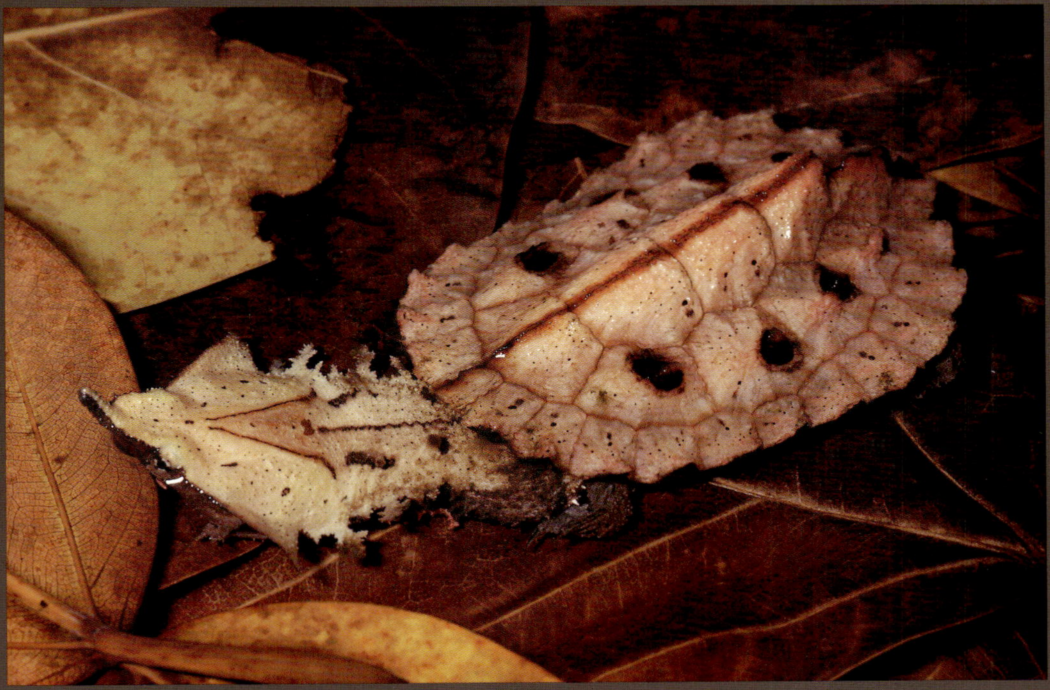

Below, a wild jaguar (Panthera onca) *photographed*
in the department of Madre de Dios in Peruvian Amazonia.
On the right, a lowland tapir (Tapirus terrestris) *crossing a river*
in the Tambopata-Candamo region of Peruvian Amazonia.

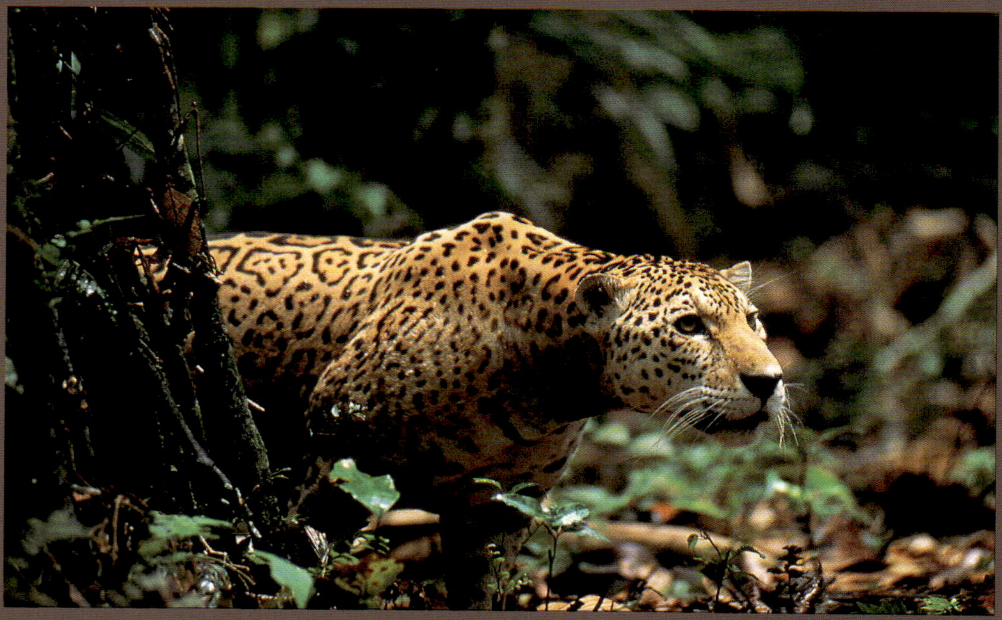

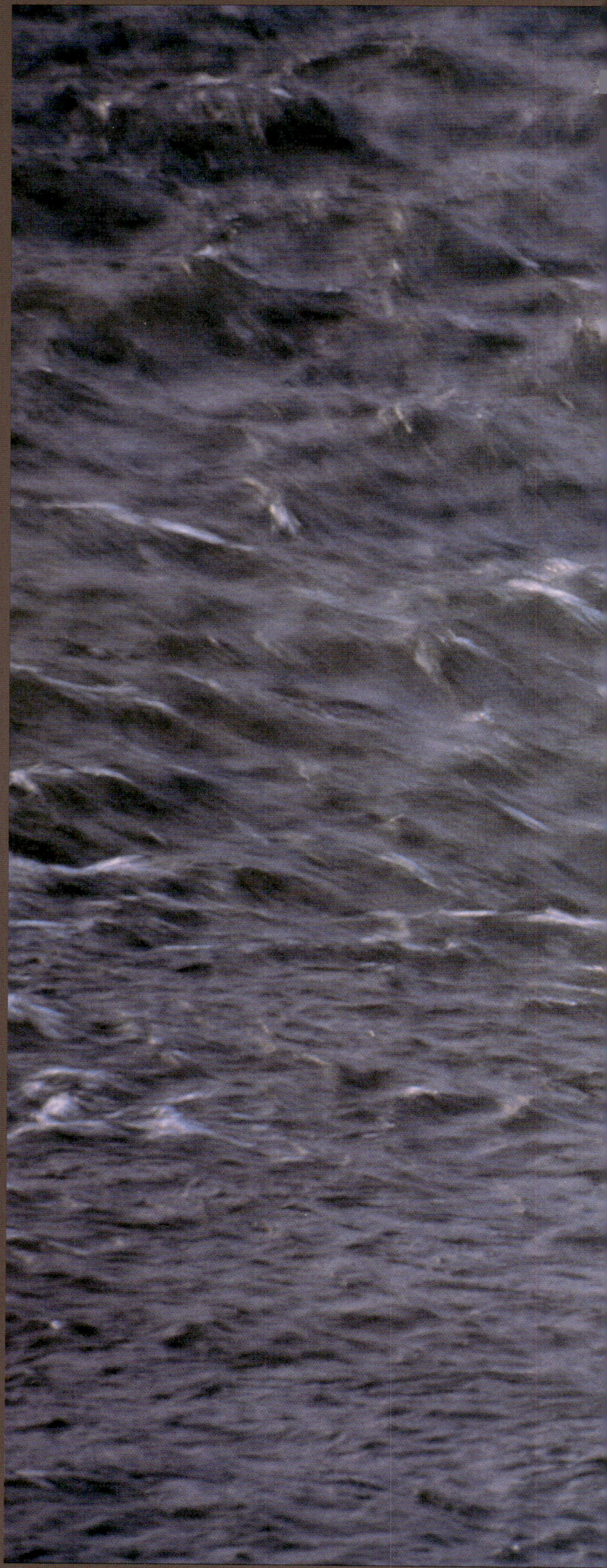

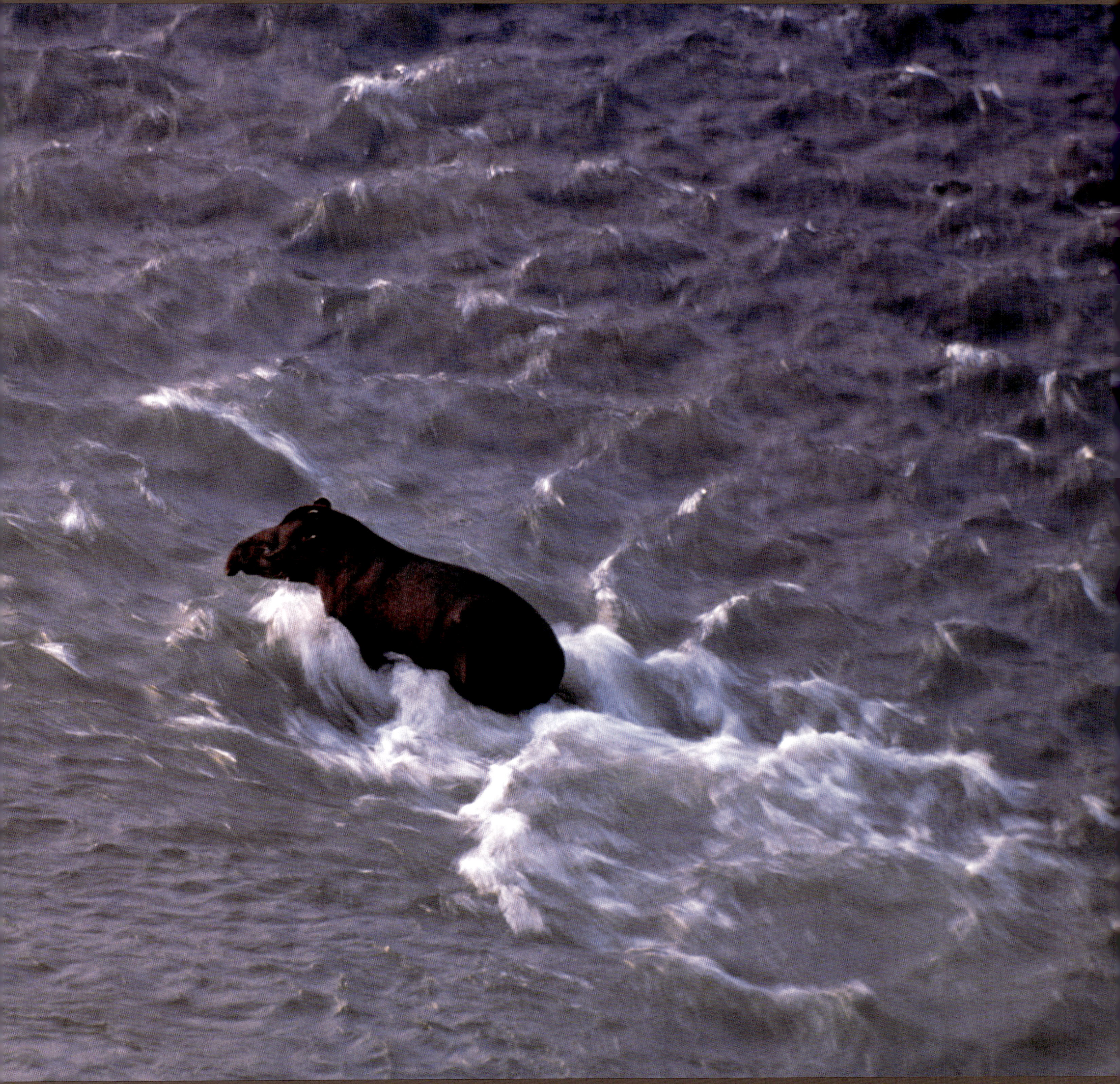

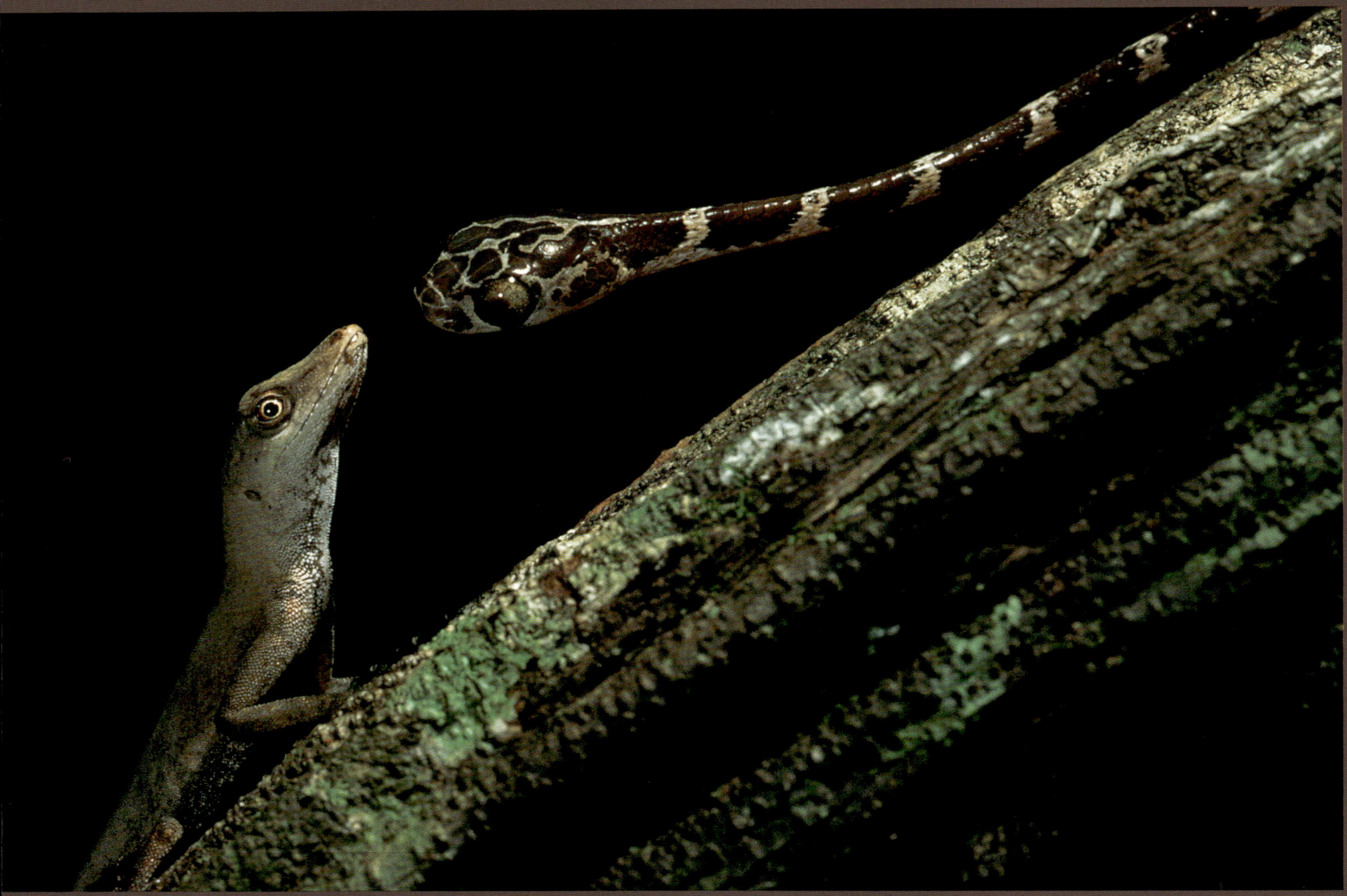

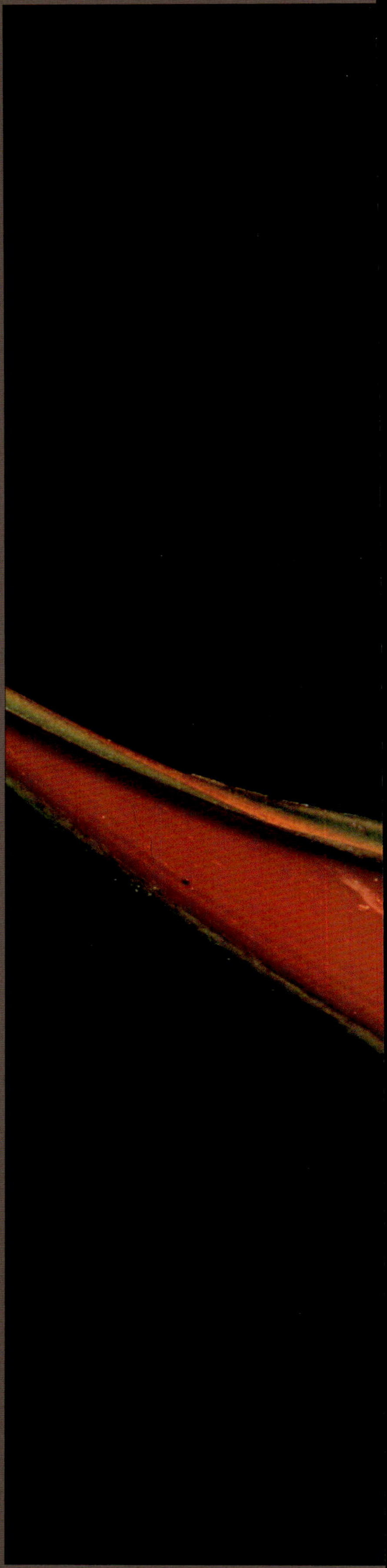

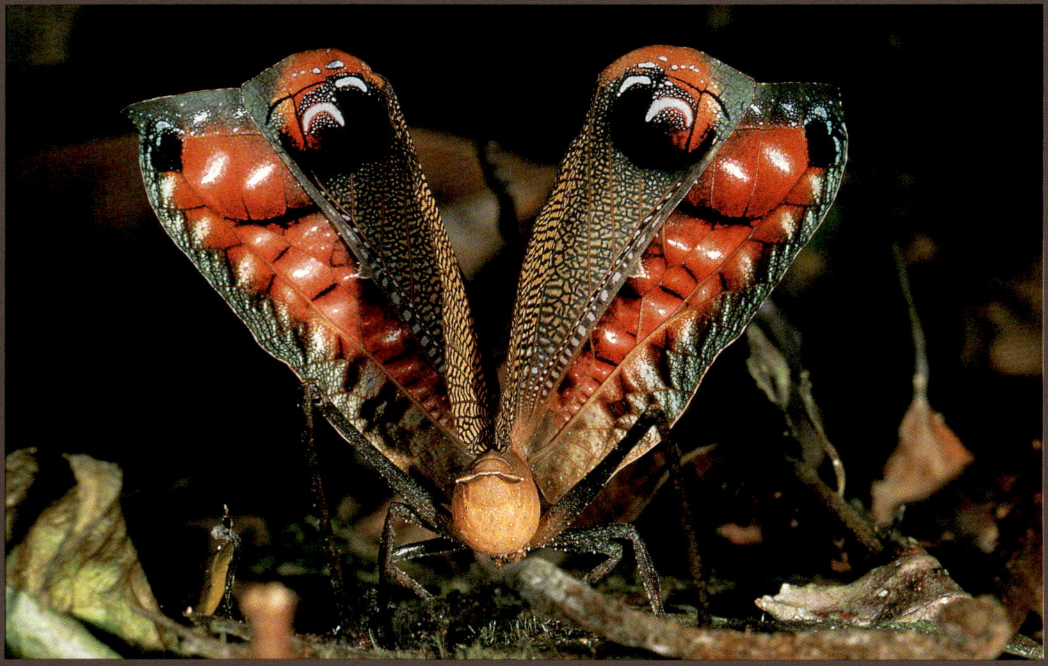

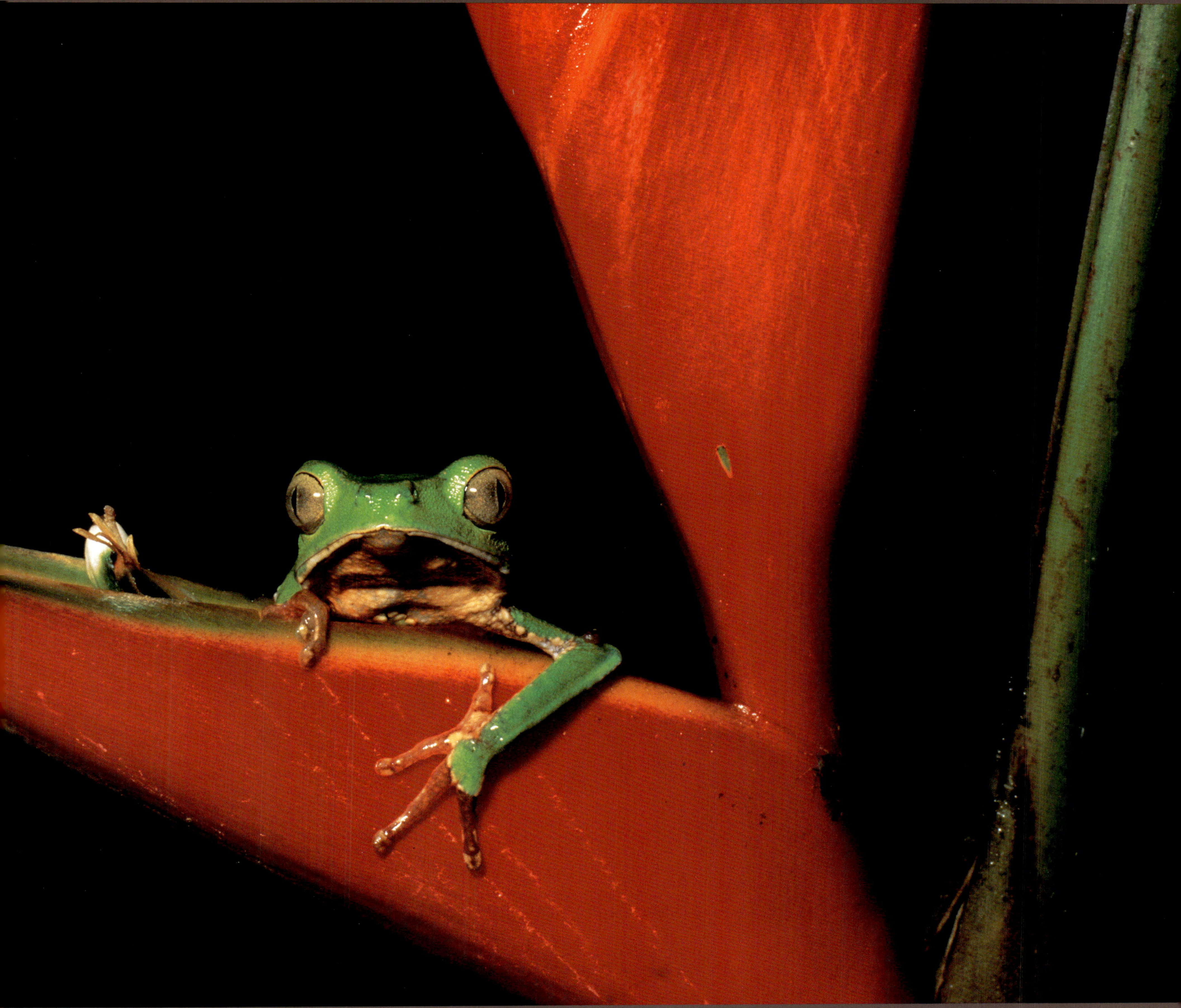

Upper left, a blunt-headed tree snake (Imantodes cenchoa) *confronting its prey, an anolis lizard* (Anolis fuscoauratus) *in Manú National Park.*
Lower left, an eye-winged lead katydid (Pterochroza ocellata) *reveals its back wings to display eyespots to potential predators.*

Katydids in Amazonia come in an incredible array of shapes and forms, many of them intended to mimic leaves and other plant parts. This species was photographed in Manú National Park.
Above, another of Peru's many tree frog species of the family Hylidae. This is Phyllomedua vaillanti *from the Tambopata-Candamo region.*

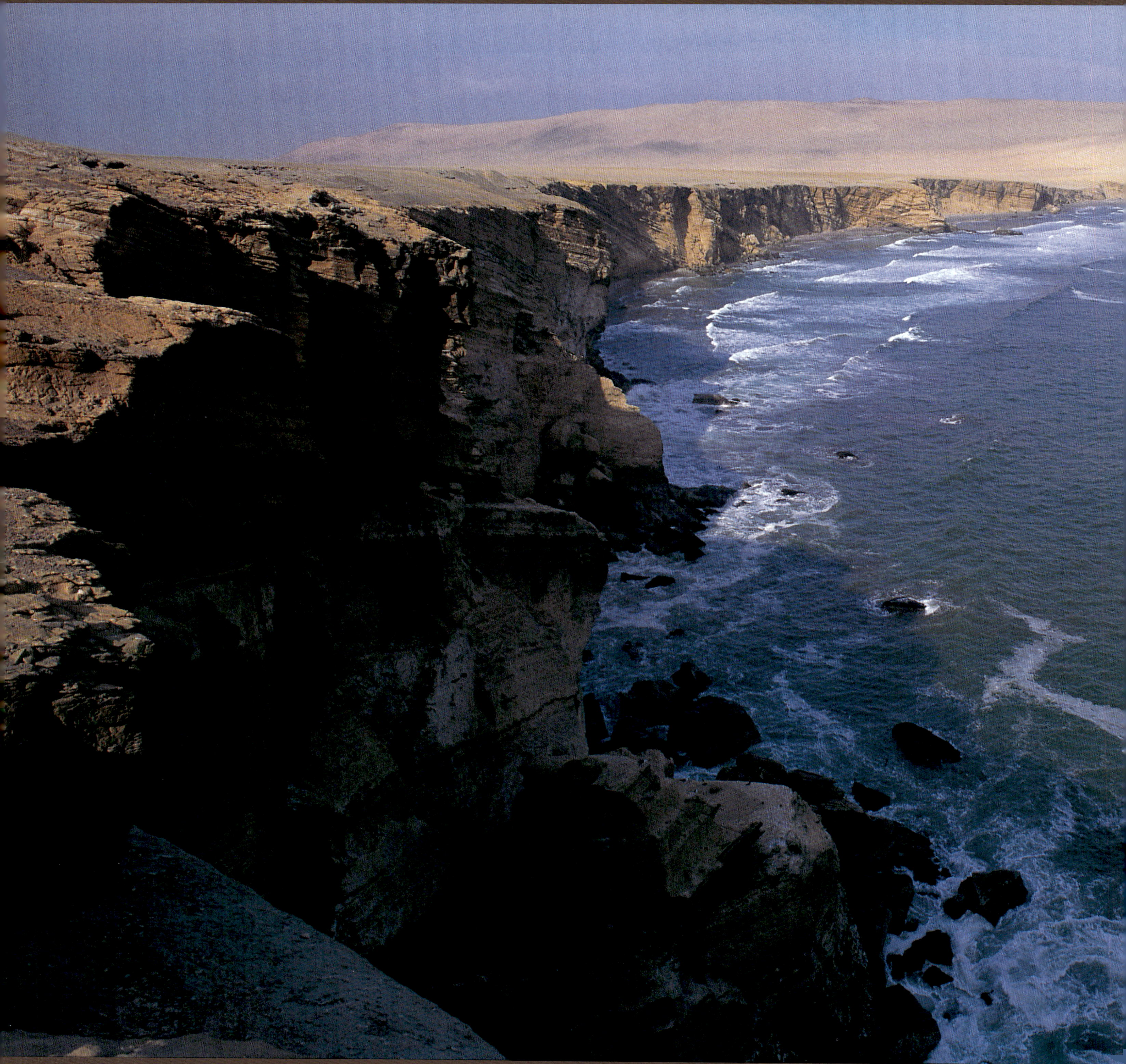

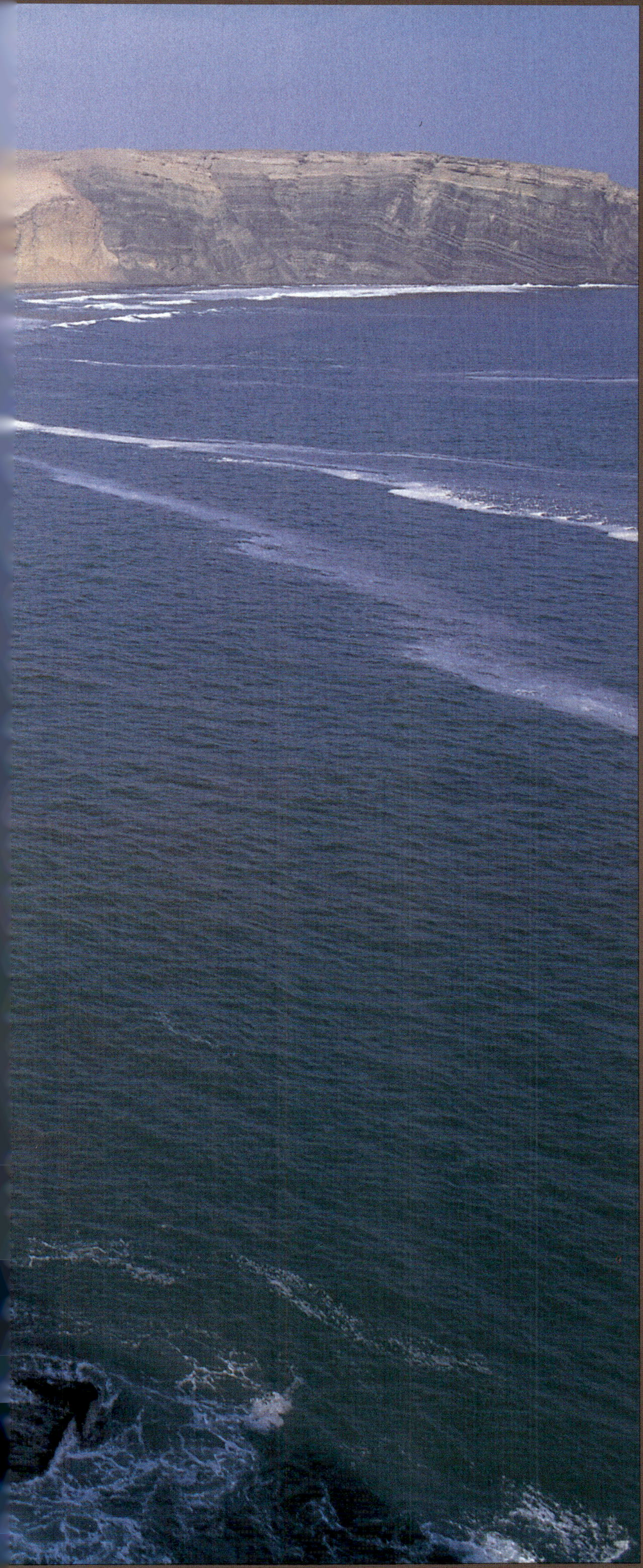

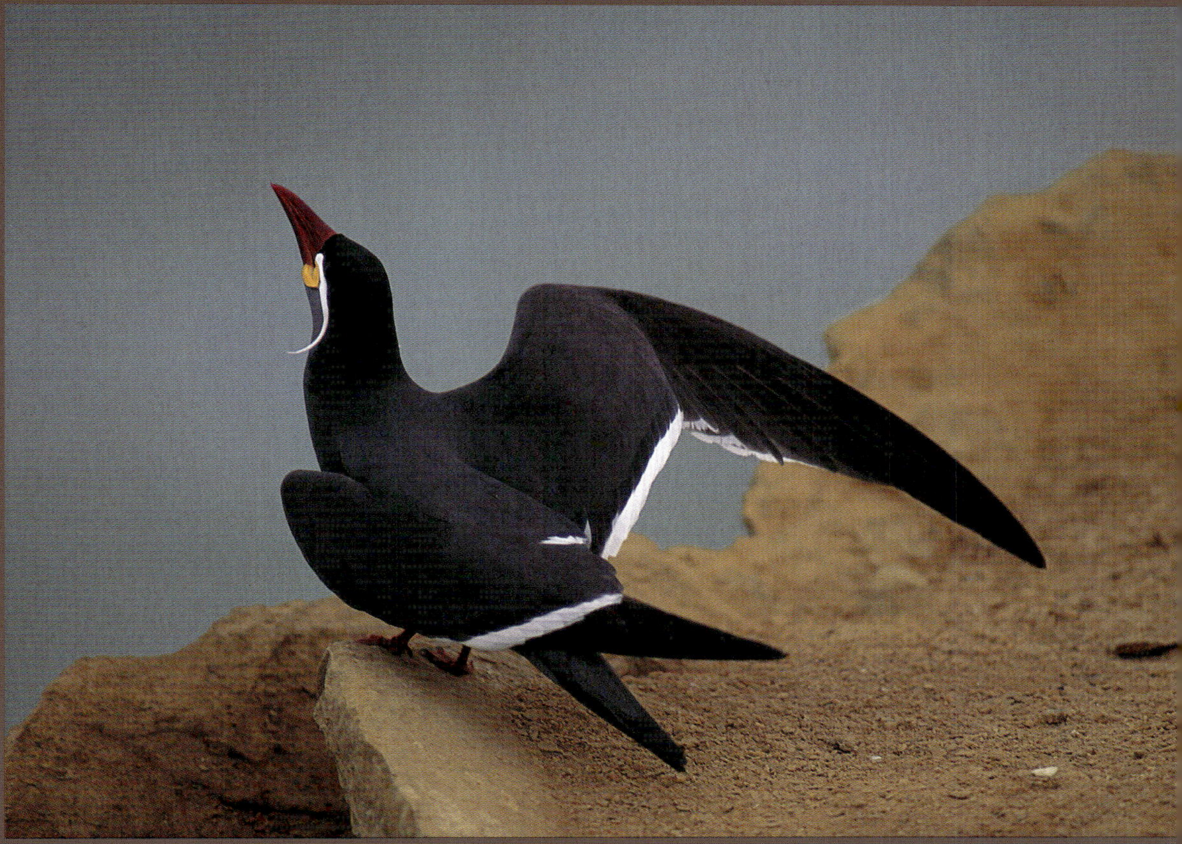

Left, the Peruvian coastal desert at Paracas.
This desert is one of the driest on Earth, and Paracas National Park
is the most important protected area in this part of Peru.
Above, an Inca tern (Larosternia inca)
photographed in Paracas National Park.

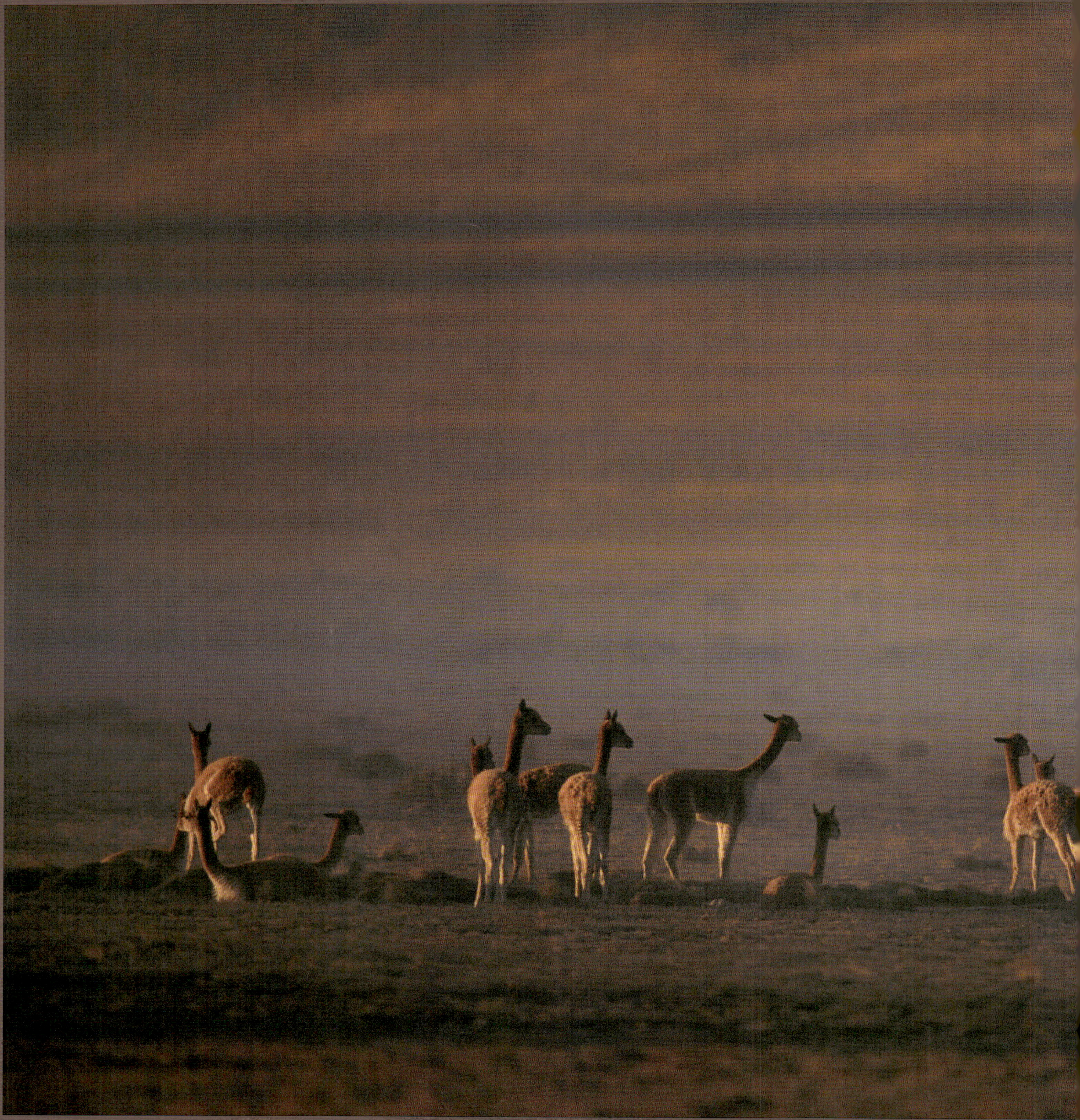

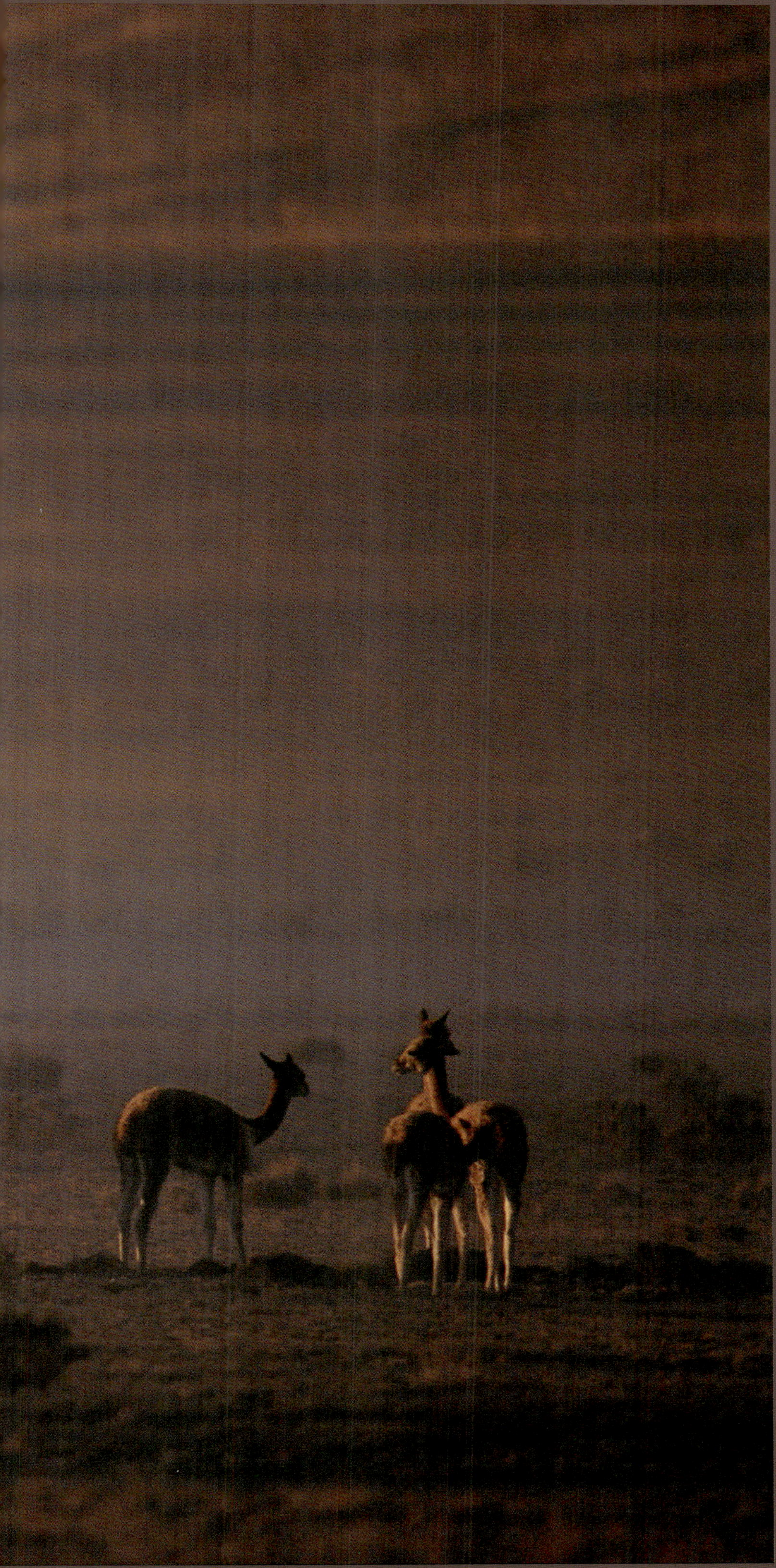

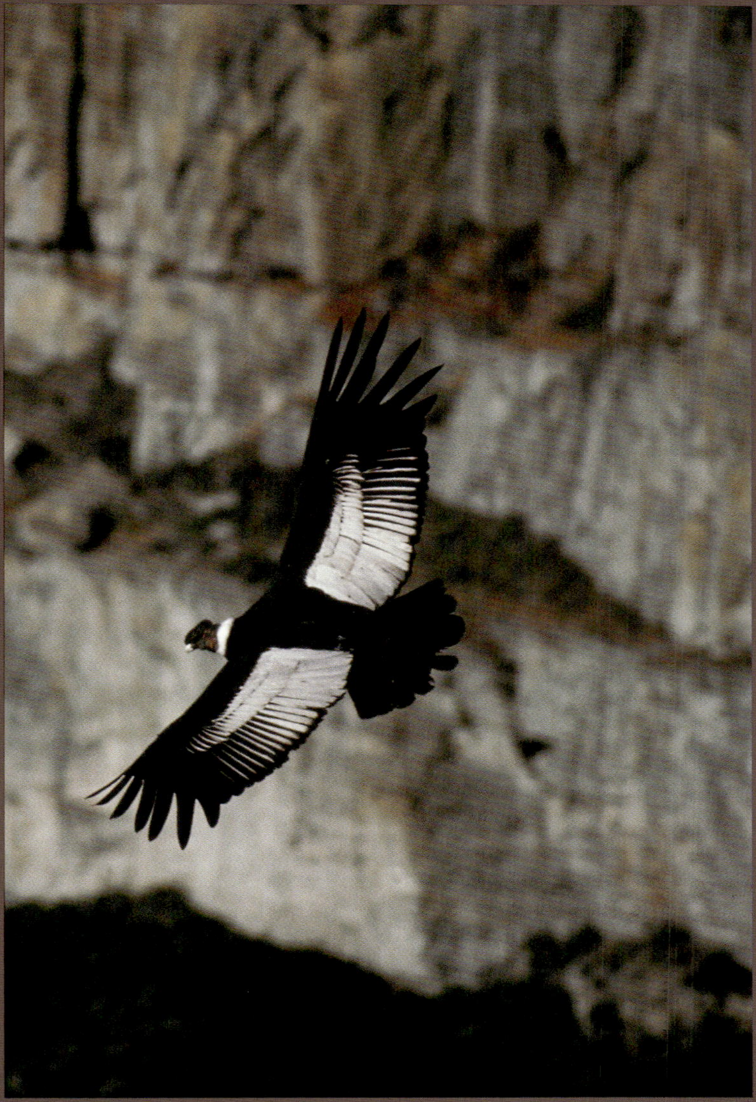

Left, a herd of vicuña (Vicugna vicugna) *in the Salinas-Aguada Blanca Reserve in the* puna *grasslands of the high Peruvian Andes. Peru has the largest remaining population of this attractive flagship species, the smallest member of the camel family. Above, the Andean condor* (Vultur gryphus), *largest member of the vulture family in South America, and the most famous avian flagship of the Andean region.*

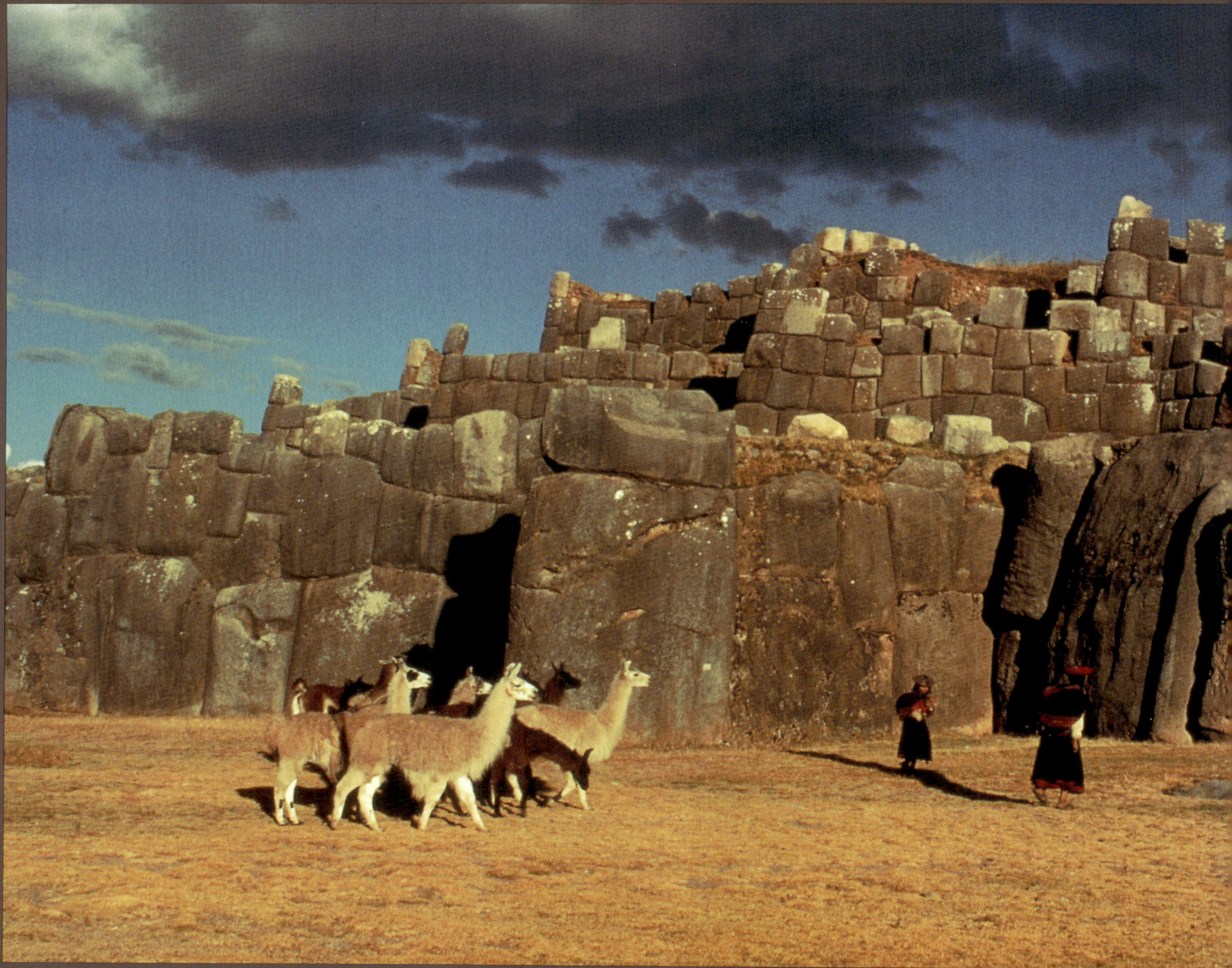

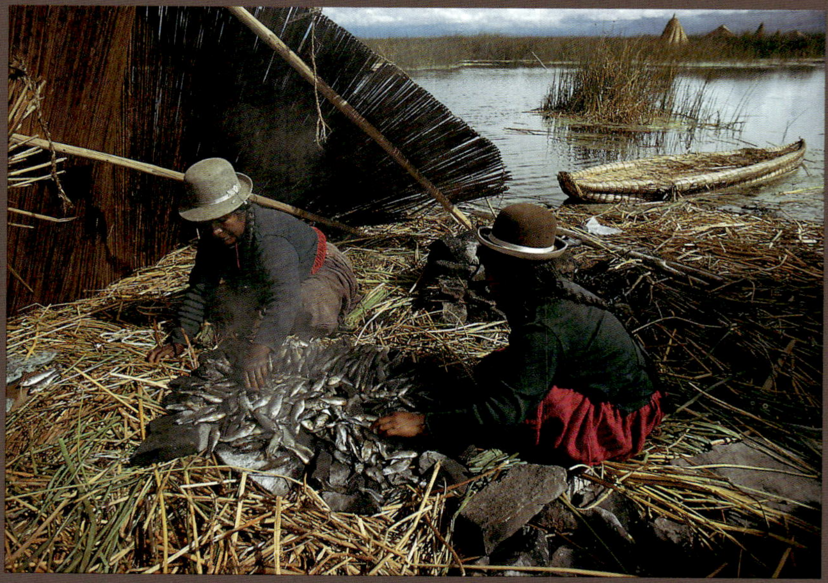

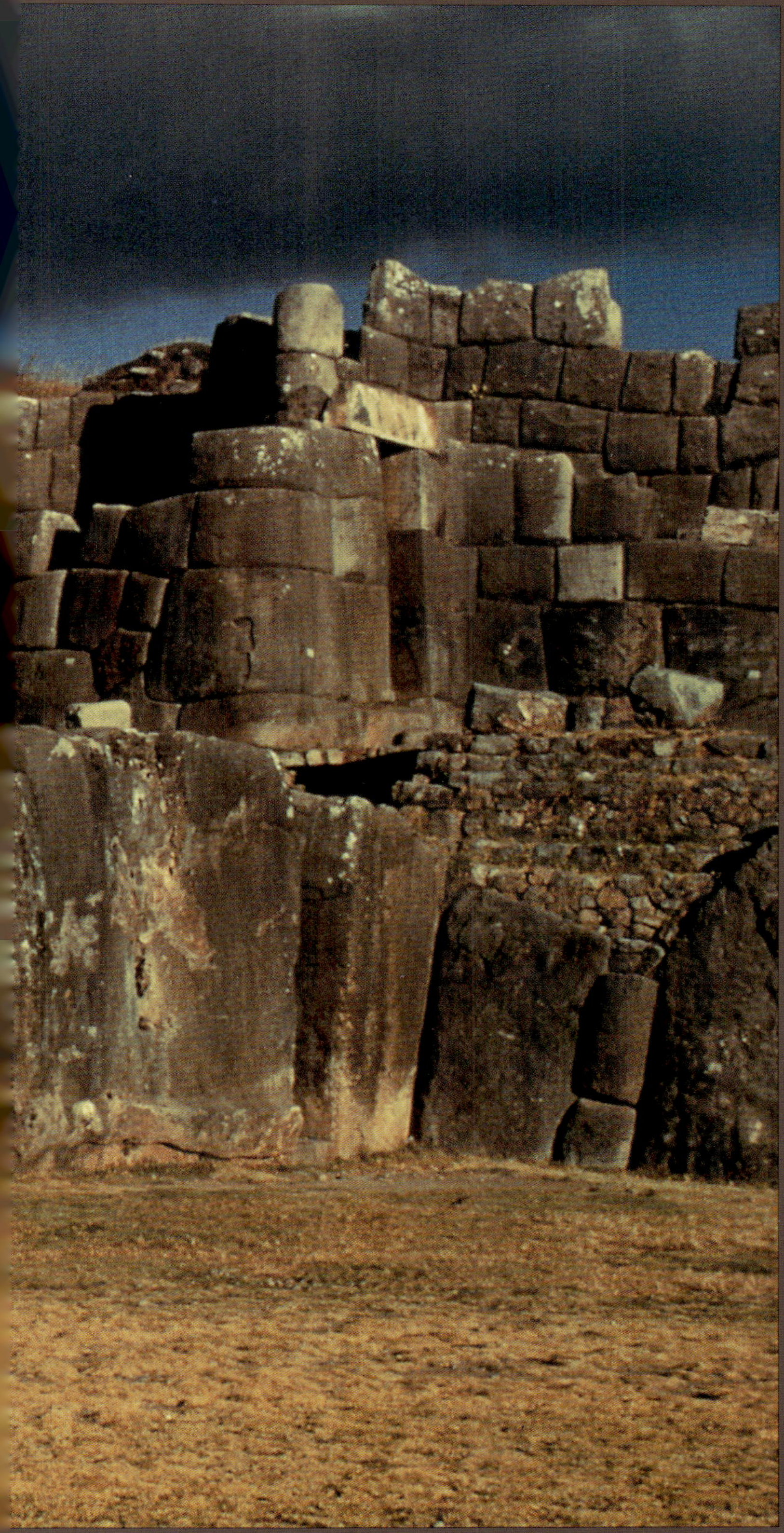

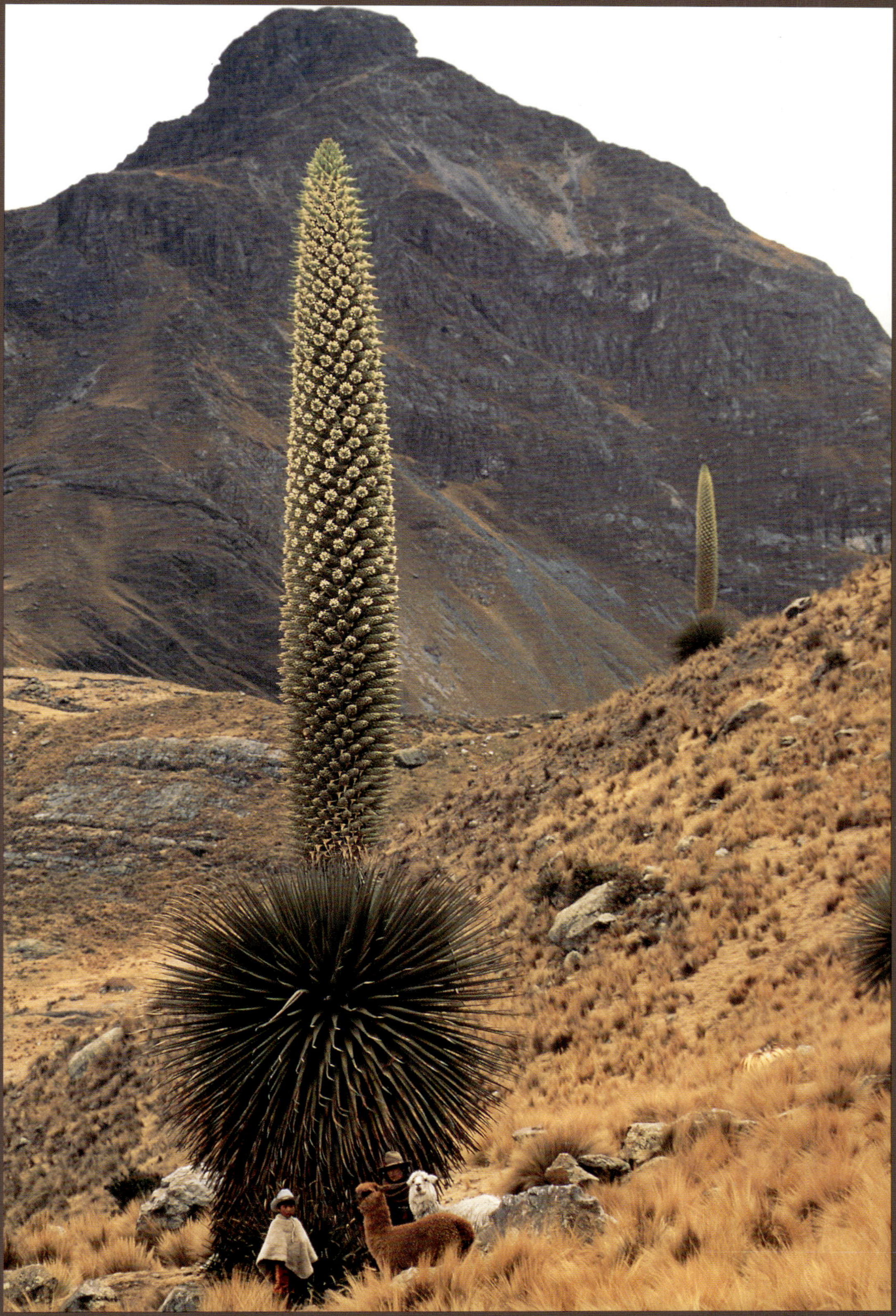

Upper left, Quechua Indians and llamas
at the Sacsahuamán ruins above Cuzco in the Peruvian Andes.
Lower left, Uru Indians cooking fish on hot stones at Lake Titicaca.
The very unusual Urus, a subgroup of the Aymara Indians,
live on floating islands of totora reed
and travel in boats made of this same reed, seen here in the background.
Above, Quechua Indians and alpacas below a spectacular example
of Puya raimondii*, the world's largest bromeliad.*
This imposing species is found in the high Andes; this example was
photographed in Huascarán National Park,
site of the highest peak in Peru.

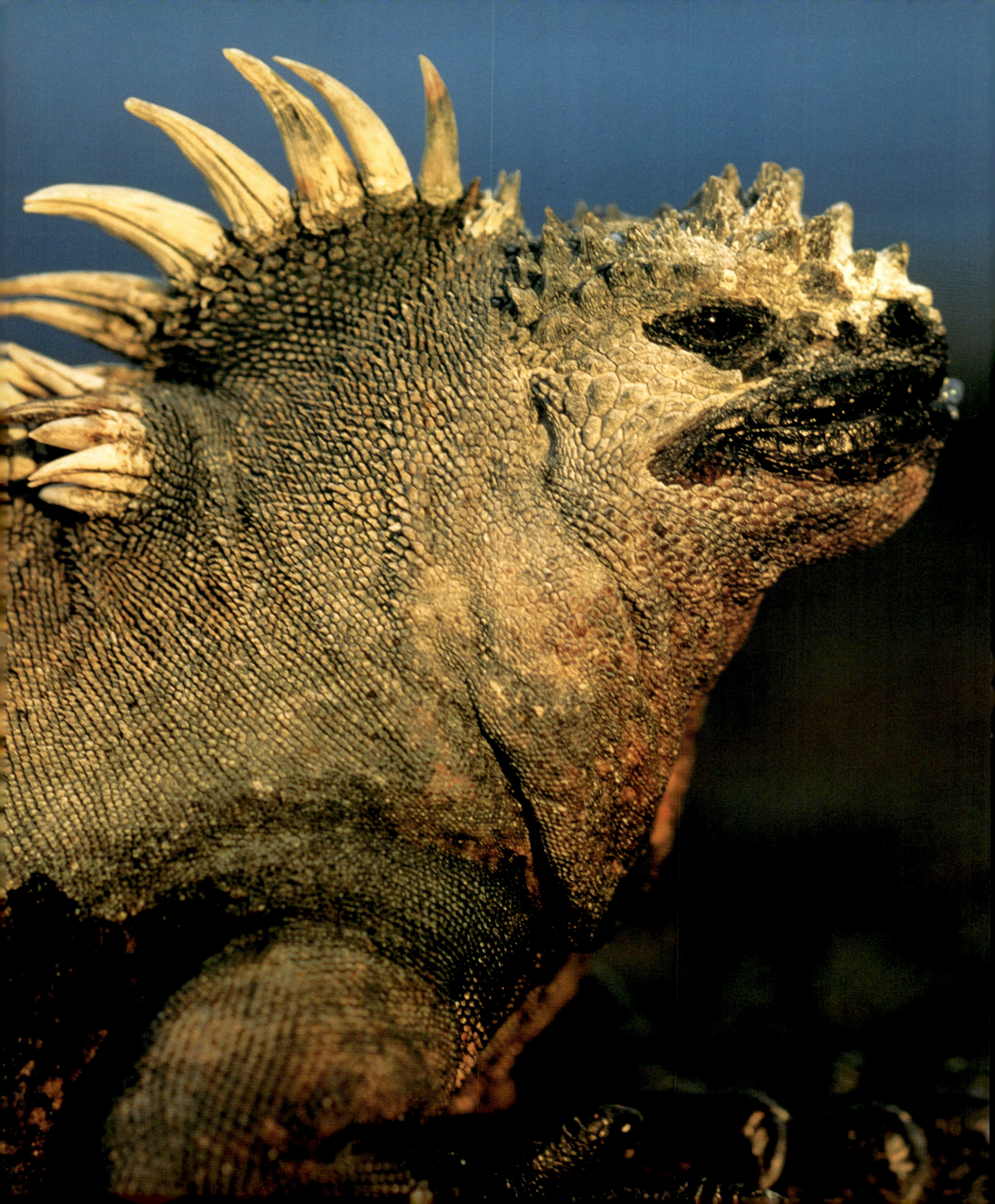

ECUADOR

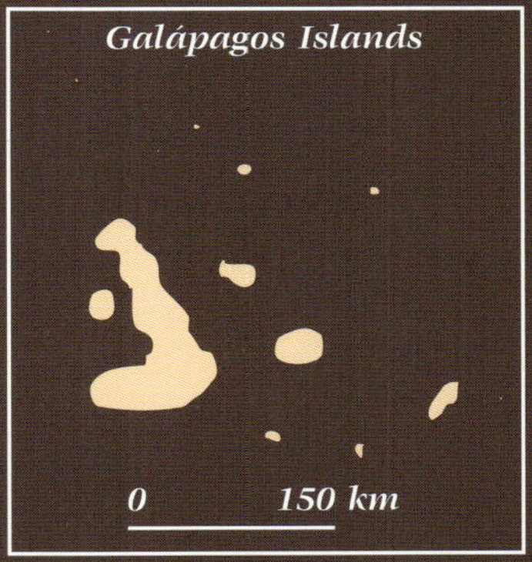

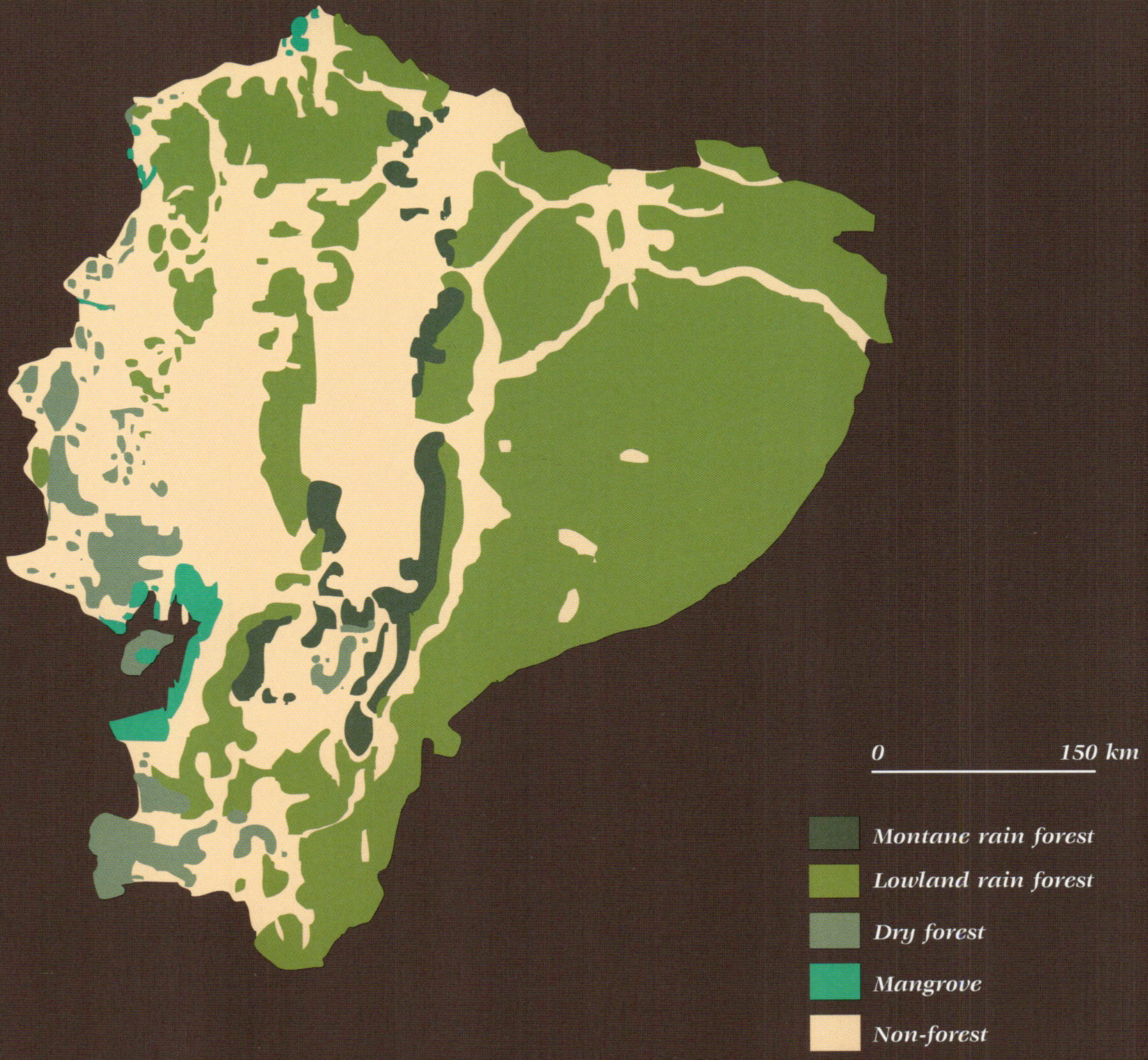

With a land area of only 283 561 km^2, Ecuador is the smallest of the megadiversity countries, occupying only 0.19% of the land surface of the planet and being about the size of the American state of Colorado. Nonetheless, as a result of a variety of geographic and climatic factors, it has surprisingly high biodiversity for such a small land area. It ranks high in virtually every category of organism and, if per unit area diversity is taken into consideration, it may well emerge at the top of the megadiversity list.

Straddling the equator (from which its name derives), Ecuador has within its borders two of the world's most important threatened biodiversity hotspots, the tropical Andes, which ranks number one on the hotspots list, and the Chocó/Darién/Western Ecuador hotspot, which in Ecuador includes both the wet forests of the province of Esmeraldas and the unique dry forest formations of Ecuador's Pacific coast. These are both highly diverse and rich in endemics, and are also among the most endangered of the hotspots. The lowland Amazonian portion of Ecuador also falls within the major tropical wilderness area of upper Amazonia. Ecuador also has within its borders 11 Endemic Bird Areas as defined by BirdLife International, four of which are first priority, and 6 Centers of Plant Diversity and Endemism as identified by WWF/IUCN. Add to this the fact that the globally unique Galápagos Archipelago is also part of Ecuador's national territory, and one finds a most impressive recipe for high biological diversity in this tiny country.

In terms of species numbers for different groups of animals and plants, Ecuador's highest ranking comes in amphibians, where it is a remarkable third globally with 402 species, exceeded only by the much larger Brazil and neighboring Colombia, and seventh in endemic amphibians (138 species, 34% endemism). In birds, Ecuador is comparably impressive, its 1 559 species (37 endemic) putting it in fourth place internationally. Translating Ecuador's diversity in these two groups into per unit area figures, Ecuador comes out with an astounding 1.42 amphibians and 5.50 birds per 1 000 km^2, compared to 0.51 amphibians and 1.59 bird species per 1 000 km^2 for Colombia, and only 0.06 amphibian species and 0.19 bird species per 1 000 km^2 for Brazil. In other words, every little piece of remaining habitat in Ecuador is of great international value, comparable to that of Madagascar and the Philippines, which also have great concentrations of per unit area diversity.

Looking at several other groups of organisms, Ecuador ranks fifth in swallowtail butterfly diversity with 69 species (3 endemic), eighth in reptile diversity with 374 species and ninth in endemic reptiles with 114 (30% endemism), sixteenth in total mam-

*On the opposite page, the Galápagos marine iguana (**Amblyrhynchus cristatus**) is the second most important flagship species for the Galápagos Islands, after the famous giant tortoises for which they are named. This unique animal, the only marine lizard species, is seen here on Fernandina Island, one of the largest and most important colonies in the archipelago.*

mal diversity with 271 species (21 endemic, 8% endemism), and possibly as high as sixth or seventh in total higher plant diversity with perhaps as many as 20 000 species. There is no consensus on higher plant endemism, with estimates ranging from 4 000-5 000. Given the high total diversity and the likelihood that much of Ecuador's flora remains poorly understood, we use the higher figure of 5 000 endemics, which qualifies Ecuador for the top 12 megadiversity countries globally.

In terms of marine diversity, the Galápagos Islands make a significant contribution to Ecuador with a number of endemics ranging from large fish like the Galápagos shark down to fully 18% of the archipelago's 1 900 reported invertebrate species. However, Ecuador does not compare in this category to some of the larger megadiversity countries like Mexico, Colombia and Indonesia, and especially to some of the smaller megadiversity countries such as Papua New Guinea and the Philippines.

Human cultural diversity is also relatively high, especially when considered, once again, in terms of per unit area diversity. At least sixteen different native ethnic groups (with numerous subgroups) exist in Ecuador, each of which considers itself a distinct nationality, with its own language and culture. All told, these make up a substantial proportion of the population and include, in the Andes, the Quechua-Quichua (of several distinct groups), and the Colorado; the Shuar, Achuar, Huaorani, Siona-Secoya and Cofán of the Amazon lowlands; the Awa (Kwaiker) and Chachi of the western lowlands; and the Afro-Ecuadorians of Esmeraldas and the Chota Valley of Imbabura province, descendants of black slaves.

Ecuador is particularly interesting amongst the megadiversity countries in that it demonstrates that a nation does not have to be large in geographic area in order to have high biodiversity. Much depends on the country's geographical position, topography and climate, and the variety of ecosystems within its borders, and less on the sheer area that it occupies. Ecuador is a very special case, and like the other small megadiversity countries (Madagascar, Papua New Guinea, Philippines) deserves special attention from the international community.

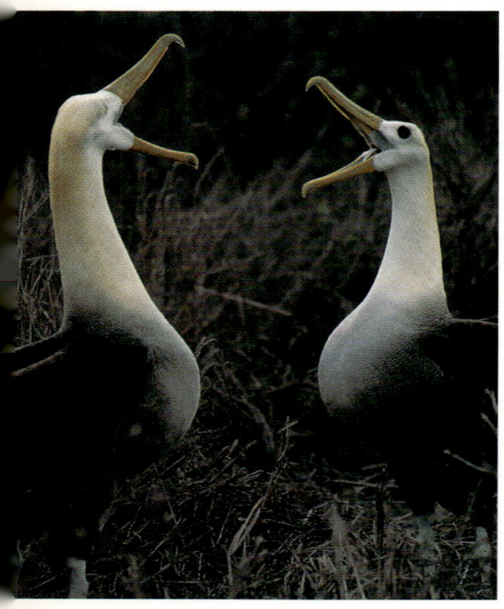

A double gape display in the courtship ritual of the waved albatross (Diomedea irrorata), *on Punta Suárez, Española Island in the Galápagos.*

Natural habitats

The country can be divided into four natural regions, as follows: the western lowlands, including the Cordillera de la Costa (Coastal Range); the high Andes, including the Cordillera del Cóndor (Condor Range); the eastern Amazonian lowlands (referred to locally as El Oriente); and the Galápagos Islands.

The western, or Pacific lowlands are bounded to the west by the Pacific and to the east by the Andes (up to approximately 900 m), with the borders of Colombia and Peru marking the northern and southern limits of the region, respectively. Spanning eight provinces (from north to south: Esmeraldas, Pichincha, Manabí, Los Ríos, Guayas, Canar, Azuay and El Oro), this region is flat except for a spine of coastal hills, the Cordillera de la Costa, paralleling the coast in the northwestern coastal provinces of Esmeraldas, Manabí and Guayas. These mountains seldom exceed 800 m in elevation yet their slopes, as well as those of the Andes to the east, create topographic variations that have resulted in high levels of floral and faunal diversity and endemism in this region. However, the vast majority of the western lowlands lies below 400 m altitude; it has a mean annual temperature of 25°C and rainfall ranging from 2 500 to 4 500 mm annually, the peak rainy months being December to June.

The coastal and foothill forests of western Ecuador are important for their high numbers of species and high levels of endemism. As many as 6 300 species of higher plants occur in this region, 1 200 (19%) of which are regional endemics (Gentry, 1986). Bird diversity is equally high in this region with more than 800 species, 40 of which are endemic to western Ecuador (Ortiz *et al.*, 1990). The region is home to approximately 142 mammals, 15 of which are also regional endemics (Albuja, 1991). Other vertebrate groups showing high regional endemism are the bats (approximately 30 of the country's 125 species; 24% endemism), as well as some 60% of the country's amphibians. Many of these endemics are limited to very small ranges such as isolated ridgetops of only a few square kilometers in size, making them particularly vulnerable to extinction.

The vegetation in the southern half of the region was originally tropical dry forest (now largely converted), while the northern half represents the southernmost extension of the Chocó biogeographic region, a lowland wet forest formation that has its greatest land area in Colombia and extends into Panama's Darién department as well. Moist forests occur in the west of the region, as well as mangrove ecosystems along the coast. The coastal dry forests in Ecuador contain many endemics, including the grotesque-trunked kapok (*Ceiba trischistandra*). Dodson and Gentry (1991) estimated that forests once covered as much as 80 000 km^2 of the western lowlands, but that of the three major lowland forest types found in this region, only some 6% of the total forest cover remains (4% in tropical moist, 1% in tropical dry, and 0.8% in tropical wet forests), making this Ecuador's most environmentally devastated region.

The Ecuadorian Andes, wedged between the Pacific lowlands to the west and the immense Amazon Basin to the east, stretch along a roughly north-south axis for the entire length of the country, splitting into two distinct cordilleras in the northern two thirds of the country, and ranging in width from 100-200 km, broadening in the north at the border with Colombia. Many of the higher peaks reach almost 6 000 m, with Chimborazo the highest at 6 310 m. The ridges connecting the mountains typically lie at an altitude of over 4 000 m above sea level, and the main Inter-Andean Valley at

2 600-2 700 m elevation is interrupted by low, transverse ridges connecting the two parallel cordilleras. The Inter-Andean Valleys provide the most hospitable environment for humans in the country, and these areas have been densely populated since pre-Columbian times.

Plants reach their highest levels of diversity and endemism in this region of Ecuador with 10 500 species, 25% of which are endemic. The montane forests, lying on the slopes of the Andes between 900- 3 000 m, contain perhaps half of the country's species while occupying only 10% of the area. The western slopes, in particular, are rich in endemics, with approximately 10% of the total Ecuadorian flora restricted to them. These montane forests, which extend north into Colombia and south into Peru, are one of the world's richest areas for bromeliad and orchid species as well. They are also botanically famous as the home of *Cinchona officinalis*, the bark of which is the source of quinine, the most important of the antimalarial drugs.

The Cordillera del Cóndor (Condor Range), straddling the Ecuador/Peru border, is a biogeographic link between the Andes mountains and the Amazonian lowlands, and an area of global conservation significance. Some of the most unique and diverse plant communities known to science can be found in this largely isolated mountain cluster which in many ways, both morphological and biological, resembles the *tepui* formations of southern Venezuela. It is isolated from the main cordillera of the Andes and protrudes into the equatorial Amazon just north of the Marañón Gap, the arid valley that separates the northern from the southern Andes.

The Amazon lowlands of Ecuador, locally referred to simply as El Oriente (the East) stretch eastward from the Andean slopes to the border with Peru (department of Loreto) and north to the border with Colombia (department of Putumayo), covering all or most of the Ecuadorian provinces of Napo, Pastaza, and Morona-Santiago. The flat plain is predominantly covered by unbroken rain forest and traversed by several rivers that drain the Andean slopes and flow east into the Amazon. This is one of the wettest parts of the Amazon Basin, and has an extraordinary biological and cultural diversity. It is the last remaining wilderness in Ecuador, occupying 50.3% of the country's territory, and is part of upper Amazonia, the richest, yet most poorly described portion of the greater Amazon Basin.

Humid lowland forests cover most of the Amazonian region. The vegetation is dominated by trees with a dense canopy, about 30 m high and very rich in lianas and epiphytes. Those forests lying closest to the base of the Andes have a distinctive and highly diverse flora, especially in the Napo region, a center of high biodiversity, where in a relatively small area some 4 000 woody plant species occur. During recent research by the Missouri Botanical Garden in this area, it was found that 70% of the tree species collected had not been previously recorded from Amazonian Ecuador, and 10% were species new to science, showing just how little is truly known of the region's vegetation. The most important tree families are Leguminocaceae, Moraceae, Lauraceae, Myristicaceae, and Meliaceae. It is estimated that overall, the Amazon region of Ecuador is home to 8 200 plants, 15% of which are endemic. This region also has the second highest recorded tree diversity on Earth, with 472 species having being found in one hectare (Valencia *et al.*, 1994). Only one site in Espírito Santo in Brazil's Atlantic Forest exceeds this total.

Lying 1 000 km west of mainland Ecuador on the equator, Galápagos is one of Earth's most curious jewels of natural history. The Galápagos (officially the Archipiélago de Colón) are true oceanic islands, having risen from the Pacific sea floor as volcanic peaks and never having been in contact with continental land masses. This isolation, combined with their relative youth (1-3 million years old), make the Galápagos a veritable laboratory of evolution, where one can see the results of the rapid adaptation of animals and plants to the unique and harsh environments of the islands, as well as the intricate mechanisms of arrival and establishment that allow some taxonomic groups (like reptiles and monoecious plants) to thrive, while others (amphibians) are absent entirely.

The Galápagos are best known for their contribution to the revolutionary thinking of a young ship's naturalist named Charles Darwin who visited the islands in 1835 aboard the survey vessel *Beagle*. Darwin changed the world with the publication of *On The Origin of Species by Means of Natural Selection* in 1859, introducing the concept of evolution to a world that then believed that the Earth had been created in seven days and that its animals and plants had been saved from a great flood by Noah. This sparked a revolution in thinking which was the birth of countless new branches of science, and more importantly an entirely new human paradigm, changing and redefining man's view of his own niche on the planet and in the universe.

The fauna and flora of the Galápagos is like nothing else on Earth, with levels of endemism that reach 100% in some taxonomic groups. As is the case in many isolated island clusters, this extremely high endemism is accompanied by low diversity overall, the complete absence or paucity of species that would find it difficult to establish themselves across large stretches of ocean (like terrestrial mammals and amphibians), and an overabundance (relative to continental biotas) of species that can readily survive a long ocean voyage (like seabirds and the plants they disperse through the seeds they eat).

The Galápagos Islands are home to 702 plant species of which 32.5% are endemic, numerous sea and land birds, and a unique variety of marine organisms. The most noteworthy among the Galápagos' array of curious inhabitants are the reptiles. These include the 14 races (three extinct) of giant tortoises, from which the islands derive their name; *galápago* is a type of saddle that resembled, in the eyes of the first Spanish visitors,

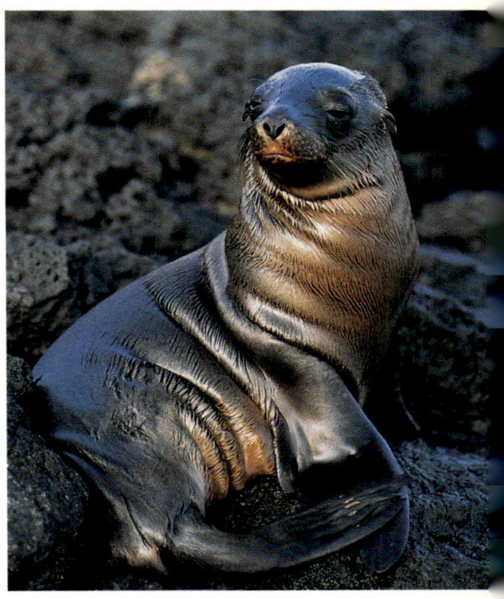

Galápagos fur seal (Arctocephalus galapagoensis), *a species endemic to the Galápagos.*

On pp. 318-319, giant tortoises (Geochelone e. elephantopus) *searching for moisture along steaming fumaroles on the caldera rim of the Alcedo Volcano on Isabela, largest of the Galápagos Islands.*

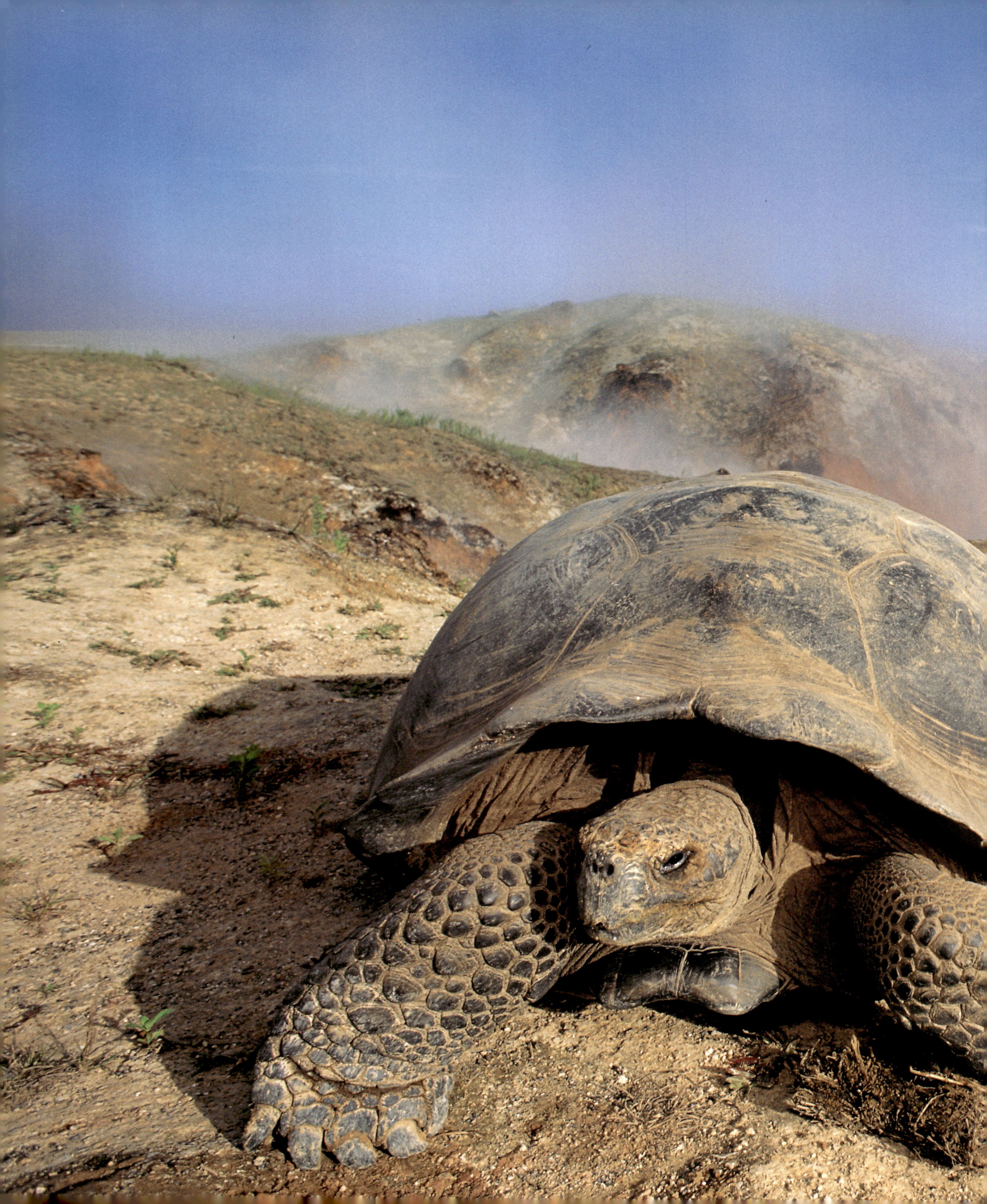

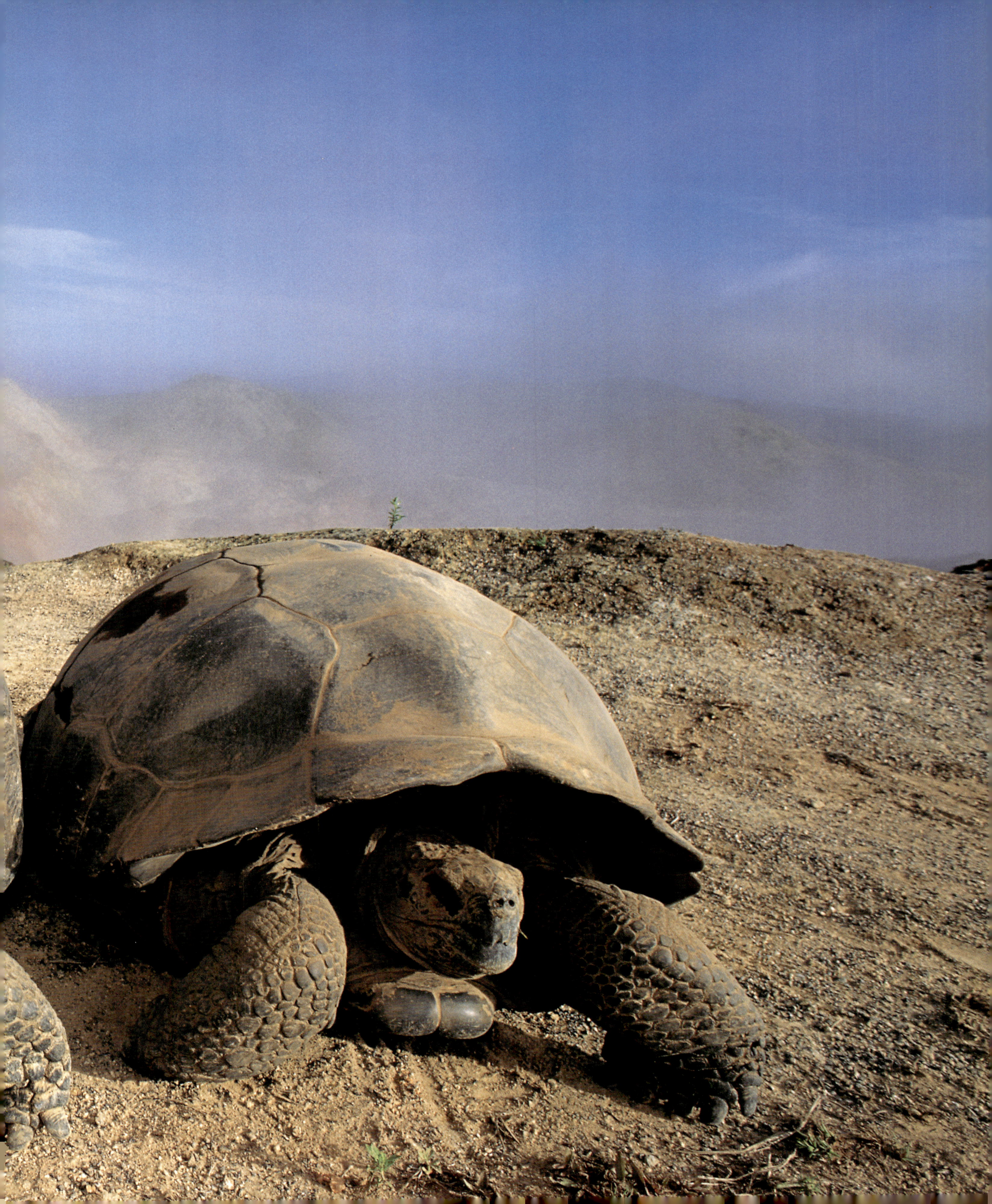

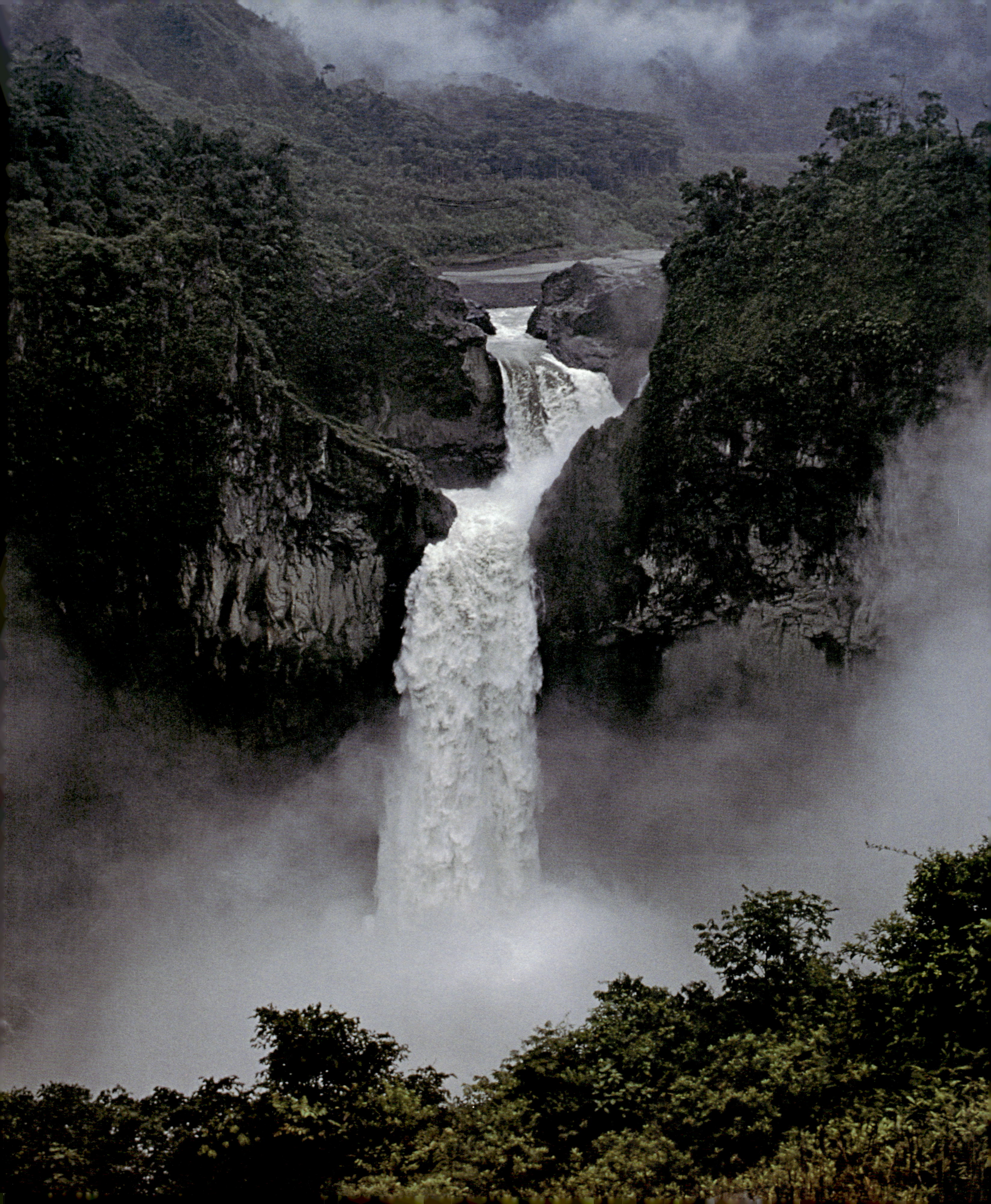

the carapaces of the saddle-backed tortoises. The Galápagos is the only place on Earth where these giant tortoises are found, though a similar species still survives on the island of Aldabra in the Seychelles. The tortoises and some of the other amazing animals of the Galápagos are discussed below in the section on flagship species.

The Galápagos' unique living heritage derives from an uncommon combination of geologic, oceanographic and climatic conditions. The thirteen large islands, six smaller islands and countless rocks and islets rise from the sea floor (3 000-4 000 m depth) as the tips of enormous volcanoes, many of which have coalesced to form a submarine shelf, the Galápagos platform, at a depth of about 400 m. Tectonically, the islands lie near the triple junction of the Cocos, Nazca and Pacific plates, and are moving at a rate of several centimeters per year in a southeasterly direction. It is likely that they were "born" by passing over a stationary plume in the Earth's molten mantle, as were the Hawaiian Islands. Galápagos is one of the most volcanically active sites on the planet today, and one of the few places where one can observe the geological processes that have formed the Earth over the millennia.

Oceanographically, the archipelago is bathed from the east by major influxes of cold Antarctic waters that travel up the Peruvian coast, such as the Humboldt Current, which flows north from Antarctica, passing the Peruvian coast and then taking a sharp westward turn at the equator to become the South Equatorial Current. From the east, the deep and cold Cromwell Countercurrent wells up against the shores of the western islands of Isabela and Fernandina. These result in cold sea temperatures year round (16-28°C) and two distinct seasons, the warm-wet season lasting from January to June, and the cold-dry season, from July to December.

Flagship species

The fauna and flora of the Galápagos is unlike anything else on the planet, and Ecuador's best known and most spectacular flagship species are those of this unique island group. In the Galápagos, the reptiles dominate the animal scene, filling the niches of large mammals elsewhere. Given the difficulties of arrival and establishment from the continent that lies across 1 000 km of open sea, naturally-occurring terrestrial mammals are nearly absent (only two bats and two native rice rats). Among the reptiles, there are two endemic species of land iguana (*Conolophus subcristatus* and *C. pallidus*), the latter of which is restricted to a single island, Barrington (Santa Fe); the three endemic species of Galápagos snakes of the genera *Alsophis* and *Philodryas*; the six species of leaf-toed geckoes (*Phyllodactylus*), five of which are endemic; and the seven endemic species of lava lizards (*Tropidurus*), which are constantly under foot and unquestionably the most ubiquitous reptiles of the arid lowlands. However, the two most outstanding reptile flagship species are the unusual marine iguanas (*Amblyrhynchus cristatus*) and the giant land tortoises (*Geochelone elephantopus* ssp.).

Galápagos tortoises (*Geochelone elephantopus* ssp.) have played a major role in the natural and human history of the islands. This is a single species with fourteen subspecific races which, until recently, were found on all major islands of the archipelago, except for Genovesa, Marchena, Culpepper, and Wenman. The races differ in the form of the carapace, some races having "domed shells," while others are "saddle-backed," and still others are "intermediate" between the two. The Galápagos tortoises are the world's largest; one incredible individual from Santa Cruz weighed in at a whopping 400 kg (Pritchard, 1996). These tortoises most likely evolved from a smaller mainland species which, upon arrival, was able to thrive in the predator-free environment, and subsequently assumed the gigantic proportions and niche of the otherwise absent large herbivores. Galápagos tortoises are strictly vegetarian, and eat in excess of 50 different plant species (Cayot, 1981), being particularly fond of the prickly-pear cactus (*Opuntia* spp.) and the manchineel, or poison guava (*Hippomane mancinella*), the latter being highly poisonous to most animals, including man.

The same processes which have shaped the evolution of this remarkable animal have also left it vulnerable. Its size and ability to survive long periods of time without food or water in holds of ships (one ship's log claimed 14 months) made it a perfect source of meat for passing sailors, and hunting took an enormous toll up until the beginning of this century. Having evolved without predation, Galápagos tortoises are also poorly equipped to withstand the more insidious threat, particularly to eggs and young, posed by competition with and predation by recently-arrived feral animals. Many subspecies are now on the verge of extinction and at least two, from Charles (Santa María) and southwestern Chatham (San Cristóbal), have already been eliminated (Pritchard, 1996). McFarland (1974) estimated that 15 000 tortoises survive in the Galápagos from an estimated original population of 250 000. Hunting is now outlawed, thus the principal threat to the tortoises comes from introduced pigs, cats, rats, and dogs, which prey upon eggs and young, and compete for scant food resources.

The Charles Darwin Research Station in the town of Puerto Ayora on Santa Cruz Island has implemented, with the help of the Galápagos National Park Service, several successful conservation programs for the tortoises. They have captive-reared and head-started several broods of the highly endangered tortoise from Hood (Española) Island, releasing the subadults once they reach a size at which they are no longer vulnerable to predation by rats. The Charles Darwin Research Station is also the home of "Lonesome George," the last surviving tortoise of the Pinta Island race. A worldwide effort has been made to locate a mate for "Lonesome

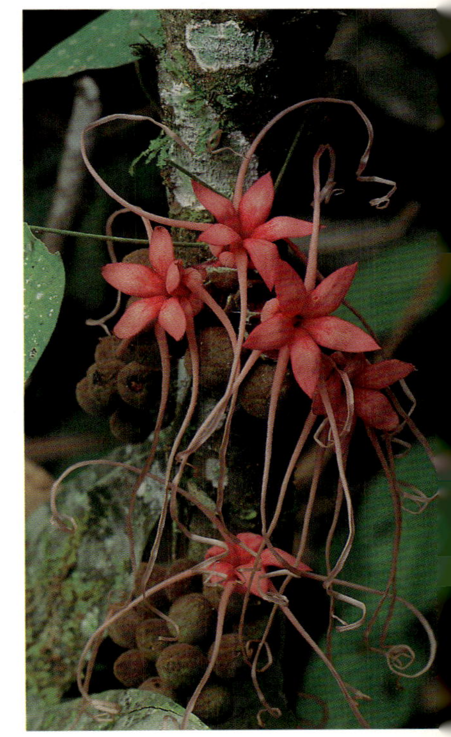

On the opposite page, the gorge of the Quiros River below the San Rafael (Coca) Falls in the eastern foothills of the Andes.

Above, the flower of a member of the cocoa family Sterculiaceae, with fruit flies as pollinators. Napo Province, Amazonian Ecuador.

George" among captive specimens that were collected by zoos in the early 1900s, but sadly it appears that he is indeed the last of his kind.

The Galápagos marine iguana (*Amblyrhynchus cristatus*) is second only to the tortoises in terms of herpetological interest. These lizards are found throughout the rocky shores of the archipelago in concentrations that can reach up to 3 000 individuals per km of coastline. The only lizards that consistently feed in the sea, these animals graze on seaweed both underwater and in the intertidal zone. They have evolved a number of unique adaptations that differentiate them from their terrestrial relatives, including a laterally-compressed tail for swimming, long claws for grasping the rock substrate in a rough surf, a gland that allows for the filtering and expulsion through the nostrils of excess salt, and a blunt snout that permits them to closely crop seaweed. Perhaps the most intriguing adaptation, however, is the ability to slow heartbeat and metabolism through a process called bradycardia that permits the animals to stay beneath the surface for as long as 60 minutes while feeding. Marine iguanas are strictly vegetarian and consume enormous quantities of marine algae of several species; it is estimated that the approximately 2 000 iguanas found at Punta Espinoza on the island of Fernandina consume on the order of 28 tons of algae each year.

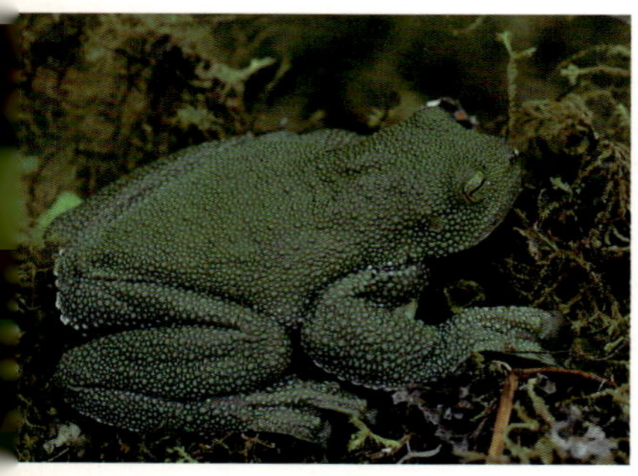

A glass frog (Centrolenella gekkodeum) *from an altitude of 2,550 m in northwestern Ecuador. Although the smallest of the megadiversity countries, Ecuador has the third highest amphibian diversity (402 species), most of it made up of frogs.*

Much of a marine iguana's time is spent basking in the sun in a variety of different postures to elevate body temperature after bathing in the cool Galápagos seas; like miniature dinosaurs, the iguanas orient themselves en masse, like patrons in a movie theater, staring at the sun or positioning themselves to maximize exposure to its warming rays. They are colonial animals, largely passive and docile except in the breeding season, when males can become aggressive. Though all of the Galápagos' marine iguanas are of the same species, there are marked differences in coloration between islands, the most distinctive being those of Hood Island, which bear striking flashes of green and red in comparison to the shades of gray and black found elsewhere in the archipelago.

The Galápagos have fittingly been termed "Islands of Birds." There are 19 resident seabirds, including such rarities as the flightless cormorant (*Phalacrocorax harrisi*), the only living cormorant that has lost the ability to fly. The Galápagos penguin (*Spheniscus mendiculus*) is the second smallest of all penguins, and thanks to the archipelago's unusually cold waters that allow it to thrive, it is the penguin species with the most northerly distribution. There is also an endemic albatross, the waved albatross (*Diomedea leptorhyncha*), which nests solely on the island of Hood, and is known to be one of the only albatrosses with a restricted home range rather than being great sea wanderers like the majority of its congeners. Galápagos seabirds are rounded out by tropic birds (*Phaëthon aethereus*), several species of storm petrels, the endemic lava gull (*Larus fuliginosus*), frigate birds, oyster catchers and the famous blue-footed booby (*Sula nebouxi*) which, unlike the other two species of booby occurring in the Galápagos —the masked booby (*Sula dactylatra*) and the red-footed booby (*Sula sula*)— seldom wanders off to sea, preferring to stay in relatively shallow coastal waters, where it is a favorite photographic subject of tourists. The dozen or so shorebirds of the Galápagos include a resident population of flamingos (*Phoenicopterus ruber*), five species of herons, and a number of other migrants.

The Galápagos is also home to 29 resident land birds, of which 22 are endemic. The world-famous Darwin's finches are a radiation of 13 species of "dull-colored birds" as described by Darwin himself; they derived from the same ancestor and have evolved different anatomical features to better use the varied ecological niches of the islands. The four species of ground finches (*Geospiza nebulosa*, *G. fuliginosa*, *G. fortis* and *G. magnirostris*) are probably the best studied of all the Darwin's finches. They are crucial to evolutionary studies because they are the most similar to the ancestral form of Darwin's finch, and therefore represent an important step in the truly extraordinary evolutionary succession that has resulted in today's radiation of Galápagos finches. Other species of Darwin's finches include the cactus finch group, the tree finch group, and the woodpecker, mangrove, sharp-billed ground or "vampire," and warbler finches. All of these birds are remarkably similar in body shape and plumage, with the greatest variation occurring in their bills, which have adapted in shape and size to allow these birds to make the most of the scarce food sources available on the islands.

Marine and coastal life abounds in the uncommonly cold and rich equatorial waters of the Galápagos, the most conspicuous coastal inhabitants being the aforementioned marine iguanas, the endemic sea lions (*Zalophus californianus wollebaeki*), and the fur seals (*Arctocephalus galapagoensis*). Adjacent seas are home to a resident population of black sea turtles (*Chelonia agassizii*), hawksbill turtles (*Eretmochelys imbricata*), leatherback turtles (*Dermochelys coriacea*), and an as-yet undescribed taxon of yellow sea turtle. A number of sharks can also be found, ranging in size from the giant whale shark (*Rhincodon typus*) down to the diminutive and unusual horn shark (*Heterodontus quoyi*), and including an endemic Galápagos shark (*Carcharhinus limbatus*). The Galápagos is also one of the few places on Earth where one can see schooling hammerhead sharks (*Sphyrna lewini* and *S. zygaena*), a rare phenomena that was first witnessed in these islands by Jacques Cousteau in 1973. Other cartilaginous fishes include the giant manta (*Manta hamiltoni*), and spotted and golden Conway rays (*Aetobatus narinari* and *Rhinoptera steindachneri*). Over three hundred species of bony fish have been reported from the Galápagos also, representing a mix of tropical and temperate taxa with some interesting endemics such as the Galápagos red-lipped batfish (*Ogcocephalus darwini*). Baleen whales, including the humpback (*Megaptera novaeangliae*), four-toothed whales, and a number of dolphins are also present in Galápagos seas.

In terms of continental flagships, northwestern South America has an extraordinary abundance of birds in global terms, and these are the leading group in terms of vertebrate diversity in Ecuador. Scientists have identified some 1 559 species of birds grouped into 21 orders, and the species count continues to grow, including such spectacular ones as the Andean hill star (*Urochroa* sp.), which is able to survive the freezing nights of the high Andes by lowering its body temperature from 40°C in the day to 15°C in the night, a remarkable feat for a warm-blooded creature. There are approximately 120 species of hummingbirds in Ecuador, and 20% are endemic. At the opposite end of the size spectrum is the Andean condor (*Vultur gryphus*), one of the largest flying birds in the world, weighing in at 13 kg.

Ecuador is home to 271 species of mammals, of which 21 are endemic. Among the noteworthy endemics are two primate taxa that are of high conservation priority given their rapidly-disappearing habitat. The first is the brown-handed spider monkey (*Ateles fusciceps fusciceps*), which has a very limited range in the northwestern interface between the lowland forests and the slopes of the Andes; here it is protected only in the Cotacachi-Cayapas Reserve, which is the largest block of intact forest in western Ecuador but is under increasing threat from logging and colonization. Another of Ecuador's primates is an endemic subspecies of the white-fronted capuchin (*Cebus albifrons aequatorialis*), which is also threatened by habitat loss in its highly-restricted range in the southwestern dry forests. Other of the large and more spectacular continental mammal species are three tapir species: the very rare mountain tapir (*Tapirus pinchaque*), the common tapir of the Amazon lowlands (*Tapirus terrestris*), and Baird's tapir or Central American tapir (*Tapirus bairdii*), which extends into northeastern Ecuador. Ecuador and Colombia are the only two countries with three species of these animals, the largest South American land mammals. The Andean spectacled bear (*Tremarctos ornatus*) and the jaguar of the Amazon lowlands (*Panthera onca*) are two other large and important mammalian flagships.

Cultural diversity

The diversity of Ecuador's human cultures is striking. Many of the same biotic and abiotic factors that have created this country's megadiversity status, such as stark contrasts in geography, temperature, and rainfall, and the extreme isolation of some areas, have contributed to a wide diversity of Amerindian groups as well. These factors, in addition to a turbulent history of conquest, introductions, and dispersal beginning in the pre-Columbian era and lasting through World War II, have created a modern radiation of cultures that is striking even in the rich and diverse context of northern South America.

Ecuador has a human population of 12 314 000. In addition to Spanish and other Old World languages spoken by the majority of its inhabitants, nine living indigenous languages are spoken by nearly three million Ecuadorians. The country's turbulent history has resulted in four distinct "classes" of people. Caucasians, living primarily in the large cities and *haciendas* of the high Sierra, make up approximately 10% of the population. Another 10% is represented by the Afro-Ecuadorians, who inhabit the coastal towns of northwestern Ecuador. The remaining approximately 80% of the population is split between the mestizo people of mixed Caucasian and Amerindian blood, and the indigenous Amerindian groups themselves.

Today, there are sixteen or more different indigenous groups (depending on how one considers the divisions among them). The largest of these groups are the Quichua Indians. The Quichuas are closely related to the Quechuas of Peru and Bolivia, and are the descendants of the Inca. The Quichua speakers are represented by a dozen or more regional communities, including the Otavaleños, who are well-known for their fine weavings sold throughout the country and the world. The Quichua predominantly inhabit the Andean region, yet there are some 60 000 lowland Amazonian Quichuas in the provinces of Pastaza and Napo between the Napo and Coca Rivers. It is uncertain as to whether the Quichua inhabited lowland areas in pre-Columbian times, and it is speculated that they may have migrated to this region in order to escape the advancing Spaniards.

The highland Quichua, although having suffered centuries of oppression, remain much more numerous and varied in cultural traditions than their lowland cousins. With communities throughout the Andes, they are found in the provinces of Imbabura, Pichincha, Cotopaxi, Tungurahua, Bolívar, Chimborazo, Cañar, Azuay, and Loja. Typically the Quichua live in small, isolated communities with their own highly-individualized social customs, religions, languages, and traditional medicines. The Andean Quichua have a surprisingly high percentage of individuals who continue to live traditional lifestyles.

The Pacific Coast Amerindian groups were nearly exterminated shortly after the arrival of the Spanish. Surviving nations today include the Colorados, or Tsáchila (1 800), the Chachilla (5 000), and a small group of the Colombian Awa (1 000).

The other major ethnic groups include the famous Shuar (40 000) and Achuar (5 000) of southern Amazonia; they are known to the outside world as the Jivaro and are renowned as blowgun hunters and producers of "shrunken heads," miniature trophies of intercultural warfare produced by shrinking the heads of enemies in hot sand before sewing them back together again. Other Amazonia groups include the Huaorani (1 000), the Siona-Secoya (650), and the Cofán (less than 500) of the

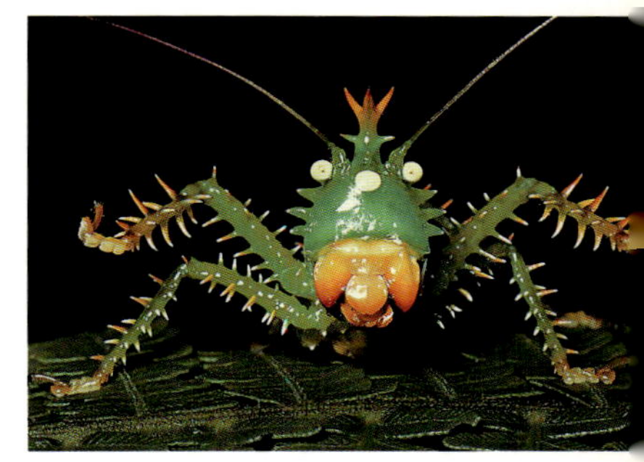

A spiky-headed katydid (Copiphora *sp.*) *in a threat display, Amazonian Ecuador.*

northern Amazon region. It is these Amazonian groups that now face the greatest challenges and receive the largest portion of worldwide media attention. Having avoided contact with the outside world until this past century, and even this century in certain groups like the Huaorani, many of these nations are now emerging into a world which is both aware of them and fighting for their rights, and advancing upon them and their traditional lands at the same time. The challenges that these people face include encroachment on their lands by settlers, both mestizo and other Amerindian groups; the invasion of large, multinational corporations looking for hydrocarbons and precious minerals; and intermittent border disputes between Ecuador and Peru.

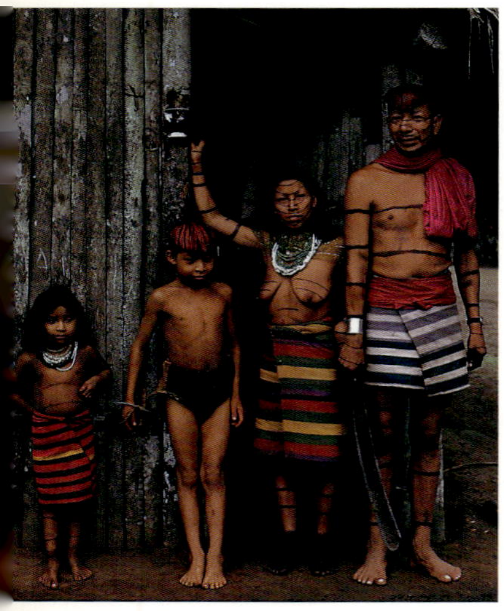

Colorados Indians from San Domingos, one of the more unusual indigenous groups in the country.

On the opposite page, a view of the Amazonian portion of Ecuador, usually referred to as El Oriente by Ecuadorians.

The Huaorani are particularly worthy of note, as their reputation as warriors has brought them international attention. The Huaorani (a word that translates in Wao to "the people") are the most nomadic and "primitive" of the Amazonian ethnic groups. They are known as Auca to neighboring groups, a name that means "savage" in Quichua, and are feared even among other indigenous forest groups. The Huaorani, in turn, refer to outsiders as "Cowode" (nonhuman). Their territory comprises some 20 000 km^2 in the province of Napo and lies in the area defined by the Napo River to the north and the Curaray River to the south, possibly extending into Peru in the far east. Within this lowland forest area, they prefer to stay away from the major rivers, as this is the territory of the "Cowode," and indeed in their conception of the afterlife they go to higher lands, leaving the "Cowode" along the river bottoms. While many of their people have begun to live in permanent settlements and have adopted some of the ways of the outside world, the Huaorani still proudly maintain their reputation for ferocity. One Huaorani subgroup called the Tagaeri (Kane), still live as nomadic hunter-gatherers in the far eastern portions of Huaorani territory and have occasional violent contacts with outsiders.

The group's ethnohistory is difficult to define, given the isolated nature of both their language, which is unrelated to other known indigenous languages, and their equally unique material culture. Sustained contact with the Huaorani was not established until 1958, at which time some anthropological studies were undertaken. The Huaorani's avoidance of river bottoms makes them a people of the interfluvial areas and nearly all of their protein is derived from rain forest animals. Waterfowl and large fish are taboo, they do not gather the abundant river turtle eggs that are an important dietary item for many other Amazonian ethnic groups, and until very recently they did not build canoes or have any knowledge of river navigation or even swimming. However, the Huaorani are master hunters with the blowgun (using curare-tipped darts) and the spear, and make the only "flat" blowguns in the world. They are also very fond of monkey meat, and consume what may be the highest percentage of monkeys of any indigenous group in the world.

The Huaorani are few, with a total population of perhaps 1 000, yet they have historically maintained the impression of far greater numbers given their custom of making sporadic and unpredictable raids on outsiders. The Huaorani live in isolated community groups divided by several days' walking, and there are still hostile interactions among their different communities. Huaorani mortality statistics reflect this lifestyle, with 40% of deaths deriving from "spearing by other Huaorani," and another 8% from being "shot by Cowode."

Times have changed for the Amerindian, and the ethnic groups of Amazonia wage a war against conquest on a much different battlefield than those that preceded them, namely within international human rights and environmental arenas. Government organizations and foreign and domestic NGOs have been active on the Indian's behalf, and Amazonian ethnic groups have established their own political action groups. These include: the Federación de Comunas-Unión de Nativos de la Amazonía Ecuatoriana (FCUNAE), the Federación de Organizaciones Indígenas del Napo (FOIN), and the Organización de Pueblos Indígenas del Pastaza (OPIP), all of them fighting to uphold Indian rights and traditional lands.

RODERIC B. MAST
CRISTINA GOETTSCH MITTERMEIER
RUSSELL A. MITTERMEIER
JOSÉ VICENTE RODRÍGUEZ
ARLO H. HEMPHILL

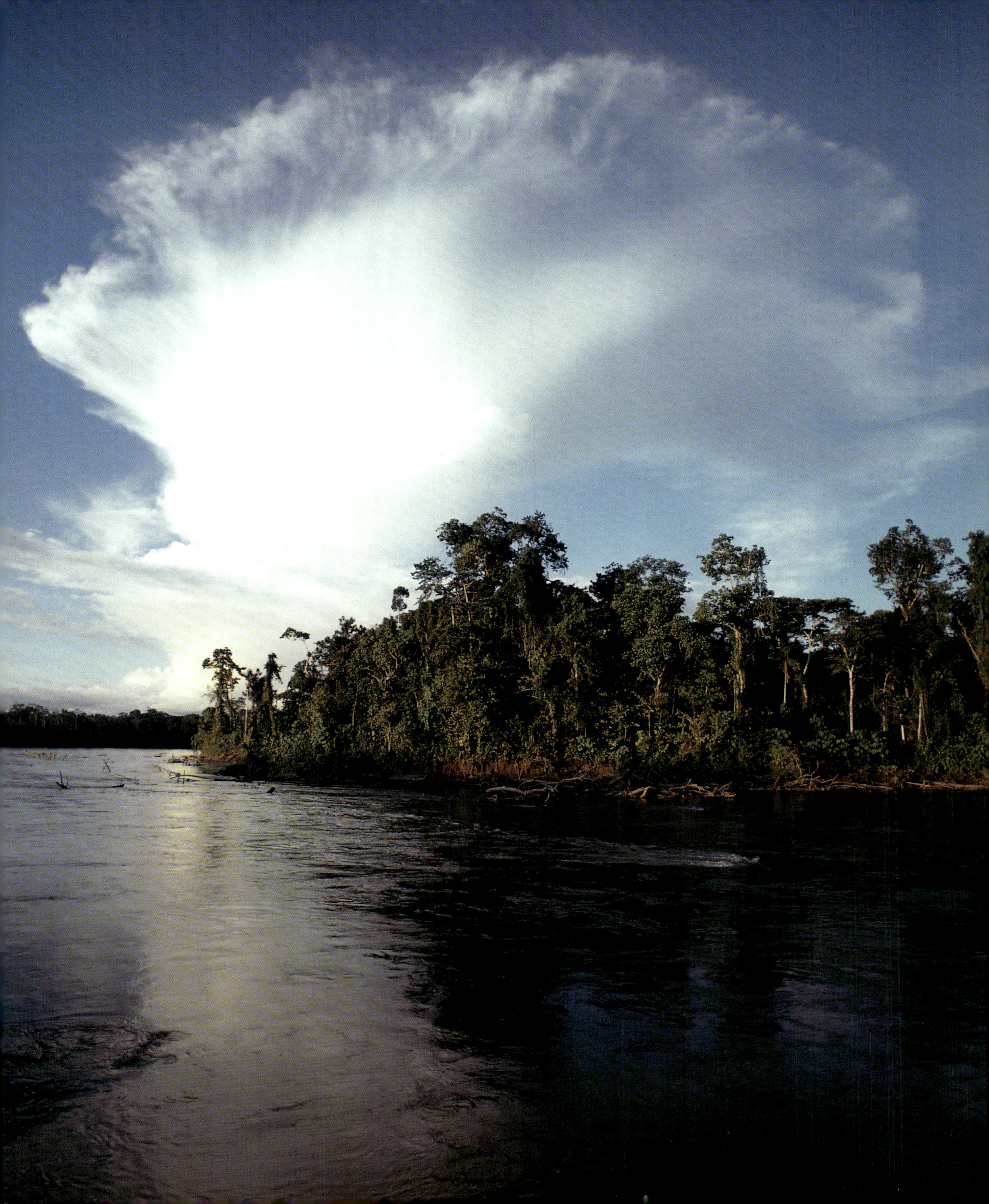

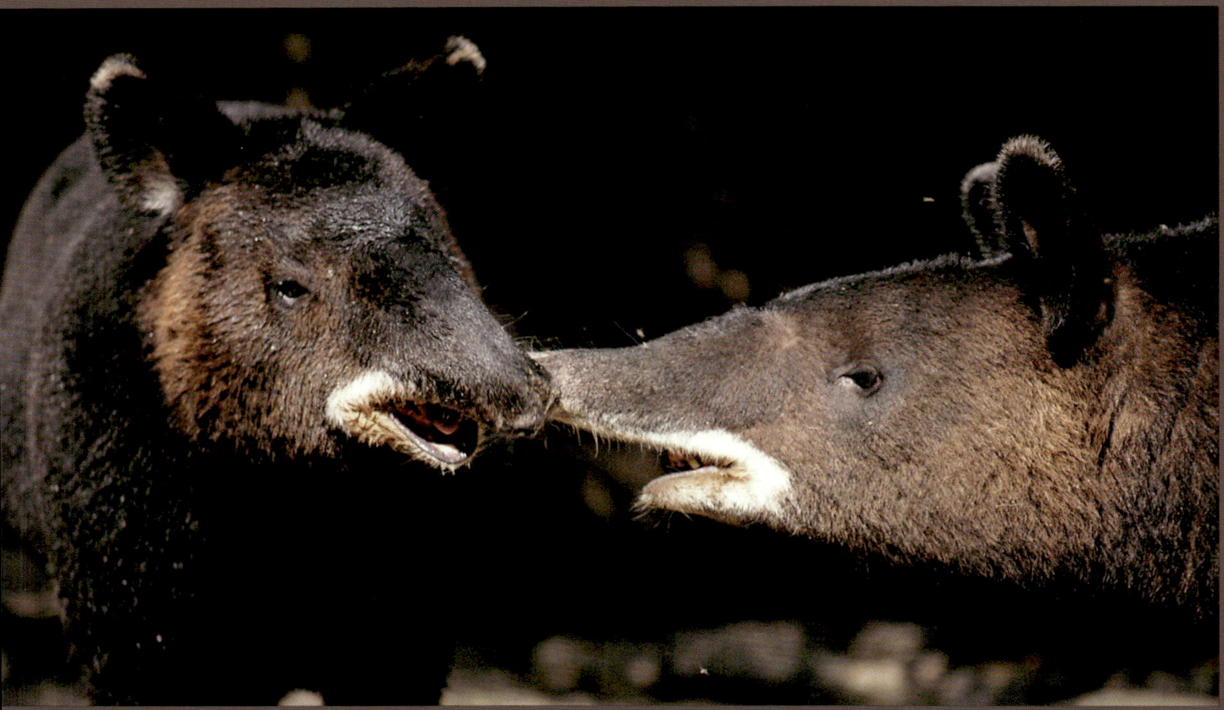

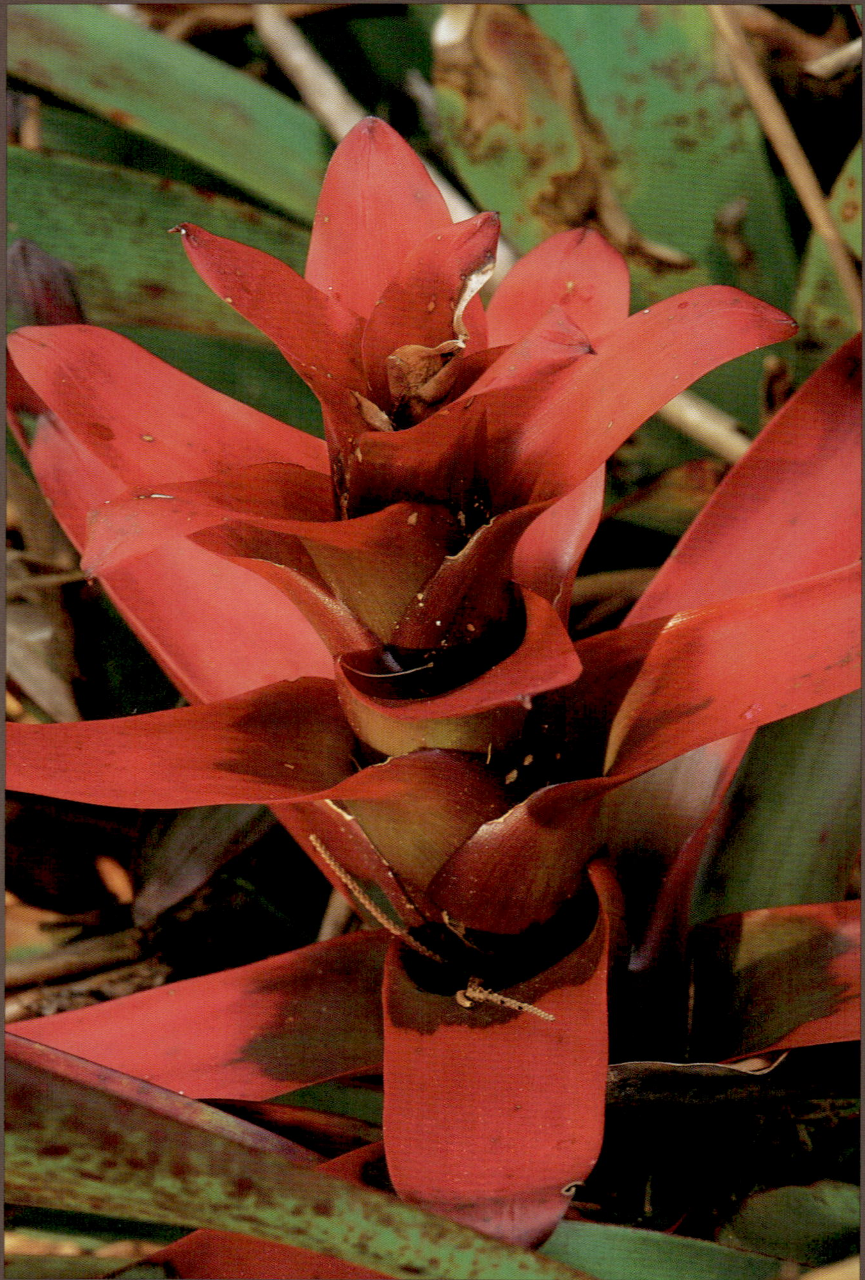

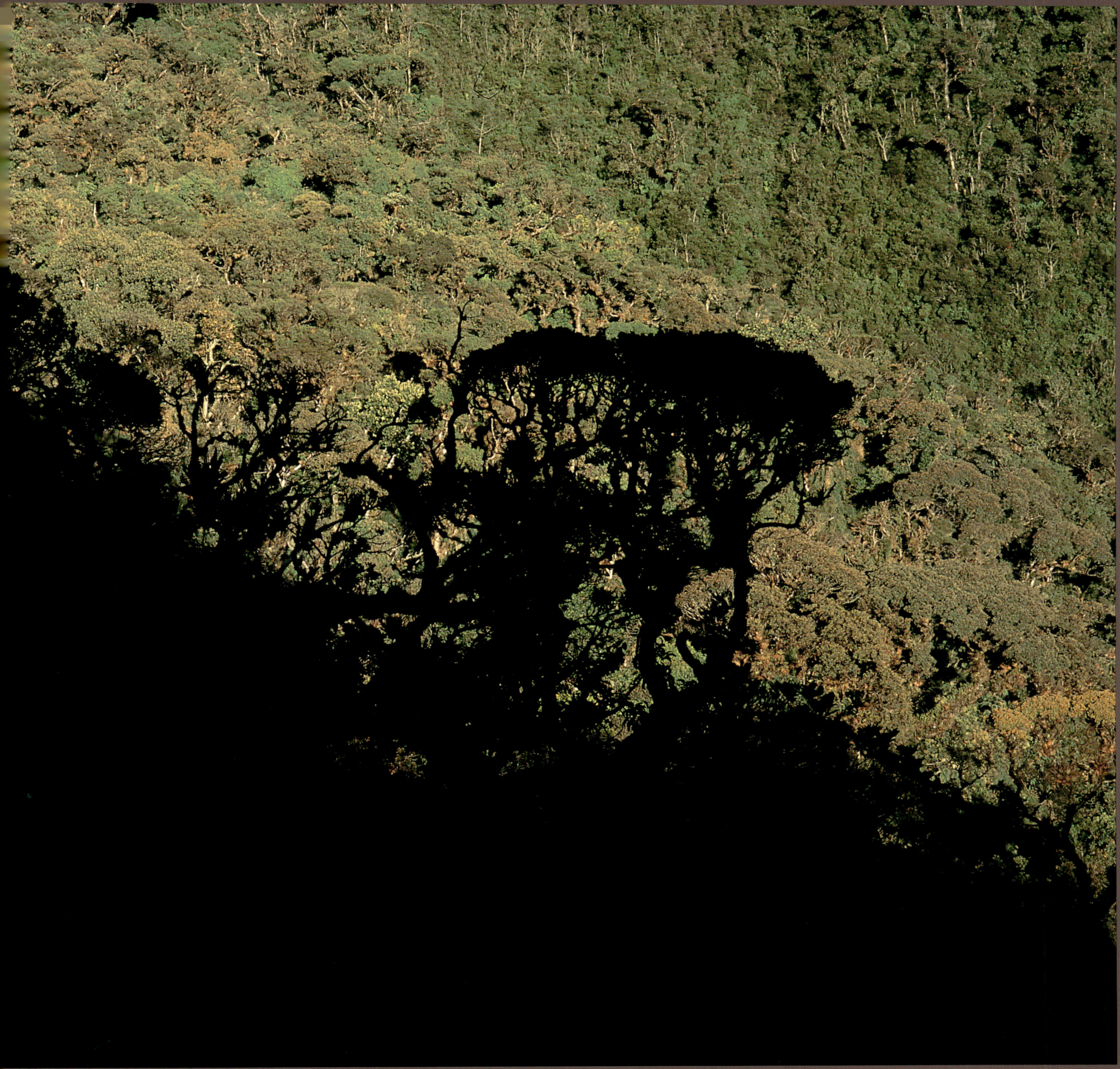

Upper left, two mountain or woolly tapirs (Tapirus pinchaque) *touch noses in a courtship ritual. This very distinctive tapir species is found only in the Andean region from Peru to Venezuela, and is the most endangered of the world's four tapirs. The Andes of Ecuador may be its most important remaining stronghold.*
Lower left, the flower of a bromeliad species. Bromeliads reach a very high diversity on the moist eastern slopes of the Andes.
Above, Podocarpus National Park in the Andes near Quito. Podocarpus National Park extends from upper Amazonia to páramo *habitats. This park protects the only large expanse of undisturbed forest left in southern Ecuador, and hosts between 600 and 800 species of birds and probably 3 000 to 4 000 species of higher plants.*

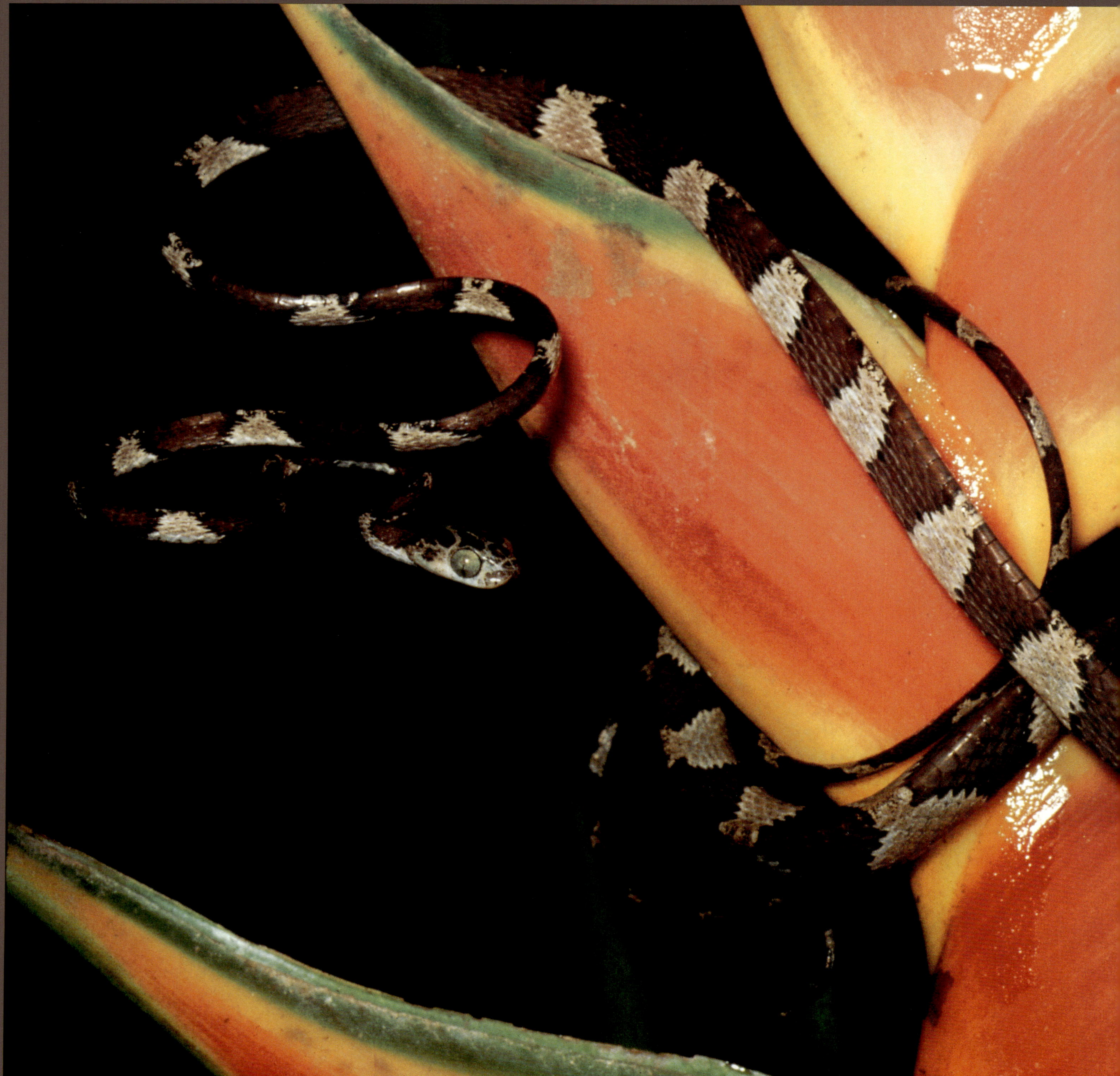

Other examples of Ecuador's great amphibian diversity, the highest in the world in diversity per unit area. Above, the blunt-headed tree snake (Imantodes cenchoa) *on the flower of a* Heliconia stricta, *Amazonian Ecuador.*
Upper right, a barred lead frog (Phyllomedua tomopterna) *from Amazonia; lower right, a harlequin poison frog* (Dendrobates histrionicus), *from the rain forests of Western Ecuador and the Chocó of Colombia.*

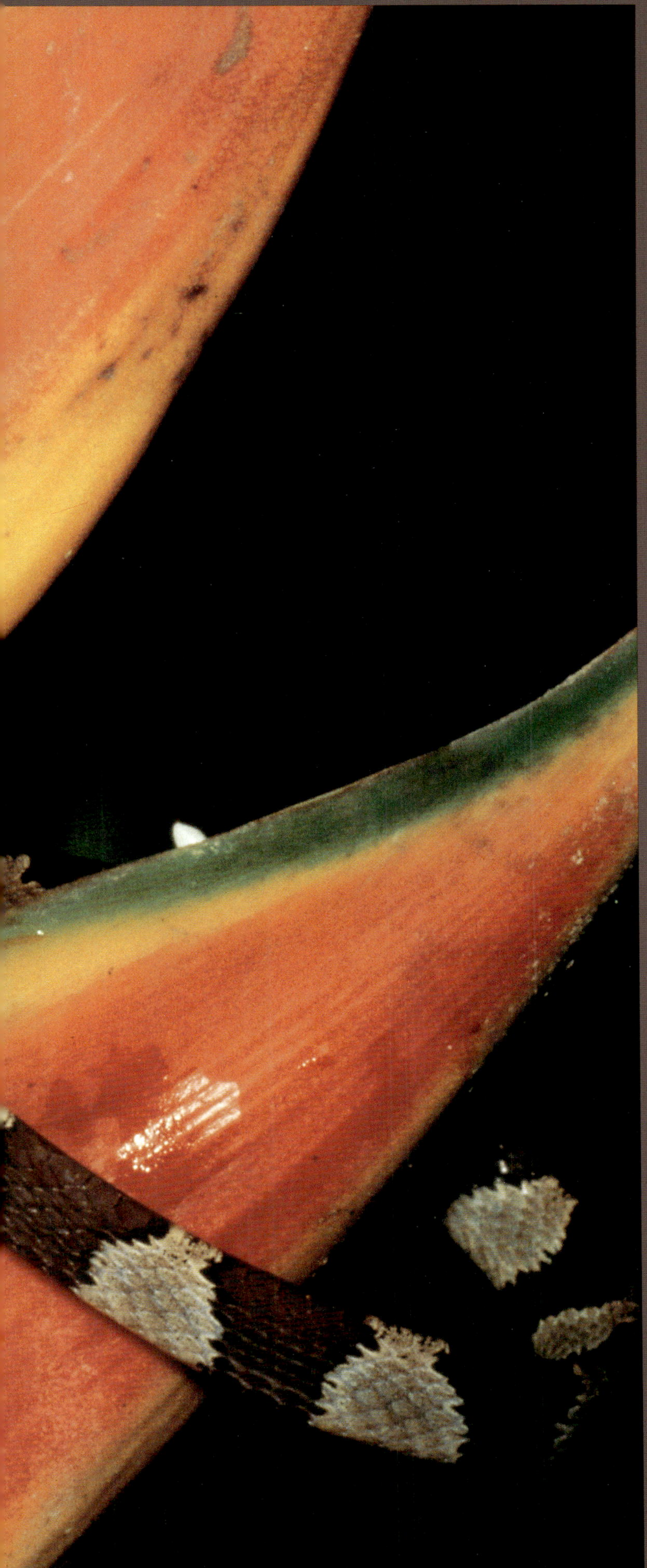

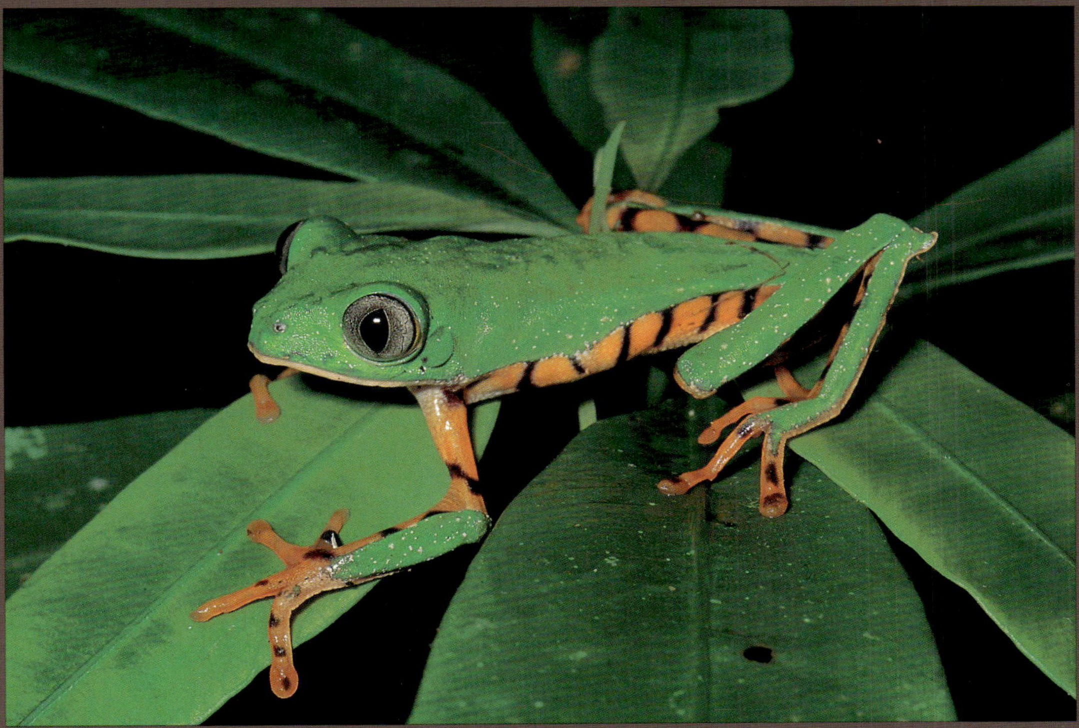

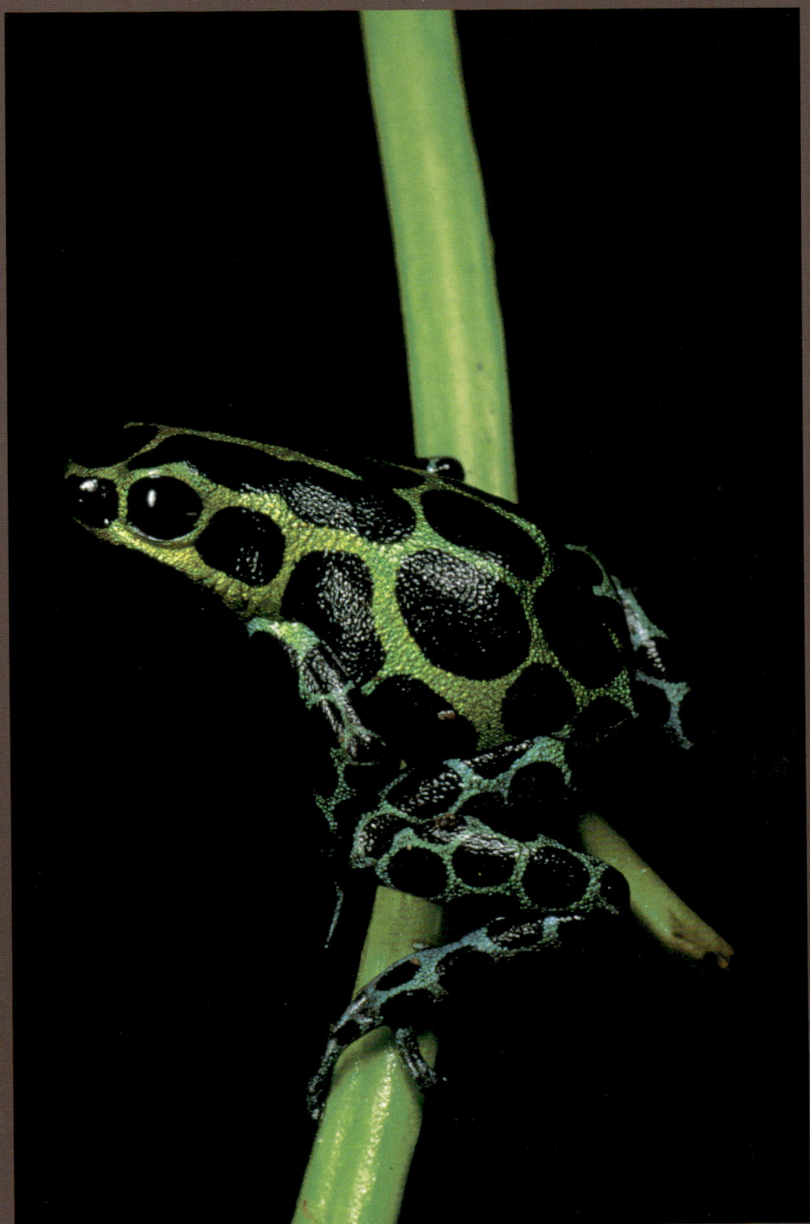

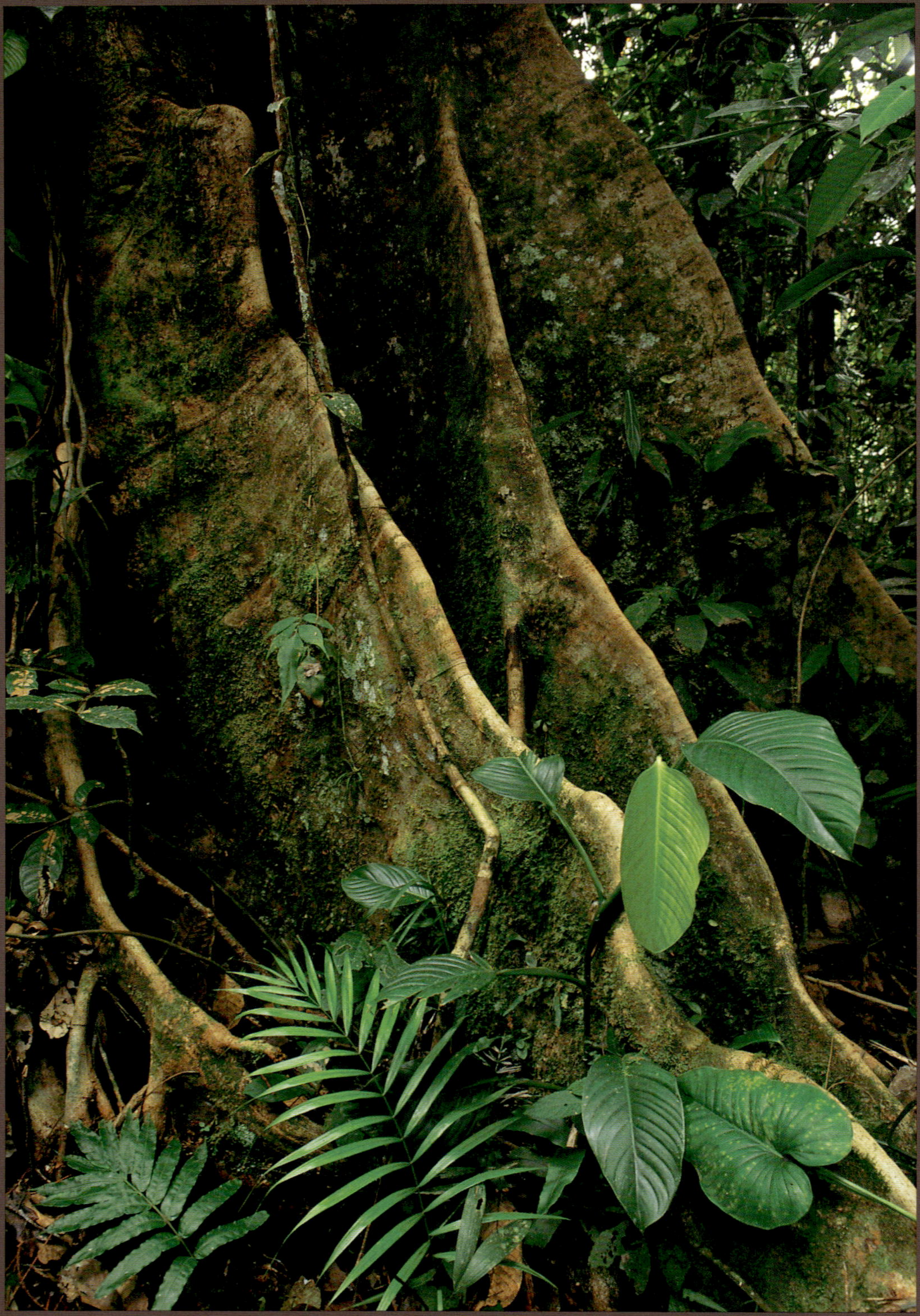

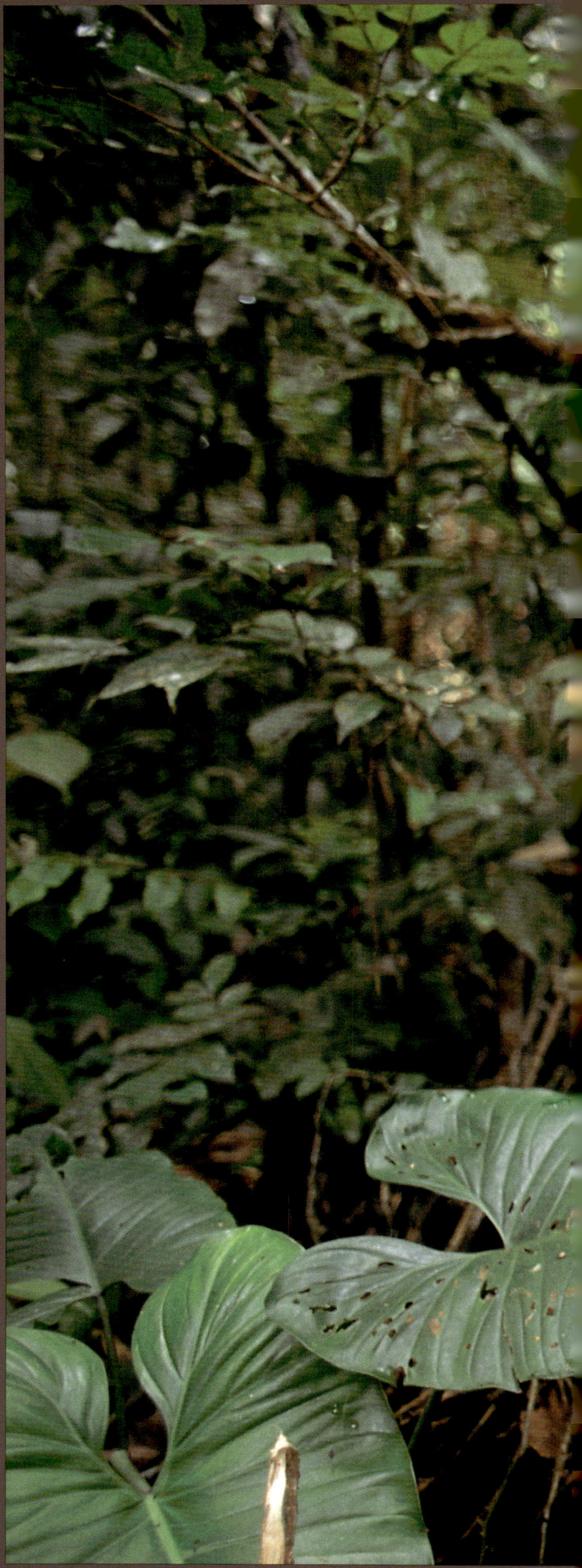

Above, the buttress roots of a large Amazonian tree. Since most of the nutrients in Amazonian forests are at the surface, roots of trees do not penetrate deeply, but rather spread across the surface of the ground. Many species also have elaborate support structures like stilts or the buttresses shown here to keep them from toppling over in storms.
Right, the gray-winged trumpeter (Psophia crepitans), *a large, forest-floor bird of Amazonia.*
The trumpeter travels in big, noisy flocks, foraging on the forest floor for whatever plant or animal food it can locate.

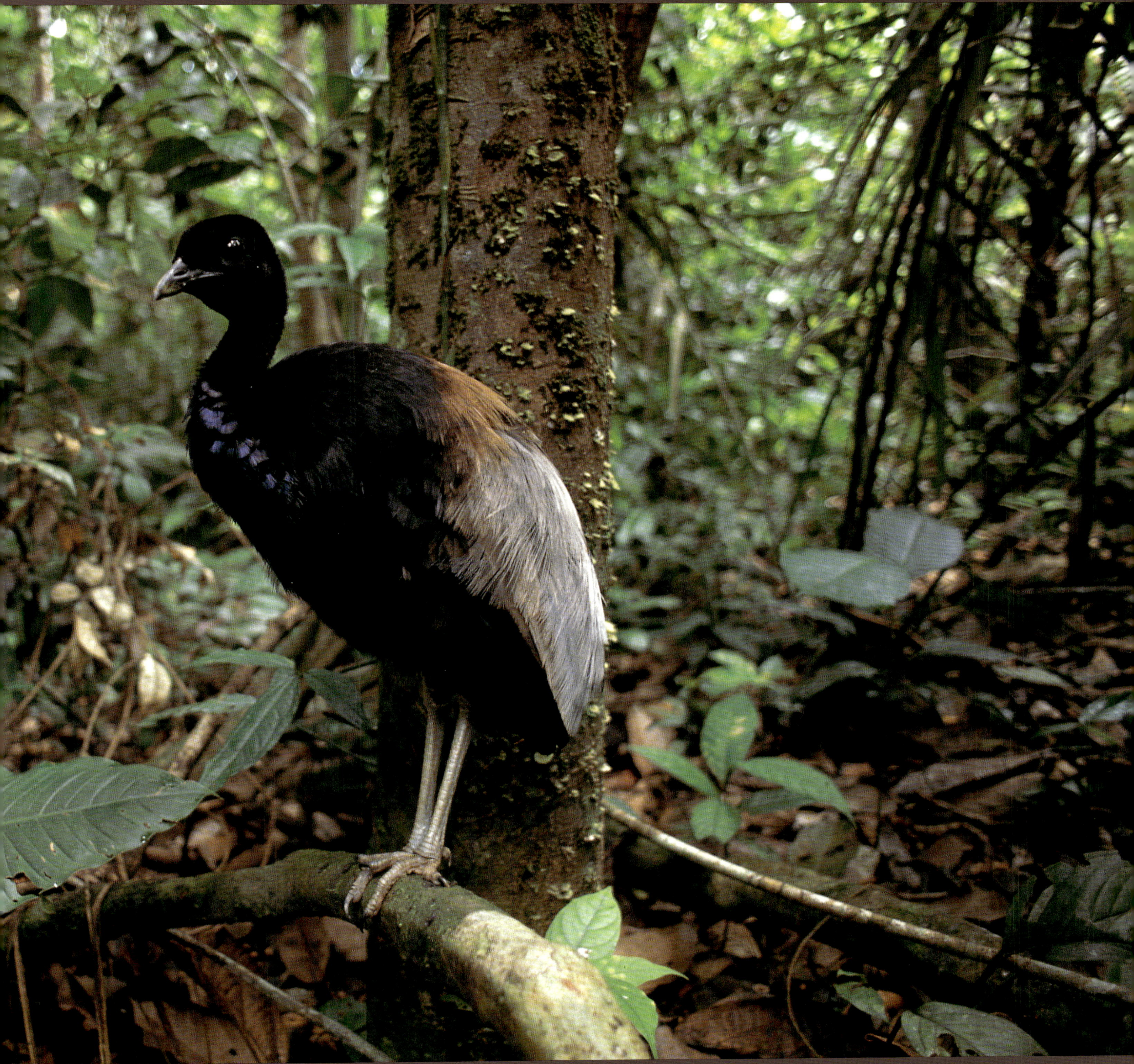

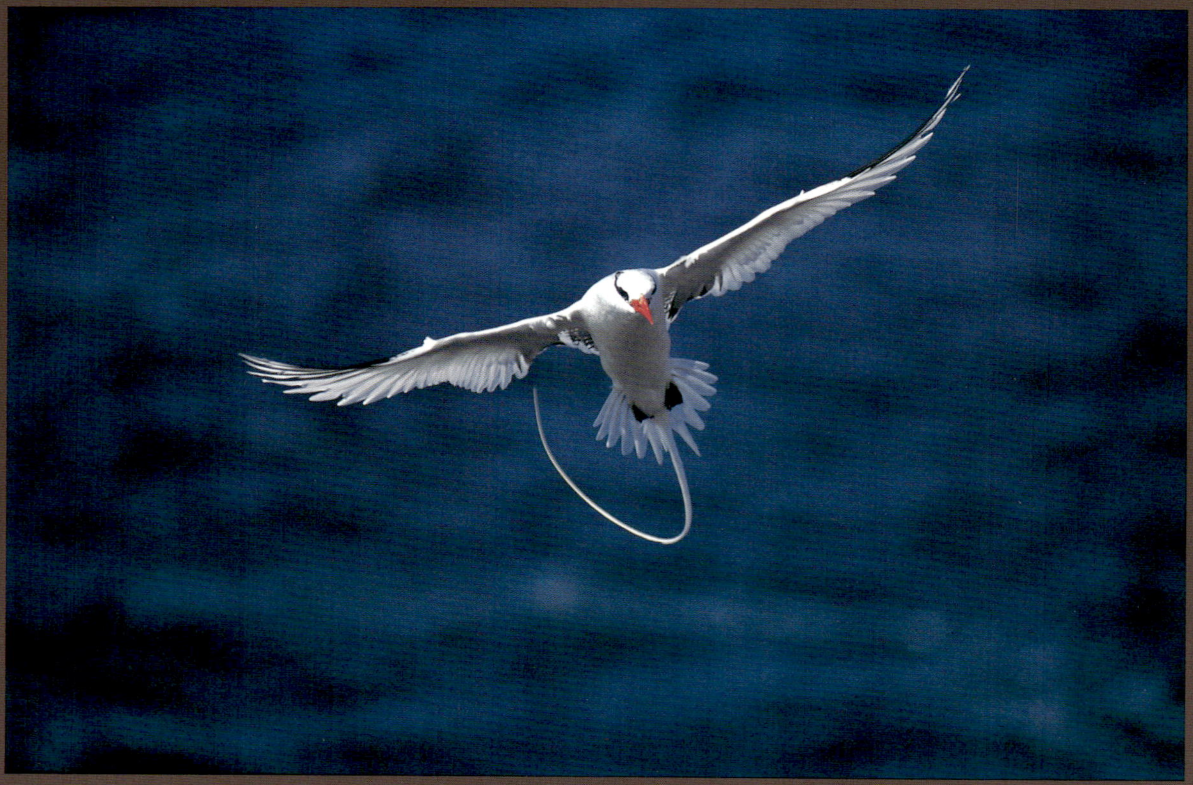

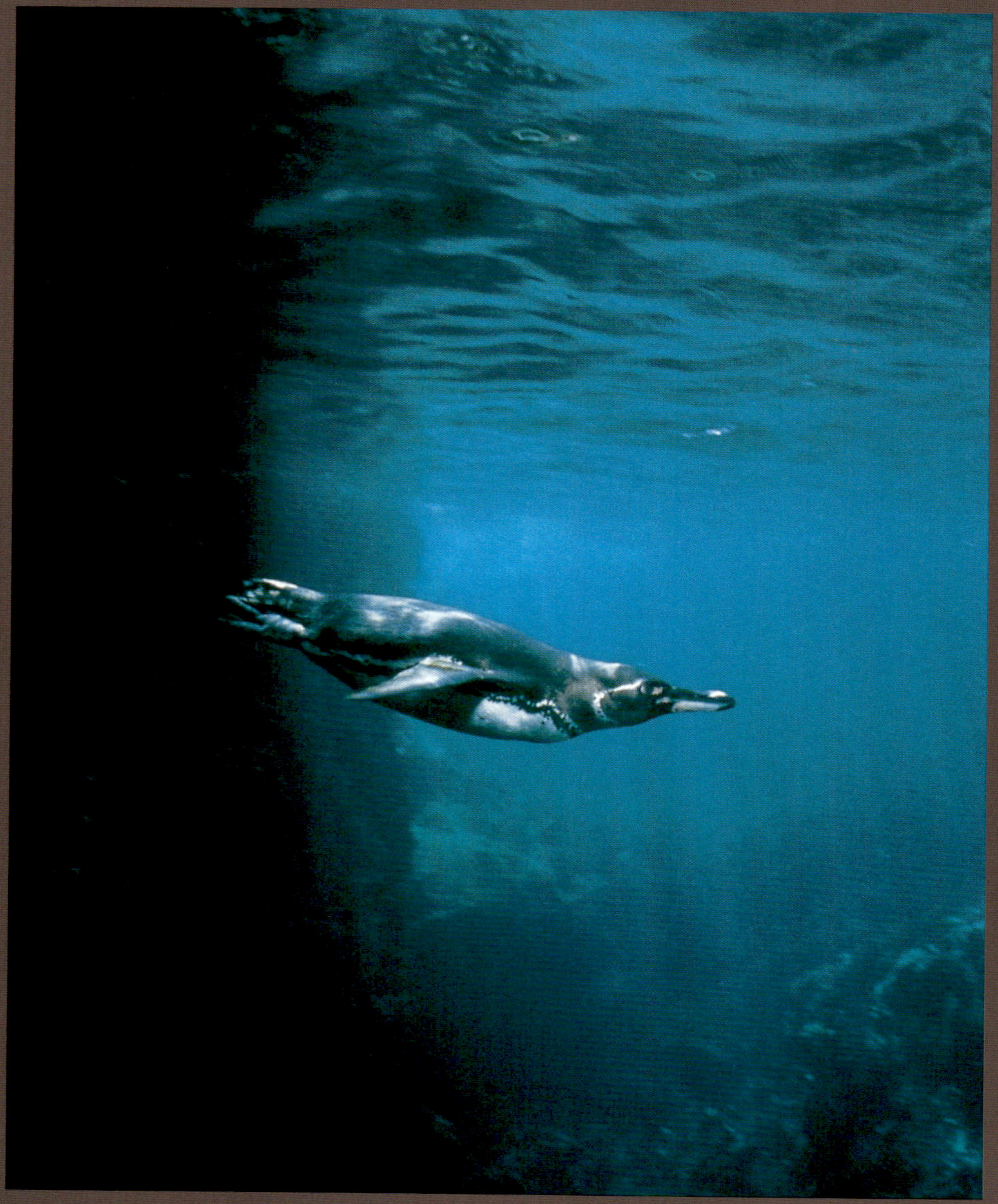

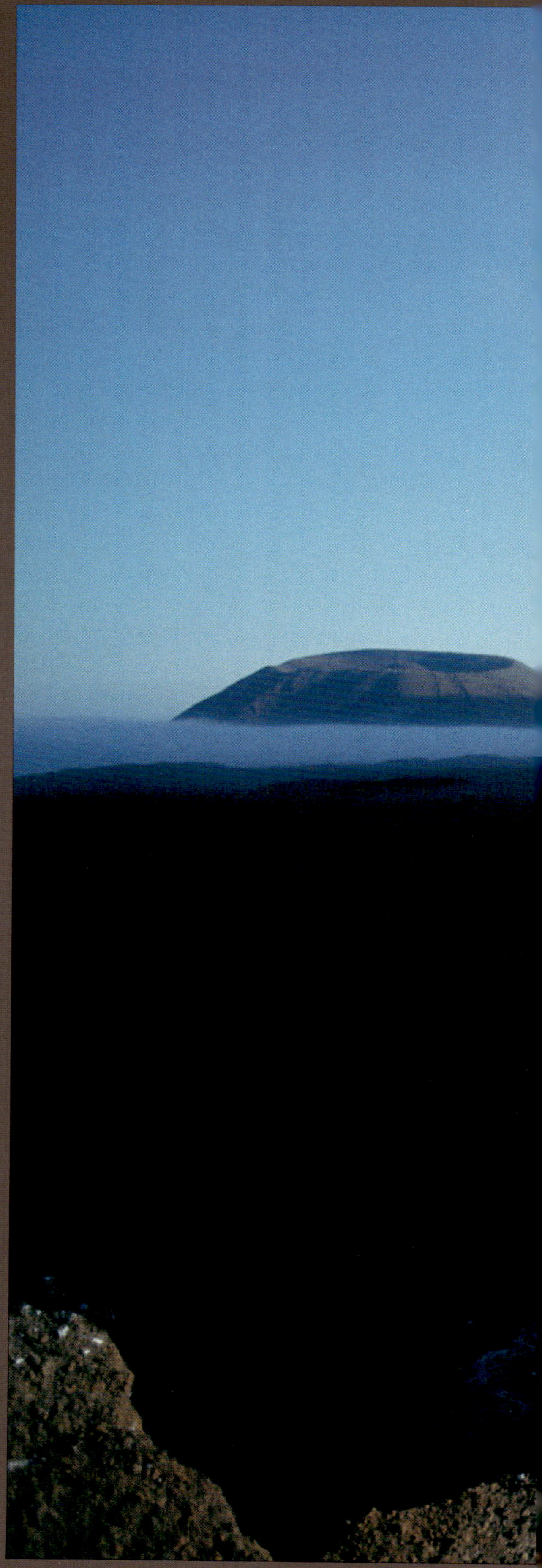

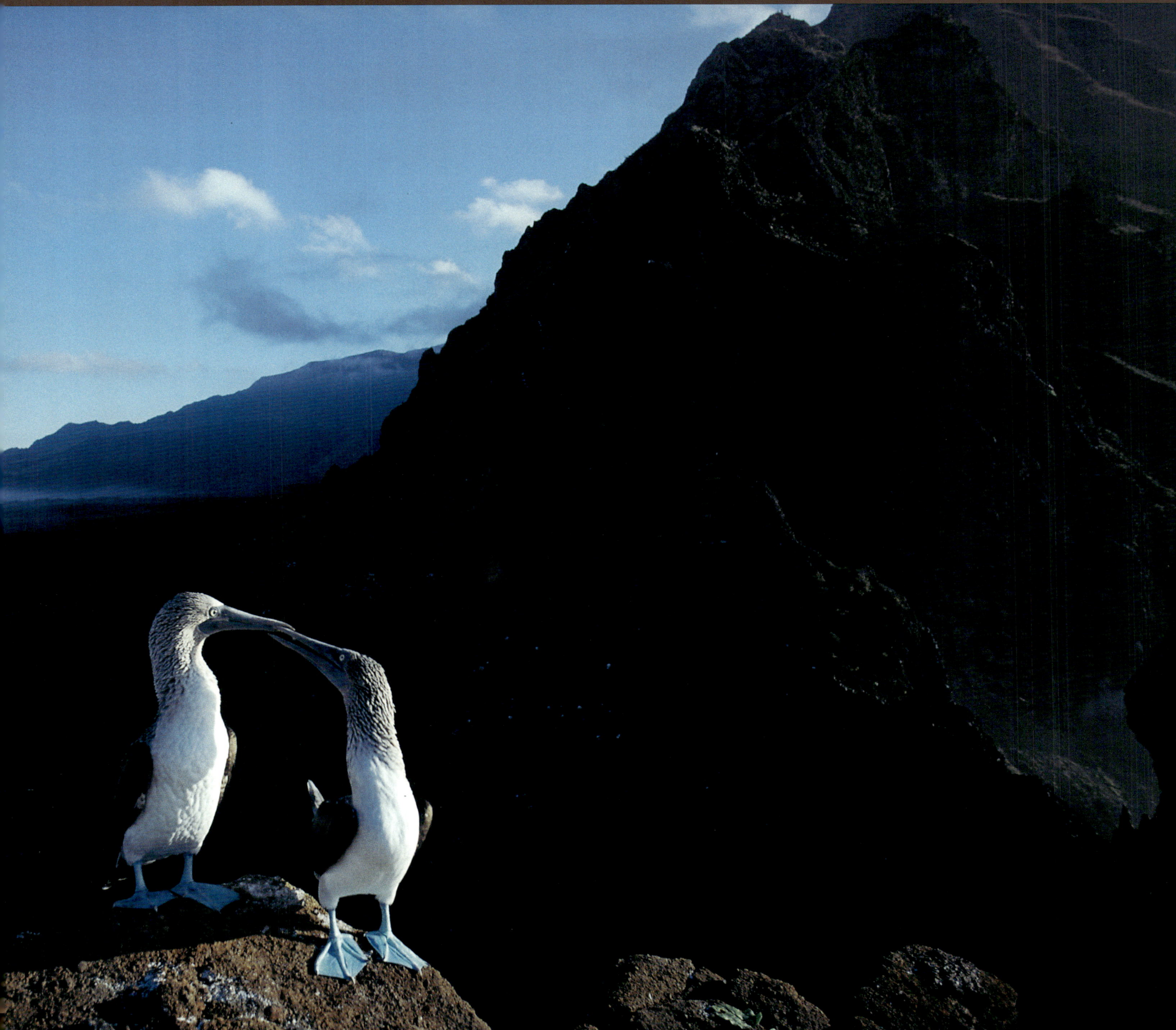

Upper left, a red-billed tropicbird (Phaëthon aethereus), *a characteristic species of the Galápagos Islands. Lower left, the Galápagos penguin* (Spheniscus mendiculus), *the only penguin species to live at the equator.*

Above, blue-footed boobies (Sula nebouxi), *with volcanic cliffs in the background, Isabela Island in the Galápagos. The blue-footed booby, with its striking baby blue feet, is one of the more popular species on the islands.*

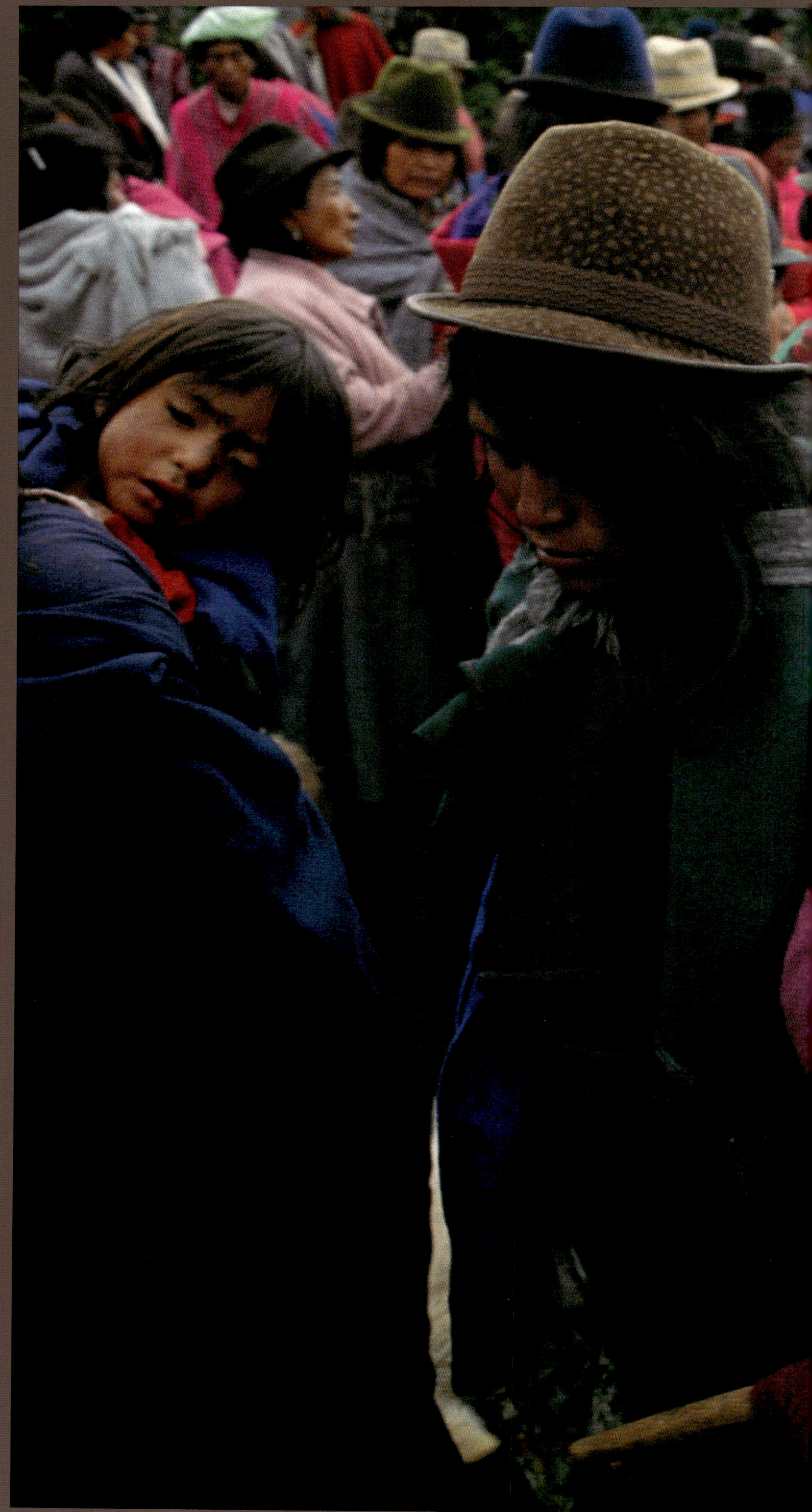

Quichua Indians in the Andes of Ecuador.
As in Peru, the Quichua are the largest indigenous group in Ecuador, and can be divided into a number of subgroups.

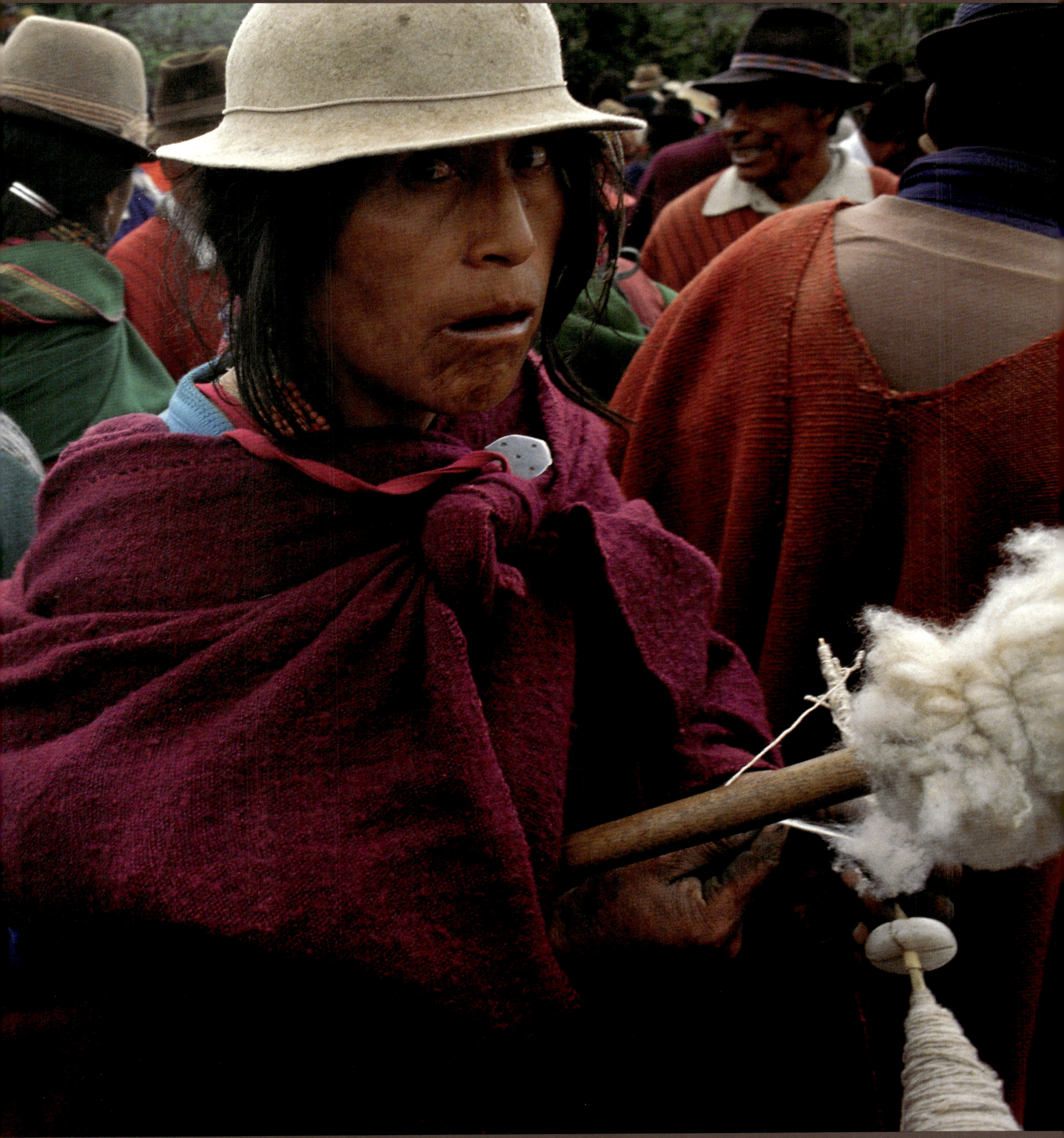

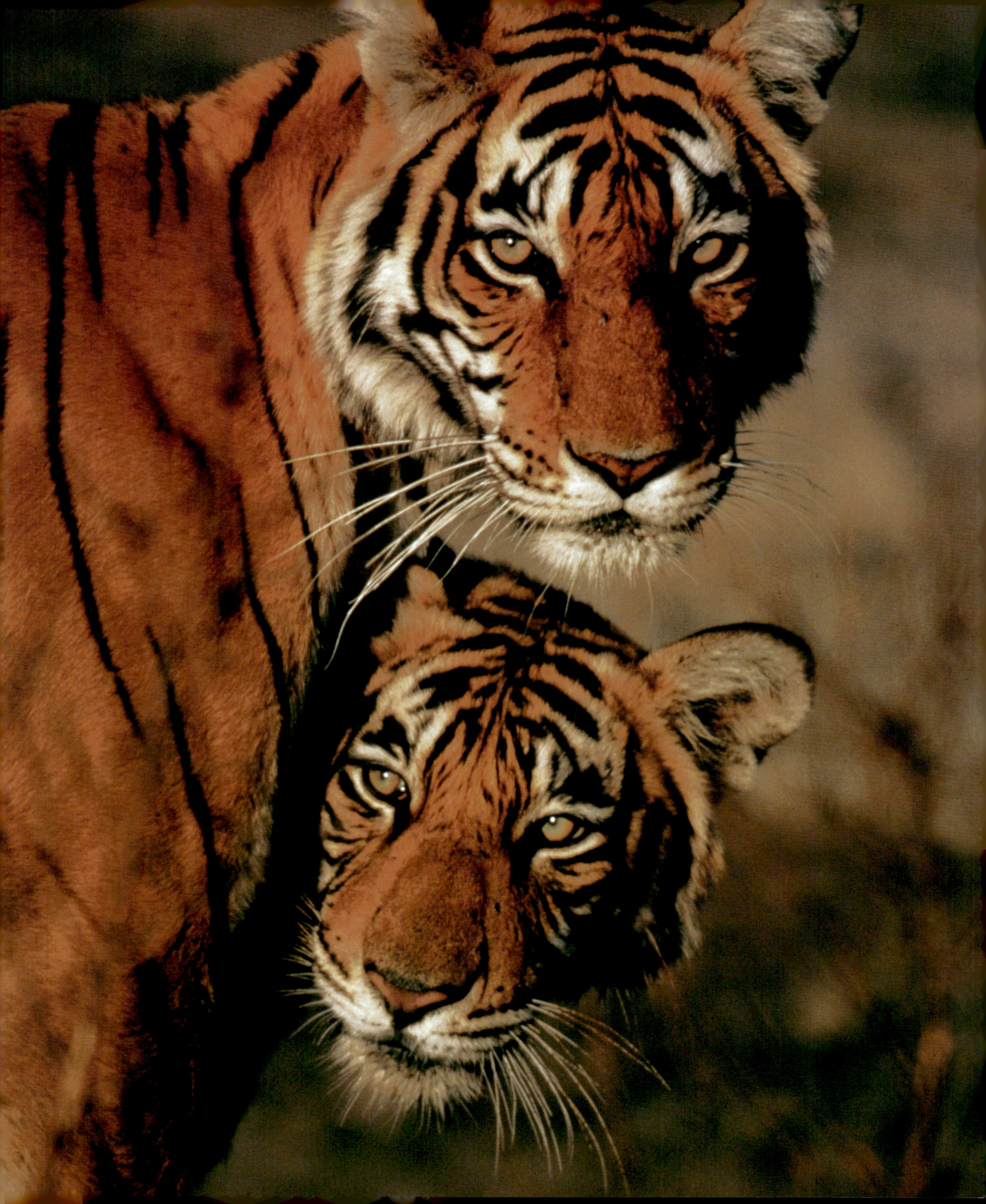

INDIA

India is a country best described in superlatives: superb mountains, expansive deserts, vast plains and plateaus, great rivers, magnificent rain forests, extensive natural resources, wonderful wildlife, and a massive human population. Its ecosystem diversity is among the highest of any nation in the world, but it is also destined to become the most populous nation in the world in 50 years. Its current human population is estimated to be 960 million —it will attain one billion near the beginning of the twenty-first century. By 2050, India is projected to surpass China in human numbers.

India's land area of 3 287 782 km^2, establishes it as the third largest country in Asia and the seventh largest in the world. Stretching from 8°N latitude in the tropics to 37°6' N in the Himalayas, this country contains a remarkable range of ecological communities from coral reefs and tropical rain forests to alpine tundra and permanent ice and snow.

India also has high species diversity and endemism, a reflection of its broad range of habitats. In mammals, it has 350 species, ranking eighth on the world list, with 44 endemics, or twelfth globally. Bird diversity is 1 258 species, or seventh highest in the world, but endemism in birds is relatively low at 52 species. Both diversity and endemism are quite high in reptiles and amphibians. India ranks sixth in reptiles with 408 species, and fourth in endemic reptiles with 187, behind only Australia, Mexico and Madagascar. In amphibians, it ranks eighth with 206 species and tenth in endemism with 110. Percent endemism in both groups is quite high, 46% (187/408) in reptiles and 53% (110/206) in amphibians. Fish diversity is also impressive, with 2 546 species in all; freshwater fish account for 750 of these, placing India in the top ten in this category as well. Based on butterflies and tiger beetles, invertebrate diversity is also quite significant in India. In total butterfly diversity, India is tied for twelfth with China, with roughly 1 200 species. However, it is very high in the well-documented swallowtail butterflies, ranking third in diversity behind Indonesia and China with 77 species, and sixth in endemism with six species. In tiger beetles, India is second with 193 species, ranking behind only Indonesia, and fourth in endemics with 82.

In higher plants, India is also in the top group, with over 17 000 species, of which a striking 7 025-7 875 are endemic, or roughly 41-46% of the entire flora. In terms of higher-level diversity, there are about 140 genera endemic to India, but no endemic families. Approximately 52 of the endemic genera are found in the Western Ghats.

India also has part of one of the world's top 19 *threatened biodiversity hotspots*, the Western Ghats/Sri Lanka, and has partly or entirely within its borders 10 Centers of Plant Diversity and Endemism as identified by WWF/IUCN. These include Kashmir, The Himalaya, Nanda Devi, Northern Sikkim and East Nepal, Namadapha, Agastyamelai Hills, Nilgiri Hills, Nallamalai Hills, Sundarbans, and the Andaman and Nicobar Islands. In addition, it has eight Endemic Bird Areas as identified by BirdLife International (ICBP): Western Himalayas, Indus Valey, Western Ghats, Eastern Himalayas, Assam Plains, Tirap Frontier, and the Andaman and Nicobar Islands.

India's human diversity is also exceptionally high, placing it in the top ranks of all nations. There are some 613 recognized tribal groups in the country, and the 1961 census listed 1 652 different languages. Although this number undoubtedly reflects some dialect variants and not necessarily full-fledged languages, the scale of linguistic and human diversity in India is enormous and near the top of the global list, along with Papua New Guinea and Indonesia.

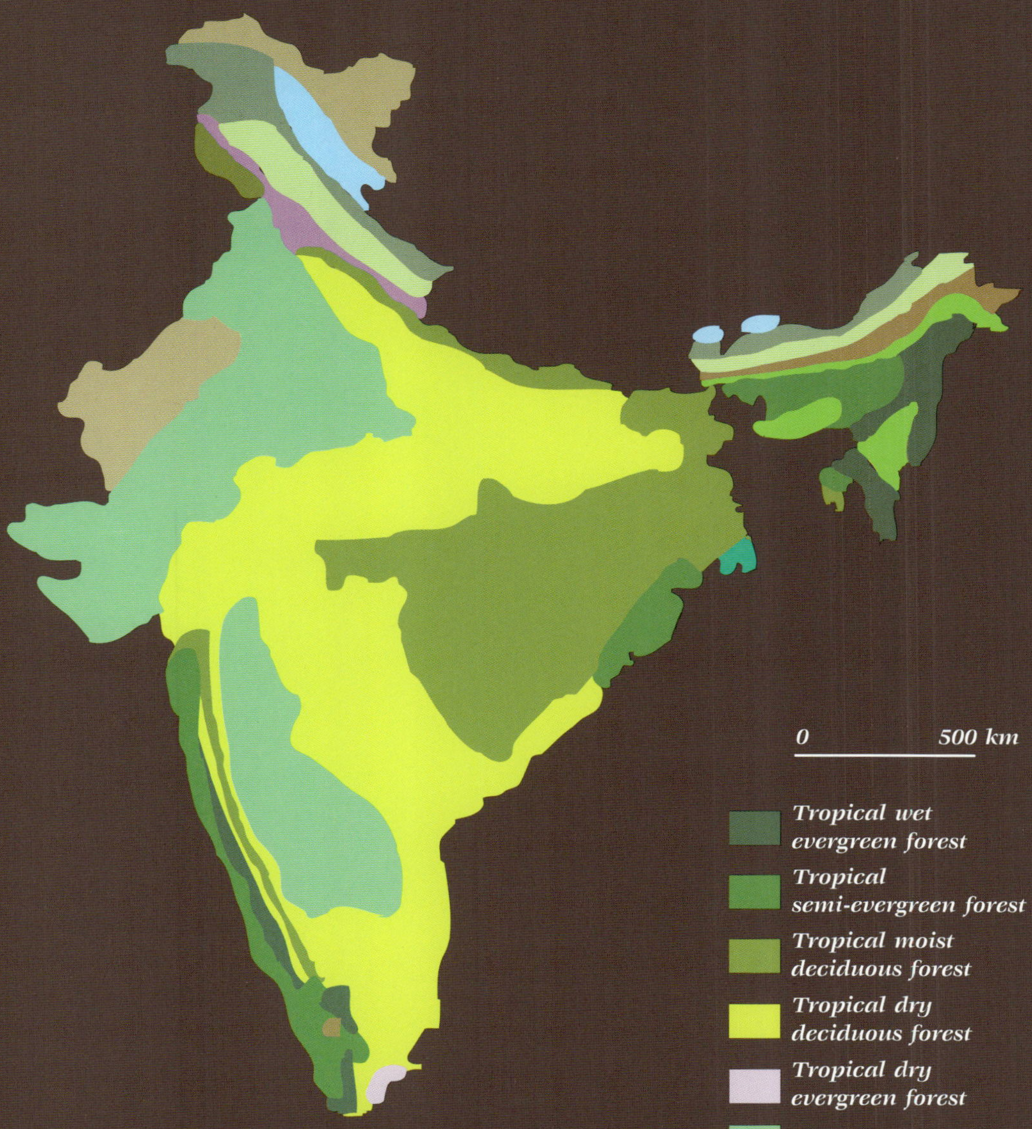

On the opposite page, the Bengal tiger (Panthera tigris tigris) is the classic flagship of India. Once relatively common throughout the country, the tiger had declined to only 1 800 animals by 1962. In the 1970s the Indian government launched "Project Tiger" to bring this magnificent animal back from the brink of extinction. By 1990 the population had increased to 4 500-6 000.

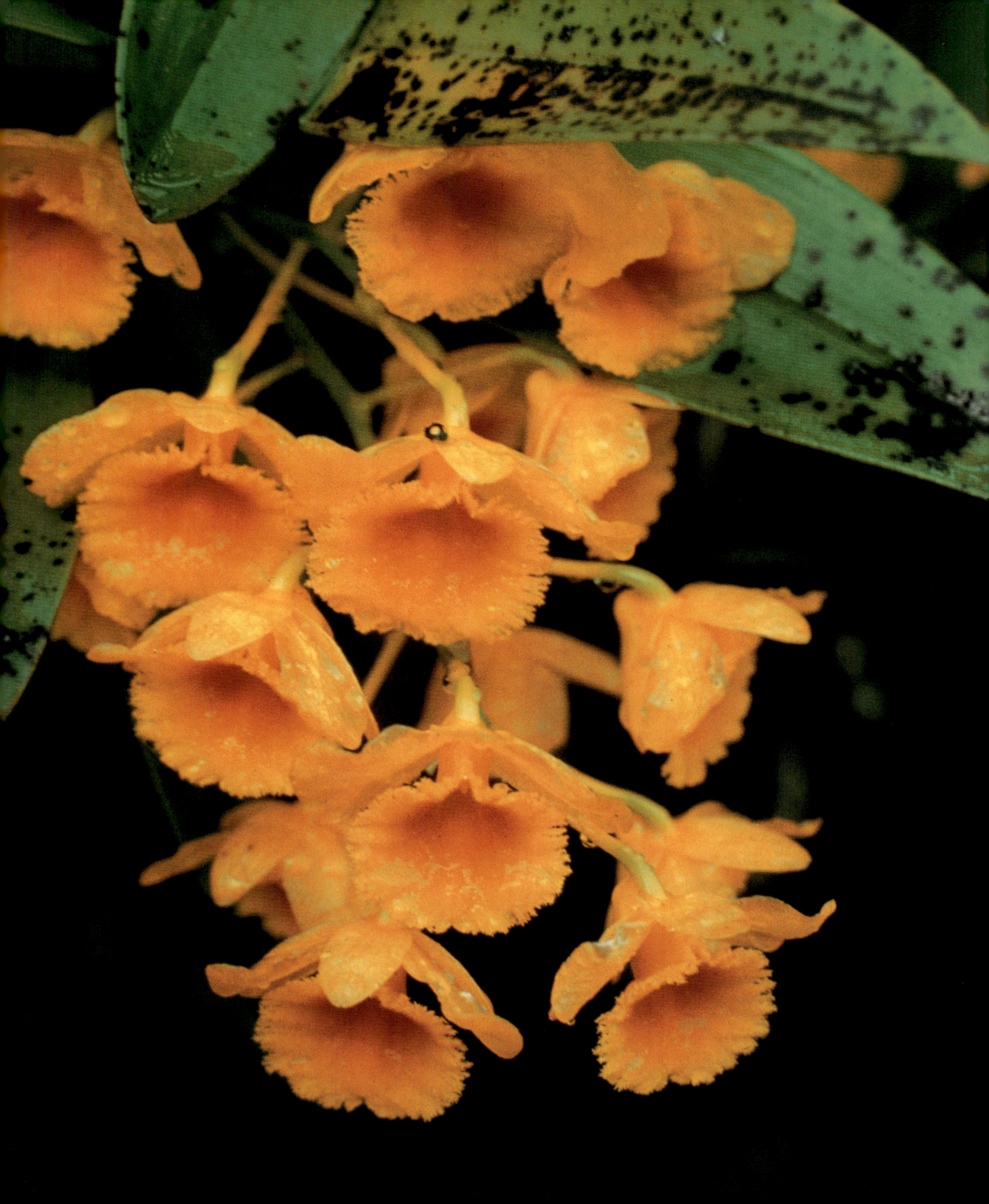

Natural habitats

In physical geography, India is divided into six rather distinct regions, and each may be subdivided further. These six are the Himalayas, the Gangetic Basin, the Thar or Great Indian Desert, the Deccan Plateau, the Western Ghat Mountains, the oceanic islands of Lakshadweep in the Arabian Sea, and the Andaman and Nicobar Islands in the Bay of Bengal.

The Himalayas stretch almost 3 000 km across the north from Arunachal Pradesh and Assam to Jammu and Kashmir. Although Mt. Everest is in Nepal, the second highest mountain (K2 or Godwin Austen) in the world is in an area claimed by both India and Pakistan, and the third (Kanchenjunga) highest is also in India. The Himalayas are bordered on the south by the foothills, the Pir Panjal Range in Jammu and Kashmir and the Dhaula Dhar in Himachal Pradesh, both rising to 3 600 to 4 600 m and the Siwaliks eastward in Uttar Pradesh, rising to 1 500 m.

Van Gruisin (1988) describes the conservation problem in the Himalayas in dramatic terms:

> Man's onslaught has rendered the Himalaya amongst the most endangered environments in the world. In the subcontinent, the Himalaya rank as the region most in need of conservation, both because of the deterioration in environment and in view of our relative ignorance of the biology and ecology of high-altitude communities. Slowly, this begins to change, but the task is immense. The Himalaya contain more endangered species of mammals than any other area of India and are remarkable in possessing almost one third of the world's mammalian species that could be called true mountain animals. M.K. Ranjitsingh has noted that.... "the outlook for wildlife in most parts of the Himalaya is grim, in some places even desperate," and G.B. Schaller warned that just as we are becoming more aware of the splendor of past and present wildlife in the Himalaya we are denying it a future. He describes mountains without wildlife as "Stones of Silence," an evocative and thought-provoking phrase.

The Gangetic Basin, 2 400 km in east-west extent from Assam to Pakistan, is a remarkably flat agricultural region with no more than 300 m in relief from west to east. It is India's breadbasket, the site of great agricultural production and advances made since the 1960s in the "Green Revolution." In the past 40 years, India has gone from a major importer of food grains (averaging 20 000 tons per day under the US P.L. 480 program) in the 1960s and early 1970s to a major exporter of food grains in the 1990s. This has been achieved through a combination of new, high-yield strains of wheat, rice, maize, and other food grains, increased use of fertilizers and pesticides, irrigation, mechanization, and agribusiness interests.

The Ganges River from the west and Brahmaputra from eastern India join in Bangladesh with some channels in India to form the Sundarbans or Gangetic Delta, one of the largest deltas in the world. Originally an area of mangrove, it is now densely populated and intensely cultivated, with only marginal mangrove habitats.

Western India is dominated by the Thar Desert bordering the alluvial plain of the Indus River in Pakistan. The Thar Desert extends throughout much of Rajasthan, southern Punjab and small portions of northern Gujarat. The western portion of the Thar Desert has been modified rapidly with the recent development of the Indira Gandhi Canal in an effort to agriculturize the arid environment and settle formerly nomadic peoples into village life.

Peninsular India is occupied by the Deccan Plateau, a hilly and eroded plateau with several ranges of mountains including the Western Ghats and Nilgiri Hills along the western coast, which capture most of the rain in southern India; the ancient Aravalli Hills, one of the oldest folded mountains of the world, stretching northwards from Gujarat to fan outward in the direction of Delhi, and the Satpuras and Vindhyas, extending from southwest to northeast.

The coastal regions and islands of India comprise over 12 067 km of shoreline and two archipelagos, Lakshadweep in the Arabian Sea, a group of approximately 25 low coral atolls, and the Andaman and Nicobar Islands, two groups of more than a hundred islands that reach elevations of 750 m above sea level in the Bay of Bengal. The Andaman and Nicobar are actually closer to Myanmar and Thailand than to India.

Virtually all landscapes of India except the tallest Himalayas are highly modified by human action. Over 40% of the total land area is cultivated (Negi, 1994). Forests occupy an estimated 639 182 km^2, or 19.4% of the total land area, but many of these are plantations or scrub secondary forests subject to grazing, fuelwood collecting, and other types of disturbance.

Forests of India have been classified into 16 major types and 221 minor types in four major categories: tropical, subtropical, temperate, and subalpine-alpine (Champion and Seth, 1968; Negi, 1989). Tropical forests include tropical wet evergreen in northeast India, in southwestern India and the Bay of Bengal islands; tropical semievergreen forests in southern and eastern India; tropical moist deciduous in certain parts of the Deccan and eastern India; tropical littoral and swamp forest in the coastal and deltaic region; tropical dry deciduous in the western and central Deccan; tropical thorn, also in the western and central Deccan; and tropical dry evergreen forest in portions of south India.

Subtropical forests include: subtropical broadleaf hill forests, in parts of the Deccan and the Terai in northern India, i.e., the lower Siwalik Hills; subtropical pine, in the Himalayan foothills; and subtropical dry evergreen forest, in northeast India.

Temperate forests include montane wet temperate forest in the Himalayas and higher elevations in the Western Ghats and Nilgiri Hills; moist temperate forest

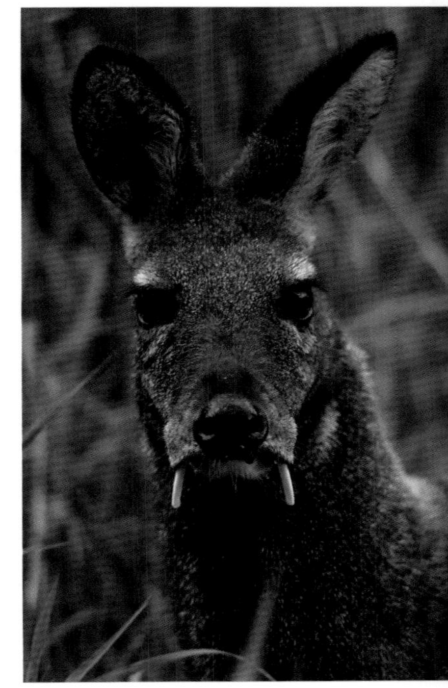

Plant diversity in India is estimated at more than 17 000 species, of which over 7 000 are endemic. Among them, orchids are very diverse. On the opposite page, Dendrobium chrysotoxum.

Above, the musk deer (Moschus moschiferus) *is an unusual member of the deer family with distinctive, protruding canine teeth and a walnut-size musk gland found under the belly of the male. Long persecuted for its glands, which are thought to cure impotence, this deer has declined or disappeared over much of its range, in spite of stringent protection.*

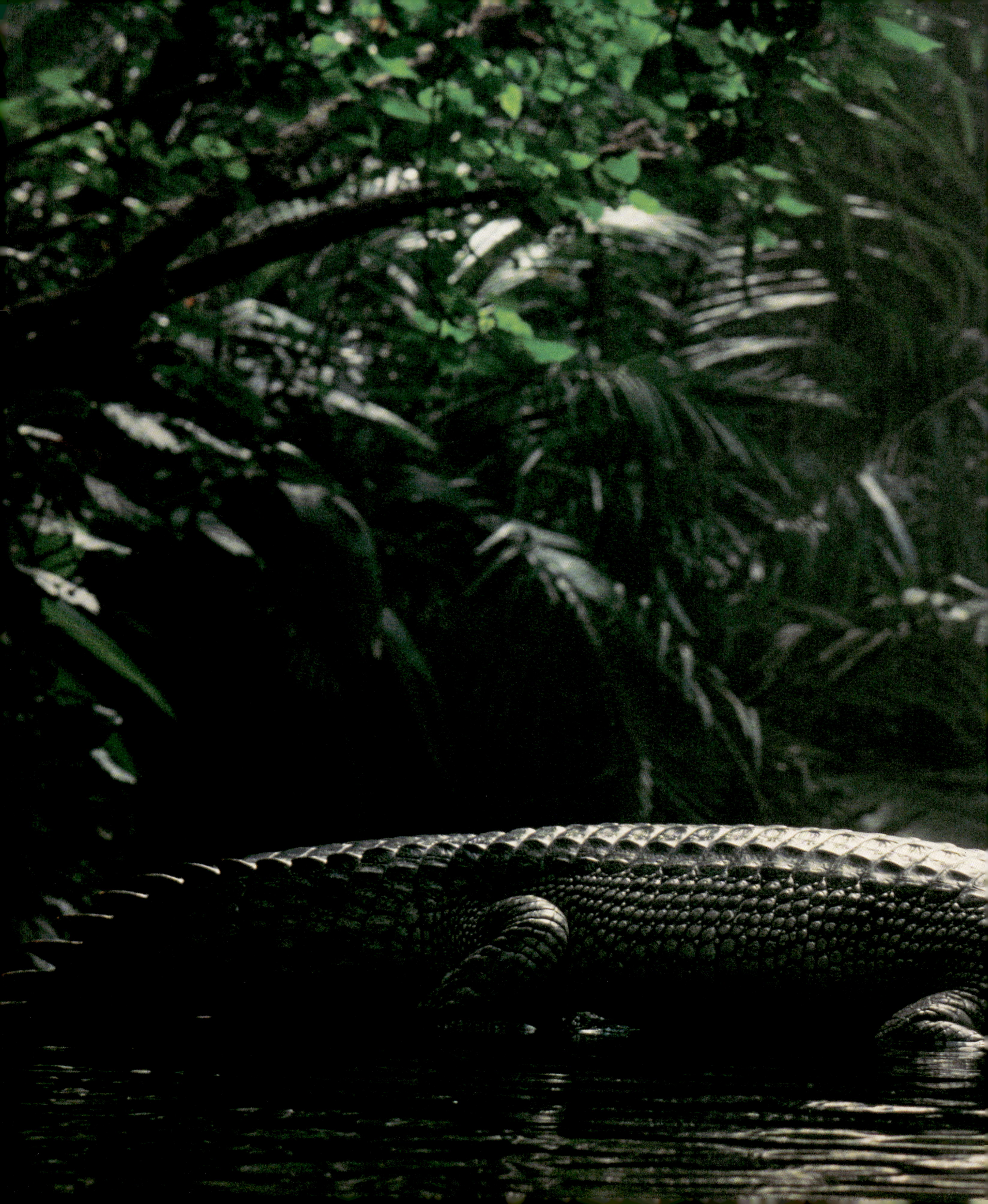

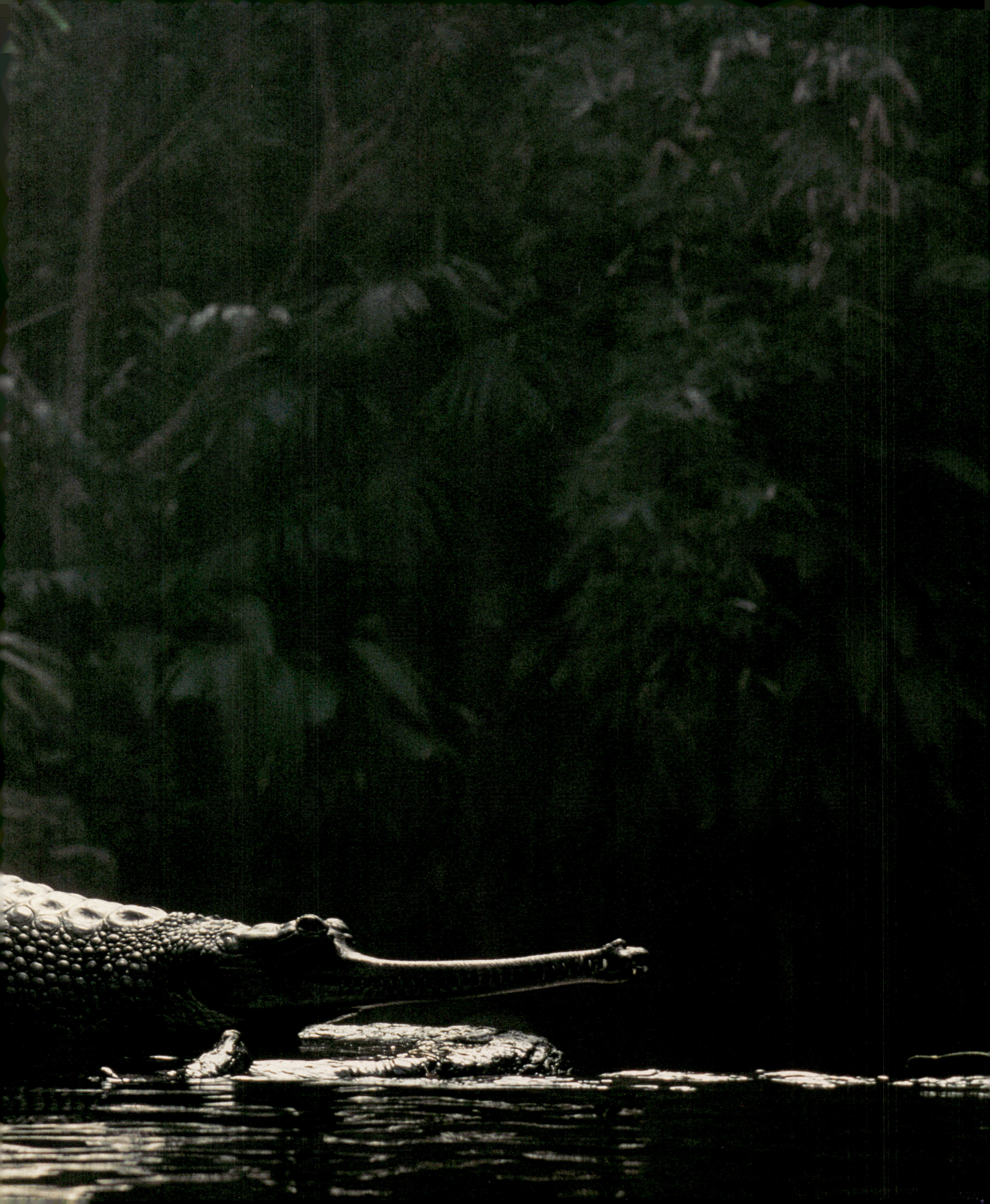

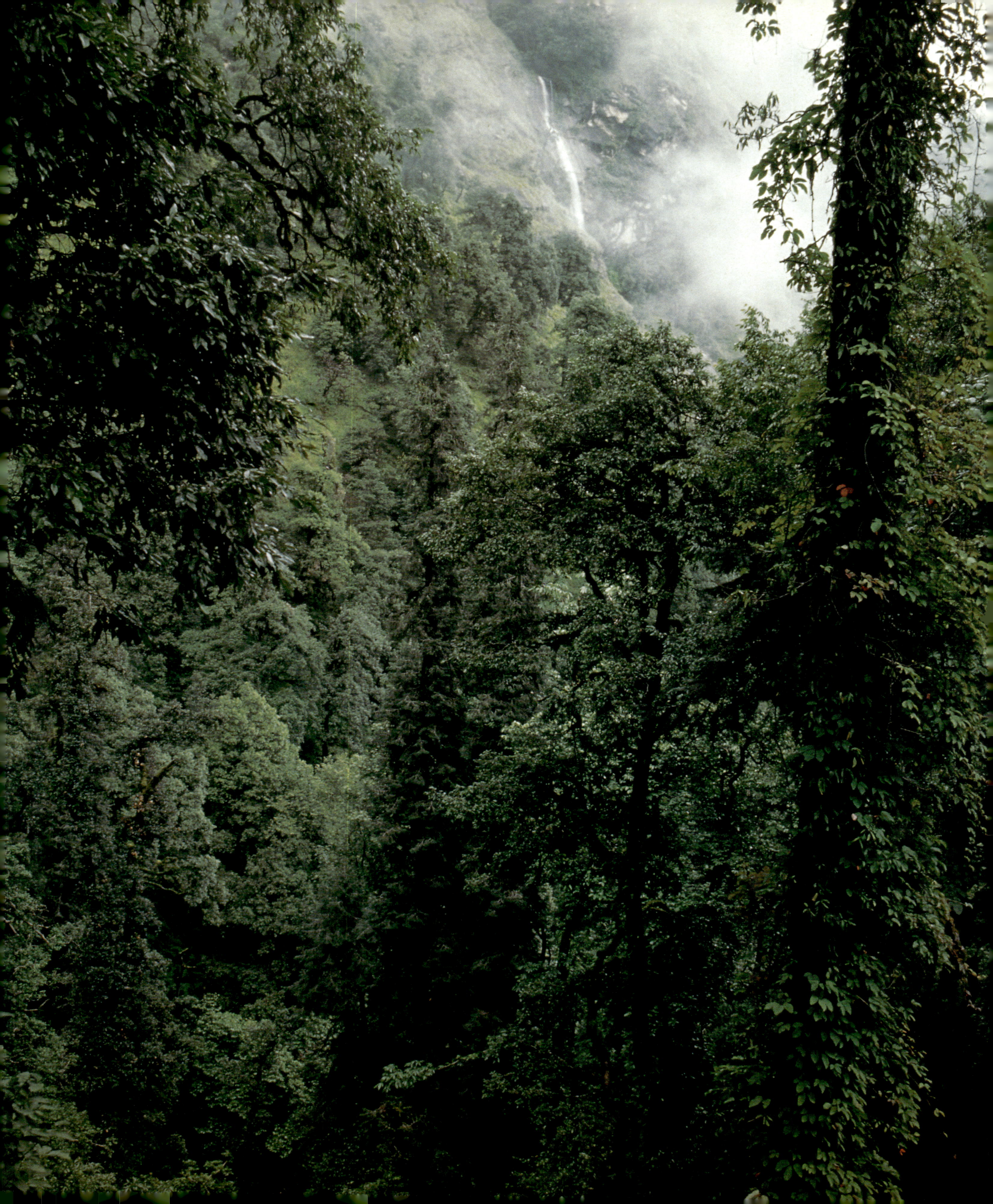

at mid-elevations in the Himalayas, and Himalayas dry temperate forest on western and northern mid-elevational slopes.

Subalpine and alpine forests include three types of Himalayan forests varying with elevation, slope and aspect, and include subalpine evergreen or birch-fir; moist alpine scrub and birch-rhododendron, and dry alpine juniper scrub.

Grassland and savanna habitats that once occupied the Gangetic Basin and portions of the Deccan Plateau have long since succumbed to agriculture and grazing. Only a few very limited areas occur in national parks, or otherwise reserved areas. These savanna areas provide an important habitat for several ungulate species, including nilgai and blackbuck (antelope).

Wetlands are an important habitat type in India that once occupied an estimated 18.4% of the country (S.K. Puri, pers. comm.), but over 70% of these have been converted to rice paddy. Natural wetlands exist in only a few sites: Chilka Lake in Orissa, Keoladeo National Park in Bharatpur, the floodplains of the Ganges, Brahmaputra and other major rivers, and the deltas of the Sundarbans and Rann of Kutch in Gujarat. These limited areas are extremely important, however, in the survival of many aquatic bird, other vertebrate, and invertebrate species.

The marine environments of India form very rich inshore coastal waters. Coral reefs occur around the Lakshadweep, Andaman, and Nicobar Islands, and the small islands and Palk Strait between India and Sri Lanka. Reef fisheries involve prawns, lobsters, oysters, and a variety of other invertebrates and bony fishes. Some coastal areas also have rich seagrass beds, important habitats for the Indian dugong and several species of marine turtles.

Flagship species

India is notable for several species which have attracted international conservation attention, among the most prominent of which are the Asiatic lion, the Bengal tiger, the Indian elephant, the Indian rhino, the snow leopard, the Asiatic cheetah, the lion-tailed macaque, the hoolock gibbon, the Nilgiri tahr, the Gangetic river dolphin, the mugger crocodile and the gharial. These dramatic animals are the merely the tip of the iceberg of threatened and endangered species in India, especially in the lower vertebrates and insects. The Threatened Plant Database maintained by the IUCN lists 96 species for India threatened with extinction, and this is certainly an underestimate.

Among the most notable flagship species, the Asiatic lion (*Panthera leo persica*), formerly the symbol of the Government of India, once had a geographic range that extended from Greece, Turkey, Syria, Iraq, Iran, and Pakistan to India, and it was widely distributed in India. In the twentieth century it has occurred only in western India and has in fact teetered on the edge of extinction, with fewer than 200 individuals in the Gir Forest of southwestern Gujarat. Under the patronage of the Nawab of Junagadh and total protection by the Government of India, the Asiatic lion has increased from 180 individuals in 1974 to 290 in 1990. Plans are underway to relocate a small group in eastern India, establishing a second population to lessen the risk of a catastrophe wiping out the only population.

Although the Asiatic lion has been the most critically endangered of the two large cats, the Indian or Bengal tiger is better known and has received more worldwide publicity. Once relatively common throughout India, the numbers and distribution of the tiger in India became reduced and scattered as forests were cut and human population increased. In the 1960s, it became apparent that tigers were seriously endangered on the subcontinent, and a national tiger census estimated that by 1962 only 1 800 of these animals remained in India. The government launched Project Tiger in the 1970s, a bold and groundbreaking attempt to set aside more tiger reserves, provide increased protection, and reduce illegal hunting and sale of tiger skins and products. With support from the international conservation community, the program achieved notable success. By 1990, the tiger population of India was estimated at 4 500 to 6 000. With continuing population growth in India (over 1.4 million people net increase per month), and intensive pressure on forests, India's tiger population will likely always remain under severe threat and in need of intensive management and monitoring.

The Indian elephant (*Elephas maximus*) is considered endangered because it now exists only in scattered remnants of its former range, which once extended from the valley of the Tigris and Euphrates Rivers in Iraq to southeast Asia. Its wild populations are now concentrated in India (estimated to number 17 000-22 000), and Myanmar (estimated at 3 000-10 000) (Santiapallai and Jackson, 1990). The Asian elephant has been domesticated for centuries as a draft animal in the timber industry. It was also used in warfare, and is still used as a ceremonial animal in India and several other countries. The total domesticated population has been estimated at 16 000; thus total numbers of Asian elephants are only one tenth those of African elephants. Although the Asian elephant breeds in captivity, most domesticated animals are tamed from wild-caught stock. The greatest threat to the Asian elephant is habitat destruction due to agricultural incursions into forest areas. Crop raiding by elephants is common, and approximately 200 villagers are killed annually in India as they attempt to interact with wild elephants. Unlike African elephants, only male Asian elephants have tusks; so poaching of males for ivory has been an additional loss. Project Elephant was launched in India in 1990 with the goals of prosecuting poachers, establishing large elephant reserves, developing corridors between existing forest areas for elephants to travel through, compensating villagers for agricultural losses, and humanely controlling or dispatching rogue ele-

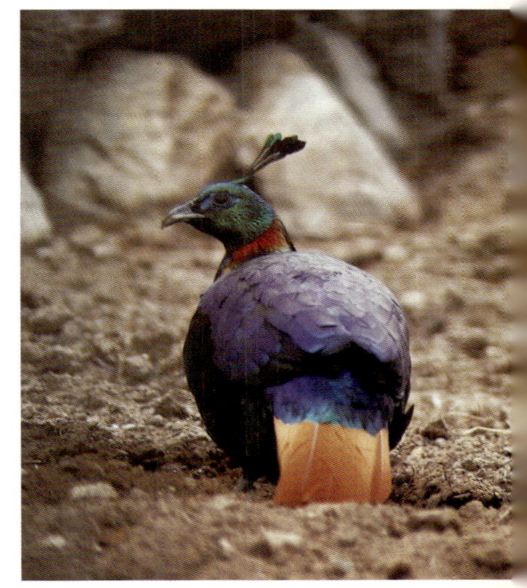

On pp. 340-341, the gharial (Gavialis gangeticus) *is one of the three largest crocodiles and indeed one of the largest living reptiles in the world. Its long, narrow snout is specialized for catching fish but, in spite of its size, it is not dangerous to humans. Although the gharial was once on the verge of extinction, conservation of key habitat areas, captive breeding and reintroduction have helped to bring it back.*

On the opposite page, Mandal Forest at 2 500 m, an example of typical broad-leaved forest in the foothills of the Himalayas.

Above, the Himalayan monal (Lophophorus impejanus), *one of the many pheasant species found in India. This animal occurs at altitudes of 2 100-4 500 m in open coniferous and mixed forest from eastern Afghanistan to Bhutan and northeast India. The individual seen here is a male.*

phants known to terrorize villages. The greatest long-term threats to elephant survival in India, as for most other wildlife, are the pressing needs of a burgeoning human population.

The Indian rhino (*Rhinoceros unicornis*) is another flagship species in northern India in which some recovery has occurred, from an estimated population of 1 000 to 1 100 in India and Nepal in 1977, to 1 950 in 1991 (Vigne and Bradley-Martin, 1991). Of these, 1 250 are found in Kaziranga National Park in Assam. Despite protection efforts, poaching remains intense, and the demand from China for rhino horn continues —58 rhinos were killed by poachers in 1990 alone.

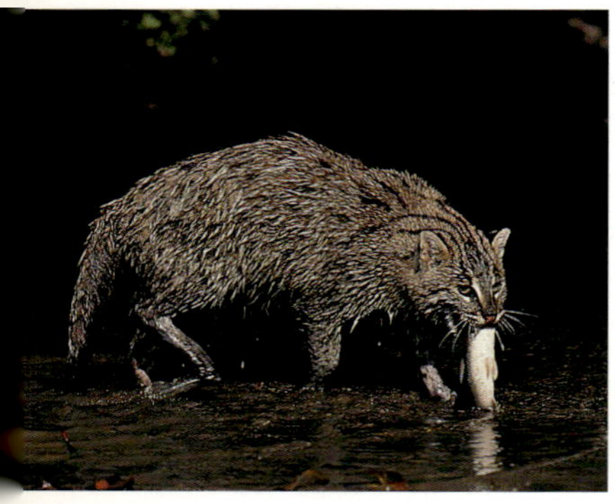

The snow leopard (*Panthera uncia*) may also be considered an endangered flagship species of India, although most of the estimated world population of 4 500 to 7 500 resides in China. India is thought to be home to 500 of these beautiful animals. A conservation effort modeled after Project Tiger has worked successfully to provide additional protection for snow leopards.

The Asiatic cheetah (*Acinonyx jubatus venaticus*), is not a conservation success story, and many reports indicate that it is already extinct in India. It may survive in very low numbers in the desert steppes of Iran, Pakistan, and Afghanistan, and one sighting was reported in November 1990 from Simlipal Wildlife Sanctuary in Orissa (Anon., 1991). Like the African cheetah, it has been shot as a livestock predator, and has also suffered from habitat destruction and decimation of its natural prey, the Indian gazelle. The total world population in 1977 was estimated at 200, and it has certainly decreased since then. The Asiatic cheetah is an example of what can inevitably happen to prominent species if they are not given adequate conservation attention.

Primates in general are very conspicuous in India, where several species of macaque (*e.g.* the rhesus macaque, *Macaca mulatta*, and the bonnet macaque, *Macaca radiata*) and the hanuman langur (*Presbytis entellus*), coexist with humans in many areas, even in cities. The best known endangered Indian primate is the lion-tailed macaque (*Macaca silenus*). An inhabitant of tropical forests in south India, primarily the Nilgiri Hills, it has received much national and international attention, including four symposiums and an international society dedicated to its protection. Field surveys in the 1970s estimated a population of less than 1 000 remaining in nature (Green and Minkowski, 1977), making it one of India's most endangered primates. Subsequent surveys, bolstered by conservation efforts in the Western Ghats, have estimated between 3 000 and 4 000 (Kumar, 1994). However, lion-tails are still vulnerable to habitat loss as more forests in southwest India are converted to teak and coffee plantations.

A fishing cat (Felis viverrina) *in Dudhwa National Park. This widespread species has a spotty, relict distribution from India to Sumatra and Java. As its name implies, it is adept at fishing and usually lives near water in brush or scrub habitat.*

On the opposite page, the Indian peafowl (Pavo cristatus) *in Bandipur National Park. This beautiful and immensely popular bird is now bred throughout the world as a domesticated species, but its original habitat is dense scrub and jungle in India, Sri Lanka, and Bangladesh.*

The hoolock gibbon (*Hylobates hoolock*), noteworthy in being India's only lesser ape, is limited in India to the northeast provinces, primarily Assam, Arunachal Pradesh, Meghalaya, Mizoram, Nagaland and Tripura. The hoolock also occurs in Bangladesh and Myanmar. Gibbon groups are small, typically only three or four individuals in a family, and are dependent upon relatively mature and intact forests. Their greatest threat is logging and related forest incursions. They are also subject to poaching for anatomical parts for the Oriental black market. Chivers (1977) estimated the total hoolock gibbon population of India to be only 1 000 individuals, but Mukherjee (1982) estimated 350 to 453 groups of gibbons in Tripura, with an average group size of 3.2, providing a population estimate of 1 120 to 1 450 for that province alone. Mukherjee noted, however, that only one third of existing forests in Tripura (1 360 km^2 out of 4 080 km^2) provided suitable gibbon habitat. More recent field surveys in 1995-1996 throughout northeast India located gibbons in only 28% of 81 reserve forests (Mohnot *et al.*,1996). The natural social behavior of gibbons, with only one adult male per group, plus their dependency upon forests with fairly intact canopy, makes them highly vulnerable to further decline.

Another important primate is the golden langur (*Presbytis geei*), which is limited to forest areas of western Assam and adjacent parts of Bhutan. Recent field studies have estimated a total wild population of less than 1 500 individuals, making it India's most endangered primate (Mukherjee *et al.*, 1997).

The Nilgiri tahr (*Hemitragus hylocrius*), the only wild goat-like ungulate south of the Himalayas, is similarly vulnerable to deforestation and poaching in the Nilgiri Hills of south India. It occurs only in the provinces of Kerala and Tamil Nadu in the woodland mosaics of the Nilgiri, Anamalai and related hills of the Western Ghats. In 1986, the total numbers were estimated at 2 000 to 2 200 (Rice, 1988), but herds were small, scattered and unusually isolated. Many of these populations are also threatened by grazing competition from domestic cattle. The tahr breeds well in captivity, and could be managed both for its wildlife values and as a potential source of protein and income for local people.

An aquatic Indian mammal that may be considered a flagship species is the Gangetic river dolphin (*Platanista gangetica*). It is native to the Gangetic and Brahmaputra river systems in India and Bangladesh, migrating as far inland as Nepal and Bhutan during seasonal high water. Once very abundant, its total populations are now estimated at no more than 5 000 individuals, with 3 000 to 4 000 of these in India (Perrin, 1989). Main threats to its survival are pollution, dams, mining, and intentional or accidental capture in fishing nets. The Ganges is seriously polluted from industrial, agricultural, and domestic sources, and both rivers are very susceptible to extreme flow variations that range from excessive flooding to drought conditions.

There are also flagship species among the reptiles. Two species of crocodiles that occur in India, the mugger (*Crocodylus palustris*), and the gharial (*Gavialis gangeticus*) are considered flagships for the country. The mugger crocodile is restricted to the Indian subcontinent, where it may be found in a number of freshwater habitats including rivers, lakes, and marshes. In India and Sri Lanka, mugger crocodiles have adapted

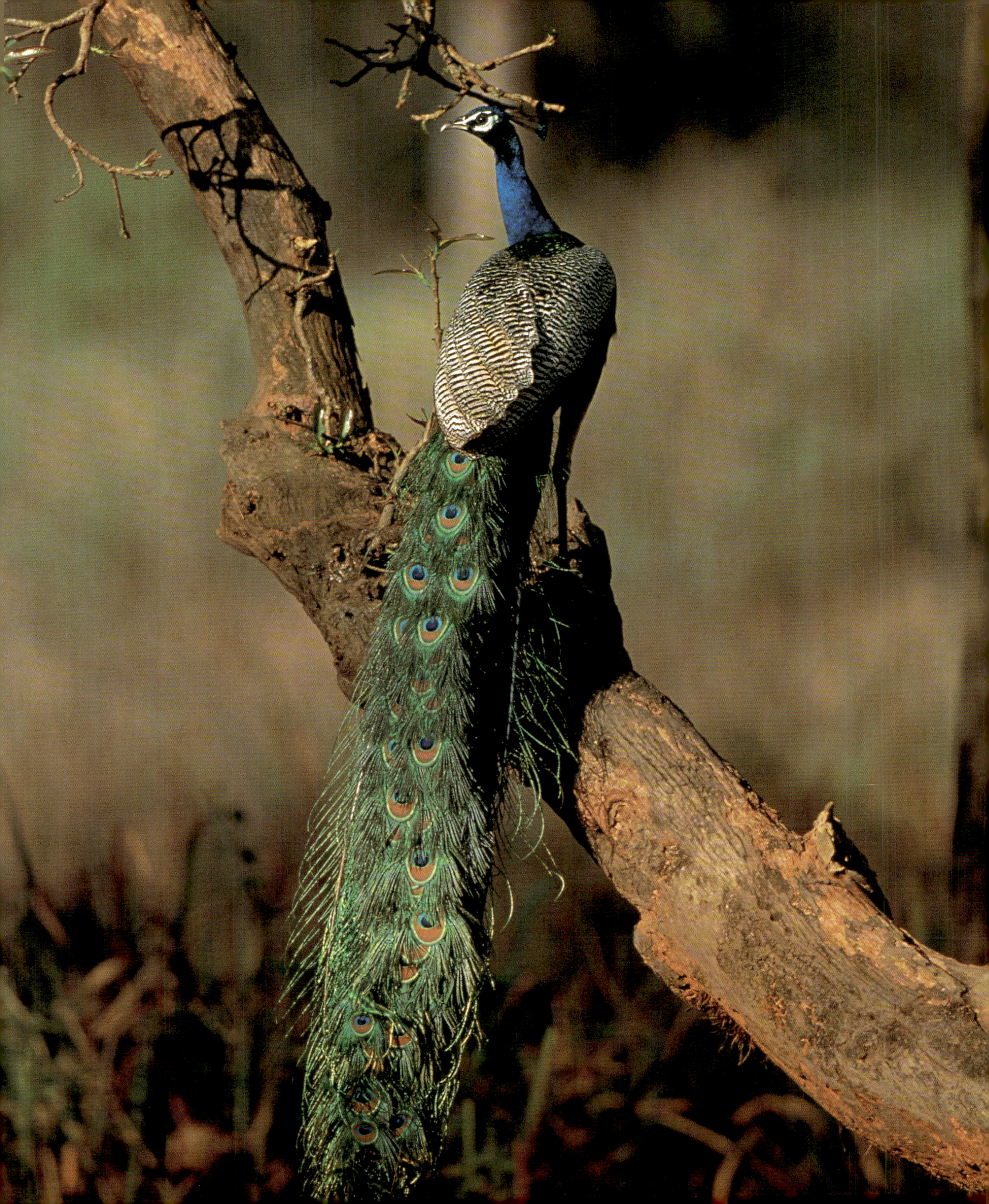

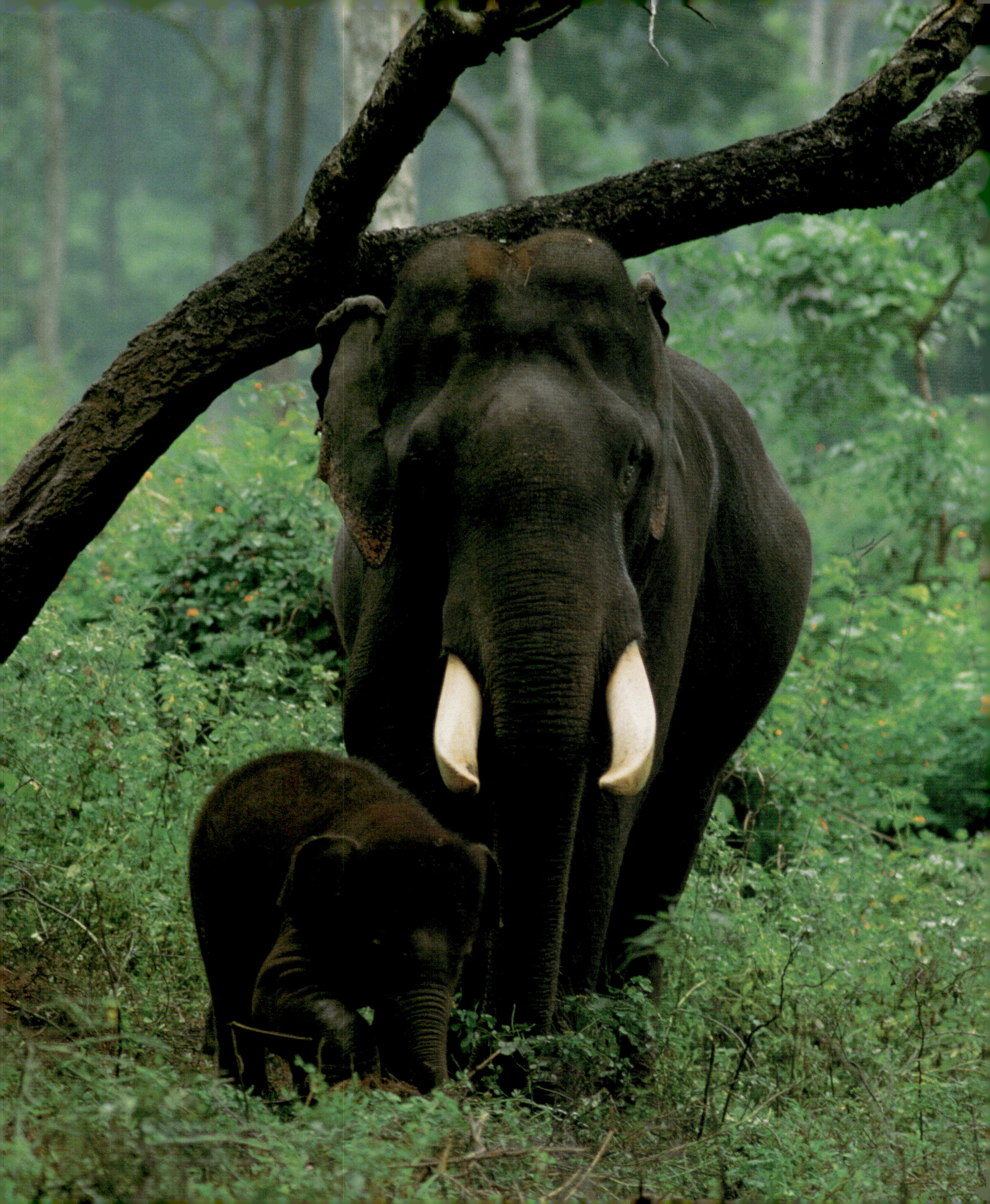

well to reservoirs, irrigation canals, and man-made ponds, and in some areas may be found in coastal saltwater lagoons (Whitaker and Whitaker, 1993). The mugger crocodile is a medium to large crocodile (maximum length ca. 4-5 m), and has the broadest snout of any living member of the genus *Crocodylus*. While illegal skin trade was a major problem in the past (1950s-1960s), the current threats to its survival include drowning in fish nets, egg predation by humans, habitat destruction, and the use of crocodile parts for medicinal purposes.

In some areas of northern India and Nepal, mugger and gharial populations are sympatric, but the two species tend to be segregated by habitat. The gharial is a resident of deep, fast-flowing rivers, but in these rivers prefers areas where the current is reduced. It has the longest snout of all crocodilians and competes with the saltwater crocodile for the title of "largest living reptile," reaching up to 6-7 m or more. The gharial is placed in a family by itself, the Gavialidae. It is arguably the most thoroughly aquatic of the extant crocodilians, and adults apparently do not have the ability to walk in a semiupright stance on land as do other crocodilians. Gharials are restricted to the northern part of the Indian subcontinent, where they are found in four river systems: the Indus (Pakistan), the Ganges (India and Nepal), the Mahanadi (India), and the Brahmaputra (Bangladesh, India, and Bhutan). The males have a strange bulbous protuberance called *ghara* at the tip of the snout; its function is not well understood, but it is apparently used as a visual indicator, as a sound resonator, or for bubbling and other sexual behaviors.

The gharial is the most unusual crocodile and a spectacular flagship species; it is also one of the most critically endangered of the crocodilians. However, unlike the case of the other seven most endangered crocodilians, good conservation programs for the gharial are now in place over much of the species' range. This animal was literally brought back from the brink of extinction by restocking programs initiated in India in 1975, and in Nepal in 1978. Since that time, over one thousand captive-reared juveniles have been released in India, and over 300 in Nepal. In India, eight protected areas have been designated for gharial management, with varying degrees of success, but despite the efforts being made, the gharial is still very much endangered; it is virtually extinct in parts of its former range (Pakistan and Bangladesh) and probably extinct in others (Bhutan and Myanmar).

The king cobra (*Ophiophagus hannah*) is an important flagship snake species for India, as are all the cobras, which figure prominently in Indian folklore and mythology. The image of the snake charmer and his entranced cobra is one of the best known Indian images. The king cobra is the largest of all poisonous snakes, reaching some 5.5 m in length, and feeds exclusively on other snakes.

There are also a number of other endangered or vulnerable mammal flagship species in India. These include the gaur or Indian bison (*Bos gaurus*), the wild yak (*Bos mutus*), the wild Asiatic water buffalo (*Bubalus arnee*), the Markhor (*Capra falconeri*), the pygmy hog (*Sus salvanius*), the gray or timber wolf (*Canis lupus*), the dhole or Asiatic wild dog (*Canis alpinus*), the Bengal fox (*Vulpes bengalensis*), the Asiatic golden cat (*Felis temmincki*), the clouded leopard (*Neofelis nebulosa*), the marbled cat (*Felis marmorata*), the rusty-spotted cat (*Felis rubiginosa*), the oriental small-clawed otter (*Aonyx cinerea*), the smooth-coated otter (*Lutra perspicillata*), the Nilgiri marten (*Martes gwatkinsi*), the sun bear (*Helarctos malayanus*), the sloth bear (*Melursus ursinus*), the Asiatic black bear (*Selenarctos thibetanus*), Jerdon's palm civet (*Paradoxurus jerdoni*), the Asiatic wild ass or onager (*Equus hemionus*), the Indian gazelle or chinkara (*Gazella gazella*), the four-horned antelope (*Tetracerus quadricornis*), the barking deer or muntjac (*Muntiacus muntjac*), the hog deer (*Axis porcinus*), the Manipur brown-antlered deer (*Cervus eldi*), the blackbuck (*Antilope cervicapra*), and the chevrotain or mouse deer (*Tragulus meminna*).

Cultural diversity

The rich biodiversity of India is mirrored in part by a complex tapestry of human culture and history. India is a land of venerable antiquity with a record of 5 000 years of civilization. This ancient history, combined with its diverse topographic and ecological setting, affords modern India a remarkable array of ethnic, cultural, linguistic, and religious groups.

The cultural history of India is as old as that of Egypt. Its roots lie in the Harappan civilization of the Indus Valley, which extended into Rajasthan and Gujarat. This was an advanced culture with well-planned cities, including the famous site of Mohenjo-Daro. Houses were spacious and built of fired bricks. There is abundant evidence of cultivated cotton and wheat; domestication of cattle, sheep, camels, and elephants; use of gold, copper, bronze and precious stones; and along the coast of Gujarat, of harbors that traded with the seaports of Sumer.

The Harappan civilization, which predated the 4th millennium BC, was destroyed by the Aryan invasions —nomadic migrations of light-skinned pastoral tribes from central Asia. They forced some of the indigenous dark-skinned people of Dravidian origins to the south. Thus began, by 2000 BC, new infusions of customs, beliefs, and racial types into Punjab and the Gangetic Basin. Hinduism emerged as an integration of the ritualistic, literary Vedas of the Aryans and the spiritual beliefs of the indigenous peoples.

The subsequent history of India is a complicated record of clans, states, kingdoms, dynasties and empires, all rocked by repeated conquests and invasions, and overlain with the development of great artistic, literary, philosophical, and religious traditions.

This brief and very incomplete overview of history in India provides a simplified view of the origins of its

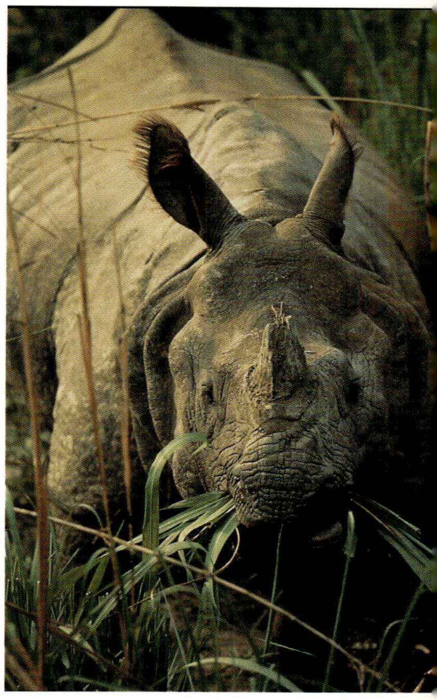

On the opposite page, the elephant (Elephas maximus) *is another of India's great flagship species, although like the tiger, it is not endemic to the country. Wild populations are now found mainly in India and Myanmar, although small numbers still occur elsewhere.*

Above, the Indian one-horned rhinoceros (Rhinoceros unicornis) *is one of three Asian rhino species and the only one that seems to be holding its own. Of some 1 950 found in India and Nepal according to the 1991 census, 1 250 were found in Kaziranga National Park in Assam. Nonetheless, poaching to serve a Chinese market for rhino horn remains a problem.*

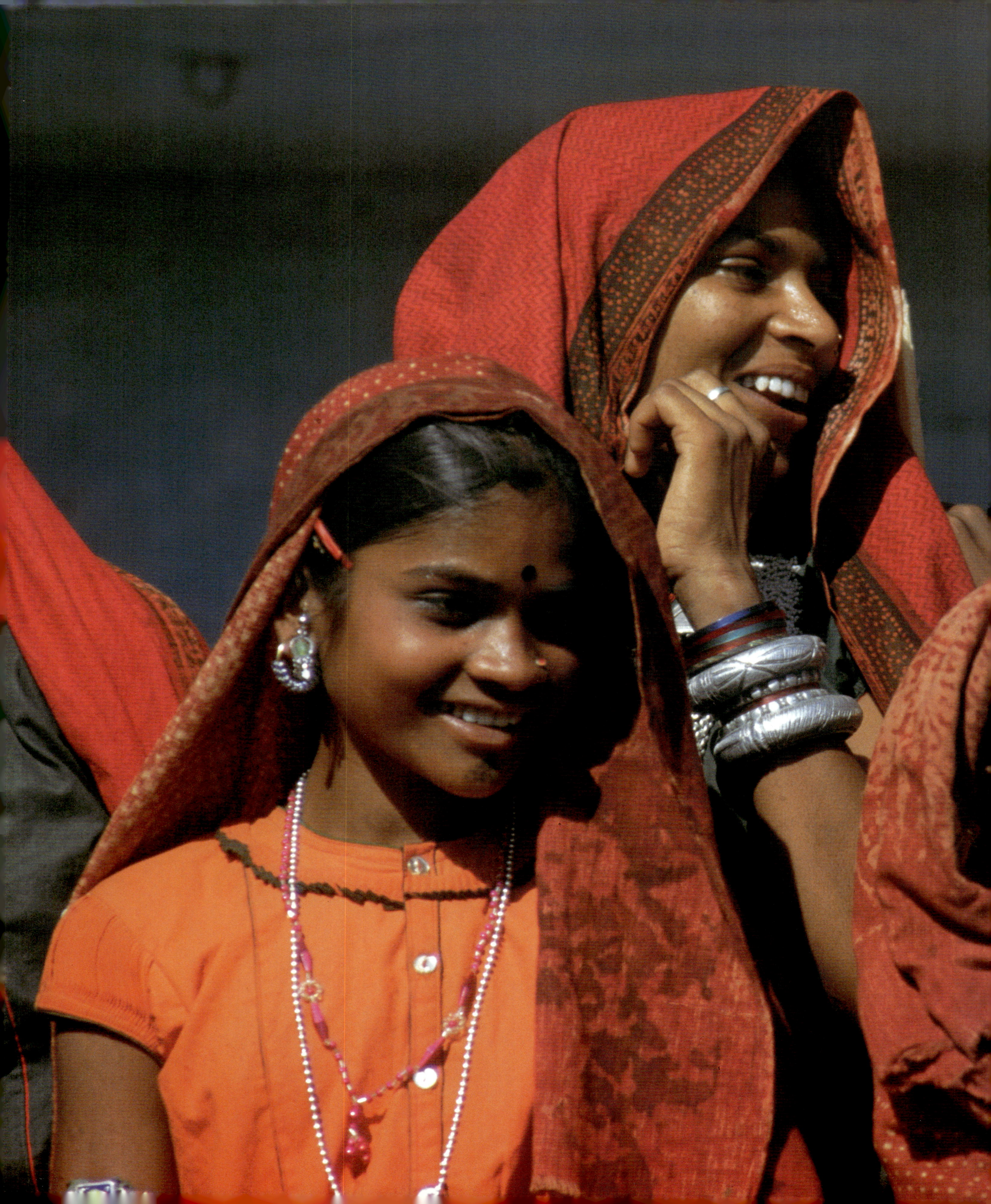

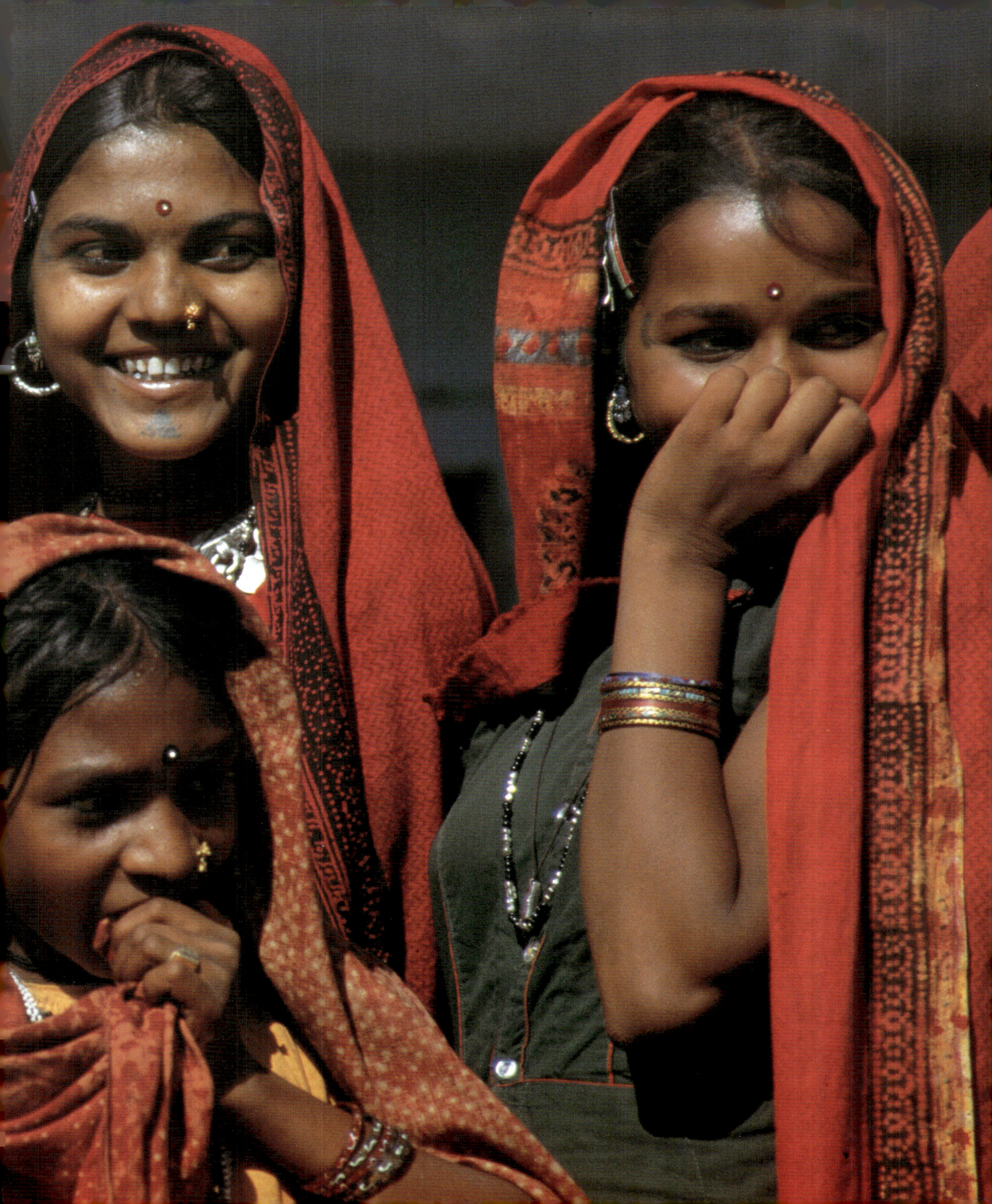

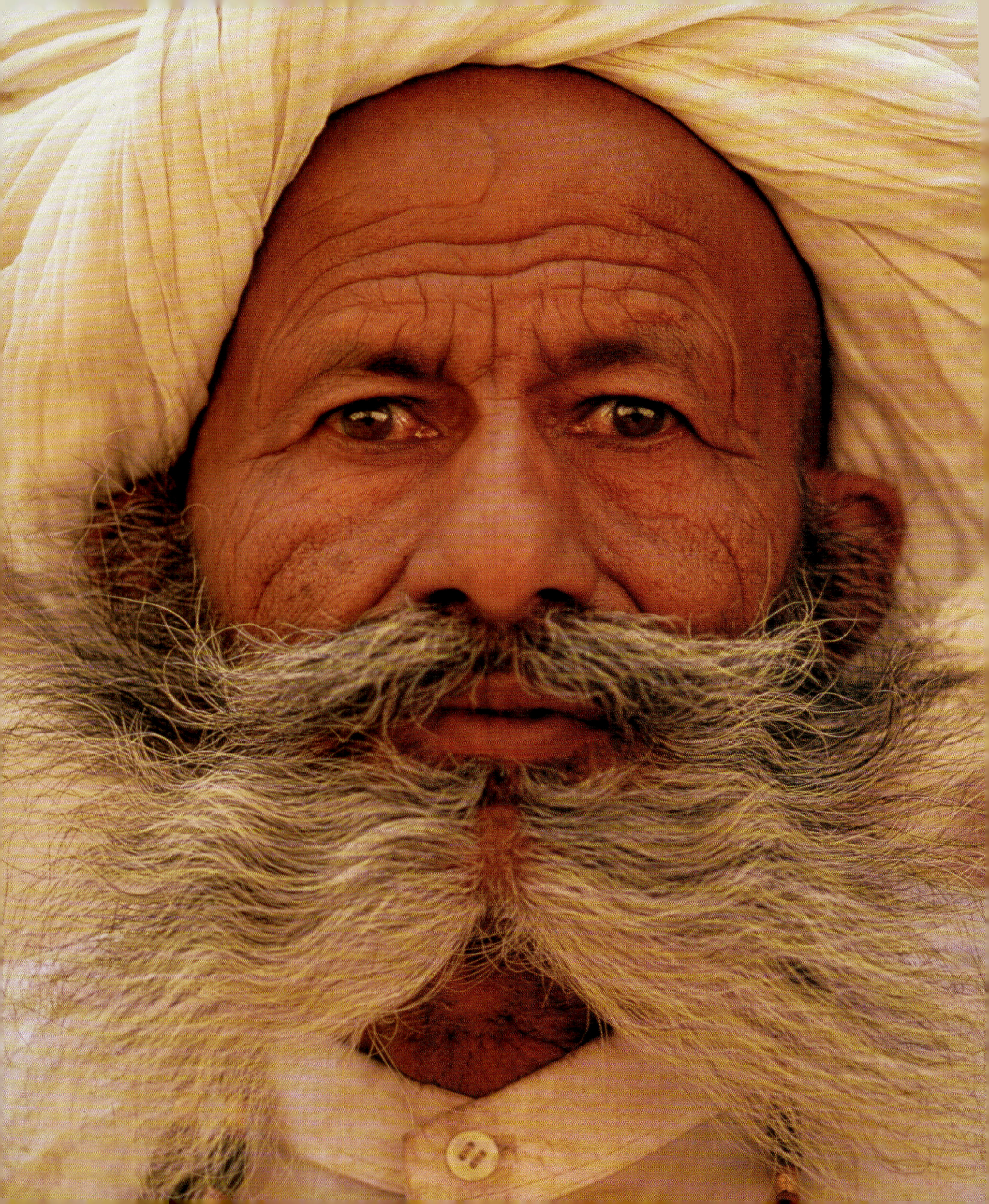

human diversity. Superimposed over 5 000 years of indigenous history were cultures from central and southeastern Asia, the Middle East, the Mediterranean, the Balkans, and all of western Europe.

Today that diversity is evident in many ethnic, linguistic, and religious groups. Modern India is approximately 82% Hindu, 12% Muslim, 2.5% Christian, 1.9% Sikh, 0.7% Buddhist, 0.5% Jain, 0.3% Zoroastrian, with smaller percentages of Parsi, and many animist religions among the tribal groups.

The numbers and diversity of tribal people in India are surprising. The tribal population of India in 1990 was approximately 102 million, about 11% of the total population, surpassing the more than 70 million "minority nationalities" in China. Thakur and Thakur (1994) list 613 recognized tribal groups in India. They define a tribe as a "...group of people speaking a common language, observing uniform rules of social organization, and working together for a common purpose." A tribe "... includes a common name, a contiguous territory, a relatively uniform culture or way of life and a tradition of common descent."

Whereas tribal groups represent about 11% of the human population in India, some provinces have over 90% of their total numbers in tribal groups, *e.g.*, Mizoram in northeast India (93.6% tribal), and Lakshadweep Islands in the Arabian Sea (93.8% tribal). Other regions with high percentages of tribal people are Nagaland (84%), Meghalaya (80.6%), and Arunachal Pradesh (69.8%). Percentages of population do not necessarily reflect tribal diversity, however. Nagaland, for example, with 84% of its total population as tribal, lists only five tribes, whereas Orissa, with only 22.2 % of its population as tribal, lists 62 distinct tribal groups; Kerala with only 1.1% tribal, lists 35 tribal groups. In general, the more peripheral provinces of India such as the northeast sector have the greatest percentages of tribal peoples; however, even Madhya Pradesh, in the central Deccan, lists 23% of its population as tribal. The lowest percentage of tribal peoples of all the states of India is Uttar Pradesh, at 0.2%.

Most tribal peoples have become integrated into the economic and cultural life of India, although some have strongly resisted. The best example of the latter are the Andaman and Nicobar Islanders: the Jarawa, Great Andamanese, Sentinelese, and Onge, who are racially Negrito peoples thought to have originated from an ancient pygmy race once widespread in Southeast Asia and Oceania (Singh, 1975).

The Andaman Islanders have been the subject of classic field studies by the British anthropologists A.R. Radcliffe-Browne (1922) and Edward Horace Mann, but in attempting to maintain their cultural integrity and isolation these people have shown hostility to Western contact. As recently as 1975, the Sentinel Islanders greeted strangers with iron-tipped arrows shot from long bows (Singh, 1975). These groups, whose life-style was once entirely hunting and gathering, have suffered greatly from outside contact, which has brought new diseases, agricultural encroachment on their hunting grounds, a bewildering array of governmental policies resulting from misguided attempts at assimilation, and even persecution, bombing and strafing in World War II. The populations of some of these tribes are now so decimated that they consist of only a few hundred people. By 1975, the Jarawa were reduced to 300 individuals on two islands, the Sentinelese to 150 individuals, the Onge to 112, and the Great Andamanese to only two dozen —all from former populations which numbered in the thousands. Thus, the tribal peoples of the Andaman Islands show the tragic, but all too frequent, pattern of modern times —the decimation and virtual extinction of indigenous people, and the loss of the ecological knowledge and traditions that have developed over millennia.

Another interesting tribal group are the Irula of Tamil Nadu State. They have become famous for their great skills in catching rats, which are one of their favorite foods, and snakes. They are especially adept at catching poisonous snakes, which they sell for venom extraction and for international trade.

Tribal people not only reflect the human diversity of India, they also impact biodiversity. Many tribal groups still subsist by active hunting and gathering combined with shifting cultivation, which affects biodiversity in many ways. At the same time, their own survival depends on the maintenance of biodiversity.

Cultural diversity is clearly evidenced by linguistic differences. The Constitution recognizes 18 major languages, but many hundreds of minor languages and dialects exist. The 1961 census of India officially listed 1 652 regional languages and dialects. Hindi is the national language, and English has become increasingly important as the language of business, government, and higher education, but great linguistic complexity exists with 13 basic languages of Indo-European origin, 4 of Dravidian origin, and 1 of Sino-Tibetan origin.

Another aspect of human diversity in India is the caste system. Caste is legally abolished in India, but still persists from a social heritage thousands of years old. Robinson (1989) noted: "The most striking characteristic of Hindu culture is the extreme emphasis placed upon the inequality of status." Caste may be considered a codified system of human diversity.

Although Mahatma Gandhi was strongly opposed to the caste system, caste continues as a pattern of social ranking that affects cultural diversity in many ways. Acquired from parents and the group into which one is born, caste greatly affects one's occupation, education, marital prospects, and economic status. The Government of India instituted certain "affirmative action" programs by which members of scheduled castes can gain educational and vocational benefits, but such programs do not completely alter basic cultural practices.

CHARLES H. SOUTHWICK
BARRY ROSENBAUM
RUSSELL A. MITTERMEIER

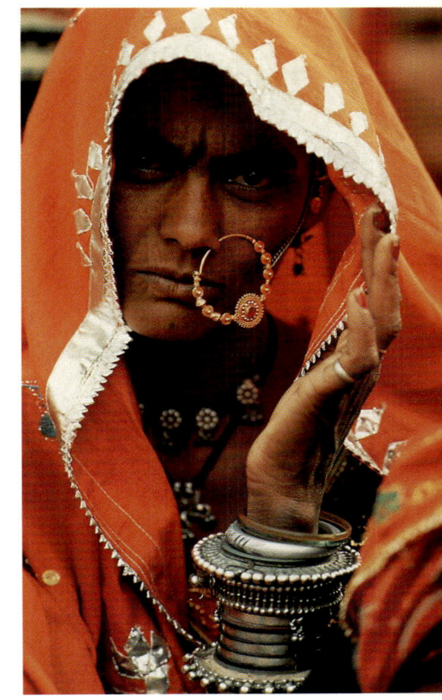

India's human population is second only to that of China, and its cultural diversity is one of the highest on Earth. An estimated 613 different ethnic groups have been recorded from India, speaking 1 652 languages and dialects.

On pp. 348-349, five Indian girls in the festival of "Bhagoriya," on the day before Holi, a Hindu spring festival.

On the opposite page, Rajasthani man from Jodhpur.

Above, a woman dressed in elaborate traditional finery in preparation for a temple visit.

India's primate fauna is one of the most diverse in Asia. It includes several endangered species, one of which is the golden langur (Presbytis geei)*, shown below. The golden langur was not described by science until 1953, and is found only in a tiny area of the India-Bhutan border. This individual was photographed in Manas National Park. Right, the hanuman langur* (Presbytis entellus) *is named after the monkey hero in the Rāmāyana, one of India's most important literary works. This langur is widespread and abundant in India, and coexists with humans in many areas. This group was photographed in Ranthambor National Park.*

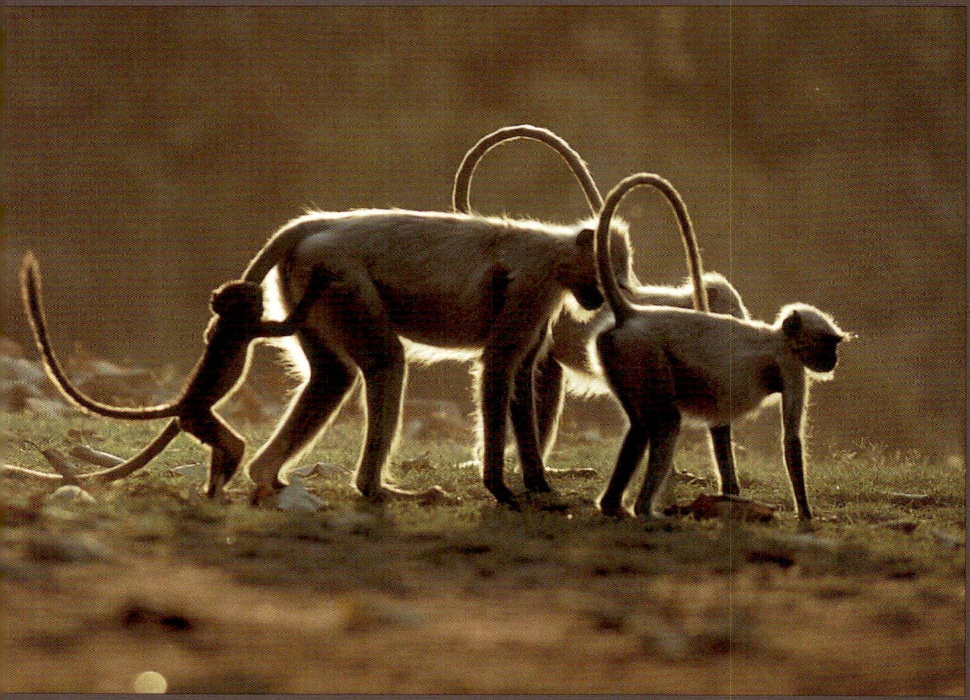

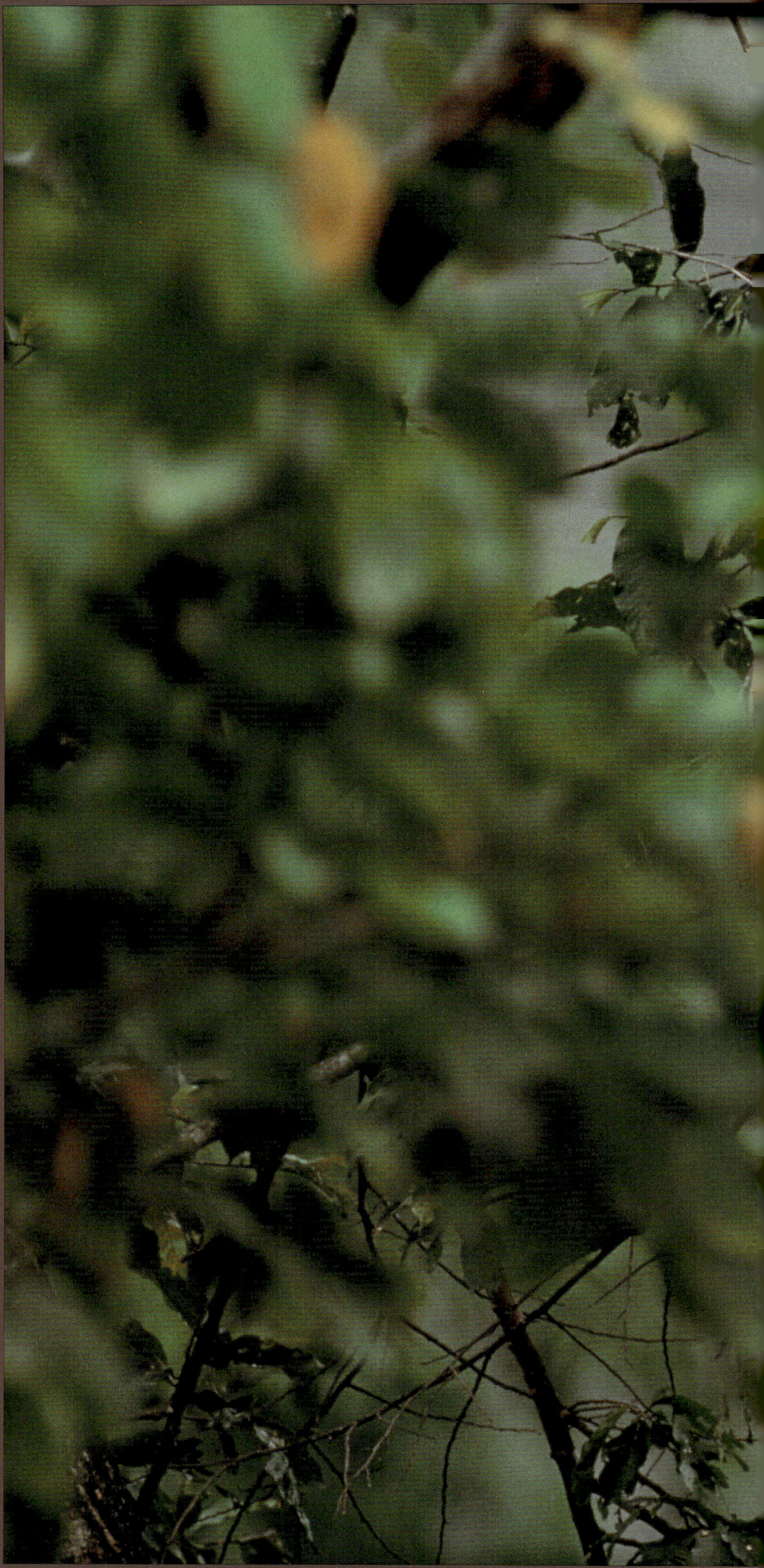

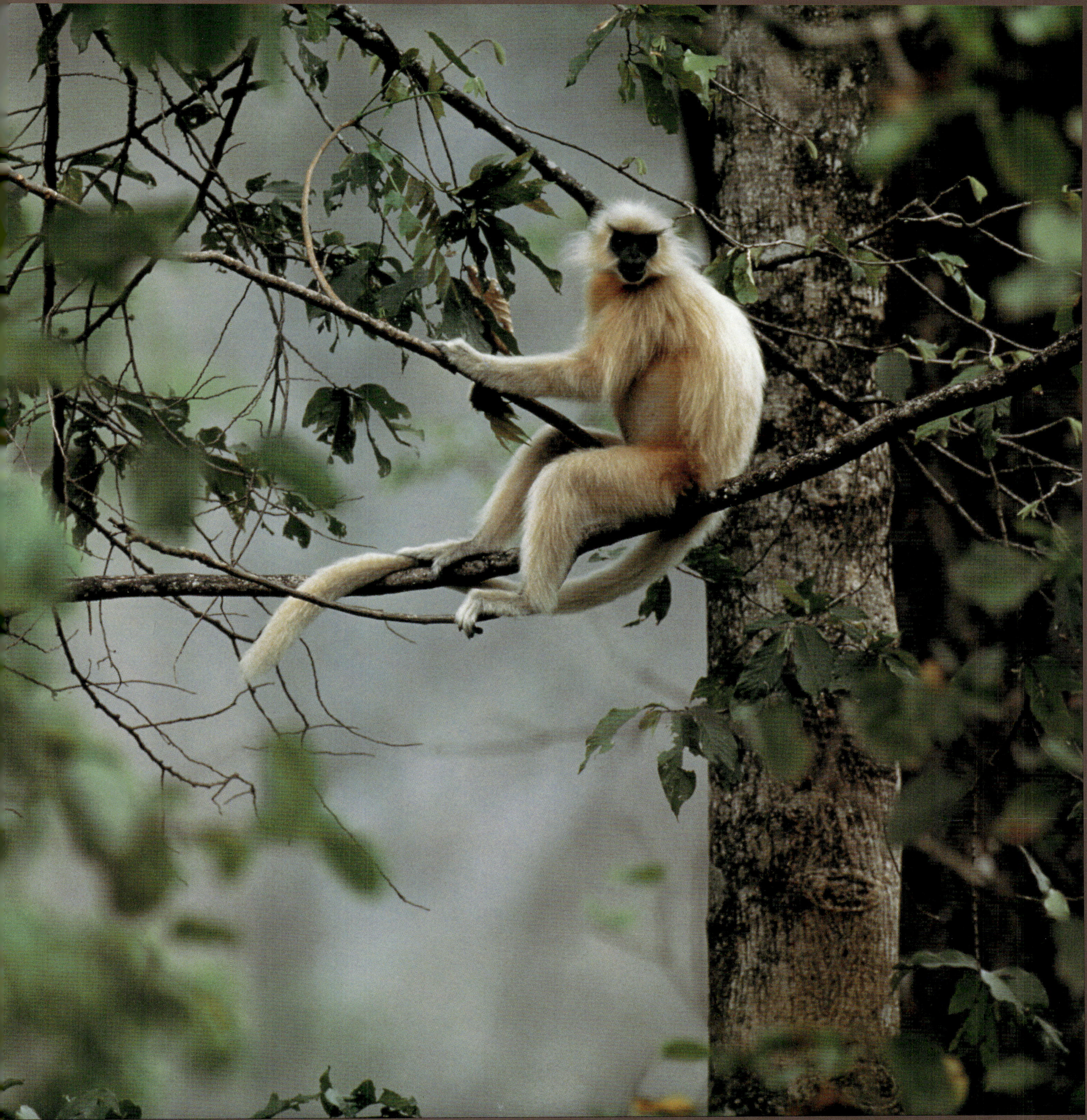

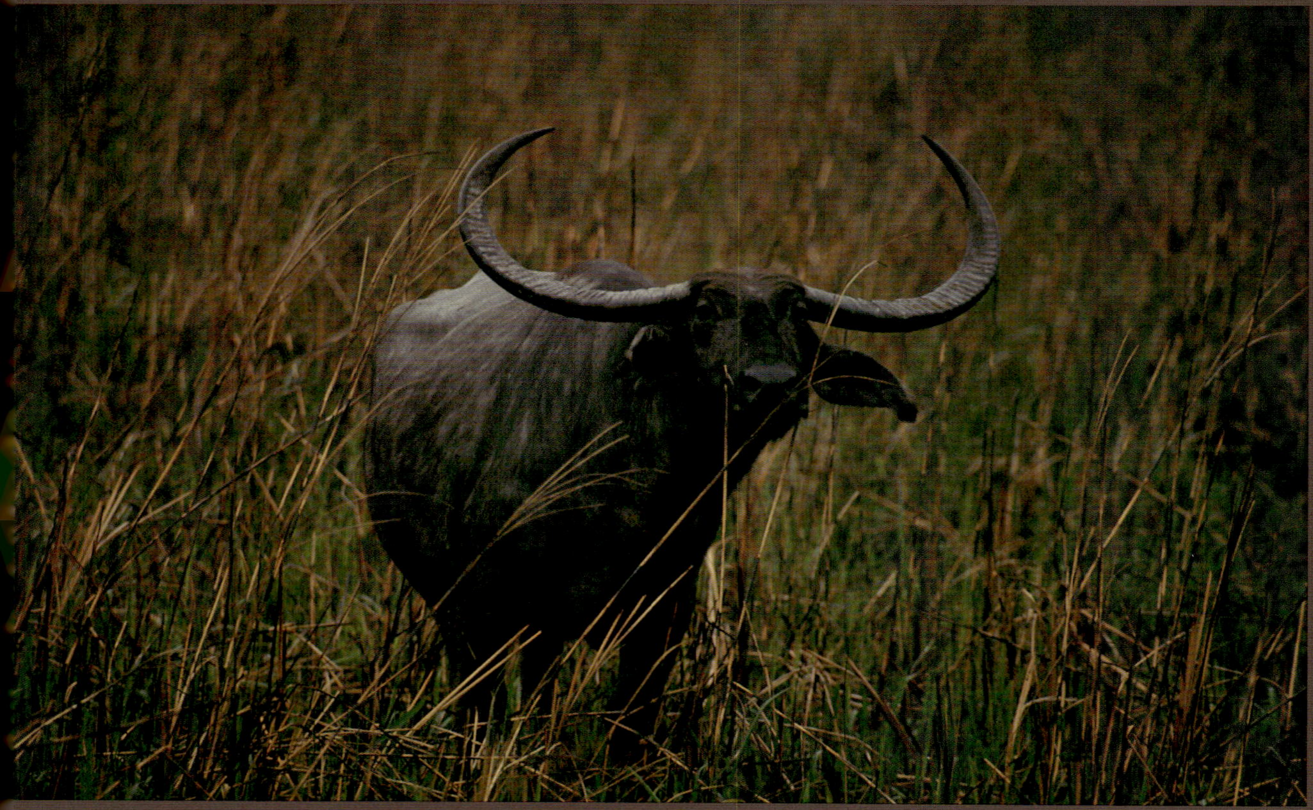

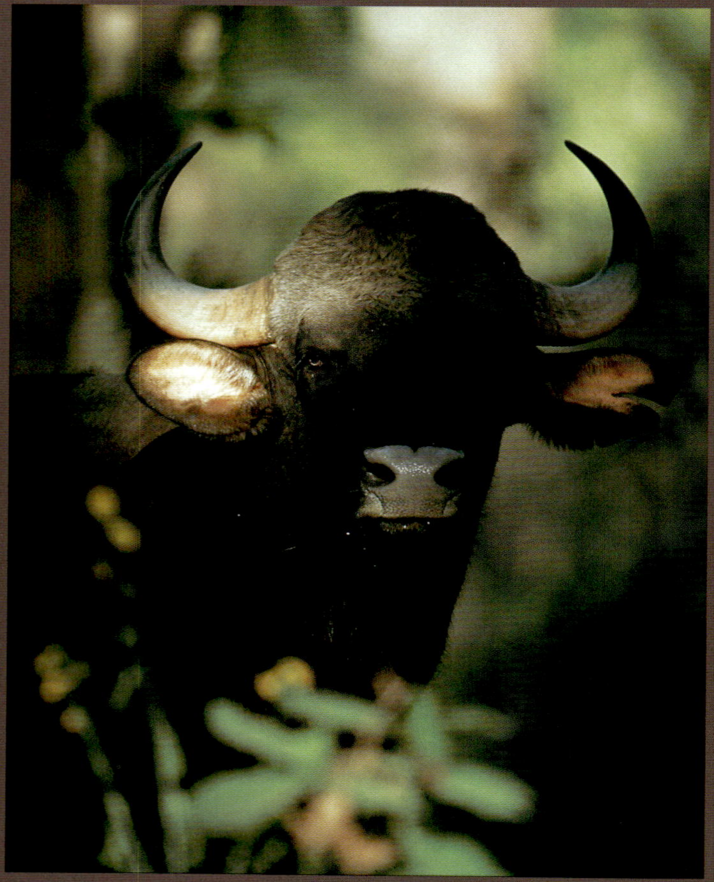

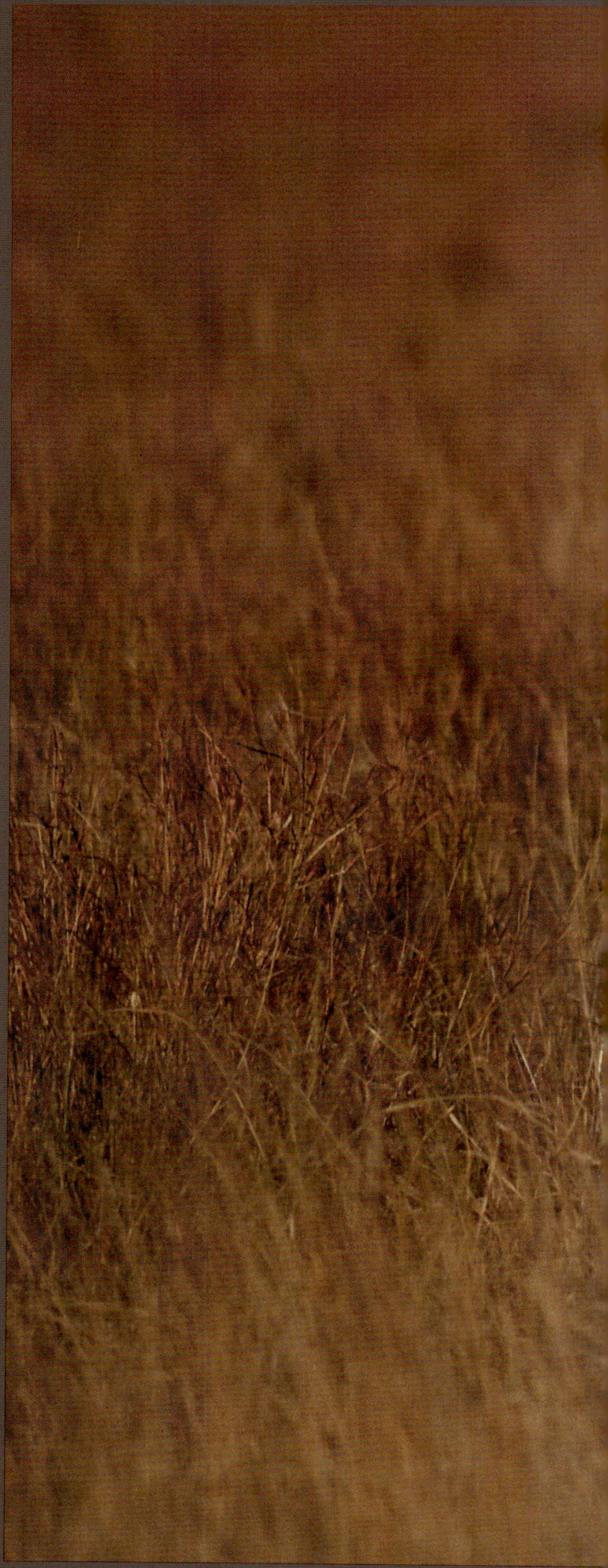

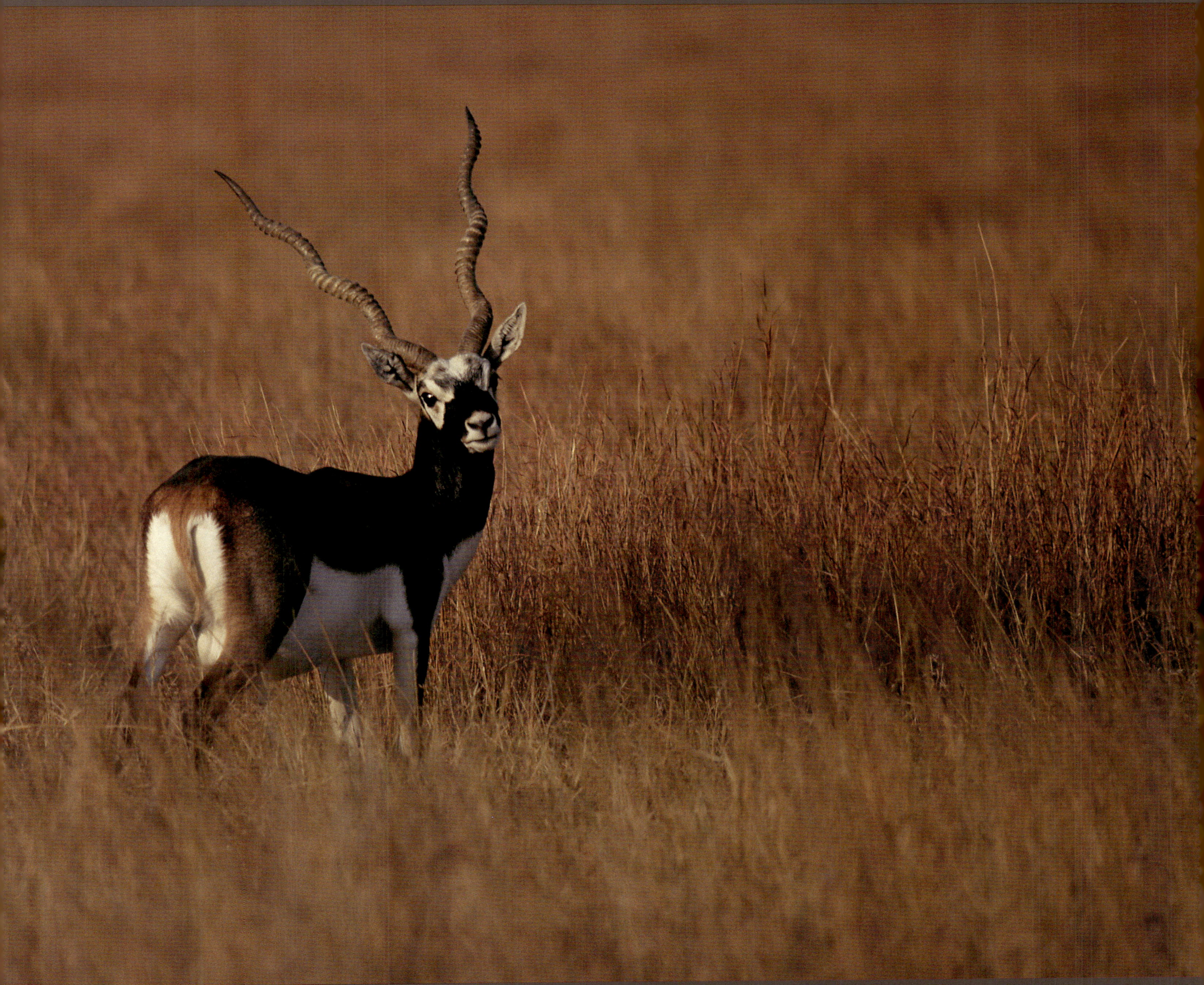

India has one of the most diverse ungulate faunas in the world, including the wild buffalo (Bubalus bubalis), *upper left; the gaur* (Bos gaurus), *lower left, and the blackbuck* (Antilope cervicapra) *above. The wild buffalo seen here in Kaziranga National Park in Assam is the ancestor of the widespread domestic buffalo, and is of great economic importance because of its close relationship to the domestic form. The gaur, seen here in Kanha National Park, is the largest of the living bovines. The male blackbuck pictured here is surely one of the most striking members of the antelope family and is especially elegant. It is also one of the fastest animals on Earth, capable of traveling at 80 km/hr for over one kilometer.*

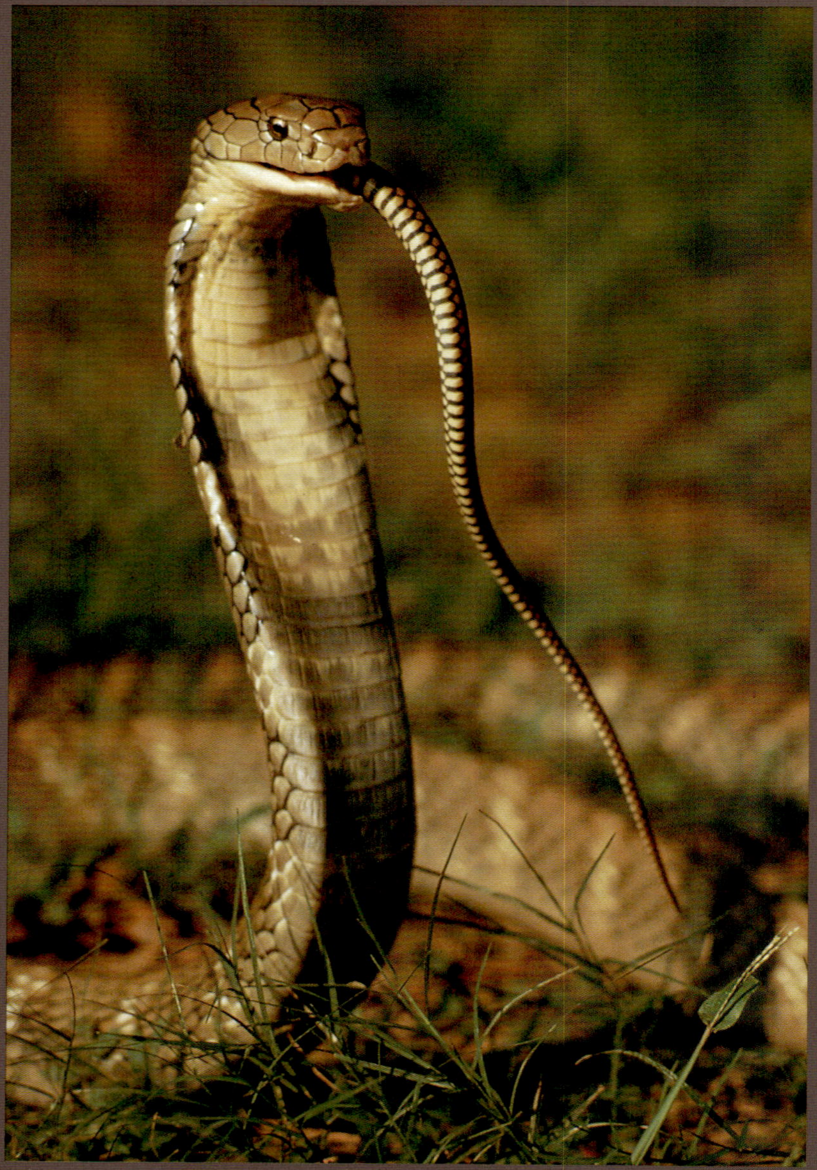

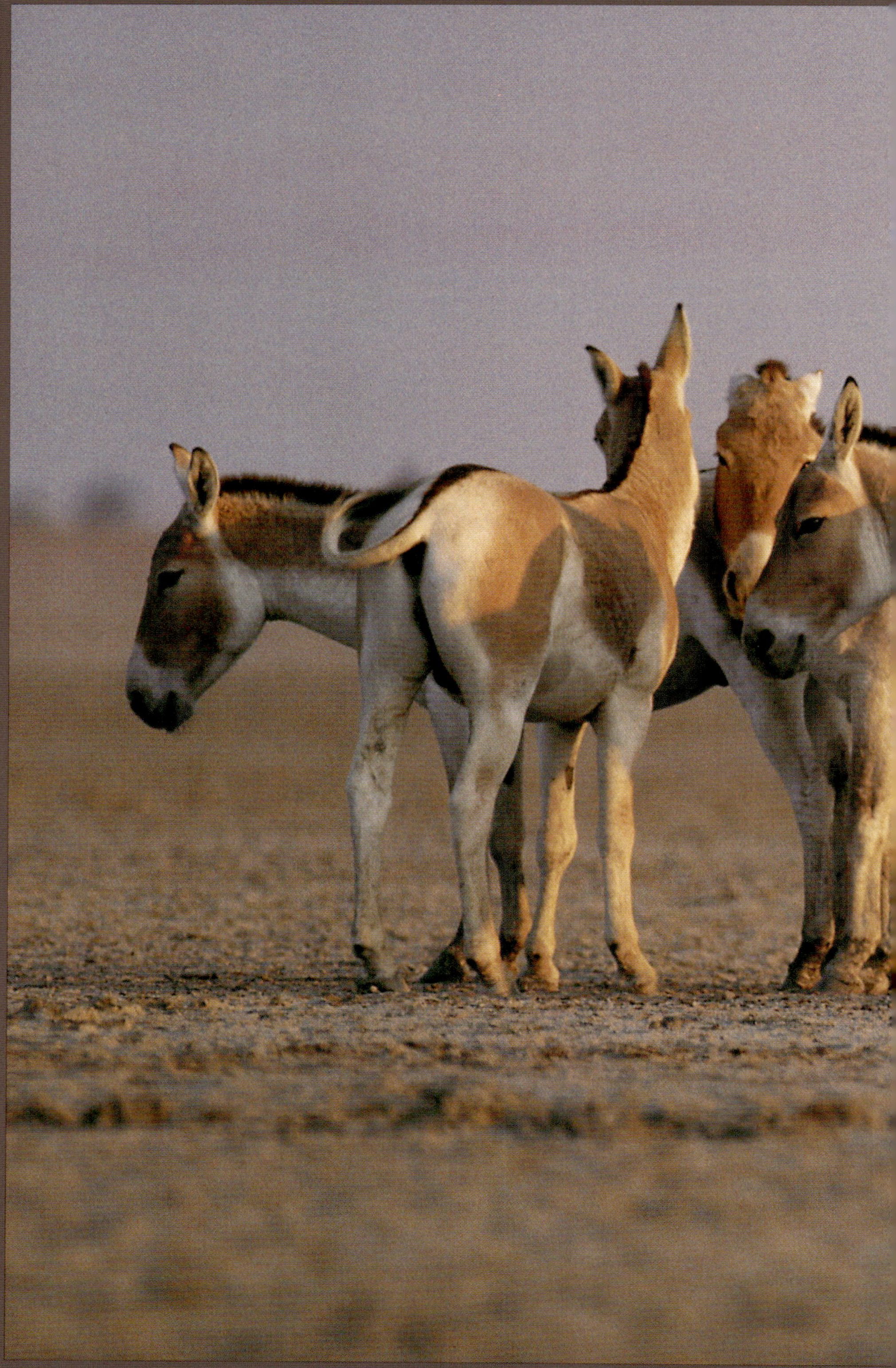

Left, the king cobra (Ophiophagus hannah) *is the longest poisonous snake in the world, reaching 5.5 m in length. It specializes on other snakes and is seen here swallowing a rat snake.*
Upper right, the Indian wild ass (Equus hemionus) *is one of the few wild equids that still survive in the modern world. Now it is found only in the Little Rann of Kutch, a scrubby desert wasteland in western India, where only about 1 000 animals survive.*
Lower right, the Asiatic lion (Panthera leo persica) *once had an enormous range from Greece and Turkey to India. It is now confined to the Gir Forest in western Gujarat, where the population was estimated at 290 in 1991.*

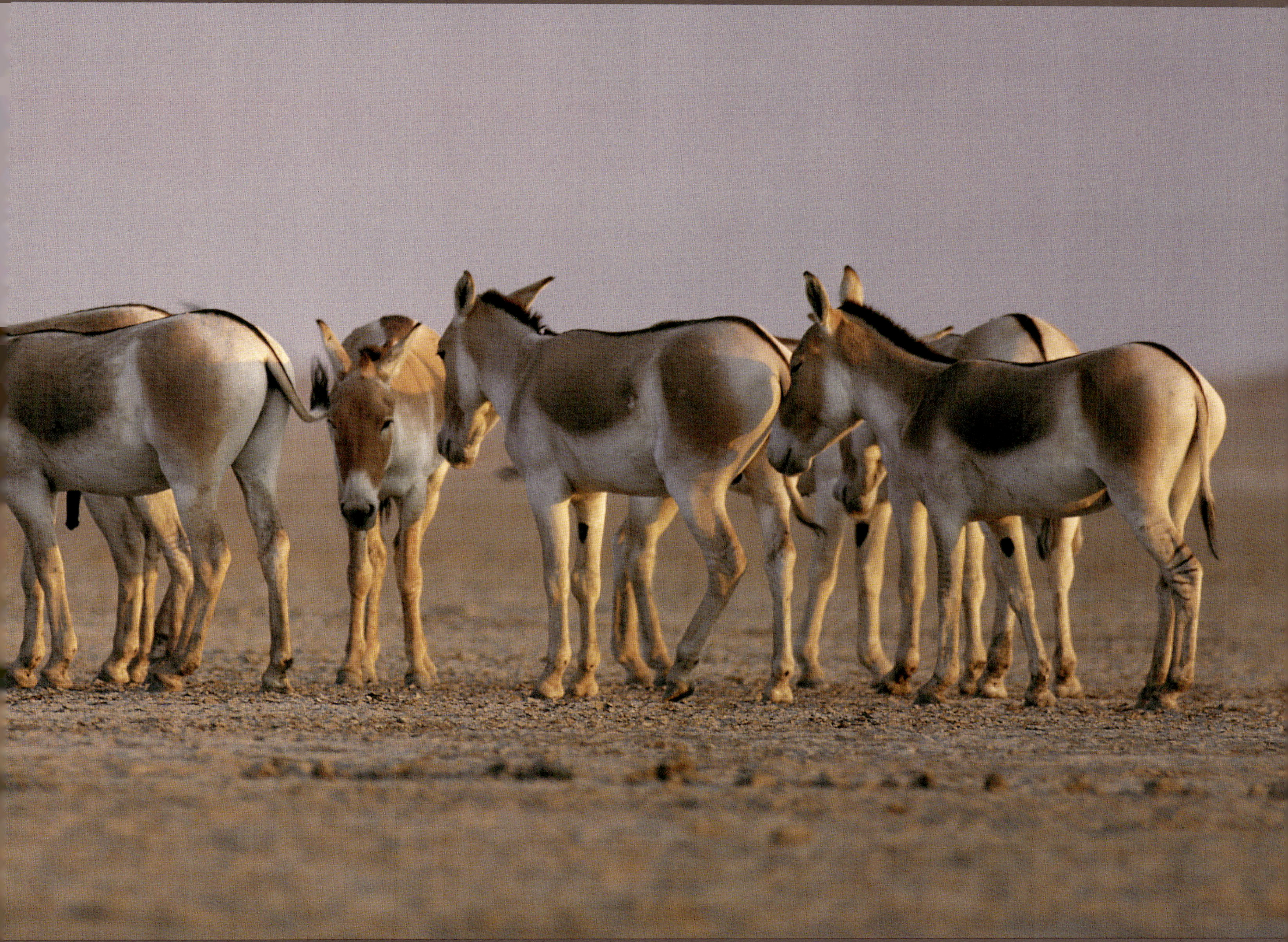

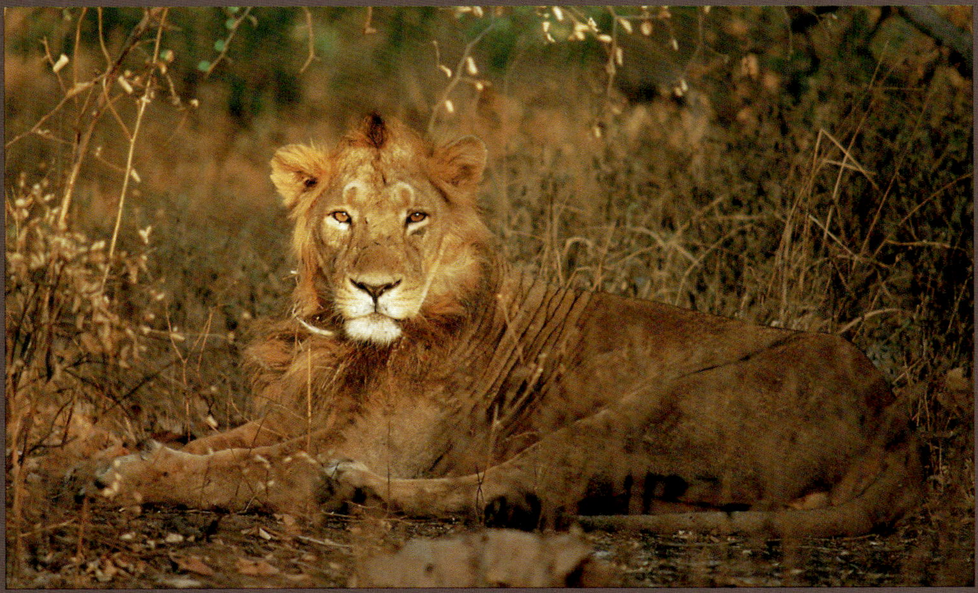

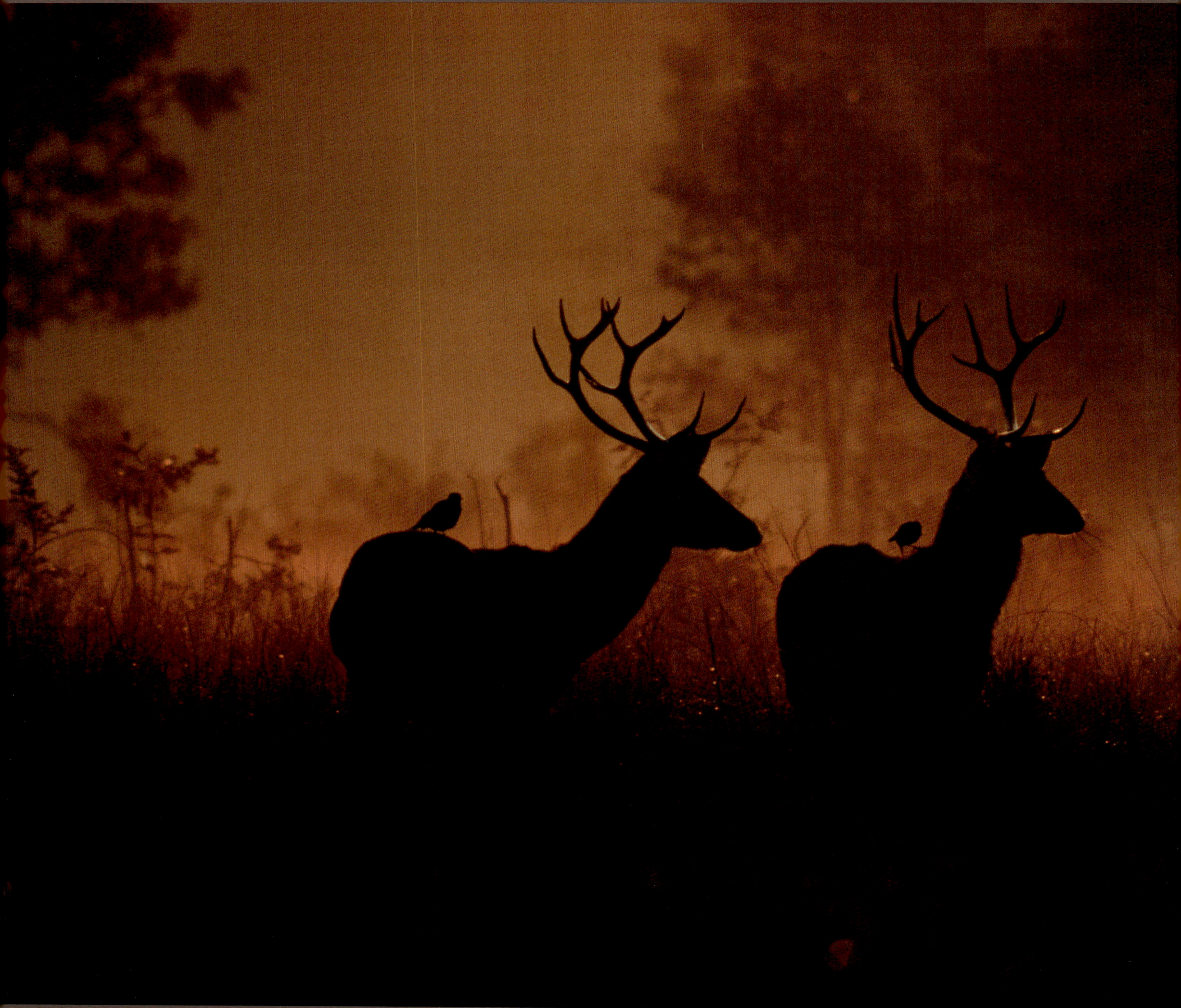

India has the greatest diversity of deer of any country. Above, the barasingha or swamp deer (Cervus duvauceli) *is adapted to alluvial and riparian grasslands that are also prime agricultural land. Consequently, it has declined dramatically in India.Upper right, the Nilgai or bluebuck* (Boselaphus tragocamelus) *is the largest of all Asian antelopes and has been protected in Hindu India because of its resemblance to the sacred cow. The individual depicted here is a calf. Lower right, the chital or spotted deer* (Axis axis) *is one of the most common and widespread wild animals in India and is also one of the most tolerant of human presence.*

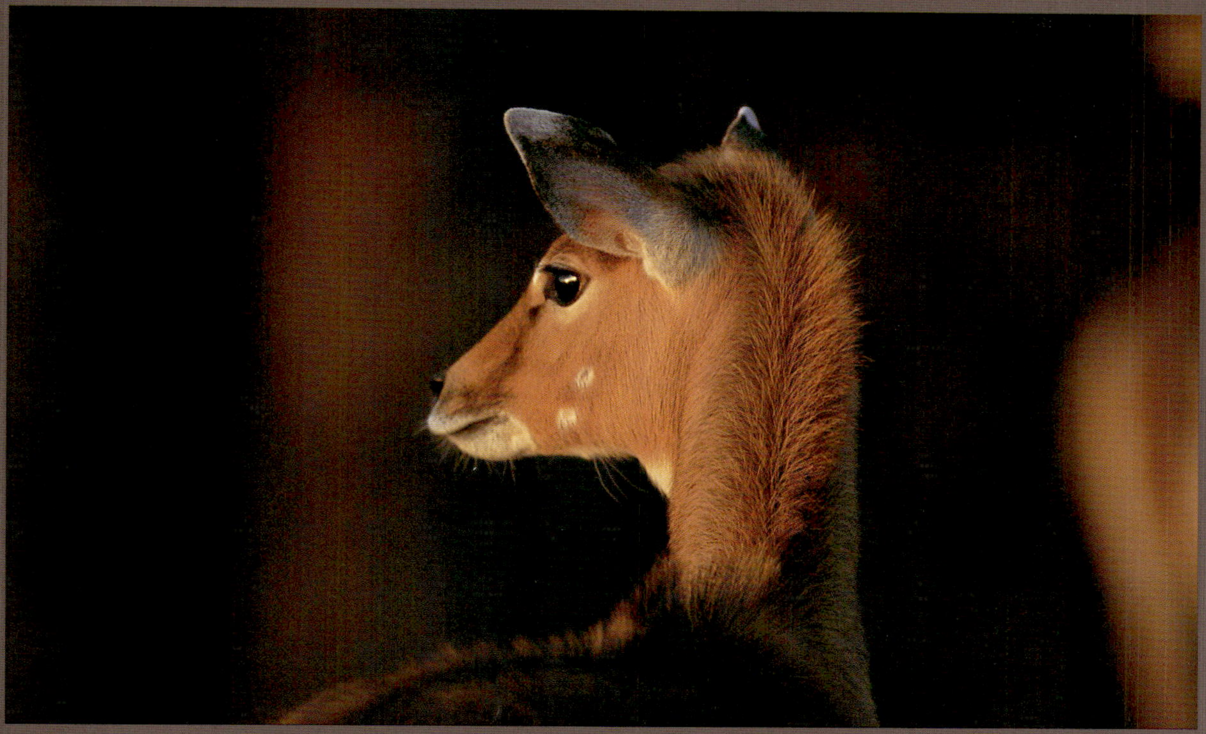

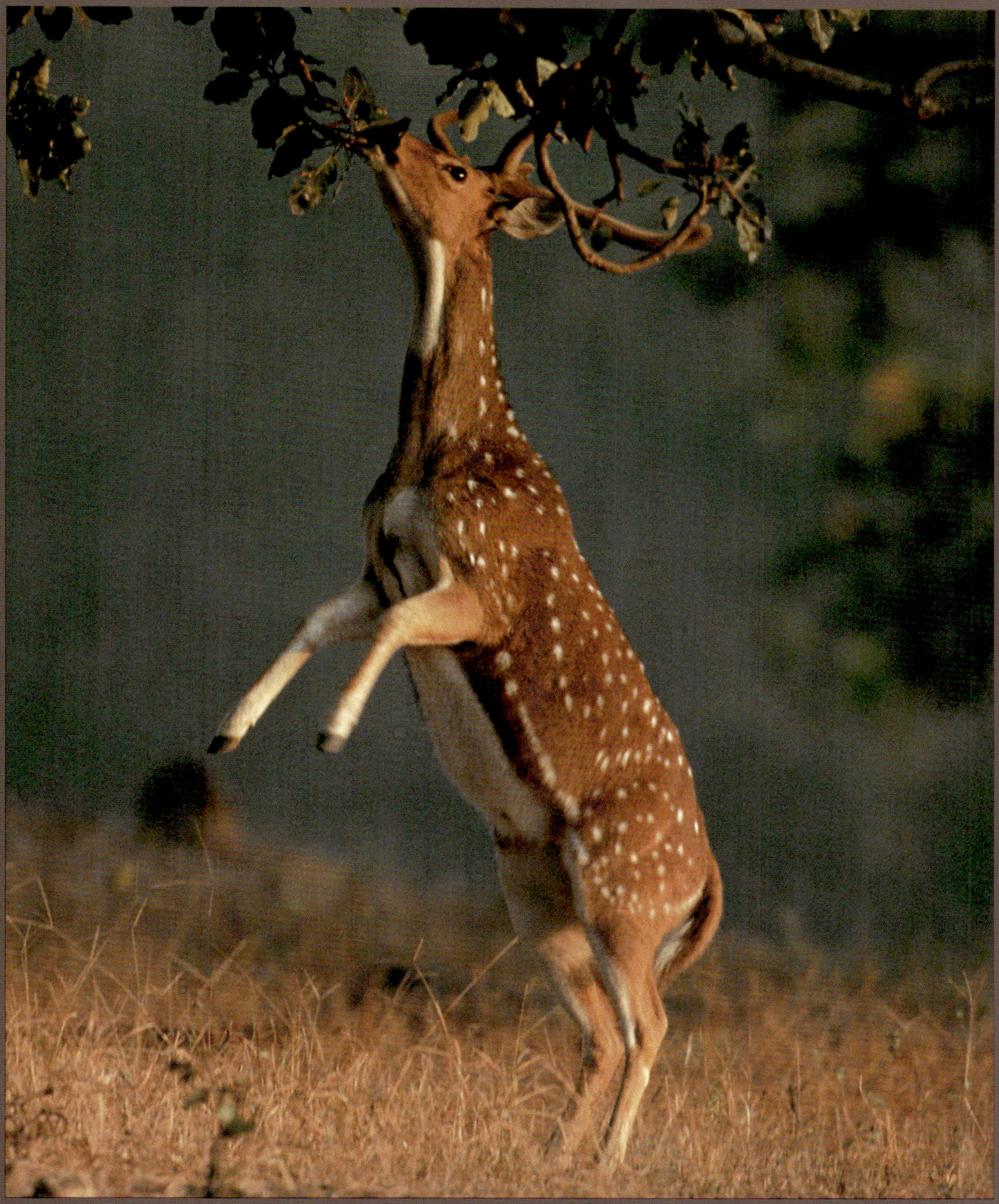

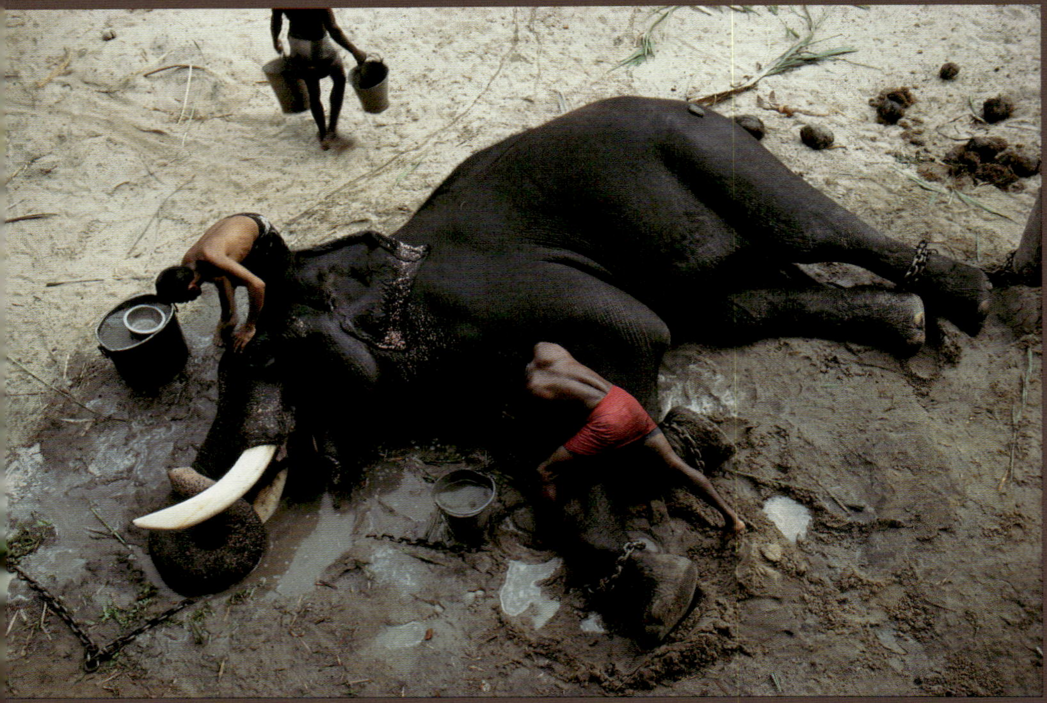

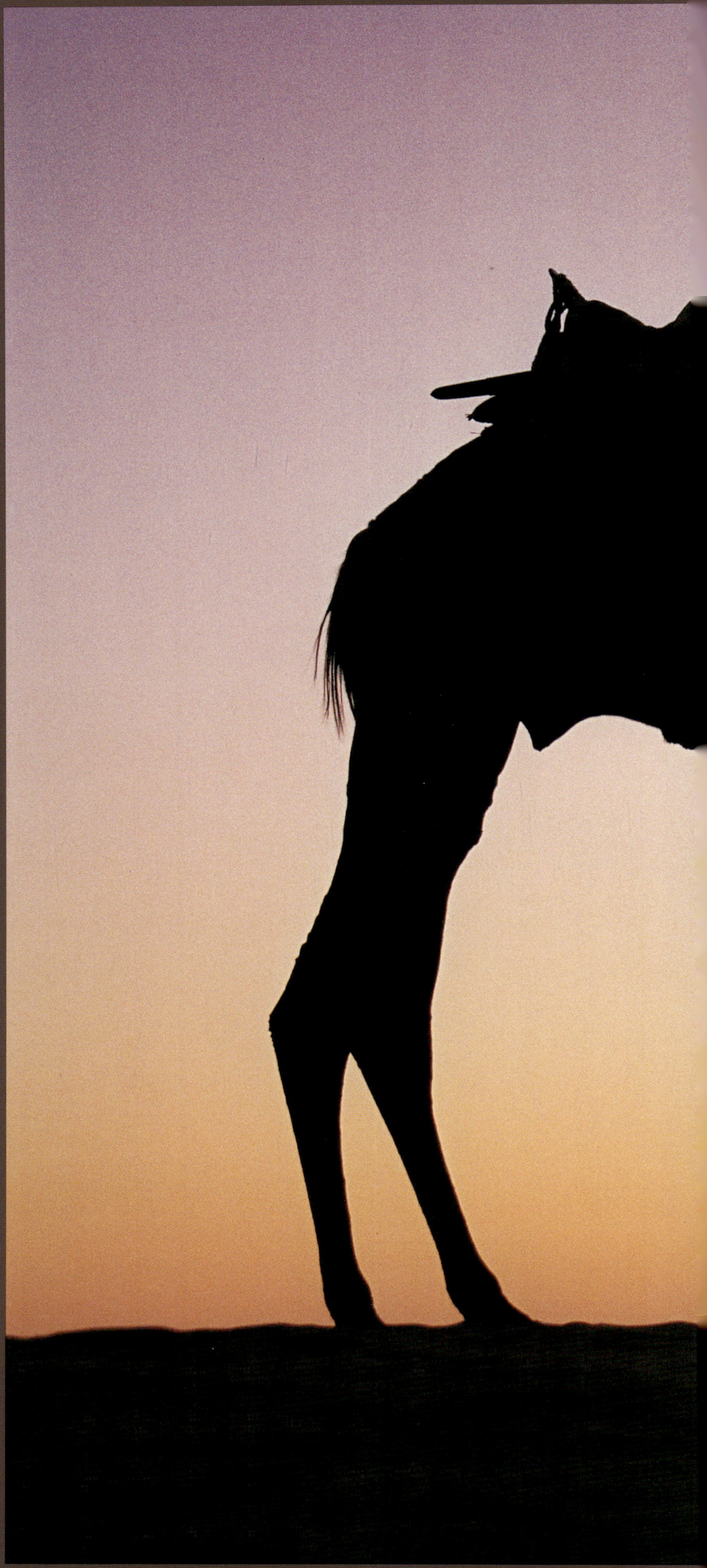

People and animals have long coexisted in India, all the more surprising given the subcontinent's tremendously high human population density. Above, the Asian elephant has been domesticated for centuries in India and is still used for ceremonial purposes. Right, the camel has been critical for human survival in arid parts of the country.

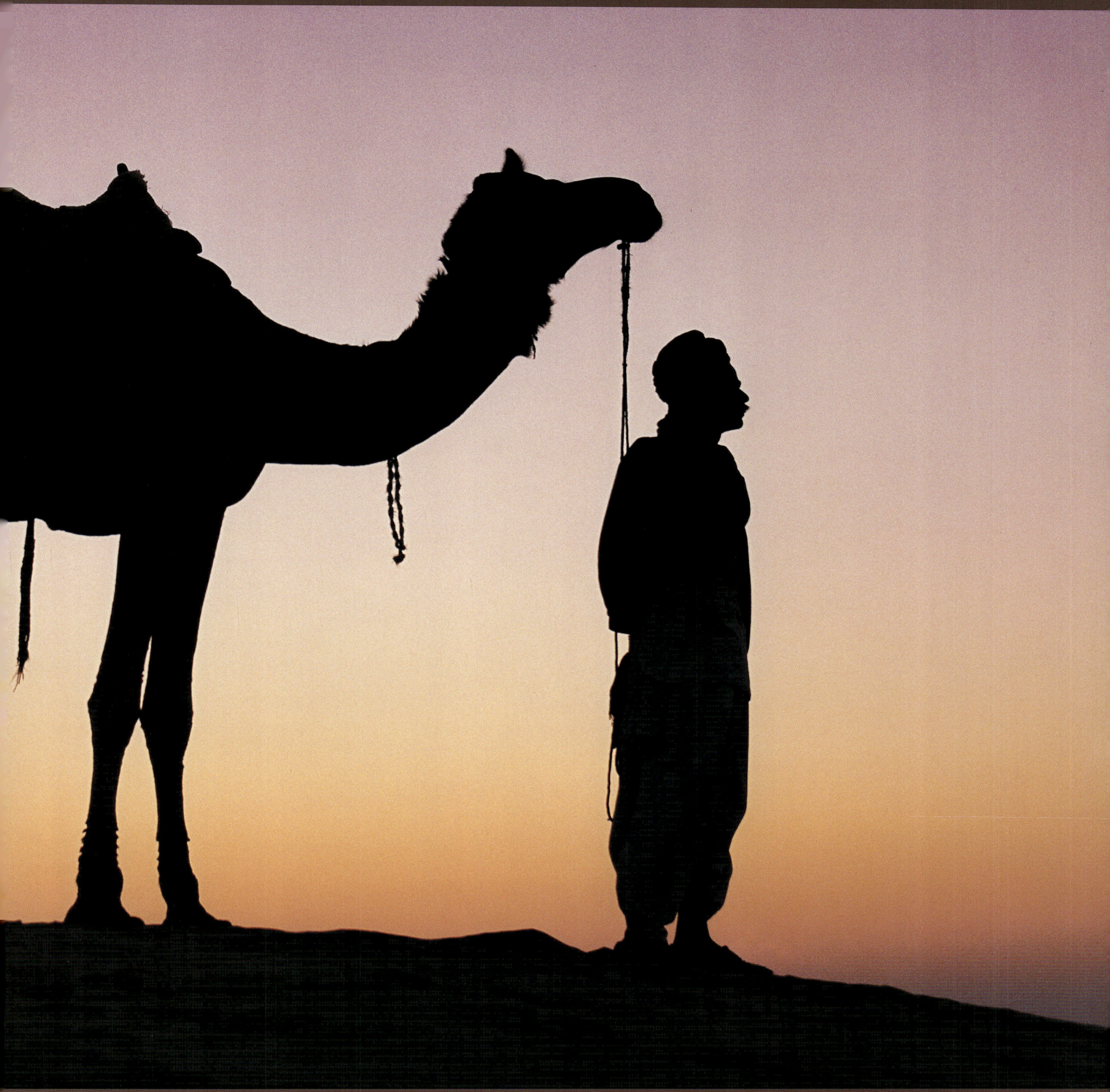

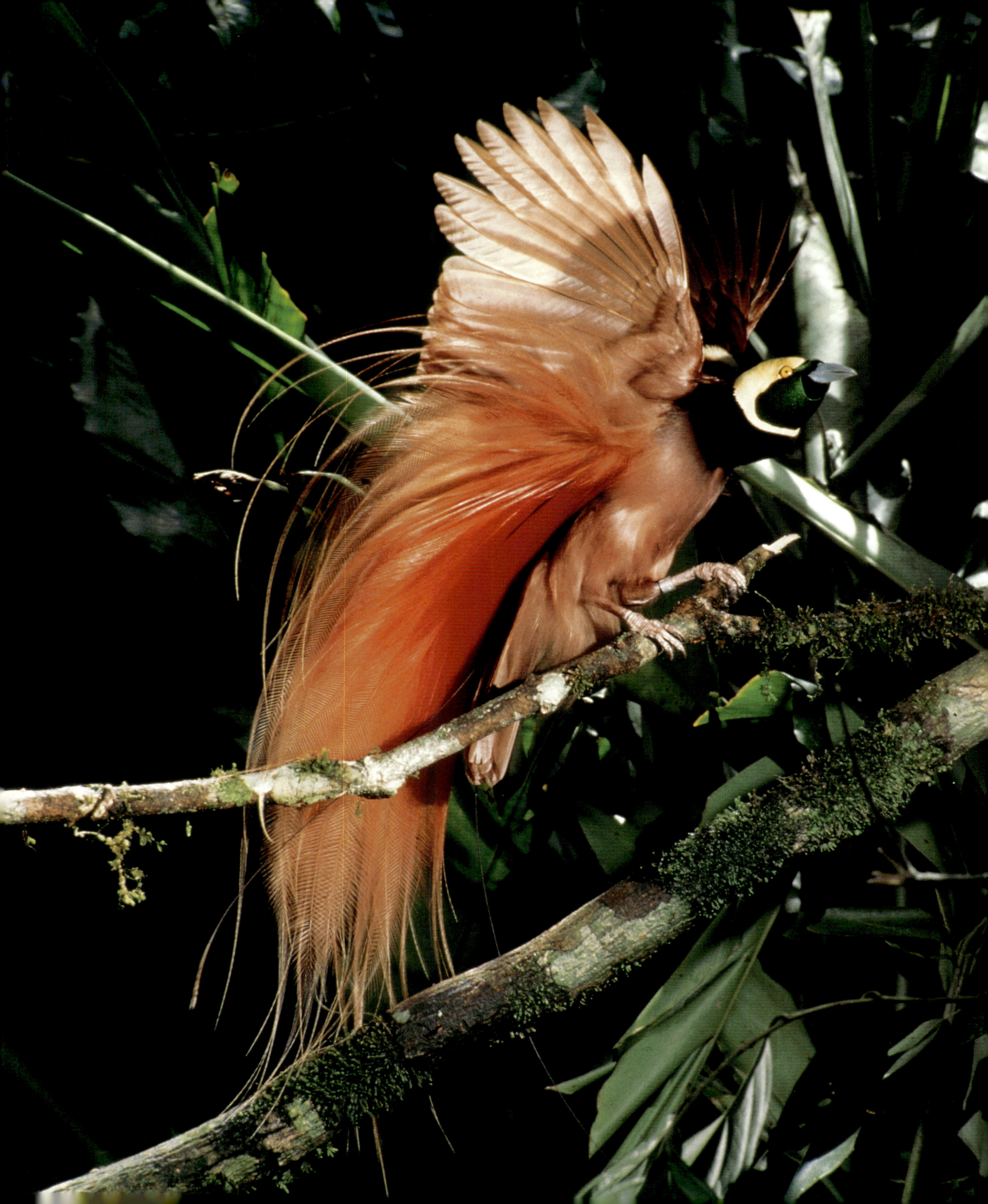

PAPUA NEW GUINEA

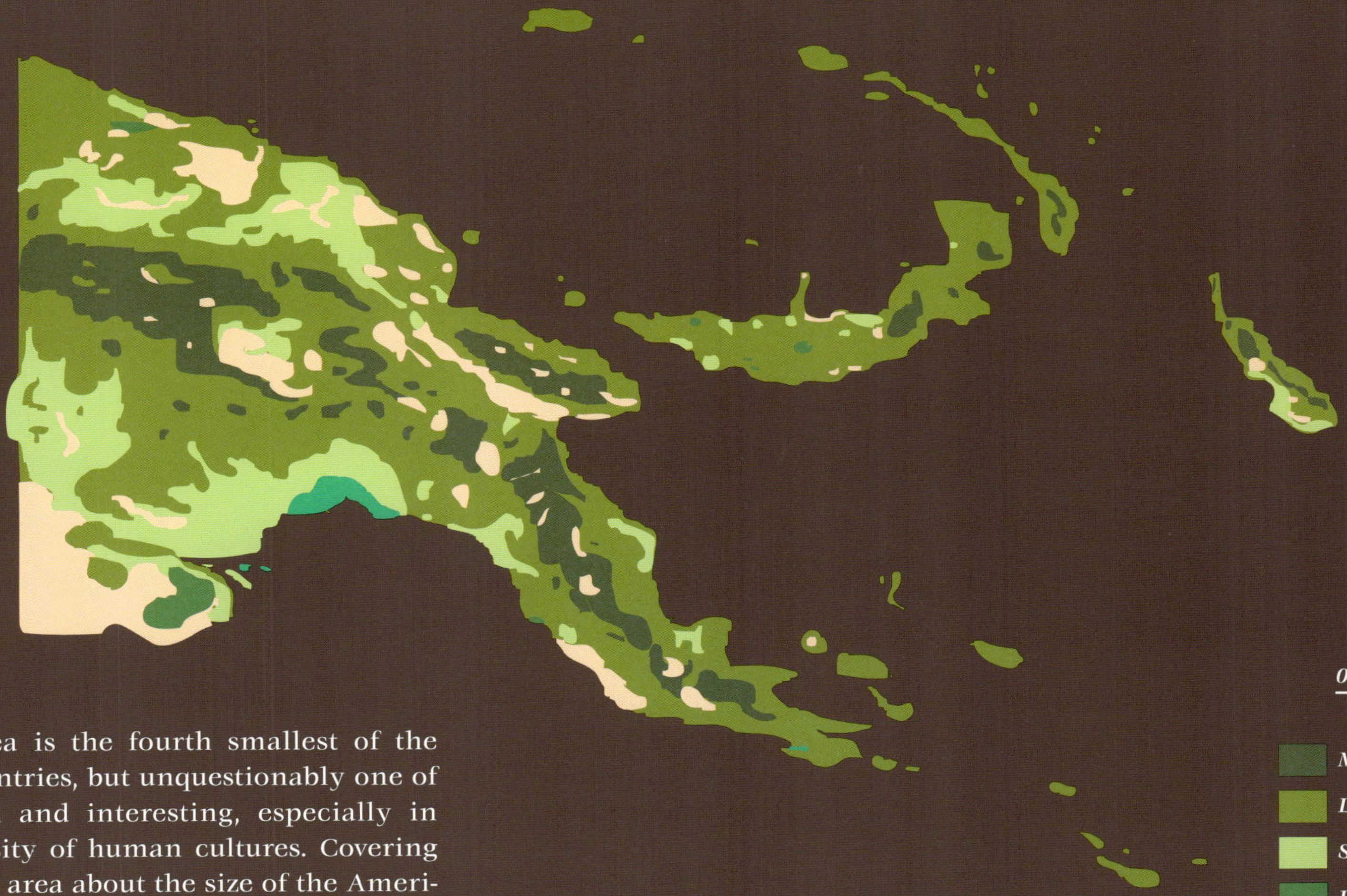

Papua New Guinea is the fourth smallest of the megadiversity countries, but unquestionably one of the most unusual and interesting, especially in terms of its diversity of human cultures. Covering 475 369 km^2, or an area about the size of the American state of California, it is composed of the eastern half of the island of New Guinea (the largest tropical island), the two northernmost islands of the Solomon chain, Bougainville and Buka, and the islands of the Bismarck and Admiralty archipelagoes, including New Britain, New Ireland, and Manus, as well as many other smaller islands. It is actually a late addition to the megadiversity list and was not recognized as a qualifier until the preparation of this book; nonetheless, it clearly belongs in this elite group.

Papua New Guinea's plant diversity is especially striking, with an estimated 15 000 to 21 000 higher plants, including at least 2 000 pteridophytes (ferns and allies) and more than 3 000 species of orchids. This represents an enormous concentration of plant diversity per unit area which can only be compared to that of Indonesia and some of the Andean countries, with their phenomenal plant richness. Plant endemism is high as well, with estimates running from 10 500 to 16 000.

Vertebrate diversity and endemism are comparably impressive. Of the five smallest megadiversity countries (Madagascar, Papua New Guinea, Malaysia, the Philippines, and Ecuador), Papua New Guinea's mammal diversity of 242 species is third behind Malaysia and Ecuador, while mammal endemism, at 57 species, is also third, but this time behind the Philippines and Madagascar.

Like in that other megadiversity island nation, Madagascar, mammal representation in Papua New Guinea is spotty, being very rich in some orders and very poor in others. Certain mammal groups are largely or entirely lacking in Papua New Guinea (*e.g.*, there are no native primates, squirrels, cats, viverrids, or ungulates). In fact, the only native placental mammals are murine rodents and bats (although there are introduced species of deer, pig, dog and cat). On the other hand, marsupial diversity is second only to Australia, and there are 92 bat species of which 32 are endemic (F. Bonaccorso, pers. comm). Among the small megadiversity countries, bird diversity again is second only to that of Ecuador, with 762 species (644 are breeding species), and endemism is moderately high at 85 species (Beehler, 1986). Reptile diversity is quite high at almost 305 species, including at least 93 snakes, 190 lizards, 14 turtles and two crocodiles. Amphibians are also high at 200 species, with 134 endemic (Allison, 1993). In terms of fish, the entire island of New Guinea has 329 freshwater species (excluding the Bismarck archi-

On the opposite page, the Raggiana bird of paradise (Paradisaea raggiana) is one of the most spectacular members of a group of birds reknowned for their magnificent plumage. It is also the national symbol of Papua New Guinea, and appears on the national flag and crest. Thirty-nine of the 43 species of birds of paradise are found on the island of New Guinea.

pelago and eastern islands but including Irian Jaya), of which 149 are endemic; 282 of these are found in the country of Papua New Guinea.

Invertebrate diversity is also quite impressive, with estimates ranging from 200 000 to 400 000 species for the entire island of New Guinea (including Irian Jaya). Data is very uneven for the different insect groups, but butterfly data indicates that Papua New Guinea has 800 species, of which 200 are endemic. In the well-documented tiger beetles, Papua New Guinea is the world's seventh richest in diversity with 111 species, and ninth in number of endemics, with 45.

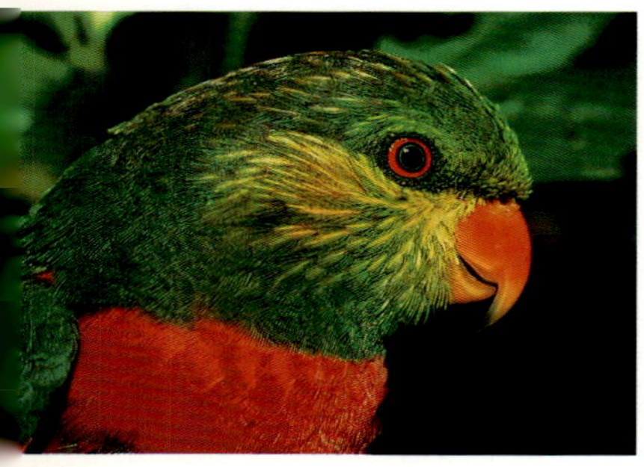

In marine diversity, Papua New Guinea is even more outstanding than in the terrestrial realm. Approximately 3 000 marine fish species are thought to occur in New Guinea waters, with roughly 2 000 inhabiting inshore reefs and associated habitats. Coral diversity is estimated at 300 species, but the total may exceed 700, making Papua New Guinea one of the most coral-rich countries in the world. Indeed, in terms of overall marine diversity, Papua New Guinea may rank in the top three countries globally, along with Indonesia (which is almost certainly first, in good part because it includes half of the island of New Guinea) and the Philippines. An example of Papua New Guinea's marine diversity is the nudibranch group. With 580 species recorded within 30 km of Madang, this is by any measure the richest site on Earth for this group (Gosliner, pers. comm.). It is also very poorly documented and merits much further study. Papua New Guinea has the richest and most diverse assemblage of tropical marine ecosystems of any area of comparable size on Earth.

In addition, Papua New Guinea is part of the three major tropical wilderness areas on Earth, since so much of it is still intact. It also has a very large number of Centers of Plant Diversity and Endemism as defined by WWF/IUCN, its total of 39 comparing to that of much larger countries like the United States and China. Furthermore, it has twelve of the globally important Endemic Bird Areas recognized by BirdLife International (ICBP): the North New Guinea Mountains, the North New Guinea Lowlands, the Adelbert and Huon Mountains, the Central New Guinea High Mountains, the Central New Guinea Mid Mountains, the Trans-Fly and Upper Fly, the Admiralty Islands, Saint Matthias Islands, New Britain and New Ireland, D'Entrecasteaux and Solomon Sea Islands, Louisiade Archipelago, and the Solomon Islands (note that the Solomons is an independent country, but the two northernmost Solomon Islands are politically part of Papua New Guinea).

Parrots and their relatives are very diverse in New Guinea, with 46 species occurring on the island as a whole. The orange-billed lorikeet (Neopsittacus pullicauda) *is found in mountainous areas over much of the island.*

Human cultural diversity in Papua New Guinea is simply phenomenal. The latest estimate is that there are some 875 different languages in the country, and most of the cultures speaking them are still intact, unlike so many other parts of the world where most tribal cultures are in rapid decline or effectively extinct. In terms of human cultures per unit area, Papua New Guinea is far and away the richest among the megadiversity countries, and only India and perhaps Indonesia can compare in terms of overall cultural diversity, the latter because it includes the western half of the island of New Guinea. Although many of these languages will persist into the next century, the orally-based cultures, belief systems, and knowledge of natural history allied to these languages are all under threat from the pervasive impacts of Western material culture. In fact, this diversity of cultures is more gravely threatened than are its rarest plants and animals.

The combination of Papua New Guinea's very high marine diversity, its impressive plant and animal diversity and endemism, its unmatched cultural diversity, and the intact condition of a large percentage of its natural ecosystems and human cultures makes this young nation one of the most interesting and important on Earth, even among the very special megadiversity countries. The international community has begun to take notice of Papua New Guinea because of its mineral and renewable natural resource wealth (mainly forests), but it should also be held in great esteem because of its very special megadiversity status.

Natural habitats

The diverse natural terrestrial habitats of Papua New Guinea can be grouped into five main categories, based mainly on vegetation and altitude, a breakdown that is adapted from Paijmans (1976), Johns (1982), and Flannery (1995). These categories are: lowland alluvial plain forest, foothill forest, lower montane forest, upper montane forest, and grassland/herbfield/woodland.

Lowland alluvial plain forest is Papua New Guinea's defining natural ecosystem —tall and floristically very rich, and botanically "Malesian"— combining Asian and Melanesian groups. Sample one-hectare plots in this habitat have been found to support from 100 to nearly 300 tree species. The canopy is usually 30-35 m with emergents sometimes reaching 50 m or more, and tree diameter can exceed 2.5 m. Palms are a common element in the shrub layer, epiphytes are plentiful, the low herb layer is patchy and there is very little leaf litter on the forest floor.

Foothill forest up to 1 000 meters is less imposing than that of the alluvial plains because of the steep slopes and shallow, unstable soils. The canopy is lower and less variable in height, closure, and crown size, and the shrub layer is reduced. The herb layer is generally denser than in the lowland forest. Stately *Araucaria* pines are emergents that can reach 50 m and which in some places form majestic stands. Oaks (*Lithocarpus* and *Cantanopsis* spp.) become dominant in the upper levels. Cloud cover may be more important than temperature in influencing changes with elevation. Some species (certain marsupials) appear to be largely restricted to this zone.

Lower montane forest occurs from 1 000-3 000 m and has a canopy of 20-30 m, a small average leaf size, and higher tree density than at lower elevations. Large accumulations of leaf litter and fallen logs are also much more evident than at lower altitudes and, at the high end of this forest type, ferns are abundant and mosses can almost completely cover the tree trunks. Tree ferns and epiphytes are common as well. Oaks predominate at lower portions of this zone, but higher up *Nothophagus*, conifers, Podocarpaceae, Myrtaceae and Elaeocarpaceae are much more common. *Pandanus* and scrambling *Nastus* bamboo form impenetrable thickets in places. A great diversity of wildlife occurs in this zone, much of it found only there. Mammals, for instance, reach their greatest diversity in lower montane forest, with the majority of New Guinea's mammal species found in this formation.

Upper montane forest occurs from 3000 m to the tree line. Canopy height varies from 6-8 m, tree ferns are common in the shrub layer, but *Pandanus* and bamboo are absent. Moss is a predominant feature covering trees, forest floor and the leaf litter, which is deep in this zone. At lower altitudes, this area is still relatively rich in species, with a few being restricted to this formation. At the higher altitudes, single tree species (*e.g.*, the conifer *Dacrycarpus compactus*) can dominate the canopy. Where this forest meets the alpine grasslands, one encounters a patchwork of closed forest and open land, often with small lakes in the grasslands —creating a beautiful environment that is called "tropic-alpine." Frost appears commonly at elevations above 3 700 m.

A variety of different types of swamp forest exist as well, especially in the Western Province and also on New Britain, southern New Ireland and Bougainville. Some of the most impressive of these are sago swamp woodland, with large stands of the edible sago palm (*Metroxylon sagu*), a staple for many of New Guinea's lowland people; pandanus swamp woodland, dominated by stands of *Pandanus*; *Nipa* palm stands; and mixed swamp forest including both sago and pandanus.

The grassland/herbfield/woodland category includes all the terrestrial non-forest environments of the country, and encompasses a number of different structural forms. One important type is the mixed savannah of monsoonal southern Papua New Guinea, in the Port Moresby region, a small patch near Raba Raba in the far southeast, and the great Trans-Fly River plains, which includes species that are found in no other New Guinean habitat but instead are shared with northern Australia (many as migrants). *Eucalyptus* and *Melaleuca* tree species are common over most of this area, fires occur often in the dry season, but much of the land is flooded in the wet season.

Man-induced grasslands, of course, occur in many areas where burning is chronic, especially where there is a long dry season. These areas are most extensive in the rain-shadow mid-montane valleys of the central and western highlands, where traditional cultures have been clearing forest for millennia. These mid-montane grasslands occur in valley bottoms mainly from 1 000 to as high as 3 000 m.

Alpine grasslands, which occur primarily above 3 500 m, also extend down to 2 800 m in frost pockets and areas with poor drainage. This tussock-grass habitat has also been burned in many sites by hunters in search of game. The tree line in areas without human disturbance is at around 3 900 m. The shrublands that abut these grasslands are rich in rhododendrons and other Ericaceae. This alpine grassland suffers extremes in weather, with frost at dawn, periods of intense solar drying when clouds are absent, and mist and rain.

Minor vegetation types include coastal vegetation, with herbaceous beach vegetation, beach scrub, beach woodland, *Casuarina* forest, and mixed littoral forest. New Guinea also has extensive mangrove communities in the tidal and brackish waters of estuaries of the larger rivers, particularly the deltas of the Purari and Fly Rivers, the Murik Lakes near the mouth of the Sepik River, and areas around the mouth of the Ramu River.

Mention should also be made of the freshwater habitats of Papua New Guinea, including the Fly River system and its huge tributary, the Strickland, in the southwestern part of the country, and the Sepik and Ramu River drainages to the north. The Fly and the Sepik are major rivers by international standards, and both have endemic faunas that are just beginning to be characterized. The largest lake in the country, Lake Kutubu, is found in the rugged hills on the fringe of the central highlands, and represents a remarkably beautiful natural feature that now lies adjacent to the base for a major oil development scheme.

Flagship species

As with the lemurs in Madagascar, one group of flagship species stands head and shoulders above the rest in Papua New Guinea —the birds of paradise (family Paradisaeidae). A group of crow-like birds in which sexual selection for spectacular male plumage has gone wild, these magnificent creatures are renowned for their nuptial plumages and elaborate courtship displays. Plumage can vary from entirely black to red, orange and iridescent green, to muted shades of brown. The many polygamous species are sexually dimorphic, with elaborately plumed males and dull and cryptic females. Males display on special courts or perches, where they attract and mate with females; in these species, only the latter construct the nest and tend the offspring. The monogamous species, on the other hand, are sexually monomorphic and more dull in coloration, and males and females share nest duties. In all, there are 43 species of birds of paradise. The 39 occurring in New Guinea can be divided into seven groups: the cnemophilines with three species, the manucodes with five species, the paradigallas with two species, the sicklebills with four species, the paradiseas with seven species, the unique McGregor's bird of paradise, and the typical birds of paradise with seventeen species. These creatures range from eastern Australia through the island of New Guinea, west to the northern Moluc-

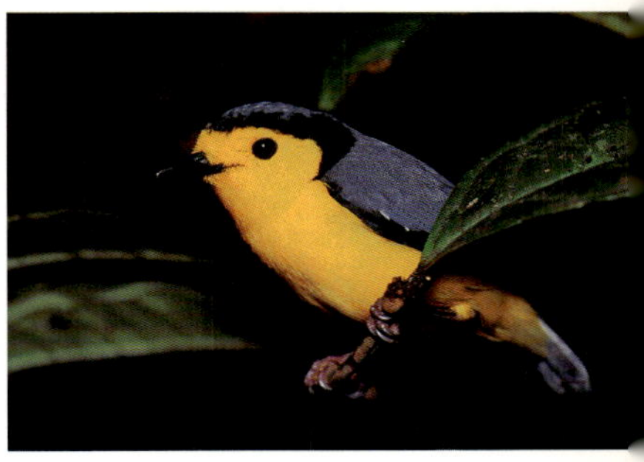

Bird diversity in Papua New Guinea is very high given the size of the country and includes 762 species of which 85 are endemic. This dwarf whistler (Pachycare flavogrisea) *is a member of the family Pachycephalidae, a group that has its center of distribution in New Guinea.*

On pp. 366-367, limestone pinnacles in the Mount Kajande area near Tari in the Southern Highlands. Much of the interior of Papua New Guinea, especially the mountains, is still covered with intact forest wilderness.

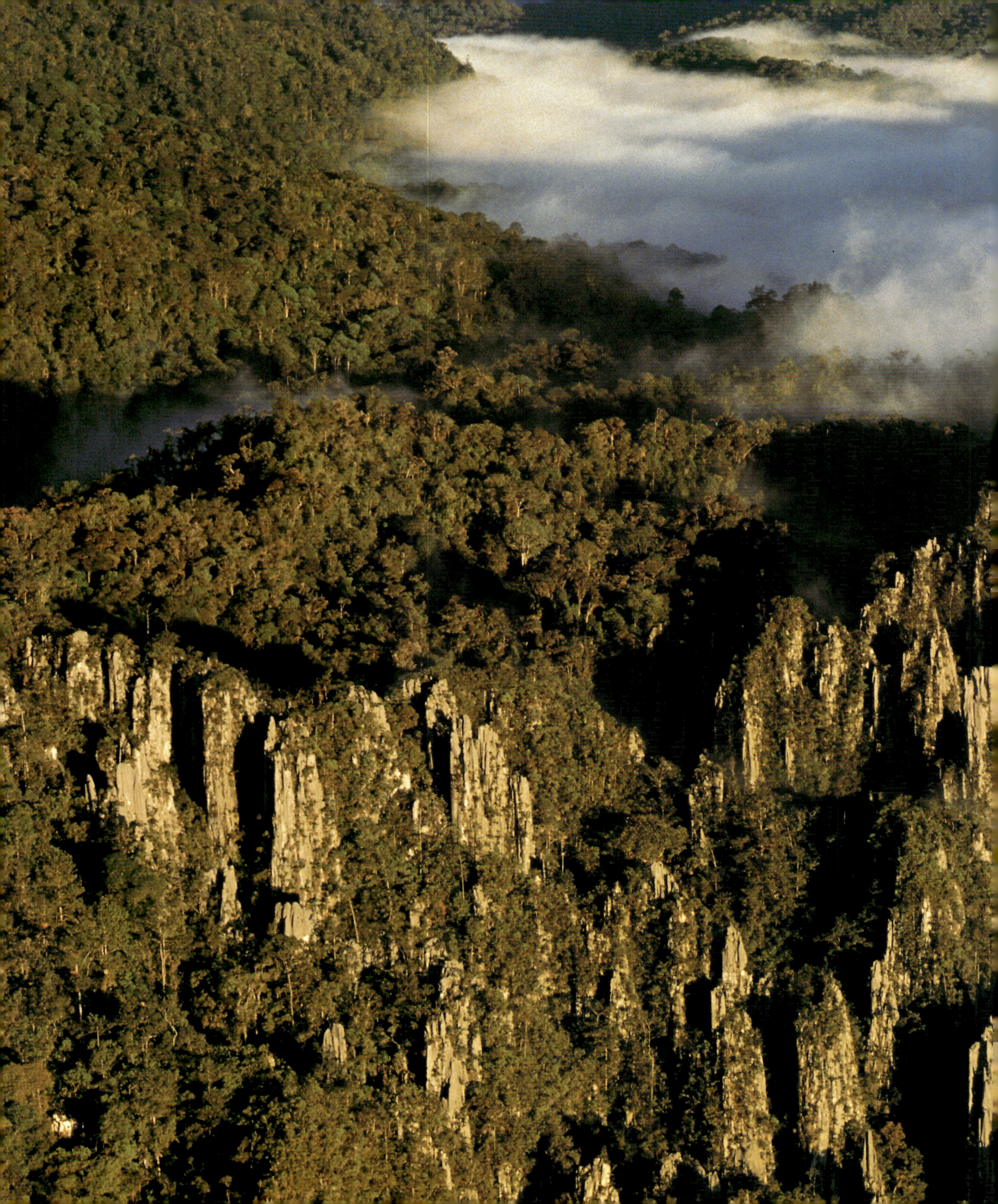

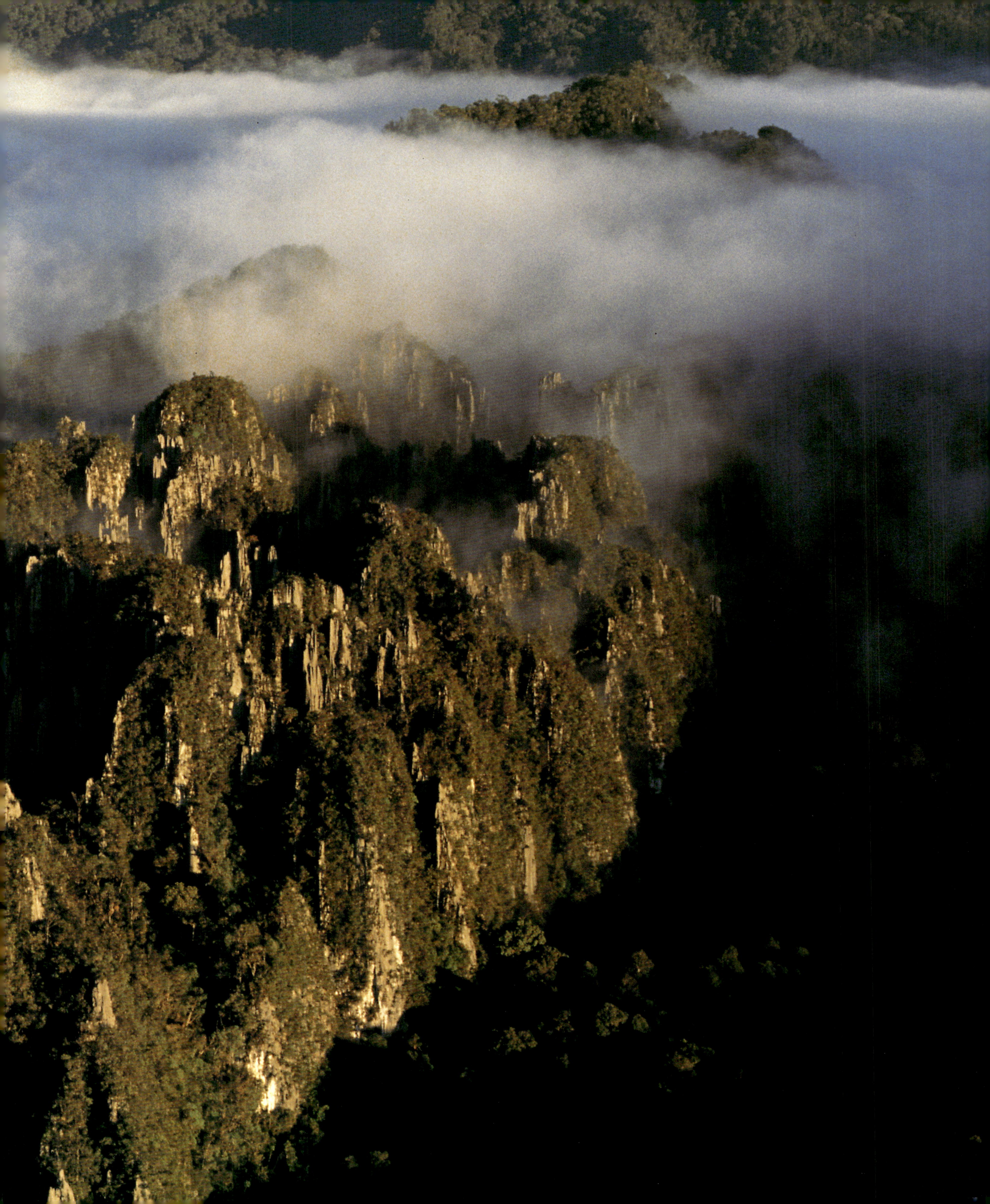

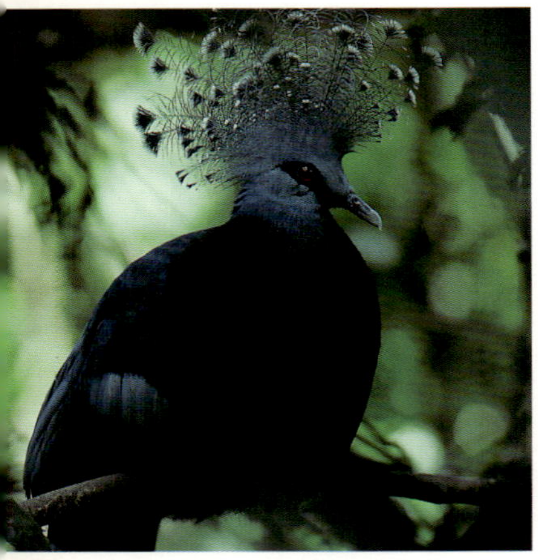

Pigeons and doves are comparable in diversity to the parrots, with 45 species on the island of New Guinea as a whole. These include the world's largest, the crowned pigeons, of which the stunning Victoria crowned pigeon (Goura victoria) *is the best known.*

The cassowaries are large, flightless, forest-dwelling birds found throughout most of New Guinea. There are three species in all, two of which are endemic to New Guinea. The third, the southern cassowary (Casuarius casuarius) *seen on the opposite page, also occurs in the forests of Queensland, Australia.*

cas in Indonesia, but are absent from the larger Papua New Guinea islands of New Britain, New Ireland, and Bougainville; they also do not occur on the other Solomon islands.

Birds of paradise have long caught the attention of people within New Guinea and the rest of the world as well. One species, the stunning Raggiana bird of paradise (*Paradisaea raggiana*), is the national symbol of Papua New Guinea. It appears both on the national flag and the national crest, and is the logo of the national airline, Air Niugini, and the country's best known beer, South Pacific. These birds are also of great significance to the human cultures of Papua New Guinea, and especially to the people of the Highlands. There, the men of many tribes feel a close bond to the birds, adorn themselves in their spectacular plumage, and mimic the birds' nuptial dances in their own dances, the famous *sing-sings*. Among the Huli wigmen of the Tari Valley in the Southern Highlands Province, for instance, men decorate elaborate wigs made of their own hair with feathers of many different species, including the ribbon-tailed astrapia (*Astrapia mayeri*), Stephanie's astrapia (*Astrapia stephaniae*), the King of Saxony bird of paradise (*Pteridophora alberti*), the Superb bird of paradise (*Lophorina superba*), the black sicklebill (*Epimachus fastuosus*), the brown sicklebill (*Epimachus meyeri*), and Lawes' parotia (*Parotia lawesii*). Although such use of these birds is widespread, it has not resulted in the loss of any species, and many of those most in demand can still be found living in the immediate vicinity of villages (*e.g.*, Superb bird of paradise and Lawes' parotia in the Tari Valley). Since only fully-plumed males are harvested (which represent only a tiny segment of the population), even heavy culling permits all females and most males to exist in peace. Fortunately, all of these birds of paradise are protected from international trade, being given the highest category of protected status in the country itself and being covered internationally by the CITES convention as well.

Birds of paradise have long held Western naturalists enthralled since they became known through tradeskins that made their way into the royal households in Europe and Asia. Four centuries ago, Jan Huyghen van Linschoten, after his voyage to the East Indies wrote:

> In these ilands onlie is found the bird, which the Portigales call passaros de Sol, that is Fowle of the Sunne, the Italians call it Manu codiatas, and the Latinists, Paradiseas, and by us called Paradise-birdes, for ye beauty of their feathers which passe al other birds. These birds are never alive, but being dead they fall on the Iland: they flie as it is said alwaies in the Sunne, and keepe themselves continually in the ayre, without lighting on the earth, for they have neither feet nor wings, but onlie head and body and the most part tayle...

Rene Lesson, the first foreigner to see these birds in the wild, was also amazed by them, stating that:

> The sight of this first bird of paradise was overwhelming. The gun remained idle in my hand for I was too astonished to shoot. It was in the virgin forest surrounding the harbor of Dorey. As I slipped carefully along the wild pigs' trails through these dusky thickets, a Paradisaea suddenly flew in graceful curves over my head. It was like a meteor whose body, cutting through the air, leaves a long trail of light.

He was referring to the lesser bird of paradise (*Paradisaea minor*), a close relative of the national symbol and comparably stunning.

Unfortunately, the outside world's fascination with the birds of paradise had its negative consequences as well, and the flintlocks of later visitors did not freeze as did those of Lesson's party on their first encounter. Indeed, bird of paradise skins were the main product sought by specialist traders visiting New Guinea as far back as 2 000 years ago and remained an important product until AD 250, when they were replaced by other products in the spice trade. Later, they again entered the trade, and by the nineteenth century, large numbers of skins were reaching Europe. Then European milliners began to use them with the feathers of other birds to decorate hats. By the turn of the century, the trade had already reached "gold rush" proportions (approximately 80 000/year), and many people began to fear that these birds would be exterminated. The trade continued for a couple of decades, but increasing concern eventually led to protective measures. Legislation was passed in Australia, Europe and the United States, as well as in the Dutch, German, and Australian portions of New Guinea itself. In the period between 1908 and 1931 different parts of New Guinea and other countries imposed controls, and the trade with the West finally ended. However, some trade continues for the Asian market to this day and, of course, there continues to be a significant trade within Papua New Guinea for feathers used by the country's many different human cultures. On the positive side, there is a growing interest in specialty ecotourism aimed specifically at the birds of paradise; today, many bird-watching groups visit the country, with birds of paradise at the top of their priority list.

Among the other spectacular birds of Papua New Guinea are the cassowaries, huge flightless birds that stand 1.3 -1.7 m tall and weigh up to 60 kg. They are forest-dwelling fruit-eaters greatly valued by the rural populace of Papua New Guinea, both for their meat and their feathers, which are widely used in masks, headdresses and other ornamentation. They are often kept in captivity or allowed to range in semicaptive conditions around villages. The cassowary is also one of the few birds in the world that can be dangerous, since it can deliver a powerful, potentially lethal kick when cornered. Three species are known, the southern cassowary (*Casuarius casuarius*), the northern cassowary (*C. unappendiculatus*), and the dwarf cassowary (*C. ben-*

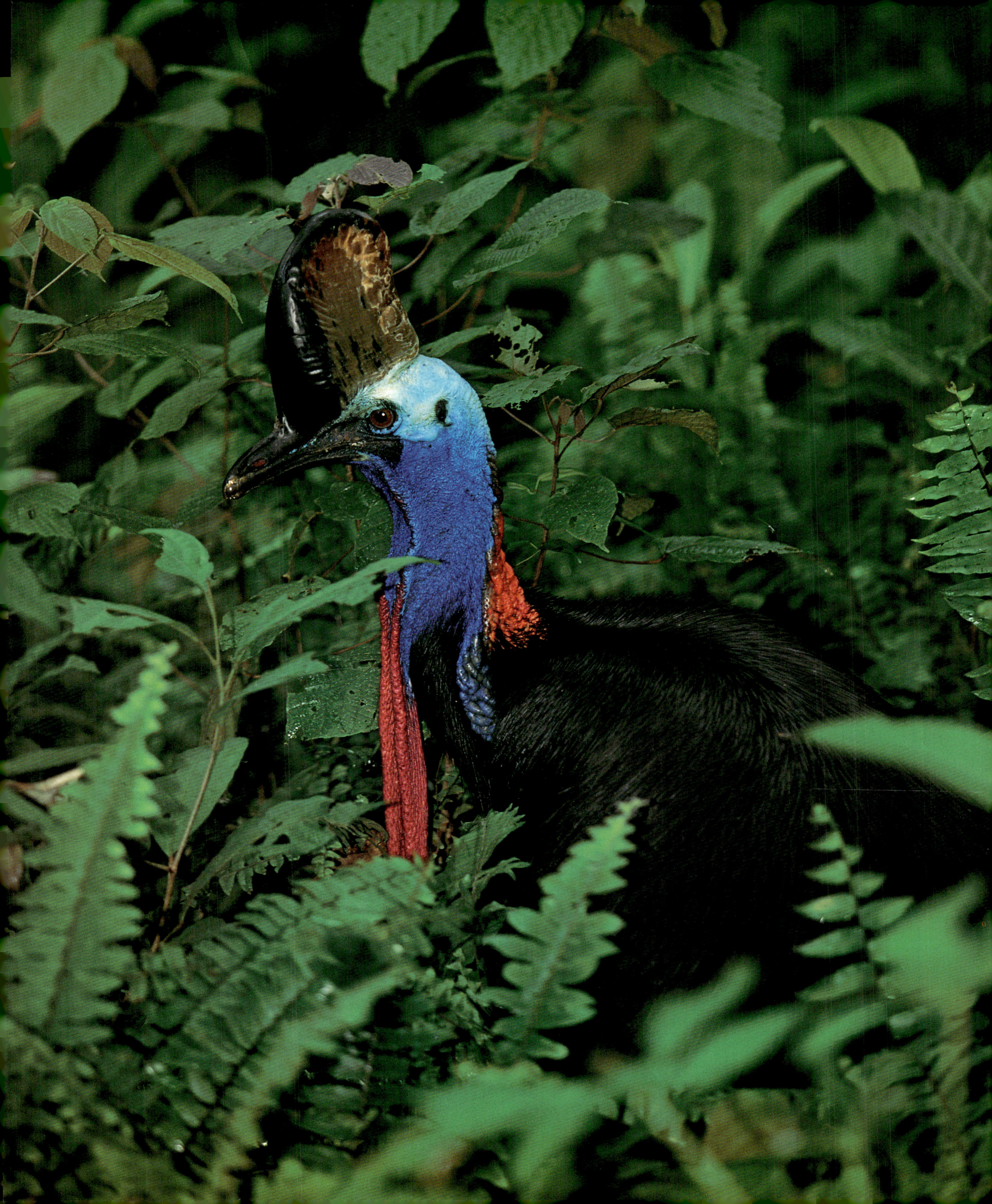

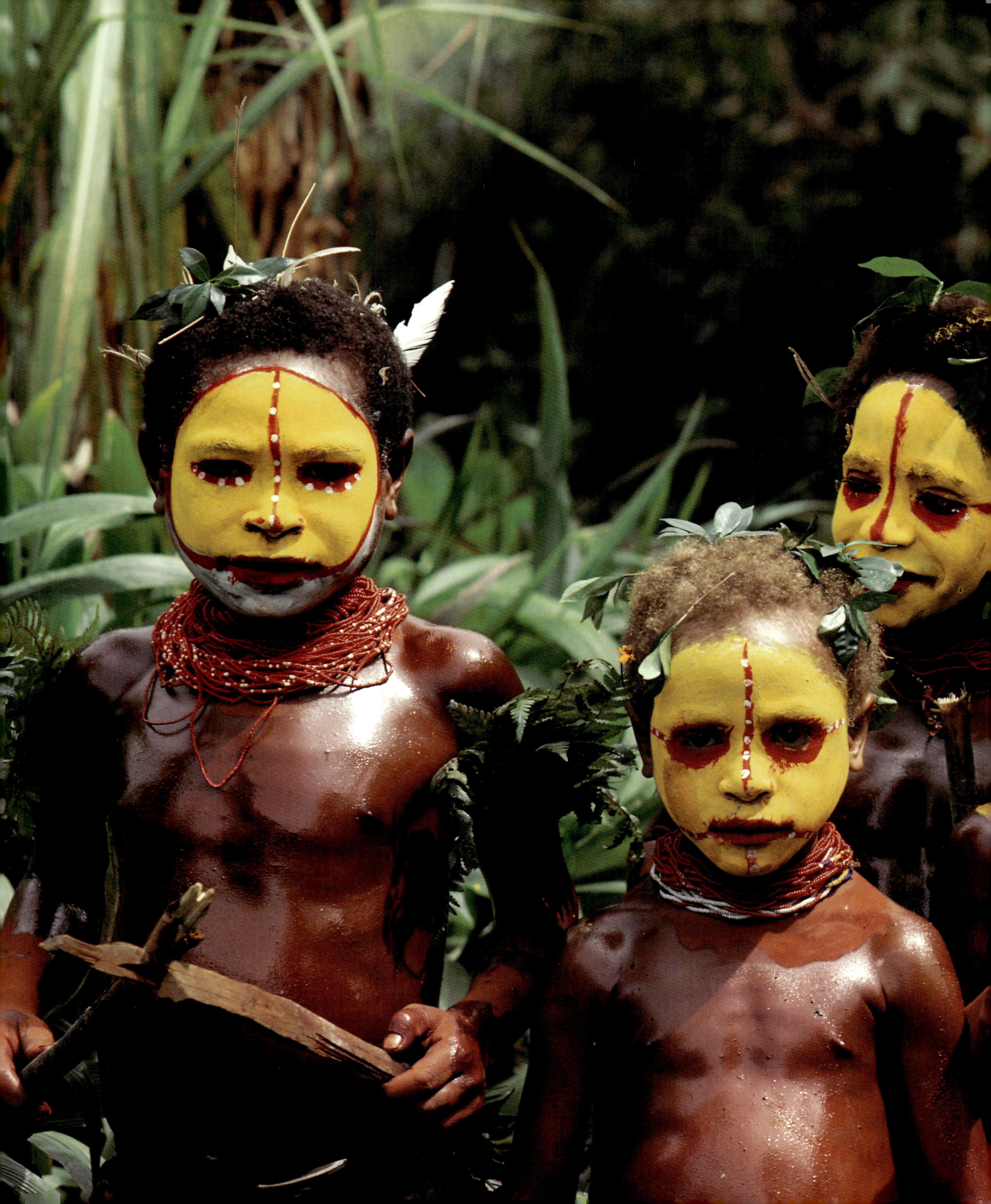

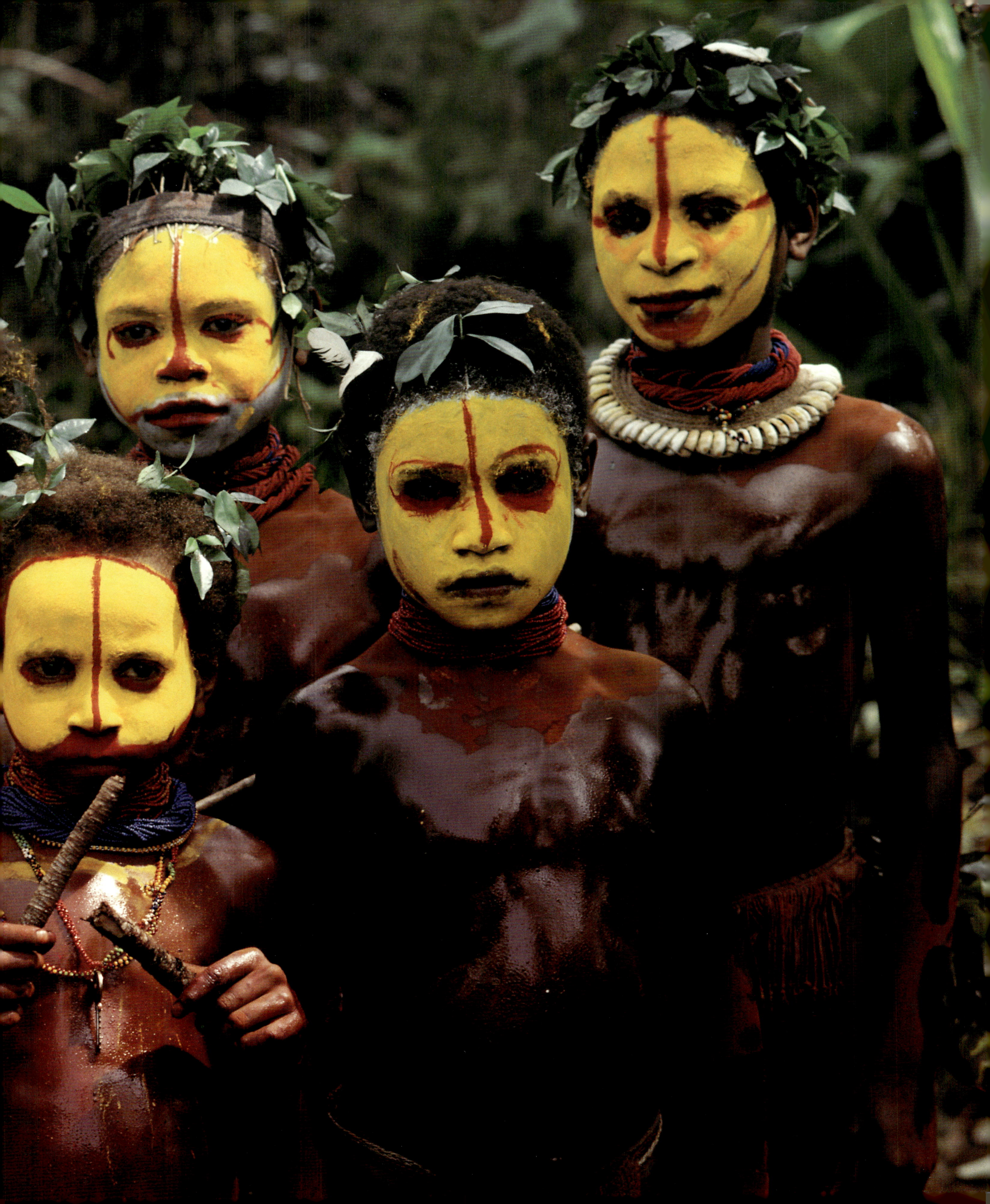

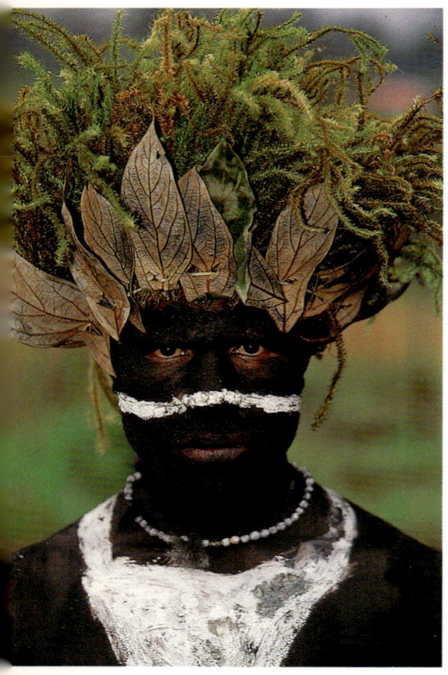

Human cultural diversity in Papua New Guinea is the highest on Earth, with no less than 875 languages still spoken.

On pp. 370-371, these Huli boys are members of one of the Southern Highlands tribes, the Huli wigmen, who are famous for their elaborate headdresses and colorful face painting. This photo was taken in the Tagali Valley.

Above, the people of Papua New Guinea use many different natural products to adorn themselves. This man from the Highlands wears a wig made of moss and leaves in preparation for a sing-sing, or traditional dance.

netti). The southern cassowary also occurs in northeastern Australia, whereas the other two are endemic to New Guinea (although cassowaries have been introduced and now range wild in Seram and New Britain).

The psittacine birds, represented by the parrots, lories and cockatoos in Australasia, are another important flagship group in Papua New Guinea. The island of New Guinea has 46 species, including lories, fig parrots, pygmy parrots, cockatoos, hanging parrots, typical parrots, and the strange vulturine parrot (*Psittrichas fulgidus*). One of the most spectacular is the huge palm cockatoo (*Probosciger aterrimus*), which can reach 64 cm and has a massive beak used to crack open large, hard seeds and nuts. The eclectus parrot (*Eclectus roratus*), though common, is especially beautiful and noteworthy, with a bright green male and a striking red-and-blue female. Most remarkable of all are the warbler-sized pygmy parrots, the world's smallest psittacines, whose diet remains a scientific mystery —apparently they harvest lichens and liverworts from the surface of tree trunks, a behavior unique in birds.

Pigeons and doves are also very diverse in New Guinea, and reach their greatest diversity there. Forty-five species of a global total of 299 (15%) are found on New Guinea, including the world's largest, the crowned pigeons, among which are the well-known Victoria crowned pigeon (*Goura victoria*) and the largest of them all, the southern crowned pigeon (*G. scheepmakeri*), which can reach 75 cm. The predominantly green fruit doves are especially attractive, some of them approaching the lorikeets and parrots in the spectacular pastel colors that adorn their heads and underparts.

New Guinea is also home to the world's only known poisonous birds, the pitohuis. Though long known to local people in New Guinea, the existence of this phenomenon was not made known to the scientific world until 1992, when a field team led by a Conservation International scientist reported on their discovery that the hooded pitohui (*Pitohui dichrous*) had strongly toxic feathers and skin —apparently used as a defense against predators and/or parasites. Most remarkable is that the toxin found in this bird is identical to that found in the poison-dart frogs of South America.

Among the mammals, the most spectacular flagship species are the tree kangaroos, family Macropodidae. Ten species and seventeen different subspecies of tree kangaroos exist in both the island of New Guinea and Queensland, Australia, of which fifteen are found on the island of New Guinea and two in Australia. Six of the eight New Guinean species are found in Papua New Guinea, five of them endemic (Flannery, 1995). Reaching a weight of up to 18 kg, the largest being Doria's tree kangaroo (*Dendrolagus d. dorianus*), they are among the largest of arboreal mammals and also among the least known. Indeed, they are so poorly known that two new species and two new subspecies have been described just in the past decade by Tim Flannery of the Australian Museum in Sydney and his colleagues; one of the seventeen forms, the Mount Menawa subspecies of Scott's tree kangaroo (*Dendrolagus scottae* subsp. indet.), known locally as the *fiwo*, has not yet been formally described. In other words, about 30% of the tree kangaroos have only become known to science in the past ten years! Another, the *wondiwoi* tree kangaroo (*Dendrolagus dorianus mayri*), discovered by the famous evolutionary biologist Ernst Mayr on the Wondiwoi Peninsula of Irian Jaya in 1928, is still known only from the type specimen. An undescribed form of *D. goodfellowi* has also been reported recently from the Foya Mountains of Irian Jaya.

This intriguing group of marsupials, which occupies at least a part of the vacant arboreal primate niche in New Guinea, is unfortunately also the most endangered group of mammals in the country. All the New Guinean species are heavily hunted for food and for their skins, and have been largely extirpated from areas inhabited by people. Tim Flannery, the world's expert on tree kangaroos, considered two of them to be among the most critically endangered of all mammals, the *tenkile* (*Dendrolagus scottae*), described in 1990, and the golden-mantled tree kangaroo or *weiman* (*Dendrolagus goodfellowi pulcherrimus*), described in 1993. The *tenkile* now occurs in an area of about 40 km^2 lying just east and west of Mount Somoro in the Torricelli Mountains. The one place where it was protected by powerful spirits, a mountain summit called Sweipini, has lost its sacred status. The golden-mantled tree kangaroo, which occurs at the extreme eastern end of the Torricelli Mountains in the vicinity of Sibilanga and possibly Weigin village, may have a remaining range as small as 20 km^2. Several of the other species, though somewhat more widespread, may also be headed in the same direction. Clearly, more attention to the conservation of these animals is urgently needed. Conservation International, in collaboration with several American zoos, has produced a poster depicting all known tree kangaroos in the Tok Pisin and Bahasa Indonesia languages, and has plans to do more work on them in the near future. A classic book on these animals has just been produced by Tim Flannery, and should do a great deal to further interest in them. A special Taxon Advisory Group on Tree Kangaroos has also been established by the American Zoo Association to assist the captive-breeding efforts, but at this time only a handful of the seventeen different kinds are kept in captivity.

Other interesting mammals include the fruit bats: there are 20 species in New Guinea, one of which appears to have the very interesting feature of lactation in males. The greater flying fox (*Pteropus neohibernicus*) may be the largest of all bats, with a wing span of 1.8 m. The egg-laying monotremes, the echidnas, are also especially noteworthy with two species occurring in New Guinea: the larger of the two, the long-beaked echidna (*Zaglossus bruijnii*) can reach 9.8 kg and is endemic. With its spines, long snout, toothless mouth, egg-laying habits and large size, it is certainly one of the strangest of all mammals. Flannery reportedly also has a new species of echidna that he is about to describe.

New Guinea's reptiles include a number of flagship species as well, among which are the beautiful green tree python (*Morelia viridis*), a perfect ecological equivalent to the emerald tree boa (*Corallus caninus*) of South America; the rare and little-known Boelen's python (*Morelia boeleni*), a high-altitude species (1000-2000 m) and the most protected reptile in Papua New Guinea (receiving protection similar to that of birds of paradise); the little-known but very attractive and unusual crocodile skink of the genus *Tribolonotus*, and the longest of all lizards, *Varanus salvadori*. This monitor lizard does not achieve the same massive bulk of the Komodo dragon (*Varanus komodoensis*) from Indonesia, but it does grow to greater length, reaching some 3.5 m. The Fly River turtle (*Carettochelys insculpta*) of the Fly and Strickland drainages of western Papua New Guinea is a large and very unusual water turtle that is found mainly in New Guinea, but also extends into extreme northern Australia.

The saltwater crocodile (*Crocodylus porosus*) and the endemic New Guinea crocodile (*Crocodylus novaguineae*) are also important to the local economy and as cult figures for many of the tribes of the Sepik River region, comparable in significance to the birds of paradise among the Highland cultures. Indeed, crocodiles achieve such a great status and importance among some of the tribes of the Karawari River region (a tributary of the middle Sepik) that they are depicted in almost every aspect of daily life. Carved crocodile heads adorn canoe prows and crocodile figures appear on drums, hooks and a variety of household items. Huge crocodile-spirit wood carvings, called *arkumban* by the people of Ambonwari village on the Konmai River, where such figures are especially elaborate and abundant, live under every house and in the *haus tambarans*, or men's spirit houses. In some tribes, men even tattoo their backs with lines of rounded scars intended to represent the scales on a crocodile's back.

The invertebrates of Papua New Guinea have their flagships as well, exemplified by the magnificent *Ornithoptera* birdwings. The birdwing butterflies are truly striking, with their giant bicolor females and smaller, more colorful males. This group includes the world's largest butterfly, Queen Alexandra's birdwing (*Ornithoptera alexandrae*), and one of the world's rarest, *Ornithoptera victoriae*, from Bougainville. There are dozens of swallowtail butterflies of all sizes, shapes, and colors. The butterfly genus *Delias*, with its remarkable color patterns on the hind wing, exemplify the diversity of Papua New Guinea's montane fauna. The Emperor or Ulysses swallowtail, *Papilio ulysses*, is Papua New Guinea's answer to the blue morphos of Latin America.

Finally, Papua New Guinea's arthropod fauna is highlighted by the world's largest moth (the *Hercules* moth), the largest walkingstick (nearly 30 cm), giant katydids, multicolored weevils of gem-like beauty, and giant bird-eating spiders. Perhaps most bizarre are the *Gymnopholus* weevils that carry a living garden on their backs —perhaps as an anti-predator form of cryptic behavior.

Cultural diversity

The human cultural diversity of Papua New Guinea can only be described as spectacular. With some 875 distinct languages spoken in the country at latest count, total human cultural diversity, both in absolute terms and certainly in per unit area, simply cannot be matched. Only a few countries like India and Indonesia come close, and the latter is especially rich because it includes the western half of the island of New Guinea (Irian Jaya). Not only do these cultures exist in Papua New Guinea, but most of them are still intact, vibrant, functioning entities, not mere remnants of what once existed, which is unfortunately the case in many other culturally rich parts of the world where tribal cultures are in serious decline. This is not to say that these societies have not changed considerably over the last century. One of the major changes relates to religion, with original belief systems largely replaced by some form of Christianity in most areas. This is just one of many adaptations that the people of Papua New Guinea have had to make to accommodate the arrival of Western civilization. The richness, diversity, and wonderful intensity of this country's cultures are best captured by photographs, as we have tried to do on the pages of this chapter, but a few of the cultures are particularly worthy of mention.

The region of Papua New Guinea that perhaps best symbolizes the uniqueness of the country's human cultures is the Highlands. The Highlands of New Guinea were among the last large land areas on Earth to be explored. When German explorers penetrated the Sepik River in the late nineteenth century, they saw high mountains to the south and a German survey team encountered Eastern Highlanders in 1914. Later the explorers Karius and Champion encountered Western Highlanders in 1927, while crossing from the Fly River to the Sepik River. However, it was not until the early 1930s that the gold-prospecting Leahy brothers put the cultures of the large valleys like the Wahgi, the Asaro, and the Tari on the map. They found a succession of wide, well-populated valleys separated by mountain barriers and sometimes with large ceremonial areas that looked like parks. Each valley had entirely different tribes with different costume and dress, and they decorated themselves in the most amazing ways imaginable, with feathers of birds of paradise, parrots and cassowaries among others, as well as dogs' teeth and shells traded from the coast. The languages of these people are also very distinctive, and fall into a loose group usually described as "Papuan" or "non-Austronesian" languages. Not only are these languages mutually unintelligible, they are so distinctive that it is not clear whether they are distant relatives in a single language family or completely distinct language families. Fortunately, in spite of considerable outside influence, including the establishment of tea and coffee plantations and other agricultural crops and the construction of the 720-km Highlands Highway, most of the Highlands tra-

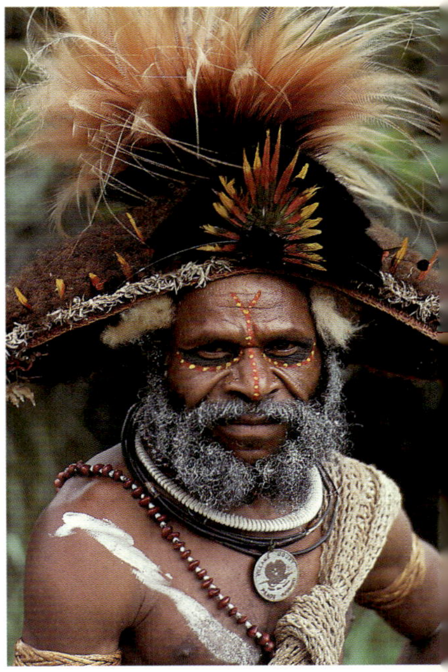

This Huli wigman from the Tari Valley is wearing a wig made of his own hair and decorated with feathers of the Raggiana bird of paradise (Paradisaea raggiana), *the Superb bird of paradise* (Lophorina superba), *and the rainbow lorikeet* (Trichoglossus haematodus). *A close spiritual relationship exists between the New Guinea Highlanders and the birds that they so admire and seek to emulate.*

ditions remain intact and are still practiced by many people. Indeed, the integrity of these Highland cultures (and many other cultures in Melanesia) is especially striking when compared to most other parts of the world (*e.g.*, North America, South America, Australia), where rapid decline and sometimes outright extinction of tribal cultures is the prevalent trend. The Highland Show, held in alternate years in Mount Hagen and Goroka, the two largest towns of the Highlands, provides a showcase of the Highlanders' different cultures; as many as 60 000 dancers from widely-separated parts of the region participate in this show every year.

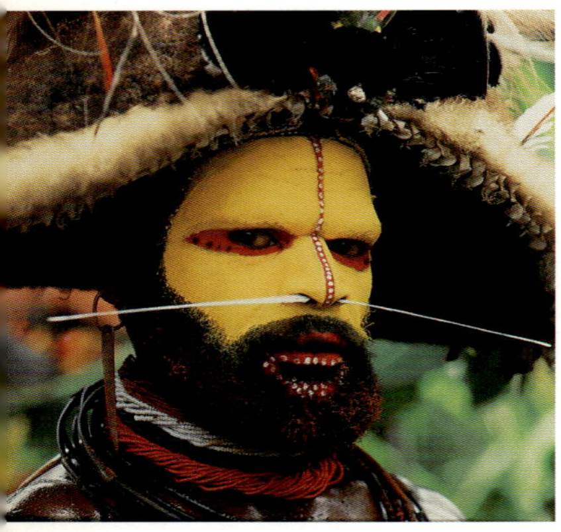

Huli wigman dressed for a dance in the Tagali Valley. His wig is adorned with cuscus fur and various bird feathers, and he is wearing necklaces made of beads and cassowary quills.

On the opposite page, Mwakiwosa Medova wears a headdress of sulphur-crested cockatoo (Cacatua galerita) *feathers, and adds dots of powdered coral mixed with water to the makeup of his face before joining the dancing, sing-sing, and feasting at the yam festival in the Trobriand Islands.*

Among the noteworthy cultures of the Highlands are the Huli of the Tari Valley, best known for their marvelous wigs made of their own hair and elaborately decorated with feathers and flowers, the neighboring Mendi, the Enga, the Melpa-speaking people of the Wahgi Valley, the Asaro people with their world famous mud-man ceremony, and the Kukukukus, long known as ferocious warriors. They are all established agriculturists, growing taro, sweet potatoes and yams in neatly organized agricultural plots, and having pigs as the principal sign of wealth. Although there has been much change in this area with missionaries, increasing influence of Western technology, and some out-migration to the larger coastal cities like Port Moresby and Lae, there is much that is still traditional in this region.

The other internationally known culture area is the Sepik River region. A lowland rain forest area centered on Papua New Guinea's longest river, the Sepik, this incredible region has almost 200 distinct languages, making it the most diverse in the country. Traditionally headhunters, the Sepik people have the sago palm as their staple food, and are renowned as some of the greatest wood-carvers in the world. Every village has its own distinctive wood-carving tradition, and some of these art forms are well-known and much sought after in international markets. These wood carvings are still used for traditional purposes, as is the case with some of the crocodile cultures of the Karawari region (*e.g.*, the Yokoim-speaking people of the Konmai and Karawari Rivers), and increasingly to attend to a growing tourist trade. Another characteristic of the Sepik cultures is the *haus tambaran* or men's spirit house, a magnificent structure that is the center of village activity. These are such a dominant feature of Papua New Guinea culture that the National Parliament building in the capital of Port Moresby is patterned after and decorated like a *haus tambaran*.

Among the singular cultures of the Sepik are the Iatmul, the Abelam, the Sawos, with their elaborate yam ceremonies, the Yokoim speakers and their heavy emphasis on the crocodile as a cult figure, the Aramblak speakers and their unusual *yipwon* spirit hooks, the Blackwater Lakes people, the Yuat River people, the Murik Lakes people at the mouth of the Sepik, and many others.

In the Western Province, the Gogodala people are particularly notable. Inhabiting the floodplain of the Aramia River north of the Fly River delta, these swamp dwellers have a very unique culture and art form symbolized by huge wooden figures of a crocodile eating a man, and also make huge, gaily-decorated canoes, which are still used in canoe races.

Off the eastern tip of the main island of New Guinea, in Milne Bay Province, are a number of interesting islands and cultures, the best known of which are the Trobriands. The Trobriand Islanders became internationally famous through the detailed studies of the Polish anthropologist Bronislaw Malinowski during the First World War. He described their complex ceremonial exchange, the Kula Ring, their sexual practices, and a number of other cultural features in a study that is now part of the classical anthropological literature. The Trobriand Islanders also have a system of hereditary chiefs, like the cultures of Polynesia and unlike much of the rest of Melanesia, where social systems tend to be more egalitarian, with headmen gaining status through personal influence and meritorious acts.

The two largest islands of the Bismarck Archipelago, New Britain and New Ireland, also have very special cultures. On New Britain, the Tolai people of the Gazelle Peninsula are famous for their *dukduk* ceremony, and the Bainengs people for their fire dance, which involves huge bark cloth masks that in days gone by reached 45 m in length! In southwest New Britain, in the vicinity of Kandrian, the Kaulong and Sengseng people are the only blowgun hunters between Sulawesi and the Amazon. Blowguns are found nowhere else in Melanesia, and are entirely absent from Polynesia and Micronesia. The Kaulong and Senseng blowguns are the largest in the world, reaching some 6 m in length, and are made of segments of bamboo glued and bound together. These people also hunt largely at night, the only blowgun culture on Earth to do so.

Finally, there are the people of New Ireland, with their *malanggan* ceremonies and intricate wood carvings. Their art form is once again completely different from anything else in the region, and represents one of the finest tribal art forms anywhere in the world. Unlike the wood carvings of the Sepik, which are made to endure, the *malanggan* carvings are for one-time use only, and are usually discarded after the funeral ceremonies in which they are employed.

In all, the complex of cultures in Papua New Guinea is truly overwhelming and can not be matched anywhere else on the planet. The fact that many of these cultures still retain a major portion of their traditional elements makes them doubly special, and contributes even more to Papua New Guinea's position among the megadiversity countries.

RUSSELL A. MITTERMEIER
BRUCE BEEHLER
GAIKOVINA KULA
CRISTINA GOETTSCH MITTERMEIER

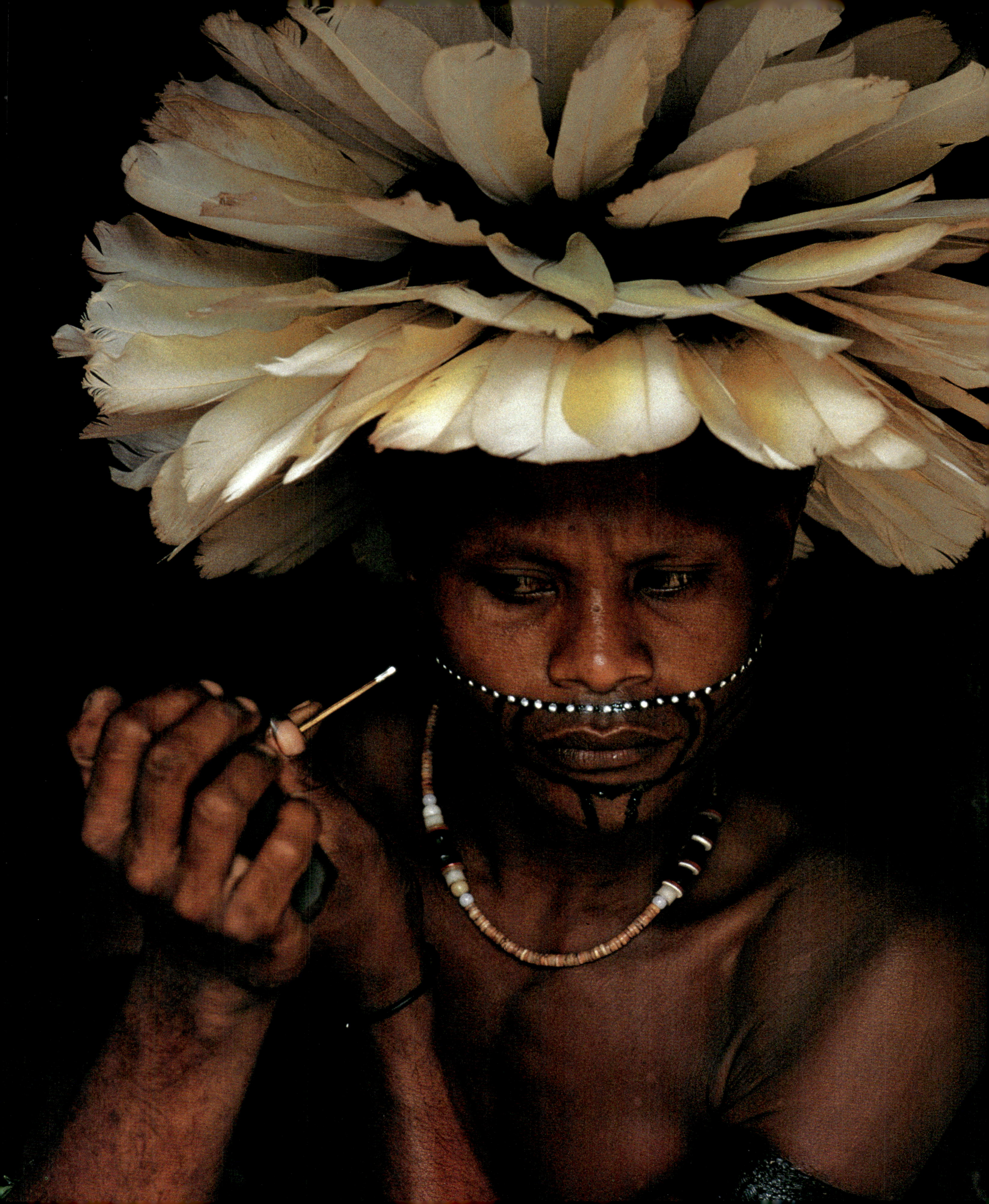

Reptile diversity in Papua New Guinea is quite high, with 305 species, most of them lizards (190 species) and snakes (93 species). Below, this crocodile skink (Tribolonotus gracilus) *from the Torricelli Mountains is one of the five species of this unusual genus of armored skinks that are endemic to New Guinea. Right, this juvenile green tree python* (Morelia viridis) *begins its life as a bright yellow animal, and gradually becomes emerald green as it matures. An arboreal species, it is an ecological equivalent of the emerald tree boa of South America.*

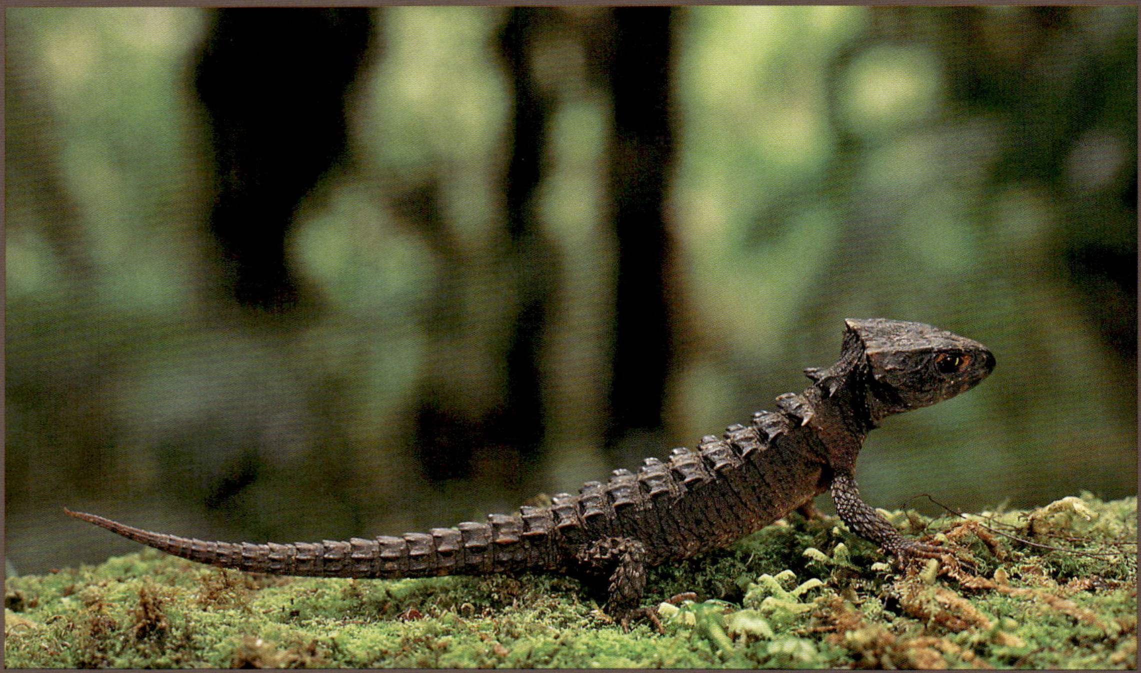

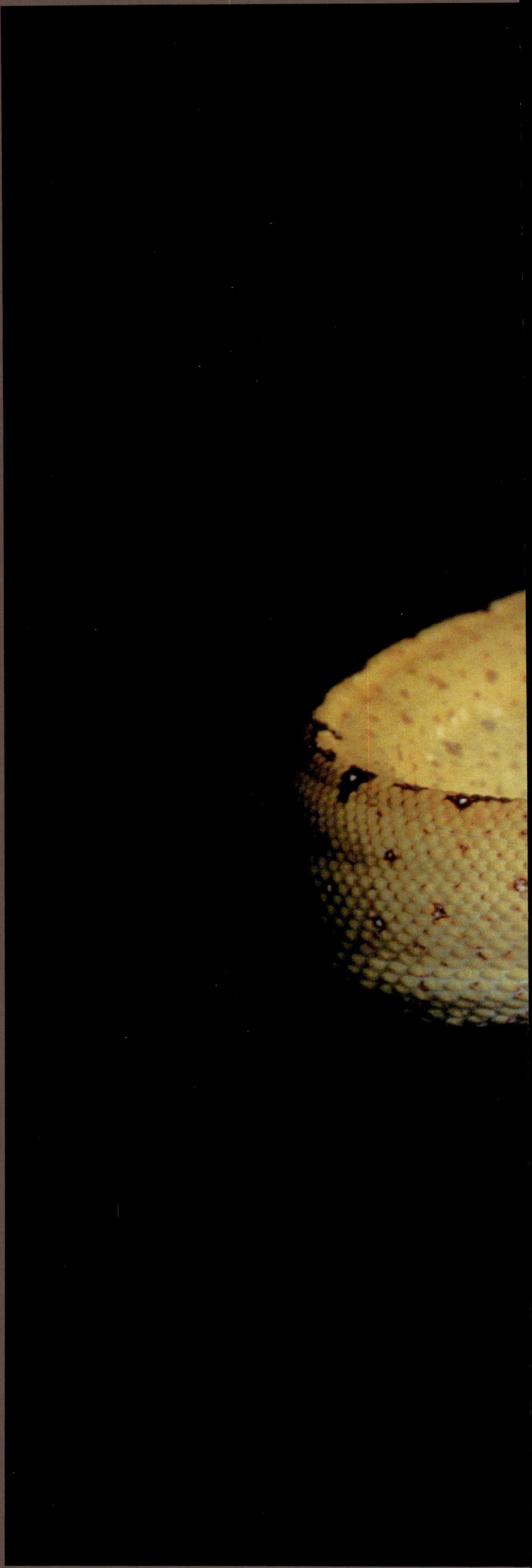

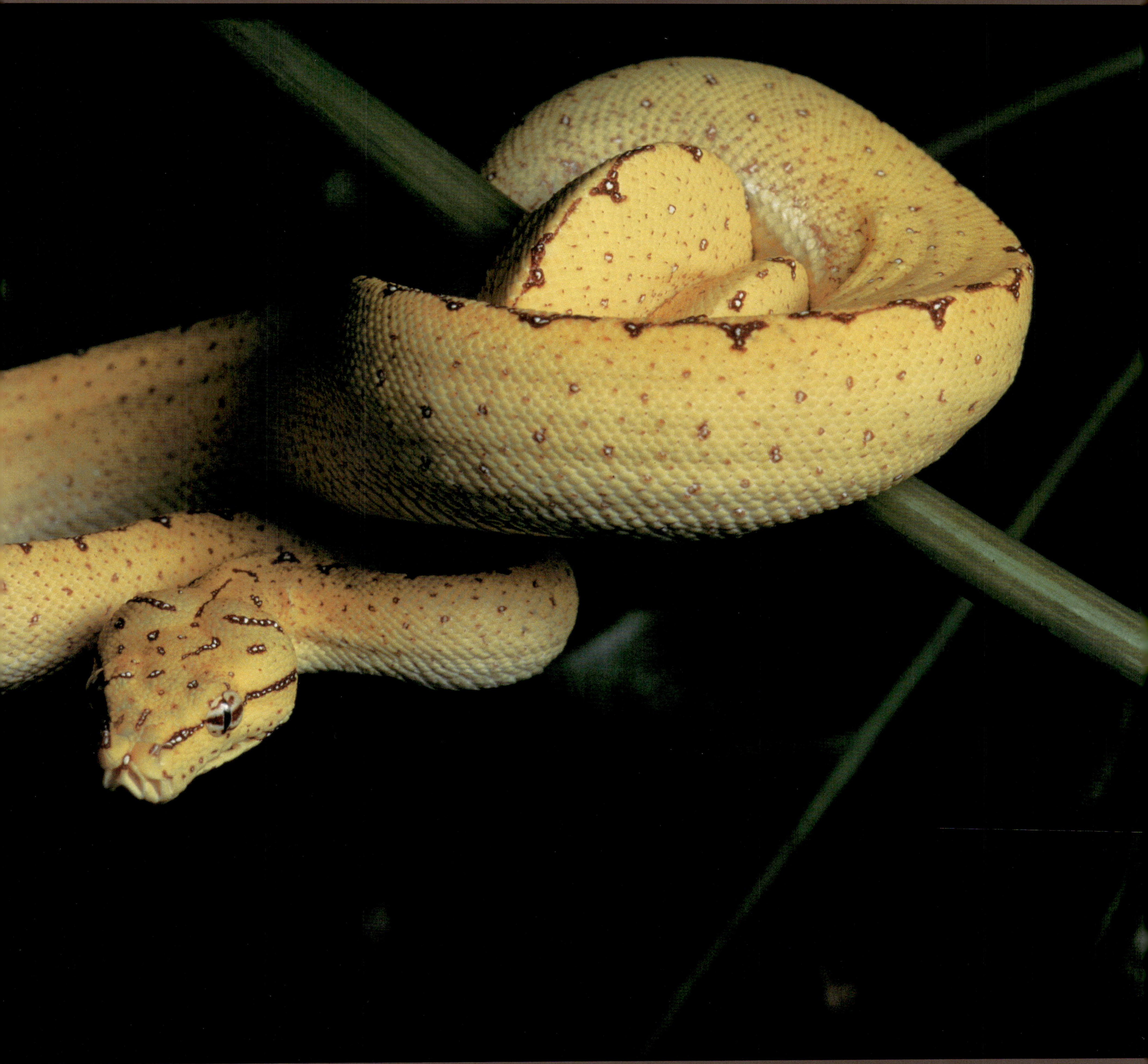

Plant diversity and endemism are very high in Papua New Guinea. Below, this species of Eucalyptus *from East New Britain has one of the most strikingly colored trunks of any tree. Right, twin falls on the upper Seribi River. The interior of the country is rich in spectacular scenery like this, much of it still unexplored and unexploited.*

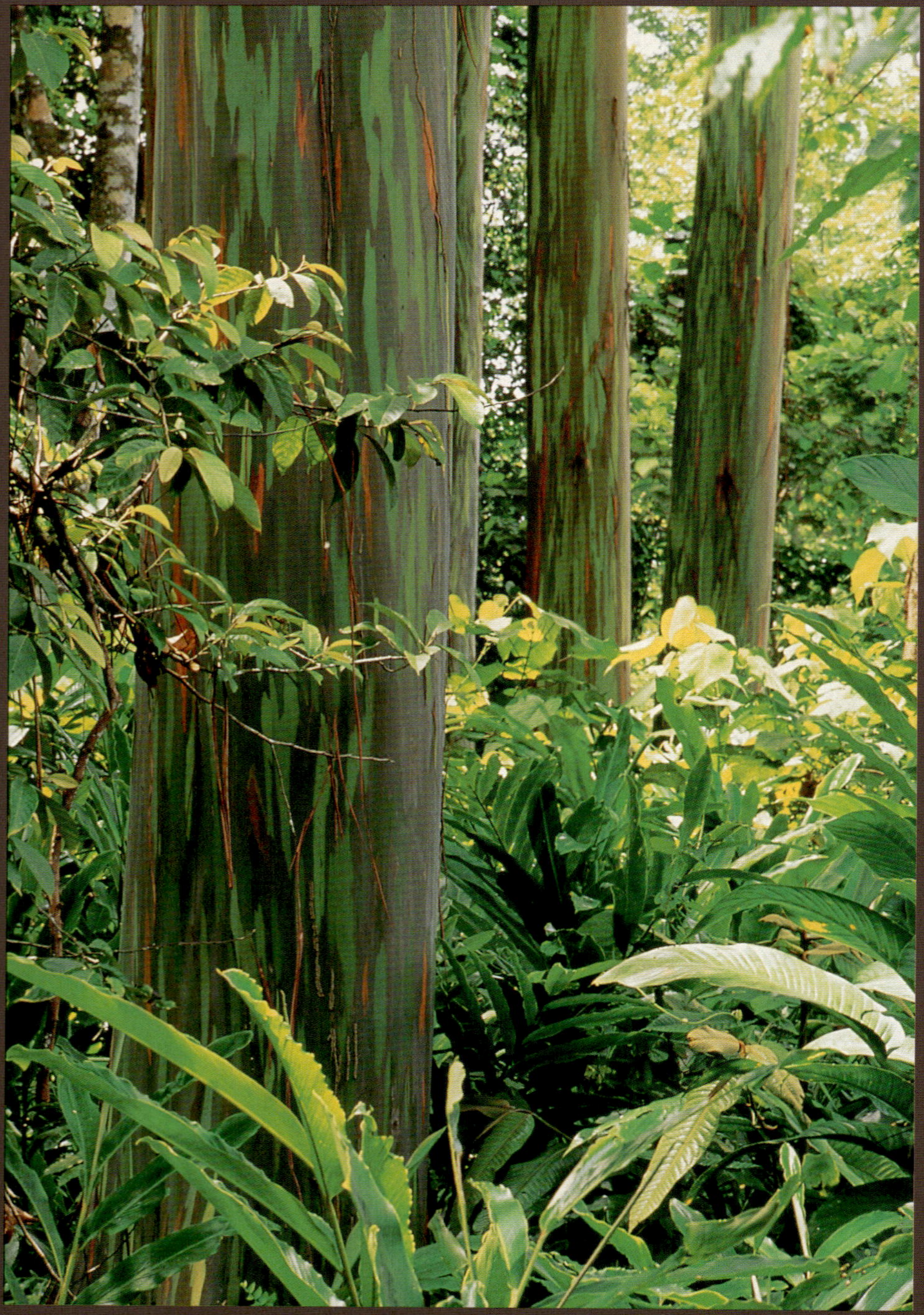

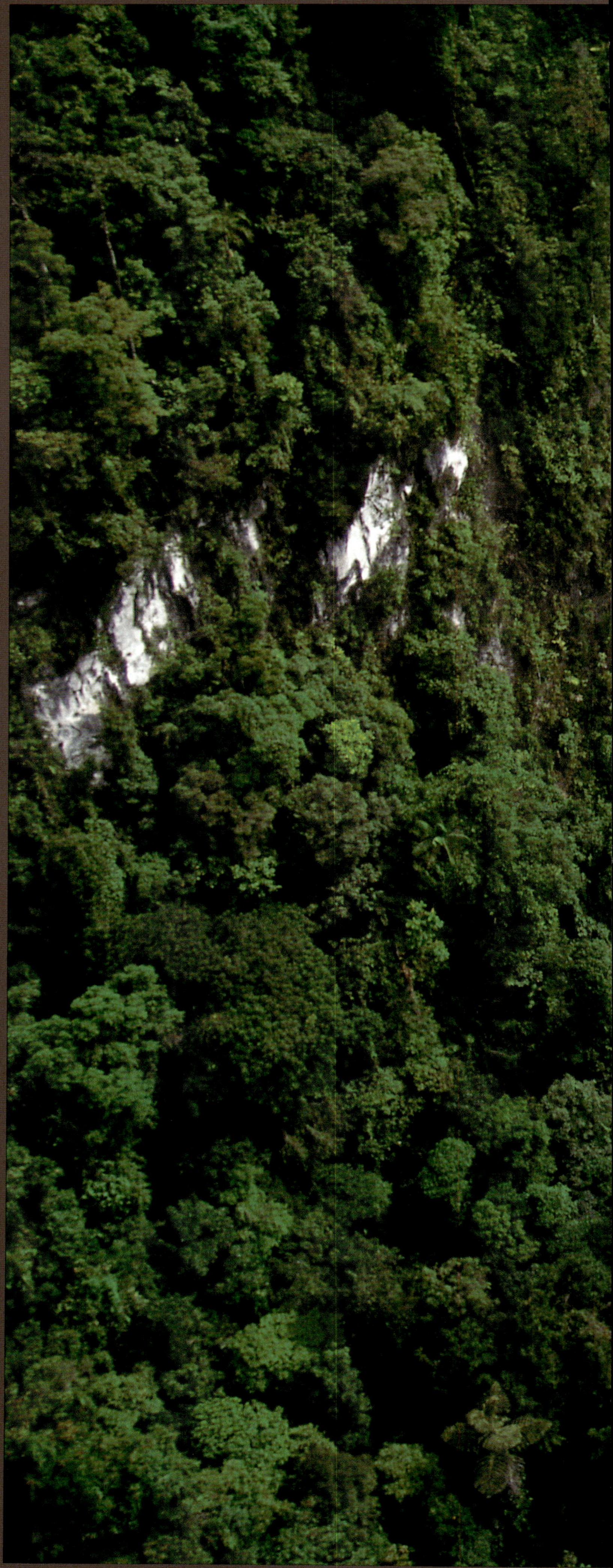

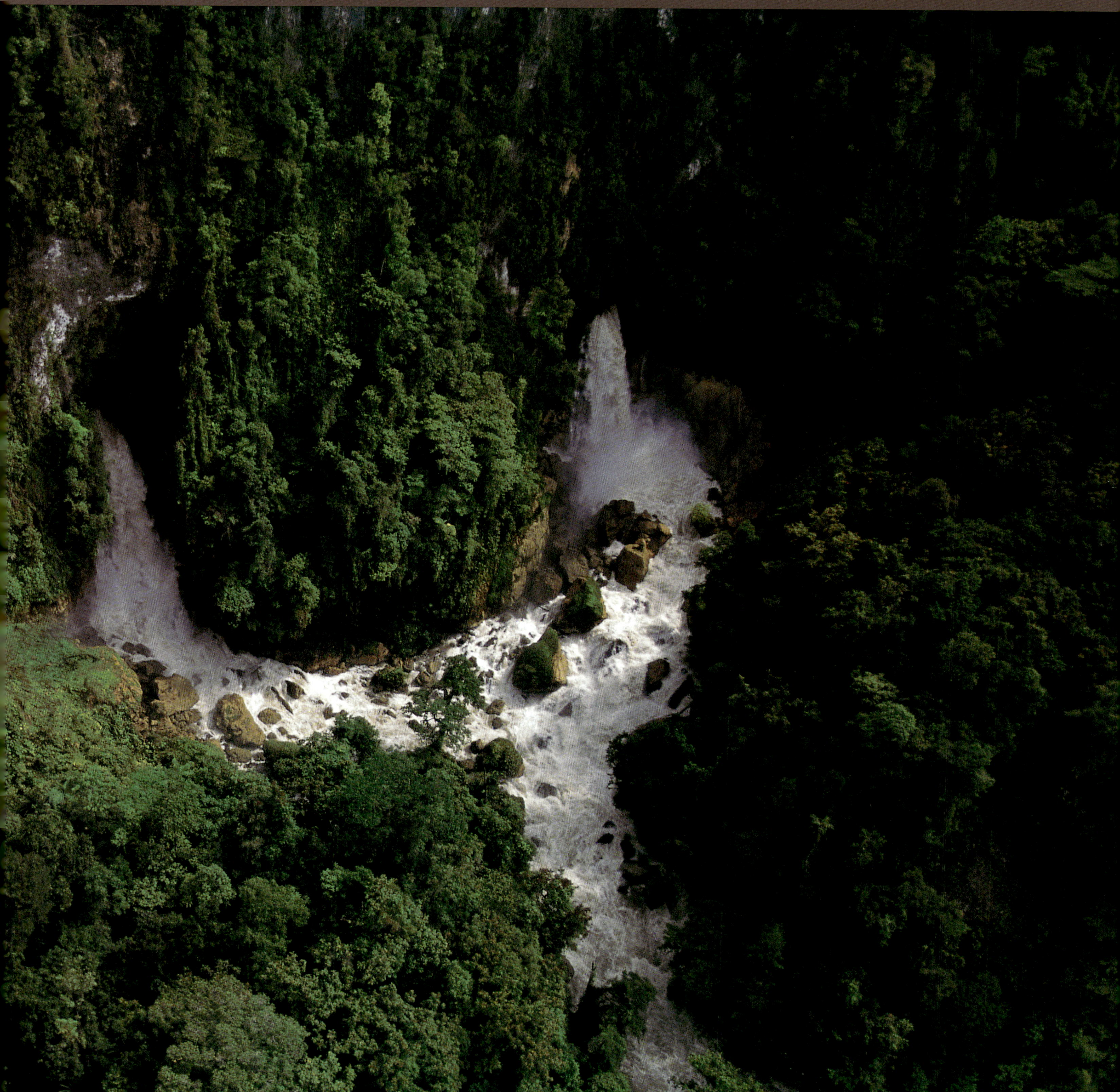

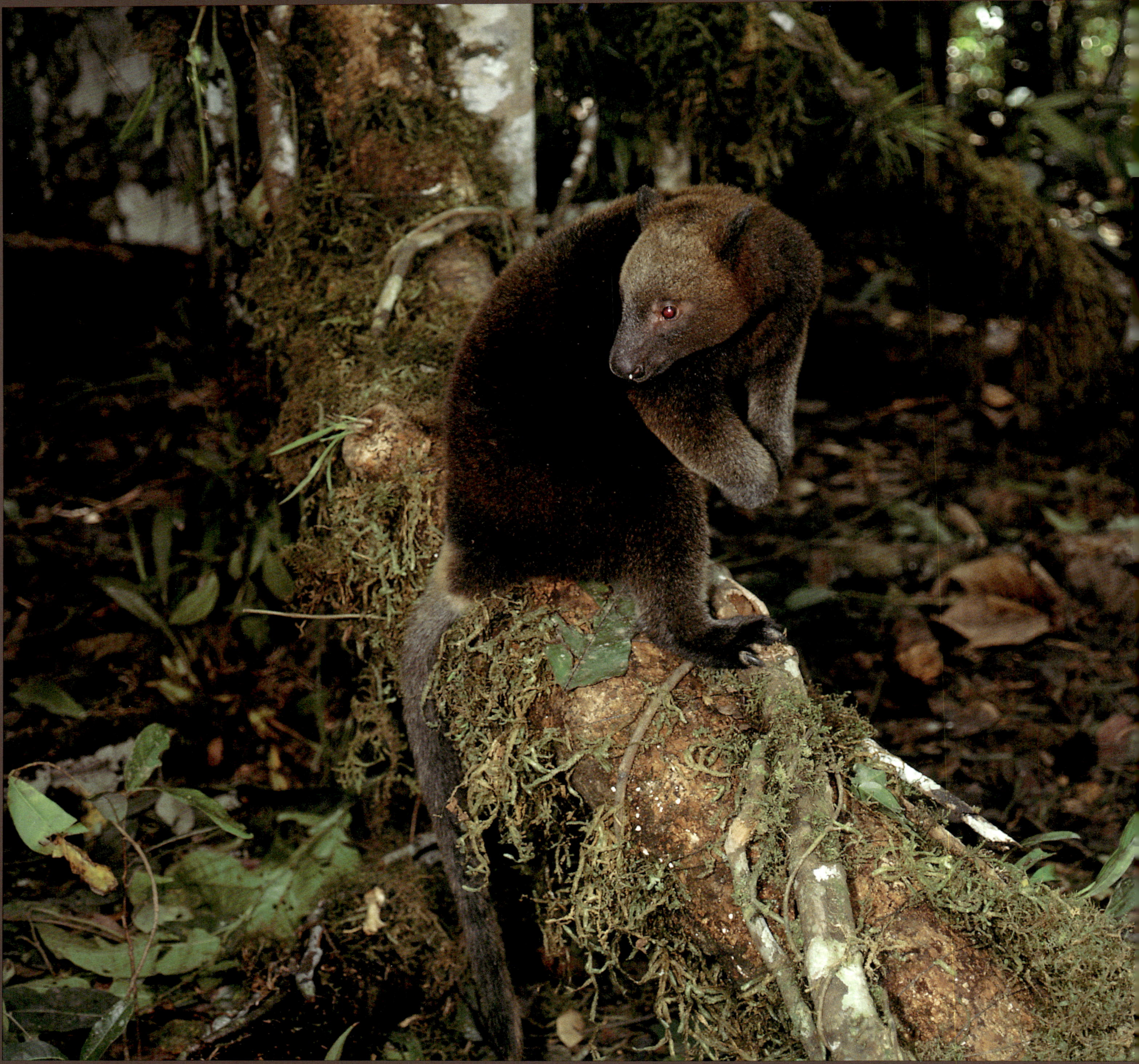

Among the mammals, the most spectacular flagship species of Papua New Guinea are the tree kangaroos. Ten species and 17 different subspecies occur on the island of New Guinea, and in Queensland, Australia. All but two are New Guinean endemics. Doria's tree kangaroo (Dendrolagus d. dorianus) *(above) is the largest and can reach 18 kg. Goodfellow's tree kangaroo* (Dendrolagus goodfellowi) *(upper right) is one of the species most often seen in captivity outside New Guinea, and also one of the most attractive. The grizzled tree kangaroo* (Dendrolagus inustus) *(lower right) is from the Torricelli Mountains in West Sepik Province.*

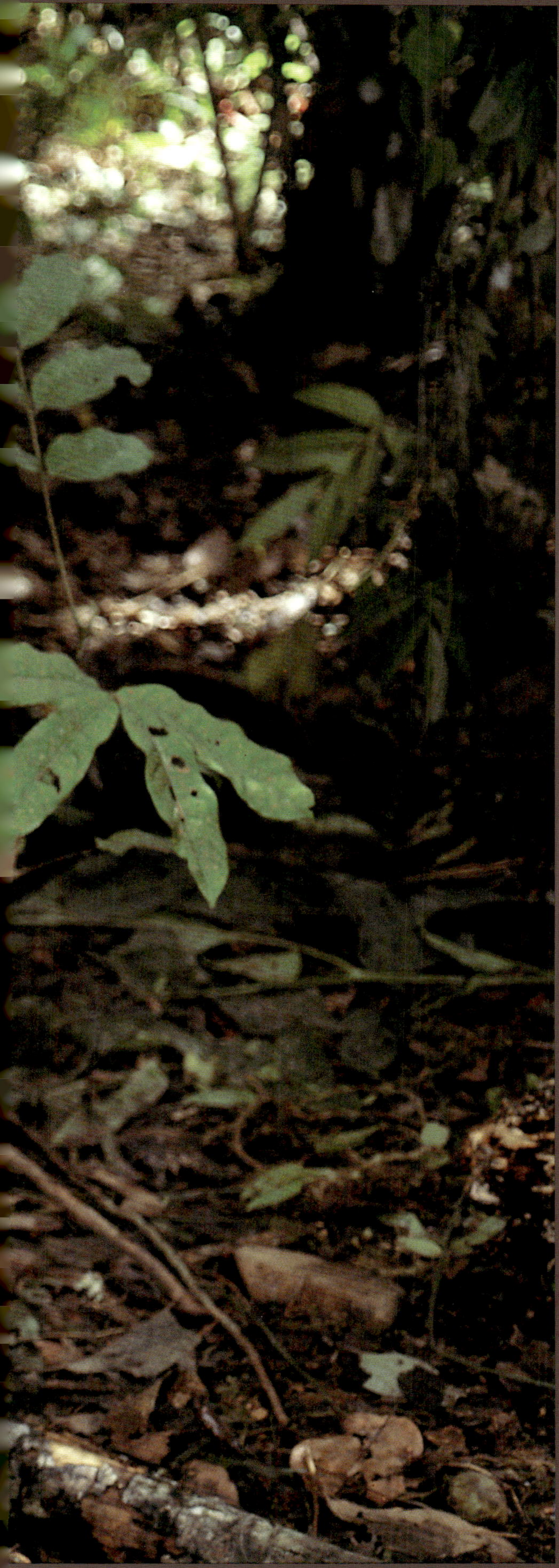

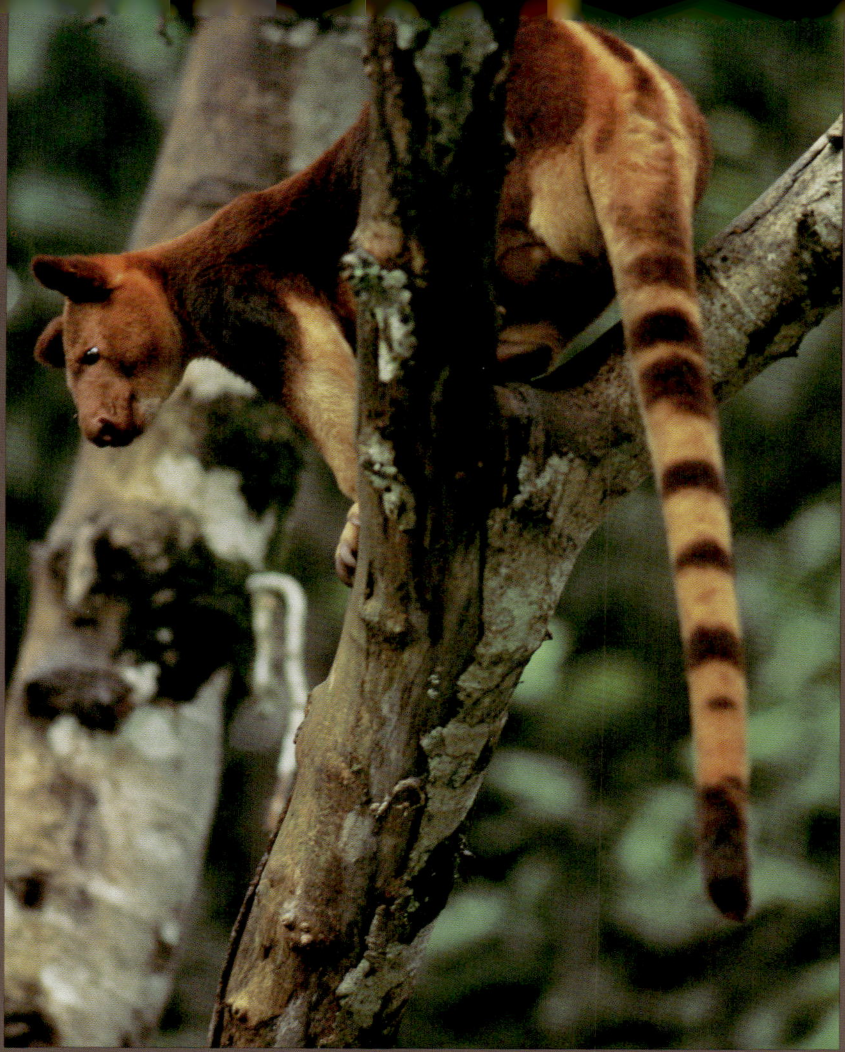

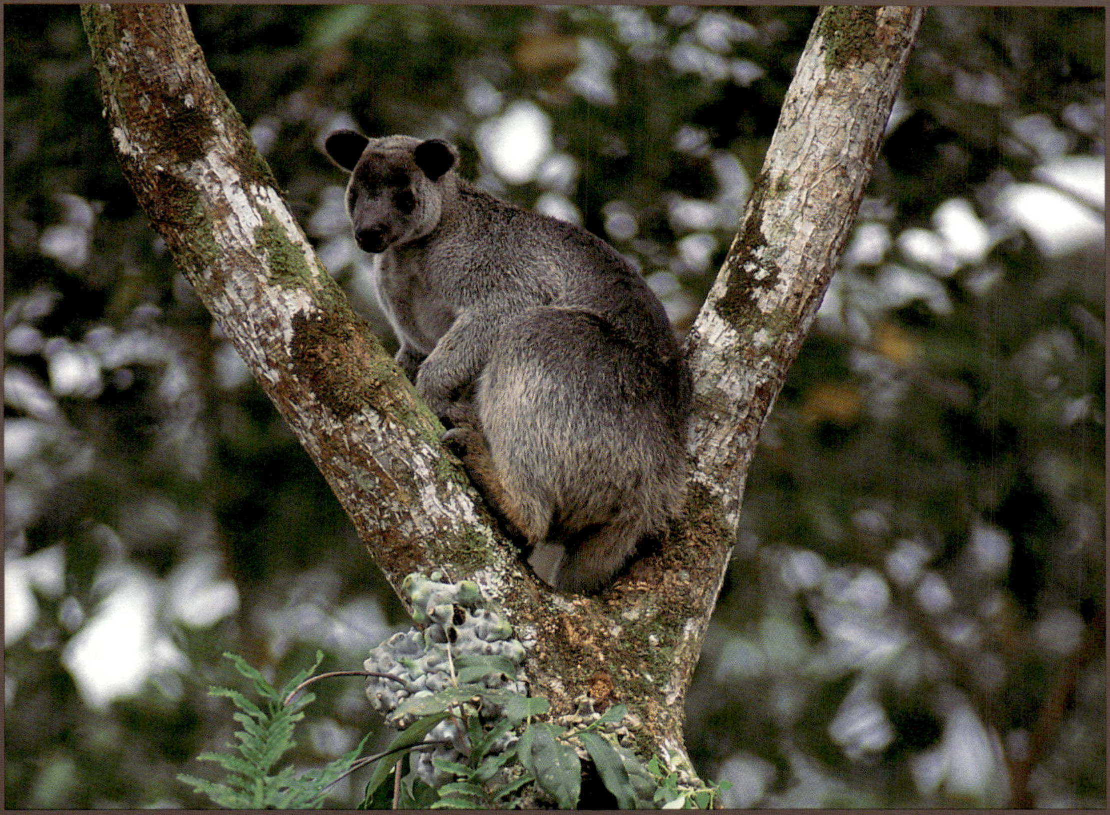

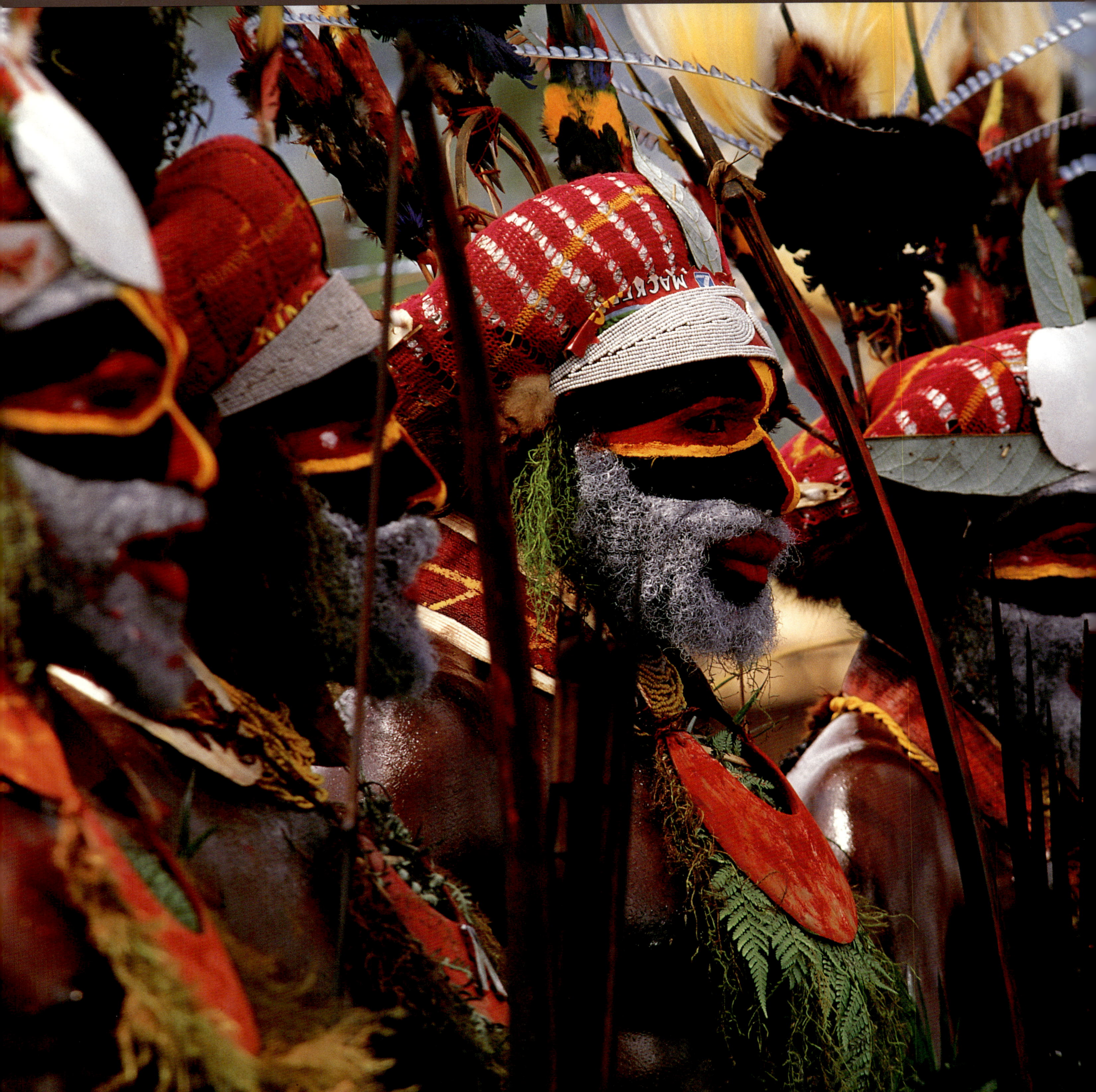

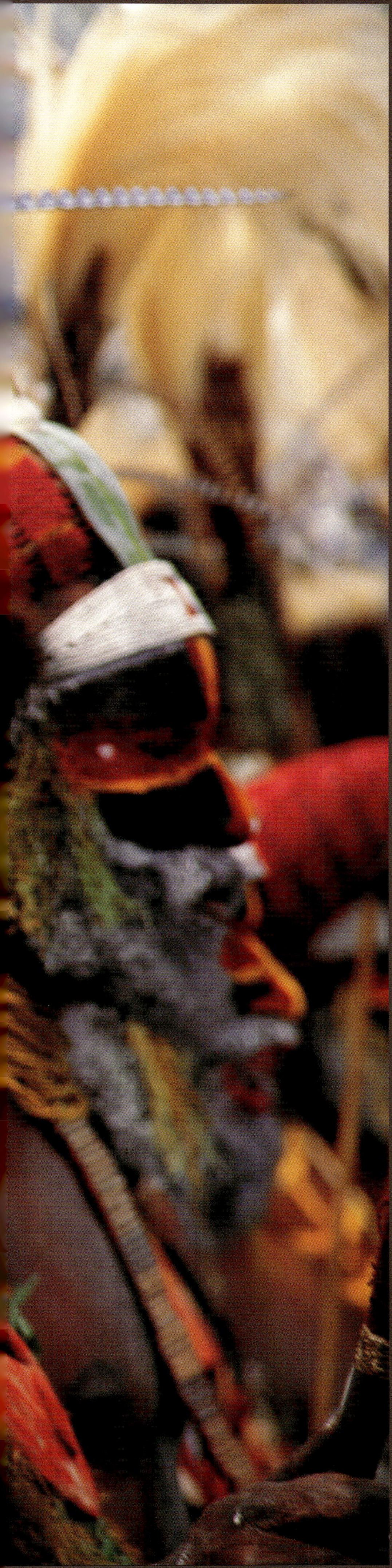

Left, a group of Mendi men dressed for a sing-sing in the Southern Highlands Province. The Mendi are neighbors of the Huli wigmen of the Tari Valley. Note again the use of bird feathers and shells. The long black and white feathers are from the male King of Saxony bird of paradise (Pteridophora alberti). *Below, Asaro mudmen from the Highlands of Papua New Guinea. The mudmen now perform this traditional dance, originally intended to frighten off enemies, for tourists.*

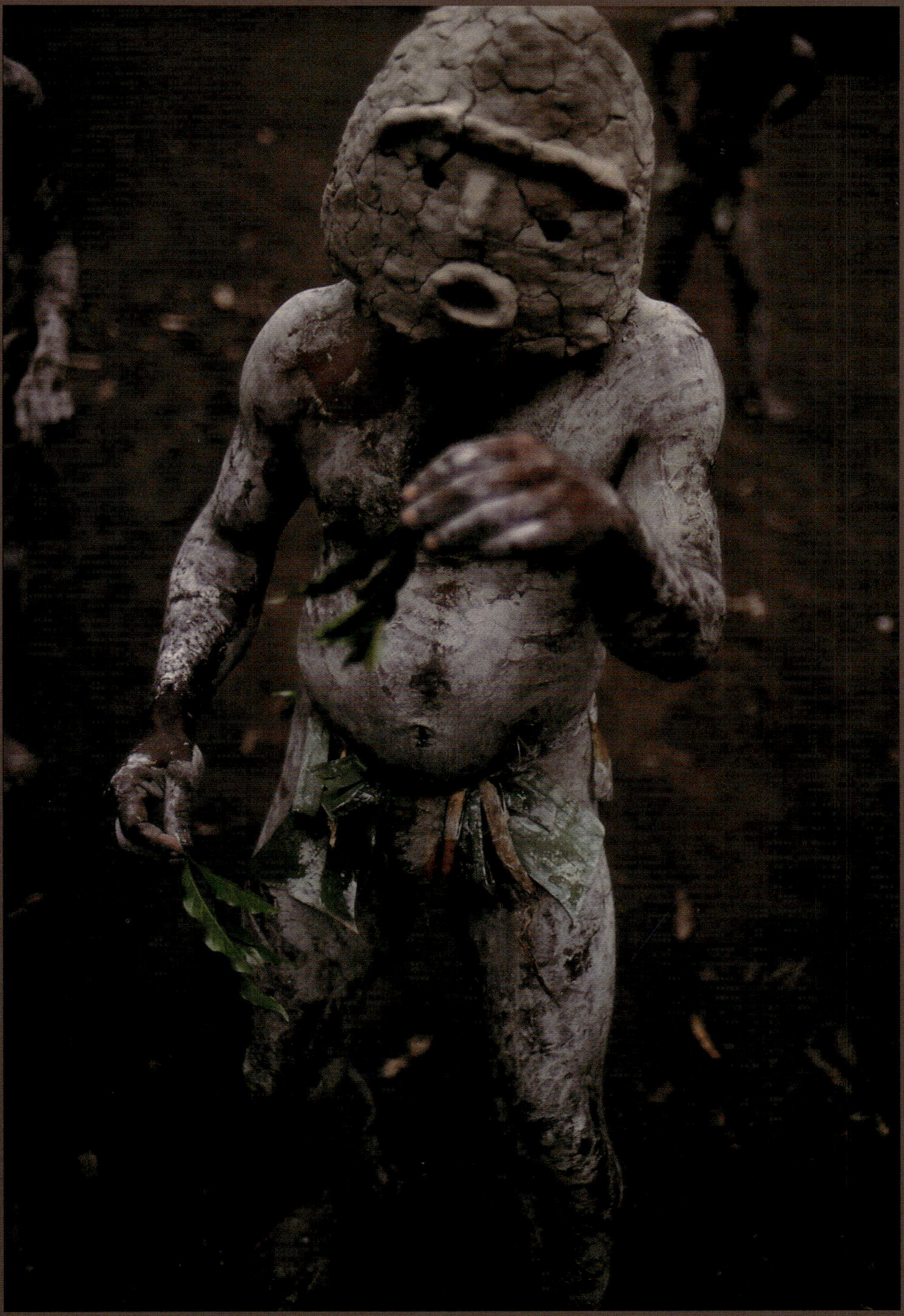

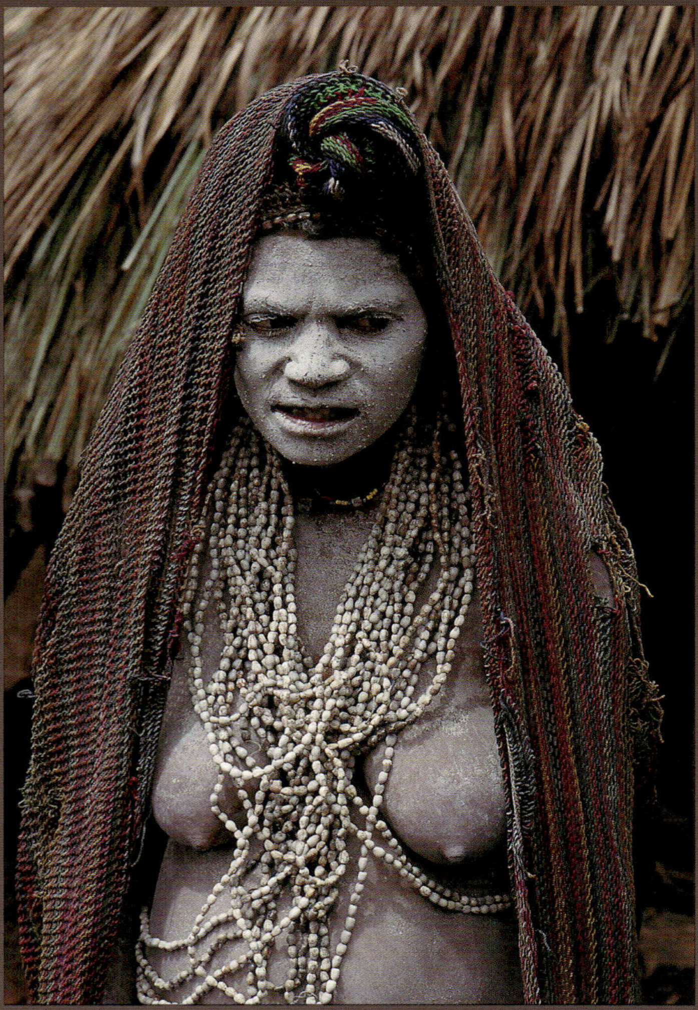

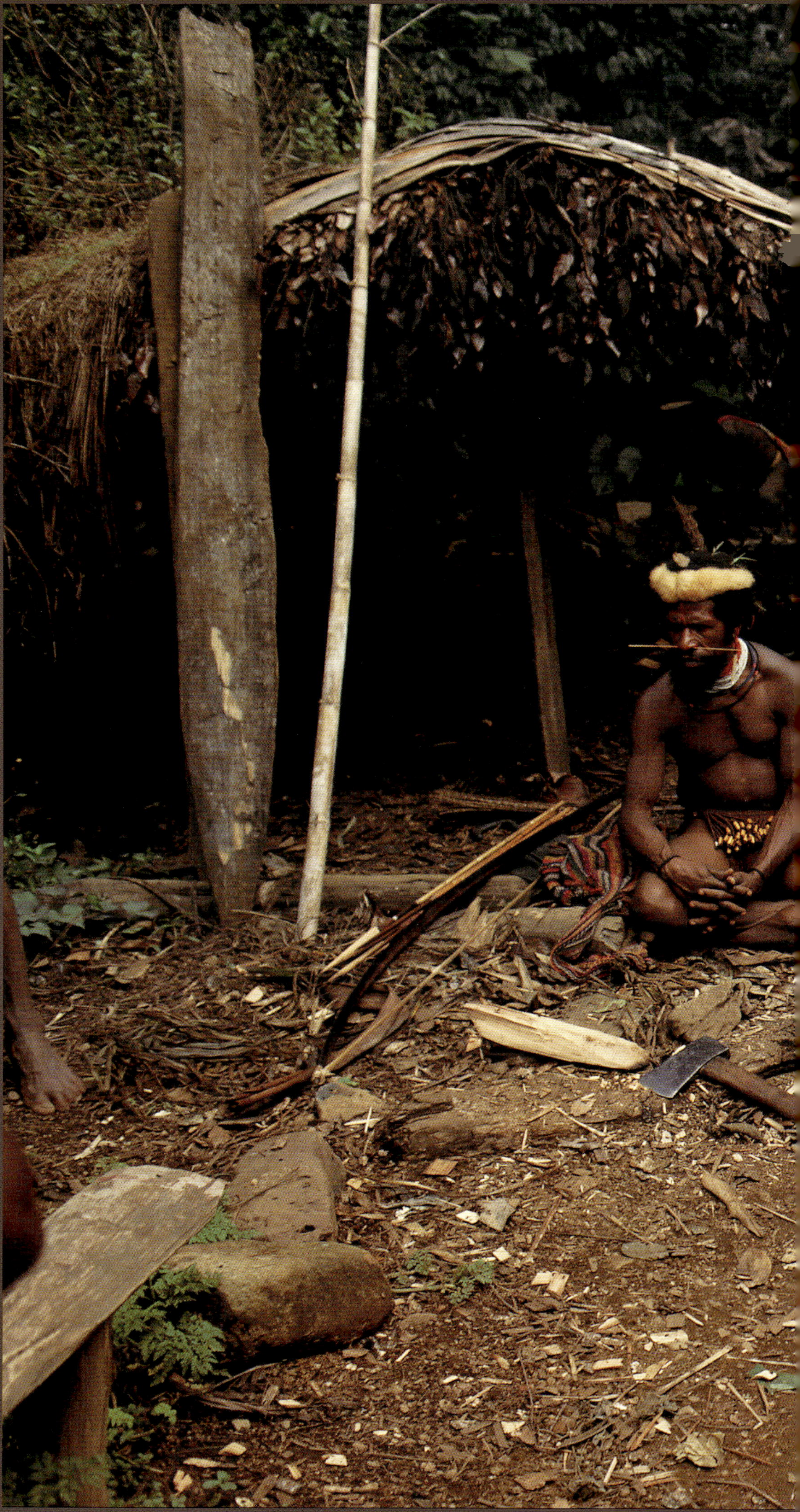

Above, Huli woman from the Tari Valley wearing her traditional bilum *bag and painted with ash as a sign of mourning. On the right, Huli wigmen in their traditional village. Note the use of the crescent-shaped day wig, and the cuscus fur headband worn by the man on the left. The man standing is wearing a skirt made of pig's tails.*

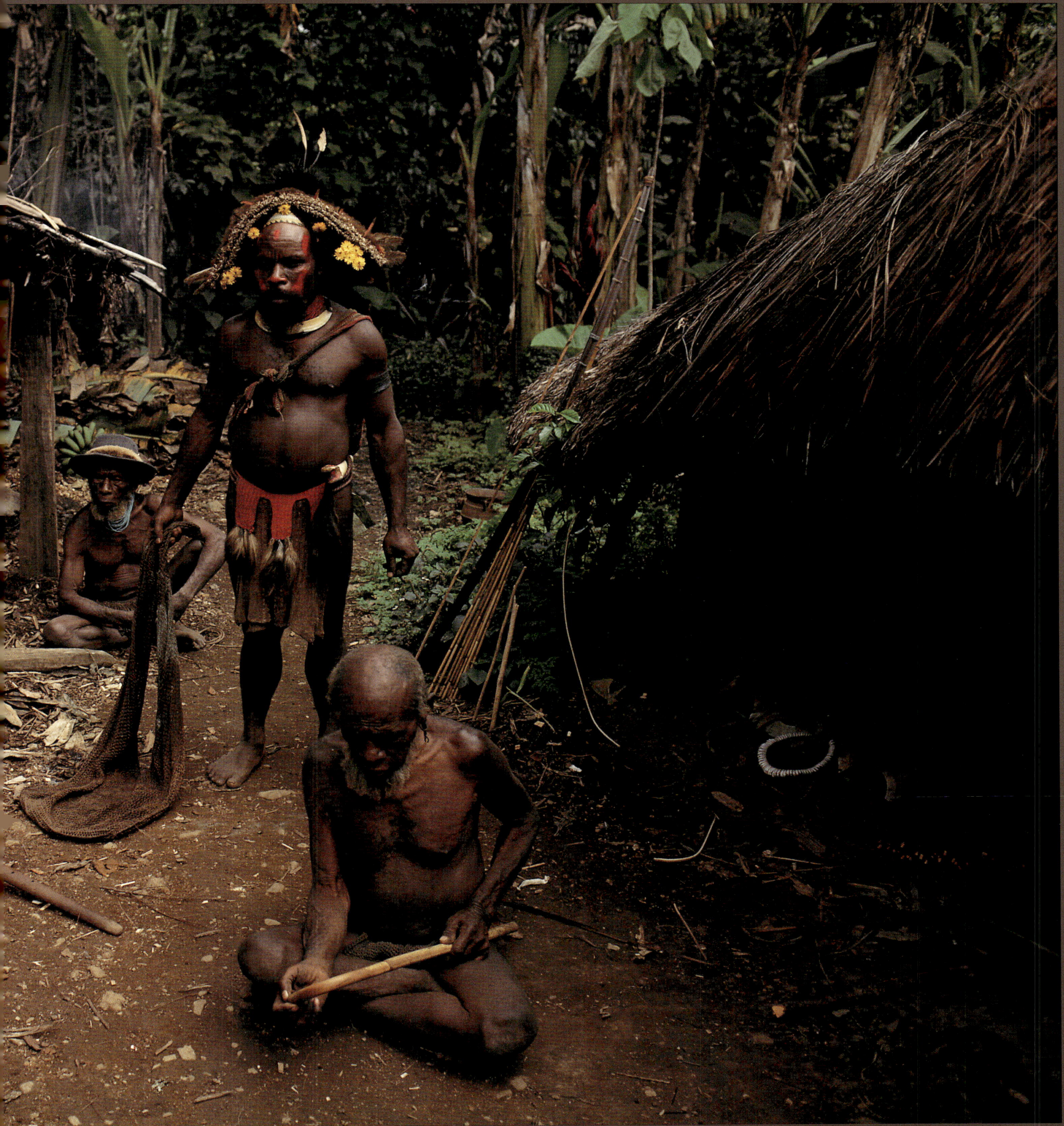

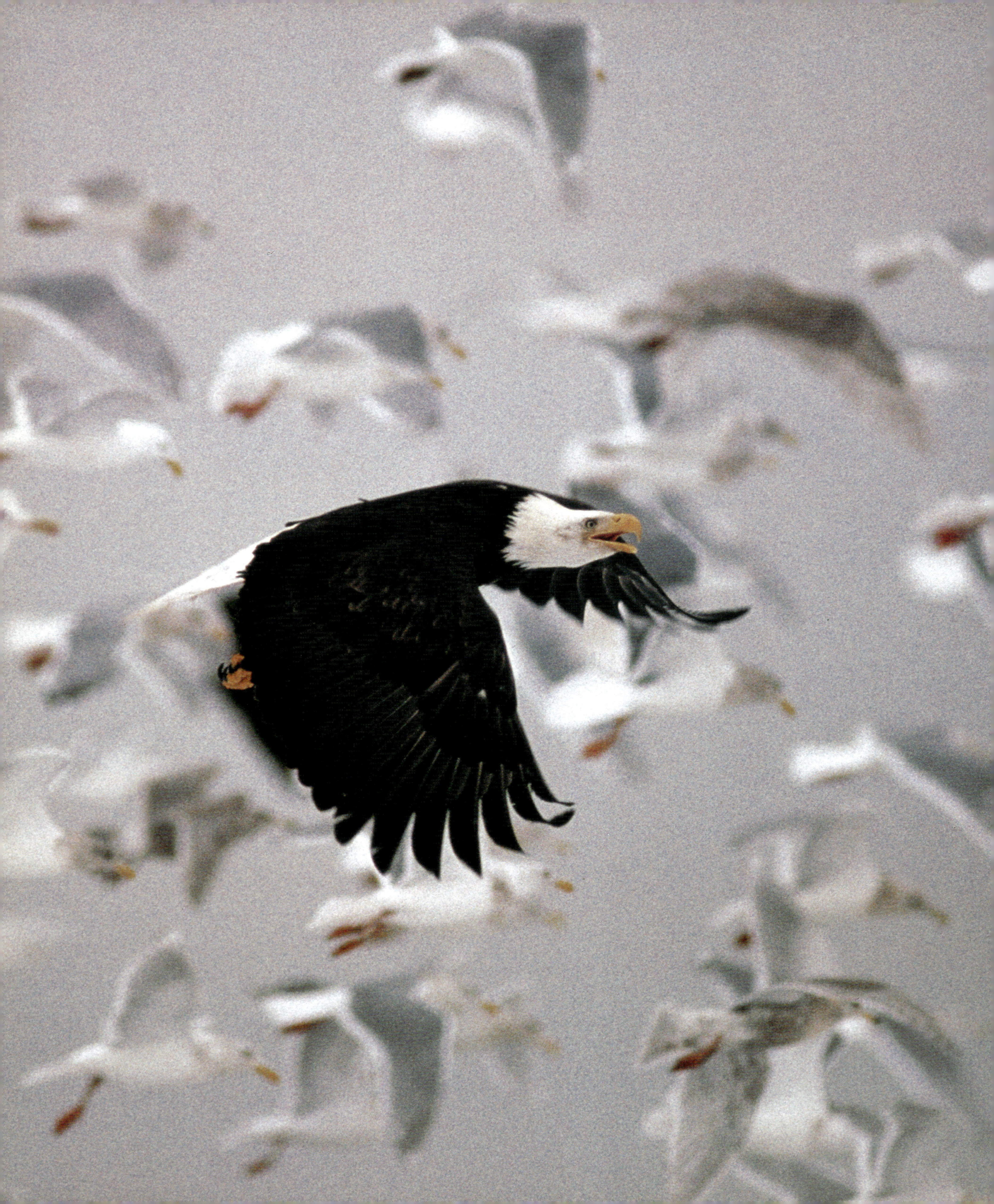

UNITED STATES OF AMERICA

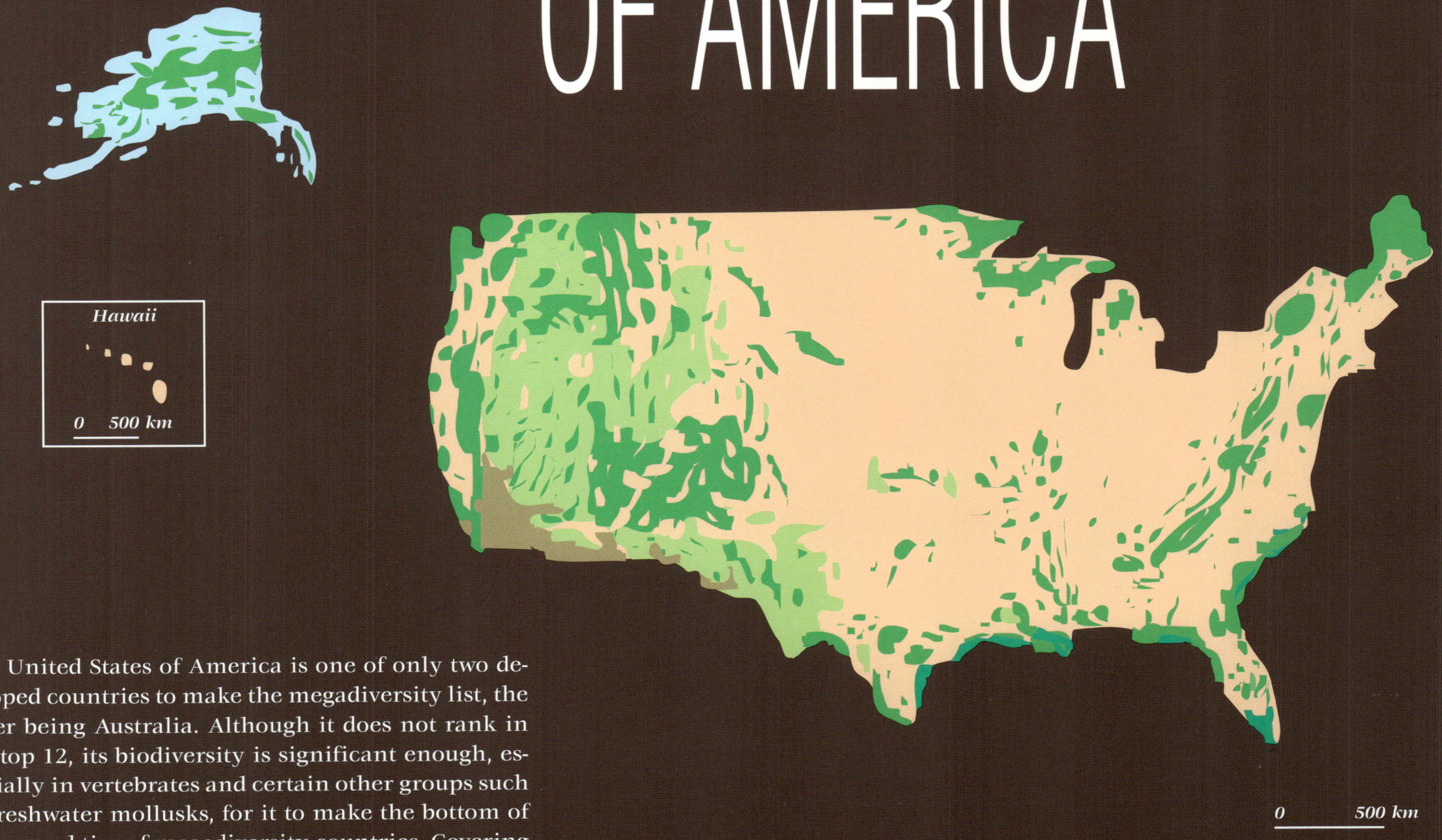

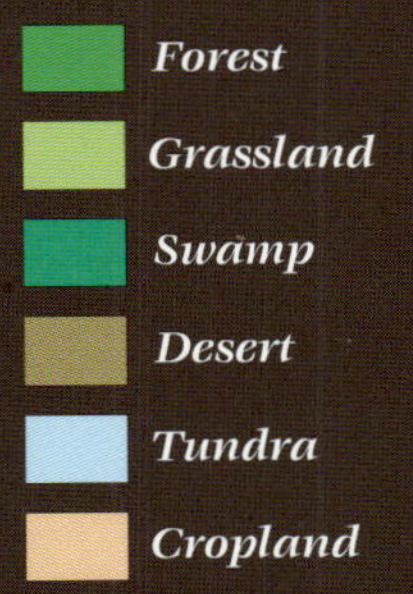

The United States of America is one of only two developed countries to make the megadiversity list, the other being Australia. Although it does not rank in the top 12, its biodiversity is significant enough, especially in vertebrates and certain other groups such as freshwater mollusks, for it to make the bottom of the second tier of megadiversity countries. Covering 9 372 143 km^2, it is the fourth largest country on Earth (after Russia, Canada and China) and fifth in population (265 000 000). Much of the United States lies within the temperate zone, with only a small tropical area and a tiny representation of tropical rain forest, most of it in the Hawaiian Islands and Puerto Rico. However, the United States extends over more than 50 degrees of latitude from north of the Arctic Circle in Alaska to southernmost Hawaii, and over 120 degrees of longitude from eastern Maine to the westernmost point of the Aleutian chain. It also ranges in altitude from Death Valley, California, at 86 m below sea level, one of the lowest points on Earth, to Mt. McKinley/Denali in Alaska, at 6 194 m, the highest point in North America. There is huge topographic relief as well, with major mountain chains like the Rockies and the Brooks Range in Alaska, and ecosystem diversity is enormous, ranging from tundra and subarctic taiga, to deserts, prairie, boreal forest, deciduous forest, temperate rain forest and tropical moist forest in the far south.

Furthermore, there are several ecoregions of global importance that help to propel the United States into the megadiversity country list. One of these is the California Floristic Province, which covers most of the state of California and extends into southern Oregon; with 2 140 endemic plant species, it accounts for more than half of the United States' total of 4 036 higher plant endemics. The other region of major consequence is the state of Hawaii, with its high diversity and endemism, especially in plants, birds and mollusks. California's Floristic Province is in the second tier of *threatened biodiversity hotspots* (having narrowly missed the first tier cutoff of 2 500 endemic plants), and the Hawaiian Islands are a major part of Polynesia/Micronesia, a first tier threatened hotspot.

Aside from these two global biodiversity hotspots, the United States also has 37 Centers of Plant Diversity and Endemism as identified by IUCN/WWF (1997); three of them are in the Pacific northeast, ten

*On the opposite page, the bald eagle (**Haliaeetus leucocephalus**) is the national symbol of the United States of America. Considered endangered for several decades, it is now making a comeback in many areas. Here, a bald eagle in southeast Alaska flies through a flock of gulls.*

in California, six in the Rocky Mountains, four in the Appalachian Mountains, one in the midwest, eight in the southeastern coastal plain, four in the southeastern deserts, and one in Hawaii. Looking at priority areas for birds, the U.S. has eight Endemic Bird Areas as identified by ICBP (1992). Four of these —the Sierra Madre Occidental, northeast Mexican Pacific slope, Sierra Madre Oriental, and northeast Mexican Gulf slope— are shared with Mexico, and four are entirely in the United States, one covering California and the other three in the Hawaiian Islands.

The United States ranks surprisingly high in diversity and endemism for several vertebrate groups. It is sixth in total mammal diversity (428 species) and sixth in mammal endemism (101 species); twelfth in total amphibian diversity (194 species) and ninth in endemic amphibians (126 species); fourteenth in total reptile diversity (261 species) and twelfth in endemic reptiles (90). Although total bird diversity is relatively low at 768 species, the 71 endemics are good enough for eleventh place on the world list in this category. The total for non-fish vertebrates is 1 651, good enough for fourteenth, but for total endemic vertebrates (388) it ranks a surprising tenth. Freshwater fish diversity is also high, the total of 790 species ranking ninth on the world list. Marine diversity is also quite high, given that the United States has long coastlines on both the Atlantic and Pacific Oceans, and that it includes marine ecosystems from the Arctic to the subtropics and the tropics; much of the country's marine diversity is concentrated in regions like Florida, the Gulf Coast, and the Hawaiian Islands, but significant numbers are contributed by the temperate and colder regions as well. Looking at invertebrates, the U.S. has relatively low butterfly diversity with only about 700 species, but in tiger beetles, it is seventh in diversity with 111 species and eighth in endemics with 45. Total higher plant diversity at 18 956 species is high at eighth or ninth on the world list, but endemism is relatively low, with at least 4 036 species.

Human cultural diversity was also once very high in the United States, with some estimates running as high as 2 000 languages spoken in the whole of North America when the Europeans first arrived in the New World. How many of these went extinct in the intervening 500 years will never be known for sure; however, some 300 languages still remain in one form or another within what is now the United States.

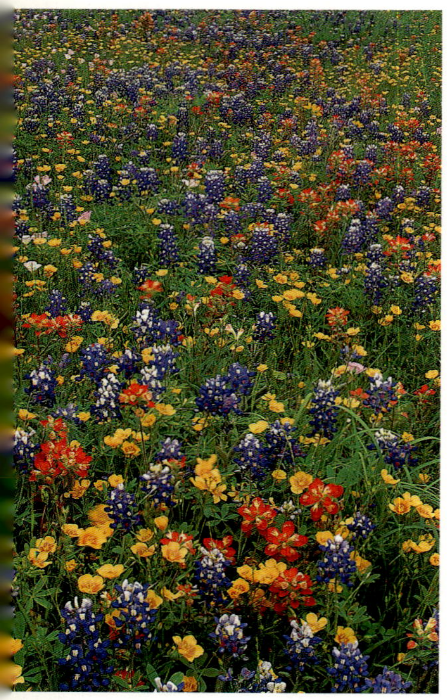

Spring wild flowers, including bluebonnets, paintbrush and flax, in a prairie in Texas. Prairie ecosystems in the United States are now down to a tiny fraction of their original extent.

Natural habitats

The natural terrestrial habitats of the United States can be divided into four domains —Polar, Humid Temperate, Dry, and Humid Tropical— on the basis of climate and vegetation. This categorization follows the work of Koeppen (1931) and Trewartha (1968), with recent revisions by Bailey (1995), who describes 14 divisions and 52 distinct provinces or ecoregions within these domains.

Located at higher latitudes, the two divisions of the Polar Domain —tundra and subarctic taiga— are found only in Alaska. Both are characterized by low average temperatures and annual precipitation, as well as by severe and prolonged winters. Together, these divisions cover 1 349 400 km^2 of the United States and account for 14.39% of the land area (Bailey, 1995).

Grasses, sedges, lichens and willow shrubs are the predominant vegetation of the northern tundra, changing to birch-lichen woodland and then to needle-leaf forest as one moves southward. Underlying these formations is permafrost, a permanently frozen layer of soil that is more than 300 m thick throughout much of this region. Needle-leaf (boreal) forests and open lichen woodland predominate in the subarctic taiga.

The Humid Temperate Domain is located within the middle latitudes (30°-60°) and is subject to pronounced seasonal cycles of temperature and precipitation. Its 21 ecoregions comprise six divisions and cover 4 205 760 km^2 within the United States, accounting for 44.9% of the national land area (Bailey, 1995).

Forests of the Warm Continental Division are found in New England, the upper mid-Atlantic states (the Adirondacks), and in the Great Lake states of Michigan, Wisconsin and Minnesota. Further south are forests of the Hot Continental Division, characterized by tall, broad-leaved deciduous trees. In the east, these forests are found as far north as southern New England and the mid-Atlantic states, and include the unique pine-oak formations known as Pine Barrens. From central Pennsylvania they extend southward throughout the Appalachian, Allegheny and Great Smoky Mountains to northern Georgia and Alabama, westward through the Ozark Highlands and beyond the Mississippi River, and northward in a thin strip through Minnesota to the Canadian border. The passenger pigeon, once abundant throughout much of this region, is now extinct. The southern Appalachian Mountains are home to 27 species of salamander —more than in any other part of North America— and they include many endemics.

The Subtropical Division begins in northern Virginia and southern Maryland, extending southward along the Piedmont Plateau to southern Florida and westward in a broad band along the Gulf of Mexico into eastern Texas and Oklahoma. Pine formations predominate along the southeastern coast, while deciduous forests are more characteristic toward the interior of the country. Due to the presence of numerous marshes, swamps and lakes, this region provides prime habitat for the American alligator, as well as significant numbers of snake, lizard, turtle, frog, toad, and salamander species.

Forests of the Marine Division comprise a relatively narrow block in the Pacific northwest, covering much of western Washington and Oregon from sea level to high mountain regions of the Coast and Cascade Ranges. These forests are characterized by abundant rainfall and mild climate, and the natural vegetation is needle-leaf forest dominated by Douglas fir, red cedar and spruce.

The Prairie Division comprises a relatively broad belt of mixed prairie and riverine forest formations that extends from the Canadian border southward to the Gulf of Mexico in southeastern Texas. Tall grasses are the dominant vegetation, with shrubs and trees almost entirely absent from some regions.

Wet winters and dry summers sustain the unique woodland and scrub habitats (also known as chaparral) of the Mediterranean Division, which is confined to California and southern Oregon. Along the Pacific coast, redwood trees grow to more than 90 m in height, as do giant sequoias further inland in the Sierra Nevada Mountains. Monterey cypress and pine, Torrey pine and Bishop pine are among the other tree species endemic to this region. More widely known as the California Floristic Province, this is a region of significant higher plant diversity and endemism, and a major reason for the United States being designated as a megadiversity country.

A characteristic common to the four divisions and 15 ecoregions of the Dry Domain is that water loss through evaporation exceeds annual levels of precipitation. Divided primarily into semiarid steppe and arid desert within the middle latitudes, these ecoregions cover approximately 3 736 980 km^2 and account for 40% of the national land area (Bailey, 1995). Tall- and short-grass prairies, montane spruce-fir and pygmy coniferous forests, chaparral, sagebrush and cacti formations typify the steppes and deserts of the Great Plains, Great Basin, Rocky Mountains, Black Hills, Grand Canyon, and the Mojave and Sonoran Deserts.

The three ecoregions that comprise the Humid Tropical Domain are all found at low latitudes, have an average temperature above 64°F (18°C), experience heavy annual rainfalls, and have no winter season. Together, the tropical savannas of the Florida Everglades and the rain forests of Puerto Rico and the Hawaiian Islands cover 46 800 km^2 and account for 0.5% of the national land area (Bailey, 1995). Cypress, mangroves, saw grass and palmettos are the predominant vegetation of the Everglades. The montane rain forests of Puerto Rico are concentrated in the Sierra de Luquillo, where mahogany, ebony, Spanish cedar, rosewood, sandalwood and mango are among the characteristic Neotropical tree species. By contrast, the rain forests of the Hawaiian archipelago, being the most isolated in the world, are populated by an exceptional array of endemic plants, land snails, flies and forest-dwelling birds, most notable among the latter being the honeycreepers (family Drepanididae). Nine of the 21 endemic species of Hawaiian forest birds have already gone extinct, and the remaining 12 are threatened with the same fate.

Flagship species

A number of species have served as flagships for conservation action in the United States, including some that were once very common or widely distributed throughout the nation, as well as others that have always been rare. Noteworthy among these are some of the larger charismatic mammals, birds, and reptiles such as the American bison (*Bison bison*), the grizzly bear (*Ursus arctos horribilis*), the gray and red wolf (*Canis lupus* and *C. rufus*), the black-footed ferret (*Mustela nigripes*), the bald eagle (*Haliaeetus leucocephalus*), the peregrine falcon (*Falco peregrinus*), the California condor (*Gymnogyps californianus*), the northern spotted owl (*Strix occidentalis caurina*), the American alligator (*Alligator mississippiensis*), as well as flagship species that are somewhat more obscure creatures such as the snail darter (*Percina tanasi*). Finally, there are the giant sequoia (*Sequoiadendron giganteum*) and coastal redwood (*Sequoia sempervirens*), the tallest trees in the United States and the largest living organisms on Earth.

North America's largest terrestrial mammal, the American bison (*Bison bison*) once ranged over a third of the continent and is believed to have numbered 60 million or more when Europeans first colonized the New World in the early sixteenth century. It also was probably the most abundant large mammal on Earth at that time. Westward-moving pioneers organized the first bison hunts in the early 1800s, initiating the slaughter that would reduce the once prodigious herds to less than one thousandth of their former size. In fact, less than 1 000 remained when national legislation to prevent the further killing of bison was passed at the turn of the present century, and many of those animals resided in zoos and on private reserves. In 1905, conservationists at the Bronx Zoo helped organize the American Bison Society and later reintroduce captive animals to the newly-created Wichita Mountains National Wildlife Refuge. Additional releases and other efforts to build the wild population have successfully established a national herd of more than 5 000 among five separate populations in and around Yellowstone National Park, Custer State Park, Badlands National Park, Theodore Roosevelt National Park and the National Bison Range. Canada's Wood Buffalo National Park supports a population of some 15 000 bison (Ricciuti, 1995).

The Spanish explorer Coronado was probably the first European to view a grizzly bear (*Ursus arctos horribilis*) during his 1540 expedition to what is now New Mexico. Although accurate estimates of the pre-colonial grizzly population do not exist, anecdotal accounts of trappers and explorers suggest that its numbers began to decline seriously beginning in the 1800s with the advent of the repeating rifle. Killed primarily because it represented a threat to human populations and was a livestock predator where its territory overlapped expanding Western ranchlands, the grizzly was driven from much of its former range and ultimately extirpated in Arizona, Colorado, New Mexico, North Dakota, Oregon, Texas, Utah and California, although it still adorns the state flag. Today, Canada and Alaska are its principal strongholds, while presumably viable populations remain in remote mountain regions of Montana,

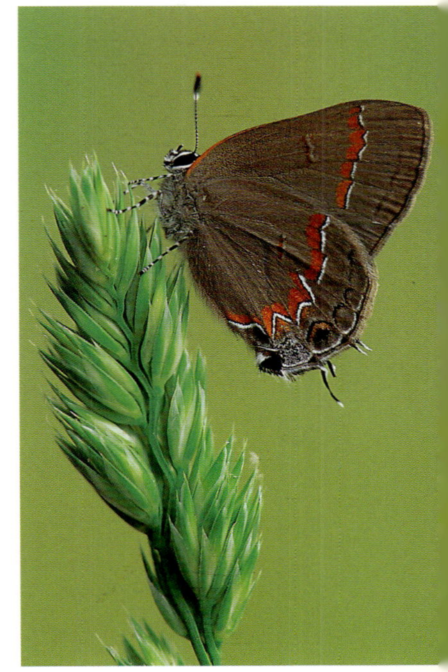

Red-banded hairstreak butterfly (Calyeopsis cecrops), *which ranges from Ohio and New Jersey south to Texas and Florida.*

On pp. 390-391, the mountain goat (Oreamnos americanus) *is one of the most spectacular large animals of western North America. Here, a mother and young feed in a meadow at Hidden Lake in Glacier National Park, Montana.*

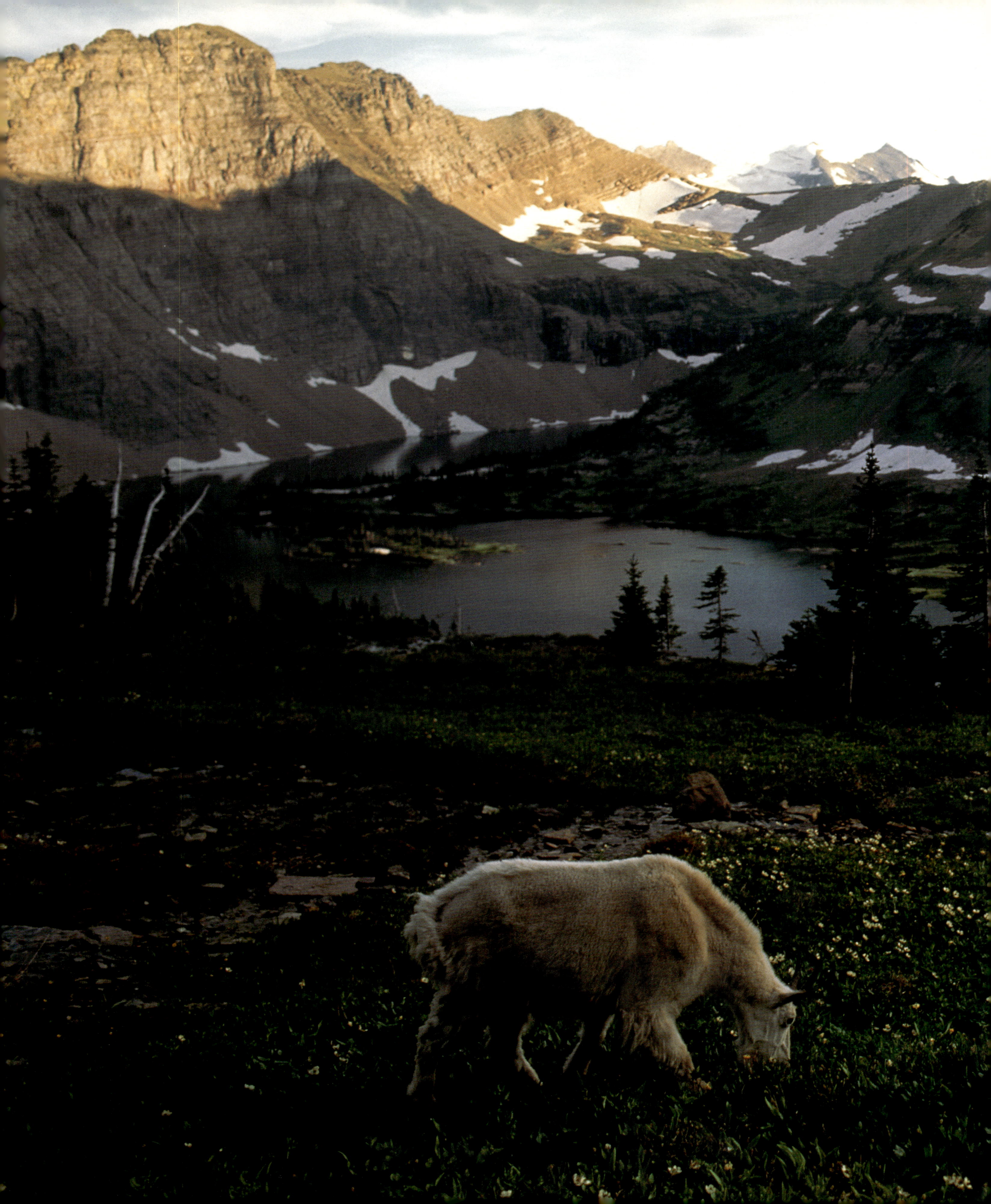

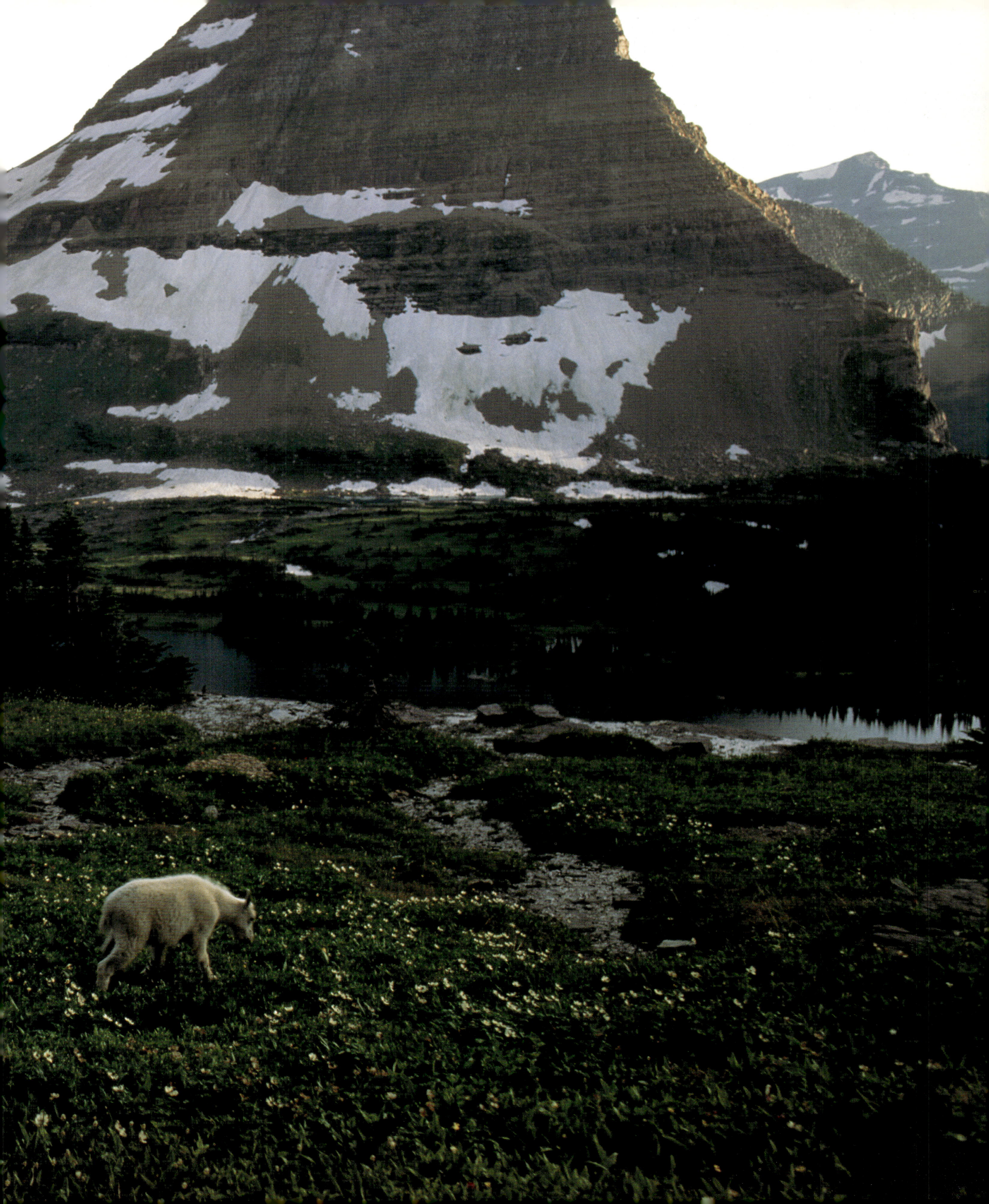

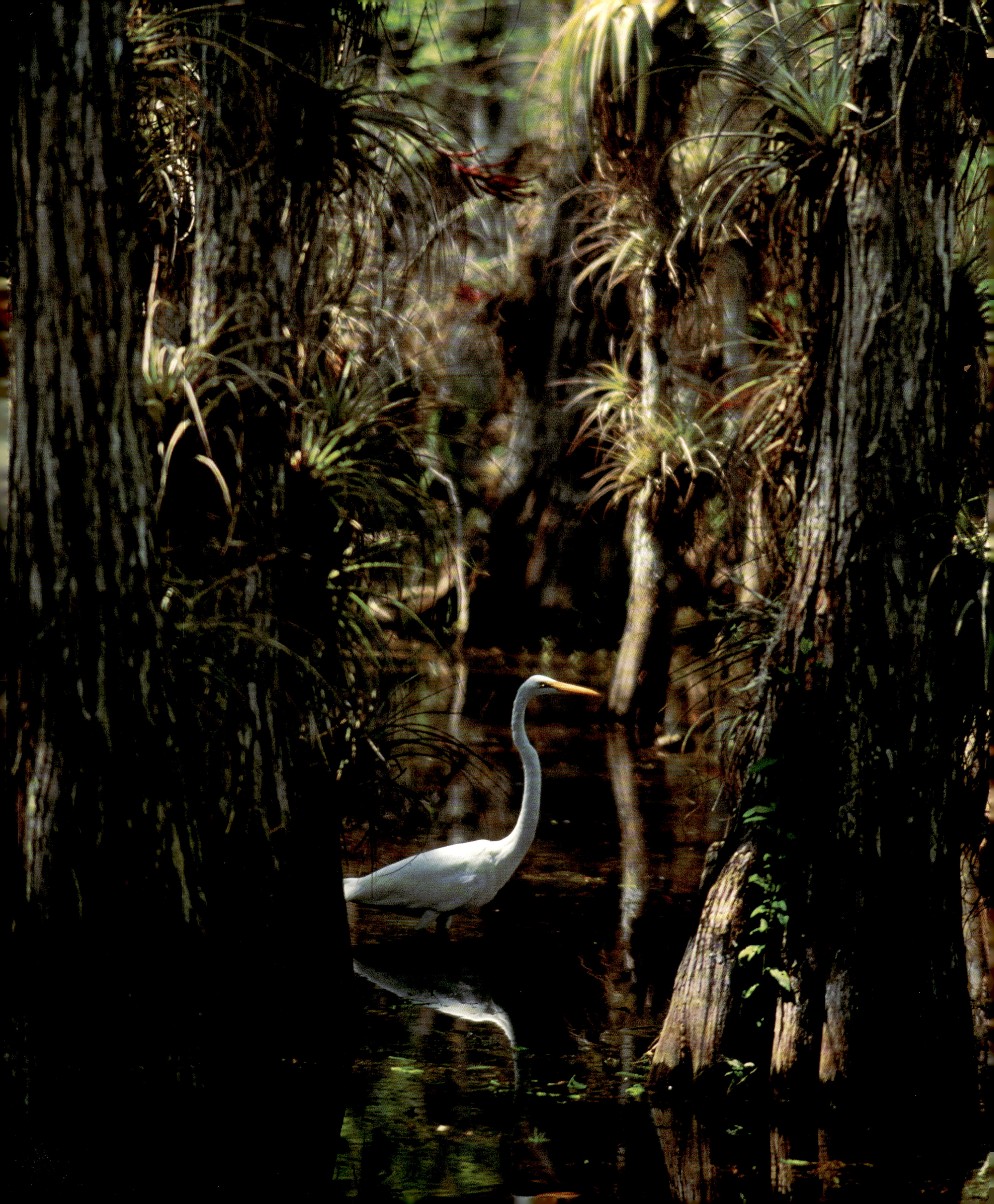

Wyoming and Idaho (Craighead, 1979). Several organizations have been created specifically to support public awareness, research and conservation efforts on behalf of grizzlies and other bears (Brown, 1993; Murray, 1995).

Historically, wolves (*Canis lupus*) have been shot, trapped and poisoned throughout the entire continental United States because, like the grizzly, they represent a threat to man and his domestic animals. The Massachusetts Bay Company established the first wolf bounty in 1630, offering a penny per hide, and the bounty system eventually spread through many states. By the mid-1900s, wolf populations had been so drastically reduced that gray wolves survived only in Michigan, Wisconsin and Minnesota; Mexican wolves occasionally wandered north into New Mexico and Arizona; and the red wolf was barely hanging on in eastern Texas, Louisiana, Oklahoma, Arkansas and Missouri (Matthiessen, 1987). In 1980, the red wolf was declared biologically extinct in the wild and only 17 animals survived in captivity. However, a reintroduction program in North Carolina's Alligator River National Wildlife Refuge has successfully reestablished a wild population of more than 60 red wolves (Chadwick, 1995). In the west, remnant gray wolf populations are also expanding and a much-debated experiment to return gray wolves to Yellowstone National Park is now underway (Chadwick, 1995; Link and Crowley, 1994).

First described by naturalists John James Audubon and John Bachman in 1851, the nocturnal and elusive black-footed ferret (*Mustela nigripes*) was not sighted again for 25 years. Black-footed ferrets once ranged throughout the Great Plains from Texas northward into southern Saskatchewan, and from as far west as the Rocky Mountains eastward through the Dakotas, Nebraska, Kansas and Oklahoma. The species has always been an obligate predator of the prairie dog, which caused the ferret's numbers to decline precipitously as a direct result of intensive efforts by western ranchers to eliminate prairie dog populations that competed with livestock for forage. By 1972, in fact, the black-footed ferret was believed to be extinct in the wild. After years of unconfirmed sightings, however, a ferret killed by a dog led to the discovery of a small remnant population in 1981 near Meeteetse, Wyoming. By 1984, biologists estimated that 130 black-footed ferrets survived at this lone site. Regrettably, the population eventually crashed due to disease outbreaks, and what are believed to be the last 19 wild ferrets were taken into captivity. Thanks to the efforts of the U.S. Fish and Wildlife Service, Wyoming Game and Fish Department, and five zoological parks, this captive population has increased to more than 500 and a reintroduction program for this species was initiated in 1991. Despite significant levels of mortality experienced in the earlier releases, the recovery program for the black-footed ferret has as its goal to establish 10 free-ranging populations. Reintroductions have been or will very likely be attempted in Wyoming, Montana, South Dakota, Utah, Colorado, and Arizona.

The bald eagle (*Haliaeetus leucocephalus*) and peregrine falcon (*Falco peregrinus*) ranged widely across the North American continent until the latter half of the twentieth century, at which time populations of both species began precipitous declines. Biologists identified organic pesticides, especially DDT, as the principal cause, the chemicals moving upward through natural food chains to accumulate at high levels in the tissues of these predatory birds. The poisons prevented proper calcification and caused thinning of the birds' egg shells to the point that they were crushed under the weight of brooding parents. Fewer and fewer chicks were produced each year as pesticide concentrations increased in populations of eagles and other affected birds of prey.

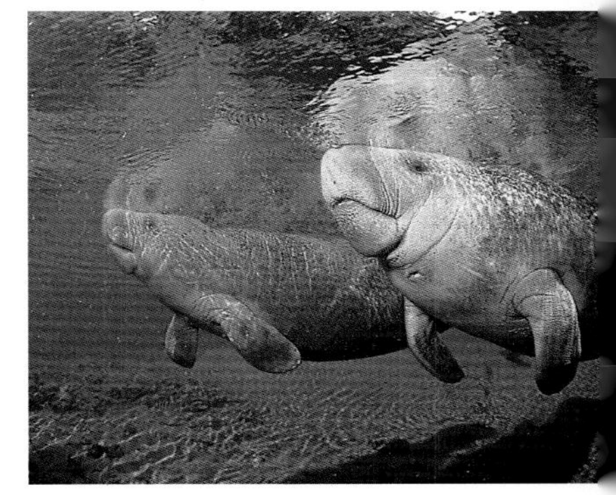

The bald eagle, already the national symbol, became the flagship of the 1973 Endangered Species Act, the nationwide effort to ban pesticides harmful to native wildlife, and programs to reestablish breeding eagle and falcon populations in a number of states. Fertile eggs and healthy young chicks from Canadian bald eagles were placed in the nests of U.S. foster parents. Under the tutelage of wildlife biologists, young birds were "hacked back" to the wild in a series of time-consuming and labor-intensive experiments. The effort has been an unqualified success. The breeding bald eagle population in the lower 48 states has increased tenfold since the 1960s to more than 4 000 pairs. In fact, the eagle's numbers have risen to the point that the species was downgraded from Endangered to Threatened status several years ago.

Although not nearly as well-known to the general public as the bald eagle, the peregrine falcon has always been a favorite of falconers, and it was this community that took up the cause of its survival. The Peregrine Fund, created in 1970 with a focus on a single species, is now an internationally-recognized conservation organization that has expanded its mission to include the protection of endangered birds of prey in many different parts of the world. In the U.S., efforts to restore declining peregrine falcon populations have included widespread reintroduction programs. In some instances, young peregrines have even been reestablished in major metropolitan areas in nests located on bridges and skyscraper ledges.

The common egret (Casmerodius alba) *(on the opposite page) and the manatee* (Trichechus manatus) *(above) are two species that find refuge in Cypress Natural Preserve in Florida. The manatee is one of the most popular flagship species in Florida. It is threatened by a variety of factors, including increasingly heavy pleasure-boat traffic in rivers in which it lives.*

The California condor (*Gymnogyps californianus*) is the largest North American bird, and one that historically ranged across much of the continent. Although a number of Native American groups considered the condor a sacred animal with supernatural powers, westward-moving settlers and ranchers typically shot these birds for target practice, collected their eggs, poisoned the animals upon which condors fed, and ultimately crowded the majestic condor out of much of its natural habitat, to the point that the species was declared endangered in 1967. By the 1970s, only 25-35 condors remained in the wild. Controversy quickly erupted over the species' recovery, until it was decided to remove the last remaining birds from the wild and attempt a

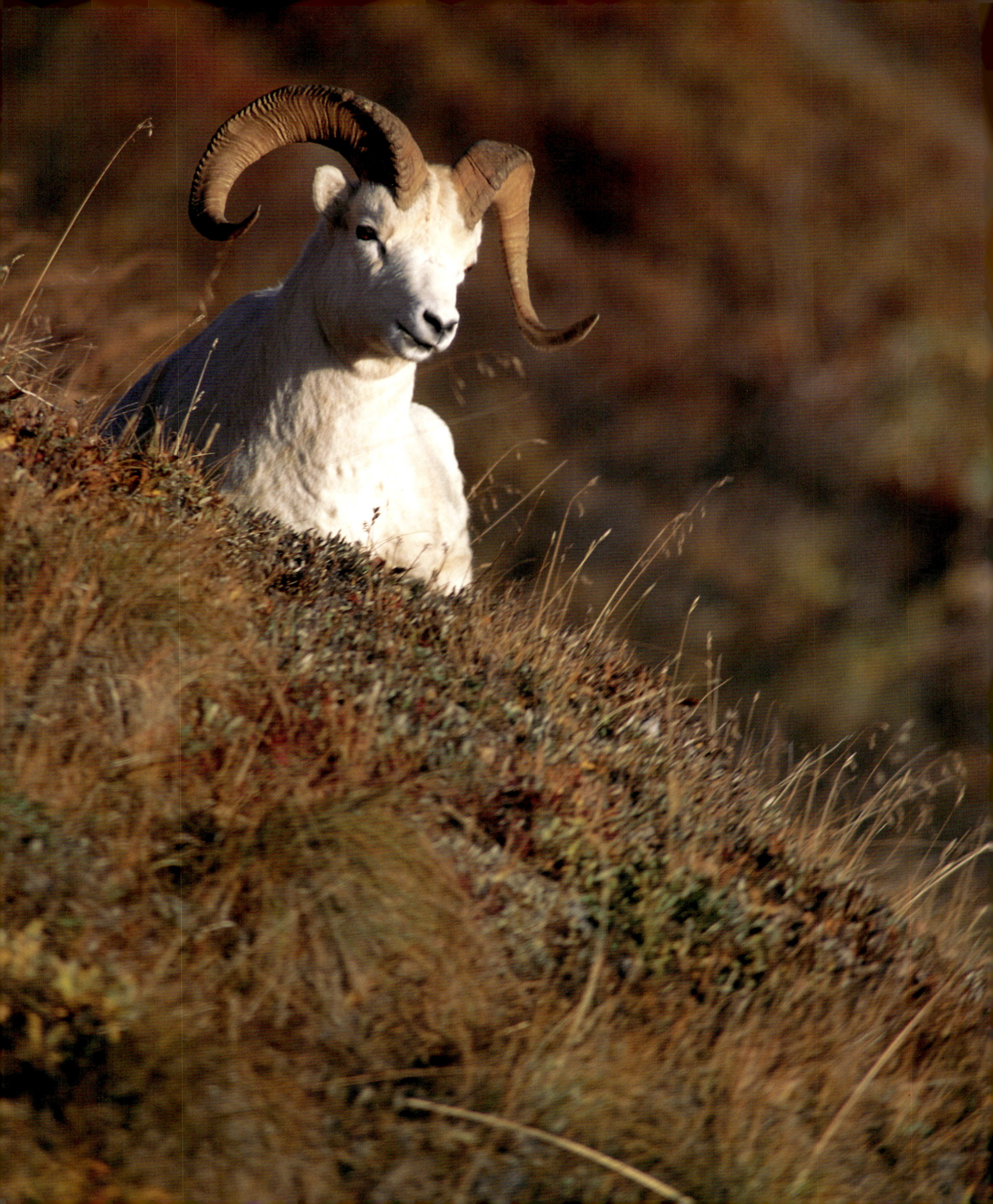

captive breeding/reintroduction program. This effort (a collaboration between the U.S. Fish and Wildlife Service, the California Department of Fish and Game, the National Audubon Society, the San Diego Wild Animal Park and the Los Angeles Zoo) resulted in the first successful breeding of the California condor in captivity (1988), an increase in its numbers to 119 (1997), and the successful return of 26 birds to the wild (21 in California and 5 in Arizona).

The northern spotted owl (*Strix occidentalis caurina*) ranges from southwestern British Columbia (Canada) southward through the Coast and Cascade Ranges into northern California. It depends on old-growth forest for its existence, each owl pair apparently requiring 800 or more hectares of forest comprised of 200 year old, 60 m-tall spruce and fir trees. Unfortunately, these same forests are prized by loggers. A conservation battle has been raging for the last decade over the fate of these forests and that of the owl, whose numbers appear to be down to the low thousands in the United States, and much lower in Canada. Approximately three fourths of its remaining habitat is public land, lying within the U.S. system of national forests. At this point, close to 3 million ha have been designated critical habitat for the northern spotted owl and are exempt from logging.

The American alligator (*Alligator mississippiensis*), the largest of North America's reptiles, gets its name from early Spanish explorers of the New World, who called it *el lagarto* —"the lizard." Alligators were for centuries widely abundant in the marshes, swamps, ponds, and drainage canals of the southeastern United States, but the value of alligator hides (and secondarily, of their meat) spawned a multimillion dollar industry and decades of illegal poaching (1940s through the 1960s) that ultimately threatened the species with extinction. Thanks to protection received under the Endangered Species Act and the Lacey Act (which made interstate shipping of illegally-taken animals or their products a federal violation), wild alligator populations began to climb again in the 1970s, and are now at the point that the species' national status has been downgraded from endangered to threatened.

A single, isolated population of the 5 cm-long snail darter (*Percina tanasi*) was discovered in a small tributary of the Tennessee River in 1973. One of approximately 150 species of darters and other members of the perch family, the snail darter is a bottom-dwelling fish that requires clean, fast-moving freshwater streams for its survival and can be considered an indicator species for this rapidly-diminishing habitat. In fact, the deterioration of freshwater river ecosystems in the United States is signaled by the fact that close to a third (47) of all native perch species are now considered endangered —all of them darters.

The plight of the snail darter goes to show how local and regional issues can be brought to national and international attention by means of a flagship species. In 1979, under the terms of the Endangered Species Act, the Tennessee Valley Authority was ordered to halt construction of the Tellico Dam —a multimillion dollar project— because it would destroy critical snail darter habitat. Anticipating that this ruling might not be heeded or that the legal decision might eventually be overturned, biologists began translocating snail darters to a number of other streams in the region. Ultimately, the U.S. Congress amended the Endangered Species Act and created a means by which projects of significant economic importance could be approved despite the negative impact they might have on an endangered species and its habitat. Fortunately, although the Tellico Dam was ultimately completed and the original snail darter habitat permanently destroyed, a number of the translocated populations survived and the species was downgraded to "threatened" status in 1984.

Abundant throughout North America in prehistoric times, redwood species and populations declined significantly during the Pleistocene ice ages and are now found only in the moist, old-growth forests along the Pacific coast. Exceeding 75 m in height and 30 m in circumference, the giant sequoia (*Sequoiadendron giganteum*) is the most massive living creature ever to inhabit the Earth. Coastal redwoods (*Sequoia sempervirens*) are even taller, with record individuals surpassing 100 m in height, but they are far less bulky than giant sequoias. Settlers pouring into northern California in the 1840s sought the termite-proof, rot-resistant redwood for construction purposes, but made relatively little impact on forest stands during the nineteenth century. However, with the mechanization of the timber industry and the advance of the railroads, west coast timber production increased dramatically during the early 1900s and reached its peaks in the 1950s and 1960s. Public outcries brought about the Save the Redwoods League and helped launch statewide and national campaigns to preserve these magnificent trees. Unfortunately, 95% of California's ancient redwood forest has already been destroyed. Most of the remaining 5% is found in officially protected areas.

A number of other species can also be considered "flagships" for conservation actions in the United States —among them, the Florida manatee (*Trichechus manatus latirostris*) and the Florida panther (*Puma concolor coryi*) in the southeast, and the Hawaiian forest bird community mentioned earlier—, each with a unique history and a significant impact upon the national effort to recognize and preserve biodiversity.

Cultural diversity

Anthropologists believe that aboriginal North Americans began emigrating from eastern Asia more than 50 000 years ago, crossing the Bering land bridge during Pleistocene glacial periods. From the northern Pacific coast region, nomadic groups moved southward along the Rocky Mountains and eastward along the river valleys of the Great Plains, eventually inhabiting the en-

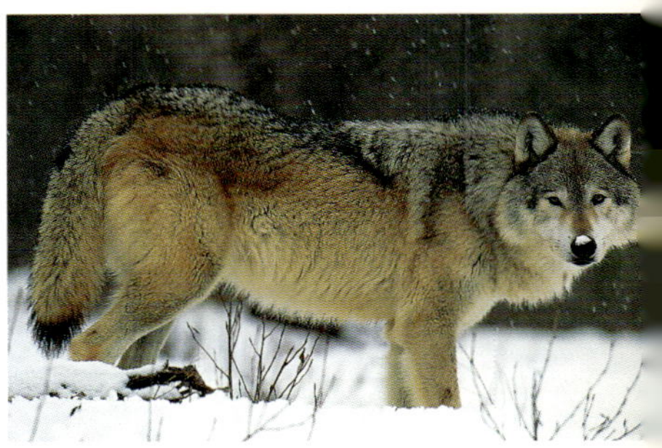

On the opposite page, the Dall sheep (Ovis dalli) *is found in rough mountain habitat over a large part of Alaska and adjacent parts of Canada. This individual was photographed in Denali (formerly Mount McKinley) National Park.*

Above, the timber wolf or gray wolf (Canis lupus) *is one of North America's best known and most maligned species. It is now the subject of a controversial, but so far successful, reintroduction program in Yellowstone National Park in Wyoming.*

tire continent. Big game hunting was the dominant way of life for the continent's early inhabitants, their prey species including woolly mammoths, camels, bison, giant beavers, giant ground sloths, giant armadillos, tapirs, musk oxen and peccaries. Many of these species appear to have gone extinct in North America between five and ten thousand years ago, a period which also coincides with archaeological evidence of increased hunting by nomadic tribes.

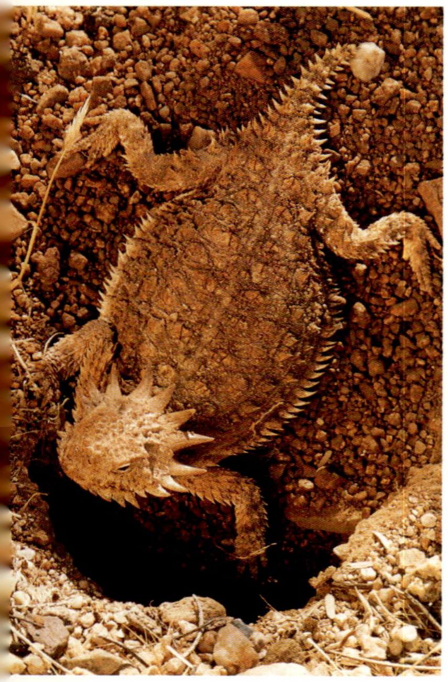

Aboriginal populations were densest where agriculture was eventually developed (largely in the southwest and southeast), along coastal areas rich in marine resources, and around rivers and lakes. Population estimates for early American cultures range widely. For example, the number of Indians inhabiting what is now the United States at the time Columbus landed in the New World has been estimated at anywhere from 750 000 to more than 10 million (Waldman, 1985) .

The terms "tribe" and "nation" have been used interchangeably by anthropologists to describe Indian groups that share a common language, culture and sociopolitical organization, but there is no agreement about how to classify these populations, extant or extinct. At the time of Columbus' arrival, however, there were believed to be more than 2 000 different language groups among native North Americans. Useful distinctions between Indian cultures can be made based upon the comparative use of wood, stone, metals, clay, shells, plant fibers and animal skins, bones and feathers to produce utensils, hunting implements, clothes and artwork. Differing designs and construction of living quarters and burial mounds, as well as religious beliefs, also help to distinguish between the many different groups (Waldman, 1985).

The regal horned lizard (Phrynosoma solare) *is found in Arizona and northern Mexico and is one of the 14 species of this unusual genus characterized by sharp spines on the head and dorsal surfaces of the body. Commonly known as "horned toads," these desert dwellers are of course lizards, not amphibians.*

On the opposite page, the Colorado River flowing through Grand Canyon National Park in Arizona. The western United States has some of the most spectacular natural scenery on Earth, and its extensive system of national parks and other protected areas conserves a large portion of the region's biodiversity.

The Algonquin- and Iroquois-speaking cultures established themselves along the entire eastern coast. The Iroquois were known as "People of the Longhouse" because of their large communal houses. The larger tribes of the southeast —Cherokee, Choctaw, Chickasaw, Creek and Seminole— were referred to by European settlers as the "Five Civilized Tribes." In the southwest, tribes such as the Hopi, Zuni, Yuma, Mojave, Apache and Navajo were heavily dependent on agriculture, while the Paiute, Ute and Shoshoni of the Great Basin region were largely hunter-gatherers. In the northwest, most tribes also relied heavily on hunting, as well as on fishing for salmon, trout and sturgeon. Indian culture evolved in the Great Plains in response to the introduction of the horse by Spanish explorers and the displacement of Native Americans by a westward movement of European settlers. The Apache, Cheyenne, Crow and Sioux were excellent horsemen and hunters who became dependent on the large herds of American bison for their existence (Waldman, 1985).

From the early 1500s through the late 1800s, European newcomers and the North American Indians engaged in what has sometimes been called the "Four-Hundred-Year War." The Spanish penetrated from the southeast, the French from the northeast, and the English from the full length of the Atlantic coast westward. As a result of European colonial expansion, Indian tribes were displaced steadily westward —ceding their lands, losing their culture, and declining drastically in number. The American Indian population reached its lowest point around 1900 when only 250 000 remained. Today, close to 1.5 million Native Americans live on more than 20 million ha in 30 states (most of it west of the Mississippi River), their land protected as federal and state reservations. Modern-day Native Americans represent approximately 0.6% of the U.S. population, and only about 300 languages survive within the borders of the U.S. from a total that 500 years ago may have been as high as 2 000 for North America as a whole (Waldman, 1985).

WILLIAM R. KONSTANT
RUSSELL A. MITTERMEIER

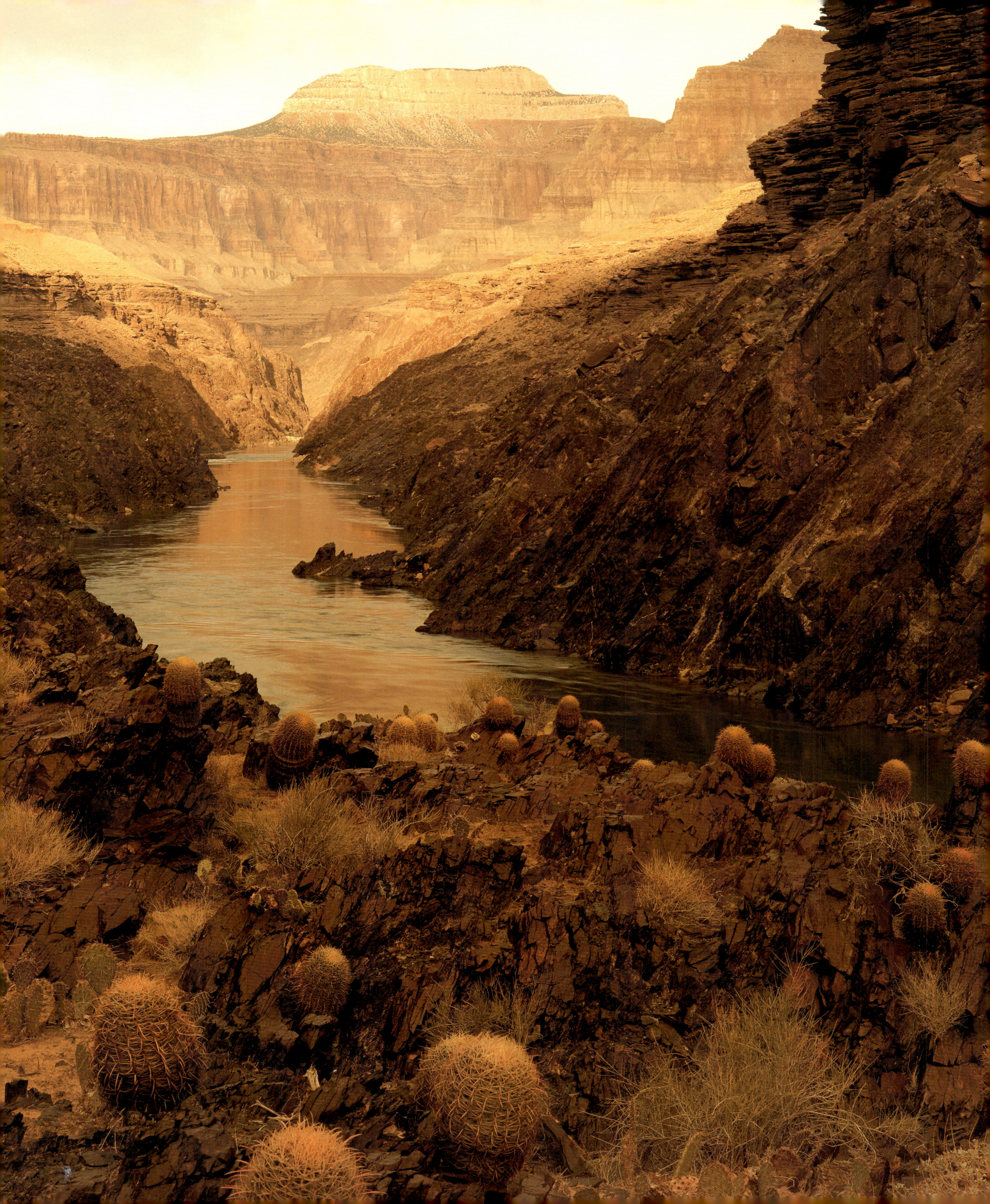

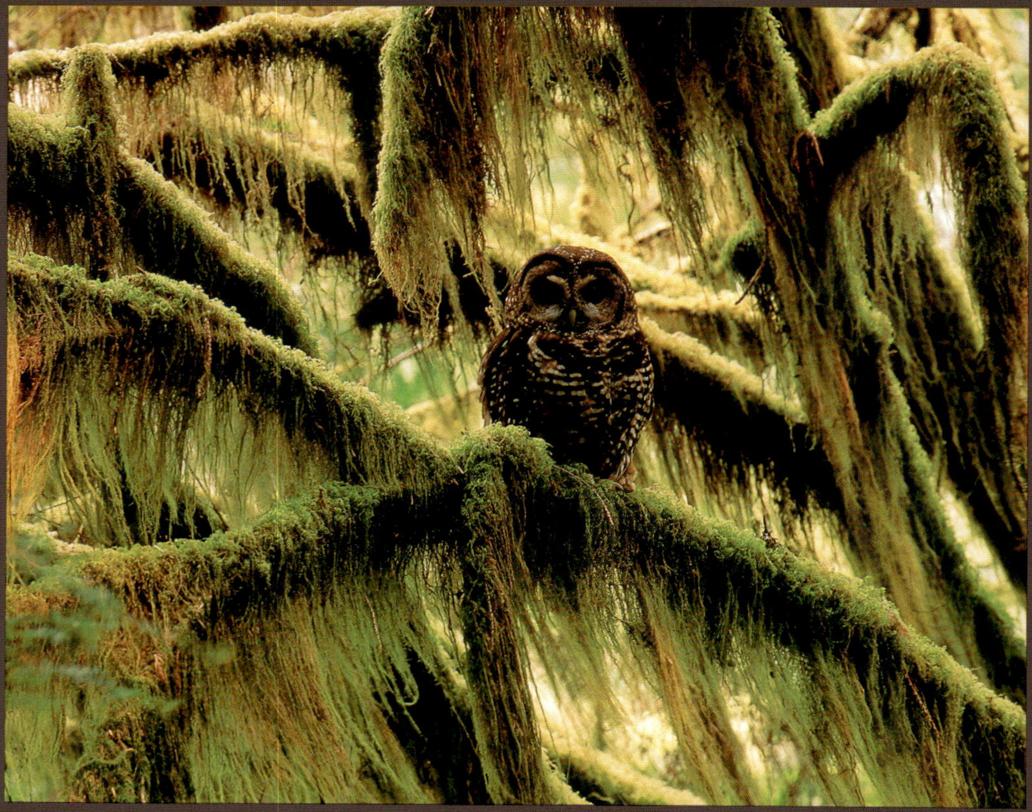

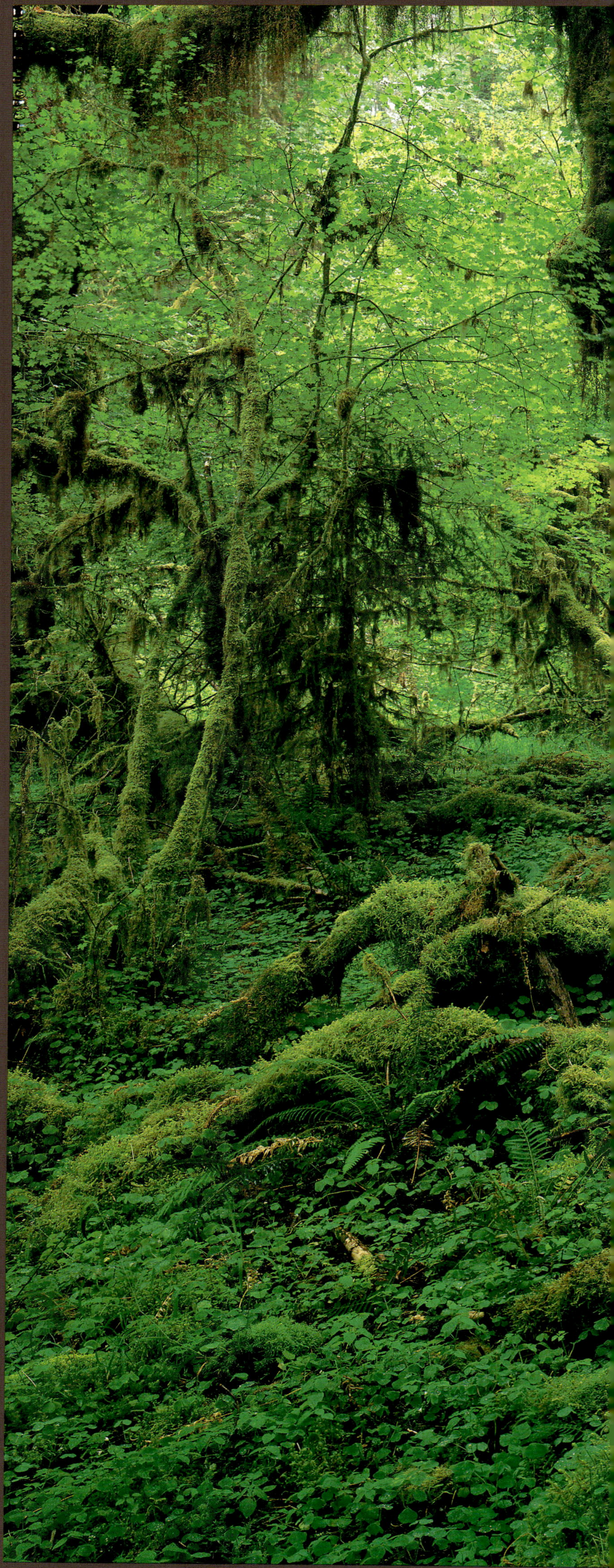

Above, temperate rain forest and spotted owl (Strix occidentalis) *in the Pacific Northwest of the United States. True temperate rain forests are even more endangered than their tropical counterparts. The spotted owl is the most famous flagship species in this unique ecosystem, which has been the source of much controversy between conservationists and loggers. The forest on the right is the Hoh Forest in Olympic National Park in Washington; club moss* (Lycopodium sp.)*, wood sorrel* (Oxalis oregana)*, and sword ferns* (Polystichum munitum) *blanket the forest floor and the lower portion of the trunks of big-leaf maples* (Acer macrophyllum).

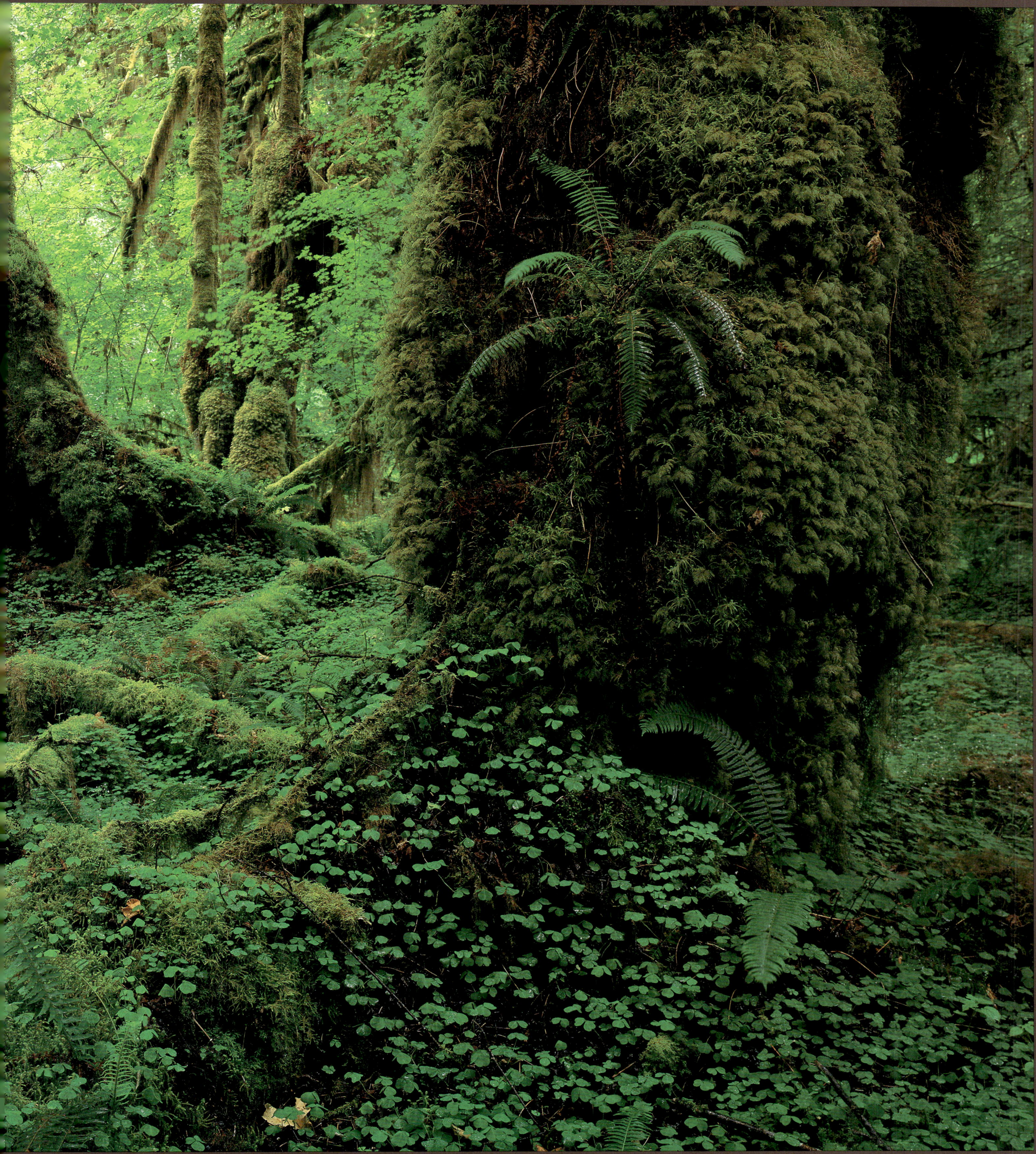

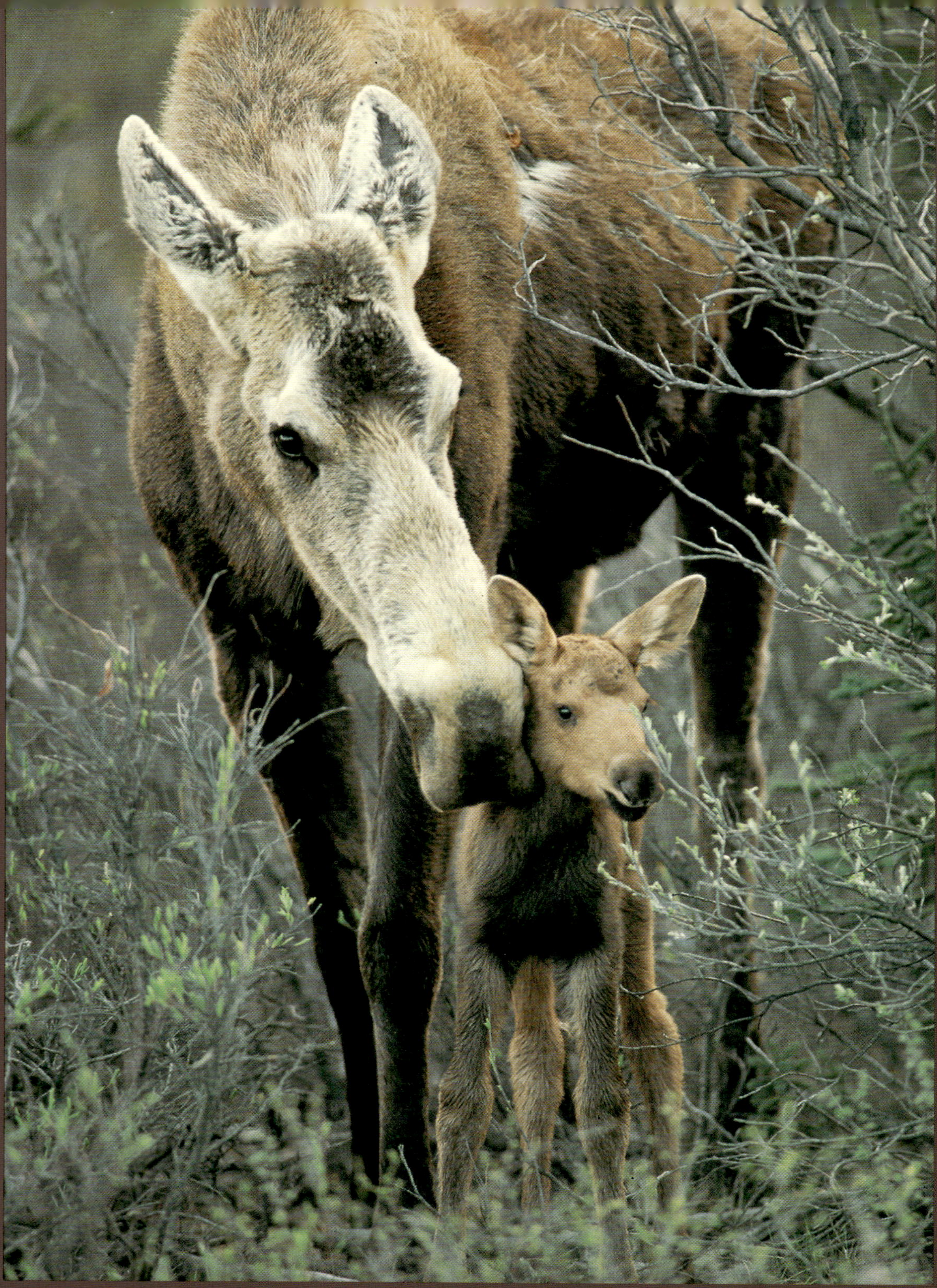

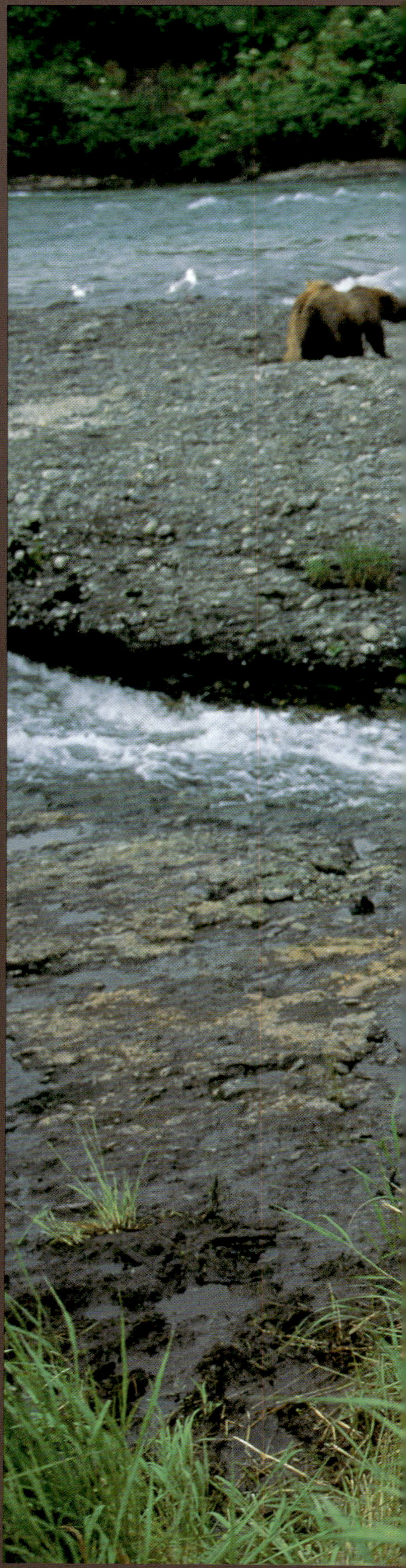

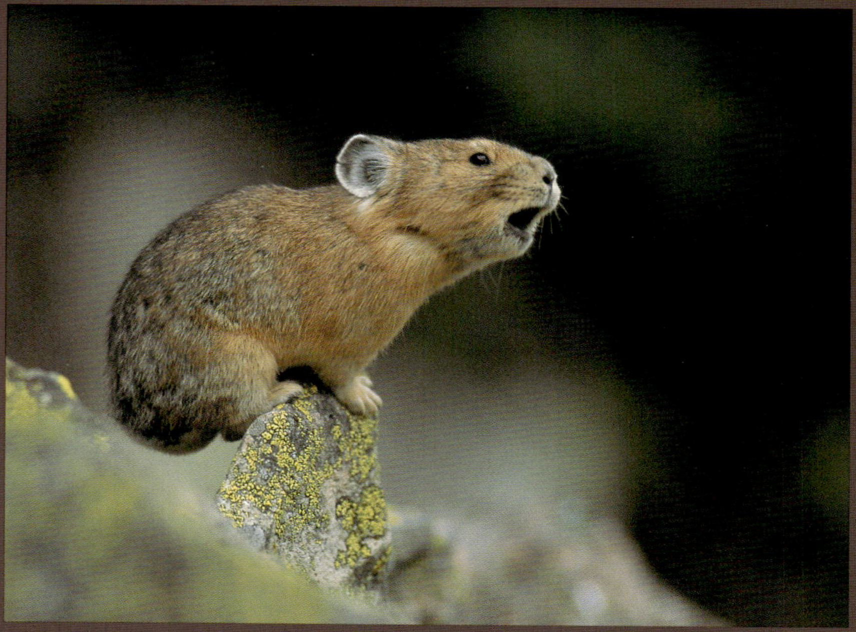

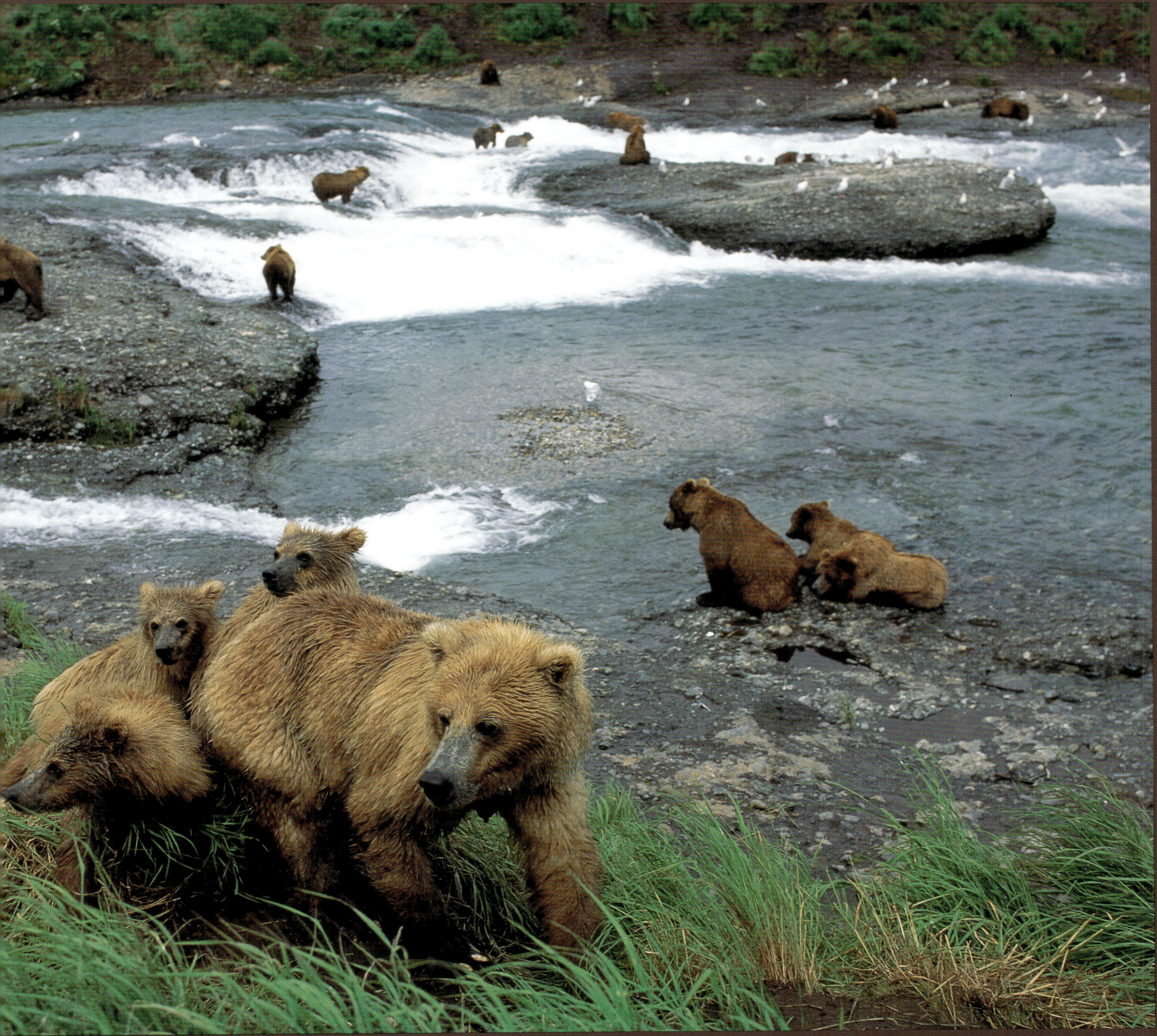

The two largest terrestrial mammals of the United States and two of the country's most important flagship mammals, the moose (Alces alces) *(upper left) and the Alaskan brown bear* (Ursus arctos) *(above). Here, a large number of brownies fish for salmon in the famous McNeil River in Alaska. Lower left, a pika* (Ochotona princeps) *calling to declare its territory in the mountains of Colorado.*

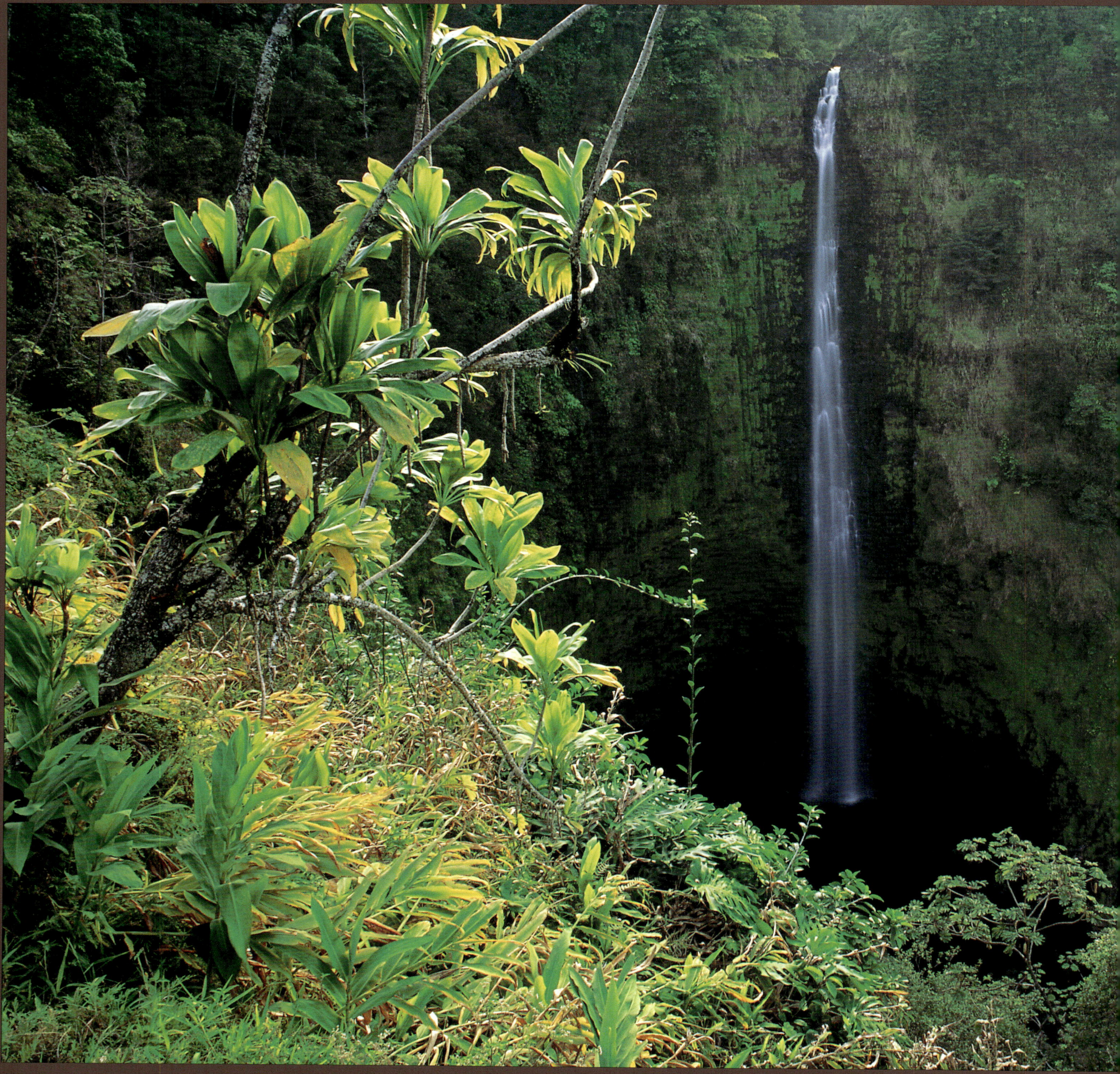

The Hawaiian Islands account for a significant portion of the biodiversity of the United States and have suffered some of the most serious recorded extinctions in recent times. Above, the Garden Isle of Kauai, with its steep volcanic slopes and its falls, is the last major stronghold for several of Hawaii's most endangered birds. Upper right, the spectacular Hawaiian silverswords include 5 species of the endemic genus Argyroxiphium. *They are restricted to the slopes of Hawaii's highest volcanoes at altitudes of 1 200-3 500 m. When they reach 10-15 years of age, they send up 2-m stalks that burst into magnificent blooms, and then they die. Lower right, the iiwi (*Vestiaria coccinea*) is one of the honeycreepers (Drepanidinae), a subfamily of birds endemic to these islands. This is a spectacular adaptive radiation of finch-like birds that once included at least 21 species of which nine are extinct, and those that remain are endangered, some of them critically.*

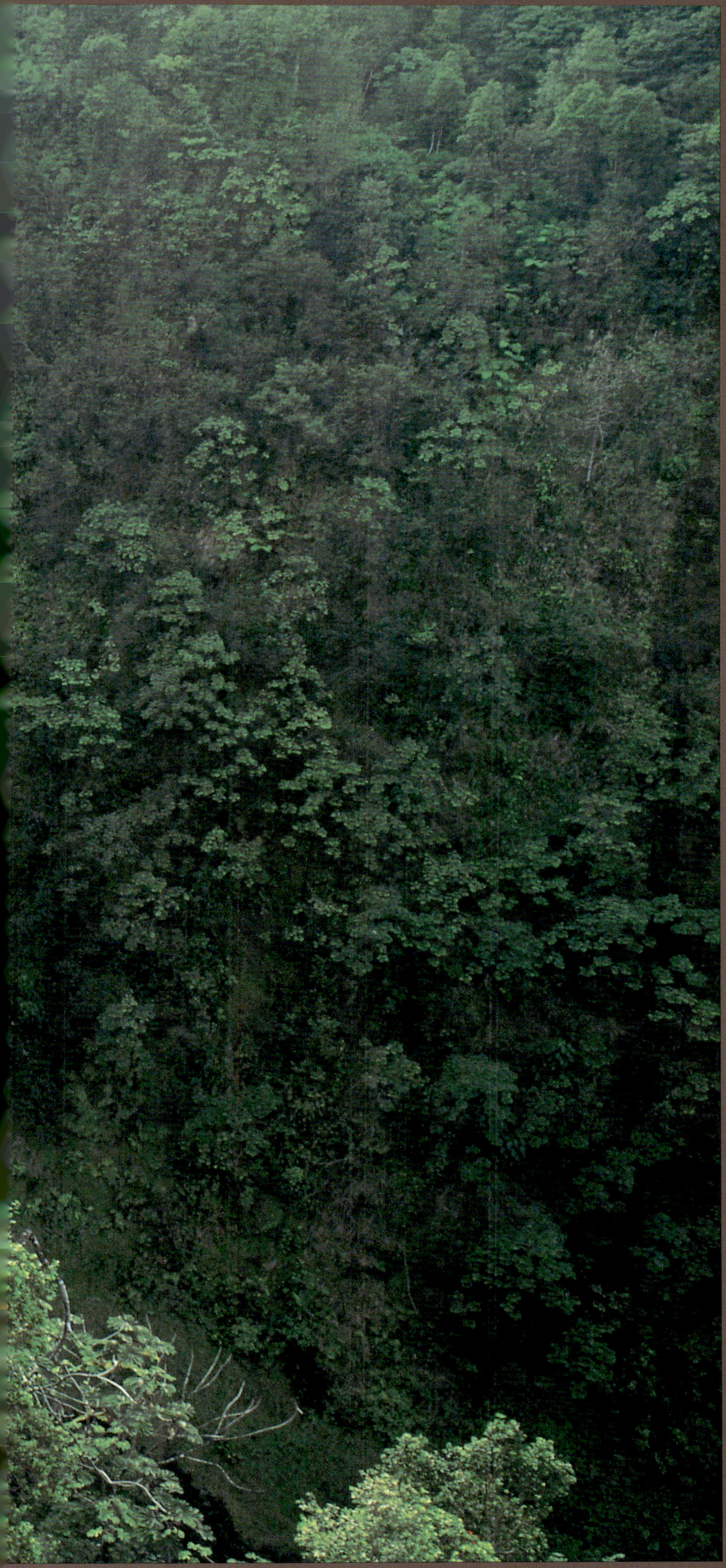

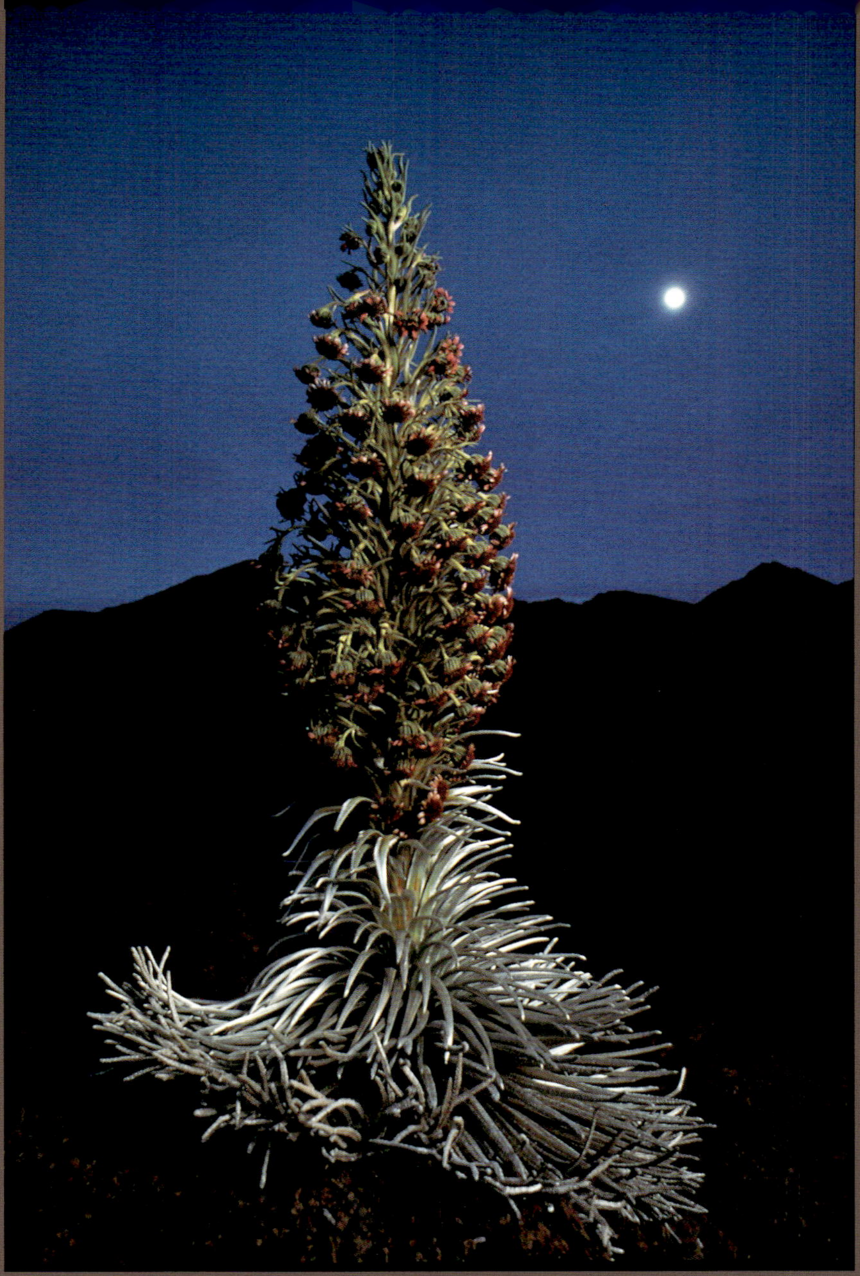

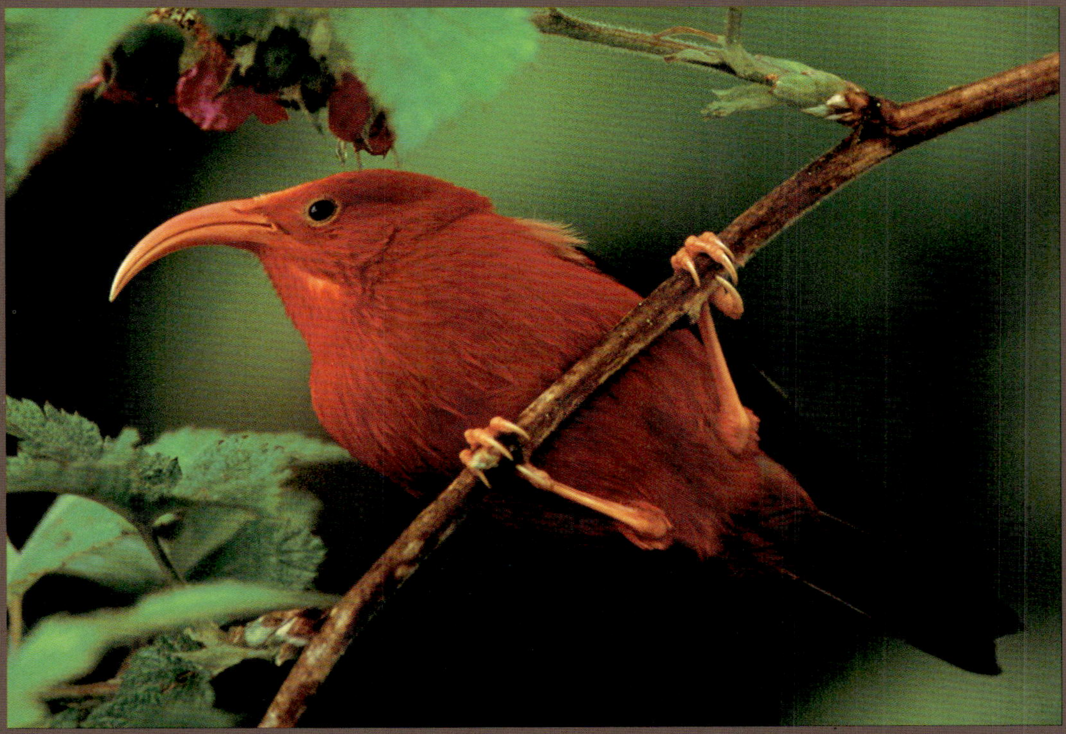

The indigenous population of the United States was and still is one of the most diverse of any country on Earth. Although a large number of cultures went extinct following the arrival of Europeans, many still exist and practice their ancestral traditions. Right, a group of Shoshone Indians, adorned with a variety of bird feathers, perform a pow-pow dance.

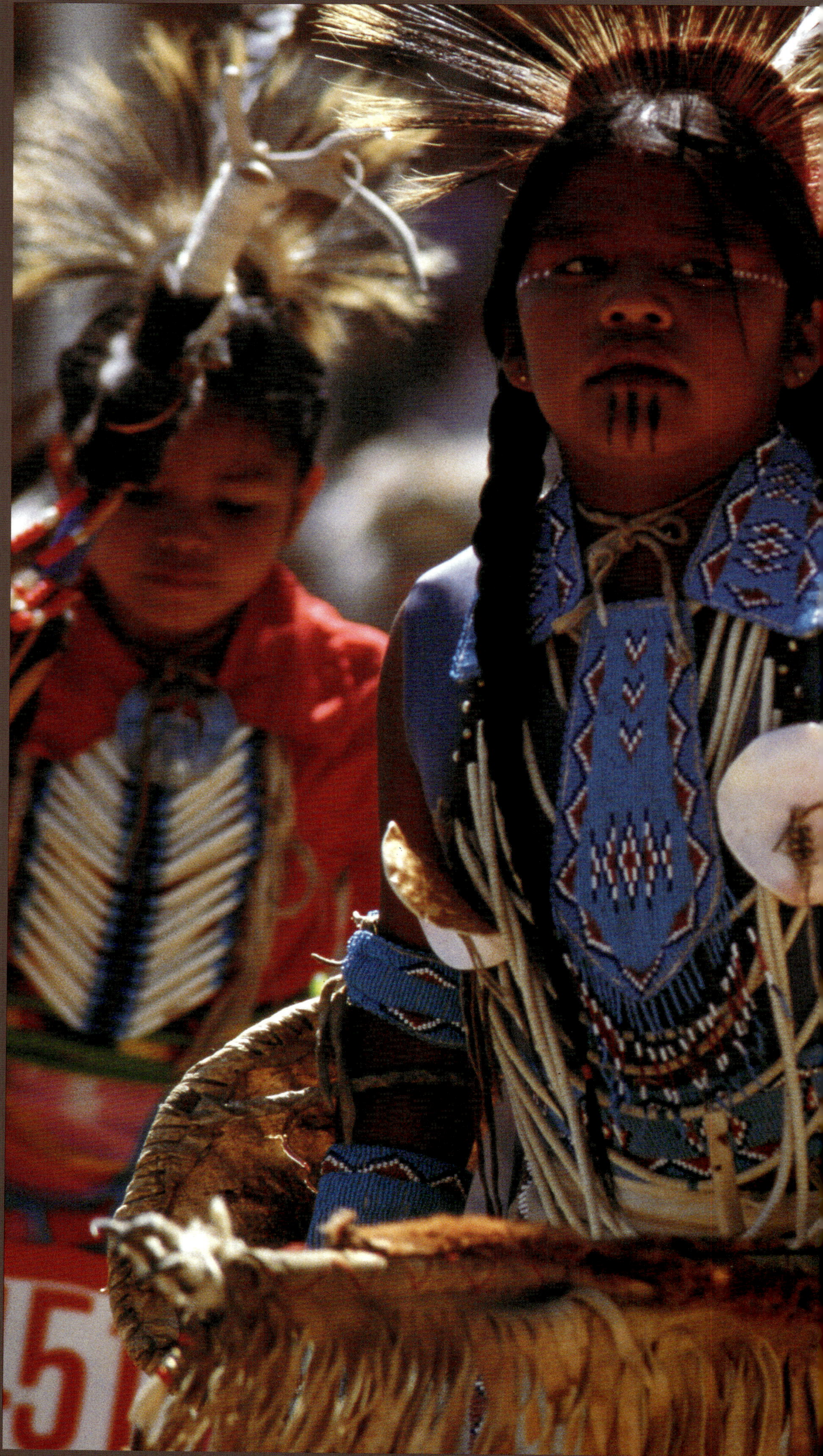

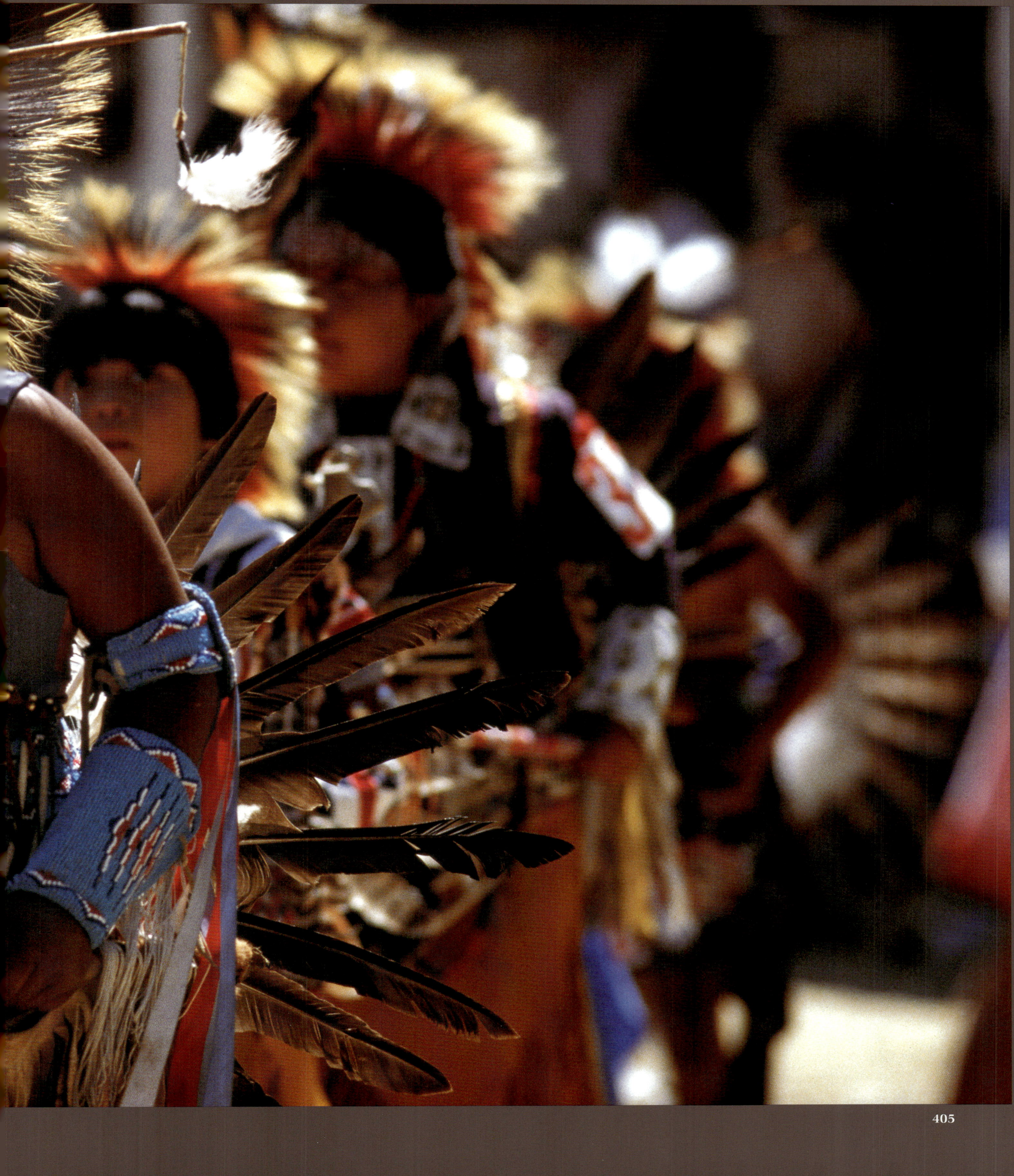

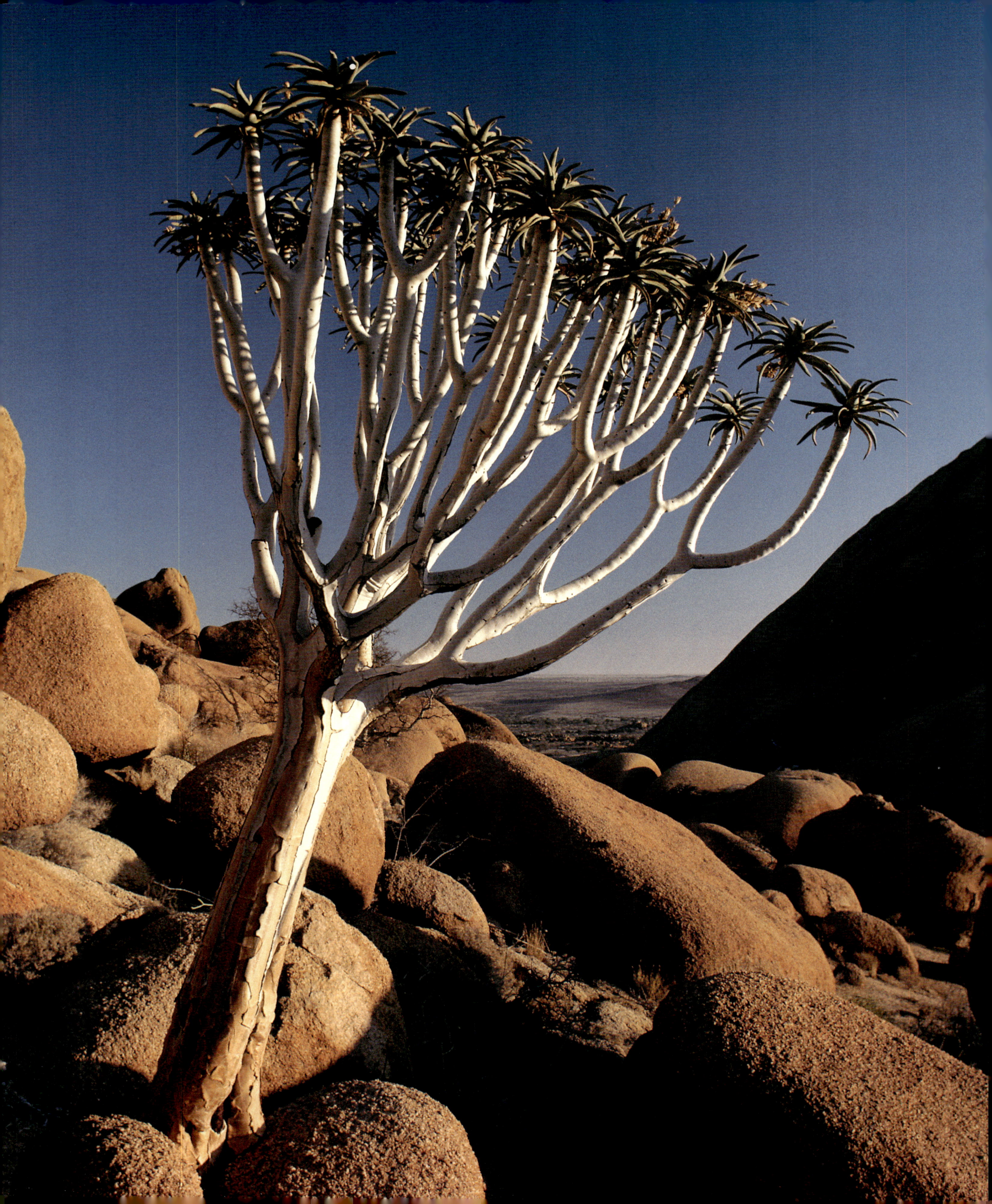

SOUTH AFRICA

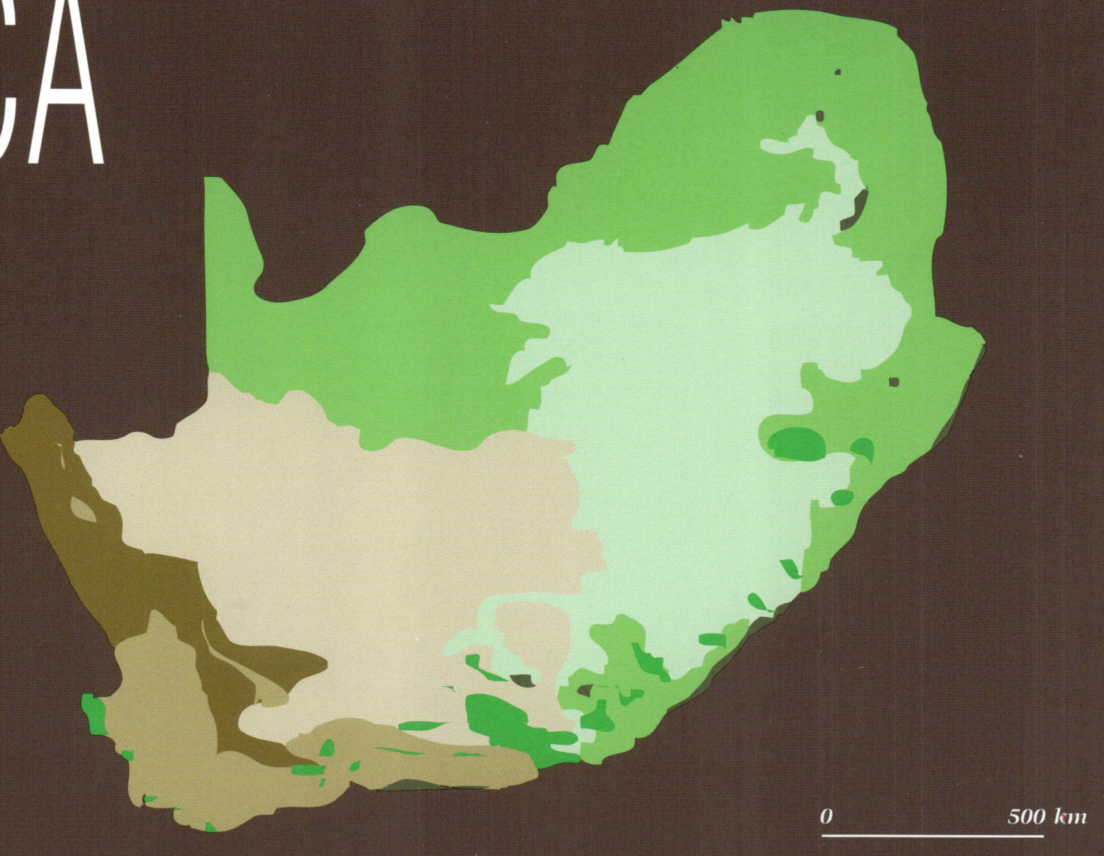

South Africa is one of the medium-sized megadiversity countries, its 1 221 037 km^2 placing it in tenth position on the list between Peru and Colombia. In many ways this country is unusual, in that it is the only one of the 17 that has no tropical rain forest systems within its borders. However, it has one of the five richest floras on Earth, and may turn out to have the highest level of plant endemism of any country. Its natural habitats range from some of the driest deserts in the world to the extensive savannas or veld formations of the interior, to the alpine habitats of the Drakensberg Mountains, and also includes the unique Cape Floristic Region.

South Africa makes the megadiversity list almost entirely on the strength of its floral diversity and endemism, and has within its borders three of the world's 19 threatened biodiversity hotspots: the Cape Floristic Region, the Western Cape or Succulent Karoo Domain and Maputaland-Pondoland. It also has six of the world's Centers of Plant Diversity and Endemism as identified by WWF/IUCN including, once again, the Cape Floristic Region, the Western Cape/Succulent Karoo Domain, and the Maputaland-Pondoland Region, and also the Drakensberg Afromontane Region, the Drakensberg Alpine Region and the Albany Center. The Cape Flora is so important that by itself, it stands as one of the world's *six major Botanical Kingdoms*, an amazing distinction for such a geographically circumscribed area. South Africa's plant diversity is estimated at 23 420 species, a huge number for a country its size, especially in the African context, and more than double the number of plants of the next African country on the list, the Democratic Republic of Congo. Its plant endemism is even more striking, with an estimated 16 500, of which some 5 870 are found as endemics just within the Fynbos of the Cape Floristic Region. This is a staggering number of endemic species by any measure, and only the world's richest countries for plants —Brazil, Colombia and Indonesia— might match it. Estimates of plant endemism in the world's top three megadiversity countries are still very rough, so it may turn out that South Africa will actually exceed them in the number of endemic plants.

Vertebrate diversity in South Africa is reasonably high, though not comparable to its plant diversity. Total non-fish vertebrate diversity stands at 1 415 species, 146 of which are endemic, a relatively low figure for megadiversity countries and the reason why South Africa doesn't make the top thirteen. The only groups in which it makes the top 12 are reptiles and tiger beetles. Its 299 reptile species place it eleventh on the world list, and 76 of these are endemic. In tiger beetles, South Africa is twelfth in diversity with 94 species, and tenth in endemism with 40 species. Bird diversity is moderately high at 774 species, a little more than the much larger United States, but only 7 bird species are endemic. Three of BirdLife International's Endemic Bird Areas are also found in South Africa. The mammal fauna includes 247 species, 27 of which are endemic, and the amphibians 95 species, of which 36 are endemic. Not surprisingly for a relatively dry country, the freshwater fish fauna is relatively small, with 153 species.

Cultural diversity in South Africa is relatively low among the megadiversity countries, but includes some important groups of African people, the oldest and largest community of white Africans of European origin, and an important group of East Indians from South Asia. The population is about 45 million, of which 76.5% are African (mainly Bantu), 12.6% Caucasian, 2.4% Asian (mostly East Indian) and 8.5% Colored (mixed African, Indian and Caucasian).

Although a recent addition to the megadiversity list, South Africa is clearly an appropriate member of this elite group and one of the most important countries on Earth for plant conservation. It is also one of only four megadiversity countries with a First World economy. In many other ways, however, it is still a developing country that was ostracized and isolated for decades because of its apartheid policy, and only reemerged as a member of the international community in 1994.

On the opposite page, a kokerboom tree **(Aloe dichotoma)**, *one of a wide variety of aloe species in the country.*

Natural habitats

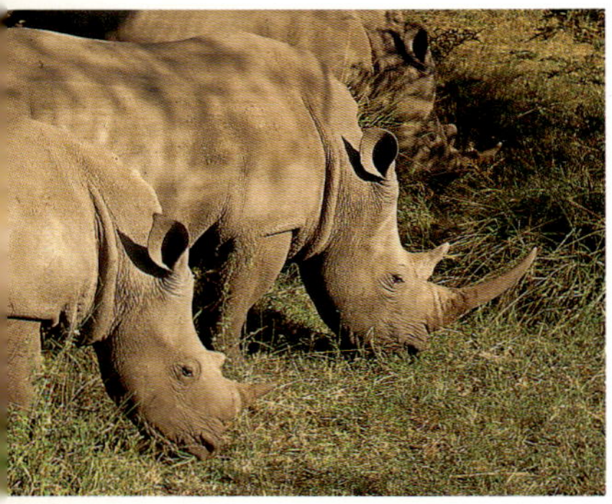

Larger than the black rhino (Diceros bicornis), *the white rhino* (Ceratotherium simum) *is found in several national parks in South Africa. This animals are still poached for their horns.*

The topography of South Africa is dominated by a central plateau of varying elevations which stretches over the country. The interior plateau or *highveld*, roughly comprising the Free State (formerly the Orange Free State) and Transvaal, is gently undulating grassland broken occasionally by highland areas. Most of the country's mineral wealth is concentrated in this region. In the western half of the plateau, the elevation is around 900 m and is dominated by a dry flatland known as the Karoo that merges with the Namib and Kalahari Deserts. On the southeastern edge of the plateau is a major mountain chain, the Drakensberg, which stretches from the eastern Transvaal southwest into the Cape Province. It is a major barrier to communication between the coastal and interior regions. In the southwestern Cape, there are ranges that rise to heights of 1 800 to 2 250 m. The interior plateau is in a wind belt, which brings rain from the Indian Ocean. In the eastern plateau region, rainfall is heaviest during the summer months, while in the western Cape rainfall peaks during the winter.

There are no navigable rivers in South Africa. Two major rivers, the Orange, and its tributary, the Vaal, flow from east to west. Hundreds of smaller rivers such as the Umzimvubu, Great Fish, Great Kei, Pongola, Umgeni and Tugela flow into the Indian Ocean. The legendary Limpopo River forms much of the northern boundary of the country.

Nine major biomes are recognized in South Africa. Starting from the east coast and moving west, the dry woodlands are dominated by mopane trees and shrubs and are typical of low-lying areas and river valleys below 1 000 m. The amazing baobab trees, with their voluminous trunks, are often common in this type of savanna, which sits on clay-based soils and commonly features seasonal flood pans. The dry woodlands have a wide diversity of species, but only a few are restricted to this biome. This is also a feature of the related, but distinct, mixed woodland biome. This type of savanna is characterized by relatively short trees, including *Acacia* and many broad-leaved species. The South Africans refer to this vegetation type as *bushveld*, which generally occurs at altitudes below 1 500 m. It includes much of South Africa's so-called *lowveld*.

The Afromontane grasslands and heaths above 1 500 m and up to 3 000 m are mostly found along the crests of the majestic Drakensberg. This is rugged countryside with erratic but generally high rainfall. Several species are restricted to this biome within South Africa, but also occur in East Africa's Rift Valley and in the Ethiopian Highlands. The *highveld* grasslands occupy much of the country's interior and are comprised of flat, hilly areas largely devoid of trees except along watercourses. This is the dominant vegetation type above 1 500 m. Cold, dry winters render this biome particularly prone to fire and frost, which prohibit the growth of most trees.

South Africa's forests are dominated by large evergreen trees and are rather limited in extent; they also contain few mammals but many birds. Coastal and lowland forests occur along the eastern seaboard, while montane forests occur in pockets above 1 200 m. The most extensive is the forested strip that extends 180 km by 16 km along the southern coast, where tall ironwood, yellowwood and stinkwood trees occur. The forest biome is characterized by high annual rainfall. Regarded productive areas for agriculture, the forests have become severely broken up in this century; today, only small fragments of true closed-canopy forest remain.

The Kalahari savanna provides the transition to the Namib Desert. This is very open terrain, dominated by seasonal grasslands with scattered trees and shrubs. Very deep sands underlie the vegetation, which occurs below 1 000 m and receives less than 400 mm of rain per year. The area is famous for its fossil dunes, dry river valleys and sand pans which give the area a desert-like appearance.

The Namib Desert occurs along the western seaboard, primarily north of the Orange River in Namibia. This area has an exceptionally arid climate and receives less than 100 mm of rainfall per year. This is a true desert, with giant sand dunes and succulent flora. Many species are restricted to this biome, including the famous "living fossil" plant *Welwitschia*.

Finally, the globally important Mediterranean-type systems of the Cape Region. The Cape Floristic Region encompasses the Cape Fold Mountains and their associated coastal plains and intermontane valleys in southern and southwestern Cape Province and several isolated outliers to the east and northwest. The region covers only 90 000 km^2, but has some 8 600 higher plant species, of which 5 870 (68%) are endemic; in terms of the number of species per unit area, this is probably the richest region on Earth for plants. About 7 000 (81%) occur within the Fynbos, which occupies 47% of the area and is the most important part of the Cape Flora. The Fynbos (Afrikaans for "fine bush," referring to the fine-leafed plant forms) is basically a sclerophyllous shrubland, fire-climax vegetation type that burns at intervals of 4-60 years; trees are typically absent, but Fynbos includes six different subtypes. Other formations in the Cape Floristic Region are *renoster* shrubland, karoo, thicket and Afromontane forest. In addition to high species endemism, there is high generic endemism (roughly 20%) and *fully six endemic families*, plus one family with only two extralimital species. This degree of endemism at the family level is even higher than that of Madagascar and New Caledonia and approaches that of Australia, which has 12 endemic plant families. Needless to say, this only further underlines the global importance of South Africa's flora.

The Cape Flora is also quite endangered, with 1 430 species or 17% of the flora of this region on the South Africa *Red List*. Threats include invasive alien species, agriculture, increasing urbanization because of a high population growth rate, and fire; although the vegetation is fire-adapted, too many fires can be a problem.

Mountain Fynbos, especially in the more species-rich west, is protected in nature reserves and mountain water catchment areas, but the lowlands, particularly the *renoster* shrubland, are very poorly protected. The region is to be proposed as a Biosphere Reserve, and should be given World Heritage status as well.

The Western Cape Domain occurs to the north of the Cape Floristic Region and is also a globally important hotspot, incorporating western parts of the Cape Province and parts of southern Namibia. It consists of low to dwarf succulent shrubland, dominated by stem and leaf succulents and fine-leafed evergreen shrubs. About 5 000 species occur here, of which some 35% are endemic, with very high diversity in succulent species. Threats to this region include overgrazing and poor land management practices leading to erosion and desertification, mining, alien plant invasion, urban development and overcollecting for the succulent trade; 900 species from this region are also on the South Africa *Red List* and less than 2% of the area is conserved.

Flagship species

Conservation in South Africa has focused on the creation of national parks and other protected areas, of which there are some 150. Over five million hectares are under some level of legal protection, of which national parks account for almost three million hectares. Largest and most famous is the Kruger National Park in the *lowveld* of the northeastern corner of the country. It covers over 1.9 million ha of dry woodland or *bushveld*, bounded on the south by the Crocodile River, on the east by Mozambique, to the north by the Limpopo River and to the west by the Transvaal Lowlands. Kruger is big game country, with 137 species of mammals, including African elephant, lion, leopard, buffalo and white rhino, in addition to 450 species of birds, 114 species of reptiles, 33 species of amphibians, 40 species of fish, and 227 species of butterflies. The vegetation varies from subtropical to tropical, with some temperate forms occurring at higher altitudes. More than 2 000 species of higher plants have been collected in this world-class national park which in peak periods accommodates over 3 000 visitors per day.

Other parks and reserves worthy of mention include: the Golden Gate Highlands National Park in the Free State; the Mountain Zebra National Park in the Great Karoo region; the Addo Elephant National Park of the eastern Cape; and the Kalahari Gemsbok National Park, which is South Africa's second largest national park, covering over 900 000 ha of wilderness, red dunes and dry riverbeds between Namibia and Botswana. Despite the general lack of water, large herds of antelopes thrive there. They have adapted their feeding habits, supplementing their diet with moisture-bearing desert plants such as tsama melon and wild cucumber.

South Africa's national parks and reserves are outstanding permanent features in a country that, in so many respects, is undergoing the trauma of rapid change. Game viewing and nature-based tourism have become important money-earners, drawing hundreds of thousands of tourists every year. Not only is South Africa fortunate to have developed a number of successful wildlife management schemes, which have brought several threatened species back from the brink of extinction, but it is also endowed with many of Africa's most beloved and charismatic species. Nowhere else on that continent is the concept of flagship species for conservation used to such perfection as in South Africa. Among many species used as vehicles for communicating the importance of biodiversity and ecosystem conservation are the rhinoceros, especially the white rhino and also the black rhino.

The white rhinoceros (*Ceratotherium simum*) has an unusual distribution, occurring in southern Africa south of the Zambezi River, and then again in northeastern Africa west of the Nile. The northern subspecies has suffered severely from poaching in recent years, and has declined to less than fifty individuals living in the DRC and perhaps the Sudan. The southern subspecies was almost exterminated during the last century, but effective protection since the 1920s resulted in a steady increase in the sole surviving population in the Hluhluwe-Umfolozi Game Reserve. By the mid-1960s, numbers had risen from perhaps 200 to nearly 2 000. As a result, the famous "Operation Rhino" was initiated by the Natal Parks Board to capture white rhinos alive for restocking other parts of their former range. This proved so successful that this subspecies could be removed from the list of endangered species. By 1987, even after translocation of many animals, there were still 1 660 white rhinos in the Hluhluwe-Umfolozi Game Reserve in South Africa, and 4 062 in the country as a whole. At this point, the main threat to the survival of the southern white rhino is posed by habitat deterioration due to the high densities attained by the species in areas like the Hluhluwe-Umfolozi Game Reserve. To help provide outlets for surplus animals, which still need to be removed annually, the Parks Board began placing animals on private ranch lands and selling them at public auction to facilitate "game ranching" as an alternative to traditional cattle ranching. In addition, some old males were sold to safari operators to be shot later by licensed hunters, generating important revenue to be plowed back into ongoing conservation efforts. Today, almost 7 000 white rhinos survive in South Africa, with an ever-increasing number on private lands —a testimony to a changing attitude towards biodiversity conservation and South Africa's innovative approach to wildlife management.

The black rhinoceros (*Diceros bicornis*) also has an important stronghold in South Africa, although this highly endangered animal is much less abundant than the white rhino. More than 50% of the world's remaining population of black rhinos occurs in Zimbabwe, South Africa and Namibia, with South Africa having a population of 577 in 11 protected areas as of 1987. In

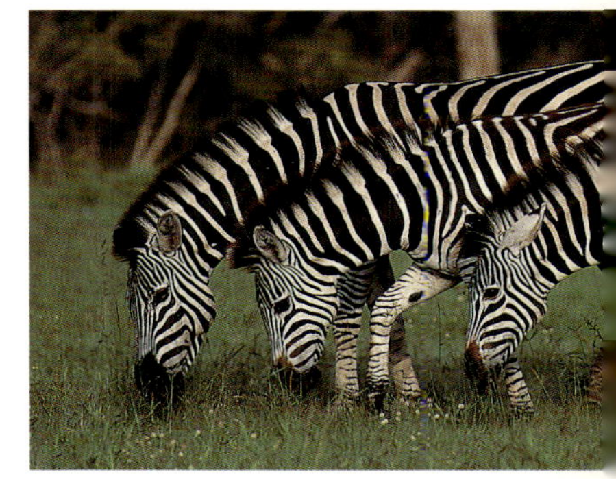

*A trio of Burchell's zebras (*Equus burchelli burchelli*) grazing in the Londolozi Reserve. When threatened, a family of this species closes up and eventually takes flight, with the lead mare at the front and the old stallion as rear guard.*

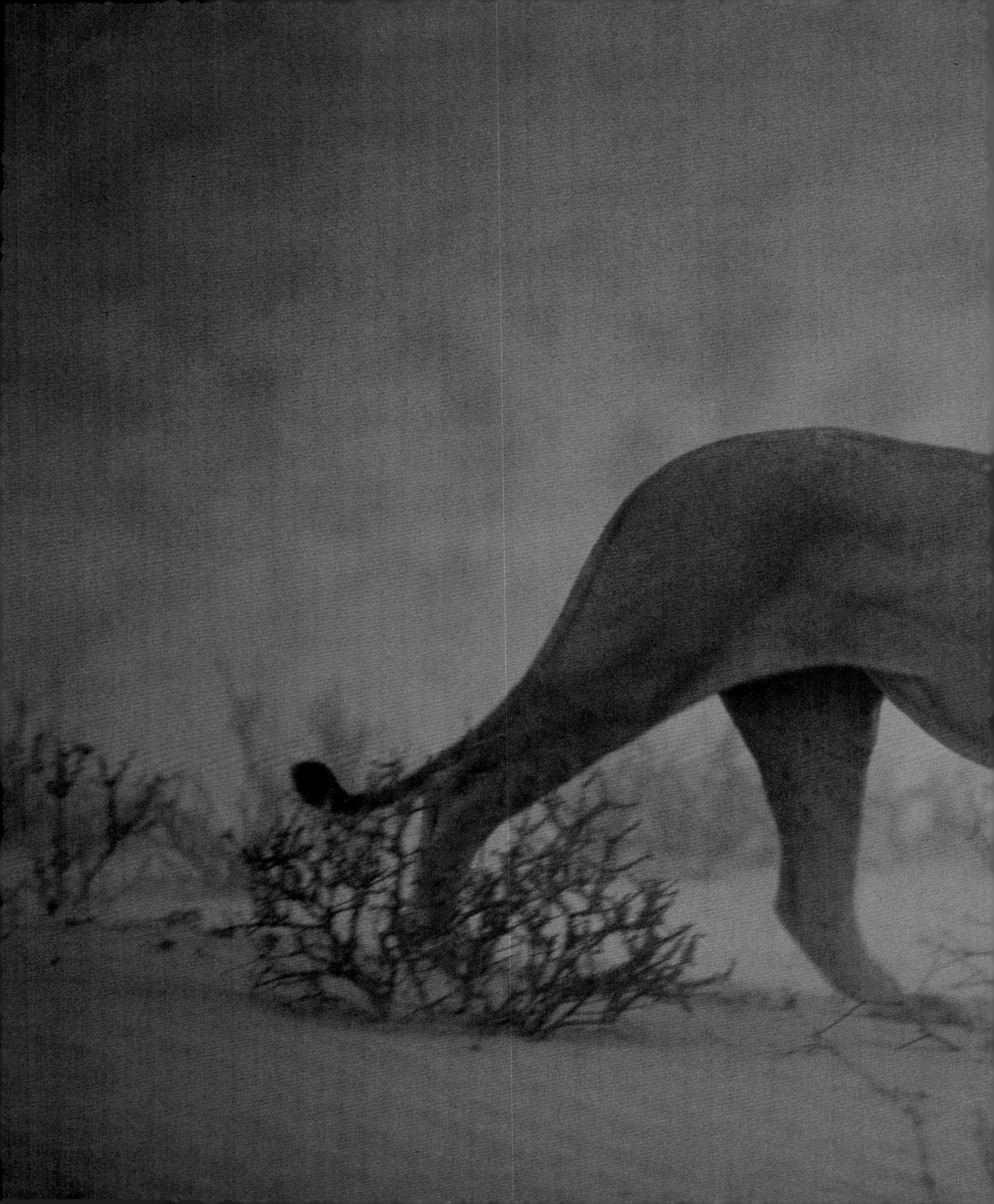

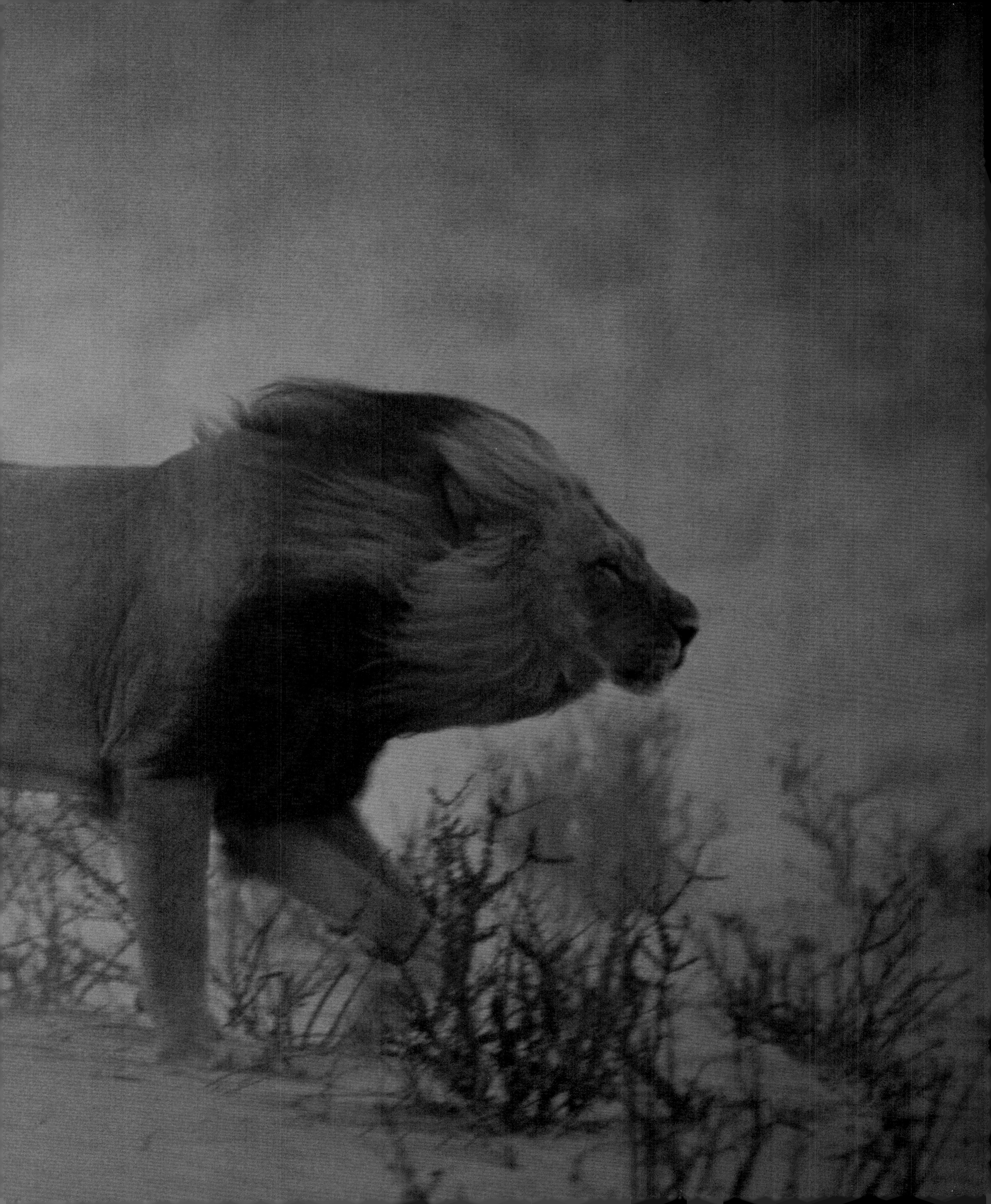

1981, the black rhino population in Africa was larger than that of the white rhino, but by 1987, the white rhino population was larger, with almost 90% of it in South Africa. Meanwhile, the black rhino population had declined from 12 753 in 1981 to 3 784 in 1987, a loss of 70.3% in six years. Only in South Africa did the population maintain some stability, declining only 7.7%, as opposed to 53.5% in the rest of southern Africa and an incredible 87.9% in West, Central and East Africa due to poaching for rhino horn (Cumming *et al.*, 1990). Obviously, South Africa must play a key role in the survival of this rhino species in the future, and could achieve the same success that it has had with the white rhino.

The zebras are another important flagship mammal group for South Africa, and part of their value derives from the fact that two of the species that were endemic to South Africa have already gone extinct. The most famous of these was the quagga (*Equus quagga*), a very distinctive relative of the wide-ranging plains zebra. The quagga was basically a brown, rather than a striped, zebra with white legs and tail. It had no distinct markings on the hind quarters, only light mottling on its back, and the only striped parts of its body were the head and neck. Although aggressive even by zebra standards, it was occasionally tamed and used in England in the 1830s as a harness animal. The Boers would also use it as a "watchdog" at night. Unfortunately, the quagga was restricted to the *veld* of the old Cape Colony, with smaller numbers in the Free State. As a result, it was an obvious source of food for the African servants of the Boers and of hides for local use and export. So the animal was killed by the thousands, and the large herds that existed in the 1840s were annihilated over the next 30 years. The last wild individual was shot in 1878, and the last captive one died in the Amsterdam Zoo in 1883.

Burchell's zebra (*Equus burchelli burchelli*), the nominate subspecies of the plains zebra and the closest relative of the quagga, unfortunately also went extinct. It was restricted to the Free State and southern Bechuanaland (now Botswana), and its range partly overlapped with that of the quagga. It was white-legged and white-tailed, and differed from other races of the plains zebra in that the stripes on its body and haunches were "shadow stripes" between its black markings. It too was originally very abundant within its range, but was heavily overexploited and had become very rare by 1850. The last individual died in 1910 in the London Zoo.

An important living zebra flagship for South Africa is the Cape Mountain zebra (*Equus zebra zebra*). The mountain zebra as a whole is found mainly in Namibia and South Africa with a small range in southern Angola, but the Cape Mountain zebra is restricted to the mountains of Cape Province. It was close to extinction in the late 1940s, when only about 100 individuals remained, but today it has increased back to nearly 600, 90% of which are found in national parks and nature reserves, and the animal occurs in a total of 21 areas (Duncan, 1992). Consequently, although its population is very small, it stands a very good chance of avoiding the fate of its *veld*-dwelling relatives mentioned above.

South Africa also has the dubious distinction of having the first African species to go extinct at the hands of Europeans. This was the unusual blue buck or *blaauwbok* (*Hippotragus leucophaeus*), a relative of the sable and roan antelopes. It lived only in the Zwellendam Province of the old Cape Colony, and apparently had an unusual bluish coloration. The early settlers referred to the animal as the "blue goat" because of its curved horns and seasonal beard. Its skin was apparently preferred to its meat, which was fed to the dogs according to a German traveler who first described the animal in 1731. The combination of intense hunting pressure and a very restricted range in one of the first areas to be populated by Europeans led to its rapid decline. By 1796, it was down to half a dozen individuals, and by 1799 or 1800 these too had been shot and their skins sent to Holland.

Less charismatic, but no less interesting as a vehicle for conservation, are the golden moles of the family Chrysochloridae. There are actually five genera and 18 species of these unusual insectivores in sub-Saharan Africa, but their stronghold is South Africa. Two genera, *Cryptochloris* and *Chrysospalax*, are endemic to South Africa, while one, *Eremitalpa*, is found in South Africa and Namibia; and two of the three species of *Chrysochloris* and 5 of the 10 species of *Amblysomus* are also South African endemics. In all, 12 of 18 species are endemic to South Africa. The giant golden mole (*Chrysospalax trevelyani*) is a forest-dweller famous in the eastern part of Cape Province. It has become a symbol for conservation in the eastern Cape, where forest fragmentation is widespread, and where remaining forest patches continue to be converted for economic development or destroyed by domestic livestock.

The giant golden mole is a solitary burrowing animal with a compact, streamlined body, short limbs and no visible tail. The backward-set fur is moisture repellent, remaining sleek and dry in muddy situations, and a dense woolly undercoat provides insulation. The eyes have been almost lost and are covered with hairy skin; the ear openings are lined with fur and the nostrils are protected by a leathery pad which assists in soil excavations. A muscular head and shoulders push and pack the soil and the strong forelimbs are equipped with digging claws. Of the four claws, the second and third are elongated and very powerful digging tools.

The giant golden mole is a prominent member of the Order Insectivora, which is rich in examples of convergent evolution, situations in which animals not closely related have adapted in similar ways to fit the demands of a specific habitat or way of life. Moles and golden moles, for example, are not closely related within the Order, yet both groups have adopted similar burrowing lifestyles and even look much alike. According to fossil records, moles developed from an animal which resembled a shrew, whereas golden moles appear to have been more closely related to the tenrecs of Madagascar. The similarity between the golden mole and the mar-

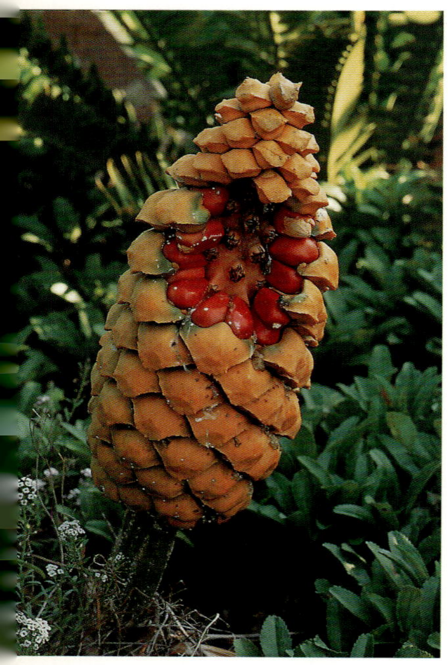

On pp. 410-411, a male lion (Panthera leo) *walks through a sandstorm in Nossob Riverbed Center, between South Africa's Kalahari Gemsbok National Park and Botswana's Gemsbok National Park.*

The cycads are quite diverse in South Africa, with 20% of the world's species occurring there. Above, the cone of the endangered cycad Encephalartos caffer *from the eastern Cape.*

supial mole of Australia is even more remarkable. In this case, the lineages have been separate for 70 million years; although one is placental and the other pouched, the golden mole is more similar in appearance to the marsupial mole than to the true mole.

Another evolutionary curiosity among South Africa's flagships is the ostrich (*Struthio camelus*), the largest of living birds. Some consider it evidence of the former existence of Gondwanaland, the great landmass that is believed to have broken apart to form Africa, Australia, Madagascar, peninsular India, South America, and the Antarctic continent, and it is certainly true that all the other giant flightless birds —rheas, emus, cassowaries, moas, and elephant birds— once inhabited these areas. Indeed, there appears to be proof of ancestral association between the ostrich and the South American rhea, which have identical bird lice. At first glance, the link is not obvious, but bird lice are very specialized, so much so that it is unlikely that they would develop separately and in parallel in two distant parts of the world.

The far smaller ancestor of the ostrich could almost certainly fly in primeval times. As evidence of this, the ostrich still possesses certain reflexes essential to flight which could never have developed in an entirely earthbound animal. But its terrestrial lifestyle and vastly increased body mass have so modified the ostrich that its atrophied wings and muscles are no longer sufficiently powerful to support it in flight; therefore it must evade its predators, not by burying its head in the sand as the legend would have it, but by running, which it does at speeds of over 50 km an hour.

At the beginning of this century, ostrich feathers enjoyed immense popularity as adornment for hats and other garments, and enormous flocks of ostriches were farmed in the Cape area of South Africa for their valuable plumage. However, the advent of the motor car changed fashions dramatically: hats with streaming feathers proved a hazard and ostrich farmers had to resign themselves to less luxurious living. All the same, a steady, if considerably smaller, demand for ostrich products has remained. Although feathers have never regained their former popularity, there is a ready and consistent market for leather products made from the attractive and unusual skin. In addition, ostrich meat, which is as red as beef and low in cholesterol, is also becoming popular, and this has once again changed the nature of ostrich farming. The eggs, which are still treasured by the Kalahari Bushmen for storing water, are also sought after as food. Each is equivalent to about 24 domestic hen's eggs and they make popular souvenirs to be sold at the farms. Indeed, the farms themselves have become an important tourist attraction with many people coming to observe these amazing birds.

Another of South Africa's extraordinary birds is the blue crane (*Anthropoides paradisea*), which is also the country's national bird. The blue crane is endemic to southern Africa and, except for an isolated population in Etosha National Park in northern Namibia and occasional sightings in Botswana and Zimbabwe, the entire population can be found in South Africa, the lowlands of Lesotho, and in western Swaziland. Together with its closest relative, the Demoiselle crane (*Anthropoides virgo*), it is less water-dependent than other crane species and is more closely associated with open grasslands. It can also be found in cultivated fields, on the edges of *vleis* and in the grassy Karoo.

Until the 1980s, blue crane populations were considered healthy throughout South Africa. Unfortunately, the situation changed drastically over the next ten years. At the first Southern African Crane Foundation Conference held in 1989, a rather grim picture was presented by speakers from different parts of South Africa. In Natal, a 90% reduction in numbers apparently took place in the period 1982-1989. A count of breeding blue cranes in Transvaal yielded only five pairs, whereas 500 had been counted only five years before. From the eastern Cape, it was reported that an estimated reduction from 8 000 to 1 000-2 000 cranes had taken place within the decade. The reasons for these frightening declines are not fully understood, but a combination of factors is probably responsible. The destruction and degradation of the grasslands once favored by blue cranes is continuing at an alarming pace. Ironically, blue cranes have adapted themselves very well to certain agricultural regions where the influence of man on the environment could not have been greater, namely the grain-growing areas of South Africa. Unfortunately, blue cranes and crowned cranes have been accused of causing crop damage by feeding on newly-sown seeds and trampling and eating newly-sprouted grain. Any damage done is probably insignificant, but this has not stopped farmers from shooting or poisoning cranes. There have also been instances, especially in the southern Cape, in which blue cranes have accidentally been killed after eating poisoned grain intended to catch guinea fowl or Egyptian geese for the pot.

Roadside counts in the Karoo and southern Cape seem to indicate that at least in these two areas the blue crane is still holding its own, and that the population could even be on the increase in the intensively-farmed southern Cape. This situation stands in stark contrast to what is happening in the rest of South Africa. For effective conservation and management purposes, it is important to know more about its life history; several key projects are now underway to ensure the survival of this important South African symbol.

While most people associate South Africa with rhinos, elephants and other charismatic megafauna, the country's flora deserves at least as much attention, if not more. Although a relatively small country, South Africa contains some 23 420 species of higher plants, or about 9% of the world total. The Fynbos biome, in the southwestern coastal strip of the Cape, is world-famous for its high plant diversity. The colorful proteas are popular flagships of this floristic kingdom, and are well known internationally.

Cycads are another exciting element of South Africa's flora. They are the surviving remnants of a

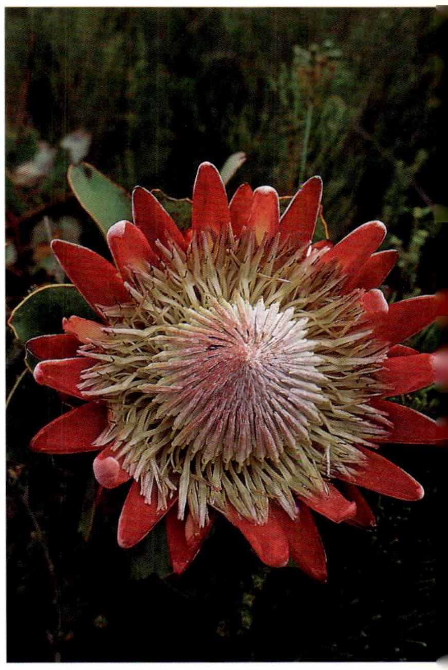

The protea is South Africa's national flower. It occurs in the Fynbos of the Cape Floristic Province, one of the world's most important regions for plants.

plant group which was relatively abundant in the world's Mesozoic flora and reached a zenith in the Jurassic Period, some 160 million years ago. Consequently, cycads are often referred to in popular terms as "living fossils." However, the layman's fascination with cycads is founded only partially on their antiquity. The size, shape and symmetry of the trunks and foliage have major decorative appeal, especially to the landscape gardener, while the curious and often "pineapple-like" fruiting structures are an added bonus. The slow growth rate and comparative scarcity of most species have resulted in an intrinsic value which has become grossly inflated by way of rumor, media publicity, and status symbol identity. However, this is all indicative of the popularity of cycads, and their potential as conservation flagships.

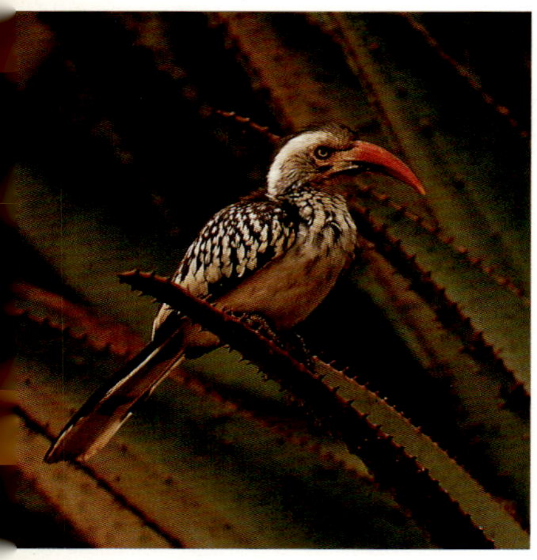

A red-billed hornbill (Tockus erythrorhynchus).

On the opposite page, the suricate or meerkat (Paracynictis selousi), *a very distinctive and appealing member of the mongoose family.*

Dr. Knut Norstog of the Fairchild Tropical Garden in Miami draws an analogy in the evolutionary sequence of plants and animals; in his comparison, cycads are the "reptiles of the plant world" since it is the reptiles which span the great gap between the amphibians, restricted to a watery environment for reproduction, and the birds and mammals which are freed from this constraint by internal fertilization. Cycads are strictly dioecious (*i.e.*, they have separate male and female individuals), with the males having the oddly archaic feature of motile spermatozoids. Species-specific pollinating insect vectors add to the complexity, and the occasional reports of cycads undergoing spontaneous and irreversible sex change are as yet without explanation by plant physiologists. In no other plant group is there such a curious mix of ancestral and derived characters.

In South Africa, there are 38 known species of *Encephalartos*, a well-known African cycad genus, and the only species of the monotypic *Stangeria*. In total, 20 per cent of all the world's cycad species are found in South Africa. These plants show a wide variety of growth forms and habitat preferences. The well-known Modjadji cycad is a tall, robust plant occurring in great stands in a few restricted localities in northern and northeastern Transvaal. *Encephalartos horridus*, found in the Karoo scrub in the eastern Cape, has viciously spiny lobes on the leaflets of its strikingly blue foliage. *Encephalartos ghellinckii* has deflexed needle-like leaflets and thrives in the high-altitude montane grasslands of the Natal Drakensberg. *Encephalartos ferox* is a fast-growing cycad bearing large, glossy green leaves and attractive scarlet cones and is found only on the stabilized sand dunes of the northern Natal coastline. Finally, *Stangeria eriopus* is a fern-like plant widely distributed in coastal grasslands and forests.

Cultural diversity

Modern South Africa is an interesting racial and cultural mix. The largest group is composed of several native African peoples, mostly of Bantu origin, who numbered some 32 million in 1991. The next largest group are the Caucasians, or white Africans of European origin, which at over 5 million represent the largest and oldest population of Europeans on the continent. They include the original Afrikaners of Dutch origin, who first arrived in 1652, and the South Africans of British origin who began to arrive after the British seized the Cape in 1795. The third largest group is the "Coloreds," the South African name for people of mixed Caucasian, African and East Indian heritage; they number nearly 4 million. Finally, there are the Asians, mostly of East Indian origin, but including some Malays, some Chinese and other small groups. They number over 1 million.

The first human inhabitants of South Africa were the Bushmen or San, a very distinctive human group living as nomadic hunter-gatherers. They are very short in stature and have yellowish-brown skin. They lived in what is now South Africa for perhaps as long as 30 000 years. The Khoikhoi or Hottentots, as they were called in South Africa, began to move into the land of the Bushmen about 2 000 years ago. They are similar in appearance, a little taller, and lived as herders of sheep and cattle. They came into conflict with the Bushmen as they moved south from their point of origin in what is today Botswana, and also eventually intermarried with the Bushmen. Both speak very distinctive languages belonging to what is called the Khoisan group and characterized by clicks of various kinds.

When the Portuguese explorer Bartolomeu Dias rounded the Cape in early 1488, he encountered the Khoikhoi, who then numbered perhaps 50 000 in the southwestern Cape; as he moved up the southeast coast, he also found well-established Bantu settlements living by agriculture and cattle-herding. These Bantu people originated in Central Africa and arrived in South Africa much later than the Khoikhoi, and probably only a few hundred years before the Europeans. Just as the Hottentots partly displaced and partly absorbed the Bushmen, the Bantu did much the same with the Hottentots.

When the Europeans began to make regular visits to South Africa in the 1590s, the country was inhabited by the San, the Khoikhoi, and Bantu populations who mainly occupied the area to the east and northeast of the Cape. The Bantu peoples were primarily Nguni, including the Xhosa and the Zulu, with Shona further to the north. All of these cultures interacted and traded to some extent with one another.

European settlement dates from the mid-seventeenth century, when the Dutch East India Company established a supply station at Table Bay in 1652 for ships rounding the Cape of Good Hope. Dutch settlers were later augmented by French and German immigrants, as well as slaves from Malaya, Madagascar and East and West Africa. They too encountered the San and the Khoikhoi, and promptly enslaved some of them to work as herdsmen; most of the others fled into the interior.

In later decades, as some of the white settlers took up cattle raising, they began to move gradually east-

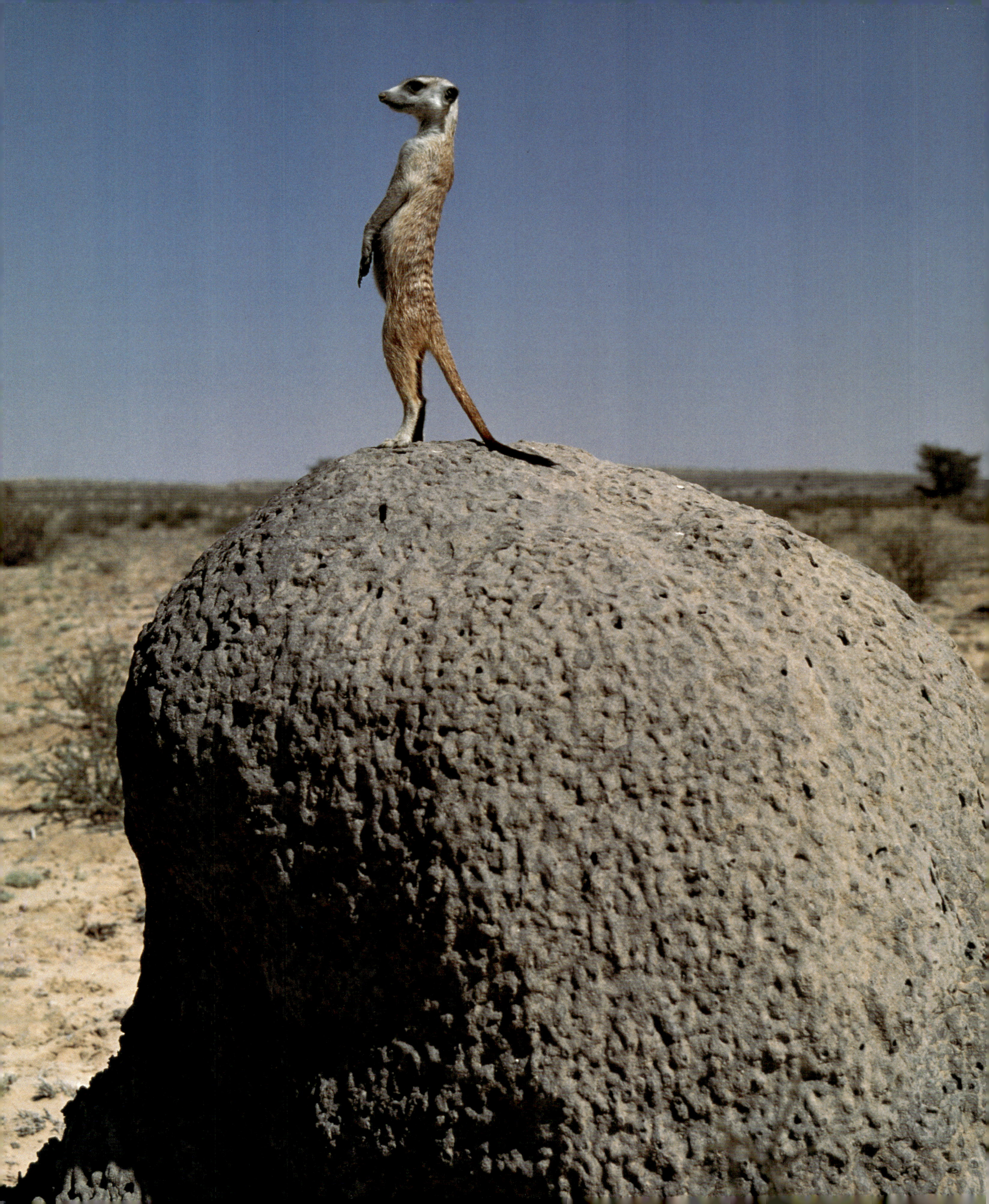

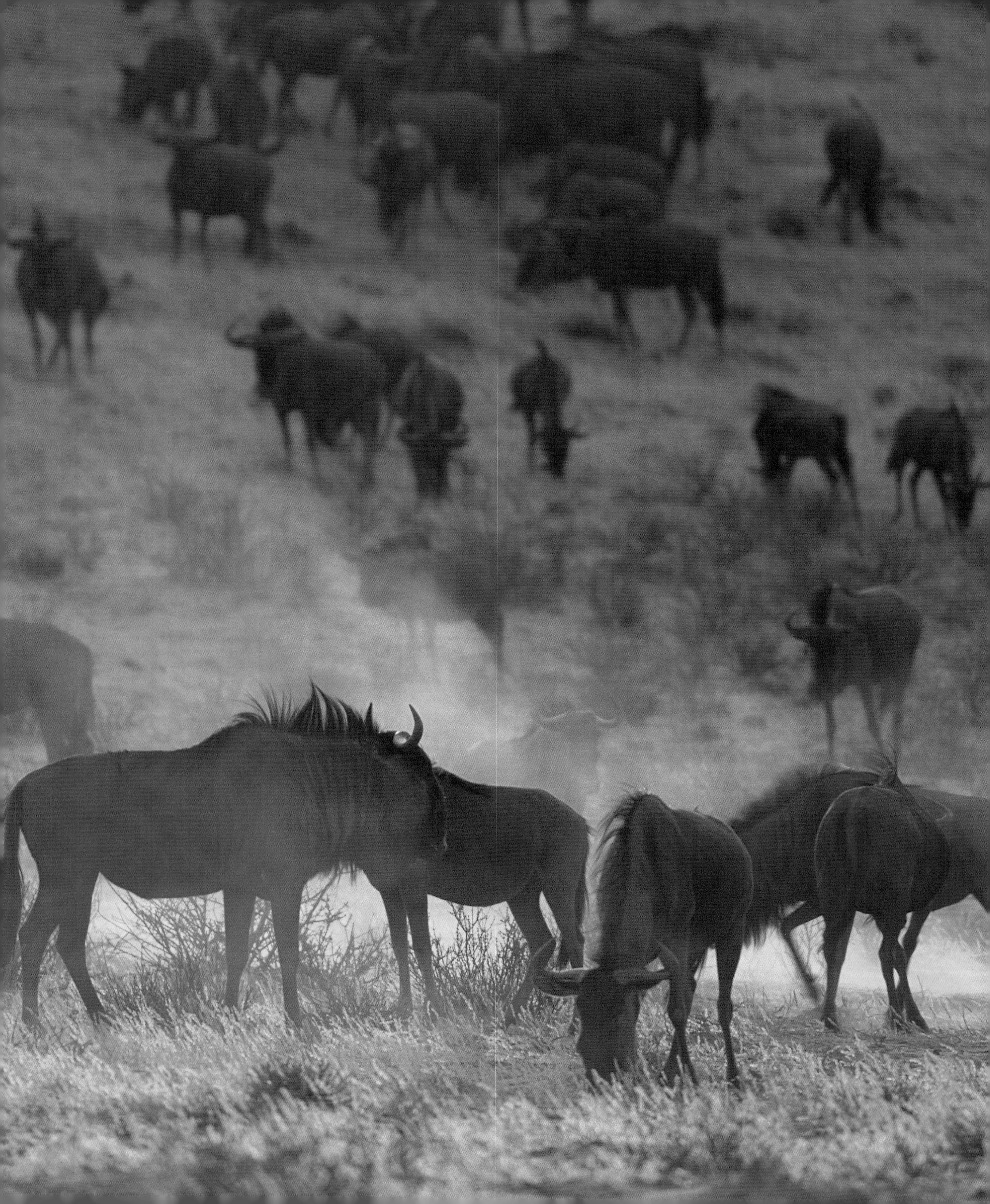

ward in search of grazing land beyond official Dutch control. In the process, they came into further contact and conflict first with Khoikhoi and San peoples, who fled further inland, and then with Xhosa and other Bantu peoples, who resisted settler encroachment on their land. Indeed, the Xhosa and the Dutch settlers fought nine wars with one another, mainly in the eighteenth and nineteenth centuries.

After England took the Cape from Holland, there began a long period of conflict with the Dutch settlers, known as Boers. In 1820, the British settled a large group of English colonists in the eastern Cape. In the 1830s, groups of Boers, reacting to the British abolition of slavery, deteriorating economic conditions, and resentment over British imperial control, moved out on what became known as the "Great Trek" into the South African interior, frequently fighting with Bantu tribes as they went. There they founded two republics, the (Orange) Free State and Transvaal. In order to prevent the Boers from laying claim to any coastline, the British annexed Natal in 1843. Following the Boer Wars in 1880-1881 and 1899-1902, the British finally defeated the Boers and made their republics into British colonies.

In 1860, the British introduced East Indians to Natal as indentured laborers on sugar plantations. At the turn of the century they were joined by a group of Indian merchants, adding to the size of this ethnic group in South Africa. Today, about 70% of Asians in South Africa are Hindu and 20% Muslim. The majority live in the Durban area.

Just prior to the time the Boers were trekking into the interior, African societies were undergoing traumatic changes because of the widespread dispersal of African peoples in the wake of a revolutionary upheaval called the Mfecane. The Mfecane originated in Natal, where the Zulu, led by their charismatic and imaginative leader, Shaka, created a major kingdom based on raiding. Zulu expansionism ignited a chain reaction of violent conflict, population dispersion, migration and devastation, which reshuffled African peoples throughout the whole of southern Africa. Despite the unsettled nature of the region, African societies resisted European expansion on a variety of fronts until the late nineteenth century, when the discovery of gold and diamonds dramatically transformed the political and economic character of southern Africa. Thereafter, the region's economy changed from one almost wholly based on agriculture to one predominantly industrial and urban.

Unfortunately, the original inhabitants of South Africa did not fare very well. In addition to being enslaved and persecuted, they had little resistance to the white man's diseases. A smallpox epidemic in 1713, for instance, wiped out a large number of them. The remaining Khoikhoi now exist mainly in southern Namibia, with the majority of those in South Africa having been absorbed into the Cape Colored population. The Bushmen still number some 60 000, but the majority live in Botswana (29 600) and Namibia (26 000), with smaller numbers in Angola (4 000) and Zambia (200). The population in South Africa itself is down to only about 25, living in desert areas of the Northwest Cape.

The indigenous African population of today is mainly Bantu in origin, and can be divided into about 10 tribes arranged in four major language groups. The largest group, the Nguni-speakers, includes Zulu, Xhosa, Ndebele and Swazi. The next largest, the Tswana-speakers, includes Tswana, Pedi and Southern Sotho. The other two major groups are the Venda and Shangaan-Tsonga.

The two biggest components of the northern Nguni are the Swazi, now largely settled in the independent kingdom of Swaziland in the northeastern part of the country, and the Zulu, a name now given to the many subgroups of the Nguni people who were forged into a single, powerful society by the great nineteenth-century leader, Shaka. The Zulu homeland is KwaZulu, adjoining Natal and led by Chief Mangosutu Buthelezi. His six million people form the largest of South Africa's ethnic groups, or about 38.5% of the black African population. The Swazi account for 4.4% of the black population of South Africa, but the majority of them are in the independent country of Swaziland.

The southern Nguni include a number of subgroups, the most numerous being the Pondo, the Tembu and, especially, the Xhosa, with whom the white settlers in the Cape came into such bloody contact. The Xhosa account for 11.6% of the black African population of South Africa.

The Ndebele people in the north are related to the Matabele of Zimbabwe, who are also of Nguni origin. The Ndebele of Transvaal are divided into a southern section, which retains its Nguni heritage, and a northern one, which has relinquished the Nguni identity and language and is now culturally part of the Sotho peoples. The Sotho, in turn, fall into three separate, though culturally-related groups: the Northern Sotho, of which the Pedi are the largest clan; the Southern Sotho, who populate the country of Lesotho; and the Tswana, whose population extends into neighboring Botswana. The Sotho account for 27.2% of the black population of South Africa and the Tswana for 6.6%.

Finally, there are the Venda and the Shangaan-Tsonga groups. The Venda, which include the distinctive but increasingly assimilated Lemba people, have a population of some 400 000, and live in a small area in northern Transvaal bordering Kruger National Park. The Shangaan-Tsonga, neighbors to the Venda and over a million strong, inhabit the South Africa-Mozambique border area.

JORGEN THOMSEN
RUSSELL A. MITTERMEIER
CRISTINA GOETTSCH MITTERMEIER

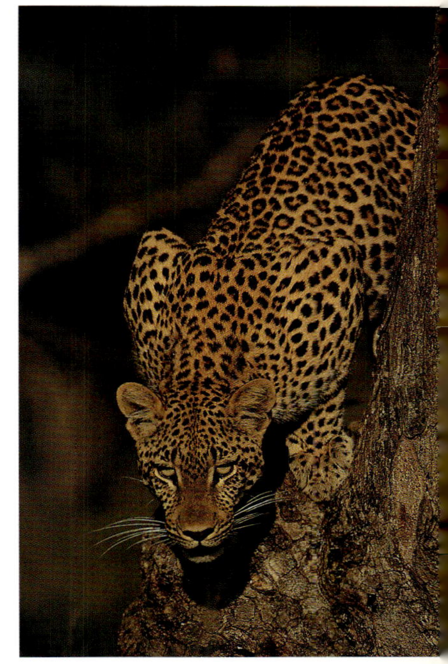

On the opposite page, a herd of blue wildebeest (Connochaetes taurinus) *at sunset, Kalahari Gemsbok National Park, northwest Cape Province.*

Above, the leopard (Panthera pardus) *is widespread in Africa; a secretive species that is rarely seen, it still survives in good numbers.*

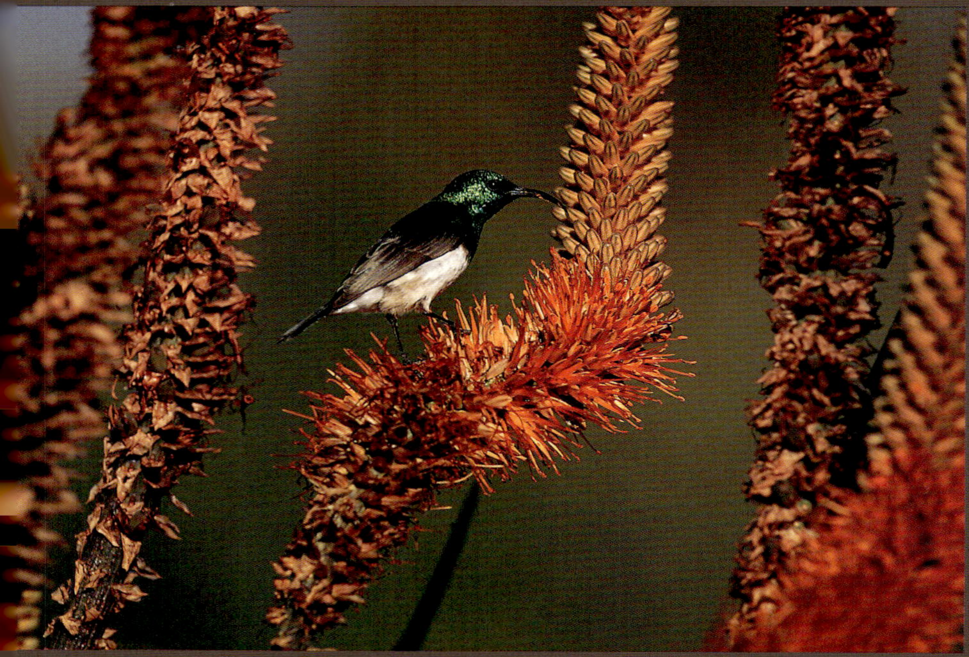

Above, a white-bellied sunbird (Nectarinia talatala)*. The sunbirds are in many ways the ecological equivalents of the New World hummingbirds. Right, a gemsbok* (Oryx gazella) *with calves in the Namib Desert. This beautiful species is characteristic of the drier regions of southern Africa. Its long, sharp horns can be used to good effect against predators like lions.*

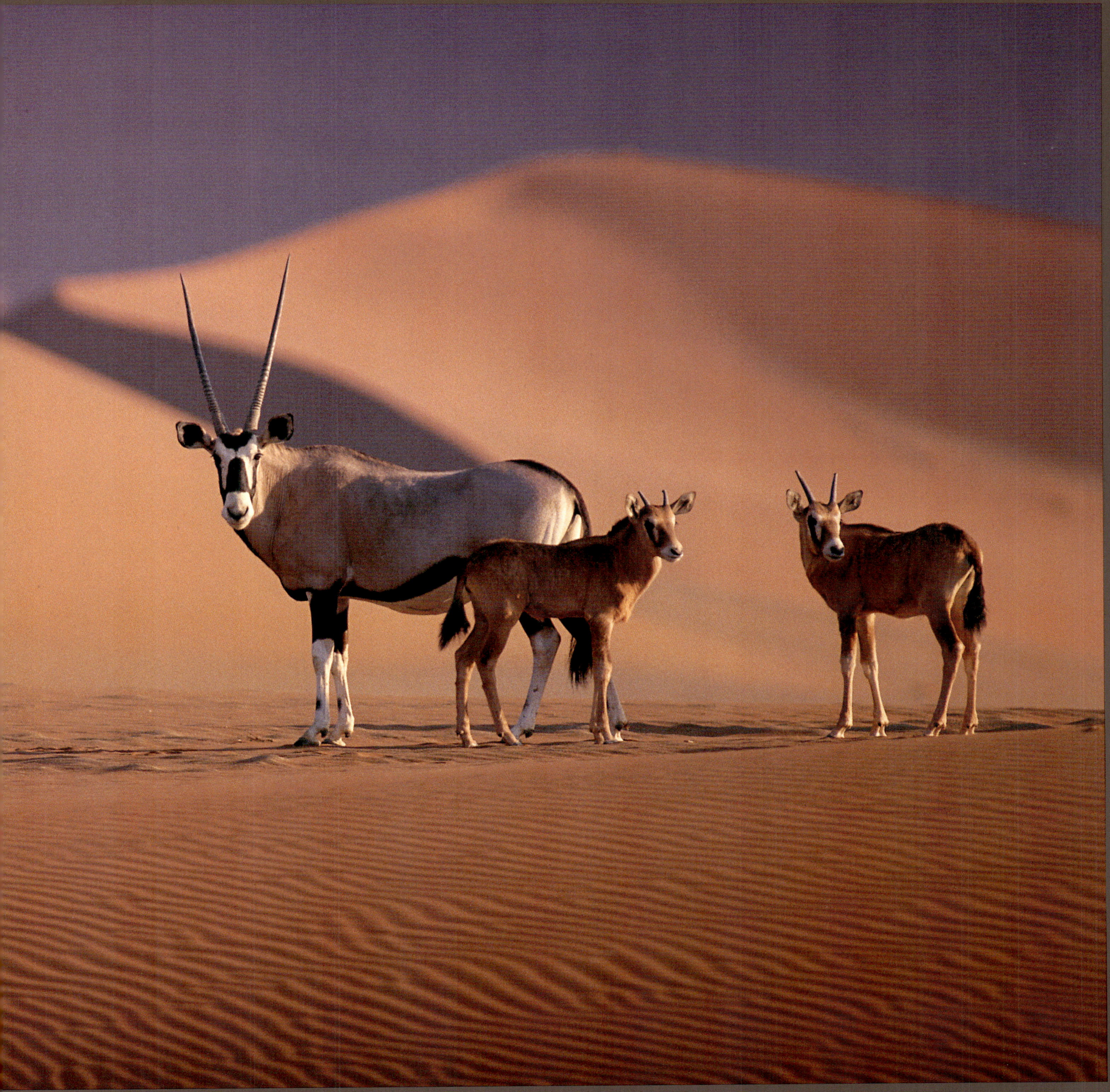

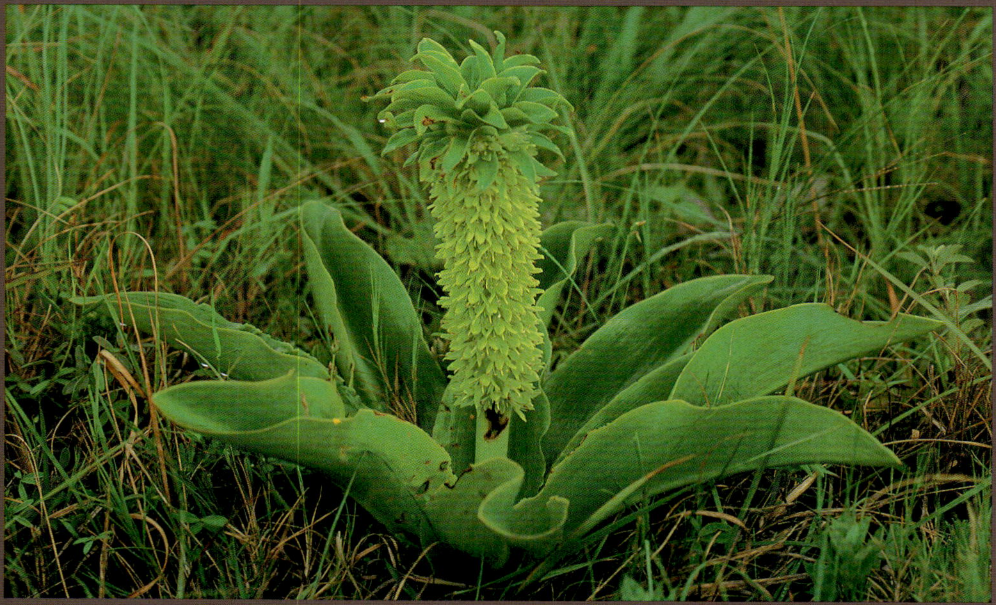

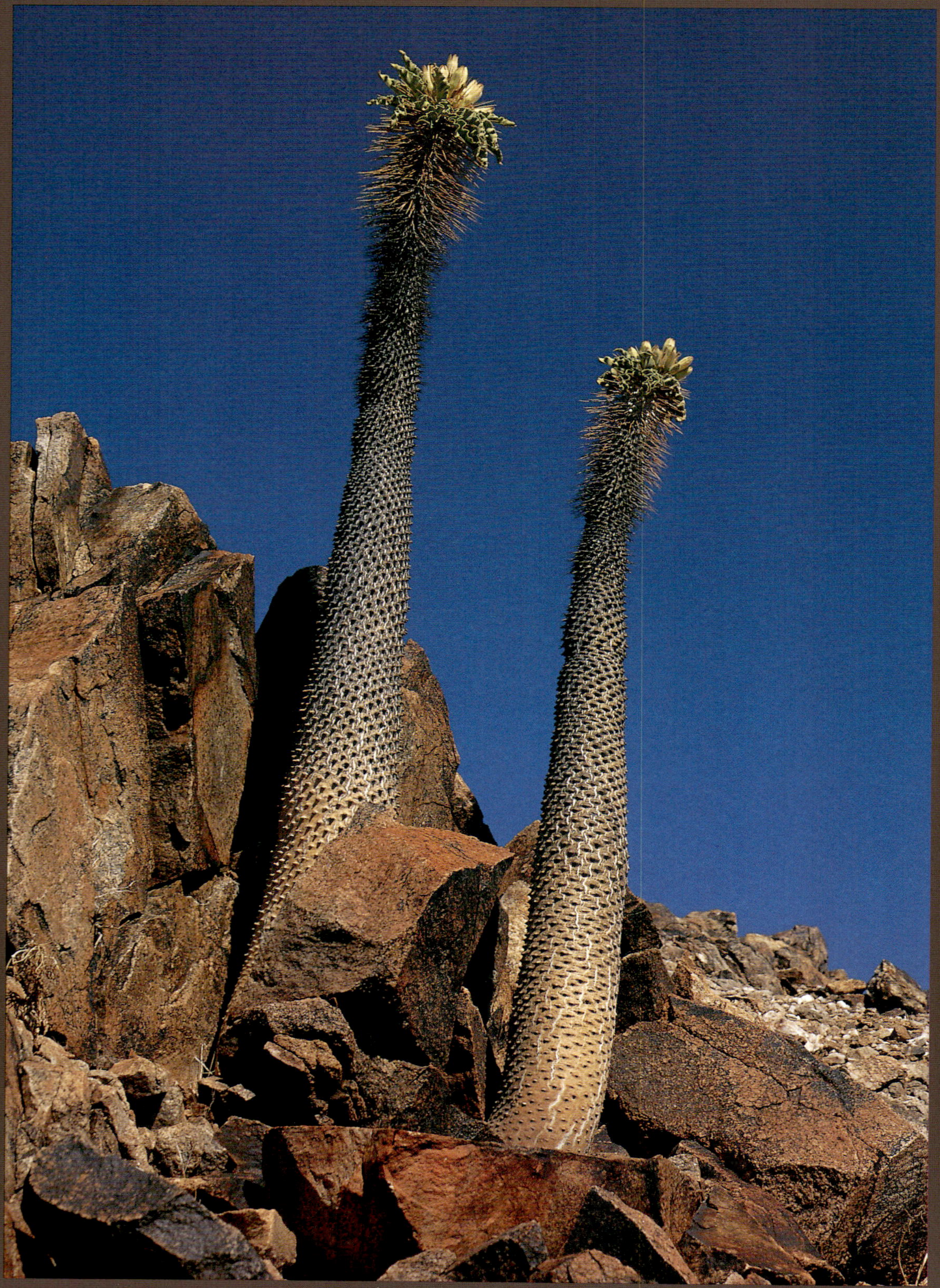

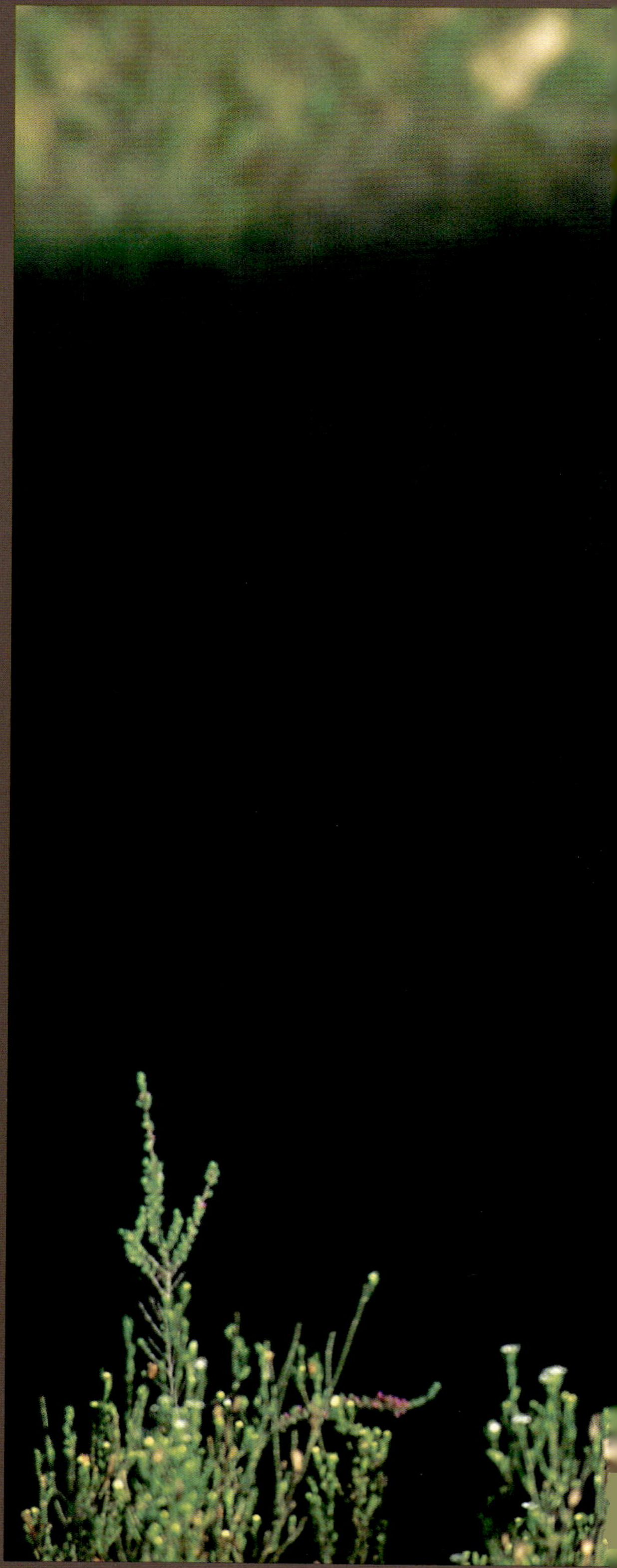

South Africa is the only megadiversity country entirely lacking in tropical rain forest. Nonetheless, its plant diversity is one of the highest on Earth. Upper left, a pineapple lily (Eucomis *sp.*) *in flower, Mkambati Game Reserve, Transkei, Cape Province. Lower left, a* Pachypodium namaquanum *(Apocynaceae) from Richtersveld National Park. Below, a Zuuberg cycad* (Encephalartos longifolius), *Zuuberg National Park, Cape Province.*

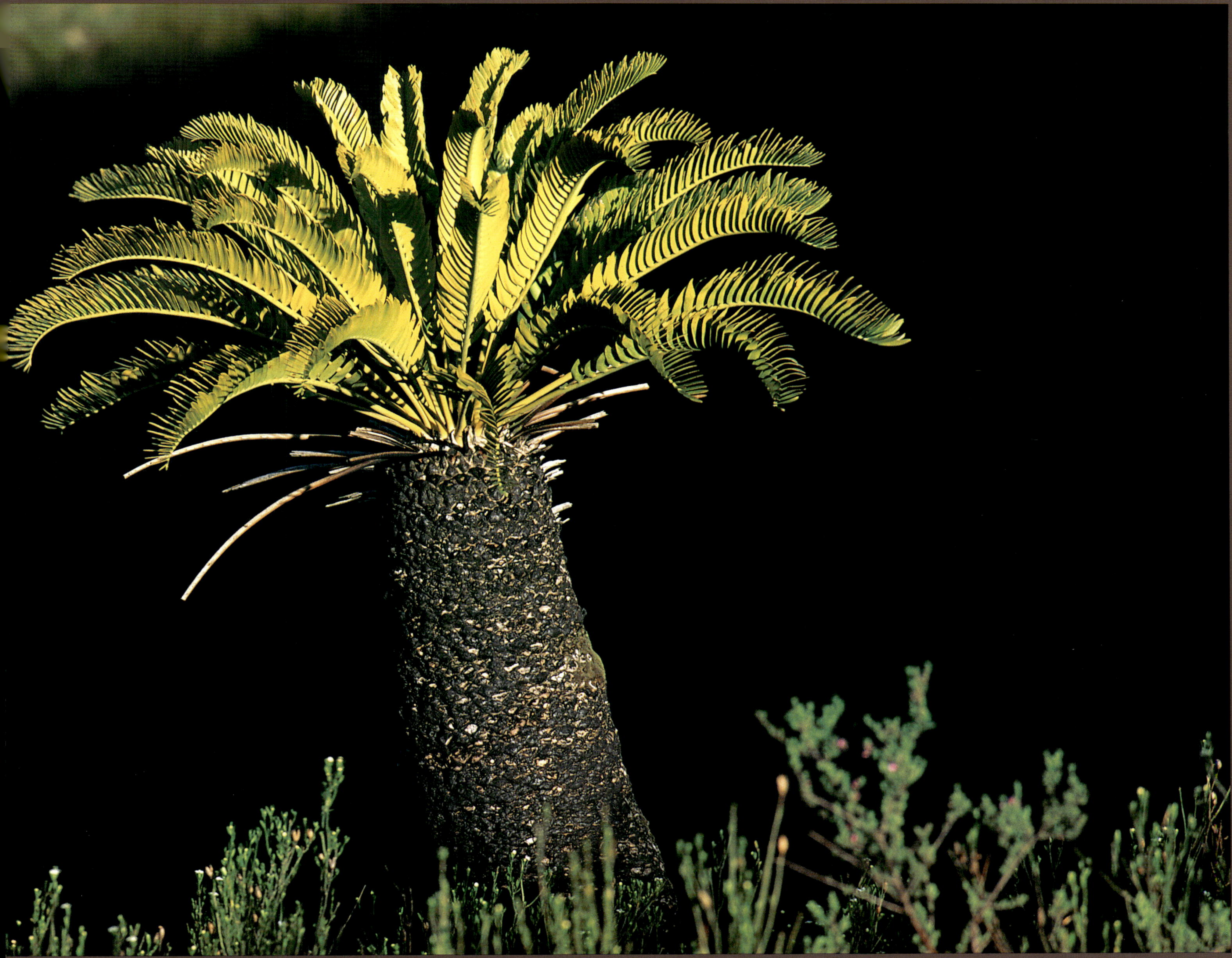

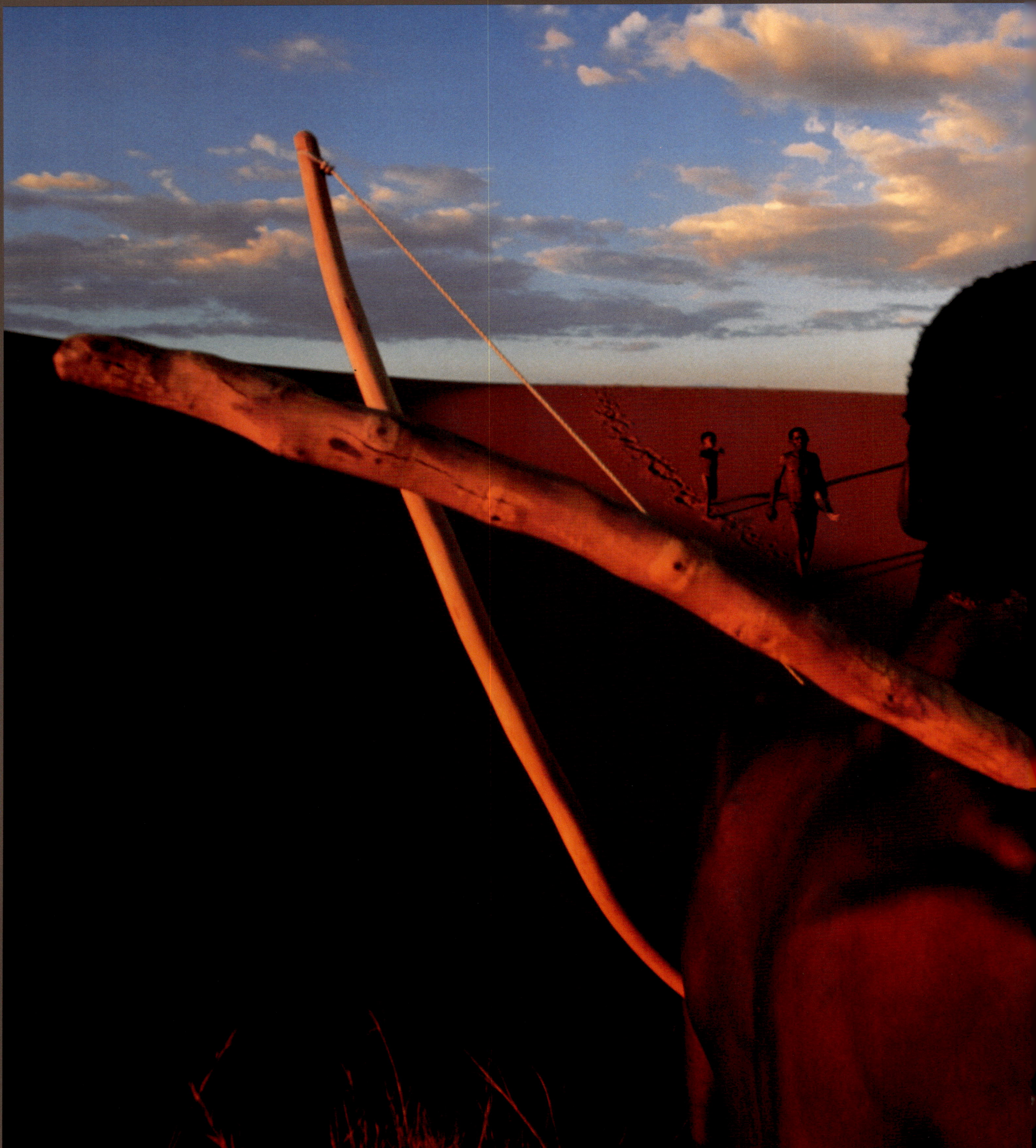

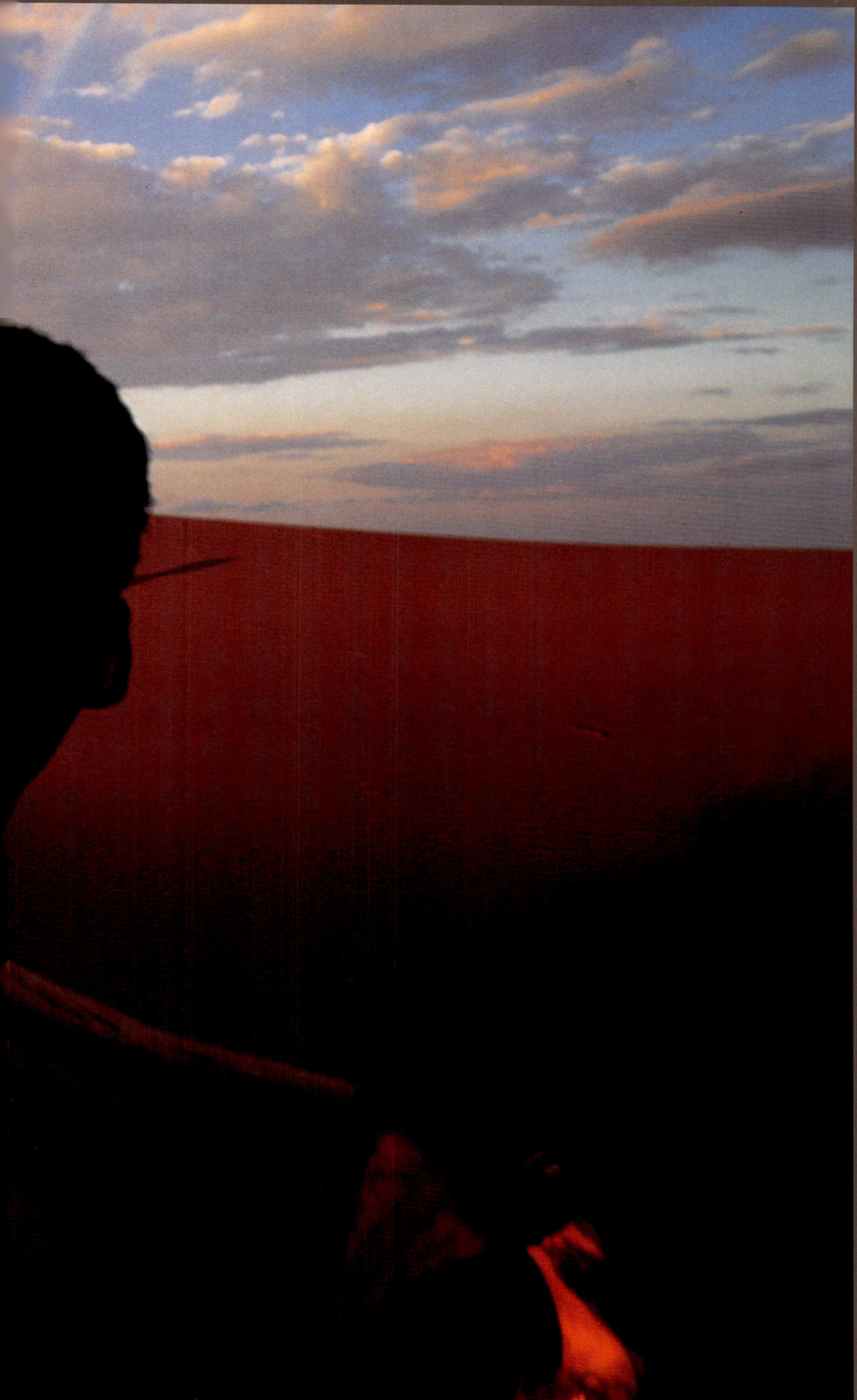

The Bushmen or San were the original inhabitants of South Africa, but were displaced, absorbed or exterminated by later waves of human groups entering the region, including the related Hottentots, the Bantu, and the Europeans. They now number no more than 25 in South Africa, most of them confined to the Northwest Cape. Here, one of the last South African Bushmen can track game for several days on the Kalahari dunes.

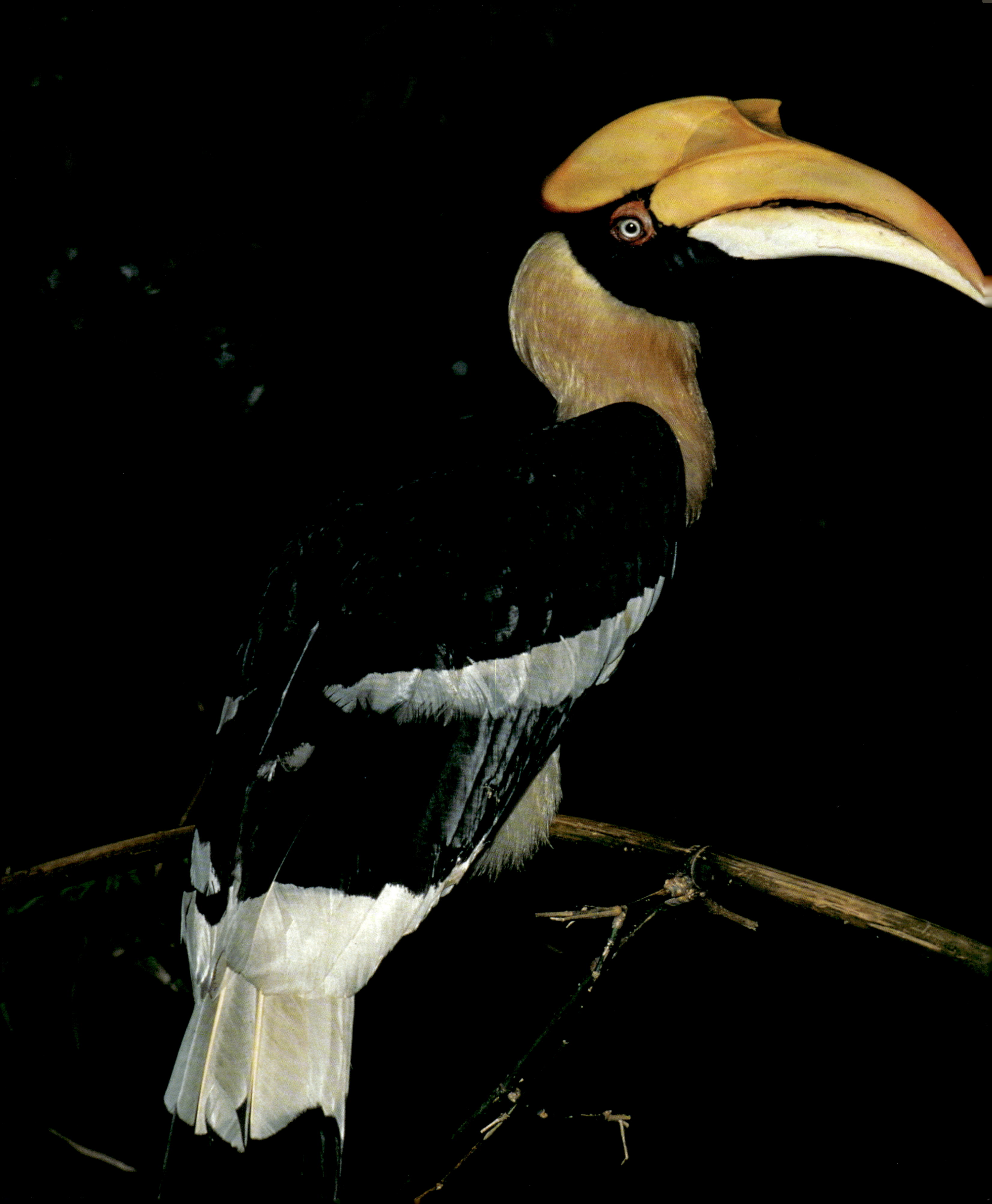

MALAYSIA

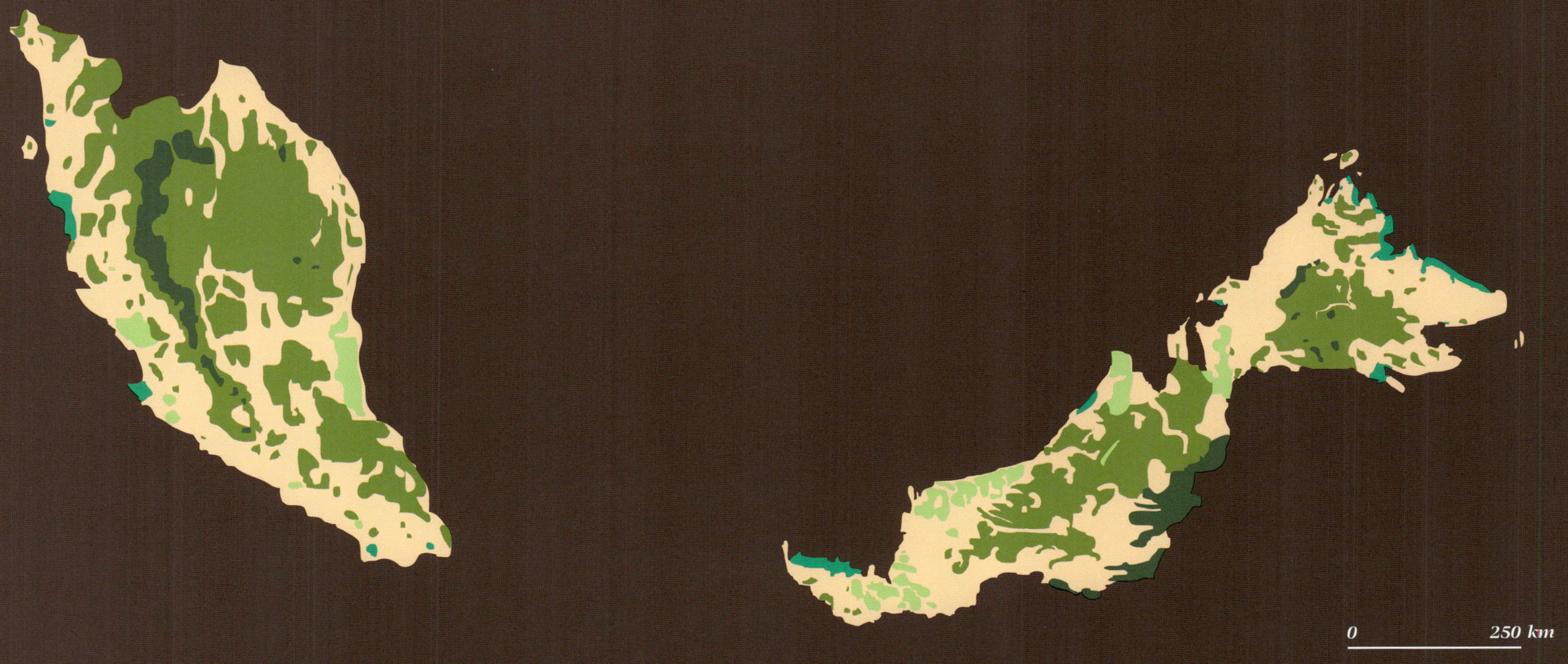

With a land area of 329 749 km^2, Malaysia is the third smallest of the megadiversity countries, with only the Philippines and Ecuador being smaller. It is a federation of 13 states, 11 of them on the Malay Peninsula and usually referred to as Peninsular Malaysia or West Malaysia, and the other two, Sarawak and Sabah, in the northern portion of the island of Borneo. Peninsular Malaysia occupies 131 598 km^2 or about 40% of the country and has more than three quarters of the population, 14.3 million out of a total 17.4 million. Sarawak and Sabah, also known as East Malaysia, have only a little more than 3 million people, but occupy 60% of the land area of the country, with Sabah contributing 73 710 km^2 and Sarawak 124 499 km^2. In spite of its small size, Malaysia has very high biodiversity, especially in plants, but in several vertebrate groups as well. This results from its tropical location, the fact that it was largely covered with tropical rain forest, that it has a large coastline and coral reef, and that it includes both the diversity of mainland southeast Asia and part of the island of Borneo. Animal endemism is relatively low because most of the animal species found in Peninsular Malaysia are also represented in Thailand and Sumatra, while most of those in Sarawak and Sabah also occur in the rest of Borneo; plant endemism, on the other hand, is quite high, and exceeds that of Ecuador and the Philippines. As with Ecuador and the Philippines, diversity and endemism per unit area in Malaysia are very high as well.

Although its landscape is mostly low-altitude, Malaysia also includes several very important montane areas. Mount Kinabalu in Sabah (4 094 m) is the highest peak in Southeast Asia and has very high levels of endemism, especially in plants. The Trus Madi and Crocker Ranges in Sabah reach as high as 2 649 m, Mount Luton in Central Sabah is a 1 657-m sandstone arc, Mount Mulu in Sarawak reaches 2 371 m and has spectacular karst formations in its vicinity, including the world's largest underground cave, and the Kelabit Highlands reach 2 348 m on Mount Murud. Peninsular Malaysia also has some highland areas, including several parallel ranges that run down the center of the country and a more dispersed mass of mountains to the east of the main ranges; the highest mountain on the peninsula is Gunung Talan at 2 189 m.

In Norman Myers' original 1988 threatened hotspots analysis, both Peninsular Malaysia and North Borneo (Sarawak, Sabah, and the independent nation of Brunei) were included as separate hotspots, mainly on the strength of their plant diversity. In the latest threatened hotspots analysis, carried out

*On the opposite page, the hornbills reach their greatest diversity on the island of Borneo, including the Malaysian states of Sarawak and Sabah. The great hornbill (**Buceros bicornis**) is one of the largest of the hornbill species.*

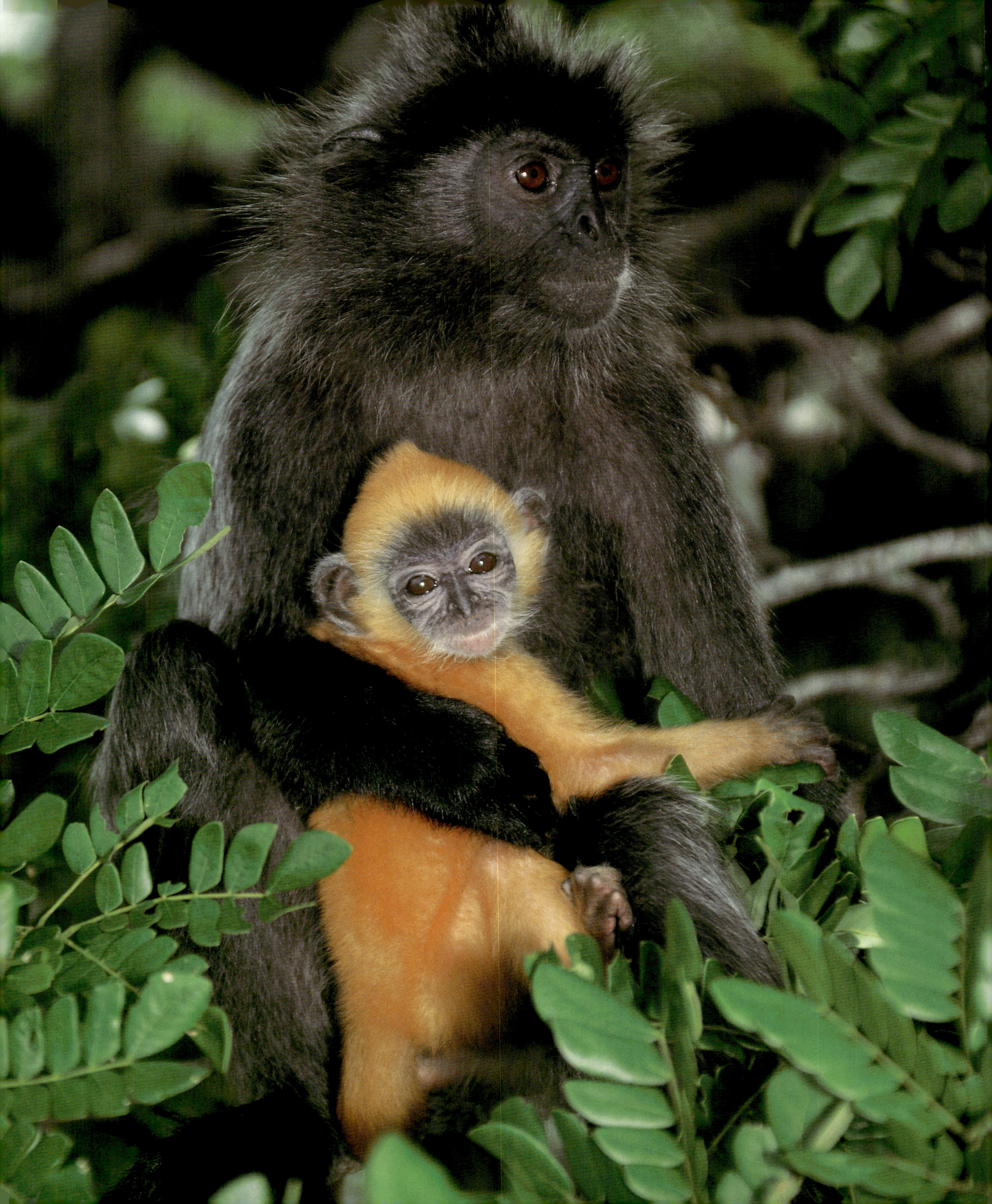

by Mittermeier, Myers, and Thomsen, both portions of Malaysia have been included in the huge Western Sunda hotspot, along with the western portions of Indonesia; this is an area that ranks among the top six hotspots on Earth.

In addition, Malaysia also has an enormous concentration of Centers of Plant Diversity and Endemism, as indicated in a recent study by IUCN/WWF: a total of 11 in Peninsular Malaysia (Dindings, Segari Melintang and Larut Hills; Endau Rompin State Park; Limestone Flora of Peninsular Malaysia; Montane Flora of Peninsular Malaysia; Pulau Tioman; Sedili Kecil Swamp Forest; Southeast Pahang Swamp Forests; Taman Negara National Park; Tasek Berah Forest Reserve; Trengganu Hills; and Ulu Belum); 12 in Sarawak (Bako National Park; Batu Laga, Linau Balui Plateau; Bukit Mersing, Anap, Tatau Region; Gunung Gading National Park; Gunung Gaharu, Gunung Apeng and Gunung Silantek, Sabal Forest Reserve; Gunung Mulu National Park; Lambir Hills National Park; Mixed Dipterocarp Forests on Humult Ultisols on Coastal Hills; Nieuwenhuis Mountains; Pueh Range (Gunung Pueh, Bukit Kanyi, Bukit Berumput to Bukit Berpayong); Pulong Tau, Gunung Murud; and the Usun Apau Plateau; and 6 in Sabah (Crocker Range/Mount Trus Madi; East Sabah Lowland and Hill Dipterocarp Forests, including Danum Valley, Sepilok Forest Reserve, Tabin Wildlife Reserve, Tawau Hills; Gunung Lotung/Maliau Basin; Kinabalu Park; Ulu Meligan/Ulu Long Pasia; and Northeast Borneo Ultramafic Flora, eastern Sabah), for a total 29 in all.

With regard to Endemic Bird Areas, as identified by BirdLife International, Malaysia has two of global importance, the Bornean Mountains on the border between Kalimantan and Sarawak and Sabah, and the whole of Peninsular Malaysia and Sumatra.

In terms of species numbers for different groups of organisms, the figures for plant diversity and endemism in Malaysia are undoubtedly the most impressive. Peninsular Malaysia alone is estimated to have more than 9 000 plant species, of which 3 000-4 500 (33-50%) are endemic, and Sarawak and Sabah combined are thought to have more than 10 000 species of higher plants, of which at least 3 500 are endemic. Adding the figures for Peninsular Malaysia and Sarawak and Sabah, there may be a total of between 6 500-8 000 endemic plant species in Malaysia as a whole, easily making the megadiversity cutoff point of 2% or 5 000 species; the total may in fact be higher, especially when endemics to the country as a whole are calculated. This figure would put Malaysia in the top 12 in plant endemism, and significantly higher than the Philippines and Ecuador, the other two smallest megadiversity countries. Indeed, Malaysia's plant endemism per unit area would compare with that of Madagascar, the highest of the megadiversity countries in this category. Total higher plant diversity for the country as a whole is more difficult to calculate because of the overlap between Peninsular Malaysia and East Malaysia, but the figure is generally thought to be on the order of 15 000. The number of genera is also very high. Peninsular Malaysia has 1 500 genera of flowering plants, of which 24 are endemic; Borneo has 61 endemic genera, many of them in Sarawak and Sabah. In terms of important plant families, Malaysia is especially rich as well. The economically important tree family Dipterocarpaceae has 156 species in Peninsular Malaysia, 26 of them endemic, and 267 in Borneo, 155 of them endemic, with a concentration of diversity in the northern part of the island.

Palm diversity is also very high, with 195 species in Peninsular Malaysia, 83 (43%) of them endemic; 213 in Sarawak, 56 (26%) of them endemic; and 131 in Sabah, 14 (11%) of them endemic. This yields a minimum of 153 endemic palms for the country as a whole, one of the highest totals in the world. Peninsular Malaysia also has about 850 orchid species, of which about 50% are endemic, and Borneo as a whole has about 1 500 orchid species in 149 genera, about 40% of them restricted to the island. These figures are truly impressive, even more so when per unit area diversity and endemism are considered.

The figures for animal diversity are also quite high, but endemism is considerably lower than for plants. Mammal diversity for the country as a whole is estimated at 286 species; some 200 occur in Peninsular Malaysia, at least 167 occur in Sabah, and 180 in Sarawak, but many of these are, of course, shared among the different components of the country; the total number of endemic mammals is 27. Bird diversity is rather high, given Malaysia's land area, with an estimated 738 species for the country as a whole, but endemism is low with only eleven.

Reptile and amphibian diversity and endemism are moderately high, with 268 reptiles, of which 68 are endemic, and 158 amphibians, of which 57 are endemic. The total of endemic non-fish vertebrates is 163, which is low compared to the other megadiversity countries, but once again high for its land area.

Little information is available on insect diversity. However, using the tiger beetles as an indicator, Malaysia ranks tenth in the world in total diversity with 95 species and fifteenth in endemism with 22 endemics.

Marine diversity in Malaysia is also significant. The country has an extensive coastline when both Peninsular Malaysia and East Malaysia are taken into consideration, and there are many important coral reef ecosystems. Indeed, Malaysia falls into the Coral Triangle, which has by far the greatest concentrations of coral reef diversity on Earth.

Human cultural diversity in Malaysia is moderately high. The modern population is racially mixed, with Bumiputras (Malays and other indigenous races) accounting for 55%, Chinese 35%, and Indians from India 10%.

In Peninsular Malaysia, the Orang Asli —composed of various pygmoid groups— are especially interesting,

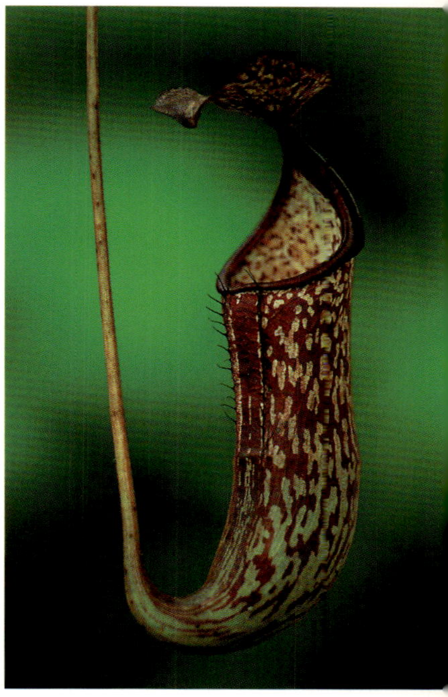

On the opposite page, silvered leaf monkey (Trachypithecus cristatus), *a common and adaptable monkey species in Malaysia. Infants have a bright orange natal coat.*

Above, the pitcher plants are very diverse on Borneo, with the greatest concentration being reached on Mount Kinabalu in Sabah. Here, a Nepenthes fusca *from the Maliau Basin in Sabah.*

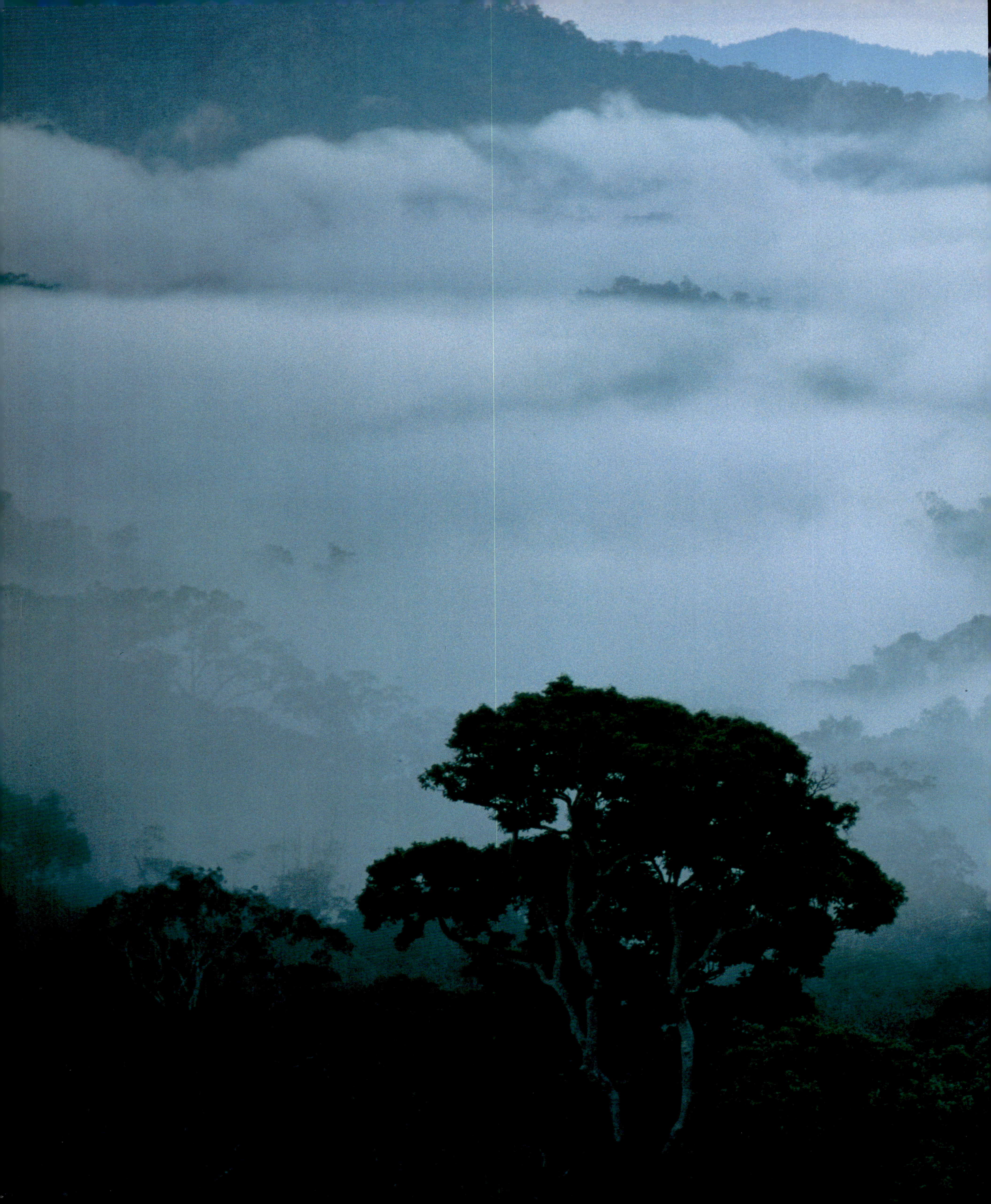

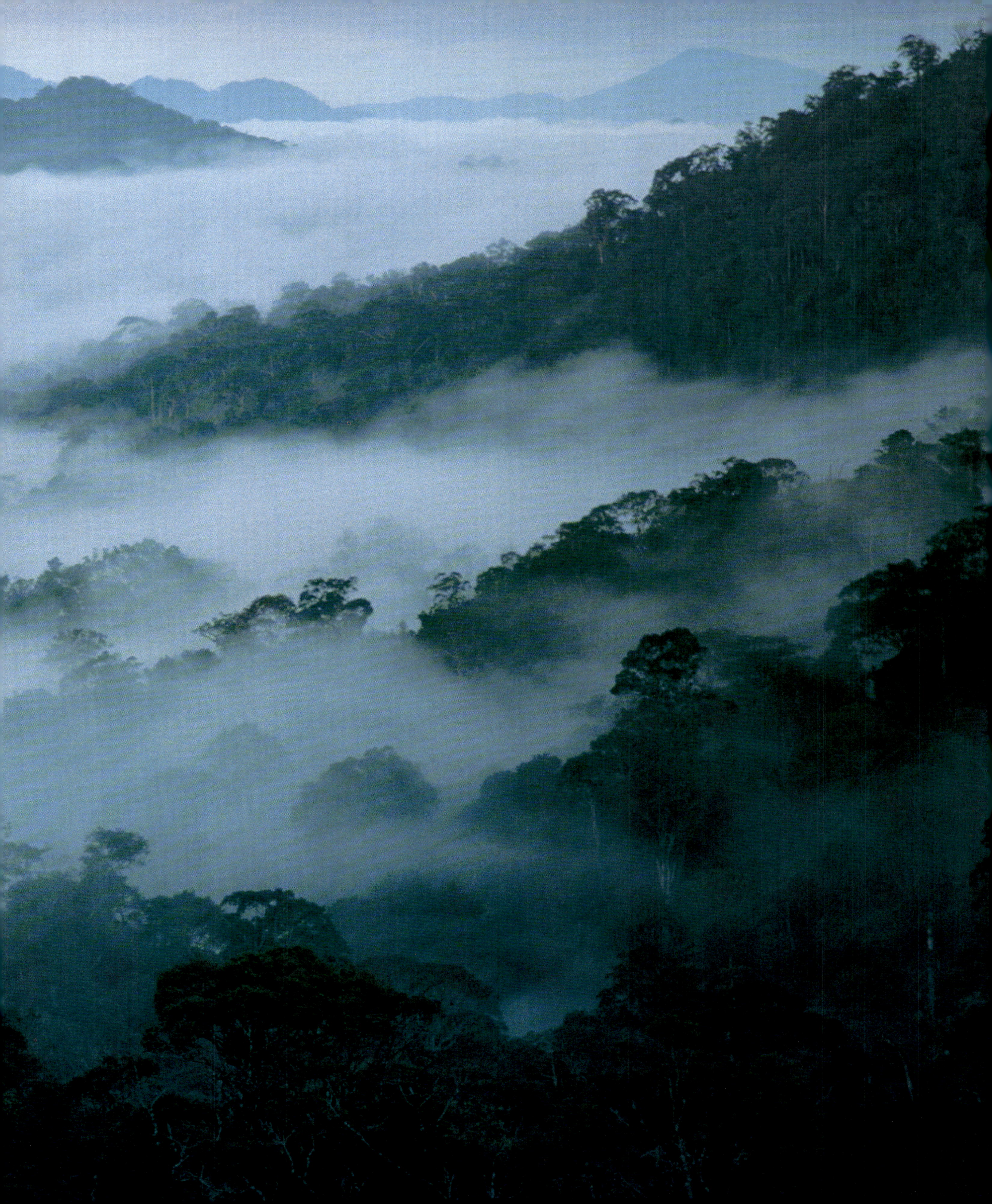

since they represent one or more early waves of human migration into Southeast Asia. In Sarawak and Sabah, there are a number of important tribal groups, with 28 in Sarawak alone.

Like Ecuador and the Philippines, Malaysia is particularly important in the context of the megadiversity countries because it demonstrates that a country does not have to be large to make the list. It is also one of the few tropical countries that has successfully made the transition from developing to developed nation, a change that has been fueled by use of its biodiversity resources, especially tropical forests, with both positive and negative consequences. As such, it provides an important model from which many other tropical countries can learn valuable lessons.

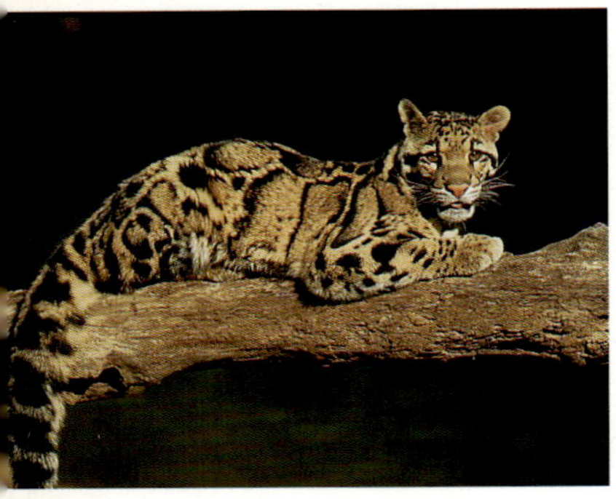

On pp. 428-429, rain forest near the border of Thailand in Perak State, West Malaysia, Malay Peninsula.

The mammal fauna of Malaysia includes a number of important flagship species. Above, the clouded leopard (Neofelis nebulosa) *is a very attractive rain forest species that is rarely seen.*

Natural habitats

Most of the natural vegetation in Malaysia, both Peninsular and East Malaysia, is tropical forest of various kinds. Indeed, the original extent of closed tropical moist forest in Malaysia as a whole is estimated to have been about 320 000 km^2 or basically the entire country; 130 000 km^2 of this was in Peninsular Malaysia, 120 000 km^2 in Sarawak, and 70 000 km^2 in Sabah. Based on estimates made in 1991, this was down to about 200 450 km^2 or 61% of the country, including 69 780 km^2 or 54% in Peninsular Malaysia, 36 000 km^2 or 51% in Sabah, and 94 670 km^2 or 79% in Sarawak.

Forest conversion has continued since these estimates were made, so the figures are likely to be somewhat lower at this time.

In Peninsular Malaysia, the dominant formation is lowland evergreen rain forest (0-300 m) and hill forest (300-1 300 m), dominated by members of the family Dipterocarpaceae; these two categories account for 44% out of the 54% that is estimated to still remain under forest. The largest area of this formation in Peninsular Malaysia is in Taman Negara National Park, with another larger area in the proposed Endau Rompin National Park. There is also a small area of semi-evergreen forest in the extreme northwest of the peninsula.

Other forest formations in Peninsular Malaysia include montane forests (5%), swamp forests (3%), and mangrove forests (1%). Hill forest grades into lower montane forest at 750-1 200 m, and is characterized by a lower canopy and an increasing number of vascular epiphytes. Above 1 500 m, formations like "elfin forest," "cloud forest" and "mossy forest" begin to appear, usually with a stunted appearance, a low canopy height of 10-20 m, and a rich abundance of epiphytes, mosses, orchids, and ericaceous species.

Lowland swamp forests are found on both the east and west coasts of Peninsular Malaysia, although many of these areas have now been cleared for agriculture. There are actually two fairly distinct types of swamp forest, one which grows on alluvial soils and is rich in nutrients and in wildlife, and the other, known as peat swamp forest, which grows on a thick layer of acidic, semi-decayed plant material known as peat. Peat is a highly infertile substance, and this forest is generally poor in wildlife, rather like the *igapó* forests of the black water rivers of the Amazon. However, peat forests do have valuable hardwood species, and most of them have been heavily logged. It is now being recognized that they also perform a valuable ecosystem maintenance service by acting as a giant sponge that absorbs flood water and releases it in dry times. Peat swamp forest is especially prevalent on the coastal plains of Sarawak.

Mangrove forests are now down to about 1 200 km^2, mostly along the west coast of Peninsular Malaysia. Most of them have been cleared for agriculture, fish ponds, and urbanization.

Small areas of heath forest also occur in Peninsular Malaysia. This is a forest type that appears on well-drained acidic soils derived from siliceous rocks; although it shares many species with lowland evergreen rain forest, there are a number of differences in species composition as well.

Families such as Myrtaceae and Casuarinaceae dominate in heath forests, as do conifers such as *Agatis*. Heath forests are characterized by low, stunted trees with small, thick leaves, and very few animals, rather like some of the poorer *terra firme* forests growing on nutrient-poor substrates of the Guayana Shield in Brazil and several other northern South American countries. A number of heath forest plants also have a commensal relationship with ants, a strategy used to protect their precious nutrients from caterpillars and other leaf predators. Pitcher plants also abound in this kind of environment, again being adapted to capture scarce nutrients by catching and digesting insects and other small animals.

Finally, there are forests growing on limestone, a feature typical of Southeast Asia, which has many distinctive limestone outcrops. In Peninsular Malaysia and Sabah, these are in the form of isolated precipitous outcrops, often in lowland areas; in Sarawak, they include both outcrops and extensive uplands, such as in Gunung Mulu National Park. Forest on limestone is generally rich in plant diversity and has high endemism as well, especially in herbs.

In Sarawak and Sabah, lowland evergreen rain forests again predominate; they make up 61 170 km^2 (65%) out of a remaining total of 94 670 km^2 of forest in Sarawak and 29 500 km^2 (82%) out of a remaining total of 36 000 km^2 in Sabah. Montane forests account for 17 060 km^2 (18%) in Sarawak and 3 100 km^2 (9%) in Sabah, and inland swamp forests cover a total of 14 080 km^2 (15%) in Sarawak but only a small area in Sabah. Mangroves cover 1 640 km^2 (2%) in Sarawak and 3 400 km^2 (9%) in Sabah. Both East Malaysian states also have some heath forest and some forest growing on limestone, although the latter is more extensive in Sarawak.

Flagship species

Malaysia has a large number of flagship species, including several of the world's most appealing charismatic megavertebrates. These include the Sumatran rhinoceros (*Dicerorhinus sumatrensis*), the Malayan tapir (*Tapirus indicus*), the Asian elephant (*Elephas maximus*), the tiger (*Panthera tigris*), the orang-utan (*Pongo pygmaeus*), the siamang (*Symphalangus syndactylus*) and other gibbons, the proboscis monkey (*Nasalis larvatus*), the reticulated python (*Python reticulatus*), the king cobra (*Ophiophagus hannah*), and a number of others. Other less known, but equally interesting animals include the Bornean tarsier (*Tarsius bancanus*), a variety of langurs and macaques, the gaur (*Bos gaurus*), the banteng (*Bos javanicus*), the Bornean earless monitor (*Lanthanotus borneensis*), and the Southeast Asian river terrapin (*Batagur baska*). All of these animals also occur in Indonesia, and are described in that chapter; some that have particularly important Malaysian populations are discussed in the following paragraphs.

The Sumatran rhinoceros (*Dicerorhinus sumatrensis*), or hairy rhinoceros as it is sometimes called, is the smallest rhinoceros species, and one of the most appealing and endangered of all the world's flagship species. It is hairier than the other larger rhinos, and lives at very low densities in the rain forest; it mainly inhabits mountains and forests at higher elevations, which may be why it has survived somewhat better than the Javan rhinoceros, which prefers coastal plains and river valleys. Once found from the foothills of the Himalayas in Bhutan and eastern India through Myanmar, Thailand and Peninsular Malaysia, and also on Sumatra and Borneo, it now survives in pockets in Myanmar, Thailand, Peninsular Malaysia, Sumatra and Borneo. Until recently, it was feared extinct in Borneo and in much of Peninsular Malaysia.

Surveys carried out since the late 1970s have indicated that it still persists, but that it is very rare and in danger of extinction over most of its range. Part of its rarity is attributable to the fact that rhino horn has long been of great value to Chinese communities throughout Asia as a form of medicine, causing it and all other rhino species to be heavily persecuted for their horns. What is worse is that the Sumatran rhino's horn is considered to be the most effective as a cure. The total world population is now estimated at between 500 and 900 individuals. In Peninsular Malaysia, there are about 100 individuals, the main rhino areas being Sungai Dusun, Ulu Selama, and Endau-Rompin. The Borneo population, recognized as a distinct subspecies, *Dicerorhinus sumatrensis harrissoni*, is critically endangered, and now found mainly in eastern Sabah, where there may be no more than 30 individuals still surviving. A small group also still exists in the Limbang catchment of Sarawak, and it is possible that some populations remain in East Kalimantan. In Sabah, the only places for it are the Tabin Wildlife Reserve (*ca.* 20 animals), the Kretam/Dent Peninsula (*ca.* 8 animals), and the Danum Valley Conservation Area (*ca.* 10 animals). As a hedge against extinction, captive-breeding programs have been established in Malaysia and Indonesia; efforts to breed these rhinos in American zoos have been notably unsuccessful, to the point that there are now only three individuals in the United States, all of them at the Cincinnati Zoo.

The Malayan tapir is the only representative of its family in the Old World, the other three living tapir species being found in South and Central America. It is very distinctive, with its long, flexible snout and striking pied, black-and-white coloration. It is more abundant and adaptable than the Sumatran rhinoceros, but within the country of Malaysia only occurs on the peninsula and not in Sarawak and Sabah. The Malayan tapir is still found in a number of forested areas, but the largest population appears to be in Taman Negara National Park.

The Asian elephant is the largest land mammal in Asia, and the second largest in the world after its African cousin, with bull elephants reaching as much as 3.5 m at the shoulder. Along with the Sumatran rhinoceros, the Asian elephant is probably the most endangered large animal in Malaysia because of the vast area that a single herd needs to survive. Much of the forest remaining in Malaysia is in fragmented blocks not large enough to support elephant herds over the long term; gene flow between herds is also impossible unless managed by man. Nonetheless, elephants have survived in scattered populations in Peninsular Malaysia and eastern Sabah, but there are no records of the species in Sarawak. The total population on the peninsula is estimated at 800-1 000 and that in Sabah is at least 500 and possibly as high as 2 000. As is the case of the Sumatran rhinoceros, the most important elephant area in Sabah is the Tabin Wildlife Reserve.

Elephants were used in the past in the timber industry, and were both status symbols and a means of transportation for families of high standing until early in this century, especially in the states of Kedah, Perak, Melaka, Terengganu, and Kelantan. There is some question as to how the elephant arrived in Sabah, since it is absent from the rest of Borneo, and there is speculation that it was introduced there several hundred years ago, possibly by the Sultan of Sulu.

The Malaysian population of the tiger once occurred throughout Peninsular Malaysia, and was estimated to be 3 500 in 1954. Since then, it has declined precipitously, as has been the case throughout the range of this animal. It is now down to perhaps 250 animals, mainly in remaining forests of Perak, Kelantan, Terengganu and Pahang. The tiger does not occur in Sarawak or Sabah, nor anywhere else in Borneo.

The orang-utan is found only on Borneo and Sumatra, and in Malaysia occurs in both Sarawak and Sabah. The largest arboreal animal, this primarily solitary ape is among the least social of all the primates, and lives in tropical forest of various kinds, feeding on durians and

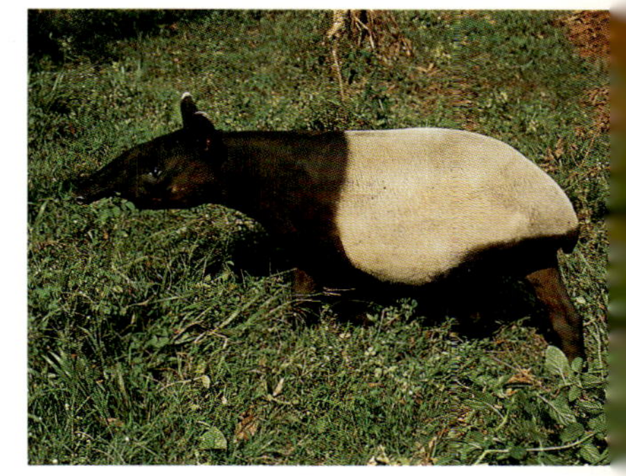

*The Malayan tapir (*Tapirus indicus*) is one of four species in the world, and the only living species found outside the Neotropics.*

*On pp. 432-433, the binturong (*Arctictis binturong*) or bear cat is a large, arboreal animal, member of the palm civet family, that inhabits the forests of several Asian countries. Females are 20 percent larger; the color of their coat varies from black to gray and has differing amounts of yellow and white at the hair tips. They have a heavily built, prehensile tail that they use to grasp a tree branch when foraging for fruit.*

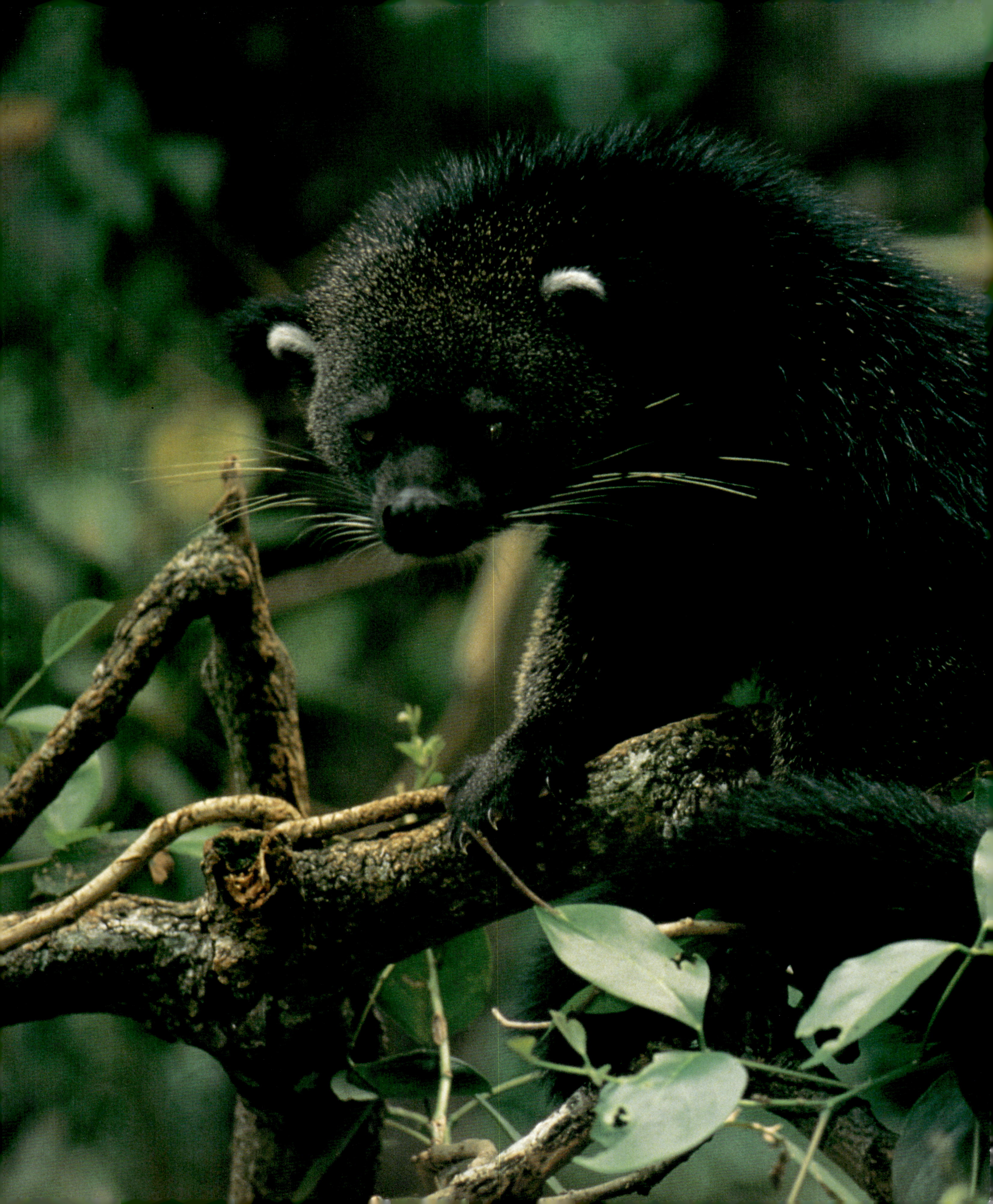

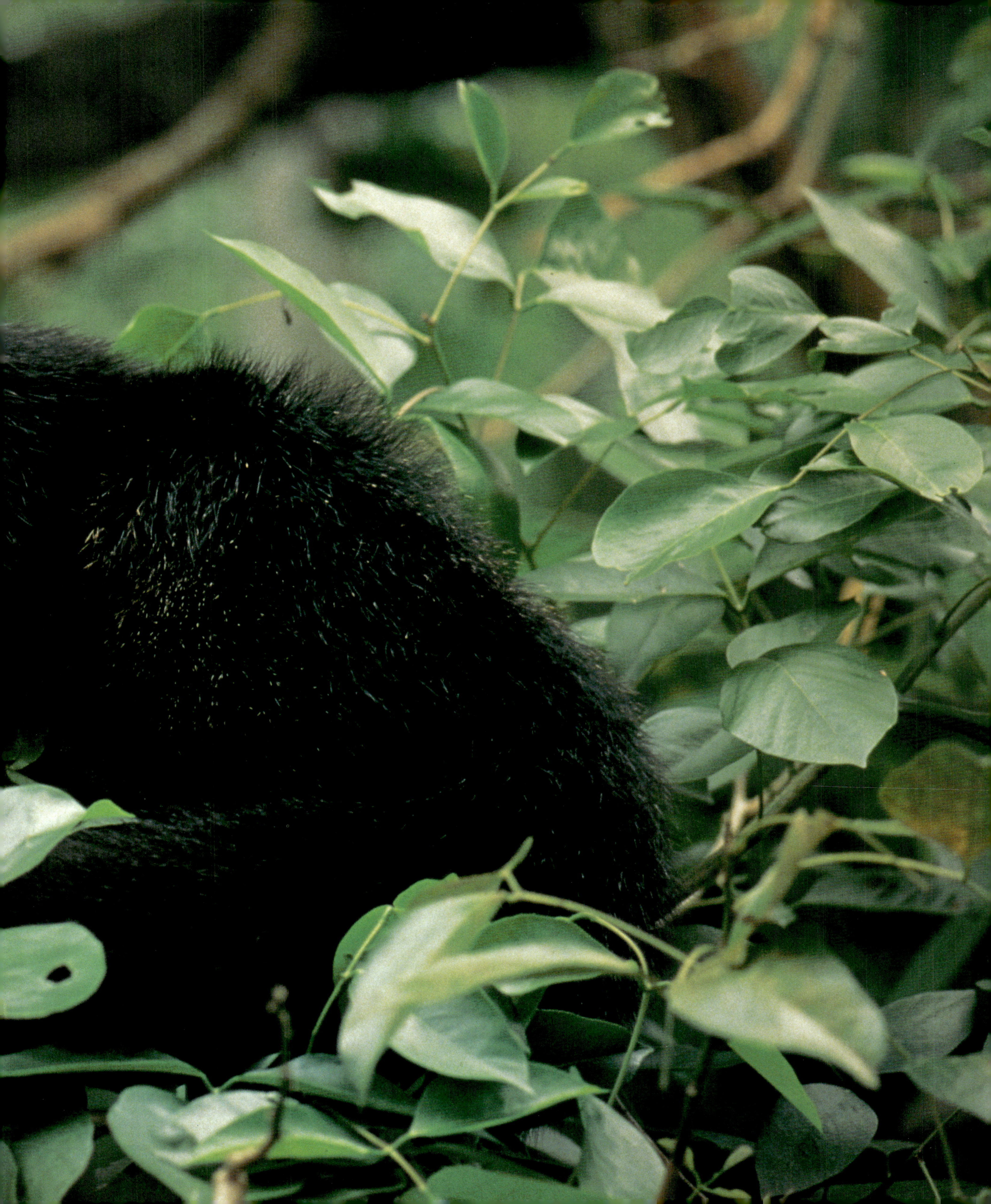

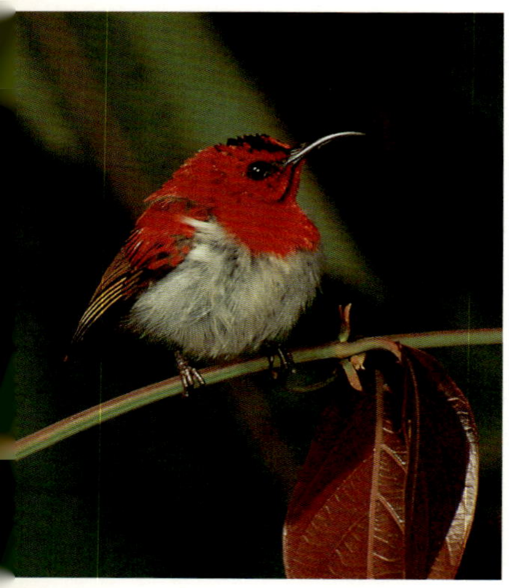

*A scarlet sunbird (*Aethopyga mytacalis*) from the forests of Malaysia.*

On the opposite page, strangler figs are common in the rain forests of Malaysia. This shot from the Danum Valley looks up from the inside, where the host tree once stood.

on a wide variety of other fruits. It also builds nests, which helps a great deal in determining whether or not it is present in a particular area, since it is otherwise difficult to locate. Much persecuted for the illegal pet trade, both internally and for export, still hunted for food in some areas, and strongly impacted by logging and other forms of forest destruction or alteration, orang-utans have declined almost everywhere, even though they are legally protected in both Malaysia and Indonesia. In Sabah, the rehabilitation station in the Sepilok Forest Reserve was one of the first to develop a program for reintroducing confiscated young orang-utans into the wild. This effort, which has been underway since 1964, has handled over 200 young orang-utans. The death rate has been rather high, with various diseases taking their toll, but health care has improved in recent years; several animals originally brought to Sepilok as infants have reached maturity, mated with wild orang-utans, and given birth to youngsters. The project has saved many young orangs that otherwise would have died, but, perhaps more important given the proximity of the station to the capital of Kota Kinabalu, Sepilok has played a very important educational role for schoolchildren, local people and international tourists.

The proboscis monkey is one of the strangest of all primates, and is characterized by the long, pendulous nose in males, which is both a visual display and a resonating chamber for loud honking vocalizations. This wonderfully grotesque animal is endemic to Borneo and is found mainly in nipa-mangrove forests all around the coast of this huge island, but in some cases it also extends far inland along major rivers as well. The species still occurs in reasonable numbers in both East Malaysian states, especially in the Samunsam Wildlife Sanctuary and Bako National Park in Sarawak and in Sepilok Forest Reserve in Sabah.

The large river terrapin (*Batagur baska*), locally known as *tuntong*, is a large species inhabiting brackish and freshwater riverine habitats over a wide but increasingly fragmented area of Southeast Asia. It has been very heavily exploited for its eggs and has declined drastically in most places. The best remaining populations appear to be those in Peninsular Malaysia, and a hatchery program has been underway there in the states of Perak, Kedah and Terengganu, together with a captive-breeding project in Perak. Another large species, the *tuntong laut* (*Callagur borneoensis*) occurs over a similar area of Southeast Asia and also in north Borneo, and requires much the same attention as the *tuntong*.

The Bornean earless monitor is a very strange lizard that is known only from a handful of specimens, and was first discovered in Sarawak. Not a monitor at all, it is apparently closest to the gila monster family, Helodermatidae. This unusual animal is a burrowing species that is rarely seen, but of considerable evolutionary interest within the lizards.

Among the snakes, two species are of particular note, the king cobra (*Ophiophagus hannah*) and the reticulated python (*Python reticulatus*). The king cobra is the largest of all poisonous snakes, reaching as much as 5.5 m in length, and one of the most dangerous. As its scientific name indicates, it feeds entirely on other snakes, and is also thought to be one of the more intelligent of snakes. The reticulated python is a beautiful animal that holds the official record as the longest snake in the world, one individual having been measured at 9.75 m; although the South American anaconda (*Eunectes murinus*) is a much bulkier animal, and has a questionable maximum length of 11.5 m, the official record for the anaconda is 8.8 m. The reticulated python has been much persecuted for its skin throughout most of its southeast Asian range, but still persists in many areas.

Cultural diversity

Human cultural diversity is moderately high in Malaysia, with some very interesting tribal people living in both Peninsular Malaysia and Sarawak and Sabah. The majority of the population is composed of the Bumiputras, the native Malays and other indigenous cultures who make up 55% of the population, with an additional 35% Chinese and 10% Indians from India who came to Malaysia during the colonial period. Islam is the official religion, but Christianity is important in East Malaysia and animist religions are still practiced by some of the tribal people. The population of 17.4 million is fairly high for a small country; however, in 1984 the Government of Malaysia announced a policy of increasing it to 70 million, and in 1985 passed a series of tax incentive measures to stimulate such growth.

Traditionally, the people of Malaysia are either hill forest rice growers, who grow "dry rice" through slash-and-burn agriculture, river people growing "wet rice" in paddies, or coastal fishermen and traders. In the past, these coastal trading communities grew into regional trading centers and, in some cases, influential sultanates that interacted with many other parts of southeast Asia. The Malay language, Bahasa Malaysia, evolved as a trade language facilitating communication among these many different peoples, and was adopted as the official language of the country after independence.

In Sarawak, there are twenty-eight different ethnic groups. Close to half the people belong to non-Muslim groups usually referred to as Dayaks, a term that is widely used but obscures the fact that it includes many different and important peoples. Most of the Dayaks are forest-dwelling peoples that still practiced headhunting until the early part of this century, and skulls continue to decorate the traditional longhouses in many places. The Ibans of the southern and central parts of the state are the largest group and account for about 30% of Sarawak's population. The Bidayuhs live

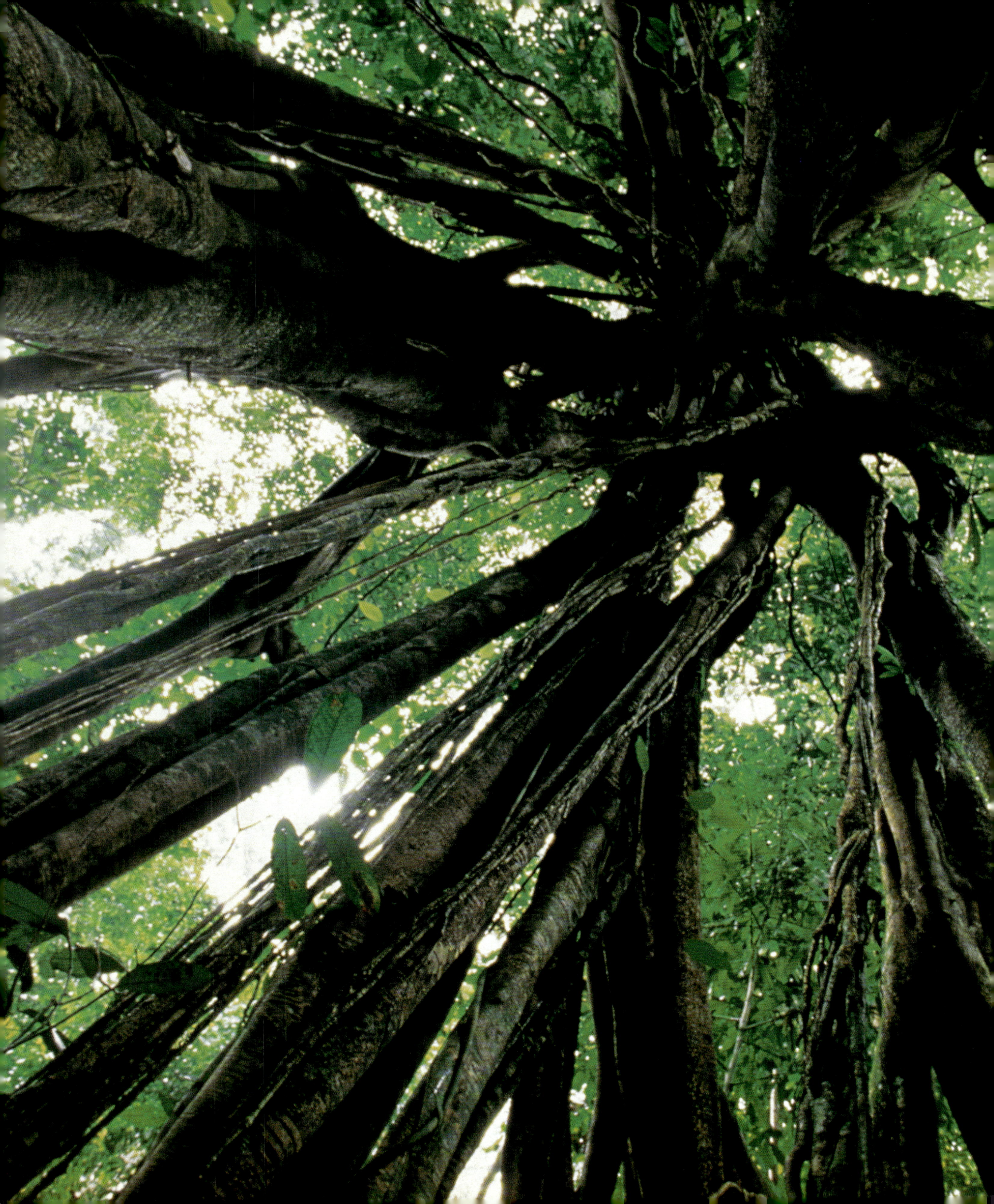

in the southern hills and plains and comprise 8% of the population, while the Kenyahs, the Kayans, the Kedayans, the Muruts, the Punans, the Kelabits, and the Penans make up an additional 5%. The Muslim Malays and Melanaus live predominantly on the coast and make up 26%, and the remainder (31%) are mostly Chinese, who live principally in the towns and larger villages.

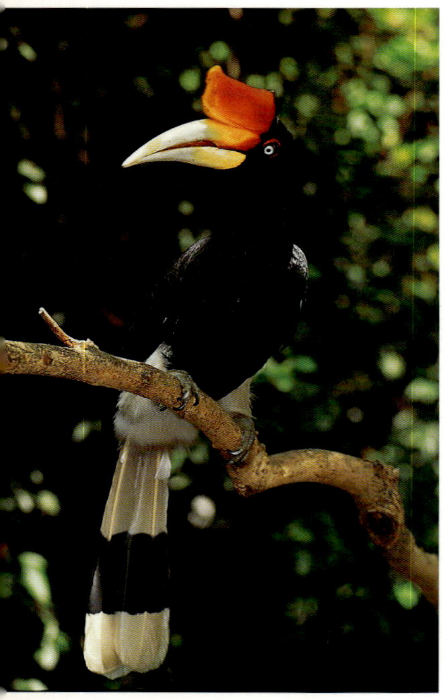

The traditionally nomadic Penans live in northeastern Sarawak, and currently number some 7 600 people, of which about 25% are settled. The remainder are either semisedentary or still nomadic. Traditionally blowgun hunters, the Penan lived mainly by hunting and gathering and some shifting agriculture. Of the roughly 100 000 indigenous people who roamed the forests of Sarawak at the turn of the century, only the Penan retain a semblance of their traditional lifestyle. Unfortunately, over the past two decades they have frequently run into conflict with logging companies, and have resisted government efforts to settle them. The treatment of these people by current Malaysian governmental authorities is reminiscent of that accorded the Indians in the United States and Brazil and aborigines in Australia, and has stimulated an international outcry on behalf of the Penan. The final outcome of this situation remains to be determined, but it appears unlikely that the Malaysian authorities will ultimately allow the Penan to continue their traditional lifestyles.

In Sabah, there are also a number of non-Muslim tribal groups. The Dusun and Kadazan peoples live by rice cultivation on the west coast, the Bisaya people occupy the mostly swampy peninsula, and the Muruts and the Lundayehs practice shifting cultivation in the interior.

From an anthropological point of view, the most interesting tribal groups in Malaysia are the Orang Asli of Peninsular Malaysia. The Orang Asli are the remnants of one or more of the early human migrations into this region, and are composed of several pygmoid groups now down to several thousand individuals. There are two main groups, the Semang, in which the men average about 149 cm and the women about 141 cm in height; these people are a true Negrito racial type, with dark skin, kinky hair and negroid features, and are related to the Negritos of the Andaman Islands and the Philippines. The second group, the Sakai, averages slightly taller, has lighter skin than the Semang, and wavy hair; they are similar in appearance to the Veddas of Sri Lanka. A third group, the straight-haired, taller, still lighter-skinned Jahun or Jakun, was believed at the turn of the century to represent the aboriginal Malay stock. Traditionally hunter-gatherers, the weapon of choice of the Semang and the Sakai was the blowgun, and this weapon is still used in certain areas. These people are now down to very small populations living in reservation-like situations in some of the remaining forests of the peninsula, with populations numbering in the low tens of thousands.

Another of Malaysia's impressive hornbill species, the rhinoceros hornbill (Buceros rhinoceros).

On the opposite page, lowland rain forest along the Segama River in Sabah.

Resettlement of Orang Asli continues to the present day. In the current Fifth Malaysia Plan, there is a comprehensive program to move 23 000 Orang Asli from the Main Range into five agriculture-based resettlement schemes. There are also indications that logging companies are impacting the lives of those Orang Asli that still live traditionally, much as is the case with the Penan in Sarawak.

RUSSELL A. MITTERMEIER
CRISTINA GOETTSCH MITTERMEIER

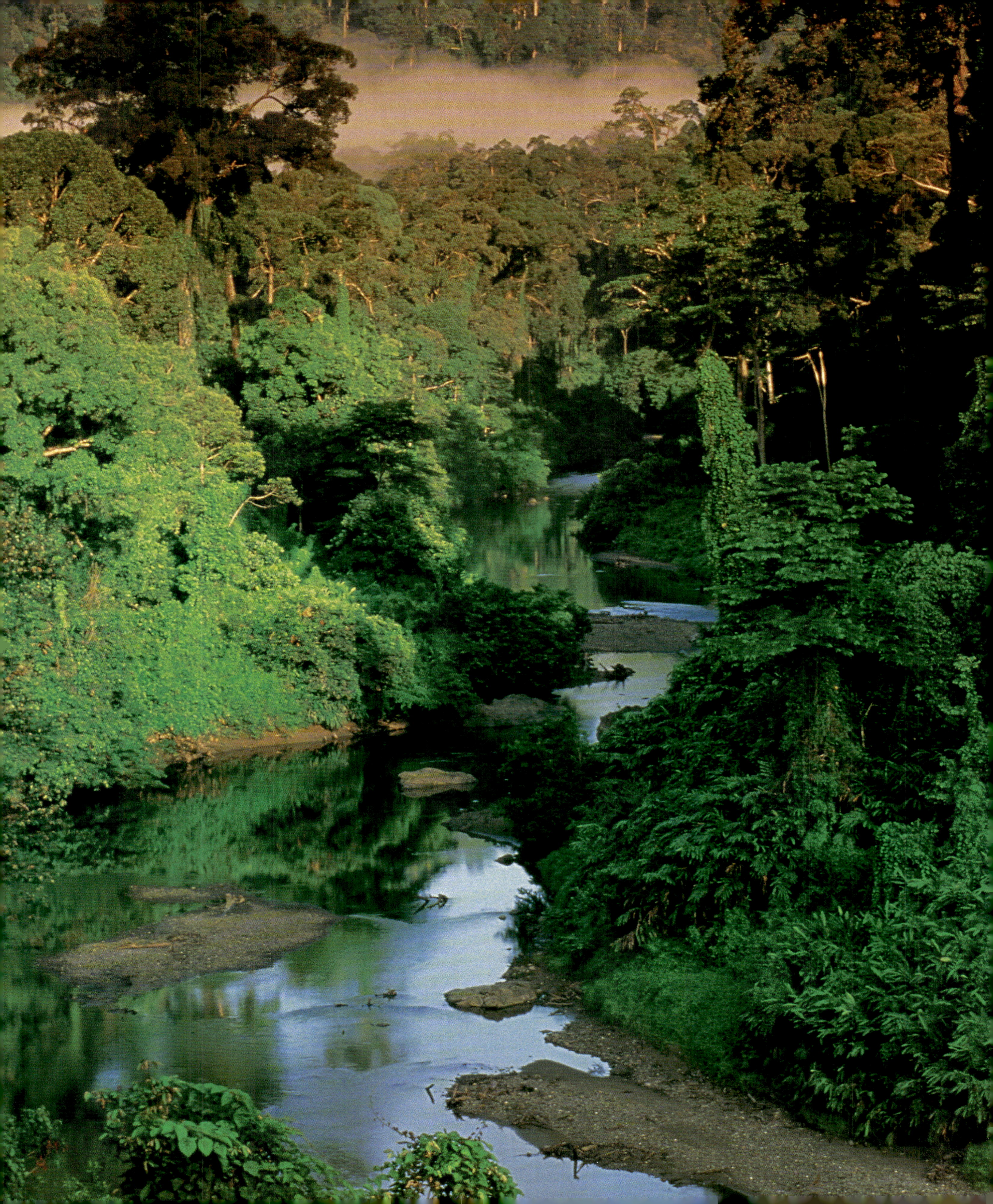

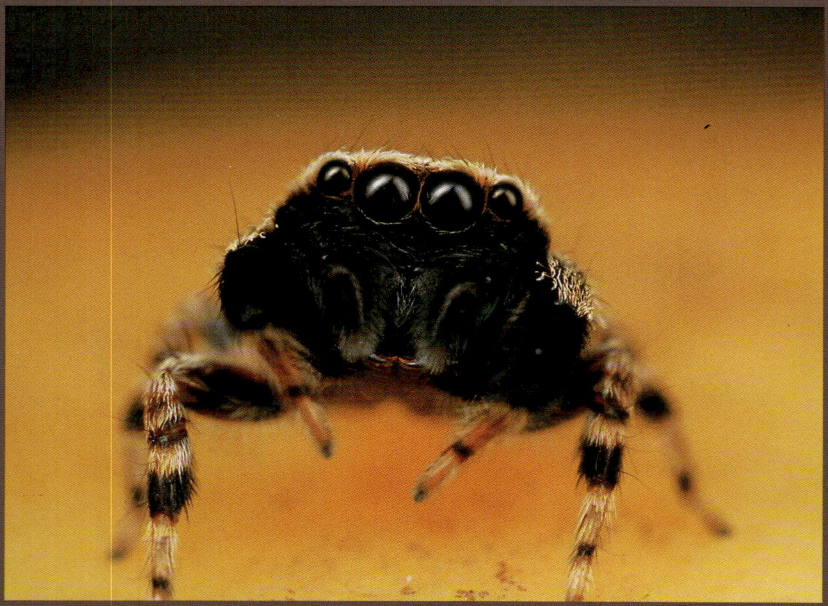

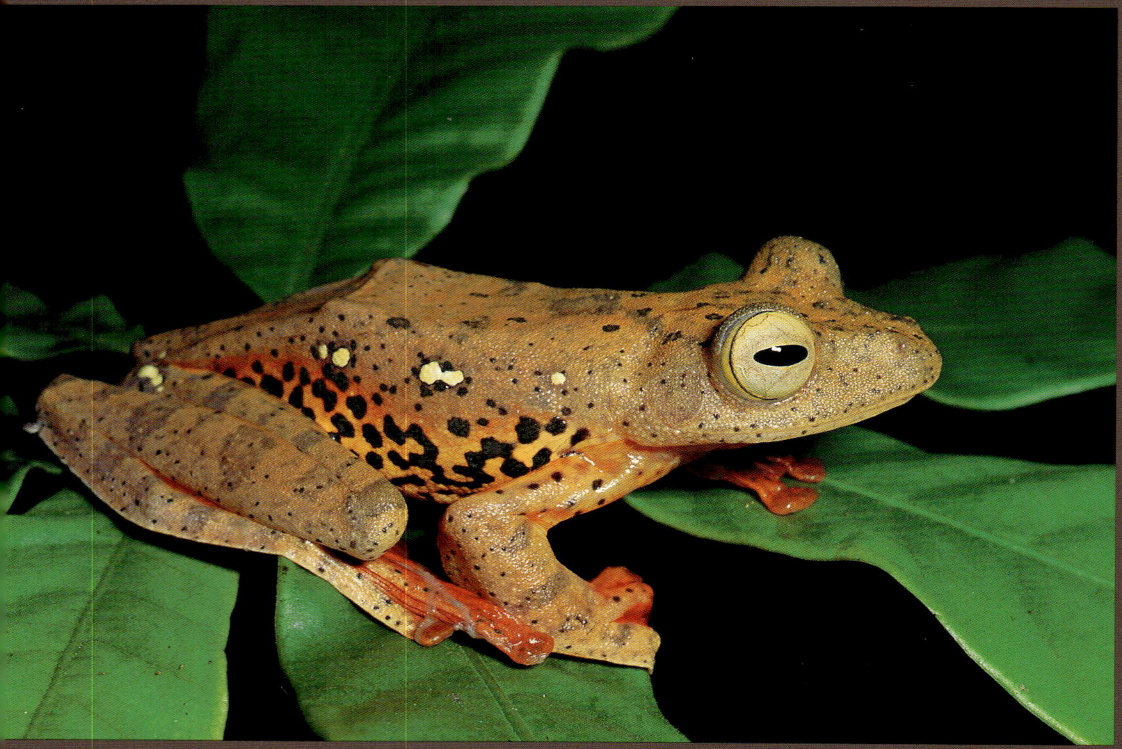

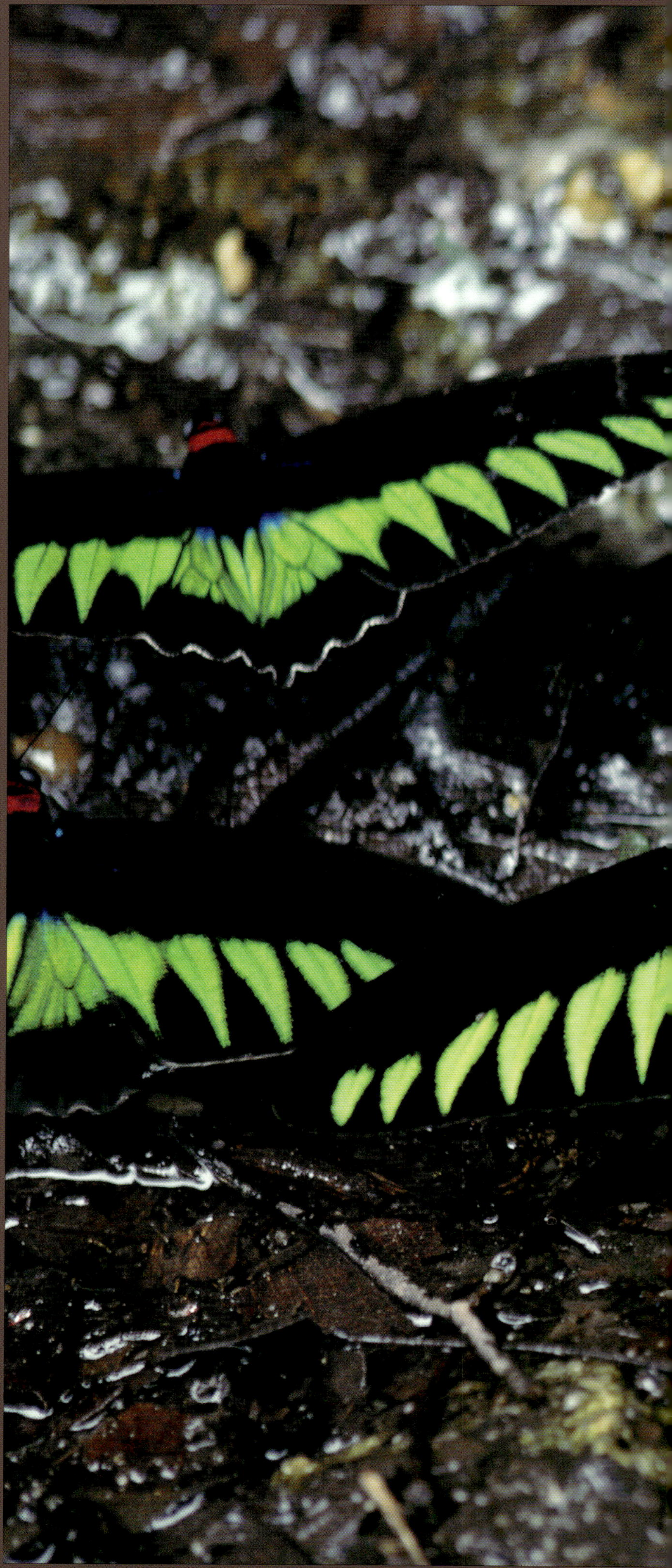

Upper left, a jumping spider. Lower left, a harlequin flying tree frog (Rhacophorus pardalis)*, one of several species capable of gliding considerable distances from tree to tree. Right, the spectacular Rajah Brooke's birdwing butterflies* (Trogonoptera brookiana).

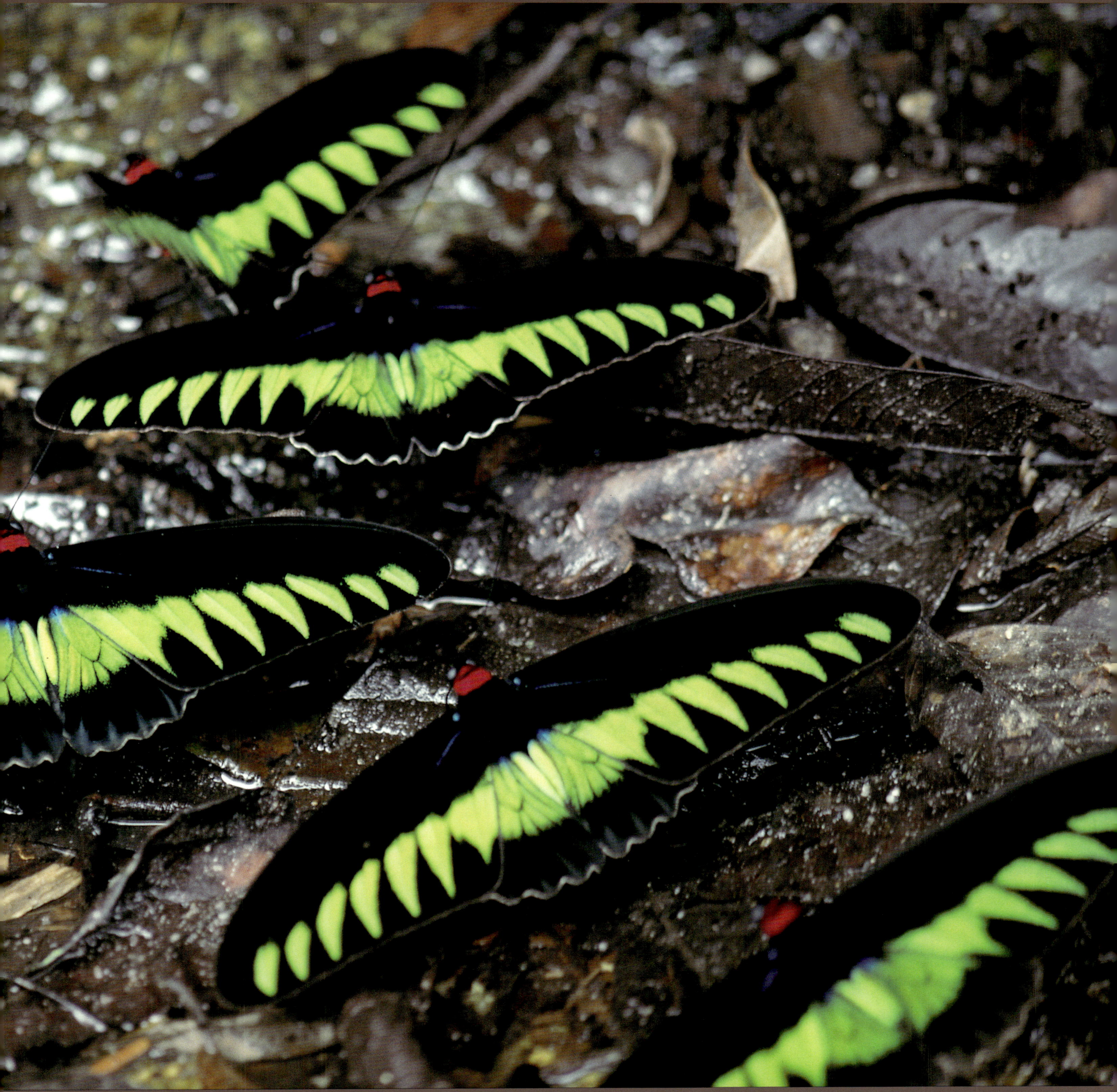

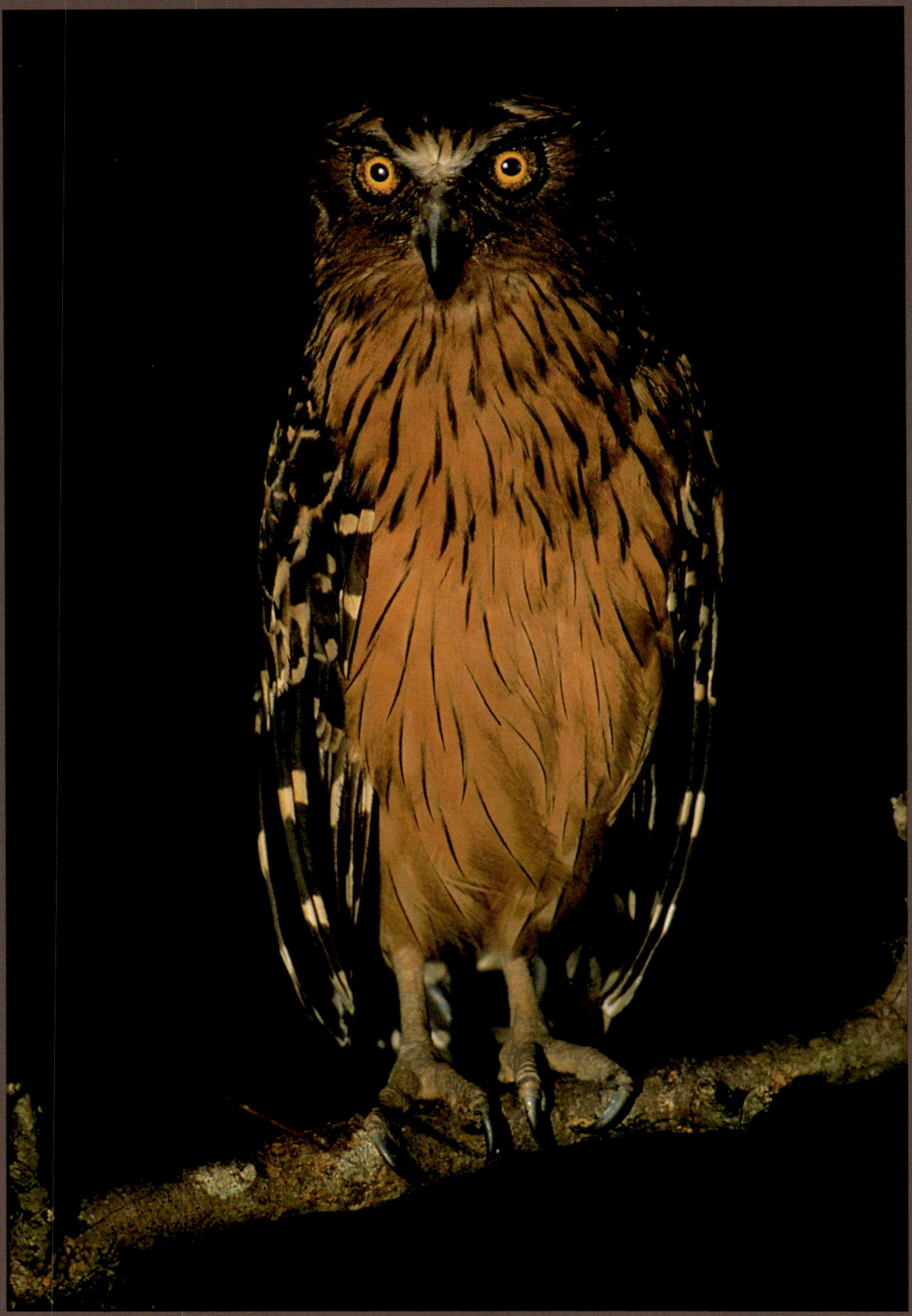

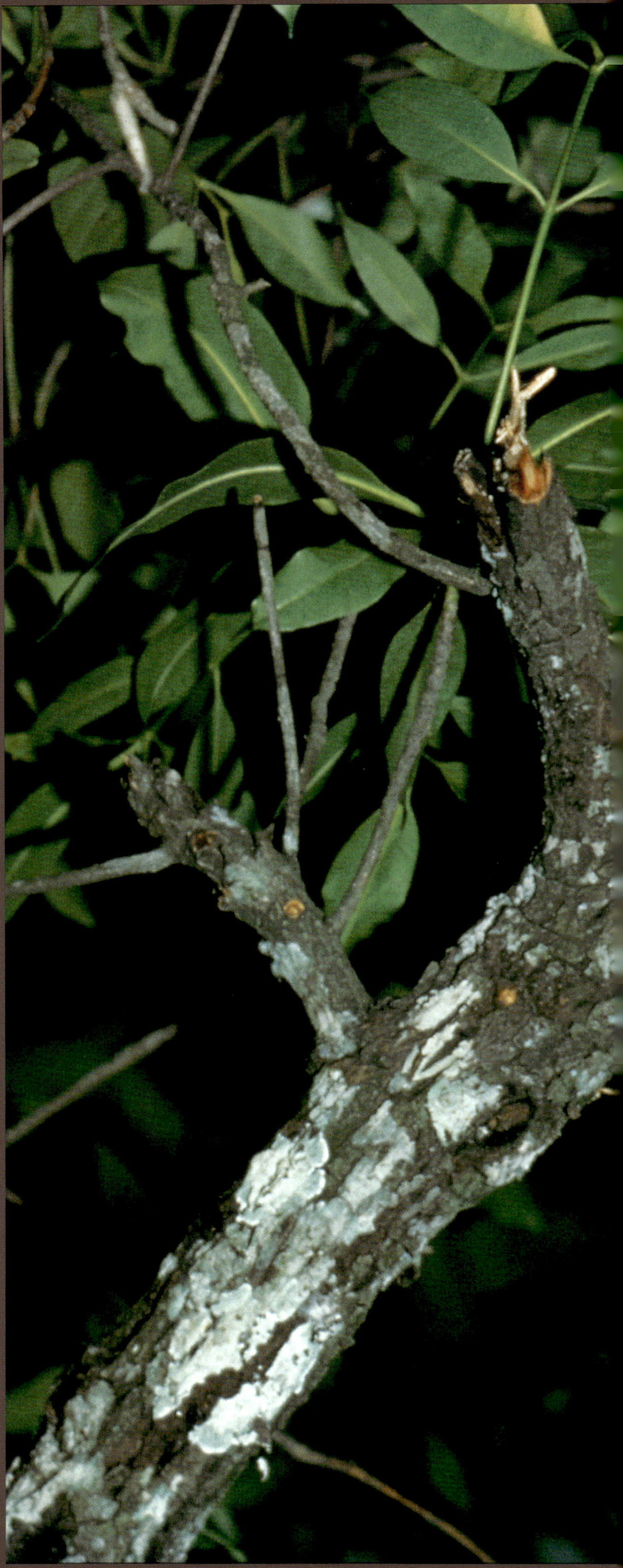

Above, a buffy fish owl (Ketupa ketupu) *from the Danum Valley in Sabah, one of the most important conservation areas in all of Borneo. More of Malaysia's mammals. Right, a flying lemur or colugo* (Cynocephalus variegatus) *mother and infant hang from a tree; not a lemur at all, this unusual creature is in its own distinctive order of mammals, the Dermoptera, which includes only one genus and two species. It does not actually fly, but glides on membranes attached to the neck and sides of the body.*

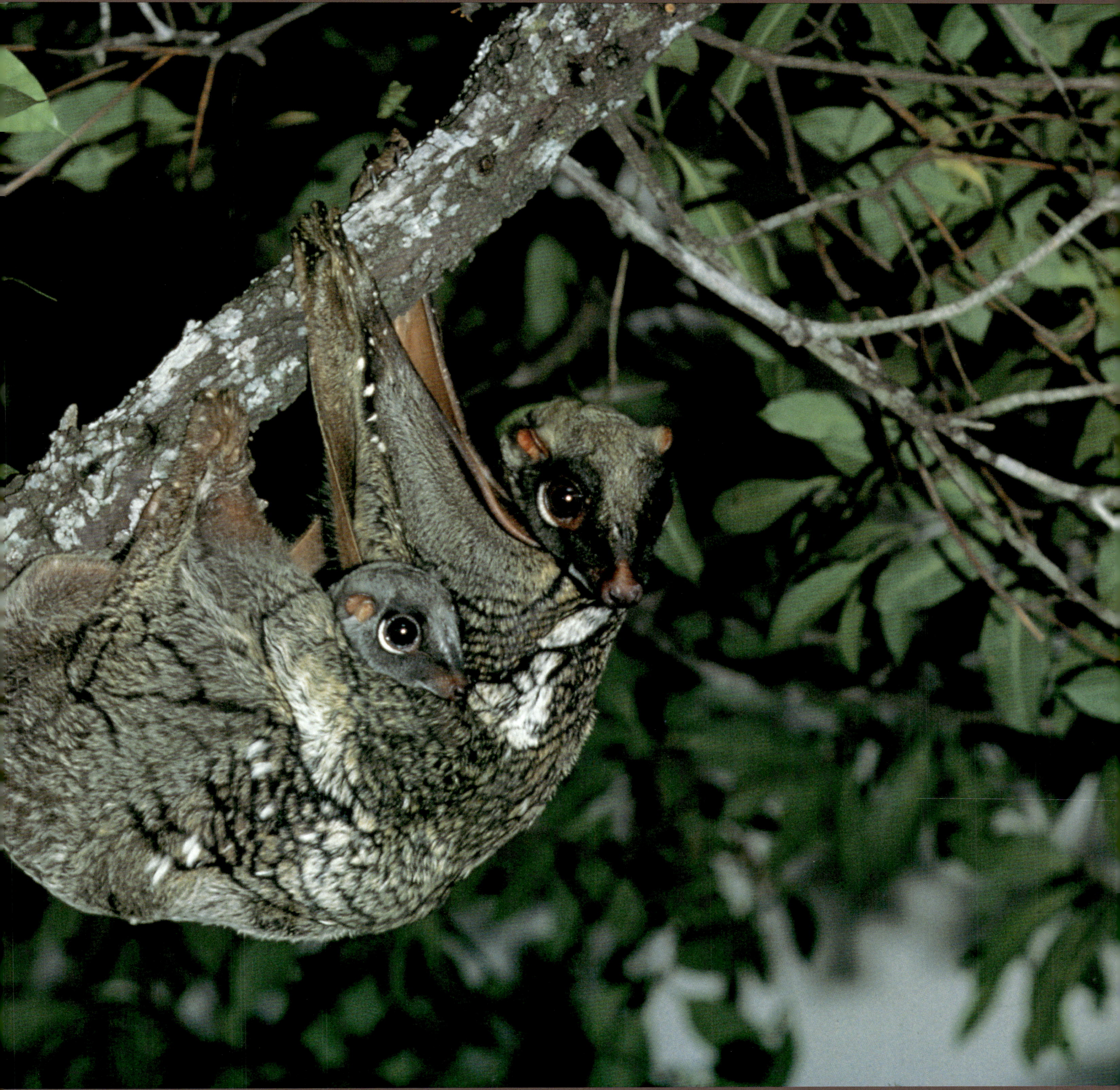

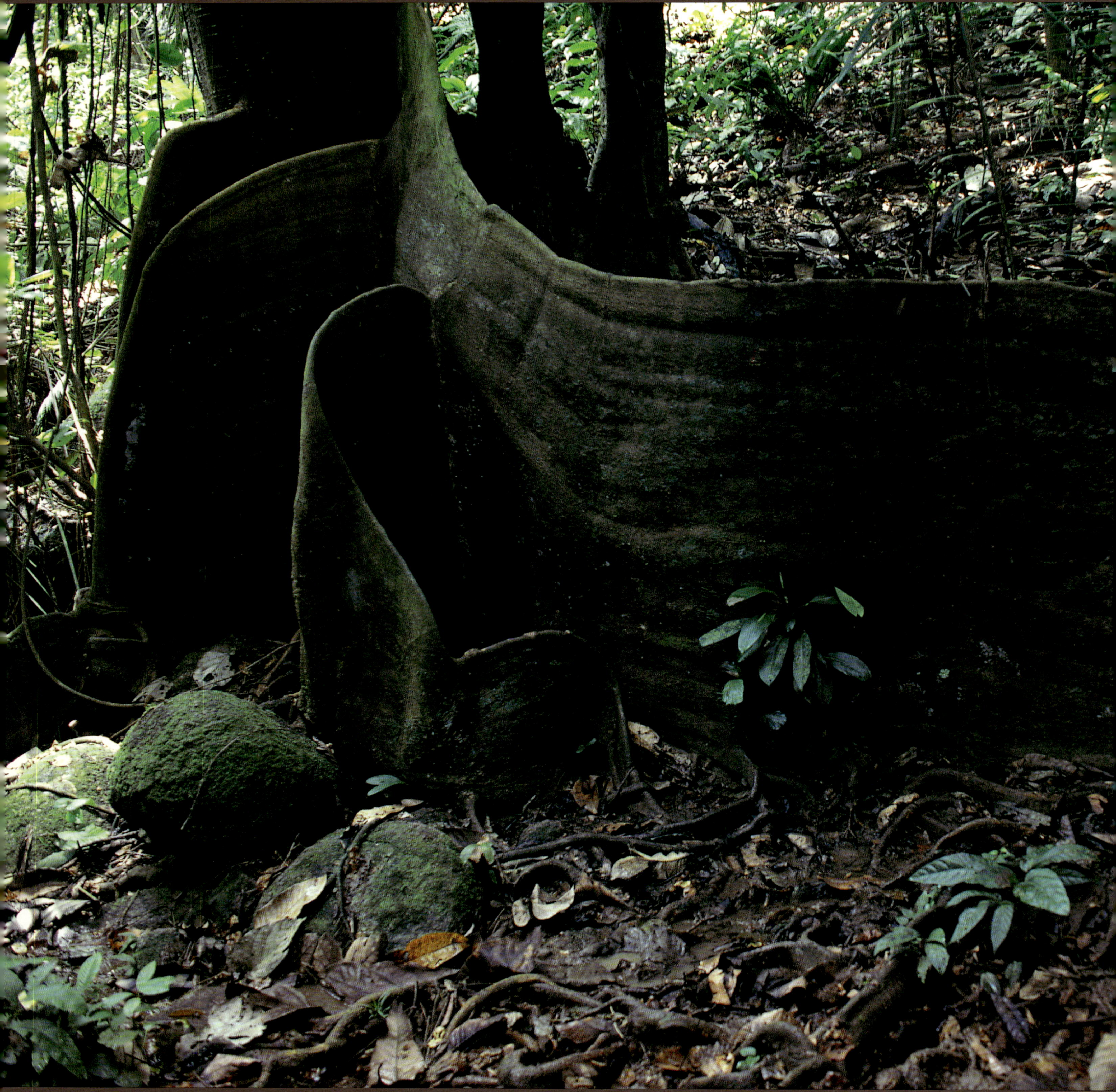

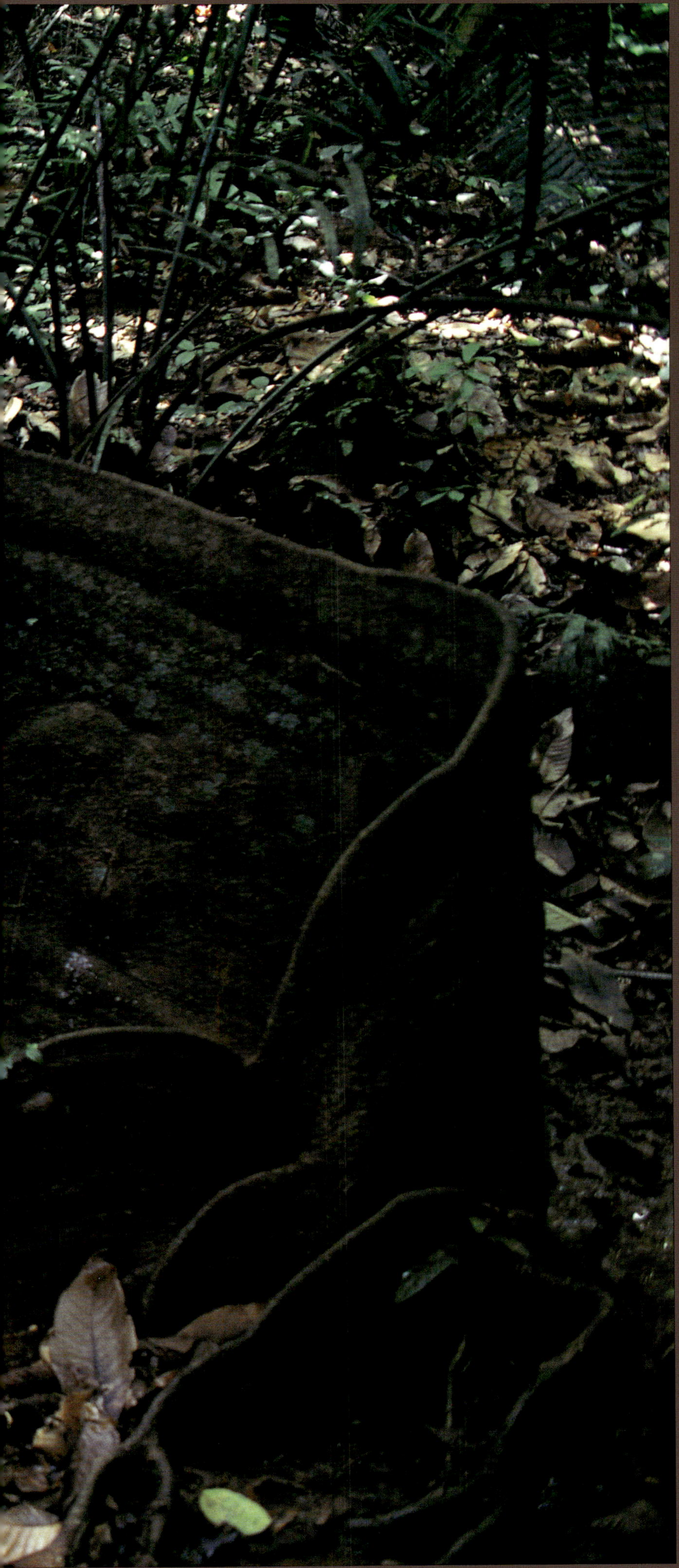

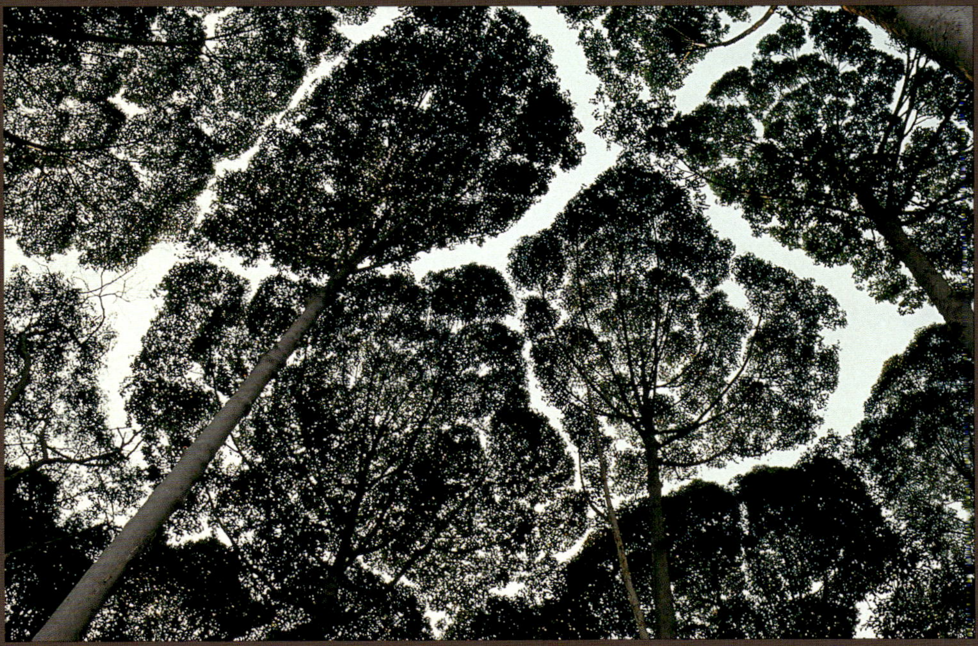

As in the rain forests of Amazonia, most of the nutrients in the Malaysian forest are at the surface and in the living biomass. Left, roots do not penetrate deeply into the substrate, leading to structures like this extensive buttress to provide stability to forest trees. Above, the unusual phenomenon of crown shyness in a forest in Peninsular Malaysia. The crowns of these trees, though quite close together, appear to be avoiding contact.

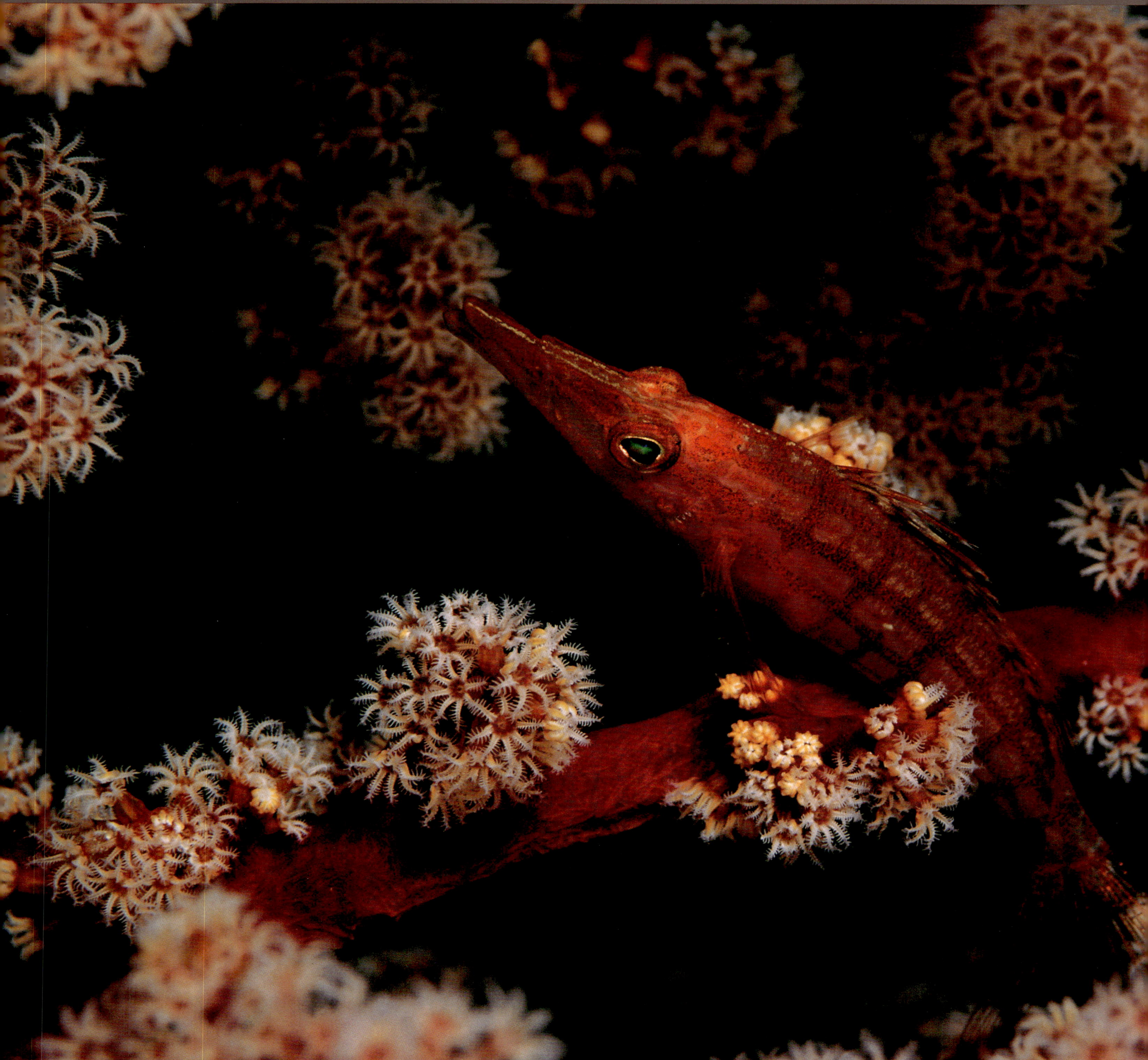

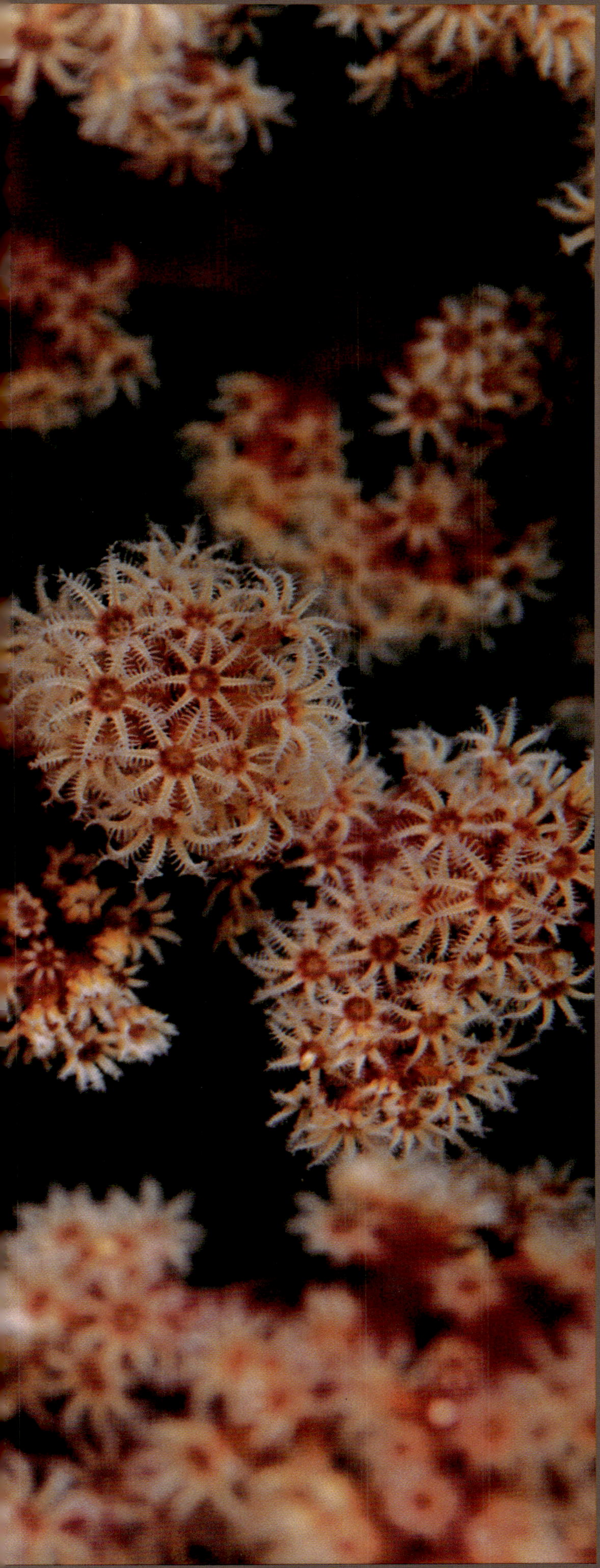

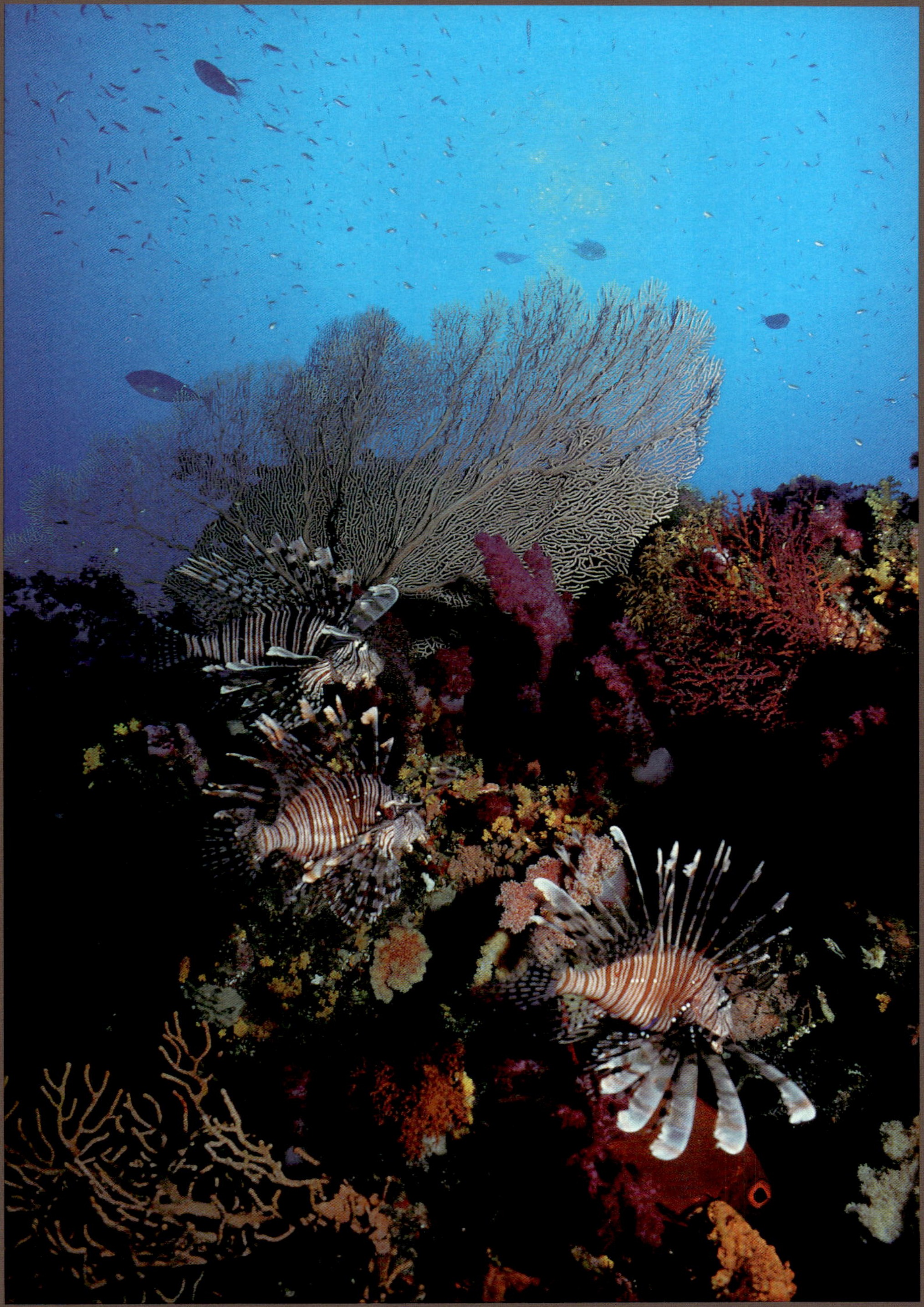

Left, a long-nosed hawkfish (Oxycirrhites typus) *on a reef in Sabah.*
Above, a wide-angle reef scene with soft corals, gorgonians and lion-fish.

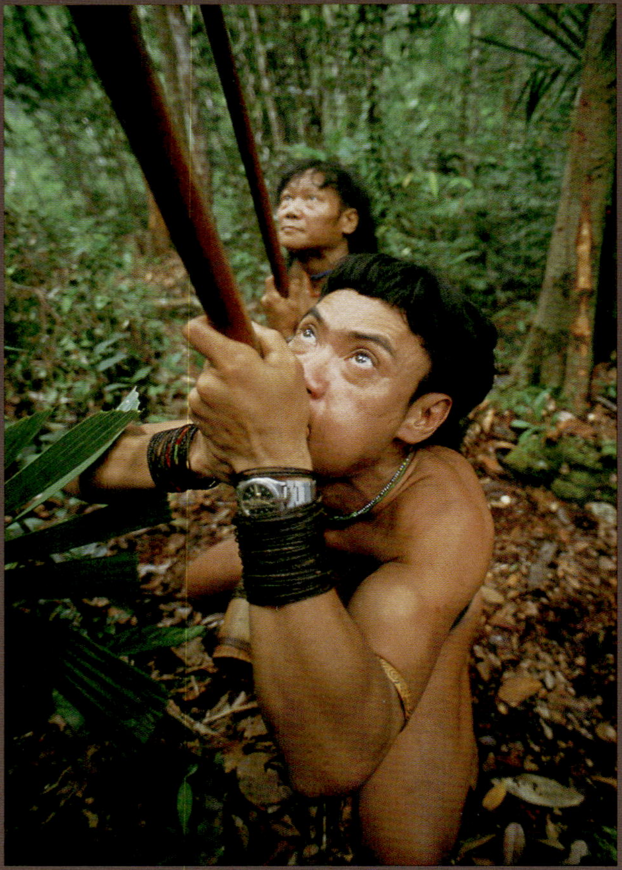

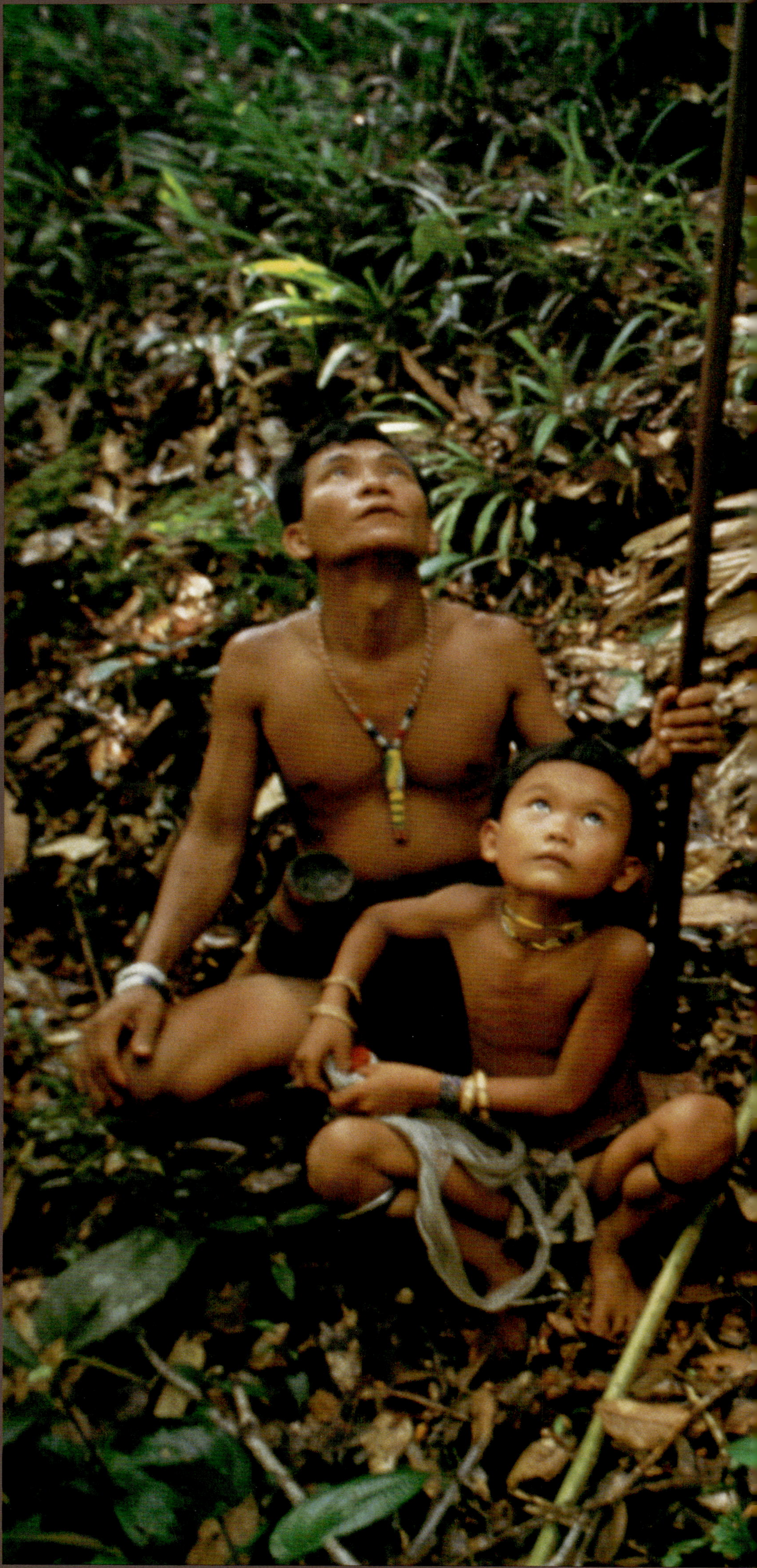

The Penan are a forest people living in the rain forests of Sarawak, where their traditional lifestyle is threatened by logging and government efforts to resettle them and integrate them into modern society. Above, Asik Nylek, an expert blowgun hunter, takes aim. The Penan live mainly by hunting and gathering in their forest home; right, a group of Penan with a wild pig shot for food.

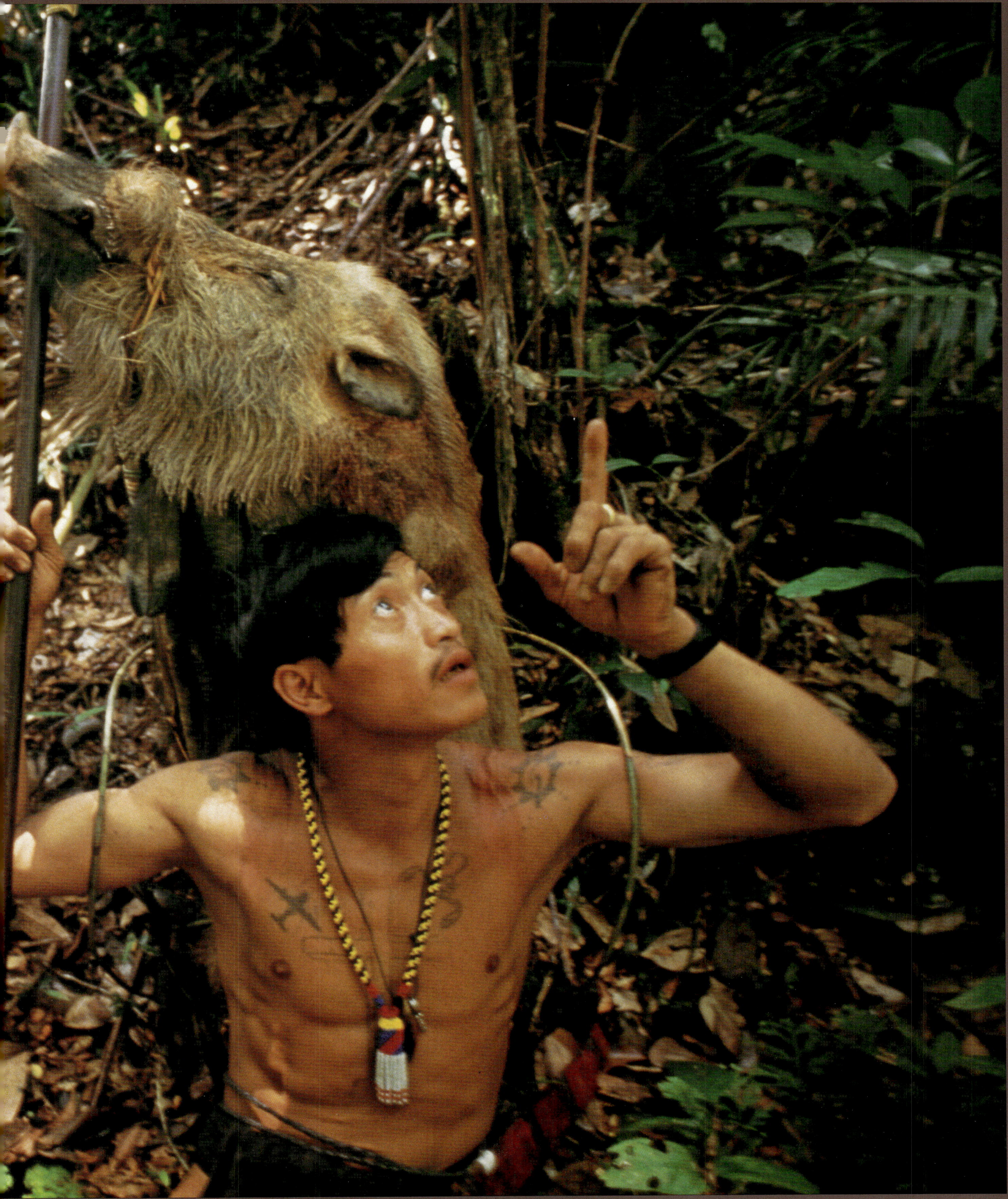

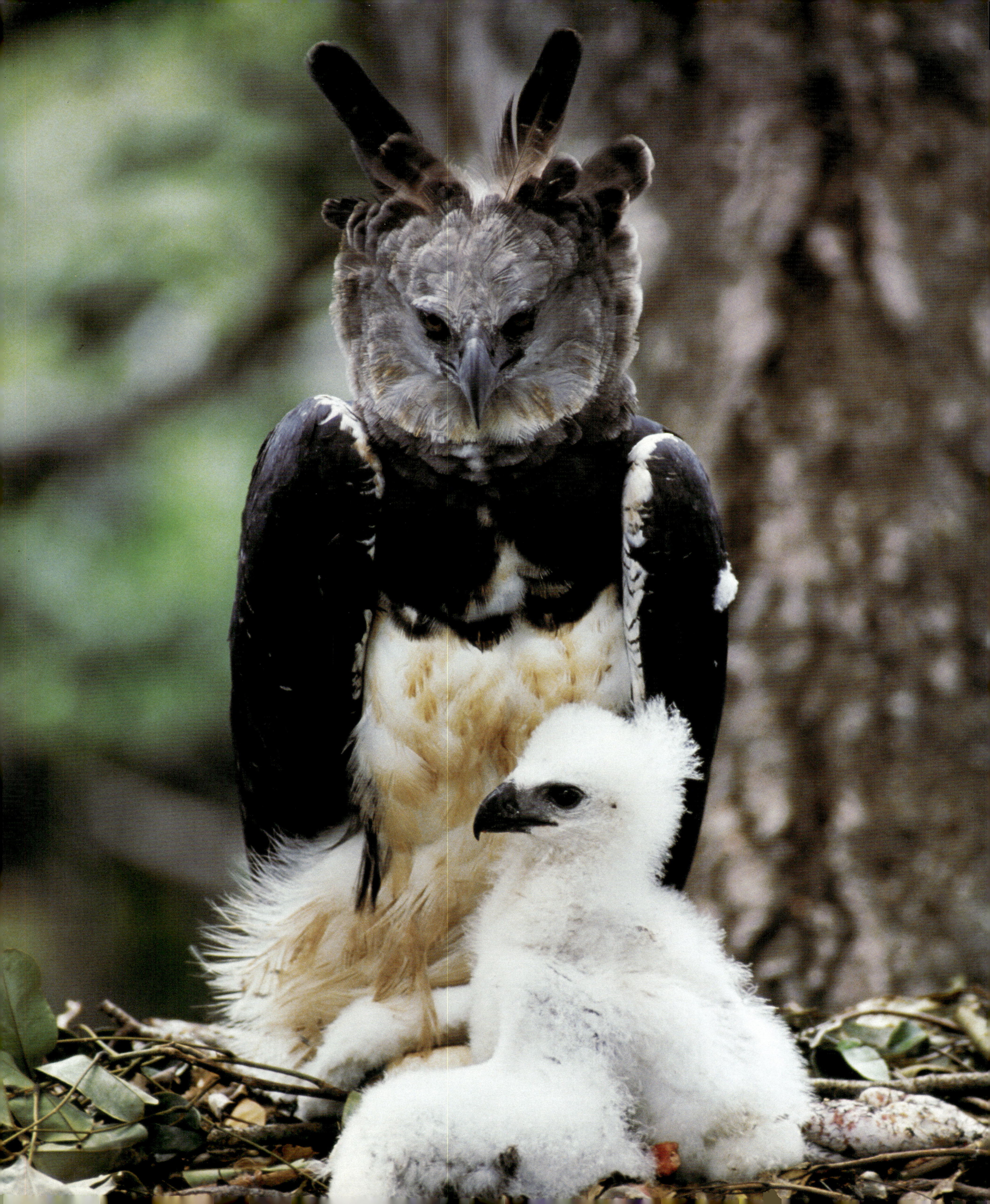

VENEZUELA

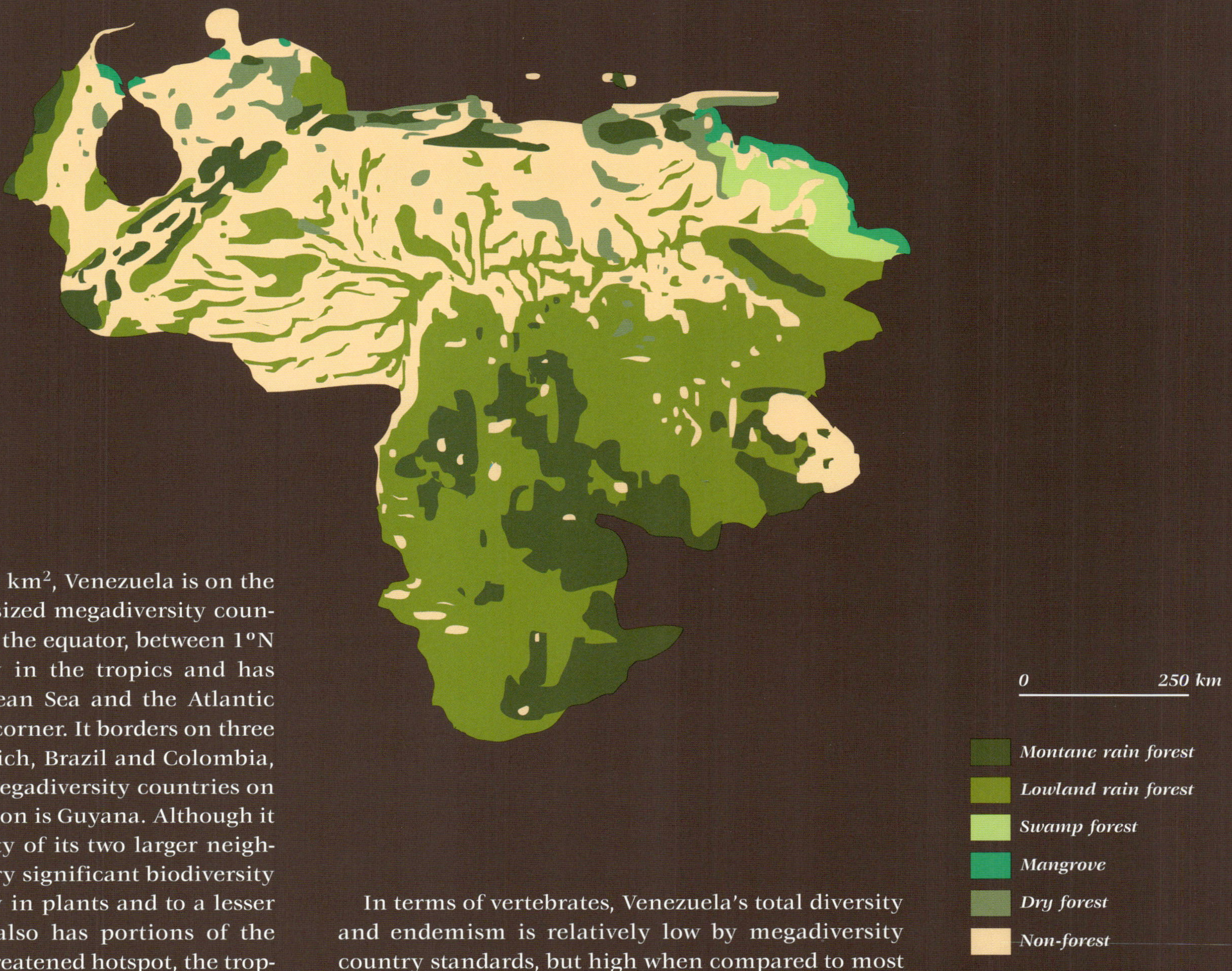

With a land area of 912 050 km^2, Venezuela is on the small end of the medium-sized megadiversity countries. Located just north of the equator, between 1°N and 12°N, it lies entirely in the tropics and has coasts in both the Caribbean Sea and the Atlantic Ocean in its northeastern corner. It borders on three other countries, two of which, Brazil and Colombia, are among the top three megadiversity countries on Earth; the third border nation is Guyana. Although it does not reach the diversity of its two larger neighbors, it nonetheless has very significant biodiversity in its own right, especially in plants and to a lesser extent in vertebrates. It also has portions of the world's highest-priority threatened hotspot, the tropical Andes, and a major portion of what is almost certainly the world's least disturbed major tropical wilderness area, the southern Guianas/southern Venezuela/northern Brazil complex.

Venezuela makes the megadiversity list mainly on the strength of its higher plant diversity, which is usually estimated at 15 000-16 000 species, with one recent estimate (WWF/IUCN, 1997) as high as 21 070. Of these, at least one third are thought to be endemic, for a minimum of 5 000 endemic species; however, the latest estimate by WWF/IUCN (1997) gives a figure of 8 000 endemics. Much of this diversity is concentrated in the southern part of the country, known as Venezuelan Guayana, which harbors at least 9 500 species. Venezuela also has three Centers of Plant Diversity and Endemism as identified by WWF/IUCN.

In terms of vertebrates, Venezuela's total diversity and endemism is relatively low by megadiversity country standards, but high when compared to most other countries on Earth. Its bird diversity is the most impressive; with 1 360 species (at least 45 of them endemic), it is in the sixth position on the world list behind only Colombia, Peru, Brazil, and Ecuador in the Neotropics and Indonesia in the Old World. Venezuela also has eight Endemic Bird Areas, as defined by BirdLife International and 26 Key Areas for Threatened Birds. The Endemic Bird Areas include the Tepuis region, the Cordillera de Caripe and the Paria Peninsula, the Cordillera de la Costa Central, the Cordillera de Mérida, the Caribbean Dry Zone, the East Andes of Venezuela/Colombia, and the upper Río Negro and Orinoco white sand forests.

Mammal diversity is moderately high at 288 species, but endemism (11 species) is low. Reptile and amphibian diversity is relatively high, and endemism in both groups is considerably higher than

On the opposite page, the harpy eagle (**Harpia harpyja**) *is the largest and most impressive bird of prey in the Neotropics. It is a rain-forest dweller that prefers undisturbed primary forest habitat, and has either disappeared or become very rare in most of its original range. The wilderness areas of southern Venezuela are one of its major strongholds.*

for mammals. Reptiles include 293 species (57 endemic), putting Venezuela in thirteenth place on the reptile diversity world list, while amphibians, with 204 species (76 endemic) rank ninth on the world diversity list. The total number of endemics for non-fish vertebrates (189) is low and does not make the cutoff of 1% of total global diversity, but the combined diversity for birds, mammals, reptiles, and amphibians (2 145) is quite high, mainly on the strength of the high number for birds, and puts Venezuela in ninth place in this category.

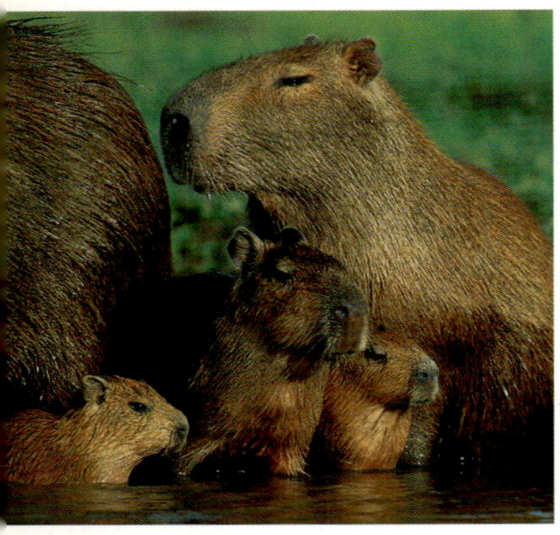

Invertebrate diversity in Venezuela is also impressive. The country ranks sixth in butterfly diversity with 2 316 species, and 14th in butterfly endemism with 115 endemics. In another representative group, the tiger beetles, Venezuela ranks relatively low, its 51 species and 6 endemics placing it in 29th position on the global list.

Freshwater diversity in Venezuela is quite impressive as well, much of it concentrated in the Orinoco, one of the world's larger rivers at 2 150 km. It divides Venezuela into two almost equal parts, each measuring some 450 000 km^2, and also has a huge delta. At least 1 250 freshwater species have been recorded from Venezuela, most of them from the Orinoco drainage, and every new expedition uncovers a host of undescribed species. Thus, Venezuela holds fourth position on the world list in this category.

Marine diversity is quite significant as well, with more than 2 000 marine fish species recorded. The country has a number of coral reef systems, especially around the islands like Margarita and the Los Roques Archipelago. Venezuela's 3 200-km coastline is also very variable and includes long, sandy beaches and steep, rocky slopes that are continuously pounded by the sea.

In global terms, the most important part of Venezuela is the Guayana region south of the Orinoco River. Not only does this area harbor the bulk of Venezuela's species diversity and the majority of its endemics, but it may be the single most intact tropical wilderness area in the world, even less altered than the largely undisturbed forests of the southern Guianas to the east. This is also the region of the spectacular *tepuis*, huge, flat-topped sandstone outcrops of the Roraima Formation that rise out of undisturbed lowland forest, and have such an otherworldly aspect to them that they served as an inspiration for Arthur Conan Doyle's famous book, *The Lost World*. Indeed, this region is so little disturbed, being interrupted only by an occasional tiny Indian settlement in a vast ocean of unbroken rain forest, that one would not be surprised to see a dinosaur suddenly appear on top of one of the sandstone mountains. Although the Roraima sandstone *tepuis* also exist in Colombia, northern Brazil, Guyana, and even Suriname, they are largely a feature of southern Venezuela, where they reach their greatest variety, and surely must rate among the top five most spectacular natural features on the entire planet. The flagship *tepui* is the famous Auyán-Tepuí, from which springs the world's tallest waterfall, Angel Falls. As spectacular as Angel Falls is, it is just one of hundreds of similar waterfalls cascading from the many *tepuis* in this region and, though it is the highest, it is surpassed in beauty by many others. A number of *tepuis* also have deep, mysterious sinkholes in them, most of which have never been explored. Immediately adjacent to the Tepuis region is the Gran Sabana, an extensive, little-modified savanna ecosystem that adds even more to the beauty of this region which, with care, could become one of the world's premier ecotourism destinations. Fortunately, a large portion of the Gran Sabana and the tepuis are conserved in the 3 million-ha Canaima National Park, one of the largest protected areas in the tropical world and the single largest in South America. All other *tepuis* not protected in Canaima or in other national parks are covered by the new Tepuyes Natural Monument created in 1991. Venezuela has taken a major step in helping to ensure the survival of this unique region, and the international community should help to implement the country's vision.

Human cultural diversity in Venezuela, though less than in most other megadiversity countries, is still impressive. Twenty-eight different tribal groups are known, and include some of the least disturbed indigenous cultures left on Earth.

The capybara (Hydrochoerus hydrochaeris) *is the world's largest rodent. It is widespread in South America, but is particularly abundant in the Llanos of Venezuela.*

Natural habitats

Venezuela has great ecosystem diversity, with virtually all Neotropical ecosystem types present and sometimes packed into narrow, steep gradients. For instance, only a few linear kilometers separate the mangroves of the hot seashore along Lake Maracaibo from the high *páramos* and permanently snow-covered Andean summits at 4 700 m. Likewise, it takes barely an hour's drive from arid cactus scrub along the Caribbean coast to magnificent, epiphyte-covered cloud forests at 1 000 m elevation, probably the most species-rich ecosystems in the southern Caribbean region.

Much of this is due to Venezuela's geological history, which is one of the oldest in the western hemisphere. Almost half of its area straddles the ancient Guayana Shield, one of the earliest landmasses of the ancestral Pangaea continent which later gave rise to Laurasia and Gondwana. It can be assumed that many of the varied ecosystems of southern Venezuela originated in Precambrian times in the western section of the early Gondwana supercontinent, before its breakup into a western and eastern portion in the early Cretaceous, approximately 120 000 000 years ago (Schubert and Huber, 1990). Since then, the continuously increasing separation from eastern Gondwana (mainly the African continent) and Laurasia has produced a pronounced isolation of the western landmasses consolidated around the nuclei of the Guayana and Brazilian Shields; this isolation, of great biogeographic significance (Simpson's "splendid isolation"), lasted for more than

100 million years. It was finally broken at the end of the Tertiary (3-5 million years ago) when the Central American land bridge consolidated and a narrow, terrestrial north-south connection between South America and the rest of the continent was established again.

In the meantime, other biogeographically important events modified the physical landscape of northern Venezuela and its associated biota. During the Eocene (some 60 million years ago), a mountain chain was uplifted along the northern and northwestern borders, giving rise to the northern Cordillera de la Costa and to the Proto-Andes in the southwest. But while the Cordillera de la Costa continued consolidating its land-forming process, the Proto-Andes underwent a complicated process marked by several heavy tectonic and volcanic events. Finally, 30 million years later (around the Miocene), the modern Andean mountains of the Eastern Cordillera began to reach their present configuration in Colombia and Venezuela, with peaks rising up to 5 000 m or more. The continuously changing montane environments, as well as the large interior lowlands, provided many new and favorable sites for biogeographic differentiation, adaptation, and speciation processes.

The great biomes —tropical forest, shrubland, grassland, and pioneer vegetation— are all very well-represented in the fascinating Venezuelan mosaic of tropical landscapes. Many of them are also found in the neighboring countries with varying species assemblages, but some are largely exclusive to Venezuela, and especially to the Guayana region.

Forest diversity in Venezuela is very pronounced at all altitudinal levels, and still covers more than 50% of the country's territory. In southern Venezuela, a still undetermined number of distinct evergreen lowland forest types is found on a wide spectrum of soils ranging from dense, multistoried, and species-rich *tierra firme* forests in the Caura plains to low, open, relatively species-poor Caatinga forests on extremely oligotrophic, often flooded, white sands in the upper Río Negro drainage. In the Orinoco delta, tall and dense mangrove forests alternate with flooded swamp forests, palm stands and semideciduous forests, creating an intricate mosaic of forest types crossed by a labyrinth of interconnected channels and temporary creeks. In northern Venezuela, magnificent evergreen lowland forests once covered the extensive alluvial plains of the southern Maracaibo Basin, but most of them have been converted into pasture lands during the last decades. The remaining lowland forests of northern Venezuela are mostly deciduous and semideciduous, covering the hills of northwestern Venezuela, the eastern piedmont region along the Andes, and the southern piedmont of the Cordillera de la Costa.

The montane forests of Venezuela harbor what are probably the most species-rich plant communities in the country, especially the many different cloud and elfin forests growing at altitudes between 1 000 and 2 500 m. The slopes of the Andes, the Cordillera de la Costa and the Guayana Highlands seem to contain an overwhelmingly large number of endemic species compared with other evergreen forest types. Unfortunately, many of these forests in the Andes and the Cordillera de la Costa have already been destroyed, but there are still some good examples surviving in national parks and natural monuments. Other peculiar forest types are found on some of the *tepui* summits in the Guayana Highlands, along creeks and in larger depressions of the sandstone plateau.

Shrublands are not as widely distributed in Venezuela as are the forests, but in some regions they harbor an incredibly high number of endemics due to their ecological adaptations to peculiar substrates. In northern Venezuela, most shrublands grow under a harsh, arid climate and are therefore made up of mainly xerophytic species of wide Caribbean distribution, whereas in the Guayana region shrublands are found on different substrates in humid and perhumid climates. Their patchy distribution in islands usually surrounded by forests has resulted in more or less isolated plant communities, which may account for their high number of endemics. The Guayana shrublands are found at all elevations, from the hot lowlands of the upper Orinoco to the rocky outcrops in the Gran Sabana uplands at 1 000 m, and even more frequently on the summits of almost all *tepuis* of the Guayana Highlands, at 1 800-2 500 m (Huber, 1989). A vivid example of the diversity of shrublands and their interesting species composition is shown on the summits of the Chimantá Massif, one of the most important *tepui* plateaus of the Guayana Highlands. There, not less than a dozen quite distinct, highly-specialized shrub communities are found growing on peat, sand, and rock outcrops, each with its own dominant species assemblage and peculiar growth forms (Huber, 1992). Most interesting is the presence of a paramoid scrub type, with a dominant genus, *Chimantaea*, that shows striking features of convergent evolution with other tropical high mountain ecosystems (*e.g.*, the high Andean *páramos* or the rosette shrub communities of the East African mountains).

The *páramos* of the Andes in Venezuela (and Colombia) are the highest-elevation shrublands growing in the Neotropics, being found between 3 000 and 4 800 m, and sometimes even higher. These impressive high mountain ecosystems are well distributed in all the Venezuelan Andes and harbor a number of endemic species of the dominant genus *Espeletia*. Its unusual growth form and beautiful yellow inflorescences characterize the numerous plant communities found within the *páramo* life zone. Small islands of *subpáramo*, a more bushy community, are also found on several peaks of the Cordillera de la Costa above the tree line at 2 200 m.

Herbaceous ecosystems cover approximately 40% of the surface of Venezuela. The most widespread ones are the savannas, which are the main vegetation type of the wide, flat interior lowlands known as the Llanos. This region is an extensive alluvial plain covering the

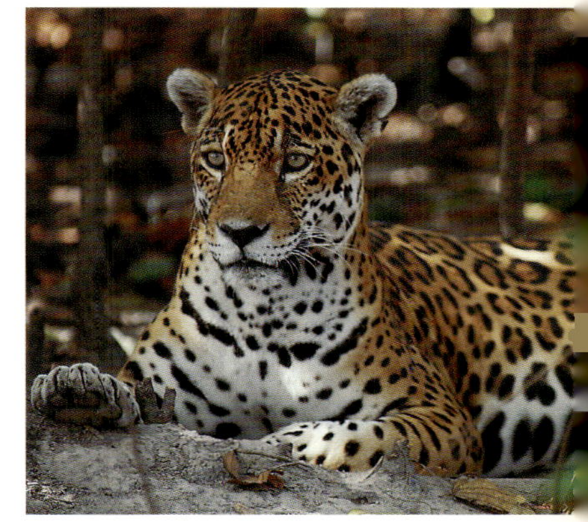

The jaguar (Panthera onca), like the harpy, is another predator at the very top of its food chain. It too prefers forested habitats, and is becoming rare over most of its range. Once again, however, the wilderness areas of southern Venezuela provide it with an important refuge.

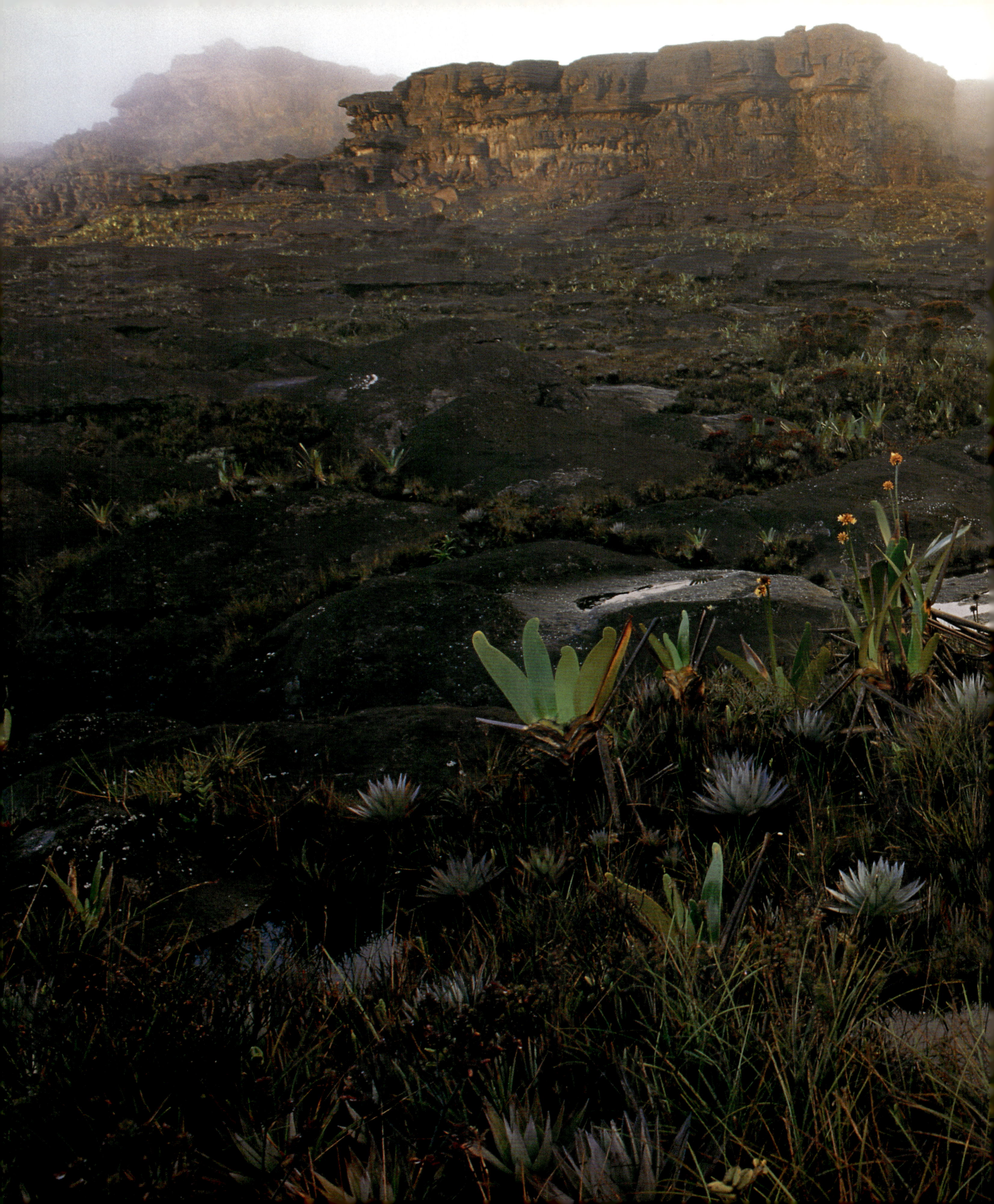

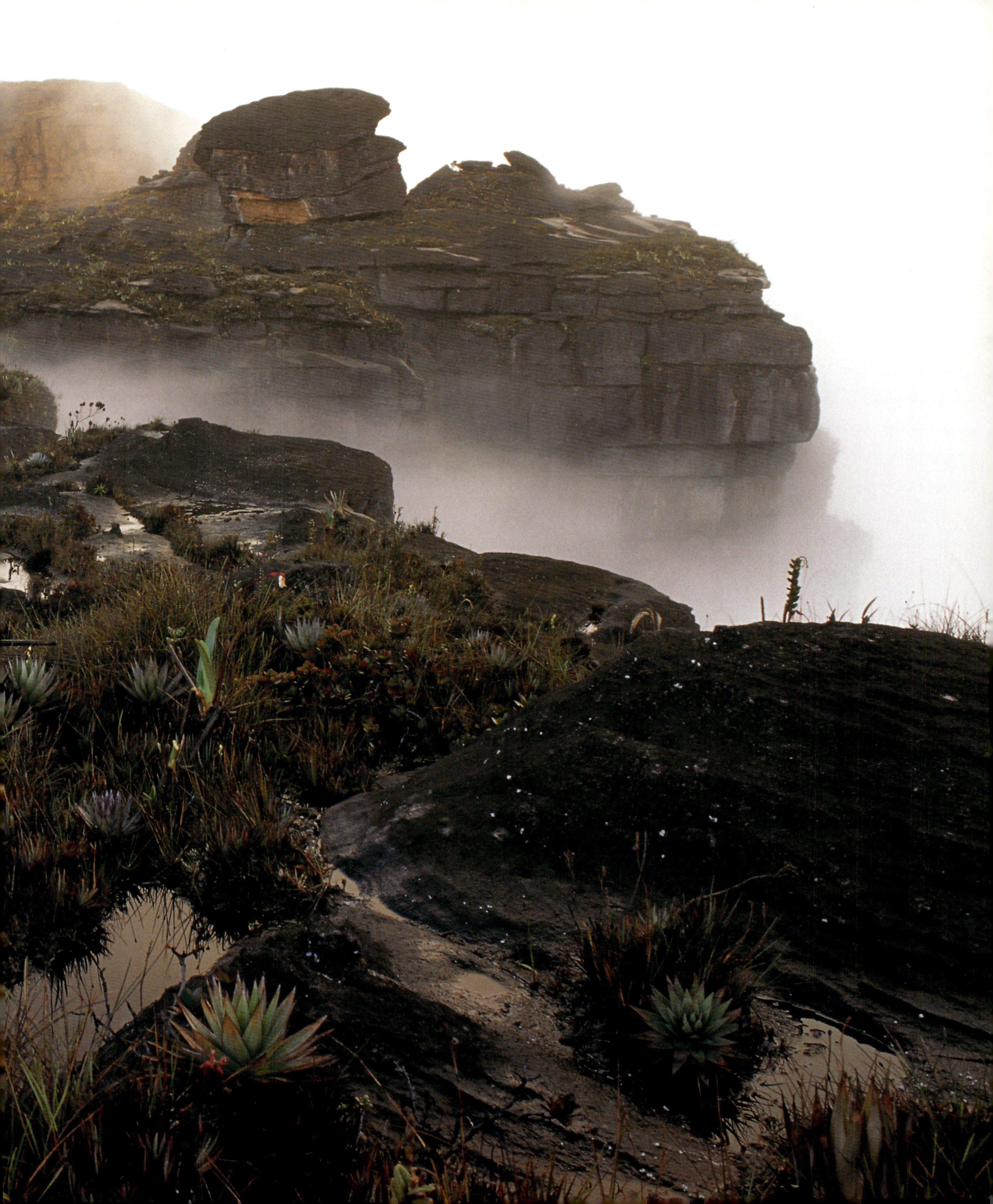

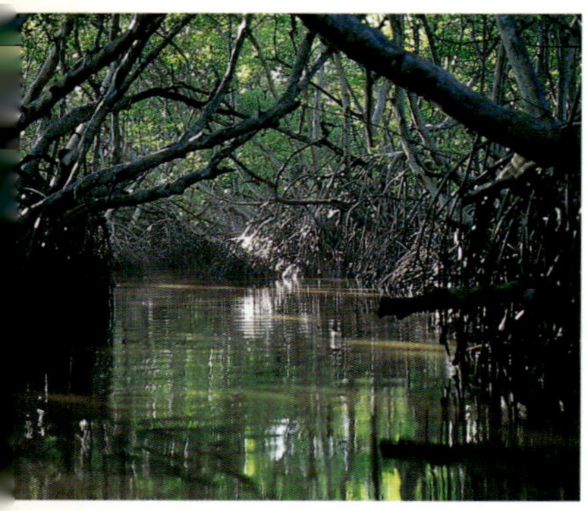

On pp. 452-453, a view of the top of Mount Roraima, at 2 772 m the highest of the tepuis, located on the border between Brazil, Guyana, and Venezuela. The Roraima Formation that includes the unique sandstone table mountains of southern Venezuela and Guyana, with outliers in Suriname, northern Brazil and Colombia, is named after this mountain.

Above, mangroves in the Golfete Cuare, Chichiviriche. Mangroves cover some 673 000 ha in Venezuela.

entire central region of Venezuela and limited to the south by the Orinoco, to the west by the Andes, and to the north by the Cordillera de la Costa. In the east, the Llanos merge with the swampy lowlands of the Orinoco delta. Although these savannas superficially appear to be homogeneous and even monotonous (as described by Alexander von Humboldt), there are a great variety of plant communities along several gradients (*e.g.*, rainfall, soil texture, soil moisture, nutrient stress, etc.). The mosaic of savanna types found in the immense Llanos region may well be comparable in its complexity and diversity to that of the forest mosaic of the southern lowlands (but at a totally different level of species complexity). Natural savannas harbor a large number of grass and sedge species, which are usually mixed with some kind of highly-specialized woody elements, such as shrubs or low trees, either standing alone or in small groups. The Venezuelan Llanos probably rank as the second richest and most extensive savannas in South America, after the Brazilian Cerrado, but above the Colombian and Bolivian savannas. Curiously enough, there are almost no endemic species in the Venezuelan Llanos, a fact that may be explained by the very recent origin of its landscapes and their associated plant communities.

In addition to the widespread savannas, which are tropical grasslands, many other herbaceous ecosystems are found in Venezuela. These are mainly in the south, where the dominant plants are not grasses, but other herbs belonging to entirely different families. This is the case of certain types of meadows growing in both the lowlands and the highlands of the Guayana region, in which members of the Rapateaceae, Xyridaceae or Eriocaulaceae families form the dominant herbaceous stratum. These meadows are usually associated with special habitats on white sand, often intermittently flooded, or on deep organic soils such as montane peats or bogs. Perhaps no other region in the Neotropics harbors more different herbaceous ecosystems than the Guayana region. It is particularly noteworthy that almost every *tepui* summit has its own complex of such meadows, with an incredibly high number of endemic species. Furthermore, in contrast to the savannas, which are either green in the rainy season or yellowish in the dry season, these Guayanan meadows are very colorful, since many of the dominant herbs and subshrubs have very showy yellow, white or red flowers. Another interesting aspect of these meadows are the curious growth forms that often predominate in the various communities. They range from dense, globose cushions to leathery, broad-leaved laminas or even peculiar tubular growth forms, such as in the famous pitcher plants (*Heliamphora*, Sarraceniaceae) or in certain endemic terrestrial bromeliads (*Brocchinia*).

The last biome with interesting ecosystems is represented by the wide range of pioneer communities which colonize rock surfaces or shallow, sandy depressions. Again, these are found mostly in the Guayana region, where they stand out both in the lowlands and the uplands of the sandstone *tepuis*. Along the middle Orinoco River, there are also numerous huge, steep-sided granite rocks (inselbergs) that range in height from 100-400 m and emerge from the surrounding forests or savannas; they are colonized by a highly specialized flora with many endemics, and were first described by Alexander von Humboldt in 1800. Likewise, the extensive, open rock surfaces of the vertical walls and the summits of the sandstone *tepuis* harbor a specialized flora which is still very poorly known. The most important plants there are many endemic species of bromeliads, orchids, and low subshrubs growing in the rock crevices and fissures.

Flagship species

Venezuela has a number of important flagship species, among them the jaguar, the giant otter, the spectacled bear, the capybara, the matamata turtle, the giant Amazon river turtle, the Orinoco crocodile, the harpy eagle and the oilbird, or *guácharo*. Although none of these are endemic to Venezuela, several of them have some of their most significant remaining populations within the borders of this country.

The spectacled bear (*Tremarctos ornatus*) is the only bear in South America, and occurs throughout the tropical Andes, from Peru, through Ecuador and Colombia to the Andean region of northwestern Venezuela. It can reach 1.75 m and 150 kg, making it one of the smallest living bears but, at the same time, one of the larger South American mammals. In Venezuela, it lives principally in the cloud forests of Mérida, Trujillo, Táchira, the Sierra de Perijá, and the high-altitude areas of the state of Lara. It is hunted throughout most of its range, in spite of being officially protected, and is considered an endangered species.

The jaguar (*Panthera onca*) is found in most South and Central American countries. Although rarely seen, it is able to persist in many areas, especially if reasonable tracts of habitat remain. In the wilderness areas of southern Venezuela, it has one of its major strongholds in northern South America, and also still occurs in reasonable numbers in the Llanos, even though it has traditionally been persecuted for its skin and because it is considered a threat to cattle.

The giant otter (*Pteronura brasiliensis*), largest of the otters and one of the most endangered large mammals in South America, once had a wide distribution in Venezuela. However, it is a diurnal, noisy, group-living, highly conspicuous animal that has a skin as valuable as that of the jaguar. Consequently, it has been much persecuted throughout its broad range, and is now restricted to certain regions of the Llanos, the Orinoco delta, and the wilderness areas of the far south in the states of Bolívar and Amazonas.

The capybara (*Hydrochoerus hydrochaeris*) is the world's largest rodent, and is common throughout much of northern South America. It is of particular interest in Venezuela because it reaches high densities in parts of the Llanos and along the Santo Domingo, Apure, and Arauca Rivers. It is eaten in large quantities during Lent and Holy Week. The church considers the capybara, the manatee and various turtles to be fish rather than meat, the turtles because they have a shell and the mammals because they are aquatic. Slaughter and processing of capybara meat during this period is a big industry for small towns in areas where the animals are especially abundant. This practice has been much studied, and appears to be sustainable.

The giant Amazon river turtle (*Podocnemis expansa*), known as *arrau* in Venezuela, is the largest freshwater turtle in South America, reaching 80 cm in length and a weight of 33 kg. It once occurred over much of the Amazon and Orinoco drainages below the large rapids, and reached incredible densities. A social nester, it hauls out on sand beaches in large numbers, making it easy prey for humans. Not surprisingly, given its size, visibility, and high productivity in terms of eggs, it has been exploited for centuries by Indians and European settlers. Alexander von Humboldt observed the Orinoco nesting beaches and the wholesale destruction of eggs during his historical voyage at the end of the eighteenth century and the first few years of the nineteenth century. He calculated that some 5 000 jars of oil were harvested annually at just three of the major Orinoco nesting beaches. Assuming that a jar contained 25 bottles of oil, each bottle being the production of 200 eggs, and allowing for wastage, he estimated that some 33 000 000 eggs had been harvested to fill those 5 000 jars —and that was just on three of the major beaches (Humboldt, 1812). The British naturalist Henry Walter Bates made similar observations between 1848 and 1859 in the Brazilian Amazon, in the vicinity of Tefé on the Solimões (=upper Amazon River). Almost 120 years later, Mittermeier was unable to find any *Podocnemis expansa* nesting in the vicinity of Tefé, although two of the smaller *Podocnemis* species, *P. unifilis* and *P. sextuberculata*, could still be found (and were also being overexploited). Not surprisingly, the *arrau* has declined throughout its range, and in fact has disappeared from many of its earlier nesting areas.

In Venezuela, the major nesting beaches for the *arrau* are seasonally inundated sandbars in the mainstream of the Orinoco River itself, downstream from Puerto Páez, Apure State. The most important of these beaches are Playa del Medio, where the species has been studied on and off since the 1960s, and Pararuma, with lesser numbers of turtles nesting on several smaller beaches. The species is now officially protected in Venezuela, but monitored populations had been dropping on both Playa del Medio and Pararuma from the 1960s to the early 1980s, so it is difficult to say how effective protection has been.

Several other relatives of the *arrau* occur in Venezuela as well, including the yellow-spotted Amazon sideneck or *taricaya* (*Podocnemis unifilis*), the red-headed sideneck or *chimpire* (*Podocnemis erythrocephala*), the Llanos sideneck or *galápago llanero* (*Podocnemis vogli*) and the big-headed sideneck or *cabezón* (*Peltocephalus dumerilianus*). This group of turtles is an ancient one, with its closest living relative in Madagascar and close fossil relatives on the African mainland.

The famous matamata (*Chelus fimbriatus*), or *caripatua* as it is called in Venezuela, is the most bizarre of all turtles. It has a broad, flat head, tiny anteriorly-situated eyes, a long snorkel-like snout unique among hard-shelled turtles, fleshy filaments on the ventral side of the head, and a laterally folding neck that is extremely thick and fleshy. The shell is flattened and very rough, with three knobby, tuberculate keels. This strange creature feeds on fish, and does so by suddenly opening its cavernous mouth and simultaneously creating an explosive expansion of the throat by means of powerful muscles and a highly developed hyoid apparatus. The expansion of the throat creates a rush of water that sucks the prey in with it. The mouth is then closed to a crack, the water is expelled and the fish swallowed. The matamata is widespread throughout the Orinoco River system, and seems to be more abundant there than anywhere else within its range, which also includes Brazil, Guyana, French Guiana, Colombia, Ecuador, Peru, Bolivia and the island of Trinidad.

The Orinoco crocodile (*Crocodylus intermedius*) is one of the most critically endangered New World crocodilians and is depleted throughout its range, which extends down the middle and lower reaches of the Orinoco River in Venezuela and Colombia. This freshwater species may be found in a wide variety of habitats, including rivers in tropical forest and piedmont streams in the foothills of the Andes, but its stronghold is definitely the Llanos region. One of the largest vertebrates in South America, it was commercially exploited from the 1930s to the 1950s, with thousands of animals being slaughtered every year for the international skin trade. Since then, there has been little recovery of wild populations, even though for years there have been efforts to captive breed it, mostly by private landowners like Tomás and Cecilia Blohm, who have kept and reproduced these animals mostly at their own expense for the last twenty years. The loss of habitat continues to be the biggest conservation challenge for this animal, which is considered dangerous to man and livestock. In Venezuela, crocodile habitat has been protected in the newly-declared Santos Luzardo National Park, along the Capanaparo and Cinaruco Rivers, but there is still no management plan for the species. A recently-declared wildlife refuge has been established along the Guaritico Channel, and this area is the site of the first releases of captive-reared young.

With 1 360 species and the sixth highest bird diversity on Earth, Venezuela has many interesting avian species. One of the most striking and interesting is the harpy eagle (*Harpia harpyja*), the heaviest raptor in the world and second in size only to the magnificent Philippine eagle (*Pithecophaga jefferyi*). The harpy is not common anywhere, since it requires a large home range to

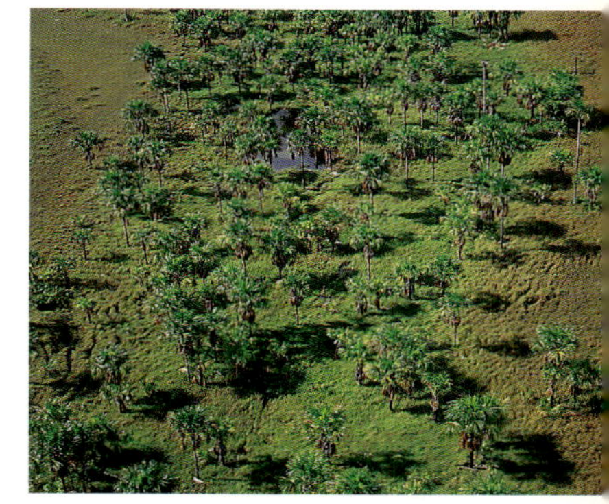

Mauritia palms (Mauritia flexuosa) *growing along rivers in the grasslands of the Gran Sabana in southern Venezuela.*

finds its prey species such as sloths, monkeys and other birds, and because it is often persecuted by hunters. However, like many other large predators, the harpy has a major stronghold in the vast wilderness areas of southern Venezuela, and stands as good a chance of surviving there as anywhere within its range.

The oilbird or *guácharo* (*Steatornis caripensis*) is a very unusual bird that is the only member of the family Steatornithidae within the Order of the nightjars and nighthawks, the Caprimulgiformes. It was first described almost 200 years ago by Alexander von Humboldt, based on specimens collected in the famous Cueva del Guácharo near Caripe, now a national park. This species lives in large colonies inside caves, nests in cavities high up in the caves, and flies out at night in search of its preferred foods, usually the fruits of certain palm species. The chicks accumulate large quantities of fat, and Humboldt reported that the local people would boil them to produce oil for cooking and lighting purposes. The oilbird has poor vision and, in an interesting parallel evolution to bats, uses echolocation to find its way in the dark.

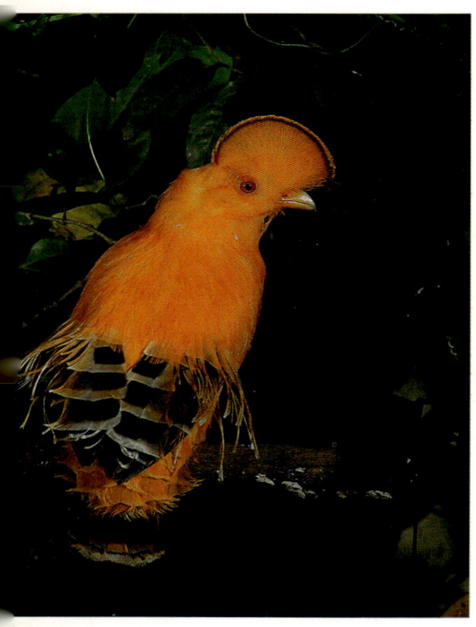

The Guianan cock of the rock (Rupicola rupicola) *is widespread in southern Venezuela and the southern Guianas and, as its name implies, is closely associated with rock formations where it builds its nest. Bright orange males display in leks to attract drab brown females.*

On the opposite page, the scarlet ibis (Eudocimus ruber) *is one of the most spectacularly-colored of all birds, and is found in selected areas along the Venezuelan coast.*

Among the plants, several families and genera could also be considered as flagships, especially some found in the Guayana region. One such group is the family Rapateaceae; it is a typically Guayanan group, and is very diverse in almost all ecosystems in this region, but especially in the high *tepui* meadows and in specialized white sand ecosystems in the lowlands. The genera *Kunhardtia*, *Marahuacaea*, and *Amphiphyllum* are found only in Venezuelan Guayana, and grow on the western and central *tepuis* of the state of Amazonas. The large genus *Stegolepis* also reaches its greatest diversity and endemism in Venezuelan Guayana, with many of its species being found as dominant herbs on almost all *tepui* summits.

Another important and highly characteristic plant group of the Guayana Highlands is the genus *Bonnetia* (Theaceae), with many, mostly endemic species in this region. Undoubtedly, this interesting shrub genus has its main center of speciation in Guayana, although a few isolated species are found as far away as Cuba and in the *restingas* of southern Brazil. Most *Bonnetias* are medium-sized shrubs with large, showy, often very beautiful flowers that vary from brilliant white to pink and dark yellow.

Mention should also be made of the Mutisieae, a tribe of the daisy family, Asteraceae. It has two main speciation centers in the Neotropics, a large one in the high central Andes and a smaller one in the Guayana mountains. The Guayana group consists of a number of characteristic, largely endemic genera found mainly in shrublands and shrubby meadows, mostly on rock or peat above 1 500 m. The most important genera are *Stenopadus*, *Gongylolepis*, *Stomatochaeta*, *Chimantaea*, *Duidaea*, and *Eurydochus*. Some of these exhibit showy capitula with beautiful red or white flowers, while others have only almost imperceptible greenish inflorescences.

Cultural diversity

It is not yet certain when the first humans arrived in Venezuela and whether they came from the south through the dense Amazonian forests and their rivers, or from the north along the arid Caribbean coast. In any case, human occupation of Venezuelan habitats seems to have occurred only during relatively recent times (approximately 15 000 years ago) (Cruxent and Rouse, 1982).

Today there are 28 ethnic groups in Venezuela with a population of approximately 316 000 people, about 1.4% of Venezuela's 22 million inhabitants (OCEI, 1993a). Twenty-four of these groups, which comprise one third of the indigenous population, live in the Venezuelan Guayana region and the Orinoco delta, whereas the remaining two thirds belong to the other four ethnic groups, and are confined to the northwestern portion of the country, *i.e.*, the Sierra de Perijá and adjacent piedmont areas in the state of Zulia.

The major indigenous groups of Venezuela are as follows: the Wayuu (53.4%), Warao (7.6%), Pemón (6%), Añú (5.5%), Yanomami (4.7 %), Guahibo (3.6%), and the Piaroa (3.0%). These 7 groups alone account for more than 84% of Venezuela's indigenous population (OCEI, 1993a). Their degree of acculturation ranges from almost completely integrated to Western society, as in the case of the Wayuu, to barely untouched groups like some Yanomami subgroups.

Some of the most important ethnic groups live in the Venezuelan Guayana region, where they are still living in close contact with their natural environment (Huber, 1995a).

The Warao Indians number about 24 000, and inhabit the intricate maze of river channels and creeks in the Orinoco delta, where they have developed a specialized lifestyle adapted to the frequently flooded environment. This important group is highly-skilled in fishing, hunting, and gathering, and belongs to an independent linguistic family. They make extensive use of the abundant forest resources, particularly palms like *moriche* (*Mauritia flexuosa*) and *manaca* (*Euterpe* spp.). Today the Warao Indians are rather acculturated to Venezuelan society, and their lifestyle has been affected strongly by outside settlers.

The Pemón Indians (approximately 19 000 people) belong to the Cariban linguistic family and occupy the southeastern region of the state of Bolívar, particularly the Gran Sabana region, the Caroní Basin, and the lower Paragua Basin. They survive by shifting agriculture, with a staple of bitter manioc, and hunting. Their villages are small, usually in open savannas, and they have an interesting, nonauthoritative, egalitarian community structure. The Pemón Indians probably arrived in the Gran Sabana uplands 500 or 600 years ago, immigrating from the adjacent savannas of the Brazilian Rio Branco to the south. During their extensive hunting trips and migrations from one site to another, they often lit great fires in the savannas, thus altering the

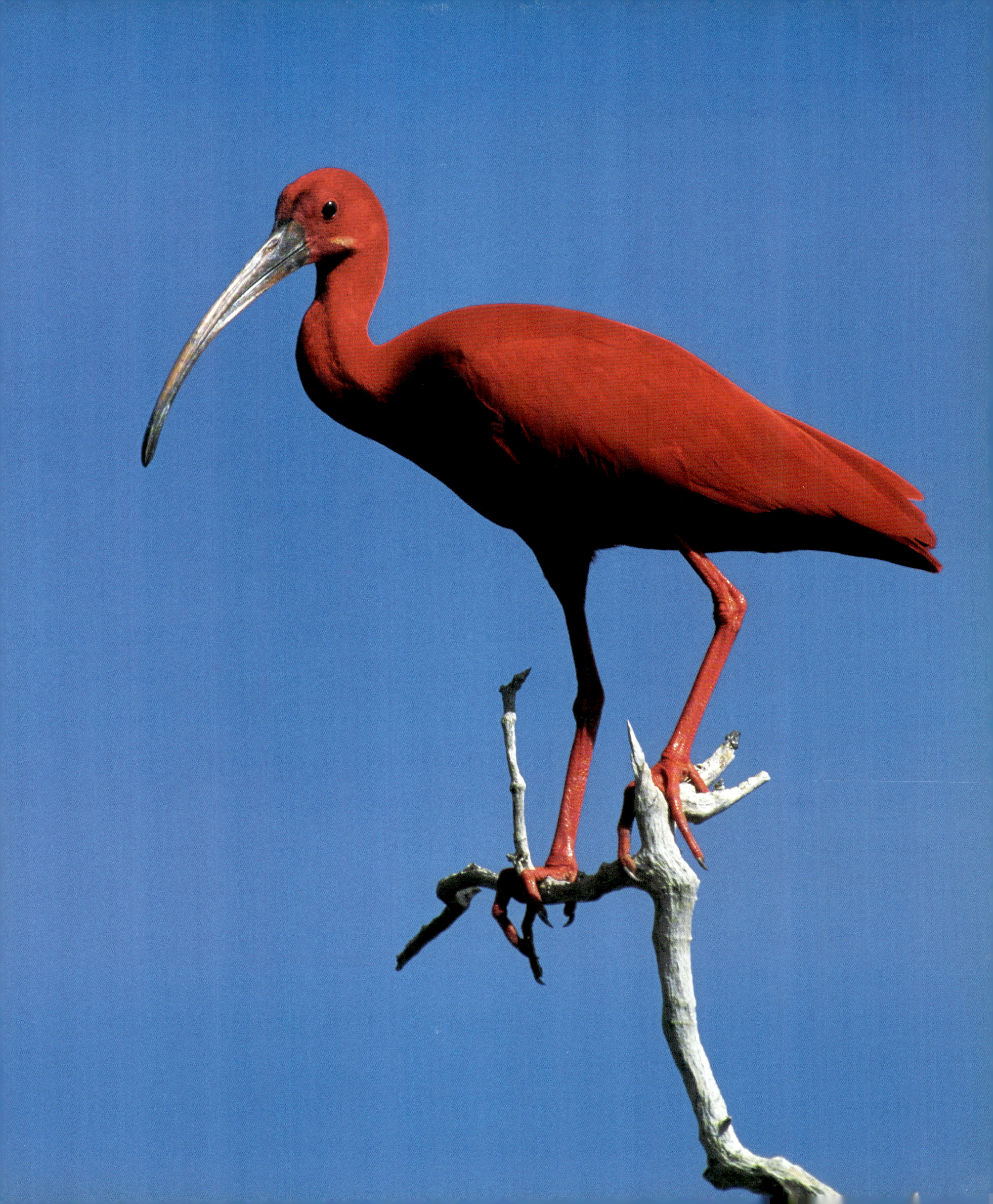

landscape. The Pemón Indians consider the summits of the large table mountains (*tepuis*) sacred sites and, therefore, avoid going to the upper montane regions.

Another important indigenous group, the Yekwana or Yekuana (also known as Makiritare), number approximately 4 400. They live in southwestern Bolívar (about 1 800), and northeastern Amazonas states (about 2 600), mainly in the headwaters of the Caura and Ventuari rivers. They belong to the Cariban linguistic family, and their traditional form of subsistence consists of slash-and-burn agriculture (with bitter manioc as the staple), hunting, and fishing. Formerly blowgun hunters, they are also skilled boat makers and navigators, with an intimate knowledge of the water courses in their territory. The Yekwana live in small settlements of 20–30 houses, loosely arranged around a larger, circular communal meeting house. In the Yekwana language, the larger mountains are called *jidi* (or *hidi*), which roughly corresponds to the Pemón term *tepui*.

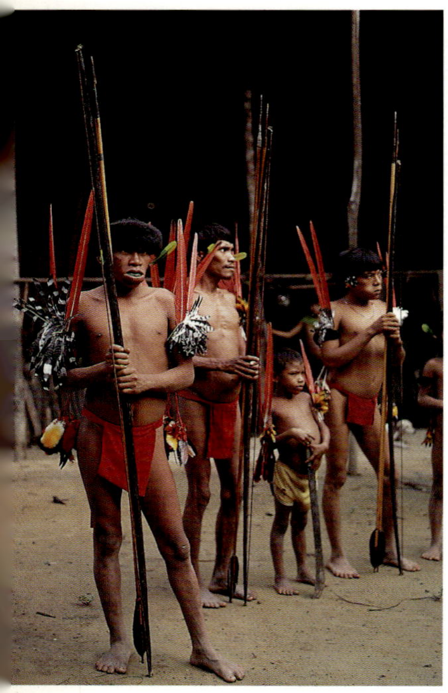

Yanomami Indian warriors in upper Orinoco, 1988. Note the arm ornaments made of macaw feathers. The Yanomami occupy an area that straddles the border between Venezuela and Brazil, and are renowned as fierce warriors. Unfortunately, after centuries of isolation, they are now being heavily impacted by Brazilian prospectors or garimpeiros *who are moving into their land to search for gold.*

On the opposite page, a Hoti Indian fishing with a spear in a rain-forest stream. The Hoti, or Ho, are a small tribe that occupies a huge domain of some 5 million hectares in the state of Bolivar in southern Venezuela. They are blowgun hunters who live in a transitional area between rain forest and savanna, and appear to still be quite traditional. However, they remain very poorly known, and their relationships to other tribes are still unclear.

Perhaps the most famous indigenous group of southern Venezuela is the Yanomami (approximately 15 000). This tribe includes four linguistic subgroups that live in the Venezuelan and Brazilian Amazon, with a small representation in the southern portion of the state of Bolívar in Venezuela. The Yanomami usually live in forest environments, where they practice shifting cultivation, hunting, gathering of forest products, and some fishing. Interestingly, one of their main staples is the banana or plantain (*Musa* spp.), which was introduced centuries ago from the Old World tropics. They build large communal houses called *shapono* or *shabono,* which are able to accommodate between 40 and 250 people. These consist of a circular roof construction surrounding a wider open plaza.

During the 1980s and 1990s, the Yanomami suffered serious cultural and health impacts from a rapid increase of immigrants into their homelands, especially miners, missionaries, and tourists. In the southern part of the Sierra Parima, numerous Yanomami families have already begun to abandon their traditional forest habitat and settle in open, artificially-created savannas.

The Piaroa, numbering some 9 370 people, are a well-organized nation of forest dwellers of the small (sometimes considered independent) Sálivan language family. Their homeland extends over the low hills and mountains of the Cuao-Sipapo Massif in the northwestern corner of Amazonas State. They live by slash-and-burn agriculture, with bitter manioc as their staple, as well as hunting, fishing, and gathering of numerous plant and animal forest products. Traditionally, their villages are small, and consist of a large, conical and circular communal house where all 15–50 village members live. The Piaroa have excellent geographic knowledge of their homeland. In their mythology, the summits of various mountains of the Cuao-Sipapo Massif, especially the spectacular, tower-like Cerro Autana, play an important role as sacred sites.

The Panare Indians, or E'ñepá as they call themselves, belong to the Cariban linguistic group. They live in the northwestern corner of the Guayana Shield, where they number about 2 700 individuals, and occupy an area of about 2 000 000 ha. Traditionally blowgun hunters, they are now in a process of acculturation.

Finally we have the Hoti or Ho, a small, enigmatic group that is one of the most traditional cultures still left in northern South America. They too live in an area to the south of Caicara on the Orinoco, and in a few places overlap with the Panare. However, they are totally unrelated, speaking a very distinctive language that may be in its own isolated family or distantly related to the Maku of the Rio Negro region far to the south in Brazil. They have a remarkable ability to maintain their culture in the face of advancing civilization. In the village of San José de Kaiamá, for example, they all still wear traditional dress, regularly hunt with their blowguns, and make dart poison, an art largely lost among other blowgun hunting groups in South America —this in spite of the fact that a Catholic mission run by nuns has been there for over a decade. Approximately 220 Hoti live in the immediate vicinity of Kaiamá, 500 live elsewhere in the region, and another 300 or so in Caño Iguana about 110-115 km away. Estimates of total population for this very poorly-known group range as high as 2 000 and they occupy a vast, otherwise uninhabited region covering some 5 000 000 km^2. Their lands have not been demarcated, but should be as soon as possible. They may be one of the few tribes that will be able to maintain an almost full assemblage of traditional culture.

OTTO HUBER
RUSSELL A. MITTERMEIER
CRISTINA GOETTSCH MITTERMEIER

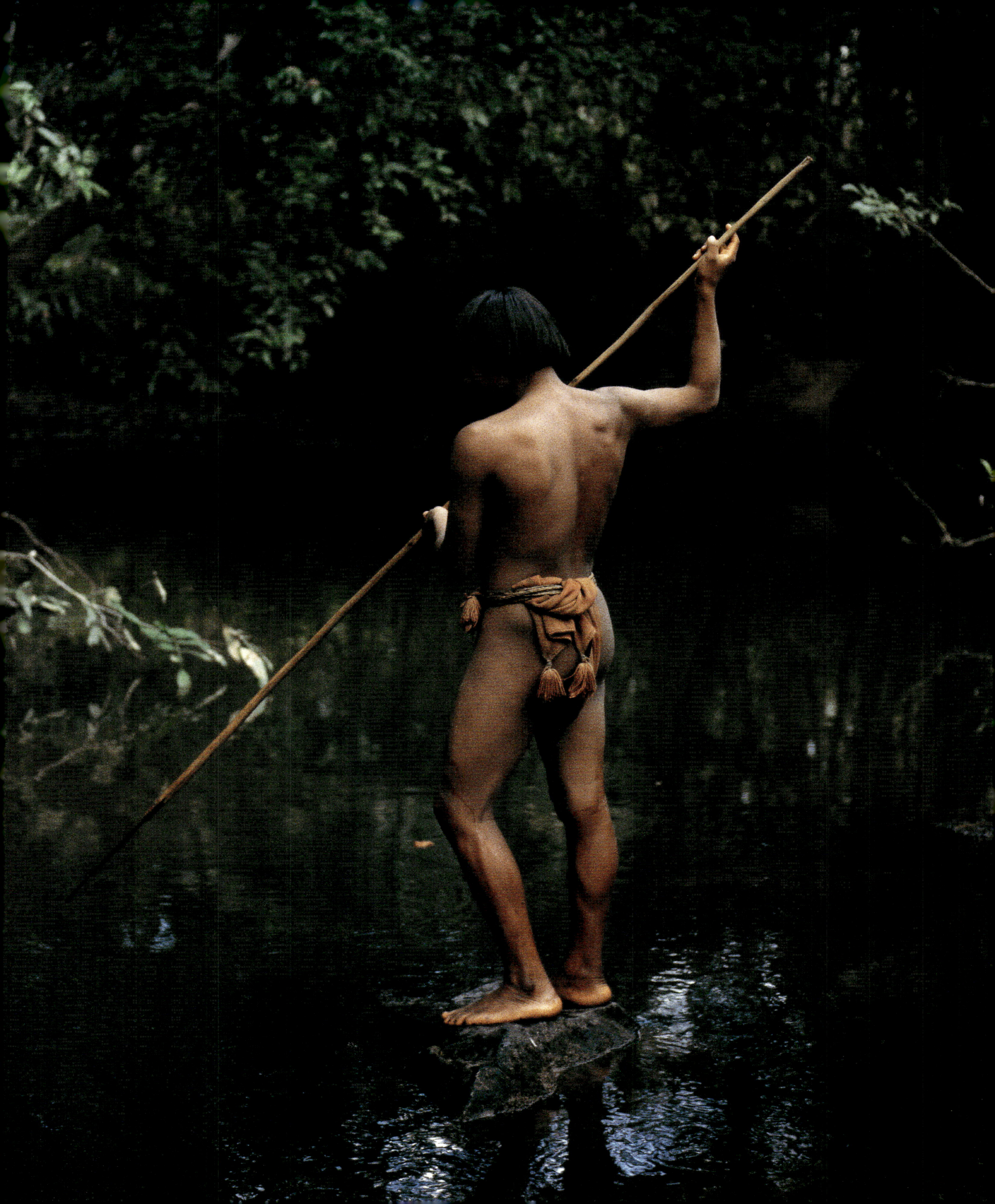

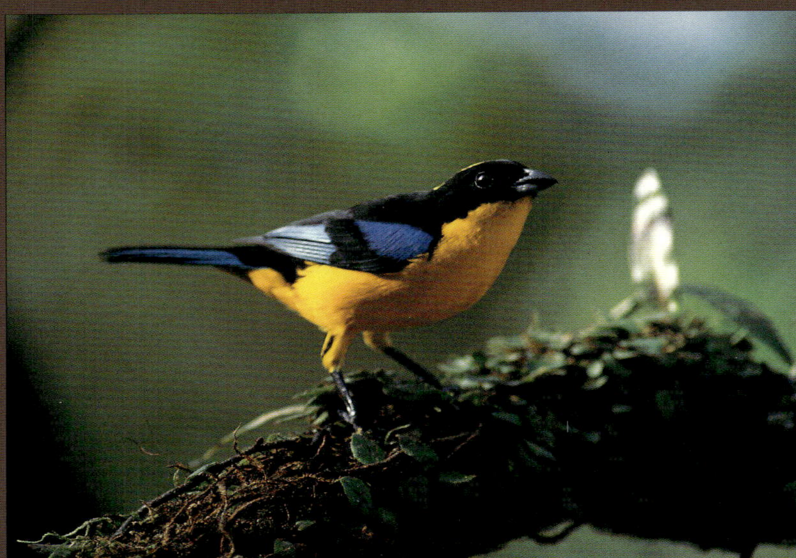

Above, a blue-winged mountain tanager (Anisognathus flavinucha) *from cloud forests in Venezuela.*
Right, cloud forest at Rancho Grande in the Henri Pittier National Park in northern Venezuela. This is the best known and oldest of Venezuela's protected areas.

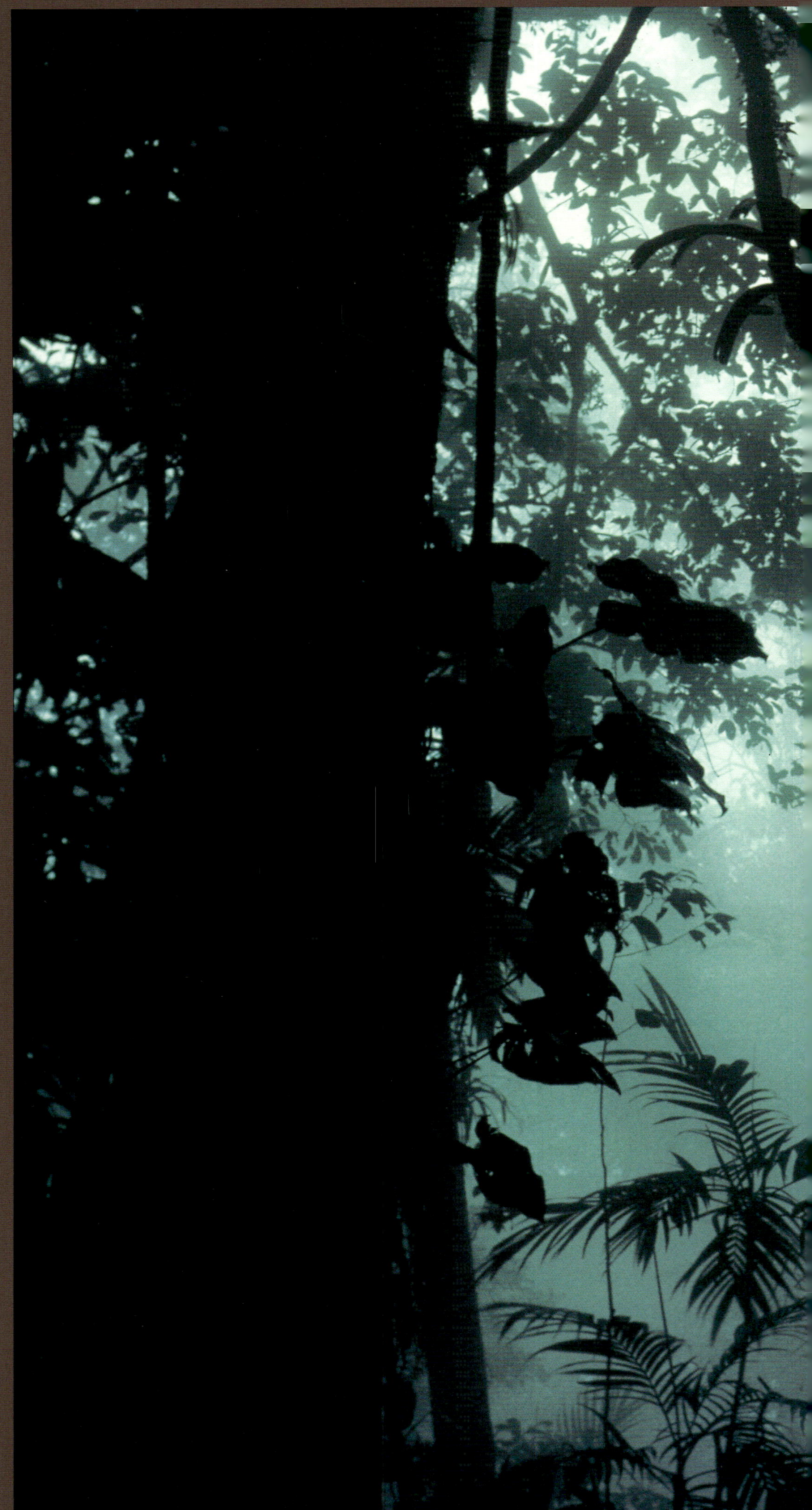

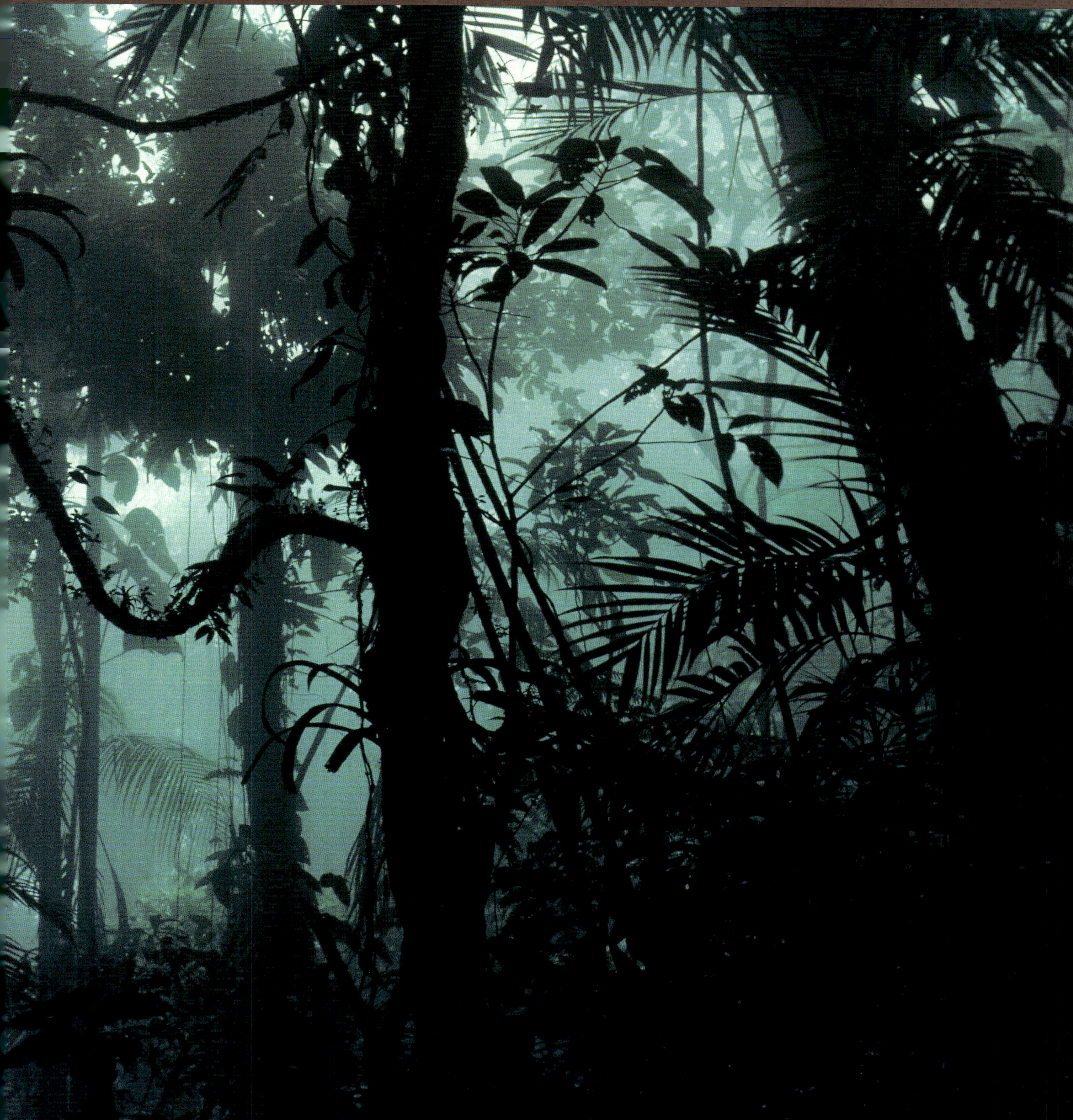

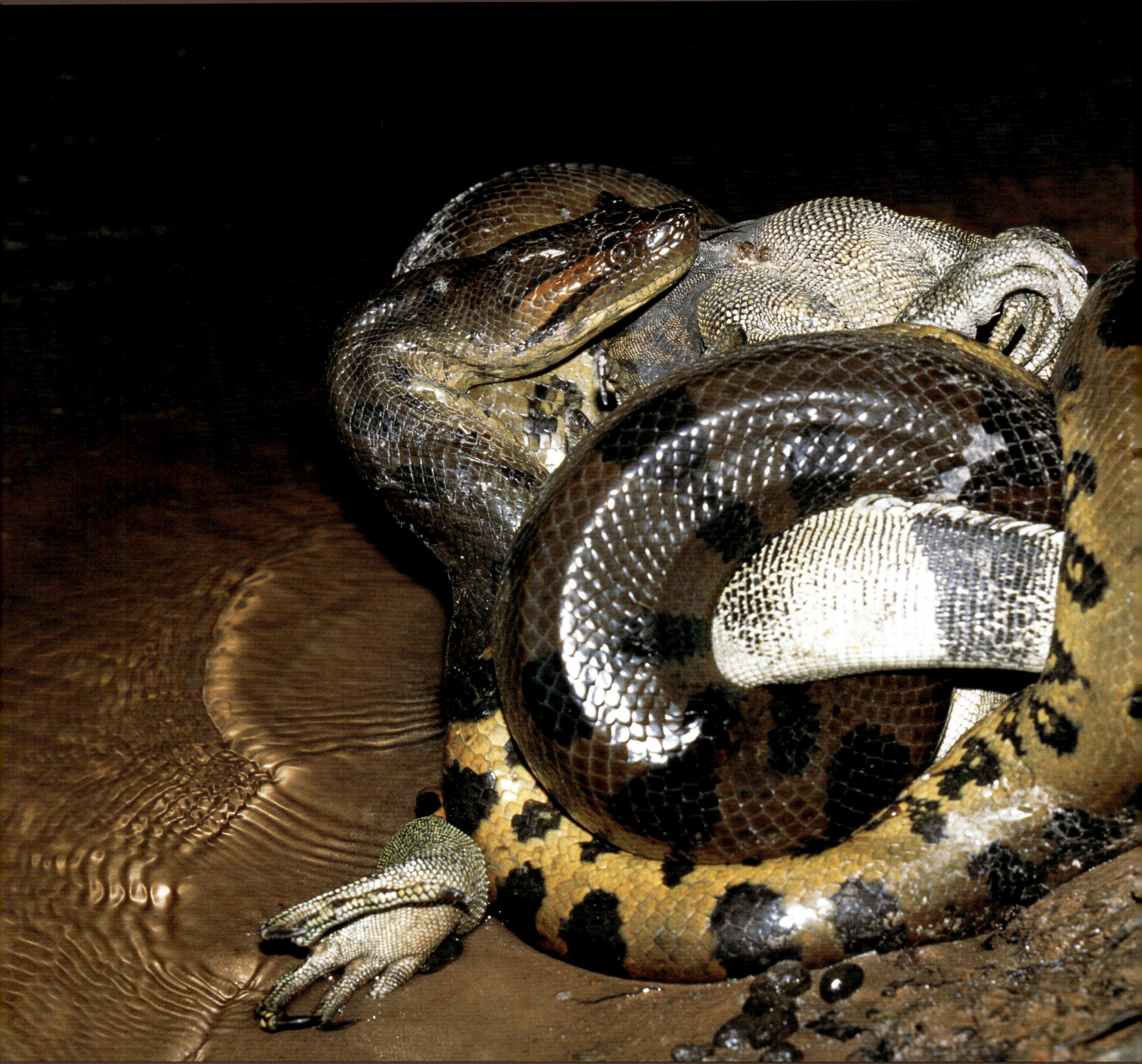

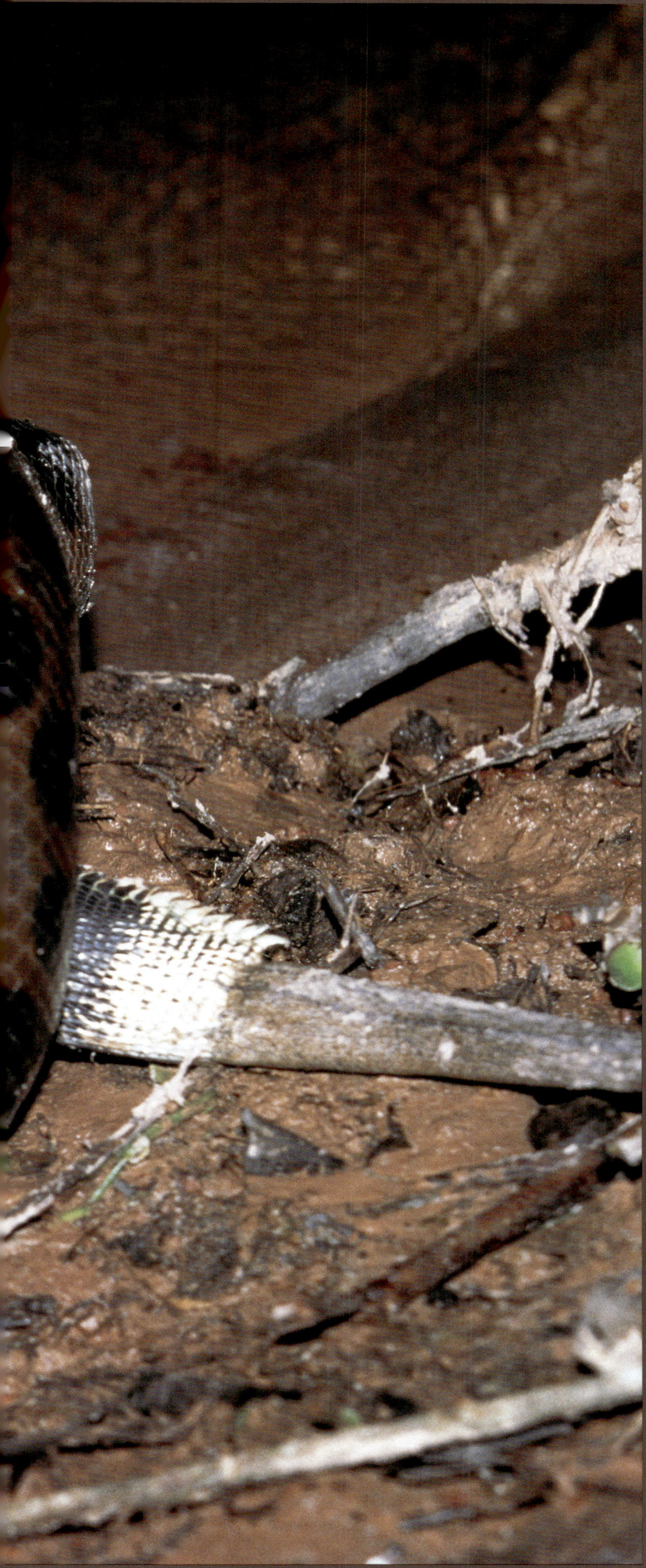

Left, an anaconda (Eunectes murinus), *the largest snake in South America and one of the two largest in the world, constricting a large male iguana* (Iguana iguana).
Below, the marsupial frog (Gastrotheca ovifera) *releasing froglets from its pouch, cloud forest habitat in Venezuela.*

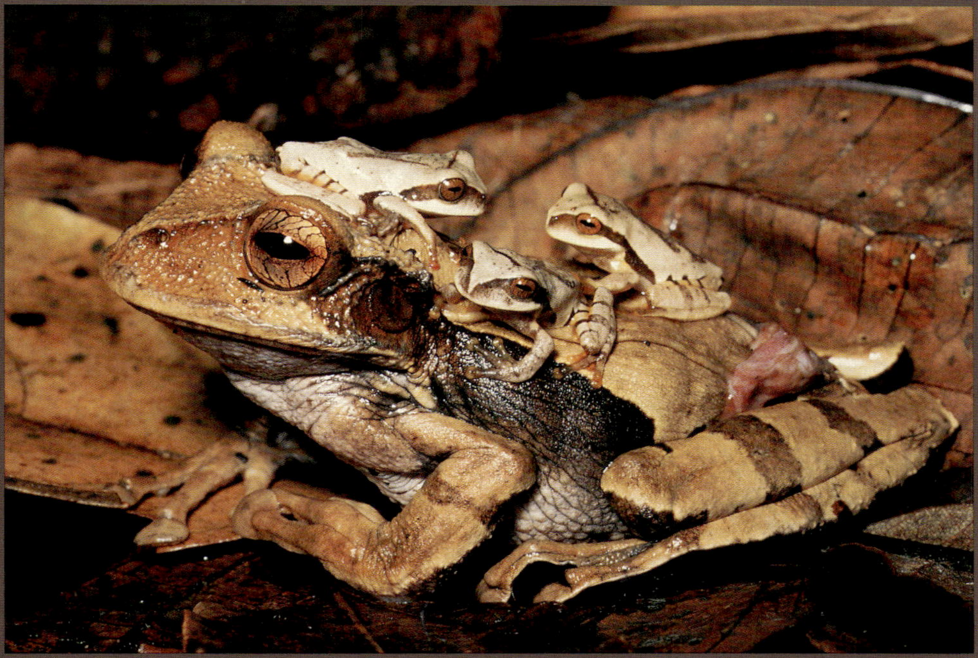

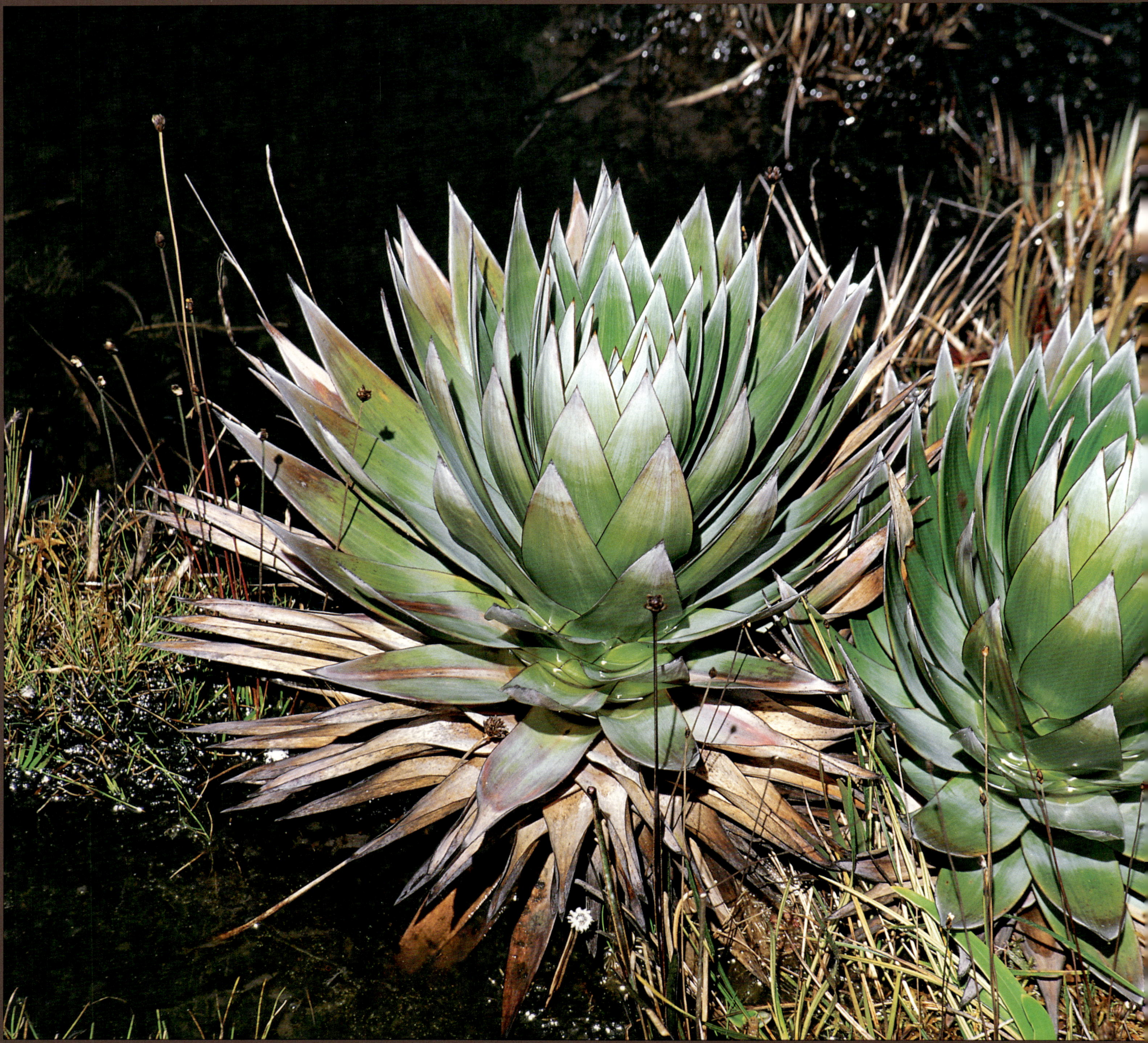

Above, a sisal, a very distinctive plant (Connellia augustae) *on top of Mount Roraima. The summits of these table mountains are sometimes quite extensive, and harbor many endemic plant species. Upper right, a small frog* (Oreophrynella quelchii) *from the top of Mount Roraima. Lower right, the imposing Auyan Tepui reflected in the quiet waters of the Carrao River. Angel Falls, the world's highest waterfall, drops from the top of this* tepui.

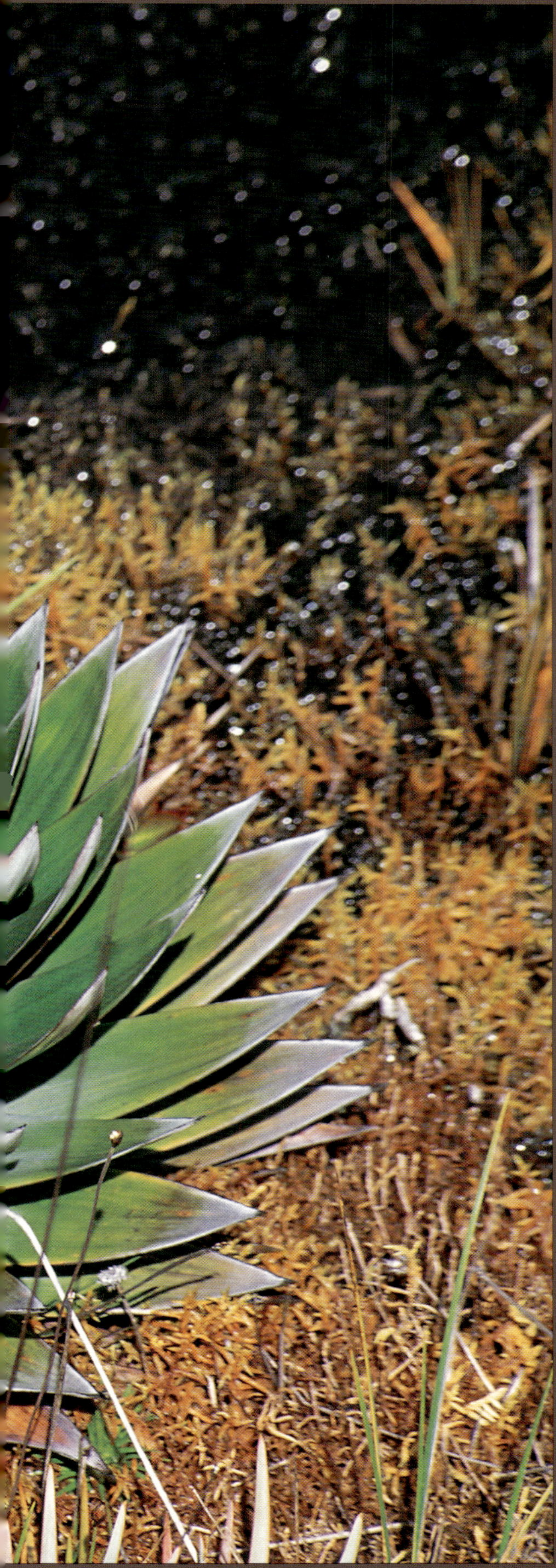
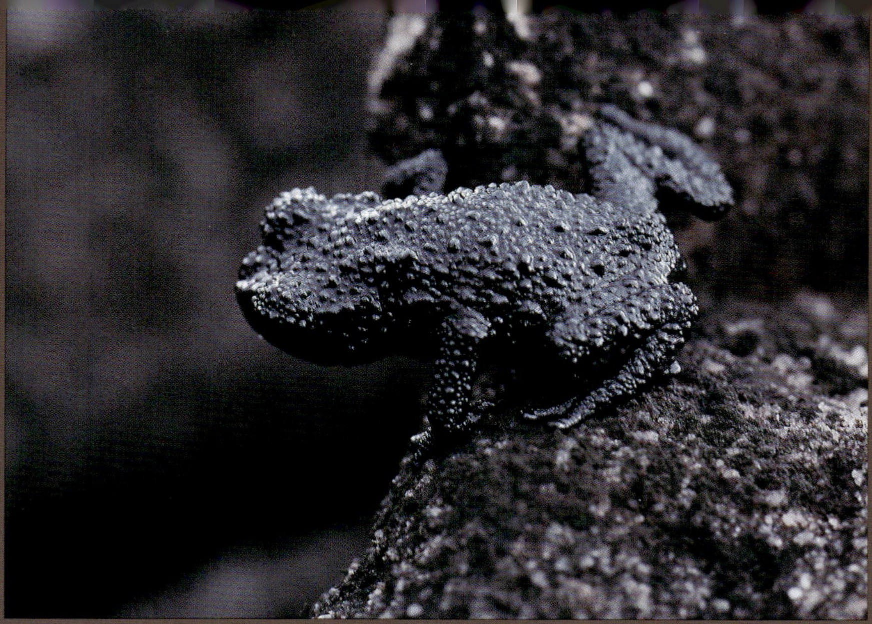
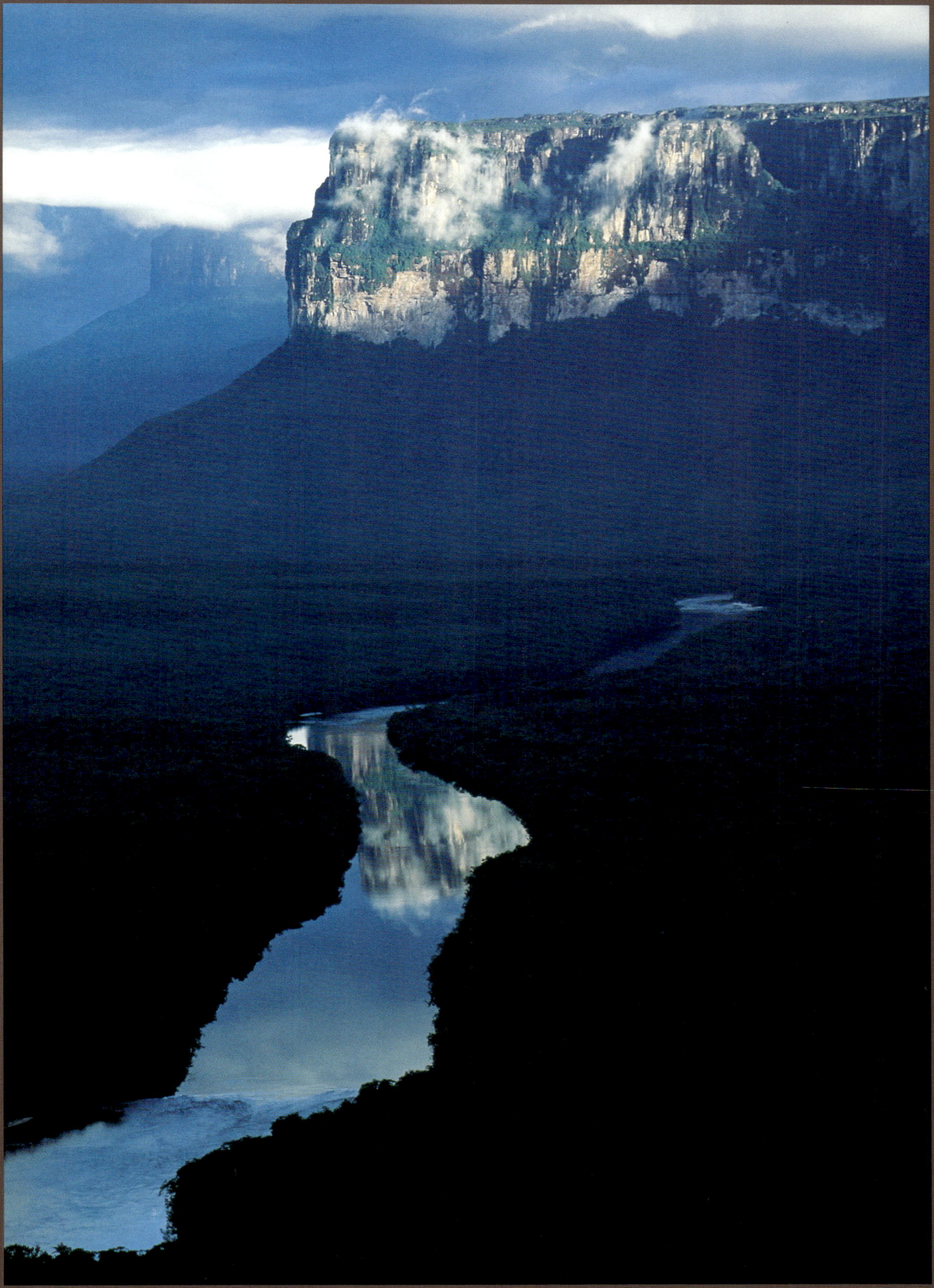

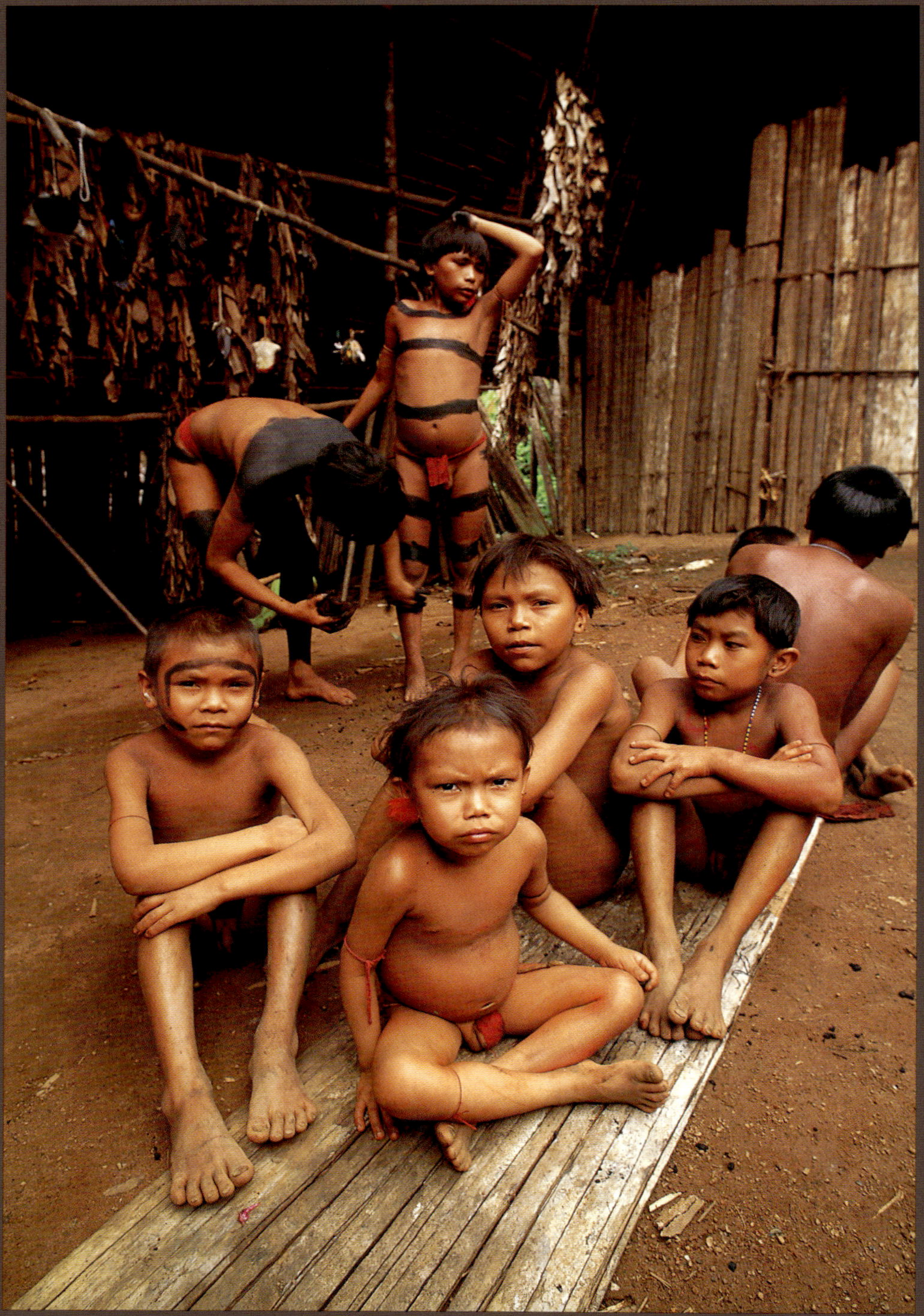
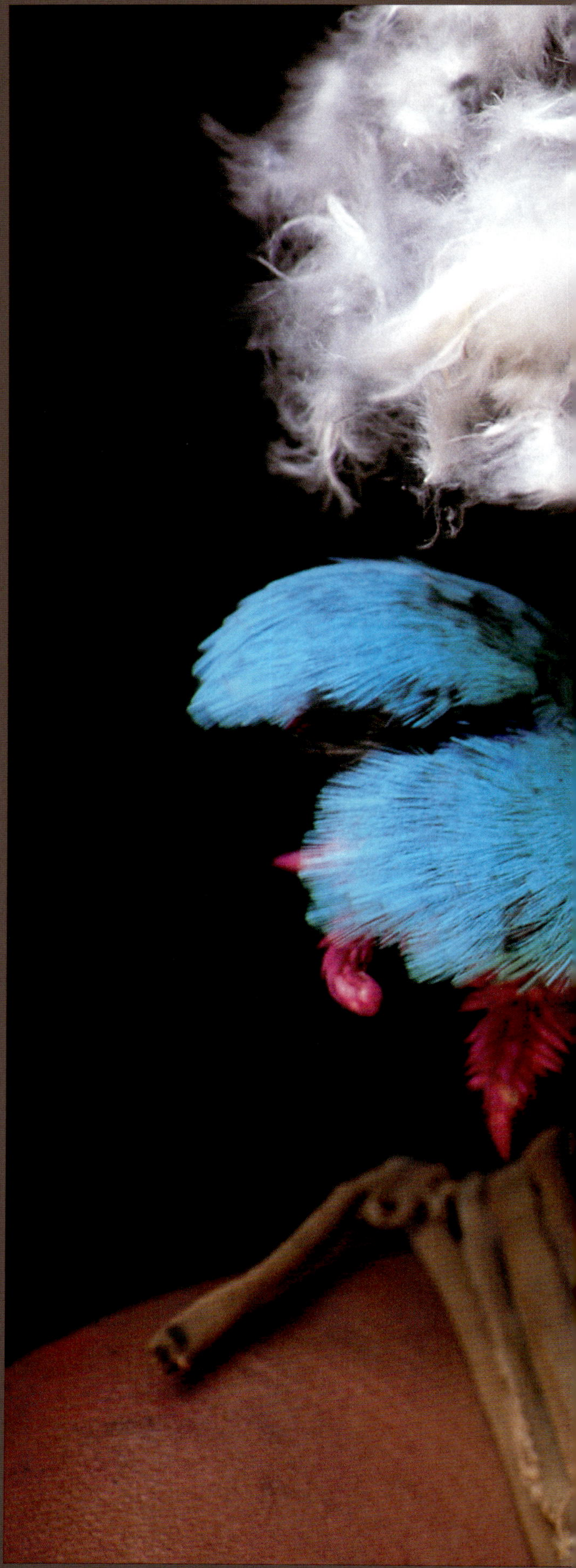
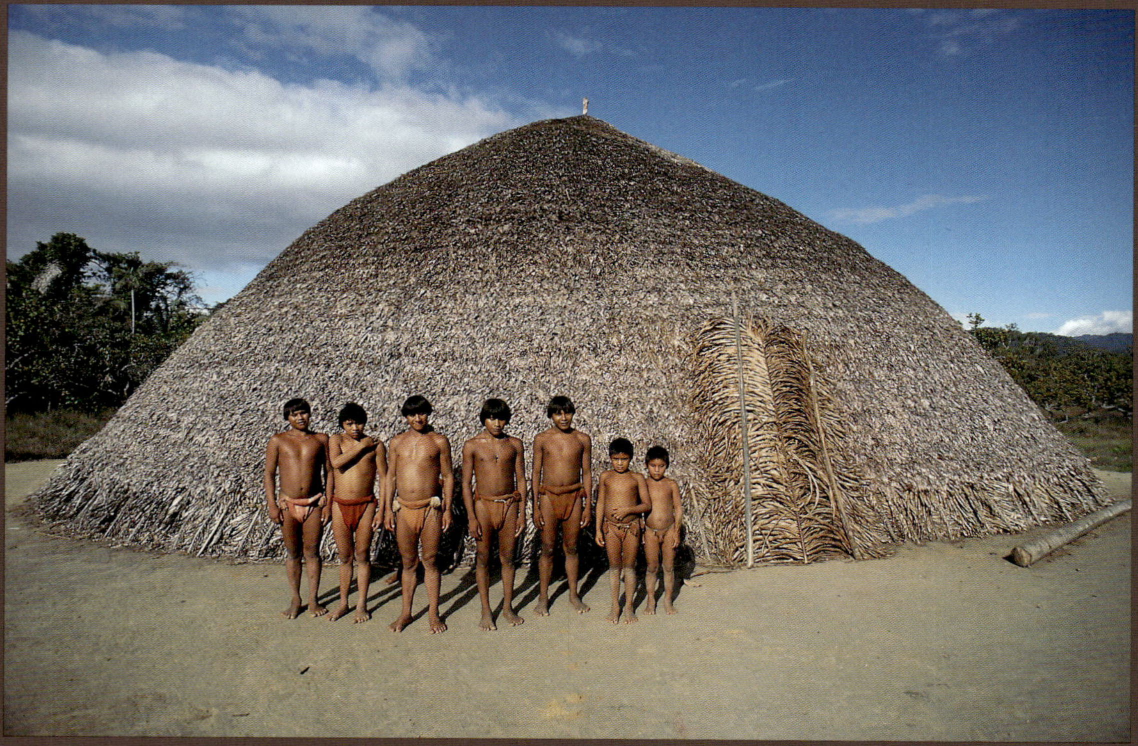

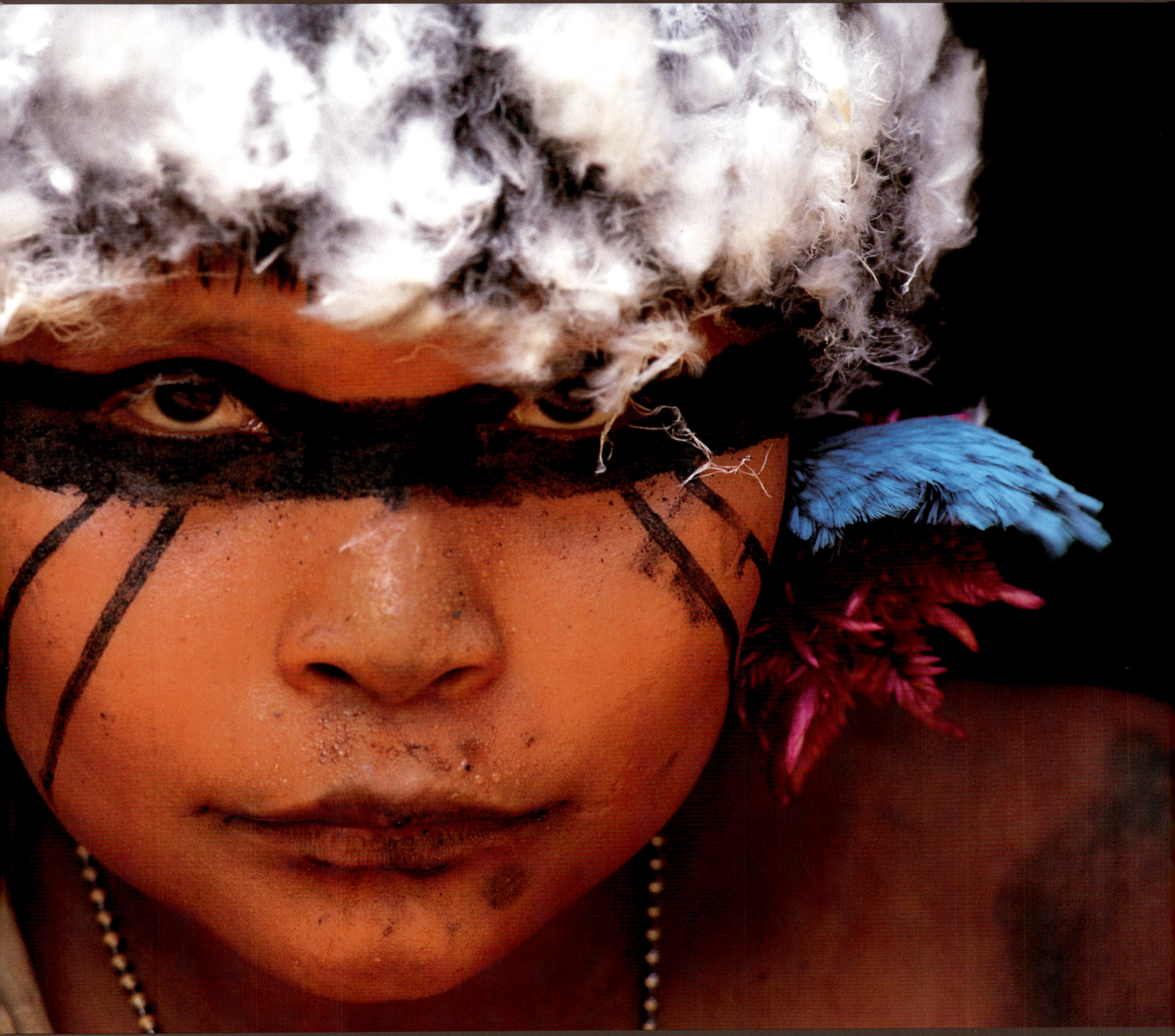

Venezuela has many distinct tribal groups. These Yanomami children, upper left, are from the Tapirapaco National Park. The Yanomami above is adorned with white down and the brilliant blue feathers of the spangled cotinga (Cotinga cayana)*. On the lower left, a group of young Hoti stand outside their very distinctive traditional house in the village of San José de Kaiamá, in the state of Bolivar.*

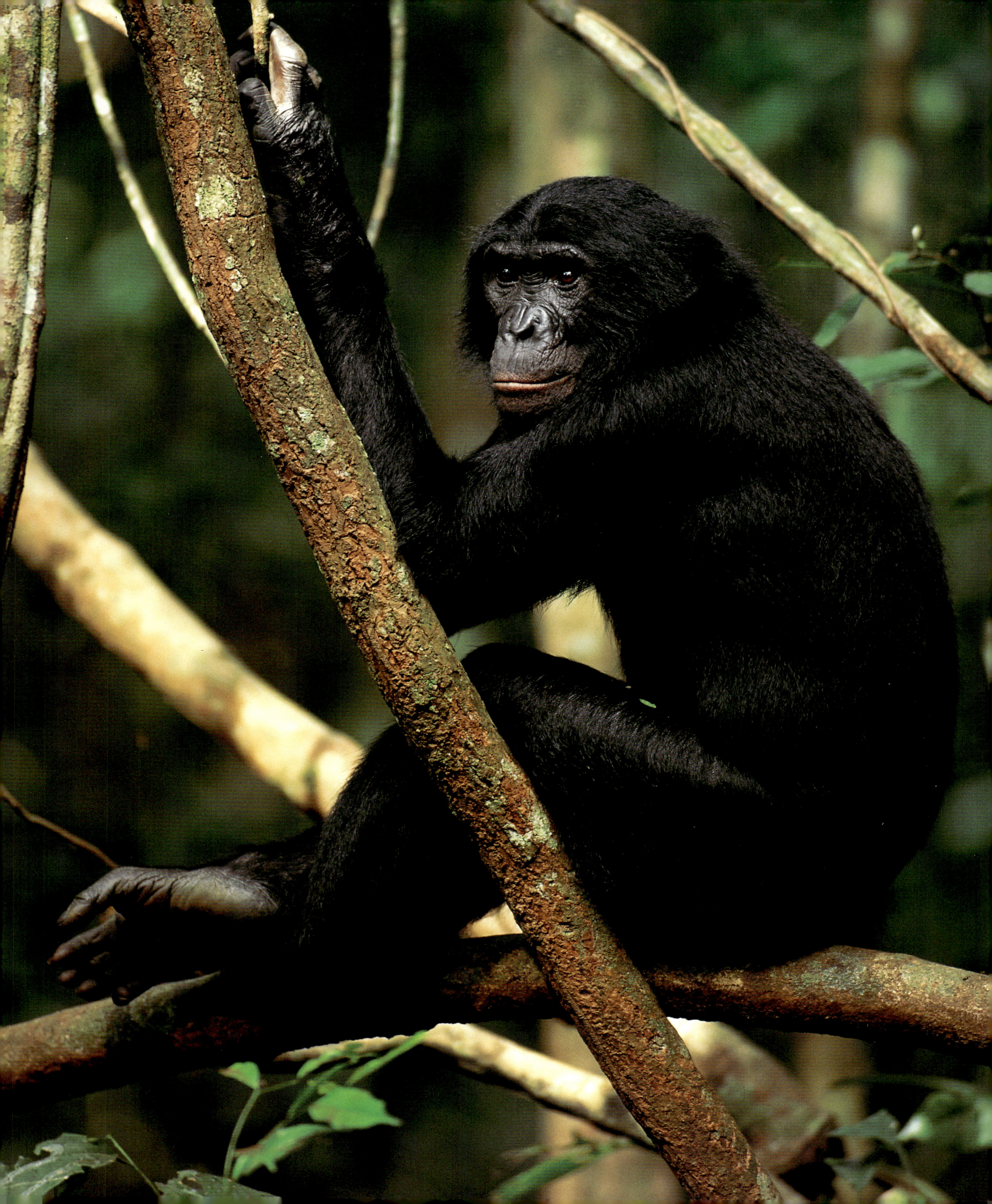

DEMOCRATIC REPUBLIC OF CONGO

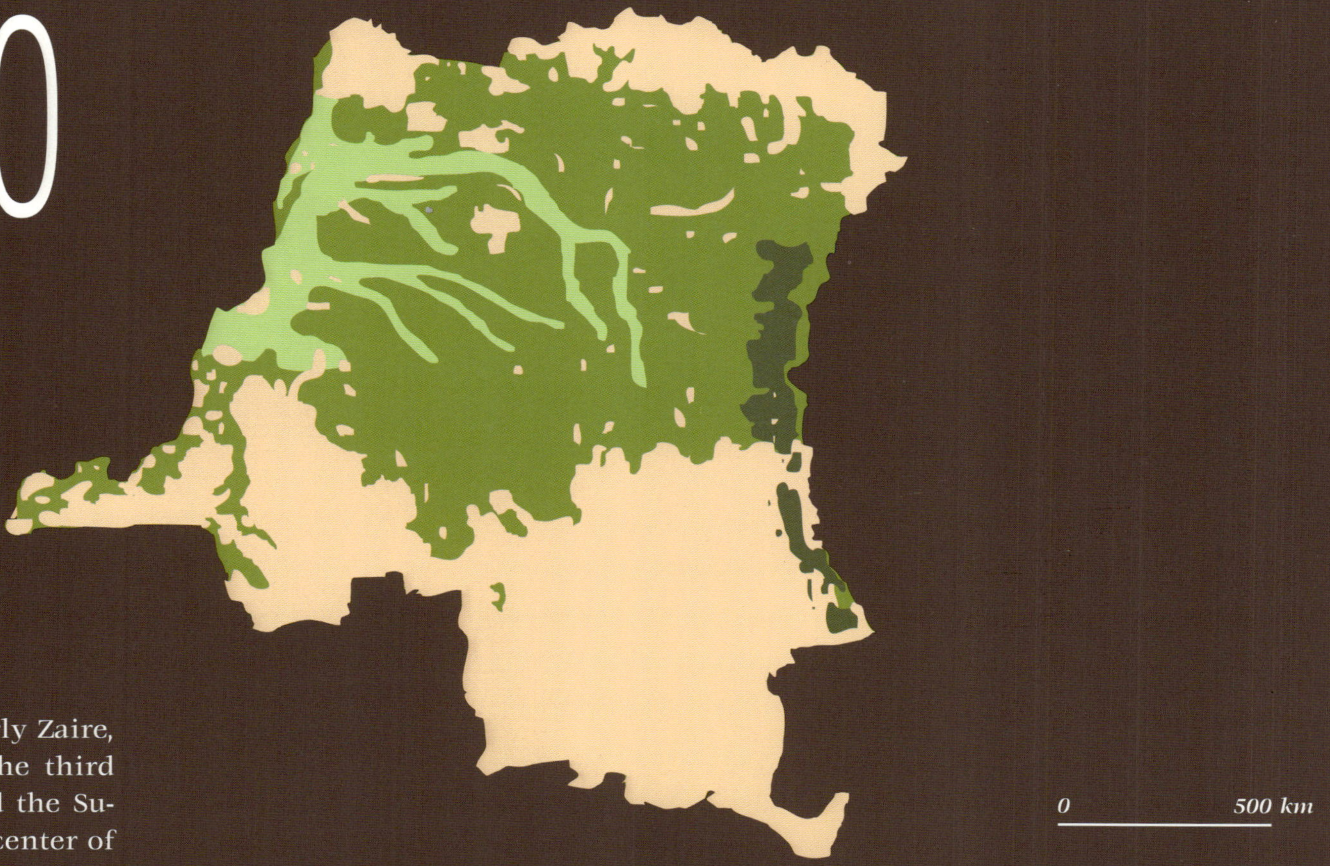

The Democratic Republic of Congo (formerly Zaire, and abbreviated hereinafter as DRC) is the third largest country in Africa after Algeria and the Sudan, and covers some 2 344 000 km^2 at the center of the continent. Overall, it is the single most biologically diverse country on the African mainland, and is at the top of the list for Africa in virtually every group of organisms except plants, in which it is second to South Africa. It also has the second highest tropical rain forest cover on Earth (being exceeded only by Brazil and slightly surpassing Indonesia), with a total of 1 190 737 km^2 of tropical rain forest. Together with Brazil and Indonesia, the DRC accounts for almost half of the tropical rain forest biome worldwide, and by itself has close to 13% of the global total and 68% of the closed tropical rain forest in Africa. The DRC also has the second largest river system on Earth, the mighty Congo, and one of the most diverse freshwater fish faunas.

Given its large land area, the DRC's human population of approximately 44 million is not that high, although the growth rate of 3.2% is among the highest in the world and population is projected to grow to 105 million by the year 2 025. About 40% of the human population lives in the cities; the country's major forested areas, especially the central basin —the Cuvette Centrale— have always been relatively sparsely populated. Consequently, this country has one of the most important major tropical wilderness areas left on Earth, second in extent only to the vast blocks remaining in northern and western Amazonia in South America.

In term of species diversity, the DRC has by far the highest mammal diversity in Africa, with 415 species (28 of them endemic), and ranks seventh in the world in this category. Bird diversity is also the highest on the continent, with 1 094 species (at least 23 endemic), and is in tenth place on the world list. Reptile diversity is moderately high, with 268 species, but only 33 are endemic. Amphibian diversity is poorly known, but appears to be low at 80 species, with 53 endemics; the actual total is likely to be higher. Freshwater fish diversity is quite high, with at least 962 species and probably many more, since there has been little research in the vast system of the Congo River. Plant diversity is high by African standards, with 11 000 species of higher plants and 3 200 endemics, but low compared to the American and Asian tropics. Overall, the DRC falls short of the megadiversity cutoff points for both plant and vertebrate endemism, and is included in the megadiversity list mainly on the strength of its tropical rain forest and its freshwater systems, its high overall vertebrate total, and because it is the most diverse country on the African continent.

As regards globally important sites for plants and birds, the DRC has 12 of Africa's 30 Centers of Plant

The Democratic Republic of Congo is the only country with five different kinds of apes, including three subspecies of the gorilla, the common chimpanzee, and the pygmy chimpanzee or bonobo (Pan paniscus) *shown on the opposite page. Not a pygmy at all, but rather a highly distinctive species of chimp, the bonobo is endemic to the rain forests of the DRC.*

Diversity and Endemism identified by IUCN/WWF (1994) lying partially or entirely within its borders, a figure matched only by Tanzania. The 12 plant centers are Mayombé, Itombwe, the Ituri lowland rain forest, Kahuzi-Biega National Park, Maiko National Park, Salonga National Park, Haut Shaba, Kundelungu National Park, Upemba National Park, the Marungu Highlands, Garamba National Park and Virunga National Park. These are mainly protected areas, and may be a reflection of research being concentrated in these areas. It also has two of the Endemic Bird Areas identified by BirdLife International (ICBP, 1992): the Albertine Rift Mountains and the East Congo lowlands. The DRC also has the largest portion of one of the three most important major tropical wilderness areas on Earth, that of the Congolian forest block of Central Africa.

Human cultural diversity is also high in the DRC, with some 250-300 languages spoken. Among the many different groups, the Pygmies of the Ituri Forest are the smallest people on Earth, and are particularly unique and interesting, since they are among the few true forest people still living a more or less traditional lifestyle.

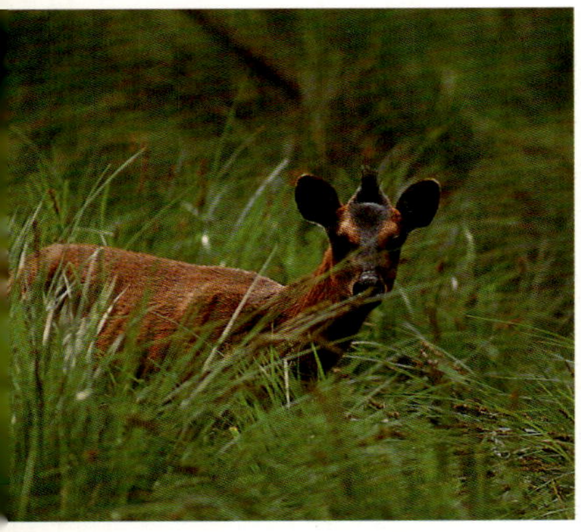

The forest antelopes known as duikers are widespread in the Guineo-Congolian forests of Central and West Africa, with eight species occurring in the DRC. This black-fronted duiker (Cephalophus nigrifrons) *was photographed in the eastern part of the country.*

On the opposite page, the potto (Perodicticus potto) *is a representative of the loris and galago family Lorisidae, which also contributes to the DRC's high primate diversity. These small, nocturnal prosimians fall into two groups, the slow-moving, short-tailed potto and ungwantibo and the highly active, leaping, long-tailed galagos.*

Natural habitats

In the context of the megadiversity countries, the DRC is moderately diverse in natural habitats, stretching as it does from the savanna woodlands of the southern fringes of the Sahara in the north to the woodlands and grasslands of Zambia and Angola in the south; from west to east, habitats range from the mangrove forests on the Atlantic coast to the mountains of the Ruwenzori, including Margherita Peak, the third highest mountain in Africa (5 119 m). However, tropical forest formations —including both closed rain forest and open forest formations— dominate the country, with 1 190 737 km^2 of rain forest , and the rest in a mosaic of Zambezian and Sudanese dry forests and savannas.

The closed evergreen forests of the DRC can be divided into three main categories: swamp and riverine forest covering some 100 000 km^2; Guineo-Congolian lowland rain forest of the Cuvette Centrale and the Bas-Zaïre, covering over 900 000 km^2; and the Afromontane forest communities on the eastern side of the country, covering more than 53 000 km^2. Few of the rain forests of the DRC are true evergreen rain forest, but rather semievergreen, becoming more deciduous towards the northern and southern fringes of the forest zone. There is usually a pronounced dry season, and total rainfall in much of the Cuvette Centrale, where most of the forest is located, is only about 2 000 mm, the lower limit of rainfall needed for moist evergreen forest. The bulk of the rain forests are found in Équateur Province, which is still 99.7% covered in forest, and Haut-Zaïre, which has 73.5% forest cover. Other important areas are Kivu, Bandundu, Kasai-Oriental and Kasai-Occidental. The forests of Bas-Zaïre are quite distinct from those of the Cuvette Centrale, and are now down to fragments on the hills of the Mayombé; they include tree species not found elsewhere in the DRC.

The rain forests of the DRC are often dominated by a single species of the family Caesalpiniaceae. In the Ituri forest, lower slope forests may be composed of almost pure stands of *mbau* or *limbali* (*Gilbertiodendron dewevrei*), and higher plateau forests of *mbau* or *bomanga* (*Brachystegia laurentii*); in other areas, the dominant species is often *mufimbi* (*Cynometra alexandri*).

To the north and south of the closed forests are the Sudanese and Zambezian dry forests. The Sudanese dry forests have been largely degraded by fire, forming open *Isoberlinia* woodlands. The Zambezi dry forest area now consists mainly of secondary miombo woodlands dominated by *Brachystegia* spp. and *Julbernardia globiflora*. Intact Zambezi dry forest now exists only as relict stands of dense forests dominated by *Entandrophragma delevoyi*, *Parinari excelsa*, and *Cryptosepalum pseudotaxus*. Although the tropical forests generally receive more attention, the Zambezian woodlands also have high diversity and endemism, especially in plants, and are also a priority. For instance, four of the centers of plant diversity and endemism in the DRC —Haut Shaba, Kundelungu National Park, the Marungu Highlands, and Upemba National Park, are within the Zambezian Regional Center of Endemism.

The montane forests of the DRC are in the Albertine Rift Mountains, which form the watershed between the Congo and Nile Rivers. They have lower plant diversity than the lowlands, but very interesting patterns of altitudinal zonation. Species of *Podocarpus*, *Prunus* and *Ocotea* occur at intermediate elevations, and then give way to elfin forests and communities dominated by bamboos, tree ferns, giant grounsels, and lobelias and, at the highest altitudes, Afroalpine moorlands with a number of ericaceous shrubs and grasses. Much mid-elevation forest has been modified by human disturbance and is covered with secondary plant communities like open *Hagenia abyssinica* woodland. Within the DRC, the Virunga National Park Center of Endemism is the most important area within the larger Afromontane Regional Center of Endemism. This area is also important for birds, and includes the Albertine Rift Mountains Endemic Bird Area. A number of globally important primate species live here as well, including the world-famous mountain gorilla (*Gorilla gorilla beringei*), which is found only in the Virunga Volcanoes and neighboring Bwindi Forest in Uganda, and the Virunga golden monkey (*Cercopithecus mitis kandti*), endemic to the Virungas and the nearby Nyungwe Forest in Rwanda.

A small area of mangrove forest also exists in the DRC, covering only 226 km^2 at the estuary of the Congo River. This area is of particular interest as a refugee for a small population of the endangered West African manatee (*Trichechus senegalensis*).

Marine systems are not very significant in the DRC, since the country has only a tiny marine area at the mouth of the Congo River, between the main part of Angola to the south and the Cabinda enclave of Angola to

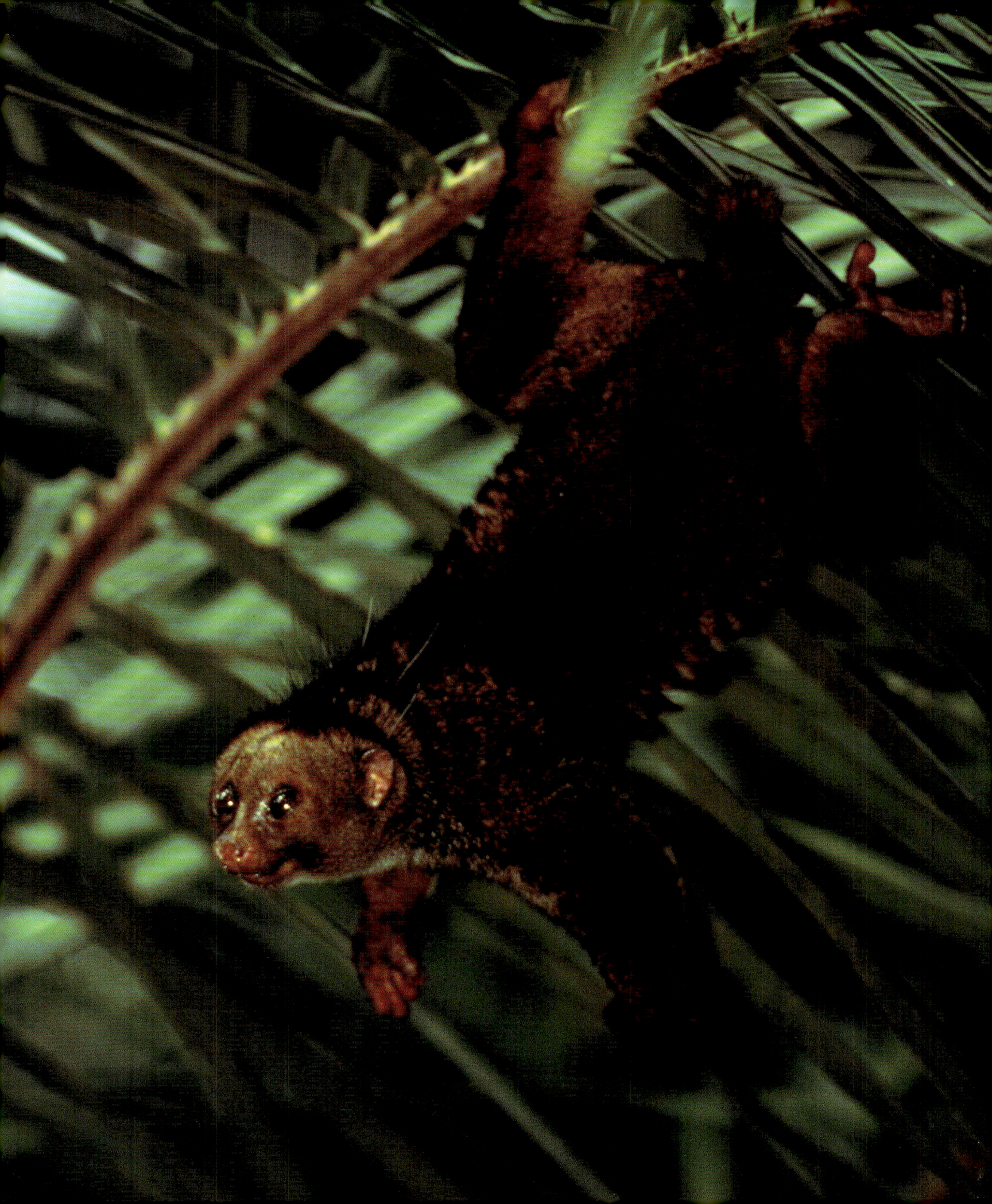

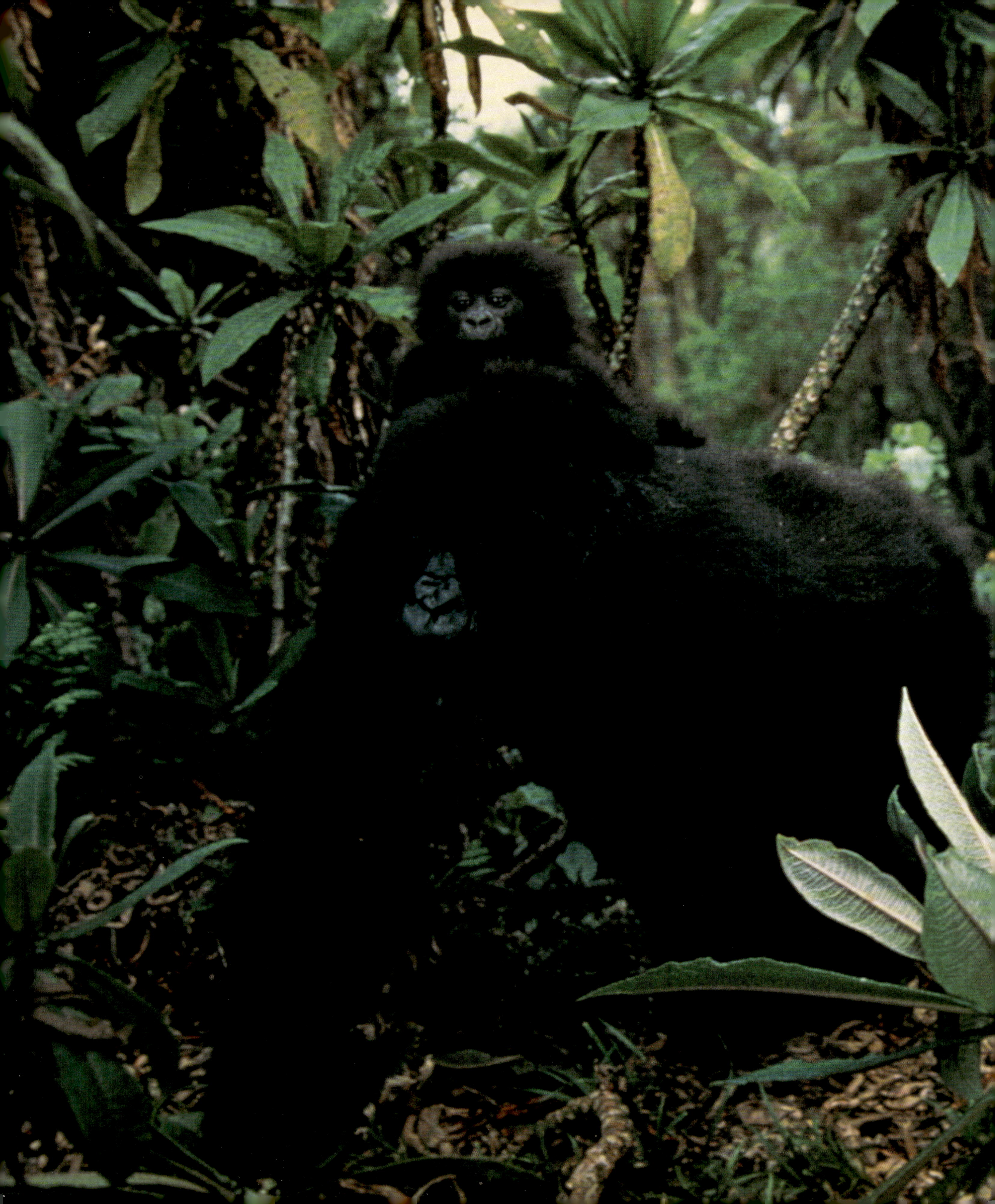

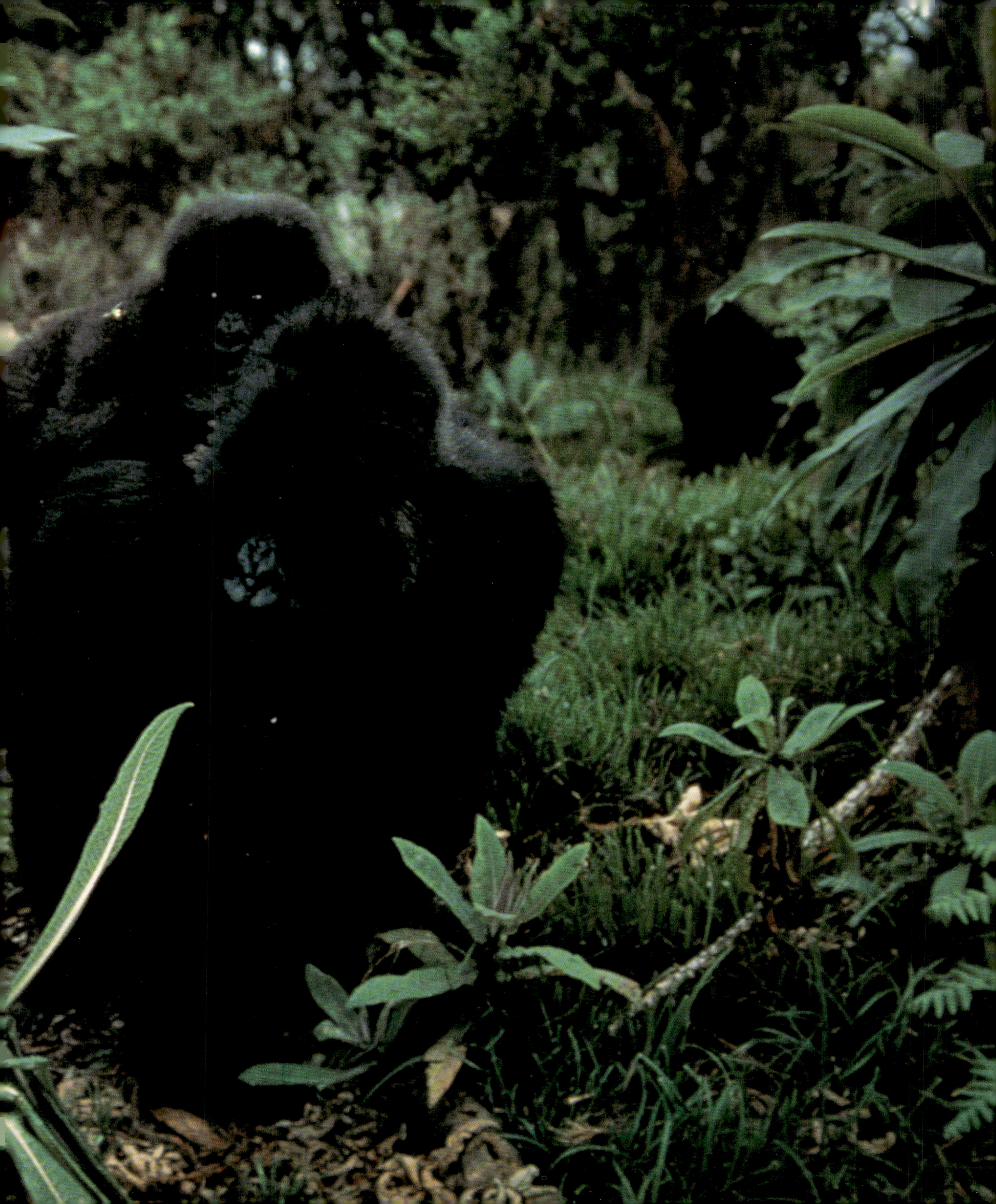

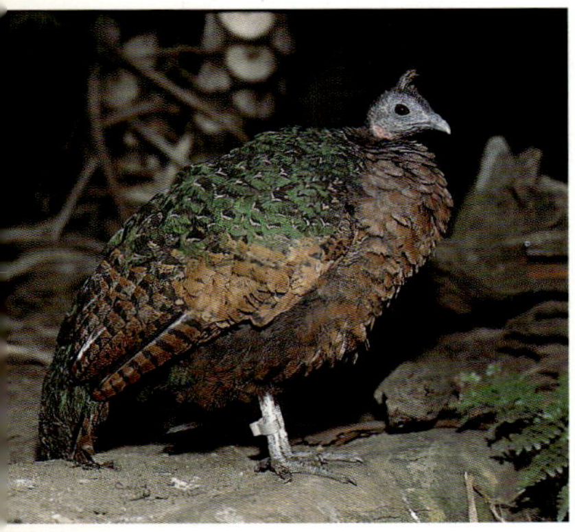

On pp. 472-473, the mountain gorilla (Gorilla gorilla beringei) *is one of the world´s best known and most critically endangered flagship species. Restricted to the Virunga Volcanoes on the border between the DRC, Rwanda and Uganda, and the nearby Bwindi (Impenetrable) Forest in Uganda, its range lies right in the middle of an area that has experienced one of the most severe social upheavals in recent decades. According to the last count, only about 400 individuals survived, but the number may now be even lower.*

Above, the Congo peacock (Afropavo congensis) *is one of the DRC's large endemic flagship species. It is also the only member of the pheasant family found outside Asia and it was not described by science until 1936. The individual shown here is a female.*

On the opposite page, another large and attractive forest antelope, the bongo (Tragelaphus euryceros) *is much more widespread than the okapi, occurring from West Africa through the Congolian Forest of Central Africa to forest outliers in East Africa.*

the north. However, its freshwater ecosystems, as mentioned above, are of great global importance. The highest diversity occurs in the Congo River drainage, where at least 790 species of fish are known to occur (probably a major underestimate), but the lakes of the Albertine Rift are of international significance, especially Lake Tanganyika which, with 214 fish species, is second in diversity only to Lake Malawi, and is followed by Lake Mweru with 88 fish species, Lake Edward, with some 55 species (35 endemic), Lake Albert with 46 species (9 endemic), and Lake Kivu with 17-33 species (6-7 endemic). Lake Tumba and Lake Mai-Ndombe in the rain forest zone also have a number of endemic species.

Flagship species

The DRC has a number of globally important flagship species, the best known being the pygmy chimpanzee or bonobo (*Pan paniscus*), the eastern gorilla (*Gorilla gorilla graueri*), the okapi (*Okapia johnstoni*), the Congo peacock (*Afropavo congensis*), and the northern white rhinoceros (*Ceratotherium simum cottoni*), all of them endemic to the DRC and of great international significance. Also of importance is the mountain gorilla (*Gorilla gorilla beringei*) and the Virunga golden monkey (*Cercopithecus mitis kandti*), both of which straddle the DRC-Uganda-Rwanda border. Primate diversity in the DRC is very high, and strongly correlated with the large extent of tropical rain forest, principal habitat of non-human primates, and the fact that the country encompasses several different regional faunal communities. With approximately 37 species, the DRC has the second highest number of primate species in the world after Brazil, but with 18 genera, its generic-level diversity is the highest in the world along with Cameroon.

The bonobo is probably our closest living relative, and shares on the order of 98% of its DNA with humans. It is not a pygmy at all, with adult males reaching 80% the size of the common chimpanzee, but is a more gracile, partly arboreal, rain forest-dwelling, very distinctive chimpanzee that was not discovered until 1935. It has a discontinuous range in the DRC, south of the Congo River and east of the Lomami River. Although the extent of its potential habitat is large, approximately 840 000 km^2 , it is thought to occur only in small isolated populations within this range. There are no precise estimates of how many remain, but the figure is probably fewer than 25 000 and more likely between 10 000 and 20 000. Only a few populations can be considered self-sustaining, with habitat destruction from commercial logging, commercial agriculture, and traditional slash-and-burn agriculture the principal threats. Bonobos are also hunted for food in some areas, but are protected by taboo in others; unfortunately, some of these taboos seem to be breaking down with cultural change and the movement of human populations. The bonobo is protected by law in the DRC, but enforcement is minimal. A small population was recently discovered in the large Salonga National Park, but some of the most important areas for this species, such as the Lomako and Wamba areas, are not yet given official protected status. The behavior of this species is very interesting, and has been the subject of much discussion among anthropologists and primatologists. Of particular note, and quite distinct from the common chimpanzee, is the ventral-ventral copulatory position and the use of sex as a displacement activity for aggression and other forms of stress. There is a captive population of bonobos in European and North American zoos, but it is small. Much more research on the species is needed, especially a survey to determine the remaining distribution of this remarkable animal. An action plan for the conservation of the bonobo was prepared in 1995, but it has been difficult to implement.

The DRC is also critically important for the survival of the gorilla, the largest living primate, and is the only country that has populations of all three recognized subspecies. The western lowland gorilla (*Gorilla gorilla gorilla*), found in equatorial West Africa and extending into the Mayombé region of the DRC, is still by far the most abundant of the gorilla subspecies, although it is declining because of habitat destruction and an increasing bushmeat trade. The total population of the western gorilla is estimated at 110 000, only a tiny portion of which occurs in the DRC. The eastern lowland gorilla (*Gorilla gorilla graueri*) is found only in the eastern DRC, and has an estimated population of 10 000 and declining; about half of these occur in national parks. Finally, the mountain gorilla (*Gorilla gorilla beringei*), one of the most endangered primates on Earth, is down to about 600 individuals in the Virunga Volcanoes on the DRC-Rwanda border and in the Bwindi (Impenetrable) Forest of Uganda. Until recently, this population had been considered stable. However, given the tremendous strife and upheaval of human populations now taking place on the border between the DRC and Rwanda, it is difficult to predict what the outcome will be for this unique animal, which was not discovered until the early years of this century.

The okapi (*Okapia johnstoni*) is perhaps the most spectacular of all the flagship species of the DRC. The only living relative of the giraffes, this beautiful animal stands about 1.5 m at the shoulder, is chocolate brown, and has alternating white and black stripes on the flanks and legs. Although there were reports of this animal as early as 1890, it was not until 1901 that Sir Harry Johnston sent a skin to the British Museum. This wonderful animal occurs mainly in the Ituri Forest, where it has been the subject of a research project supported by the Wildlife Conservation Society for over a decade.

The Ituri Forest itself is an important flagship, and one of the most important rain forest areas left on Earth, both in biological and cultural terms. The area is comprised of the 60 000-km^2 watershed of the Ituri River in the eastern DRC, and includes a number of different forest communities, among which are monospecific forests, mixed forests of varying composition, and

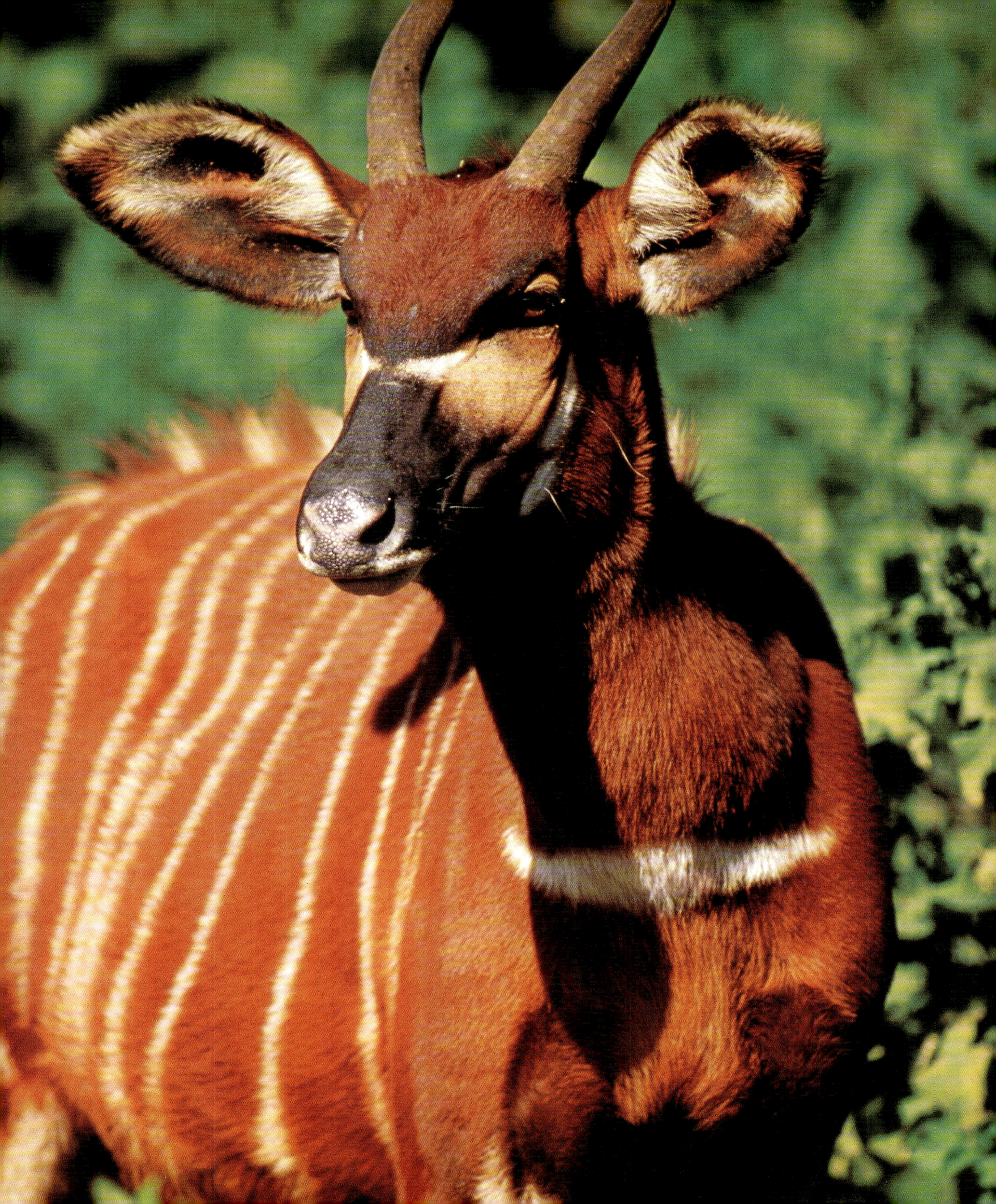

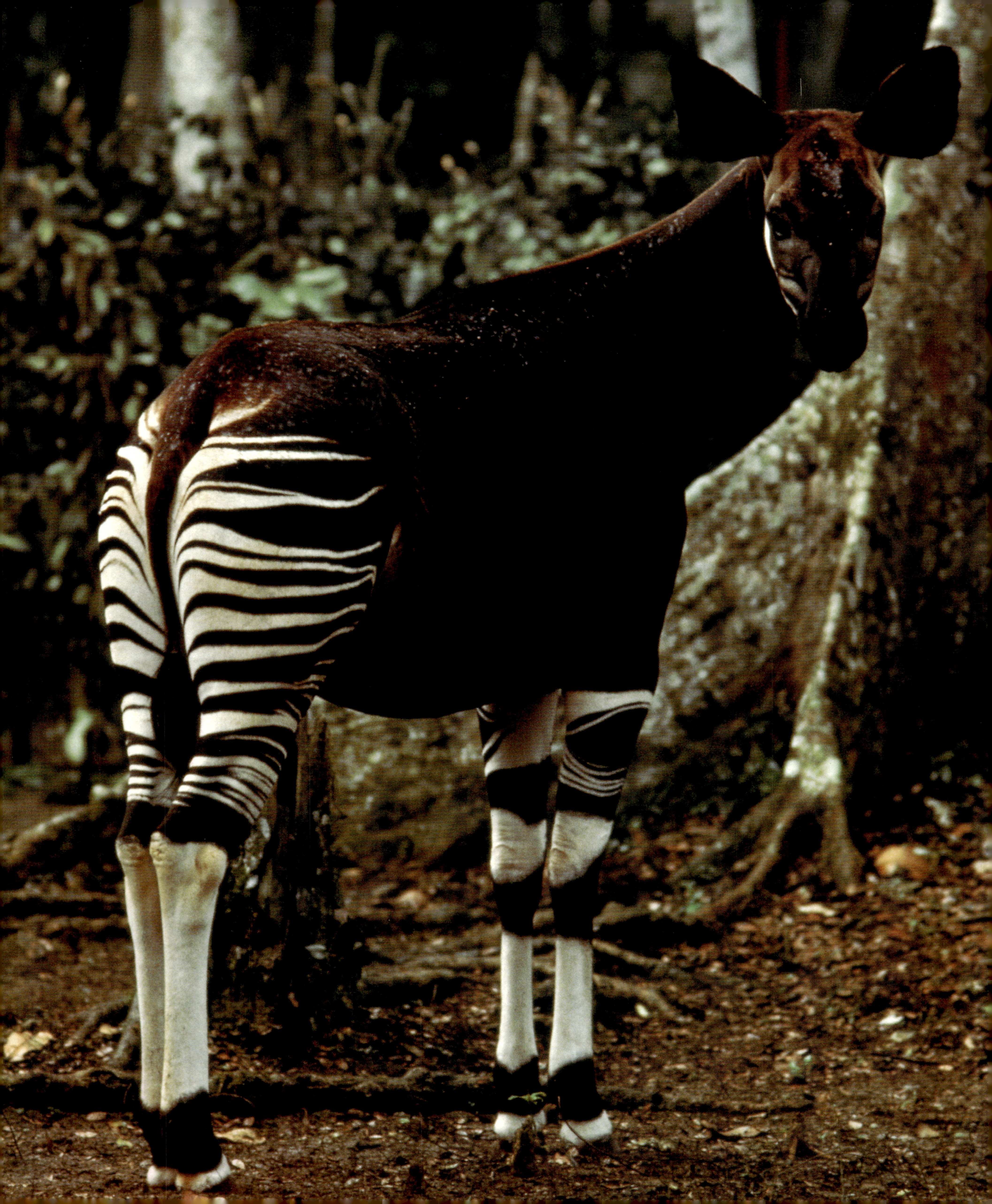

unique xerophylic communities on bare granite. Nearly 15% of its mammal species are endemic, it has the highest diversity of monkeys in Africa, and is the main stronghold for the okapi, one of the DRC's two most important endemic species. During the 1950s, a government station was established in Epulu in the central Ituri to capture okapi. It adapted indigenous capture techniques, and employed local people, especially the Mbuti and Efe Pygmies, to capture, habituate and feed the okapis. A system of locally-controlled forest reserves was established along the road for 70 km on either side of Epulu to serve as okapi capture zones, and each reserve was associated with a particular village which worked on the okapi capture project. The economic return and the status these villagers obtained by working with the okapi provided a strong incentive to protect the capture areas, and the reserves were also used for traditional hunting and gathering. Village tradition protected the forest in the reserves from agricultural incursions and excessive hunting, with the result that there are large stands of mature forest along the road running through the central Ituri.

Unfortunately, this traditional system is now being threatened as a result of logging, gold mining and other activities. This threatens both the forest and the indigenous people living within it. To counter these trends, and in recognition of the great biological and cultural significance of the Ituri Forest, a national park proposal was developed in 1986 with World Wildlife Fund (WWF) and the Institut Zairois pour la Conservation de la Nature (IZCN). As it currently stands, the proposal calls for a reserve of about 1 350 000 ha, with multiple-use buffer zones centered on the road, with the more remote portions of the park being given total protection. This area has an estimated okapi population of 4 500-6 500. If the national park proposal succeeds, it will be largely the result of the attention given to one flagship species, the okapi. Hopefully, it will be possible to make this national park a reality in the face of the many different social and economic crises now faced by this country.

The DRC also has very high overall antelope diversity, with 30 species occurring in the country. Among these are the bongo, the sitatunga, and at least eight species of forest duiker.

The northern white rhino (*Ceratotherium simum cottoni*) is the most endangered of all the living taxa of rhinos, although it is outwardly very similar to the southern white rhino of southern Africa, which is the most abundant remaining form. Formerly more widespread, the northern white rhino is now down to about 25 individuals, all of them restricted to the savannas of Garamba National Park in the extreme northeastern DRC.

The most important flagship bird species in the DRC is the Congo peacock (*Afropavo congensis*). Endemic to the central forests of the country, it was not discovered until 1936, when it created a real sensation in the ornithological world. It has only two close relatives living in Southeast Asia. This species is very difficult to observe in the wild, in part because it appears to be largely inactive during the day, but its very distinctive call is one of the characteristic sounds of the DRC rain forest.

Cultural diversity

Like most of the megadiversity countries, the DRC has striking cultural diversity, with some 250-300 different languages spoken. However, there are four main languages spoken there, in addition to French, which is the official language. Lingala is spoken mainly in the capital and largest city, Kinshasa, and in Équateur Province; Kikongo, the language of the sixteenth- and seventeenth-century Kongo Kingdom, is spoken in the provinces of Bas-Zaïre and Bandundu; Tshiluba is the principal language in Kasai-Oriental and Kasai-Occidental provinces; and Swahili is the language of the provinces of Haut-Zaïre, Kivu and Shaba in the east. Most of the people are of Bantu origin, but there are also Sudanese and Nilotic groups in the north and northeast.

Probably the most interesting and most famous of all of the people of the DRC are the Pygmies. Although they are not restricted to the DRC, being found in most of the other equatorial African countries as well, the groups living in the DRC, and especially the Mbuti Pygmies living in the Ituri forest, have been the most studied and also appear to be the most traditional. They are the smallest people on Earth, with men averaging only four feet eight and a half inches (1.44 m) and women four feet six inches (1.37 m), and number between 20 000 and 40 000 according to recent estimates. They inhabit an area of about 100 000 km^2 in the Ituri forest, and are divided into two different groups: the net hunters (Mbuti Swa and Bakango) and the archers (Mbuti Efe). The archers hunt with bows and arrows and the net hunters drive game into long nets that are stretched through the forest. These people are mainly hunter-gatherers and deep forest dwellers, but they have evolved an unusual symbiotic relationship with neighboring Bantu agriculturists who live in clearings in the forest and open areas. They trade game and other forest products to the Bantu for iron goods, tobacco, and cultivated food, and the Bantu people tend to think of the Pygmies as their "servants"; the Pygmies' view of the situation, of course, is quite different. The Pygmies have lost their own language and tend to speak the language of the Bantu groups with whom they are in closest contact, although they speak it with strong dialectal changes and their own unique Pygmy phonetics. They also have a different philosophy regarding the forest, for they consider it generous and friendly, and feel themselves to be "the children of the forest." In contrast, the Bantu and other villagers consider it a hostile environment.

RUSSELL A. MITTERMEIER
CRISTINA GOETTSCH MITTERMEIER

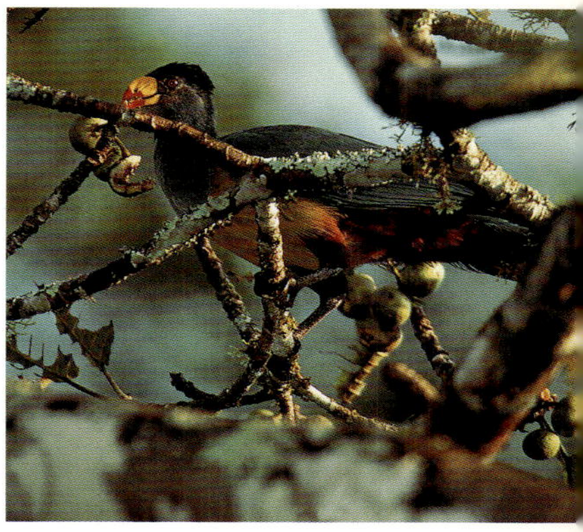

On the opposite page, the okapi (Okapia johnstoni) *is a large, forest-dwelling relative of the giraffe and one of the DRC's three most important flagship species. It is found mainly in the Ituri Forest in the eastern part of the country and has been the subject of an international conservation program for over a decade.*

The DRC has the highest bird diversity in the African region. Above, this blue plantain eater (Corythaeola cristata) *is a widespread member of the turaco family Musophagidae, which are large, brightly-colored, arboreal, fruit-eating birds.*

On pp. 478-479, the Virunga Volcanoes on the DRC/Rwanda/Uganda border, home of the mountain gorilla. Mt. Mikeno is seen in the background. In the foreground is the crater of Mt. Visoke, with its subalpine vegetation.

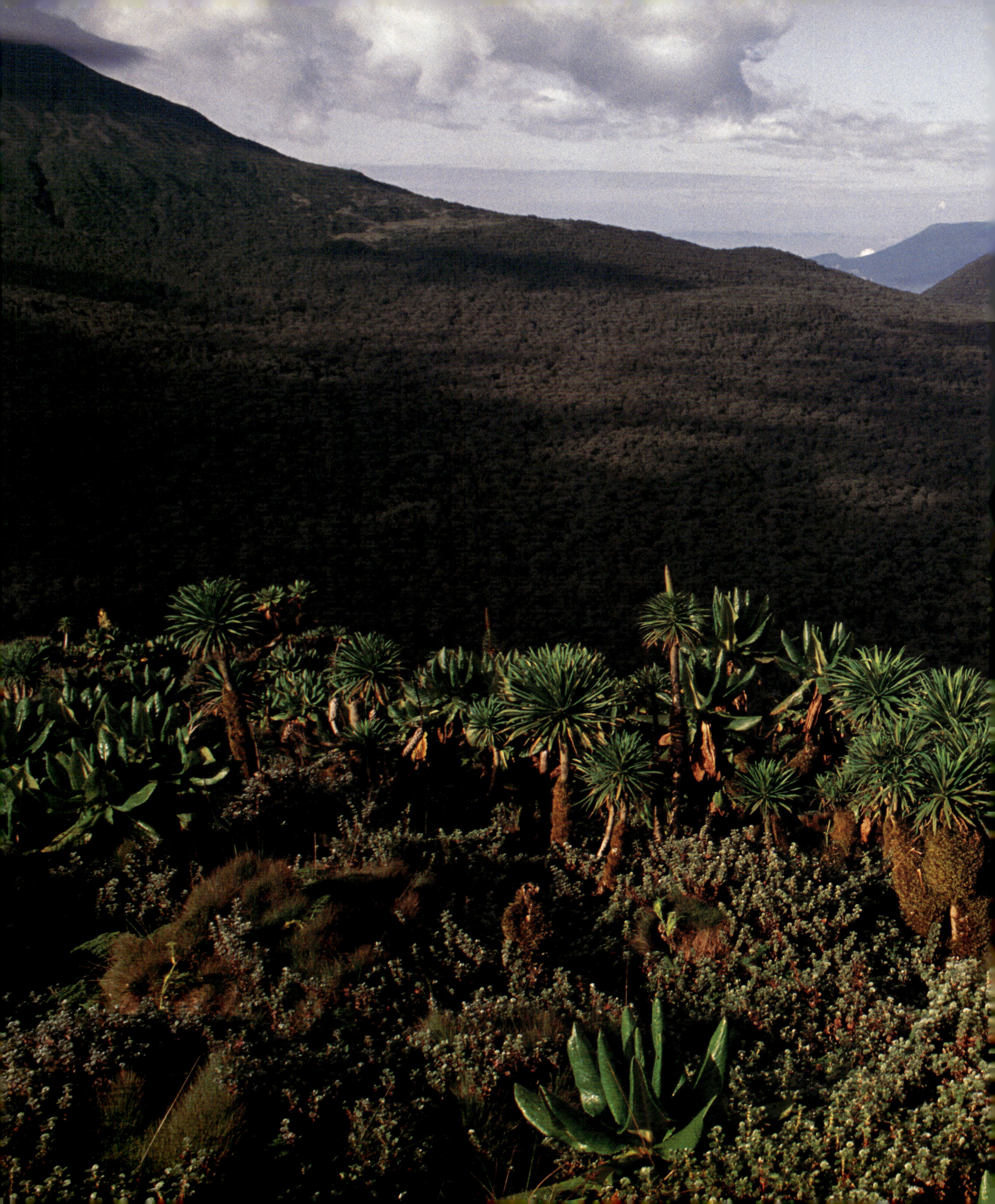

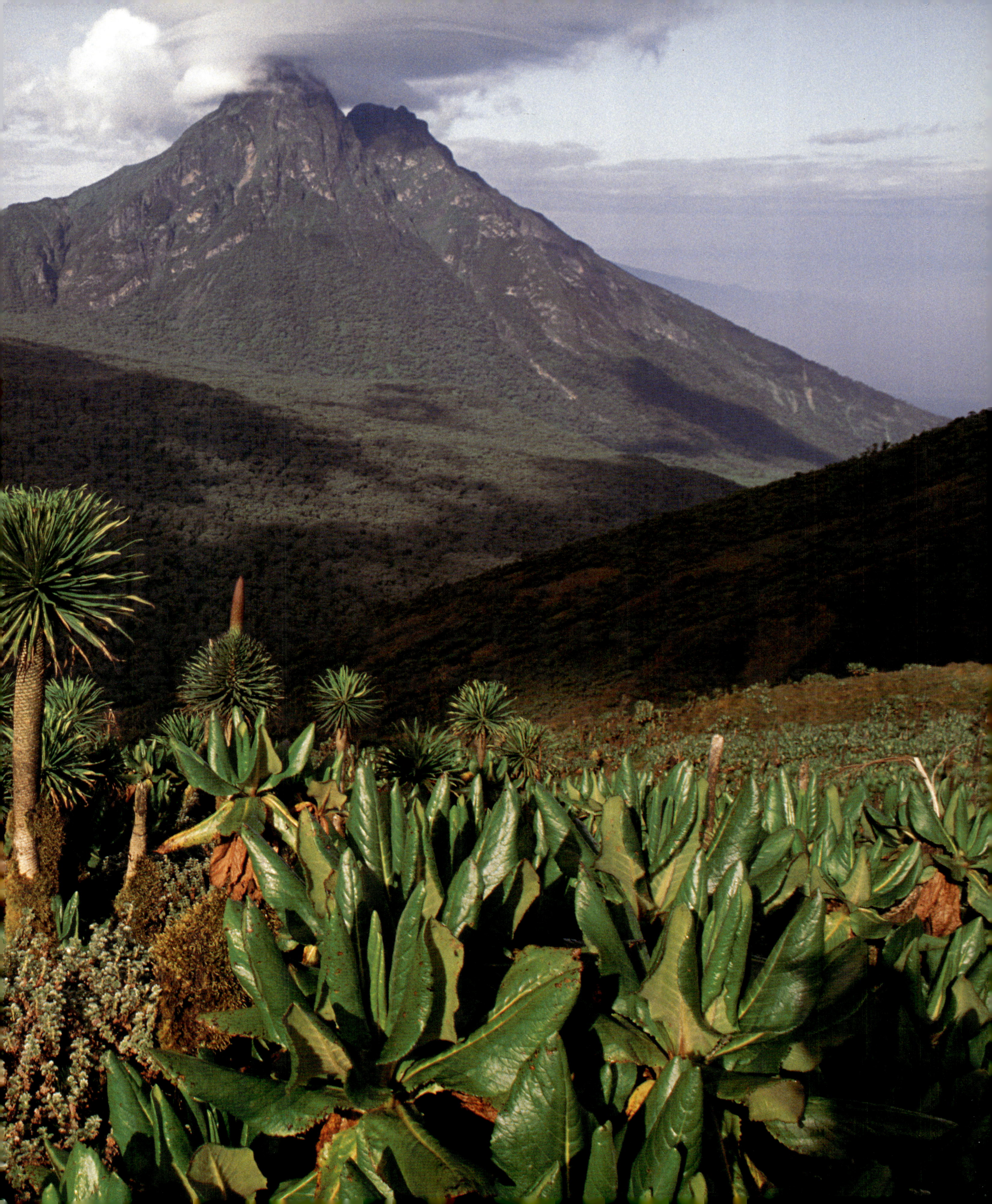

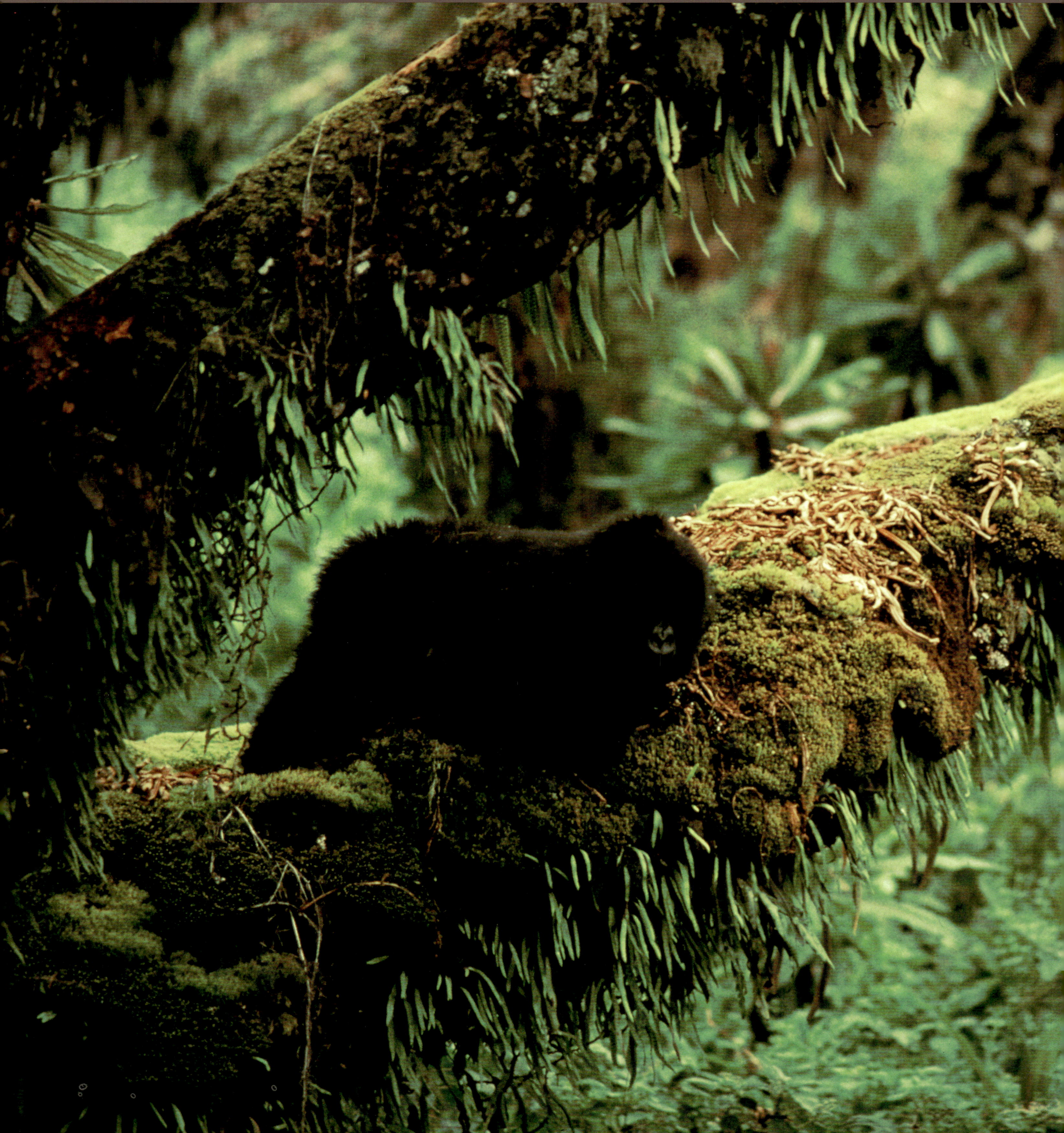

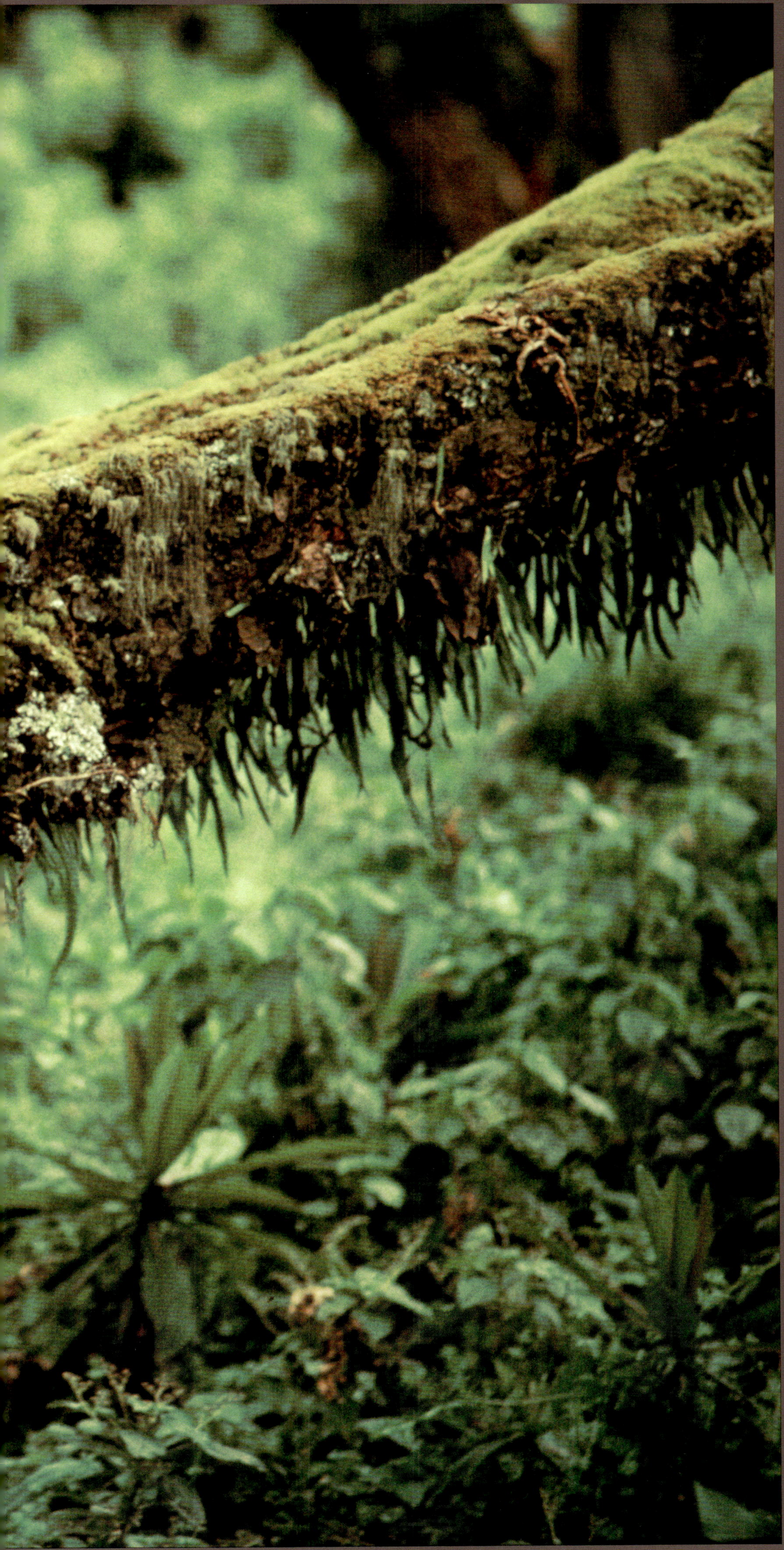

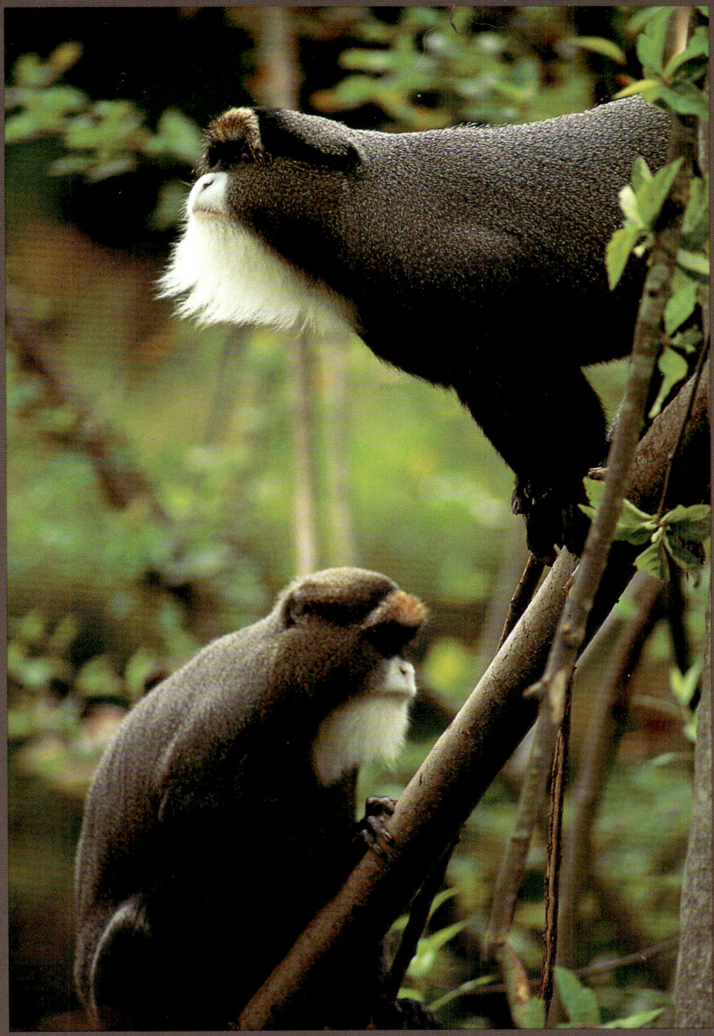

*The DRC's primate diversity is the second highest on Earth after that of Brazil, with 18 genera and approximately 37 species. The colorful guenons, represented above by De Brazza's monkey (*Cercopithecus neglectus*), constitute a particularly large primate radiation that reaches its peak in the DRC. Left, a mountain gorilla rests on a tree branch.*

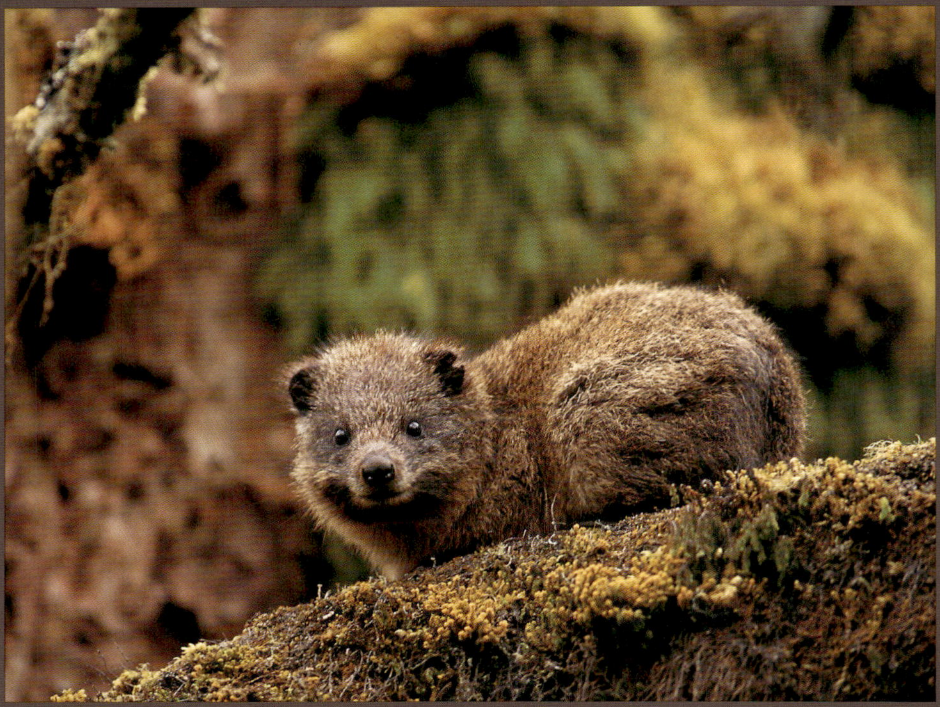

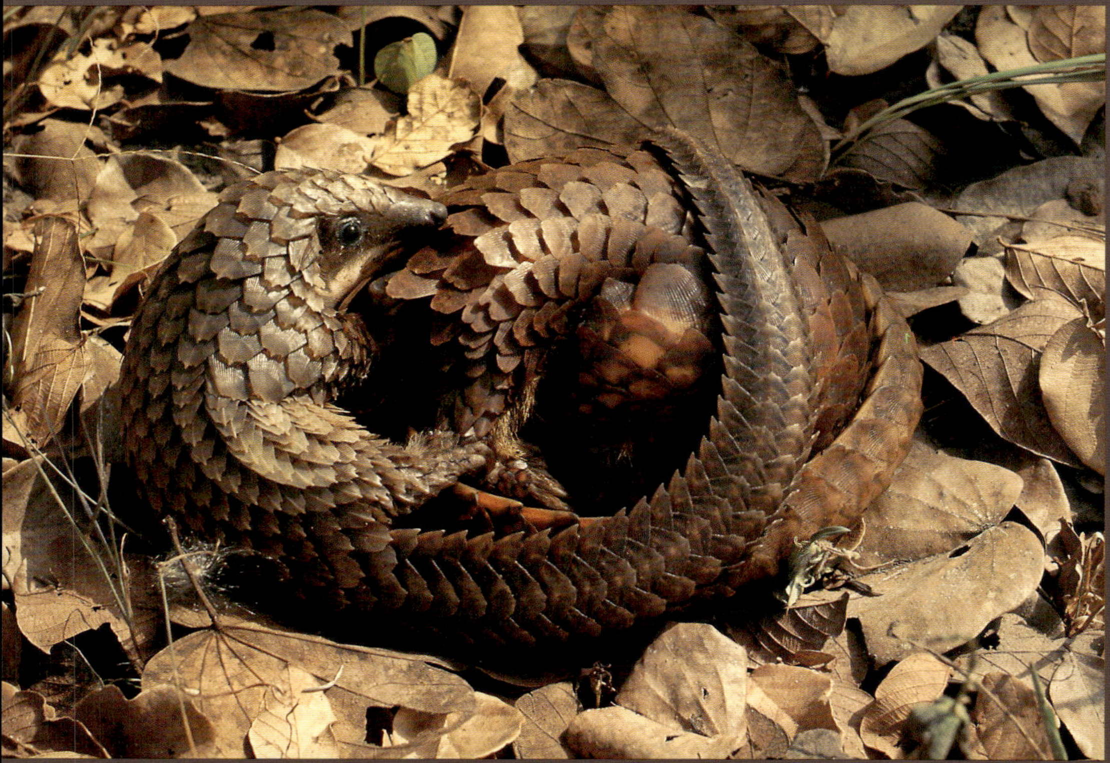

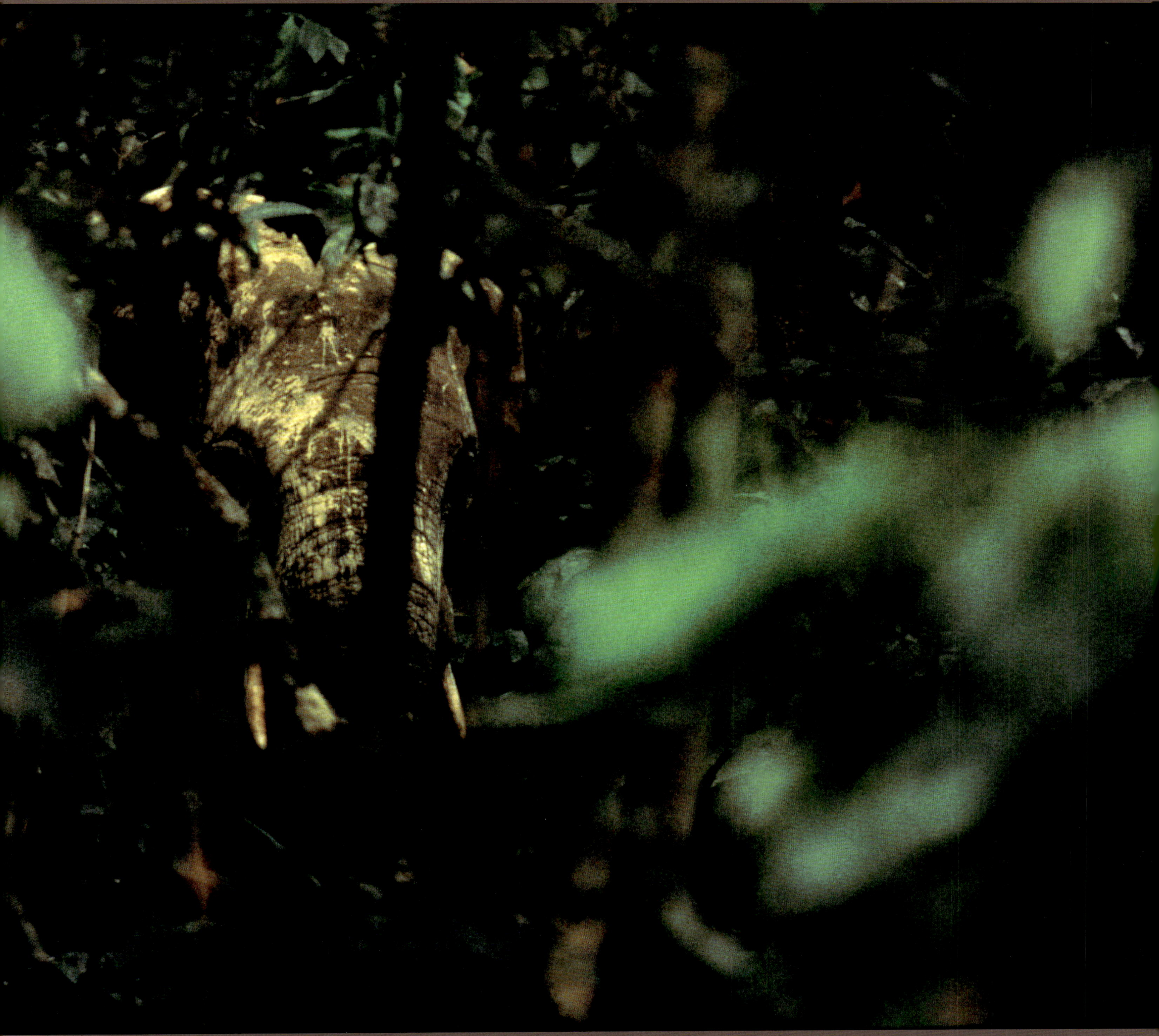

Above, the forest elephant (Loxodonta africana cyclotis) *is generally somewhat smaller than its savanna-dwelling cousin, and has more rounded ears. It was abundant in the DRC, but has declined precipitously because of large-scale, uncontrolled ivory poaching.*
Upper left, the tree hyrax or tree dassie (Dendrohyrax arboreus) *is a tree-dwelling counterpart of the rock hyrax.*

Amazingly, this small animal is the closest living relative of the elephant. This individual was photographed in the Virunga Volcanoes. The DRC has by far the richest mammal fauna in Africa with 415 species. Of the four pangolin species found in Africa, three are found in the DRC, including the white-bellied pangolin (Manis tricuspis), *shown lower left.*

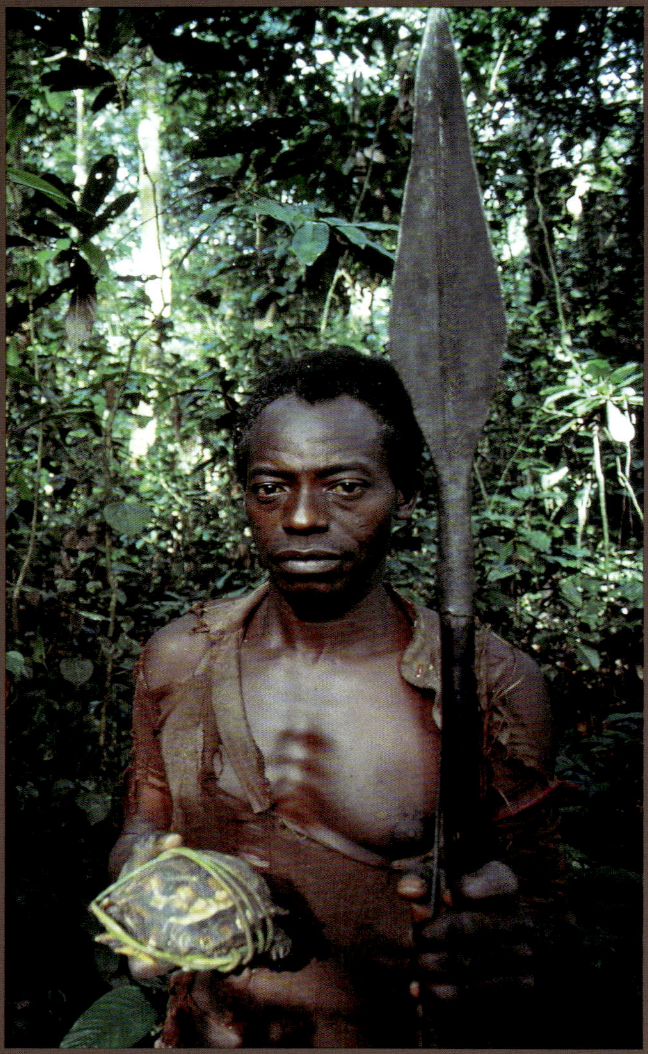

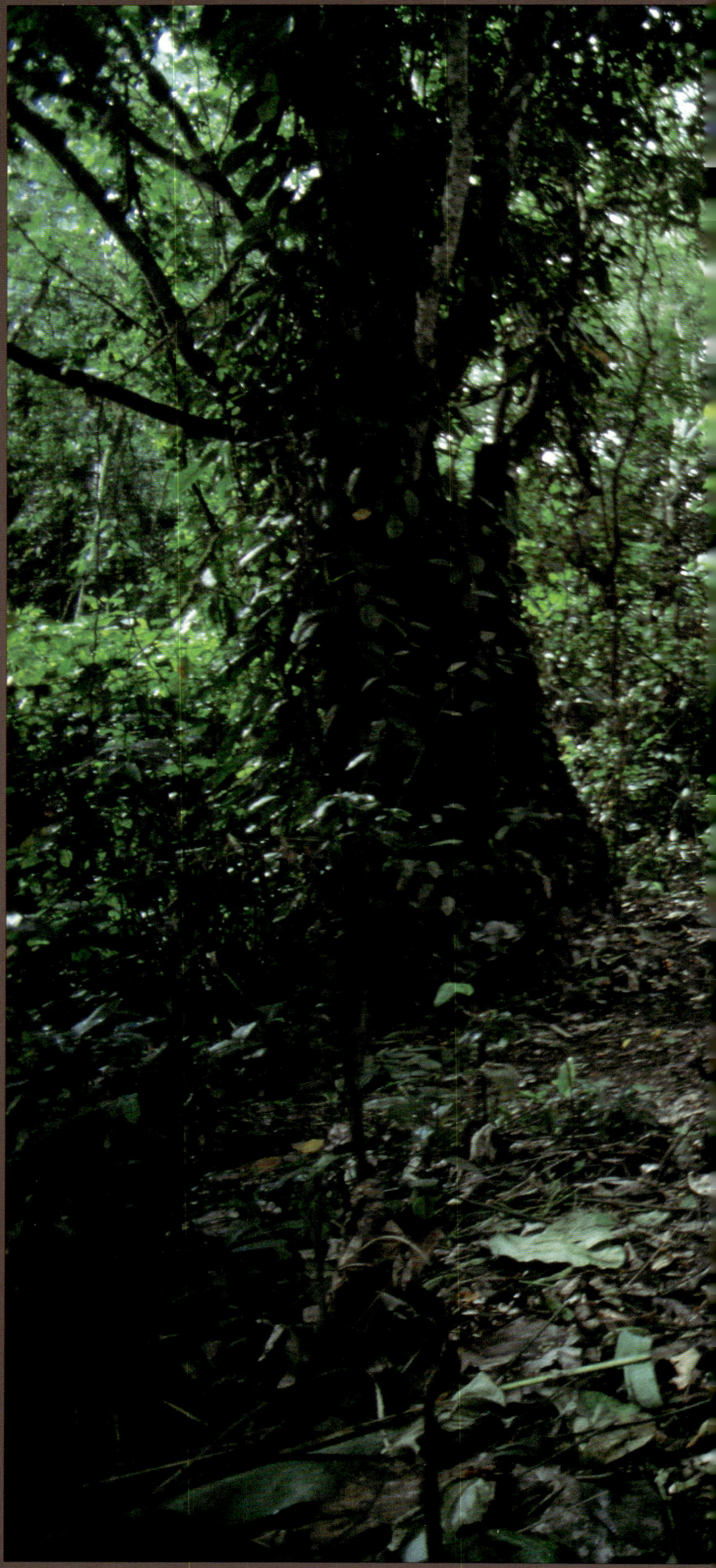

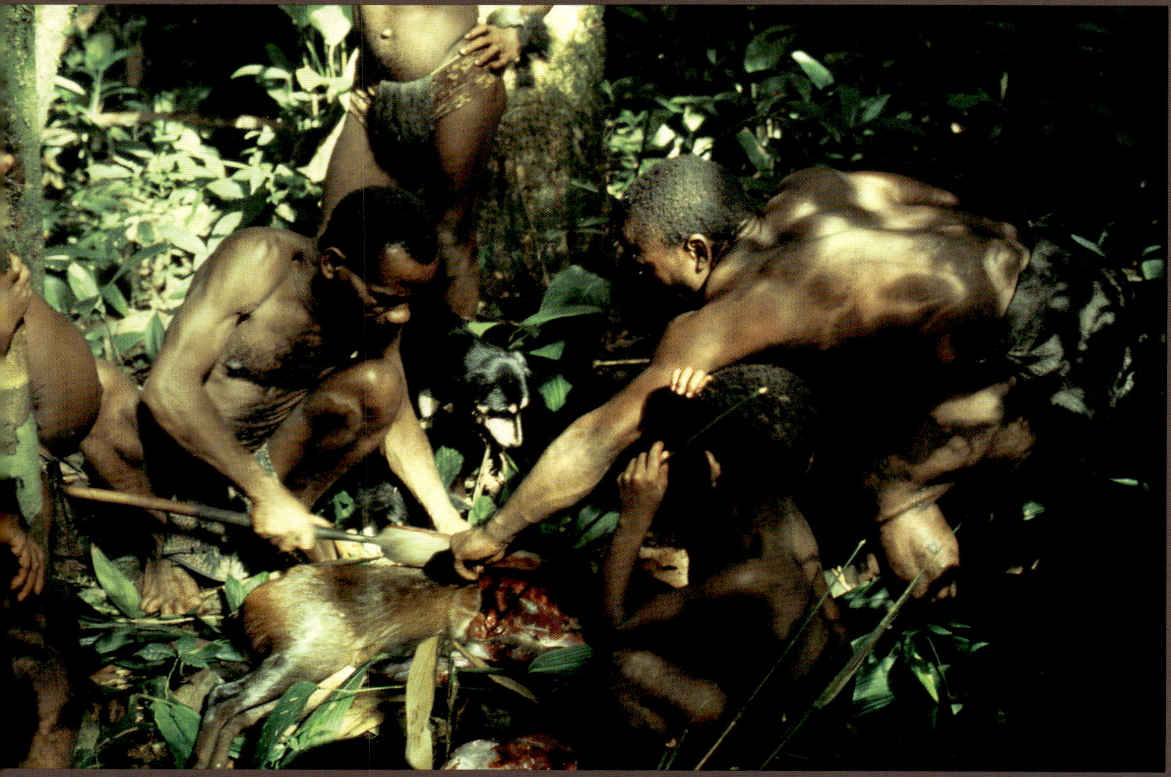

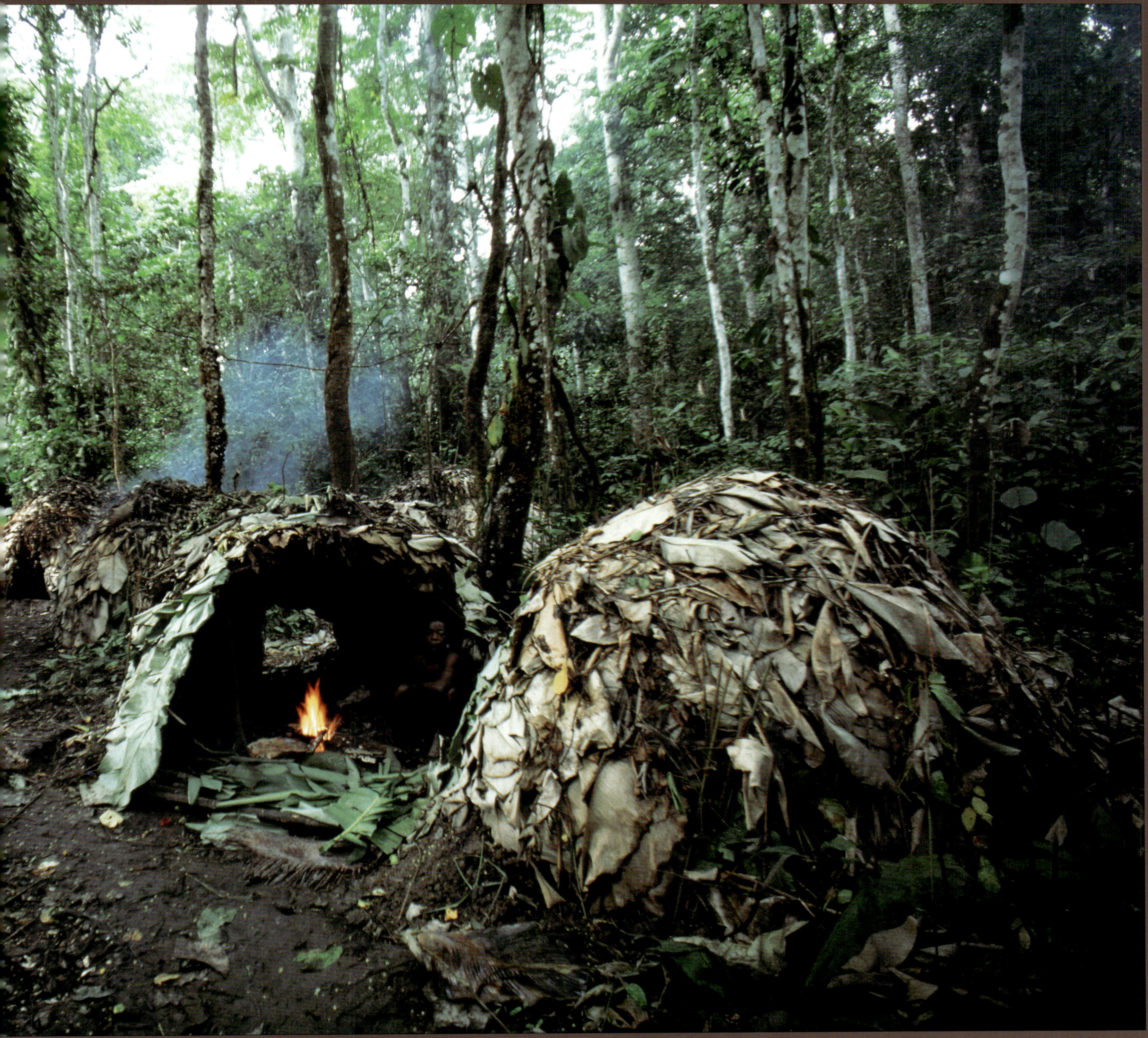

The Pygmies of Central Africa's rain forest are the smallest humans on Earth and one of the few remaining true forest-dwelling peoples. They are not restricted to the DRC, but are found in the greatest numbers in that country. They live in bee hive-shaped huts (above) and are hunter-gatherers, living off the plant and animal products of the forest and whatever they trade with their agricultural neighbors. Here, a man (upper left) displays a hinged-back tortoise (Kinixys sp.) collected in the forest, and a group of hunters butchers a duiker (lower left).

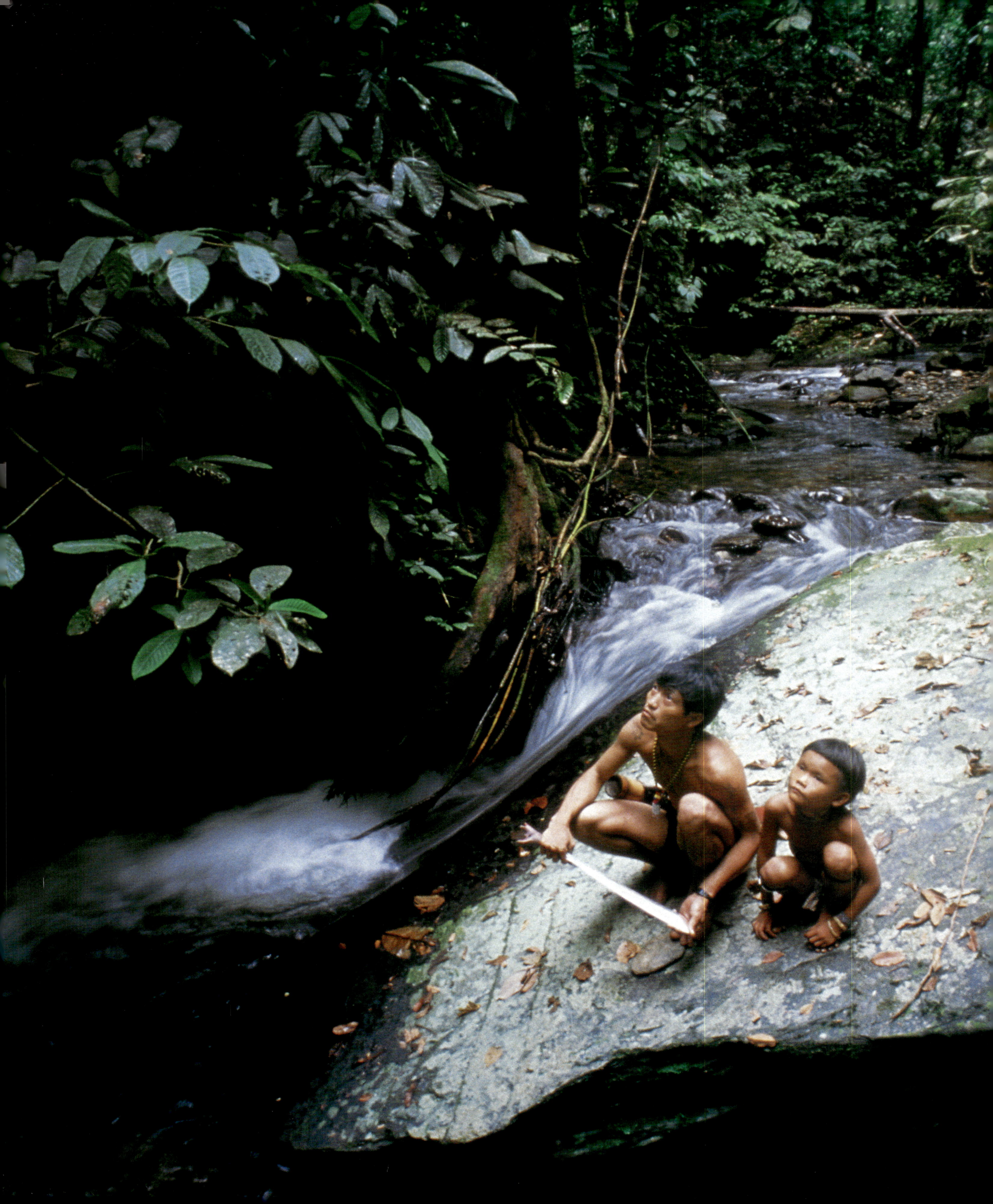

EPILOGUE

In this book, we have dealt mainly with the wealth of diversity that is found in the megadiversity countries. We have emphasized how special these countries are, both in terms of their biological resources and in terms of human cultures, and we have given special attention to their flagship species. However, for a variety of reasons, we have not focused on the threats to biodiversity, the impending extinction spasms that we face, and the many different conservation programs underway to ensure that biodiversity is maintained through the critical twenty-first century. In the next book in this series, which will cover the 27 most threatened biodiversity "hotspots," we will shift the focus substantially and concentrate on those areas at greatest risk and those in which the first major extinction spasms are likely to occur if appropriate measures are not taken in the last few years of this millennium. The biodiversity hotspots are those areas in the terrestrial realm that have the highest concentrations of plant and animal species and, at the same time, are at greatest risk. With few exceptions, they are down to less than 25% of their original natural vegetation, and in quite a few cases, less than 5% remains. Clearly, they must be the principal theaters of operation for the international conservation community over the next five, ten, or twenty years, or our children and grandchildren will inherit a much impoverished planet and quite possibly one on which life as we know it cannot be sustained.

Not surprisingly, fully two thirds of the threatened biodiversity hotspots occur partly or entirely within the borders of the 17 megadiversity countries and, in some cases, they are one and the same. It is especially informative to examine two in which the overlap between megadiversity countries and threatened hotspots is complete, namely the Philippines and Madagascar. They are, in many ways, microcosms of what is happening around the world, with the only difference being that in these two countries the problems are even more acute and the concentrations of biodiversity at risk are even greater than virtually anywhere else.

Threats to the magnificent biodiversity of Madagascar are many, but, without a doubt, habitat destruction in its many different manifestations is the most important. Areas such as the central plateau region of the country are now almost entirely deforested, with most of the remaining natural habitats being found in the coastal areas of the east, west and south. Estimates vary, but roughly 80 to over 90% of Madagascar's natural forest cover has already been lost. The underpinnings of forest destruction in Madagascar are complex and historical. The human population is nearly 14 million people in an area of 587 045 km^2, not very high compared to many other developing countries, and roughly comparable to the population of greater Los Angeles in an area slightly smaller than the State of Texas. Nonetheless, it is growing annually by 3.1%, with a doubling time of 25 years. So the population issue, though not as severe as in many parts of the tropics, is definitely an issue in Madagascar as well.

Another factor is that the people of Madagascar are relative newcomers to their island, arriving there from southeast Asia, Africa, and intermediate points as recently as 1 500 to 2 000 years ago. Rather than evolving more appropriate land-use practices adapted to the delicate Malagasy ecosystems, they imported a mix of traditions such as rice-growing from Southeast Asia and cattle-raising from East Africa that have had a devastating impact on the country's biodiversity.

The rates of forest loss in Madagascar have been enormous, with an estimated 111 000 ha lost per year in the eastern rain forest region from 1950 to 1985 (Green and Sussman, 1990). Slash-and-burn agriculture is a problem wherever natural habitats remain, and is probably the single greatest cause of forest loss. Madagascar's lateritic soils, with their low fertility, are not suited to agricultural uses like slash-and-burn agriculture. Exposing them to the ravages of erosion in this way has already turned much of the country's central plateau into a lifeless moonscape of baked red earth, giving rise to one of Madagascar's popular, albeit tragic, nicknames, *La Grande Île Rouge* (The Great Red Island). The results of this misuse of land have been disastrous. According to a recent World Bank estimate, some 100-300 million dollars of future potential is being lost each year in Madagascar due to widespread erosion. And the quality of life is not improving; a *Human Development Report* for 1993, published by the United Nations Development Programme (UNDP) ranks Madagascar 123rd out of 173 countries on a human development index that incorporates data on poverty, quality of life and GNP.

The case of rice helps illustrate some of Madagascar's critical land-use problems. Rice is the mainstay of Madagascar, and its cultivation is the primary livelihood of 70% of the country's population; indeed, the Malagasy have the greatest per capita rice consumption of any country on Earth. Paddy rice is grown in the valleys and low plains, whereas dry rice is cultivated mostly on deforested hillsides. Traditionally, about 70% of Madagascar's rice is grown in paddies, which can be

On the opposite page, Penan people in the rain forest in Mulu National Park in Sarawak. The Penan are nomadic hunter-gatherers whose lifestyle is being severely threatened by large-scale logging operations and government efforts to integrate them into modern Malaysian society.

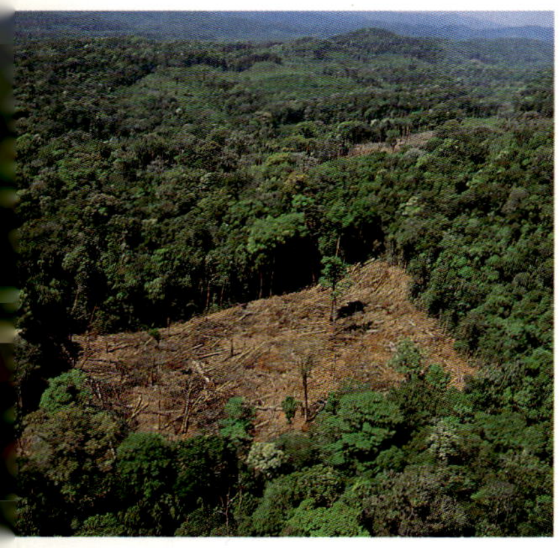

We have already lost more than 60% of the Earth's primary tropical rain forest, the most diverse ecosystem on the planet.

maintained year after year and have relatively high productivity. Dry rice, on the other hand, is grown by slash-and-burn methods and a deforested slope will produce for only two to three years. This practice is not only disastrous for the forest and its wildlife, but also exacerbates erosion and destroys the watershed for paddy rice in the valleys below.

Another subsistence-level land-use issue is fuelwood. Every year, large areas of natural forest are cut down to provide firewood and charcoal for cooking. This problem is especially severe in the southern spiny desert region, where very poor Antandroy tribesmen convert large areas of slow-growing Didiereaceae bush into bags of charcoal and stacks of firewood for sale to townspeople. Needless to say, such use of these globally unique forests, with 95% plant endemism at the species level and close to 50% at the generic level, is very sad indeed.

The cattle-raising tradition is yet another problem. There are as many cattle in Madagascar as there are people. The effects of overgrazing, clearance of forest for cattle pasture, and periodic burning to provide the "green bite" of new grass have been at least as bad as on the savannas of East Africa. As in East Africa, the cattle-raising people of Madagascar are very closely tied to their herds, and changing this destructive tradition will be most difficult.

Hunting is an issue as well. As in most other poor countries, bushmeat is part of the diet of local people, and certain species are under particularly heavy pressure. In Madagascar, animals that are heavily impacted by hunting include many of the lemur species, a variety of birds, and some of the larger reptiles like tortoises, freshwater turtles, and marine turtles.

Major international development programs funded by a variety of bilateral and multilateral donors have also wreaked havoc on Madagascar's environment. These have been grand in scope and short on careful planning, and have paid little or no attention to the environment or to the local people they were theoretically intended to help.

Mining is also an issue. Madagascar is rich in minerals, including some very valuable ones like titanium and gold. In the past, mining activities have paid little attention to the health of the environment, and the impacts have at times been disastrous. Some of the international mining companies have started to become more environmentally responsible, but a great deal more needs to be done.

In spite of what little remains, certain predatory international logging companies are still trying to obtain major concessions in Madagascar. Just this past year, a Malaysian conglomerate attempted to obtain four major logging concessions in different parts of the country. Fortunately, one of these was rejected and a 300 000-ha national park on the critically-important Masoala Peninsula was created instead, but the other three are still pending. This sort of exploitation of the remnants of globally-important megadiversity countries and hotspot ecosystems has no place at the end of the twentieth century.

Due to these negative impacts on the environment, Madagascar has already suffered some of the most severe extinction spasms in recent times. Since the arrival of humans on the island, eight genera and at least 15 species of lemur have gone extinct, representing almost a third of all known species (living and extinct) and 36% (8 of 22) of the known lemur genera. Disappearing with them were several other creatures, including at least one species of giant tortoise, a pygmy hippopotamus, and an aardvark, and the spectacular elephant birds, which included one species, *Aepyornis maximus*, that reached 450-500 kg and was the largest bird that ever lived. And the extinction spasms have not ended, since many of Madagascar's living species are in the endangered or critically endangered categories.

Unfortunately, the situation in the Philippines is even worse than in Madagascar. The Philippines at 300 780 km^2 is only 51% the size of Madagascar, yet it has a population of over 73 million, more than five times that of Madagascar. Furthermore, consumption patterns are increasing more rapidly in the Philippines, meaning that the average impact of each Filipino is greater than that of his or her Malagasy counterpart.

The practice most damaging to the Philippine environment has been massive logging in the recent past. As late as 1945, around 60 to 65% of the country was still covered by old-growth forest, but the rate of logging accelerated quickly after World War II, with forest cover dropping to 55% in 1950, 30% by 1975, 20% by 1988, and no more than about 8% today, and probably much less if one eliminates secondary forest formation from the equation. From 1969 to 1988, forest loss averaged 210 000 km^2 per year, almost double that of Madagascar in the comparable period and three times the global average for tropical rain forest. As if the logging itself were not bad enough, the secondary effect of building roads that opened up previously inaccessible areas was at least as important, since landless subsistence farmers quickly poured into such areas.

Subsistence farmers typically have low-protein diets and, as in Madagascar, rarely pass up an opportunity to catch and eat any wild animal that comes along. Although there is no quantitative data on this in the Philippines, it must have great cumulative impact, especially given how restricted and fragmented remaining wildlife habitat is. Commercial hunting plays a smaller, but nonetheless significant role, and is directed primarily towards catching animals for export for the pet trade or for medical research.

The Philippines also has some of the richest marine ecosystems on Earth, and these are exploited by some 10 million subsistence fishermen. These people lived sustainably when their numbers were much smaller, but even the high productivity and large areas of coral reefs and other productive waters are no longer able to support so many. And heavy overfishing by commercial boats, often owned by foreigners, has depleted the fish resources even more. On top of this, some of the capture techniques in use are extremely destructive and include dynamiting of reefs and the use of poisons.

As a result of these many pressures, the Philippines and Madagascar rank very high in number of threatened species (Introduction, Tables 26-27), and are clearly among the highest-priority countries for biodiversity conservation. When their extremely high levels of endemism and diversity per unit area are superimposed onto the tiny natural areas that remain, it becomes clear that every little bit of what is left is critically important both in national and in global terms. Although both of these countries have been a focus of international conservation activity over the past 10-15 years, much is yet to be done. Only 1.9% of Madagascar's land area is in protected areas, and in the Philippines it is only 1.3%. In addition, only a small portion of this already very restricted area is adequately protected, and many of the parks and reserves in both countries receive no significant protection whatsoever. Given the fundamental importance of these areas in global efforts to conserve biodiversity, there needs to be much more of an effort to increase the area under official protection and to more effectively implement protective legislation already in place.

We have highlighted two of the most threatened megadiversity countries because of their great importance and their high risk factor. Nonetheless, what is happening to them is also taking place in most of the other megadiversity countries, and especially in the hotspot ecosystems within them. Among other megadiversity country-based hotspots of particular concern are the slopes of the tropical Andes in Colombia, Venezuela, Ecuador and Peru; the Chocó/Darién/Western Ecuador region of Colombia and Ecuador; the Atlantic forest region and the Cerrado of Brazil; the rain forests and dry forests of Mexico; the California Floristic Province of the United States and northern Mexico; the Cape Floristic Province and Western Cape of South Africa; the Western Ghats of India; virtually all of Malaysia and Indonesia west of New Guinea; and the southwestern and southeastern corners of Australia. Although for the most part these are somewhat better off than Madagascar and the Philippines, there are portions of several of these hotspots that are at least as endangered as these two countries and perhaps more so.

Countering the many negative trends that we see around the world are quite a few positive ones as well. Over the past few years, and especially since the Earth Summit in Rio de Janeiro in 1992, the issue of biodiversity conservation has assumed greater importance in international affairs. Although much of the self-appraisal taking place right now, five years after the Rio Summit, is negative, this gathering of international leaders, the largest of its kind in history, had significant impact and introduced the term "biodiversity" into the geopolitical lexicon. Since that time, there has been a significant increase in support for biodiversity conservation activities from institutions like the World Bank, the Inter-American Development Bank, the European Union, the United Nations Development Programme (UNDP), and many bilateral agencies like USAID, German GTZ and KfW, British ODA, Dutch foreign assistance and the aid agencies of the Scandinavian countries. For their part, the Japanese have become more involved and are now starting to incorporate concepts like hotspots into their Overseas Development Assistance. The creation of the Global Environment Facility (GEF), a direct output of the Earth Summit and a joint effort of the World Bank, UNDP and the United Nations Environment Programme (UNEP), has also been significant. The GEF's delivery mechanisms left much to be desired early on and are still in need of improvement, but it has invested well over a billion dollars in biodiversity-related activities over the past five years.

Actions of these international donors have been supplemented by a number of private foundations, especially in the United States, with some like the MacArthur Foundation and the W. Alton Jones Foundation playing a particularly important role. Foundations interested in biodiversity are especially catalytic, since they invest their funds with far less bureaucracy than their government counterparts, and consequently get much more impact for each dollar invested. Finally, and perhaps most important, has been the increasing role of the private sector since 1992. Many corporations are beginning to recognize that the environment and especially biodiversity conservation are fundamentally important to economic growth and to their own bottom line, and are supporting more and more international projects. Some are doing this purely for image, in recognition of the fact that being green can be good for corporate revenues; others are doing it because their leaders believe that it is simply the right thing to do. Companies like the latter, which have shown both great vision and a responsibility to society, are still few in number, but they are increasing and are perhaps the greatest source of optimism for the future. The corporate sponsor of this book, CEMEX, is one such foresighted company. The fact that some of these companies are based in the tropical megadiversity countries is especially gratifying and a very positive sign for the future. Once again, we hope that this book will help to increase awareness, and stimulate this growing interest in and movement on behalf of biodiversity still further.

In the next book in this series, entitled *Hotspots: Earth's Most Endangered Ecosystems*, we will examine the threatened hotspot ecosystems, go into more detail on the problems that they face, and begin the process of outlining strategies for their long-term survival. Thanks again to CEMEX for making this important series on global biodiversity possible, and for ensuring that it is accessible to so many people.

RUSSELL A. MITTERMEIER
CRISTINA GOETTSCH MITTERMEIER
PATRICIO ROBLES GIL

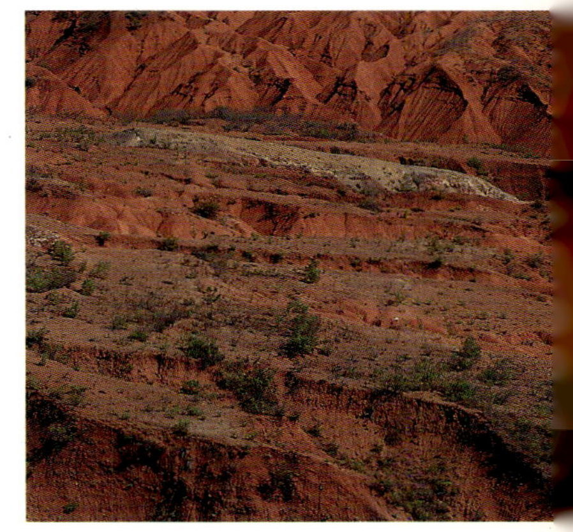

In the forests that have lost their vegetation, the rains wash away the soil, causing severe erosion. Thus the soil is left completely infertile.

On pp. 490-491, a terraced rice field in Bali. Much of the Earth's most fertile land has been converted for agriculture and other human uses.

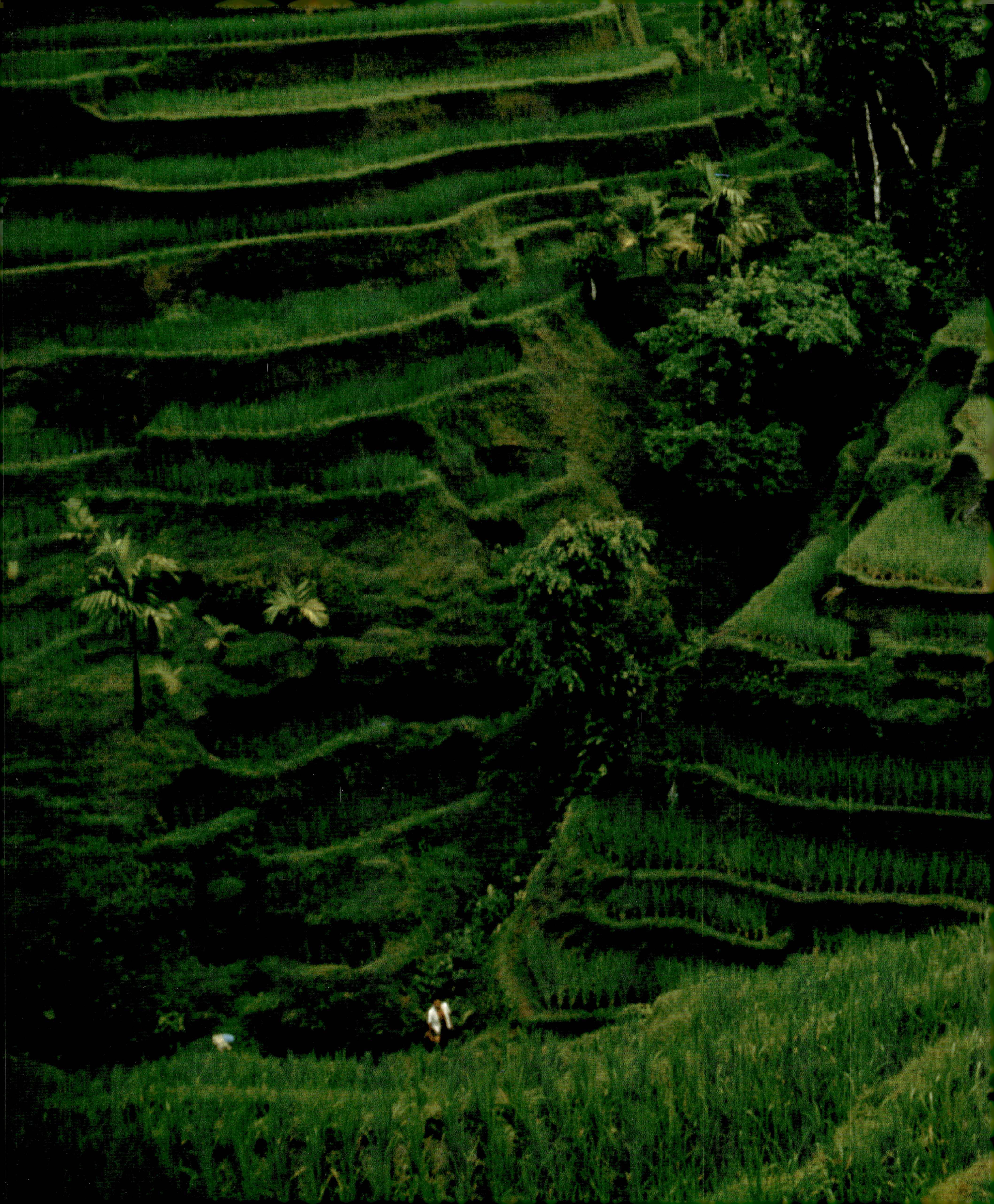

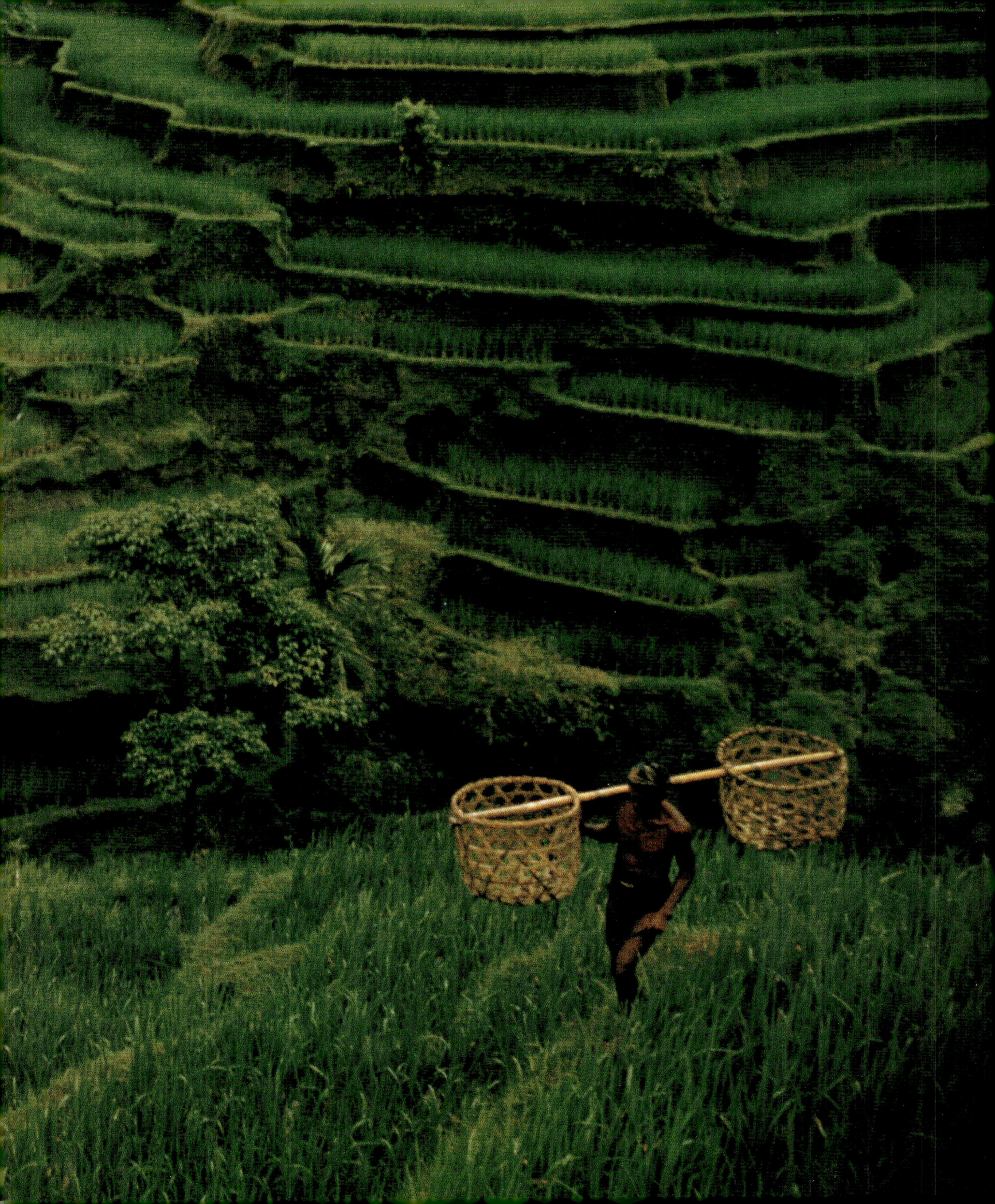

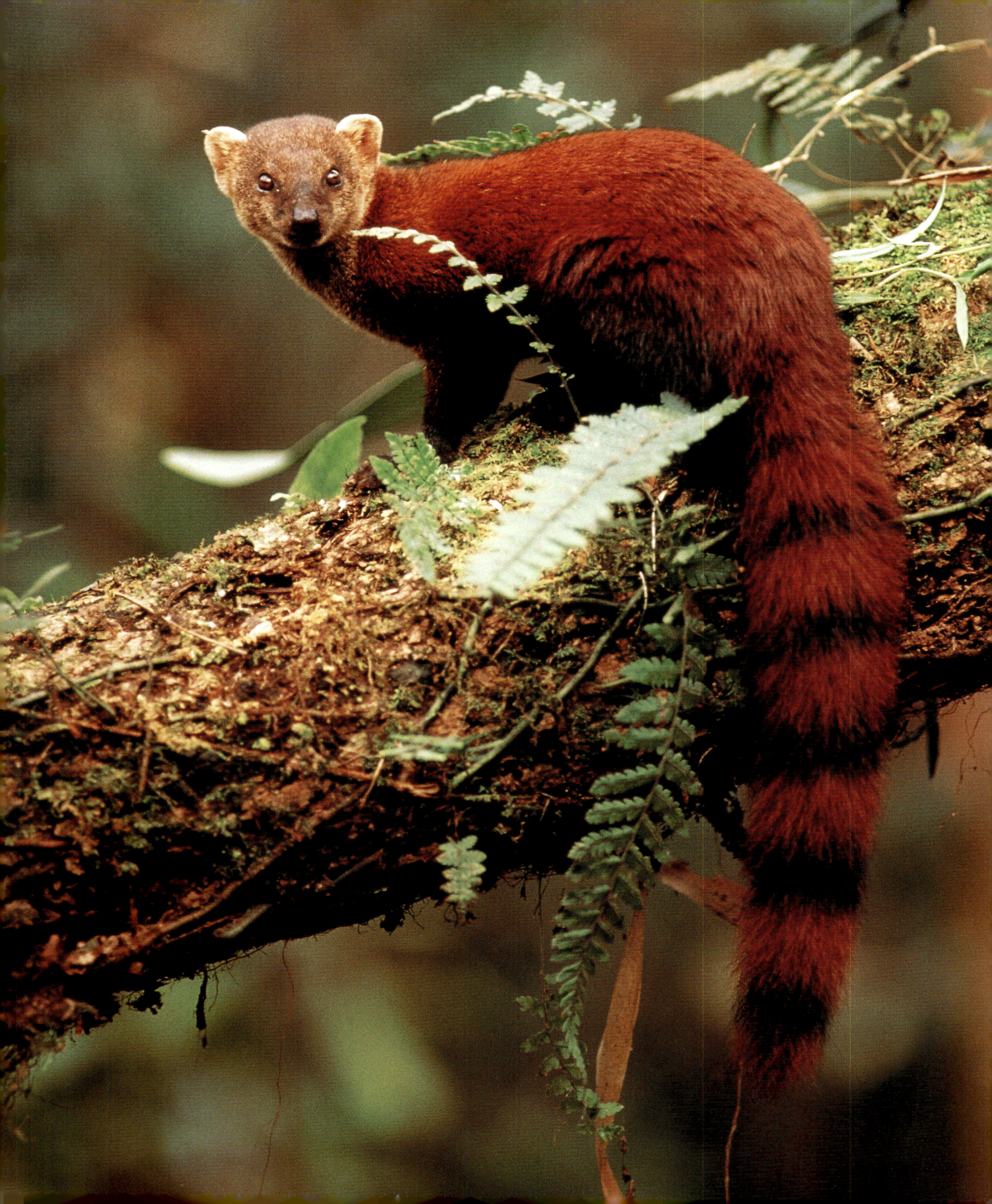

BIBLIOGRAPHY

ALBUJA, V.L. 1991. Lista de vertebrados del Ecuador: mamíferos. *Politécnica* 16:163-203.

ALBUJA, V.L. and B.D. PATTERSON. 1996. A new species of northern shrew-opossum (Paucituberculata: Caenolestidae) from the Cordillera del Cóndor, Ecuador. *Journal of Mammology* 77:41-53.

ALCALÁ, A.C. 1986. *Guide to the Philippine Flora and Fauna*. Vol. 10, Amphibians and Reptiles. Natural Resources Management Center and University of the Philippines, Manila.

ALCALÁ, A.C. and C.C. CUSTODIO. 1987. The status of endemic Philippine amphibians. *Sylvatrop* 5:72-86.

ALCORN, B.J. 1993. *Papua New Guinea Conservation Needs Assessment*. Government of Papua New Guinea, Department of Environment and Conservation, Boroko.

ALLAN, T. and A. WARREN. 1993. *Deserts. The Encroaching Wilderness. A World Conservation Atlas*. Oxford University Press, New York.

ALLEN, G.R. 1991. *Field Guide to the Freshwater Fishes of New Guinea*. Christensen Research Institute, Madang.

ALLEN, G.R. and N.J. CROSS. 1982. *Rainbow Fishes of Papua New Guinea and Australia*. T.F.H. Publications, New Jersey.

ALLEN, G.R. and R. SWAINSTON. 1993. *Reef Fishes of New Guinea. A Field Guide for Divers, Anglers and Naturalists*. Publication No. 8, Christensen Research Institute, Madang.

ALLISON, A. 1993. Biodiversity and conservation of fishes, amphibians, and reptiles of Papua New Guinea. In: B. BEEHLER. (Ed.), *A Biodiversity Analysis for Papua New Guinea*. Biodiversity Support Program, Washington, D.C.

ANONYMOUS. 1991. Cheetah sighting in India. *Oryx* 25:69.

ANONYMOUS. 1994. *China's Agenda 21*. China Environmental Science Press, Beijing.

ANDREW, P. 1992. *The Birds of Indonesia: A Checklist (Peter´s Sequence)*. Kulia Checklist No. 1, Indonesian Ornithological Society, Jakarta.

ASCORRA, C.F., D.L. WILSON, and M. ROMO. 1991. *Lista anotada de los quirópteros del Parque Nacional Manú, Perú*. Publicaciones del Museo de Historia Natural, Universidad Nacional Mayor de San Marcos, Lima, 42:1-14.

ATSIC. 1992. *Indigenous Australia*. Informational kit produced by Aboriginal and Torres Strait Islander Commission, Canberra.

BAILEY, R.G. 1995. *Descriptions of the Ecoregions of the United States*. Miscellaneous Publication 1391, Forest Service, U.S. Department of Agriculture, Washington, D.C.

BANCO DE OCCIDENTE (Colombia). 1991. *El bosque nublado de Colombia*. Gráficas Bogotá, Bogotá.

BANCROFT, N. 1995. *Warriors, Warfare and the Native American Indian*. Salamander, London.

BARLOW, C. and J. HARDJONO (Eds.). 1995. *Indonesia Assessment 1995. Development in Eastern Indonesia*. Research School of Pacific and Asian Studies, Australian National Univ., Canberra.

BARLOW, H.S. 1982. *An Introduction to the Moths of Southeast Asia*. Malayan Nature Society, Kuala Lumpur.

BARRERA-GUEVARA, J.C. 1992. Biología reproductiva de la totoaba, *Totoaba macdonaldi* (Gilbert, 1891) (Teleostei: Sciaenidae). Instituto Tecnológico y de Estudios Superiores de Monterrey, Guaymas (Sonora, Mexico).

BARRY, T. (Ed.). 1992. *Mexico: A Country Guide*. Inter-Hemispheric Education Resource Center, Albuquerque.

BEATTIE, A.J. (Ed.). 1995. *Australia's Biodiversity Living Wealth*. Reed Books, Chatswood.

BEAZLEY, M. 1981. *International Book of the Forest*. Simon & Schuster, New York.

BEEHLER, B.M. 1991. *A Naturalist in New Guinea*. University of Texas Press, Austin.

BEEHLER, B.M., T.K. PRATT, and D.A. ZIMMERMAN. 1986. *Birds of New Guinea*. Princeton University Press, Princeton.

BENÍTEZ, L. and A. GARCÉS. 1993. *Culturas ecuatorianas de ayer y hoy*. Ediciones Abya-Yala, Cayambe.

BERNARDES, A.T., A.B.M. MACHADO, and A.B. RYLANDS. 1990. *Fauna brasileira ameaçada de extinção*. Fundação Biodiversitas/IBAMA, Belo Horizonte.

BIANCHI y AA. VV., C. 1982. *Artesanías y técnicas shuar. Jivaro Indians from Ecuador*. Ediciones Mundo Shuar, Quito.

BIBBY, C.J., M.J. CROSBY, M.F. HEATH, T.H. JOHNSON, A.J. LONG, A.J. STATTERSFIELD, and S.J. THIRGOOD. 1992. *Putting Biodiversity on the Map. Global Priorities for Conservation*. ICBP, Cambridge.

BLAKERS, M., S.J.J.F. DAVIES, and P.N. REILLY. 1984. *The Atlas of Australian Birds*. Melbourne University Press, Melbourne.

BOGERT, C.M. and M. DEL CAMPO. 1993. *The Gila Monster and Its Allies* (and a retrospective essay by Daniel D. Beck). Society for the Study of Amphibians and Reptiles, Oxford.

BOWANDER, B. 1982. Deforestation in India. *International Journal of Environmental Studies* 18: 223-236.

BRAKO, L. and J.L. ZARUCCHI. 1993. Catalogue of the flowering plants and gymnosperms of Peru. *Monographs in Systematic Botany* 45: 1286.

BRAKO, L. and J.L. ZARUCCHI. 1996. *Catalogue of the Flowering Plants and Gymnosperms of Peru. Catálogo de las angiospermas y gimnospermas del Perú*. Missouri Botanical Garden, Saint Louis.

BRAUN, B., P.G. ROE, and A. MEKLER (Eds.). 1995. *Arts of the Amazon*. Thames and Hudson, London.

BROAD, R. and J. CAVANAGH. 1993. *Plundering Paradise. The Struggle for the Environment in the Philippines*. University of California Press, Berkeley.

BROCKELMAN, W.Y. and D.J. CHIVERS. 1984. Gibbon conservation: Looking to the future. In: H. PREUSCHOFT, D.J. CHIVERS, W.Y. BROCKELMAN, and N. CREEL (Eds.), *The Lesser Apes: Evolution and Behavioral Biology*. Edinburgh University Press, Edinburgh.

BROKENSHA, P. 1975. *The Pitjantjatjara and Their Crafts*. The Aboriginal Arts Board, Alice Springs (Australia).

BROWN, G. 1993. *The Great Bear Almanac*. Lyons and Burford, New York.

BROWN, K. 1979. *Ecologia geográfica e evolução nas florestas neotropicais*. Universidade Estadual de Campinas, Brasil.

BUSTARD, R. 1972. *Sea Turtles. Their Natural History and Conservation*. Collins, London.

BYE, R. 1993. The role of humans in the diversification of plants in Mexico. In: T.P. RAMAMOORTHY, R. BYE, A. LOT, and J. FA (Eds.), *Biological Diversity of Mexico: Origins and Distribution*. Oxford University Press, New York, pp. 707-731.

CALDECOTT, J.O., M.D. JENKINS, T.H. JOHNSON, and B. GROOMBRIDGE. 1996. Priorities for conserving global species richness and endemism. *Biodiversity and Conservation* 5:699-728.

CAMPBELL, D.G. and H.D. HAMMOND (Eds.). 1989. *Floristic Inventory of Tropical Countries*. New York Botanical Garden/World Wildlife Fund, New York.

CAMPBELL, J.A. and W.W. LAMAR. 1989. *The Venomous Reptiles of Latin America*. Comstock Publishing Associates, New York.

CAPELL, A. 1969. *A Survey of New Guinea Languages*. Sydney University Press, Sydney.

CARPANETO, G.M. and F.P. GERMI. 1989. The mammals in the zoological culture of the Mbuti pygmies in north-eastern Zaire. *Hystrix* 1:1-83.

CARRIZOSA-UMAÑA, J. and J. HERNÁNDEZ-CAMACHO. 1990. *Selva y futuro. Colombia*. Sello Editorial, Bogotá.

CARVALHO FILHO, A. 1994. *Peixes da costa brasileira*. Editora Marca d'Agua, São Paulo.

CAYOT, L. 1981. *Tortoise Plant Interactions on Santa Cruz*. Charles Darwin Research Station, annual report, Quito, pp. 94-95.

CDC-UNALM. 1991. *Plan director del Sistema Nacional de Unidades de Conservación (SINUC), una aproximación desde la diversidad biológica (propuesta CDC-UNALM)*. Centro de Datos para la Conservación, Universidad Nacional Agraria, La Molina, Lima.

CDC-UNALM. 1992. *Estado de conservación de la diversidad natural de la región noroeste del Perú*. Centro de Datos para la Conservación, Universidad Nacional Agraria, La Molina, Lima.

CEBALLOS, G. and F. ECCARDI. 1993. In: P. ROBLES GIL (Ed.), *Mexican Diversity of Fauna*. Cemex, Mexico.

CHADWICK, D.H. 1995. Dead or alive: The Endangered Species Act. *National Geographic* 187: 2-41.

CHAMPION, H.G. and S.K. SETH. 1968. *A Revised Survey of the Forest Types of India*. Government of India Press, Delhi.

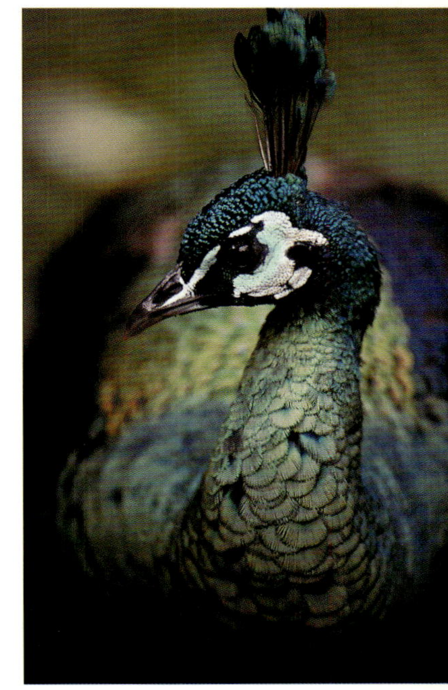

On the opposite page, the ring-tailed mongoose (Galidia elegans) *from Madagascar, one of the world's most threatened biodiversity hotspots.*

Above, black-winged peafowl (Pavo cristatus *mut.* nigripennis). *All peacocks' calls are loud and unpleasant, in contrast with their beautiful courtship dances, where the male displays his long, round tail.*

CHAPMAN, J.A. and J.E.C. FLUX (Eds.). 1990. *Rabbits, Hares and Pikas. Status Survey and Conservation Action Plan.* IUCN/SSC Lagomorph Specialist Group, Gland (Switzerland).

CHEN, L., X. WANG, and S. WANG (Eds.). 1992. *Chinese Biodiversity. Status and Conservation Strategy.* Chinese Science Press, Beijing.

CHINA WILDLIFE CONSERVATION ASSOCIATION. n.d. *Wildlife of China.* China Forestry Publishing House, Beijing.

CHIVERS, D.J. 1977. The lesser apes. In: H.S.H. PRINCE RANIER and G. BOURNE (Eds.), *Primate Conservation.* Academic Press, New York.

COATES, B.J. 1990. *The Birds of Papua New Guinea.* Dove Publications, Alderley (Australia).

COGGER, G.H. 1992. *Reptiles and Amphibians of Australia.* 5th ed., Reed Books, Sydney.

COIMBRA-FILHO, A.F. and I. DE GUSMÃO-CÂMARA. 1996. *Os limites originais do bioma Mata Atlântica na região nordeste do Brasil.* Fundacão Brasileira para a Conservação da Natureza, Brasil.

COLBURN, T.D., J.P. DUMANOSKI, and J.P. MYERS. 1996. *Our Stolen Future.* Penguin Books, New York.

COLLAR, N.J., M.J. CROSBY, and A.J. STATTERSFIELD. 1994. *Birds to Watch 2.* BirdLife Conservation, series 4: 407.

COLLINS, N.M. and M.G. MORRIS. 1985. *Threatened Swallowtail Butterflies of the World. The IUCN Red Data Book.* IUCN, Gland (Switzerland) and Cambridge (U.K.).

COLLINS, N.M., J.A. SAYER, and T.C. WHITMORE. 1991. *The Conservation Atlas of Tropical Forests: Asia and the Pacific.* MacMillan, London.

COMMON, I.F.B. and D.F. WATERHOUSE. 1981. *Butterflies of Australia.* Angus and Robertson, East Melbourne.

CONNIF, R. 1996. Tarantulas. *National Geographic* 190: 98-115.

CONSERVATION INTERNATIONAL. 1990. *Biodiversity at Risk.* A Map produced by Conservation International, Washington, D.C.

CONSERVATION INTERNATIONAL. 1990. *The Rain Forest Imperative.* Conservation International, Washington, D.C.

COOMBER, J. 1991. *The Orchids of Java.* Royal Botanic Gardens, Kew.

COOPER ORNITHOLOGICAL SOCIETY. 1957. *Distributional Checklist of the Birds of Mexico.* The Cooper Ornithological Society, Berkeley.

CORBETT, A.S. and H.M. PENDLEBURY. 1978. *The Butterflies of the Malay Peninsula.* 3rd ed. (revised by J.N. Eliot), Malayan Nature Society, Kuala Lumpur.

COVENEY, J. 1993. *Australia's Conservation Reserves.* Cambridge University Press, Melbourne.

CRAIGHEAD, JR., F.C. 1979. *Track of the Grizzly.* Sierra Club, San Francisco.

CRUXENT, J.M. and I. ROUSE. 1982. *Arqueología cronológica de Venezuela.* Armitano, Caracas.

CUATRECASAS, J. 1958. Aspectos de la vegetación natural de Colombia. *Revista de la Academia Colombiana de Ciencias Exactas, Físicas y Naturales* 10:221-264.

CUMMING, D.H.M., R.F. DU TOIT, and S.N. STUART (Eds.). 1990. *African Elephants and Rhinos. Status Survey and Conservation Action Plan.* IUCN-The World Conservation Union/SSC African Elephant and Rhino Specialist Group, Gland (Switzerland).

CUSTODIO, C.C., M.V. LEPITEN, and L.R. HEANEY. 1996. *Bubalus mindorensis. Mammalian Species* 520:1-5.

D'ABRERA, B. 1977. *Birdwing Butterflies of the Australian Region.* Lansdowne, Melbourne.

DALTON, B. 1995. *Indonesia Handbook.* Moon Publications, Chico, CA.

DAVIS, I.L. 1972. *A Field Guide to the Birds of Mexico and Central America.* University of Texas Press, Austin.

DAVIS, W. and T. HENLEY. 1990. *Penan Voice for the Borneo Rainforest.* WCWC/WILD, Vancouver.

DAY, D. 1989. *Vanished Species.* Gallery Books, New York.

DEFLER, T. 1995. La conservación de los primates en Colombia. *Trianea* 5:255-287.

DEKKER, R.W.R.J., P.J.K. MCGOWAN, and WPA/BIRDLIFE/SSC MEGAPODE SPECIALIST GROUP (Eds.). 1995. *Megapodes. An Action Plan for Their Conservation 1995-1999.* IUCN-The World Conservation Union, Gland (Switzerland).

DENEVAN, W.M. 1992. The pristine myth: The landscape of the Americas in 1492. *Annals of the Association of American Geographers* 82: 369-385.

DEVELOPMENTAL ALTERNATIVES. 1992. *An Aerial Reconnaissance of Closed Canopy Forests.* Natural Resources Management Program, Manila.

DHARMA, B. 1988. *Siput dan Kerang Indonesia (Indonesian Shells).* Sarana Graha, Jakarta.

DHARMA, B. 1992. *Siput dan Kerang Indonesia II (Indonesian Shells II).* Christa Hemmen, Wiesbaden.

DINERSTEIN, E., D.M. OLSON, D.J. GRAHAM, A.L. WEBSTER, S.A. PRIMM, M.P. BOOKBINDER, and G. LEDEC. 1995. *A Conservation Assessment of Latin America and the Caribbean.* The World Bank/The World Wildlife Fund, Washington, D.C.

DIRZO, R. 1994. In: P. ROBLES GIL (Ed.), *Mexican Diversity of Flora.* Cemex, Mexico.

DODSON, C.H. and A.H. GENTRY. 1991. Biological extinction in western Ecuador. *Annals Missouri Botanical Garden* 78:273-295.

DOUROJEANNI, M.J. 1988. *Si el árbol de la quina hablara.* Fundación Peruana para la Conservación de la Naturaleza, Lima.

DOUROJEANNI, M.J. and C.F. PONCE. 1978. *Los parques nacionales de Perú.* INCAFO, Madrid.

DUFFY, K. 1985. *Children of the Forest.* Dodd Mead, New York.

DUGAN, P. 1993. *Wetlands in Danger: A World Conservation Atlas.* Oxford University Press, New York.

DUNCAN, P. (Ed.). 1992. *Zebras, Asses, and Horses: An Action Plan for Their Conservation.* IUCN-The World Conservation Union/SSC Equid Specialist Group, Gland (Switzerland).

DURNING, A.T. 1993. Supporting indigenous peoples. In: LESTER BROWN *et al.*, *State of the World: A Worldwatch Institute Report on Progress Toward a Sustainable Society.* W.W. Norton, New York.

DYNE, G.R. and D.W. WALTON (Eds.). 1987. *Fauna of Australia.* General articles, Australian Government Publishing Service, Canberra.

ELLIS, J.A. 1995. *Australia's Aboriginal Heritage.* Harper Schools, North Blackburn.

EMMONS, L.H. and F. FEER. 1990. *Neotropical Rainforest Mammals: A Field Guide.* The University of Chicago Press, Chicago.

ERWIN, T.L. 1991. An evolutionary basis for conservation strategies. *Science* 253:750-752.

EUDEY, A.A. (Ed.). 1987. *IUCN/SSC Primate Specialist Group. Action Plan for Asian Primate Conservation: 1987-1991.* IUCN-The World Conservation Union, Lock Haven, PA.

FAO. 1990. *Efectos ecológicos de los* Eucalyptus. Núm. 59, Colección Estudios FAO, Roma.

FELGER, R.S. and M.B. MOSER. 1992. *People of the Desert and Sea: Ethnobotany of the Seri Indians.* The University of Arizona Press, Tucson.

FINN, F. 1980. *Sterndale's Mammalia of India.* Bishen Singh Mahendra Pal Singh, Dehra Dun.

FLANNERY, T.F. 1995. *Mammals of New Guinea.* Reed Books, Chatswood (Australia).

FLANNERY, T.F. 1996a. *The Future Eaters. An Ecological History of the Australasian Land and People.* Reed Books, Chatswood (Australia).

FLANNERY, T.F. 1996b. *Mammals of the South-West Pacific and Moluccan Islands.* Reed, Sydney.

FLANNERY, T.F., R. MARTIN, and A. SZALAY. 1996. *Tree Kangaroos. A Curious Natural History.* Reed Books, Melbourne.

FOLEY, W.A. 1991. *The Yimas Language of New Guinea.* Stanford University Press, Stanford.

FONSECA, G.A.B., DA. 1985. The vanishing Brazilian Atlantic forest. *Biological Conservation* 33: 1-18.

FONSECA, G.A.B., DA, G. HERRMANN, Y.L.R. LEITE, R.A. MITTERMEIER, A.B. RYLANDS, and J.L. PATTON. 1996. *Lista anotada dos mamíferos do Brasil.* Conservation International and Fundação Biodiversitas, Belo Horizonte.

FONTANA, B.L. 1979. *Tarahumara: Where Night is the Day of the Moon.* Photographs by John P. Shaefer. Northland Press, Flagstaff, AZ.

FORESTA, R.A. 1984. *America's National Parks and Their Keepers.* Resources for the Future, Washington, DC.

FREITAG, S., A.O. NICHOLLS, and A.S. VAN JAARSVELD. 1996. Nature reserve selection in the Transvaal, South Africa: What data should we be using? *Biodiversity and Conservation* 5:685-698.

FRITH, C. and D. FRITH. 1995. *Australia's Wet Tropics Rainforest Life.* 2nd ed., Frith and Frith, Queensland.

FUNDACIÓN AMBIENTAL E INPARQUES. 1983. *Los parques nacionales de Venezuela.* Instituto Nacional de Parques (Inparques), Caracas.

GALLINA-T., M.P. and L. SANGRI. 1979. *Bellezas naturales de México.* Centro Iberoamericano de Cooperación, Madrid.

GENTRY, A.H. 1982. Phytogeographic patterns in northwest South America and southern Central America for a Chocó refugium. In: G.T. PRANCE (Ed.), *Biological Diversification in the Tropics.* Columbia University Press, New York, pp. 112-136.

GENTRY, A.H. 1986. Endemism in tropical versus temperate plant communities. In: M.A. SOULÉ (Ed.), *Conservation Biology. The Science of Scarcity and Diversity.* Sinauer Associates, Sunderland, MA.

GENTRY, A.H. 1990. La región amazónica. In: C.C. URIBE *et al.*, *Selva húmeda de Colombia.* Villegas, Bogotá.

GENTRY, A.H. 1991. El bosque nublado de Colombia. In: BANCO DE OCCIDENTE (Ed.), *Bosques de niebla de Colombia.* OP Gráficas Bogotá, Bogotá, pp. 23-52.

GENTRY, A.H. 1992. *Exactra* (Bignoniaceae), a genus from the Chocó region of Ecuador and Colombia. *Systematic Botany* 17:503-507.

GIBBS, R.M. 1986. *The Aborigines.* 2nd ed., Longman Cheshire, Melbourne.

GLAW, F. and M. VENCES. 1994. *A Fieldguide to the Amphibians and Reptiles of Madagascar.* 2nd ed., Zoologisches Forschungsinstitut und Museum Alexander Koenig, Bonn.

GONDELLES, R. 1977. *Los parques nacionales de Venezuela.* INCAFO, Madrid.

GONZALES, P.C. and C.P. REES. 1988. *Birds of the Philippines.* Haribon Foundation for the Conservation of Natural Resources, Manila.

GOULD, H. 1973. *Kangaroos.* Macmillan, Australia.

GREEN, G.M.G. and R.W. SUSSMAN. 1990. Deforestation history of the eastern rain forests of Madagascar from satellite images. *Science* 248: 212-215.

GREEN, S. and K. MINKOWSKI. 1977. The lion-tailed monkey and its south Indian rain forest habitat. In: H.S.H PRINCE RAINIER and G. BOURNE (Eds.), *Primate Conservation*. Academic Press, New York, pp. 290-334.

GREMONE, C., F. CERVIGÓN, S. GORZULA, G. MEDINA, and D. NOVOA. n.d. *Fauna de Venezuela. Vertebrados*. Biosfera, Caracas.

GRESSITT, J.L. 1982. *Biogeography and Ecology of New Guinea*. W. Junk, The Hague.

GREWAL, B. (Ed.). 1988. *Indian Wildlife*. Insight Guides, APA Publications, Singapore.

GROVES, R.H. (Ed.). 1994. *Australian Vegetation*. 2nd ed., Cambridge University Press, Melbourne.

HALLINAN, J.P. 1990. *Revelation of the Malagans. The Ritual Art of New Ireland*. The Gold Coast City Art Gallery, Garry Anderson Gallery, and Collectors Gallery, Sydney.

HAMILTON, L., J. STEYERMARK, J.P. VEILLON, and E. MONDOLFI. 1976. *Conservación de los bosques húmedos de Venezuela*. Sierra Club, Caracas.

HARCOURT, C.S. and J.A. SAYER (Eds.). 1996. *The Conservation Atlas of Tropical Forests: The Americas*. IUCN-The World Conservation Union, Simon & Schuster, New York.

HARPER, P. and E.S. PEPLOW. 1991. *Philippines Handbook*. Moon Publications, Chico, CA.

HEANEY, L.R. 1986. Biogeography of mammals in Southeast Asia: Estimates of rates of colonization, extinction and speciation. *Biological Journal of the Linnean Society* 28:127-165.

HEANEY, L.R. 1991. A synopsis of climatic and vegetational change in Southeast Asia. *Climatic Change* 19:53-61.

HEANEY, L.R. 1993. Biodiversity patterns and the conservation of mammals in the Philippines. *Asia Life Sciences* 2:261-274.

HEANEY, L.R. and P.D. HEIDEMAN. 1987. Philippine fruit bats, endangered and extinct. *Bats* 5:3-5.

HEANEY, L.R. and E.A. RICKART. 1990. Correlations of clades and clines: Geographic, elevational, and phylogenetic distribution patterns among Philippine mammals. In: G. PETERS and R. HUTTERER (Eds.), *Vertebrates in the Tropics*, Museum Alexander Koenig, Bonn, pp. 321-332.

HEANEY, L.R. and R.C.B. UTZURRUM. 1992. A review of the conservation status of Philippine land mammals. *Association of Systematic Biologists of the Philippines Communications* 3:1-13.

HERMES, N. 1990. *Australia's Endangered Wildlife*. Child & Associates, Frenchs Forest.

HERNÁNDEZ-CAMACHO, J.I., A. HURTADO-GUERRA, R. ORTIZ-QUIJANO, and T. WALSBURGER. 1992. Unidades biogeográficas de Colombia. In: G. HALFFTER (Ed.), *La diversidad biológica de Latinoamérica, Acta Zoológica Mexicana*, volumen especial, México.

HESSE, K. and T. AERTS. 1996. *Baining Life and Lore*. University of Papua New Guinea Press, Port Moresby.

HIGGINS, G., L. CRONIN, and J. McDONALD (Eds.). 1988. *Presenting Australia's National Parks*. 2nd ed., Child & Associates, Frenchs Forest.

HOLMES, D. and S.V. NASH. 1989. *Birds of Java and Bali*. Oxford University Press, Kuala Lumpur.

HOLMES, D. and S.V. NASH. 1990. *Birds of Sumatra*. Oxford University Press, Kuala Lumpur.

HOWELL, S.N.G. and S. WEBB. 1995. *A Guide to the Birds of Mexico and Northern Central America*. Oxford University Press, New York.

HUBER, O. 1989. Shrublands of the Venezuelan Guayana. In: L.B. HOLM-NIELSEN, I.C. NIELSEN, and H. BALSLEV (Eds.), *Tropical Forests: Botanical Dynamics, Speciation and Diversity*. Academic Press, London, pp. 271-285.

HUBER, O. 1992. La vegetación. In: O. HUBER (Ed.), *El macizo del Chimantá, Escudo de Guayana, Venezuela: un ensayo ecológico tepuyano*. Oscar Todtmann, Caracas, pp. 161-178.

HUBER, O. 1995a. Geographical and physical features. In: P.E. BERRY, B.K. HOLST, and K. YATSKIEVYCH (Eds.), *Flora of the Venezuelan Guayana*. Missouri Botanical Garden, St. Louis, and Timber Press, Portland, pp. 1-61.

HUBER, O. 1995b. Vegetation. In: P.E. BERRY, B.K. HOLST, and K. YATSKIEVYCH (Eds.), *Flora of the Venezuelan Guayana*. Missouri Botanical Garden, St. Louis, and Timber Press, Portland, pp. 97-160.

HUMBOLDT, A. VON and A. BONPLAND. 1812. *Recueil d'observations de zoologie et d'anatomie comparée*. Vol. 1, F. Schoell and G. Dufour, Paris.

ICBP. 1992. *Putting Biodiversity on the Map: Priority Areas for Global Conservation*. International Council for Bird Preservation, Cambridge (U.K.).

INPARQUES. 1982. *Guía de los parques nacionales y monumentos naturales de Venezuela*. Fundación de Educación Ambiental, Caracas.

INSTITUTE OF BOTANY, C.A.S. 1992. *China Plant Red Data Book. Rare and Endangered Plants*. Science Press China, Beijing.

INSTITUTO GEOGRÁFICO NACIONAL. 1989. *Atlas del Perú*. Lima.

ISAACS, J. 1984. *Australia's Living Heritage*. Lansdowne, Sydney.

IUCN. 1996. *1996 IUCN Red List of Threatened Animals*. J. BAILLIE and B. GROOMBRIDGE (Eds.), IUCN-The World Conservation Union/Species Survival Commission, Gland (Switzerland).

JAIRAJPURI, S.M. 1991. *Animal Resources of India*. Zoological Survey on India, Calcutta.

JI, Z., Z. GUANGMEI, W. HUADONG, and X. JIALIN (Eds.). 1990. *The Natural History of China*. McGraw-Hill, New York.

JOHNS, R.J. 1977. *The Vegetation of Papua New Guinea. Training Manual for the Forestry College*. Forestry College, Bulolo.

JOHNS, R.J. 1982. Plant zonation. In: J. GRESSIT (Ed.), *Biogeography and Ecology of New Guinea*. Part 1. General and physical background; Part 2. Man and his impact on the environment; Part 3. Vegetation and flora. W. Junk, The Hague.

JORGE-PADUA, M.T. and A.F. COIMBRA-FILHO. 1979. *Os parques nacionais do Brasil*. INCAFO, Madrid.

KANE, J. 1996. *Savages*. Random House, Vintage Books, New York.

KARAN, P.P. and M. WEINER. 1996. *India*. World Book Press, Chicago.

KARNOW, S. 1989. *In Our Image: America's Empire in the Philippines*. Random House, New York.

KEAST, A., R.L. CROCKER, and C.S. CHRISTIAN (Eds.). 1959. *Biogeography and Ecology of Australia*. W. Junk, The Hague.

KENNEDY, M. (Ed.) 1992. *Australasian Marsupials and Monotremes*. IUCN-The World Conservation Union, Gland (Switzerland).

KHAN, M.KBM. 1989. *Asian Rhinos. An Action Plan for Their Conservation*. IUCN-The World Conservation Union, Gland (Switzerland).

KIRK, M. 1987. *Man as Art. New Guinea Body Decoration*. Taco Verlagsgesellschaft, Berlin.

KOEPPEN, W. 1931. *Grundriss der Klimakinde*. Walter de Gruyter, Berlin.

KOTTELAT, M., A.J. WHITTEN, S.N. KARTIKASARI, and S. WIRJOATMODJO. 1993. *Freshwater Fishes of Western Indonesia and Sulawesi*. Periplus Editions, Singapore.

KOTTELAT, M. and T. WHITTEN. 1996. *Freshwater Biodiversity in Asia with Special Reference to Fish*. World Bank Technical Paper No. 343, The World Bank, Washington, D.C.

KRUTCH, J.W. 1961. *The Forgotten Peninsula: A Naturalist in Baja California*. The University of Arizona Press, Tucson.

KUITER, R.H. 1992. *Tropical Reef-Fishes of the Western Pacific: Indonesia and Adjacent Waters*. Gramedia, Jakarta.

KUMAR, A. 1994. Lion-tailed macaque (*Macaca silenus*). Population and habitat viability assessment. IVth International Symposium on the Lion-tailed Macaque, Madras, India, October 1993. Executive Summary, Workshop Report.

KUMMER, D.M. 1992. *Deforestation in the Postwar Philippines*. University of Chicago Press, Chicago.

LAYCOCK, G. 1966. *The Alien Animals: The Story of Imported Wildlife*. Natural History Press, New York.

LEAHY, M.J. 1994. *Explorations into Highland New Guinea*. Crawford House Press, Bathurst (Australia).

LEATHERWOOD, S. and R.R. REEVES. 1983. *The Sierra Club Handbook of Whales and Dolphins*. Sierra Club, San Francisco.

LENTINO, M. and M.L. GOODWIN. 1993. *Lista de las aves del Parque Nacional Henri Pittier, Estado Aragua, Venezuela*. Sociedad Conservacionista Audubon de Venezuela, Caracas.

LINARES, O. 1987. *Murciélagos de Venezuela*. Cuadernos Lagoven, Caracas.

LINCOLN, L. 1987. *An Assemblage of Spirits: Idea and Image in New Ireland*. George Braziller, New York.

LINDBLAD, S.O. and L. LINDBLAD (Eds.). 1987. *Baja California*. Rizzoli, New York.

LINK, M. and K. CROWLEY. 1994. *Following the Pack: The World of Wolf Research*. Voyageur Press, Stillwater, MN.

LIPPENS, L. and H. WILLE. n.d. *Les oiseaux du Zaïre*. La présidence de la République du Zaïre. Belgique.

LOURENÇO, W.R. and E. FLORES. 1990. Scorpions (Chelicerata) from Colombia. The scorpiofauna of the Pacific Region (Chocó), with some biogeographic considerations. *Amazoniana* XI:119-133.

LOVELOCK, C. 1993. *Field Guide to the Mangroves of Queensland*. Australian Institute of Marine Science. Cairns.

MACKINNON, J. 1974. *In Search of the Red Ape*. Collins, London.

MACKINNON, J. 1996. *Wild China*. New Holland, London.

MACKINNON, J. *et al.* 1996. *A Biodiversity Review of China*. WWF China Programme, WWF-International, Hong Kong.

MACKINNON, J. and K. PHILLIPS. 1993. *Field Guide to the Birds of Sumatra, Borneo, Java and Bali*. Oxford University Press, Oxford.

MACKINNON, K.S. 1992. *Nature's Treasurehouse: The Wildlife of Indonesia*. Gramedia, Jakarta.

MACKINNON, K.S., G. HATTA, H. HALIM, and A. MANGALIK. 1996. *The Ecology of Kalimantan*. Periplus Editions, Singapore.

MAGSALAY, P.M. 1993. Rediscovery of four Cebu endemic birds (Philippines). *Asia Life Sciences* 2:141-148.

MAJUPURIA, C.T. 1986. *Wildlife Wealth of India*. Tecpress Service, LP, Bangkok.

MARGULIS, L., K.V. SCHWARTZ, and M. DOLAN. 1994. *The Illustrated Five Kingdoms. A Guide to the Diversity of Life on Earth*. HarperCollins College, New York.

MARNR. 1992. *Áreas naturales protegidas de Venezuela*. Dirección General Sectorial de Planificación y Ordenación del Ambiente, Caracas.

MARTÍNEZ, R.J. 1986. *Diversidad monolingüe de México en 1970*. UNAM, México.

MASON, V. and F. JARVIS. 1989. *Birds of Bali*. Periplus Editions, Singapore.

MATTHIESSEN, P. 1987. *Wildlife in America*. Viking Penguin, New York.

MAYR, E. and W.H. PHELPS. 1967. The origin of the bird fauna of the south Venezuelan highlands. *Bulletin of the American Museum of Natural History* 136:269-328.

MAYUGA, S. and A. YUSON. 1989. *Insight Guides. Philippines*. Apa Publications, Quezon City.

MCALLISTER, D.E., A.L. HAMILTON, and B. HARVEY. 1997. *Working Draft: Global Freshwater Biodiversity* (For SIFR/IDCR).

MCFARLAND, C.G., J. VILLA, and B. TORO. 1974. The Galápagos giant tortoises (*Geochelone elephantopus*). Part I: Status of surviving populations. *Biological Conservation* 6:118-133.

MCGOWAN, P.J.K., P.J. GARSON, and WPA/BIRDLIFE/SSC PHEASANT SPECIALIST GROUP (Eds.). 1995. *Pheasants. Status Survey and Conservation Action Plan 1995-1999*. IUCN, The World Conservation Union, Gland (Switzerland).

MCNEELY, J.A., K.R. MILLER, W.V. REID, R.A. MITTERMEIER, and T.B. WERNER (Eds.). 1990. *Conserving the World's Biological Diversity*. International Union for Conservation of Nature and Natural Resources, World Resources Institute, Conservation International, World Wildlife Fund-U.S. and the World Bank, Gland (Switzerland) and Washington, D.C.

MEDELLÍN, R.A. 1994. Mammal diversity and conservation in the Selva Lacandona, Chiapas, Mexico. *Conservation Biology* 8: 780-799.

MEDWAY, L. 1969. *The Wild Animals of Malaya*. Oxford University Press, Kuala Lumpur.

MEDWAY, L. 1977. *Mammals of Borneo: Field Keys and an Annotated Checklist*. Monographs of the Malaysian Branch of the Royal Asiatic Society, Kuala Lumpur.

MENA, P.A. and M.L. SUÁREZ (Eds.). 1993. *La investigación para la conservación de la diversidad biológica en el Ecuador*. Memorias del simposio llevado a cabo del 10 al 12 de junio de 1992. EcoCiencia. Fundación Ecuatoriana de Estudios Ecológicos, Quito.

MENZIES, J. 1991. *A Handbook of New Guinea Marsupials and Monotremes*. Kristen Press, Madang.

MESSEL, H., F.W. KING, and J.P. ROSS (Eds.). 1992. *Crocodiles. An Action Plan for Their Conservation*. Compiled by John Thorbjarnarson. International Union for Conservation of Nature and Natural Resources, Gland (Switzerland).

MEYER DE SCHAUENSEE, R. 1964. *The Birds of Colombia and Adjacent Areas of South America and Central America*. Livingston, Wynnewood, PA.

MEYER DE SCHAUENSEE, R. and J.W.H. PHELPS. 1978. *A Guide to the Birds of Venezuela*. Princeton University Press, Princeton, NJ.

MINTER-FUNAI. 1989. *Terras Indígenas: Legislação-Situação Fundiária*. Ministério do Interior (MINTER), Fundação Nacional do Índio (FUNAI), Superintendência de Asuntos Fundiários, Documentação Interna Indígena, Brasília.

MITCHELL, J.G. 1997. In the line of fire: Our national forests. *National Geographic* 191:58-87.

MITTERMEIER, R.A. 1988. Primate diversity and the tropical forest: Case studies from Brazil and Madagascar, and the importance of the megadiversity countries. In: E.O. WILSON (Ed.), *Biodiversity*. National Academy Press, Washington, D.C.

MITTERMEIER, R.A. and T.B. WERNER. 1990. Wealth of plants and animals unites "megadiversity" countries. *Tropicus* 4(1):1, 4-5.

MITTERMEIER, R.A. and C. GOETTSCH MITTERMEIER. 1992. La importancia de la diversidad biológica en México. In: J. SARUKHÁN and R. DIRZO (Eds.), *Mexico Confronts the Challenges of Biodiversity*. Comisión Nacional para el Conocimiento y Uso de la Biodiversidad, México.

MITTERMEIER, R.A., T. WERNER, J.M. AYRES, and G.A.B. DA FONSECA. 1992. O país da megadiversidade. *Ciencia Hoje* 14:20-27.

MITTERMEIER, R.A., I. TATTERSALL, W.R. KONSTANT, D.M. MEYERS, and R.B. MAST. 1994. *Lemurs of Madagascar*. Conservation International Tropical Field Guide Series. Conservation International, Washington, D.C.

MITTERMEIER, R.A, N. MYERS, and J. THOMSEN (in preparation). *Hotspots*. Conservation International, Washington, D.C.

MOCHI, U. and T.D. CARTER. 1971. *Hoofed Animals of the World*. Charles Scribner's Sons, New York.

MOHNOT, S.M., C.M. SOUTHWICK, and D. FERGUSON. 1996. *Annual Report of the Indo-U.S. Primate Project, Jodhpur*. Ministry of International Affairs, U.S. Fish and Wildlife Service, Ministry of Environment and Forests, Government of India, Washington, D.C. and New Delhi.

MONK, K.A., Y. DE FRETES, and G. REKSODIHARDJO-LILLEY. 1997. *The Ecology of the Nusa Tenggara and Maluku*. Periplus, Singapore.

MONTULET, J.P. 1984. *Les cervidés du monde entier*. Lechevalier, Paris.

MUKHERJEE, R.P. 1982. Survey of non-human primates in Tripura, India. *Zool. Soc. of India* 34: 70-81.

MUKHERJEE, R.P., A. SRIVASTAVA, S.M. MONHOT, I. BERNSTEIN, and C.H. SOUTHWICK. 1997. Population status of the golden langur (*Presbytis geei*) in Assam. *American Journal of Primatology* (Abstract in press).

MULLER, K. 1992. *Underwater Indonesia: A Guide to the World's Best Diving*. Periplus, Singapore.

MURRAY, J.A. 1995. *Grizzly Bears: An Illustrated Field Guide*. Roberts Rinehart, Boulder.

MYERS, N. 1988. Threatened biotas: Hotspots in tropical forests. *The Environmentalist* 8:118-208.

MYERS, N. 1992. *The Primary Source: Tropical Forests and Our Future*. W.W. Norton, New York.

MYERS, N. 1994. *Ultimate Security: The Environmental Basis of Political Stability*. W.W. Norton, New York.

NAYAR, M.P. and R.K. SHASTRI (Eds.). 1987. *Red Data Book of Indian Plants*. Botanical Survey of India, Calcutta.

NEGI, S.S. 1989. *Forest Types of India, Nepal and Bhutan*. PEBA, Delhi.

NEGI, S.S. 1994. *Biodiversity and Its Conservation in India*. Indus Publication Co., New Delhi.

NEGI, S.S., J. BROWNELL, and Z. JIANKANG. 1989. *Biology and Conservation of the River Dolphins*. Proceedings of the Workshop on the Biology and Conservation of the Platanistoid Dolphins, Whuan (China), IUCN, Gland (Switzerland).

NEPA. 1994. *China Biodiversity Conservation Action Plan*. Environment Science Press, Beijing.

NEW, T.R. and N.M. COLLINS. 1991. *Swallowtail Butterflies. An Action Plan for Their Conservation*. IUCN/SSC Lepidoptera Specialist Group, Gland (Switzerland).

NORSE, E.A., K.L. ROSENBAUM, D.S. WILCOVE, B.A. WILCOX, W.H. ROMME, D.W. JOHNSON, and M.L. STOUT. 1986. *Conserving Biological Diversity in Our National Parks*. The Wilderness Society, Washington, D.C.

NOSS, R.F. and R.L. PETERS. 1995. *Endangered Ecosystems: A Status Report on America's Vanishing Habitat and Wildlife. Defenders of Wildlife*, Washington, D.C.

NOWAK, R.M. and J.L. PARADISO. 1983. *Walker's Mammals of the World*. 4th ed., The Johns Hopkins University Press, Baltimore.

OCEI. 1993a. *Censo Indígena de Venezuela 1992*. Oficina Central de Estadística e Informática, Caracas.

OCEI. 1993b. *El Censo 90 en Venezuela*. Oficina Central de Estadística e Informática, Caracas.

OLIVER, W.L.R. (Ed.) 1993. *Pigs, Peccaries and Hippos*. IUCN-The World Conservation Union, Gland (Switzerland).

OLIVER, W.L.R., C.R. COX, P.C. GONZALES, and L.R. HEANEY. 1993. Cloud rats in the Philippines. Preliminary report on distribution and status. *Oryx* 27:41-48.

OLIVER, W.L.R. and L.R. HEANEY. 1996. Biodiversity and conservation in the Philippines. *International Zoo News* 43:329-337.

ONERN. 1976. *Mapa ecológico de Perú: guía explicativa*. Oficina Nacional de Evaluación de Recursos Naturales, Lima.

ORTIZ-CRESPO, F., P.J. GREENFIELD, and J.C. MATHEUS. 1990. *Aves del Ecuador*. Feprotur, Quito.

O'SHEA, M. 1996. *A Guide to the Snakes of Papua New Guinea*. Independent Publishing, Port Moresby.

PADUA, C. and J. BALLOU. 1997. *Leontopithecus chrysopygus*. Metapopulation Management Action Plan. 5/9/97, Draft.

PAIJMANS, S.K. (Ed.). 1976. *New Guinea Vegetation*. Australia National University Press, Canberra.

PANWAR, H.S. 1982. Project Tiger. In: V.B. SAHARIA (Ed.), *Wildlife in India*. Natray Publications, Dehra Dun.

PARKER, T.A., S.A. PARKER, and M.A. PLENGE. 1982. *An Annotated Checklist of Peruvian Birds*. Buteo Books, Vermillion, SD.

PAYMAL, N. and C. SOSA. 1993. *Mundos amazónicos, pueblos y culturas de la Amazonia ecuatoriana*. Fundación Sinchi Sacha, Quito.

PAYNE, J. 1990. *Wild Malaysia. The Wildlife and Scenery of Peninsular Malaysia, Sarawak and Sabah*. WWF, London.

PAYNE, J., C.M. FRANCIS, and K. PHILLIPS. 1985. *A Field Guide to the Mammals of Borneo*. Sabah Society, Kota Kinabalu and WWK Malaysia, Kuala Lumpur.

PEARSON, D.L. and F. CASSOLA. 1992. World-wide species richness patterns of tiger beetles (Coleoptera: Cicindelidae): Indicator taxon for biodiversity and conservation studies. *Conservation Biology* 6:376-391.

PECKOVER, S.W. 1990. *Birds of Paradise*. Robert Brown, Carina (Australia).

PERRIN, W.F. (Ed.) 1989. *Dolphins, Porpoises, and Whales. An Action Plan for the Conservation of Biological Diversity: 1988-1992*. 2nd ed., IUCN/Fisheries Service, NOAA, Gland (Switzerland).

PERÚ (Government of). 1992. *Informe Nacional de Perú a la Conferencia de las Naciones Unidas sobre el Medio Ambiente y el Desarrollo*. CNUMAD 92, Lima.

PÉRUSSE, Y. 1993. *Bushwalking in Papua New Guinea*. 2nd. ed., Second Lonely Planet, Hawthorn (Australia).

PETOCZ, R.G. 1989. *Conservation and Development in Irian Jaya*. Brill, Leiden.

PIZZEY, G. 1988. *A Field Guide to the Birds of Australia*. Illustrated by Ron Doyle, Princeton University Press, Princeton.

PORTER, G. and D.J. GANAPIN, JR. 1988. *Resources, Population and the Philippines' Future. A Case Study*. World Resources Institute, Washington, D.C.

POSEY, D.A. *et al*. 1987. *Alternatives to Destruction: The Science of the Mêbêngôkre*. Museu Paraense Emilio Goeldi, Belém.

PRITCHARD, P.C.H. 1996. *The Galápagos tortoises, nomenclatural and survival status*. Chelonian Research Monographs No. 1, Chelonian Research Foundation, Lunenburg, MA.

PRITCHARD, P.C.H. and P. TREBBAU. 1984. *The Turtles of Venezuela*. Contributions to Herpetology, No. 2, Society for the Study of Amphibians and Reptiles, Oxford, OH.

PULIDO, V. 1991. *El libro rojo de la fauna silvestre del Perú*. P. Aguilar, Lima.

PULLIAM, H.R. and B. BABBIT. 1997. Science and the protection of endangered species. *Science* 275:499-500.

RABOR, D.S. 1977. *Philippine Birds and Mammals*. U.P. Science Education Center, University of the Philippines Press, Quezon City.

RABOR, D.S. 1981. *Philippine Reptiles and Amphibians*. Pundasyon sa Pagpapaunlad sa Pagtuturo ng Agham, Quezon City.

RADCLIFFE-BROWNE, A.R. 1922. *The Andaman Islanders: A Study in Social Anthropology*. Cambridge University Press, Cambridge.

RAVEN, P.H. 1995. Biodiversity and the future of China. *Pacific Science Association. Information Bulletin* 47 (1-2):1-37.

REEVES, R.R. and S. LEATHERWOOD (Eds.). 1994. *Dolphins, Porpoises and Whales*. 1994-1998 Action Plan for the Conservation of Cetaceans, IUCN-The World Conservation Union, Oxford.

REICHEL-DOLMATOFF, G. 1991. *Indians of Colombia: Experience and Cognition*. Villegas, Santa Fe de Bogotá.

REPETTO, R. 1988. *The Forest for the Trees? Government Policies and the Misuse of Forest Resources*. World Resources Institute, Washington, D.C.

RHANDARE, S.S. and J.K. MUKHOPADHYA (Eds.). 1995. *Statistical Outline of India, 1995-1996*. Dept. of Economics and Statistics, Tata Services, Bombay.

RIBEIRO, D. 1957. *Culturas e linguas indígenas do Brasil*. Centro Brasileiro de Pesquisas Educacionais, Rio de Janeiro.

RICARDO, C.A. (Ed.). 1996. *Povos indígenas do Brasil: 1991-1995*. Instituto Socioambiental, São Paulo/Brasília.

RICCIARDI, M. 1991. *Vanishing Amazon*. Abrams, New York.

RICCIUTI, E. 1995. *The Natural History of North America*. Crescent Books, New York.

RICE, C.G. 1988. Reproductive biology of Nilgiri tahr (*Hemitragus hylocrius*) (Mammalia: Bovidae). *Journal of the Zoological Society of London* 214: 269-284.

RICKART, E.A. 1993. Diversity patterns of mammals along elevational and diversity gradients in the Philippines: Implications for conservation. *Asia Life Sciences* 2:251-260.

RICKART, E.A., L.R. HEANEY, and R.B. UTZURRUM. 1991. Distribution and ecology of small mammals along an elevation transect in southeastern Luzon, Philippines. *Journal of Mammology* 72:458-469.

RÍOS, M.A. 1993. *La diversidad natural del Perú y la situación actual de su protección*. MSc. Thesis, Universidad Nacional Agraria, La Molina, Lima.

RITCHIE, R. 1989. *Seeing the Rainforests in 19th-century Australia*. Rainforest Publishing, Sydney.

ROBINSON, F. (Ed.). 1989. *The Cambridge Encyclopedia of India*. Cambridge University Press, Cambridge.

ROBINSON, M. 1994. *A Field Guide to Frogs of Australia*. Australian Museum/Reed Books, Chatswood.

ROBLES GIL P. and F. ECCARDI (Eds.). 1991. *Lacandonia, el último refugio*. UNAM, Mexico.

RODRÍGUEZ, L.O. (Ed.). 1996. *Diversidad biológica de Perú. Zonas prioritarias para su conservación*. Franpe GTZ-Instituto Nacional de los Recursos Naturales, Perú.

RODRÍGUEZ, L.O., J.H. CÓRDOVA, and J. ICOCHEA. 1993. *Lista preliminar de los anfibios del Perú*. Publicaciones del Museo de Historia Natural, Universidad Nacional Mayor de San Marcos, Lima.

RODRÍGUEZ-M., J.V., J.I. HERNÁNDEZ-CAMACHO, J.V. RUEDA-A., and J. MORALES-S. 1986. *Colombia. Fauna en peligro*. Procam, Instituto Nacional de los Recursos Naturales Renovables y del Ambiente, Bogotá.

RODRÍGUEZ-M., J.V., J.I. HERNÁNDEZ-CAMACHO, T.R. DEFLER, M. ALBERICO, R.B. MAST, R.A. MITTERMEIER, and A. CADENA. 1995. *Mamíferos colombianos: sus nombres comunes e indígenas*. Occasional Papers in Conservation Biology, Dec. 1995.

ROMERO, D.A. 1992. *Venezuela: mágico país de la biodiversidad*. Ediciones Co-Bo, Caracas.

ROMERO, F.J. and A. VELÁZQUEZ. 1994. *El conejo zacatuche. Tan lejos de Dios y tan cerca de la Ciudad de México*. Instituto Nacional de Ecología/Consejo Nacional de la Fauna, México.

ROTH, L.H. 1968. *The Natives of Sarawak and British North Borneo*. University of Malaya Press, Singapore.

RUEDA-A., J.V., J.E. MAYORGA, and G. ULLOA. 1992. Observaciones sobre la captura comercial de tortugas marinas en la península de La Guajira, Colombia. In: J.V. RODRÍGUEZ and H. SÁNCHEZ PÁEZ (Eds.),*Contribución al conocimiento de las tortugas de Colombia*. Libro 4. Biblioteca Andrés Posada Arango, Bogotá, pp. 133-190.

RUIZ-CARRANZA, P.M., M.C. ARDILA-ROBAYO, and J.D. LYNCH. 1996. Lista actualizada de la fauna de Amphibia de Colombia. *Revista de la Academia Colombiana de Ciencias Exactas, Físicas y Naturales* 20:365-413.

RYLANDS, A.B. 1991. *The Status of Conservation Areas in the Brazilian Amazon*. World Wildlife Fund Publications, Washington, D.C.

RYLANDS, A.B. 1995. Áreas protegidas na Amazônia brasileira. In: G.A. FONSECA, M. SCHMINK, L.P. de S. PINTO, and F. BRITO (Eds.), *Abordagens interdisciplinares para a conservação da biodiversidade e dinâmica do uso da terra no Novo Mundo*, Conservation International do Brasil, Universidade Federal de Minas Gerais and University of Florida, Gainesville.

RYLANDS, A.B. and L.P. DE S. PINTO. In press. *Conservação da biodiversidade na Amazônia brasileira: uma análise do sistema de unidades de conservação*. Cadernos FBDS, No. 1, Fundação Brasileira para o Desenvolvimento Sustentável, Rio de Janeiro.

SALAZAR-VALLEJO, S.I. and N.E. GONZÁLEZ (Eds.). 1993. *Biodiversidad marina y costera de México*. Comisión Nacional para el Conocimiento y Uso de la Biodiversidad/Centro de Investigaciones de Quintana Roo, México.

SALVADORI, F. and P. FLORIO. 1977. *Rare and Beautiful Animals of the World*. Optimum, Milan.

SÁNCHEZ-PÁEZ, H., J.I. HERNÁNDEZ-CAMACHO, J.V. RODRÍGUEZ MAHECHA, and C.C. URIBE. 1990. *Nuevos parques nacionales, Colombia*. Instituto Nacional de los Recursos Naturales Renovables y del Ambiente, Bogotá.

SANTIAPALLAI, C. and P. JACKSON. 1990. *The Asian Elephant: An Action Plan for Its Conservation*. IUCN. The World Conservation Union/SSC Asian Elephant Specialist Group, Gland (Switzerland).

SARUKHÁN, J. and R. DIRZO (Eds.). 1992. *Mexico Confronts the Challenges of Biodiversity*. Comisión Nacional para el Conocimiento y Uso de la Biodiversidad, Mexico.

SCALATER, A. 1981. *Catalogue of Mammalia*. Cosmo, New Delhi.

SCHUBERT, C. and O. HUBER. 1990. *The Gran Sabana: Panorama of a Region*. Lagoven Booklets, Caracas.

SCHULTES, R.E. 1988. *Where the Gods Reign: Plants and Peoples of the Colombian Amazon*. Synergetic Press, London.

SCHWARTZ, J.T. 1987. *The Red Ape*. Houghton Mifflin, Boston.

SEDESOL. 1993. *The Protected Areas of Mexico*. Secretaría de Desarrollo Social, Mexico.

SEKHRAN, N. and S. MILLER. 1994. *Papua New Guinea Country Study on Biological Diversity*. The Department of Environment and Conservation, Conservation Resource Center, Africa Centre for Resources and Environment, Waigani.

SERVENTY, V. 1988. *Saving Australia*. Child & Associates, Frenchs Forest.

SESHADRI, B. 1986. *India's Wildlife and Wildlife Reserves*. Sterling, New Delhi.

SHARP, I. 1994. *Green Indonesia*. Oxford University Press, Oxford.

SICK, H. 1985. *Ornitologia Brasileira. Uma introdução*. Editora Universidade de Brasília, Brasília.

SIERRA, R. 1996. *La deforestación en el norte del Ecuador. 1983-1993*. EcoCiencia, Quito.

SIMPSON, K., N. DAY, and P. TRUSLER. 1996. *Field Guide to the Birds of Australia*. Viking Penguin, Ringwood.

SINGH, R. 1975. The last Andaman Islanders. *National Geographic* 148: 66-91.

SKEAT, W.W. and C.O. BLAGDEN. 1906. *Pagan Races of the Malay Peninsula*. Macmillan, London.

SMIDT, D. (Ed.). 1993. *Asmat Art. Woodcarvings of Southwest New Guinea*. George Braziller, New York.

SMITH, H. 1996. *Tiwi. The Life and Art of Australia's Tiwi People*. JB Gifts, Sydney.

SMITH, H.M. and R.B. SMITH. 1971-1980. *Synopsis of the Herpetofauna of Mexico*. Eric Lundberg, Augusta, WV.

SOUTHWICK, C.H. 1976. *Ecology and Quality of Our Environment*. Willard Grant Press, Boston.

SOUTHWICK, C.H. 1996. *Global Ecology in Human Perspective*. Oxford University Press, Oxford.

STEADMAN, D.W. and S. ZOUSMER. 1988. *Galápagos: Discovery on Darwin's Islands*. Smithsonian Institution Press, Washington, D.C.

STEENIS, C.G.G.J. VAN 1972. *The Mountain Flora of Java*. Brill, Leiden.

STEIN, B.A. and S.R. FLACK. 1997. *1997 Species Report Card: The State of U.S. Plants and Animals*. The Nature Conservancy, Arlington.

STEWARD, J.H. (Ed.). 1948. *Handbook of South American Indians*. Smithsonian Institution, Bureau of American Ethnology, Washington, D.C.

STEWART, D. (Ed.). 1988. *Burnum Burnum's Aboriginal Australia. A Traveler's Guide*. Angus & Robertson, London.

STEYERMARK, J.A., P.E. BERRY, and B.K. HOLST (Eds.). 1995. *Flora of the Venezuelan Guayana*. Missouri Botanical Garden, St. Louis and Timber Press, Portland, OR.

STIRLING, M.W. 1955. *Indians of the Americas*. Na-

On the opposite page,
Huli wigman warrior
in traditional attire.
Tari Valley region,
Highlands of Papua New Guinea.

tional Geographic Society, Washington, D.C.

STONE, D. 1994. *Tanah Air. Indonesia's Biodiversity*. Archipelago Press, Singapore.

STORR, G.M. and R.E. JOHNSTONE. 1990. *A Field Guide to the Birds of Australia*. Western Australian Museum, Perth.

STRAHAN, R. (Ed.) 1983. *The Australian Museum Complete Book of Australian Mammals*. Angus and Robertson, London.

STRAHAN, R. (Ed.). 1995. *The Mammals of Australia*. Reed Books, Chatswood.

STRIER, K.B. and G.A.B. DA FONSECA. 1996. *The Endangered Muriquis in Brazil's Atlantic Forest* (A version of this paper was presented at the meeting of the International Primatological Society in Madison, Wisconsin, U.S.A., in August, 1996).

STUART, S.N., R.J. ADAMS, and M.D. JENKINS. 1990. *Biodiversity in Sub-Saharan Africa and Its Islands*. Occasional Papers of the IUCN Species Survival Commission, The World Conservation Union, No. 6, Gland (Switzerland).

SUNQUIST, F. 1977. Where cats and herders mix. *International Wildlife* 27:26-33.

SWADLING, P. 1996. *Plumes from Paradise*. Papua New Guinea National Museum, Boroko.

TELLO, J. 1979. *Mamíferos de Venezuela*. Fundación La Salle de Ciencias Naturales, Caracas.

TERBORGH, J. and B. WINTER. 1980. Some causes of extinction. In: M. SOULÉ and B. WILCOX (Eds.), *Conservation Biology: An Evolutionary-Ecological Perspective*. Sinauer Associates, Sunderland, MA, pp. 119-133.

THAKUR, D. and D.N. THAKUR (Eds.). 1994. *Tribal Life in India*. Deep and Deep Publications, Delhi.

TINDALE, N.B. 1974. *Aboriginal Tribes of Australia*. University of California Press, London.

TOLEDO, V.M. 1994. Biodiversity and cultural diversity in Mexico. *Different Drummer* 1: 16-19.

TOLEDO, V.M. 1995. In: P. ROBLES GIL (Ed.), *Mexico. Diversity of Cultures*. Cemex, Mexico.

TORRE, J. and L.T. FINDLEY. 1994. Osteology of the vaquita, *Phocoena sinus*. Final Report presented to Conservation International, Centro de Investigación y Desarrollo de Sonora/Centro de Conservación y Aprovechamiento de los Recursos Naturales/Instituto Tecnológico y de Estudios Superiores de Monterrey.

TORRES, H. (Ed.). 1992. *South American Camelids. An Action Plan for Their Conservation*. IUCN/Species Survival Commission, South American Camelid Specialist Group, Gland (Switzerland).

TREWARTHA, G.T. 1968. *An Introduction to Weather and Climate*. 4th ed., McGraw-Hill, New York.

TURNBULL, C.M. 1965. *Wayward Servants: The Two Worlds of the African Pygmies*. The Natural History Press, Garden City, NY.

VALENCIA, R., H. BALSLEV, and G. PAZ y MIÑO C. 1994. High tree alpha-diversity in Amazonian Ecuador. *Biodiversity and Conservation* 3:21-28.

VAN DER HAMMEN and C. CASTAÑO. 1995. Tras las huellas del pasado prehistórico. *Chiribiquete Maloca de los Dioses* 1:18-21.

VAN DER ZON, A.P.M. 1979. *Mammals of Indonesia*. FO/INS/78/061. Special Report to the Food and Agriculture Organization of the United Nations, Bogor.

VAN GRUISIN, J. 1988. Wildlife of the Himalaya. In: B. GREWAL (Ed.), *Indian Wildlife*. Insight Guides, APA Publications, Singapore.

VARESCHI, V. 1992. *Ecología de la vegetación tropical*. Sociedad Venezolana de Ciencias Naturales, Caracas.

VARGAS, H.V. 1991. *Sistema nacional de áreas protegidas y la vida silvestre del Ecuador*. Subsecretaría Forestal y de Recursos Naturales Renovables, Quito.

VEILLON, J. 1989. *Los bosques naturales de Venezuela*. Instituto de Silvicultura, Universidad de los Andes, Mérida/Todtmann, Caracas.

VERSWIJVER, G. 1992. *Kaiapó Amazonia*. Royal Museum for Central Africa, Tervuren (Belgium).

VIGNE, L. and E. BRADLEY-MARTIN. 1991. Assam's rhinos face poaching threats. *Oryx* 25: 215-221.

VITUG, M.D. 1993. *Power from the Forest: The Politics of Logging*. Philippine Center for Creative Journalism, Manila.

WAGNER, W.L., D.R. HERBST, and S.H. SOHMER. 1990. *Manual of the Flowering Plants of Hawaii*. Bishop Museum Special Publication No. 83, University of Hawaii Press/Bishop Museum Press, Honolulu.

WALDMAN, C. 1985. *Atlas of the North American Indian*. Facts on File, New York.

WALLACE, A.R. 1962. *The Malay Archipelago*. Dover, New York.

WANG SUNG (Ed.). 1990. *Proceedings of the Symposium on Biological Diversity, 29-31 March 1990, Beijing*. Biodiversity Working Group/Bur, Bio-Sci and Tech. CAS, Beijing.

WANG SUNG. 1995. *The Chinese Diversity Action Plan*. National Biodiversity Planning, guidelines based on early experiences around the world, pp. 111-114.

WANG SUNG and BIODIVERSITY COMMITTEE/CAS (Eds.). 1992. *Biodiversity in China*. Science Press China, Beijing.

WANG SUNG *et al.* (in preparation). *China's Red Data Book of Threatened Animals*.

WARREN, W., A.J. WHITTEN, M. JENKINS, A. VICKERS, and T. HUNTER. 1995. *Balinese Gardens*. Periplus, Singapore.

WCMC (WORLD CONSERVATION MONITORING CENTRE). 1992. *Global Diversity*. Chapman and Hall, London.

WEBB, G. and C. MANOLIS. 1989. *Grahame Webb's Crocodiles of Australia*. Reed Books, Frenchs Forest.

WEGE, D.C. and A.J. LONG. 1995. *Key Areas for Threatened Birds in the Neotropics*. BirdLife Conservation Series No. 5, BirdLife International, Cambridge (U.K.).

WHEELER, T. and J. MURRAY. 1993. *Papua New Guinea: A Travel Survival Kit*. 5th ed., Lonely Planet, Hawthorn (Australia).

WHITAKER, R., Z. WHITAKER, and D. INDRANEIL. 1993. *The World of Turtles and Crocodiles*. National Book Trust, New Delhi.

WHITE, F. 1983. *The Vegetation of Africa*. UNESCO, La Chaux de Fonds.

WHITMORE, T.C. 1984. *Tropical Rain Forests of the Far East*. 2nd ed., Clarendon, Oxford.

WHITMORE, T.C. 1990. *An Introduction to Tropical Rain Forests*. Clarendon, Oxford.

WHITTEN, T. 1982. *The Gibbons of Siberut*. Dent, London.

WHITTEN, T., S.J. DAMANIK, and J. ANWAR. 1987. *The Ecology of Sumatra*. 2nd ed., Gadjah Mada University Press, Yogyakarta.

WHITTEN, T., M. MUSTAFA, and G.S. HENDERSON. 1987. *The Ecology of Sulawesi*. Gadjah Mada University Press, Yogyakarta.

WHITTEN, T. and J.E.J. WHITTEN. 1992. *Wild Indonesia*. New Holland, London.

WHITTEN, T., R.E. SOERIAATMADJA, and S. AFIFF. 1996. *The Ecology of Java and Bali*. Periplus Editions, Singapore.

WHITTEN, T. and J.E.J. WHITTEN. 1996. *Indonesia Heritage Encyclopedia*. Wildlife Archipelago, Singapore.

WOOD, J.D., O.C. SHARMA, H.M. MEKIJIAN, D.W. STONE, J.I. HERNÁNDEZ-CAMACHO, and J.V. RODRÍGUEZ-M. 1990. *Incidence of colitis and colon cancer in feral cotton-top tamarins* (Saguinus oedipus). American Association for the Study of Liver Diseases/American Gastroenterological Association, Washington, D.C.

WORLD BANK. 1989. *Philippines: Environment and Natural Resource Management Study*. The World Bank, Washington, D.C.

WRI (WORLD RESOURCES INSTITUTE). 1997. *World Resources, 1996-1997. A Guide to Global Environment*. Oxford University Press, Oxford.

WWF and IUCN. 1994-1997. *Centers of Plant Diversity. A Guide and Strategy for Their Conservation*. 3 volumes. IUCN Publications Unit, Cambridge (U.K.).

ZHAO, ER-MI. and K. ADLER. 1993. *Herpetology of China*. Society for the Study of Amphibians and Reptiles, Oxford, OH.

ZSLOWSKY, D. and WILDERNESS SOCIETY. 1986. *These American Lands*. Henry Holt, New York.

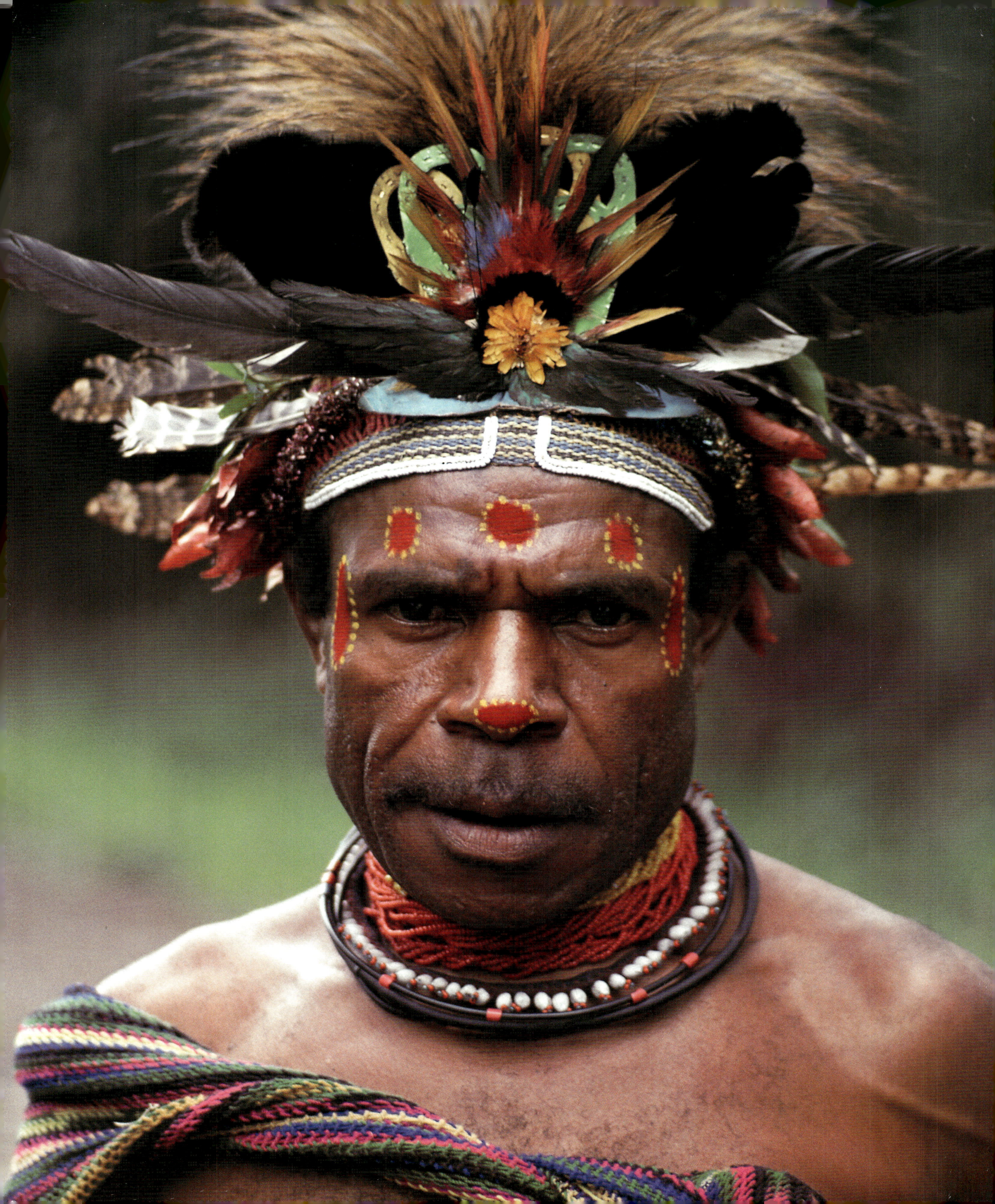

PHOTOGRAPHS

PRELIMINARY PAGES AND INTRODUCTION

© Fred Bavendam/Tony Stone Images/Mexico: 16
© John Cancalosi: 33
© Gerry Ellis/Ellis Nature Photography: 2
© François Gohier: 30
© Frans Lanting, 13
© Frans Lanting/Minden Pictures: 4-5
© Michael K. Nichols/National Geographic Society: 6
© Patricio Robles Gil/Sierra Madre: 25, 34-35
© Mike Severns/Tony Stone Images/Mexico: 20-21
© Darryl Torckler /Tony Stone Images/Mexico: 10-11
© Michael P. Turco: 19
© Art Wolfe: 8, 27, 36

BRAZIL

© Erwin & Peggy Bauer/Bruce Coleman, Inc.: 62-63
© Wolfgang Bayer/ Bruce Coleman, Inc.: 65 top
© Stephen Dalton/NHPA/Silvestris: 64-65
© Nicholas DeVore III/National Geographic Society/ Photographers Aspen: 50-51
© Michael Durham/Ellis Nature Photography: 45
© Victor Englebert: 70, 70-71
© John Giustina/The Wildlife Collection: 66 bottom
© François Gohier: 56 bottom, 56-57
© Frans Lanting/Minden Pictures: 62 top
© Luiz Claudio Marigo: 42-43, 46, 47, 48, 49, 58 bottom, 61 top, 66 top
© Loren McIntyre: 55, 58 top, 58-59, 68-69, 69, 72, 72-73
© Russell A. Mittermeier: 38
© Patricio Robles Gil/Sierra Madre: 40, 41, 44, 56 top, 60-61, 61 bottom, 65 bottom
© Rod Williams/Bruce Coleman, Inc.: 62 bottom
© Art Wolfe: 52, 53, 54
© Andrew L. Young/Archipelago Films, Inc.: 66-67

INDONESIA

© Josef Beck/Silvestris: 78-79
© Tom Brakefield/Bruce Coleman, Inc.: 102 bottom
© John Cancalosi/DRK Photo: 88
© Alain Compost/Peter Arnold, Inc.: 95
© Gerald Cubitt: 83
© Tui De Roy/Auscape: 82
© Tui De Roy/The Roving Tortoise: 84-85, 102-103
© Gerry Ellis/Ellis Nature Photography: 76, 86
© Victor Englebert: 96, 97
© Jean-Paul Ferrero/Auscape: 81, 87, 98-99, 100-101, 102 top
© Rio Helmi/Auscape: 92, 94
© Mitsuhiko Imamori/Nature Production: 100
© Mark Jones/The Roving Tortoise: 77, 80
© Frans Lanting/Minden Pictures: 99 top
© Wayne Lawler/Auscape: 99 bottom
© Michael K. Nichols/National Geographic Society: 89
© Charles O'Rear/VM Internacional: 106 top
© George Steinmetz: 106 bottom, 106-107
© Art Wolfe: 93
© Art Wolfe/Tony Stone Images/Mexico: 74
© Norbert Wu: 90, 91, 104
© Norbert Wu/Tony Stone Images/Mexico: 104-105

COLOMBIA

© Sebastián Beláustegui: 127, 136-137, 137
© James P. Blair/National Geographic Society: 134-135
© Santiago Carrizosa: 114
© Loren McIntyre: 138
© Mark Moffett/Minden Pictures: 108, 132-133, 134 top, 134 bottom
© Leo Postl/Silvestris: 112-113
© Juan Manuel Renjifo: 110, 121, 122, 124, 126, 128-129, 133 top
© Patricio Robles Gil/Sierra Madre: 111, 116, 117, 119, 120, 123, 125, 128, 129 bottom, 130-131, 131 top, 131 bottom, 133 bottom
© Carlos Adolfo Sastoque: 115
© Michael P. Turco: 118
© Jonathan Wright/Bruce Coleman, Inc.: 138-139

MEXICO

© Guillermo Aldana: 158, 177
© Lorenzo Armendáriz: 176-177
© Sebastián Beláustegui: 174, 174-175
© Thomas C. Boyden: 140
© John Cancalosi: 153
© Pablo Cervantes/Sierra Madre: 143
© Claudio Contreras/Sierra Madre: 162-163, 166 top, 166-167
© Jack Dykinga: 144
© Fulvio Eccardi/Sierra Madre: 146, 170-171
© Ernesto C. Enkerlin: 145
© François Gohier: 170, 173
© George Grall/National Geographic Society: 160 top, 160 bottom, 160-161
© Pablo Ortiz Monasterio: 156-157, 159
© Antonio Ramírez/Sierra Madre: 166 bottom, 169 bottom
© Patricio Robles Gil/Sierra Madre: 142, 147, 148-149, 152, 154, 155, 163 top, 163 bottom, 164, 164-165, 168-169, 169 top, 172-173
© Adolfo de Sostoa y Xavier Ferrer: 151
© Merlin D. Tuttle/Bat Conservation International: 150

AUSTRALIA

© Stanley Breeden/DRK Photo: 182, 184
© John Cancalosi: 188, 189, 193, 194 bottom, 200 top, 200-201, 202
© John Cancalosi/Auscape: 194 top
© Chuck Davis/Tony Stone Images/Mexico: 190
© Gerry Ellis/Ellis Nature Photography: 178, 192, 196 bottom
© Jean-Paul Ferrero/Auscape: 194-195, 196 top, 204-205
© Ferrero-Labat/Auscape: 198-199
© Michael & Patricia Fodgen: 196-197
© David Alan Harvey/National Geographic Society: 202-203
© J. M. La Roque/Auscape: 186-187
© D. Parer-E. Parer Cook/Auscape: 185, 191, 199
© Jamie Plaza van Roon/Auscape: 183
© Becca Saunders/Auscape: 205 bottom
© John Shaw: 200 bottom
© Medford Taylor/National Geographic Society: 206-207
© Darryl Torckler/Tony Stone Images/Mexico: 205 top
© Glen Trelfo/Auscape: 180, 181
© Belinda Wright/National Geographic Society: 206

MADAGASCAR

© Erwin & Peggy Bauer: 211, 215
© Dani/Jeske/Silvestris: 227
© John Giustina/The Wildlife Collection: 231 top
© Masahiro Iijima/Nature Production: 232-233
© Mitsuhiko Imamori/Nature Production: 214, 222, 230-231, 232
© Frans Lanting: 212-213, 218-219, 224, 226-227, 228, 234-235, 235
© Frans Lanting/Minden Pictures: 208, 210, 216, 221, 224-225, 225 bottom
© Tom & Pat Leeson: 228-229
© Bill Love/Blue Chameleon Ventures: 220, 231 bottom
© Russell A. Mittermeier: 217, 223, 229 bottom

PHILIPPINES

© Bob Bowdey/La Mer Majik: 244
© Sisse Brimberg/National Geographic Society: 248
© Roger Brown/Vireo: 254
© John Cancalosi: 252
© Haroldo Castro/Conservation International: 243
© Andrew Drake/Tony Stone Images/Mexico: 245
© Jean-Paul Ferrero/Auscape: 247, 251, 252-253
© Michael Freeman/Bruce Coleman, Inc.: 249
© Neil L. Rettig, F.R.E.E., Ltd.: 250-251
© Doug Wechsler/Vireo: 236, 238, 239, 240-241, 242, 246, 254-255

CHINA

© Heather Angel/Biophotos: 263, 272-273
© James P. Blair/National Geographic Society: 270
© Jim Brandenburg/Minden Pictures: 274 top
© John Cancalosi: 274-275
© John Chellman/Animals Animals: 262
© Nigel J. Dennis/NHPA/Silvestris: 278-279
© Eric Dragesco/Jacana: 277
© Tom & Pat Leeson: 267
© Judy Y. Mundy: 271
© Richard Packwood/Animals Animals: 274 bottom
© Fritz Pölking: 264-265

© Hans Reinhard/Bruce Coleman, Inc.: 266
© Patricio Robles Gil/Sierra Madre: 268-269
© Galen Rowell: 276-277, 278, 280-281, 281
© George Schaller: 258, 259, 273
© Lynn M. Stone/Bruce Coleman, Inc.: 256
© TH Fotowerbung/Silvestris: 260-261

PERU

© Frank S. Balthis: 297
© André Bärtschi: 285, 293, 300-301, 301, 304, 306 top, 306 bottom
© John Cancalosi: 303
© James Carmichael/NHPA/Silvestris: 292
© Martyn Colbeck/Oxford Scientific Films: 282
© Victor Englebert: 312 bottom
© Michael Fogden/Oxford Scientific Films: 284
© Michael & Patricia Fogden: 295, 303 bottom
© Frans Lanting/Minden Pictures: 289, 291, 298-299
© Loren McIntyre: 312-313
© Mark Moffett/Minden Pictures: 302-303
© Patricio Robles Gil/Sierra Madre: 286-287, 288, 290, 294, 298 top, 298 bottom, 308, 309, 310-311
© Galen Rowell: 313
© Kevin Schaffer: 304-305, 306-307
© Art Wolfe: 296
© Gunter Ziesler/Peter Arnold, Inc.: 311

ECUADOR

© Frank S. Balthis: 316
© Sebastián Beláustegui: 334-335
© Tui De Roy/The Roving Tortoise: 318-319, 332-333
© Michael & Patricia Fogden: 320, 323, 328-329, 329 top, 330, 330-331
© Russell Mittermeier: 326 top
© Mark Moffett/Minden Pictures: 329 bottom
© M. Timothy O'Keefe/Bruce Coleman, Inc.: 325
© Patricio Robles Gil/Sierra Madre: 326 bottom, 326-327
© Galen Rowell: 317
© Kevin Schaffer: 314
© Stella Snead/Bruce Coleman, Inc.: 324
© Tom L. Ulrich: 332 top
© Doug Wechsler/Vireo: 321, 322
© Norbert Wu: 332 bottom

INDIA

© James P. Blair/National Geographic Society: 348-349
© Stanley Breeden/DRK Photo: 338
© Stanley Breeden/National Geographic Society: 339
© Gerald Cubitt: 342, 347
© Nicholas DeVore III/National Geographic Society/Photographers Aspen: 343
© Michael Durham/Ellis Nature Photography: 359 top
© Gerry Ellis/Ellis Nature Photography: 340-341
© E. Hanumantha/BIF/Photo Researchers, Inc.: 354-355
© Joe McDonald/Natural Selection: 351
© Kevin Schaffer: 350
© Anup & Manoj Shah: 336, 345, 352, 352-353, 354 bottom, 356-357, 357 bottom, 359 bottom
© Amrit P. Singh: 354 top, 360-361
© James L. Stanfield/National Geographic Society: 360
© Konrad Wothe/Ellis Nature Photography: 346
© Belinda Wright/DRK Photo: 344, 356, 358-359

PAPUA NEW GUINEA

© Bruce Beehler: 364
© Gerry Ellis/Ellis Nature Photography: 365, 368, 370-371, 374, 376-377, 378-379, 381 top
© Peter Essick/Aurora & Quanta Productions: 375
© Jean-Paul Ferrero/Auscape: 366-367, 376, 381 bottom
© Russell A. Mittermeier: 373, 384, 384-385
© D. Parer- E. Parer-Cook/Auscape: 362, 380-381
© Ulli Seer/LOOK: 383
© Art Wolfe: 372, 378, 382-383
© Konrad Wothe/Ellis Nature Photography: 369

UNITED STATES OF AMERICA

© James L. Amos/National Geographic Society: 404-405
© John Cancalosi: 396, 400 bottom
© Jack Dykinga: 398-399
© Gerry Ellis/Ellis Nature Photography: 392
© Ralph Lee Hopkins/The Wildlife Collection: 397
© Chris Johns/National Geographic Society: 403 top, 403 bottom
© Stephen J. Kraseman/DRK Photo: 390-391
© Frans Lanting: 402-403
© Thomas Mangelsen: 386, 398
© Doug Perrine/DRK Photo: 393
© Patricio Robles Gil/Sierra Madre: 394
© John Shaw: 388, 389
© Kennan Ward: 400 top, 400-401
© Art Wolfe: 395

SOUTH AFRICA

© Gerald Cubitt: 412
© Michael Fogden/Animals Animals: 406
© Michael Fodgen/Oxford Scientific Films: 420 bottom
© Raúl Gil: 408
© Lex Hes: 409, 413, 416, 417, 418, 420 top, 420-421
© Chris Johns/National Geographic Society: 410-411, 422-423
© M. P. Kahl/Vireo: 414
© David McDonald/Oxford Scientific Films: 415
© Rob Nunnington/Oxford Scientific Films: 418-419

MALAYSIA

© Bob Bowdey/La Mer Majik: 444-445, 445
© Tom Brakefield/DRK PHOTO: 430
© Daniel J. Cox/Oxford Scientific Films: 436
© Gerald Cubitt: 424, 431, 432-433
© Fletcher & Baylis/Wildside Photography: 426, 438-439, 442-443
© Michael Fogden/Oxford Scientific Films: 438 bottom
© David A. Harvey/National Geographic Society: 428-429
© David Hiser/Photographers Aspen: 446, 446-447
© Frans Lanting: 427
© Frans Lanting/Minden Pictures: 437, 440
© Mark Moffet/Minden Pictures: 438 top, 443
© Peter Ward/Bruce Coleman, Inc.: 440-441
© Doug Wechsler/Vireo: 434
© Gunter Ziesler/Peter Arnold, Inc.: 435

VENEZUELA

© John Cancalosi: 451, 460
© Jay Dickman/National Geographic Society: 465 bottom
© Michael & Patricia Fogden: 463
© François Gohier: 457, 462-463
© Gavriel Jecan/Art Wolfe, Inc.: 466 top
© Luiz Claudio Marigo: 456
© Russell A. Mittermeier: 458, 459
© Russell A. Mittermeier/Bruce Coleman, Inc.: 466 bottom
© Neil L. Rettig, F.R.E.E., Ltd.: 448
© Patricio Robles Gil: 452-453, 454, 455, 464-465, 465 top
© Brian Rogers/Biophotos: 460-461
© Kevin Schaffer: 450
© Art Wolfe: 466-467

DEMOCRATIC REPUBLIC OF CONGO

© José Azel/Aurora & Quanta Productions: 484-485
© Robert C. Bailey: 484 bottom
© James P. Blair/National Geographic Society: 484 top
© Martyn Colbeck/Oxford Scientific Films: 468
© Gerard/Jacana: 482 bottom
© Gerry Ellis/Ellis Nature Photography: 470, 478-479, 481, 482 top
© George Holton/BIF/Photo Researchers, Inc.: 476
© Rich Kirchner/Natural Selection: 475
© Tom McHugh/BIF/Photo Researchers, Inc.: 474
© Michael K. Nichols/National Geographic Society: 471, 472-473, 477, 480-481, 482-483

EPILOGUE AND BIBLIOGRAPHY

© Gilbert Grosvenor/National Geographic Society: 490-491
© David Hiser/Photographers Aspen: 486
© Frans Lanting/Minden Pictures: 492
© Cristina Goettsch Mittermeier: 499
© Patricio Robles Gil/Sierra Madre: 488, 489, 493

ACKNOWLEDGEMENTS

Brazil: *Associação Mico-Leão-Dourado, Centro de Primatologia do Rio de Janeiro, Fazenda Kaiman, IBAMA, Parque Nacional das Emas, Parque Nacional Serra dos Órgãos;*
Colombia: *Fundación Felca, INDERENA, Ministerio del Medio Ambiente, Parque Nacional Amacuyacu, Reserva La Planada, Reserva Natural Río Niambi, Unidad Administrativa Especial del Sistema de Parques Nacionales Naturales;*
Ecuador: *Fundación Arco Iris, Instituto Ecuatoriano Forestal y de Áreas Naturales y Vida Silvestre, Jardín Botánico de la Universidad Nacional de Loja "Reinaldo Espinosa", Parque Nacional Podocarpus;*
Mexico: *Secretaría de Medio Ambiente, Recursos Naturales y Pesca;*
Peru: *INRENA, Ministerio de Agricultura, Parque Nacional de Manú, Reserva Nacional de Paracas, Reserva Nacional de Salinas y Aguada Blanca;*
Venezuela: *Fundación para la Defensa de la Naturaleza, INPARQUES.*

Thomas Akre, Patricia Almada-Villela, Pablo Andrade, Salvador Arias Montes, John Arvin, Samuel Marcial Bisbikus, Alicia Calaf, Santos Calderón, Guillermo Cantillo, Alex Cortés, Juan Cuazo, María Elfis Chávez, Luis Fernando Dias da Silva, Diego Díaz Martín, Fernando Fernández, Ana Luisa Figueroa, Lloyd T. Finley, Juan Manuel García, Luis Guzmán, Débora Herzog, Eugenia Huerta, Carlos Jaramillo, María Fernanda Jaramillo, Arturo Jiménez, Gabriela Lara, Elías León, Fausto López, Armando Luis, Juan Carlos Luna, Jorge Llorente, Rodrigo Medellín, Luis Medina, Alfredo Mendoza, Olga Lucía Mesa, María Teresa Miozzi, Alfonso Ocampo, Yolanda Padilla, Alcides Pissinatti, Carlos Ponce, Jorge Quiñónez, Denis Marçal Rambaldi, Antonio Ramírez, Constanza Ríos, Guillermo Rivera, Nancy Rojas, José Andrés Rojo, María Isabel de Rojo, Patricia Rojo, Anthony Rylands, Kristin Salaman, Paul Salaman, Cristian Samper, Teresa Segovia, Michael Small, Renán Valega, Juárez Vieira Santos, Carlos G. Zárate.

PRODUCTION
Agrupación Sierra Madre, S.C.
Redacta, S.A. de C.V.
Impetus Comunicación, S.A. de C.V.

EDITORIAL DIRECTION
Antonio Bolívar

EDITORIAL REVISION AND CORRECTION
Susan Beth Kapilian

GRAPHIC DESIGN AND PHOTO EDITING
Patricio Robles Gil
Martín Jon García-Urtiaga
Lucila Flores de Clavé

TECHNICAL SUPPORT
Martín Jon García-Urtiaga

COORDINATION OF PRODUCTION
Blanca Luz Pulido
Eugenia Pallares
Elena León
María Luisa Madrazo
Georgina Ferrer

TYPESETTING
Rocío Moreno
Daniel Olivares

MAPS
Stephen Nash
Yvonne Udaeta

PRINTING LIAISON
Ignacio Narro

First English edition, 1997

ISBN 968-6397-50-7

Printed in Canada
by Quebecor Printing Inc.